College Physics for AP® Courses

Part 2: Chapters 18 - 34

Senior Contributing Authors

Irina Lyublinskaya, Cuny College Of Staten Island
Gregg Wolfe, Avonworth High School
Douglas Ingram, Texas Christian University
Liza Pujji, Manukau Institute Of Technology
Sudhi Oberoi, Raman Research Institute
Nathan Czuba, Sabio Academy

This book does not contain the chapters 1-17 or the Answers for chapters 1-17 of the original book *College Physics for AP® Courses* by OpenStax.

Chapters 1-17 and the chapter Answers of the original book *The College Physics for AP® Courses* by OpenStax may be found in a book with the ISBN of 9781680920765.

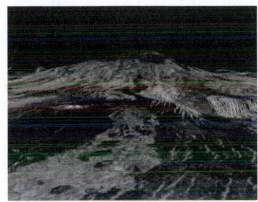

Figure 24.16 An image of Sif Mons with lava flows on Venus, based on Magellan synthetic aperture radar data combined with radar altimetry to produce a three-dimensional map of the surface. The Venusian atmosphere is opaque to visible light, but not to the microwaves that were used to create this image. (credit: NSSDC, NASA/JPL)

Heating with Microwaves

How does the ubiquitous microwave oven produce microwaves electronically, and why does food absorb them preferentially? Microwaves at a frequency of 2.45 GHz are produced by accelerating electrons. The microwaves are then used to induce an alternating electric field in the oven.

Water and some other constituents of food have a slightly negative charge at one end and a slightly positive charge at one end (called polar molecules). The range of microwave frequencies is specially selected so that the polar molecules, in trying to keep orienting themselves with the electric field, absorb these energies and increase their temperatures—called dielectric heating.

The energy thereby absorbed results in thermal agitation heating food and not the plate, which does not contain water. Hot spots in the food are related to constructive and destructive interference patterns. Rotating antennas and food turntables help spread out the hot spots.

Another use of microwaves for heating is within the human body. Microwaves will penetrate more than shorter wavelengths into tissue and so can accomplish "deep heating" (called microwave diathermy). This is used for treating muscular pains, spasms, tendonitis, and rheumatoid arthritis.

Making Connections: Take-Home Experiment—Microwave Ovens

1. Look at the door of a microwave oven. Describe the structure of the door. Why is there a metal grid on the door? How does the size of the holes in the grid compare with the wavelengths of microwaves used in microwave ovens? What is this wavelength?

2. Place a glass of water (about 250 ml) in the microwave and heat it for 30 seconds. Measure the temperature gain (the ΔT). Assuming that the power output of the oven is 1000 W, calculate the efficiency of the heat-transfer process.

3. Remove the rotating turntable or moving plate and place a cup of water in several places along a line parallel with the opening. Heat for 30 seconds and measure the ΔT for each position. Do you see cases of destructive interference?

Microwaves generated by atoms and molecules far away in time and space can be received and detected by electronic circuits. Deep space acts like a blackbody with a 2.7 K temperature, radiating most of its energy in the microwave frequency range. In 1964, Penzias and Wilson detected this radiation and eventually recognized that it was the radiation of the Big Bang's cooled remnants.

Infrared Radiation

The microwave and infrared regions of the electromagnetic spectrum overlap (see **Figure 24.10**). **Infrared radiation** is generally produced by thermal motion and the vibration and rotation of atoms and molecules. Electronic transitions in atoms and molecules can also produce infrared radiation.

The range of infrared frequencies extends up to the lower limit of visible light, just below red. In fact, infrared means "below red." Frequencies at its upper limit are too high to be produced by accelerating electrons in circuits, but small systems, such as atoms and molecules, can vibrate fast enough to produce these waves.

Water molecules rotate and vibrate particularly well at infrared frequencies, emitting and absorbing them so efficiently that the emissivity for skin is $e = 0.97$ in the infrared. Night-vision scopes can detect the infrared emitted by various warm objects, including humans, and convert it to visible light.

We can examine radiant heat transfer from a house by using a camera capable of detecting infrared radiation. Reconnaissance satellites can detect buildings, vehicles, and even individual humans by their infrared emissions, whose power radiation is proportional to the fourth power of the absolute temperature. More mundanely, we use infrared lamps, some of which are called quartz heaters, to preferentially warm us because we absorb infrared better than our surroundings.

The Sun radiates like a nearly perfect blackbody (that is, it has $e = 1$), with a 6000 K surface temperature. About half of the solar energy arriving at the Earth is in the infrared region, with most of the rest in the visible part of the spectrum, and a relatively small amount in the ultraviolet. On average, 50 percent of the incident solar energy is absorbed by the Earth.

The relatively constant temperature of the Earth is a result of the energy balance between the incoming solar radiation and the energy radiated from the Earth. Most of the infrared radiation emitted from the Earth is absorbed by CO_2 and H_2O in the atmosphere and then radiated back to Earth or into outer space. This radiation back to Earth is known as the greenhouse effect, and it maintains the surface temperature of the Earth about $40°C$ higher than it would be if there is no absorption. Some scientists think that the increased concentration of CO_2 and other greenhouse gases in the atmosphere, resulting from increases in fossil fuel burning, has increased global average temperatures.

Visible Light

Visible light is the narrow segment of the electromagnetic spectrum to which the normal human eye responds. Visible light is produced by vibrations and rotations of atoms and molecules, as well as by electronic transitions within atoms and molecules. The receivers or detectors of light largely utilize electronic transitions. We say the atoms and molecules are excited when they absorb and relax when they emit through electronic transitions.

Figure 24.17 shows this part of the spectrum, together with the colors associated with particular pure wavelengths. We usually refer to visible light as having wavelengths of between 400 nm and 750 nm. (The retina of the eye actually responds to the lowest ultraviolet frequencies, but these do not normally reach the retina because they are absorbed by the cornea and lens of the eye.)

Red light has the lowest frequencies and longest wavelengths, while violet has the highest frequencies and shortest wavelengths. Blackbody radiation from the Sun peaks in the visible part of the spectrum but is more intense in the red than in the violet, making the Sun yellowish in appearance.

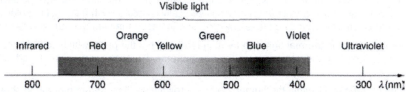

Figure 24.17 A small part of the electromagnetic spectrum that includes its visible components. The divisions between infrared, visible, and ultraviolet are not perfectly distinct, nor are those between the seven rainbow colors.

Living things—plants and animals—have evolved to utilize and respond to parts of the electromagnetic spectrum they are embedded in. Visible light is the most predominant and we enjoy the beauty of nature through visible light. Plants are more selective. Photosynthesis makes use of parts of the visible spectrum to make sugars.

Example 24.3 Integrated Concept Problem: Correcting Vision with Lasers

During laser vision correction, a brief burst of 193-nm ultraviolet light is projected onto the cornea of a patient. It makes a spot 0.80 mm in diameter and evaporates a layer of cornea $0.30 \ \mu m$ thick. Calculate the energy absorbed, assuming the corneal tissue has the same properties as water; it is initially at $34°C$. Assume the evaporated tissue leaves at a temperature of $100°C$.

Strategy

The energy from the laser light goes toward raising the temperature of the tissue and also toward evaporating it. Thus we have two amounts of heat to add together. Also, we need to find the mass of corneal tissue involved.

Solution

To figure out the heat required to raise the temperature of the tissue to $100°C$, we can apply concepts of thermal energy. We know that

$$Q = mc\Delta T, \qquad\qquad (24.11)$$

where Q is the heat required to raise the temperature, ΔT is the desired change in temperature, m is the mass of tissue to be heated, and c is the specific heat of water equal to 4186 J/kg/K.

Without knowing the mass m at this point, we have

$$Q = m(4186 \text{ J/kg/K})(100°C - 34°C) = m(276{,}276 \text{ J/kg}) = m(276 \text{ kJ/kg}). \qquad\qquad (24.12)$$

The latent heat of vaporization of water is 2256 kJ/kg, so that the energy needed to evaporate mass m is

$$Q_v = mL_v = m(2256 \text{ kJ/kg}). \qquad\qquad (24.13)$$

To find the mass m, we use the equation $\rho = m/V$, where ρ is the density of the tissue and V is its volume. For this case,

$$m = \rho V \tag{24.14}$$
$$= (1000 \text{ kg/m}^3)(\text{area} \times \text{thickness}(\text{m}^3))$$
$$= (1000 \text{ kg/m}^3)(\pi(0.80 \times 10^{-3} \text{ m})^2 / 4)(0.30 \times 10^{-6} \text{ m})$$
$$= 0.151 \times 10^{-9} \text{ kg}.$$

Therefore, the total energy absorbed by the tissue in the eye is the sum of Q and Q_v:

$$Q_{tot} = m(c\Delta T + L_v) = (0.151 \times 10^{-9} \text{ kg})(276 \text{ kJ/kg} + 2256 \text{ kJ/kg}) = 382 \times 10^{-9} \text{ kJ}. \tag{24.15}$$

Discussion

The lasers used for this eye surgery are excimer lasers, whose light is well absorbed by biological tissue. They evaporate rather than burn the tissue, and can be used for precision work. Most lasers used for this type of eye surgery have an average power rating of about one watt. For our example, if we assume that each laser burst from this pulsed laser lasts for 10 ns, and there are 400 bursts per second, then the average power is $Q_{tot} \times 400 = 150 \text{ mW}$.

Optics is the study of the behavior of visible light and other forms of electromagnetic waves. Optics falls into two distinct categories. When electromagnetic radiation, such as visible light, interacts with objects that are large compared with its wavelength, its motion can be represented by straight lines like rays. Ray optics is the study of such situations and includes lenses and mirrors.

When electromagnetic radiation interacts with objects about the same size as the wavelength or smaller, its wave nature becomes apparent. For example, observable detail is limited by the wavelength, and so visible light can never detect individual atoms, because they are so much smaller than its wavelength. Physical or wave optics is the study of such situations and includes all wave characteristics.

Take-Home Experiment: Colors That Match

When you light a match you see largely orange light; when you light a gas stove you see blue light. Why are the colors different? What other colors are present in these?

Ultraviolet Radiation

Ultraviolet means "above violet." The electromagnetic frequencies of **ultraviolet radiation (UV)** extend upward from violet, the highest-frequency visible light. Ultraviolet is also produced by atomic and molecular motions and electronic transitions. The wavelengths of ultraviolet extend from 400 nm down to about 10 nm at its highest frequencies, which overlap with the lowest X-ray frequencies. It was recognized as early as 1801 by Johann Ritter that the solar spectrum had an invisible component beyond the violet range.

Solar UV radiation is broadly subdivided into three regions: UV-A (320–400 nm), UV-B (290–320 nm), and UV-C (220–290 nm), ranked from long to shorter wavelengths (from smaller to larger energies). Most UV-B and all UV-C is absorbed by ozone (O_3) molecules in the upper atmosphere. Consequently, 99% of the solar UV radiation reaching the Earth's surface is UV-A.

Human Exposure to UV Radiation

It is largely exposure to UV-B that causes skin cancer. It is estimated that as many as 20% of adults will develop skin cancer over the course of their lifetime. Again, treatment is often successful if caught early. Despite very little UV-B reaching the Earth's surface, there are substantial increases in skin-cancer rates in countries such as Australia, indicating how important it is that UV-B and UV-C continue to be absorbed by the upper atmosphere.

All UV radiation can damage collagen fibers, resulting in an acceleration of the aging process of skin and the formation of wrinkles. Because there is so little UV-B and UV-C reaching the Earth's surface, sunburn is caused by large exposures, and skin cancer from repeated exposure. Some studies indicate a link between overexposure to the Sun when young and melanoma later in life.

The tanning response is a defense mechanism in which the body produces pigments to absorb future exposures in inert skin layers above living cells. Basically UV-B radiation excites DNA molecules, distorting the DNA helix, leading to mutations and the possible formation of cancerous cells.

Repeated exposure to UV-B may also lead to the formation of cataracts in the eyes—a cause of blindness among people living in the equatorial belt where medical treatment is limited. Cataracts, clouding in the eye's lens and a loss of vision, are age related; 60% of those between the ages of 65 and 74 will develop cataracts. However, treatment is easy and successful, as one replaces the lens of the eye with a plastic lens. Prevention is important. Eye protection from UV is more effective with plastic sunglasses than those made of glass.

A major acute effect of extreme UV exposure is the suppression of the immune system, both locally and throughout the body.

Low-intensity ultraviolet is used to sterilize haircutting implements, implying that the energy associated with ultraviolet is deposited in a manner different from lower-frequency electromagnetic waves. (Actually this is true for all electromagnetic waves with frequencies greater than visible light.)

Flash photography is generally not allowed of precious artworks and colored prints because the UV radiation from the flash can cause photo-degradation in the artworks. Often artworks will have an extra-thick layer of glass in front of them, which is especially designed to absorb UV radiation.

UV Light and the Ozone Layer

If all of the Sun's ultraviolet radiation reached the Earth's surface, there would be extremely grave effects on the biosphere from the severe cell damage it causes. However, the layer of ozone (O_3) in our upper atmosphere (10 to 50 km above the Earth) protects life by absorbing most of the dangerous UV radiation.

Unfortunately, today we are observing a depletion in ozone concentrations in the upper atmosphere. This depletion has led to the formation of an "ozone hole" in the upper atmosphere. The hole is more centered over the southern hemisphere, and changes with the seasons, being largest in the spring. This depletion is attributed to the breakdown of ozone molecules by refrigerant gases called chlorofluorocarbons (CFCs).

The UV radiation helps dissociate the CFC's, releasing highly reactive chlorine (Cl) atoms, which catalyze the destruction of the ozone layer. For example, the reaction of $CFCl_3$ with a photon of light $(h\nu)$ can be written as:

$$CFCl_3 + h\nu \rightarrow CFCl_2 + Cl. \tag{24.16}$$

The Cl atom then catalyzes the breakdown of ozone as follows:

$$Cl + O_3 \rightarrow ClO + O_2 \text{ and } ClO + O_3 \rightarrow Cl + 2O_2. \tag{24.17}$$

A single chlorine atom could destroy ozone molecules for up to two years before being transported down to the surface. The CFCs are relatively stable and will contribute to ozone depletion for years to come. CFCs are found in refrigerants, air conditioning systems, foams, and aerosols.

International concern over this problem led to the establishment of the "Montreal Protocol" agreement (1987) to phase out CFC production in most countries. However, developing-country participation is needed if worldwide production and elimination of CFCs is to be achieved. Probably the largest contributor to CFC emissions today is India. But the protocol seems to be working, as there are signs of an ozone recovery. (See **Figure 24.18**.)

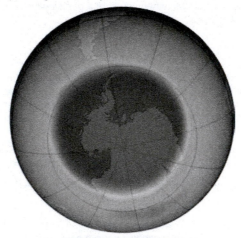

0   100 200 300 400 500 600 700
Total Ozone (Dobson units)

Figure 24.18 This map of ozone concentration over Antarctica in October 2011 shows severe depletion suspected to be caused by CFCs. Less dramatic but more general depletion has been observed over northern latitudes, suggesting the effect is global. With less ozone, more ultraviolet radiation from the Sun reaches the surface, causing more damage. (credit: NASA Ozone Watch)

Benefits of UV Light

Besides the adverse effects of ultraviolet radiation, there are also benefits of exposure in nature and uses in technology. Vitamin D production in the skin (epidermis) results from exposure to UVB radiation, generally from sunlight. A number of studies indicate lack of vitamin D can result in the development of a range of cancers (prostate, breast, colon), so a certain amount of UV exposure is helpful. Lack of vitamin D is also linked to osteoporosis. Exposures (with no sunscreen) of 10 minutes a day to arms, face, and legs might be sufficient to provide the accepted dietary level. However, in the winter time north of about $37°$ latitude, most UVB gets blocked by the atmosphere.

UV radiation is used in the treatment of infantile jaundice and in some skin conditions. It is also used in sterilizing workspaces and tools, and killing germs in a wide range of applications. It is also used as an analytical tool to identify substances.

When exposed to ultraviolet, some substances, such as minerals, glow in characteristic visible wavelengths, a process called fluorescence. So-called black lights emit ultraviolet to cause posters and clothing to fluoresce in the visible. Ultraviolet is also used in special microscopes to detect details smaller than those observable with longer-wavelength visible-light microscopes.

Things Great and Small: A Submicroscopic View of X-Ray Production

X-rays can be created in a high-voltage discharge. They are emitted in the material struck by electrons in the discharge current. There are two mechanisms by which the electrons create X-rays.

The first method is illustrated in **Figure 24.19**. An electron is accelerated in an evacuated tube by a high positive voltage. The electron strikes a metal plate (e.g., copper) and produces X-rays. Since this is a high-voltage discharge, the electron gains sufficient energy to ionize the atom.

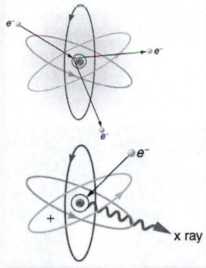

Figure 24.19 Artist's conception of an electron ionizing an atom followed by the recapture of an electron and emission of an X-ray. An energetic electron strikes an atom and knocks an electron out of one of the orbits closest to the nucleus. Later, the atom captures another electron, and the energy released by its fall into a low orbit generates a high-energy EM wave called an X-ray.

In the case shown, an inner-shell electron (one in an orbit relatively close to and tightly bound to the nucleus) is ejected. A short time later, another electron is captured and falls into the orbit in a single great plunge. The energy released by this fall is given to an EM wave known as an X-ray. Since the orbits of the atom are unique to the type of atom, the energy of the X-ray is characteristic of the atom, hence the name characteristic X-ray.

The second method by which an energetic electron creates an X-ray when it strikes a material is illustrated in **Figure 24.20**. The electron interacts with charges in the material as it penetrates. These collisions transfer kinetic energy from the electron to the electrons and atoms in the material.

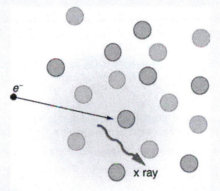

Figure 24.20 Artist's conception of an electron being slowed by collisions in a material and emitting X-ray radiation. This energetic electron makes numerous collisions with electrons and atoms in a material it penetrates. An accelerated charge radiates EM waves, a second method by which X-rays are created.

A loss of kinetic energy implies an acceleration, in this case decreasing the electron's velocity. Whenever a charge is accelerated, it radiates EM waves. Given the high energy of the electron, these EM waves can have high energy. We call them X-rays. Since the process is random, a broad spectrum of X-ray energy is emitted that is more characteristic of the electron energy than the type of material the electron encounters. Such EM radiation is called "bremsstrahlung" (German for "braking radiation").

X-Rays

In the 1850s, scientists (such as Faraday) began experimenting with high-voltage electrical discharges in tubes filled with rarefied gases. It was later found that these discharges created an invisible, penetrating form of very high frequency electromagnetic radiation. This radiation was called an **X-ray**, because its identity and nature were unknown.

As described in **Things Great and Small**, there are two methods by which X-rays are created—both are submicroscopic processes and can be caused by high-voltage discharges. While the low-frequency end of the X-ray range overlaps with the ultraviolet, X-rays extend to much higher frequencies (and energies).

X-rays have adverse effects on living cells similar to those of ultraviolet radiation, and they have the additional liability of being more penetrating, affecting more than the surface layers of cells. Cancer and genetic defects can be induced by exposure to X-rays. Because of their effect on rapidly dividing cells, X-rays can also be used to treat and even cure cancer.

The widest use of X-rays is for imaging objects that are opaque to visible light, such as the human body or aircraft parts. In humans, the risk of cell damage is weighed carefully against the benefit of the diagnostic information obtained. However, questions have risen in recent years as to accidental overexposure of some people during CT scans—a mistake at least in part due to poor monitoring of radiation dose.

The ability of X-rays to penetrate matter depends on density, and so an X-ray image can reveal very detailed density information. **Figure 24.21** shows an example of the simplest type of X-ray image, an X-ray shadow on film. The amount of information in a simple X-ray image is impressive, but more sophisticated techniques, such as CT scans, can reveal three-dimensional information with details smaller than a millimeter.

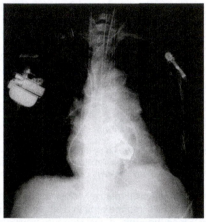

Figure 24.21 This shadow X-ray image shows many interesting features, such as artificial heart valves, a pacemaker, and the wires used to close the sternum. (credit: P. P. Urone)

The use of X-ray technology in medicine is called radiology—an established and relatively cheap tool in comparison to more sophisticated technologies. Consequently, X-rays are widely available and used extensively in medical diagnostics. During World War I, mobile X-ray units, advocated by Madame Marie Curie, were used to diagnose soldiers.

Because they can have wavelengths less than 0.01 nm, X-rays can be scattered (a process called X-ray diffraction) to detect the shape of molecules and the structure of crystals. X-ray diffraction was crucial to Crick, Watson, and Wilkins in the determination of the shape of the double-helix DNA molecule.

X-rays are also used as a precise tool for trace-metal analysis in X-ray induced fluorescence, in which the energy of the X-ray emissions are related to the specific types of elements and amounts of materials present.

Gamma Rays

Soon after nuclear radioactivity was first detected in 1896, it was found that at least three distinct types of radiation were being emitted. The most penetrating nuclear radiation was called a **gamma ray (γ ray)** (again a name given because its identity and character were unknown), and it was later found to be an extremely high frequency electromagnetic wave.

In fact, γ rays are any electromagnetic radiation emitted by a nucleus. This can be from natural nuclear decay or induced nuclear processes in nuclear reactors and weapons. The lower end of the γ-ray frequency range overlaps the upper end of the X-ray range, but γ rays can have the highest frequency of any electromagnetic radiation.

Gamma rays have characteristics identical to X-rays of the same frequency—they differ only in source. At higher frequencies, γ rays are more penetrating and more damaging to living tissue. They have many of the same uses as X-rays, including cancer therapy. Gamma radiation from radioactive materials is used in nuclear medicine.

Figure 24.22 shows a medical image based on γ rays. Food spoilage can be greatly inhibited by exposing it to large doses of γ radiation, thereby obliterating responsible microorganisms. Damage to food cells through irradiation occurs as well, and the long-term hazards of consuming radiation-preserved food are unknown and controversial for some groups. Both X-ray and γ-ray technologies are also used in scanning luggage at airports.

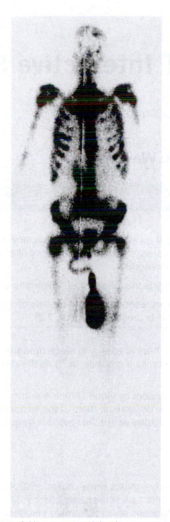

Figure 24.22 This is an image of the γ rays emitted by nuclei in a compound that is concentrated in the bones and eliminated through the kidneys. Bone cancer is evidenced by nonuniform concentration in similar structures. For example, some ribs are darker than others. (credit: P. P. Urone)

Detecting Electromagnetic Waves from Space

A final note on star gazing. The entire electromagnetic spectrum is used by researchers for investigating stars, space, and time. As noted earlier, Penzias and Wilson detected microwaves to identify the background radiation originating from the Big Bang. Radio telescopes such as the Arecibo Radio Telescope in Puerto Rico and Parkes Observatory in Australia were designed to detect radio waves.

Infrared telescopes need to have their detectors cooled by liquid nitrogen to be able to gather useful signals. Since infrared radiation is predominantly from thermal agitation, if the detectors were not cooled, the vibrations of the molecules in the antenna would be stronger than the signal being collected.

The most famous of these infrared sensitive telescopes is the James Clerk Maxwell Telescope in Hawaii. The earliest telescopes, developed in the seventeenth century, were optical telescopes, collecting visible light. Telescopes in the ultraviolet, X-ray, and γ-ray regions are placed outside the atmosphere on satellites orbiting the Earth.

The Hubble Space Telescope (launched in 1990) gathers ultraviolet radiation as well as visible light. In the X-ray region, there is the Chandra X-ray Observatory (launched in 1999), and in the γ-ray region, there is the new Fermi Gamma-ray Space

Telescope (launched in 2008—taking the place of the Compton Gamma Ray Observatory, 1991–2000.).

PhET Explorations: Color Vision

Make a whole rainbow by mixing red, green, and blue light. Change the wavelength of a monochromatic beam or filter white light. View the light as a solid beam, or see the individual photons.

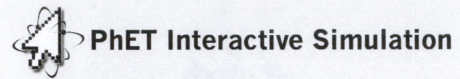

Figure 24.23 Color Vision (http://cnx.org/content/m55432/1.3/color-vision_en.jar)

24.4 Energy in Electromagnetic Waves

Learning Objectives

By the end of this section, you will be able to:

- Explain how the energy and amplitude of an electromagnetic wave are related.
- Given its power output and the heating area, calculate the intensity of a microwave oven's electromagnetic field, as well as its peak electric and magnetic field strengths.

The information presented in this section supports the following AP® learning objectives and science practices:

- **6.F.2.1** The student is able to describe representations and models of electromagnetic waves that explain the transmission of energy when no medium is present. **(S.P. 6.4, 7.2)**

Anyone who has used a microwave oven knows there is energy in **electromagnetic waves**. Sometimes this energy is obvious, such as in the warmth of the summer sun. Other times it is subtle, such as the unfelt energy of gamma rays, which can destroy living cells.

Electromagnetic waves can bring energy into a system by virtue of their **electric and magnetic fields**. These fields can exert forces and move charges in the system and, thus, do work on them. If the frequency of the electromagnetic wave is the same as the natural frequencies of the system (such as microwaves at the resonant frequency of water molecules), the transfer of energy is much more efficient.

Connections: Waves and Particles

The behavior of electromagnetic radiation clearly exhibits wave characteristics. But we shall find in later modules that at high frequencies, electromagnetic radiation also exhibits particle characteristics. These particle characteristics will be used to explain more of the properties of the electromagnetic spectrum and to introduce the formal study of modern physics.

Another startling discovery of modern physics is that particles, such as electrons and protons, exhibit wave characteristics. This simultaneous sharing of wave and particle properties for all submicroscopic entities is one of the great symmetries in nature.

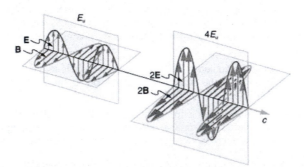

Figure 24.24 Energy carried by a wave is proportional to its amplitude squared. With electromagnetic waves, larger E-fields and B-fields exert larger forces and can do more work.

But there is energy in an electromagnetic wave, whether it is absorbed or not. Once created, the fields carry energy away from a source. If absorbed, the field strengths are diminished and anything left travels on. Clearly, the larger the strength of the electric and magnetic fields, the more work they can do and the greater the energy the electromagnetic wave carries.

A wave's energy is proportional to its **amplitude** squared (E^2 or B^2). This is true for waves on guitar strings, for water waves, and for sound waves, where amplitude is proportional to pressure. In electromagnetic waves, the amplitude is the **maximum field strength** of the electric and magnetic fields. (See **Figure 24.24**.)

Thus the energy carried and the **intensity** I of an electromagnetic wave is proportional to E^2 and B^2. In fact, for a continuous sinusoidal electromagnetic wave, the average intensity I_{ave} is given by

$$I_{ave} = \frac{c\varepsilon_0 E_0^2}{2},$$ (24.18)

where c is the speed of light, ε_0 is the permittivity of free space, and E_0 is the maximum electric field strength; intensity, as always, is power per unit area (here in W/m^2).

The average intensity of an electromagnetic wave I_{ave} can also be expressed in terms of the magnetic field strength by using the relationship $B = E/c$, and the fact that $\varepsilon_0 = 1/\mu_0 c^2$, where μ_0 is the permeability of free space. Algebraic manipulation produces the relationship

$$I_{ave} = \frac{cB_0^2}{2\mu_0},$$ (24.19)

where B_0 is the maximum magnetic field strength.

One more expression for I_{ave} in terms of both electric and magnetic field strengths is useful. Substituting the fact that $c \cdot B_0 = E_0$, the previous expression becomes

$$I_{ave} = \frac{E_0 B_0}{2\mu_0}.$$ (24.20)

Whichever of the three preceding equations is most convenient can be used, since they are really just different versions of the same principle: Energy in a wave is related to amplitude squared. Furthermore, since these equations are based on the assumption that the electromagnetic waves are sinusoidal, peak intensity is twice the average; that is, $I_0 = 2I_{ave}$.

Example 24.4 Calculate Microwave Intensities and Fields

On its highest power setting, a certain microwave oven projects 1.00 kW of microwaves onto a 30.0 by 40.0 cm area. (a) What is the intensity in W/m^2? (b) Calculate the peak electric field strength E_0 in these waves. (c) What is the peak magnetic field strength B_0?

Strategy

In part (a), we can find intensity from its definition as power per unit area. Once the intensity is known, we can use the equations below to find the field strengths asked for in parts (b) and (c).

Solution for (a)

Entering the given power into the definition of intensity, and noting the area is 0.300 by 0.400 m, yields

$$I = \frac{P}{A} = \frac{1.00 \text{ kW}}{0.300 \text{ m} \times 0.400 \text{ m}}.$$ (24.21)

Here $I = I_{ave}$, so that

$$I_{ave} = \frac{1000 \text{ W}}{0.120 \text{ m}^2} = 8.33 \times 10^3 \text{ W/m}^2.$$ (24.22)

Note that the peak intensity is twice the average:

$$I_0 = 2I_{ave} = 1.67 \times 10^4 \text{ W/m}^2.$$ (24.23)

Solution for (b)

To find E_0, we can rearrange the first equation given above for I_{ave} to give

$$E_0 = \left(\frac{2I_{ave}}{c\varepsilon_0}\right)^{1/2}.$$ (24.24)

Entering known values gives

$$\begin{aligned} E_0 &= \sqrt{\frac{2(8.33 \times 10^3 \text{ W/m}^2)}{(3.00 \times 10^8 \text{ m/s})(8.85 \times 10^{-12} \text{ C}^2/\text{N} \cdot \text{m}^2)}} \\ &= 2.51 \times 10^3 \text{ V/m}. \end{aligned}$$ (24.25)

Solution for (c)

Perhaps the easiest way to find magnetic field strength, now that the electric field strength is known, is to use the relationship given by

$$B_0 = \frac{E_0}{c}. \tag{24.26}$$

Entering known values gives

$$\begin{aligned} B_0 &= \frac{2.51 \times 10^3 \text{ V/m}}{3.0 \times 10^8 \text{ m/s}} \\ &= 8.35 \times 10^{-6} \text{ T}. \end{aligned} \tag{24.27}$$

Discussion

As before, a relatively strong electric field is accompanied by a relatively weak magnetic field in an electromagnetic wave, since $B = E/c$, and c is a large number.

Glossary

amplitude: the height, or magnitude, of an electromagnetic wave

amplitude modulation (AM): a method for placing information on electromagnetic waves by modulating the amplitude of a carrier wave with an audio signal, resulting in a wave with constant frequency but varying amplitude

carrier wave: an electromagnetic wave that carries a signal by modulation of its amplitude or frequency

electric field: a vector quantity (**E**); the lines of electric force per unit charge, moving radially outward from a positive charge and in toward a negative charge

electric field lines: a pattern of imaginary lines that extend between an electric source and charged objects in the surrounding area, with arrows pointed away from positively charged objects and toward negatively charged objects. The more lines in the pattern, the stronger the electric field in that region

electric field strength: the magnitude of the electric field, denoted E-field

electromagnetic spectrum: the full range of wavelengths or frequencies of electromagnetic radiation

electromagnetic waves: radiation in the form of waves of electric and magnetic energy

electromotive force (emf): energy produced per unit charge, drawn from a source that produces an electrical current

extremely low frequency (ELF): electromagnetic radiation with wavelengths usually in the range of 0 to 300 Hz, but also about 1kHz

frequency: the number of complete wave cycles (up-down-up) passing a given point within one second (cycles/second)

frequency modulation (FM): a method of placing information on electromagnetic waves by modulating the frequency of a carrier wave with an audio signal, producing a wave of constant amplitude but varying frequency

gamma ray: (γ ray); extremely high frequency electromagnetic radiation emitted by the nucleus of an atom, either from natural nuclear decay or induced nuclear processes in nuclear reactors and weapons. The lower end of the γ-ray frequency range overlaps the upper end of the X-ray range, but γ rays can have the highest frequency of any electromagnetic radiation

hertz: an SI unit denoting the frequency of an electromagnetic wave, in cycles per second

infrared radiation (IR): a region of the electromagnetic spectrum with a frequency range that extends from just below the red region of the visible light spectrum up to the microwave region, or from $0.74 \ \mu m$ to $300 \ \mu m$

intensity: the power of an electric or magnetic field per unit area, for example, Watts per square meter

magnetic field: a vector quantity (**B**); can be used to determine the magnetic force on a moving charged particle

magnetic field lines: a pattern of continuous, imaginary lines that emerge from and enter into opposite magnetic poles. The density of the lines indicates the magnitude of the magnetic field

magnetic field strength: the magnitude of the magnetic field, denoted B-field

maximum field strength: the maximum amplitude an electromagnetic wave can reach, representing the maximum amount of electric force and/or magnetic flux that the wave can exert

Maxwell's equations: a set of four equations that comprise a complete, overarching theory of electromagnetism

microwaves: electromagnetic waves with wavelengths in the range from 1 mm to 1 m; they can be produced by currents in macroscopic circuits and devices

oscillate: to fluctuate back and forth in a steady beat

radar: a common application of microwaves. Radar can determine the distance to objects as diverse as clouds and aircraft, as well as determine the speed of a car or the intensity of a rainstorm

radio waves: electromagnetic waves with wavelengths in the range from 1 mm to 100 km; they are produced by currents in wires and circuits and by astronomical phenomena

resonant: a system that displays enhanced oscillation when subjected to a periodic disturbance of the same frequency as its natural frequency

RLC circuit: an electric circuit that includes a resistor, capacitor and inductor

speed of light: in a vacuum, such as space, the speed of light is a constant 3×10^8 m/s

standing wave: a wave that oscillates in place, with nodes where no motion happens

thermal agitation: the thermal motion of atoms and molecules in any object at a temperature above absolute zero, which causes them to emit and absorb radiation

transverse wave: a wave, such as an electromagnetic wave, which oscillates perpendicular to the axis along the line of travel

TV: video and audio signals broadcast on electromagnetic waves

ultra-high frequency (UHF): TV channels in an even higher frequency range than VHF, of 470 to 1000 MHz

ultraviolet radiation (UV): electromagnetic radiation in the range extending upward in frequency from violet light and overlapping with the lowest X-ray frequencies, with wavelengths from 400 nm down to about 10 nm

very high frequency (VHF): TV channels utilizing frequencies in the two ranges of 54 to 88 MHz and 174 to 222 MHz

visible light: the narrow segment of the electromagnetic spectrum to which the normal human eye responds

wavelength: the distance from one peak to the next in a wave

X-ray: invisible, penetrating form of very high frequency electromagnetic radiation, overlapping both the ultraviolet range and the γ-ray range

Section Summary

24.1 Maxwell's Equations: Electromagnetic Waves Predicted and Observed

- Electromagnetic waves consist of oscillating electric and magnetic fields and propagate at the speed of light c. They were predicted by Maxwell, who also showed that

$$c = \frac{1}{\sqrt{\mu_0 \varepsilon_0}},$$

where μ_0 is the permeability of free space and ε_0 is the permittivity of free space.

- Maxwell's prediction of electromagnetic waves resulted from his formulation of a complete and symmetric theory of electricity and magnetism, known as Maxwell's equations.
- These four equations are paraphrased in this text, rather than presented numerically, and encompass the major laws of electricity and magnetism. First is Gauss's law for electricity, second is Gauss's law for magnetism, third is Faraday's law of induction, including Lenz's law, and fourth is Ampere's law in a symmetric formulation that adds another source of magnetism—changing electric fields.

24.2 Production of Electromagnetic Waves

- Electromagnetic waves are created by oscillating charges (which radiate whenever accelerated) and have the same frequency as the oscillation.
- Since the electric and magnetic fields in most electromagnetic waves are perpendicular to the direction in which the wave moves, it is ordinarily a transverse wave.
- The strengths of the electric and magnetic parts of the wave are related by

$$\frac{E}{B} = c,$$

which implies that the magnetic field B is very weak relative to the electric field E.

24.3 The Electromagnetic Spectrum

- The relationship among the speed of propagation, wavelength, and frequency for any wave is given by $v_W = f\lambda$, so that for electromagnetic waves,

$$c = f\lambda,$$

where f is the frequency, λ is the wavelength, and c is the speed of light.

- The electromagnetic spectrum is separated into many categories and subcategories, based on the frequency and wavelength, source, and uses of the electromagnetic waves.
- Any electromagnetic wave produced by currents in wires is classified as a radio wave, the lowest frequency electromagnetic waves. Radio waves are divided into many types, depending on their applications, ranging up to microwaves at their highest frequencies.
- Infrared radiation lies below visible light in frequency and is produced by thermal motion and the vibration and rotation of atoms and molecules. Infrared's lower frequencies overlap with the highest-frequency microwaves.
- Visible light is largely produced by electronic transitions in atoms and molecules, and is defined as being detectable by the human eye. Its colors vary with frequency, from red at the lowest to violet at the highest.
- Ultraviolet radiation starts with frequencies just above violet in the visible range and is produced primarily by electronic transitions in atoms and molecules.
- X-rays are created in high-voltage discharges and by electron bombardment of metal targets. Their lowest frequencies overlap the ultraviolet range but extend to much higher values, overlapping at the high end with gamma rays.
- Gamma rays are nuclear in origin and are defined to include the highest-frequency electromagnetic radiation of any type.

24.4 Energy in Electromagnetic Waves

- The energy carried by any wave is proportional to its amplitude squared. For electromagnetic waves, this means intensity can be expressed as

$$I_{ave} = \frac{c\varepsilon_0 E_0^2}{2},$$

where I_{ave} is the average intensity in W/m^2, and E_0 is the maximum electric field strength of a continuous sinusoidal wave.

- This can also be expressed in terms of the maximum magnetic field strength B_0 as

$$I_{ave} = \frac{cB_0^2}{2\mu_0}$$

and in terms of both electric and magnetic fields as

$$I_{ave} = \frac{E_0 B_0}{2\mu_0}.$$

- The three expressions for I_{ave} are all equivalent.

Conceptual Questions

24.2 Production of Electromagnetic Waves

1. The direction of the electric field shown in each part of **Figure 24.5** is that produced by the charge distribution in the wire. Justify the direction shown in each part, using the Coulomb force law and the definition of $\mathbf{E} = \mathbf{F}/q$, where q is a positive test charge.

2. Is the direction of the magnetic field shown in **Figure 24.6** (a) consistent with the right-hand rule for current (RHR-2) in the direction shown in the figure?

3. Why is the direction of the current shown in each part of **Figure 24.6** opposite to the electric field produced by the wire's charge separation?

4. In which situation shown in **Figure 24.25** will the electromagnetic wave be more successful in inducing a current in the wire? Explain.

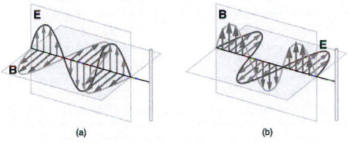

(a)   (b)

Figure 24.25 Electromagnetic waves approaching long straight wires.

5. In which situation shown in **Figure 24.26** will the electromagnetic wave be more successful in inducing a current in the loop? Explain.

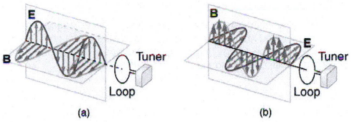

(a)   (b)

Figure 24.26 Electromagnetic waves approaching a wire loop.

6. Should the straight wire antenna of a radio be vertical or horizontal to best receive radio waves broadcast by a vertical transmitter antenna? How should a loop antenna be aligned to best receive the signals? (Note that the direction of the loop that produces the best reception can be used to determine the location of the source. It is used for that purpose in tracking tagged animals in nature studies, for example.)

7. Under what conditions might wires in a DC circuit emit electromagnetic waves?

8. Give an example of interference of electromagnetic waves.

9. Figure 24.27 shows the interference pattern of two radio antennas broadcasting the same signal. Explain how this is analogous to the interference pattern for sound produced by two speakers. Could this be used to make a directional antenna system that broadcasts preferentially in certain directions? Explain.

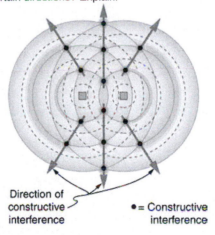

Direction of constructive interference

• = Constructive interference

Figure 24.27 An overhead view of two radio broadcast antennas sending the same signal, and the interference pattern they produce.

10. Can an antenna be any length? Explain your answer.

24.3 The Electromagnetic Spectrum

11. If you live in a region that has a particular TV station, you can sometimes pick up some of its audio portion on your FM radio receiver. Explain how this is possible. Does it imply that TV audio is broadcast as FM?

12. Explain why people who have the lens of their eye removed because of cataracts are able to see low-frequency ultraviolet.

13. How do fluorescent soap residues make clothing look "brighter and whiter" in outdoor light? Would this be effective in candlelight?

14. Give an example of resonance in the reception of electromagnetic waves.

15. Illustrate that the size of details of an object that can be detected with electromagnetic waves is related to their wavelength, by comparing details observable with two different types (for example, radar and visible light or infrared and X-rays).

16. Why don't buildings block radio waves as completely as they do visible light?

17. Make a list of some everyday objects and decide whether they are transparent or opaque to each of the types of electromagnetic waves.

18. Your friend says that more patterns and colors can be seen on the wings of birds if viewed in ultraviolet light. Would you agree with your friend? Explain your answer.

19. The rate at which information can be transmitted on an electromagnetic wave is proportional to the frequency of the wave. Is this consistent with the fact that laser telephone transmission at visible frequencies carries far more conversations per optical fiber than conventional electronic transmission in a wire? What is the implication for ELF radio communication with submarines?

20. Give an example of energy carried by an electromagnetic wave.

21. In an MRI scan, a higher magnetic field requires higher frequency radio waves to resonate with the nuclear type whose density and location is being imaged. What effect does going to a larger magnetic field have on the most efficient antenna to broadcast those radio waves? Does it favor a smaller or larger antenna?

22. Laser vision correction often uses an excimer laser that produces 193-nm electromagnetic radiation. This wavelength is extremely strongly absorbed by the cornea and ablates it in a manner that reshapes the cornea to correct vision defects. Explain how the strong absorption helps concentrate the energy in a thin layer and thus give greater accuracy in shaping the cornea. Also explain how this strong absorption limits damage to the lens and retina of the eye.

Problems & Exercises

24.1 Maxwell's Equations: Electromagnetic Waves Predicted and Observed

1. Verify that the correct value for the speed of light c is obtained when numerical values for the permeability and permittivity of free space (μ_0 and ε_0) are entered into the equation $c = \dfrac{1}{\sqrt{\mu_0 \varepsilon_0}}$.

2. Show that, when SI units for μ_0 and ε_0 are entered, the units given by the right-hand side of the equation in the problem above are m/s.

24.2 Production of Electromagnetic Waves

3. What is the maximum electric field strength in an electromagnetic wave that has a maximum magnetic field strength of 5.00×10^{-4} T (about 10 times the Earth's)?

4. The maximum magnetic field strength of an electromagnetic field is 5×10^{-6} T. Calculate the maximum electric field strength if the wave is traveling in a medium in which the speed of the wave is $0.75c$.

5. Verify the units obtained for magnetic field strength B in Example 24.1 (using the equation $B = \dfrac{E}{c}$) are in fact teslas (T).

24.3 The Electromagnetic Spectrum

6. (a) Two microwave frequencies are authorized for use in microwave ovens: 900 and 2560 MHz. Calculate the wavelength of each. (b) Which frequency would produce smaller hot spots in foods due to interference effects?

7. (a) Calculate the range of wavelengths for AM radio given its frequency range is 540 to 1600 kHz. (b) Do the same for the FM frequency range of 88.0 to 108 MHz.

8. A radio station utilizes frequencies between commercial AM and FM. What is the frequency of a 11.12-m-wavelength channel?

9. Find the frequency range of visible light, given that it encompasses wavelengths from 380 to 760 nm.

10. Combing your hair leads to excess electrons on the comb. How fast would you have to move the comb up and down to produce red light?

11. Electromagnetic radiation having a $15.0 - \mu$m wavelength is classified as infrared radiation. What is its frequency?

12. Approximately what is the smallest detail observable with a microscope that uses ultraviolet light of frequency 1.20×10^{15} Hz ?

13. A radar used to detect the presence of aircraft receives a pulse that has reflected off an object 6×10^{-5} s after it was transmitted. What is the distance from the radar station to the reflecting object?

14. Some radar systems detect the size and shape of objects such as aircraft and geological terrain. Approximately what is the smallest observable detail utilizing 500-MHz radar?

15. Determine the amount of time it takes for X-rays of frequency 3×10^{18} Hz to travel (a) 1 mm and (b) 1 cm.

16. If you wish to detect details of the size of atoms (about 1×10^{-10} m) with electromagnetic radiation, it must have a wavelength of about this size. (a) What is its frequency? (b) What type of electromagnetic radiation might this be?

17. If the Sun suddenly turned off, we would not know it until its light stopped coming. How long would that be, given that the Sun is 1.50×10^{11} m away?

18. Distances in space are often quoted in units of light years, the distance light travels in one year. (a) How many meters is a light year? (b) How many meters is it to Andromeda, the nearest large galaxy, given that it is 2.00×10^6 light years away? (c) The most distant galaxy yet discovered is 12.0×10^9 light years away. How far is this in meters?

19. A certain 50.0-Hz AC power line radiates an electromagnetic wave having a maximum electric field strength of 13.0 kV/m. (a) What is the wavelength of this very low frequency electromagnetic wave? (b) What is its maximum magnetic field strength?

20. During normal beating, the heart creates a maximum 4.00-mV potential across 0.300 m of a person's chest, creating a 1.00-Hz electromagnetic wave. (a) What is the maximum electric field strength created? (b) What is the corresponding maximum magnetic field strength in the electromagnetic wave? (c) What is the wavelength of the electromagnetic wave?

21. (a) The ideal size (most efficient) for a broadcast antenna with one end on the ground is one-fourth the wavelength ($\lambda/4$) of the electromagnetic radiation being sent out. If a new radio station has such an antenna that is 50.0 m high, what frequency does it broadcast most efficiently? Is this in the AM or FM band? (b) Discuss the analogy of the fundamental resonant mode of an air column closed at one end to the resonance of currents on an antenna that is one-fourth their wavelength.

22. (a) What is the wavelength of 100-MHz radio waves used in an MRI unit? (b) If the frequencies are swept over a ± 1.00 range centered on 100 MHz, what is the range of wavelengths broadcast?

23. (a) What is the frequency of the 193-nm ultraviolet radiation used in laser eye surgery? (b) Assuming the accuracy with which this EM radiation can ablate the cornea is directly proportional to wavelength, how much more accurate can this UV be than the shortest visible wavelength of light?

24. TV-reception antennas for VHF are constructed with cross wires supported at their centers, as shown in **Figure 24.28**. The ideal length for the cross wires is one-half the wavelength to be received, with the more expensive antennas having one for each channel. Suppose you measure the lengths of the wires for particular channels and find them to be 1.94 and 0.753 m long, respectively. What are the frequencies for these channels?

Figure 24.28 A television reception antenna has cross wires of various lengths to most efficiently receive different wavelengths.

25. Conversations with astronauts on lunar walks had an echo that was used to estimate the distance to the Moon. The sound spoken by the person on Earth was transformed into a radio signal sent to the Moon, and transformed back into sound on a speaker inside the astronaut's space suit. This sound was picked up by the microphone in the space suit (intended for the astronaut's voice) and sent back to Earth as a radio echo of sorts. If the round-trip time was 2.60 s, what was the approximate distance to the Moon, neglecting any delays in the electronics?

26. Lunar astronauts placed a reflector on the Moon's surface, off which a laser beam is periodically reflected. The distance to the Moon is calculated from the round-trip time. (a) To what accuracy in meters can the distance to the Moon be determined, if this time can be measured to 0.100 ns? (b) What percent accuracy is this, given the average distance to the Moon is 3.84×10^8 m ?

27. Radar is used to determine distances to various objects by measuring the round-trip time for an echo from the object. (a) How far away is the planet Venus if the echo time is 1000 s? (b) What is the echo time for a car 75.0 m from a Highway Police radar unit? (c) How accurately (in nanoseconds) must you be able to measure the echo time to an airplane 12.0 km away to determine its distance within 10.0 m?

28. Integrated Concepts

(a) Calculate the ratio of the highest to lowest frequencies of electromagnetic waves the eye can see, given the wavelength range of visible light is from 380 to 760 nm. (b) Compare this with the ratio of highest to lowest frequencies the ear can hear.

29. Integrated Concepts

(a) Calculate the rate in watts at which heat transfer through radiation occurs (almost entirely in the infrared) from 1.0 m^2 of the Earth's surface at night. Assume the emissivity is 0.90, the temperature of the Earth is $15°C$, and that of outer space is 2.7 K. (b) Compare the intensity of this radiation with that coming to the Earth from the Sun during the day, which averages about 800 W/m^2 , only half of which is absorbed. (c) What is the maximum magnetic field strength in the outgoing radiation, assuming it is a continuous wave?

24.4 Energy in Electromagnetic Waves

30. What is the intensity of an electromagnetic wave with a peak electric field strength of 125 V/m?

31. Find the intensity of an electromagnetic wave having a peak magnetic field strength of 4.00×10^{-9} T .

32. Assume the helium-neon lasers commonly used in student physics laboratories have power outputs of 0.250 mW. (a) If such a laser beam is projected onto a circular spot 1.00 mm in diameter, what is its intensity? (b) Find the peak magnetic field strength. (c) Find the peak electric field strength.

33. An AM radio transmitter broadcasts 50.0 kW of power uniformly in all directions. (a) Assuming all of the radio waves that strike the ground are completely absorbed, and that there is no absorption by the atmosphere or other objects, what is the intensity 30.0 km away? (Hint: Half the power will be spread over the area of a hemisphere.) (b) What is the maximum electric field strength at this distance?

34. Suppose the maximum safe intensity of microwaves for human exposure is taken to be 1.00 W/m^2 . (a) If a radar unit leaks 10.0 W of microwaves (other than those sent by its antenna) uniformly in all directions, how far away must you be to be exposed to an intensity considered to be safe? Assume that the power spreads uniformly over the area of a sphere with no complications from absorption or reflection. (b) What is the maximum electric field strength at the safe intensity? (Note that early radar units leaked more than modern ones do. This caused identifiable health problems, such as cataracts, for people who worked near them.)

35. A 2.50-m-diameter university communications satellite dish receives TV signals that have a maximum electric field strength (for one channel) of $7.50~\mu V/m$. (See **Figure 24.29**.) (a) What is the intensity of this wave? (b) What is the power received by the antenna? (c) If the orbiting satellite broadcasts uniformly over an area of $1.50\times10^{13}~m^2$ (a large fraction of North America), how much power does it radiate?

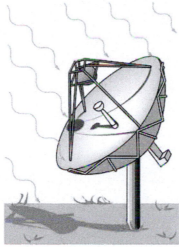

Figure 24.29 Satellite dishes receive TV signals sent from orbit. Although the signals are quite weak, the receiver can detect them by being tuned to resonate at their frequency.

36. Lasers can be constructed that produce an extremely high intensity electromagnetic wave for a brief time—called pulsed lasers. They are used to ignite nuclear fusion, for example. Such a laser may produce an electromagnetic wave with a maximum electric field strength of $1.00\times10^{11}~V/m$ for a time of 1.00 ns. (a) What is the maximum magnetic field strength in the wave? (b) What is the intensity of the beam? (c) What energy does it deliver on a 1.00-mm^2 area?

37. Show that for a continuous sinusoidal electromagnetic wave, the peak intensity is twice the average intensity ($I_0 = 2I_{ave}$), using either the fact that $E_0 = \sqrt{2}E_{rms}$, or $B_0 = \sqrt{2}B_{rms}$, where rms means average (actually root mean square, a type of average).

38. Suppose a source of electromagnetic waves radiates uniformly in all directions in empty space where there are no absorption or interference effects. (a) Show that the intensity is inversely proportional to r^2, the distance from the source squared. (b) Show that the magnitudes of the electric and magnetic fields are inversely proportional to r.

39. Integrated Concepts

An LC circuit with a 5.00-pF capacitor oscillates in such a manner as to radiate at a wavelength of 3.30 m. (a) What is the resonant frequency? (b) What inductance is in series with the capacitor?

40. Integrated Concepts

What capacitance is needed in series with an $800 - \mu H$ inductor to form a circuit that radiates a wavelength of 196 m?

41. Integrated Concepts

Police radar determines the speed of motor vehicles using the same Doppler-shift technique employed for ultrasound in medical diagnostics. Beats are produced by mixing the double Doppler-shifted echo with the original frequency. If 1.50×10^9-Hz microwaves are used and a beat frequency of 150 Hz is produced, what is the speed of the vehicle? (Assume the same Doppler-shift formulas are valid with the speed of sound replaced by the speed of light.)

42. Integrated Concepts

Assume the mostly infrared radiation from a heat lamp acts like a continuous wave with wavelength $1.50~\mu m$. (a) If the lamp's 200-W output is focused on a person's shoulder, over a circular area 25.0 cm in diameter, what is the intensity in W/m^2? (b) What is the peak electric field strength? (c) Find the peak magnetic field strength. (d) How long will it take to increase the temperature of the 4.00-kg shoulder by $2.00°~C$, assuming no other heat transfer and given that its specific heat is $3.47\times10^3~J/kg\cdot°C$?

43. Integrated Concepts

On its highest power setting, a microwave oven increases the temperature of 0.400 kg of spaghetti by $45.0°C$ in 120 s. (a) What was the rate of power absorption by the spaghetti, given that its specific heat is $3.76\times10^3~J/kg\cdot°C$? (b) Find the average intensity of the microwaves, given that they are absorbed over a circular area 20.0 cm in diameter. (c) What is the peak electric field strength of the microwave? (d) What is its peak magnetic field strength?

44. Integrated Concepts

Electromagnetic radiation from a 5.00-mW laser is concentrated on a 1.00-mm^2 area. (a) What is the intensity in W/m^2? (b) Suppose a 2.00-nC static charge is in the beam. What is the maximum electric force it experiences? (c) If the static charge moves at 400 m/s, what maximum magnetic force can it feel?

45. Integrated Concepts

A 200-turn flat coil of wire 30.0 cm in diameter acts as an antenna for FM radio at a frequency of 100 MHz. The magnetic field of the incoming electromagnetic wave is perpendicular to the coil and has a maximum strength of $1.00\times10^{-12}~T$. (a) What power is incident on the coil? (b) What average emf is induced in the coil over one-fourth of a cycle? (c) If the radio receiver has an inductance of $2.50~\mu H$, what capacitance must it have to resonate at 100 MHz?

46. Integrated Concepts

If electric and magnetic field strengths vary sinusoidally in time, being zero at $t = 0$, then $E = E_0 \sin 2\pi ft$ and $B = B_0 \sin 2\pi ft$. Let $f = 1.00~GHz$ here. (a) When are the field strengths first zero? (b) When do they reach their most negative value? (c) How much time is needed for them to complete one cycle?

47. Unreasonable Results

A researcher measures the wavelength of a 1.20-GHz electromagnetic wave to be 0.500 m. (a) Calculate the speed at which this wave propagates. (b) What is unreasonable about this result? (c) Which assumptions are unreasonable or inconsistent?

48. Unreasonable Results

The peak magnetic field strength in a residential microwave oven is 9.20×10^{-5} T . (a) What is the intensity of the microwave? (b) What is unreasonable about this result? (c) What is wrong about the premise?

49. Unreasonable Results

An LC circuit containing a 2.00-H inductor oscillates at such a frequency that it radiates at a 1.00-m wavelength. (a) What is the capacitance of the circuit? (b) What is unreasonable about this result? (c) Which assumptions are unreasonable or inconsistent?

50. Unreasonable Results

An LC circuit containing a 1.00-pF capacitor oscillates at such a frequency that it radiates at a 300-nm wavelength. (a) What is the inductance of the circuit? (b) What is unreasonable about this result? (c) Which assumptions are unreasonable or inconsistent?

51. Create Your Own Problem

Consider electromagnetic fields produced by high voltage power lines. Construct a problem in which you calculate the intensity of this electromagnetic radiation in W/m^2 based on the measured magnetic field strength of the radiation in a home near the power lines. Assume these magnetic field strengths are known to average less than a μT . The intensity is small enough that it is difficult to imagine mechanisms for biological damage due to it. Discuss how much energy may be radiating from a section of power line several hundred meters long and compare this to the power likely to be carried by the lines. An idea of how much power this is can be obtained by calculating the approximate current responsible for μT fields at distances of tens of meters.

52. Create Your Own Problem

Consider the most recent generation of residential satellite dishes that are a little less than half a meter in diameter. Construct a problem in which you calculate the power received by the dish and the maximum electric field strength of the microwave signals for a single channel received by the dish. Among the things to be considered are the power broadcast by the satellite and the area over which the power is spread, as well as the area of the receiving dish.

Test Prep for AP® Courses

24.2 Production of Electromagnetic Waves

1. If an electromagnetic wave is described as having a frequency of 3 GHz, what are its period and wavelength (in a vacuum)?
   a. 3.0×10^9s, 10 cm
   b. 3.3×10^{-10}s, 10 cm
   c. 3.3×10^{-10}s, 10 m
   d. 3.0×10^9s, 10 m

2. Describe the outcome if you attempt to produce a longitudinal electromagnetic wave.

3. A wave is travelling through a medium until it hits the end of the medium and there is nothing but vacuum beyond. What happens to a mechanical wave? Electromagnetic wave?
   a. reflects backward, continues on
   b. reflects backward, reflects backward
   c. continues on, continues on
   d. stops, continues on

4. You're on the moon, skipping around, and your radio breaks. What would be the best way to communicate this problem to your friend, who is also skipping around on the moon: yelling or flashing a light? Why?

5. Given the waveform in **Figure 24.5**(d), if $T = 3.0 \times 10^{-9}$ s and $E = 5.0 \times 10^5$ N/C, which of the following is the correct equation for the wave at the antenna?

a. $\left(5.0 \times 10^5 \text{ N/C}\right) \sin\left(2\pi\left(3.0 \times 10^{-9} \text{ s}\right)t\right)$

b. $\left(5.0 \times 10^5 \text{ N/C}\right) \sin\left(\dfrac{2\pi t}{\left(3.0 \times 10^{-9} \text{ s}\right)}\right)$

c. $\left(5.0 \times 10^5 \text{ N/C}\right) \cos\left(2\pi\left(3.0 \times 10^{-9} \text{ s}\right)t\right)$

d. $\left(5.0 \times 10^5 \text{ N/C}\right) \cos\left(\dfrac{2\pi t}{\left(3.0 \times 10^{-9} \text{ s}\right)}\right)$

6. Given the waveform in **Figure 24.5**(d), if $f = 2.0$ GHz and $E = 6.0 \times 10^5$ N/C, what is the correct equation for the magnetic field wave at the antenna?

7. In Heinrich Hertz's spark gap experiment (**Figure 24.4**), how will the induced sparks in Loop 2 compare to those created in Loop 1?
a. Stronger
b. Weaker
c. Need to know the tuner settings to tell
d. Weaker, but how much depends on the tuner settings.

8. The sun is far away from the Earth, and the intervening space is very close to empty. Yet the tilt of the Earth's axis of rotation relative to the sun results in seasons. Explain why, given what you have learned in this section.

24.3 The Electromagnetic Spectrum

9. The correct ordering from least to greatest wavelength is:
a. ELF, FM radio, microwaves, infrared, red, green, ultraviolet, x-ray, gamma ray
b. ELF, FM radio, microwaves, infrared, green, red, ultraviolet, x-ray, gamma ray
c. gamma ray, x-ray, ultraviolet, red, green, infrared, microwaves, FM radio, ELF
d. gamma ray, x-ray, ultraviolet, green, red, infrared, microwaves, FM radio, ELF

10. Describe how our vision would be different if we could see energy in what we define as the radio spectrum.

24.4 Energy in Electromagnetic Waves

11. An old microwave oven outputs only half the electric field it used to. How much longer does it take to cook things in this microwave oven?
a. Four times as long
b. Twice as long
c. Half the time
d. One fourth the time

12. Describe at least two improvements you could make to a radar set to make it more sensitive (able to detect things at longer ranges). Explain why these would work.

25 GEOMETRIC OPTICS

Figure 25.1 Image seen as a result of reflection of light on a plane smooth surface. (credit: NASA Goddard Photo and Video, via Flickr)

Chapter Outline

25.1. The Ray Aspect of Light

25.2. The Law of Reflection

25.3. The Law of Refraction

25.4. Total Internal Reflection

25.5. Dispersion: The Rainbow and Prisms

25.6. Image Formation by Lenses

25.7. Image Formation by Mirrors

Connection for AP® Courses

Many visual aspects of light result from the transfer of energy in the form of electromagnetic waves (Big Idea 6). Light from this page or screen is formed into an image by the lens of your eye, much like the lens of the camera that make a photograph. Mirrors, like lenses, can also form images that in turn are captured by your eye (Essential Knowledge 6.E.2, Essential Knowledge 6.E.4). In this chapter, you will explore the behavior of light as an electromagnetic wave and learn:

- what makes a diamond sparkle (Essential Knowledge 6.E.3),
- how images are formed by lenses for the purposes of magnification or photography (Essential Knowledge 6.E.5),
- why objects in some mirrors are closer than they appear (Essential Knowledge 6.E.2), and
- why clear mountain streams are always a little bit deeper than they appear to be.

You will examine different ways of thinking about and modeling light and when each method is most appropriate (Enduring Understanding 6.F, Essential Knowledge 6.F.4). You will also learn how to use simple geometry to predict how light will move when crossing from one medium to another, or when passing through a lens, or when reflecting off a curved surface (Enduring Understanding 6.E, Essential Knowledge 6.E.1). With this knowledge, you will be able to predict what kind of image will form when light interacts with matter.

Big Idea 6 Waves can transfer energy and momentum from one location to another without the permanent transfer of mass and serve as a mathematical model for the description of other phenomena.

Enduring Understanding 6.E The direction of propagation of a wave such as light may be changed when the wave encounters an interface between two media.

Essential Knowledge 6.E.1 When light travels from one medium to another, some of the light is transmitted, some is reflected, and some is absorbed. (Qualitative understanding only.)

Essential Knowledge 6.E.2 When light hits a smooth reflecting surface at an angle, it reflects at the same angle on the other side of the line perpendicular to the surface (specular reflection); and this law of reflection accounts for the size and location of images seen in plane mirrors.

Essential Knowledge 6.E.3 When light travels across a boundary from one transparent material to another, the speed of propagation changes. At a non–normal incident angle, the path of the light ray bends closer to the perpendicular in the optically slower substance. This is called refraction.

Essential Knowledge 6.E.4 The reflection of light from surfaces can be used to form images.

Essential Knowledge 6.E.5 The refraction of light as it travels from one transparent medium to another can be used to form images.

Enduring Understanding 6.F Electromagnetic radiation can be modeled as waves or as fundamental particles.

Essential Knowledge 6.F.4 The nature of light requires that different models of light are most appropriate at different scales.

25.1 The Ray Aspect of Light

Learning Objectives

By the end of this section, you will be able to:

- List the ways by which light travels from a source to another location.

The information presented in this section supports the following AP® learning objectives and science practices:

- **6.F.4.1** The student is able to select a model of radiant energy that is appropriate to the spatial or temporal scale of an interaction with matter. **(S.P. 6.4, 7.1)**

There are three ways in which light can travel from a source to another location. (See **Figure 25.2**.) It can come directly from the source through empty space, such as from the Sun to Earth. Or light can travel through various media, such as air and glass, to the person. Light can also arrive after being reflected, such as by a mirror. In all of these cases, light is modeled as traveling in straight lines called rays. Light may change direction when it encounters objects (such as a mirror) or in passing from one material to another (such as in passing from air to glass), but it then continues in a straight line or as a ray. The word **ray** comes from mathematics and here means a straight line that originates at some point. It is acceptable to visualize light rays as laser rays (or even science fiction depictions of ray guns).

Ray

The word "ray" comes from mathematics and here means a straight line that originates at some point.

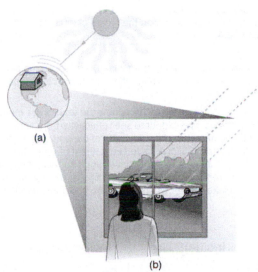

Figure 25.2 Three methods for light to travel from a source to another location. (a) Light reaches the upper atmosphere of Earth traveling through empty space directly from the source. (b) Light can reach a person in one of two ways. It can travel through media like air and glass. It can also reflect from an object like a mirror. In the situations shown here, light interacts with objects large enough that it travels in straight lines, like a ray.

Experiments, as well as our own experiences, show that when light interacts with objects several times as large as its wavelength, it travels in straight lines and acts like a ray. Its wave characteristics are not pronounced in such situations. Since the wavelength of light is less than a micron (a thousandth of a millimeter), it acts like a ray in the many common situations in which it encounters objects larger than a micron. For example, when light encounters anything we can observe with unaided eyes, such as a mirror, it acts like a ray, with only subtle wave characteristics. We will concentrate on the ray characteristics in this chapter.

Since light moves in straight lines, changing directions when it interacts with materials, it is described by geometry and simple trigonometry. This part of optics, where the ray aspect of light dominates, is therefore called **geometric optics**. There are two laws that govern how light changes direction when it interacts with matter. These are the law of reflection, for situations in which light bounces off matter, and the law of refraction, for situations in which light passes through matter.

Making Connections: Models of Light

There are three different ways of thinking about or modeling light. Our earliest understanding of light dates back at least to the ancient Greeks, who recorded their observations of the behavior of light as a ray. These philosophers noted that reflection, refraction, and formation of images can be explained by assuming objects emit and/or reflect light rays that travel in straight lines until they encounter an object or surface.

By the end of the 17th century, scientists came to understand that light also behaves like a wave. It exhibits phenomena associated with waves, such as diffraction and interference (which we will study in later chapters). Two hundred years later, scientists studying the smallest structures in nature showed that light can also be thought of as a stream of particles we call "photons," each carrying its own individual portion (or "quantum") of energy.

In this chapter, we will be discussing the behavior of light as it interacts with surfaces that are much larger than the wavelength of the light. In such cases, the light is very well modeled as a ray. When light interacts with smaller surfaces or openings (with sizes comparable to or smaller than the wavelength of light), the wavelike properties of light manifest more clearly—with profoundly interesting and useful results. When light interacts with individual atoms, the particle nature of light becomes more clearly apparent. We will study those situations in later chapters.

Geometric Optics

The part of optics dealing with the ray aspect of light is called geometric optics.

25.2 The Law of Reflection

Learning Objectives

By the end of this section, you will be able to:

- Explain reflection of light from polished and rough surfaces.

The information presented in this section supports the following AP® learning objectives and science practices:

- **6.E.1.1** The student is able to make claims using connections across concepts about the behavior of light as the wave travels from one medium into another, as some is transmitted, some is reflected, and some is absorbed. **(S.P. 6.4, 7.2)**
- **6.E.2.1** The student is able to make predictions about the locations of object and image relative to the location of a

> reflecting surface. The prediction should be based on the model of specular reflection with all angles measured relative to the normal to the surface. **(S.P. 6.4, 7.2)**

Whenever we look into a mirror, or squint at sunlight glinting from a lake, we are seeing a reflection. When you look at this page, too, you are seeing light reflected from it. Large telescopes use reflection to form an image of stars and other astronomical objects.

The law of reflection is illustrated in **Figure 25.3**, which also shows how the angles are measured relative to the perpendicular to the surface at the point where the light ray strikes. We expect to see reflections from smooth surfaces, but **Figure 25.4** illustrates how a rough surface reflects light. Since the light strikes different parts of the surface at different angles, it is reflected in many different directions, or diffused. Diffused light is what allows us to see a sheet of paper from any angle, as illustrated in **Figure 25.5**. Many objects, such as people, clothing, leaves, and walls, have rough surfaces and can be seen from all sides. A mirror, on the other hand, has a smooth surface (compared with the wavelength of light) and reflects light at specific angles, as illustrated in **Figure 25.6**. When the moon reflects from a lake, as shown in **Figure 25.7**, a combination of these effects takes place.

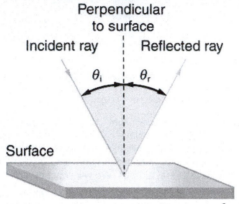

Figure 25.3 The law of reflection states that the angle of reflection equals the angle of incidence— $\theta_r = \theta_i$. The angles are measured relative to the perpendicular to the surface at the point where the ray strikes the surface.

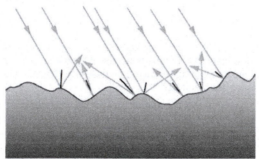

Figure 25.4 Light is diffused when it reflects from a rough surface. Here many parallel rays are incident, but they are reflected at many different angles since the surface is rough.

Figure 25.5 When a sheet of paper is illuminated with many parallel incident rays, it can be seen at many different angles, because its surface is rough and diffuses the light.

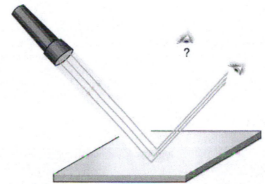

Figure 25.6 A mirror illuminated by many parallel rays reflects them in only one direction, since its surface is very smooth. Only the observer at a particular angle will see the reflected light.

Figure 25.7 Moonlight is spread out when it is reflected by the lake, since the surface is shiny but uneven. (credit: Diego Torres Silvestre, Flickr)

The law of reflection is very simple: The angle of reflection equals the angle of incidence.

The Law of Reflection

The angle of reflection equals the angle of incidence.

When we see ourselves in a mirror, it appears that our image is actually behind the mirror. This is illustrated in **Figure 25.8**. We see the light coming from a direction determined by the law of reflection. The angles are such that our image is exactly the same distance behind the mirror as we stand away from the mirror. If the mirror is on the wall of a room, the images in it are all behind the mirror, which can make the room seem bigger. Although these mirror images make objects appear to be where they cannot be (like behind a solid wall), the images are not figments of our imagination. Mirror images can be photographed and videotaped by instruments and look just as they do with our eyes (optical instruments themselves). The precise manner in which images are formed by mirrors and lenses will be treated in later sections of this chapter.

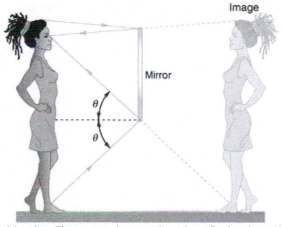

Figure 25.8 Our image in a mirror is behind the mirror. The two rays shown are those that strike the mirror at just the correct angles to be reflected into the eyes of the person. The image appears to be in the direction the rays are coming from when they enter the eyes.

Take-Home Experiment: Law of Reflection

Take a piece of paper and shine a flashlight at an angle at the paper, as shown in **Figure 25.5**. Now shine the flashlight at a mirror at an angle. Do your observations confirm the predictions in **Figure 25.5** and **Figure 25.6**? Shine the flashlight on various surfaces and determine whether the reflected light is diffuse or not. You can choose a shiny metallic lid of a pot or your skin. Using the mirror and flashlight, can you confirm the law of reflection? You will need to draw lines on a piece of paper showing the incident and reflected rays. (This part works even better if you use a laser pencil.)

25.3 The Law of Refraction

Learning Objectives

By the end of this section, you will be able to:

- Determine the index of refraction, given the speed of light in a medium.

The information presented in this section supports the following AP® learning objectives and science practices:

- **6.E.1.1** The student is able to make claims using connections across concepts about the behavior of light as the wave travels from one medium into another, as some is transmitted, some is reflected, and some is absorbed. **(S.P. 6.4, 7.2)**
- **6.E.3.1** The student is able to describe models of light traveling across a boundary from one transparent material to another when the speed of propagation changes, causing a change in the path of the light ray at the boundary of the two media. **(S.P. 1.1, 1.4)**
- **6.E.3.2** The student is able to plan data collection strategies as well as perform data analysis and evaluation of the evidence for finding the relationship between the angle of incidence and the angle of refraction for light crossing boundaries from one transparent material to another (Snell's law). **(S.P. 4.1, 5.1, 5.2, 5.3)**
- **6.E.3.3** The student is able to make claims and predictions about path changes for light traveling across a boundary from one transparent material to another at non-normal angles resulting from changes in the speed of propagation. **(S.P. 6.4, 7.2)**

It is easy to notice some odd things when looking into a fish tank. For example, you may see the same fish appearing to be in two different places. (See **Figure 25.9**.) This is because light coming from the fish to us changes direction when it leaves the tank, and in this case, it can travel two different paths to get to our eyes. The changing of a light ray's direction (loosely called bending) when it passes through variations in matter is called **refraction**. Refraction is responsible for a tremendous range of optical phenomena, from the action of lenses to voice transmission through optical fibers.

Refraction

The changing of a light ray's direction (loosely called bending) when it passes through variations in matter is called refraction.

Speed of Light

The speed of light c not only affects refraction, it is one of the central concepts of Einstein's theory of relativity. As the accuracy of the measurements of the speed of light were improved, c was found not to depend on the velocity of the source or the observer. However, the speed of light does vary in a precise manner with the material it traverses. These facts have far-reaching implications, as we will see in **Special Relativity**. It makes connections between space and time and alters our expectations that all observers measure the same time for the same event, for example. The speed of light is so important that its value in a vacuum is one of the most fundamental constants in nature as well as being one of the four fundamental SI units.

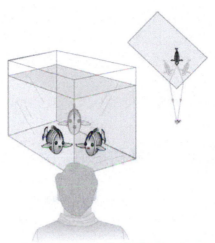

Figure 25.9 Looking at the fish tank as shown, we can see the same fish in two different locations, because light changes directions when it passes from water to air. In this case, the light can reach the observer by two different paths, and so the fish seems to be in two different places. This bending of light is called refraction and is responsible for many optical phenomena.

Why does light change direction when passing from one material (medium) to another? It is because light changes speed when going from one material to another. So before we study the law of refraction, it is useful to discuss the speed of light and how it varies in different media.

The Speed of Light

Early attempts to measure the speed of light, such as those made by Galileo, determined that light moved extremely fast, perhaps instantaneously. The first real evidence that light traveled at a finite speed came from the Danish astronomer Ole Roemer in the late 17th century. Roemer had noted that the average orbital period of one of Jupiter's moons, as measured from Earth, varied depending on whether Earth was moving toward or away from Jupiter. He correctly concluded that the apparent change in period was due to the change in distance between Earth and Jupiter and the time it took light to travel this distance.

From his 1676 data, a value of the speed of light was calculated to be 2.26×10^8 m/s (only 25% different from today's accepted value). In more recent times, physicists have measured the speed of light in numerous ways and with increasing accuracy. One particularly direct method, used in 1887 by the American physicist Albert Michelson (1852–1931), is illustrated in Figure 25.10. Light reflected from a rotating set of mirrors was reflected from a stationary mirror 35 km away and returned to the rotating mirrors. The time for the light to travel can be determined by how fast the mirrors must rotate for the light to be returned to the observer's eye.

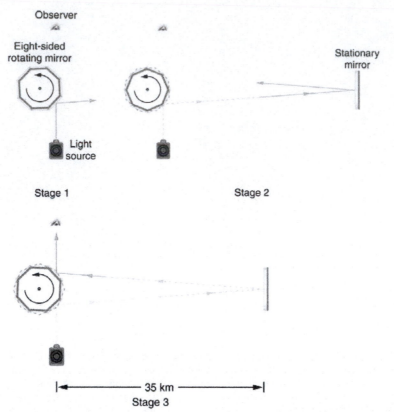

Figure 25.10 A schematic of early apparatus used by Michelson and others to determine the speed of light. As the mirrors rotate, the reflected ray is only briefly directed at the stationary mirror. The returning ray will be reflected into the observer's eye only if the next mirror has rotated into the correct position just as the ray returns. By measuring the correct rotation rate, the time for the round trip can be measured and the speed of light calculated. Michelson's calculated value of the speed of light was only 0.04% different from the value used today.

The speed of light is now known to great precision. In fact, the speed of light in a vacuum c is so important that it is accepted as one of the basic physical quantities and has the fixed value

$$c = 2.9972458 \times 10^8 \text{ m/s} \approx 3.00 \times 10^8 \text{ m/s,} \tag{25.1}$$

where the approximate value of 3.00×10^8 m/s is used whenever three-digit accuracy is sufficient. The speed of light through matter is less than it is in a vacuum, because light interacts with atoms in a material. The speed of light depends strongly on the type of material, since its interaction with different atoms, crystal lattices, and other substructures varies. We define the **index of refraction** n of a material to be

$$n = \frac{c}{v}, \tag{25.2}$$

where v is the observed speed of light in the material. Since the speed of light is always less than c in matter and equals c only in a vacuum, the index of refraction is always greater than or equal to one.

Value of the Speed of Light

$$c = 2.9972458 \times 10^8 \text{ m/s} \approx 3.00 \times 10^8 \text{ m/s} \tag{25.3}$$

Index of Refraction

$$n = \frac{c}{v} \tag{25.4}$$

That is, $n \geq 1$. Table 25.1 gives the indices of refraction for some representative substances. The values are listed for a particular wavelength of light, because they vary slightly with wavelength. (This can have important effects, such as colors produced by a prism.) Note that for gases, n is close to 1.0. This seems reasonable, since atoms in gases are widely separated and light travels at c in the vacuum between atoms. It is common to take $n = 1$ for gases unless great precision is needed. Although the speed of light v in a medium varies considerably from its value c in a vacuum, it is still a large speed.

Table 25.1 Index of Refraction in Various Media

Medium	n
Gases at $0°C$, 1 atm	
Air	1.000293
Carbon dioxide	1.00045
Hydrogen	1.000139
Oxygen	1.000271
Liquids at $20°C$	
Benzene	1.501
Carbon disulfide	1.628
Carbon tetrachloride	1.461
Ethanol	1.361
Glycerine	1.473
Water, fresh	1.333
Solids at $20°C$	
Diamond	2.419
Fluorite	1.434
Glass, crown	1.52
Glass, flint	1.66
Ice at $20°C$	1.309
Polystyrene	1.49
Plexiglas	1.51
Quartz, crystalline	1.544
Quartz, fused	1.458
Sodium chloride	1.544
Zircon	1.923

Example 25.1 Speed of Light in Matter

Calculate the speed of light in zircon, a material used in jewelry to imitate diamond.

Strategy

The speed of light in a material, v, can be calculated from the index of refraction n of the material using the equation $n = c/v$.

Solution

The equation for index of refraction states that $n = c/v$. Rearranging this to determine v gives

$$v = \frac{c}{n}. \tag{25.5}$$

The index of refraction for zircon is given as 1.923 in **Table 25.1**, and c is given in the equation for speed of light. Entering these values in the last expression gives

$$
\begin{aligned}
v &= \frac{3.00 \times 10^8 \text{ m/s}}{1.923} \\
&= 1.56 \times 10^8 \text{ m/s}.
\end{aligned}
\tag{25.6}
$$

Discussion

This speed is slightly larger than half the speed of light in a vacuum and is still high compared with speeds we normally experience. The only substance listed in **Table 25.1** that has a greater index of refraction than zircon is diamond. We shall see later that the large index of refraction for zircon makes it sparkle more than glass, but less than diamond.

Law of Refraction

Figure 25.11 shows how a ray of light changes direction when it passes from one medium to another. As before, the angles are measured relative to a perpendicular to the surface at the point where the light ray crosses it. (Some of the incident light will be reflected from the surface, but for now we will concentrate on the light that is transmitted.) The change in direction of the light ray depends on how the speed of light changes. The change in the speed of light is related to the indices of refraction of the media involved. In the situations shown in **Figure 25.11**, medium 2 has a greater index of refraction than medium 1. This means that the speed of light is less in medium 2 than in medium 1. Note that as shown in **Figure 25.11**(a), the direction of the ray moves closer to the perpendicular when it slows down. Conversely, as shown in **Figure 25.11**(b), the direction of the ray moves away from the perpendicular when it speeds up. The path is exactly reversible. In both cases, you can imagine what happens by thinking about pushing a lawn mower from a footpath onto grass, and vice versa. Going from the footpath to grass, the front wheels are slowed and pulled to the side as shown. This is the same change in direction as for light when it goes from a fast medium to a slow one. When going from the grass to the footpath, the front wheels can move faster and the mower changes direction as shown. This, too, is the same change in direction as for light going from slow to fast.

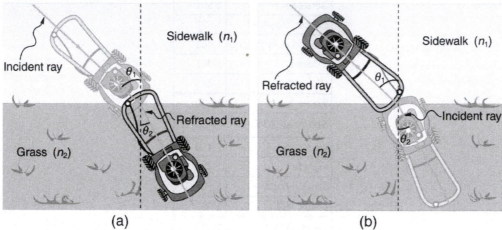

Figure 25.11 The change in direction of a light ray depends on how the speed of light changes when it crosses from one medium to another. The speed of light is greater in medium 1 than in medium 2 in the situations shown here. (a) A ray of light moves closer to the perpendicular when it slows down. This is analogous to what happens when a lawn mower goes from a footpath to grass. (b) A ray of light moves away from the perpendicular when it speeds up. This is analogous to what happens when a lawn mower goes from grass to footpath. The paths are exactly reversible.

The amount that a light ray changes its direction depends both on the incident angle and the amount that the speed changes. For a ray at a given incident angle, a large change in speed causes a large change in direction, and thus a large change in angle. The exact mathematical relationship is the **law of refraction**, or "Snell's Law," which is stated in equation form as

$$n_1 \sin \theta_1 = n_2 \sin \theta_2. \tag{25.7}$$

Here n_1 and n_2 are the indices of refraction for medium 1 and 2, and θ_1 and θ_2 are the angles between the rays and the perpendicular in medium 1 and 2, as shown in **Figure 25.11**. The incoming ray is called the incident ray and the outgoing ray the refracted ray, and the associated angles the incident angle and the refracted angle. The law of refraction is also called Snell's law after the Dutch mathematician Willebrord Snell (1591–1626), who discovered it in 1621. Snell's experiments showed that the law of refraction was obeyed and that a characteristic index of refraction n could be assigned to a given medium. Snell was not aware that the speed of light varied in different media, but through experiments he was able to determine indices of refraction from the way light rays changed direction.

The Law of Refraction

$$n_1 \sin \theta_1 = n_2 \sin \theta_2 \tag{25.8}$$

Take-Home Experiment: A Broken Pencil

A classic observation of refraction occurs when a pencil is placed in a glass half filled with water. Do this and observe the shape of the pencil when you look at the pencil sideways, that is, through air, glass, water. Explain your observations. Draw ray diagrams for the situation.

Example 25.2 Determine the Index of Refraction from Refraction Data

Find the index of refraction for medium 2 in **Figure 25.11**(a), assuming medium 1 is air and given the incident angle is $30.0°$ and the angle of refraction is $22.0°$.

Strategy

The index of refraction for air is taken to be 1 in most cases (and up to four significant figures, it is 1.000). Thus $n_1 = 1.00$ here. From the given information, $\theta_1 = 30.0°$ and $\theta_2 = 22.0°$. With this information, the only unknown in Snell's law is n_2, so that it can be used to find this unknown.

Solution

Snell's law is

$$n_1 \sin \theta_1 = n_2 \sin \theta_2. \qquad (25.9)$$

Rearranging to isolate n_2 gives

$$n_2 = n_1 \frac{\sin \theta_1}{\sin \theta_2}. \qquad (25.10)$$

Entering known values,

$$
\begin{aligned}
n_2 &= 1.00 \frac{\sin 30.0°}{\sin 22.0°} = \frac{0.500}{0.375} \qquad (25.11) \\
&= 1.33.
\end{aligned}
$$

Discussion

This is the index of refraction for water, and Snell could have determined it by measuring the angles and performing this calculation. He would then have found 1.33 to be the appropriate index of refraction for water in all other situations, such as when a ray passes from water to glass. Today we can verify that the index of refraction is related to the speed of light in a medium by measuring that speed directly.

Example 25.3 A Larger Change in Direction

Suppose that in a situation like that in **Example 25.2**, light goes from air to diamond and that the incident angle is $30.0°$. Calculate the angle of refraction θ_2 in the diamond.

Strategy

Again the index of refraction for air is taken to be $n_1 = 1.00$, and we are given $\theta_1 = 30.0°$. We can look up the index of refraction for diamond in **Table 25.1**, finding $n_2 = 2.419$. The only unknown in Snell's law is θ_2, which we wish to determine.

Solution

Solving Snell's law for $\sin \theta_2$ yields

$$\sin \theta_2 = \frac{n_1}{n_2} \sin \theta_1. \qquad (25.12)$$

Entering known values,

$$\sin \theta_2 = \frac{1.00}{2.419} \sin 30.0° = (0.413)(0.500) = 0.207. \qquad (25.13)$$

The angle is thus

$$\theta_2 = \sin^{-1} 0.207 = 11.9°. \qquad (25.14)$$

Discussion

For the same $30°$ angle of incidence, the angle of refraction in diamond is significantly smaller than in water ($11.9°$ rather than $22°$ —see the preceding example). This means there is a larger change in direction in diamond. The cause of a large change in direction is a large change in the index of refraction (or speed). In general, the larger the change in speed, the greater the effect on the direction of the ray.

25.4 Total Internal Reflection

Learning Objectives

By the end of this section, you will be able to:

- Explain the phenomenon of total internal reflection.
- Describe the workings and uses of fiber optics.
- Analyze the reason for the sparkle of diamonds.

The information presented in this section supports the following AP® learning objectives and science practices:

- **6.E.1.1** The student is able to make claims using connections across concepts about the behavior of light as the wave travels from one medium into another, as some is transmitted, some is reflected, and some is absorbed. **(S.P. 6.4, 7.2)**

A good-quality mirror may reflect more than 90% of the light that falls on it, absorbing the rest. But it would be useful to have a mirror that reflects all of the light that falls on it. Interestingly, we can produce *total reflection* using an aspect of *refraction*.

Consider what happens when a ray of light strikes the surface between two materials, such as is shown in **Figure 25.12**(a). Part of the light crosses the boundary and is refracted; the rest is reflected. If, as shown in the figure, the index of refraction for the second medium is less than for the first, the ray bends away from the perpendicular. (Since $n_1 > n_2$, the angle of refraction is greater than the angle of incidence—that is, $\theta_2 > \theta_1$.) Now imagine what happens as the incident angle is increased. This causes θ_2 to increase also. The largest the angle of refraction θ_2 can be is 90°, as shown in **Figure 25.12**(b).The **critical angle** θ_c for a combination of materials is defined to be the incident angle θ_1 that produces an angle of refraction of 90°. That is, θ_c is the incident angle for which $\theta_2 = 90^\circ$. If the incident angle θ_1 is greater than the critical angle, as shown in **Figure 25.12**(c), then all of the light is reflected back into medium 1, a condition called **total internal reflection**.

Critical Angle

The incident angle θ_1 that produces an angle of refraction of 90° is called the critical angle, θ_c.

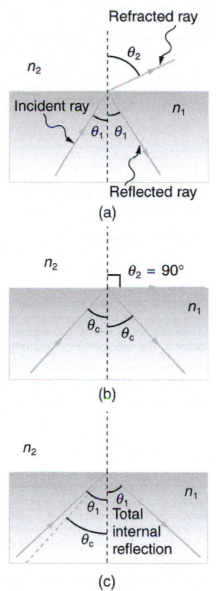

Figure 25.12 (a) A ray of light crosses a boundary where the speed of light increases and the index of refraction decreases. That is, $n_2 < n_1$. The ray bends away from the perpendicular. (b) The critical angle θ_c is the one for which the angle of refraction is . (c) Total internal reflection occurs when the incident angle is greater than the critical angle.

Snell's law states the relationship between angles and indices of refraction. It is given by

$$n_1 \sin \theta_1 = n_2 \sin \theta_2. \tag{25.15}$$

When the incident angle equals the critical angle ($\theta_1 = \theta_c$), the angle of refraction is 90° ($\theta_2 = 90^\circ$). Noting that $\sin 90^\circ = 1$, Snell's law in this case becomes

$$n_1 \sin \theta_1 = n_2. \tag{25.16}$$

The critical angle θ_c for a given combination of materials is thus

$$\theta_c = \sin^{-1}(n_2/n_1) \text{ for } n_1 > n_2. \tag{25.17}$$

Total internal reflection occurs for any incident angle greater than the critical angle θ_c, and it can only occur when the second medium has an index of refraction less than the first. Note the above equation is written for a light ray that travels in medium 1 and reflects from medium 2, as shown in the figure.

Example 25.4 How Big is the Critical Angle Here?

What is the critical angle for light traveling in a polystyrene (a type of plastic) pipe surrounded by air?

Strategy

The index of refraction for polystyrene is found to be 1.49 in **Figure 25.13**, and the index of refraction of air can be taken to be 1.00, as before. Thus, the condition that the second medium (air) has an index of refraction less than the first (plastic) is satisfied, and the equation $\theta_c = \sin^{-1}(n_2/n_1)$ can be used to find the critical angle θ_c. Here, then, $n_2 = 1.00$ and $n_1 = 1.49$.

Solution

The critical angle is given by

$$\theta_c = \sin^{-1}(n_2/n_1). \tag{25.18}$$

Substituting the identified values gives

$$\theta_c = \sin^{-1}(1.00/1.49) = \sin^{-1}(0.671) \tag{25.19}$$
$$42.2°.$$

Discussion

This means that any ray of light inside the plastic that strikes the surface at an angle greater than $42.2°$ will be totally reflected. This will make the inside surface of the clear plastic a perfect mirror for such rays without any need for the silvering used on common mirrors. Different combinations of materials have different critical angles, but any combination with $n_1 > n_2$ can produce total internal reflection. The same calculation as made here shows that the critical angle for a ray going from water to air is $48.6°$, while that from diamond to air is $24.4°$, and that from flint glass to crown glass is $66.3°$. There is no total reflection for rays going in the other direction—for example, from air to water—since the condition that the second medium must have a smaller index of refraction is not satisfied. A number of interesting applications of total internal reflection follow.

Fiber Optics: Endoscopes to Telephones

Fiber optics is one application of total internal reflection that is in wide use. In communications, it is used to transmit telephone, internet, and cable TV signals. **Fiber optics** employs the transmission of light down fibers of plastic or glass. Because the fibers are thin, light entering one is likely to strike the inside surface at an angle greater than the critical angle and, thus, be totally reflected (See **Figure 25.13**.) The index of refraction outside the fiber must be smaller than inside, a condition that is easily satisfied by coating the outside of the fiber with a material having an appropriate refractive index. In fact, most fibers have a varying refractive index to allow more light to be guided along the fiber through total internal refraction. Rays are reflected around corners as shown, making the fibers into tiny light pipes.

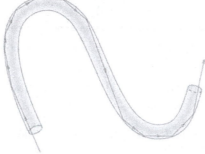

Figure 25.13 Light entering a thin fiber may strike the inside surface at large or grazing angles and is completely reflected if these angles exceed the critical angle. Such rays continue down the fiber, even following it around corners, since the angles of reflection and incidence remain large.

Bundles of fibers can be used to transmit an image without a lens, as illustrated in **Figure 25.14**. The output of a device called an **endoscope** is shown in **Figure 25.14**(b). Endoscopes are used to explore the body through various orifices or minor incisions. Light is transmitted down one fiber bundle to illuminate internal parts, and the reflected light is transmitted back out through another to be observed. Surgery can be performed, such as arthroscopic surgery on the knee joint, employing cutting tools attached to and observed with the endoscope. Samples can also be obtained, such as by lassoing an intestinal polyp for external examination.

Fiber optics has revolutionized surgical techniques and observations within the body. There are a host of medical diagnostic and therapeutic uses. The flexibility of the fiber optic bundle allows it to navigate around difficult and small regions in the body, such as the intestines, the heart, blood vessels, and joints. Transmission of an intense laser beam to burn away obstructing plaques in major arteries as well as delivering light to activate chemotherapy drugs are becoming commonplace. Optical fibers have in fact enabled microsurgery and remote surgery where the incisions are small and the surgeon's fingers do not need to touch the diseased tissue.

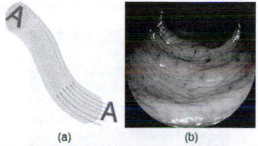

(a) (b)

Figure 25.14 (a) An image is transmitted by a bundle of fibers that have fixed neighbors. (b) An endoscope is used to probe the body, both transmitting light to the interior and returning an image such as the one shown. (credit: Med_Chaos, Wikimedia Commons)

Fibers in bundles are surrounded by a cladding material that has a lower index of refraction than the core. (See **Figure 25.15.**) The cladding prevents light from being transmitted between fibers in a bundle. Without cladding, light could pass between fibers in contact, since their indices of refraction are identical. Since no light gets into the cladding (there is total internal reflection back into the core), none can be transmitted between clad fibers that are in contact with one another. The cladding prevents light from escaping out of the fiber; instead most of the light is propagated along the length of the fiber, minimizing the loss of signal and ensuring that a quality image is formed at the other end. The cladding and an additional protective layer make optical fibers flexible and durable.

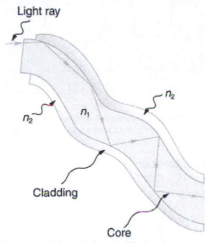

Figure 25.15 Fibers in bundles are clad by a material that has a lower index of refraction than the core to ensure total internal reflection, even when fibers are in contact with one another. This shows a single fiber with its cladding.

Cladding

The cladding prevents light from being transmitted between fibers in a bundle.

Special tiny lenses that can be attached to the ends of bundles of fibers are being designed and fabricated. Light emerging from a fiber bundle can be focused and a tiny spot can be imaged. In some cases the spot can be scanned, allowing quality imaging of a region inside the body. Special minute optical filters inserted at the end of the fiber bundle have the capacity to image tens of microns below the surface without cutting the surface—non-intrusive diagnostics. This is particularly useful for determining the extent of cancers in the stomach and bowel.

Most telephone conversations and Internet communications are now carried by laser signals along optical fibers. Extensive optical fiber cables have been placed on the ocean floor and underground to enable optical communications. Optical fiber communication systems offer several advantages over electrical (copper) based systems, particularly for long distances. The fibers can be made so transparent that light can travel many kilometers before it becomes dim enough to require amplification—much superior to copper conductors. This property of optical fibers is called *low loss*. Lasers emit light with characteristics that allow far more conversations in one fiber than are possible with electric signals on a single conductor. This property of optical fibers is called *high bandwidth*. Optical signals in one fiber do not produce undesirable effects in other adjacent fibers. This property of optical fibers is called *reduced crosstalk*. We shall explore the unique characteristics of laser radiation in a later chapter.

Corner Reflectors and Diamonds

A light ray that strikes an object consisting of two mutually perpendicular reflecting surfaces is reflected back exactly parallel to the direction from which it came. This is true whenever the reflecting surfaces are perpendicular, and it is independent of the angle of incidence. Such an object, shown in **m55438 (http://cnx.org/content/m55438/1.2/#Figure 25.16)** , is called a **corner reflector**, since the light bounces from its inside corner. Many inexpensive reflector buttons on bicycles, cars, and warning signs have corner reflectors designed to return light in the direction from which it originated. It was more expensive for astronauts to place one on the moon. Laser signals can be bounced from that corner reflector to measure the gradually increasing distance to the moon with great precision.

(a)

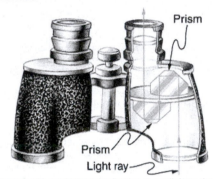

(b)

Figure 25.16 (a) Astronauts placed a corner reflector on the moon to measure its gradually increasing orbital distance. (credit: NASA) (b) The bright spots on these bicycle safety reflectors are reflections of the flash of the camera that took this picture on a dark night. (credit: Julo, Wikimedia Commons)

Corner reflectors are perfectly efficient when the conditions for total internal reflection are satisfied. With common materials, it is easy to obtain a critical angle that is less than $45°$. One use of these perfect mirrors is in binoculars, as shown in **Figure 25.17**. Another use is in periscopes found in submarines.

Figure 25.17 These binoculars employ corner reflectors with total internal reflection to get light to the observer's eyes.

The Sparkle of Diamonds

Total internal reflection, coupled with a large index of refraction, explains why diamonds sparkle more than other materials. The critical angle for a diamond-to-air surface is only $24.4°$, and so when light enters a diamond, it has trouble getting back out. (See **Figure 25.18**.) Although light freely enters the diamond, it can exit only if it makes an angle less than $24.4°$. Facets on diamonds are specifically intended to make this unlikely, so that the light can exit only in certain places. Good diamonds are very clear, so that the light makes many internal reflections and is concentrated at the few places it can exit—hence the sparkle. (Zircon is a natural gemstone that has an exceptionally large index of refraction, but not as large as diamond, so it is not as highly prized. Cubic zirconia is manufactured and has an even higher index of refraction (≈ 2.17), but still less than that of diamond.) The colors you see emerging from a sparkling diamond are not due to the diamond's color, which is usually nearly colorless. Those colors result from dispersion, the topic of **Dispersion: The Rainbow and Prisms**. Colored diamonds get their color from structural defects of the crystal lattice and the inclusion of minute quantities of graphite and other materials. The Argyle Mine in Western Australia produces around 90% of the world's pink, red, champagne, and cognac diamonds, while around 50% of the world's clear diamonds come from central and southern Africa.

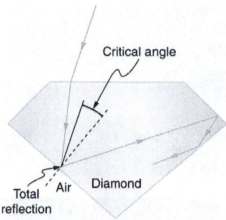

Figure 25.18 Light cannot easily escape a diamond, because its critical angle with air is so small. Most reflections are total, and the facets are placed so that light can exit only in particular ways—thus concentrating the light and making the diamond sparkle.

PhET Explorations: Bending Light

Explore bending of light between two media with different indices of refraction. See how changing from air to water to glass changes the bending angle. Play with prisms of different shapes and make rainbows.

Figure 25.19 Bending Light (http://cnx.org/content/m55440/1.4/bending-light_en.jar)

25.5 Dispersion: The Rainbow and Prisms

Learning Objectives
By the end of this section, you will be able to: • Explain the phenomenon of dispersion and discuss its advantages and disadvantages.

Everyone enjoys the spectacle of a rainbow glimmering against a dark stormy sky. How does sunlight falling on clear drops of rain get broken into the rainbow of colors we see? The same process causes white light to be broken into colors by a clear glass prism or a diamond. (See **Figure 25.20.**)

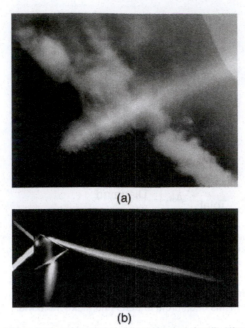

(a)

(b)

Figure 25.20 The colors of the rainbow (a) and those produced by a prism (b) are identical. (credit: Alfredo55, Wikimedia Commons; NASA)

We see about six colors in a rainbow—red, orange, yellow, green, blue, and violet; sometimes indigo is listed, too. Those colors are associated with different wavelengths of light, as shown in Figure 25.21. When our eye receives pure-wavelength light, we tend to see only one of the six colors, depending on wavelength. The thousands of other hues we can sense in other situations are our eye's response to various mixtures of wavelengths. White light, in particular, is a fairly uniform mixture of all visible wavelengths. Sunlight, considered to be white, actually appears to be a bit yellow because of its mixture of wavelengths, but it does contain all visible wavelengths. The sequence of colors in rainbows is the same sequence as the colors plotted versus wavelength in Figure 25.21. What this implies is that white light is spread out according to wavelength in a rainbow. **Dispersion** is defined as the spreading of white light into its full spectrum of wavelengths. More technically, dispersion occurs whenever there is a process that changes the direction of light in a manner that depends on wavelength. Dispersion, as a general phenomenon, can occur for any type of wave and always involves wavelength-dependent processes.

Dispersion

Dispersion is defined to be the spreading of white light into its full spectrum of wavelengths.

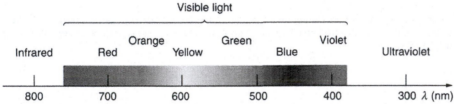

Figure 25.21 Even though rainbows are associated with seven colors, the rainbow is a continuous distribution of colors according to wavelengths.

Refraction is responsible for dispersion in rainbows and many other situations. The angle of refraction depends on the index of refraction, as we saw in The Law of Refraction. We know that the index of refraction n depends on the medium. But for a given medium, n also depends on wavelength. (See Table 25.2. Note that, for a given medium, n increases as wavelength decreases and is greatest for violet light. Thus violet light is bent more than red light, as shown for a prism in Figure 25.22(b), and the light is dispersed into the same sequence of wavelengths as seen in Figure 25.20 and Figure 25.21.

Making Connections: Dispersion

Any type of wave can exhibit dispersion. Sound waves, all types of electromagnetic waves, and water waves can be dispersed according to wavelength. Dispersion occurs whenever the speed of propagation depends on wavelength, thus separating and spreading out various wavelengths. Dispersion may require special circumstances and can result in spectacular displays such as in the production of a rainbow. This is also true for sound, since all frequencies ordinarily travel at the same speed. If you listen to sound through a long tube, such as a vacuum cleaner hose, you can easily hear it is dispersed by interaction with the tube. Dispersion, in fact, can reveal a great deal about what the wave has encountered that disperses its wavelengths. The dispersion of electromagnetic radiation from outer space, for example, has revealed much about what exists between the stars—the so-called empty space.

Table 25.2 Index of Refraction n in Selected Media at Various Wavelengths

Medium	Red (660 nm)	Orange (610 nm)	Yellow (580 nm)	Green (550 nm)	Blue (470 nm)	Violet (410 nm)
Water	1.331	1.332	1.333	1.335	1.338	1.342
Diamond	2.410	2.415	2.417	2.426	2.444	2.458
Glass, crown	1.512	1.514	1.518	1.519	1.524	1.530
Glass, flint	1.662	1.665	1.667	1.674	1.684	1.698
Polystyrene	1.488	1.490	1.492	1.493	1.499	1.506
Quartz, fused	1.455	1.456	1.458	1.459	1.462	1.468

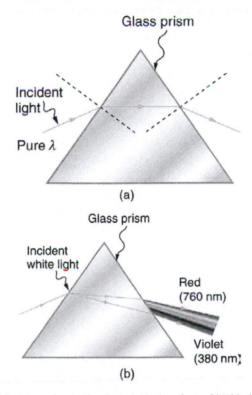

Figure 25.22 (a) A pure wavelength of light falls onto a prism and is refracted at both surfaces. (b) White light is dispersed by the prism (shown exaggerated). Since the index of refraction varies with wavelength, the angles of refraction vary with wavelength. A sequence of red to violet is produced, because the index of refraction increases steadily with decreasing wavelength.

Rainbows are produced by a combination of refraction and reflection. You may have noticed that you see a rainbow only when you look away from the sun. Light enters a drop of water and is reflected from the back of the drop, as shown in **Figure 25.23**. The light is refracted both as it enters and as it leaves the drop. Since the index of refraction of water varies with wavelength, the light is dispersed, and a rainbow is observed, as shown in **Figure 25.24** (a). (There is no dispersion caused by reflection at the back surface, since the law of reflection does not depend on wavelength.) The actual rainbow of colors seen by an observer depends on the myriad of rays being refracted and reflected toward the observer's eyes from numerous drops of water. The effect is most spectacular when the background is dark, as in stormy weather, but can also be observed in waterfalls and lawn sprinklers. The arc of a rainbow comes from the need to be looking at a specific angle relative to the direction of the sun, as illustrated in **Figure 25.24** (b). (If there are two reflections of light within the water drop, another "secondary" rainbow is produced. This rare event produces an arc that lies above the primary rainbow arc—see **Figure 25.24** (c).)

Rainbows

Rainbows are produced by a combination of refraction and reflection.

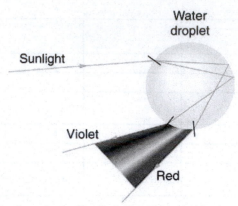

Figure 25.23 Part of the light falling on this water drop enters and is reflected from the back of the drop. This light is refracted and dispersed both as it enters and as it leaves the drop.

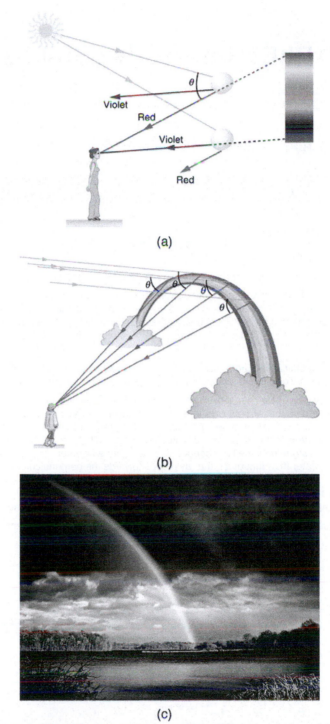

Violet

Red

Violet

Red

(a)

(b)

(c)

Figure 25.24 (a) Different colors emerge in different directions, and so you must look at different locations to see the various colors of a rainbow. (b) The arc of a rainbow results from the fact that a line between the observer and any point on the arc must make the correct angle with the parallel rays of sunlight to receive the refracted rays. (c) Double rainbow. (credit: Nicholas, Wikimedia Commons)

Dispersion may produce beautiful rainbows, but it can cause problems in optical systems. White light used to transmit messages in a fiber is dispersed, spreading out in time and eventually overlapping with other messages. Since a laser produces a nearly pure wavelength, its light experiences little dispersion, an advantage over white light for transmission of information. In contrast, dispersion of electromagnetic waves coming to us from outer space can be used to determine the amount of matter they pass through. As with many phenomena, dispersion can be useful or a nuisance, depending on the situation and our human goals.

PhET Explorations: Geometric Optics

How does a lens form an image? See how light rays are refracted by a lens. Watch how the image changes when you adjust the focal length of the lens, move the object, move the lens, or move the screen.

25.6 Image Formation by Lenses

Learning Objectives

By the end of this section, you will be able to:

- List the rules for ray tracking for thin lenses.
- Illustrate the formation of images using the technique of ray tracing.
- Determine power of a lens given the focal length.

The information presented in this section supports the following AP® learning objectives and science practices:

- **6.E.5.1** The student is able to use quantitative and qualitative representations and models to analyze situations and solve problems about image formation occurring due to the refraction of light through thin lenses. **(S.P. 1.4, 2.2)**
- **6.E.5.2** The student is able to plan data collection strategies, perform data analysis and evaluation of evidence, and refine scientific questions about the formation of images due to refraction for thin lenses. **(S.P. 3.2, 4.1, 5.1, 5.2, 5.3)**

Lenses are found in a huge array of optical instruments, ranging from a simple magnifying glass to the eye to a camera's zoom lens. In this section, we will use the law of refraction to explore the properties of lenses and how they form images.

The word *lens* derives from the Latin word for a lentil bean, the shape of which is similar to the convex lens in **Figure 25.26**. The convex lens shown has been shaped so that all light rays that enter it parallel to its axis cross one another at a single point on the opposite side of the lens. (The axis is defined to be a line normal to the lens at its center, as shown in **Figure 25.26**.) Such a lens is called a **converging (or convex) lens** for the converging effect it has on light rays. An expanded view of the path of one ray through the lens is shown, to illustrate how the ray changes direction both as it enters and as it leaves the lens. Since the index of refraction of the lens is greater than that of air, the ray moves towards the perpendicular as it enters and away from the perpendicular as it leaves. (This is in accordance with the law of refraction.) Due to the lens's shape, light is thus bent toward the axis at both surfaces. The point at which the rays cross is defined to be the **focal point** F of the lens. The distance from the center of the lens to its focal point is defined to be the **focal length** f of the lens. **Figure 25.27** shows how a converging lens, such as that in a magnifying glass, can converge the nearly parallel light rays from the sun to a small spot.

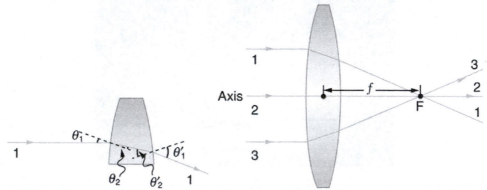

Figure 25.26 Rays of light entering a converging lens parallel to its axis converge at its focal point F. (Ray 2 lies on the axis of the lens.) The distance from the center of the lens to the focal point is the lens's focal length f. An expanded view of the path taken by ray 1 shows the perpendiculars and the angles of incidence and refraction at both surfaces.

Converging or Convex Lens

The lens in which light rays that enter it parallel to its axis cross one another at a single point on the opposite side with a converging effect is called converging lens.

Focal Point F

The point at which the light rays cross is called the focal point F of the lens.

Focal Length f

The distance from the center of the lens to its focal point is called focal length f.

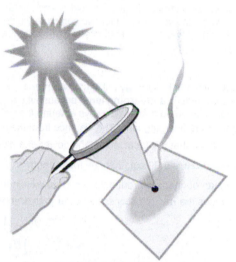

Figure 25.27 Sunlight focused by a converging magnifying glass can burn paper. Light rays from the sun are nearly parallel and cross at the focal point of the lens. The more powerful the lens, the closer to the lens the rays will cross.

The greater effect a lens has on light rays, the more powerful it is said to be. For example, a powerful converging lens will focus parallel light rays closer to itself and will have a smaller focal length than a weak lens. The light will also focus into a smaller and more intense spot for a more powerful lens. The **power** P of a lens is defined to be the inverse of its focal length. In equation form, this is

$$P = \frac{1}{f}. \tag{25.20}$$

Power P

The **power** P of a lens is defined to be the inverse of its focal length. In equation form, this is

$$P = \frac{1}{f}. \tag{25.21}$$

where f is the focal length of the lens, which must be given in meters (and not cm or mm). The power of a lens P has the unit diopters (D), provided that the focal length is given in meters. That is, $1\ \mathrm{D} = 1\,/\,\mathrm{m}$, or $1\ \mathrm{m}^{-1}$. (Note that this power (optical power, actually) is not the same as power in watts defined in **Work, Energy, and Energy Resources**. It is a concept related to the effect of optical devices on light.) Optometrists prescribe common spectacles and contact lenses in units of diopters.

Example 25.5 What is the Power of a Common Magnifying Glass?

Suppose you take a magnifying glass out on a sunny day and you find that it concentrates sunlight to a small spot 8.00 cm away from the lens. What are the focal length and power of the lens?

Strategy

The situation here is the same as those shown in **Figure 25.26** and **Figure 25.27**. The Sun is so far away that the Sun's rays are nearly parallel when they reach Earth. The magnifying glass is a convex (or converging) lens, focusing the nearly parallel rays of sunlight. Thus the focal length of the lens is the distance from the lens to the spot, and its power is the inverse of this distance (in m).

Solution

The focal length of the lens is the distance from the center of the lens to the spot, given to be 8.00 cm. Thus,

$$f = 8.00\ \mathrm{cm}. \tag{25.22}$$

To find the power of the lens, we must first convert the focal length to meters; then, we substitute this value into the equation for power. This gives

$$P = \frac{1}{f} = \frac{1}{0.0800 \text{ m}} = 12.5 \text{ D}. \qquad\qquad (25.23)$$

Discussion

This is a relatively powerful lens. The power of a lens in diopters should not be confused with the familiar concept of power in watts. It is an unfortunate fact that the word "power" is used for two completely different concepts. If you examine a prescription for eyeglasses, you will note lens powers given in diopters. If you examine the label on a motor, you will note energy consumption rate given as a power in watts.

Figure 25.28 shows a concave lens and the effect it has on rays of light that enter it parallel to its axis (the path taken by ray 2 in the figure is the axis of the lens). The concave lens is a **diverging lens**, because it causes the light rays to bend away (diverge) from its axis. In this case, the lens has been shaped so that all light rays entering it parallel to its axis appear to originate from the same point, F, defined to be the focal point of a diverging lens. The distance from the center of the lens to the focal point is again called the focal length f of the lens. Note that the focal length and power of a diverging lens are defined to be negative.

For example, if the distance to F in **Figure 25.28** is 5.00 cm, then the focal length is $f = -5.00 \text{ cm}$ and the power of the lens is $P = -20 \text{ D}$. An expanded view of the path of one ray through the lens is shown in the figure to illustrate how the shape of the lens, together with the law of refraction, causes the ray to follow its particular path and be diverged.

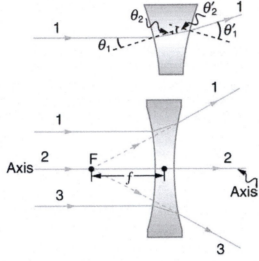

Figure 25.28 Rays of light entering a diverging lens parallel to its axis are diverged, and all appear to originate at its focal point F. The dashed lines are not rays—they indicate the directions from which the rays appear to come. The focal length f of a diverging lens is negative. An expanded view of the path taken by ray 1 shows the perpendiculars and the angles of incidence and refraction at both surfaces.

Diverging Lens

A lens that causes the light rays to bend away from its axis is called a diverging lens.

As noted in the initial discussion of the law of refraction in **The Law of Refraction**, the paths of light rays are exactly reversible. This means that the direction of the arrows could be reversed for all of the rays in **Figure 25.26** and **Figure 25.28**. For example, if a point light source is placed at the focal point of a convex lens, as shown in **Figure 25.29**, parallel light rays emerge from the other side.

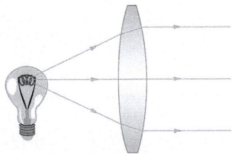

Figure 25.29 A small light source, like a light bulb filament, placed at the focal point of a convex lens, results in parallel rays of light emerging from the other side. The paths are exactly the reverse of those shown in **Figure 25.26**. This technique is used in lighthouses and sometimes in traffic lights to produce a directional beam of light from a source that emits light in all directions.

Ray Tracing and Thin Lenses

Ray tracing is the technique of determining or following (tracing) the paths that light rays take. For rays passing through matter, the law of refraction is used to trace the paths. Here we use ray tracing to help us understand the action of lenses in situations ranging from forming images on film to magnifying small print to correcting nearsightedness. While ray tracing for complicated lenses, such as those found in sophisticated cameras, may require computer techniques, there is a set of simple rules for tracing rays through thin lenses. A **thin lens** is defined to be one whose thickness allows rays to refract, as illustrated in **Figure 25.26**, but does not allow properties such as dispersion and aberrations. An ideal thin lens has two refracting surfaces but the lens is thin enough to assume that light rays bend only once. A thin symmetrical lens has two focal points, one on either side and both at the same distance from the lens. (See **Figure 25.30**.) Another important characteristic of a thin lens is that light rays through its center are deflected by a negligible amount, as seen in **Figure 25.31**.

Thin Lens

A thin lens is defined to be one whose thickness allows rays to refract but does not allow properties such as dispersion and aberrations.

Take-Home Experiment: A Visit to the Optician

Look through your eyeglasses (or those of a friend) backward and forward and comment on whether they act like thin lenses.

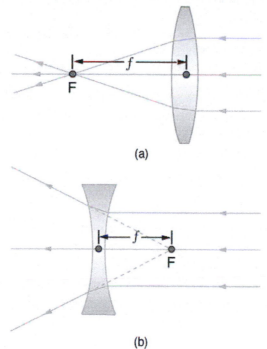

(a)

(b)

Figure 25.30 Thin lenses have the same focal length on either side. (a) Parallel light rays entering a converging lens from the right cross at its focal point on the left. (b) Parallel light rays entering a diverging lens from the right seem to come from the focal point on the right.

Figure 25.31 The light ray through the center of a thin lens is deflected by a negligible amount and is assumed to emerge parallel to its original path (shown as a shaded line).

Using paper, pencil, and a straight edge, ray tracing can accurately describe the operation of a lens. The rules for ray tracing for thin lenses are based on the illustrations already discussed:

1. A ray entering a converging lens parallel to its axis passes through the focal point F of the lens on the other side. (See rays 1 and 3 in **Figure 25.26**.)

2. A ray entering a diverging lens parallel to its axis seems to come from the focal point F. (See rays 1 and 3 in **Figure 25.28**.)

3. A ray passing through the center of either a converging or a diverging lens does not change direction. (See **Figure 25.31**, and see ray 2 in **Figure 25.26** and **Figure 25.28**.)

4. A ray entering a converging lens through its focal point exits parallel to its axis. (The reverse of rays 1 and 3 in **Figure 25.26**.)

5. A ray that enters a diverging lens by heading toward the focal point on the opposite side exits parallel to the axis. (The reverse of rays 1 and 3 in **Figure 25.28**.)

Rules for Ray Tracing
1. A ray entering a converging lens parallel to its axis passes through the focal point F of the lens on the other side.

2. A ray entering a diverging lens parallel to its axis seems to come from the focal point F.

3. A ray passing through the center of either a converging or a diverging lens does not change direction.

4. A ray entering a converging lens through its focal point exits parallel to its axis.

5. A ray that enters a diverging lens by heading toward the focal point on the opposite side exits parallel to the axis.

Image Formation by Thin Lenses

In some circumstances, a lens forms an obvious image, such as when a movie projector casts an image onto a screen. In other cases, the image is less obvious. Where, for example, is the image formed by eyeglasses? We will use ray tracing for thin lenses to illustrate how they form images, and we will develop equations to describe the image formation quantitatively.

Consider an object some distance away from a converging lens, as shown in **Figure 25.32**. To find the location and size of the image formed, we trace the paths of selected light rays originating from one point on the object, in this case the top of the person's head. The figure shows three rays from the top of the object that can be traced using the ray tracing rules given above. (Rays leave this point going in many directions, but we concentrate on only a few with paths that are easy to trace.) The first ray is one that enters the lens parallel to its axis and passes through the focal point on the other side (rule 1). The second ray passes through the center of the lens without changing direction (rule 3). The third ray passes through the nearer focal point on its way into the lens and leaves the lens parallel to its axis (rule 4). The three rays cross at the same point on the other side of the lens. The image of the top of the person's head is located at this point. All rays that come from the same point on the top of the person's head are refracted in such a way as to cross at the point shown. Rays from another point on the object, such as her belt buckle, will also cross at another common point, forming a complete image, as shown. Although three rays are traced in **Figure 25.32**, only two are necessary to locate the image. It is best to trace rays for which there are simple ray tracing rules. Before applying ray tracing to other situations, let us consider the example shown in **Figure 25.32** in more detail.

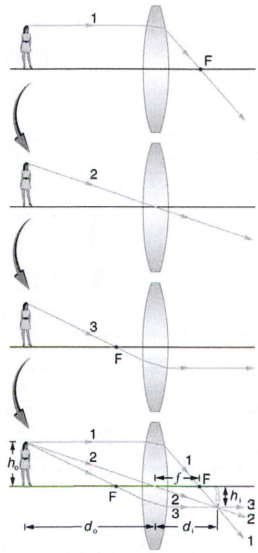

Figure 25.32 Ray tracing is used to locate the image formed by a lens. Rays originating from the same point on the object are traced—the three chosen rays each follow one of the rules for ray tracing, so that their paths are easy to determine. The image is located at the point where the rays cross. In this case, a real image—one that can be projected on a screen—is formed.

The image formed in **Figure 25.32** is a **real image**, meaning that it can be projected. That is, light rays from one point on the object actually cross at the location of the image and can be projected onto a screen, a piece of film, or the retina of an eye, for example. **Figure 25.33** shows how such an image would be projected onto film by a camera lens. This figure also shows how a real image is projected onto the retina by the lens of an eye. Note that the image is there whether it is projected onto a screen or not.

Real Image

The image in which light rays from one point on the object actually cross at the location of the image and can be projected onto a screen, a piece of film, or the retina of an eye is called a real image.

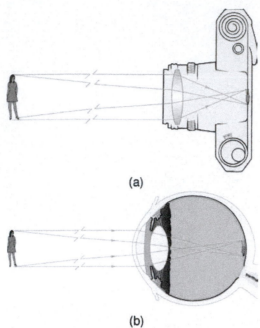

(a)

(b)

Figure 25.33 Real images can be projected. (a) A real image of the person is projected onto film. (b) The converging nature of the multiple surfaces that make up the eye result in the projection of a real image on the retina.

Several important distances appear in **Figure 25.32**. We define d_o to be the object distance, the distance of an object from the center of a lens. **Image distance** d_i is defined to be the distance of the image from the center of a lens. The height of the object and height of the image are given the symbols h_o and h_i, respectively. Images that appear upright relative to the object have heights that are positive and those that are inverted have negative heights. Using the rules of ray tracing and making a scale drawing with paper and pencil, like that in **Figure 25.32**, we can accurately describe the location and size of an image. But the real benefit of ray tracing is in visualizing how images are formed in a variety of situations. To obtain numerical information, we use a pair of equations that can be derived from a geometric analysis of ray tracing for thin lenses. The **thin lens equations** are

$$\frac{1}{d_o} + \frac{1}{d_i} = \frac{1}{f} \tag{25.24}$$

and

$$\frac{h_i}{h_o} = -\frac{d_i}{d_o} = m. \tag{25.25}$$

We define the ratio of image height to object height (h_i / h_o) to be the **magnification** m. (The minus sign in the equation above will be discussed shortly.) The thin lens equations are broadly applicable to all situations involving thin lenses (and "thin" mirrors, as we will see later). We will explore many features of image formation in the following worked examples.

Image Distance

The distance of the image from the center of the lens is called image distance.

Thin Lens Equations and Magnification

$$\frac{1}{d_o} + \frac{1}{d_i} = \frac{1}{f} \tag{25.26}$$

$$\frac{h_i}{h_o} = -\frac{d_i}{d_o} = m \tag{25.27}$$

Example 25.6 Finding the Image of a Light Bulb Filament by Ray Tracing and by the Thin Lens Equations

A clear glass light bulb is placed 0.750 m from a convex lens having a 0.500 m focal length, as shown in **Figure 25.34**. Use ray tracing to get an approximate location for the image. Then use the thin lens equations to calculate (a) the location of the image and (b) its magnification. Verify that ray tracing and the thin lens equations produce consistent results.

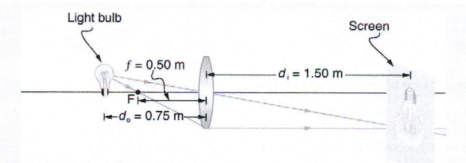

Figure 25.34 A light bulb placed 0.750 m from a lens having a 0.500 m focal length produces a real image on a poster board as discussed in the example above. Ray tracing predicts the image location and size.

Strategy and Concept

Since the object is placed farther away from a converging lens than the focal length of the lens, this situation is analogous to those illustrated in **Figure 25.32** and **Figure 25.33**. Ray tracing to scale should produce similar results for d_i. Numerical solutions for d_i and m can be obtained using the thin lens equations, noting that $d_o = 0.750$ m and $f = 0.500$ m.

Solutions (Ray tracing)

The ray tracing to scale in **Figure 25.34** shows two rays from a point on the bulb's filament crossing about 1.50 m on the far side of the lens. Thus the image distance d_i is about 1.50 m. Similarly, the image height based on ray tracing is greater than the object height by about a factor of 2, and the image is inverted. Thus m is about –2. The minus sign indicates that the image is inverted.

The thin lens equations can be used to find d_i from the given information:

$$\frac{1}{d_o} + \frac{1}{d_i} = \frac{1}{f}.$$

(25.28)

Rearranging to isolate d_i gives

$$\frac{1}{d_i} = \frac{1}{f} - \frac{1}{d_o}.$$

(25.29)

Entering known quantities gives a value for $1/d_i$:

$$\frac{1}{d_i} = \frac{1}{0.500 \text{ m}} - \frac{1}{0.750 \text{ m}} = \frac{0.667}{\text{m}}.$$

(25.30)

This must be inverted to find d_i:

$$d_i = \frac{\text{m}}{0.667} = 1.50 \text{ m}.$$

(25.31)

Note that another way to find d_i is to rearrange the equation:

$$\frac{1}{d_i} = \frac{1}{f} - \frac{1}{d_o}.$$

(25.32)

This yields the equation for the image distance as:

$$d_i = \frac{fd_o}{d_o - f}.$$

(25.33)

Note that there is no inverting here.

The thin lens equations can be used to find the magnification m, since both d_i and d_o are known. Entering their values gives

$$m = -\frac{d_i}{d_o} = -\frac{1.50 \text{ m}}{0.750 \text{ m}} = -2.00.$$

(25.34)

Discussion

Note that the minus sign causes the magnification to be negative when the image is inverted. Ray tracing and the use of the thin lens equations produce consistent results. The thin lens equations give the most precise results, being limited only by the accuracy of the given information. Ray tracing is limited by the accuracy with which you can draw, but it is highly useful both conceptually and visually.

Real images, such as the one considered in the previous example, are formed by converging lenses whenever an object is farther from the lens than its focal length. This is true for movie projectors, cameras, and the eye. We shall refer to these as *case 1* images. A case 1 image is formed when $d_o > f$ and f is positive, as in **Figure 25.35**(a). (A summary of the three cases or types of image formation appears at the end of this section.)

A different type of image is formed when an object, such as a person's face, is held close to a convex lens. The image is upright and larger than the object, as seen in **Figure 25.35**(b), and so the lens is called a magnifier. If you slowly pull the magnifier away from the face, you will see that the magnification steadily increases until the image begins to blur. Pulling the magnifier even farther away produces an inverted image as seen in **Figure 25.35**(a). The distance at which the image blurs, and beyond which it inverts, is the focal length of the lens. To use a convex lens as a magnifier, the object must be closer to the converging lens than its focal length. This is called a *case 2* image. A case 2 image is formed when $d_o < f$ and f is positive.

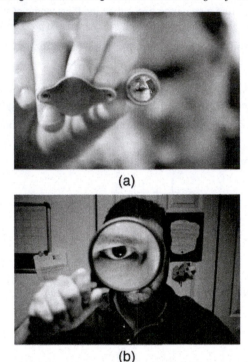

(a)

(b)

Figure 25.35 (a) When a converging lens is held farther away from the face than the lens's focal length, an inverted image is formed. This is a case 1 image. Note that the image is in focus but the face is not, because the image is much closer to the camera taking this photograph than the face. (credit: DaMongMan, Flickr) (b) A magnified image of a face is produced by placing it closer to the converging lens than its focal length. This is a case 2 image. (credit: Casey Fleser, Flickr)

Figure 25.36 uses ray tracing to show how an image is formed when an object is held closer to a converging lens than its focal length. Rays coming from a common point on the object continue to diverge after passing through the lens, but all appear to originate from a point at the location of the image. The image is on the same side of the lens as the object and is farther away from the lens than the object. This image, like all case 2 images, cannot be projected and, hence, is called a **virtual image**. Light rays only appear to originate at a virtual image; they do not actually pass through that location in space. A screen placed at the location of a virtual image will receive only diffuse light from the object, not focused rays from the lens. Additionally, a screen placed on the opposite side of the lens will receive rays that are still diverging, and so no image will be projected on it. We can see the magnified image with our eyes, because the lens of the eye converges the rays into a real image projected on our retina. Finally, we note that a virtual image is upright and larger than the object, meaning that the magnification is positive and greater than 1.

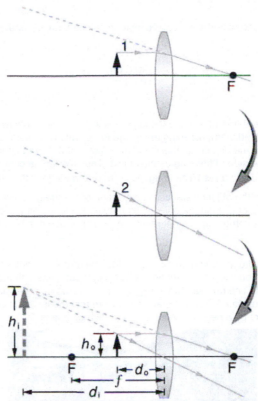

Figure 25.36 Ray tracing predicts the image location and size for an object held closer to a converging lens than its focal length. Ray 1 enters parallel to the axis and exits through the focal point on the opposite side, while ray 2 passes through the center of the lens without changing path. The two rays continue to diverge on the other side of the lens, but both appear to come from a common point, locating the upright, magnified, virtual image. This is a case 2 image.

Virtual Image

An image that is on the same side of the lens as the object and cannot be projected on a screen is called a virtual image.

Example 25.7 Image Produced by a Magnifying Glass

Suppose the book page in **Figure 25.36** (a) is held 7.50 cm from a convex lens of focal length 10.0 cm, such as a typical magnifying glass might have. What magnification is produced?

Strategy and Concept

We are given that $d_o = 7.50 \text{ cm}$ and $f = 10.0 \text{ cm}$, so we have a situation where the object is placed closer to the lens than its focal length. We therefore expect to get a case 2 virtual image with a positive magnification that is greater than 1. Ray tracing produces an image like that shown in **Figure 25.36**, but we will use the thin lens equations to get numerical solutions in this example.

Solution

To find the magnification m, we try to use magnification equation, $m = -d_i / d_o$. We do not have a value for d_i, so that we must first find the location of the image using lens equation. (The procedure is the same as followed in the preceding example, where d_o and f were known.) Rearranging the magnification equation to isolate d_i gives

$$\frac{1}{d_i} = \frac{1}{f} - \frac{1}{d_o}. \tag{25.35}$$

Entering known values, we obtain a value for $1/d_i$:

$$\frac{1}{d_i} = \frac{1}{10.0 \text{ cm}} - \frac{1}{7.50 \text{ cm}} = \frac{-0.0333}{\text{cm}}. \tag{25.36}$$

This must be inverted to find d_i:

$$d_i = -\frac{\text{cm}}{0.0333} = -30.0 \text{ cm}. \tag{25.37}$$

Now the thin lens equation can be used to find the magnification m, since both d_i and d_o are known. Entering their values gives

$$m = -\frac{d_i}{d_o} = -\frac{-30.0 \text{ cm}}{10.0 \text{ cm}} = 3.00. \tag{25.38}$$

Discussion

A number of results in this example are true of all case 2 images, as well as being consistent with **Figure 25.36**. Magnification is indeed positive (as predicted), meaning the image is upright. The magnification is also greater than 1, meaning that the image is larger than the object—in this case, by a factor of 3. Note that the image distance is negative. This means the image is on the same side of the lens as the object. Thus the image cannot be projected and is virtual. (Negative values of d_i occur for virtual images.) The image is farther from the lens than the object, since the image

distance is greater in magnitude than the object distance. The location of the image is not obvious when you look through a magnifier. In fact, since the image is bigger than the object, you may think the image is closer than the object. But the image is farther away, a fact that is useful in correcting farsightedness, as we shall see in a later section.

A third type of image is formed by a diverging or concave lens. Try looking through eyeglasses meant to correct nearsightedness. (See **Figure 25.37**.) You will see an image that is upright but smaller than the object. This means that the magnification is positive but less than 1. The ray diagram in **Figure 25.38** shows that the image is on the same side of the lens as the object and, hence, cannot be projected—it is a virtual image. Note that the image is closer to the lens than the object. This is a *case 3* image, formed for any object by a negative focal length or diverging lens.

Figure 25.37 A car viewed through a concave or diverging lens looks upright. This is a case 3 image. (credit: Daniel Oines, Flickr)

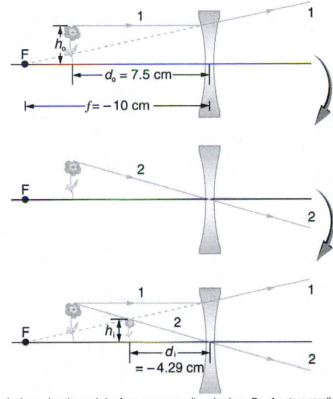

Figure 25.38 Ray tracing predicts the image location and size for a concave or diverging lens. Ray 1 enters parallel to the axis and is bent so that it appears to originate from the focal point. Ray 2 passes through the center of the lens without changing path. The two rays appear to come from a common point, locating the upright image. This is a case 3 image, which is closer to the lens than the object and smaller in height.

Example 25.8 Image Produced by a Concave Lens

Suppose an object such as a book page is held 7.50 cm from a concave lens of focal length –10.0 cm. Such a lens could be used in eyeglasses to correct pronounced nearsightedness. What magnification is produced?

Strategy and Concept

This example is identical to the preceding one, except that the focal length is negative for a concave or diverging lens. The method of solution is thus the same, but the results are different in important ways.

Solution

To find the magnification m, we must first find the image distance d_i using thin lens equation

$$\frac{1}{d_i} = \frac{1}{f} - \frac{1}{d_o},$$ (25.39)

or its alternative rearrangement

$$d_i = \frac{fd_o}{d_o - f}.$$ (25.40)

We are given that $f = -10.0 \text{ cm}$ and $d_o = 7.50 \text{ cm}$. Entering these yields a value for $1/d_i$:

$$\frac{1}{d_i} = \frac{1}{-10.0 \text{ cm}} - \frac{1}{7.50 \text{ cm}} = \frac{-0.2333}{\text{cm}}.$$ (25.41)

This must be inverted to find d_i:

$$d_i = -\frac{\text{cm}}{0.2333} = -4.29 \text{ cm}.$$ (25.42)

Or

$$d_i = \frac{(7.5)(-10)}{(7.5 - (-10))} = -75/17.5 = -4.29 \text{ cm}.$$ (25.43)

Now the magnification equation can be used to find the magnification m, since both d_i and d_o are known. Entering their values gives

$$m = -\frac{d_i}{d_o} = -\frac{-4.29 \text{ cm}}{7.50 \text{ cm}} = 0.571.$$

(25.44)

Discussion

A number of results in this example are true of all case 3 images, as well as being consistent with **Figure 25.38**. Magnification is positive (as predicted), meaning the image is upright. The magnification is also less than 1, meaning the image is smaller than the object—in this case, a little over half its size. The image distance is negative, meaning the image is on the same side of the lens as the object. (The image is virtual.) The image is closer to the lens than the object, since the image distance is smaller in magnitude than the object distance. The location of the image is not obvious when you look through a concave lens. In fact, since the image is smaller than the object, you may think it is farther away. But the image is closer than the object, a fact that is useful in correcting nearsightedness, as we shall see in a later section.

Table 25.3 summarizes the three types of images formed by single thin lenses. These are referred to as case 1, 2, and 3 images. Convex (converging) lenses can form either real or virtual images (cases 1 and 2, respectively), whereas concave (diverging) lenses can form only virtual images (always case 3). Real images are always inverted, but they can be either larger or smaller than the object. For example, a slide projector forms an image larger than the slide, whereas a camera makes an image smaller than the object being photographed. Virtual images are always upright and cannot be projected. Virtual images are larger than the object only in case 2, where a convex lens is used. The virtual image produced by a concave lens is always smaller than the object—a case 3 image. We can see and photograph virtual images only by using an additional lens to form a real image.

Table 25.3 Three Types of Images Formed By Thin Lenses

Type	Formed when	Image type	d_i	m
Case 1	f positive, $d_o > f$	real	positive	negative
Case 2	f positive, $d_o < f$	virtual	negative	positive $m > 1$
Case 3	f negative	virtual	negative	positive $m < 1$

In **Image Formation by Mirrors**, we shall see that mirrors can form exactly the same types of images as lenses.

Take-Home Experiment: Concentrating Sunlight

Find several lenses and determine whether they are converging or diverging. In general those that are thicker near the edges are diverging and those that are thicker near the center are converging. On a bright sunny day take the converging lenses outside and try focusing the sunlight onto a piece of paper. Determine the focal lengths of the lenses. Be careful because the paper may start to burn, depending on the type of lens you have selected.

Problem-Solving Strategies for Lenses

Step 1. Examine the situation to determine that image formation by a lens is involved.

Step 2. Determine whether ray tracing, the thin lens equations, or both are to be employed. A sketch is very useful even if ray tracing is not specifically required by the problem. Write symbols and values on the sketch.

Step 3. Identify exactly what needs to be determined in the problem (identify the unknowns).

Step 4. Make a list of what is given or can be inferred from the problem as stated (identify the knowns). It is helpful to determine whether the situation involves a case 1, 2, or 3 image. While these are just names for types of images, they have certain characteristics (given in **Table 25.3**) that can be of great use in solving problems.

Step 5. If ray tracing is required, use the ray tracing rules listed near the beginning of this section.

Step 6. Most quantitative problems require the use of the thin lens equations. These are solved in the usual manner by substituting knowns and solving for unknowns. Several worked examples serve as guides.

Step 7. Check to see if the answer is reasonable: Does it make sense? If you have identified the type of image (case 1, 2, or 3), you should assess whether your answer is consistent with the type of image, magnification, and so on.

Misconception Alert

We do not realize that light rays are coming from every part of the object, passing through every part of the lens, and all can be used to form the final image.

We generally feel the entire lens, or mirror, is needed to form an image. Actually, half a lens will form the same, though a fainter, image.

25.7 Image Formation by Mirrors

We only have to look as far as the nearest bathroom to find an example of an image formed by a mirror. Images in flat mirrors are the same size as the object and are located behind the mirror. Like lenses, mirrors can form a variety of images. For example, dental mirrors may produce a magnified image, just as makeup mirrors do. Security mirrors in shops, on the other hand, form images that are smaller than the object. We will use the law of reflection to understand how mirrors form images, and we will find that mirror images are analogous to those formed by lenses.

Figure 25.39 helps illustrate how a flat mirror forms an image. Two rays are shown emerging from the same point, striking the mirror, and being reflected into the observer's eye. The rays can diverge slightly, and both still get into the eye. If the rays are extrapolated backward, they seem to originate from a common point behind the mirror, locating the image. (The paths of the reflected rays into the eye are the same as if they had come directly from that point behind the mirror.) Using the law of reflection—the angle of reflection equals the angle of incidence—we can see that the image and object are the same distance from the mirror. This is a virtual image, since it cannot be projected—the rays only appear to originate from a common point behind the mirror. Obviously, if you walk behind the mirror, you cannot see the image, since the rays do not go there. But in front of the mirror, the rays behave exactly as if they had come from behind the mirror, so that is where the image is situated.

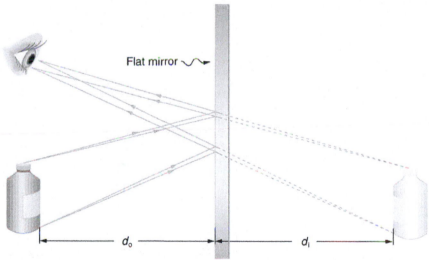

Figure 25.39 Two sets of rays from common points on an object are reflected by a flat mirror into the eye of an observer. The reflected rays seem to originate from behind the mirror, locating the virtual image.

Now let us consider the focal length of a mirror—for example, the concave spherical mirrors in **Figure 25.40**. Rays of light that strike the surface follow the law of reflection. For a mirror that is large compared with its radius of curvature, as in **Figure 25.40**(a), we see that the reflected rays do not cross at the same point, and the mirror does not have a well-defined focal point. If the mirror had the shape of a parabola, the rays would all cross at a single point, and the mirror would have a well-defined focal point. But parabolic mirrors are much more expensive to make than spherical mirrors. The solution is to use a mirror that is small compared with its radius of curvature, as shown in **Figure 25.40**(b). (This is the mirror equivalent of the thin lens approximation.) To a very good approximation, this mirror has a well-defined focal point at F that is the focal distance f from the center of the mirror. The focal length f of a concave mirror is positive, since it is a converging mirror.

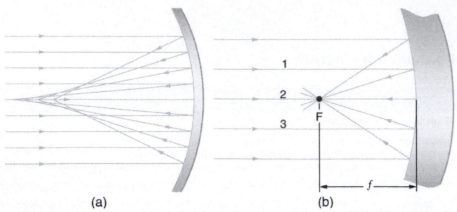

Figure 25.40 (a) Parallel rays reflected from a large spherical mirror do not all cross at a common point. (b) If a spherical mirror is small compared with its radius of curvature, parallel rays are focused to a common point. The distance of the focal point from the center of the mirror is its focal length f. Since this mirror is converging, it has a positive focal length.

Just as for lenses, the shorter the focal length, the more powerful the mirror; thus, $P = 1/f$ for a mirror, too. A more strongly curved mirror has a shorter focal length and a greater power. Using the law of reflection and some simple trigonometry, it can be shown that the focal length is half the radius of curvature, or

$$f = \frac{R}{2},$$ (25.45)

where R is the radius of curvature of a spherical mirror. The smaller the radius of curvature, the smaller the focal length and, thus, the more powerful the mirror.

The convex mirror shown in **Figure 25.41** also has a focal point. Parallel rays of light reflected from the mirror seem to originate from the point F at the focal distance f behind the mirror. The focal length and power of a convex mirror are negative, since it is a diverging mirror.

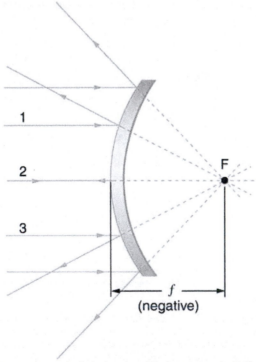

Figure 25.41 Parallel rays of light reflected from a convex spherical mirror (small in size compared with its radius of curvature) seem to originate from a well-defined focal point at the focal distance f behind the mirror. Convex mirrors diverge light rays and, thus, have a negative focal length.

Ray tracing is as useful for mirrors as for lenses. The rules for ray tracing for mirrors are based on the illustrations just discussed:

1. A ray approaching a concave converging mirror parallel to its axis is reflected through the focal point F of the mirror on the same side. (See rays 1 and 3 in **Figure 25.40**(b).)

2. A ray approaching a convex diverging mirror parallel to its axis is reflected so that it seems to come from the focal point F

behind the mirror. (See rays 1 and 3 in **Figure 25.41**.)

3.   Any ray striking the center of a mirror is followed by applying the law of reflection; it makes the same angle with the axis when leaving as when approaching. (See ray 2 in **Figure 25.42**.)

4.   A ray approaching a concave converging mirror through its focal point is reflected parallel to its axis. (The reverse of rays 1 and 3 in **Figure 25.40**.)

5.   A ray approaching a convex diverging mirror by heading toward its focal point on the opposite side is reflected parallel to the axis. (The reverse of rays 1 and 3 in **Figure 25.41**.)

We will use ray tracing to illustrate how images are formed by mirrors, and we can use ray tracing quantitatively to obtain numerical information. But since we assume each mirror is small compared with its radius of curvature, we can use the thin lens equations for mirrors just as we did for lenses.

Consider the situation shown in **Figure 25.42**, concave spherical mirror reflection, in which an object is placed farther from a concave (converging) mirror than its focal length. That is, f is positive and $d_o > f$, so that we may expect an image similar to the case 1 real image formed by a converging lens. Ray tracing in **Figure 25.42** shows that the rays from a common point on the object all cross at a point on the same side of the mirror as the object. Thus a real image can be projected onto a screen placed at this location. The image distance is positive, and the image is inverted, so its magnification is negative. This is a *case 1 image for mirrors*. It differs from the case 1 image for lenses only in that the image is on the same side of the mirror as the object. It is otherwise identical.

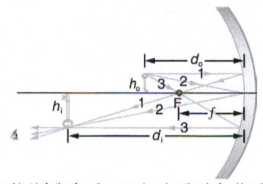

Figure 25.42 A case 1 image for a mirror. An object is farther from the converging mirror than its focal length. Rays from a common point on the object are traced using the rules in the text. Ray 1 approaches parallel to the axis, ray 2 strikes the center of the mirror, and ray 3 goes through the focal point on the way toward the mirror. All three rays cross at the same point after being reflected, locating the inverted real image. Although three rays are shown, only two of the three are needed to locate the image and determine its height.

Example 25.9 A Concave Reflector

Electric room heaters use a concave mirror to reflect infrared (IR) radiation from hot coils. Note that IR follows the same law of reflection as visible light. Given that the mirror has a radius of curvature of 50.0 cm and produces an image of the coils 3.00 m away from the mirror, where are the coils?

Strategy and Concept

We are given that the concave mirror projects a real image of the coils at an image distance $d_i = 3.00 \text{ m}$. The coils are the object, and we are asked to find their location—that is, to find the object distance d_o. We are also given the radius of curvature of the mirror, so that its focal length is $f = R/2 = 25.0 \text{ cm}$ (positive since the mirror is concave or converging). Assuming the mirror is small compared with its radius of curvature, we can use the thin lens equations, to solve this problem.

Solution

Since d_i and f are known, thin lens equation can be used to find d_o:

$$\frac{1}{d_o} + \frac{1}{d_i} = \frac{1}{f}. \tag{25.46}$$

Rearranging to isolate d_o gives

$$\frac{1}{d_o} = \frac{1}{f} - \frac{1}{d_i}. \tag{25.47}$$

Entering known quantities gives a value for $1/d_o$:

$$\frac{1}{d_o} = \frac{1}{0.250 \text{ m}} - \frac{1}{3.00 \text{ m}} = \frac{3.667}{\text{m}}. \tag{25.48}$$

This must be inverted to find d_o :

$$d_o = \frac{1\,\text{m}}{3.667} = 27.3\text{ cm.} \tag{25.49}$$

Discussion

Note that the object (the filament) is farther from the mirror than the mirror's focal length. This is a case 1 image ($d_o > f$ and f positive), consistent with the fact that a real image is formed. You will get the most concentrated thermal energy directly in front of the mirror and 3.00 m away from it. Generally, this is not desirable, since it could cause burns. Usually, you want the rays to emerge parallel, and this is accomplished by having the filament at the focal point of the mirror.

Note that the filament here is not much farther from the mirror than its focal length and that the image produced is considerably farther away. This is exactly analogous to a slide projector. Placing a slide only slightly farther away from the projector lens than its focal length produces an image significantly farther away. As the object gets closer to the focal distance, the image gets farther away. In fact, as the object distance approaches the focal length, the image distance approaches infinity and the rays are sent out parallel to one another.

Example 25.10 Solar Electric Generating System

One of the solar technologies used today for generating electricity is a device (called a parabolic trough or concentrating collector) that concentrates the sunlight onto a blackened pipe that contains a fluid. This heated fluid is pumped to a heat exchanger, where its heat energy is transferred to another system that is used to generate steam—and so generate electricity through a conventional steam cycle. **Figure 25.43** shows such a working system in southern California. Concave mirrors are used to concentrate the sunlight onto the pipe. The mirror has the approximate shape of a section of a cylinder. For the problem, assume that the mirror is exactly one-quarter of a full cylinder.

a. If we wish to place the fluid-carrying pipe 40.0 cm from the concave mirror at the mirror's focal point, what will be the radius of curvature of the mirror?

b. Per meter of pipe, what will be the amount of sunlight concentrated onto the pipe, assuming the insolation (incident solar radiation) is 0.900 kW/m^2 ?

c. If the fluid-carrying pipe has a 2.00-cm diameter, what will be the temperature increase of the fluid per meter of pipe over a period of one minute? Assume all the solar radiation incident on the reflector is absorbed by the pipe, and that the fluid is mineral oil.

Strategy

To solve an *Integrated Concept Problem* we must first identify the physical principles involved. Part (a) is related to the current topic. Part (b) involves a little math, primarily geometry. Part (c) requires an understanding of heat and density.

Solution to (a)

To a good approximation for a concave or semi-spherical surface, the point where the parallel rays from the sun converge will be at the focal point, so $R = 2f = 80.0\text{ cm}$.

Solution to (b)

The insolation is 900 W/m^2 . We must find the cross-sectional area A of the concave mirror, since the power delivered is $900\text{ W/m}^2 \times A$. The mirror in this case is a quarter-section of a cylinder, so the area for a length L of the mirror is $A = \frac{1}{4}(2\pi R)L$. The area for a length of 1.00 m is then

$$A = \frac{\pi}{2}R(1.00\text{ m}) = \frac{(3.14)}{2}(0.800\text{ m})(1.00\text{ m}) = 1.26\text{ m}^2. \tag{25.50}$$

The insolation on the 1.00-m length of pipe is then

$$\left(9.00\times10^2\,\frac{\text{W}}{\text{m}^2}\right)\left(1.26\text{ m}^2\right) = 1130\text{ W}. \tag{25.51}$$

Solution to (c)

The increase in temperature is given by $Q = mc\Delta T$. The mass m of the mineral oil in the one-meter section of pipe is

$$m \; = \; \rho V = \rho \pi \left(\frac{d}{2}\right)^2 (1.00 \text{ m}) \tag{25.52}$$
$$= \; \left(8.00 \times 10^2 \text{ kg/m}^3\right)(3.14)(0.0100 \text{ m})^2 (1.00 \text{ m})$$
$$= \; 0.251 \text{ kg}.$$

Therefore, the increase in temperature in one minute is

$$\Delta T \; = \; Q/mc \tag{25.53}$$
$$= \; \frac{(1130 \text{ W})(60.0 \text{ s})}{(0.251 \text{ kg})(1670 \text{ J·kg/°C})}$$
$$= \; 162°C.$$

Discussion for (c)

An array of such pipes in the California desert can provide a thermal output of 250 MW on a sunny day, with fluids reaching temperatures as high as $400°C$. We are considering only one meter of pipe here, and ignoring heat losses along the pipe.

Figure 25.43 Parabolic trough collectors are used to generate electricity in southern California. (credit: kjkolb, Wikimedia Commons)

What happens if an object is closer to a concave mirror than its focal length? This is analogous to a case 2 image for lenses ($d_o < f$ and f positive), which is a magnifier. In fact, this is how makeup mirrors act as magnifiers. **Figure 25.44**(a) uses ray tracing to locate the image of an object placed close to a concave mirror. Rays from a common point on the object are reflected in such a manner that they appear to be coming from behind the mirror, meaning that the image is virtual and cannot be projected. As with a magnifying glass, the image is upright and larger than the object. This is a *case 2 image for mirrors* and is exactly analogous to that for lenses.

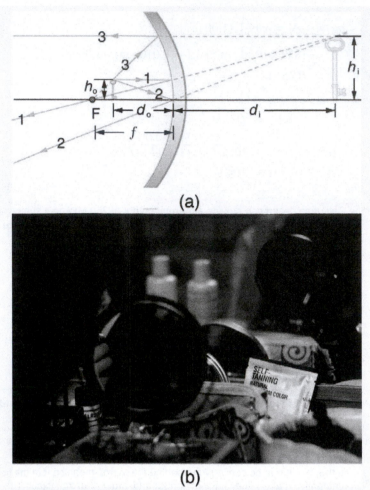

(a)

(b)

Figure 25.44 (a) Case 2 images for mirrors are formed when a converging mirror has an object closer to it than its focal length. Ray 1 approaches parallel to the axis, ray 2 strikes the center of the mirror, and ray 3 approaches the mirror as if it came from the focal point. (b) A magnifying mirror showing the reflection. (credit: Mike Melrose, Flickr)

All three rays appear to originate from the same point after being reflected, locating the upright virtual image behind the mirror and showing it to be larger than the object. (b) Makeup mirrors are perhaps the most common use of a concave mirror to produce a larger, upright image.

A convex mirror is a diverging mirror (f is negative) and forms only one type of image. It is a *case 3* image—one that is upright and smaller than the object, just as for diverging lenses. **Figure 25.45**(a) uses ray tracing to illustrate the location and size of the case 3 image for mirrors. Since the image is behind the mirror, it cannot be projected and is thus a virtual image. It is also seen to be smaller than the object.

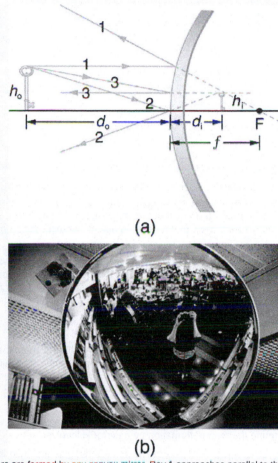

(a)

(b)

Figure 25.45 Case 3 images for mirrors are formed by any convex mirror. Ray 1 approaches parallel to the axis, ray 2 strikes the center of the mirror, and ray 3 approaches toward the focal point. All three rays appear to originate from the same point after being reflected, locating the upright virtual image behind the mirror and showing it to be smaller than the object. (b) Security mirrors are convex, producing a smaller, upright image. Because the image is smaller, a larger area is imaged compared to what would be observed for a flat mirror (and hence security is improved). (credit: Laura D'Alessandro, Flickr)

Example 25.11 Image in a Convex Mirror

A keratometer is a device used to measure the curvature of the cornea, particularly for fitting contact lenses. Light is reflected from the cornea, which acts like a convex mirror, and the keratometer measures the magnification of the image. The smaller the magnification, the smaller the radius of curvature of the cornea. If the light source is 12.0 cm from the cornea and the image's magnification is 0.0320, what is the cornea's radius of curvature?

Strategy

If we can find the focal length of the convex mirror formed by the cornea, we can find its radius of curvature (the radius of curvature is twice the focal length of a spherical mirror). We are given that the object distance is $d_o = 12.0$ cm and that $m = 0.0320$. We first solve for the image distance d_i, and then for f.

Solution

$m = -d_i / d_o$. Solving this expression for d_i gives

$$d_i = -m d_o. \tag{25.54}$$

Entering known values yields

$$d_i = -(0.0320)(12.0 \text{ cm}) = -0.384 \text{ cm}. \tag{25.55}$$

$$\frac{1}{f} = \frac{1}{d_o} + \frac{1}{d_i} \tag{25.56}$$

Substituting known values,

$$\frac{1}{f} = \frac{1}{12.0 \text{ cm}} + \frac{1}{-0.384 \text{ cm}} = \frac{-2.52}{\text{cm}}.$$ (25.57)

This must be inverted to find f :

$$f = \frac{\text{cm}}{-2.52} = -0.400 \text{ cm}.$$ (25.58)

The radius of curvature is twice the focal length, so that

$$R = 2 \mid f \mid = 0.800 \text{ cm}.$$ (25.59)

Discussion

Although the focal length f of a convex mirror is defined to be negative, we take the absolute value to give us a positive value for R . The radius of curvature found here is reasonable for a cornea. The distance from cornea to retina in an adult eye is about 2.0 cm. In practice, many corneas are not spherical, complicating the job of fitting contact lenses. Note that the image distance here is negative, consistent with the fact that the image is behind the mirror, where it cannot be projected. In this section's Problems and Exercises, you will show that for a fixed object distance, the smaller the radius of curvature, the smaller the magnification.

The three types of images formed by mirrors (cases 1, 2, and 3) are exactly analogous to those formed by lenses, as summarized in the table at the end of **Image Formation by Lenses**. It is easiest to concentrate on only three types of images—then remember that concave mirrors act like convex lenses, whereas convex mirrors act like concave lenses.

Take-Home Experiment: Concave Mirrors Close to Home

Find a flashlight and identify the curved mirror used in it. Find another flashlight and shine the first flashlight onto the second one, which is turned off. Estimate the focal length of the mirror. You might try shining a flashlight on the curved mirror behind the headlight of a car, keeping the headlight switched off, and determine its focal length.

Problem-Solving Strategy for Mirrors

Step 1. Examine the situation to determine that image formation by a mirror is involved.

Step 2. Refer to the **Problem-Solving Strategies for Lenses**. The same strategies are valid for mirrors as for lenses with one qualification—use the ray tracing rules for mirrors listed earlier in this section.

Glossary

converging lens: a convex lens in which light rays that enter it parallel to its axis converge at a single point on the opposite side

converging mirror: a concave mirror in which light rays that strike it parallel to its axis converge at one or more points along the axis

corner reflector: an object consisting of two mutually perpendicular reflecting surfaces, so that the light that enters is reflected back exactly parallel to the direction from which it came

critical angle: incident angle that produces an angle of refraction of $90°$

dispersion: spreading of white light into its full spectrum of wavelengths

diverging lens: a concave lens in which light rays that enter it parallel to its axis bend away (diverge) from its axis

diverging mirror: a convex mirror in which light rays that strike it parallel to its axis bend away (diverge) from its axis

fiber optics: transmission of light down fibers of plastic or glass, applying the principle of total internal reflection

focal length: distance from the center of a lens or curved mirror to its focal point

focal point: for a converging lens or mirror, the point at which converging light rays cross; for a diverging lens or mirror, the point from which diverging light rays appear to originate

geometric optics: part of optics dealing with the ray aspect of light

index of refraction: for a material, the ratio of the speed of light in vacuum to that in the material

law of reflection: angle of reflection equals the angle of incidence

law of reflection: angle of reflection equals the angle of incidence

magnification: ratio of image height to object height

mirror: smooth surface that reflects light at specific angles, forming an image of the person or object in front of it

power: inverse of focal length

rainbow: dispersion of sunlight into a continuous distribution of colors according to wavelength, produced by the refraction and reflection of sunlight by water droplets in the sky

ray: straight line that originates at some point

real image: image that can be projected

refraction: changing of a light ray's direction when it passes through variations in matter

virtual image: image that cannot be projected

zircon: natural gemstone with a large index of refraction

Section Summary

25.1 The Ray Aspect of Light
- A straight line that originates at some point is called a ray.
- The part of optics dealing with the ray aspect of light is called geometric optics.
- Light can travel in three ways from a source to another location: (1) directly from the source through empty space; (2) through various media; (3) after being reflected from a mirror.

25.2 The Law of Reflection
- The angle of reflection equals the angle of incidence.
- A mirror has a smooth surface and reflects light at specific angles.
- Light is diffused when it reflects from a rough surface.
- Mirror images can be photographed and videotaped by instruments.

25.3 The Law of Refraction
- The changing of a light ray's direction when it passes through variations in matter is called refraction.
- The speed of light in vacuum $c = 2.9972458 \times 10^8$ m/s $\approx 3.00 \times 10^8$ m/s.
- Index of refraction $n = \frac{c}{v}$, where v is the speed of light in the material, c is the speed of light in vacuum, and n is the index of refraction.
- Snell's law, the law of refraction, is stated in equation form as $n_1 \sin \theta_1 = n_2 \sin \theta_2$.

25.4 Total Internal Reflection
- The incident angle that produces an angle of refraction of $90°$ is called critical angle.
- Total internal reflection is a phenomenon that occurs at the boundary between two mediums, such that if the incident angle in the first medium is greater than the critical angle, then all the light is reflected back into that medium.
- Fiber optics involves the transmission of light down fibers of plastic or glass, applying the principle of total internal reflection.
- Endoscopes are used to explore the body through various orifices or minor incisions, based on the transmission of light through optical fibers.
- Cladding prevents light from being transmitted between fibers in a bundle.
- Diamonds sparkle due to total internal reflection coupled with a large index of refraction.

25.5 Dispersion: The Rainbow and Prisms
- The spreading of white light into its full spectrum of wavelengths is called dispersion.
- Rainbows are produced by a combination of refraction and reflection and involve the dispersion of sunlight into a continuous distribution of colors.
- Dispersion produces beautiful rainbows but also causes problems in certain optical systems.

25.6 Image Formation by Lenses
- Light rays entering a converging lens parallel to its axis cross one another at a single point on the opposite side.
- For a converging lens, the focal point is the point at which converging light rays cross; for a diverging lens, the focal point is the point from which diverging light rays appear to originate.
- The distance from the center of the lens to its focal point is called the focal length f.

- Power P of a lens is defined to be the inverse of its focal length, $P = \frac{1}{f}$.

- A lens that causes the light rays to bend away from its axis is called a diverging lens.
- Ray tracing is the technique of graphically determining the paths that light rays take.
- The image in which light rays from one point on the object actually cross at the location of the image and can be projected onto a screen, a piece of film, or the retina of an eye is called a real image.

- Thin lens equations are $\frac{1}{d_o} + \frac{1}{d_i} = \frac{1}{f}$ and $\frac{h_i}{h_o} = -\frac{d_i}{d_o} = m$ (magnification).

- The distance of the image from the center of the lens is called image distance.
- An image that is on the same side of the lens as the object and cannot be projected on a screen is called a virtual image.

25.7 Image Formation by Mirrors
- The characteristics of an image formed by a flat mirror are: (a) The image and object are the same distance from the mirror, (b) The image is a virtual image, and (c) The image is situated behind the mirror.
- Image length is half the radius of curvature.

$$f = \frac{R}{2}$$

- A convex mirror is a diverging mirror and forms only one type of image, namely a virtual image.

Conceptual Questions

25.2 The Law of Reflection

1. Using the law of reflection, explain how powder takes the shine off of a person's nose. What is the name of the optical effect?

25.3 The Law of Refraction

2. Diffusion by reflection from a rough surface is described in this chapter. Light can also be diffused by refraction. Describe how this occurs in a specific situation, such as light interacting with crushed ice.

3. Why is the index of refraction always greater than or equal to 1?

4. Does the fact that the light flash from lightning reaches you before its sound prove that the speed of light is extremely large or simply that it is greater than the speed of sound? Discuss how you could use this effect to get an estimate of the speed of light.

5. Will light change direction toward or away from the perpendicular when it goes from air to water? Water to glass? Glass to air?

6. Explain why an object in water always appears to be at a depth shallower than it actually is? Why do people sometimes sustain neck and spinal injuries when diving into unfamiliar ponds or waters?

7. Explain why a person's legs appear very short when wading in a pool. Justify your explanation with a ray diagram showing the path of rays from the feet to the eye of an observer who is out of the water.

8. Why is the front surface of a thermometer curved as shown?

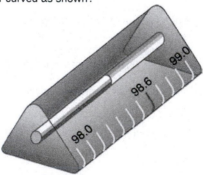

Figure 25.46 The curved surface of the thermometer serves a purpose.

9. Suppose light were incident from air onto a material that had a negative index of refraction, say −1.3; where does the refracted light ray go?

25.4 Total Internal Reflection

10. A ring with a colorless gemstone is dropped into water. The gemstone becomes invisible when submerged. Can it be a diamond? Explain.

11. A high-quality diamond may be quite clear and colorless, transmitting all visible wavelengths with little absorption. Explain how it can sparkle with flashes of brilliant color when illuminated by white light.

12. Is it possible that total internal reflection plays a role in rainbows? Explain in terms of indices of refraction and angles, perhaps referring to **Figure 25.47**. Some of us have seen the formation of a double rainbow. Is it physically possible to observe a triple rainbow?

Figure 25.47 Double rainbows are not a very common observance. (credit: InvictusOU812, Flickr)

13. The most common type of mirage is an illusion that light from faraway objects is reflected by a pool of water that is not really there. Mirages are generally observed in deserts, when there is a hot layer of air near the ground. Given that the refractive index of air is lower for air at higher temperatures, explain how mirages can be formed.

25.6 Image Formation by Lenses

14. It can be argued that a flat piece of glass, such as in a window, is like a lens with an infinite focal length. If so, where does it form an image? That is, how are d_i and d_o related?

15. You can often see a reflection when looking at a sheet of glass, particularly if it is darker on the other side. Explain why you can often see a double image in such circumstances.

16. When you focus a camera, you adjust the distance of the lens from the film. If the camera lens acts like a thin lens, why can it not be a fixed distance from the film for both near and distant objects?

17. A thin lens has two focal points, one on either side, at equal distances from its center, and should behave the same for light entering from either side. Look through your eyeglasses (or those of a friend) backward and forward and comment on whether they are thin lenses.

18. Will the focal length of a lens change when it is submerged in water? Explain.

25.7 Image Formation by Mirrors

19. What are the differences between real and virtual images? How can you tell (by looking) whether an image formed by a single lens or mirror is real or virtual?

20. Can you see a virtual image? Can you photograph one? Can one be projected onto a screen with additional lenses or mirrors? Explain your responses.

21. Is it necessary to project a real image onto a screen for it to exist?

22. At what distance is an image *always* located—at d_o, d_i, or f?

23. Under what circumstances will an image be located at the focal point of a lens or mirror?

24. What is meant by a negative magnification? What is meant by a magnification that is less than 1 in magnitude?

25. Can a case 1 image be larger than the object even though its magnification is always negative? Explain.

26. Figure 25.48 shows a light bulb between two mirrors. One mirror produces a beam of light with parallel rays; the other keeps light from escaping without being put into the beam. Where is the filament of the light in relation to the focal point or radius of curvature of each mirror?

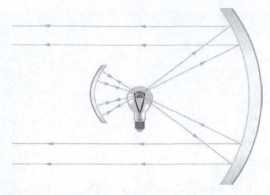

Figure 25.48 The two mirrors trap most of the bulb's light and form a directional beam as in a headlight.

27. Devise an arrangement of mirrors allowing you to see the back of your head. What is the minimum number of mirrors needed for this task?

28. If you wish to see your entire body in a flat mirror (from head to toe), how tall should the mirror be? Does its size depend upon your distance away from the mirror? Provide a sketch.

29. It can be argued that a flat mirror has an infinite focal length. If so, where does it form an image? That is, how are d_i and d_o related?

30. Why are diverging mirrors often used for rear-view mirrors in vehicles? What is the main disadvantage of using such a mirror compared with a flat one?

Problems & Exercises

25.1 The Ray Aspect of Light

1. Suppose a man stands in front of a mirror as shown in **Figure 25.49**. His eyes are 1.65 m above the floor, and the top of his head is 0.13 m higher. Find the height above the floor of the top and bottom of the smallest mirror in which he can see both the top of his head and his feet. How is this distance related to the man's height?

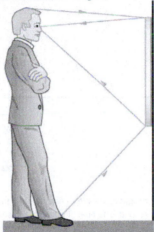

Figure 25.49 A full-length mirror is one in which you can see all of yourself. It need not be as big as you, and its size is independent of your distance from it.

25.2 The Law of Reflection

2. Show that when light reflects from two mirrors that meet each other at a right angle, the outgoing ray is parallel to the incoming ray, as illustrated in the following figure.

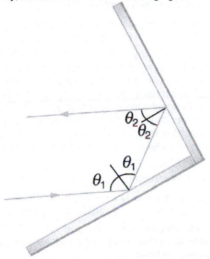

Figure 25.50 A corner reflector sends the reflected ray back in a direction parallel to the incident ray, independent of incoming direction.

3. Light shows staged with lasers use moving mirrors to swing beams and create colorful effects. Show that a light ray reflected from a mirror changes direction by 2θ when the mirror is rotated by an angle θ.

4. A flat mirror is neither converging nor diverging. To prove this, consider two rays originating from the same point and diverging at an angle θ. Show that after striking a plane mirror, the angle between their directions remains θ.

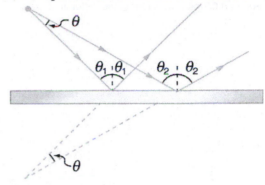

Figure 25.51 A flat mirror neither converges nor diverges light rays. Two rays continue to diverge at the same angle after reflection.

25.3 The Law of Refraction

5. What is the speed of light in water? In glycerine?

6. What is the speed of light in air? In crown glass?

7. Calculate the index of refraction for a medium in which the speed of light is 2.012×10^8 m/s , and identify the most likely substance based on **Table 25.1**.

8. In what substance in **Table 25.1** is the speed of light 2.290×10^8 m/s ?

9. There was a major collision of an asteroid with the Moon in medieval times. It was described by monks at Canterbury Cathedral in England as a red glow on and around the Moon. How long after the asteroid hit the Moon, which is 3.84×10^5 km away, would the light first arrive on Earth?

10. A scuba diver training in a pool looks at his instructor as shown in **Figure 25.52**. What angle does the ray from the instructor's face make with the perpendicular to the water at the point where the ray enters? The angle between the ray in the water and the perpendicular to the water is $25.0°$.

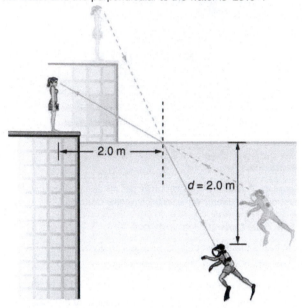

Figure 25.52 A scuba diver in a pool and his trainer look at each other.

11. Components of some computers communicate with each other through optical fibers having an index of refraction $n = 1.55$. What time in nanoseconds is required for a signal to travel 0.200 m through such a fiber?

12. (a) Using information in **Figure 25.52**, find the height of the instructor's head above the water, noting that you will first have to calculate the angle of incidence. (b) Find the apparent depth of the diver's head below water as seen by the instructor.

13. Suppose you have an unknown clear substance immersed in water, and you wish to identify it by finding its index of refraction. You arrange to have a beam of light enter it at an angle of $45.0°$, and you observe the angle of refraction to be $40.3°$. What is the index of refraction of the substance and its likely identity?

14. On the Moon's surface, lunar astronauts placed a corner reflector, off which a laser beam is periodically reflected. The distance to the Moon is calculated from the round-trip time. What percent correction is needed to account for the delay in time due to the slowing of light in Earth's atmosphere? Assume the distance to the Moon is precisely 3.84×10^8 m, and Earth's atmosphere (which varies in density with altitude) is equivalent to a layer 30.0 km thick with a constant index of refraction $n = 1.000293$.

15. Suppose **Figure 25.53** represents a ray of light going from air through crown glass into water, such as going into a fish tank. Calculate the amount the ray is displaced by the glass (Δx), given that the incident angle is $40.0°$ and the glass is 1.00 cm thick.

16. Figure 25.53 shows a ray of light passing from one medium into a second and then a third. Show that θ_3 is the same as it would be if the second medium were not present (provided total internal reflection does not occur).

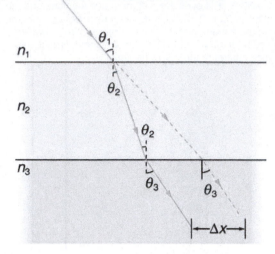

Figure 25.53 A ray of light passes from one medium to a third by traveling through a second. The final direction is the same as if the second medium were not present, but the ray is displaced by Δx (shown exaggerated).

17. Unreasonable Results

Suppose light travels from water to another substance, with an angle of incidence of $10.0°$ and an angle of refraction of $14.9°$. (a) What is the index of refraction of the other substance? (b) What is unreasonable about this result? (c) Which assumptions are unreasonable or inconsistent?

18. Construct Your Own Problem

Consider sunlight entering the Earth's atmosphere at sunrise and sunset—that is, at a $90°$ incident angle. Taking the boundary between nearly empty space and the atmosphere to be sudden, calculate the angle of refraction for sunlight. This lengthens the time the Sun appears to be above the horizon, both at sunrise and sunset. Now construct a problem in which you determine the angle of refraction for different models of the atmosphere, such as various layers of varying density. Your instructor may wish to guide you on the level of complexity to consider and on how the index of refraction varies with air density.

19. Unreasonable Results

Light traveling from water to a gemstone strikes the surface at an angle of $80.0°$ and has an angle of refraction of $15.2°$. (a) What is the speed of light in the gemstone? (b) What is unreasonable about this result? (c) Which assumptions are unreasonable or inconsistent?

25.4 Total Internal Reflection

20. Verify that the critical angle for light going from water to air is $48.6°$, as discussed at the end of **Example 25.4**, regarding the critical angle for light traveling in a polystyrene (a type of plastic) pipe surrounded by air.

21. (a) At the end of **Example 25.4**, it was stated that the critical angle for light going from diamond to air is $24.4°$. Verify this. (b) What is the critical angle for light going from zircon to air?

22. An optical fiber uses flint glass clad with crown glass. What is the critical angle?

23. At what minimum angle will you get total internal reflection of light traveling in water and reflected from ice?

24. Suppose you are using total internal reflection to make an efficient corner reflector. If there is air outside and the incident angle is $45.0°$, what must be the minimum index of refraction of the material from which the reflector is made?

25. You can determine the index of refraction of a substance by determining its critical angle. (a) What is the index of refraction of a substance that has a critical angle of $68.4°$ when submerged in water? What is the substance, based on **Table 25.1**? (b) What would the critical angle be for this substance in air?

26. A ray of light, emitted beneath the surface of an unknown liquid with air above it, undergoes total internal reflection as shown in **Figure 25.54**. What is the index of refraction for the liquid and its likely identification?

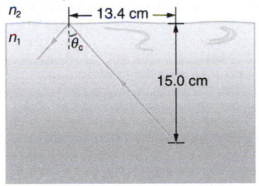

Figure 25.54 A light ray inside a liquid strikes the surface at the critical angle and undergoes total internal reflection.

27. A light ray entering an optical fiber surrounded by air is first refracted and then reflected as shown in **Figure 25.55**. Show that if the fiber is made from crown glass, any incident ray will be totally internally reflected.

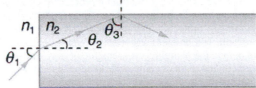

Figure 25.55 A light ray enters the end of a fiber, the surface of which is perpendicular to its sides. Examine the conditions under which it may be totally internally reflected.

25.5 Dispersion: The Rainbow and Prisms

28. (a) What is the ratio of the speed of red light to violet light in diamond, based on **Table 25.2**? (b) What is this ratio in polystyrene? (c) Which is more dispersive?

29. A beam of white light goes from air into water at an incident angle of $75.0°$. At what angles are the red (660 nm) and violet (410 nm) parts of the light refracted?

30. By how much do the critical angles for red (660 nm) and violet (410 nm) light differ in a diamond surrounded by air?

31. (a) A narrow beam of light containing yellow (580 nm) and green (550 nm) wavelengths goes from polystyrene to air, striking the surface at a $30.0°$ incident angle. What is the angle between the colors when they emerge? (b) How far would they have to travel to be separated by 1.00 mm?

32. A parallel beam of light containing orange (610 nm) and violet (410 nm) wavelengths goes from fused quartz to water, striking the surface between them at a $60.0°$ incident angle. What is the angle between the two colors in water?

33. A ray of 610 nm light goes from air into fused quartz at an incident angle of $55.0°$. At what incident angle must 470 nm light enter flint glass to have the same angle of refraction?

34. A narrow beam of light containing red (660 nm) and blue (470 nm) wavelengths travels from air through a 1.00 cm thick flat piece of crown glass and back to air again. The beam strikes at a $30.0°$ incident angle. (a) At what angles do the two colors emerge? (b) By what distance are the red and blue separated when they emerge?

35. A narrow beam of white light enters a prism made of crown glass at a $45.0°$ incident angle, as shown in **Figure 25.56**. At what angles, θ_R and θ_V, do the red (660 nm) and violet (410 nm) components of the light emerge from the prism?

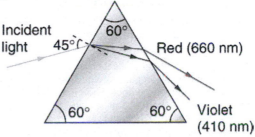

Figure 25.56 This prism will disperse the white light into a rainbow of colors. The incident angle is $45.0°$, and the angles at which the red and violet light emerge are θ_R and θ_V.

25.6 Image Formation by Lenses

36. What is the power in diopters of a camera lens that has a 50.0 mm focal length?

37. Your camera's zoom lens has an adjustable focal length ranging from 80.0 to 200 mm. What is its range of powers?

38. What is the focal length of 1.75 D reading glasses found on the rack in a pharmacy?

39. You note that your prescription for new eyeglasses is −4.50 D. What will their focal length be?

40. How far from the lens must the film in a camera be, if the lens has a 35.0 mm focal length and is being used to photograph a flower 75.0 cm away? Explicitly show how you follow the steps in the Problem-Solving Strategy for lenses.

41. A certain slide projector has a 100 mm focal length lens. (a) How far away is the screen, if a slide is placed 103 mm from the lens and produces a sharp image? (b) If the slide is 24.0 by 36.0 mm, what are the dimensions of the image? Explicitly show how you follow the steps in the Problem-Solving Strategy for lenses.

42. A doctor examines a mole with a 15.0 cm focal length magnifying glass held 13.5 cm from the mole (a) Where is the image? (b) What is its magnification? (c) How big is the image of a 5.00 mm diameter mole?

43. How far from a piece of paper must you hold your father's 2.25 D reading glasses to try to burn a hole in the paper with sunlight?

44. A camera with a 50.0 mm focal length lens is being used to photograph a person standing 3.00 m away. (a) How far from the lens must the film be? (b) If the film is 36.0 mm high, what fraction of a 1.75 m tall person will fit on it? (c) Discuss how reasonable this seems, based on your experience in taking or posing for photographs.

45. A camera lens used for taking close-up photographs has a focal length of 22.0 mm. The farthest it can be placed from the film is 33.0 mm. (a) What is the closest object that can be photographed? (b) What is the magnification of this closest object?

46. Suppose your 50.0 mm focal length camera lens is 51.0 mm away from the film in the camera. (a) How far away is an object that is in focus? (b) What is the height of the object if its image is 2.00 cm high?

47. (a) What is the focal length of a magnifying glass that produces a magnification of 3.00 when held 5.00 cm from an object, such as a rare coin? (b) Calculate the power of the magnifier in diopters. (c) Discuss how this power compares to those for store-bought reading glasses (typically 1.0 to 4.0 D). Is the magnifier's power greater, and should it be?

48. What magnification will be produced by a lens of power −4.00 D (such as might be used to correct myopia) if an object is held 25.0 cm away?

49. In Example 25.7, the magnification of a book held 7.50 cm from a 10.0 cm focal length lens was found to be 3.00. (a) Find the magnification for the book when it is held 8.50 cm from the magnifier. (b) Do the same for when it is held 9.50 cm from the magnifier. (c) Comment on the trend in m as the object distance increases as in these two calculations.

50. Suppose a 200 mm focal length telephoto lens is being used to photograph mountains 10.0 km away. (a) Where is the image? (b) What is the height of the image of a 1000 m high cliff on one of the mountains?

51. A camera with a 100 mm focal length lens is used to photograph the sun and moon. What is the height of the image of the sun on the film, given the sun is 1.40×10^6 km in diameter and is 1.50×10^8 km away?

52. Combine thin lens equations to show that the magnification for a thin lens is determined by its focal length and the object distance and is given by $m = f / (f - d_o)$.

25.7 Image Formation by Mirrors

53. What is the focal length of a makeup mirror that has a power of 1.50 D?

54. Some telephoto cameras use a mirror rather than a lens. What radius of curvature mirror is needed to replace a 800 mm focal length telephoto lens?

55. (a) Calculate the focal length of the mirror formed by the shiny back of a spoon that has a 3.00 cm radius of curvature. (b) What is its power in diopters?

56. Find the magnification of the heater element in Example 25.9. Note that its large magnitude helps spread out the reflected energy.

57. What is the focal length of a makeup mirror that produces a magnification of 1.50 when a person's face is 12.0 cm away? Explicitly show how you follow the steps in the **Problem-Solving Strategy for Mirrors**.

58. A shopper standing 3.00 m from a convex security mirror sees his image with a magnification of 0.250. (a) Where is his image? (b) What is the focal length of the mirror? (c) What is its radius of curvature? Explicitly show how you follow the steps in the **Problem-Solving Strategy for Mirrors**.

59. An object 1.50 cm high is held 3.00 cm from a person's cornea, and its reflected image is measured to be 0.167 cm high. (a) What is the magnification? (b) Where is the image? (c) Find the radius of curvature of the convex mirror formed by the cornea. (Note that this technique is used by optometrists to measure the curvature of the cornea for contact lens fitting. The instrument used is called a keratometer, or curve measurer.)

60. Ray tracing for a flat mirror shows that the image is located a distance behind the mirror equal to the distance of the object from the mirror. This is stated $d_i = -d_o$, since this is a negative image distance (it is a virtual image). (a) What is the focal length of a flat mirror? (b) What is its power?

61. Show that for a flat mirror $h_i = h_o$, knowing that the image is a distance behind the mirror equal in magnitude to the distance of the object from the mirror.

62. Use the law of reflection to prove that the focal length of a mirror is half its radius of curvature. That is, prove that $f = R / 2$. Note this is true for a spherical mirror only if its diameter is small compared with its radius of curvature.

63. Referring to the electric room heater considered in the first example in this section, calculate the intensity of IR radiation in $\mathrm{W/m}^2$ projected by the concave mirror on a person 3.00 m away. Assume that the heating element radiates 1500 W and has an area of $100 \ \mathrm{cm}^2$, and that half of the radiated power is reflected and focused by the mirror.

64. Consider a 250-W heat lamp fixed to the ceiling in a bathroom. If the filament in one light burns out then the remaining three still work. Construct a problem in which you determine the resistance of each filament in order to obtain a certain intensity projected on the bathroom floor. The ceiling is 3.0 m high. The problem will need to involve concave mirrors behind the filaments. Your instructor may wish to guide you on the level of complexity to consider in the electrical components.

25.1 The Ray Aspect of Light

1. When light from a distant object reflects off of a concave mirror and comes to a focus some distance in front of the mirror, we model light as a _____ to explain and predict the behavior of light and the formation of an image.
   a. wave
   b. particle
   c. ray
   d. all of the above

2. Light of wavelength 500 nm is incident on a narrow slit of width 150 nm. Which model of light most accurately predicts the behavior of the light after it passes through the slit? Explain your answer.

25.2 The Law of Reflection

3. An object is 2 meters in front of a flat mirror. Ray 1 from the object travels in a direction toward the mirror and normal to the mirror's surface. Ray 2 from the object travels at an angle of 5° from the direction of ray 1, and it also reflects off the mirror's surface. At what distance behind the mirror do these two reflected rays appear to converge to form an image?
   a. 0.2 m
   b. 0.5 m
   c. 2 m
   d. 4 m

4. Two light rays originate from object A, at a distance of 50 cm in front of a flat mirror, diverging at an angle of 10°. Both of the rays strike a flat mirror and reflect. Two light rays originate from object B, at a distance of 50 cm in front of a convex mirror, diverging at an angle of 10°. Both of the rays strike the convex mirror and reflect. For which object do the reflected rays appear to converge behind the mirror closer to the surface of the mirror, thus forming a closer (larger) image? Explain with the help of a sketch or diagram.

25.3 The Law of Refraction

5. When light travels from air into water, which of the following statements is accurate?
   a. The wavelength decreases, and the speed decreases.
   b. The wavelength decreases, and the speed increases.
   c. The wavelength increases, and the speed decreases.
   d. The wavelength increases, and the speed increases.

6. When a light ray travels from air into glass, which of the following statements is accurate after the light enters the glass?
   a. The ray bends away from the normal, and the speed decreases.
   b. The ray bends away from the normal, and the speed increases.
   c. The ray bends toward the normal, and the speed increases.
   d. The ray bends toward the normal, and the speed decreases.

7.

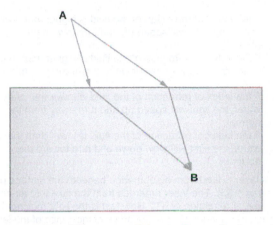

Figure 25.57 Two different potential paths from point A to point B are shown. Point A is in the air, and point B is in water. For which of these paths (upper or lower) would light travel from point A to point B faster? Which of the paths more accurately represents how a light ray would travel from point A to point B? Explain.

8. Students in a lab group are given a plastic cube with a hollow cube-shaped space in the middle that fills about half the volume of the cube. The index of refraction of the plastic is known. The hollow space is filled with a gas, and the students are asked to collect the data needed to find the index of refraction of the gas. The students take the following set of measurements:

Angle of incidence of the light in the air above the plastic block: 30°

Angle of refraction of the beam as it enters the plastic from the air: 45°

Angle of refraction of the beam as it enters the plastic from the gas: 45°

The three measurements are shared with a second lab group. Can the second group determine a value for the index of refraction of the gas from only this data?

   a. Yes, because they have information about the beam in air and in the plastic above the gas.
   b. Yes, because they have information about the beam on both sides of the gas.
   c. No, because they need additional information to determine the angle of the beam in the gas.
   d. No, because they do not have multiple data points to analyze.

9. Students in a lab group are given a plastic cube with a hollow cube-shaped space in the middle that fills about half the volume of the cube. The index of refraction of the plastic is known. The hollow space is filled with a gas, and the students are asked to collect the data needed to find the index of refraction of the gas. What information would you need to collect, and how would you use this information in order to deduce the index of refraction of the gas in the cube?

10. Light travels through water and crosses a boundary at a non-normal angle into a different fluid with an unknown index of refraction. Which of the following is true about the path of the light after crossing the boundary?

a. If the index of refraction of the fluid is higher than that of water, the light will speed up and turn toward the normal.
b. If the index of refraction of the fluid is higher than that of water, the light will slow down and turn away from the normal.
c. If the index of refraction of the fluid is lower than that of water, the light will speed up and turn away from the normal.
d. If the index of refraction of the fluid is lower than that of water, the light will slow down and turn toward the normal.

11. A laser is fired from a submarine beneath the surface of a lake ($n = 1.33$). The laser emerges from the lake into air with an angle of refraction of 67°. How fast is the light moving through the water? What is the angle of incidence of the laser light when it crosses the boundary between the lake and the air?

25.4 Total Internal Reflection

12. As light travels from air into water, what happens to the frequency of the light? Consider how the wavelength and speed of light change; then use the relationship between speed, wavelength, and frequency for a wave. What about light that is reflected off the surface of water? What happens to its wavelength, speed, and frequency?

25.6 Image Formation by Lenses

13. An object is 25 cm in front of a converging lens with a focal length of 25 cm. Where will the resulting image be located?
a. 25 cm in front of the lens
b. 25 cm behind the lens
c. 50 cm behind the lens
d. at infinity (either in front of or behind the lens)

14. A detective holds a magnifying glass 5.0 cm above an object he is studying, creating an upright image twice as large as the object. What is the focal length of the lens used for the magnifying glass?

15. A student wishes to predict the magnification of an image given the distance from the object to a converging lens with an unknown index of refraction. What data must the student collect in order to make such a prediction for any object distance?
a. A specific object distance and the image distance associated with that object distance.
b. A specific image distance and a determination of whether the image formed is upright or inverted.
c. The diameter and index of refraction of the lens.
d. The radius of curvature of each side of the lens.

16. Given a converging lens of unknown focal length and unknown index of refraction, explain what materials you would need and what procedure you would follow in order to experimentally determine the focal length of the lens.

25.7 Image Formation by Mirrors

17. A student is testing the properties of a mirror with an unknown radius of curvature. The student notices that no matter how far an object is placed from the mirror, the image seen in the mirror is always upright and smaller than the object. What can the student deduce about this mirror?

a. The mirror is convex.
b. The mirror is flat.
c. The mirror is concave.
d. More information is required to deduce the shape of the mirror.

18. A student notices a small printed sentence at the bottom of the driver's side mirror on her car. It reads, "Objects in the mirror are closer than they appear." Which type of mirror is this (convex, concave, or flat)? How could you confirm the shape of the mirror experimentally?

19. A mirror shows an upright image twice as large as the object when the object is 10 cm away from the mirror. What is the focal length of the mirror?
a. -10 cm
b. 10 cm
c. 20 cm
d. 40 cm

20. A mirror shows an inverted image that is equal in size to the object when the object is 20 cm away from the mirror. Describe the image that will be formed if this object is moved to a distance of 5 cm away from the mirror.

26   VISION AND OPTICAL INSTRUMENTS

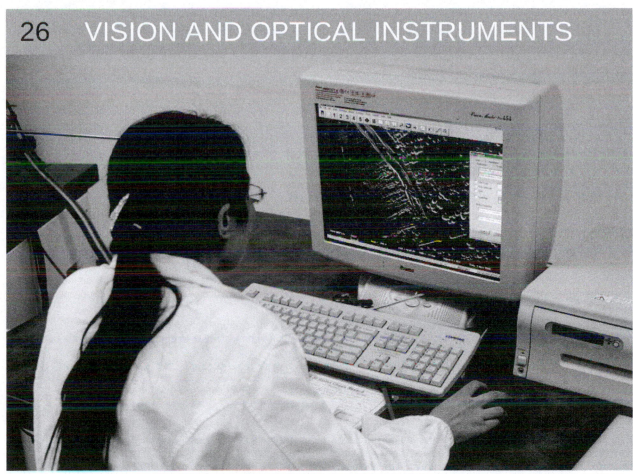

Figure 26.1 A scientist examines minute details on the surface of a disk drive at a magnification of 100,000 times. The image was produced using an electron microscope. (credit: Robert Scoble)

Chapter Outline

26.1. Physics of the Eye

26.2. Vision Correction

26.3. Color and Color Vision

26.4. Microscopes

26.5. Telescopes

26.6. Aberrations

Connection for AP® Courses

Seeing faces and objects we love and cherish—one's favorite teddy bear, a picture on the wall, or the sun rising over the mountains—is a delight. Intricate images help us understand nature and are invaluable for developing techniques and technologies in order to improve the quality of life. The image of a red blood cell that almost fills the cross-sectional area of a tiny capillary makes us wonder how blood makes it through and does not get stuck. We are able to see bacteria and viruses and understand their structure. It is the knowledge of physics that provides the fundamental understanding and the models required to develop new techniques and instruments. Therefore, physics is called an *enabling science*—it enables development and advancement in other areas. It is through optics and imaging that physics enables advancement in major areas of biosciences.

This chapter builds an understanding of vision and optical instruments on the idea that waves can transfer energy and momentum without the transfer of matter. In support of Big Idea 6, the way light waves travel is addressed using both conceptual and mathematical models. Throughout this unit, the direction of this travel is manipulated through the use of instruments like microscopes and telescopes, in support of Enduring Understanding 6.E.

When light enters a new transparent medium, like the crystalline lens of your eye or the glass lens of a microscope, it is bent either away or toward the line perpendicular to the boundary surface. This process is called "refraction," as outlined in Essential Knowledge 6.E.3. In both the eye and the microscope, lenses use refraction in order to redirect light and form images. These images, alluded to by Essential Knowledge 6.E.4, can be magnified, shrunk, or inverted, depending upon the lens arrangement.

When a new medium is not fully transparent, the incident light may be reflected or absorbed, and some light may be transmitted. This idea, referenced in Essential Knowledge 6.E.1, is utilized in the construction of telescopes. By relying on the law of reflection and the idea that reflective surfaces can be used to form images, telescopes can be constructed using mirrors to distort the path of light. This distortion allows the person using the telescope to see objects at great distance. While household telescopes utilize wavelengths in the visible light range, telescopes like the Chandra X-ray Observatory and Square Kilometre Array are capable of collecting wavelengths of considerably different size. Essential Knowledge 6.E.2, 6.E.4, and 6.F.1 are all addressed within this telescope discussion.

While ray tracing may easily predict the images formed by lenses and mirrors, only the wave model can be used to describe observations of color. This concept, covered in Section 26.3, underlines Essential Knowledge 6.F.4, the idea that different models of light are appropriate at different scales. The understanding and utilization of both the particle and wave models of light, as described in Enduring Understanding 6.F, is critical to success throughout this chapter.

Big Idea 6 Waves can transfer energy and momentum from one location to another without the permanent transfer of mass and serve as a mathematical model for the description of other phenomena.

Enduring Understanding 6.E The direction of propagation of a wave such as light may be changed when the wave encounters an interface between two media.

Essential Knowledge 6.E.1 When light travels from one medium to another, some of the light is transmitted, some is reflected, and some is absorbed.

Essential Knowledge 6.E.2 When light hits a smooth reflecting surface at an angle, it reflects at the same angle on the other side of the line perpendicular to the surface (specular reflection); and this law of reflection accounts for the size and location of images seen in plane mirrors.

Essential Knowledge 6.E.3 When light travels across a boundary from one transparent material to another, the speed of propagation changes. At a non-normal incident angle, the path of the light ray bends closer to the perpendicular in the optically slower substance. This is called refraction.

Essential Knowledge 6.E.4 The reflection of light from surfaces can be used to form images.

Essential Knowledge 6.E.5 The refraction of light as it travels from one transparent medium to another can be used to form images.

Enduring Understanding 6.F Electromagnetic radiation can be modeled as waves or as fundamental particles.

Essential Knowledge 6.F.1 Types of electromagnetic radiation are characterized by their wavelengths, and certain ranges of wavelength have been given specific names. These include (in order of increasing wavelength spanning a range from picometers to kilometers) gamma rays, x-rays, ultraviolet, visible light, infrared, microwaves, and radio waves.

Essential Knowledge 6.F.4 The nature of light requires that different models of light are most appropriate at different scales.

26.1 **Physics of the Eye**

Learning Objectives

By the end of this section, you will be able to:

- Explain the image formation by the eye.
- Explain why peripheral images lack detail and color.
- Define refractive indices.
- Analyze the accommodation of the eye for distant and near vision.

The information presented in this section supports the following AP® learning objectives and science practices:

- **6.E.5.1** The student is able to use quantitative and qualitative representations and models to analyze situations and solve problems about image formation occurring due to the refraction of light through thin lenses. **(S.P. 1.4, 2.2)**

The eye is perhaps the most interesting of all optical instruments. The eye is remarkable in how it forms images and in the richness of detail and color it can detect. However, our eyes commonly need some correction, to reach what is called "normal" vision, but should be called ideal rather than normal. Image formation by our eyes and common vision correction are easy to analyze with the optics discussed in **Geometric Optics**.

Figure 26.2 shows the basic anatomy of the eye. The cornea and lens form a system that, to a good approximation, acts as a single thin lens. For clear vision, a real image must be projected onto the light-sensitive retina, which lies at a fixed distance from the lens. The lens of the eye adjusts its power to produce an image on the retina for objects at different distances. The center of the image falls on the fovea, which has the greatest density of light receptors and the greatest acuity (sharpness) in the visual field. The variable opening (or pupil) of the eye along with chemical adaptation allows the eye to detect light intensities from the lowest observable to 10^{10} times greater (without damage). This is an incredible range of detection. Our eyes perform a vast number of functions, such as sense direction, movement, sophisticated colors, and distance. Processing of visual nerve impulses begins with interconnections in the retina and continues in the brain. The optic nerve conveys signals received by the eye to the brain.

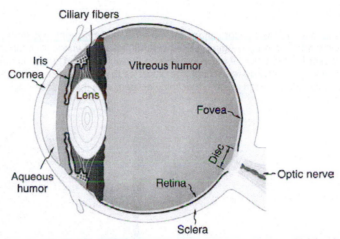

Figure 26.2 The cornea and lens of an eye act together to form a real image on the light-sensing retina, which has its densest concentration of receptors in the fovea and a blind spot over the optic nerve. The power of the lens of an eye is adjustable to provide an image on the retina for varying object distances. Layers of tissues with varying indices of refraction in the lens are shown here. However, they have been omitted from other pictures for clarity.

Refractive indices are crucial to image formation using lenses. **Table 26.1** shows refractive indices relevant to the eye. The biggest change in the refractive index, and bending of rays, occurs at the cornea rather than the lens. The ray diagram in **Figure 26.3** shows image formation by the cornea and lens of the eye. The rays bend according to the refractive indices provided in **Table 26.1**. The cornea provides about two-thirds of the power of the eye, owing to the fact that speed of light changes considerably while traveling from air into cornea. The lens provides the remaining power needed to produce an image on the retina. The cornea and lens can be treated as a single thin lens, even though the light rays pass through several layers of material (such as cornea, aqueous humor, several layers in the lens, and vitreous humor), changing direction at each interface. The image formed is much like the one produced by a single convex lens. This is a case 1 image. Images formed in the eye are inverted but the brain inverts them once more to make them seem upright.

Table 26.1 Refractive Indices Relevant to the Eye

Material	Index of Refraction
Water	1.33
Air	1.0
Cornea	1.38
Aqueous humor	1.34
Lens	1.41 average (varies throughout the lens, greatest in center)
Vitreous humor	1.34

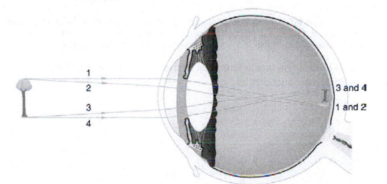

Figure 26.3 An image is formed on the retina with light rays converging most at the cornea and upon entering and exiting the lens. Rays from the top and bottom of the object are traced and produce an inverted real image on the retina. The distance to the object is drawn smaller than scale.

As noted, the image must fall precisely on the retina to produce clear vision — that is, the image distance d_i must equal the lens-to-retina distance. Because the lens-to-retina distance does not change, the image distance d_i must be the same for objects at all distances. The eye manages this by varying the power (and focal length) of the lens to accommodate for objects at various distances. The process of adjusting the eye's focal length is called **accommodation**. A person with normal (ideal) vision can see objects clearly at distances ranging from 25 cm to essentially infinity. However, although the near point (the shortest distance at which a sharp focus can be obtained) increases with age (becoming meters for some older people), we will consider it to be 25 cm in our treatment here.

Figure 26.4 shows the accommodation of the eye for distant and near vision. Since light rays from a nearby object can diverge and still enter the eye, the lens must be more converging (more powerful) for close vision than for distant vision. To be more converging, the lens is made thicker by the action of the ciliary muscle surrounding it. The eye is most relaxed when viewing distant objects, one reason that microscopes and telescopes are designed to produce distant images. Vision of very distant objects is called *totally relaxed*, while close vision is termed *accommodated*, with the closest vision being *fully accommodated*.

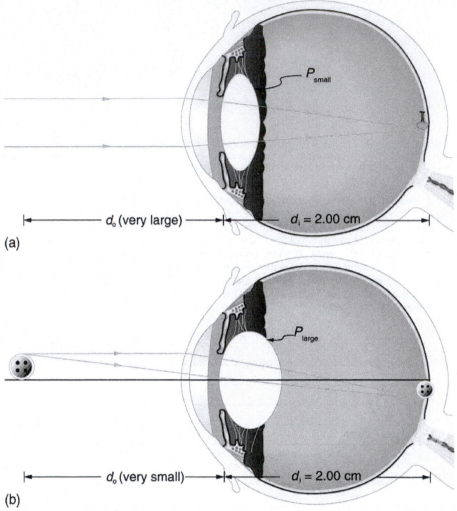

Figure 26.4 Relaxed and accommodated vision for distant and close objects. (a) Light rays from the same point on a distant object must be nearly parallel while entering the eye and more easily converge to produce an image on the retina. (b) Light rays from a nearby object can diverge more and still enter the eye. A more powerful lens is needed to converge them on the retina than if they were parallel.

We will use the thin lens equations to examine image formation by the eye quantitatively. First, note the power of a lens is given as $p = 1/f$, so we rewrite the thin lens equations as

$$P = \frac{1}{d_\mathrm{o}} + \frac{1}{d_\mathrm{i}}$$

(26.1)

and

$$\frac{h_\mathrm{i}}{h_\mathrm{o}} = -\frac{d_\mathrm{i}}{d_\mathrm{o}} = m.$$

(26.2)

We understand that d_i must equal the lens-to-retina distance to obtain clear vision, and that normal vision is possible for objects at distances $d_\mathrm{o} = 25$ cm to infinity.

Take-Home Experiment: The Pupil

Look at the central transparent area of someone's eye, the pupil, in normal room light. Estimate the diameter of the pupil. Now turn off the lights and darken the room. After a few minutes turn on the lights and promptly estimate the diameter of the pupil. What happens to the pupil as the eye adjusts to the room light? Explain your observations.

The eye can detect an impressive amount of detail, considering how small the image is on the retina. To get some idea of how small the image can be, consider the following example.

Example 26.1 Size of Image on Retina

What is the size of the image on the retina of a 1.20×10^{-2} cm diameter human hair, held at arm's length (60.0 cm) away? Take the lens-to-retina distance to be 2.00 cm.

Strategy

We want to find the height of the image h_i, given the height of the object is $h_o = 1.20 \times 10^{-2}$ cm. We also know that the object is 60.0 cm away, so that $d_o = 60.0$ cm . For clear vision, the image distance must equal the lens-to-retina distance, and so $d_i = 2.00$ cm . The equation $\frac{h_i}{h_o} = -\frac{d_i}{d_o} = m$ can be used to find h_i with the known information.

Solution

The only unknown variable in the equation $\frac{h_i}{h_o} = -\frac{d_i}{d_o} = m$ is h_i:

$$\frac{h_i}{h_o} = -\frac{d_i}{d_o}. \tag{26.3}$$

Rearranging to isolate h_i yields

$$h_i = -h_o \cdot \frac{d_i}{d_o}. \tag{26.4}$$

Substituting the known values gives

$$\begin{aligned} h_i &= -(1.20 \times 10^{-2} \text{ cm}) \frac{2.00 \text{ cm}}{60.0 \text{ cm}} \\ &= -4.00 \times 10^{-4} \text{ cm}. \end{aligned} \tag{26.5}$$

Discussion

This truly small image is not the smallest discernible—that is, the limit to visual acuity is even smaller than this. Limitations on visual acuity have to do with the wave properties of light and will be discussed in the next chapter. Some limitation is also due to the inherent anatomy of the eye and processing that occurs in our brain.

Example 26.2 Power Range of the Eye

Calculate the power of the eye when viewing objects at the greatest and smallest distances possible with normal vision, assuming a lens-to-retina distance of 2.00 cm (a typical value).

Strategy

For clear vision, the image must be on the retina, and so $d_i = 2.00$ cm here. For distant vision, $d_o \approx \infty$, and for close vision, $d_o = 25.0$ cm , as discussed earlier. The equation $P = \frac{1}{d_o} + \frac{1}{d_i}$ as written just above, can be used directly to solve for P in both cases, since we know d_i and d_o . Power has units of diopters, where $1 \text{ D} = 1/\text{m}$, and so we should express all distances in meters.

Solution

For distant vision,

$$P = \frac{1}{d_o} + \frac{1}{d_i} = \frac{1}{\infty} + \frac{1}{0.0200 \text{ m}}. \tag{26.6}$$

Since $1 / \infty = 0$, this gives

$$P = 0 + 50.0 / \text{m} = 50.0 \text{ D (distant vision)}. \tag{26.7}$$

Now, for close vision,

$$
\begin{aligned}
P &= \frac{1}{d_o} + \frac{1}{d_i} = \frac{1}{0.250 \text{ m}} + \frac{1}{0.0200 \text{ m}} \\
&= \frac{4.00}{\text{m}} + \frac{50.0}{\text{m}} = 4.00 \text{ D} + 50.0 \text{ D} \\
&= 54.0 \text{ D (close vision)}.
\end{aligned}
$$

(26.8)

Discussion

For an eye with this typical 2.00 cm lens-to-retina distance, the power of the eye ranges from 50.0 D (for distant totally relaxed vision) to 54.0 D (for close fully accommodated vision), which is an 8% increase. This increase in power for close vision is consistent with the preceding discussion and the ray tracing in **Figure 26.4**. An 8% ability to accommodate is considered normal but is typical for people who are about 40 years old. Younger people have greater accommodation ability, whereas older people gradually lose the ability to accommodate. When an optometrist identifies accommodation as a problem in elder people, it is most likely due to stiffening of the lens. The lens of the eye changes with age in ways that tend to preserve the ability to see distant objects clearly but do not allow the eye to accommodate for close vision, a condition called **presbyopia** (literally, elder eye). To correct this vision defect, we place a converging, positive power lens in front of the eye, such as found in reading glasses. Commonly available reading glasses are rated by their power in diopters, typically ranging from 1.0 to 3.5 D.

26.2 Vision Correction

Learning Objectives

By the end of this section, you will be able to:

- Identify and discuss common vision defects.
- Explain nearsightedness and farsightedness corrections.
- Explain laser vision correction.

The information presented in this section supports the following AP® learning objectives and science practices:

- **6.F.1.1** The student is able to make qualitative comparisons of the wavelengths of types of electromagnetic radiation. **(S.P. 6.4, 7.2)**

The need for some type of vision correction is very common. Common vision defects are easy to understand, and some are simple to correct. **Figure 26.5** illustrates two common vision defects. **Nearsightedness**, or **myopia**, is the inability to see distant objects clearly while close objects are clear. The eye overconverges the nearly parallel rays from a distant object, and the rays cross in front of the retina. More divergent rays from a close object are converged on the retina for a clear image. The distance to the farthest object that can be seen clearly is called the **far point** of the eye (normally infinity). **Farsightedness**, or **hyperopia**, is the inability to see close objects clearly while distant objects may be clear. A farsighted eye does not converge sufficient rays from a close object to make the rays meet on the retina. Less diverging rays from a distant object can be converged for a clear image. The distance to the closest object that can be seen clearly is called the **near point** of the eye (normally 25 cm).

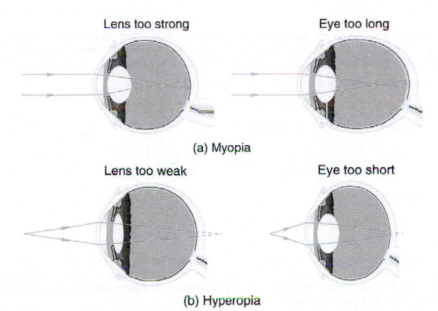

(a) Myopia

(b) Hyperopia

Figure 26.5 (a) The nearsighted (myopic) eye converges rays from a distant object in front of the retina; thus, they are diverging when they strike the retina, producing a blurry image. This can be caused by the lens of the eye being too powerful or the length of the eye being too great. (b) The farsighted (hyperopic) eye is unable to converge the rays from a close object by the time they strike the retina, producing blurry close vision. This can be caused by insufficient power in the lens or by the eye being too short.

Since the nearsighted eye over converges light rays, the correction for nearsightedness is to place a diverging spectacle lens in front of the eye. This reduces the power of an eye that is too powerful. Another way of thinking about this is that a diverging spectacle lens produces a case 3 image, which is closer to the eye than the object (see **Figure 26.6**). To determine the spectacle power needed for correction, you must know the person's far point—that is, you must know the greatest distance at which the person can see clearly. Then the image produced by a spectacle lens must be at this distance or closer for the nearsighted person to be able to see it clearly. It is worth noting that wearing glasses does not change the eye in any way. The eyeglass lens is simply used to create an image of the object at a distance where the nearsighted person can see it clearly. Whereas someone not wearing glasses can see clearly *objects* that fall between their near point and their far point, someone wearing glasses can see *images* that fall between their near point and their far point.

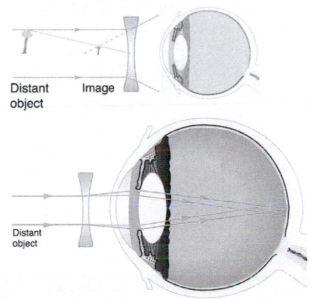

Figure 26.6 Correction of nearsightedness requires a diverging lens that compensates for the overconvergence by the eye. The diverging lens produces an image closer to the eye than the object, so that the nearsighted person can see it clearly.

Example 26.3 Correcting Nearsightedness

What power of spectacle lens is needed to correct the vision of a nearsighted person whose far point is 30.0 cm? Assume the spectacle (corrective) lens is held 1.50 cm away from the eye by eyeglass frames.

Strategy

You want this nearsighted person to be able to see very distant objects clearly. That means the spectacle lens must produce an image 30.0 cm from the eye for an object very far away. An image 30.0 cm from the eye will be 28.5 cm to the left of the spectacle lens (see Figure 26.6). Therefore, we must get $d_i = -28.5 \text{ cm}$ when $d_o \approx \infty$. The image distance is negative, because it is on the same side of the spectacle as the object.

Solution

Since d_i and d_o are known, the power of the spectacle lens can be found using $P = \frac{1}{d_o} + \frac{1}{d_i}$ as written earlier:

$$P = \frac{1}{d_o} + \frac{1}{d_i} = \frac{1}{\infty} + \frac{1}{-0.285 \text{ m}}. \tag{26.9}$$

Since $1/\infty = 0$, we obtain:

$$P = 0 - 3.51 / \text{m} = -3.51 \text{ D}. \tag{26.10}$$

Discussion

The negative power indicates a diverging (or concave) lens, as expected. The spectacle produces a case 3 image closer to the eye, where the person can see it. If you examine eyeglasses for nearsighted people, you will find the lenses are thinnest in the center. Additionally, if you examine a prescription for eyeglasses for nearsighted people, you will find that the prescribed power is negative and given in units of diopters.

Since the farsighted eye under converges light rays, the correction for farsightedness is to place a converging spectacle lens in front of the eye. This increases the power of an eye that is too weak. Another way of thinking about this is that a converging spectacle lens produces a case 2 image, which is farther from the eye than the object (see Figure 26.7). To determine the spectacle power needed for correction, you must know the person's near point—that is, you must know the smallest distance at which the person can see clearly. Then the image produced by a spectacle lens must be at this distance or farther for the farsighted person to be able to see it clearly.

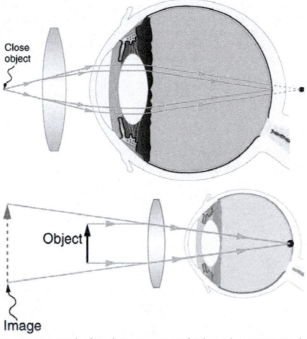

Figure 26.7 Correction of farsightedness uses a converging lens that compensates for the under convergence by the eye. The converging lens produces an image farther from the eye than the object, so that the farsighted person can see it clearly.

Example 26.4 Correcting Farsightedness

What power of spectacle lens is needed to allow a farsighted person, whose near point is 1.00 m, to see an object clearly that is 25.0 cm away? Assume the spectacle (corrective) lens is held 1.50 cm away from the eye by eyeglass frames.

Strategy

When an object is held 25.0 cm from the person's eyes, the spectacle lens must produce an image 1.00 m away (the near point). An image 1.00 m from the eye will be 98.5 cm to the left of the spectacle lens because the spectacle lens is 1.50 cm from the eye (see **Figure 26.7**). Therefore, $d_i = -98.5 \text{ cm}$. The image distance is negative, because it is on the same side of the spectacle as the object. The object is 23.5 cm to the left of the spectacle, so that $d_o = 23.5 \text{ cm}$.

Solution

Since d_i and d_o are known, the power of the spectacle lens can be found using $P = \frac{1}{d_o} + \frac{1}{d_i}$:

$$P = \frac{1}{d_o} + \frac{1}{d_i} = \frac{1}{0.235 \text{ m}} + \frac{1}{-0.985 \text{ m}} \tag{26.11}$$
$$= 4.26 \text{ D} - 1.02 \text{ D} = 3.24 \text{ D}.$$

Discussion

The positive power indicates a converging (convex) lens, as expected. The convex spectacle produces a case 2 image farther from the eye, where the person can see it. If you examine eyeglasses of farsighted people, you will find the lenses to be thickest in the center. In addition, a prescription of eyeglasses for farsighted people has a prescribed power that is positive.

Another common vision defect is **astigmatism**, an unevenness or asymmetry in the focus of the eye. For example, rays passing through a vertical region of the eye may focus closer than rays passing through a horizontal region, resulting in the image appearing elongated. This is mostly due to irregularities in the shape of the cornea but can also be due to lens irregularities or unevenness in the retina. Because of these irregularities, different parts of the lens system produce images at different locations. The eye-brain system can compensate for some of these irregularities, but they generally manifest themselves as less distinct vision or sharper images along certain axes. **Figure 26.8** shows a chart used to detect astigmatism. Astigmatism can be at least partially corrected with a spectacle having the opposite irregularity of the eye. If an eyeglass prescription has a cylindrical correction, it is there to correct astigmatism. The normal corrections for short- or farsightedness are spherical corrections, uniform along all axes.

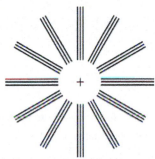

Figure 26.8 This chart can detect astigmatism, unevenness in the focus of the eye. Check each of your eyes separately by looking at the center cross (without spectacles if you wear them). If lines along some axes appear darker or clearer than others, you have an astigmatism.

Contact lenses have advantages over glasses beyond their cosmetic aspects. One problem with glasses is that as the eye moves, it is not at a fixed distance from the spectacle lens. Contacts rest on and move with the eye, eliminating this problem. Because contacts cover a significant portion of the cornea, they provide superior peripheral vision compared with eyeglasses. Contacts also correct some corneal astigmatism caused by surface irregularities. The tear layer between the smooth contact and the cornea fills in the irregularities. Since the index of refraction of the tear layer and the cornea are very similar, you now have a regular optical surface in place of an irregular one. If the curvature of a contact lens is not the same as the cornea (as may be necessary with some individuals to obtain a comfortable fit), the tear layer between the contact and cornea acts as a lens. If the tear layer is thinner in the center than at the edges, it has a negative power, for example. Skilled optometrists will adjust the power of the contact to compensate.

Laser vision correction has progressed rapidly in the last few years. It is the latest and by far the most successful in a series of procedures that correct vision by reshaping the cornea. As noted at the beginning of this section, the cornea accounts for about two-thirds of the power of the eye. Thus, small adjustments of its curvature have the same effect as putting a lens in front of the eye. To a reasonable approximation, the power of multiple lenses placed close together equals the sum of their powers. For example, a concave spectacle lens (for nearsightedness) having $P = -3.00 \text{ D}$ has the same effect on vision as reducing the power of the eye itself by 3.00 D. So to correct the eye for nearsightedness, the cornea is flattened to reduce its power. Similarly, to correct for farsightedness, the curvature of the cornea is enhanced to increase the power of the eye—the same effect as the positive power spectacle lens used for farsightedness. Laser vision correction uses high intensity electromagnetic radiation to ablate (to remove material from the surface) and reshape the corneal surfaces.

Today, the most commonly used laser vision correction procedure is *Laser in situ Keratomileusis (LASIK)*. The top layer of the cornea is surgically peeled back and the underlying tissue ablated by multiple bursts of finely controlled ultraviolet radiation produced by an excimer laser. Lasers are used because they not only produce well-focused intense light, but they also emit very pure wavelength electromagnetic radiation that can be controlled more accurately than mixed wavelength light. The 193 nm wavelength UV commonly used is extremely and strongly absorbed by corneal tissue, allowing precise evaporation of very thin layers. A computer controlled program applies more bursts, usually at a rate of 10 per second, to the areas that require deeper removal. Typically a spot less than 1 mm in diameter and about $0.3~\mu m$ in thickness is removed by each burst.

Nearsightedness, farsightedness, and astigmatism can be corrected with an accuracy that produces normal distant vision in more than 90% of the patients, in many cases right away. The corneal flap is replaced; healing takes place rapidly and is nearly painless. More than 1 million Americans per year undergo LASIK (see **Figure 26.9**).

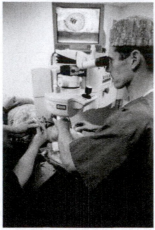

Figure 26.9 Laser vision correction is being performed using the LASIK procedure. Reshaping of the cornea by laser ablation is based on a careful assessment of the patient's vision and is computer controlled. The upper corneal layer is temporarily peeled back and minimally disturbed in LASIK, providing for more rapid and less painful healing of the less sensitive tissues below. (credit: U.S. Navy photo by Mass Communication Specialist 1st Class Brien Aho)

26.3 Color and Color Vision

Learning Objectives

By the end of this section, you will be able to:

- Explain the simple theory of color vision.
- Outline the coloring properties of light sources.
- Describe the retinex theory of color vision.

The information presented in this section supports the following AP® learning objectives and science practices:

- **6.F.4.1** The student is able to select a model of radiant energy that is appropriate to the spatial or temporal scale of an interaction with matter. **(S.P. 6.4, 7.1)**

The gift of vision is made richer by the existence of color. Objects and lights abound with thousands of hues that stimulate our eyes, brains, and emotions. Two basic questions are addressed in this brief treatment—what does color mean in scientific terms, and how do we, as humans, perceive it?

Simple Theory of Color Vision

We have already noted that color is associated with the wavelength of visible electromagnetic radiation. When our eyes receive pure-wavelength light, we tend to see only a few colors. Six of these (most often listed) are red, orange, yellow, green, blue, and violet. These are the rainbow of colors produced when white light is dispersed according to different wavelengths. There are thousands of other **hues** that we can perceive. These include brown, teal, gold, pink, and white. One simple theory of color vision implies that all these hues are our eye's response to different combinations of wavelengths. This is true to an extent, but we find that color perception is even subtler than our eye's response for various wavelengths of light.

The two major types of light-sensing cells (photoreceptors) in the retina are **rods and cones**. Rods are more sensitive than cones by a factor of about 1000 and are solely responsible for peripheral vision as well as vision in very dark environments. They are also important for motion detection. There are about 120 million rods in the human retina. Rods do not yield color information. You may notice that you lose color vision when it is very dark, but you retain the ability to discern grey scales.

Take-Home Experiment: Rods and Cones

1. Go into a darkened room from a brightly lit room, or from outside in the Sun. How long did it take to start seeing shapes more clearly? What about color? Return to the bright room. Did it take a few minutes before you could see things clearly?

2. Demonstrate the sensitivity of foveal vision. Look at the letter G in the word ROGERS. What about the clarity of the letters on either side of G?

Cones are most concentrated in the fovea, the central region of the retina. There are no rods here. The fovea is at the center of the macula, a 5 mm diameter region responsible for our central vision. The cones work best in bright light and are responsible for high resolution vision. There are about 6 million cones in the human retina. There are three types of cones, and each type is sensitive to different ranges of wavelengths, as illustrated in **Figure 26.10**. A **simplified theory of color vision** is that there are three *primary colors* corresponding to the three types of cones. The thousands of other hues that we can distinguish among are created by various combinations of stimulations of the three types of cones. Color television uses a three-color system in which the screen is covered with equal numbers of red, green, and blue phosphor dots. The broad range of hues a viewer sees is produced by various combinations of these three colors. For example, you will perceive yellow when red and green are illuminated with the correct ratio of intensities. White may be sensed when all three are illuminated. Then, it would seem that all hues can be produced by adding three primary colors in various proportions. But there is an indication that color vision is more sophisticated. There is no unique set of three primary colors. Another set that works is yellow, green, and blue. A further indication of the need for a more complex theory of color vision is that various different combinations can produce the same hue. Yellow can be sensed with yellow light, or with a combination of red and green, and also with white light from which violet has been removed. The three-primary-colors aspect of color vision is well established; more sophisticated theories expand on it rather than deny it.

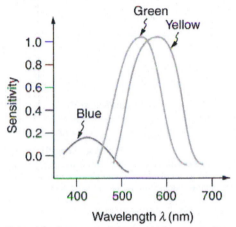

Figure 26.10 The image shows the relative sensitivity of the three types of cones, which are named according to wavelengths of greatest sensitivity. Rods are about 1000 times more sensitive, and their curve peaks at about 500 nm. Evidence for the three types of cones comes from direct measurements in animal and human eyes and testing of color blind people.

Consider why various objects display color—that is, why are feathers blue and red in a crimson rosella? The *true color of an object* is defined by its absorptive or reflective characteristics. **Figure 26.11** shows white light falling on three different objects, one pure blue, one pure red, and one black, as well as pure red light falling on a white object. Other hues are created by more complex absorption characteristics. Pink, for example on a galah cockatoo, can be due to weak absorption of all colors except red. An object can appear a different color under non-white illumination. For example, a pure blue object illuminated with pure red light will *appear* black, because it absorbs all the red light falling on it. But, the true color of the object is blue, which is independent of illumination.

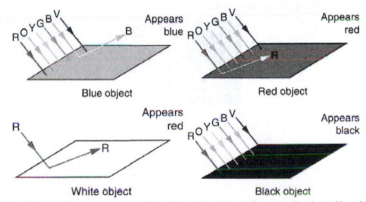

Figure 26.11 Absorption characteristics determine the true color of an object. Here, three objects are illuminated by white light, and one by pure red light. White is the equal mixture of all visible wavelengths; black is the absence of light.

Similarly, *light sources have colors* that are defined by the wavelengths they produce. A helium-neon laser emits pure red light. In fact, the phrase "pure red light" is defined by having a sharp constrained spectrum, a characteristic of laser light. The Sun produces a broad yellowish spectrum, fluorescent lights emit bluish-white light, and incandescent lights emit reddish-white hues as seen in **Figure 26.12**. As you would expect, you sense these colors when viewing the light source directly or when illuminating a white object with them. All of this fits neatly into the simplified theory that a combination of wavelengths produces various hues.

This activity is best done with plastic sheets of different colors as they allow more light to pass through to our eyes. However, thin sheets of paper and fabric can also be used. Overlay different colors of the material and hold them up to a white light. Using the theory described above, explain the colors you observe. You could also try mixing different crayon colors.

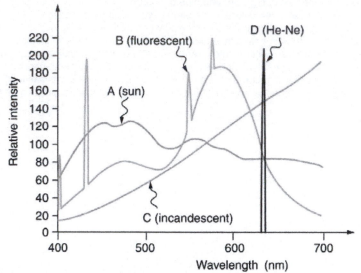

Figure 26.12 Emission spectra for various light sources are shown. Curve A is average sunlight at Earth's surface, curve B is light from a fluorescent lamp, and curve C is the output of an incandescent light. The spike for a helium-neon laser (curve D) is due to its pure wavelength emission. The spikes in the fluorescent output are due to atomic spectra—a topic that will be explored later.

Color Constancy and a Modified Theory of Color Vision

The eye-brain color-sensing system can, by comparing various objects in its view, perceive the true color of an object under varying lighting conditions—an ability that is called **color constancy**. We can sense that a white tablecloth, for example, is white whether it is illuminated by sunlight, fluorescent light, or candlelight. The wavelengths entering the eye are quite different in each case, as the graphs in **Figure 26.12** imply, but our color vision can detect the true color by comparing the tablecloth with its surroundings.

Theories that take color constancy into account are based on a large body of anatomical evidence as well as perceptual studies. There are nerve connections among the light receptors on the retina, and there are far fewer nerve connections to the brain than there are rods and cones. This means that there is signal processing in the eye before information is sent to the brain. For example, the eye makes comparisons between adjacent light receptors and is very sensitive to edges as seen in **Figure 26.13**. Rather than responding simply to the light entering the eye, which is uniform in the various rectangles in this figure, the eye responds to the edges and senses false darkness variations.

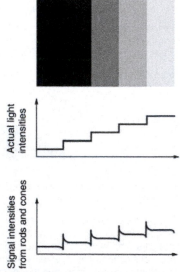

Figure 26.13 The importance of edges is shown. Although the grey strips are uniformly shaded, as indicated by the graph immediately below them, they do not appear uniform at all. Instead, they are perceived darker on the dark side and lighter on the light side of the edge, as shown in the bottom graph. This is due to nerve impulse processing in the eye.

One theory that takes various factors into account was advanced by Edwin Land (1909 – 1991), the creative founder of the Polaroid Corporation. Land proposed, based partly on his many elegant experiments, that the three types of cones are organized into systems called **retinexes**. Each retinex forms an image that is compared with the others, and the eye-brain system thus can compare a candle-illuminated white table cloth with its generally reddish surroundings and determine that it is actually white. This **retinex theory of color vision** is an example of modified theories of color vision that attempt to account for its subtleties. One striking experiment performed by Land demonstrates that some type of image comparison may produce color vision. Two pictures are taken of a scene on black-and-white film, one using a red filter, the other a blue filter. Resulting black-and-white slides are then projected and superimposed on a screen, producing a black-and-white image, as expected. Then a red filter is placed in front of the slide taken with a red filter, and the images are again superimposed on a screen. You would expect an image in various shades of pink, but instead, the image appears to humans in full color with all the hues of the original scene. This implies that color vision can be induced by comparison of the black-and-white and red images. Color vision is not completely understood or explained, and the retinex theory is not totally accepted. It is apparent that color vision is much subtler than what a first look might imply.

PhET Explorations: Color Vision

Make a whole rainbow by mixing red, green, and blue light. Change the wavelength of a monochromatic beam or filter white light. View the light as a solid beam, or see the individual photons.

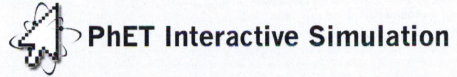

Figure 26.14 Color Vision (http://cnx.org/content/m55447/1.2/color-vision_en.jar)

26.4 Microscopes

Learning Objectives
By the end of this section, you will be able to: • Investigate different types of microscopes. • Learn how an image is formed in a compound microscope. The information presented in this section supports the following AP® learning objectives and science practices: • **6.E.5.1** The student is able to use quantitative and qualitative representations and models to analyze situations and solve problems about image formation occurring due to the refraction of light through thin lenses. **(S.P. 1.4, 2.2)**

Although the eye is marvelous in its ability to see objects large and small, it obviously has limitations to the smallest details it can detect. Human desire to see beyond what is possible with the naked eye led to the use of optical instruments. In this section we will examine microscopes, instruments for enlarging the detail that we cannot see with the unaided eye. The microscope is a multiple-element system having more than a single lens or mirror. (See **Figure 26.15**) A microscope can be made from two convex lenses. The image formed by the first element becomes the object for the second element. The second element forms its own image, which is the object for the third element, and so on. Ray tracing helps to visualize the image formed. If the device is composed of thin lenses and mirrors that obey the thin lens equations, then it is not difficult to describe their behavior numerically.

Figure 26.15 Multiple lenses and mirrors are used in this microscope. (credit: U.S. Navy photo by Tom Watanabe)

Microscopes were first developed in the early 1600s by eyeglass makers in The Netherlands and Denmark. The simplest **compound microscope** is constructed from two convex lenses as shown schematically in **Figure 26.16**. The first lens is called the **objective lens**, and has typical magnification values from $5\times$ to $100\times$. In standard microscopes, the objectives are mounted such that when you switch between objectives, the sample remains in focus. Objectives arranged in this way are described as parfocal. The second, the **eyepiece**, also referred to as the ocular, has several lenses which slide inside a cylindrical barrel. The focusing ability is provided by the movement of both the objective lens and the eyepiece. The purpose of a microscope is to magnify small objects, and both lenses contribute to the final magnification. Additionally, the final enlarged image is produced in a location far enough from the observer to be easily viewed, since the eye cannot focus on objects or images that are too close.

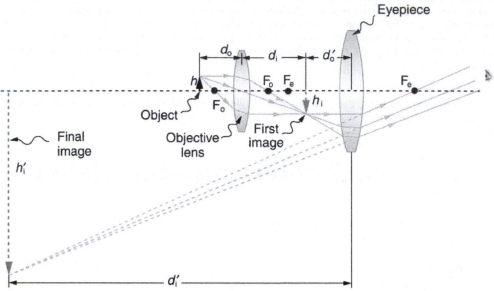

Figure 26.16 A compound microscope composed of two lenses, an objective and an eyepiece. The objective forms a case 1 image that is larger than the object. This first image is the object for the eyepiece. The eyepiece forms a case 2 final image that is further magnified.

To see how the microscope in **Figure 26.16** forms an image, we consider its two lenses in succession. The object is slightly farther away from the objective lens than its focal length f_o, producing a case 1 image that is larger than the object. This first image is the object for the second lens, or eyepiece. The eyepiece is intentionally located so it can further magnify the image. The eyepiece is placed so that the first image is closer to it than its focal length f_e. Thus the eyepiece acts as a magnifying glass, and the final image is made even larger. The final image remains inverted, but it is farther from the observer, making it easy to view (the eye is most relaxed when viewing distant objects and normally cannot focus closer than 25 cm). Since each lens produces a magnification that multiplies the height of the image, it is apparent that the overall magnification m is the product of the individual magnifications:

$$m = m_o m_e, \tag{26.12}$$

where m_o is the magnification of the objective and m_e is the magnification of the eyepiece. This equation can be generalized for any combination of thin lenses and mirrors that obey the thin lens equations.

Overall Magnification

The overall magnification of a multiple-element system is the product of the individual magnifications of its elements.

Example 26.5 Microscope Magnification

Calculate the magnification of an object placed 6.20 mm from a compound microscope that has a 6.00 mm focal length objective and a 50.0 mm focal length eyepiece. The objective and eyepiece are separated by 23.0 cm.

Strategy and Concept

This situation is similar to that shown in **Figure 26.16**. To find the overall magnification, we must find the magnification of the objective, then the magnification of the eyepiece. This involves using the thin lens equation.

Solution

The magnification of the objective lens is given as

$$m_o = -\frac{d_i}{d_o}, \tag{26.13}$$

where d_o and d_i are the object and image distances, respectively, for the objective lens as labeled in **Figure 26.16**. The object distance is given to be $d_o = 6.20$ mm , but the image distance d_i is not known. Isolating d_i , we have

$$\frac{1}{d_i} = \frac{1}{f_o} - \frac{1}{d_o},$$

(26.14)

where f_o is the focal length of the objective lens. Substituting known values gives

$$\frac{1}{d_i} = \frac{1}{6.00 \text{ mm}} - \frac{1}{6.20 \text{ mm}} = \frac{0.00538}{\text{mm}}.$$

(26.15)

We invert this to find d_i :

$$d_i = 186 \text{ mm}.$$

(26.16)

Substituting this into the expression for m_o gives

$$m_o = -\frac{d_i}{d_o} = -\frac{186 \text{ mm}}{6.20 \text{ mm}} = -30.0.$$

(26.17)

Now we must find the magnification of the eyepiece, which is given by

$$m_e = -\frac{d_i'}{d_o''},$$

(26.18)

where d_i' and d_o' are the image and object distances for the eyepiece (see **Figure 26.16**). The object distance is the distance of the first image from the eyepiece. Since the first image is 186 mm to the right of the objective and the eyepiece is 230 mm to the right of the objective, the object distance is $d_o' = 230 \text{ mm} - 186 \text{ mm} = 44.0 \text{ mm}$. This places the first image closer to the eyepiece than its focal length, so that the eyepiece will form a case 2 image as shown in the figure. We still need to find the location of the final image d_i' in order to find the magnification. This is done as before to obtain a value for $1 / d_i'$:

$$\frac{1}{d_i'} = \frac{1}{f_e} - \frac{1}{d_o'} = \frac{1}{50.0 \text{ mm}} - \frac{1}{44.0 \text{ mm}} = -\frac{0.00273}{\text{mm}}.$$

(26.19)

Inverting gives

$$d_i' = -\frac{\text{mm}}{0.00273} = -367 \text{ mm}.$$

(26.20)

The eyepiece's magnification is thus

$$m_e = -\frac{d_i'}{d_o'} = -\frac{-367 \text{ mm}}{44.0 \text{ mm}} = 8.33.$$

(26.21)

So the overall magnification is

$$m = m_o m_e = (-30.0)(8.33) = -250.$$

(26.22)

Discussion

Both the objective and the eyepiece contribute to the overall magnification, which is large and negative, consistent with **Figure 26.16**, where the image is seen to be large and inverted. In this case, the image is virtual and inverted, which cannot happen for a single element (case 2 and case 3 images for single elements are virtual and upright). The final image is 367 mm (0.367 m) to the left of the eyepiece. Had the eyepiece been placed farther from the objective, it could have formed a case 1 image to the right. Such an image could be projected on a screen, but it would be behind the head of the person in the figure and not appropriate for direct viewing. The procedure used to solve this example is applicable in any multiple-element system. Each element is treated in turn, with each forming an image that becomes the object for the next element. The process is not more difficult than for single lenses or mirrors, only lengthier.

Normal optical microscopes can magnify up to $1500\times$ with a theoretical resolution of -0.2 µm . The lenses can be quite complicated and are composed of multiple elements to reduce aberrations. Microscope objective lenses are particularly important as they primarily gather light from the specimen. Three parameters describe microscope objectives: the **numerical aperture** (NA) , the magnification (m) , and the working distance. The NA is related to the light gathering ability of a lens and is obtained using the angle of acceptance θ formed by the maximum cone of rays focusing on the specimen (see **Figure 26.17(a)**) and is given by

$$NA = n \sin \alpha, \tag{26.23}$$

where n is the refractive index of the medium between the lens and the specimen and $\alpha = \theta/2$. As the angle of acceptance given by θ increases, NA becomes larger and more light is gathered from a smaller focal region giving higher resolution. A $0.75NA$ objective gives more detail than a $0.10NA$ objective.

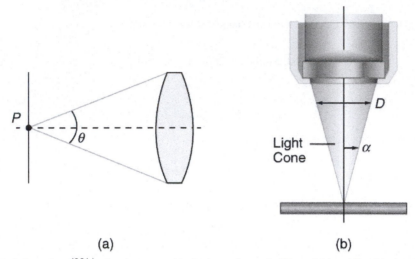

(a)   (b)

Figure 26.17 (a) The numerical aperture (NA) of a microscope objective lens refers to the light-gathering ability of the lens and is calculated using half the angle of acceptance θ. (b) Here, α is half the acceptance angle for light rays from a specimen entering a camera lens, and D is the diameter of the aperture that controls the light entering the lens.

While the numerical aperture can be used to compare resolutions of various objectives, it does not indicate how far the lens could be from the specimen. This is specified by the "working distance," which is the distance (in mm usually) from the front lens element of the objective to the specimen, or cover glass. The higher the NA the closer the lens will be to the specimen and the more chances there are of breaking the cover slip and damaging both the specimen and the lens. The focal length of an objective lens is different than the working distance. This is because objective lenses are made of a combination of lenses and the focal length is measured from inside the barrel. The working distance is a parameter that microscopists can use more readily as it is measured from the outermost lens. The working distance decreases as the NA and magnification both increase.

The term $f/\#$ in general is called the f-number and is used to denote the light per unit area reaching the image plane. In photography, an image of an object at infinity is formed at the focal point and the f-number is given by the ratio of the focal length f of the lens and the diameter D of the aperture controlling the light into the lens (see **Figure 26.17**(b)). If the acceptance angle is small the NA of the lens can also be used as given below.

$$f/\# = \frac{f}{D} \approx \frac{1}{2NA}. \tag{26.24}$$

As the f-number decreases, the camera is able to gather light from a larger angle, giving wide-angle photography. As usual there is a trade-off. A greater $f/\#$ means less light reaches the image plane. A setting of $f/16$ usually allows one to take pictures in bright sunlight as the aperture diameter is small. In optical fibers, light needs to be focused into the fiber. **Figure 26.18** shows the angle used in calculating the NA of an optical fiber.

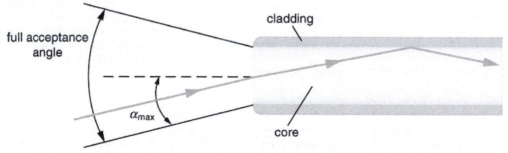

Figure 26.18 Light rays enter an optical fiber. The numerical aperture of the optical fiber can be determined by using the angle α_{max}.

Can the NA be larger than 1.00? The answer is 'yes' if we use immersion lenses in which a medium such as oil, glycerine or water is placed between the objective and the microscope cover slip. This minimizes the mismatch in refractive indices as light rays go through different media, generally providing a greater light-gathering ability and an increase in resolution. **Figure 26.19** shows light rays when using air and immersion lenses.

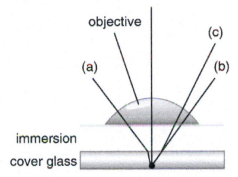

Figure 26.19 Light rays from a specimen entering the objective. Paths for immersion medium of air (a), water (b) $(n = 1.33)$, and oil (c) $(n = 1.51)$ are shown. The water and oil immersions allow more rays to enter the objective, increasing the resolution.

When using a microscope we do not see the entire extent of the sample. Depending on the eyepiece and objective lens we see a restricted region which we say is the field of view. The objective is then manipulated in two-dimensions above the sample to view other regions of the sample. Electronic scanning of either the objective or the sample is used in scanning microscopy. The image formed at each point during the scanning is combined using a computer to generate an image of a larger region of the sample at a selected magnification.

When using a microscope, we rely on gathering light to form an image. Hence most specimens need to be illuminated, particularly at higher magnifications, when observing details that are so small that they reflect only small amounts of light. To make such objects easily visible, the intensity of light falling on them needs to be increased. Special illuminating systems called condensers are used for this purpose. The type of condenser that is suitable for an application depends on how the specimen is examined, whether by transmission, scattering or reflecting. See **Figure 26.20** for an example of each. White light sources are common and lasers are often used. Laser light illumination tends to be quite intense and it is important to ensure that the light does not result in the degradation of the specimen.

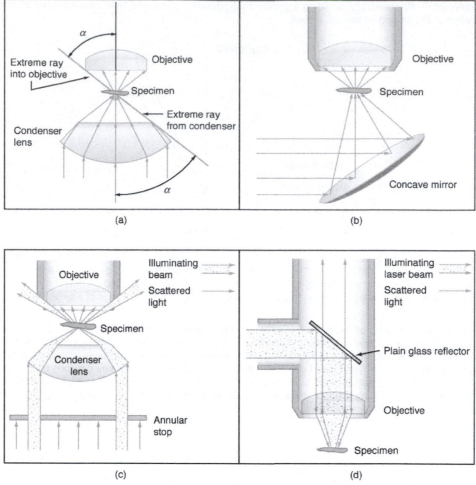

Figure 26.20 Illumination of a specimen in a microscope. (a) Transmitted light from a condenser lens. (b) Transmitted light from a mirror condenser. (c) Dark field illumination by scattering (the illuminating beam misses the objective lens). (d) High magnification illumination with reflected light – normally laser light.

We normally associate microscopes with visible light but x ray and electron microscopes provide greater resolution. The focusing and basic physics is the same as that just described, even though the lenses require different technology. The electron microscope requires vacuum chambers so that the electrons can proceed unheeded. Magnifications of 50 million times provide the ability to determine positions of individual atoms within materials. An electron microscope is shown in **Figure 26.21**. We do not use our eyes to form images; rather images are recorded electronically and displayed on computers. In fact observing and saving images formed by optical microscopes on computers is now done routinely. Video recordings of what occurs in a microscope can be made for viewing by many people at later dates. Physics provides the science and tools needed to generate the sequence of time-lapse images of meiosis similar to the sequence sketched in **Figure 26.22**.

Figure 26.21 An electron microscope has the capability to image individual atoms on a material. The microscope uses vacuum technology, sophisticated detectors and state of the art image processing software. (credit: Dave Pape)

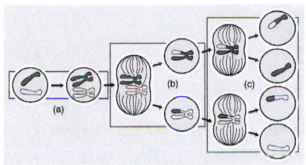

Figure 26.22 The image shows a sequence of events that takes place during meiosis. (credit: PatríciaR, Wikimedia Commons; National Center for Biotechnology Information)

Take-Home Experiment: Make a Lens

Look through a clear glass or plastic bottle and describe what you see. Now fill the bottle with water and describe what you see. Use the water bottle as a lens to produce the image of a bright object and estimate the focal length of the water bottle lens. How is the focal length a function of the depth of water in the bottle?

26.5 Telescopes

Learning Objectives

By the end of this section, you will be able to:

- Outline the invention of the telescope.
- Describe the working of a telescope.

The information presented in this section supports the following AP® learning objectives and science practices:

- **6.E.4.1** The student is able to plan data collection strategies and perform data analysis and evaluation of evidence about the formation of images due to reflection of light from curved spherical mirrors. **(S.P. 3.2, 4.1, 5.1, 5.2, 5.3)**
- **6.E.5.1** The student is able to use quantitative and qualitative representations and models to analyze situations and solve problems about image formation occurring due to the refraction of light through thin lenses. **(S.P. 1.4, 2.2)**

Telescopes are meant for viewing distant objects, producing an image that is larger than the image that can be seen with the unaided eye. Telescopes gather far more light than the eye, allowing dim objects to be observed with greater magnification and better resolution. Although Galileo is often credited with inventing the telescope, he actually did not. What he did was more important. He constructed several early telescopes, was the first to study the heavens with them, and made monumental discoveries using them. Among these are the moons of Jupiter, the craters and mountains on the Moon, the details of sunspots, and the fact that the Milky Way is composed of vast numbers of individual stars.

Figure 26.23(a) shows a telescope made of two lenses, the convex objective and the concave eyepiece, the same construction used by Galileo. Such an arrangement produces an upright image and is used in spyglasses and opera glasses.

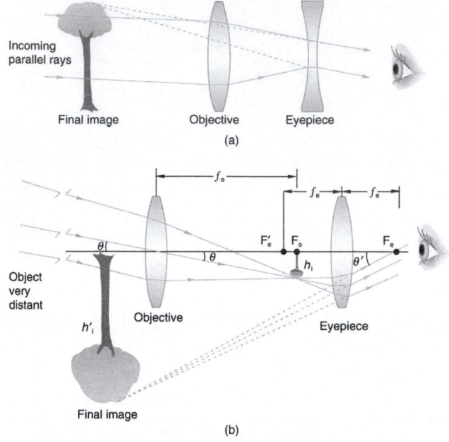

Figure 26.23 (a) Galileo made telescopes with a convex objective and a concave eyepiece. These produce an upright image and are used in spyglasses. (b) Most simple telescopes have two convex lenses. The objective forms a case 1 image that is the object for the eyepiece. The eyepiece forms a case 2 final image that is magnified.

The most common two-lens telescope, like the simple microscope, uses two convex lenses and is shown in **Figure 26.23**(b). The object is so far away from the telescope that it is essentially at infinity compared with the focal lengths of the lenses ($d_o \approx \infty$). The first image is thus produced at $d_i = f_o$, as shown in the figure. To prove this, note that

$$\frac{1}{d_i} = \frac{1}{f_o} - \frac{1}{d_o} = \frac{1}{f_o} - \frac{1}{\infty}. \qquad (26.25)$$

Because $1/\infty = 0$, this simplifies to

$$\frac{1}{d_i} = \frac{1}{f_o}, \qquad (26.26)$$

which implies that $d_i = f_o$, as claimed. It is true that for any distant object and any lens or mirror, the image is at the focal length.

The first image formed by a telescope objective as seen in **Figure 26.23**(b) will not be large compared with what you might see by looking at the object directly. For example, the spot formed by sunlight focused on a piece of paper by a magnifying glass is the image of the Sun, and it is small. The telescope eyepiece (like the microscope eyepiece) magnifies this first image. The distance between the eyepiece and the objective lens is made slightly less than the sum of their focal lengths so that the first image is closer to the eyepiece than its focal length. That is, d_o' is less than f_e , and so the eyepiece forms a case 2 image that is large and to the left for easy viewing. If the angle subtended by an object as viewed by the unaided eye is θ , and the angle subtended by the telescope image is θ' , then the **angular magnification** M is defined to be their ratio. That is, $M = \theta'/\theta$. It can be shown that the angular magnification of a telescope is related to the focal lengths of the objective and eyepiece; and is given by

$$M = \frac{\theta'}{\theta} = -\frac{f_o}{f_e}. \qquad (26.27)$$

The minus sign indicates the image is inverted. To obtain the greatest angular magnification, it is best to have a long focal length objective and a short focal length eyepiece. The greater the angular magnification M , the larger an object will appear when viewed through a telescope, making more details visible. Limits to observable details are imposed by many factors, including lens quality and atmospheric disturbance.

The image in most telescopes is inverted, which is unimportant for observing the stars but a real problem for other applications, such as telescopes on ships or telescopic gun sights. If an upright image is needed, Galileo's arrangement in **Figure 26.23**(a) can be used. But a more common arrangement is to use a third convex lens as an eyepiece, increasing the distance between the first two and inverting the image once again as seen in **Figure 26.24**.

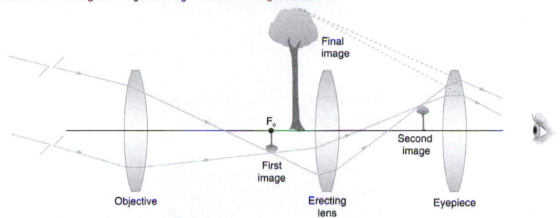

Figure 26.24 This arrangement of three lenses in a telescope produces an upright final image. The first two lenses are far enough apart that the second lens inverts the image of the first one more time. The third lens acts as a magnifier and keeps the image upright and in a location that is easy to view.

A telescope can also be made with a concave mirror as its first element or objective, since a concave mirror acts like a convex lens as seen in **Figure 26.25**. Flat mirrors are often employed in optical instruments to make them more compact or to send light to cameras and other sensing devices. There are many advantages to using mirrors rather than lenses for telescope objectives. Mirrors can be constructed much larger than lenses and can, thus, gather large amounts of light, as needed to view distant galaxies, for example. Large and relatively flat mirrors have very long focal lengths, so that great angular magnification is possible.

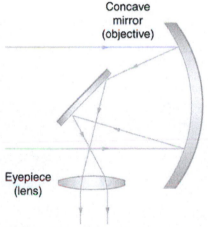

Figure 26.25 A two-element telescope composed of a mirror as the objective and a lens for the eyepiece is shown. This telescope forms an image in the same manner as the two-convex-lens telescope already discussed, but it does not suffer from chromatic aberrations. Such telescopes can gather more light, since larger mirrors than lenses can be constructed.

Telescopes, like microscopes, can utilize a range of frequencies from the electromagnetic spectrum. **Figure 26.26**(a) shows the Australia Telescope Compact Array, which uses six 22-m antennas for mapping the southern skies using radio waves. **Figure 26.26**(b) shows the focusing of x rays on the Chandra X-ray Observatory—a satellite orbiting earth since 1999 and looking at high temperature events as exploding stars, quasars, and black holes. X rays, with much more energy and shorter wavelengths than RF and light, are mainly absorbed and not reflected when incident perpendicular to the medium. But they can be reflected when incident at small glancing angles, much like a rock will skip on a lake if thrown at a small angle. The mirrors for the Chandra consist of a long barrelled pathway and 4 pairs of mirrors to focus the rays at a point 10 meters away from the entrance. The mirrors are extremely smooth and consist of a glass ceramic base with a thin coating of metal (iridium). Four pairs of precision manufactured mirrors are exquisitely shaped and aligned so that x rays ricochet off the mirrors like bullets off a wall, focusing on a spot.

Figure 26.26 (a) The Australia Telescope Compact Array at Narrabri (500 km NW of Sydney). (credit: Ian Bailey) (b) The focusing of x rays on the Chandra Observatory, a satellite orbiting earth. X rays ricochet off 4 pairs of mirrors forming a barrelled pathway leading to the focus point. (credit: NASA)

A current exciting development is a collaborative effort involving 17 countries to construct a Square Kilometre Array (SKA) of telescopes capable of covering from 80 MHz to 2 GHz. The initial stage of the project is the construction of the Australian Square Kilometre Array Pathfinder in Western Australia (see Figure 26.27). The project will use cutting-edge technologies such as **adaptive optics** in which the lens or mirror is constructed from lots of carefully aligned tiny lenses and mirrors that can be manipulated using computers. A range of rapidly changing distortions can be minimized by deforming or tilting the tiny lenses and mirrors. The use of adaptive optics in vision correction is a current area of research.

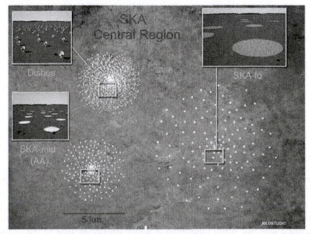

Figure 26.27 An artist's impression of the Australian Square Kilometre Array Pathfinder in Western Australia is displayed. (credit: SPDO, XILOSTUDIOS)

26.6 Aberrations

Learning Objectives
By the end of this section, you will be able to: • Describe optical aberration.

Real lenses behave somewhat differently from how they are modeled using the thin lens equations, producing **aberrations**. An aberration is a distortion in an image. There are a variety of aberrations due to a lens size, material, thickness, and position of the object. One common type of aberration is chromatic aberration, which is related to color. Since the index of refraction of lenses depends on color or wavelength, images are produced at different places and with different magnifications for different colors. (The law of reflection is independent of wavelength, and so mirrors do not have this problem. This is another advantage for mirrors in optical systems such as telescopes.) Figure 26.28(a) shows chromatic aberration for a single convex lens and its partial correction with a two-lens system. Violet rays are bent more than red, since they have a higher index of refraction and are thus focused closer to the lens. The diverging lens partially corrects this, although it is usually not possible to do so completely. Lenses of different materials and having different dispersions may be used. For example an achromatic doublet consisting of a converging lens made of crown glass and a diverging lens made of flint glass in contact can dramatically reduce chromatic aberration (see Figure 26.28(b)).

Quite often in an imaging system the object is off-center. Consequently, different parts of a lens or mirror do not refract or reflect the image to the same point. This type of aberration is called a coma and is shown in Figure 26.29. The image in this case often appears pear-shaped. Another common aberration is spherical aberration where rays converging from the outer edges of a lens converge to a focus closer to the lens and rays closer to the axis focus further (see Figure 26.30). Aberrations due to astigmatism in the lenses of the eyes are discussed in Vision Correction, and a chart used to detect astigmatism is shown in Figure 26.8. Such aberrations and can also be an issue with manufactured lenses.

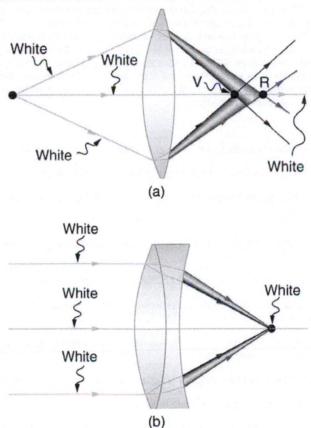

(a)

(b)

Figure 26.28 (a) Chromatic aberration is caused by the dependence of a lens's index of refraction on color (wavelength). The lens is more powerful for violet (V) than for red (R), producing images with different locations and magnifications. (b) Multiple-lens systems can partially correct chromatic aberrations, but they may require lenses of different materials and add to the expense of optical systems such as cameras.

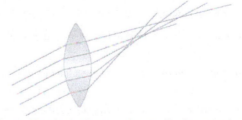

Figure 26.29 A coma is an aberration caused by an object that is off-center, often resulting in a pear-shaped image. The rays originate from points that are not on the optical axis and they do not converge at one common focal point.

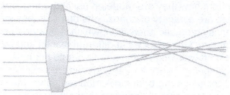

Figure 26.30 Spherical aberration is caused by rays focusing at different distances from the lens.

The image produced by an optical system needs to be bright enough to be discerned. It is often a challenge to obtain a sufficiently bright image. The brightness is determined by the amount of light passing through the optical system. The optical components determining the brightness are the diameter of the lens and the diameter of pupils, diaphragms or aperture stops placed in front of lenses. Optical systems often have entrance and exit pupils to specifically reduce aberrations but they inevitably reduce brightness as well. Consequently, optical systems need to strike a balance between the various components used. The iris in the eye dilates and constricts, acting as an entrance pupil. You can see objects more clearly by looking through a small hole made with your hand in the shape of a fist. Squinting, or using a small hole in a piece of paper, also will make the object sharper.

So how are aberrations corrected? The lenses may also have specially shaped surfaces, as opposed to the simple spherical shape that is relatively easy to produce. Expensive camera lenses are large in diameter, so that they can gather more light, and need several elements to correct for various aberrations. Further, advances in materials science have resulted in lenses with a range of refractive indices—technically referred to as graded index (GRIN) lenses. Spectacles often have the ability to provide a range of focusing ability using similar techniques. GRIN lenses are particularly important at the end of optical fibers in endoscopes. Advanced computing techniques allow for a range of corrections on images after the image has been collected and certain characteristics of the optical system are known. Some of these techniques are sophisticated versions of what are available on commercial packages like Adobe Photoshop.

Glossary

aberration: failure of rays to converge at one focus because of limitations or defects in a lens or mirror

accommodation: the ability of the eye to adjust its focal length is known as accommodation

adaptive optics: optical technology in which computers adjust the lenses and mirrors in a device to correct for image distortions

angular magnification: a ratio related to the focal lengths of the objective and eyepiece and given as $M = -\dfrac{f_o}{f_e}$

astigmatism: the result of an inability of the cornea to properly focus an image onto the retina

color constancy: a part of the visual perception system that allows people to perceive color in a variety of conditions and to see some consistency in the color

compound microscope: a microscope constructed from two convex lenses, the first serving as the ocular lens(close to the eye) and the second serving as the objective lens

eyepiece: the lens or combination of lenses in an optical instrument nearest to the eye of the observer

far point: the object point imaged by the eye onto the retina in an unaccommodated eye

farsightedness: another term for hyperopia, the condition of an eye where incoming rays of light reach the retina before they converge into a focused image

hues: identity of a color as it relates specifically to the spectrum

hyperopia: the condition of an eye where incoming rays of light reach the retina before they converge into a focused image

laser vision correction: a medical procedure used to correct astigmatism and eyesight deficiencies such as myopia and hyperopia

myopia: a visual defect in which distant objects appear blurred because their images are focused in front of the retina rather than being focused on the retina

near point: the point nearest the eye at which an object is accurately focused on the retina at full accommodation

nearsightedness: another term for myopia, a visual defect in which distant objects appear blurred because their images are focused in front of the retina rather than being focused on the retina

numerical aperture: a number or measure that expresses the ability of a lens to resolve fine detail in an object being observed. Derived by mathematical formula

$$NA = n \sin \alpha,$$

where n is the refractive index of the medium between the lens and the specimen and $\alpha = \theta / 2$

objective lens: the lens nearest to the object being examined

presbyopia: a condition in which the lens of the eye becomes progressively unable to focus on objects close to the viewer

retinex: a theory proposed to explain color and brightness perception and constancies; is a combination of the words retina and cortex, which are the two areas responsible for the processing of visual information

retinex theory of color vision: the ability to perceive color in an ambient-colored environment

rods and cones: two types of photoreceptors in the human retina; rods are responsible for vision at low light levels, while cones are active at higher light levels

simplified theory of color vision: a theory that states that there are three primary colors, which correspond to the three types of cones

Section Summary

26.1 Physics of the Eye

- Image formation by the eye is adequately described by the thin lens equations:

$$P = \frac{1}{d_o} + \frac{1}{d_i} \text{ and } \frac{h_i}{h_o} = -\frac{d_i}{d_o} = m.$$

- The eye produces a real image on the retina by adjusting its focal length and power in a process called accommodation.
- For close vision, the eye is fully accommodated and has its greatest power, whereas for distant vision, it is totally relaxed and has its smallest power.
- The loss of the ability to accommodate with age is called presbyopia, which is corrected by the use of a converging lens to add power for close vision.

26.2 Vision Correction

- Nearsightedness, or myopia, is the inability to see distant objects and is corrected with a diverging lens to reduce power.
- Farsightedness, or hyperopia, is the inability to see close objects and is corrected with a converging lens to increase power.
- In myopia and hyperopia, the corrective lenses produce images at a distance that the person can see clearly—the far point and near point, respectively.

26.3 Color and Color Vision

- The eye has four types of light receptors—rods and three types of color-sensitive cones.
- The rods are good for night vision, peripheral vision, and motion changes, while the cones are responsible for central vision and color.
- We perceive many hues, from light having mixtures of wavelengths.
- A simplified theory of color vision states that there are three primary colors, which correspond to the three types of cones, and that various combinations of the primary colors produce all the hues.
- The true color of an object is related to its relative absorption of various wavelengths of light. The color of a light source is related to the wavelengths it produces.
- Color constancy is the ability of the eye-brain system to discern the true color of an object illuminated by various light sources.
- The retinex theory of color vision explains color constancy by postulating the existence of three retinexes or image systems, associated with the three types of cones that are compared to obtain sophisticated information.

26.4 Microscopes

- The microscope is a multiple-element system having more than a single lens or mirror.
- Many optical devices contain more than a single lens or mirror. These are analysed by considering each element sequentially. The image formed by the first is the object for the second, and so on. The same ray tracing and thin lens techniques apply to each lens element.
- The overall magnification of a multiple-element system is the product of the magnifications of its individual elements. For a two-element system with an objective and an eyepiece, this is

$$m = m_o m_e,$$

where m_o is the magnification of the objective and m_e is the magnification of the eyepiece, such as for a microscope.

- Microscopes are instruments for allowing us to see detail we would not be able to see with the unaided eye and consist of a range of components.
- The eyepiece and objective contribute to the magnification. The numerical aperture (NA) of an objective is given by

$$NA = n \sin \alpha$$

where n is the refractive index and α the angle of acceptance.

- Immersion techniques are often used to improve the light gathering ability of microscopes. The specimen is illuminated by transmitted, scattered or reflected light though a condenser.
- The $f/\#$ describes the light gathering ability of a lens. It is given by

$$f/\# = \frac{f}{D} \approx \frac{1}{2NA}.$$

26.5 Telescopes
- Simple telescopes can be made with two lenses. They are used for viewing objects at large distances and utilize the entire range of the electromagnetic spectrum.
- The angular magnification M for a telescope is given by

$$M = \frac{\theta'}{\theta} = -\frac{f_o}{f_e},$$

where θ is the angle subtended by an object viewed by the unaided eye, θ' is the angle subtended by a magnified image, and f_o and f_e are the focal lengths of the objective and the eyepiece.

26.6 Aberrations
- Aberrations or image distortions can arise due to the finite thickness of optical instruments, imperfections in the optical components, and limitations on the ways in which the components are used.
- The means for correcting aberrations range from better components to computational techniques.

Conceptual Questions

26.1 Physics of the Eye

1. If the lens of a person's eye is removed because of cataracts (as has been done since ancient times), why would you expect a spectacle lens of about 16 D to be prescribed?

2. A cataract is cloudiness in the lens of the eye. Is light dispersed or diffused by it?

3. When laser light is shone into a relaxed normal-vision eye to repair a tear by spot-welding the retina to the back of the eye, the rays entering the eye must be parallel. Why?

4. How does the power of a dry contact lens compare with its power when resting on the tear layer of the eye? Explain.

5. Why is your vision so blurry when you open your eyes while swimming under water? How does a face mask enable clear vision?

26.2 Vision Correction

6. It has become common to replace the cataract-clouded lens of the eye with an internal lens. This intraocular lens can be chosen so that the person has perfect distant vision. Will the person be able to read without glasses? If the person was nearsighted, is the power of the intraocular lens greater or less than the removed lens?

7. If the cornea is to be reshaped (this can be done surgically or with contact lenses) to correct myopia, should its curvature be made greater or smaller? Explain. Also explain how hyperopia can be corrected.

8. If there is a fixed percent uncertainty in LASIK reshaping of the cornea, why would you expect those people with the greatest correction to have a poorer chance of normal distant vision after the procedure?

9. A person with presbyopia has lost some or all of the ability to accommodate the power of the eye. If such a person's distant vision is corrected with LASIK, will she still need reading glasses? Explain.

26.3 Color and Color Vision

10. A pure red object on a black background seems to disappear when illuminated with pure green light. Explain why.

11. What is color constancy, and what are its limitations?

12. There are different types of color blindness related to the malfunction of different types of cones. Why would it be particularly useful to study those rare individuals who are color blind only in one eye or who have a different type of color blindness in each eye?

13. Propose a way to study the function of the rods alone, given they can sense light about 1000 times dimmer than the cones.

26.4 Microscopes

14. Geometric optics describes the interaction of light with macroscopic objects. Why, then, is it correct to use geometric optics to analyse a microscope's image?

15. The image produced by the microscope in **Figure 26.16** cannot be projected. Could extra lenses or mirrors project it? Explain.

16. Why not have the objective of a microscope form a case 2 image with a large magnification? (Hint: Consider the location of that image and the difficulty that would pose for using the eyepiece as a magnifier.)

17. What advantages do oil immersion objectives offer?

18. How does the NA of a microscope compare with the NA of an optical fiber?

26.5 Telescopes

19. If you want your microscope or telescope to project a real image onto a screen, how would you change the placement of the eyepiece relative to the objective?

26.6 Aberrations

20. List the various types of aberrations. What causes them and how can each be reduced?

Problems & Exercises

26.1 Physics of the Eye

Unless otherwise stated, the lens-to-retina distance is 2.00 cm.

1. What is the power of the eye when viewing an object 50.0 cm away?

2. Calculate the power of the eye when viewing an object 3.00 m away.

3. (a) The print in many books averages 3.50 mm in height. How high is the image of the print on the retina when the book is held 30.0 cm from the eye?

(b) Compare the size of the print to the sizes of rods and cones in the fovea and discuss the possible details observable in the letters. (The eye-brain system can perform better because of interconnections and higher order image processing.)

4. Suppose a certain person's visual acuity is such that he can see objects clearly that form an image $4.00 \ \mu m$ high on his retina. What is the maximum distance at which he can read the 75.0 cm high letters on the side of an airplane?

5. People who do very detailed work close up, such as jewellers, often can see objects clearly at much closer distance than the normal 25 cm.

(a) What is the power of the eyes of a woman who can see an object clearly at a distance of only 8.00 cm?

(b) What is the size of an image of a 1.00 mm object, such as lettering inside a ring, held at this distance?

(c) What would the size of the image be if the object were held at the normal 25.0 cm distance?

26.2 Vision Correction

6. What is the far point of a person whose eyes have a relaxed power of 50.5 D?

7. What is the near point of a person whose eyes have an accommodated power of 53.5 D?

8. (a) A laser vision correction reshaping the cornea of a myopic patient reduces the power of his eye by 9.00 D, with a $\pm 5.0\%$ uncertainty in the final correction. What is the range of diopters for spectacle lenses that this person might need after LASIK procedure? (b) Was the person nearsighted or farsighted before the procedure? How do you know?

9. In a LASIK vision correction, the power of a patient's eye is increased by 3.00 D. Assuming this produces normal close vision, what was the patient's near point before the procedure?

10. What was the previous far point of a patient who had laser vision correction that reduced the power of her eye by 7.00 D, producing normal distant vision for her?

11. A severely myopic patient has a far point of 5.00 cm. By how many diopters should the power of his eye be reduced in laser vision correction to obtain normal distant vision for him?

12. A student's eyes, while reading the blackboard, have a power of 51.0 D. How far is the board from his eyes?

13. The power of a physician's eyes is 53.0 D while examining a patient. How far from her eyes is the feature being examined?

14. A young woman with normal distant vision has a 10.0% ability to accommodate (that is, increase) the power of her eyes. What is the closest object she can see clearly?

15. The far point of a myopic administrator is 50.0 cm. (a) What is the relaxed power of his eyes? (b) If he has the normal 8.00% ability to accommodate, what is the closest object he can see clearly?

16. A very myopic man has a far point of 20.0 cm. What power contact lens (when on the eye) will correct his distant vision?

17. Repeat the previous problem for eyeglasses held 1.50 cm from the eyes.

18. A myopic person sees that her contact lens prescription is $-4.00 \ D$. What is her far point?

19. Repeat the previous problem for glasses that are 1.75 cm from the eyes.

20. The contact lens prescription for a mildly farsighted person is 0.750 D, and the person has a near point of 29.0 cm. What is the power of the tear layer between the cornea and the lens if the correction is ideal, taking the tear layer into account?

21. A nearsighted man cannot see objects clearly beyond 20 cm from his eyes. How close must he stand to a mirror in order to see what he is doing when he shaves?

22. A mother sees that her child's contact lens prescription is 0.750 D. What is the child's near point?

23. Repeat the previous problem for glasses that are 2.20 cm from the eyes.

24. The contact lens prescription for a nearsighted person is $-4.00 \ D$ and the person has a far point of 22.5 cm. What is the power of the tear layer between the cornea and the lens if the correction is ideal, taking the tear layer into account?

25. Unreasonable Results

A boy has a near point of 50 cm and a far point of 500 cm. Will a $-4.00 \ D$ lens correct his far point to infinity?

26.4 Microscopes

26. A microscope with an overall magnification of 800 has an objective that magnifies by 200. (a) What is the magnification of the eyepiece? (b) If there are two other objectives that can be used, having magnifications of 100 and 400, what other total magnifications are possible?

27. (a) What magnification is produced by a 0.150 cm focal length microscope objective that is 0.155 cm from the object being viewed? (b) What is the overall magnification if an $8\times$ eyepiece (one that produces a magnification of 8.00) is used?

28. (a) Where does an object need to be placed relative to a microscope for its 0.500 cm focal length objective to produce a magnification of -400? (b) Where should the 5.00 cm focal length eyepiece be placed to produce a further fourfold (4.00) magnification?

29. You switch from a $1.40NA \ 60\times$ oil immersion objective to a $1.40NA \ 60\times$ oil immersion objective. What are the acceptance angles for each? Compare and comment on the values. Which would you use first to locate the target area on your specimen?

30. An amoeba is 0.305 cm away from the 0.300 cm focal length objective lens of a microscope. (a) Where is the image formed by the objective lens? (b) What is this image's magnification? (c) An eyepiece with a 2.00 cm focal length is placed 20.0 cm from the objective. Where is the final image? (d) What magnification is produced by the eyepiece? (e) What is the overall magnification? (See **Figure 26.16**.)

31. You are using a standard microscope with a $0.10NA$ $4\times$ objective and switch to a $0.65NA$ $40\times$ objective. What are the acceptance angles for each? Compare and comment on the values. Which would you use first to locate the target area on of your specimen? (See **Figure 26.17**.)

32. Unreasonable Results

Your friends show you an image through a microscope. They tell you that the microscope has an objective with a 0.500 cm focal length and an eyepiece with a 5.00 cm focal length. The resulting overall magnification is 250,000. Are these viable values for a microscope?

26.5 Telescopes

Unless otherwise stated, the lens-to-retina distance is 2.00 cm.

33. What is the angular magnification of a telescope that has a 100 cm focal length objective and a 2.50 cm focal length eyepiece?

34. Find the distance between the objective and eyepiece lenses in the telescope in the above problem needed to produce a final image very far from the observer, where vision is most relaxed. Note that a telescope is normally used to view very distant objects.

35. A large reflecting telescope has an objective mirror with a 10.0 m radius of curvature. What angular magnification does it produce when a 3.00 m focal length eyepiece is used?

36. A small telescope has a concave mirror with a 2.00 m radius of curvature for its objective. Its eyepiece is a 4.00 cm focal length lens. (a) What is the telescope's angular magnification? (b) What angle is subtended by a 25,000 km diameter sunspot? (c) What is the angle of its telescopic image?

37. A $7.5\times$ binocular produces an angular magnification of -7.50, acting like a telescope. (Mirrors are used to make the image upright.) If the binoculars have objective lenses with a 75.0 cm focal length, what is the focal length of the eyepiece lenses?

38. Construct Your Own Problem

Consider a telescope of the type used by Galileo, having a convex objective and a concave eyepiece as illustrated in **Figure 26.23**(a). Construct a problem in which you calculate the location and size of the image produced. Among the things to be considered are the focal lengths of the lenses and their relative placements as well as the size and location of the object. Verify that the angular magnification is greater than one. That is, the angle subtended at the eye by the image is greater than the angle subtended by the object.

26.6 Aberrations

39. Integrated Concepts

(a) During laser vision correction, a brief burst of 193 nm ultraviolet light is projected onto the cornea of the patient. It makes a spot 1.00 mm in diameter and deposits 0.500 mJ of energy. Calculate the depth of the layer ablated, assuming the corneal tissue has the same properties as water and is initially at $34.0°C$. The tissue's temperature is increased to $100°C$ and evaporated without further temperature increase.

(b) Does your answer imply that the shape of the cornea can be finely controlled?

Test Prep for AP® Courses

26.1 Physics of the Eye

1. A tree that is 3 m tall is viewed from a distance of 25 m. If the cornea-to-retina distance of an ideal eye is 2 cm, how tall is the image of the tree on the observer's retina?
a. 0.24 cm
b. 0.5 cm
c. 0.5 m
d. 0.08 cm

2. Often people with lens-to-retina distances smaller than 2 cm purchase glasses to place in front of their eyes.
a. Explain why people with lens-to-retina distances smaller than 2 cm need glasses.
b. Explain whether the glasses should be composed of converging or diverging lenses.
c. Draw a ray diagram demonstrating the ability to see with and without the glasses.

26.2 Vision Correction

3. Which of the following types of light have a wavelength greater than that of visible light?
I.   gamma rays
II.   infrared
III. radio
IV.   ultraviolet
a. I, II, and III
b. I and IV only
c. II and III only
d. III only

4. In LASIK surgery, a coherent UV light of 193 nm is focused on the corneal tissue.
a. Explain the importance of using light that is all the same wavelength.
b. Explain why UV light is more effective than infrared light at evaporating the corneal tissue.

26.3 Color and Color Vision

5. A student sees a piece of paper sitting on a table. Which of the following would not result in the student observing the paper as yellow?
a. Yellow light shines on a black paper.
b. White light shines on a yellow paper.
c. Yellow light shines on a white paper.
d. Red and green lights shine on a white paper.

6. A white light is projected onto a tablecloth. Using the light reflecting off the tablecloth, an observer determines that the color of the tablecloth is blue.
a. Using the wave model of light, explain how the observer is capable of making this judgment.
b. Describe how using the particle model of light limits our explanation of the observer's judgment.

26.4 Microscopes

7. Which of the following correctly describes the image created by a microscope?
a. The image is real, inverted, and magnified.
b. The image is virtual, inverted, and magnified.
c. The image is real, upright, and magnified.
d. The image is virtual, upright, and magnified.

8. Use the diagram shown below to answer the following questions.

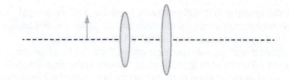

Figure 26.31

Draw two rays leaving the arrow shown to the left of both lenses. Use ray tracing to draw the images created by the objective and eyepiece lenses. Label the images as i_o and i_e.

26.5 Telescopes

9. Which of the following is an advantage to using a concave mirror in the construction of a telescope?
I.   The telescope can gather more light than a telescope using lenses.
II.   The telescope does not suffer from chromatic aberration.
III. The telescope can provide greater magnification than a telescope using lenses.
a. I and III only
b. II only
c. I and II only
d. I, II, and III

10. A spherical mirror is used to construct a telescope.
a. Using the picture below, draw two rays incident on the object mirror and continue their path through the eye lens.

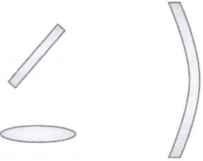

Figure 26.32
b. The plane mirror is replaced with a concave lens. Using the picture below, draw the path of two incident rays.

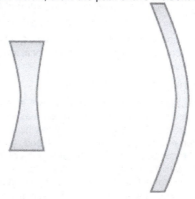

Figure 26.33
c. Using the concave lens setup, describe the final image created by the concave lens.

11. Two concave lenses, of focal lengths 500 mm and 20 mm, are used in the construction of a telescope. Given any potential arrangement, what is the largest possible

magnification the telescope may have?
   a. 100×
   b. 10,000×
   c. 25×
   d. 4×

27 WAVE OPTICS

Figure 27.1 The colors reflected by this compact disc vary with angle and are not caused by pigments. Colors such as these are direct evidence of the wave character of light. (credit: Infopro, Wikimedia Commons)

Chapter Outline

Connection for AP® Courses

If you have ever looked at the reds, blues, and greens in a sunlit soap bubble and wondered how straw-colored soapy water could produce them, you have hit upon one of the many phenomena that can only be explained by the wave character of light. The same is true for the colors seen in an oil slick or in the light reflected from an optical data disk. These and other interesting phenomena, such as the dispersion of white light into a rainbow of colors when passed through a narrow slit, cannot be explained fully by geometric optics. In these cases, light interacts with objects and exhibits a number of wave characteristics. The branch of optics that considers the behavior of light when it exhibits wave characteristics is called "wave optics" (or sometimes "physical optics").

Figure 27.2 Multicolored soap bubbles (credit: Scott Robinson, Flickr).

These soap bubbles exhibit brilliant colors when exposed to sunlight. How are the colors produced if they are not pigments in the soap?

This chapter supports Big Idea 6 in its coverage of wave optics by presenting explanations and examples of many phenomena that can only be explained by the wave aspect of light. You will learn how only waves can exhibit diffraction and interference patterns that we observe in light (Enduring Understanding 6.C). As explained by Huygens's principle, diffraction is the bending of waves around the edges of a nontransparent object or after passing through an opening (Essential Knowledge 6.C.4). Interference results from the superposition of two or more traveling waves (Enduring Understanding 6.D, Enduring Understanding 6.D.1). Superposition causes variations in the resultant wave amplitude (Essential Knowledge 6.D.2). The interference can be described as constructive interference, which increases amplitude, and destructive interference, which decreases amplitude. Based on an understanding of diffraction and interference of light, this chapter also explains experimental observations that occur when light passes through an opening or set of openings with dimensions comparable to the wavelength of the light – specifically the effects of double-slit, multiple-slit (Essential Knowledge 6.C.3), and single-slit (Essential Knowledge 6.C.2) openings. Another aspect of light waves that you will learn about in this chapter is polarization, a phenomenon in which light waves all vibrate in a single plane. The explanation for this phenomenon is based on the fact that light is a traveling electromagnetic wave (Enduring Understanding 6.A) that propagates via transverse oscillations of both electric and magnetic field vectors (Enduring Understanding 6.A.1). Light waves can be polarized by passing through filters. Many sunglasses contain polarizing filters to reduce glare, and certain types of 3-D glasses use polarization to create an effect of depth on the movie screen.

Big Idea 6 Waves can transfer energy and momentum from one location to another without the permanent transfer of mass and serve as a mathematical model for the description of other phenomena.

Enduring Understanding 6.A A wave is a traveling disturbance that transfers energy and momentum.

Essential Knowledge 6.A.1 Waves can propagate via different oscillation modes such as transverse and longitudinal.

Enduring Understanding 6.C Only waves exhibit interference and diffraction.

Essential Knowledge 6.C.2 When waves pass through an opening whose dimensions are comparable to the wavelength, a diffraction pattern can be observed.

Essential Knowledge 6.C.3 When waves pass through a set of openings whose spacing is comparable to the wavelength, an interference pattern can be observed. Examples should include monochromatic double-slit interference.

Essential Knowledge 6.C.4 When waves pass by an edge, they can diffract into the "shadow region" behind the edge. Examples should include hearing around corners, but not seeing around them, and water waves bending around obstacles.

Enduring Understanding 6.D Interference and superposition lead to standing waves and beats.

Essential Knowledge 6.D.1 Two or more wave pulses can interact in such a way as to produce amplitude variations in the resultant wave. When two pulses cross, they travel through each other; they do not bounce off each other. Where the pulses overlap, the resulting displacement can be determined by adding the displacements of the two pulses. This is called superposition.

Essential Knowledge 6.D.2 Two or more traveling waves can interact in such a way as to produce amplitude variations in the resultant wave.

27.1 The Wave Aspect of Light: Interference

Learning Objectives

By the end of this section, you will be able to:

> - Discuss the wave character of light.
> - Identify the changes when light enters a medium.

We know that visible light is the type of electromagnetic wave to which our eyes respond. Like all other electromagnetic waves, it obeys the equation

$$c = f\lambda, \tag{27.1}$$

where $c = 3\times10^8$ m/s is the speed of light in vacuum, f is the frequency of the electromagnetic waves, and λ is its wavelength. The range of visible wavelengths is approximately 380 to 760 nm. As is true for all waves, light travels in straight lines and acts like a ray when it interacts with objects several times as large as its wavelength. However, when it interacts with smaller objects, it displays its wave characteristics prominently. Interference is the hallmark of a wave, and in **Figure 27.3** both the ray and wave characteristics of light can be seen. The laser beam emitted by the observatory epitomizes a ray, traveling in a straight line. However, passing a pure-wavelength beam through vertical slits with a size close to the wavelength of the beam reveals the wave character of light, as the beam spreads out horizontally into a pattern of bright and dark regions caused by systematic constructive and destructive interference. Rather than spreading out, a ray would continue traveling straight ahead after passing through slits.

Making Connections: Waves

The most certain indication of a wave is interference. This wave characteristic is most prominent when the wave interacts with an object that is not large compared with the wavelength. Interference is observed for water waves, sound waves, light waves, and (as we will see in **Special Relativity**) for matter waves, such as electrons scattered from a crystal.

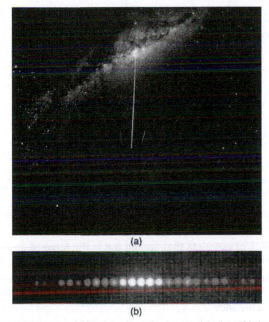

(a)

(b)

Figure 27.3 (a) The laser beam emitted by an observatory acts like a ray, traveling in a straight line. This laser beam is from the Paranal Observatory of the European Southern Observatory. (credit: Yuri Beletsky, European Southern Observatory) (b) A laser beam passing through a grid of vertical slits produces an interference pattern—characteristic of a wave. (credit: Shim'on and Slava Rybka, Wikimedia Commons)

Light has wave characteristics in various media as well as in a vacuum. When light goes from a vacuum to some medium, like water, its speed and wavelength change, but its frequency f remains the same. (We can think of light as a forced oscillation that must have the frequency of the original source.) The speed of light in a medium is $v = c/n$, where n is its index of refraction. If we divide both sides of equation $c = f\lambda$ by n, we get $c/n = v = f\lambda/n$. This implies that $v = f\lambda_n$, where λ_n is the **wavelength in a medium** and that

$$\lambda_n = \frac{\lambda}{n}, \tag{27.2}$$

where λ is the wavelength in vacuum and n is the medium's index of refraction. Therefore, the wavelength of light is smaller in any medium than it is in vacuum. In water, for example, which has $n = 1.333$, the range of visible wavelengths is $(380 \text{ nm})/1.333$ to $(760 \text{ nm})/1.333$, or $\lambda_n = 285$ to 570 nm. Although wavelengths change while traveling from one medium to another, colors do not, since colors are associated with frequency.

27.2 Huygens's Principle: Diffraction

Figure 27.4 shows how a transverse wave looks as viewed from above and from the side. A light wave can be imagined to propagate like this, although we do not actually see it wiggling through space. From above, we view the wavefronts (or wave crests) as we would by looking down on the ocean waves. The side view would be a graph of the electric or magnetic field. The view from above is perhaps the most useful in developing concepts about wave optics.

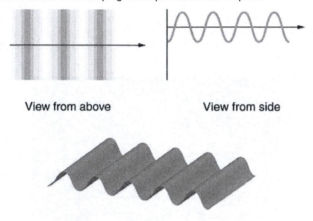

View from above   View from side

Overall view

Figure 27.4 A transverse wave, such as an electromagnetic wave like light, as viewed from above and from the side. The direction of propagation is perpendicular to the wavefronts (or wave crests) and is represented by an arrow like a ray.

The Dutch scientist Christiaan Huygens (1629–1695) developed a useful technique for determining in detail how and where waves propagate. Starting from some known position, **Huygens's principle** states that:

Every point on a wavefront is a source of wavelets that spread out in the forward direction at the same speed as the wave itself. The new wavefront is a line tangent to all of the wavelets.

Figure 27.5 shows how Huygens's principle is applied. A wavefront is the long edge that moves, for example, the crest or the trough. Each point on the wavefront emits a semicircular wave that moves at the propagation speed v. These are drawn at a time t later, so that they have moved a distance $s = vt$. The new wavefront is a line tangent to the wavelets and is where we would expect the wave to be a time t later. Huygens's principle works for all types of waves, including water waves, sound waves, and light waves. We will find it useful not only in describing how light waves propagate, but also in explaining the laws of reflection and refraction. In addition, we will see that Huygens's principle tells us how and where light rays interfere.

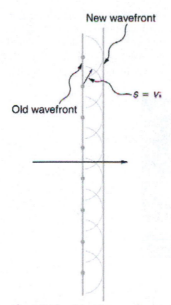

Figure 27.5 Huygens's principle applied to a straight wavefront. Each point on the wavefront emits a semicircular wavelet that moves a distance $s = vt$. The new wavefront is a line tangent to the wavelets.

Figure 27.6 shows how a mirror reflects an incoming wave at an angle equal to the incident angle, verifying the law of reflection. As the wavefront strikes the mirror, wavelets are first emitted from the left part of the mirror and then the right. The wavelets closer to the left have had time to travel farther, producing a wavefront traveling in the direction shown.

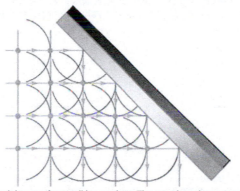

Figure 27.6 Huygens's principle applied to a straight wavefront striking a mirror. The wavelets shown were emitted as each point on the wavefront struck the mirror. The tangent to these wavelets shows that the new wavefront has been reflected at an angle equal to the incident angle. The direction of propagation is perpendicular to the wavefront, as shown by the downward-pointing arrows.

The law of refraction can be explained by applying Huygens's principle to a wavefront passing from one medium to another (see Figure 27.7). Each wavelet in the figure was emitted when the wavefront crossed the interface between the media. Since the speed of light is smaller in the second medium, the waves do not travel as far in a given time, and the new wavefront changes direction as shown. This explains why a ray changes direction to become closer to the perpendicular when light slows down. Snell's law can be derived from the geometry in Figure 27.7, but this is left as an exercise for ambitious readers.

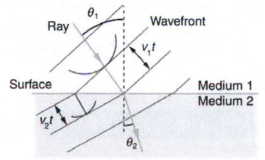

Figure 27.7 Huygens's principle applied to a straight wavefront traveling from one medium to another where its speed is less. The ray bends toward the perpendicular, since the wavelets have a lower speed in the second medium.

What happens when a wave passes through an opening, such as light shining through an open door into a dark room? For light, we expect to see a sharp shadow of the doorway on the floor of the room, and we expect no light to bend around corners into other parts of the room. When sound passes through a door, we expect to hear it everywhere in the room and, thus, expect that sound spreads out when passing through such an opening (see Figure 27.8). What is the difference between the behavior of sound waves and light waves in this case? The answer is that light has very short wavelengths and acts like a ray. Sound has wavelengths on the order of the size of the door and bends around corners (for frequency of 1000 Hz,

$$\lambda = c / f = (330 \text{ m} / \text{s}) / (1000 \text{ s}^{-1}) = 0.33 \text{ m}$$, about three times smaller than the width of the doorway).

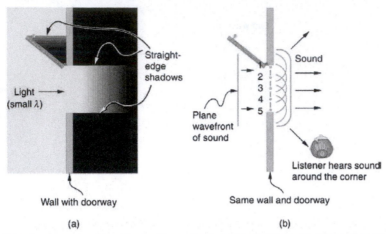

(a)   (b)

Figure 27.8 (a) Light passing through a doorway makes a sharp outline on the floor. Since light's wavelength is very small compared with the size of the door, it acts like a ray. (b) Sound waves bend into all parts of the room, a wave effect, because their wavelength is similar to the size of the door.

If we pass light through smaller openings, often called slits, we can use Huygens's principle to see that light bends as sound does (see Figure 27.9). The bending of a wave around the edges of an opening or an obstacle is called **diffraction**. Diffraction is a wave characteristic and occurs for all types of waves. If diffraction is observed for some phenomenon, it is evidence that the phenomenon is a wave. Thus the horizontal diffraction of the laser beam after it passes through slits in Figure 27.3 is evidence that light is a wave.

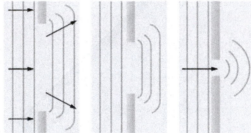

Figure 27.9 Huygens's principle applied to a straight wavefront striking an opening. The edges of the wavefront bend after passing through the opening, a process called diffraction. The amount of bending is more extreme for a small opening, consistent with the fact that wave characteristics are most noticeable for interactions with objects about the same size as the wavelength.

Making Connections: Diffraction

Diffraction of light waves passing though openings is illustrated in Figure 27.9. But the phenomenon of diffraction occurs in all waves, including sound and water waves. We are able to hear sounds from nearby rooms as a result of diffraction of sound waves around obstacles and corners. The diffraction of water waves can be visually seen when waves bend around boats.

As shown in Figure 27.8, the wavelengths of the different types of waves affect their behavior and diffraction. In fact, no observable diffraction occurs if the wave's wavelength is much smaller than the obstacle or slit. For example, light waves diffract around extremely small objects but cannot diffract around large obstacles, as their wavelength is very small. On the other hand, sound waves have long wavelengths and hence can diffract around large objects.

27.3 Young's Double Slit Experiment

Learning Objectives

By the end of this section, you will be able to:

- Explain the phenomena of interference.
- Define constructive interference for a double slit and destructive interference for a double slit.

The information presented in this section supports the following AP® learning objectives and science practices:

- **6.C.3.1** The student is able to qualitatively apply the wave model to quantities that describe the generation of

interference patterns to make predictions about interference patterns that form when waves pass through a set of openings whose spacing and widths are small, but larger than the wavelength. **(S.P. 1.4, 6.4)**
- **6.D.1.1** The student is able to use representations of individual pulses and construct representations to model the interaction of two wave pulses to analyze the superposition of two pulses. **(S.P. 1.1, 1.4)**
- **6.D.1.3** The student is able to design a plan for collecting data to quantify the amplitude variations when two or more traveling waves or wave pulses interact in a given medium. **(S.P. 4.2)**
- **6.D.2.1** The student is able to analyze data or observations or evaluate evidence of the interaction of two or more traveling waves in one or two dimensions (i.e., circular wave fronts) to evaluate the variations in resultant amplitudes. **(S.P. 5.1)**

Although Christiaan Huygens thought that light was a wave, Isaac Newton did not. Newton felt that there were other explanations for color, and for the interference and diffraction effects that were observable at the time. Owing to Newton's tremendous stature, his view generally prevailed. The fact that Huygens's principle worked was not considered evidence that was direct enough to prove that light is a wave. The acceptance of the wave character of light came many years later when, in 1801, the English physicist and physician Thomas Young (1773–1829) did his now-classic double slit experiment (see **Figure 27.10**).

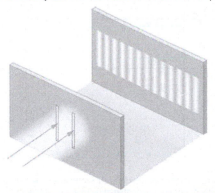

Figure 27.10 Young's double slit experiment. Here pure-wavelength light sent through a pair of vertical slits is diffracted into a pattern on the screen of numerous vertical lines spread out horizontally. Without diffraction and interference, the light would simply make two lines on the screen.

Why do we not ordinarily observe wave behavior for light, such as observed in Young's double slit experiment? First, light must interact with something small, such as the closely spaced slits used by Young, to show pronounced wave effects. Furthermore, Young first passed light from a single source (the Sun) through a single slit to make the light somewhat coherent. By **coherent**, we mean waves are in phase or have a definite phase relationship. **Incoherent** means the waves have random phase relationships. Why did Young then pass the light through a double slit? The answer to this question is that two slits provide two coherent light sources that then interfere constructively or destructively. Young used sunlight, where each wavelength forms its own pattern, making the effect more difficult to see. We illustrate the double slit experiment with monochromatic (single λ) light to clarify the effect. **Figure 27.11** shows the pure constructive and destructive interference of two waves having the same wavelength and amplitude.

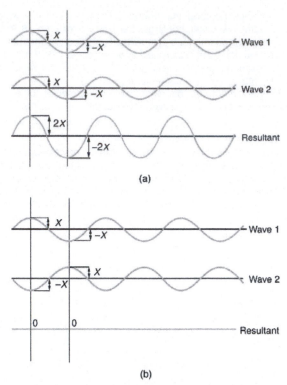

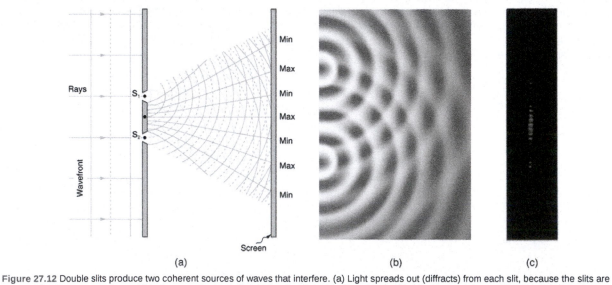

Figure 27.11 The amplitudes of waves add. (a) Pure constructive interference is obtained when identical waves are in phase. (b) Pure destructive interference occurs when identical waves are exactly out of phase, or shifted by half a wavelength.

When light passes through narrow slits, it is diffracted into semicircular waves, as shown in **Figure 27.12**(a). Pure constructive interference occurs where the waves are crest to crest or trough to trough. Pure destructive interference occurs where they are crest to trough. The light must fall on a screen and be scattered into our eyes for us to see the pattern. An analogous pattern for water waves is shown in **Figure 27.12**(b). Note that regions of constructive and destructive interference move out from the slits at well-defined angles to the original beam. These angles depend on wavelength and the distance between the slits, as we shall see below.

Figure 27.12 Double slits produce two coherent sources of waves that interfere. (a) Light spreads out (diffracts) from each slit, because the slits are narrow. These waves overlap and interfere constructively (bright lines) and destructively (dark regions). We can only see this if the light falls onto a screen and is scattered into our eyes. (b) Double slit interference pattern for water waves are nearly identical to that for light. Wave action is greatest in regions of constructive interference and least in regions of destructive interference. (c) When light that has passed through double slits falls on a screen, we see a pattern such as this. (credit: PASCO)

Making Connections: Interference

In addition to light waves, the phenomenon of interference also occurs in other waves, including water and sound waves. You will observe patterns of constructive and destructive interference if you throw two stones in a lake simultaneously. The crests (and troughs) of the two waves interfere constructively whereas the crest of a wave interferes destructively with the trough of the other wave. Similarly, sound waves traveling in the same medium interfere with each other. Their amplitudes add if they interfere constructively or subtract if there is destructive interference.

To understand the double slit interference pattern, we consider how two waves travel from the slits to the screen, as illustrated in **Figure 27.13**. Each slit is a different distance from a given point on the screen. Thus different numbers of wavelengths fit into each path. Waves start out from the slits in phase (crest to crest), but they may end up out of phase (crest to trough) at the screen if the paths differ in length by half a wavelength, interfering destructively as shown in **Figure 27.13**(a). If the paths differ by a whole wavelength, then the waves arrive in phase (crest to crest) at the screen, interfering constructively as shown in **Figure 27.13**(b). More generally, if the paths taken by the two waves differ by any half-integral number of wavelengths [$(1/2)\lambda$, $(3/2)\lambda$, $(5/2)\lambda$, etc.], then destructive interference occurs. Similarly, if the paths taken by the two waves differ by any integral number of wavelengths (λ, 2λ, 3λ, etc.), then constructive interference occurs.

Take-Home Experiment: Using Fingers as Slits

Look at a light, such as a street lamp or incandescent bulb, through the narrow gap between two fingers held close together. What type of pattern do you see? How does it change when you allow the fingers to move a little farther apart? Is it more distinct for a monochromatic source, such as the yellow light from a sodium vapor lamp, than for an incandescent bulb?

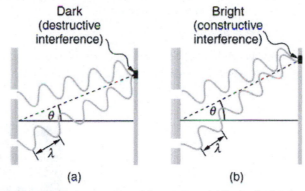

Figure 27.13 Waves follow different paths from the slits to a common point on a screen. (a) Destructive interference occurs here, because one path is a half wavelength longer than the other. The waves start in phase but arrive out of phase. (b) Constructive interference occurs here because one path is a whole wavelength longer than the other. The waves start out and arrive in phase.

Figure 27.14 shows how to determine the path length difference for waves traveling from two slits to a common point on a screen. If the screen is a large distance away compared with the distance between the slits, then the angle θ between the path and a line from the slits to the screen (see the figure) is nearly the same for each path. The difference between the paths is shown in the figure; simple trigonometry shows it to be $d \sin \theta$, where d is the distance between the slits. To obtain **constructive interference for a double slit**, the path length difference must be an integral multiple of the wavelength, or

$$d \, \sin \theta = m\lambda, \ \text{ for } m = 0, \, 1, \, -1, \, 2, \, -2, \ \ldots \ \text{(constructive)}. \tag{27.3}$$

Similarly, to obtain **destructive interference for a double slit**, the path length difference must be a half-integral multiple of the wavelength, or

$$d \, \sin \theta = \left(m + \frac{1}{2}\right)\!\lambda, \ \text{ for } m = 0, \, 1, \, -1, \, 2, \, -2, \ \ldots \ \text{(destructive)}, \tag{27.4}$$

where λ is the wavelength of the light, d is the distance between slits, and θ is the angle from the original direction of the beam as discussed above. We call m the **order** of the interference. For example, $m = 4$ is fourth-order interference.

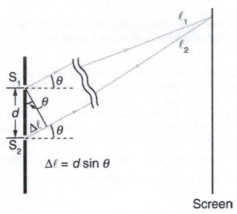

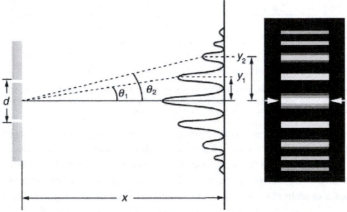

Figure 27.14 The paths from each slit to a common point on the screen differ by an amount $d \sin \theta$, assuming the distance to the screen is much greater than the distance between slits (not to scale here).

The equations for double slit interference imply that a series of bright and dark lines are formed. For vertical slits, the light spreads out horizontally on either side of the incident beam into a pattern called interference fringes, illustrated in **Figure 27.15**. The intensity of the bright fringes falls off on either side, being brightest at the center. The closer the slits are, the more is the spreading of the bright fringes. We can see this by examining the equation

$$d \sin \theta = m\lambda, \text{ for } m = 0, 1, -1, 2, -2, \dots . \tag{27.5}$$

For fixed λ and m, the smaller d is, the larger θ must be, since $\sin \theta = m\lambda / d$. This is consistent with our contention that wave effects are most noticeable when the object the wave encounters (here, slits a distance d apart) is small. Small d gives large θ, hence a large effect.

Figure 27.15 The interference pattern for a double slit has an intensity that falls off with angle. The photograph shows multiple bright and dark lines, or fringes, formed by light passing through a double slit.

Making Connections: Amplitude of Interference Fringe

The amplitude of the interference fringe at a point depends on the amplitudes of the two coherent waves (A_1 and A_2) arriving at that point and can be found using the relationship

$$A^2 = A_1^2 + A_2^2 + 2A_1A_2 \cos\delta,$$

where δ is the phase difference between the arriving waves.

This equation is also applicable for Young's double slit experiment. If the two waves come from the same source or two sources with the same amplitude, then $A_1 = A_2$, and the amplitude of the interference fringe can be calculated using

$$A^2 = 2A_1^2 (1 + \cos\delta).$$

The amplitude will be maximum when $\cos\delta = 1$ or $\delta = 0$. This means the central fringe has the maximum amplitude. Also the intensity of a wave is directly proportional to its amplitude (i.e., $I \propto A^2$) and consequently the central fringe also has the maximum intensity.

Example 27.1 Finding a Wavelength from an Interference Pattern

Suppose you pass light from a He-Ne laser through two slits separated by 0.0100 mm and find that the third bright line on a screen is formed at an angle of $10.95°$ relative to the incident beam. What is the wavelength of the light?

Strategy

The third bright line is due to third-order constructive interference, which means that $m = 3$. We are given $d = 0.0100$ mm and $\theta = 10.95°$. The wavelength can thus be found using the equation $d \sin \theta = m\lambda$ for constructive interference.

Solution

The equation is $d \sin \theta = m\lambda$. Solving for the wavelength λ gives

$$\lambda = \frac{d \sin \theta}{m}. \tag{27.6}$$

Substituting known values yields

$$\lambda = \frac{(0.0100 \text{ mm})(\sin 10.95°)}{3} \tag{27.7}$$
$$= 6.33 \times 10^{-4} \text{ mm} = 633 \text{ nm}.$$

Discussion

To three digits, this is the wavelength of light emitted by the common He-Ne laser. Not by coincidence, this red color is similar to that emitted by neon lights. More important, however, is the fact that interference patterns can be used to measure wavelength. Young did this for visible wavelengths. This analytical technique is still widely used to measure electromagnetic spectra. For a given order, the angle for constructive interference increases with λ, so that spectra (measurements of intensity versus wavelength) can be obtained.

Example 27.2 Calculating Highest Order Possible

Interference patterns do not have an infinite number of lines, since there is a limit to how big m can be. What is the highest-order constructive interference possible with the system described in the preceding example?

Strategy and Concept

The equation $d \sin \theta = m\lambda$ (for $m = 0, 1, -1, 2, -2, \ldots$) describes constructive interference. For fixed values of d and λ, the larger m is, the larger $\sin \theta$ is. However, the maximum value that $\sin \theta$ can have is 1, for an angle of $90°$. (Larger angles imply that light goes backward and does not reach the screen at all.) Let us find which m corresponds to this maximum diffraction angle.

Solution

Solving the equation $d \sin \theta = m\lambda$ for m gives

$$m = \frac{d \sin \theta}{\lambda}. \tag{27.8}$$

Taking $\sin \theta = 1$ and substituting the values of d and λ from the preceding example gives

$$m = \frac{(0.0100 \text{ mm})(1)}{633 \text{ nm}} \approx 15.8. \tag{27.9}$$

Therefore, the largest integer m can be is 15, or

$$m = 15. \tag{27.10}$$

Discussion

The number of fringes depends on the wavelength and slit separation. The number of fringes will be very large for large slit separations. However, if the slit separation becomes much greater than the wavelength, the intensity of the interference pattern changes so that the screen has two bright lines cast by the slits, as expected when light behaves like a ray. We also note that the fringes get fainter further away from the center. Consequently, not all 15 fringes may be observable.

Applying the Science Practices: Double Slit Experiment

Design an Experiment

Design a double slit experiment to find the wavelength of a He-Ne laser light. Your setup may include the He-Ne laser, a glass plate with two slits, paper, measurement apparatus, and a light intensity recorder. Write a step-by-step procedure for the experiment, draw a diagram of the set-up, and describe the steps followed to calculate the wavelength of the laser light.

Analyze Data

A double slit experiment is performed using three lasers. The table below shows the locations of the bright fringes that are recorded (in meters) on a screen.

Table 27.1

Fringe	Location for Laser 1	Location for Laser 2	Location for Laser 3
3	0.371	0.344	0.395
2	0.314	0.296	0.330
1	0.257	0.248	0.265
0	0.200	0.200	0.200
-1	0.143	0.152	0.135
-2	0.086	0.104	0.070
-3	0.029	0.056	0.005

a. Assuming the screen is 2.00 m away from the slits, find the angles for the first, second, and third bright fringes for each laser.

b. If the distance between the slits is 0.02 mm, calculate the wavelengths of the three lasers used in the experiment.

c. If the amplitudes of the three lasers are in the ratio 1:2:3, find the ratio of intensities of the central bright fringes formed by the three lasers.

27.4 Multiple Slit Diffraction

Learning Objectives

By the end of this section, you will be able to:

-   Discuss the pattern obtained from diffraction grating.
-   Explain diffraction grating effects.

The information presented in this section supports the following AP® learning objectives and science practices:

-   **6.C.3.1** The student is able to qualitatively apply the wave model to quantities that describe the generation of interference patterns to make predictions about interference patterns that form when waves pass through a set of openings whose spacing and widths are small, but larger than the wavelength. **(S.P. 1.4, 6.4)**

An interesting thing happens if you pass light through a large number of evenly spaced parallel slits, called a **diffraction grating**. An interference pattern is created that is very similar to the one formed by a double slit (see **Figure 27.16**). A diffraction grating can be manufactured by scratching glass with a sharp tool in a number of precisely positioned parallel lines, with the untouched regions acting like slits. These can be photographically mass produced rather cheaply. Diffraction gratings work both for transmission of light, as in **Figure 27.16**, and for reflection of light, as on butterfly wings and the Australian opal in **Figure 27.17** or the CD pictured in the opening photograph of this chapter, **Figure 27.1**. In addition to their use as novelty items, diffraction gratings are commonly used for spectroscopic dispersion and analysis of light. What makes them particularly useful is the fact that they form a sharper pattern than double slits do. That is, their bright regions are narrower and brighter, while their dark regions are darker. **Figure 27.18** shows idealized graphs demonstrating the sharper pattern. Natural diffraction gratings occur in the feathers of certain birds. Tiny, finger-like structures in regular patterns act as reflection gratings, producing constructive interference that gives the feathers colors not solely due to their pigmentation. This is called iridescence.

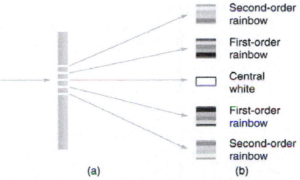

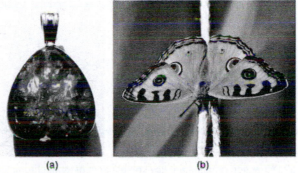

Figure 27.16 A diffraction grating is a large number of evenly spaced parallel slits. (a) Light passing through is diffracted in a pattern similar to a double slit, with bright regions at various angles. (b) The pattern obtained for white light incident on a grating. The central maximum is white, and the higher-order maxima disperse white light into a rainbow of colors.

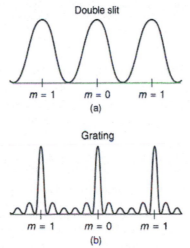

Figure 27.17 (a) This Australian opal and (b) the butterfly wings have rows of reflectors that act like reflection gratings, reflecting different colors at different angles. (credits: (a) Opals-On-Black.com, via Flickr (b) whologwhy, Flickr)

Double slit

Grating

Figure 27.18 Idealized graphs of the intensity of light passing through a double slit (a) and a diffraction grating (b) for monochromatic light. Maxima can be produced at the same angles, but those for the diffraction grating are narrower and hence sharper. The maxima become narrower and the regions between darker as the number of slits is increased.

The analysis of a diffraction grating is very similar to that for a double slit (see **Figure 27.19**). As we know from our discussion of double slits in **Young's Double Slit Experiment**, light is diffracted by each slit and spreads out after passing through. Rays traveling in the same direction (at an angle θ relative to the incident direction) are shown in the figure. Each of these rays travels a different distance to a common point on a screen far away. The rays start in phase, and they can be in or out of phase when they reach a screen, depending on the difference in the path lengths traveled. As seen in the figure, each ray travels a distance $d \sin \theta$ different from that of its neighbor, where d is the distance between slits. If this distance equals an integral number of wavelengths, the rays all arrive in phase, and constructive interference (a maximum) is obtained. Thus, the condition necessary to obtain **constructive interference for a diffraction grating** is

$$d \sin \theta = m\lambda, \text{ for } m = 0, 1, -1, 2, -2, \ldots \text{ (constructive)}, \tag{27.11}$$

where d is the distance between slits in the grating, λ is the wavelength of light, and m is the order of the maximum. Note that this is exactly the same equation as for double slits separated by d. However, the slits are usually closer in diffraction gratings than in double slits, producing fewer maxima at larger angles.

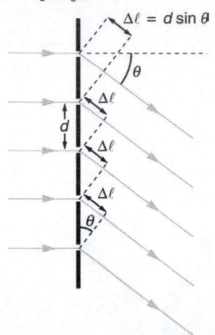

Figure 27.19 Diffraction grating showing light rays from each slit traveling in the same direction. Each ray travels a different distance to reach a common point on a screen (not shown). Each ray travels a distance $d \sin \theta$ different from that of its neighbor.

Where are diffraction gratings used? Diffraction gratings are key components of monochromators used, for example, in optical imaging of particular wavelengths from biological or medical samples. A diffraction grating can be chosen to specifically analyze a wavelength emitted by molecules in diseased cells in a biopsy sample or to help excite strategic molecules in the sample with a selected frequency of light. Another vital use is in optical fiber technologies where fibers are designed to provide optimum performance at specific wavelengths. A range of diffraction gratings are available for selecting specific wavelengths for such use.

Take-Home Experiment: Rainbows on a CD

The spacing d of the grooves in a CD or DVD can be well determined by using a laser and the equation $d \sin \theta = m\lambda$, for $m = 0, 1, -1, 2, -2, \ldots$. However, we can still make a good estimate of this spacing by using white light and the rainbow of colors that comes from the interference. Reflect sunlight from a CD onto a wall and use your best judgment of the location of a strongly diffracted color to find the separation d.

Example 27.3 Calculating Typical Diffraction Grating Effects

Diffraction gratings with 10,000 lines per centimeter are readily available. Suppose you have one, and you send a beam of white light through it to a screen 2.00 m away. (a) Find the angles for the first-order diffraction of the shortest and longest wavelengths of visible light (380 and 760 nm). (b) What is the distance between the ends of the rainbow of visible light produced on the screen for first-order interference? (See **Figure 27.20**.)

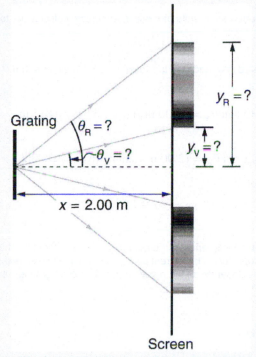

Figure 27.20 The diffraction grating considered in this example produces a rainbow of colors on a screen a distance $x = 2.00 \text{ m}$ from the grating. The distances along the screen are measured perpendicular to the x-direction. In other words, the rainbow pattern extends out of the page.

Strategy

The angles can be found using the equation

$$d \, \sin \theta = m\lambda \text{ (for } m = 0, 1, -1, 2, -2, \ldots) \tag{27.12}$$

once a value for the slit spacing d has been determined. Since there are 10,000 lines per centimeter, each line is separated by $1/10{,}000$ of a centimeter. Once the angles are found, the distances along the screen can be found using simple trigonometry.

Solution for (a)

The distance between slits is $d = (1 \text{ cm}) / 10{,}000 = 1.00 \times 10^{-4} \text{ cm}$ or $1.00 \times 10^{-6} \text{ m}$. Let us call the two angles θ_V for violet (380 nm) and θ_R for red (760 nm). Solving the equation $d \sin \theta_V = m\lambda$ for $\sin \theta_V$,

$$\sin \theta_V = \frac{m\lambda_V}{d}, \tag{27.13}$$

where $m = 1$ for first order and $\lambda_V = 380 \text{ nm} = 3.80 \times 10^{-7} \text{ m}$. Substituting these values gives

$$\sin \theta_V = \frac{3.80 \times 10^{-7} \text{ m}}{1.00 \times 10^{-6} \text{ m}} = 0.380. \tag{27.14}$$

Thus the angle θ_V is

$$\theta_V = \sin^{-1} 0.380 = 22.33°. \tag{27.15}$$

Similarly,

$$\sin \theta_R = \frac{7.60 \times 10^{-7} \text{ m}}{1.00 \times 10^{-6} \text{ m}}. \tag{27.16}$$

Thus the angle θ_R is

$$\theta_R = \sin^{-1} 0.760 = 49.46°. \tag{27.17}$$

Notice that in both equations, we reported the results of these intermediate calculations to four significant figures to use with the calculation in part (b).

Solution for (b)

The distances on the screen are labeled y_V and y_R in **Figure 27.20**. Noting that $\tan \theta = y / x$, we can solve for y_V and y_R. That is,

$$y_V = x \tan \theta_V = (2.00 \text{ m})(\tan 22.33°) = 0.815 \text{ m} \tag{27.18}$$

and

$$y_R = x \tan \theta_R = (2.00 \text{ m})(\tan 49.46°) = 2.338 \text{ m}. \tag{27.19}$$

The distance between them is therefore

$$y_R - y_V = 1.52 \text{ m}. \tag{27.20}$$

Discussion

The large distance between the red and violet ends of the rainbow produced from the white light indicates the potential this diffraction grating has as a spectroscopic tool. The more it can spread out the wavelengths (greater dispersion), the more detail can be seen in a spectrum. This depends on the quality of the diffraction grating—it must be very precisely made in addition to having closely spaced lines.

27.5 Single Slit Diffraction

Light passing through a single slit forms a diffraction pattern somewhat different from those formed by double slits or diffraction gratings. **Figure 27.21** shows a single slit diffraction pattern. Note that the central maximum is larger than those on either side, and that the intensity decreases rapidly on either side. In contrast, a diffraction grating produces evenly spaced lines that dim slowly on either side of center.

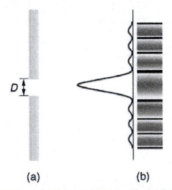

(a)   (b)

Figure 27.21 (a) Single slit diffraction pattern. Monochromatic light passing through a single slit has a central maximum and many smaller and dimmer maxima on either side. The central maximum is six times higher than shown. (b) The drawing shows the bright central maximum and dimmer and thinner maxima on either side.

The analysis of single slit diffraction is illustrated in **Figure 27.22**. Here we consider light coming from different parts of the *same* slit. According to Huygens's principle, every part of the wavefront in the slit emits wavelets. These are like rays that start out in phase and head in all directions. (Each ray is perpendicular to the wavefront of a wavelet.) Assuming the screen is very far away compared with the size of the slit, rays heading toward a common destination are nearly parallel. When they travel straight ahead, as in **Figure 27.22**(a), they remain in phase, and a central maximum is obtained. However, when rays travel at an angle θ relative to the original direction of the beam, each travels a different distance to a common location, and they can arrive in or out of phase. In **Figure 27.22**(b), the ray from the bottom travels a distance of one wavelength λ farther than the ray from the top. Thus a ray from the center travels a distance $\lambda/2$ farther than the one on the left, arrives out of phase, and interferes destructively. A ray from slightly above the center and one from slightly above the bottom will also cancel one another. In fact, each ray from the slit will have another to interfere destructively, and a minimum in intensity will occur at this angle. There will be another minimum at the same angle to the right of the incident direction of the light.

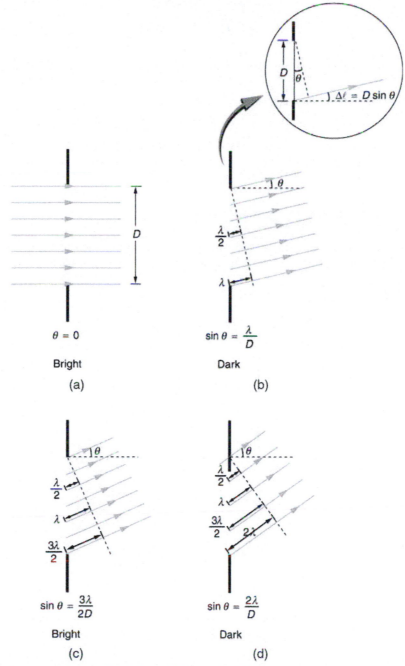

Figure 27.22 Light passing through a single slit is diffracted in all directions and may interfere constructively or destructively, depending on the angle. The difference in path length for rays from either side of the slit is seen to be $D \sin \theta$.

At the larger angle shown in Figure 27.22(c), the path lengths differ by $3\lambda/2$ for rays from the top and bottom of the slit. One ray travels a distance λ different from the ray from the bottom and arrives in phase, interfering constructively. Two rays, each from slightly above those two, will also add constructively. Most rays from the slit will have another to interfere with constructively, and a maximum in intensity will occur at this angle. However, all rays do not interfere constructively for this situation, and so the maximum is not as intense as the central maximum. Finally, in Figure 27.22(d), the angle shown is large enough to produce a second minimum. As seen in the figure, the difference in path length for rays from either side of the slit is $D \sin \theta$, and we see that a destructive minimum is obtained when this distance is an integral multiple of the wavelength.

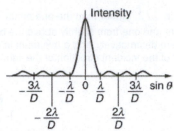

Figure 27.23 A graph of single slit diffraction intensity showing the central maximum to be wider and much more intense than those to the sides. In fact the central maximum is six times higher than shown here.

Thus, to obtain **destructive interference for a single slit**,

$$D \sin \theta = m\lambda, \text{ for } m = 1, -1, 2, -2, 3, \; \dots \text{ (destructive)}, \qquad (27.21)$$

where D is the slit width, λ is the light's wavelength, θ is the angle relative to the original direction of the light, and m is the order of the minimum. **Figure 27.23** shows a graph of intensity for single slit interference, and it is apparent that the maxima on either side of the central maximum are much less intense and not as wide. This is consistent with the illustration in **Figure 27.21**(b).

Example 27.4 Calculating Single Slit Diffraction

Visible light of wavelength 550 nm falls on a single slit and produces its second diffraction minimum at an angle of $45.0°$ relative to the incident direction of the light. (a) What is the width of the slit? (b) At what angle is the first minimum produced?

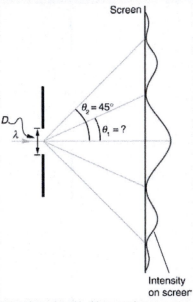

Figure 27.24 A graph of the single slit diffraction pattern is analyzed in this example.

Strategy

From the given information, and assuming the screen is far away from the slit, we can use the equation $D \sin \theta = m\lambda$ first to find D, and again to find the angle for the first minimum θ_1.

Solution for (a)

We are given that $\lambda = 550$ nm , $m = 2$, and $\theta_2 = 45.0°$. Solving the equation $D \sin \theta = m\lambda$ for D and substituting known values gives

$$
\begin{aligned}
D &= \frac{m\lambda}{\sin \theta_2} = \frac{2(550 \text{ nm})}{\sin 45.0°} \\
&= \frac{1100 \times 10^{-9}}{0.707} \\
&= 1.56 \times 10^{-6}.
\end{aligned}
\qquad (27.22)
$$

Solution for (b)

Solving the equation $D \sin \theta = m\lambda$ for $\sin \theta_1$ and substituting the known values gives

$$\sin \theta_1 = \frac{m\lambda}{D} = \frac{1\left(550 \times 10^{-9} \text{ m}\right)}{1.56 \times 10^{-6} \text{ m}}. \tag{27.23}$$

Thus the angle θ_1 is

$$\theta_1 = \sin^{-1} 0.354 = 20.7°. \tag{27.24}$$

Discussion

We see that the slit is narrow (it is only a few times greater than the wavelength of light). This is consistent with the fact that light must interact with an object comparable in size to its wavelength in order to exhibit significant wave effects such as this single slit diffraction pattern. We also see that the central maximum extends $20.7°$ on either side of the original beam, for a width of about $41°$. The angle between the first and second minima is only about $24°$ $(45.0° - 20.7°)$. Thus the second maximum is only about half as wide as the central maximum.

27.6 Limits of Resolution: The Rayleigh Criterion

Learning Objectives
By the end of this section, you will be able to: • Discuss the Rayleigh criterion. The information presented in this section supports the following AP® learning objectives and science practices: • **6.C.2.1** The student is able to make claims about the diffraction pattern produced when a wave passes through a small opening and to qualitatively apply the wave model to quantities that describe the generation of a diffraction pattern when a wave passes through an opening whose dimensions are comparable to the wavelength of the wave.

Light diffracts as it moves through space, bending around obstacles, interfering constructively and destructively. While this can be used as a spectroscopic tool—a diffraction grating disperses light according to wavelength, for example, and is used to produce spectra—diffraction also limits the detail we can obtain in images. **Figure 27.25**(a) shows the effect of passing light through a small circular aperture. Instead of a bright spot with sharp edges, a spot with a fuzzy edge surrounded by circles of light is obtained. This pattern is caused by diffraction similar to that produced by a single slit. Light from different parts of the circular aperture interferes constructively and destructively. The effect is most noticeable when the aperture is small, but the effect is there for large apertures, too.

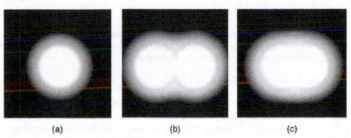

(a) (b) (c)

Figure 27.25 (a) Monochromatic light passed through a small circular aperture produces this diffraction pattern. (b) Two point light sources that are close to one another produce overlapping images because of diffraction. (c) If they are closer together, they cannot be resolved or distinguished.

How does diffraction affect the detail that can be observed when light passes through an aperture? **Figure 27.25**(b) shows the diffraction pattern produced by two point light sources that are close to one another. The pattern is similar to that for a single point source, and it is just barely possible to tell that there are two light sources rather than one. If they were closer together, as in **Figure 27.25**(c), we could not distinguish them, thus limiting the detail or resolution we can obtain. This limit is an inescapable consequence of the wave nature of light.

There are many situations in which diffraction limits the resolution. The acuity of our vision is limited because light passes through the pupil, the circular aperture of our eye. Be aware that the diffraction-like spreading of light is due to the limited diameter of a light beam, not the interaction with an aperture. Thus light passing through a lens with a diameter D shows this effect and spreads, blurring the image, just as light passing through an aperture of diameter D does. So diffraction limits the resolution of any system having a lens or mirror. Telescopes are also limited by diffraction, because of the finite diameter D of their primary mirror.

Draw two lines on a white sheet of paper (several mm apart). How far away can you be and still distinguish the two lines? What does this tell you about the size of the eye's pupil? Can you be quantitative? (The size of an adult's pupil is discussed in **Physics of the Eye**.)

Just what is the limit? To answer that question, consider the diffraction pattern for a circular aperture, which has a central maximum that is wider and brighter than the maxima surrounding it (similar to a slit) [see **Figure 27.26**(a)]. It can be shown that, for a circular aperture of diameter D, the first minimum in the diffraction pattern occurs at $\theta = 1.22\,\lambda/D$ (providing the aperture is large compared with the wavelength of light, which is the case for most optical instruments). The accepted criterion for determining the diffraction limit to resolution based on this angle was developed by Lord Rayleigh in the 19th century. The **Rayleigh criterion** for the diffraction limit to resolution states that *two images are just resolvable when the center of the diffraction pattern of one is directly over the first minimum of the diffraction pattern of the other*. See **Figure 27.26**(b). The first minimum is at an angle of $\theta = 1.22\,\lambda/D$, so that two point objects are just resolvable if they are separated by the angle

$$\theta = 1.22\frac{\lambda}{D},\qquad(27.25)$$

where λ is the wavelength of light (or other electromagnetic radiation) and D is the diameter of the aperture, lens, mirror, etc., with which the two objects are observed. In this expression, θ has units of radians.

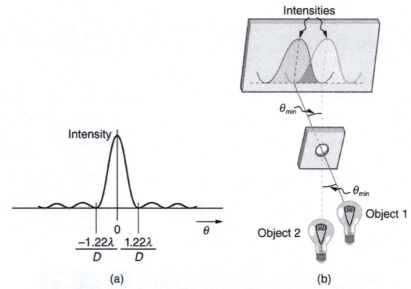

Figure 27.26 (a) Graph of intensity of the diffraction pattern for a circular aperture. Note that, similar to a single slit, the central maximum is wider and brighter than those to the sides. (b) Two point objects produce overlapping diffraction patterns. Shown here is the Rayleigh criterion for being just resolvable. The central maximum of one pattern lies on the first minimum of the other.

Connections: Limits to Knowledge

All attempts to observe the size and shape of objects are limited by the wavelength of the probe. Even the small wavelength of light prohibits exact precision. When extremely small wavelength probes as with an electron microscope are used, the system is disturbed, still limiting our knowledge, much as making an electrical measurement alters a circuit. Heisenberg's uncertainty principle asserts that this limit is fundamental and inescapable, as we shall see in quantum mechanics.

Example 27.5 Calculating Diffraction Limits of the Hubble Space Telescope

The primary mirror of the orbiting Hubble Space Telescope has a diameter of 2.40 m. Being in orbit, this telescope avoids the degrading effects of atmospheric distortion on its resolution. (a) What is the angle between two just-resolvable point light sources (perhaps two stars)? Assume an average light wavelength of 550 nm. (b) If these two stars are at the 2 million light year distance of the Andromeda galaxy, how close together can they be and still be resolved? (A light year, or ly, is the distance light travels in 1 year.)

Strategy

The Rayleigh criterion stated in the equation $\theta = 1.22\frac{\lambda}{D}$ gives the smallest possible angle θ between point sources, or

the best obtainable resolution. Once this angle is found, the distance between stars can be calculated, since we are given how far away they are.

Solution for (a)

The Rayleigh criterion for the minimum resolvable angle is

$$\theta = 1.22\frac{\lambda}{D}. \tag{27.26}$$

Entering known values gives

$$\theta = 1.22\frac{550\times10^{-9}\text{ m}}{2.40\text{ m}} \tag{27.27}$$
$$= 2.80\times10^{-7}\text{ rad.}$$

Solution for (b)

The distance s between two objects a distance r away and separated by an angle θ is $s = r\theta$.

Substituting known values gives

$$\begin{aligned} s &= (2.0\times10^6\text{ ly})(2.80\times10^{-7}\text{ rad}) \\ &= 0.56\text{ ly.} \end{aligned} \tag{27.28}$$

Discussion

The angle found in part (a) is extraordinarily small (less than 1/50,000 of a degree), because the primary mirror is so large compared with the wavelength of light. As noticed, diffraction effects are most noticeable when light interacts with objects having sizes on the order of the wavelength of light. However, the effect is still there, and there is a diffraction limit to what is observable. The actual resolution of the Hubble Telescope is not quite as good as that found here. As with all instruments, there are other effects, such as non-uniformities in mirrors or aberrations in lenses that further limit resolution. However, **Figure 27.27** gives an indication of the extent of the detail observable with the Hubble because of its size and quality and especially because it is above the Earth's atmosphere.

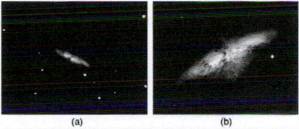

(a)   (b)

Figure 27.27 These two photographs of the M82 galaxy give an idea of the observable detail using the Hubble Space Telescope compared with that using a ground-based telescope. (a) On the left is a ground-based image. (credit: Ricnun, Wikimedia Commons) (b) The photo on the right was captured by Hubble. (credit: NASA, ESA, and the Hubble Heritage Team (STScI/AURA))

The answer in part (b) indicates that two stars separated by about half a light year can be resolved. The average distance between stars in a galaxy is on the order of 5 light years in the outer parts and about 1 light year near the galactic center. Therefore, the Hubble can resolve most of the individual stars in Andromeda galaxy, even though it lies at such a huge distance that its light takes 2 million years for its light to reach us. **Figure 27.28** shows another mirror used to observe radio waves from outer space.

Figure 27.28 A 305-m-diameter natural bowl at Arecibo in Puerto Rico is lined with reflective material, making it into a radio telescope. It is the largest curved focusing dish in the world. Although D for Arecibo is much larger than for the Hubble Telescope, it detects much longer wavelength radiation and its diffraction limit is significantly poorer than Hubble's. Arecibo is still very useful, because important information is carried by radio waves that is not carried by visible light. (credit: Tatyana Temirbulatova, Flickr)

Diffraction is not only a problem for optical instruments but also for the electromagnetic radiation itself. Any beam of light having a finite diameter D and a wavelength λ exhibits diffraction spreading. The beam spreads out with an angle θ given by the equation $\theta = 1.22\frac{\lambda}{D}$. Take, for example, a laser beam made of rays as parallel as possible (angles between rays as close to

$\theta = 0°$ as possible) instead spreads out at an angle $\theta = 1.22\,\lambda/D$, where D is the diameter of the beam and λ is its wavelength. This spreading is impossible to observe for a flashlight, because its beam is not very parallel to start with. However, for long-distance transmission of laser beams or microwave signals, diffraction spreading can be significant (see **Figure 27.29**). To avoid this, we can increase D. This is done for laser light sent to the Moon to measure its distance from the Earth. The laser beam is expanded through a telescope to make D much larger and θ smaller.

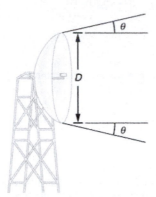

Figure 27.29 The beam produced by this microwave transmission antenna will spread out at a minimum angle $\theta = 1.22\,\lambda/D$ due to diffraction. It is impossible to produce a near-parallel beam, because the beam has a limited diameter.

In most biology laboratories, resolution is presented when the use of the microscope is introduced. The ability of a lens to produce sharp images of two closely spaced point objects is called resolution. The smaller the distance x by which two objects can be separated and still be seen as distinct, the greater the resolution. The resolving power of a lens is defined as that distance x. An expression for resolving power is obtained from the Rayleigh criterion. In **Figure 27.30**(a) we have two point objects separated by a distance x. According to the Rayleigh criterion, resolution is possible when the minimum angular separation is

$$\theta = 1.22\frac{\lambda}{D} = \frac{x}{d}, \tag{27.29}$$

where d is the distance between the specimen and the objective lens, and we have used the small angle approximation (i.e., we have assumed that x is much smaller than d), so that $\tan\theta \approx \sin\theta \approx \theta$.

Therefore, the resolving power is

$$x = 1.22\frac{\lambda d}{D}. \tag{27.30}$$

Another way to look at this is by re-examining the concept of Numerical Aperture (NA) discussed in **Microscopes**. There, NA is a measure of the maximum acceptance angle at which the fiber will take light and still contain it within the fiber. **Figure 27.30**(b) shows a lens and an object at point P. The NA here is a measure of the ability of the lens to gather light and resolve fine detail. The angle subtended by the lens at its focus is defined to be $\theta = 2\alpha$. From the figure and again using the small angle approximation, we can write

$$\sin\alpha = \frac{D/2}{d} = \frac{D}{2d}. \tag{27.31}$$

The NA for a lens is $NA = n\sin\alpha$, where n is the index of refraction of the medium between the objective lens and the object at point P.

From this definition for NA, we can see that

$$x = 1.22\frac{\lambda d}{D} = 1.22\frac{\lambda}{2\sin\alpha} = 0.61\frac{\lambda n}{NA}. \tag{27.32}$$

In a microscope, NA is important because it relates to the resolving power of a lens. A lens with a large NA will be able to resolve finer details. Lenses with larger NA will also be able to collect more light and so give a brighter image. Another way to describe this situation is that the larger the NA, the larger the cone of light that can be brought into the lens, and so more of the diffraction modes will be collected. Thus the microscope has more information to form a clear image, and so its resolving power will be higher.

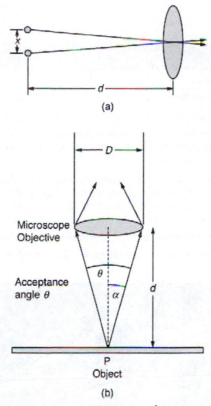

Figure 27.30 (a) Two points separated by at distance x and a positioned a distance d away from the objective. (credit: Infopro, Wikimedia Commons) (b) Terms and symbols used in discussion of resolving power for a lens and an object at point P. (credit: Infopro, Wikimedia Commons)

One of the consequences of diffraction is that the focal point of a beam has a finite width and intensity distribution. Consider focusing when only considering geometric optics, shown in **Figure 27.31**(a). The focal point is infinitely small with a huge intensity and the capacity to incinerate most samples irrespective of the NA of the objective lens. For wave optics, due to diffraction, the focal point spreads to become a focal spot (see **Figure 27.31**(b)) with the size of the spot decreasing with increasing NA. Consequently, the intensity in the focal spot increases with increasing NA. The higher the NA, the greater the chances of photodegrading the specimen. However, the spot never becomes a true point.

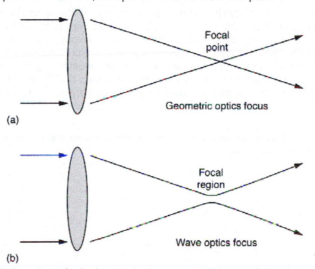

Figure 27.31 (a) In geometric optics, the focus is a point, but it is not physically possible to produce such a point because it implies infinite intensity. (b) In wave optics, the focus is an extended region.

27.7 Thin Film Interference

The bright colors seen in an oil slick floating on water or in a sunlit soap bubble are caused by interference. The brightest colors are those that interfere constructively. This interference is between light reflected from different surfaces of a thin film; thus, the effect is known as **thin film interference**. As noticed before, interference effects are most prominent when light interacts with something having a size similar to its wavelength. A thin film is one having a thickness t smaller than a few times the wavelength of light, λ. Since color is associated indirectly with λ and since all interference depends in some way on the ratio of λ to the size of the object involved, we should expect to see different colors for different thicknesses of a film, as in **Figure 27.32**.

Figure 27.32 These soap bubbles exhibit brilliant colors when exposed to sunlight. (credit: Scott Robinson, Flickr)

What causes thin film interference? **Figure 27.33** shows how light reflected from the top and bottom surfaces of a film can interfere. Incident light is only partially reflected from the top surface of the film (ray 1). The remainder enters the film and is itself partially reflected from the bottom surface. Part of the light reflected from the bottom surface can emerge from the top of the film (ray 2) and interfere with light reflected from the top (ray 1). Since the ray that enters the film travels a greater distance, it may be in or out of phase with the ray reflected from the top. However, consider for a moment, again, the bubbles in **Figure 27.32**. The bubbles are darkest where they are thinnest. Furthermore, if you observe a soap bubble carefully, you will note it gets dark at the point where it breaks. For very thin films, the difference in path lengths of ray 1 and ray 2 in **Figure 27.33** is negligible; so why should they interfere destructively and not constructively? The answer is that a phase change can occur upon reflection. The rule is as follows:

When light reflects from a medium having an index of refraction greater than that of the medium in which it is traveling, a $180°$ phase change (or a $\lambda/2$ shift) occurs.

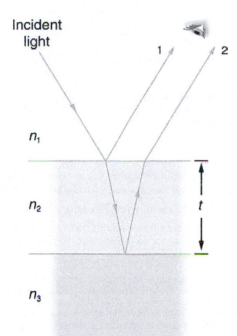

Figure 27.33 Light striking a thin film is partially reflected (ray 1) and partially refracted at the top surface. The refracted ray is partially reflected at the bottom surface and emerges as ray 2. These rays will interfere in a way that depends on the thickness of the film and the indices of refraction of the various media.

If the film in **Figure 27.33** is a soap bubble (essentially water with air on both sides), then there is a $\lambda/2$ shift for ray 1 and none for ray 2. Thus, when the film is very thin, the path length difference between the two rays is negligible, they are exactly out of phase, and destructive interference will occur at all wavelengths and so the soap bubble will be dark here.

The thickness of the film relative to the wavelength of light is the other crucial factor in thin film interference. Ray 2 in **Figure 27.33** travels a greater distance than ray 1. For light incident perpendicular to the surface, ray 2 travels a distance approximately $2t$ farther than ray 1. When this distance is an integral or half-integral multiple of the wavelength in the medium ($\lambda_n = \lambda/n$, where λ is the wavelength in vacuum and n is the index of refraction), constructive or destructive interference occurs, depending also on whether there is a phase change in either ray.

Example 27.6 Calculating Non-reflective Lens Coating Using Thin Film Interference

Sophisticated cameras use a series of several lenses. Light can reflect from the surfaces of these various lenses and degrade image clarity. To limit these reflections, lenses are coated with a thin layer of magnesium fluoride that causes destructive thin film interference. What is the thinnest this film can be, if its index of refraction is 1.38 and it is designed to limit the reflection of 550-nm light, normally the most intense visible wavelength? The index of refraction of glass is 1.52.

Strategy

Refer to **Figure 27.33** and use $n_1 = 100$ for air, $n_2 = 1.38$, and $n_3 = 1.52$. Both ray 1 and ray 2 will have a $\lambda/2$ shift upon reflection. Thus, to obtain destructive interference, ray 2 will need to travel a half wavelength farther than ray 1. For rays incident perpendicularly, the path length difference is $2t$.

Solution

To obtain destructive interference here,

$$2t = \frac{\lambda_{n_2}}{2},$$

(27.33)

where λ_{n_2} is the wavelength in the film and is given by $\lambda_{n_2} = \frac{\lambda}{n_2}$.

Thus,

$$2t = \frac{\lambda/n_2}{2}.$$

(27.34)

Solving for t and entering known values yields

$$t = \frac{\lambda/n_2}{4} = \frac{(550 \text{ nm})/1.38}{4} \tag{27.35}$$
$$= 99.6 \text{ nm}.$$

Discussion

Films such as the one in this example are most effective in producing destructive interference when the thinnest layer is used, since light over a broader range of incident angles will be reduced in intensity. These films are called non-reflective coatings; this is only an approximately correct description, though, since other wavelengths will only be partially cancelled. Non-reflective coatings are used in car windows and sunglasses.

Thin film interference is most constructive or most destructive when the path length difference for the two rays is an integral or half-integral wavelength, respectively. That is, for rays incident perpendicularly, $2t = \lambda_n, 2\lambda_n, 3\lambda_n, \dots$ or $2t = \lambda_n/2, 3\lambda_n/2, 5\lambda_n/2, \dots$. To know whether interference is constructive or destructive, you must also determine if there is a phase change upon reflection. Thin film interference thus depends on film thickness, the wavelength of light, and the refractive indices. For white light incident on a film that varies in thickness, you will observe rainbow colors of constructive interference for various wavelengths as the thickness varies.

Example 27.7 Soap Bubbles: More Than One Thickness can be Constructive

(a) What are the three smallest thicknesses of a soap bubble that produce constructive interference for red light with a wavelength of 650 nm? The index of refraction of soap is taken to be the same as that of water. (b) What three smallest thicknesses will give destructive interference?

Strategy and Concept

Use **Figure 27.33** to visualize the bubble. Note that $n_1 = n_3 = 1.00$ for air, and $n_2 = 1.333$ for soap (equivalent to water). There is a $\lambda/2$ shift for ray 1 reflected from the top surface of the bubble, and no shift for ray 2 reflected from the bottom surface. To get constructive interference, then, the path length difference ($2t$) must be a half-integral multiple of the wavelength—the first three being $\lambda_n/2$, $3\lambda_n/2$, and $5\lambda_n/2$. To get destructive interference, the path length difference must be an integral multiple of the wavelength—the first three being 0, λ_n, and $2\lambda_n$.

Solution for (a)

Constructive interference occurs here when

$$2t_c = \frac{\lambda_n}{2}, \frac{3\lambda_n}{2}, \frac{5\lambda_n}{2}, \dots . \tag{27.36}$$

The smallest constructive thickness t_c thus is

$$t_c = \frac{\lambda_n}{4} = \frac{\lambda/n}{4} = \frac{(650 \text{ nm})/1.333}{4} \tag{27.37}$$
$$= 122 \text{ nm}.$$

The next thickness that gives constructive interference is $t'_c = 3\lambda_n/4$, so that

$$t'_c = 366 \text{ nm}. \tag{27.38}$$

Finally, the third thickness producing constructive interference is $t''_c \le 5\lambda_n/4$, so that

$$t''_c = 610 \text{ nm}. \tag{27.39}$$

Solution for (b)

For *destructive interference*, the path length difference here is an integral multiple of the wavelength. The first occurs for zero thickness, since there is a phase change at the top surface. That is,

$$t_d = 0. \tag{27.40}$$

The first non-zero thickness producing destructive interference is

$$2t'_d = \lambda_n. \tag{27.41}$$

Substituting known values gives

$$t'_d = \frac{\lambda}{2} = \frac{\lambda/n}{2} = \frac{(650 \text{ nm})/1.333}{2} \tag{27.42}$$
$$= 244 \text{ nm}.$$

Finally, the third destructive thickness is $2t''_d = 2\lambda_n$, so that

$$t''_d = \lambda_n = \frac{\lambda}{n} = \frac{650 \text{ nm}}{1.333} \tag{27.43}$$
$$= 488 \text{ nm}.$$

Discussion

If the bubble was illuminated with pure red light, we would see bright and dark bands at very uniform increases in thickness. First would be a dark band at 0 thickness, then bright at 122 nm thickness, then dark at 244 nm, bright at 366 nm, dark at 488 nm, and bright at 610 nm. If the bubble varied smoothly in thickness, like a smooth wedge, then the bands would be evenly spaced.

Another example of thin film interference can be seen when microscope slides are separated (see **Figure 27.34**). The slides are very flat, so that the wedge of air between them increases in thickness very uniformly. A phase change occurs at the second surface but not the first, and so there is a dark band where the slides touch. The rainbow colors of constructive interference repeat, going from violet to red again and again as the distance between the slides increases. As the layer of air increases, the bands become more difficult to see, because slight changes in incident angle have greater effects on path length differences. If pure-wavelength light instead of white light is used, then bright and dark bands are obtained rather than repeating rainbow colors.

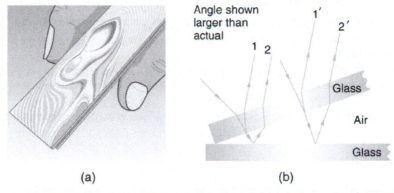

(a)   (b)

Figure 27.34 (a) The rainbow color bands are produced by thin film interference in the air between the two glass slides. (b) Schematic of the paths taken by rays in the wedge of air between the slides.

An important application of thin film interference is found in the manufacturing of optical instruments. A lens or mirror can be compared with a master as it is being ground, allowing it to be shaped to an accuracy of less than a wavelength over its entire surface. **Figure 27.35** illustrates the phenomenon called Newton's rings, which occurs when the plane surfaces of two lenses are placed together. (The circular bands are called Newton's rings because Isaac Newton described them and their use in detail. Newton did not discover them; Robert Hooke did, and Newton did not believe they were due to the wave character of light.) Each successive ring of a given color indicates an increase of only one wavelength in the distance between the lens and the blank, so that great precision can be obtained. Once the lens is perfect, there will be no rings.

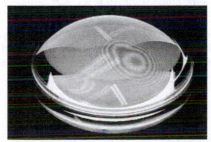

Figure 27.35 "Newton's rings" interference fringes are produced when two plano-convex lenses are placed together with their plane surfaces in contact. The rings are created by interference between the light reflected off the two surfaces as a result of a slight gap between them, indicating that these surfaces are not precisely plane but are slightly convex. (credit: Ulf Seifert, Wikimedia Commons)

The wings of certain moths and butterflies have nearly iridescent colors due to thin film interference. In addition to pigmentation, the wing's color is affected greatly by constructive interference of certain wavelengths reflected from its film-coated surface. Car manufacturers are offering special paint jobs that use thin film interference to produce colors that change with angle. This expensive option is based on variation of thin film path length differences with angle. Security features on credit cards, banknotes, driving licenses and similar items prone to forgery use thin film interference, diffraction gratings, or holograms.

Australia led the way with dollar bills printed on polymer with a diffraction grating security feature making the currency difficult to forge. Other countries such as New Zealand and Taiwan are using similar technologies, while the United States currency includes a thin film interference effect.

Making Connections: Take-Home Experiment—Thin Film Interference

One feature of thin film interference and diffraction gratings is that the pattern shifts as you change the angle at which you look or move your head. Find examples of thin film interference and gratings around you. Explain how the patterns change for each specific example. Find examples where the thickness changes giving rise to changing colors. If you can find two microscope slides, then try observing the effect shown in **Figure 27.34**. Try separating one end of the two slides with a hair or maybe a thin piece of paper and observe the effect.

Problem-Solving Strategies for Wave Optics

Step 1. *Examine the situation to determine that interference is involved*. Identify whether slits or thin film interference are considered in the problem.

Step 2. *If slits are involved*, note that diffraction gratings and double slits produce very similar interference patterns, but that gratings have narrower (sharper) maxima. Single slit patterns are characterized by a large central maximum and smaller maxima to the sides.

Step 3. *If thin film interference is involved, take note of the path length difference between the two rays that interfere*. Be certain to use the wavelength in the medium involved, since it differs from the wavelength in vacuum. Note also that there is an additional $\lambda / 2$ phase shift when light reflects from a medium with a greater index of refraction.

Step 4. *Identify exactly what needs to be determined in the problem (identify the unknowns)*. A written list is useful. Draw a diagram of the situation. Labeling the diagram is useful.

Step 5. *Make a list of what is given or can be inferred from the problem as stated (identify the knowns)*.

Step 6. *Solve the appropriate equation for the quantity to be determined (the unknown), and enter the knowns*. Slits, gratings, and the Rayleigh limit involve equations.

Step 7. *For thin film interference, you will have constructive interference for a total shift that is an integral number of wavelengths*. *You will have destructive interference for a total shift of a half-integral number of wavelengths*. Always keep in mind that crest to crest is constructive whereas crest to trough is destructive.

Step 8. *Check to see if the answer is reasonable: Does it make sense?* Angles in interference patterns cannot be greater than $90°$, for example.

27.8 Polarization

Learning Objectives

By the end of this section, you will be able to:

- Discuss the meaning of polarization.
- Discuss the property of optical activity of certain materials.

The information presented in this section supports the following AP® learning objectives and science practices:

- **6.A.1.3** The student is able to analyze data (or a visual representation) to identify patterns that indicate that a particular mechanical wave is polarized and construct an explanation of the fact that the wave must have a vibration perpendicular to the direction of energy propagation. **(S.P. 5.1, 6.2)**

Polaroid sunglasses are familiar to most of us. They have a special ability to cut the glare of light reflected from water or glass (see **Figure 27.36**). Polaroids have this ability because of a wave characteristic of light called polarization. What is polarization? How is it produced? What are some of its uses? The answers to these questions are related to the wave character of light.

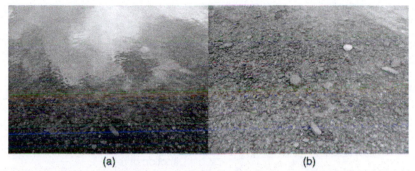

Figure 27.36 These two photographs of a river show the effect of a polarizing filter in reducing glare in light reflected from the surface of water. Part (b) of this figure was taken with a polarizing filter and part (a) was not. As a result, the reflection of clouds and sky observed in part (a) is not observed in part (b). Polarizing sunglasses are particularly useful on snow and water. (credit: Amithshs, Wikimedia Commons)

Light is one type of electromagnetic (EM) wave. As noted earlier, EM waves are *transverse waves* consisting of varying electric and magnetic fields that oscillate perpendicular to the direction of propagation (see **Figure 27.37**). There are specific directions for the oscillations of the electric and magnetic fields. **Polarization** is the attribute that a wave's oscillations have a definite direction relative to the direction of propagation of the wave. (This is not the same type of polarization as that discussed for the separation of charges.) Waves having such a direction are said to be **polarized**. For an EM wave, we define the **direction of polarization** to be the direction parallel to the electric field. Thus we can think of the electric field arrows as showing the direction of polarization, as in **Figure 27.37**.

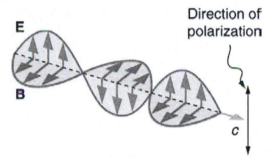

Figure 27.37 An EM wave, such as light, is a transverse wave. The electric and magnetic fields are perpendicular to the direction of propagation.

To examine this further, consider the transverse waves in the ropes shown in **Figure 27.38**. The oscillations in one rope are in a vertical plane and are said to be **vertically polarized**. Those in the other rope are in a horizontal plane and are **horizontally polarized**. If a vertical slit is placed on the first rope, the waves pass through. However, a vertical slit blocks the horizontally polarized waves. For EM waves, the direction of the electric field is analogous to the disturbances on the ropes.

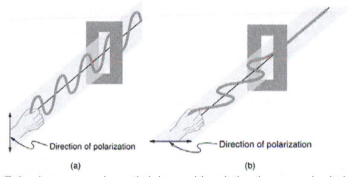

Figure 27.38 The transverse oscillations in one rope are in a vertical plane, and those in the other rope are in a horizontal plane. The first is said to be vertically polarized, and the other is said to be horizontally polarized. Vertical slits pass vertically polarized waves and block horizontally polarized waves.

The Sun and many other light sources produce waves that are randomly polarized (see **Figure 27.39**). Such light is said to be **unpolarized** because it is composed of many waves with all possible directions of polarization. Polaroid materials, invented by the founder of Polaroid Corporation, Edwin Land, act as a *polarizing* slit for light, allowing only polarization in one direction to pass through. Polarizing filters are composed of long molecules aligned in one direction. Thinking of the molecules as many slits, analogous to those for the oscillating ropes, we can understand why only light with a specific polarization can get through. The **axis of a polarizing filter** is the direction along which the filter passes the electric field of an EM wave (see **Figure 27.40**).

Random polarization

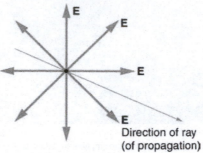

Direction of ray
(of propagation)

Figure 27.39 The slender arrow represents a ray of unpolarized light. The bold arrows represent the direction of polarization of the individual waves composing the ray. Since the light is unpolarized, the arrows point in all directions.

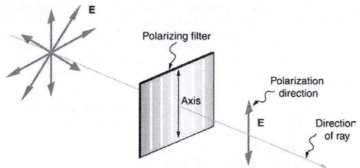

Figure 27.40 A polarizing filter has a polarization axis that acts as a slit passing through electric fields parallel to its direction. The direction of polarization of an EM wave is defined to be the direction of its electric field.

Figure 27.41 shows the effect of two polarizing filters on originally unpolarized light. The first filter polarizes the light along its axis. When the axes of the first and second filters are aligned (parallel), then all of the polarized light passed by the first filter is also passed by the second. If the second polarizing filter is rotated, only the component of the light parallel to the second filter's axis is passed. When the axes are perpendicular, no light is passed by the second.

Only the component of the EM wave parallel to the axis of a filter is passed. Let us call the angle between the direction of polarization and the axis of a filter θ. If the electric field has an amplitude E, then the transmitted part of the wave has an amplitude $E \cos \theta$ (see **Figure 27.42**). Since the intensity of a wave is proportional to its amplitude squared, the intensity I of the transmitted wave is related to the incident wave by

$$I = I_0 \cos^2 \theta, \tag{27.44}$$

where I_0 is the intensity of the polarized wave before passing through the filter. (The above equation is known as Malus's law.)

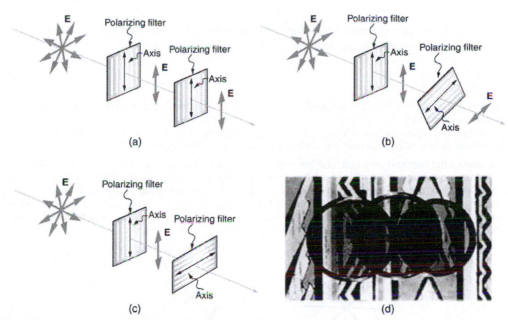

Figure 27.41 The effect of rotating two polarizing filters, where the first polarizes the light. (a) All of the polarized light is passed by the second polarizing filter, because its axis is parallel to the first. (b) As the second is rotated, only part of the light is passed. (c) When the second is perpendicular to the first, no light is passed. (d) In this photograph, a polarizing filter is placed above two others. Its axis is perpendicular to the filter on the right (dark area) and parallel to the filter on the left (lighter area). (credit: P.P. Urone)

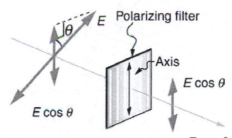

Figure 27.42 A polarizing filter transmits only the component of the wave parallel to its axis, $E \cos \theta$, reducing the intensity of any light not polarized parallel to its axis.

Example 27.8 Calculating Intensity Reduction by a Polarizing Filter

What angle is needed between the direction of polarized light and the axis of a polarizing filter to reduce its intensity by 90.0%?

Strategy

When the intensity is reduced by 90.0%, it is 10.0% or 0.100 times its original value. That is, $I = 0.100 I_0$. Using this information, the equation $I = I_0 \cos^2 \theta$ can be used to solve for the needed angle.

Solution

Solving the equation $I = I_0 \cos^2 \theta$ for $\cos \theta$ and substituting with the relationship between I and I_0 gives

$$\cos \theta = \sqrt{\frac{I}{I_0}} = \sqrt{\frac{0.100 I_0}{I_0}} = 0.3162. \tag{27.45}$$

Solving for θ yields

$$\theta = \cos^{-1} 0.3162 = 71.6°. \tag{27.46}$$

Discussion

A fairly large angle between the direction of polarization and the filter axis is needed to reduce the intensity to 10.0% of its original value. This seems reasonable based on experimenting with polarizing films. It is interesting that, at an angle of $45°$, the intensity is reduced to 50% of its original value (as you will show in this section's Problems & Exercises). Note that

$71.6°$ is $18.4°$ from reducing the intensity to zero, and that at an angle of $18.4°$ the intensity is reduced to 90.0% of its original value (as you will also show in Problems & Exercises), giving evidence of symmetry.

Polarization by Reflection

By now you can probably guess that Polaroid sunglasses cut the glare in reflected light because that light is polarized. You can check this for yourself by holding Polaroid sunglasses in front of you and rotating them while looking at light reflected from water or glass. As you rotate the sunglasses, you will notice the light gets bright and dim, but not completely black. This implies the reflected light is partially polarized and cannot be completely blocked by a polarizing filter.

Figure 27.43 illustrates what happens when unpolarized light is reflected from a surface. Vertically polarized light is preferentially refracted at the surface, so that *the reflected light is left more horizontally polarized*. The reasons for this phenomenon are beyond the scope of this text, but a convenient mnemonic for remembering this is to imagine the polarization direction to be like an arrow. Vertical polarization would be like an arrow perpendicular to the surface and would be more likely to stick and not be reflected. Horizontal polarization is like an arrow bouncing on its side and would be more likely to be reflected. Sunglasses with vertical axes would then block more reflected light than unpolarized light from other sources.

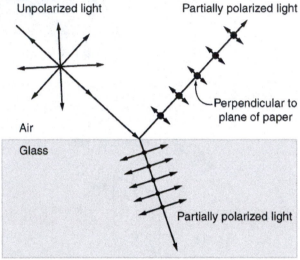

Figure 27.43 Polarization by reflection. Unpolarized light has equal amounts of vertical and horizontal polarization. After interaction with a surface, the vertical components are preferentially absorbed or refracted, leaving the reflected light more horizontally polarized. This is akin to arrows striking on their sides bouncing off, whereas arrows striking on their tips go into the surface.

Since the part of the light that is not reflected is refracted, the amount of polarization depends on the indices of refraction of the media involved. It can be shown that **reflected light is completely polarized** at a angle of reflection θ_b, given by

$$\tan \theta_b = \frac{n_2}{n_1},$$

(27.47)

where n_1 is the medium in which the incident and reflected light travel and n_2 is the index of refraction of the medium that forms the interface that reflects the light. This equation is known as **Brewster's law**, and θ_b is known as **Brewster's angle**, named after the 19th-century Scottish physicist who discovered them.

Things Great and Small: Atomic Explanation of Polarizing Filters

Polarizing filters have a polarization axis that acts as a slit. This slit passes electromagnetic waves (often visible light) that have an electric field parallel to the axis. This is accomplished with long molecules aligned perpendicular to the axis as shown in **Figure 27.44**.

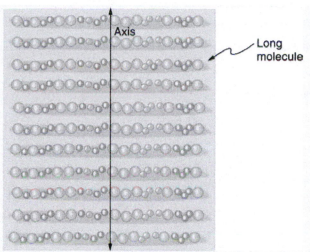

Figure 27.44 Long molecules are aligned perpendicular to the axis of a polarizing filter. The component of the electric field in an EM wave perpendicular to these molecules passes through the filter, while the component parallel to the molecules is absorbed.

Figure 27.45 illustrates how the component of the electric field parallel to the long molecules is absorbed. An electromagnetic wave is composed of oscillating electric and magnetic fields. The electric field is strong compared with the magnetic field and is more effective in exerting force on charges in the molecules. The most affected charged particles are the electrons in the molecules, since electron masses are small. If the electron is forced to oscillate, it can absorb energy from the EM wave. This reduces the fields in the wave and, hence, reduces its intensity. In long molecules, electrons can more easily oscillate parallel to the molecule than in the perpendicular direction. The electrons are bound to the molecule and are more restricted in their movement perpendicular to the molecule. Thus, the electrons can absorb EM waves that have a component of their electric field parallel to the molecule. The electrons are much less responsive to electric fields perpendicular to the molecule and will allow those fields to pass. Thus the axis of the polarizing filter is perpendicular to the length of the molecule.

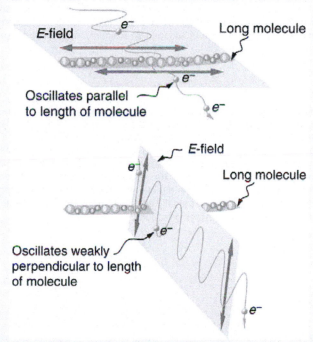

Figure 27.45 Artist's conception of an electron in a long molecule oscillating parallel to the molecule. The oscillation of the electron absorbs energy and reduces the intensity of the component of the EM wave that is parallel to the molecule.

Example 27.9 Calculating Polarization by Reflection

(a) At what angle will light traveling in air be completely polarized horizontally when reflected from water? (b) From glass?

Strategy

All we need to solve these problems are the indices of refraction. Air has $n_1 = 1.00$, water has $n_2 = 1.333$, and crown glass has $n'_2 = 1.520$. The equation $\tan \theta_b = \frac{n_2}{n_1}$ can be directly applied to find θ_b in each case.

Solution for (a)

Putting the known quantities into the equation

$$\tan \theta_b = \frac{n_2}{n_1} \tag{27.48}$$

gives

$$\tan \theta_b = \frac{n_2}{n_1} = \frac{1.333}{1.00} = 1.333. \tag{27.49}$$

Solving for the angle θ_b yields

$$\theta_b = \tan^{-1} 1.333 = 53.1°. \tag{27.50}$$

Solution for (b)

Similarly, for crown glass and air,

$$\tan \theta'_b = \frac{n'_2}{n_1} = \frac{1.520}{1.00} = 1.52. \tag{27.51}$$

Thus,

$$\theta'_b = \tan^{-1} 1.52 = 56.7°. \tag{27.52}$$

Discussion

Light reflected at these angles could be completely blocked by a good polarizing filter held with its *axis vertical*. Brewster's angle for water and air are similar to those for glass and air, so that sunglasses are equally effective for light reflected from either water or glass under similar circumstances. Light not reflected is refracted into these media. So at an incident angle equal to Brewster's angle, the refracted light will be slightly polarized vertically. It will not be completely polarized vertically, because only a small fraction of the incident light is reflected, and so a significant amount of horizontally polarized light is refracted.

Polarization by Scattering

If you hold your Polaroid sunglasses in front of you and rotate them while looking at blue sky, you will see the sky get bright and dim. This is a clear indication that light scattered by air is partially polarized. **Figure 27.46** helps illustrate how this happens. Since light is a transverse EM wave, it vibrates the electrons of air molecules perpendicular to the direction it is traveling. The electrons then radiate like small antennae. Since they are oscillating perpendicular to the direction of the light ray, they produce EM radiation that is polarized perpendicular to the direction of the ray. When viewing the light along a line perpendicular to the original ray, as in **Figure 27.46**, there can be no polarization in the scattered light parallel to the original ray, because that would require the original ray to be a longitudinal wave. Along other directions, a component of the other polarization can be projected along the line of sight, and the scattered light will only be partially polarized. Furthermore, multiple scattering can bring light to your eyes from other directions and can contain different polarizations.

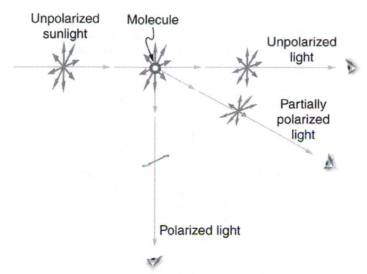

Figure 27.46 Polarization by scattering. Unpolarized light scattering from air molecules shakes their electrons perpendicular to the direction of the original ray. The scattered light therefore has a polarization perpendicular to the original direction and none parallel to the original direction.

Photographs of the sky can be darkened by polarizing filters, a trick used by many photographers to make clouds brighter by contrast. Scattering from other particles, such as smoke or dust, can also polarize light. Detecting polarization in scattered EM waves can be a useful analytical tool in determining the scattering source.

There is a range of optical effects used in sunglasses. Besides being Polaroid, other sunglasses have colored pigments embedded in them, while others use non-reflective or even reflective coatings. A recent development is photochromic lenses, which darken in the sunlight and become clear indoors. Photochromic lenses are embedded with organic microcrystalline molecules that change their properties when exposed to UV in sunlight, but become clear in artificial lighting with no UV.

Take-Home Experiment: Polarization

Find Polaroid sunglasses and rotate one while holding the other still and look at different surfaces and objects. Explain your observations. What is the difference in angle from when you see a maximum intensity to when you see a minimum intensity? Find a reflective glass surface and do the same. At what angle does the glass need to be oriented to give minimum glare?

Liquid Crystals and Other Polarization Effects in Materials

While you are undoubtedly aware of liquid crystal displays (LCDs) found in watches, calculators, computer screens, cellphones, flat screen televisions, and other myriad places, you may not be aware that they are based on polarization. Liquid crystals are so named because their molecules can be aligned even though they are in a liquid. Liquid crystals have the property that they can rotate the polarization of light passing through them by $90°$. Furthermore, this property can be turned off by the application of a voltage, as illustrated in **Figure 27.47**. It is possible to manipulate this characteristic quickly and in small well-defined regions to create the contrast patterns we see in so many LCD devices.

In flat screen LCD televisions, there is a large light at the back of the TV. The light travels to the front screen through millions of tiny units called pixels (picture elements). One of these is shown in **Figure 27.47** (a) and (b). Each unit has three cells, with red, blue, or green filters, each controlled independently. When the voltage across a liquid crystal is switched off, the liquid crystal passes the light through the particular filter. One can vary the picture contrast by varying the strength of the voltage applied to the liquid crystal.

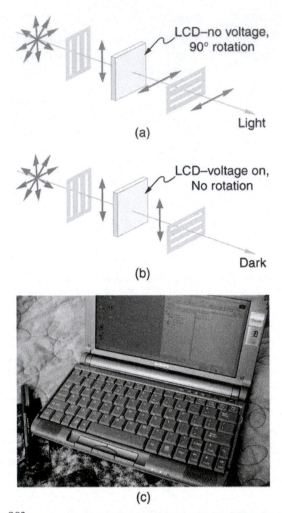

(a)

(b)

(c)

Figure 27.47 (a) Polarized light is rotated $90°$ by a liquid crystal and then passed by a polarizing filter that has its axis perpendicular to the original polarization direction. (b) When a voltage is applied to the liquid crystal, the polarized light is not rotated and is blocked by the filter, making the region dark in comparison with its surroundings. (c) LCDs can be made color specific, small, and fast enough to use in laptop computers and TVs. (credit: Jon Sullivan)

Many crystals and solutions rotate the plane of polarization of light passing through them. Such substances are said to be **optically active**. Examples include sugar water, insulin, and collagen (see **Figure 27.48**). In addition to depending on the type of substance, the amount and direction of rotation depends on a number of factors. Among these is the concentration of the substance, the distance the light travels through it, and the wavelength of light. Optical activity is due to the asymmetric shape of molecules in the substance, such as being helical. Measurements of the rotation of polarized light passing through substances can thus be used to measure concentrations, a standard technique for sugars. It can also give information on the shapes of molecules, such as proteins, and factors that affect their shapes, such as temperature and pH.

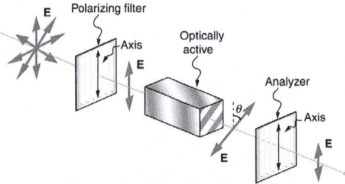

Figure 27.48 Optical activity is the ability of some substances to rotate the plane of polarization of light passing through them. The rotation is detected with a polarizing filter or analyzer.

Glass and plastic become optically active when stressed; the greater the stress, the greater the effect. Optical stress analysis on complicated shapes can be performed by making plastic models of them and observing them through crossed filters, as seen in

Figure 27.49. It is apparent that the effect depends on wavelength as well as stress. The wavelength dependence is sometimes also used for artistic purposes.

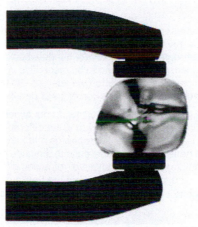

Figure 27.49 Optical stress analysis of a plastic lens placed between crossed polarizers. (credit: Infopro, Wikimedia Commons)

Another interesting phenomenon associated with polarized light is the ability of some crystals to split an unpolarized beam of light into two. Such crystals are said to be **birefringent** (see **Figure 27.50**). Each of the separated rays has a specific polarization. One behaves normally and is called the ordinary ray, whereas the other does not obey Snell's law and is called the extraordinary ray. Birefringent crystals can be used to produce polarized beams from unpolarized light. Some birefringent materials preferentially absorb one of the polarizations. These materials are called dichroic and can produce polarization by this preferential absorption. This is fundamentally how polarizing filters and other polarizers work. The interested reader is invited to further pursue the numerous properties of materials related to polarization.

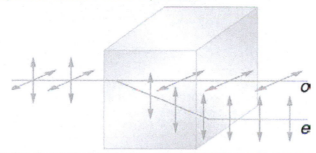

Figure 27.50 Birefringent materials, such as the common mineral calcite, split unpolarized beams of light into two. The ordinary ray behaves as expected, but the extraordinary ray does not obey Snell's law.

27.9 *Extended Topic* Microscopy Enhanced by the Wave Characteristics of Light

Learning Objectives
By the end of this section, you will be able to: • Discuss the different types of microscopes.

Physics research underpins the advancement of developments in microscopy. As we gain knowledge of the wave nature of electromagnetic waves and methods to analyze and interpret signals, new microscopes that enable us to "see" more are being developed. It is the evolution and newer generation of microscopes that are described in this section.

The use of microscopes (microscopy) to observe small details is limited by the wave nature of light. Owing to the fact that light diffracts significantly around small objects, it becomes impossible to observe details significantly smaller than the wavelength of light. One rule of thumb has it that all details smaller than about λ are difficult to observe. Radar, for example, can detect the size of an aircraft, but not its individual rivets, since the wavelength of most radar is several centimeters or greater. Similarly, visible light cannot detect individual atoms, since atoms are about 0.1 nm in size and visible wavelengths range from 380 to 760 nm. Ironically, special techniques used to obtain the best possible resolution with microscopes take advantage of the same wave characteristics of light that ultimately limit the detail.

Making Connections: Waves

All attempts to observe the size and shape of objects are limited by the wavelength of the probe. Sonar and medical ultrasound are limited by the wavelength of sound they employ. We shall see that this is also true in electron microscopy, since electrons have a wavelength. Heisenberg's uncertainty principle asserts that this limit is fundamental and inescapable, as we shall see in quantum mechanics.

The most obvious method of obtaining better detail is to utilize shorter wavelengths. **Ultraviolet (UV) microscopes** have been constructed with special lenses that transmit UV rays and utilize photographic or electronic techniques to record images. The shorter UV wavelengths allow somewhat greater detail to be observed, but drawbacks, such as the hazard of UV to living tissue and the need for special detection devices and lenses (which tend to be dispersive in the UV), severely limit the use of UV microscopes. Elsewhere, we will explore practical uses of very short wavelength EM waves, such as x rays, and other short-wavelength probes, such as electrons in electron microscopes, to detect small details.

Another difficulty in microscopy is the fact that many microscopic objects do not absorb much of the light passing through them. The lack of contrast makes image interpretation very difficult. **Contrast** is the difference in intensity between objects and the background on which they are observed. Stains (such as dyes, fluorophores, etc.) are commonly employed to enhance contrast, but these tend to be application specific. More general wave interference techniques can be used to produce contrast. **Figure 27.51** shows the passage of light through a sample. Since the indices of refraction differ, the number of wavelengths in the paths differs. Light emerging from the object is thus out of phase with light from the background and will interfere differently, producing enhanced contrast, especially if the light is coherent and monochromatic—as in laser light.

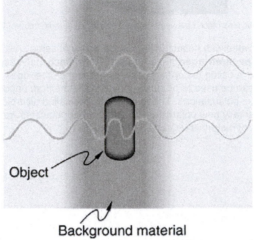

Figure 27.51 Light rays passing through a sample under a microscope will emerge with different phases depending on their paths. The object shown has a greater index of refraction than the background, and so the wavelength decreases as the ray passes through it. Superimposing these rays produces interference that varies with path, enhancing contrast between the object and background.

Interference microscopes enhance contrast between objects and background by superimposing a reference beam of light upon the light emerging from the sample. Since light from the background and objects differ in phase, there will be different amounts of constructive and destructive interference, producing the desired contrast in final intensity. **Figure 27.52** shows schematically how this is done. Parallel rays of light from a source are split into two beams by a half-silvered mirror. These beams are called the object and reference beams. Each beam passes through identical optical elements, except that the object beam passes through the object we wish to observe microscopically. The light beams are recombined by another half-silvered mirror and interfere. Since the light rays passing through different parts of the object have different phases, interference will be significantly different and, hence, have greater contrast between them.

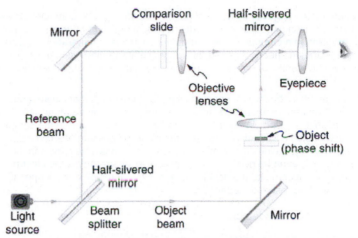

Figure 27.52 An interference microscope utilizes interference between the reference and object beam to enhance contrast. The two beams are split by a half-silvered mirror; the object beam is sent through the object, and the reference beam is sent through otherwise identical optical elements. The beams are recombined by another half-silvered mirror, and the interference depends on the various phases emerging from different parts of the object, enhancing contrast.

Another type of microscope utilizing wave interference and differences in phases to enhance contrast is called the **phase-contrast microscope**. While its principle is the same as the interference microscope, the phase-contrast microscope is simpler to use and construct. Its impact (and the principle upon which it is based) was so important that its developer, the Dutch physicist Frits Zernike (1888–1966), was awarded the Nobel Prize in 1953. **Figure 27.53** shows the basic construction of a phase-contrast microscope. Phase differences between light passing through the object and background are produced by passing the rays through different parts of a phase plate (so called because it shifts the phase of the light passing through it). These two light rays are superimposed in the image plane, producing contrast due to their interference.

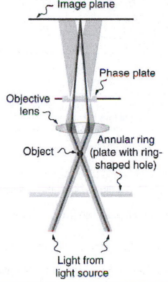

Figure 27.53 Simplified construction of a phase-contrast microscope. Phase differences between light passing through the object and background are produced by passing the rays through different parts of a phase plate. The light rays are superimposed in the image plane, producing contrast due to their interference.

A **polarization microscope** also enhances contrast by utilizing a wave characteristic of light. Polarization microscopes are useful for objects that are optically active or birefringent, particularly if those characteristics vary from place to place in the object. Polarized light is sent through the object and then observed through a polarizing filter that is perpendicular to the original polarization direction. Nearly transparent objects can then appear with strong color and in high contrast. Many polarization effects are wavelength dependent, producing color in the processed image. Contrast results from the action of the polarizing filter in passing only components parallel to its axis.

Apart from the UV microscope, the variations of microscopy discussed so far in this section are available as attachments to fairly standard microscopes or as slight variations. The next level of sophistication is provided by commercial **confocal microscopes**, which use the extended focal region shown in **Figure 27.31**(b) to obtain three-dimensional images rather than two-dimensional images. Here, only a single plane or region of focus is identified; out-of-focus regions above and below this plane are subtracted out by a computer so the image quality is much better. This type of microscope makes use of fluorescence, where a laser provides the excitation light. Laser light passing through a tiny aperture called a pinhole forms an extended focal region within the specimen. The reflected light passes through the objective lens to a second pinhole and the photomultiplier detector, see **Figure 27.54**. The second pinhole is the key here and serves to block much of the light from points that are not at the focal point of the objective lens. The pinhole is conjugate (coupled) to the focal point of the lens. The second pinhole and detector are scanned, allowing reflected light from a small region or section of the extended focal region to be imaged at any one time. The out-of-focus light is excluded. Each image is stored in a computer, and a full scanned image is generated in a short time. Live cell processes can also be imaged at adequate scanning speeds allowing the imaging of three-dimensional microscopic movement. Confocal microscopy enhances images over conventional optical microscopy, especially for thicker specimens, and so has become quite popular.

The next level of sophistication is provided by microscopes attached to instruments that isolate and detect only a small wavelength band of light—monochromators and spectral analyzers. Here, the monochromatic light from a laser is scattered from the specimen. This scattered light shifts up or down as it excites particular energy levels in the sample. The uniqueness of the observed scattered light can give detailed information about the chemical composition of a given spot on the sample with high contrast—like molecular fingerprints. Applications are in materials science, nanotechnology, and the biomedical field. Fine details in biochemical processes over time can even be detected. The ultimate in microscopy is the electron microscope—to be discussed later. Research is being conducted into the development of new prototype microscopes that can become commercially available, providing better diagnostic and research capacities.

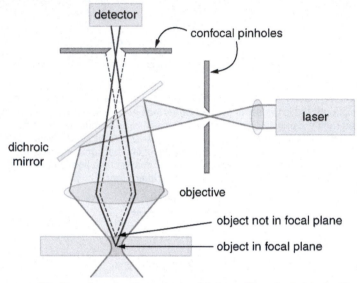

Figure 27.54 A confocal microscope provides three-dimensional images using pinholes and the extended depth of focus as described by wave optics. The right pinhole illuminates a tiny region of the sample in the focal plane. In-focus light rays from this tiny region pass through the dichroic mirror and the second pinhole to a detector and a computer. Out-of-focus light rays are blocked. The pinhole is scanned sideways to form an image of the entire focal plane. The pinhole can then be scanned up and down to gather images from different focal planes. The result is a three-dimensional image of the specimen.

Glossary

axis of a polarizing filter: the direction along which the filter passes the electric field of an EM wave

birefringent: crystals that split an unpolarized beam of light into two beams

Brewster's angle: $\theta_b = \tan^{-1}\left(\dfrac{n_2}{n_1}\right)$, where n_2 is the index of refraction of the medium from which the light is reflected and n_1 is the index of refraction of the medium in which the reflected light travels

Brewster's law: $\tan\theta_b = \dfrac{n_2}{n_1}$, where n_1 is the medium in which the incident and reflected light travel and n_2 is the index of refraction of the medium that forms the interface that reflects the light

coherent: waves are in phase or have a definite phase relationship

confocal microscopes: microscopes that use the extended focal region to obtain three-dimensional images rather than two-dimensional images

constructive interference for a diffraction grating: occurs when the condition $d\ \sin\theta = m\lambda$ (for $m = 0,\ 1,\ -1,\ 2,\ -2,\ ...$) is satisfied, where d is the distance between slits in the grating, λ is the wavelength of light, and m is the order of the maximum

constructive interference for a double slit: the path length difference must be an integral multiple of the wavelength

contrast: the difference in intensity between objects and the background on which they are observed

destructive interference for a double slit: the path length difference must be a half-integral multiple of the wavelength

destructive interference for a single slit: occurs when $D \sin \theta = m\lambda$, (for $m = 1, -1, 2, -2, 3, \ldots$), where D is the slit width, λ is the light's wavelength, θ is the angle relative to the original direction of the light, and m is the order of the minimum

diffraction: the bending of a wave around the edges of an opening or an obstacle

diffraction grating: a large number of evenly spaced parallel slits

direction of polarization: the direction parallel to the electric field for EM waves

horizontally polarized: the oscillations are in a horizontal plane

Huygens's principle: every point on a wavefront is a source of wavelets that spread out in the forward direction at the same speed as the wave itself. The new wavefront is a line tangent to all of the wavelets

incoherent: waves have random phase relationships

interference microscopes: microscopes that enhance contrast between objects and background by superimposing a reference beam of light upon the light emerging from the sample

optically active: substances that rotate the plane of polarization of light passing through them

order: the integer m used in the equations for constructive and destructive interference for a double slit

phase-contrast microscope: microscope utilizing wave interference and differences in phases to enhance contrast

polarization: the attribute that wave oscillations have a definite direction relative to the direction of propagation of the wave

polarization microscope: microscope that enhances contrast by utilizing a wave characteristic of light, useful for objects that are optically active

polarized: waves having the electric and magnetic field oscillations in a definite direction

Rayleigh criterion: two images are just resolvable when the center of the diffraction pattern of one is directly over the first minimum of the diffraction pattern of the other

reflected light that is completely polarized: light reflected at the angle of reflection θ_b, known as Brewster's angle

thin film interference: interference between light reflected from different surfaces of a thin film

ultraviolet (UV) microscopes: microscopes constructed with special lenses that transmit UV rays and utilize photographic or electronic techniques to record images

unpolarized: waves that are randomly polarized

vertically polarized: the oscillations are in a vertical plane

wavelength in a medium: $\lambda_n = \lambda / n$, where λ is the wavelength in vacuum, and n is the index of refraction of the medium

Section Summary

27.1 The Wave Aspect of Light: Interference
- Wave optics is the branch of optics that must be used when light interacts with small objects or whenever the wave characteristics of light are considered.
- Wave characteristics are those associated with interference and diffraction.
- Visible light is the type of electromagnetic wave to which our eyes respond and has a wavelength in the range of 380 to 760 nm.
- Like all EM waves, the following relationship is valid in vacuum: $c = f\lambda$, where $c = 3 \times 10^8$ m/s is the speed of light, f is the frequency of the electromagnetic wave, and λ is its wavelength in vacuum.
- The wavelength λ_n of light in a medium with index of refraction n is $\lambda_n = \lambda / n$. Its frequency is the same as in vacuum.

27.2 Huygens's Principle: Diffraction
- An accurate technique for determining how and where waves propagate is given by Huygens's principle: Every point on a wavefront is a source of wavelets that spread out in the forward direction at the same speed as the wave itself. The new

wavefront is a line tangent to all of the wavelets.
- Diffraction is the bending of a wave around the edges of an opening or other obstacle.

27.3 Young's Double Slit Experiment
- Young's double slit experiment gave definitive proof of the wave character of light.
- An interference pattern is obtained by the superposition of light from two slits.
- There is constructive interference when $d \sin \theta = m\lambda$ (for $m = 0, 1, -1, 2, -2, \ldots$), where d is the distance between the slits, θ is the angle relative to the incident direction, and m is the order of the interference.
- There is destructive interference when $d \sin \theta = \left(m + \frac{1}{2}\right)\lambda$ (for $m = 0, 1, -1, 2, -2, \ldots$).

27.4 Multiple Slit Diffraction
- A diffraction grating is a large collection of evenly spaced parallel slits that produces an interference pattern similar to but sharper than that of a double slit.
- There is constructive interference for a diffraction grating when $d \sin \theta = m\lambda$ (for $m = 0, 1, -1, 2, -2, \ldots$), where d is the distance between slits in the grating, λ is the wavelength of light, and m is the order of the maximum.

27.5 Single Slit Diffraction
- A single slit produces an interference pattern characterized by a broad central maximum with narrower and dimmer maxima to the sides.
- There is destructive interference for a single slit when $D \sin \theta = m\lambda$, (for $m = 1, -1, 2, -2, 3, \ldots$), where D is the slit width, λ is the light's wavelength, θ is the angle relative to the original direction of the light, and m is the order of the minimum. Note that there is no $m = 0$ minimum.

27.6 Limits of Resolution: The Rayleigh Criterion
- Diffraction limits resolution.
- For a circular aperture, lens, or mirror, the Rayleigh criterion states that two images are just resolvable when the center of the diffraction pattern of one is directly over the first minimum of the diffraction pattern of the other.
- This occurs for two point objects separated by the angle $\theta = 1.22\frac{\lambda}{D}$, where λ is the wavelength of light (or other electromagnetic radiation) and D is the diameter of the aperture, lens, mirror, etc. This equation also gives the angular spreading of a source of light having a diameter D.

27.7 Thin Film Interference
- Thin film interference occurs between the light reflected from the top and bottom surfaces of a film. In addition to the path length difference, there can be a phase change.
- When light reflects from a medium having an index of refraction greater than that of the medium in which it is traveling, a $180°$ phase change (or a $\lambda/2$ shift) occurs.

27.8 Polarization
- Polarization is the attribute that wave oscillations have a definite direction relative to the direction of propagation of the wave.
- EM waves are transverse waves that may be polarized.
- The direction of polarization is defined to be the direction parallel to the electric field of the EM wave.
- Unpolarized light is composed of many rays having random polarization directions.
- Light can be polarized by passing it through a polarizing filter or other polarizing material. The intensity I of polarized light after passing through a polarizing filter is $I = I_0 \cos^2 \theta$, where I_0 is the original intensity and θ is the angle between the direction of polarization and the axis of the filter.
- Polarization is also produced by reflection.
- Brewster's law states that reflected light will be completely polarized at the angle of reflection θ_b, known as Brewster's angle, given by a statement known as Brewster's law: $\tan \theta_b = \frac{n_2}{n_1}$, where n_1 is the medium in which the incident and reflected light travel and n_2 is the index of refraction of the medium that forms the interface that reflects the light.
- Polarization can also be produced by scattering.
- There are a number of types of optically active substances that rotate the direction of polarization of light passing through them.

27.9 *Extended Topic* Microscopy Enhanced by the Wave Characteristics of Light

- To improve microscope images, various techniques utilizing the wave characteristics of light have been developed. Many of these enhance contrast with interference effects.

Conceptual Questions

27.1 The Wave Aspect of Light: Interference

1. What type of experimental evidence indicates that light is a wave?

2. Give an example of a wave characteristic of light that is easily observed outside the laboratory.

27.2 Huygens's Principle: Diffraction

3. How do wave effects depend on the size of the object with which the wave interacts? For example, why does sound bend around the corner of a building while light does not?

4. Under what conditions can light be modeled like a ray? Like a wave?

5. Go outside in the sunlight and observe your shadow. It has fuzzy edges even if you do not. Is this a diffraction effect? Explain.

6. Why does the wavelength of light decrease when it passes from vacuum into a medium? State which attributes change and which stay the same and, thus, require the wavelength to decrease.

7. Does Huygens's principle apply to all types of waves?

27.3 Young's Double Slit Experiment

8. Young's double slit experiment breaks a single light beam into two sources. Would the same pattern be obtained for two independent sources of light, such as the headlights of a distant car? Explain.

9. Suppose you use the same double slit to perform Young's double slit experiment in air and then repeat the experiment in water. Do the angles to the same parts of the interference pattern get larger or smaller? Does the color of the light change? Explain.

10. Is it possible to create a situation in which there is only destructive interference? Explain.

11. Figure 27.55 shows the central part of the interference pattern for a pure wavelength of red light projected onto a double slit. The pattern is actually a combination of single slit and double slit interference. Note that the bright spots are evenly spaced. Is this a double slit or single slit characteristic? Note that some of the bright spots are dim on either side of the center. Is this a single slit or double slit characteristic? Which is smaller, the slit width or the separation between slits? Explain your responses.

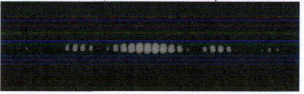

Figure 27.55 This double slit interference pattern also shows signs of single slit interference. (credit: PASCO)

27.4 Multiple Slit Diffraction

12. What is the advantage of a diffraction grating over a double slit in dispersing light into a spectrum?

13. What are the advantages of a diffraction grating over a prism in dispersing light for spectral analysis?

14. Can the lines in a diffraction grating be too close together to be useful as a spectroscopic tool for visible light? If so, what type of EM radiation would the grating be suitable for? Explain.

15. If a beam of white light passes through a diffraction grating with vertical lines, the light is dispersed into rainbow colors on the right and left. If a glass prism disperses white light to the right into a rainbow, how does the sequence of colors compare with that produced on the right by a diffraction grating?

16. Suppose pure-wavelength light falls on a diffraction grating. What happens to the interference pattern if the same light falls on a grating that has more lines per centimeter? What happens to the interference pattern if a longer-wavelength light falls on the same grating? Explain how these two effects are consistent in terms of the relationship of wavelength to the distance between slits.

17. Suppose a feather appears green but has no green pigment. Explain in terms of diffraction.

18. It is possible that there is no minimum in the interference pattern of a single slit. Explain why. Is the same true of double slits and diffraction gratings?

27.5 Single Slit Diffraction

19. As the width of the slit producing a single-slit diffraction pattern is reduced, how will the diffraction pattern produced change?

27.6 Limits of Resolution: The Rayleigh Criterion

20. A beam of light always spreads out. Why can a beam not be created with parallel rays to prevent spreading? Why can lenses, mirrors, or apertures not be used to correct the spreading?

27.7 Thin Film Interference

21. What effect does increasing the wedge angle have on the spacing of interference fringes? If the wedge angle is too large, fringes are not observed. Why?

22. How is the difference in paths taken by two originally in-phase light waves related to whether they interfere constructively or destructively? How can this be affected by reflection? By refraction?

23. Is there a phase change in the light reflected from either surface of a contact lens floating on a person's tear layer? The index of refraction of the lens is about 1.5, and its top surface is dry.

24. In placing a sample on a microscope slide, a glass cover is placed over a water drop on the glass slide. Light incident from above can reflect from the top and bottom of the glass cover and from the glass slide below the water drop. At which surfaces will there be a phase change in the reflected light?

25. Answer the above question if the fluid between the two pieces of crown glass is carbon disulfide.

26. While contemplating the food value of a slice of ham, you notice a rainbow of color reflected from its moist surface. Explain its origin.

27. An inventor notices that a soap bubble is dark at its thinnest and realizes that destructive interference is taking place for all wavelengths. How could she use this knowledge to make a non-reflective coating for lenses that is effective at all wavelengths? That is, what limits would there be on the index of refraction and thickness of the coating? How might this be impractical?

28. A non-reflective coating like the one described in **Example 27.6** works ideally for a single wavelength and for perpendicular incidence. What happens for other wavelengths and other incident directions? Be specific.

29. Why is it much more difficult to see interference fringes for light reflected from a thick piece of glass than from a thin film? Would it be easier if monochromatic light were used?

27.8 Polarization

30. Under what circumstances is the phase of light changed by reflection? Is the phase related to polarization?

31. Can a sound wave in air be polarized? Explain.

32. No light passes through two perfect polarizing filters with perpendicular axes. However, if a third polarizing filter is placed between the original two, some light can pass. Why is this? Under what circumstances does most of the light pass?

33. Explain what happens to the energy carried by light that it is dimmed by passing it through two crossed polarizing filters.

34. When particles scattering light are much smaller than its wavelength, the amount of scattering is proportional to $1/\lambda^4$. Does this mean there is more scattering for small λ than large λ? How does this relate to the fact that the sky is blue?

35. Using the information given in the preceding question, explain why sunsets are red.

36. When light is reflected at Brewster's angle from a smooth surface, it is 100% polarized parallel to the surface. Part of the light will be refracted into the surface. Describe how you would do an experiment to determine the polarization of the refracted light. What direction would you expect the polarization to have and would you expect it to be 100%?

27.9 *Extended Topic* Microscopy Enhanced by the Wave Characteristics of Light

37. Explain how microscopes can use wave optics to improve contrast and why this is important.

38. A bright white light under water is collimated and directed upon a prism. What range of colors does one see emerging?

Problems & Exercises

27.1 The Wave Aspect of Light: Interference

1. Show that when light passes from air to water, its wavelength decreases to 0.750 times its original value.

2. Find the range of visible wavelengths of light in crown glass.

3. What is the index of refraction of a material for which the wavelength of light is 0.671 times its value in a vacuum? Identify the likely substance.

4. Analysis of an interference effect in a clear solid shows that the wavelength of light in the solid is 329 nm. Knowing this light comes from a He-Ne laser and has a wavelength of 633 nm in air, is the substance zircon or diamond?

5. What is the ratio of thicknesses of crown glass and water that would contain the same number of wavelengths of light?

27.3 Young's Double Slit Experiment

6. At what angle is the first-order maximum for 450-nm wavelength blue light falling on double slits separated by 0.0500 mm?

7. Calculate the angle for the third-order maximum of 580-nm wavelength yellow light falling on double slits separated by 0.100 mm.

8. What is the separation between two slits for which 610-nm orange light has its first maximum at an angle of $30.0°$?

9. Find the distance between two slits that produces the first minimum for 410-nm violet light at an angle of $45.0°$.

10. Calculate the wavelength of light that has its third minimum at an angle of $30.0°$ when falling on double slits separated by $3.00 \, \mu m$. Explicitly, show how you follow the steps in **Problem-Solving Strategies for Wave Optics**.

11. What is the wavelength of light falling on double slits separated by $2.00 \, \mu m$ if the third-order maximum is at an angle of $60.0°$?

12. At what angle is the fourth-order maximum for the situation in **Exercise 27.6**?

13. What is the highest-order maximum for 400-nm light falling on double slits separated by $25.0 \, \mu m$?

14. Find the largest wavelength of light falling on double slits separated by $1.20 \, \mu m$ for which there is a first-order maximum. Is this in the visible part of the spectrum?

15. What is the smallest separation between two slits that will produce a second-order maximum for 720-nm red light?

16. (a) What is the smallest separation between two slits that will produce a second-order maximum for any visible light? (b) For all visible light?

17. (a) If the first-order maximum for pure-wavelength light falling on a double slit is at an angle of $10.0°$, at what angle is the second-order maximum? (b) What is the angle of the first minimum? (c) What is the highest-order maximum possible here?

18. **Figure 27.56** shows a double slit located a distance x from a screen, with the distance from the center of the screen given by y. When the distance d between the slits is relatively large, there will be numerous bright spots, called fringes. Show that, for small angles (where $\sin \theta \approx \theta$, with θ in radians), the distance between fringes is given by $\Delta y = x\lambda / d$.

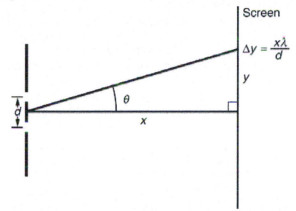

Figure 27.56 The distance between adjacent fringes is $\Delta y = x\lambda / d$, assuming the slit separation d is large compared with λ.

19. Using the result of the problem above, calculate the distance between fringes for 633-nm light falling on double slits separated by 0.0800 mm, located 3.00 m from a screen as in **Figure 27.56**.

20. Using the result of the problem two problems prior, find the wavelength of light that produces fringes 7.50 mm apart on a screen 2.00 m from double slits separated by 0.120 mm (see **Figure 27.56**).

27.4 Multiple Slit Diffraction

21. A diffraction grating has 2000 lines per centimeter. At what angle will the first-order maximum be for 520-nm-wavelength green light?

22. Find the angle for the third-order maximum for 580-nm-wavelength yellow light falling on a diffraction grating having 1500 lines per centimeter.

23. How many lines per centimeter are there on a diffraction grating that gives a first-order maximum for 470-nm blue light at an angle of $25.0°$?

24. What is the distance between lines on a diffraction grating that produces a second-order maximum for 760-nm red light at an angle of $60.0°$?

25. Calculate the wavelength of light that has its second-order maximum at $45.0°$ when falling on a diffraction grating that has 5000 lines per centimeter.

26. An electric current through hydrogen gas produces several distinct wavelengths of visible light. What are the wavelengths of the hydrogen spectrum, if they form first-order maxima at angles of $24.2°$, $25.7°$, $29.1°$, and $41.0°$ when projected on a diffraction grating having 10,000 lines per centimeter? Explicitly show how you follow the steps in **Problem-Solving Strategies for Wave Optics**

27. (a) What do the four angles in the above problem become if a 5000-line-per-centimeter diffraction grating is used? (b) Using this grating, what would the angles be for the second-order maxima? (c) Discuss the relationship between integral reductions in lines per centimeter and the new angles of various order maxima.

28. What is the maximum number of lines per centimeter a diffraction grating can have and produce a complete first-order spectrum for visible light?

29. The yellow light from a sodium vapor lamp *seems* to be of pure wavelength, but it produces two first-order maxima at $36.093°$ and $36.129°$ when projected on a 10,000 line per centimeter diffraction grating. What are the two wavelengths to an accuracy of 0.1 nm?

30. What is the spacing between structures in a feather that acts as a reflection grating, given that they produce a first-order maximum for 525-nm light at a $30.0°$ angle?

31. Structures on a bird feather act like a reflection grating having 8000 lines per centimeter. What is the angle of the first-order maximum for 600-nm light?

32. An opal such as that shown in **Figure 27.17** acts like a reflection grating with rows separated by about $8~\mu\text{m}$. If the opal is illuminated normally, (a) at what angle will red light be seen and (b) at what angle will blue light be seen?

33. At what angle does a diffraction grating produces a second-order maximum for light having a first-order maximum at $20.0°$?

34. Show that a diffraction grating cannot produce a second-order maximum for a given wavelength of light unless the first-order maximum is at an angle less than $30.0°$.

35. If a diffraction grating produces a first-order maximum for the shortest wavelength of visible light at $30.0°$, at what angle will the first-order maximum be for the longest wavelength of visible light?

36. (a) Find the maximum number of lines per centimeter a diffraction grating can have and produce a maximum for the smallest wavelength of visible light. (b) Would such a grating be useful for ultraviolet spectra? (c) For infrared spectra?

37. (a) Show that a 30,000-line-per-centimeter grating will not produce a maximum for visible light. (b) What is the longest wavelength for which it does produce a first-order maximum? (c) What is the greatest number of lines per centimeter a diffraction grating can have and produce a complete second-order spectrum for visible light?

38. A He–Ne laser beam is reflected from the surface of a CD onto a wall. The brightest spot is the reflected beam at an angle equal to the angle of incidence. However, fringes are also observed. If the wall is 1.50 m from the CD, and the first fringe is 0.600 m from the central maximum, what is the spacing of grooves on the CD?

39. The analysis shown in the figure below also applies to diffraction gratings with lines separated by a distance d. What is the distance between fringes produced by a diffraction grating having 125 lines per centimeter for 600-nm light, if the screen is 1.50 m away?

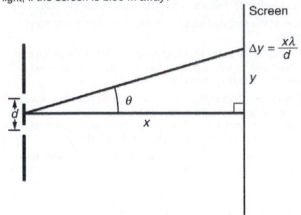

Figure 27.57 The distance between adjacent fringes is $\Delta y = x\lambda / d$, assuming the slit separation d is large compared with λ.

40. Unreasonable Results

Red light of wavelength of 700 nm falls on a double slit separated by 400 nm. (a) At what angle is the first-order maximum in the diffraction pattern? (b) What is unreasonable about this result? (c) Which assumptions are unreasonable or inconsistent?

41. Unreasonable Results

(a) What visible wavelength has its fourth-order maximum at an angle of $25.0°$ when projected on a 25,000-line-per-centimeter diffraction grating? (b) What is unreasonable about this result? (c) Which assumptions are unreasonable or inconsistent?

42. Construct Your Own Problem

Consider a spectrometer based on a diffraction grating. Construct a problem in which you calculate the distance between two wavelengths of electromagnetic radiation in your spectrometer. Among the things to be considered are the wavelengths you wish to be able to distinguish, the number of lines per meter on the diffraction grating, and the distance from the grating to the screen or detector. Discuss the practicality of the device in terms of being able to discern between wavelengths of interest.

27.5 Single Slit Diffraction

43. (a) At what angle is the first minimum for 550-nm light falling on a single slit of width $1.00~\mu\text{m}$? (b) Will there be a second minimum?

44. (a) Calculate the angle at which a $2.00\text{-}\mu\text{m}$ -wide slit produces its first minimum for 410-nm violet light. (b) Where is the first minimum for 700-nm red light?

45. (a) How wide is a single slit that produces its first minimum for 633-nm light at an angle of $28.0°$? (b) At what angle will the second minimum be?

46. (a) What is the width of a single slit that produces its first minimum at $60.0°$ for 600-nm light? (b) Find the wavelength of light that has its first minimum at $62.0°$.

47. Find the wavelength of light that has its third minimum at an angle of $48.6°$ when it falls on a single slit of width $3.00 \, \mu m$.

48. Calculate the wavelength of light that produces its first minimum at an angle of $36.9°$ when falling on a single slit of width $1.00 \, \mu m$.

49. (a) Sodium vapor light averaging 589 nm in wavelength falls on a single slit of width $7.50 \, \mu m$. At what angle does it produces its second minimum? (b) What is the highest-order minimum produced?

50. (a) Find the angle of the third diffraction minimum for 633-nm light falling on a slit of width $20.0 \, \mu m$. (b) What slit width would place this minimum at $85.0°$? Explicitly show how you follow the steps in **Problem-Solving Strategies for Wave Optics**

51. (a) Find the angle between the first minima for the two sodium vapor lines, which have wavelengths of 589.1 and 589.6 nm, when they fall upon a single slit of width $2.00 \, \mu m$. (b) What is the distance between these minima if the diffraction pattern falls on a screen 1.00 m from the slit? (c) Discuss the ease or difficulty of measuring such a distance.

52. (a) What is the minimum width of a single slit (in multiples of λ) that will produce a first minimum for a wavelength λ? (b) What is its minimum width if it produces 50 minima? (c) 1000 minima?

53. (a) If a single slit produces a first minimum at $14.5°$, at what angle is the second-order minimum? (b) What is the angle of the third-order minimum? (c) Is there a fourth-order minimum? (d) Use your answers to illustrate how the angular width of the central maximum is about twice the angular width of the next maximum (which is the angle between the first and second minima).

54. A double slit produces a diffraction pattern that is a combination of single and double slit interference. Find the ratio of the width of the slits to the separation between them, if the first minimum of the single slit pattern falls on the fifth maximum of the double slit pattern. (This will greatly reduce the intensity of the fifth maximum.)

55. Integrated Concepts

A water break at the entrance to a harbor consists of a rock barrier with a 50.0-m-wide opening. Ocean waves of 20.0-m wavelength approach the opening straight on. At what angle to the incident direction are the boats inside the harbor most protected against wave action?

56. Integrated Concepts

An aircraft maintenance technician walks past a tall hangar door that acts like a single slit for sound entering the hangar. Outside the door, on a line perpendicular to the opening in the door, a jet engine makes a 600-Hz sound. At what angle with the door will the technician observe the first minimum in sound intensity if the vertical opening is 0.800 m wide and the speed of sound is 340 m/s?

27.6 Limits of Resolution: The Rayleigh

Criterion

57. The 300-m-diameter Arecibo radio telescope pictured in **Figure 27.28** detects radio waves with a 4.00 cm average wavelength.

(a) What is the angle between two just-resolvable point sources for this telescope?

(b) How close together could these point sources be at the 2 million light year distance of the Andromeda galaxy?

58. Assuming the angular resolution found for the Hubble Telescope in **Example 27.5**, what is the smallest detail that could be observed on the Moon?

59. Diffraction spreading for a flashlight is insignificant compared with other limitations in its optics, such as spherical aberrations in its mirror. To show this, calculate the minimum angular spreading of a flashlight beam that is originally 5.00 cm in diameter with an average wavelength of 600 nm.

60. (a) What is the minimum angular spread of a 633-nm wavelength He-Ne laser beam that is originally 1.00 mm in diameter?

(b) If this laser is aimed at a mountain cliff 15.0 km away, how big will the illuminated spot be?

(c) How big a spot would be illuminated on the Moon, neglecting atmospheric effects? (This might be done to hit a corner reflector to measure the round-trip time and, hence, distance.) Explicitly show how you follow the steps in **Problem-Solving Strategies for Wave Optics.**

61. A telescope can be used to enlarge the diameter of a laser beam and limit diffraction spreading. The laser beam is sent through the telescope in opposite the normal direction and can then be projected onto a satellite or the Moon.

(a) If this is done with the Mount Wilson telescope, producing a 2.54-m-diameter beam of 633-nm light, what is the minimum angular spread of the beam?

(b) Neglecting atmospheric effects, what is the size of the spot this beam would make on the Moon, assuming a lunar distance of $3.84 \times 10^8 \, m$?

62. The limit to the eye's acuity is actually related to diffraction by the pupil.

(a) What is the angle between two just-resolvable points of light for a 3.00-mm-diameter pupil, assuming an average wavelength of 550 nm?

(b) Take your result to be the practical limit for the eye. What is the greatest possible distance a car can be from you if you can resolve its two headlights, given they are 1.30 m apart?

(c) What is the distance between two just-resolvable points held at an arm's length (0.800 m) from your eye?

(d) How does your answer to (c) compare to details you normally observe in everyday circumstances?

63. What is the minimum diameter mirror on a telescope that would allow you to see details as small as 5.00 km on the Moon some 384,000 km away? Assume an average wavelength of 550 nm for the light received.

64. You are told not to shoot until you see the whites of their eyes. If the eyes are separated by 6.5 cm and the diameter of your pupil is 5.0 mm, at what distance can you resolve the two eyes using light of wavelength 555 nm?

65. (a) The planet Pluto and its Moon Charon are separated by 19,600 km. Neglecting atmospheric effects, should the 5.08-m-diameter Mount Palomar telescope be able to resolve these bodies when they are 4.50×10^9 km from Earth? Assume an average wavelength of 550 nm.

(b) In actuality, it is just barely possible to discern that Pluto and Charon are separate bodies using an Earth-based telescope. What are the reasons for this?

66. The headlights of a car are 1.3 m apart. What is the maximum distance at which the eye can resolve these two headlights? Take the pupil diameter to be 0.40 cm.

67. When dots are placed on a page from a laser printer, they must be close enough so that you do not see the individual dots of ink. To do this, the separation of the dots must be less than Raleigh's criterion. Take the pupil of the eye to be 3.0 mm and the distance from the paper to the eye of 35 cm; find the minimum separation of two dots such that they cannot be resolved. How many dots per inch (dpi) does this correspond to?

68. Unreasonable Results

An amateur astronomer wants to build a telescope with a diffraction limit that will allow him to see if there are people on the moons of Jupiter.

(a) What diameter mirror is needed to be able to see 1.00 m detail on a Jovian Moon at a distance of 7.50×10^8 km from Earth? The wavelength of light averages 600 nm.

(b) What is unreasonable about this result?

(c) Which assumptions are unreasonable or inconsistent?

69. Construct Your Own Problem

Consider diffraction limits for an electromagnetic wave interacting with a circular object. Construct a problem in which you calculate the limit of angular resolution with a device, using this circular object (such as a lens, mirror, or antenna) to make observations. Also calculate the limit to spatial resolution (such as the size of features observable on the Moon) for observations at a specific distance from the device. Among the things to be considered are the wavelength of electromagnetic radiation used, the size of the circular object, and the distance to the system or phenomenon being observed.

27.7 Thin Film Interference

70. A soap bubble is 100 nm thick and illuminated by white light incident perpendicular to its surface. What wavelength and color of visible light is most constructively reflected, assuming the same index of refraction as water?

71. An oil slick on water is 120 nm thick and illuminated by white light incident perpendicular to its surface. What color does the oil appear (what is the most constructively reflected wavelength), given its index of refraction is 1.40?

72. Calculate the minimum thickness of an oil slick on water that appears blue when illuminated by white light perpendicular to its surface. Take the blue wavelength to be 470 nm and the index of refraction of oil to be 1.40.

73. Find the minimum thickness of a soap bubble that appears red when illuminated by white light perpendicular to its surface. Take the wavelength to be 680 nm, and assume the same index of refraction as water.

74. A film of soapy water ($n = 1.33$) on top of a plastic cutting board has a thickness of 233 nm. What color is most strongly reflected if it is illuminated perpendicular to its surface?

75. What are the three smallest non-zero thicknesses of soapy water ($n = 1.33$) on Plexiglas if it appears green (constructively reflecting 520-nm light) when illuminated perpendicularly by white light? Explicitly show how you follow the steps in **Problem Solving Strategies for Wave Optics.**

76. Suppose you have a lens system that is to be used primarily for 700-nm red light. What is the second thinnest coating of fluorite (magnesium fluoride) that would be non-reflective for this wavelength?

77. (a) As a soap bubble thins it becomes dark, because the path length difference becomes small compared with the wavelength of light and there is a phase shift at the top surface. If it becomes dark when the path length difference is less than one-fourth the wavelength, what is the thickest the bubble can be and appear dark at all visible wavelengths? Assume the same index of refraction as water. (b) Discuss the fragility of the film considering the thickness found.

78. A film of oil on water will appear dark when it is very thin, because the path length difference becomes small compared with the wavelength of light and there is a phase shift at the top surface. If it becomes dark when the path length difference is less than one-fourth the wavelength, what is the thickest the oil can be and appear dark at all visible wavelengths? Oil has an index of refraction of 1.40.

79. Figure 27.34 shows two glass slides illuminated by pure-wavelength light incident perpendicularly. The top slide touches the bottom slide at one end and rests on a 0.100-mm-diameter hair at the other end, forming a wedge of air. (a) How far apart are the dark bands, if the slides are 7.50 cm long and 589-nm light is used? (b) Is there any difference if the slides are made from crown or flint glass? Explain.

80. Figure 27.34 shows two 7.50-cm-long glass slides illuminated by pure 589-nm wavelength light incident perpendicularly. The top slide touches the bottom slide at one end and rests on some debris at the other end, forming a wedge of air. How thick is the debris, if the dark bands are 1.00 mm apart?

81. Repeat **Exercise 27.70**, but take the light to be incident at a $45°$ angle.

82. Repeat **Exercise 27.71**, but take the light to be incident at a $45°$ angle.

83. Unreasonable Results

To save money on making military aircraft invisible to radar, an inventor decides to coat them with a non-reflective material having an index of refraction of 1.20, which is between that of air and the surface of the plane. This, he reasons, should be much cheaper than designing Stealth bombers. (a) What thickness should the coating be to inhibit the reflection of 4.00-cm wavelength radar? (b) What is unreasonable about this result? (c) Which assumptions are unreasonable or inconsistent?

27.8 Polarization

84. What angle is needed between the direction of polarized light and the axis of a polarizing filter to cut its intensity in half?

85. The angle between the axes of two polarizing filters is $45.0°$. By how much does the second filter reduce the intensity of the light coming through the first?

86. If you have completely polarized light of intensity $150 \text{ W} / \text{m}^2$, what will its intensity be after passing through a polarizing filter with its axis at an $89.0°$ angle to the light's polarization direction?

87. What angle would the axis of a polarizing filter need to make with the direction of polarized light of intensity 1.00 kW/m^2 to reduce the intensity to 10.0 W/m^2?

88. At the end of **Example 27.8**, it was stated that the intensity of polarized light is reduced to 90.0% of its original value by passing through a polarizing filter with its axis at an angle of $18.4°$ to the direction of polarization. Verify this statement.

89. Show that if you have three polarizing filters, with the second at an angle of $45°$ to the first and the third at an angle of $90.0°$ to the first, the intensity of light passed by the first will be reduced to 25.0% of its value. (This is in contrast to having only the first and third, which reduces the intensity to zero, so that placing the second between them increases the intensity of the transmitted light.)

90. Prove that, if I is the intensity of light transmitted by two polarizing filters with axes at an angle θ and I' is the intensity when the axes are at an angle $90.0°-\theta$, then $I + I' = I_0$, the original intensity. (Hint: Use the trigonometric identities $\cos(90.0°-\theta) = \sin \theta$ and $\cos^2 \theta + \sin^2 \theta = 1$.)

91. At what angle will light reflected from diamond be completely polarized?

92. What is Brewster's angle for light traveling in water that is reflected from crown glass?

93. A scuba diver sees light reflected from the water's surface. At what angle will this light be completely polarized?

94. At what angle is light inside crown glass completely polarized when reflected from water, as in a fish tank?

95. Light reflected at $55.6°$ from a window is completely polarized. What is the window's index of refraction and the likely substance of which it is made?

96. (a) Light reflected at $62.5°$ from a gemstone in a ring is completely polarized. Can the gem be a diamond? (b) At what angle would the light be completely polarized if the gem was in water?

97. If θ_b is Brewster's angle for light reflected from the top of an interface between two substances, and θ'_b is Brewster's angle for light reflected from below, prove that $\theta_b + \theta'_b = 90.0°$.

98. Integrated Concepts

If a polarizing filter reduces the intensity of polarized light to 50.0% of its original value, by how much are the electric and magnetic fields reduced?

99. Integrated Concepts

Suppose you put on two pairs of Polaroid sunglasses with their axes at an angle of $15.0°$. How much longer will it take the light to deposit a given amount of energy in your eye compared with a single pair of sunglasses? Assume the lenses are clear except for their polarizing characteristics.

100. Integrated Concepts

(a) On a day when the intensity of sunlight is $1.00 \text{ kW} / \text{m}^2$, a circular lens 0.200 m in diameter focuses light onto water in a black beaker. Two polarizing sheets of plastic are placed in front of the lens with their axes at an angle of $20.0°$. Assuming the sunlight is unpolarized and the polarizers are 100% efficient, what is the initial rate of heating of the water in $°\text{C}/\text{s}$, assuming it is 80.0% absorbed? The aluminum beaker has a mass of 30.0 grams and contains 250 grams of water. (b) Do the polarizing filters get hot? Explain.

27.2 Huygens's Principle: Diffraction

1. Which of the following statements is true about Huygens's principle of secondary wavelets?
   a. It can be used to explain the particle behavior of waves.
   b. It states that each point on a wavefront can be considered a new wave source.
   c. It can be used to find the velocity of a wave.
   d. All of the above.

2. Explain why the amount of bending that occurs during diffraction depends on the width of the opening through which light passes.

27.3 Young's Double Slit Experiment

3. Superposition of which of the following light waves may produce interference fringes? Select *two* answers.

$Wave_1 = A_1\sin(2\omega t)$

$Wave_2 = A_2\sin(4\omega t)$

$Wave_3 = A_3\sin(2\omega t + \theta)$

$Wave_4 = A_4\sin(4\omega t + \theta)$.

   a. $Wave_1$ and $Wave_2$
   b. $Wave_2$ and $Wave_4$
   c. $Wave_3$ and $Wave_1$
   d. $Wave_4$ and $Wave_3$

4. In a double slit experiment with monochromatic light, the separation between the slits is 2 mm. If the screen is moved by 100 mm toward the slits, the distance between the central bright line and the second bright line changes by 32 μm. Calculate the wavelength of the light used for the experiment.

5. In a double slit experiment, a student measures the maximum and minimum intensities when two waves with equal amplitudes are used. The student then doubles the amplitudes of the two waves and performs the measurements again. Which of the following will remain unchanged?
   a. The intensity of the bright fringe
   b. The intensity of the dark fringe
   c. The difference in the intensities of consecutive bright and dark fringes
   d. None of the above

6. Draw a figure to show the resultant wave produced when two coherent waves (with equal amplitudes x) interact in phase. What is the amplitude of the resultant wave? If the phase difference between the coherent waves is changed to 60º, what will be new amplitude?

7. What will be the amplitude of the central fringe if the amplitudes of the two waves in a double slit experiment are a and $3a$?
   a. $2a$
   b. $4a$
   c. $8a^2$
   d. $16a^2$

8. If the ratio of amplitudes of the two waves in a double slit experiment is 3:4, calculate the ratio of minimum intensity (dark fringe) to maximum intensity (bright fringe).

27.4 Multiple Slit Diffraction

9. Which of the following cannot be a possible outcome of passing white light through several evenly spaced parallel slits?

   a. The central maximum will be white but the higher-order maxima will disperse into a rainbow of colors.
   b. The central maximum and higher-order maxima will be of equal widths.
   c. The lower wavelength components of light will have less diffraction compared to higher wavelength components for all maxima except the central one.
   d. None of the above.

10. White light is passed through a diffraction grating to a screen some distance away. The nth-order diffraction angle for the longest wavelength (760 nm) is 53.13º. Find the nth-order diffraction angle for the shortest wavelength (380 nm). What will be the change in the two angles if the distance between the screen and the grating is doubled?

27.5 Single Slit Diffraction

11. A diffraction pattern is formed on a screen when light of wavelength 410 nm is passed through a single slit of width 1 μm. If the source light is replaced by another light of wavelength 700 nm, what should be the width of the slit so that the new light produces a pattern with the same spacing?
   a. 0.6 μm
   b. 1 μm
   c. 1.4 μm
   d. 1.7 μm

12. Monochromatic light passing through a single slit forms a diffraction pattern on a screen. If the second minimum occurs at an angle of 15º, find the angle for the fourth minimum.

27.6 Limits of Resolution: The Rayleigh Criterion

13. What is the relationship between the width (W) of the central diffraction maximum formed through a circular aperture and the size (S) of the aperture?
   a. W increases as S increases.
   b. W decreases as S increases.
   c. W can increase or decrease as S decreases.
   d. W can neither increase nor decrease as S decreases.

14. Light from two sources passes through a circular aperture to form images on a screen. State the Rayleigh criterion for the images to be just resolvable and draw a figure to visually explain it.

27.7 Thin Film Interference

15. Which of the following best describes the cause of thin film interference?
   a. Light reflecting from a medium having an index of refraction less than that of the medium in which it is traveling.
   b. Light reflecting from a medium having an index of refraction greater than that of the medium in which it is traveling.
   c. Light changing its wavelength and speed after reflection.
   d. Light reflecting from the top and bottom surfaces of a film.

16. A film of magnesium fluoride ($n = 1.38$) is used to coat a glass camera lens ($n = 1.52$). If the thickness of the film is 105 nm, calculate the wavelength of visible light that will have the most limited reflection.

27.8 Polarization

17. Which of the following statements is true for the direction of polarization for a polarized light wave?

a. It is parallel to the direction of propagation and perpendicular to the direction of the electric field.
b. It is perpendicular to the direction of propagation and parallel to the direction of the electric field.
c. It is parallel to the directions of propagation and the electric field.
d. It is perpendicular to the directions of propagation and the electric field.

18. In an experiment, light is passed through two polarizing filters. The image below shows the first filter and axis of polarization.

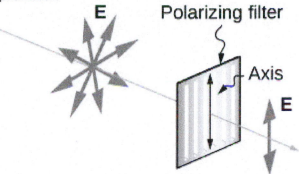

Figure 27.58

The intensity of the resulting light (after the first filter) is recorded as *I*. Three configurations (at different angles) are set up for the second filter, and the intensity of light is recorded for each configuration. The results are shown in the table below:

Table 27.2

Set up	Angle of second filter compared to first filter	Intensity of light after second filter
Configuration A	θ_1	*I*
Configuration B	θ_2	0.5*I*
Configuration C	θ_3	0

Complete the table by calculating θ_1, θ_2, and θ_3.

28 SPECIAL RELATIVITY

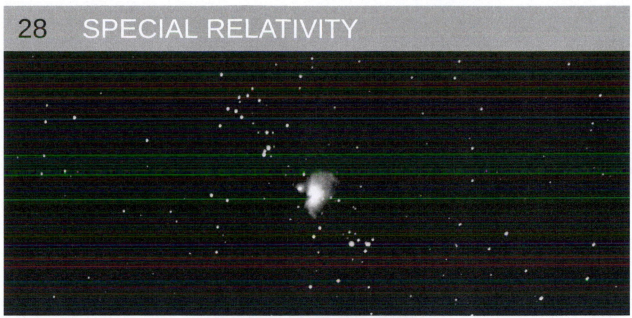

Figure 28.1 Special relativity explains why traveling to other star systems, such as these in the Orion Nebula, is unreasonable using our current level of technology. (credit: s58y, Flickr)

Chapter Outline

28.1. Einstein's Postulates

28.2. Simultaneity And Time Dilation

28.3. Length Contraction

28.4. Relativistic Addition of Velocities

28.5. Relativistic Momentum

28.6. Relativistic Energy

Connection for AP® Courses

In this chapter you will be introduced to the theory of special relativity, which was first described by Albert Einstein in the year 1905. The chapter opens with a discussion of Einstein's postulates that form the basis of special relativity. You will learn about an essential physics framework that is used to describe the observations and measurements made by an observer in what is called the "inertial frame of reference" (Enduring Understanding 3.A). Special relativity is a universally accepted theory that defines a relationship between space and time (Essential Knowledge 1.D.3). When the speed of an object approaches the speed of light, Newton's laws no longer hold, which means that classical (Newtonian) mechanics (Enduring Understanding 1.D) is not sufficient to define the physical properties of such a system. This is where special relativity comes into play. Many interesting and counterintuitive physical results follow from the theory of special relativity. In this chapter we will explore the concepts of simultaneity, time dilation, and length contraction.

Further into the chapter you will find information that supports the concepts of relativistic velocity addition, relativistic momentum, and energy (Enduring Understanding 4.C). Learning these concepts will help you understand how the mass (Enduring Understanding 1.C and Essential Knowledge 4.C.4) of an object can appear to be different for different observers and how matter can be converted into energy and then back to matter so that the energy of the system remains conserved. (Essential Knowledge 1.C.4 and Enduring Understanding 5.B). The information and examples presented in the chapter support Big Ideas 1, 3, 4, and 5 of the AP® Physics Curriculum Framework.

The content of this chapter supports:

Big Idea 1 Objects and systems have properties such as mass and charge. Systems may have internal structure.

Enduring Understanding 1.C Objects and systems have properties of inertial mass and gravitational mass that are experimentally verified to be the same and that satisfy conservation principles.

Essential Knowledge 1.C.4 In certain processes, mass can be converted to energy and energy can be converted to mass according to $E = mc^2$, the equation derived from the theory of special relativity.

Enduring Understanding 1.D Classical mechanics cannot describe all properties of objects.

Essential Knowledge 1.D.3 Properties of space and time cannot always be treated as absolute.

Big Idea 3 The interactions of an object with other objects can be described by forces.

Enduring Understanding 3.A All forces share certain common characteristics when considered by observers in inertial reference frames.

Essential Knowledge 3.A.1 An observer in a particular reference frame can describe the motion of an object using such quantities as position, displacement, distance, velocity, speed, and acceleration.

Big Idea 4 Interactions between systems can result in changes in those systems.

Enduring Understanding 4.C Interactions with other objects or systems can change the total energy of a system.

Essential Knowledge 4.C.4 Mass can be converted into energy and energy can be converted into mass.

Big Idea 5 Changes that occur as a result of interactions are constrained by conservation laws.

Enduring Understanding 5.B The energy of a system is conserved.

Essential Knowledge 5.B.11 Beyond the classical approximation, mass is actually part of the internal energy of an object or system with $E = mc^2$.

Figure 28.2 Many people think that Albert Einstein (1879–1955) was the greatest physicist of the 20th century. Not only did he develop modern relativity, thus revolutionizing our concept of the universe, he also made fundamental contributions to the foundations of quantum mechanics. (credit: The Library of Congress)

It is important to note that although classical mechanic, in general, and classical relativity, in particular, are limited, they are extremely good approximations for large, slow-moving objects. Otherwise, we could not use classical physics to launch satellites or build bridges. In the classical limit (objects larger than submicroscopic and moving slower than about 1% of the speed of light), relativistic mechanics becomes the same as classical mechanics. This fact will be noted at appropriate places throughout this chapter.

28.1 Einstein's Postulates

Learning Objectives

By the end of this section, you will be able to:

- State and explain both of Einstein's postulates.
- Explain what an inertial frame of reference is.
- Describe one way the speed of light can be changed.

The information presented in this section supports the following AP® learning objectives and science practices:

- **1.D.3.1** The student is able to articulate the reasons that classical mechanics must be replaced by special relativity to describe the experimental results and theoretical predictions that show that the properties of space and time are not absolute. [Students will be expected to recognize situations in which nonrelativistic classical physics breaks down and to explain how relativity addresses that breakdown, but students will not be expected to know in which of two reference frames a given series of events corresponds to a greater or lesser time interval, or a greater or lesser spatial distance; they will just need to know that observers in the two reference frames can "disagree" about some time and distance intervals.] **(SP 6.3, 7.1)**

Figure 28.3 Special relativity resembles trigonometry in that both are reliable because they are based on postulates that flow one from another in a logical way. (credit: Jon Oakley, Flickr)

Have you ever used the Pythagorean Theorem and gotten a wrong answer? Probably not, unless you made a mistake in either your algebra or your arithmetic. Each time you perform the same calculation, you know that the answer will be the same. Trigonometry is reliable because of the certainty that one part always flows from another in a logical way. Each part is based on a set of postulates, and you can always connect the parts by applying those postulates. Physics is the same way with the exception that *all* parts must describe nature. If we are careful to choose the correct postulates, then our theory will follow and will be verified by experiment.

Einstein essentially did the theoretical aspect of this method for **relativity**. With two deceptively simple postulates and a careful consideration of how measurements are made, he produced the theory of **special relativity.**

Einstein's First Postulate

The first postulate upon which Einstein based the theory of special relativity relates to reference frames. All velocities are measured relative to some frame of reference. For example, a car's motion is measured relative to its starting point or the road it is moving over, a projectile's motion is measured relative to the surface it was launched from, and a planet's orbit is measured relative to the star it is orbiting around. The simplest frames of reference are those that are not accelerated and are not rotating. Newton's first law, the law of inertia, holds exactly in such a frame.

Inertial Reference Frame

An **inertial frame of reference** is a reference frame in which a body at rest remains at rest and a body in motion moves at a constant speed in a straight line unless acted on by an outside force.

The laws of physics seem to be simplest in inertial frames. For example, when you are in a plane flying at a constant altitude and speed, physics seems to work exactly the same as if you were standing on the surface of the Earth. However, in a plane that is taking off, matters are somewhat more complicated. In these cases, the net force on an object, F, is not equal to the product of mass and acceleration, ma. Instead, F is equal to ma plus a fictitious force. This situation is not as simple as in an inertial frame. Not only are laws of physics simplest in inertial frames, but they should be the same in all inertial frames, since there is no preferred frame and no absolute motion. Einstein incorporated these ideas into his **first postulate of special relativity**.

First Postulate of Special Relativity

The laws of physics are the same and can be stated in their simplest form in all inertial frames of reference.

As with many fundamental statements, there is more to this postulate than meets the eye. The laws of physics include only those that satisfy this postulate. We shall find that the definitions of relativistic momentum and energy must be altered to fit. Another outcome of this postulate is the famous equation $E = mc^2$.

Einstein's Second Postulate

The second postulate upon which Einstein based his theory of special relativity deals with the speed of light. Late in the 19th century, the major tenets of classical physics were well established. Two of the most important were the laws of electricity and magnetism and Newton's laws. In particular, the laws of electricity and magnetism predict that light travels at

$c = 3.00 \times 10^8$ m/s in a vacuum, but they do not specify the frame of reference in which light has this speed.

There was a contradiction between this prediction and Newton's laws, in which velocities add like simple vectors. If the latter were true, then two observers moving at different speeds would see light traveling at different speeds. Imagine what a light wave would look like to a person traveling along with it at a speed c. If such a motion were possible then the wave would be stationary relative to the observer. It would have electric and magnetic fields that varied in strength at various distances from the observer but were constant in time. This is not allowed by Maxwell's equations. So either Maxwell's equations are wrong, or an object with mass cannot travel at speed c. Einstein concluded that the latter is true. An object with mass cannot travel at speed c. This conclusion implies that light in a vacuum must always travel at speed c relative to any observer. Maxwell's equations are correct, and Newton's addition of velocities is not correct for light.

Investigations such as Young's double slit experiment in the early-1800s had convincingly demonstrated that light is a wave. Many types of waves were known, and all travelled in some medium. Scientists therefore assumed that a medium carried light, even in a vacuum, and light travelled at a speed c relative to that medium. Starting in the mid-1880s, the American physicist A. A. Michelson, later aided by E. W. Morley, made a series of direct measurements of the speed of light. The results of their measurements were startling.

Michelson-Morley Experiment

The **Michelson-Morley experiment** demonstrated that the speed of light in a vacuum is independent of the motion of the Earth about the Sun.

The eventual conclusion derived from this result is that light, unlike mechanical waves such as sound, does not need a medium to carry it. Furthermore, the Michelson-Morley results implied that the speed of light c is independent of the motion of the source relative to the observer. That is, everyone observes light to move at speed c regardless of how they move relative to the source or one another. For a number of years, many scientists tried unsuccessfully to explain these results and still retain the general applicability of Newton's laws.

It was not until 1905, when Einstein published his first paper on special relativity, that the currently accepted conclusion was reached. Based mostly on his analysis that the laws of electricity and magnetism would not allow another speed for light, and only slightly aware of the Michelson-Morley experiment, Einstein detailed his **second postulate of special relativity**.

Second Postulate of Special Relativity

The speed of light c is a constant, independent of the relative motion of the source.

Deceptively simple and counterintuitive, this and the first postulate leave all else open for change. Some fundamental concepts do change. Among the changes are the loss of agreement on the elapsed time for an event, the variation of distance with speed, and the realization that matter and energy can be converted into one another. You will read about these concepts in the following sections.

Misconception Alert: Constancy of the Speed of Light

The speed of light is a constant $c = 3.00 \times 10^8$ m/s *in a vacuum*. If you remember the effect of the index of refraction from **The Law of Refraction**, the speed of light is lower in matter.

Check Your Understanding

Explain how special relativity differs from general relativity.

Solution

Special relativity applies only to unaccelerated motion, but general relativity applies to accelerated motion.

28.2 Simultaneity And Time Dilation

Learning Objectives

By the end of this section, you will be able to:

- Describe simultaneity.
- Describe time dilation.
- Calculate γ.

- Compare proper time and the observer's measured time.
- Explain why the twin paradox is a false paradox.

Figure 28.4 Elapsed time for a foot race is the same for all observers, but at relativistic speeds, elapsed time depends on the relative motion of the observer and the event that is observed. (credit: Jason Edward Scott Bain, Flickr)

Do time intervals depend on who observes them? Intuitively, we expect the time for a process, such as the elapsed time for a foot race, to be the same for all observers. Our experience has been that disagreements over elapsed time have to do with the accuracy of measuring time. When we carefully consider just how time is measured, however, we will find that elapsed time depends on the relative motion of an observer with respect to the process being measured.

Simultaneity

Consider how we measure elapsed time. If we use a stopwatch, for example, how do we know when to start and stop the watch? One method is to use the arrival of light from the event, such as observing a light turning green to start a drag race. The timing will be more accurate if some sort of electronic detection is used, avoiding human reaction times and other complications.

Now suppose we use this method to measure the time interval between two flashes of light produced by flash lamps. (See Figure 28.5.) Two flash lamps with observer A midway between them are on a rail car that moves to the right relative to observer B. In the frame of reference of observer B, the light flashes are emitted just as A passes B, so that both A and B are equidistant from the lamps when the light is emitted. Observer B measures the time interval between the arrival of the light flashes. According to postulate 2, the speed of light is not affected by the motion of the lamps relative to B. Therefore, light travels equal distances to him at equal speeds. Thus observer B measures the flashes to be simultaneous.

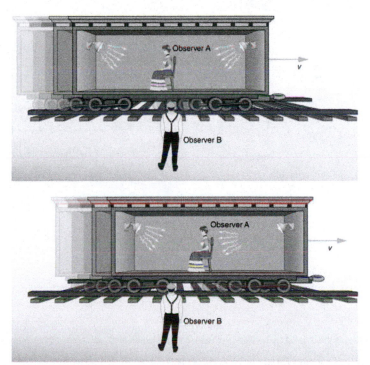

Figure 28.5 Observer B measures the elapsed time between the arrival of light flashes as described in the text. Observer A moves with the lamps on a rail car. Observer B perceives that the light flashes occurred simultaneously. Observer A perceives that the light on the right flashes before the light on the left.

Now consider what observer B sees happen to observer A. Observer B perceives light from the right reaching observer A before light from the left, because she has moved toward that flash lamp, lessening the distance the light must travel and reducing the time it takes to get to her. Light travels at speed c relative to both observers, but observer B remains equidistant between the points where the flashes were emitted, while A gets closer to the emission point on the right. From observer B's point of view, then, there is a time interval between the arrival of the flashes to observer A. Observer B measures the flashes to arrive simultaneously relative to him but not relative to A.

Now consider what observer A sees happening. She sees the light from the right arriving before the right to the left. Since both lamps are the same distance from her in her reference frame, from her perspective, the right flash occurred before the left flash. Here a relative velocity between observers affects whether two events are observed to be simultaneous. *Simultaneity is not absolute.*

This illustrates the power of clear thinking. We might have guessed incorrectly that if light is emitted simultaneously, then two observers halfway between the sources would see the flashes simultaneously. But careful analysis shows this not to be the case. Einstein was brilliant at this type of *thought experiment* (in German, "Gedankenexperiment"). He very carefully considered how an observation is made and disregarded what might seem obvious. The validity of thought experiments, of course, is determined by actual observation. The genius of Einstein is evidenced by the fact that experiments have repeatedly confirmed his theory of relativity.

In summary: Two events are defined to be simultaneous if an observer measures them as occurring at the same time (such as by receiving light from the events). Two events are not necessarily simultaneous to all observers.

Time Dilation

The consideration of the measurement of elapsed time and simultaneity leads to an important relativistic effect.

Time dilation

Time dilation is the phenomenon of time passing slower for an observer who is moving relative to another observer.

Suppose, for example, an astronaut measures the time it takes for light to cross her ship, bounce off a mirror, and return. (See **Figure 28.6**.) How does the elapsed time the astronaut measures compare with the elapsed time measured for the same event by a person on the Earth? Asking this question (another thought experiment) produces a profound result. We find that the elapsed time for a process depends on who is measuring it. In this case, the time measured by the astronaut is smaller than the time measured by the Earth-bound observer. The passage of time is different for the observers because the distance the light travels in the astronaut's frame is smaller than in the Earth-bound frame. Light travels at the same speed in each frame, and so it will take longer to travel the greater distance in the Earth-bound frame.

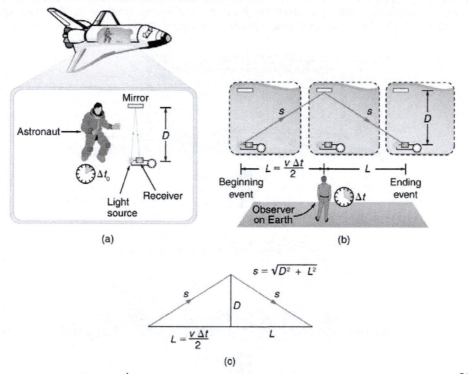

Figure 28.6 (a) An astronaut measures the time Δt_0 for light to cross her ship using an electronic timer. Light travels a distance $2D$ in the astronaut's frame. (b) A person on the Earth sees the light follow the longer path $2s$ and take a longer time Δt. (c) These triangles are used to find the relationship between the two distances $2D$ and $2s$.

To quantitatively verify that time depends on the observer, consider the paths followed by light as seen by each observer. (See **Figure 28.6**(c).) The astronaut sees the light travel straight across and back for a total distance of $2D$, twice the width of her ship. The Earth-bound observer sees the light travel a total distance $2s$. Since the ship is moving at speed v to the right relative to the Earth, light moving to the right hits the mirror in this frame. Light travels at a speed c in both frames, and because time is the distance divided by speed, the time measured by the astronaut is

$$\Delta t_0 = \frac{2D}{c}.$$ (28.1)

This time has a separate name to distinguish it from the time measured by the Earth-bound observer.

Making Connections: GPS Navigation

For GPS navigation to work properly, satellites have to take into account the effects of both special relativity and general relativity. GPS satellites move at speeds of a few miles per second, and although these speeds are just tiny fractions of the speed of light, the accuracy of timing that is needed to pinpoint a position requires that we account for the effects of *special relativity* (that is, the slower motion of satellite time relative to an observer on Earth). Additionally, GPS satellites are in orbit roughly ten thousand miles above the Earth, where the gravitational force is weaker. From the theory of *general relativity*, the weaker gravitational force means that time on the satellite is ticking faster. If these two relativistic effects were not accounted for, GPS units would lose their accuracy in a matter of minutes.

Proper Time

Proper time Δt_0 is the time measured by an observer at rest relative to the event being observed.

In the case of the astronaut observe the reflecting light, the astronaut measures proper time. The time measured by the Earth-bound observer is

$$\Delta t = \frac{2s}{c}.$$ (28.2)

To find the relationship between Δt_0 and Δt, consider the triangles formed by D and s. (See **Figure 28.6**(c).) The third side of these similar triangles is L, the distance the astronaut moves as the light goes across her ship. In the frame of the Earth-bound observer,

$$L = \frac{v\Delta t}{2}.$$ (28.3)

Using the Pythagorean Theorem, the distance s is found to be

$$s = \sqrt{D^2 + \left(\frac{v\Delta t}{2}\right)^2}.$$ (28.4)

Substituting s into the expression for the time interval Δt gives

$$\Delta t = \frac{2s}{c} = \frac{2\sqrt{D^2 + \left(\frac{v\Delta t}{2}\right)^2}}{c}.$$ (28.5)

We square this equation, which yields

$$(\Delta t)^2 = \frac{4\left(D^2 + \frac{v^2(\Delta t)^2}{4}\right)}{c^2} = \frac{4D^2}{c^2} + \frac{v^2}{c^2}(\Delta t)^2.$$ (28.6)

Note that if we square the first expression we had for Δt_0, we get $(\Delta t_0)^2 = \frac{4D^2}{c^2}$. This term appears in the preceding equation, giving us a means to relate the two time intervals. Thus,

$$(\Delta t)^2 = (\Delta t_0)^2 + \frac{v^2}{c^2}(\Delta t)^2.$$ (28.7)

Gathering terms, we solve for Δt:

$$(\Delta t)^2\left(1 - \frac{v^2}{c^2}\right) = (\Delta t_0)^2.$$ (28.8)

Thus,

$$(\Delta t)^2 = \frac{(\Delta t_0)^2}{1 - \frac{v^2}{c^2}}.$$ (28.9)

Taking the square root yields an important relationship between elapsed times:

$$\Delta t = \frac{\Delta t_0}{\sqrt{1 - \frac{v^2}{c^2}}} = \gamma \Delta t_0, \tag{28.10}$$

where

$$\gamma = \frac{1}{\sqrt{1 - \frac{v^2}{c^2}}}. \tag{28.11}$$

This equation for Δt is truly remarkable. First, as contended, elapsed time is not the same for different observers moving relative to one another, even though both are in inertial frames. Proper time Δt_0 measured by an observer, like the astronaut moving with the apparatus, is smaller than time measured by other observers. Since those other observers measure a longer time Δt, the effect is called time dilation. The Earth-bound observer sees time dilate (get longer) for a system moving relative to the Earth. Alternatively, according to the Earth-bound observer, time slows in the moving frame, since less time passes there. All clocks moving relative to an observer, including biological clocks such as aging, are observed to run slow compared with a clock stationary relative to the observer.

Note that if the relative velocity is much less than the speed of light ($v \ll c$), then $\frac{v^2}{c^2}$ is extremely small, and the elapsed times Δt and Δt_0 are nearly equal. At low velocities, modern relativity approaches classical physics—our everyday experiences have very small relativistic effects.

The equation $\Delta t = \gamma \Delta t_0$ also implies that relative velocity cannot exceed the speed of light. As v approaches c, Δt approaches infinity. This would imply that time in the astronaut's frame stops at the speed of light. If v exceeded c, then we would be taking the square root of a negative number, producing an imaginary value for Δt.

There is considerable experimental evidence that the equation $\Delta t = \gamma \Delta t_0$ is correct. One example is found in cosmic ray particles that continuously rain down on the Earth from deep space. Some collisions of these particles with nuclei in the upper atmosphere result in short-lived particles called muons. The half-life (amount of time for half of a material to decay) of a muon is $1.52 \ \mu s$ when it is at rest relative to the observer who measures the half-life. This is the proper time Δt_0. Muons produced by cosmic ray particles have a range of velocities, with some moving near the speed of light. It has been found that the muon's half-life as measured by an Earth-bound observer (Δt) varies with velocity exactly as predicted by the equation $\Delta t = \gamma \Delta t_0$. The faster the muon moves, the longer it lives. We on the Earth see the muon's half-life time dilated—as viewed from our frame, the muon decays more slowly than it does when at rest relative to us.

Example 28.1 Calculating Δt for a Relativistic Event: How Long Does a Speedy Muon Live?

Suppose a cosmic ray colliding with a nucleus in the Earth's upper atmosphere produces a muon that has a velocity $v = 0.950c$. The muon then travels at constant velocity and lives $1.52 \ \mu s$ as measured in the muon's frame of reference. (You can imagine this as the muon's internal clock.) How long does the muon live as measured by an Earth-bound observer? (See **Figure 28.7**.)

Figure 28.7 A muon in the Earth's atmosphere lives longer as measured by an Earth-bound observer than measured by the muon's internal clock.

Strategy

A clock moving with the system being measured observes the proper time, so the time we are given is $\Delta t_0 = 1.52 \ \mu s$. The Earth-bound observer measures Δt as given by the equation $\Delta t = \gamma \Delta t_0$. Since we know the velocity, the calculation is straightforward.

Solution

1) Identify the knowns. $v = 0.950c$, $\Delta t_0 = 1.52 \ \mu s$

2) Identify the unknown. Δt

3) Choose the appropriate equation.

Use,

$$\Delta t = \gamma \Delta t_0, \tag{28.12}$$

where

$$\gamma = \frac{1}{\sqrt{1 - \frac{v^2}{c^2}}}. \tag{28.13}$$

4) Plug the knowns into the equation.

First find γ.

$$
\begin{aligned}
\gamma &= \frac{1}{\sqrt{1 - \frac{v^2}{c^2}}} \\
&= \frac{1}{\sqrt{1 - \frac{(0.950c)^2}{c^2}}} \\
&= \frac{1}{\sqrt{1 - (0.950)^2}} \\
&= 3.20.
\end{aligned}
\tag{28.14}
$$

Use the calculated value of γ to determine Δt.

$$
\begin{aligned}
\Delta t &= \gamma \Delta t_0 \\
&= (3.20)(1.52 \ \mu s) \\
&= 4.87 \ \mu s
\end{aligned}
\tag{28.15}
$$

Discussion

One implication of this example is that since $\gamma = 3.20$ at 95.0% of the speed of light ($v = 0.950c$), the relativistic effects are significant. The two time intervals differ by this factor of 3.20, where classically they would be the same. Something moving at $0.950c$ is said to be highly relativistic.

Another implication of the preceding example is that everything an astronaut does when moving at 95.0% of the speed of light relative to the Earth takes 3.20 times longer when observed from the Earth. Does the astronaut sense this? Only if she looks outside her spaceship. All methods of measuring time in her frame will be affected by the same factor of 3.20. This includes her wristwatch, heart rate, cell metabolism rate, nerve impulse rate, and so on. She will have no way of telling, since all of her clocks will agree with one another because their relative velocities are zero. Motion is relative, not absolute. But what if she does look out the window?

Real-World Connections

It may seem that special relativity has little effect on your life, but it is probably more important than you realize. One of the most common effects is through the Global Positioning System (GPS). Emergency vehicles, package delivery services, electronic maps, and communications devices are just a few of the common uses of GPS, and the GPS system could not work without taking into account relativistic effects. GPS satellites rely on precise time measurements to communicate. The signals travel at relativistic speeds. Without corrections for time dilation, the satellites could not communicate, and the GPS system would fail within minutes.

The Twin Paradox

An intriguing consequence of time dilation is that a space traveler moving at a high velocity relative to the Earth would age less than her Earth-bound twin. Imagine the astronaut moving at such a velocity that $\gamma = 30.0$, as in **Figure 28.8**. A trip that takes 2.00 years in her frame would take 60.0 years in her Earth-bound twin's frame. Suppose the astronaut traveled 1.00 year to another star system. She briefly explored the area, and then traveled 1.00 year back. If the astronaut was 40 years old when she left, she would be 42 upon her return. Everything on the Earth, however, would have aged 60.0 years. Her twin, if still alive, would be 100 years old.

The situation would seem different to the astronaut. Because motion is relative, the spaceship would seem to be stationary and the Earth would appear to move. (This is the sensation you have when flying in a jet.) If the astronaut looks out the window of the spaceship, she will see time slow down on the Earth by a factor of $\gamma = 30.0$. To her, the Earth-bound sister will have aged only 2/30 (1/15) of a year, while she aged 2.00 years. The two sisters cannot both be correct.

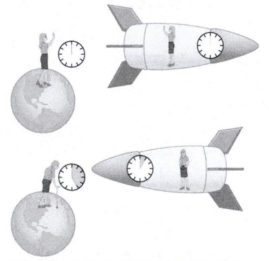

Figure 28.8 The twin paradox asks why the traveling twin ages less than the Earth-bound twin. That is the prediction we obtain if we consider the Earth-bound twin's frame. In the astronaut's frame, however, the Earth is moving and time runs slower there. Who is correct?

As with all paradoxes, the premise is faulty and leads to contradictory conclusions. In fact, the astronaut's motion is significantly different from that of the Earth-bound twin. The astronaut accelerates to a high velocity and then decelerates to view the star system. To return to the Earth, she again accelerates and decelerates. The Earth-bound twin does not experience these accelerations. So the situation is not symmetric, and it is not correct to claim that the astronaut will observe the same effects as her Earth-bound twin. If you use special relativity to examine the twin paradox, you must keep in mind that the theory is expressly based on inertial frames, which by definition are not accelerated or rotating. Einstein developed general relativity to deal with accelerated frames and with gravity, a prime source of acceleration. You can also use general relativity to address the twin paradox and, according to general relativity, the astronaut will age less. Some important conceptual aspects of general relativity are discussed in **General Relativity and Quantum Gravity** of this course.

In 1971, American physicists Joseph Hafele and Richard Keating verified time dilation at low relative velocities by flying extremely accurate atomic clocks around the Earth on commercial aircraft. They measured elapsed time to an accuracy of a few nanoseconds and compared it with the time measured by clocks left behind. Hafele and Keating's results were within experimental uncertainties of the predictions of relativity. Both special and general relativity had to be taken into account, since gravity and accelerations were involved as well as relative motion.

Check Your Understanding

1. What is γ if $v = 0.650c$?

Solution

$$\gamma = \frac{1}{\sqrt{1 - \frac{v^2}{c^2}}} = \frac{1}{\sqrt{1 - \frac{(0.650c)^2}{c^2}}} = 1.32$$

2. A particle travels at 1.90×10^8 m/s and lives 2.10×10^{-8} s when at rest relative to an observer. How long does the particle live as viewed in the laboratory?

Solution

$$\Delta t = \frac{\Delta_t}{\sqrt{1 - \frac{v^2}{c^2}}} = \frac{2.10 \times 10^{-8} \text{ s}}{\sqrt{1 - \frac{(1.90 \times 10^8 \text{ m/s})^2}{(3.00 \times 10^8 \text{ m/s})^2}}} = 2.71 \times 10^{-8} \text{ s}$$

28.3 Length Contraction

Learning Objectives

By the end of this section, you will be able to:

- Describe proper length.
- Calculate length contraction.
- Explain why we do not notice these effects at everyday scales.

Figure 28.9 People might describe distances differently, but at relativistic speeds, the distances really are different. (credit: Corey Leopold, Flickr)

Have you ever driven on a road that seems like it goes on forever? If you look ahead, you might say you have about 10 km left to go. Another traveler might say the road ahead looks like it's about 15 km long. If you both measured the road, however, you would agree. Traveling at everyday speeds, the distance you both measure would be the same. You will read in this section, however, that this is not true at relativistic speeds. Close to the speed of light, distances measured are not the same when measured by different observers.

Proper Length

One thing all observers agree upon is relative speed. Even though clocks measure different elapsed times for the same process, they still agree that relative speed, which is distance divided by elapsed time, is the same. This implies that distance, too, depends on the observer's relative motion. If two observers see different times, then they must also see different distances for relative speed to be the same to each of them.

The muon discussed in **Example 28.1** illustrates this concept. To an observer on the Earth, the muon travels at $0.950c$ for $7.05 \ \mu s$ from the time it is produced until it decays. Thus it travels a distance

$$L_0 = v\Delta t = (0.950)(3.00\times10^8 \text{ m/s})(7.05\times10^{-6} \text{ s}) = 2.01 \text{ km} \tag{28.16}$$

relative to the Earth. In the muon's frame of reference, its lifetime is only $2.20 \ \mu s$. It has enough time to travel only

$$L = v\Delta t_0 = (0.950)(3.00\times10^8 \text{ m/s})(2.20\times10^{-6} \text{ s}) = 0.627 \text{ km}. \tag{28.17}$$

The distance between the same two events (production and decay of a muon) depends on who measures it and how they are moving relative to it.

Proper Length

Proper length L_0 is the distance between two points measured by an observer who is at rest relative to both of the points.

The Earth-bound observer measures the proper length L_0, because the points at which the muon is produced and decays are stationary relative to the Earth. To the muon, the Earth, air, and clouds are moving, and so the distance L it sees is not the proper length.

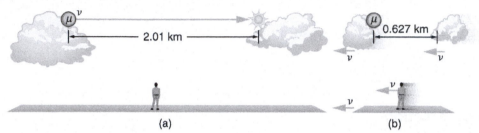

Figure 28.10 (a) The Earth-bound observer sees the muon travel 2.01 km between clouds. (b) The muon sees itself travel the same path, but only a distance of 0.627 km. The Earth, air, and clouds are moving relative to the muon in its frame, and all appear to have smaller lengths along the direction of travel.

Length Contraction

To develop an equation relating distances measured by different observers, we note that the velocity relative to the Earth-bound observer in our muon example is given by

$$v = \frac{L_0}{\Delta t}. \tag{28.18}$$

The time relative to the Earth-bound observer is Δt, since the object being timed is moving relative to this observer. The velocity relative to the moving observer is given by

$$v = \frac{L}{\Delta t_0}. \tag{28.19}$$

The moving observer travels with the muon and therefore observes the proper time Δt_0. The two velocities are identical; thus,

$$\frac{L_0}{\Delta t} = \frac{L}{\Delta t_0}. \tag{28.20}$$

We know that $\Delta t = \gamma\Delta t_0$. Substituting this equation into the relationship above gives

$$L = \frac{L_0}{\gamma}. \tag{28.21}$$

Substituting for γ gives an equation relating the distances measured by different observers.

Length Contraction

Length contraction L is the shortening of the measured length of an object moving relative to the observer's frame.

$$L = L_0\sqrt{1 - \frac{v^2}{c^2}}. \tag{28.22}$$

If we measure the length of anything moving relative to our frame, we find its length L to be smaller than the proper length L_0 that would be measured if the object were stationary. For example, in the muon's reference frame, the distance between the points where it was produced and where it decayed is shorter. Those points are fixed relative to the Earth but moving relative to the muon. Clouds and other objects are also contracted along the direction of motion in the muon's reference frame.

Making Connections: Length Contraction

One of the consequences of Einstein's theory of special relativity is the concept of length contraction. Consider a 10-cm stick. If this stick is traveling past you at a speed close to the speed of light, its length will no longer appear to be 10 cm. The length measured when the stick is at rest is calledits proper length. The length measured when the stick is in motion close to the speed of light will always be less than the proper length. This is what is known as length contraction. But the effect of length contraction can only be observed if the stick moves really fast—close to the speed of light. In principle, when the speed of the stick is equal to the speed of light,the stick should have no length.

Example 28.2 Calculating Length Contraction: The Distance between Stars Contracts when You Travel at High Velocity

Suppose an astronaut, such as the twin discussed in **Simultaneity and Time Dilation**, travels so fast that $\gamma = 30.00$. (a) She travels from the Earth to the nearest star system, Alpha Centauri, 4.300 light years (ly) away as measured by an Earth-bound observer. How far apart are the Earth and Alpha Centauri as measured by the astronaut? (b) In terms of c, what is her velocity relative to the Earth? You may neglect the motion of the Earth relative to the Sun. (See **Figure 28.11**.)

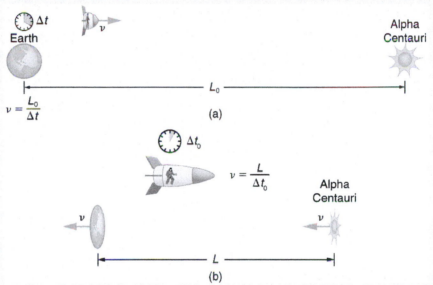

Figure 28.11 (a) The Earth-bound observer measures the proper distance between the Earth and the Alpha Centauri. (b) The astronaut observes a length contraction, since the Earth and the Alpha Centauri move relative to her ship. She can travel this shorter distance in a smaller time (her proper time) without exceeding the speed of light.

Strategy

First note that a light year (ly) is a convenient unit of distance on an astronomical scale—it is the distance light travels in a year. For part (a), note that the 4.300 ly distance between the Alpha Centauri and the Earth is the proper distance L_0, because it is measured by an Earth-bound observer to whom both stars are (approximately) stationary. To the astronaut, the Earth and the Alpha Centauri are moving by at the same velocity, and so the distance between them is the contracted length L. In part (b), we are given γ, and so we can find v by rearranging the definition of γ to express v in terms of c.

Solution for (a)

1. Identify the knowns. $L_0 - 4.300$ ly ; $\gamma = 30.00$

2. Identify the unknown. L

3. Choose the appropriate equation. $L = \dfrac{L_0}{\gamma}$

4. Rearrange the equation to solve for the unknown.

$$L = \frac{L_0}{\gamma}$$ (28.23)

$$= \frac{4.300 \text{ ly}}{30.00}$$

$$= 0.1433 \text{ ly}$$

Solution for (b)

1. Identify the known. $\gamma = 30.00$

2. Identify the unknown. v in terms of c

3. Choose the appropriate equation. $\gamma = \dfrac{1}{\sqrt{1 - \dfrac{v^2}{c^2}}}$

4. Rearrange the equation to solve for the unknown.

$$\gamma = \frac{1}{\sqrt{1 - \frac{v^2}{c^2}}}$$ (28.24)

$$30.00 = \frac{1}{\sqrt{1 - \frac{v^2}{c^2}}}$$

Squaring both sides of the equation and rearranging terms gives

$$900.0 = \frac{1}{1 - \frac{v^2}{c^2}}$$ (28.25)

so that

$$1 - \frac{v^2}{c^2} = \frac{1}{900.0}$$ (28.26)

and

$$\frac{v^2}{c^2} = 1 - \frac{1}{900.0} = 0.99888....$$ (28.27)

Taking the square root, we find

$$\frac{v}{c} = 0.99944,$$ (28.28)

which is rearranged to produce a value for the velocity

$$v = 0.9994c.$$ (28.29)

Discussion

First, remember that you should not round off calculations until the final result is obtained, or you could get erroneous results. This is especially true for special relativity calculations, where the differences might only be revealed after several decimal places. The relativistic effect is large here ($\gamma = 30.00$), and we see that v is approaching (not equaling) the speed of light. Since the distance as measured by the astronaut is so much smaller, the astronaut can travel it in much less time in her frame.

People could be sent very large distances (thousands or even millions of light years) and age only a few years on the way if they traveled at extremely high velocities. But, like emigrants of centuries past, they would leave the Earth they know forever. Even if they returned, thousands to millions of years would have passed on the Earth, obliterating most of what now exists. There is also a more serious practical obstacle to traveling at such velocities; immensely greater energies than classical physics predicts would be needed to achieve such high velocities. This will be discussed in **Relatavistic Energy**.

Why don't we notice length contraction in everyday life? The distance to the grocery shop does not seem to depend on whether we are moving or not. Examining the equation $L = L_0\sqrt{1 - \dfrac{v^2}{c^2}}$, we see that at low velocities ($v \ll c$) the lengths are nearly equal, the classical expectation. But length contraction is real, if not commonly experienced. For example, a charged particle, like an electron, traveling at relativistic velocity has electric field lines that are compressed along the direction of motion as seen by a stationary observer. (See **Figure 28.12**.) As the electron passes a detector, such as a coil of wire, its field interacts much more briefly, an effect observed at particle accelerators such as the 3 km long Stanford Linear Accelerator (SLAC). In fact, to an electron traveling down the beam pipe at SLAC, the accelerator and the Earth are all moving by and are length contracted. The relativistic effect is so great than the accelerator is only 0.5 m long to the electron. It is actually easier to get the electron beam down the pipe, since the beam does not have to be as precisely aimed to get down a short pipe as it would down one 3 km long. This, again, is an experimental verification of the Special Theory of Relativity.

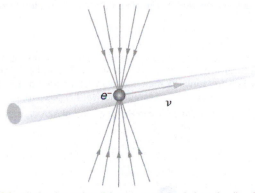

Figure 28.12 The electric field lines of a high-velocity charged particle are compressed along the direction of motion by length contraction. This produces a different signal when the particle goes through a coil, an experimentally verified effect of length contraction.

Check Your Understanding

A particle is traveling through the Earth's atmosphere at a speed of $0.750c$. To an Earth-bound observer, the distance it travels is 2.50 km. How far does the particle travel in the particle's frame of reference?

Solution

$$L=L_0\sqrt{1-\frac{v^2}{c^2}} = (2.50 \text{ km})\sqrt{1-\frac{(0.750c)^2}{c^2}} = 1.65 \text{ km} \qquad (28.30)$$

28.4 Relativistic Addition of Velocities

Learning Objectives

By the end of this section, you will be able to:

- Calculate relativistic velocity addition.
- Explain when relativistic velocity addition should be used instead of classical addition of velocities.
- Calculate relativistic Doppler shift.

The information presented in this section supports the following AP® learning objectives and science practices:

- **1.D.3.1** The student is able to articulate the reasons that classical mechanics must be replaced by special relativity to describe the experimental results and theoretical predictions that show that the properties of space and time are not absolute. [Students will be expected to recognize situations in which nonrelativistic classical physics breaks down and to explain how relativity addresses that breakdown, but students will not be expected to know in which of two reference frames a given series of events corresponds to a greater or lesser time interval, or a greater or lesser spatial distance; they will just need to know that observers in the two reference frames can "disagree" about some time and distance intervals.] **(SP 6.3, 7.1)**

Figure 28.13 The total velocity of a kayak, like this one on the Deerfield River in Massachusetts, is its velocity relative to the water as well as the water's velocity relative to the riverbank. (credit: abkfenris, Flickr)

If you've ever seen a kayak move down a fast-moving river, you know that remaining in the same place would be hard. The river current pulls the kayak along. Pushing the oars back against the water can move the kayak forward in the water, but that only accounts for part of the velocity. The kayak's motion is an example of classical addition of velocities. In classical physics, velocities add as vectors. The kayak's velocity is the vector sum of its velocity relative to the water and the water's velocity relative to the riverbank.

Classical Velocity Addition

For simplicity, we restrict our consideration of velocity addition to one-dimensional motion. Classically, velocities add like regular numbers in one-dimensional motion. (See **Figure 28.14**.) Suppose, for example, a girl is riding in a sled at a speed 1.0 m/s relative to an observer. She throws a snowball first forward, then backward at a speed of 1.5 m/s relative to the sled. We denote direction with plus and minus signs in one dimension; in this example, forward is positive. Let v be the velocity of the sled relative to the Earth, u the velocity of the snowball relative to the Earth-bound observer, and u' the velocity of the snowball relative to the sled.

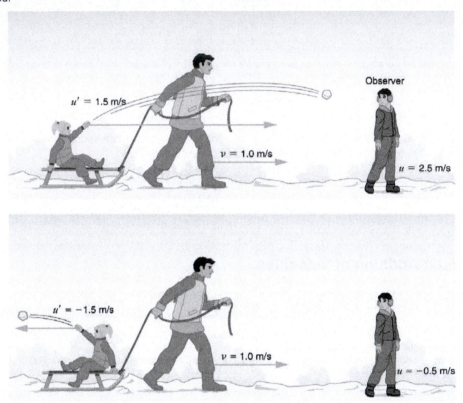

Figure 28.14 Classically, velocities add like ordinary numbers in one-dimensional motion. Here the girl throws a snowball forward and then backward from a sled. The velocity of the sled relative to the Earth is $v = 1.0 \text{ m/s}$. The velocity of the snowball relative to the truck is u', while its velocity relative to the Earth is u. Classically, $u = v + u'$.

Classical Velocity Addition
$$u = v + u'$$
(28.31)

Thus, when the girl throws the snowball forward, $u = 1.0 \text{ m/s} + 1.5 \text{ m/s} = 2.5 \text{ m/s}$. It makes good intuitive sense that the snowball will head towards the Earth-bound observer faster, because it is thrown forward from a moving vehicle. When the girl throws the snowball backward, $u = 1.0 \text{ m/s} + (-1.5 \text{ m/s}) = -0.5 \text{ m/s}$. The minus sign means the snowball moves away from the Earth-bound observer.

Relativistic Velocity Addition

The second postulate of relativity (verified by extensive experimental observation) says that classical velocity addition does not apply to light. Imagine a car traveling at night along a straight road, as in **Figure 28.15**. If classical velocity addition applied to light, then the light from the car's headlights would approach the observer on the sidewalk at a speed $u = v + c$. But we know that light will move away from the car at speed c relative to the driver of the car, and light will move towards the observer on the sidewalk at speed c, too.

Figure 28.15 According to experiment and the second postulate of relativity, light from the car's headlights moves away from the car at speed c and towards the observer on the sidewalk at speed c. Classical velocity addition is not valid.

Relativistic Velocity Addition

Either light is an exception, or the classical velocity addition formula only works at low velocities. The latter is the case. The correct formula for one-dimensional **relativistic velocity addition** is

$$u = \frac{v+u'}{1 + \frac{vu'}{c^2}},$$ (28.32)

where v is the relative velocity between two observers, u is the velocity of an object relative to one observer, and u' is the velocity relative to the other observer. (For ease of visualization, we often choose to measure u in our reference frame, while someone moving at v relative to us measures u'.) Note that the term $\frac{vu'}{c^2}$ becomes very small at low velocities, and $u = \frac{v+u'}{1 + \frac{vu'}{c^2}}$ gives a result very close to classical velocity addition. As before, we see that classical velocity addition is an excellent approximation to the correct relativistic formula for small velocities. No wonder that it seems correct in our experience.

Example 28.3 Showing that the Speed of Light towards an Observer is Constant (in a Vacuum): The Speed of Light is the Speed of Light

Suppose a spaceship heading directly towards the Earth at half the speed of light sends a signal to us on a laser-produced beam of light. Given that the light leaves the ship at speed c as observed from the ship, calculate the speed at which it approaches the Earth.

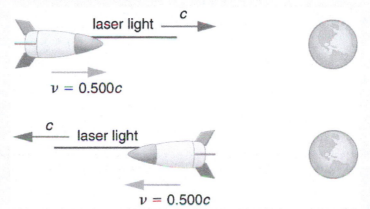

Figure 28.16

Strategy

Because the light and the spaceship are moving at relativistic speeds, we cannot use simple velocity addition. Instead, we can determine the speed at which the light approaches the Earth using relativistic velocity addition.

Solution

1. Identify the knowns. $v=0.500c$; $u' = c$

2. Identify the unknown. u

3. Choose the appropriate equation. $u = \dfrac{v+u'}{1 + \frac{vu'}{c^2}}$

4. Plug the knowns into the equation.

$$
\begin{aligned}
u &= \frac{v+u'}{1 + \frac{vu'}{c^2}} \\
&= \frac{0.500c + c}{1 + \frac{(0.500c)(c)}{c^2}} \\
&= \frac{(0.500 + 1)c}{1 + \frac{0.500c^2}{c^2}} \\
&= \frac{1.500c}{1 + 0.500} \\
&= \frac{1.500c}{1.500} \\
&= c
\end{aligned}
$$

(28.33)

Discussion

Relativistic velocity addition gives the correct result. Light leaves the ship at speed c and approaches the Earth at speed c . The speed of light is independent of the relative motion of source and observer, whether the observer is on the ship or Earth-bound.

Velocities cannot add to greater than the speed of light, provided that v is less than c and u' does not exceed c . The following example illustrates that relativistic velocity addition is not as symmetric as classical velocity addition.

Example 28.4 Comparing the Speed of Light towards and away from an Observer: Relativistic Package Delivery

Suppose the spaceship in the previous example is approaching the Earth at half the speed of light and shoots a canister at a speed of $0.750c$. (a) At what velocity will an Earth-bound observer see the canister if it is shot directly towards the Earth? (b) If it is shot directly away from the Earth? (See **Figure 28.17**.)

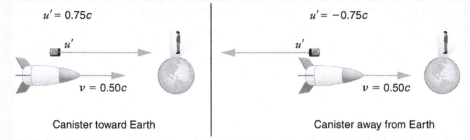

Canister toward Earth   Canister away from Earth

Figure 28.17

Strategy

Because the canister and the spaceship are moving at relativistic speeds, we must determine the speed of the canister by an Earth-bound observer using relativistic velocity addition instead of simple velocity addition.

Solution for (a)

1. Identify the knowns. $v=0.500c$; $u' = 0.750c$

2. Identify the unknown. u

3. Choose the appropriate equation. $u = \dfrac{v + u'}{1 + \dfrac{vu'}{c^2}}$

4. Plug the knowns into the equation.

$$
\begin{aligned}
u &= \frac{v + u'}{1 + \dfrac{vu'}{c^2}} \\[2mm]
&= \frac{0.500c + 0.750c}{1 + \dfrac{(0.500c)(0.750c)}{c^2}} \\[2mm]
&= \frac{1.250c}{1 + 0.375} \\[2mm]
&= 0.909c
\end{aligned}
\tag{28.34}
$$

Solution for (b)

1. Identify the knowns. $v = 0.500c$; $u' = -0.750c$

2. Identify the unknown. u

3. Choose the appropriate equation. $u = \dfrac{v + u'}{1 + \dfrac{vu'}{c^2}}$

4. Plug the knowns into the equation.

$$
\begin{aligned}
u &= \frac{v + u'}{1 + \dfrac{vu'}{c^2}} \\[2mm]
&= \frac{0.500c + (-0.750c)}{1 + \dfrac{(0.500c)(-0.750c)}{c^2}} \\[2mm]
&= \frac{-0.250c}{1 - 0.375} \\[2mm]
&= -0.400c
\end{aligned}
\tag{28.35}
$$

Discussion

The minus sign indicates velocity away from the Earth (in the opposite direction from v), which means the canister is heading towards the Earth in part (a) and away in part (b), as expected. But relativistic velocities do not add as simply as they do classically. In part (a), the canister does approach the Earth faster, but not at the simple sum of $1.250c$. The total velocity is less than you would get classically. And in part (b), the canister moves away from the Earth at a velocity of $-0.400c$, which is *faster* than the $-0.250c$ you would expect classically. The velocities are not even symmetric. In part (a) the canister moves $0.409c$ faster than the ship relative to the Earth, whereas in part (b) it moves $0.900c$ slower than the ship.

Doppler Shift

Although the speed of light does not change with relative velocity, the frequencies and wavelengths of light do. First discussed for sound waves, a Doppler shift occurs in any wave when there is relative motion between source and observer.

Relativistic Doppler Effects

The observed wavelength of electromagnetic radiation is longer (called a red shift) than that emitted by the source when the source moves away from the observer and shorter (called a blue shift) when the source moves towards the observer.

$$
= \lambda_{\text{obs}} = \lambda_s \sqrt{\frac{1 + \frac{u}{c}}{1 - \frac{u}{c}}}.
\tag{28.36}
$$

In the Doppler equation, λ_{obs} is the observed wavelength, λ_s is the source wavelength, and u is the relative velocity of the source to the observer. The velocity u is positive for motion away from an observer and negative for motion toward an observer. In terms of source frequency and observed frequency, this equation can be written

$$f_{\text{obs}} = f_s \sqrt{\frac{1 - \frac{u}{c}}{1 + \frac{u}{c}}}. \tag{28.37}$$

Notice that the − and + signs are different than in the wavelength equation.

Career Connection: Astronomer

If you are interested in a career that requires a knowledge of special relativity, there's probably no better connection than astronomy. Astronomers must take into account relativistic effects when they calculate distances, times, and speeds of black holes, galaxies, quasars, and all other astronomical objects. To have a career in astronomy, you need at least an undergraduate degree in either physics or astronomy, but a Master's or doctoral degree is often required. You also need a good background in high-level mathematics.

Example 28.5 Calculating a Doppler Shift: Radio Waves from a Receding Galaxy

Suppose a galaxy is moving away from the Earth at a speed $0.825c$. It emits radio waves with a wavelength of 0.525 m. What wavelength would we detect on the Earth?

Strategy

Because the galaxy is moving at a relativistic speed, we must determine the Doppler shift of the radio waves using the relativistic Doppler shift instead of the classical Doppler shift.

Solution

1. Identify the knowns. $u = 0.825c$; $\lambda_s = 0.525\ m$

2. Identify the unknown. λ_{obs}

3. Choose the appropriate equation. $\lambda_{\text{obs}} = \lambda_s \sqrt{\frac{1 + \frac{u}{c}}{1 - \frac{u}{c}}}$

4. Plug the knowns into the equation.

$$\begin{aligned}
\lambda_{\text{obs}} &= \lambda_s \sqrt{\frac{1 + \frac{u}{c}}{1 - \frac{u}{c}}} \\[6pt]
&= (0.525\ \text{m}) \sqrt{\frac{1 + \frac{0.825c}{c}}{1 - \frac{0.825c}{c}}} \\[6pt]
&= 1.70\ \text{m.}
\end{aligned} \tag{28.38}$$

Discussion

Because the galaxy is moving away from the Earth, we expect the wavelengths of radiation it emits to be redshifted. The wavelength we calculated is 1.70 m, which is redshifted from the original wavelength of 0.525 m.

The relativistic Doppler shift is easy to observe. This equation has everyday applications ranging from Doppler-shifted radar velocity measurements of transportation to Doppler-radar storm monitoring. In astronomical observations, the relativistic Doppler shift provides velocity information such as the motion and distance of stars.

Check Your Understanding

Suppose a space probe moves away from the Earth at a speed $0.350c$. It sends a radio wave message back to the Earth at a frequency of 1.50 GHz. At what frequency is the message received on the Earth?

Solution

$$f_{\text{obs}} = f_s \sqrt{\frac{1 - \frac{u}{c}}{1 + \frac{u}{c}}} = (1.50\ \text{GHz}) \sqrt{\frac{1 - \frac{0.350c}{c}}{1 + \frac{0.350c}{c}}} = 1.04\ \text{GHz} \tag{28.39}$$

28.5 Relativistic Momentum

Learning Objectives

By the end of this section, you will be able to:

- Calculate relativistic momentum.
- Explain why the only mass it makes sense to talk about is rest mass.

Figure 28.18 Momentum is an important concept for these football players from the University of California at Berkeley and the University of California at Davis. Players with more mass often have a larger impact because their momentum is larger. For objects moving at relativistic speeds, the effect is even greater. (credit: John Martinez Pavliga)

In classical physics, momentum is a simple product of mass and velocity. However, we saw in the last section that when special relativity is taken into account, massive objects have a speed limit. What effect do you think mass and velocity have on the momentum of objects moving at relativistic speeds?

Momentum is one of the most important concepts in physics. The broadest form of Newton's second law is stated in terms of momentum. Momentum is conserved whenever the net external force on a system is zero. This makes momentum conservation a fundamental tool for analyzing collisions. All of **Work, Energy, and Energy Resources** is devoted to momentum, and momentum has been important for many other topics as well, particularly where collisions were involved. We will see that momentum has the same importance in modern physics. Relativistic momentum is conserved, and much of what we know about subatomic structure comes from the analysis of collisions of accelerator-produced relativistic particles.

The first postulate of relativity states that the laws of physics are the same in all inertial frames. Does the law of conservation of momentum survive this requirement at high velocities? The answer is yes, provided that the momentum is defined as follows.

Relativistic Momentum

Relativistic momentum p is classical momentum multiplied by the relativistic factor γ.

$$p = \gamma m u, \tag{28.40}$$

where m is the **rest mass** of the object, u is its velocity relative to an observer, and the relativistic factor

$$\gamma = \frac{1}{\sqrt{1 - \frac{u^2}{c^2}}}. \tag{28.41}$$

Note that we use u for velocity here to distinguish it from relative velocity v between observers. Only one observer is being considered here. With p defined in this way, total momentum p_{tot} is conserved whenever the net external force is zero, just as in classical physics. Again we see that the relativistic quantity becomes virtually the same as the classical at low velocities. That is, relativistic momentum $\gamma m u$ becomes the classical mu at low velocities, because γ is very nearly equal to 1 at low velocities.

Relativistic momentum has the same intuitive feel as classical momentum. It is greatest for large masses moving at high velocities, but, because of the factor γ, relativistic momentum approaches infinity as u approaches c. (See **Figure 28.19**.)

This is another indication that an object with mass cannot reach the speed of light. If it did, its momentum would become infinite, an unreasonable value.

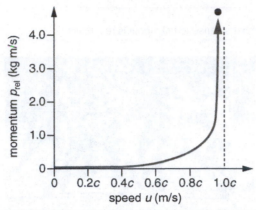

Figure 28.19 Relativistic momentum approaches infinity as the velocity of an object approaches the speed of light.

Misconception Alert: Relativistic Mass and Momentum

The relativistically correct definition of momentum as $p = \gamma m u$ is sometimes taken to imply that mass varies with velocity: $m_{var} = \gamma m$, particularly in older textbooks. However, note that m is the mass of the object as measured by a person at rest relative to the object. Thus, m is defined to be the rest mass, which could be measured at rest, perhaps using gravity. When a mass is moving relative to an observer, the only way that its mass can be determined is through collisions or other means in which momentum is involved. Since the mass of a moving object cannot be determined independently of momentum, the only meaningful mass is rest mass. Thus, when we use the term mass, assume it to be identical to rest mass.

Relativistic momentum is defined in such a way that the conservation of momentum will hold in all inertial frames. Whenever the net external force on a system is zero, relativistic momentum is conserved, just as is the case for classical momentum. This has been verified in numerous experiments.

In **Relativistic Energy**, the relationship of relativistic momentum to energy is explored. That subject will produce our first inkling that objects without mass may also have momentum.

Check Your Understanding

What is the momentum of an electron traveling at a speed $0.985c$? The rest mass of the electron is 9.11×10^{-31} kg .

Solution

$$p = \gamma m u = \frac{mu}{\sqrt{1 - \frac{u^2}{c^2}}} = \frac{(9.11 \times 10^{-31} \text{ kg})(0.985)(3.00 \times 10^8 \text{ m/s})}{\sqrt{1 - \frac{(0.985c)^2}{c^2}}} = 1.56 \times 10^{-21} \text{ kg} \cdot \text{m/s} \qquad (28.42)$$

28.6 Relativistic Energy

Learning Objectives

By the end of this section, you will be able to:

- Compute the total energy of a relativistic object.
- Compute the kinetic energy of a relativistic object.
- Describe rest energy, and explain how it can be converted to other forms.
- Explain why objects with mass cannot travel at c, the speed of light.

The information presented in this section supports the following AP® learning objectives and science practices:

- **5.B.11.1** The student is able to apply conservation of mass and conservation of energy concepts to a natural phenomenon and use the equation $E = mc^2$ to make a related calculation. **(S.P. 2.2, 7.2)**

Figure 28.20 The National Spherical Torus Experiment (NSTX) has a fusion reactor in which hydrogen isotopes undergo fusion to produce helium. In this process, a relatively small mass of fuel is converted into a large amount of energy. (credit: Princeton Plasma Physics Laboratory)

A tokamak is a form of experimental fusion reactor, which can change mass to energy. Accomplishing this requires an understanding of relativistic energy. Nuclear reactors are proof of the conservation of relativistic energy.

Conservation of energy is one of the most important laws in physics. Not only does energy have many important forms, but each form can be converted to any other. We know that classically the total amount of energy in a system remains constant. Relativistically, energy is still conserved, provided its definition is altered to include the possibility of mass changing to energy, as in the reactions that occur within a nuclear reactor. Relativistic energy is intentionally defined so that it will be conserved in all inertial frames, just as is the case for relativistic momentum. As a consequence, we learn that several fundamental quantities are related in ways not known in classical physics. All of these relationships are verified by experiment and have fundamental consequences. The altered definition of energy contains some of the most fundamental and spectacular new insights into nature found in recent history.

Total Energy and Rest Energy

The first postulate of relativity states that the laws of physics are the same in all inertial frames. Einstein showed that the law of conservation of energy is valid relativistically, if we define energy to include a relativistic factor.

Total Energy

Total energy E is defined to be

$$E = \gamma mc^2,$$

(28.43)

where m is mass, c is the speed of light, $\gamma = \dfrac{1}{\sqrt{1 - \dfrac{v^2}{c^2}}}$, and v is the velocity of the mass relative to an observer. There

are many aspects of the total energy E that we will discuss—among them are how kinetic and potential energies are included in E, and how E is related to relativistic momentum. But first, note that at rest, total energy is not zero. Rather, when $v = 0$, we have $\gamma = 1$, and an object has rest energy.

Rest Energy

Rest energy is

$$E_0 = mc^2.$$

(28.44)

This is the correct form of Einstein's most famous equation, which for the first time showed that energy is related to the mass of an object at rest. For example, if energy is stored in the object, its rest mass increases. This also implies that mass can be destroyed to release energy. The implications of these first two equations regarding relativistic energy are so broad that they were not completely recognized for some years after Einstein published them in 1907, nor was the experimental proof that they are correct widely recognized at first. Einstein, it should be noted, did understand and describe the meanings and implications of his theory.

Example 28.6 Calculating Rest Energy: Rest Energy is Very Large

Calculate the rest energy of a 1.00-g mass.

Strategy

One gram is a small mass—less than half the mass of a penny. We can multiply this mass, in SI units, by the speed of light squared to find the equivalent rest energy.

Solution

1. Identify the knowns. $m = 1.00 \times 10^{-3}$ kg ; $c = 3.00 \times 10^8$ m/s

2. Identify the unknown. E_0

3. Choose the appropriate equation. $E_0 = mc^2$

4. Plug the knowns into the equation.

$$
\begin{aligned}
E_0 &= mc^2 = (1.00 \times 10^{-3} \text{ kg})(3.00 \times 10^8 \text{ m/s})^2 \\
&= 9.00 \times 10^{13} \text{ kg} \cdot \text{m}^2/\text{s}^2
\end{aligned}
\tag{28.45}
$$

5. Convert units.

   Noting that $1 \text{ kg} \cdot \text{m}^2/\text{s}^2 = 1 \text{ J}$, we see the rest mass energy is

$$
E_0 = 9.00 \times 10^{13} \text{ J}.
\tag{28.46}
$$

Discussion

This is an enormous amount of energy for a 1.00-g mass. We do not notice this energy, because it is generally not available. Rest energy is large because the speed of light c is a large number and c^2 is a very large number, so that mc^2 is huge for any macroscopic mass. The 9.00×10^{13} J rest mass energy for 1.00 g is about twice the energy released by the Hiroshima atomic bomb and about 10,000 times the kinetic energy of a large aircraft carrier. If a way can be found to convert rest mass energy into some other form (and all forms of energy can be converted into one another), then huge amounts of energy can be obtained from the destruction of mass.

Today, the practical applications of *the conversion of mass into another form of energy*, such as in nuclear weapons and nuclear power plants, are well known. But examples also existed when Einstein first proposed the correct form of relativistic energy, and he did describe some of them. Nuclear radiation had been discovered in the previous decade, and it had been a mystery as to where its energy originated. The explanation was that, in certain nuclear processes, a small amount of mass is destroyed and energy is released and carried by nuclear radiation. But the amount of mass destroyed is so small that it is difficult to detect that any is missing. Although Einstein proposed this as the source of energy in the radioactive salts then being studied, it was many years before there was broad recognition that mass could be and, in fact, commonly is converted to energy. (See **Figure 28.21**.)

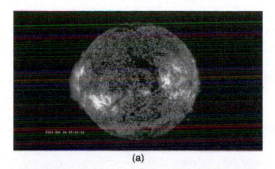

(a)

(b)

Figure 28.21 The Sun (a) and the Susquehanna Steam Electric Station (b) both convert mass into energy—the Sun via nuclear fusion, the electric station via nuclear fission. (credits: (a) NASA/Goddard Space Flight Center, Scientific Visualization Studio; (b) U.S. government)

Because of the relationship of rest energy to mass, we now consider mass to be a form of energy rather than something separate. There had not even been a hint of this prior to Einstein's work. Such conversion is now known to be the source of the Sun's energy, the energy of nuclear decay, and even the source of energy keeping Earth's interior hot.

Making Connections: Mass-Energy Conservation

Nuclear power plants and nuclear weapons are practical examples of conversion of mass into energy. In nuclear processes, a small amount of mass is destroyed and converted into energy, which is released in the form of nuclear radiation. The amount of mass destroyed is, however, very small and cannot be easily detected. Mass-energy equivalence is very important in this regard. The famous equation $E = mc^2$, where c is the speed of light, tells us how much energy is equivalent to how much mass. The speed of light is a large number, and the square of that is even larger. This implies that a small mass when destroyed has the capability of producing a very large amount of energy. To summarize: mass and energy are really the same quantities, and we can calculate the conversion of one into the other using the speed of light. Mass conservation has to take energy into account and vice versa. This is mass-energy conservation.

Stored Energy and Potential Energy

What happens to energy stored in an object at rest, such as the energy put into a battery by charging it, or the energy stored in a toy gun's compressed spring? The energy input becomes part of the total energy of the object and, thus, increases its rest mass. All stored and potential energy becomes mass in a system. Why is it we don't ordinarily notice this? In fact, conservation of mass (meaning total mass is constant) was one of the great laws verified by 19th-century science. Why was it not noticed to be incorrect? The following example helps answer these questions.

Example 28.7 Calculating Rest Mass: A Small Mass Increase due to Energy Input

A car battery is rated to be able to move 600 ampere-hours (A·h) of charge at 12.0 V. (a) Calculate the increase in rest mass of such a battery when it is taken from being fully depleted to being fully charged. (b) What percent increase is this, given the battery's mass is 20.0 kg?

Strategy

In part (a), we first must find the energy stored in the battery, which equals what the battery can supply in the form of electrical potential energy. Since $\text{PE}_{\text{elec}} = qV$, we have to calculate the charge q in 600 A·h, which is the product of the current I and the time t. We then multiply the result by 12.0 V. We can then calculate the battery's increase in mass using

$\Delta E = \mathrm{PE}_{elec} = (\Delta m)c^2$. Part (b) is a simple ratio converted to a percentage.

Solution for (a)

1. Identify the knowns. $I \cdot t = 600 \, \mathrm{A \cdot h}$; $V = 12.0 \, \mathrm{V}$; $c = 3.00 \times 10^8 \, \mathrm{m/s}$

2. Identify the unknown. Δm

3. Choose the appropriate equation. $\mathrm{PE}_{elec} = (\Delta m)c^2$

4. Rearrange the equation to solve for the unknown. $\Delta m = \dfrac{\mathrm{PE}_{elec}}{c^2}$

5. Plug the knowns into the equation.

$$
\begin{aligned}
\Delta m &= \frac{\mathrm{PE}_{elec}}{c^2} \\
&= \frac{qV}{c^2} \\
&= \frac{(It)V}{c^2} \\
&= \frac{(600 \, \mathrm{A \cdot h})(12.0 \, \mathrm{V})}{(3.00 \times 10^8)^2}.
\end{aligned}
$$
(28.47)

Write amperes A as coulombs per second (C/s), and convert hours to seconds.

$$
\begin{aligned}
\Delta m &= \frac{(600 \, \mathrm{C/s \cdot h}\left(\frac{3600 \, \mathrm{s}}{1 \, \mathrm{h}}\right)(12.0 \, \mathrm{J/C})}{(3.00 \times 10^8 \, \mathrm{m/s})^2} \\
&= \frac{(2.16 \times 10^6 \, \mathrm{C})(12.0 \, \mathrm{J/C})}{(3.00 \times 10^8 \, \mathrm{m/s})^2}
\end{aligned}
$$
(28.48)

Using the conversion $1 \, \mathrm{kg \cdot m^2/s^2} = 1 \, \mathrm{J}$, we can write the mass as

$\Delta m = 2.88 \times 10^{-10} \, \mathrm{kg}$.

Solution for (b)

1. Identify the knowns. $\Delta m = 2.88 \times 10^{-10} \, \mathrm{kg}$; $m = 20.0 \, \mathrm{kg}$

2. Identify the unknown. % change

3. Choose the appropriate equation. % increase $= \dfrac{\Delta m}{m} \times 100\%$

4. Plug the knowns into the equation.

$$
\begin{aligned}
\% \text{ increase} &= \frac{\Delta m}{m} \times 100\% \\
&= \frac{2.88 \times 10^{-10} \, \mathrm{kg}}{20.0 \, \mathrm{kg}} \times 100\% \\
&= 1.44 \times 10^{-9} \, \%.
\end{aligned}
$$
(28.49)

Discussion

Both the actual increase in mass and the percent increase are very small, since energy is divided by c^2, a very large number. We would have to be able to measure the mass of the battery to a precision of a billionth of a percent, or 1 part in 10^{11}, to notice this increase. It is no wonder that the mass variation is not readily observed. In fact, this change in mass is so small that we may question how you could verify it is real. The answer is found in nuclear processes in which the percentage of mass destroyed is large enough to be measured. The mass of the fuel of a nuclear reactor, for example, is measurably smaller when its energy has been used. In that case, stored energy has been released (converted mostly to heat and electricity) and the rest mass has decreased. This is also the case when you use the energy stored in a battery, except that the stored energy is much greater in nuclear processes, making the change in mass measurable in practice as well as in theory.

Kinetic Energy and the Ultimate Speed Limit

Kinetic energy is energy of motion. Classically, kinetic energy has the familiar expression $\frac{1}{2}mv^2$. The relativistic expression for kinetic energy is obtained from the work-energy theorem. This theorem states that the net work on a system goes into kinetic energy. If our system starts from rest, then the work-energy theorem is

$$W_{\text{net}} = \text{KE}. \tag{28.50}$$

Relativistically, at rest we have rest energy $E_0 = mc^2$. The work increases this to the total energy $E = \gamma mc^2$. Thus,

$$W_{\text{net}} = E - E_0 = \gamma mc^2 - mc^2 = (\gamma - 1)mc^2. \tag{28.51}$$

Relativistically, we have $W_{\text{net}} = \text{KE}_{\text{rel}}$.

Relativistic Kinetic Energy

Relativistic kinetic energy is

$$\text{KE}_{\text{rel}} = (\gamma - 1)mc^2. \tag{28.52}$$

When motionless, we have $v = 0$ and

$$\gamma = \frac{1}{\sqrt{1 - \frac{v^2}{c^2}}} = 1, \tag{28.53}$$

so that $\text{KE}_{\text{rel}} = 0$ at rest, as expected. But the expression for relativistic kinetic energy (such as total energy and rest energy) does not look much like the classical $\frac{1}{2}mv^2$. To show that the classical expression for kinetic energy is obtained at low velocities, we note that the binomial expansion for γ at low velocities gives

$$\gamma = 1 + \frac{1}{2}\frac{v^2}{c^2}. \tag{28.54}$$

A binomial expansion is a way of expressing an algebraic quantity as a sum of an infinite series of terms. In some cases, as in the limit of small velocity here, most terms are very small. Thus the expression derived for γ here is not exact, but it is a very accurate approximation. Thus, at low velocities,

$$\gamma - 1 = \frac{1}{2}\frac{v^2}{c^2}. \tag{28.55}$$

Entering this into the expression for relativistic kinetic energy gives

$$\text{KE}_{\text{rel}} = \left[\frac{1}{2}\frac{v^2}{c^2}\right]mc^2 = \frac{1}{2}mv^2 = \text{KE}_{\text{class}}. \tag{28.56}$$

So, in fact, relativistic kinetic energy does become the same as classical kinetic energy when $v \ll c$.

It is even more interesting to investigate what happens to kinetic energy when the velocity of an object approaches the speed of light. We know that γ becomes infinite as v approaches c, so that KE_{rel} also becomes infinite as the velocity approaches the speed of light. (See **Figure 28.22**.) An infinite amount of work (and, hence, an infinite amount of energy input) is required to accelerate a mass to the speed of light.

The Speed of Light

No object with mass can attain the speed of light.

So the speed of light is the ultimate speed limit for any particle having mass. All of this is consistent with the fact that velocities less than c always add to less than c. Both the relativistic form for kinetic energy and the ultimate speed limit being c have been confirmed in detail in numerous experiments. No matter how much energy is put into accelerating a mass, its velocity can only approach—not reach—the speed of light.

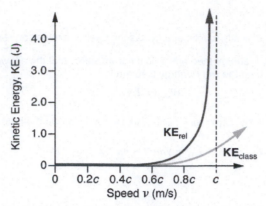

Figure 28.22 This graph of KE_{rel} versus velocity shows how kinetic energy approaches infinity as velocity approaches the speed of light. It is thus not possible for an object having mass to reach the speed of light. Also shown is KE_{class} , the classical kinetic energy, which is similar to relativistic kinetic energy at low velocities. Note that much more energy is required to reach high velocities than predicted classically.

Example 28.8 Comparing Kinetic Energy: Relativistic Energy Versus Classical Kinetic Energy

An electron has a velocity $v = 0.990c$. (a) Calculate the kinetic energy in MeV of the electron. (b) Compare this with the classical value for kinetic energy at this velocity. (The mass of an electron is 9.11×10^{-31} kg .)

Strategy

The expression for relativistic kinetic energy is always correct, but for (a) it must be used since the velocity is highly relativistic (close to c). First, we will calculate the relativistic factor γ , and then use it to determine the relativistic kinetic energy. For (b), we will calculate the classical kinetic energy (which would be close to the relativistic value if v were less than a few percent of c) and see that it is not the same.

Solution for (a)

1. Identify the knowns. $v = 0.990c$; $m = 9.11 \times 10^{-31}$ kg

2. Identify the unknown. KE_{rel}

3. Choose the appropriate equation. $\mathrm{KE}_{rel} = (\gamma - 1)mc^2$

4. Plug the knowns into the equation.
    First calculate γ . We will carry extra digits because this is an intermediate calculation.

$$\gamma = \frac{1}{\sqrt{1 - \frac{v^2}{c^2}}} \tag{28.57}$$

$$= \frac{1}{\sqrt{1 - \frac{(0.990c)^2}{c^2}}}$$

$$= \frac{1}{\sqrt{1 - (0.990)^2}}$$

$$= 7.0888$$

Next, we use this value to calculate the kinetic energy.

$$\mathrm{KE}_{rel} = (\gamma - 1)mc^2 \tag{28.58}$$

$$= (7.0888 - 1)(9.11 \times 10^{-31} \text{ kg})(3.00 \times 10^8 \text{ m/s})^2$$

$$= 4.99 \times 10^{-13} \text{ J}$$

5. Convert units.

$$\mathrm{KE}_{rel} = (4.99 \times 10^{-13} \text{ J})\left(\frac{1 \text{ MeV}}{1.60 \times 10^{-13} \text{ J}}\right) \tag{28.59}$$

$$= 3.12 \text{ MeV}$$

Solution for (b)

1. List the knowns. $v = 0.990c$; $m = 9.11 \times 10^{-31}$ kg

2. List the unknown. KE_{class}

3. Choose the appropriate equation. $\text{KE}_{\text{class}} = \frac{1}{2}mv^2$

4. Plug the knowns into the equation.

$$
\begin{aligned}
\text{KE}_{\text{class}} &= \frac{1}{2}mv^2 \\
&= \frac{1}{2}(9.00 \times 10^{-31} \text{ kg})(0.990)^2(3.00 \times 10^8 \text{ m/s})^2 \\
&= 4.02 \times 10^{-14} \text{ J}
\end{aligned}
$$

(28.60)

5. Convert units.

$$
\begin{aligned}
\text{KE}_{\text{class}} &= 4.02 \times 10^{-14} \text{ J}\left(\frac{1 \text{ MeV}}{1.60 \times 10^{-13} \text{ J}}\right) \\
&= 0.251 \text{ MeV}
\end{aligned}
$$

(28.61)

Discussion

As might be expected, since the velocity is 99.0% of the speed of light, the classical kinetic energy is significantly off from the correct relativistic value. Note also that the classical value is much smaller than the relativistic value. In fact, $\text{KE}_{\text{rel}}/\text{KE}_{\text{class}} = 12.4$ here. This is some indication of how difficult it is to get a mass moving close to the speed of light.

Much more energy is required than predicted classically. Some people interpret this extra energy as going into increasing the mass of the system, but, as discussed in **Relativistic Momentum**, this cannot be verified unambiguously. What is certain is that ever-increasing amounts of energy are needed to get the velocity of a mass a little closer to that of light. An energy of 3 MeV is a very small amount for an electron, and it can be achieved with present-day particle accelerators. SLAC, for example, can accelerate electrons to over 50×10^9 eV $= 50,000$ MeV .

Is there any point in getting v a little closer to c than 99.0% or 99.9%? The answer is yes. We learn a great deal by doing this. The energy that goes into a high-velocity mass can be converted to any other form, including into entirely new masses. (See **Figure 28.23**.) Most of what we know about the substructure of matter and the collection of exotic short-lived particles in nature has been learned this way. Particles are accelerated to extremely relativistic energies and made to collide with other particles, producing totally new species of particles. Patterns in the characteristics of these previously unknown particles hint at a basic substructure for all matter. These particles and some of their characteristics will be covered in **Particle Physics**.

Figure 28.23 The Fermi National Accelerator Laboratory, near Batavia, Illinois, was a subatomic particle collider that accelerated protons and antiprotons to attain energies up to 1 Tev (a trillion electronvolts). The circular ponds near the rings were built to dissipate waste heat. This accelerator was shut down in September 2011. (credit: Fermilab, Reidar Hahn)

Relativistic Energy and Momentum

We know classically that kinetic energy and momentum are related to each other, since

$$
\text{KE}_{\text{class}} = \frac{p^2}{2m} = \frac{(mv)^2}{2m} = \frac{1}{2}mv^2.
$$

(28.62)

Relativistically, we can obtain a relationship between energy and momentum by algebraically manipulating their definitions. This produces

$$E^2 = (pc)^2 + (mc^2)^2, \tag{28.63}$$

where E is the relativistic total energy and p is the relativistic momentum. This relationship between relativistic energy and relativistic momentum is more complicated than the classical, but we can gain some interesting new insights by examining it. First, total energy is related to momentum and rest mass. At rest, momentum is zero, and the equation gives the total energy to be the rest energy mc^2 (so this equation is consistent with the discussion of rest energy above). However, as the mass is accelerated, its momentum p increases, thus increasing the total energy. At sufficiently high velocities, the rest energy term $(mc^2)^2$ becomes negligible compared with the momentum term $(pc)^2$; thus, $E = pc$ at extremely relativistic velocities.

If we consider momentum p to be distinct from mass, we can determine the implications of the equation $E^2 = (pc)^2 + (mc^2)^2,$ for a particle that has no mass. If we take m to be zero in this equation, then $E = pc$, or $p = E/c$. Massless particles have this momentum. There are several massless particles found in nature, including photons (these are quanta of electromagnetic radiation). Another implication is that a massless particle must travel at speed c and only at speed c. While it is beyond the scope of this text to examine the relationship in the equation $E^2 = (pc)^2 + (mc^2)^2,$ in detail, we can see that the relationship has important implications in special relativity.

Problem-Solving Strategies for Relativity

1. *Examine the situation to determine that it is necessary to use relativity*. Relativistic effects are related to $\gamma = \dfrac{1}{\sqrt{1 - \frac{v^2}{c^2}}}$, the quantitative relativistic factor. If γ is very close to 1, then relativistic effects are small and differ very little from the usually easier classical calculations.

2. *Identify exactly what needs to be determined in the problem (identify the unknowns)*.

3. *Make a list of what is given or can be inferred from the problem as stated (identify the knowns)*. Look in particular for information on relative velocity v.

4. *Make certain you understand the conceptual aspects of the problem before making any calculations*. Decide, for example, which observer sees time dilated or length contracted before plugging into equations. If you have thought about who sees what, who is moving with the event being observed, who sees proper time, and so on, you will find it much easier to determine if your calculation is reasonable.

5. *Determine the primary type of calculation to be done to find the unknowns identified above*. You will find the section summary helpful in determining whether a length contraction, relativistic kinetic energy, or some other concept is involved.

6. *Do not round off during the calculation*. As noted in the text, you must often perform your calculations to many digits to see the desired effect. You may round off at the very end of the problem, but do not use a rounded number in a subsequent calculation.

7. *Check the answer to see if it is reasonable: Does it make sense?* This may be more difficult for relativity, since we do not encounter it directly. But you can look for velocities greater than c or relativistic effects that are in the wrong direction (such as a time contraction where a dilation was expected).

Check Your Understanding

A photon decays into an electron-positron pair. What is the kinetic energy of the electron if its speed is $0.992c$?

Solution

$$
\begin{aligned}
\mathrm{KE_{rel}} &= (\gamma - 1)mc^2 = \left(\frac{1}{\sqrt{1 - \frac{v^2}{c^2}}} - 1 \right) mc^2 \\[2mm]
&= \left(\frac{1}{\sqrt{1 - \frac{(0.992c)^2}{c^2}}} - 1 \right)(9.11 \times 10^{-31}\ \mathrm{kg})(3.00 \times 10^8\ \mathrm{m/s})^2 = 5.67 \times 10^{-13}\ \mathrm{J}
\end{aligned} \tag{28.64}
$$

Glossary

classical velocity addition: the method of adding velocities when $v \ll c$; velocities add like regular numbers in one-dimensional motion: $u = v + u'$, where v is the velocity between two observers, u is the velocity of an object relative to one observer, and u' is the velocity relative to the other observer

first postulate of special relativity: the idea that the laws of physics are the same and can be stated in their simplest form in all inertial frames of reference

inertial frame of reference: a reference frame in which a body at rest remains at rest and a body in motion moves at a constant speed in a straight line unless acted on by an outside force

length contraction: L , the shortening of the measured length of an object moving relative to the observer's frame:

$$L = L_0 \sqrt{1 - \frac{v^2}{c^2}} = \frac{L_0}{\gamma}$$

Michelson-Morley experiment: an investigation performed in 1887 that proved that the speed of light in a vacuum is the same in all frames of reference from which it is viewed

proper length: L_0 ; the distance between two points measured by an observer who is at rest relative to both of the points; Earth-bound observers measure proper length when measuring the distance between two points that are stationary relative to the Earth

proper time: Δt_0 . the time measured by an observer at rest relative to the event being observed: $\Delta t = \dfrac{\Delta t_0}{\sqrt{1 - \frac{v^2}{c^2}}} = \gamma \Delta t_0$,

where $\gamma = \dfrac{1}{\sqrt{1 - \frac{v^2}{c^2}}}$

relativistic Doppler effects: a change in wavelength of radiation that is moving relative to the observer; the wavelength of the radiation is longer (called a red shift) than that emitted by the source when the source moves away from the observer and shorter (called a blue shift) when the source moves toward the observer; the shifted wavelength is described by the equation

$$\lambda_{obs} = \lambda_s \sqrt{\frac{1 + \frac{u}{c}}{1 - \frac{u}{c}}}$$

where λ_{obs} is the observed wavelength, λ_s is the source wavelength, and u is the velocity of the source to the observer

relativistic kinetic energy: the kinetic energy of an object moving at relativistic speeds: $KE_{rel} = (\gamma - 1)mc^2$, where

$$\gamma = \frac{1}{\sqrt{1 - \frac{v^2}{c^2}}}$$

relativistic momentum: p , the momentum of an object moving at relativistic velocity; $p = \gamma m u$, where m is the rest mass

of the object, u is its velocity relative to an observer, and the relativistic factor $\gamma = \dfrac{1}{\sqrt{1 - \frac{u^2}{c^2}}}$

relativistic velocity addition: the method of adding velocities of an object moving at a relativistic speed: $u = \dfrac{v + u'}{1 + \frac{vu'}{c^2}}$, where

v is the relative velocity between two observers, u is the velocity of an object relative to one observer, and u' is the velocity relative to the other observer

relativity: the study of how different observers measure the same event

rest energy: the energy stored in an object at rest: $E_0 = mc^2$

rest mass: the mass of an object as measured by a person at rest relative to the object

second postulate of special relativity: the idea that the speed of light c is a constant, independent of the source

special relativity: the theory that, in an inertial frame of reference, the motion of an object is relative to the frame from which it is viewed or measured

time dilation: the phenomenon of time passing slower to an observer who is moving relative to another observer

total energy: defined as $E = \gamma mc^2$, where $\gamma = \dfrac{1}{\sqrt{1 - \dfrac{v^2}{c^2}}}$

twin paradox: this asks why a twin traveling at a relativistic speed away and then back towards the Earth ages less than the Earth-bound twin. The premise to the paradox is faulty because the traveling twin is accelerating, and special relativity does not apply to accelerating frames of reference

Section Summary

28.1 Einstein's Postulates

- Relativity is the study of how different observers measure the same event.
- Modern relativity is divided into two parts. Special relativity deals with observers who are in uniform (unaccelerated) motion, whereas general relativity includes accelerated relative motion and gravity. Modern relativity is correct in all circumstances and, in the limit of low velocity and weak gravitation, gives the same predictions as classical relativity.
- An inertial frame of reference is a reference frame in which a body at rest remains at rest and a body in motion moves at a constant speed in a straight line unless acted on by an outside force.
- Modern relativity is based on Einstein's two postulates. The first postulate of special relativity is the idea that the laws of physics are the same and can be stated in their simplest form in all inertial frames of reference. The second postulate of special relativity is the idea that the speed of light c is a constant, independent of the relative motion of the source.
- The Michelson-Morley experiment demonstrated that the speed of light in a vacuum is independent of the motion of the Earth about the Sun.

28.2 Simultaneity And Time Dilation

- Two events are defined to be simultaneous if an observer measures them as occurring at the same time. They are not necessarily simultaneous to all observers—simultaneity is not absolute.
- Time dilation is the phenomenon of time passing slower for an observer who is moving relative to another observer.
- Observers moving at a relative velocity v do not measure the same elapsed time for an event. Proper time Δt_0 is the time measured by an observer at rest relative to the event being observed. Proper time is related to the time Δt measured by an Earth-bound observer by the equation

$$\Delta t = \frac{\Delta t_0}{\sqrt{1 - \dfrac{v^2}{c^2}}} = \gamma \Delta t_0,$$

where

$$\gamma = \frac{1}{\sqrt{1 - \dfrac{v^2}{c^2}}}.$$

- The equation relating proper time and time measured by an Earth-bound observer implies that relative velocity cannot exceed the speed of light.
- The twin paradox asks why a twin traveling at a relativistic speed away and then back towards the Earth ages less than the Earth-bound twin. The premise to the paradox is faulty because the traveling twin is accelerating. Special relativity does not apply to accelerating frames of reference.
- Time dilation is usually negligible at low relative velocities, but it does occur, and it has been verified by experiment.

28.3 Length Contraction

- All observers agree upon relative speed.
- Distance depends on an observer's motion. Proper length L_0 is the distance between two points measured by an observer who is at rest relative to both of the points. Earth-bound observers measure proper length when measuring the distance between two points that are stationary relative to the Earth.
- Length contraction L is the shortening of the measured length of an object moving relative to the observer's frame:

$$L = L_0 \sqrt{1 - \frac{v^2}{c^2}} = \frac{L_0}{\gamma}.$$

28.4 Relativistic Addition of Velocities

- With classical velocity addition, velocities add like regular numbers in one-dimensional motion: $u = v + u'$, where v is the velocity between two observers, u is the velocity of an object relative to one observer, and u' is the velocity relative to the other observer.
- Velocities cannot add to be greater than the speed of light. Relativistic velocity addition describes the velocities of an object moving at a relativistic speed:

$$u = \frac{v + u'}{1 + \frac{vu'}{c^2}}$$

- An observer of electromagnetic radiation sees **relativistic Doppler effects** if the source of the radiation is moving relative to the observer. The wavelength of the radiation is longer (called a red shift) than that emitted by the source when the source moves away from the observer and shorter (called a blue shift) when the source moves toward the observer. The shifted wavelength is described by the equation

$$\lambda_{obs} = \lambda_s \sqrt{\frac{1 + \frac{u}{c}}{1 - \frac{u}{c}}}$$

λ_{obs} is the observed wavelength, λ_s is the source wavelength, and u is the relative velocity of the source to the observer.

28.5 Relativistic Momentum

- The law of conservation of momentum is valid whenever the net external force is zero and for relativistic momentum. Relativistic momentum p is classical momentum multiplied by the relativistic factor γ.
- $p = \gamma m u$, where m is the rest mass of the object, u is its velocity relative to an observer, and the relativistic factor

$$\gamma = \frac{1}{\sqrt{1 - \frac{u^2}{c^2}}}.$$

- At low velocities, relativistic momentum is equivalent to classical momentum.
- Relativistic momentum approaches infinity as u approaches c. This implies that an object with mass cannot reach the speed of light.
- Relativistic momentum is conserved, just as classical momentum is conserved.

28.6 Relativistic Energy

- Relativistic energy is conserved as long as we define it to include the possibility of mass changing to energy.
- Total Energy is defined as: $E = \gamma mc^2$, where $\gamma = \frac{1}{\sqrt{1 - \frac{v^2}{c^2}}}$.

- Rest energy is $E_0 = mc^2$, meaning that mass is a form of energy. If energy is stored in an object, its mass increases. Mass can be destroyed to release energy.
- We do not ordinarily notice the increase or decrease in mass of an object because the change in mass is so small for a large increase in energy.
- The relativistic work-energy theorem is $W_{net} = E - E_0 = \gamma mc^2 - mc^2 = (\gamma - 1)mc^2$.
- Relativistically, $W_{net} = KE_{rel}$, where KE_{rel} is the relativistic kinetic energy.
- Relativistic kinetic energy is $KE_{rel} = (\gamma - 1)mc^2$, where $\gamma = \frac{1}{\sqrt{1 - \frac{v^2}{c^2}}}$. At low velocities, relativistic kinetic energy reduces to classical kinetic energy.
- **No object with mass can attain the speed of light** because an infinite amount of work and an infinite amount of energy input is required to accelerate a mass to the speed of light.
- The equation $E^2 = (pc)^2 + (mc^2)^2$ relates the relativistic total energy E and the relativistic momentum p. At extremely high velocities, the rest energy mc^2 becomes negligible, and $E = pc$.

Conceptual Questions

28.1 Einstein's Postulates

1. Which of Einstein's postulates of special relativity includes a concept that does not fit with the ideas of classical physics? Explain.

2. Is Earth an inertial frame of reference? Is the Sun? Justify your response.

3. When you are flying in a commercial jet, it may appear to you that the airplane is stationary and the Earth is moving beneath you. Is this point of view valid? Discuss briefly.

28.2 Simultaneity And Time Dilation

4. Does motion affect the rate of a clock as measured by an observer moving with it? Does motion affect how an observer moving relative to a clock measures its rate?

5. To whom does the elapsed time for a process seem to be longer, an observer moving relative to the process or an observer moving with the process? Which observer measures proper time?

6. How could you travel far into the future without aging significantly? Could this method also allow you to travel into the past?

28.3 Length Contraction

7. To whom does an object seem greater in length, an observer moving with the object or an observer moving relative to the object? Which observer measures the object's proper length?

8. Relativistic effects such as time dilation and length contraction are present for cars and airplanes. Why do these effects seem strange to us?

9. Suppose an astronaut is moving relative to the Earth at a significant fraction of the speed of light. (a) Does he observe the rate of his clocks to have slowed? (b) What change in the rate of Earth-bound clocks does he see? (c) Does his ship seem to him to shorten? (d) What about the distance between stars that lie on lines parallel to his motion? (e) Do he and an Earth-bound observer agree on his velocity relative to the Earth?

28.4 Relativistic Addition of Velocities

10. Explain the meaning of the terms "red shift" and "blue shift" as they relate to the relativistic Doppler effect.

11. What happens to the relativistic Doppler effect when relative velocity is zero? Is this the expected result?

12. Is the relativistic Doppler effect consistent with the classical Doppler effect in the respect that λ_{obs} is larger for motion away?

13. All galaxies farther away than about 50×10^6 ly exhibit a red shift in their emitted light that is proportional to distance, with those farther and farther away having progressively greater red shifts. What does this imply, assuming that the only source of red shift is relative motion? (Hint: At these large distances, it is space itself that is expanding, but the effect on light is the same.)

28.5 Relativistic Momentum

14. How does modern relativity modify the law of conservation of momentum?

15. Is it possible for an external force to be acting on a system and relativistic momentum to be conserved? Explain.

28.6 Relativistic Energy

16. How are the classical laws of conservation of energy and conservation of mass modified by modern relativity?

17. What happens to the mass of water in a pot when it cools, assuming no molecules escape or are added? Is this observable in practice? Explain.

18. Consider a thought experiment. You place an expanded balloon of air on weighing scales outside in the early morning. The balloon stays on the scales and you are able to measure changes in its mass. Does the mass of the balloon change as the day progresses? Discuss the difficulties in carrying out this experiment.

19. The mass of the fuel in a nuclear reactor decreases by an observable amount as it puts out energy. Is the same true for the coal and oxygen combined in a conventional power plant? If so, is this observable in practice for the coal and oxygen? Explain.

20. We know that the velocity of an object with mass has an upper limit of c. Is there an upper limit on its momentum? Its energy? Explain.

21. Given the fact that light travels at c, can it have mass? Explain.

22. If you use an Earth-based telescope to project a laser beam onto the Moon, you can move the spot across the Moon's surface at a velocity greater than the speed of light. Does this violate modern relativity? (Note that light is being sent from the Earth to the Moon, not across the surface of the Moon.)

Problems & Exercises

28.2 Simultaneity And Time Dilation

1. (a) What is γ if $v = 0.250c$? (b) If $v = 0.500c$?

2. (a) What is γ if $v = 0.100c$? (b) If $v = 0.900c$?

3. Particles called π -mesons are produced by accelerator beams. If these particles travel at 2.70×10^8 m/s and live 2.60×10^{-8} s when at rest relative to an observer, how long do they live as viewed in the laboratory?

4. Suppose a particle called a kaon is created by cosmic radiation striking the atmosphere. It moves by you at $0.980c$, and it lives 1.24×10^{-8} s when at rest relative to an observer. How long does it live as you observe it?

5. A neutral π -meson is a particle that can be created by accelerator beams. If one such particle lives 1.40×10^{-16} s as measured in the laboratory, and 0.840×10^{-16} s when at rest relative to an observer, what is its velocity relative to the laboratory?

6. A neutron lives 900 s when at rest relative to an observer. How fast is the neutron moving relative to an observer who measures its life span to be 2065 s?

7. If relativistic effects are to be less than 1%, then γ must be less than 1.01. At what relative velocity is $\gamma = 1.01$?

8. If relativistic effects are to be less than 3%, then γ must be less than 1.03. At what relative velocity is $\gamma = 1.03$?

9. (a) At what relative velocity is $\gamma = 1.50$? (b) At what relative velocity is $\gamma = 100$?

10. (a) At what relative velocity is $\gamma = 2.00$? (b) At what relative velocity is $\gamma = 10.0$?

11. Unreasonable Results

(a) Find the value of γ for the following situation. An Earth-bound observer measures 23.9 h to have passed while signals from a high-velocity space probe indicate that 24.0 h have passed on board. (b) What is unreasonable about this result? (c) Which assumptions are unreasonable or inconsistent?

28.3 Length Contraction

12. A spaceship, 200 m long as seen on board, moves by the Earth at $0.970c$. What is its length as measured by an Earth-bound observer?

13. How fast would a 6.0 m-long sports car have to be going past you in order for it to appear only 5.5 m long?

14. (a) How far does the muon in **Example 28.1** travel according to the Earth-bound observer? (b) How far does it travel as viewed by an observer moving with it? Base your calculation on its velocity relative to the Earth and the time it lives (proper time). (c) Verify that these two distances are related through length contraction $\gamma = 3.20$.

15. (a) How long would the muon in **Example 28.1** have lived as observed on the Earth if its velocity was $0.0500c$? (b) How far would it have traveled as observed on the Earth? (c) What distance is this in the muon's frame?

16. (a) How long does it take the astronaut in **Example 28.2** to travel 4.30 ly at $0.99944c$ (as measured by the Earth-bound observer)? (b) How long does it take according to the astronaut? (c) Verify that these two times are related through time dilation with $\gamma = 30.00$ as given.

17. (a) How fast would an athlete need to be running for a 100-m race to look 100 yd long? (b) Is the answer consistent with the fact that relativistic effects are difficult to observe in ordinary circumstances? Explain.

18. Unreasonable Results

(a) Find the value of γ for the following situation. An astronaut measures the length of her spaceship to be 25.0 m, while an Earth-bound observer measures it to be 100 m. (b) What is unreasonable about this result? (c) Which assumptions are unreasonable or inconsistent?

19. Unreasonable Results

A spaceship is heading directly toward the Earth at a velocity of $0.800c$. The astronaut on board claims that he can send a canister toward the Earth at $1.20c$ relative to the Earth. (a) Calculate the velocity the canister must have relative to the spaceship. (b) What is unreasonable about this result? (c) Which assumptions are unreasonable or inconsistent?

28.4 Relativistic Addition of Velocities

20. Suppose a spaceship heading straight towards the Earth at $0.750c$ can shoot a canister at $0.500c$ relative to the ship. (a) What is the velocity of the canister relative to the Earth, if it is shot directly at the Earth? (b) If it is shot directly away from the Earth?

21. Repeat the previous problem with the ship heading directly away from the Earth.

22. If a spaceship is approaching the Earth at $0.100c$ and a message capsule is sent toward it at $0.100c$ relative to the Earth, what is the speed of the capsule relative to the ship?

23. (a) Suppose the speed of light were only 3000 m/s . A jet fighter moving toward a target on the ground at 800 m/s shoots bullets, each having a muzzle velocity of 1000 m/s . What are the bullets' velocity relative to the target? (b) If the speed of light was this small, would you observe relativistic effects in everyday life? Discuss.

24. If a galaxy moving away from the Earth has a speed of 1000 km/s and emits 656 nm light characteristic of hydrogen (the most common element in the universe). (a) What wavelength would we observe on the Earth? (b) What type of electromagnetic radiation is this? (c) Why is the speed of the Earth in its orbit negligible here?

25. A space probe speeding towards the nearest star moves at $0.250c$ and sends radio information at a broadcast frequency of 1.00 GHz. What frequency is received on the Earth?

26. If two spaceships are heading directly towards each other at $0.800c$, at what speed must a canister be shot from the first ship to approach the other at $0.999c$ as seen by the second ship?

27. Two planets are on a collision course, heading directly towards each other at $0.250c$. A spaceship sent from one planet approaches the second at $0.750c$ as seen by the second planet. What is the velocity of the ship relative to the first planet?

28. When a missile is shot from one spaceship towards another, it leaves the first at $0.950c$ and approaches the other at $0.750c$. What is the relative velocity of the two ships?

29. What is the relative velocity of two spaceships if one fires a missile at the other at $0.750c$ and the other observes it to approach at $0.950c$?

30. Near the center of our galaxy, hydrogen gas is moving directly away from us in its orbit about a black hole. We receive 1900 nm electromagnetic radiation and know that it was 1875 nm when emitted by the hydrogen gas. What is the speed of the gas?

31. A highway patrol officer uses a device that measures the speed of vehicles by bouncing radar off them and measuring the Doppler shift. The outgoing radar has a frequency of 100 GHz and the returning echo has a frequency 15.0 kHz higher. What is the velocity of the vehicle? Note that there are two Doppler shifts in echoes. Be certain not to round off until the end of the problem, because the effect is small.

32. Prove that for any relative velocity v between two observers, a beam of light sent from one to the other will approach at speed c (provided that v is less than c, of course).

33. Show that for any relative velocity v between two observers, a beam of light projected by one directly away from the other will move away at the speed of light (provided that v is less than c, of course).

34. (a) All but the closest galaxies are receding from our own Milky Way Galaxy. If a galaxy 12.0×10^9 ly ly away is receding from us at 0. $0.900c$, at what velocity relative to us must we send an exploratory probe to approach the other galaxy at $0.990c$, as measured from that galaxy? (b) How long will it take the probe to reach the other galaxy as measured from the Earth? You may assume that the velocity of the other galaxy remains constant. (c) How long will it then take for a radio signal to be beamed back? (All of this is possible in principle, but not practical.)

28.5 Relativistic Momentum

35. Find the momentum of a helium nucleus having a mass of 6.68×10^{-27} kg that is moving at $0.200c$.

36. What is the momentum of an electron traveling at $0.980c$?

37. (a) Find the momentum of a 1.00×10^9 kg asteroid heading towards the Earth at 30.0 km/s. (b) Find the ratio of this momentum to the classical momentum. (Hint: Use the approximation that $\gamma = 1 + (1/2)v^2/c^2$ at low velocities.)

38. (a) What is the momentum of a 2000 kg satellite orbiting at 4.00 km/s? (b) Find the ratio of this momentum to the classical momentum. (Hint: Use the approximation that $\gamma = 1 + (1/2)v^2/c^2$ at low velocities.)

39. What is the velocity of an electron that has a momentum of 3.04×10^{-21} kg·m/s? Note that you must calculate the velocity to at least four digits to see the difference from c.

40. Find the velocity of a proton that has a momentum of 4.48×-10^{-19} kg·m/s.

41. (a) Calculate the speed of a 1.00-μg particle of dust that has the same momentum as a proton moving at $0.999c$. (b) What does the small speed tell us about the mass of a proton compared to even a tiny amount of macroscopic matter?

42. (a) Calculate γ for a proton that has a momentum of 1.00 kg·m/s. (b) What is its speed? Such protons form a rare component of cosmic radiation with uncertain origins.

28.6 Relativistic Energy

43. What is the rest energy of an electron, given its mass is 9.11×10^{-31} kg? Give your answer in joules and MeV.

44. Find the rest energy in joules and MeV of a proton, given its mass is 1.67×10^{-27} kg.

45. If the rest energies of a proton and a neutron (the two constituents of nuclei) are 938.3 and 939.6 MeV respectively, what is the difference in their masses in kilograms?

46. The Big Bang that began the universe is estimated to have released 10^{68} J of energy. How many stars could half this energy create, assuming the average star's mass is 4.00×10^{30} kg?

47. A supernova explosion of a 2.00×10^{31} kg star produces 1.00×10^{44} kg of energy. (a) How many kilograms of mass are converted to energy in the explosion? (b) What is the ratio $\Delta m/m$ of mass destroyed to the original mass of the star?

48. (a) Using data from Table 7.1, calculate the mass converted to energy by the fission of 1.00 kg of uranium. (b) What is the ratio of mass destroyed to the original mass, $\Delta m/m$?

49. (a) Using data from Table 7.1, calculate the amount of mass converted to energy by the fusion of 1.00 kg of hydrogen. (b) What is the ratio of mass destroyed to the original mass, $\Delta m/m$? (c) How does this compare with $\Delta m/m$ for the fission of 1.00 kg of uranium?

50. There is approximately 10^{34} J of energy available from fusion of hydrogen in the world's oceans. (a) If 10^{33} J of this energy were utilized, what would be the decrease in mass of the oceans? Assume that 0.08% of the mass of a water molecule is converted to energy during the fusion of hydrogen. (b) How great a volume of water does this correspond to? (c) Comment on whether this is a significant fraction of the total mass of the oceans.

51. A muon has a rest mass energy of 105.7 MeV, and it decays into an electron and a massless particle. (a) If all the lost mass is converted into the electron's kinetic energy, find γ for the electron. (b) What is the electron's velocity?

52. A π-meson is a particle that decays into a muon and a massless particle. The π-meson has a rest mass energy of 139.6 MeV, and the muon has a rest mass energy of 105.7 MeV. Suppose the π-meson is at rest and all of the missing mass goes into the muon's kinetic energy. How fast will the muon move?

53. (a) Calculate the relativistic kinetic energy of a 1000-kg car moving at 30.0 m/s if the speed of light were only 45.0 m/s. (b) Find the ratio of the relativistic kinetic energy to classical.

54. Alpha decay is nuclear decay in which a helium nucleus is emitted. If the helium nucleus has a mass of 6.80×10^{-27} kg and is given 5.00 MeV of kinetic energy, what is its velocity?

55. (a) Beta decay is nuclear decay in which an electron is emitted. If the electron is given 0.750 MeV of kinetic energy, what is its velocity? (b) Comment on how the high velocity is consistent with the kinetic energy as it compares to the rest mass energy of the electron.

56. A positron is an antimatter version of the electron, having exactly the same mass. When a positron and an electron meet, they annihilate, converting all of their mass into energy. (a) Find the energy released, assuming negligible kinetic energy before the annihilation. (b) If this energy is given to a proton in the form of kinetic energy, what is its velocity? (c) If this energy is given to another electron in the form of kinetic energy, what is its velocity?

57. What is the kinetic energy in MeV of a π-meson that lives 1.40×10^{-16} s as measured in the laboratory, and 0.840×10^{-16} s when at rest relative to an observer, given that its rest energy is 135 MeV?

58. Find the kinetic energy in MeV of a neutron with a measured life span of 2065 s, given its rest energy is 939.6 MeV, and rest life span is 900s.

59. (a) Show that $(pc)^2/(mc^2)^2 = \gamma^2 - 1$. This means that at large velocities $pc \gg mc^2$. (b) Is $E \approx pc$ when $\gamma = 30.0$, as for the astronaut discussed in the twin paradox?

60. One cosmic ray neutron has a velocity of $0.250c$ relative to the Earth. (a) What is the neutron's total energy in MeV? (b) Find its momentum. (c) Is $E \approx pc$ in this situation? Discuss in terms of the equation given in part (a) of the previous problem.

61. What is γ for a proton having a mass energy of 938.3 MeV accelerated through an effective potential of 1.0 TV (teravolt) at Fermilab outside Chicago?

62. (a) What is the effective accelerating potential for electrons at the Stanford Linear Accelerator, if $\gamma = 1.00 \times 10^5$ for them? (b) What is their total energy (nearly the same as kinetic in this case) in GeV?

63. (a) Using data from Table 7.1, find the mass destroyed when the energy in a barrel of crude oil is released. (b) Given these barrels contain 200 liters and assuming the density of crude oil is 750 kg/m^3, what is the ratio of mass destroyed to original mass, $\Delta m / m$?

64. (a) Calculate the energy released by the destruction of 1.00 kg of mass. (b) How many kilograms could be lifted to a 10.0 km height by this amount of energy?

65. A Van de Graaff accelerator utilizes a 50.0 MV potential difference to accelerate charged particles such as protons. (a) What is the velocity of a proton accelerated by such a potential? (b) An electron?

66. Suppose you use an average of 500 kW·h of electric energy per month in your home. (a) How long would 1.00 g of mass converted to electric energy with an efficiency of 38.0% last you? (b) How many homes could be supplied at the 500 kW·h per month rate for one year by the energy from the described mass conversion?

67. (a) A nuclear power plant converts energy from nuclear fission into electricity with an efficiency of 35.0%. How much mass is destroyed in one year to produce a continuous 1000 MW of electric power? (b) Do you think it would be possible to observe this mass loss if the total mass of the fuel is 10^4 kg?

68. Nuclear-powered rockets were researched for some years before safety concerns became paramount. (a) What fraction of a rocket's mass would have to be destroyed to get it into a low Earth orbit, neglecting the decrease in gravity? (Assume an orbital altitude of 250 km, and calculate both the kinetic energy (classical) and the gravitational potential energy needed.) (b) If the ship has a mass of 1.00×10^5 kg (100 tons), what total yield nuclear explosion in tons of TNT is needed?

69. The Sun produces energy at a rate of 4.00×10^{26} W by the fusion of hydrogen. (a) How many kilograms of hydrogen undergo fusion each second? (b) If the Sun is 90.0% hydrogen and half of this can undergo fusion before the Sun changes character, how long could it produce energy at its current rate? (c) How many kilograms of mass is the Sun losing per second? (d) What fraction of its mass will it have lost in the time found in part (b)?

70. Unreasonable Results

A proton has a mass of 1.67×10^{-27} kg. A physicist measures the proton's total energy to be 50.0 MeV. (a) What is the proton's kinetic energy? (b) What is unreasonable about this result? (c) Which assumptions are unreasonable or inconsistent?

71. Construct Your Own Problem

Consider a highly relativistic particle. Discuss what is meant by the term "highly relativistic." (Note that, in part, it means that the particle cannot be massless.) Construct a problem in which you calculate the wavelength of such a particle and show that it is very nearly the same as the wavelength of a massless particle, such as a photon, with the same energy. Among the things to be considered are the rest energy of the particle (it should be a known particle) and its total energy, which should be large compared to its rest energy.

72. Construct Your Own Problem

Consider an astronaut traveling to another star at a relativistic velocity. Construct a problem in which you calculate the time for the trip as observed on the Earth and as observed by the astronaut. Also calculate the amount of mass that must be converted to energy to get the astronaut and ship to the velocity travelled. Among the things to be considered are the distance to the star, the velocity, and the mass of the astronaut and ship. Unless your instructor directs you otherwise, do not include any energy given to other masses, such as rocket propellants.

Test Prep for AP® Courses

28.1 Einstein's Postulates

1. Which of the following statements describes the Michelson-Morley experiment?
   a. The speed of light is independent of the motion of the source relative to the observer.
   b. The speed of light is different in different frames of reference.
   c. The speed of light changes with changes in the observer.
   d. The speed of light is dependent on the motion of the source.

28.4 Relativistic Addition of Velocities

2. What happens when velocities comparable to the speed of light are involved in an observation?
   a. Newton's second law of motion, $F = ma$, governs the motion of the object.
   b. Newton's second law of motion, $F = ma$, no longer governs the dynamics of the object.
   c. Such velocities cannot be determined mathematically.
   d. None of the above

3. How is the relativistic Doppler effect different from the classical Doppler effect?

28.6 Relativistic Energy

4. A mass of 50 g is completely converted into energy. What is the energy that will be obtained when such a conversion takes place?

5. Show that relativistic kinetic energy becomes the same as classical kinetic energy when $v = c$.

6. The relativistic energy of a particle in terms of momentum is given by:

   a. $E = \sqrt{p^2 c^2 + m_0^2 c^4}$

   b. $E = \sqrt{p^2 c^2 + m_0^4 c^4}$

   c. $E = \sqrt{p^2 c^2 + m_0^2 c^2}$

   d. $E = \sqrt{p^2 c^4 + m_0^2 c^2}$

29 INTRODUCTION TO QUANTUM PHYSICS

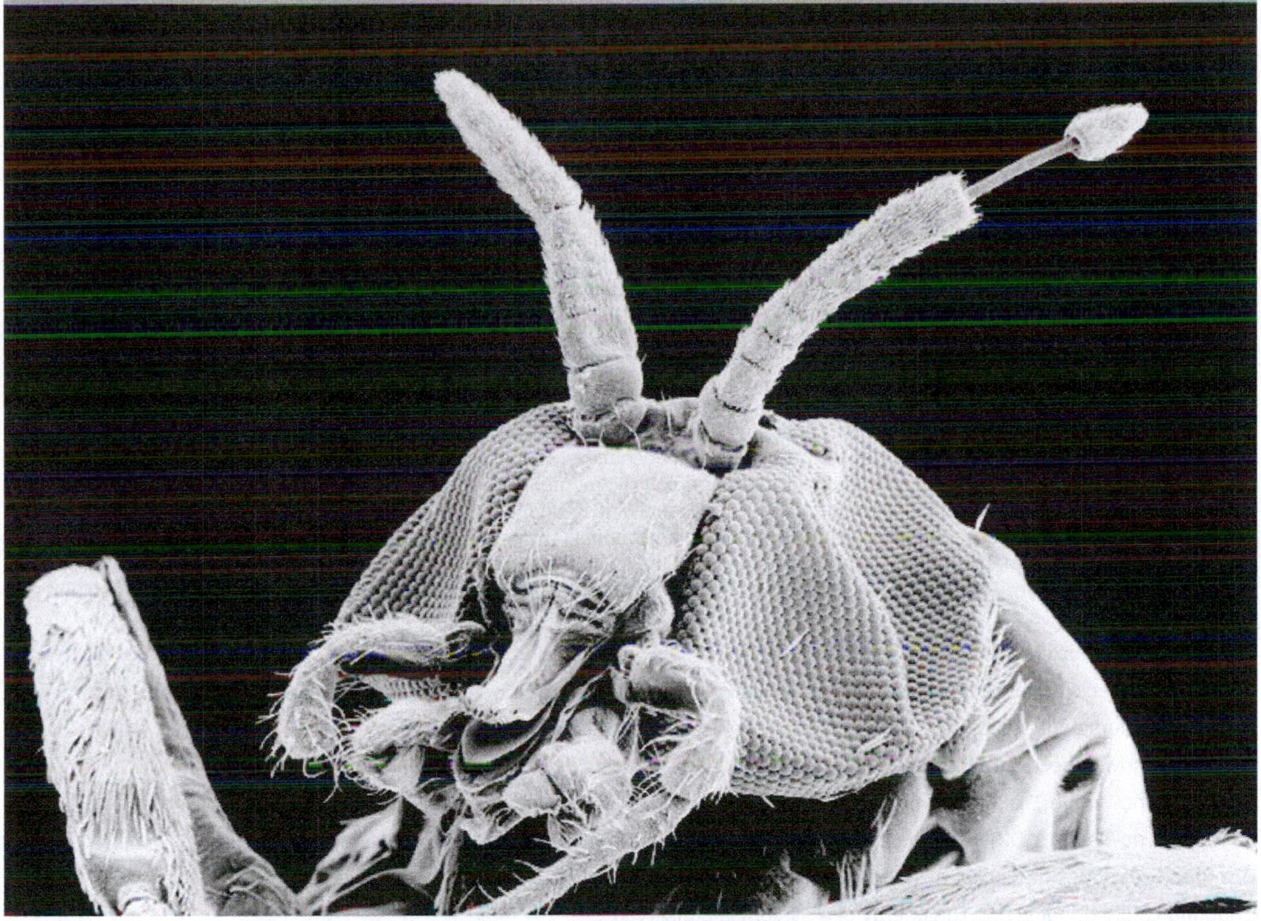

Figure 29.1 A black fly imaged by an electron microscope is as monstrous as any science-fiction creature. (credit: U.S. Department of Agriculture via Wikimedia Commons)

Chapter Outline

Connection for AP® Courses

In this chapter, the basic principles of **quantum mechanics** are introduced. Quantum mechanics is the branch of physics needed to deal with submicroscopic objects. Because these objects are smaller than those, such as computers, books, or cars, that we can observe directly with our senses, and so generally must be observed with the aid of instruments, parts of quantum mechanics seem as foreign and bizarre as the effects of relative motion near the speed of light. Yet through experimental results, quantum mechanics has been shown to be valid. Truth is often stranger than fiction.

Quantum theory was developed initially to explain the behavior of electromagnetic energy in certain situations, such as **blackbody radiation** or the **photoelectric effect**, which could not be understood in terms of classical electrodynamics (Essential Knowledge 1.D.2). In the quantum model, light is treated as a packet of energy called a **photon**, which has both the properties of a wave and a particle (Essential Knowledge 6.F.3). The energy of a photon is directly proportional to its frequency.

This new model for light provided the foundation for one of the most important ideas in quantum theory: wave-particle duality. Just as light has properties of both waves and particles, matter also has the properties of waves and particles (Essential Knowledge 1.D.1). This interpretation of matter and energy explained observations at the atomic level that could not be explained by classical mechanics or electromagnetic theory (Enduring Understanding 1.D). The quantum interpretation of energy and matter at the atomic level, most notably the internal structure of atoms, supports Big Idea 1 of the AP Physics Curriculum Framework.

Big Idea 1 is also supported by the **correspondence principle**. Classical mechanics cannot accurately describe systems at the atomic level, whereas quantum mechanics is able to describe systems at both levels. However, the properties of matter that are described by waves become insignificant at the macroscopic level, so that for large systems of matter, the quantum description closely approaches, or *corresponds to*, the classical description (Essential Knowledge 6.G.1, Essential Knowledge 6.G.2, Essential Knowledge 6.F.3).

Big Ideas 5 and 6 are supported by the descriptions of energy and momentum transfer at the quantum level. Although quantum mechanics overturned a number of fundamental ideas of classical physics, the most important principles, such as energy conservation and momentum conservation, remained intact (Enduring Understanding 5.B, Enduring Understanding 5.D). Quantum mechanics expands on these principles, so that the particle-like behavior of electromagnetic energy describes momentum transfer, while the wave-like behavior of matter accounts for why electrons produce diffraction patterns when they pass through the atomic lattices of crystals.

At the quantum level, the effects of measurement are very different from those at the macroscopic level. Because the wave properties of matter are more prominent for small particles, such as electrons, and a wave does not have a specific location, the position and momentum of matter cannot be measured with absolute precision (Essential Knowledge 1.D.3). Rather, the particle has a certain probability of being in a location interval for a specific momentum, or being located within a particular interval of time for a specific energy (Enduring Understanding 7.C, Essential Knowledge 7.C.1). These probabilistic limits on measurement are described by **Heisenberg's uncertainty principle**, which connects wave-particle duality to the non-absolute properties of space and time. At the quantum level, measurements affect the system being measured, and so restrict the degree to which properties can be known. The discussion of this probabilistic interpretation supports Big Idea 7 of the AP Physics Curriculum Framework.

The concepts in this chapter support:

Big Idea 1 Objects and systems have properties such as mass and charge. Systems may have internal structure.

Enduring Understanding 1.D Classical mechanics cannot describe all properties of objects.

Essential Knowledge 1.D.1 Objects classically thought of as particles can exhibit properties of waves.

Essential Knowledge 1.D.2 Certain phenomena classically thought of as waves can exhibit properties of particles.

Essential Knowledge 1.D.3 Properties of space and time cannot always be treated as absolute.

Big Idea 5 Changes that occur as a result of interactions are constrained by conservation laws.

Enduring Understanding 5.B The energy of a system is conserved.

Essential Knowledge 5.B.8 Energy transfer occurs when photons are absorbed or emitted, for example, by atoms or nuclei.

Enduring Understanding 5.D The linear momentum of a system is conserved.

Essential Knowledge 5.D.1 In a collision between objects, linear momentum is conserved. In an elastic collision, kinetic energy is the same before and after.

Big Idea 6 Waves can transfer energy and momentum from one location to another without the permanent transfer of mass and serve as a mathematical model for the description of other phenomena.

Enduring Understanding 6.F Electromagnetic radiation can be modeled as waves or as fundamental particles.

Essential Knowledge 6.F.3 Photons are individual energy packets of electromagnetic waves, with $E_{photon} = hf$, where h is Planck's constant and f is the frequency of the associated light wave.

Essential Knowledge 6.F.4 The nature of light requires that different models of light are most appropriate at different scales.

Enduring Understanding 6.G All matter can be modeled as waves or as particles.

Essential Knowledge 6.G.1 Under certain regimes of energy or distance, matter can be modeled as a classical particle.

Essential Knowledge 6.G.2 Under certain regimes of energy or distance, matter can be modeled as a wave. The behavior in these regimes is described by quantum mechanics.

Big Idea 7. The mathematics of probability can be used to describe the behavior of complex systems and to interpret the behavior of quantum mechanical systems.

Enduring Understanding 7.C At the quantum scale, matter is described by a wave function, which leads to a probabilistic description of the microscopic world.

Essential Knowledge 7.C.1 The probabilistic description of matter is modeled by a wave function, which can be assigned to an object and used to describe its motion and interactions. The absolute value of the wave function is related to the probability of finding a particle in some spatial region. (Qualitative treatment only, using graphical analysis.)

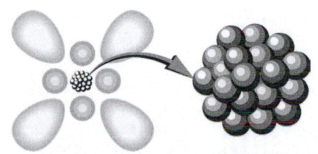

Figure 29.2 Atoms and their substructure are familiar examples of objects that require quantum mechanics to be fully explained. Certain of their characteristics, such as the discrete electron shells, are classical physics explanations. In quantum mechanics we conceptualize discrete "electron clouds" around the nucleus.

Making Connections: Realms of Physics
_____

Classical physics is a good approximation of modern physics under conditions first discussed in the **The Nature of Science and Physics**. Quantum mechanics is valid in general, and it must be used rather than classical physics to describe small objects, such as atoms.

Atoms, molecules, and fundamental electron and proton charges are all examples of physical entities that are **quantized**—that is, they appear only in certain discrete values and do not have every conceivable value. Quantized is the opposite of continuous. We cannot have a fraction of an atom, or part of an electron's charge, or 14-1/3 cents, for example. Rather, everything is built of integral multiples of these substructures. Quantum physics is the branch of physics that deals with small objects and the quantization of various entities, including energy and angular momentum. Just as with classical physics, quantum physics has several subfields, such as mechanics and the study of electromagnetic forces. The **correspondence principle** states that in the classical limit (large, slow-moving objects), **quantum mechanics** becomes the same as classical physics. In this chapter, we begin the development of quantum mechanics and its description of the strange submicroscopic world. In later chapters, we will examine many areas, such as atomic and nuclear physics, in which quantum mechanics is crucial.

29.1 Quantization of Energy

Learning Objectives

By the end of this section, you will be able to:

- Explain Max Planck's contribution to the development of quantum mechanics.
- Explain why atomic spectra indicate quantization.

The information presented in this section supports the following AP® learning objectives and science practices:

- **5.B.8.1** The student is able to describe emission or absorption spectra associated with electronic or nuclear transitions as transitions between allowed energy states of the atom in terms of the principle of energy conservation, including characterization of the frequency of radiation emitted or absorbed. **(S.P. 1.2, 7.2)**

Planck's Contribution

Energy is quantized in some systems, meaning that the system can have only certain energies and not a continuum of energies, unlike the classical case. This would be like having only certain speeds at which a car can travel because its kinetic energy can have only certain values. We also find that some forms of energy transfer take place with discrete lumps of energy. While most of us are familiar with the quantization of matter into lumps called atoms, molecules, and the like, we are less aware that energy, too, can be quantized. Some of the earliest clues about the necessity of quantum mechanics over classical physics came from the quantization of energy.

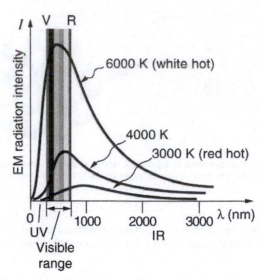

Figure 29.3 Graphs of blackbody radiation (from an ideal radiator) at three different radiator temperatures. The intensity or rate of radiation emission increases dramatically with temperature, and the peak of the spectrum shifts toward the visible and ultraviolet parts of the spectrum. The shape of the spectrum cannot be described with classical physics.

Where is the quantization of energy observed? Let us begin by considering the emission and absorption of electromagnetic (EM) radiation. The EM spectrum radiated by a hot solid is linked directly to the solid's temperature. (See Figure 29.3.) An ideal radiator is one that has an emissivity of 1 at all wavelengths and, thus, is jet black. Ideal radiators are therefore called **blackbodies**, and their EM radiation is called **blackbody radiation**. It was discussed that the total intensity of the radiation varies as T^4, the fourth power of the absolute temperature of the body, and that the peak of the spectrum shifts to shorter wavelengths at higher temperatures. All of this seems quite continuous, but it was the curve of the spectrum of intensity versus wavelength that gave a clue that the energies of the atoms in the solid are quantized. In fact, providing a theoretical explanation for the experimentally measured shape of the spectrum was a mystery at the turn of the century. When this "ultraviolet catastrophe" was eventually solved, the answers led to new technologies such as computers and the sophisticated imaging techniques described in earlier chapters. Once again, physics as an enabling science changed the way we live.

The German physicist Max Planck (1858–1947) used the idea that atoms and molecules in a body act like oscillators to absorb and emit radiation. The energies of the oscillating atoms and molecules had to be quantized to correctly describe the shape of the blackbody spectrum. Planck deduced that the energy of an oscillator having a frequency f is given by

$$E = \left(n + \frac{1}{2}\right)hf.$$ (29.1)

Here n is any nonnegative integer (0, 1, 2, 3, ...). The symbol h stands for **Planck's constant**, given by

$$h = 6.626 \times 10^{-34} \text{ J} \cdot \text{s.}$$ (29.2)

The equation $E = \left(n + \frac{1}{2}\right)hf$ means that an oscillator having a frequency f (emitting and absorbing EM radiation of frequency f) can have its energy increase or decrease only in *discrete* steps of size

$$\Delta E = hf.$$ (29.3)

It might be helpful to mention some macroscopic analogies of this quantization of energy phenomena. This is like a pendulum that has a characteristic oscillation frequency but can swing with only certain amplitudes. Quantization of energy also resembles a standing wave on a string that allows only particular harmonics described by integers. It is also similar to going up and down a hill using discrete stair steps rather than being able to move up and down a continuous slope. Your potential energy takes on discrete values as you move from step to step.

Using the quantization of oscillators, Planck was able to correctly describe the experimentally known shape of the blackbody spectrum. This was the first indication that energy is sometimes quantized on a small scale and earned him the Nobel Prize in Physics in 1918. Although Planck's theory comes from observations of a macroscopic object, its analysis is based on atoms and molecules. It was such a revolutionary departure from classical physics that Planck himself was reluctant to accept his own idea that energy states are not continuous. The general acceptance of Planck's energy quantization was greatly enhanced by Einstein's explanation of the photoelectric effect (discussed in the next section), which took energy quantization a step further. Planck was fully involved in the development of both early quantum mechanics and relativity. He quickly embraced Einstein's special relativity, published in 1905, and in 1906 Planck was the first to suggest the correct formula for relativistic momentum, $p = \gamma mu$.

Figure 29.4 The German physicist Max Planck had a major influence on the early development of quantum mechanics, being the first to recognize that energy is sometimes quantized. Planck also made important contributions to special relativity and classical physics. (credit: Library of Congress, Prints and Photographs Division via Wikimedia Commons)

Note that Planck's constant h is a very small number. So for an infrared frequency of 10^{14} Hz being emitted by a blackbody, for example, the difference between energy levels is only $\Delta E = hf = (6.63 \times 10^{-34} \text{ J·s})(10^{14} \text{ Hz}) = 6.63 \times 10^{-20}$ J, or about 0.4 eV. This 0.4 eV of energy is significant compared with typical atomic energies, which are on the order of an electron volt, or thermal energies, which are typically fractions of an electron volt. But on a macroscopic or classical scale, energies are typically on the order of joules. Even if macroscopic energies are quantized, the quantum steps are too small to be noticed. This is an example of the correspondence principle. For a large object, quantum mechanics produces results indistinguishable from those of classical physics.

Atomic Spectra

Now let us turn our attention to the *emission and absorption of EM radiation by gases*. The Sun is the most common example of a body containing gases emitting an EM spectrum that includes visible light. We also see examples in neon signs and candle flames. Studies of emissions of hot gases began more than two centuries ago, and it was soon recognized that these emission spectra contained huge amounts of information. The type of gas and its temperature, for example, could be determined. We now know that these EM emissions come from electrons transitioning between energy levels in individual atoms and molecules; thus, they are called **atomic spectra**. Atomic spectra remain an important analytical tool today. **Figure 29.5** shows an example of an emission spectrum obtained by passing an electric discharge through a material. One of the most important characteristics of these spectra is that they are discrete. By this we mean that only certain wavelengths, and hence frequencies, are emitted. This is called a line spectrum. If frequency and energy are associated as $\Delta E = hf$, the energies of the electrons in the emitting atoms and molecules are quantized. This is discussed in more detail later in this chapter.

Figure 29.5 Emission spectrum of oxygen. When an electrical discharge is passed through a substance, its atoms and molecules absorb energy, which is reemitted as EM radiation. The discrete nature of these emissions implies that the energy states of the atoms and molecules are quantized. Such atomic spectra were used as analytical tools for many decades before it was understood why they are quantized. (credit: Teravolt, Wikimedia Commons)

It was a major puzzle that atomic spectra are quantized. Some of the best minds of 19th-century science failed to explain why this might be. Not until the second decade of the 20th century did an answer based on quantum mechanics begin to emerge. Again a macroscopic or classical body of gas was involved in the studies, but the effect, as we shall see, is due to individual atoms and molecules.

PhET Explorations: Models of the Hydrogen Atom

How did scientists figure out the structure of atoms without looking at them? Try out different models by shooting light at the atom. Check how the prediction of the model matches the experimental results.

Figure 29.6 Models of the Hydrogen Atom (http://cnx.org/content/m55023/1.2/hydrogen-atom_en.jar)

29.2 The Photoelectric Effect

When light strikes materials, it can eject electrons from them. This is called the **photoelectric effect**, meaning that light (*photo*) produces electricity. One common use of the photoelectric effect is in light meters, such as those that adjust the automatic iris on various types of cameras. In a similar way, another use is in solar cells, as you probably have in your calculator or have seen on a roof top or a roadside sign. These make use of the photoelectric effect to convert light into electricity for running different devices.

Figure 29.7 The photoelectric effect can be observed by allowing light to fall on the metal plate in this evacuated tube. Electrons ejected by the light are collected on the collector wire and measured as a current. A retarding voltage between the collector wire and plate can then be adjusted so as to determine the energy of the ejected electrons. For example, if it is sufficiently negative, no electrons will reach the wire. (credit: P.P. Urone)

This effect has been known for more than a century and can be studied using a device such as that shown in **Figure 29.7**. This figure shows an evacuated tube with a metal plate and a collector wire that are connected by a variable voltage source, with the collector more negative than the plate. When light (or other EM radiation) strikes the plate in the evacuated tube, it may eject electrons. If the electrons have energy in electron volts (eV) greater than the potential difference between the plate and the wire in volts, some electrons will be collected on the wire. Since the electron energy in eV is qV, where q is the electron charge and V is the potential difference, the electron energy can be measured by adjusting the retarding voltage between the wire and the plate. The voltage that stops the electrons from reaching the wire equals the energy in eV. For example, if -3.00 V barely stops the electrons, their energy is 3.00 eV. The number of electrons ejected can be determined by measuring the current between the wire and plate. The more light, the more electrons; a little circuitry allows this device to be used as a light meter.

What is really important about the photoelectric effect is what Albert Einstein deduced from it. Einstein realized that there were several characteristics of the photoelectric effect that could be explained only if *EM radiation is itself quantized*: the apparently continuous stream of energy in an EM wave is actually composed of energy quanta called photons. In his explanation of the photoelectric effect, Einstein defined a quantized unit or quantum of EM energy, which we now call a **photon**, with an energy proportional to the frequency of EM radiation. In equation form, the **photon energy** is

$$E = hf, \tag{29.4}$$

where E is the energy of a photon of frequency f and h is Planck's constant. This revolutionary idea looks similar to Planck's quantization of energy states in blackbody oscillators, but it is quite different. It is the quantization of EM radiation itself. EM waves are composed of photons and are not continuous smooth waves as described in previous chapters on optics. Their energy is absorbed and emitted in lumps, not continuously. This is exactly consistent with Planck's quantization of energy levels in blackbody oscillators, since these oscillators increase and decrease their energy in steps of hf by absorbing and emitting photons having $E = hf$. We do not observe this with our eyes, because there are so many photons in common light sources that individual photons go unnoticed. (See **Figure 29.8**.) The next section of the text (**Photon Energies and the Electromagnetic Spectrum**) is devoted to a discussion of photons and some of their characteristics and implications. For now, we will use the photon concept to explain the photoelectric effect, much as Einstein did.

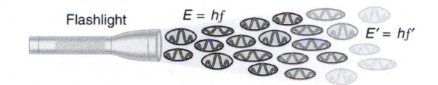

Figure 29.8 An EM wave of frequency f is composed of photons, or individual quanta of EM radiation. The energy of each photon is $E = hf$, where h is Planck's constant and f is the frequency of the EM radiation. Higher intensity means more photons per unit area. The flashlight emits large numbers of photons of many different frequencies, hence others have energy $E' = hf'$, and so on.

The photoelectric effect has the properties discussed below. All these properties are consistent with the idea that individual photons of EM radiation are absorbed by individual electrons in a material, with the electron gaining the photon's energy. Some of these properties are inconsistent with the idea that EM radiation is a simple wave. For simplicity, let us consider what happens with monochromatic EM radiation in which all photons have the same energy hf.

1. If we vary the frequency of the EM radiation falling on a material, we find the following: For a given material, there is a threshold frequency f_0 for the EM radiation below which no electrons are ejected, regardless of intensity. Individual photons interact with individual electrons. Thus if the photon energy is too small to break an electron away, no electrons will be ejected. If EM radiation was a simple wave, sufficient energy could be obtained by increasing the intensity.

2. *Once EM radiation falls on a material, electrons are ejected without delay.* As soon as an individual photon of a sufficiently high frequency is absorbed by an individual electron, the electron is ejected. If the EM radiation were a simple wave, several minutes would be required for sufficient energy to be deposited to the metal surface to eject an electron.

3. The number of electrons ejected per unit time is proportional to the intensity of the EM radiation and to no other characteristic. High-intensity EM radiation consists of large numbers of photons per unit area, with all photons having the same characteristic energy hf.

4. If we vary the intensity of the EM radiation and measure the energy of ejected electrons, we find the following: *The maximum kinetic energy of ejected electrons is independent of the intensity of the EM radiation.* Since there are so many electrons in a material, it is extremely unlikely that two photons will interact with the same electron at the same time, thereby increasing the energy given it. Instead (as noted in 3 above), increased intensity results in more electrons of the same energy being ejected. If EM radiation were a simple wave, a higher intensity could give more energy, and higher-energy electrons would be ejected.

5. The kinetic energy of an ejected electron equals the photon energy minus the binding energy of the electron in the specific material. An individual photon can give all of its energy to an electron. The photon's energy is partly used to break the electron away from the material. The remainder goes into the ejected electron's kinetic energy. In equation form, this is given by

$$\mathrm{KE}_e = hf - \mathrm{BE},\qquad(29.5)$$

where KE_e is the maximum kinetic energy of the ejected electron, hf is the photon's energy, and BE is the **binding energy** of the electron to the particular material. (BE is sometimes called the *work function* of the material.) This equation, due to Einstein in 1905, explains the properties of the photoelectric effect quantitatively. An individual photon of EM radiation (it does not come any other way) interacts with an individual electron, supplying enough energy, BE, to break it away, with the remainder going to kinetic energy. The binding energy is $\mathrm{BE} = hf_0$, where f_0 is the threshold frequency for the particular material. **Figure 29.9** shows a graph of maximum KE_e versus the frequency of incident EM radiation falling on a particular material.

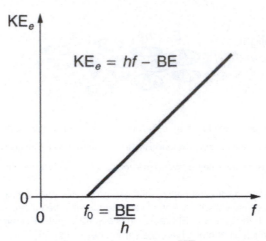

Figure 29.9 Photoelectric effect. A graph of the kinetic energy of an ejected electron, KE_e, versus the frequency of EM radiation impinging on a certain material. There is a threshold frequency below which no electrons are ejected, because the individual photon interacting with an individual electron has insufficient energy to break it away. Above the threshold energy, KE_e increases linearly with f, consistent with $\mathrm{KE}_e = hf - \mathrm{BE}$.

The slope of this line is h —the data can be used to determine Planck's constant experimentally. Einstein gave the first successful explanation of such data by proposing the idea of photons—quanta of EM radiation.

Einstein's idea that EM radiation is quantized was crucial to the beginnings of quantum mechanics. It is a far more general concept than its explanation of the photoelectric effect might imply. All EM radiation can also be modeled in the form of photons, and the characteristics of EM radiation are entirely consistent with this fact. (As we will see in the next section, many aspects of EM radiation, such as the hazards of ultraviolet (UV) radiation, can be explained *only* by photon properties.) More famous for modern relativity, Einstein planted an important seed for quantum mechanics in 1905, the same year he published his first paper on special relativity. His explanation of the photoelectric effect was the basis for the Nobel Prize awarded to him in 1921. Although his other contributions to theoretical physics were also noted in that award, special and general relativity were not fully recognized in spite of having been partially verified by experiment by 1921. Although hero-worshipped, this great man never received Nobel recognition for his most famous work—relativity.

Example 29.1 Calculating Photon Energy and the Photoelectric Effect: A Violet Light

(a) What is the energy in joules and electron volts of a photon of 420-nm violet light? (b) What is the maximum kinetic energy of electrons ejected from calcium by 420-nm violet light, given that the binding energy (or work function) of electrons for calcium metal is 2.71 eV?

Strategy

To solve part (a), note that the energy of a photon is given by $E = hf$. For part (b), once the energy of the photon is calculated, it is a straightforward application of $\mathrm{KE}_e = hf - \mathrm{BE}$ to find the ejected electron's maximum kinetic energy, since BE is given.

Solution for (a)

Photon energy is given by

$$E = hf \tag{29.6}$$

Since we are given the wavelength rather than the frequency, we solve the familiar relationship $c = f\lambda$ for the frequency, yielding

$$f = \frac{c}{\lambda}. \tag{29.7}$$

Combining these two equations gives the useful relationship

$$E = \frac{hc}{\lambda}. \tag{29.8}$$

Now substituting known values yields

$$E = \frac{\left(6.63\times10^{-34}\ \mathrm{J\cdot s}\right)\left(3.00\times10^8\ \mathrm{m/s}\right)}{420\times10^{-9}\ \mathrm{m}} = 4.74\times10^{-19}\ \mathrm{J}. \tag{29.9}$$

Converting to eV, the energy of the photon is

$$E = \left(4.74 \times 10^{-19} \text{ J}\right) \frac{1 \text{ eV}}{1.6 \times 10^{-19} \text{ J}} = 2.96 \text{ eV}. \qquad (29.10)$$

Solution for (b)

Finding the kinetic energy of the ejected electron is now a simple application of the equation $\text{KE}_e = hf - \text{BE}$. Substituting the photon energy and binding energy yields

$$\text{KE}_e = hf - \text{BE} = 2.96 \text{ eV} - 2.71 \text{ eV} = 0.246 \text{ eV}. \qquad (29.11)$$

Discussion

The energy of this 420-nm photon of violet light is a tiny fraction of a joule, and so it is no wonder that a single photon would be difficult for us to sense directly—humans are more attuned to energies on the order of joules. But looking at the energy in electron volts, we can see that this photon has enough energy to affect atoms and molecules. A DNA molecule can be broken with about 1 eV of energy, for example, and typical atomic and molecular energies are on the order of eV, so that the UV photon in this example could have biological effects. The ejected electron (called a *photoelectron*) has a rather low energy, and it would not travel far, except in a vacuum. The electron would be stopped by a retarding potential of but 0.26 eV. In fact, if the photon wavelength were longer and its energy less than 2.71 eV, then the formula would give a negative kinetic energy, an impossibility. This simply means that the 420-nm photons with their 2.96-eV energy are not much above the frequency threshold. You can show for yourself that the threshold wavelength is 459 nm (blue light). This means that if calcium metal is used in a light meter, the meter will be insensitive to wavelengths longer than those of blue light. Such a light meter would be completely insensitive to red light, for example.

PhET Explorations: Photoelectric Effect

See how light knocks electrons off a metal target, and recreate the experiment that spawned the field of quantum mechanics.

Figure 29.10 Photoelectric Effect (http://cnx.org/content/m55064/1.2/photoelectric_en.jar)

29.3 Photon Energies and the Electromagnetic Spectrum

Learning Objectives
By the end of this section, you will be able to: • Explain the relationship between the energy of a photon in joules or electron volts and its wavelength or frequency. • Calculate the number of photons per second emitted by a monochromatic source of specific wavelength and power. The information presented in this section supports the following AP® learning objectives and science practices: • **6.F.3.1** The student is able to support the photon model of radiant energy with evidence provided by the photoelectric effect. **(S.P. 6.4)**

Ionizing Radiation

A photon is a quantum of EM radiation. Its energy is given by $E = hf$ and is related to the frequency f and wavelength λ of the radiation by

$$E = hf = \frac{hc}{\lambda} \text{(energy of a photon)}, \qquad (29.12)$$

where E is the energy of a single photon and c is the speed of light. When working with small systems, energy in eV is often useful. Note that Planck's constant in these units is

$$h = 4.14 \times 10^{-15} \text{ eV} \cdot \text{s}. \qquad (29.13)$$

Since many wavelengths are stated in nanometers (nm), it is also useful to know that

$$hc = 1240 \text{ eV} \cdot \text{nm}. \qquad (29.14)$$

These will make many calculations a little easier.

All EM radiation is composed of photons. **Figure 29.11** shows various divisions of the EM spectrum plotted against wavelength, frequency, and photon energy. Previously in this book, photon characteristics were alluded to in the discussion of some of the characteristics of UV, x rays, and γ rays, the first of which start with frequencies just above violet in the visible spectrum. It was noted that these types of EM radiation have characteristics much different than visible light. We can now see that such properties arise because photon energy is larger at high frequencies.

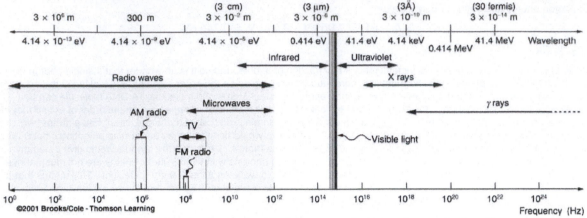

Figure 29.11 The EM spectrum, showing major categories as a function of photon energy in eV, as well as wavelength and frequency. Certain characteristics of EM radiation are directly attributable to photon energy alone.

Table 29.1 Representative Energies for Submicroscopic Effects (Order of Magnitude Only)

Rotational energies of molecules	10^{-5} eV
Vibrational energies of molecules	0.1 eV
Energy between outer electron shells in atoms	1 eV
Binding energy of a weakly bound molecule	1 eV
Energy of red light	2 eV
Binding energy of a tightly bound molecule	10 eV
Energy to ionize atom or molecule	10 to 1000 eV

Photons act as individual quanta and interact with individual electrons, atoms, molecules, and so on. The energy a photon carries is, thus, crucial to the effects it has. **Table 29.1** lists representative submicroscopic energies in eV. When we compare photon energies from the EM spectrum in **Figure 29.11** with energies in the table, we can see how effects vary with the type of EM radiation.

Gamma rays, a form of nuclear and cosmic EM radiation, can have the highest frequencies and, hence, the highest photon energies in the EM spectrum. For example, a γ-ray photon with $f = 10^{21}$ Hz has an energy

$E = hf = 6.63 \times 10^{-13}$ J $= 4.14$ MeV. This is sufficient energy to ionize thousands of atoms and molecules, since only 10 to 1000 eV are needed per ionization. In fact, γ rays are one type of **ionizing radiation**, as are x rays and UV, because they produce ionization in materials that absorb them. Because so much ionization can be produced, a single γ-ray photon can cause significant damage to biological tissue, killing cells or damaging their ability to properly reproduce. When cell reproduction is disrupted, the result can be cancer, one of the known effects of exposure to ionizing radiation. Since cancer cells are rapidly reproducing, they are exceptionally sensitive to the disruption produced by ionizing radiation. This means that ionizing radiation has positive uses in cancer treatment as well as risks in producing cancer.

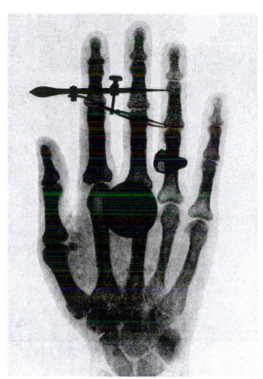

Figure 29.12 One of the first x-ray images, taken by Röentgen himself. The hand belongs to Bertha Röentgen, his wife. (credit: Wilhelm Conrad Röntgen, via Wikimedia Commons)

High photon energy also enables γ rays to penetrate materials, since a collision with a single atom or molecule is unlikely to absorb all the γ ray's energy. This can make γ rays useful as a probe, and they are sometimes used in medical imaging. **x rays**, as you can see in **Figure 29.11**, overlap with the low-frequency end of the γ ray range. Since x rays have energies of keV and up, individual x-ray photons also can produce large amounts of ionization. At lower photon energies, x rays are not as penetrating as γ rays and are slightly less hazardous. X rays are ideal for medical imaging, their most common use, and a fact that was recognized immediately upon their discovery in 1895 by the German physicist W. C. Roentgen (1845–1923). (See **Figure 29.12**.) Within one year of their discovery, x rays (for a time called Roentgen rays) were used for medical diagnostics. Roentgen received the 1901 Nobel Prize for the discovery of x rays.

Connections: Conservation of Energy

Once again, we find that conservation of energy allows us to consider the initial and final forms that energy takes, without having to make detailed calculations of the intermediate steps. **Example 29.2** is solved by considering only the initial and final forms of energy.

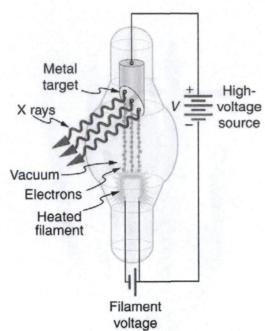

Figure 29.13 X rays are produced when energetic electrons strike the copper anode of this cathode ray tube (CRT). Electrons (shown here as separate particles) interact individually with the material they strike, sometimes producing photons of EM radiation.

While γ rays originate in nuclear decay, x rays are produced by the process shown in **Figure 29.13**. Electrons ejected by thermal agitation from a hot filament in a vacuum tube are accelerated through a high voltage, gaining kinetic energy from the electrical potential energy. When they strike the anode, the electrons convert their kinetic energy to a variety of forms, including thermal energy. But since an accelerated charge radiates EM waves, and since the electrons act individually, photons are also produced. Some of these x-ray photons obtain the kinetic energy of the electron. The accelerated electrons originate at the cathode, so such a tube is called a cathode ray tube (CRT), and various versions of them are found in older TV and computer screens as well as in x-ray machines.

Example 29.2 X-ray Photon Energy and X-ray Tube Voltage

Find the maximum energy in eV of an x-ray photon produced by electrons accelerated through a potential difference of 50.0 kV in a CRT like the one in **Figure 29.13**.

Strategy

Electrons can give all of their kinetic energy to a single photon when they strike the anode of a CRT. (This is something like the photoelectric effect in reverse.) The kinetic energy of the electron comes from electrical potential energy. Thus we can simply equate the maximum photon energy to the electrical potential energy—that is, $hf = qV$. (We do not have to calculate each step from beginning to end if we know that all of the starting energy qV is converted to the final form hf.)

Solution

The maximum photon energy is $hf = qV$, where q is the charge of the electron and V is the accelerating voltage. Thus,

$$hf = (1.60 \times 10^{-19} \text{ C})(50.0 \times 10^3 \text{ V}).$$

(29.15)

From the definition of the electron volt, we know $1 \text{ eV} = 1.60 \times 10^{-19} \text{ J}$, where $1 \text{ J} = 1 \text{ C} \cdot \text{V}$. Gathering factors and converting energy to eV yields

$$hf = (50.0 \times 10^3)(1.60 \times 10^{-19} \text{ C} \cdot \text{V})\left(\frac{1 \text{ eV}}{1.60 \times 10^{-19} \text{ C} \cdot \text{V}}\right) = (50.0 \times 10^3)(1 \text{ eV}) = 50.0 \text{ keV}.$$

(29.16)

Discussion

This example produces a result that can be applied to many similar situations. If you accelerate a single elementary charge, like that of an electron, through a potential given in volts, then its energy in eV has the same numerical value. Thus a 50.0-kV potential generates 50.0 keV electrons, which in turn can produce photons with a maximum energy of 50 keV. Similarly, a 100-kV potential in an x-ray tube can generate up to 100-keV x-ray photons. Many x-ray tubes have adjustable voltages so that various energy x rays with differing energies, and therefore differing abilities to penetrate, can be generated.

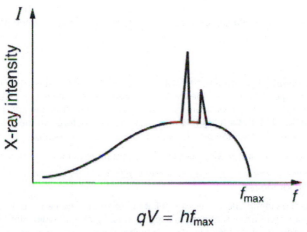

Figure 29.14 X-ray spectrum obtained when energetic electrons strike a material. The smooth part of the spectrum is bremsstrahlung, while the peaks are characteristic of the anode material. Both are atomic processes that produce energetic photons known as x-ray photons.

Figure 29.14 shows the spectrum of x rays obtained from an x-ray tube. There are two distinct features to the spectrum. First, the smooth distribution results from electrons being decelerated in the anode material. A curve like this is obtained by detecting many photons, and it is apparent that the maximum energy is unlikely. This decelerating process produces radiation that is called **bremsstrahlung** (German for *braking radiation*). The second feature is the existence of sharp peaks in the spectrum; these are called **characteristic x rays**, since they are characteristic of the anode material. Characteristic x rays come from atomic excitations unique to a given type of anode material. They are akin to lines in atomic spectra, implying the energy levels of atoms are quantized. Phenomena such as discrete atomic spectra and characteristic x rays are explored further in **Atomic Physics**.

Ultraviolet radiation (approximately 4 eV to 300 eV) overlaps with the low end of the energy range of x rays, but UV is typically lower in energy. UV comes from the de-excitation of atoms that may be part of a hot solid or gas. These atoms can be given energy that they later release as UV by numerous processes, including electric discharge, nuclear explosion, thermal agitation, and exposure to x rays. A UV photon has sufficient energy to ionize atoms and molecules, which makes its effects different from those of visible light. UV thus has some of the same biological effects as γ rays and x rays. For example, it can cause skin cancer and is used as a sterilizer. The major difference is that several UV photons are required to disrupt cell reproduction or kill a bacterium, whereas single γ-ray and X-ray photons can do the same damage. But since UV does have the energy to alter molecules, it can do what visible light cannot. One of the beneficial aspects of UV is that it triggers the production of vitamin D in the skin, whereas visible light has insufficient energy per photon to alter the molecules that trigger this production. Infantile jaundice is treated by exposing the baby to UV (with eye protection), called phototherapy, the beneficial effects of which are thought to be related to its ability to help prevent the buildup of potentially toxic bilirubin in the blood.

Example 29.3 Photon Energy and Effects for UV

Short-wavelength UV is sometimes called vacuum UV, because it is strongly absorbed by air and must be studied in a vacuum. Calculate the photon energy in eV for 100-nm vacuum UV, and estimate the number of molecules it could ionize or break apart.

Strategy

Using the equation $E = hf$ and appropriate constants, we can find the photon energy and compare it with energy information in **Table 29.1**.

Solution

The energy of a photon is given by

$$E = hf = \frac{hc}{\lambda}.$$ (29.17)

Using $hc = 1240 \text{ eV} \cdot \text{nm}$, we find that

$$E = \frac{hc}{\lambda} = \frac{1240 \text{ eV} \cdot \text{nm}}{100 \text{ nm}} = 12.4 \text{ eV}.$$ (29.18)

Discussion

According to **Table 29.1**, this photon energy might be able to ionize an atom or molecule, and it is about what is needed to break up a tightly bound molecule, since they are bound by approximately 10 eV. This photon energy could destroy about a dozen weakly bound molecules. Because of its high photon energy, UV disrupts atoms and molecules it interacts with. One good consequence is that all but the longest-wavelength UV is strongly absorbed and is easily blocked by sunglasses. In fact, most of the Sun's UV is absorbed by a thin layer of ozone in the upper atmosphere, protecting sensitive organisms on

Earth. Damage to our ozone layer by the addition of such chemicals as CFC's has reduced this protection for us.

Visible Light

The range of photon energies for **visible light** from red to violet is 1.63 to 3.26 eV, respectively (left for this chapter's Problems and Exercises to verify). These energies are on the order of those between outer electron shells in atoms and molecules. This means that these photons can be absorbed by atoms and molecules. A *single* photon can actually stimulate the retina, for example, by altering a receptor molecule that then triggers a nerve impulse. Photons can be absorbed or emitted only by atoms and molecules that have precisely the correct quantized energy step to do so. For example, if a red photon of frequency f encounters a molecule that has an energy step, $\Delta E,$ equal to $hf,$ then the photon can be absorbed. Violet flowers absorb red and reflect violet; this implies there is no energy step between levels in the receptor molecule equal to the violet photon's energy, but there is an energy step for the red.

There are some noticeable differences in the characteristics of light between the two ends of the visible spectrum that are due to photon energies. Red light has insufficient photon energy to expose most black-and-white film, and it is thus used to illuminate darkrooms where such film is developed. Since violet light has a higher photon energy, dyes that absorb violet tend to fade more quickly than those that do not. (See **Figure 29.15**.) Take a look at some faded color posters in a storefront some time, and you will notice that the blues and violets are the last to fade. This is because other dyes, such as red and green dyes, absorb blue and violet photons, the higher energies of which break up their weakly bound molecules. (Complex molecules such as those in dyes and DNA tend to be weakly bound.) Blue and violet dyes reflect those colors and, therefore, do not absorb these more energetic photons, thus suffering less molecular damage.

Figure 29.15 Why do the reds, yellows, and greens fade before the blues and violets when exposed to the Sun, as with this poster? The answer is related to photon energy. (credit: Deb Collins, Flickr)

Transparent materials, such as some glasses, do not absorb any visible light, because there is no energy step in the atoms or molecules that could absorb the light. Since individual photons interact with individual atoms, it is nearly impossible to have two photons absorbed simultaneously to reach a large energy step. Because of its lower photon energy, visible light can sometimes pass through many kilometers of a substance, while higher frequencies like UV, x ray, and γ rays are absorbed, because they have sufficient photon energy to ionize the material.

Example 29.4 How Many Photons per Second Does a Typical Light Bulb Produce?

Assuming that 10.0% of a 100-W light bulb's energy output is in the visible range (typical for incandescent bulbs) with an average wavelength of 580 nm, calculate the number of visible photons emitted per second.

Strategy

Power is energy per unit time, and so if we can find the energy per photon, we can determine the number of photons per second. This will best be done in joules, since power is given in watts, which are joules per second.

Solution

The power in visible light production is 10.0% of 100 W, or 10.0 J/s. The energy of the average visible photon is found by substituting the given average wavelength into the formula

$$E = \frac{hc}{\lambda}. \tag{29.19}$$

This produces

$$E = \frac{(6.63 \times 10^{-34} \text{ J} \cdot \text{s})(3.00 \times 10^8 \text{ m/s})}{580 \times 10^{-9} \text{ m}} = 3.43 \times 10^{-19} \text{ J}. \tag{29.20}$$

The number of visible photons per second is thus

$$\text{photon/s} = \frac{10.0 \text{ J/s}}{3.43 \times 10^{-19} \text{ J/photon}} = 2.92 \times 10^{19} \text{ photon/s}. \tag{29.21}$$

Discussion

This incredible number of photons per second is verification that individual photons are insignificant in ordinary human experience. It is also a verification of the correspondence principle—on the macroscopic scale, quantization becomes essentially continuous or classical. Finally, there are so many photons emitted by a 100-W lightbulb that it can be seen by the unaided eye many kilometers away.

Lower-Energy Photons

Infrared radiation (IR) has even lower photon energies than visible light and cannot significantly alter atoms and molecules. IR can be absorbed and emitted by atoms and molecules, particularly between closely spaced states. IR is extremely strongly absorbed by water, for example, because water molecules have many states separated by energies on the order of 10^{-5} eV to 10^{-2} eV, well within the IR and microwave energy ranges. This is why in the IR range, skin is almost jet black, with an emissivity near 1—there are many states in water molecules in the skin that can absorb a large range of IR photon energies. Not all molecules have this property. Air, for example, is nearly transparent to many IR frequencies.

Microwaves are the highest frequencies that can be produced by electronic circuits, although they are also produced naturally. Thus microwaves are similar to IR but do not extend to as high frequencies. There are states in water and other molecules that have the same frequency and energy as microwaves, typically about 10^{-5} eV. This is one reason why food absorbs microwaves more strongly than many other materials, making microwave ovens an efficient way of putting energy directly into food.

Photon energies for both IR and microwaves are so low that huge numbers of photons are involved in any significant energy transfer by IR or microwaves (such as warming yourself with a heat lamp or cooking pizza in the microwave). Visible light, IR, microwaves, and all lower frequencies cannot produce ionization with single photons and do not ordinarily have the hazards of higher frequencies. When visible, IR, or microwave radiation *is* hazardous, such as the inducement of cataracts by microwaves, the hazard is due to huge numbers of photons acting together (not to an accumulation of photons, such as sterilization by weak UV). The negative effects of visible, IR, or microwave radiation can be thermal effects, which could be produced by any heat source. But one difference is that at very high intensity, strong electric and magnetic fields can be produced by photons acting together. Such electromagnetic fields (EMF) can actually ionize materials.

Misconception Alert: High-Voltage Power Lines

Although some people think that living near high-voltage power lines is hazardous to one's health, ongoing studies of the transient field effects produced by these lines show their strengths to be insufficient to cause damage. Demographic studies also fail to show significant correlation of ill effects with high-voltage power lines. The American Physical Society issued a report over 10 years ago on power-line fields, which concluded that the scientific literature and reviews of panels show no consistent, significant link between cancer and power-line fields. They also felt that the "diversion of resources to eliminate a threat which has no persuasive scientific basis is disturbing."

It is virtually impossible to detect individual photons having frequencies below microwave frequencies, because of their low photon energy. But the photons are there. A continuous EM wave can be modeled as photons. At low frequencies, EM waves are generally treated as time- and position-varying electric and magnetic fields with no discernible quantization. This is another example of the correspondence principle in situations involving huge numbers of photons.

PhET Explorations: Color Vision

Make a whole rainbow by mixing red, green, and blue light. Change the wavelength of a monochromatic beam or filter white light. View the light as a solid beam, or see the individual photons.

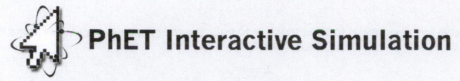

Figure 29.16 Color Vision (http://cnx.org/content/m55036/1.3/color-vision_en.jar)

29.4 Photon Momentum

Learning Objectives

By the end of this section, you will be able to:

- Relate the linear momentum of a photon to its energy or wavelength, and apply linear momentum conservation to simple processes involving the emission, absorption, or reflection of photons.
- Account qualitatively for the increase of photon wavelength that is observed, and explain the significance of the Compton wavelength.

The information presented in this section supports the following AP® learning objectives and science practices:

- **5.D.1.6** The student is able to make predictions of the dynamical properties of a system undergoing a collision by application of the principle of linear momentum conservation and the principle of the conservation of energy in situations in which an elastic collision may also be assumed. **(S.P. 6.4)**
- **5.D.1.7** The student is able to classify a given collision situation as elastic or inelastic, justify the selection of conservation of linear momentum and restoration of kinetic energy as the appropriate principles for analyzing an elastic collision, solve for missing variables, and calculate their values. **(S.P. 2.1, 2.2)**

Measuring Photon Momentum

The quantum of EM radiation we call a **photon** has properties analogous to those of particles we can see, such as grains of sand. A photon interacts as a unit in collisions or when absorbed, rather than as an extensive wave. Massive quanta, like electrons, also act like macroscopic particles—something we expect, because they are the smallest units of matter. Particles carry momentum as well as energy. Despite photons having no mass, there has long been evidence that EM radiation carries momentum. (Maxwell and others who studied EM waves predicted that they would carry momentum.) It is now a well-established fact that photons *do* have momentum. In fact, photon momentum is suggested by the photoelectric effect, where photons knock electrons out of a substance. **Figure 29.17** shows macroscopic evidence of photon momentum.

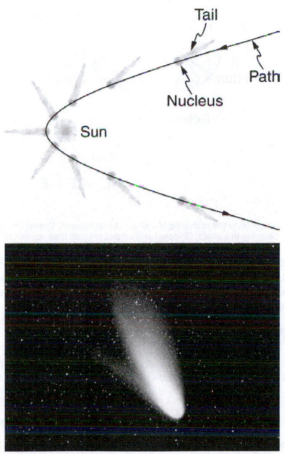

Figure 29.17 The tails of the Hale-Bopp comet point away from the Sun, evidence that light has momentum. Dust emanating from the body of the comet forms this tail. Particles of dust are pushed away from the Sun by light reflecting from them. The blue ionized gas tail is also produced by photons interacting with atoms in the comet material. (credit: Geoff Chester, U.S. Navy, via Wikimedia Commons)

Figure 29.17 shows a comet with two prominent tails. What most people do not know about the tails is that they always point *away* from the Sun rather than trailing behind the comet (like the tail of Bo Peep's sheep). Comet tails are composed of gases and dust evaporated from the body of the comet and ionized gas. The dust particles recoil away from the Sun when photons scatter from them. Evidently, photons carry momentum in the direction of their motion (away from the Sun), and some of this momentum is transferred to dust particles in collisions. Gas atoms and molecules in the blue tail are most affected by other particles of radiation, such as protons and electrons emanating from the Sun, rather than by the momentum of photons.

Connections: Conservation of Momentum

Not only is momentum conserved in all realms of physics, but all types of particles are found to have momentum. We expect particles with mass to have momentum, but now we see that massless particles including photons also carry momentum.

Momentum is conserved in quantum mechanics just as it is in relativity and classical physics. Some of the earliest direct experimental evidence of this came from scattering of x-ray photons by electrons in substances, named Compton scattering after the American physicist Arthur H. Compton (1892–1962). Around 1923, Compton observed that x rays scattered from materials had a decreased energy and correctly analyzed this as being due to the scattering of photons from electrons. This phenomenon could be handled as a collision between two particles—a photon and an electron at rest in the material. Energy and momentum are conserved in the collision. (See **Figure 29.18**) He won a Nobel Prize in 1929 for the discovery of this scattering, now called the **Compton effect**, because it helped prove that **photon momentum** is given by

$$p = \frac{h}{\lambda},$$ (29.22)

where h is Planck's constant and λ is the photon wavelength. (Note that relativistic momentum given as $p = \gamma mu$ is valid only for particles having mass.)

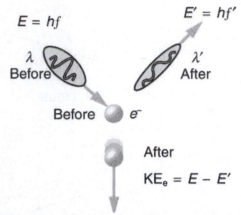

Figure 29.18 The Compton effect is the name given to the scattering of a photon by an electron. Energy and momentum are conserved, resulting in a reduction of both for the scattered photon. Studying this effect, Compton verified that photons have momentum.

We can see that photon momentum is small, since $p = h / \lambda$ and h is very small. It is for this reason that we do not ordinarily observe photon momentum. Our mirrors do not recoil when light reflects from them (except perhaps in cartoons). Compton saw the effects of photon momentum because he was observing x rays, which have a small wavelength and a relatively large momentum, interacting with the lightest of particles, the electron.

Example 29.5 Electron and Photon Momentum Compared

(a) Calculate the momentum of a visible photon that has a wavelength of 500 nm. (b) Find the velocity of an electron having the same momentum. (c) What is the energy of the electron, and how does it compare with the energy of the photon?

Strategy

Finding the photon momentum is a straightforward application of its definition: $p = \frac{h}{\lambda}$. If we find the photon momentum is small, then we can assume that an electron with the same momentum will be nonrelativistic, making it easy to find its velocity and kinetic energy from the classical formulas.

Solution for (a)

Photon momentum is given by the equation:

$$p = \frac{h}{\lambda}.$$

(29.23)

Entering the given photon wavelength yields

$$p = \frac{6.63 \times 10^{-34} \text{ J} \cdot \text{s}}{500 \times 10^{-9} \text{ m}} = 1.33 \times 10^{-27} \text{ kg} \cdot \text{m/s}.$$

(29.24)

Solution for (b)

Since this momentum is indeed small, we will use the classical expression $p = mv$ to find the velocity of an electron with this momentum. Solving for v and using the known value for the mass of an electron gives

$$v = \frac{p}{m} = \frac{1.33 \times 10^{-27} \text{ kg} \cdot \text{m/s}}{9.11 \times 10^{-31} \text{ kg}} = 1460 \text{ m/s} \approx 1460 \text{ m/s}.$$

(29.25)

Solution for (c)

The electron has kinetic energy, which is classically given by

$$KE_e = \frac{1}{2} m v^2.$$

(29.26)

Thus,

$$KE_e = \frac{1}{2}(9.11 \times 10^{-3} \text{ kg})(1455 \text{ m/s})^2 = 9.64 \times 10^{-25} \text{ J}.$$

(29.27)

Converting this to eV by multiplying by $(1 \text{ eV}) / (1.602 \times 10^{-19} \text{ J})$ yields

$$KE_e = 6.02 \times 10^{-6} \text{ eV}.$$

(29.28)

The photon energy E is

$$E = \frac{hc}{\lambda} = \frac{1240 \text{ eV} \cdot \text{nm}}{500 \text{ nm}} = 2.48 \text{ eV}, \quad\quad (29.29)$$

which is about five orders of magnitude greater.

Discussion

Photon momentum is indeed small. Even if we have huge numbers of them, the total momentum they carry is small. An electron with the same momentum has a 1460 m/s velocity, which is clearly nonrelativistic. A more massive particle with the same momentum would have an even smaller velocity. This is borne out by the fact that it takes far less energy to give an electron the same momentum as a photon. But on a quantum-mechanical scale, especially for high-energy photons interacting with small masses, photon momentum is significant. Even on a large scale, photon momentum can have an effect if there are enough of them and if there is nothing to prevent the slow recoil of matter. Comet tails are one example, but there are also proposals to build space sails that use huge low-mass mirrors (made of aluminized Mylar) to reflect sunlight. In the vacuum of space, the mirrors would gradually recoil and could actually take spacecraft from place to place in the solar system. (See **Figure 29.19**.)

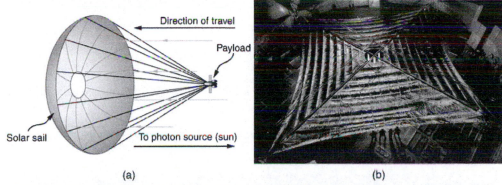

(a)   (b)

Figure 29.19 (a) Space sails have been proposed that use the momentum of sunlight reflecting from gigantic low-mass sails to propel spacecraft about the solar system. A Russian test model of this (the Cosmos 1) was launched in 2005, but did not make it into orbit due to a rocket failure. (b) A U.S. version of this, labeled LightSail-1, is scheduled for trial launches in the first part of this decade. It will have a 40-m^2 sail. (credit: Kim Newton/NASA)

Relativistic Photon Momentum

There is a relationship between photon momentum p and photon energy E that is consistent with the relation given previously for the relativistic total energy of a particle as $E^2 = (pc)^2 + (mc)^2$. We know m is zero for a photon, but p is not, so that $E^2 = (pc)^2 + (mc)^2$ becomes

$$E = pc, \quad\quad (29.30)$$

or

$$p = \frac{E}{c} \text{(photons)}. \quad\quad (29.31)$$

To check the validity of this relation, note that $E = hc / \lambda$ for a photon. Substituting this into $p = E/c$ yields

$$p = (hc / \lambda) / c = \frac{h}{\lambda}, \quad\quad (29.32)$$

as determined experimentally and discussed above. Thus, $p = E/c$ is equivalent to Compton's result $p = h / \lambda$. For a further verification of the relationship between photon energy and momentum, see **Example 29.6**.

Photon Detectors

Almost all detection systems talked about thus far—eyes, photographic plates, photomultiplier tubes in microscopes, and CCD cameras—rely on particle-like properties of photons interacting with a sensitive area. A change is caused and either the change is cascaded or zillions of points are recorded to form an image we detect. These detectors are used in biomedical imaging systems, and there is ongoing research into improving the efficiency of receiving photons, particularly by cooling detection systems and reducing thermal effects.

Example 29.6 Photon Energy and Momentum

Show that $p = E/c$ for the photon considered in the **Example 29.5**.

Strategy

We will take the energy E found in **Example 29.5**, divide it by the speed of light, and see if the same momentum is obtained as before.

Solution

Given that the energy of the photon is 2.48 eV and converting this to joules, we get

$$p = \frac{E}{c} = \frac{(2.48 \text{ eV})(1.60 \times 10^{-19} \text{ J/eV})}{3.00 \times 10^8 \text{ m/s}} = 1.33 \times 10^{-27} \text{ kg} \cdot \text{m/s}. \qquad (29.33)$$

Discussion

This value for momentum is the same as found before (note that unrounded values are used in all calculations to avoid even small rounding errors), an expected verification of the relationship $p = E/c$. This also means the relationship between energy, momentum, and mass given by $E^2 = (pc)^2 + (mc)^2$ applies to both matter and photons. Once again, note that p is not zero, even when m is.

Problem-Solving Suggestion

Note that the forms of the constants $h = 4.14 \times 10^{-15}$ eV $\cdot$ s and $hc = 1240$ eV $\cdot$ nm may be particularly useful for this section's Problems and Exercises.

29.5 The Particle-Wave Duality

Learning Objectives

By the end of this section, you will be able to:

- Explain what the term particle-wave duality means, and why it is applied to EM radiation.

The information presented in this section supports the following AP® learning objectives and science practices:

- **1.D.1.1** The student is able to explain why classical mechanics cannot describe all properties of objects by articulating the reasons that classical mechanics must be refined and an alternative explanation developed when classical particles display wave properties. **(S.P. 6.3)**

We have long known that EM radiation is a wave, capable of interference and diffraction. We now see that light can be modeled as photons, which are massless particles. This may seem contradictory, since we ordinarily deal with large objects that never act like both wave and particle. An ocean wave, for example, looks nothing like a rock. To understand small-scale phenomena, we make analogies with the large-scale phenomena we observe directly. When we say something behaves like a wave, we mean it shows interference effects analogous to those seen in overlapping water waves. (See **Figure 29.20**.) Two examples of waves are sound and EM radiation. When we say something behaves like a particle, we mean that it interacts as a discrete unit with no interference effects. Examples of particles include electrons, atoms, and photons of EM radiation. How do we talk about a phenomenon that acts like both a particle and a wave?

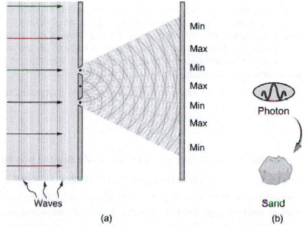

Figure 29.20 (a) The interference pattern for light through a double slit is a wave property understood by analogy to water waves. (b) The properties of photons having quantized energy and momentum and acting as a concentrated unit are understood by analogy to macroscopic particles.

There is no doubt that EM radiation interferes and has the properties of wavelength and frequency. There is also no doubt that it behaves as particles—photons with discrete energy. We call this twofold nature the **particle-wave duality**, meaning that EM radiation has both particle and wave properties. This so-called duality is simply a term for properties of the photon analogous to phenomena we can observe directly, on a macroscopic scale. If this term seems strange, it is because we do not ordinarily observe details on the quantum level directly, and our observations yield either particle *or* wavelike properties, but never both simultaneously.

Since we have a particle-wave duality for photons, and since we have seen connections between photons and matter in that both have momentum, it is reasonable to ask whether there is a particle-wave duality for matter as well. If the EM radiation we once thought to be a pure wave has particle properties, is it possible that matter has wave properties? The answer is yes. The consequences are tremendous, as we will begin to see in the next section.

PhET Explorations: Quantum Wave Interference

When do photons, electrons, and atoms behave like particles and when do they behave like waves? Watch waves spread out and interfere as they pass through a double slit, then get detected on a screen as tiny dots. Use quantum detectors to explore how measurements change the waves and the patterns they produce on the screen.

Figure 29.21 Quantum Wave Interference (http://cnx.org/content/m55042/1.2/quantum-wave-interference_en.jar)

29.6 The Wave Nature of Matter

Learning Objectives
By the end of this section, you will be able to: • Describe the Davisson-Germer experiment, and explain how it provides evidence for the wave nature of electrons. The information presented in this section supports the following AP® learning objectives and science practices: • **1.D.1.1** The student is able to explain why classical mechanics cannot describe all properties of objects by articulating the reasons that classical mechanics must be refined and an alternative explanation developed when classical particles display wave properties. **(S.P. 6.3)** • **6.G.1.1** The student is able to make predictions about using the scale of the problem to determine at what regimes a particle or wave model is more appropriate. **(S.P. 6.4, 7.1)** • **6.G.2.1** The student is able to articulate the evidence supporting the claim that a wave model of matter is appropriate to explain the diffraction of matter interacting with a crystal, given conditions where a particle of matter has momentum corresponding to a de Broglie wavelength smaller than the separation between adjacent atoms in the crystal. **(S.P. 6.1)**

De Broglie Wavelength

In 1923 a French physics graduate student named Prince Louis-Victor de Broglie (1892–1987) made a radical proposal based on the hope that nature is symmetric. If EM radiation has both particle and wave properties, then nature would be symmetric if matter also had both particle and wave properties. If what we once thought of as an unequivocal wave (EM radiation) is also a particle, then what we think of as an unequivocal particle (matter) may also be a wave. De Broglie's suggestion, made as part of his doctoral thesis, was so radical that it was greeted with some skepticism. A copy of his thesis was sent to Einstein, who said it was not only probably correct, but that it might be of fundamental importance. With the support of Einstein and a few other prominent physicists, de Broglie was awarded his doctorate.

De Broglie took both relativity and quantum mechanics into account to develop the proposal that *all particles have a wavelength*, given by

$$\lambda = \frac{h}{p} \text{(matter and photons)}, \tag{29.34}$$

where h is Planck's constant and p is momentum. This is defined to be the **de Broglie wavelength**. (Note that we already have this for photons, from the equation $p = h/\lambda$.) The hallmark of a wave is interference. If matter is a wave, then it must exhibit constructive and destructive interference. Why isn't this ordinarily observed? The answer is that in order to see significant interference effects, a wave must interact with an object about the same size as its wavelength. Since h is very small, λ is also small, especially for macroscopic objects. A 3-kg bowling ball moving at 10 m/s, for example, has

$$\lambda = h/p = (6.63 \times 10^{-34} \text{ J·s}) / [(3 \text{ kg})(10 \text{ m/s})] = 2 \times 10^{-35} \text{ m}. \tag{29.35}$$

This means that to see its wave characteristics, the bowling ball would have to interact with something about 10^{-35} m in size—far smaller than anything known. When waves interact with objects much larger than their wavelength, they show negligible interference effects and move in straight lines (such as light rays in geometric optics). To get easily observed interference effects from particles of matter, the longest wavelength and hence smallest mass possible would be useful. Therefore, this effect was first observed with electrons.

American physicists Clinton J. Davisson and Lester H. Germer in 1925 and, independently, British physicist G. P. Thomson (son of J. J. Thomson, discoverer of the electron) in 1926 scattered electrons from crystals and found diffraction patterns. These patterns are exactly consistent with interference of electrons having the de Broglie wavelength and are somewhat analogous to light interacting with a diffraction grating. (See **Figure 29.22**.)

Connections: Waves

All microscopic particles, whether massless, like photons, or having mass, like electrons, have wave properties. The relationship between momentum and wavelength is fundamental for all particles.

De Broglie's proposal of a wave nature for all particles initiated a remarkably productive era in which the foundations for quantum mechanics were laid. In 1926, the Austrian physicist Erwin Schrödinger (1887–1961) published four papers in which the wave nature of particles was treated explicitly with wave equations. At the same time, many others began important work. Among them was German physicist Werner Heisenberg (1901–1976) who, among many other contributions to quantum mechanics, formulated a mathematical treatment of the wave nature of matter that used matrices rather than wave equations. We will deal with some specifics in later sections, but it is worth noting that de Broglie's work was a watershed for the development of quantum mechanics. De Broglie was awarded the Nobel Prize in 1929 for his vision, as were Davisson and G. P. Thomson in 1937 for their experimental verification of de Broglie's hypothesis.

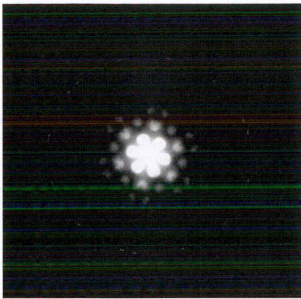

Figure 29.22 This diffraction pattern was obtained for electrons diffracted by crystalline silicon. Bright regions are those of constructive interference, while dark regions are those of destructive interference. (credit: Ndthe, Wikimedia Commons)

Example 29.7 Electron Wavelength versus Velocity and Energy

For an electron having a de Broglie wavelength of 0.167 nm (appropriate for interacting with crystal lattice structures that are about this size): (a) Calculate the electron's velocity, assuming it is nonrelativistic. (b) Calculate the electron's kinetic energy in eV.

Strategy

For part (a), since the de Broglie wavelength is given, the electron's velocity can be obtained from $\lambda = h / p$ by using the nonrelativistic formula for momentum, $p = mv$. For part (b), once v is obtained (and it has been verified that v is nonrelativistic), the classical kinetic energy is simply $(1/2)mv^2$.

Solution for (a)

Substituting the nonrelativistic formula for momentum ($p = mv$) into the de Broglie wavelength gives

$$\lambda = \frac{h}{p} = \frac{h}{mv}. \tag{29.36}$$

Solving for v gives

$$v = \frac{h}{m\lambda}. \tag{29.37}$$

Substituting known values yields

$$v = \frac{6.63 \times 10^{-34} \text{ J} \cdot \text{s}}{(9.11 \times 10^{-31} \text{ kg})(0.167 \times 10^{-9} \text{ m})} = 4.36 \times 10^6 \text{ m/s}. \tag{29.38}$$

Solution for (b)

While fast compared with a car, this electron's speed is not highly relativistic, and so we can comfortably use the classical formula to find the electron's kinetic energy and convert it to eV as requested.

$$
\begin{aligned}
\text{KE} &= \tfrac{1}{2}mv^2 \\
&= \tfrac{1}{2}(9.11 \times 10^{-31} \text{ kg})(4.36 \times 10^6 \text{ m/s})^2 \\
&= (86.4 \times 10^{-18} \text{ J})\left(\frac{1 \text{ eV}}{1.602 \times 10^{-19} \text{ J}}\right) \\
&= 54.0 \text{ eV}
\end{aligned}
\tag{29.39}
$$

Discussion

This low energy means that these 0.167-nm electrons could be obtained by accelerating them through a 54.0-V electrostatic potential, an easy task. The results also confirm the assumption that the electrons are nonrelativistic, since their velocity is just over 1% of the speed of light and the kinetic energy is about 0.01% of the rest energy of an electron (0.511 MeV). If the electrons had turned out to be relativistic, we would have had to use more involved calculations employing relativistic formulas.

Electron Microscopes

One consequence or use of the wave nature of matter is found in the electron microscope. As we have discussed, there is a limit to the detail observed with any probe having a wavelength. Resolution, or observable detail, is limited to about one wavelength. Since a potential of only 54 V can produce electrons with sub-nanometer wavelengths, it is easy to get electrons with much smaller wavelengths than those of visible light (hundreds of nanometers). Electron microscopes can, thus, be constructed to detect much smaller details than optical microscopes. (See **Figure 29.23**.)

There are basically two types of electron microscopes. The transmission electron microscope (TEM) accelerates electrons that are emitted from a hot filament (the cathode). The beam is broadened and then passes through the sample. A magnetic lens focuses the beam image onto a fluorescent screen, a photographic plate, or (most probably) a CCD (light sensitive camera), from which it is transferred to a computer. The TEM is similar to the optical microscope, but it requires a thin sample examined in a vacuum. However it can resolve details as small as 0.1 nm (10^{-10} m), providing magnifications of 100 million times the size of the original object. The TEM has allowed us to see individual atoms and structure of cell nuclei.

The scanning electron microscope (SEM) provides images by using secondary electrons produced by the primary beam interacting with the surface of the sample (see **Figure 29.23**). The SEM also uses magnetic lenses to focus the beam onto the sample. However, it moves the beam around electrically to "scan" the sample in the x and y directions. A CCD detector is used to process the data for each electron position, producing images like the one at the beginning of this chapter. The SEM has the advantage of not requiring a thin sample and of providing a 3-D view. However, its resolution is about ten times less than a TEM.

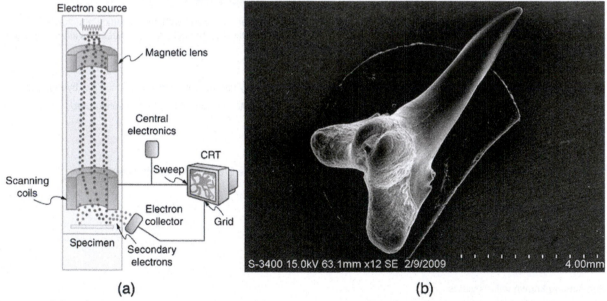

Figure 29.23 Schematic of a scanning electron microscope (SEM) (a) used to observe small details, such as those seen in this image of a tooth of a *Himipristis*, a type of shark (b). (credit: Dallas Krentzel, Flickr)

Electrons were the first particles with mass to be directly confirmed to have the wavelength proposed by de Broglie. Subsequently, protons, helium nuclei, neutrons, and many others have been observed to exhibit interference when they interact with objects having sizes similar to their de Broglie wavelength. The de Broglie wavelength for massless particles was well established in the 1920s for photons, and it has since been observed that all massless particles have a de Broglie wavelength $\lambda = h / p$. The wave nature of all particles is a universal characteristic of nature. We shall see in following sections that implications of the de Broglie wavelength include the quantization of energy in atoms and molecules, and an alteration of our basic view of nature on the microscopic scale. The next section, for example, shows that there are limits to the precision with which we may make predictions, regardless of how hard we try. There are even limits to the precision with which we may measure an object's location or energy.

Making Connections: A Submicroscopic Diffraction Grating

The wave nature of matter allows it to exhibit all the characteristics of other, more familiar, waves. Diffraction gratings, for example, produce diffraction patterns for light that depend on grating spacing and the wavelength of the light. This effect, as with most wave phenomena, is most pronounced when the wave interacts with objects having a size similar to its wavelength. For gratings, this is the spacing between multiple slits.) When electrons interact with a system having a spacing similar to the electron wavelength, they show the same types of interference patterns as light does for diffraction gratings, as shown at top left in **Figure 29.24**.

Atoms are spaced at regular intervals in a crystal as parallel planes, as shown in the bottom part of **Figure 29.24**. The spacings between these planes act like the openings in a diffraction grating. At certain incident angles, the paths of electrons scattering from successive planes differ by one wavelength and, thus, interfere constructively. At other angles, the path length differences are not an integral wavelength, and there is partial to total destructive interference. This type of scattering from a large crystal with well-defined lattice planes can produce dramatic interference patterns. It is called *Bragg reflection*, for the father-and-son team who first explored and analyzed it in some detail. The expanded view also shows the path-length differences and indicates how these depend on incident angle θ in a manner similar to the diffraction patterns for x rays reflecting from a crystal.

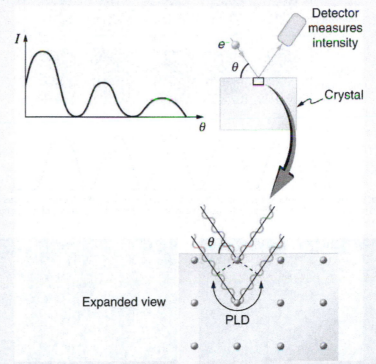

Figure 29.24 The diffraction pattern at top left is produced by scattering electrons from a crystal and is graphed as a function of incident angle relative to the regular array of atoms in a crystal, as shown at bottom. Electrons scattering from the second layer of atoms travel farther than those scattered from the top layer. If the path length difference (PLD) is an integral wavelength, there is constructive interference.

Let us take the spacing between parallel planes of atoms in the crystal to be d. As mentioned, if the path length difference (PLD) for the electrons is a whole number of wavelengths, there will be constructive interference—that is,

$\text{PLD} = n\lambda (n = 1, 2, 3, \dots)$. Because $\text{AB} = \text{BC} = d \sin\theta$, we have constructive interference when $n\lambda = 2d \sin\theta$.

This relationship is called the *Bragg equation* and applies not only to electrons but also to x rays.

The wavelength of matter is a submicroscopic characteristic that explains a macroscopic phenomenon such as Bragg reflection. Similarly, the wavelength of light is a submicroscopic characteristic that explains the macroscopic phenomenon of diffraction patterns.

29.7 Probability: The Heisenberg Uncertainty Principle

Learning Objectives
By the end of this section, you will be able to:
• Use both versions of Heisenberg's uncertainty principle in calculations.
• Explain the implications of Heisenberg's uncertainty principle for measurements.
The information presented in this section supports the following AP® learning objectives and science practices:
• **7.C.1.1** The student is able to use a graphical wave function representation of a particle to predict qualitatively the probability of finding a particle in a specific spatial region. **(S.P. 1.4)**

Probability Distribution

Matter and photons are waves, implying they are spread out over some distance. What is the position of a particle, such as an electron? Is it at the center of the wave? The answer lies in how you measure the position of an electron. Experiments show that you will find the electron at some definite location, unlike a wave. But if you set up exactly the same situation and measure it again, you will find the electron in a different location, often far outside any experimental uncertainty in your measurement. Repeated measurements will display a statistical distribution of locations that appears wavelike. (See **Figure 29.25**.)

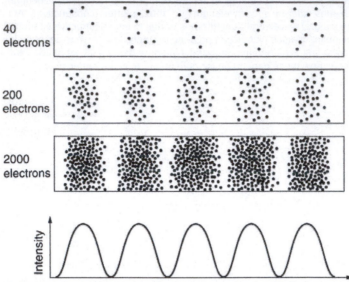

Figure 29.25 The building up of the diffraction pattern of electrons scattered from a crystal surface. Each electron arrives at a definite location, which cannot be precisely predicted. The overall distribution shown at the bottom can be predicted as the diffraction of waves having the de Broglie wavelength of the electrons.

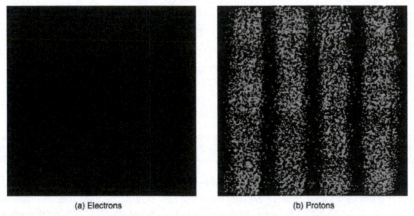

(a) Electrons   (b) Protons

Figure 29.26 Double-slit interference for electrons (a) and photons (b) is identical for equal wavelengths and equal slit separations. Both patterns are probability distributions in the sense that they are built up by individual particles traversing the apparatus, the paths of which are not individually predictable.

After de Broglie proposed the wave nature of matter, many physicists, including Schrödinger and Heisenberg, explored the consequences. The idea quickly emerged that, *because of its wave character, a particle's trajectory and destination cannot be precisely predicted for each particle individually*. However, each particle goes to a definite place (as illustrated in **Figure 29.25**). After compiling enough data, you get a distribution related to the particle's wavelength and diffraction pattern. There is a certain *probability* of finding the particle at a given location, and the overall pattern is called a **probability distribution**. Those who developed quantum mechanics devised equations that predicted the probability distribution in various circumstances.

It is somewhat disquieting to think that you cannot predict exactly where an individual particle will go, or even follow it to its destination. Let us explore what happens if we try to follow a particle. Consider the double-slit patterns obtained for electrons and photons in **Figure 29.26**. First, we note that these patterns are identical, following $d \sin \theta = m\lambda$, the equation for double-slit constructive interference developed in **Photon Energies and the Electromagnetic Spectrum**, where d is the slit separation and λ is the electron or photon wavelength.

Both patterns build up statistically as individual particles fall on the detector. This can be observed for photons or electrons—for now, let us concentrate on electrons. You might imagine that the electrons are interfering with one another as any waves do. To test this, you can lower the intensity until there is never more than one electron between the slits and the screen. The same interference pattern builds up! This implies that a particle's probability distribution spans both slits, and the particles actually interfere with themselves. Does this also mean that the electron goes through both slits? An electron is a basic unit of matter that is not divisible. But it is a fair question, and so we should look to see if the electron traverses one slit or the other, or both. One possibility is to have coils around the slits that detect charges moving through them. What is observed is that an electron always goes through one slit or the other; it does not split to go through both. But there is a catch. If you determine that the electron went through one of the slits, you no longer get a double slit pattern—instead, you get single slit interference. There is no escape by using another method of determining which slit the electron went through. Knowing the particle went through one slit forces a single-slit pattern. If you do not observe which slit the electron goes through, you obtain a double-slit pattern.

Heisenberg Uncertainty

How does knowing which slit the electron passed through change the pattern? The answer is fundamentally important—*measurement affects the system being observed*. Information can be lost, and in some cases it is impossible to measure two physical quantities simultaneously to exact precision. For example, you can measure the position of a moving electron by scattering light or other electrons from it. Those probes have momentum themselves, and by scattering from the electron, they change its momentum *in a manner that loses information*. There is a limit to absolute knowledge, even in principle.

Figure 29.27 Werner Heisenberg was one of the best of those physicists who developed early quantum mechanics. Not only did his work enable a description of nature on the very small scale, it also changed our view of the availability of knowledge. Although he is universally recognized for his brilliance and the importance of his work (he received the Nobel Prize in 1932, for example), Heisenberg remained in Germany during World War II and headed the German effort to build a nuclear bomb, permanently alienating himself from most of the scientific community. (credit: Author Unknown, via Wikimedia Commons)

It was Werner Heisenberg who first stated this limit to knowledge in 1929 as a result of his work on quantum mechanics and the wave characteristics of all particles. (See **Figure 29.27**). Specifically, consider simultaneously measuring the position and momentum of an electron (it could be any particle). There is an **uncertainty in position** Δx that is approximately equal to the wavelength of the particle. That is,

$$\Delta x \approx \lambda. \tag{29.40}$$

As discussed above, a wave is not located at one point in space. If the electron's position is measured repeatedly, a spread in locations will be observed, implying an uncertainty in position Δx. To detect the position of the particle, we must interact with it, such as having it collide with a detector. In the collision, the particle will lose momentum. This change in momentum could be anywhere from close to zero to the total momentum of the particle, $p = h/\lambda$. It is not possible to tell how much momentum will be transferred to a detector, and so there is an **uncertainty in momentum** Δp, too. In fact, the uncertainty in momentum may be as large as the momentum itself, which in equation form means that

$$\Delta p \approx \frac{h}{\lambda}. \tag{29.41}$$

The uncertainty in position can be reduced by using a shorter-wavelength electron, since $\Delta x \approx \lambda$. But shortening the wavelength increases the uncertainty in momentum, since $\Delta p \approx h/\lambda$. Conversely, the uncertainty in momentum can be reduced by using a longer-wavelength electron, but this increases the uncertainty in position. Mathematically, you can express this trade-off by multiplying the uncertainties. The wavelength cancels, leaving

$$\Delta x \Delta p \approx h. \tag{29.42}$$

So if one uncertainty is reduced, the other must increase so that their product is $\approx h$.

With the use of advanced mathematics, Heisenberg showed that the best that can be done in a *simultaneous measurement of position and momentum* is

$$\Delta x \Delta p \geq \frac{h}{4\pi}. \tag{29.43}$$

This is known as the **Heisenberg uncertainty principle**. It is impossible to measure position x and momentum p simultaneously with uncertainties Δx and Δp that multiply to be less than $h/4\pi$. Neither uncertainty can be zero. Neither uncertainty can become small without the other becoming large. A small wavelength allows accurate position measurement, but it increases the momentum of the probe to the point that it further disturbs the momentum of a system being measured. For example, if an electron is scattered from an atom and has a wavelength small enough to detect the position of electrons in the atom, its momentum can knock the electrons from their orbits in a manner that loses information about their original motion. It is therefore impossible to follow an electron in its orbit around an atom. If you measure the electron's position, you will find it in a definite location, but the atom will be disrupted. Repeated measurements on identical atoms will produce interesting probability distributions for electrons around the atom, but they will not produce motion information. The probability distributions are referred to as electron clouds or orbitals. The shapes of these orbitals are often shown in general chemistry texts and are discussed in **The Wave Nature of Matter Causes Quantization**.

Example 29.8 Heisenberg Uncertainty Principle in Position and Momentum for an Atom

(a) If the position of an electron in an atom is measured to an accuracy of 0.0100 nm, what is the electron's uncertainty in velocity? (b) If the electron has this velocity, what is its kinetic energy in eV?

Strategy

The uncertainty in position is the accuracy of the measurement, or $\Delta x = 0.0100$ nm. Thus the smallest uncertainty in momentum Δp can be calculated using $\Delta x \Delta p \geq h/4\pi$. Once the uncertainty in momentum Δp is found, the uncertainty in velocity can be found from $\Delta p = m\Delta v$.

Solution for (a)

Using the equals sign in the uncertainty principle to express the minimum uncertainty, we have

$$\Delta x \Delta p = \frac{h}{4\pi}. \tag{29.44}$$

Solving for Δp and substituting known values gives

$$\Delta p = \frac{h}{4\pi \Delta x} = \frac{6.63 \times 10^{-34} \text{ J} \cdot \text{s}}{4\pi(1.00 \times 10^{-11} \text{ m})} = 5.28 \times 10^{-24} \text{ kg} \cdot \text{m/s}. \tag{29.45}$$

Thus,

$$\Delta p = 5.28 \times 10^{-24} \text{ kg} \cdot \text{m/s} = m\Delta v. \tag{29.46}$$

Solving for Δv and substituting the mass of an electron gives

$$\Delta v = \frac{\Delta p}{m} = \frac{5.28 \times 10^{-24} \text{ kg} \cdot \text{m/s}}{9.11 \times 10^{-31} \text{ kg}} = 5.79 \times 10^{6} \text{ m/s}. \tag{29.47}$$

Solution for (b)

Although large, this velocity is not highly relativistic, and so the electron's kinetic energy is

$$
\begin{aligned}
\text{KE}_e &= \tfrac{1}{2}mv^2 \\
&= \tfrac{1}{2}(9.11 \times 10^{-31} \text{ kg})(5.79 \times 10^{6} \text{ m/s})^2 \\
&= (1.53 \times 10^{-17} \text{ J})\left(\frac{1 \text{ eV}}{1.60 \times 10^{-19} \text{ J}}\right) = 95.5 \text{ eV}.
\end{aligned} \tag{29.48}
$$

Discussion

Since atoms are roughly 0.1 nm in size, knowing the position of an electron to 0.0100 nm localizes it reasonably well inside the atom. This would be like being able to see details one-tenth the size of the atom. But the consequent uncertainty in velocity is large. You certainly could not follow it very well if its velocity is so uncertain. To get a further idea of how large the uncertainty in velocity is, we assumed the velocity of the electron was equal to its uncertainty and found this gave a kinetic energy of 95.5 eV. This is significantly greater than the typical energy difference between levels in atoms (see **Table 29.1**), so that it is impossible to get a meaningful energy for the electron if we know its position even moderately well.

Why don't we notice Heisenberg's uncertainty principle in everyday life? The answer is that Planck's constant is very small. Thus the lower limit in the uncertainty of measuring the position and momentum of large objects is negligible. We can detect sunlight reflected from Jupiter and follow the planet in its orbit around the Sun. The reflected sunlight alters the momentum of Jupiter and creates an uncertainty in its momentum, but this is totally negligible compared with Jupiter's huge momentum. The correspondence principle tells us that the predictions of quantum mechanics become indistinguishable from classical physics for large objects, which is the case here.

Heisenberg Uncertainty for Energy and Time

There is another form of **Heisenberg's uncertainty principle** for *simultaneous measurements of energy and time*. In equation form,

$$\Delta E \Delta t \geq \frac{h}{4\pi}, \tag{29.49}$$

where ΔE is the **uncertainty in energy** and Δt is the **uncertainty in time**. This means that within a time interval Δt, it is not possible to measure energy precisely—there will be an uncertainty ΔE in the measurement. In order to measure energy more precisely (to make ΔE smaller), we must increase Δt. This time interval may be the amount of time we take to make the measurement, or it could be the amount of time a particular state exists, as in the next **Example 29.9**.

Example 29.9 Heisenberg Uncertainty Principle for Energy and Time for an Atom

An atom in an excited state temporarily stores energy. If the lifetime of this excited state is measured to be 1.0×10^{-10} s , what is the minimum uncertainty in the energy of the state in eV?

Strategy

The minimum uncertainty in energy ΔE is found by using the equals sign in $\Delta E \Delta t \geq h/4\pi$ and corresponds to a reasonable choice for the uncertainty in time. The largest the uncertainty in time can be is the full lifetime of the excited state, or $\Delta t = 1.0 \times 10^{-10}$ s .

Solution

Solving the uncertainty principle for ΔE and substituting known values gives

$$\Delta E = \frac{h}{4\pi \Delta t} = \frac{6.63 \times 10^{-34} \text{ J} \cdot \text{s}}{4\pi (1.0 \times 10^{-10} \text{ s})} = 5.3 \times 10^{-25} \text{ J}. \tag{29.50}$$

Now converting to eV yields

$$\Delta E = (5.3 \times 10^{-25} \text{ J}) \left(\frac{1 \text{ eV}}{1.6 \times 10^{-19} \text{ J}} \right) = 3.3 \times 10^{-6} \text{ eV}. \tag{29.51}$$

Discussion

The lifetime of 10^{-10} s is typical of excited states in atoms—on human time scales, they quickly emit their stored energy. An uncertainty in energy of only a few millionths of an eV results. This uncertainty is small compared with typical excitation energies in atoms, which are on the order of 1 eV. So here the uncertainty principle limits the accuracy with which we can measure the lifetime and energy of such states, but not very significantly.

The uncertainty principle for energy and time can be of great significance if the lifetime of a system is very short. Then Δt is very small, and ΔE is consequently very large. Some nuclei and exotic particles have extremely short lifetimes (as small as 10^{-25} s), causing uncertainties in energy as great as many GeV (10^{9} eV). Stored energy appears as increased rest mass, and so this means that there is significant uncertainty in the rest mass of short-lived particles. When measured repeatedly, a spread of masses or decay energies are obtained. The spread is ΔE. You might ask whether this uncertainty in energy could be avoided by not measuring the lifetime. The answer is no. Nature knows the lifetime, and so its brevity affects the energy of the particle. This is so well established experimentally that the uncertainty in decay energy is used to calculate the lifetime of short-lived states. Some nuclei and particles are so short-lived that it is difficult to measure their lifetime. But if their decay energy can be measured, its spread is ΔE, and this is used in the uncertainty principle ($\Delta E \Delta t \geq h/4\pi$) to calculate the lifetime Δt.

There is another consequence of the uncertainty principle for energy and time. If energy is uncertain by ΔE, then conservation of energy can be violated by ΔE for a time Δt. Neither the physicist nor nature can tell that conservation of energy has been violated, if the violation is temporary and smaller than the uncertainty in energy. While this sounds innocuous enough, we shall see in later chapters that it allows the temporary creation of matter from nothing and has implications for how nature transmits forces over very small distances.

Finally, note that in the discussion of particles and waves, we have stated that individual measurements produce precise or particle-like results. A definite position is determined each time we observe an electron, for example. But repeated measurements produce a spread in values consistent with wave characteristics. The great theoretical physicist Richard Feynman

(1918–1988) commented, "What there are, are particles." When you observe enough of them, they distribute themselves as you would expect for a wave phenomenon. However, what there are as they travel we cannot tell because, when we do try to measure, we affect the traveling.

29.8 The Particle-Wave Duality Reviewed

Particle-wave duality—the fact that all particles have wave properties—is one of the cornerstones of quantum mechanics. We first came across it in the treatment of photons, those particles of EM radiation that exhibit both particle and wave properties, but not at the same time. Later it was noted that particles of matter have wave properties as well. The dual properties of particles and waves are found for all particles, whether massless like photons, or having a mass like electrons. (See **Figure 29.28**.)

Figure 29.28 On a quantum-mechanical scale (i.e., very small), particles with and without mass have wave properties. For example, both electrons and photons have wavelengths but also behave as particles.

There are many submicroscopic particles in nature. Most have mass and are expected to act as particles, or the smallest units of matter. All these masses have wave properties, with wavelengths given by the de Broglie relationship $\lambda = h/p$. So, too, do combinations of these particles, such as nuclei, atoms, and molecules. As a combination of masses becomes large, particularly if it is large enough to be called macroscopic, its wave nature becomes difficult to observe. This is consistent with our common experience with matter.

Some particles in nature are massless. We have only treated the photon so far, but all massless entities travel at the speed of light, have a wavelength, and exhibit particle and wave behaviors. They have momentum given by a rearrangement of the de Broglie relationship, $p = h/\lambda$. In large combinations of these massless particles (such large combinations are common only for photons or EM waves), there is mostly wave behavior upon detection, and the particle nature becomes difficult to observe. This is also consistent with experience. (See **Figure 29.29**.)

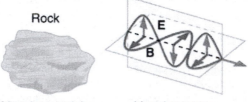

Rock

Massive particle   Massless wave

Figure 29.29 On a classical scale (macroscopic), particles with mass behave as particles and not as waves. Particles without mass act as waves and not as particles.

The particle-wave duality is a universal attribute. It is another connection between matter and energy. Not only has modern physics been able to describe nature for high speeds and small sizes, it has also discovered new connections and symmetries. There is greater unity and symmetry in nature than was known in the classical era—but they were dreamt of. A beautiful poem written by the English poet William Blake some two centuries ago contains the following four lines:

To see the World in a Grain of Sand

And a Heaven in a Wild Flower

Hold Infinity in the palm of your hand

And Eternity in an hour

Integrated Concepts

The problem set for this section involves concepts from this chapter and several others. Physics is most interesting when applied to general situations involving more than a narrow set of physical principles. For example, photons have momentum, hence the relevance of **Linear Momentum and Collisions**. The following topics are involved in some or all of the problems in this section:

- **Dynamics: Newton's Laws of Motion**
- **Work, Energy, and Energy Resources**
- **Linear Momentum and Collisions**
- **Heat and Heat Transfer Methods**
- **Electric Potential and Electric Field**
- **Electric Current, Resistance, and Ohm's Law**
- **Wave Optics**
- **Special Relativity**

Problem-Solving Strategy

1. Identify which physical principles are involved.

2. Solve the problem using strategies outlined in the text.

Example 29.10 illustrates how these strategies are applied to an integrated-concept problem.

Example 29.10 Recoil of a Dust Particle after Absorbing a Photon

The following topics are involved in this integrated concepts worked example:

Table 29.2 Topics

Photons (quantum mechanics)
Linear Momentum

A 550-nm photon (visible light) is absorbed by a 1.00-μg particle of dust in outer space. (a) Find the momentum of such a photon. (b) What is the recoil velocity of the particle of dust, assuming it is initially at rest?

Strategy Step 1

To solve an *integrated-concept problem*, such as those following this example, we must first identify the physical principles involved and identify the chapters in which they are found. Part (a) of this example asks for the *momentum of a photon*, a topic of the present chapter. Part (b) considers *recoil following a collision*, a topic of **Linear Momentum and Collisions**.

Strategy Step 2

The following solutions to each part of the example illustrate how specific problem-solving strategies are applied. These involve identifying knowns and unknowns, checking to see if the answer is reasonable, and so on.

Solution for (a)

The momentum of a photon is related to its wavelength by the equation:

$$p = \frac{h}{\lambda}. \tag{29.52}$$

Entering the known value for Planck's constant h and given the wavelength λ, we obtain

$$\begin{aligned} p &= \frac{6.63 \times 10^{-34} \text{ J} \cdot \text{s}}{550 \times 10^{-9} \text{ m}} \\ &= 1.21 \times 10^{-27} \text{ kg} \cdot \text{m/s}. \end{aligned} \tag{29.53}$$

Discussion for (a)

This momentum is small, as expected from discussions in the text and the fact that photons of visible light carry small amounts of energy and momentum compared with those carried by macroscopic objects.

Solution for (b)

Conservation of momentum in the absorption of this photon by a grain of dust can be analyzed using the equation:

$$p_1 + p_2 = p'_1 + p'_2 (F_{\text{net}} = 0). \tag{29.54}$$

The net external force is zero, since the dust is in outer space. Let 1 represent the photon and 2 the dust particle. Before the collision, the dust is at rest (relative to some observer); after the collision, there is no photon (it is absorbed). So conservation of momentum can be written

$$p_1 = p'_2 = mv, \tag{29.55}$$

where p_1 is the photon momentum before the collision and p'_2 is the dust momentum after the collision. The mass and recoil velocity of the dust are m and v, respectively. Solving this for v, the requested quantity, yields

$$v = \frac{p}{m}, \tag{29.56}$$

where p is the photon momentum found in part (a). Entering known values (noting that a microgram is 10^{-9} kg) gives

$$\begin{aligned} v &= \frac{1.21 \times 10^{-27} \text{ kg} \cdot \text{m/s}}{1.00 \times 10^{-9} \text{ kg}} \\ &= 1.21 \times 10^{-18} \text{ m/s}. \end{aligned} \tag{29.57}$$

Discussion

The recoil velocity of the particle of dust is extremely small. As we have noted, however, there are immense numbers of photons in sunlight and other macroscopic sources. In time, collisions and absorption of many photons could cause a significant recoil of the dust, as observed in comet tails.

Glossary

atomic spectra: the electromagnetic emission from atoms and molecules

binding energy: also called the *work function*; the amount of energy necessary to eject an electron from a material

blackbody: an ideal radiator, which can radiate equally well at all wavelengths

blackbody radiation: the electromagnetic radiation from a blackbody

bremsstrahlung: German for *braking radiation*; produced when electrons are decelerated

characteristic x rays: x rays whose energy depends on the material they were produced in

Compton effect: the phenomenon whereby x rays scattered from materials have decreased energy

correspondence principle: in the classical limit (large, slow-moving objects), quantum mechanics becomes the same as classical physics

de Broglie wavelength: the wavelength possessed by a particle of matter, calculated by $\lambda = h/p$

gamma ray: also γ-ray; highest-energy photon in the EM spectrum

Heisenberg's uncertainty principle: a fundamental limit to the precision with which pairs of quantities (momentum and position, and energy and time) can be measured

infrared radiation: photons with energies slightly less than red light

ionizing radiation: radiation that ionizes materials that absorb it

microwaves: photons with wavelengths on the order of a micron (μm)

particle-wave duality: the property of behaving like either a particle or a wave; the term for the phenomenon that all particles have wave characteristics

photoelectric effect: the phenomenon whereby some materials eject electrons when light is shined on them

photon: a quantum, or particle, of electromagnetic radiation

photon energy: the amount of energy a photon has; $E = hf$

photon momentum: the amount of momentum a photon has, calculated by $p = \frac{h}{\lambda} = \frac{E}{c}$

Planck's constant: $h = 6.626 \times 10^{-34}$ J · s

probability distribution: the overall spatial distribution of probabilities to find a particle at a given location

quantized: the fact that certain physical entities exist only with particular discrete values and not every conceivable value

quantum mechanics: the branch of physics that deals with small objects and with the quantization of various entities, especially energy

ultraviolet radiation: UV; ionizing photons slightly more energetic than violet light

uncertainty in energy: lack of precision or lack of knowledge of precise results in measurements of energy

uncertainty in momentum: lack of precision or lack of knowledge of precise results in measurements of momentum

uncertainty in position: lack of precision or lack of knowledge of precise results in measurements of position

uncertainty in time: lack of precision or lack of knowledge of precise results in measurements of time

visible light: the range of photon energies the human eye can detect

x ray: EM photon between γ-ray and UV in energy

Section Summary

29.1 Quantization of Energy
- The first indication that energy is sometimes quantized came from blackbody radiation, which is the emission of EM radiation by an object with an emissivity of 1.
- Planck recognized that the energy levels of the emitting atoms and molecules were quantized, with only the allowed values of $E = \left(n + \frac{1}{2}\right)hf$, where n is any non-negative integer (0, 1, 2, 3, ...).
- h is Planck's constant, whose value is $h = 6.626 \times 10^{-34}\ \text{J} \cdot \text{s}$.
- Thus, the oscillatory absorption and emission energies of atoms and molecules in a blackbody could increase or decrease only in steps of size $\Delta E = hf$ where f is the frequency of the oscillatory nature of the absorption and emission of EM radiation.
- Another indication of energy levels being quantized in atoms and molecules comes from the lines in atomic spectra, which are the EM emissions of individual atoms and molecules.

29.2 The Photoelectric Effect
- The photoelectric effect is the process in which EM radiation ejects electrons from a material.
- Einstein proposed photons to be quanta of EM radiation having energy $E = hf$, where f is the frequency of the radiation.
- All EM radiation is composed of photons. As Einstein explained, all characteristics of the photoelectric effect are due to the interaction of individual photons with individual electrons.
- The maximum kinetic energy KE_e of ejected electrons (photoelectrons) is given by $\text{KE}_e = hf - \text{BE}$, where hf is the photon energy and BE is the binding energy (or work function) of the electron to the particular material.

29.3 Photon Energies and the Electromagnetic Spectrum
- Photon energy is responsible for many characteristics of EM radiation, being particularly noticeable at high frequencies.
- Photons have both wave and particle characteristics.

29.4 Photon Momentum
- Photons have momentum, given by $p = \frac{h}{\lambda}$, where λ is the photon wavelength.
- Photon energy and momentum are related by $p = \frac{E}{c}$, where $E = hf = hc / \lambda$ for a photon.

29.5 The Particle-Wave Duality
- EM radiation can behave like either a particle or a wave.
- This is termed particle-wave duality.

29.6 The Wave Nature of Matter
- Particles of matter also have a wavelength, called the de Broglie wavelength, given by $\lambda = \frac{h}{p}$, where p is momentum.

• Matter is found to have the same *interference characteristics* as any other wave.

29.7 Probability: The Heisenberg Uncertainty Principle

• Matter is found to have the same interference characteristics as any other wave.
• There is now a probability distribution for the location of a particle rather than a definite position.
• Another consequence of the wave character of all particles is the Heisenberg uncertainty principle, which limits the precision with which certain physical quantities can be known simultaneously. For position and momentum, the uncertainty principle is $\Delta x \Delta p \geq \frac{h}{4\pi}$, where Δx is the uncertainty in position and Δp is the uncertainty in momentum.

• For energy and time, the uncertainty principle is $\Delta E \Delta t \geq \frac{h}{4\pi}$ where ΔE is the uncertainty in energy and Δt is the uncertainty in time.
• These small limits are fundamentally important on the quantum-mechanical scale.

29.8 The Particle-Wave Duality Reviewed

• The particle-wave duality refers to the fact that all particles—those with mass and those without mass—have wave characteristics.
• This is a further connection between mass and energy.

Conceptual Questions

29.1 Quantization of Energy

1. Give an example of a physical entity that is quantized. State specifically what the entity is and what the limits are on its values.

2. Give an example of a physical entity that is not quantized, in that it is continuous and may have a continuous range of values.

3. What aspect of the blackbody spectrum forced Planck to propose quantization of energy levels in its atoms and molecules?

4. If Planck's constant were large, say 10^{34} times greater than it is, we would observe macroscopic entities to be quantized. Describe the motions of a child's swing under such circumstances.

5. Why don't we notice quantization in everyday events?

29.2 The Photoelectric Effect

6. Is visible light the only type of EM radiation that can cause the photoelectric effect?

7. Which aspects of the photoelectric effect cannot be explained without photons? Which can be explained without photons? Are the latter inconsistent with the existence of photons?

8. Is the photoelectric effect a direct consequence of the wave character of EM radiation or of the particle character of EM radiation? Explain briefly.

9. Insulators (nonmetals) have a higher BE than metals, and it is more difficult for photons to eject electrons from insulators. Discuss how this relates to the free charges in metals that make them good conductors.

10. If you pick up and shake a piece of metal that has electrons in it free to move as a current, no electrons fall out. Yet if you heat the metal, electrons can be boiled off. Explain both of these facts as they relate to the amount and distribution of energy involved with shaking the object as compared with heating it.

29.3 Photon Energies and the Electromagnetic Spectrum

11. Why are UV, x rays, and γ rays called ionizing radiation?

12. How can treating food with ionizing radiation help keep it from spoiling? UV is not very penetrating. What else could be used?

13. Some television tubes are CRTs. They use an approximately 30-kV accelerating potential to send electrons to the screen, where the electrons stimulate phosphors to emit the light that forms the pictures we watch. Would you expect x rays also to be created?

14. Tanning salons use "safe" UV with a longer wavelength than some of the UV in sunlight. This "safe" UV has enough photon energy to trigger the tanning mechanism. Is it likely to be able to cause cell damage and induce cancer with prolonged exposure?

15. Your pupils dilate when visible light intensity is reduced. Does wearing sunglasses that lack UV blockers increase or decrease the UV hazard to your eyes? Explain.

16. One could feel heat transfer in the form of infrared radiation from a large nuclear bomb detonated in the atmosphere 75 km from you. However, none of the profusely emitted x rays or γ rays reaches you. Explain.

17. Can a single microwave photon cause cell damage? Explain.

18. In an x-ray tube, the maximum photon energy is given by $hf = qV.$ Would it be technically more correct to say $hf = qV + \mathrm{BE},$ where BE is the binding energy of electrons in the target anode? Why isn't the energy stated the latter way?

29.4 Photon Momentum

19. Which formula may be used for the momentum of all particles, with or without mass?

20. Is there any measurable difference between the momentum of a photon and the momentum of matter?

21. Why don't we feel the momentum of sunlight when we are on the beach?

29.6 The Wave Nature of Matter

22. How does the interference of water waves differ from the interference of electrons? How are they analogous?

23. Describe one type of evidence for the wave nature of matter.

24. Describe one type of evidence for the particle nature of EM radiation.

29.7 Probability: The Heisenberg Uncertainty Principle

25. What is the Heisenberg uncertainty principle? Does it place limits on what can be known?

29.8 The Particle-Wave Duality Reviewed

26. In what ways are matter and energy related that were not known before the development of relativity and quantum mechanics?

In what ways are matter and energy related that were not known before the development of relativity and quantum mechanics?

Problems & Exercises

29.1 Quantization of Energy

1. A LiBr molecule oscillates with a frequency of 1.7×10^{13} Hz. (a) What is the difference in energy in eV between allowed oscillator states? (b) What is the approximate value of n for a state having an energy of 1.0 eV?

2. The difference in energy between allowed oscillator states in HBr molecules is 0.330 eV. What is the oscillation frequency of this molecule?

3. A physicist is watching a 15-kg orangutan at a zoo swing lazily in a tire at the end of a rope. He (the physicist) notices that each oscillation takes 3.00 s and hypothesizes that the energy is quantized. (a) What is the difference in energy in joules between allowed oscillator states? (b) What is the value of n for a state where the energy is 5.00 J? (c) Can the quantization be observed?

29.2 The Photoelectric Effect

4. What is the longest-wavelength EM radiation that can eject a photoelectron from silver, given that the binding energy is 4.73 eV? Is this in the visible range?

5. Find the longest-wavelength photon that can eject an electron from potassium, given that the binding energy is 2.24 eV. Is this visible EM radiation?

6. What is the binding energy in eV of electrons in magnesium, if the longest-wavelength photon that can eject electrons is 337 nm?

7. Calculate the binding energy in eV of electrons in aluminum, if the longest-wavelength photon that can eject them is 304 nm.

8. What is the maximum kinetic energy in eV of electrons ejected from sodium metal by 450-nm EM radiation, given that the binding energy is 2.28 eV?

9. UV radiation having a wavelength of 120 nm falls on gold metal, to which electrons are bound by 4.82 eV. What is the maximum kinetic energy of the ejected photoelectrons?

10. Violet light of wavelength 400 nm ejects electrons with a maximum kinetic energy of 0.860 eV from sodium metal. What is the binding energy of electrons to sodium metal?

11. UV radiation having a 300-nm wavelength falls on uranium metal, ejecting 0.500-eV electrons. What is the binding energy of electrons to uranium metal?

12. What is the wavelength of EM radiation that ejects 2.00-eV electrons from calcium metal, given that the binding energy is 2.71 eV? What type of EM radiation is this?

13. Find the wavelength of photons that eject 0.100-eV electrons from potassium, given that the binding energy is 2.24 eV. Are these photons visible?

14. What is the maximum velocity of electrons ejected from a material by 80-nm photons, if they are bound to the material by 4.73 eV?

15. Photoelectrons from a material with a binding energy of 2.71 eV are ejected by 420-nm photons. Once ejected, how long does it take these electrons to travel 2.50 cm to a detection device?

16. A laser with a power output of 2.00 mW at a wavelength of 400 nm is projected onto calcium metal. (a) How many electrons per second are ejected? (b) What power is carried away by the electrons, given that the binding energy is 2.71 eV?

17. (a) Calculate the number of photoelectrons per second ejected from a 1.00-mm^2 area of sodium metal by 500-nm EM radiation having an intensity of 1.30 kW/m^2 (the intensity of sunlight above the Earth's atmosphere). (b) Given that the binding energy is 2.28 eV, what power is carried away by the electrons? (c) The electrons carry away less power than brought in by the photons. Where does the other power go? How can it be recovered?

18. Unreasonable Results

Red light having a wavelength of 700 nm is projected onto magnesium metal to which electrons are bound by 3.68 eV. (a) Use $KE_e = hf - BE$ to calculate the kinetic energy of the ejected electrons. (b) What is unreasonable about this result? (c) Which assumptions are unreasonable or inconsistent?

19. Unreasonable Results

(a) What is the binding energy of electrons to a material from which 4.00-eV electrons are ejected by 400-nm EM radiation? (b) What is unreasonable about this result? (c) Which assumptions are unreasonable or inconsistent?

29.3 Photon Energies and the Electromagnetic Spectrum

20. What is the energy in joules and eV of a photon in a radio wave from an AM station that has a 1530-kHz broadcast frequency?

21. (a) Find the energy in joules and eV of photons in radio waves from an FM station that has a 90.0-MHz broadcast frequency. (b) What does this imply about the number of photons per second that the radio station must broadcast?

22. Calculate the frequency in hertz of a 1.00-MeV γ-ray photon.

23. (a) What is the wavelength of a 1.00-eV photon? (b) Find its frequency in hertz. (c) Identify the type of EM radiation.

24. Do the unit conversions necessary to show that $hc = 1240$ eV $\cdot$ nm, as stated in the text.

25. Confirm the statement in the text that the range of photon energies for visible light is 1.63 to 3.26 eV, given that the range of visible wavelengths is 380 to 760 nm.

26. (a) Calculate the energy in eV of an IR photon of frequency 2.00×10^{13} Hz. (b) How many of these photons would need to be absorbed simultaneously by a tightly bound molecule to break it apart? (c) What is the energy in eV of a γ ray of frequency 3.00×10^{20} Hz? (d) How many tightly bound molecules could a single such γ ray break apart?

27. Prove that, to three-digit accuracy, $h = 4.14 \times 10^{-15}$ eV $\cdot$ s, as stated in the text.

28. (a) What is the maximum energy in eV of photons produced in a CRT using a 25.0-kV accelerating potential, such as a color TV? (b) What is their frequency?

29. What is the accelerating voltage of an x-ray tube that produces x rays with a shortest wavelength of 0.0103 nm?

30. (a) What is the ratio of power outputs by two microwave ovens having frequencies of 950 and 2560 MHz, if they emit the same number of photons per second? (b) What is the ratio of photons per second if they have the same power output?

31. How many photons per second are emitted by the antenna of a microwave oven, if its power output is 1.00 kW at a frequency of 2560 MHz?

32. Some satellites use nuclear power. (a) If such a satellite emits a 1.00-W flux of γ rays having an average energy of 0.500 MeV, how many are emitted per second? (b) These γ rays affect other satellites. How far away must another satellite be to only receive one γ ray per second per square meter?

33. (a) If the power output of a 650-kHz radio station is 50.0 kW, how many photons per second are produced? (b) If the radio waves are broadcast uniformly in all directions, find the number of photons per second per square meter at a distance of 100 km. Assume no reflection from the ground or absorption by the air.

34. How many x-ray photons per second are created by an x-ray tube that produces a flux of x rays having a power of 1.00 W? Assume the average energy per photon is 75.0 keV.

35. (a) How far away must you be from a 650-kHz radio station with power 50.0 kW for there to be only one photon per second per square meter? Assume no reflections or absorption, as if you were in deep outer space. (b) Discuss the implications for detecting intelligent life in other solar systems by detecting their radio broadcasts.

36. Assuming that 10.0% of a 100-W light bulb's energy output is in the visible range (typical for incandescent bulbs) with an average wavelength of 580 nm, and that the photons spread out uniformly and are not absorbed by the atmosphere, how far away would you be if 500 photons per second enter the 3.00-mm diameter pupil of your eye? (This number easily stimulates the retina.)

37. Construct Your Own Problem

Consider a laser pen. Construct a problem in which you calculate the number of photons per second emitted by the pen. Among the things to be considered are the laser pen's wavelength and power output. Your instructor may also wish for you to determine the minimum diffraction spreading in the beam and the number of photons per square centimeter the pen can project at some large distance. In this latter case, you will also need to consider the output size of the laser beam, the distance to the object being illuminated, and any absorption or scattering along the way.

29.4 Photon Momentum

38. (a) Find the momentum of a 4.00-cm-wavelength microwave photon. (b) Discuss why you expect the answer to (a) to be very small.

39. (a) What is the momentum of a 0.0100-nm-wavelength photon that could detect details of an atom? (b) What is its energy in MeV?

40. (a) What is the wavelength of a photon that has a momentum of $5.00 \times 10^{-29} \ \text{kg} \cdot \text{m/s}$? (b) Find its energy in eV.

41. (a) A γ-ray photon has a momentum of $8.00 \times 10^{-21} \ \text{kg} \cdot \text{m/s}$. What is its wavelength? (b) Calculate its energy in MeV.

42. (a) Calculate the momentum of a photon having a wavelength of $2.50 \ \mu\text{m}$. (b) Find the velocity of an electron having the same momentum. (c) What is the kinetic energy of the electron, and how does it compare with that of the photon?

43. Repeat the previous problem for a 10.0-nm-wavelength photon.

44. (a) Calculate the wavelength of a photon that has the same momentum as a proton moving at 1.00% of the speed of light. (b) What is the energy of the photon in MeV? (c) What is the kinetic energy of the proton in MeV?

45. (a) Find the momentum of a 100-keV x-ray photon. (b) Find the equivalent velocity of a neutron with the same momentum. (c) What is the neutron's kinetic energy in keV?

46. Take the ratio of relativistic rest energy, $E = \gamma m c^2$, to relativistic momentum, $p = \gamma m u$, and show that in the limit that mass approaches zero, you find $E / p = c$.

47. Construct Your Own Problem

Consider a space sail such as mentioned in **Example 29.5**. Construct a problem in which you calculate the light pressure on the sail in N/m^2 produced by reflecting sunlight. Also calculate the force that could be produced and how much effect that would have on a spacecraft. Among the things to be considered are the intensity of sunlight, its average wavelength, the number of photons per square meter this implies, the area of the space sail, and the mass of the system being accelerated.

48. Unreasonable Results

A car feels a small force due to the light it sends out from its headlights, equal to the momentum of the light divided by the time in which it is emitted. (a) Calculate the power of each headlight, if they exert a total force of $2.00 \times 10^{-2} \ \text{N}$ backward on the car. (b) What is unreasonable about this result? (c) Which assumptions are unreasonable or inconsistent?

29.6 The Wave Nature of Matter

49. At what velocity will an electron have a wavelength of 1.00 m?

50. What is the wavelength of an electron moving at 3.00% of the speed of light?

51. At what velocity does a proton have a 6.00-fm wavelength (about the size of a nucleus)? Assume the proton is nonrelativistic. (1 femtometer = 10^{-15} m.)

52. What is the velocity of a 0.400-kg billiard ball if its wavelength is 7.50 cm (large enough for it to interfere with other billiard balls)?

53. Find the wavelength of a proton moving at 1.00% of the speed of light.

54. Experiments are performed with ultracold neutrons having velocities as small as 1.00 m/s. (a) What is the wavelength of such a neutron? (b) What is its kinetic energy in eV?

55. (a) Find the velocity of a neutron that has a 6.00-fm wavelength (about the size of a nucleus). Assume the neutron is nonrelativistic. (b) What is the neutron's kinetic energy in MeV?

56. What is the wavelength of an electron accelerated through a 30.0-kV potential, as in a TV tube?

57. What is the kinetic energy of an electron in a TEM having a 0.0100-nm wavelength?

58. (a) Calculate the velocity of an electron that has a wavelength of 1.00 μm. (b) Through what voltage must the electron be accelerated to have this velocity?

59. The velocity of a proton emerging from a Van de Graaff accelerator is 25.0% of the speed of light. (a) What is the proton's wavelength? (b) What is its kinetic energy, assuming it is nonrelativistic? (c) What was the equivalent voltage through which it was accelerated?

60. The kinetic energy of an electron accelerated in an x-ray tube is 100 keV. Assuming it is nonrelativistic, what is its wavelength?

61. Unreasonable Results

(a) Assuming it is nonrelativistic, calculate the velocity of an electron with a 0.100-fm wavelength (small enough to detect details of a nucleus). (b) What is unreasonable about this result? (c) Which assumptions are unreasonable or inconsistent?

29.7 Probability: The Heisenberg Uncertainty Principle

62. (a) If the position of an electron in a membrane is measured to an accuracy of 1.00 μm, what is the electron's minimum uncertainty in velocity? (b) If the electron has this velocity, what is its kinetic energy in eV? (c) What are the implications of this energy, comparing it to typical molecular binding energies?

63. (a) If the position of a chlorine ion in a membrane is measured to an accuracy of 1.00 μm, what is its minimum uncertainty in velocity, given its mass is 5.86×10^{-26} kg ?

(b) If the ion has this velocity, what is its kinetic energy in eV, and how does this compare with typical molecular binding energies?

64. Suppose the velocity of an electron in an atom is known to an accuracy of 2.0×10^3 m/s (reasonably accurate compared with orbital velocities). What is the electron's minimum uncertainty in position, and how does this compare with the approximate 0.1-nm size of the atom?

65. The velocity of a proton in an accelerator is known to an accuracy of 0.250% of the speed of light. (This could be small compared with its velocity.) What is the smallest possible uncertainty in its position?

66. A relatively long-lived excited state of an atom has a lifetime of 3.00 ms. What is the minimum uncertainty in its energy?

67. (a) The lifetime of a highly unstable nucleus is 10^{-20} s . What is the smallest uncertainty in its decay energy? (b) Compare this with the rest energy of an electron.

68. The decay energy of a short-lived particle has an uncertainty of 1.0 MeV due to its short lifetime. What is the smallest lifetime it can have?

69. The decay energy of a short-lived nuclear excited state has an uncertainty of 2.0 eV due to its short lifetime. What is the smallest lifetime it can have?

70. What is the approximate uncertainty in the mass of a muon, as determined from its decay lifetime?

71. Derive the approximate form of Heisenberg's uncertainty principle for energy and time, $\Delta E \Delta t \approx h$, using the following arguments: Since the position of a particle is uncertain by $\Delta x \approx \lambda$, where λ is the wavelength of the photon used to examine it, there is an uncertainty in the time the photon takes to traverse Δx. Furthermore, the photon has an energy related to its wavelength, and it can transfer some or all of this energy to the object being examined. Thus the uncertainty in the energy of the object is also related to λ. Find Δt and ΔE; then multiply them to give the approximate uncertainty principle.

29.8 The Particle-Wave Duality Reviewed

72. Integrated Concepts

The 54.0-eV electron in **Example 29.7** has a 0.167-nm wavelength. If such electrons are passed through a double slit and have their first maximum at an angle of $25.0°$, what is the slit separation d ?

73. Integrated Concepts

An electron microscope produces electrons with a 2.00-pm wavelength. If these are passed through a 1.00-nm single slit, at what angle will the first diffraction minimum be found?

74. Integrated Concepts

A certain heat lamp emits 200 W of mostly IR radiation averaging 1500 nm in wavelength. (a) What is the average photon energy in joules? (b) How many of these photons are required to increase the temperature of a person's shoulder by $2.0°C$, assuming the affected mass is 4.0 kg with a specific heat of 0.83 kcal/kg · °C. Also assume no other significant heat transfer. (c) How long does this take?

75. Integrated Concepts

On its high power setting, a microwave oven produces 900 W of 2560 MHz microwaves. (a) How many photons per second is this? (b) How many photons are required to increase the temperature of a 0.500-kg mass of pasta by $45.0°C$, assuming a specific heat of 0.900 kcal/kg · °C ? Neglect all other heat transfer. (c) How long must the microwave operator wait for their pasta to be ready?

76. Integrated Concepts

(a) Calculate the amount of microwave energy in joules needed to raise the temperature of 1.00 kg of soup from $20.0°C$ to $100°C$. (b) What is the total momentum of all the microwave photons it takes to do this? (c) Calculate the velocity of a 1.00-kg mass with the same momentum. (d) What is the kinetic energy of this mass?

77. Integrated Concepts

(a) What is γ for an electron emerging from the Stanford Linear Accelerator with a total energy of 50.0 GeV? (b) Find its momentum. (c) What is the electron's wavelength?

78. Integrated Concepts

(a) What is γ for a proton having an energy of 1.00 TeV, produced by the Fermilab accelerator? (b) Find its momentum. (c) What is the proton's wavelength?

79. Integrated Concepts

An electron microscope passes 1.00-pm-wavelength electrons through a circular aperture $2.00\ \mu\text{m}$ in diameter. What is the angle between two just-resolvable point sources for this microscope?

80. Integrated Concepts

(a) Calculate the velocity of electrons that form the same pattern as 450-nm light when passed through a double slit. (b) Calculate the kinetic energy of each and compare them. (c) Would either be easier to generate than the other? Explain.

81. Integrated Concepts

(a) What is the separation between double slits that produces a second-order minimum at $45.0°$ for 650-nm light? (b) What slit separation is needed to produce the same pattern for 1.00-keV protons.

82. Integrated Concepts

A laser with a power output of 2.00 mW at a wavelength of 400 nm is projected onto calcium metal. (a) How many electrons per second are ejected? (b) What power is carried away by the electrons, given that the binding energy is 2.71 eV? (c) Calculate the current of ejected electrons. (d) If the photoelectric material is electrically insulated and acts like a 2.00-pF capacitor, how long will current flow before the capacitor voltage stops it?

83. Integrated Concepts

One problem with x rays is that they are not sensed. Calculate the temperature increase of a researcher exposed in a few seconds to a nearly fatal accidental dose of x rays under the following conditions. The energy of the x-ray photons is 200 keV, and 4.00×10^{13} of them are absorbed per kilogram of tissue, the specific heat of which is $0.830\ \text{kcal/kg} \cdot °\text{C}$. (Note that medical diagnostic x-ray machines *cannot* produce an intensity this great.)

84. Integrated Concepts

A 1.00-fm photon has a wavelength short enough to detect some information about nuclei. (a) What is the photon momentum? (b) What is its energy in joules and MeV? (c) What is the (relativistic) velocity of an electron with the same momentum? (d) Calculate the electron's kinetic energy.

85. Integrated Concepts

The momentum of light is exactly reversed when reflected straight back from a mirror, assuming negligible recoil of the mirror. Thus the change in momentum is twice the photon momentum. Suppose light of intensity $1.00\ \text{kW/m}^2$ reflects from a mirror of area $2.00\ \text{m}^2$. (a) Calculate the energy reflected in 1.00 s. (b) What is the momentum imparted to the mirror? (c) Using the most general form of Newton's second law, what is the force on the mirror? (d) Does the assumption of no mirror recoil seem reasonable?

86. Integrated Concepts

Sunlight above the Earth's atmosphere has an intensity of $1.30\ \text{kW/m}^2$. If this is reflected straight back from a mirror that has only a small recoil, the light's momentum is exactly reversed, giving the mirror twice the incident momentum. (a) Calculate the force per square meter of mirror. (b) Very low mass mirrors can be constructed in the near weightlessness of space, and attached to a spaceship to sail it. Once done, the average mass per square meter of the spaceship is 0.100 kg. Find the acceleration of the spaceship if all other forces are balanced. (c) How fast is it moving 24 hours later?

Test Prep for AP® Courses

29.1 Quantization of Energy

1. The visible spectrum of sunlight shows a range of colors from red to violet. This spectrum has numerous dark lines spread throughout it. Noting that the surface of the Sun is much cooler than the interior, so that the surface is comparable to a cool gas through which light passes, which of the following statements correctly explains the dark lines?

a. The cooler, denser surface material scatters certain wavelengths of light, forming dark lines.
b. The atoms at the surface absorb certain wavelengths of light, causing the dark lines at those wavelengths.
c. The atoms in the Sun's interior emit light of specific wavelength, so that parts of the spectrum are dark.
d. The atoms at the surface are excited by the high interior temperatures, so that the dark lines are merely wavelengths at which those atoms don't emit energy.

2. A log in a fireplace burns for nearly an hour, at which point it consists mostly of small, hot embers. These embers glow a bright orange and whitish-yellow color. Describe the characteristics of the energy of this system, both in terms of energy transfer and the quantum behavior of blackbodies.

29.2 The Photoelectric Effect

3. A metal exposed to a beam of light with a wavelength equal to or shorter than a specific wavelength emits electrons. What property of light, as described in the quantum explanation of blackbody radiation, accounts for this photoelectric process?
a. The energy of light increases as its speed increases.
b. The energy of light increases as its intensity increases.
c. The energy of light increases as its frequency increases.
d. The energy of light increases as its wavelength increases.

4. During his experiments that confirmed the existence of electromagnetic waves, Heinrich Hertz used a spark across a gap between two electrodes to provide the rapidly changing electric current that produced electromagnetic waves. He noticed, however, that production of the spark required a lower voltage in a well-lighted laboratory than when the room was dark. Describe how this curious event can be explained in terms of the quantum interpretation of the photoelectric effect.

29.3 Photon Energies and the Electromagnetic Spectrum

5. A microwave oven produces electromagnetic radiation in the radio portion of the spectrum. These microwave photons are absorbed by water molecules, resulting in an increase in the molecules' rotational energies. This added energy is transferred by heat to the surrounding food, which as a result becomes hot very quickly. If the energy absorbed by a water molecule is 1.0×10^{-5} eV, what is the corresponding wavelength of the microwave photons?
a. 1.22 GHz
b. 2.45 GHz
c. 4.90 GHz
d. 9.80 Hz

6. In the intensity versus frequency curve for x rays (**Figure 29.14**), the intensity is mostly a smooth curve associated with *bremsstrahlung* ("breaking radiation"). However, there are two spikes (characteristic x rays) that exhibit high-intensity output. Explain how the smooth curve can be described by classical electrodynamics, whereas the peaks require a quantum mechanical interpretation. (Recall that the acceleration or deceleration of electric charges causes the emission of electromagnetic radiation.)

29.4 Photon Momentum

7. The mass of a proton is 1.67×10^{-27} kg. If a proton has the same momentum as a photon with a wavelength of 325 nm, what is its speed?

a. 2.73×10^{-3} m/s
b. 0.819 m/s
c. 1.22 m/s
d. 2.71×10^{4} m/s

8. A strip of metal foil with a mass of 5.00×10^{-7} kg is suspended in a vacuum and exposed to a pulse of light. The velocity of the foil changes from zero to 1.00×10^{-3} m/s in the same direction as the initial light pulse, and the light pulse is entirely reflected from the surface of the foil. Given that the wavelength of the light is 450 nm, and assuming that this wavelength is the same before and after the collision, how many photons in the pulse collide with the foil?

9. In an experiment in which the Compton effect is observed, a "gamma ray" photon with a wavelength of 5.00×10^{-13} m scatters from an electron. If the change in the electron energy is 1.60×10^{-15} J, what is the wavelength of the photon after the collision with the electron?
a. 4.95×10^{-13} m
b. 4.98×10^{-13} m
c. 5.02×10^{-13} m
d. 5.05×10^{-13} m

10. Consider two experiments involving a metal sphere with a radius of 2.00 μm that is suspended in a vacuum. In one experiment, a pulse of N photons reflects from the surface of the sphere, causing the sphere to acquire momentum. In a second experiment, an identical pulse of photons is completely absorbed by the sphere, so that the sphere acquires momentum. Identify each type of collision as either elastic or inelastic, and, assuming that the change in the photon wavelength can be ignored, use linear momentum conservation to derive the expression for the momentum of the sphere in each experiment.

29.5 The Particle-Wave Duality

11. The ground state of a certain type of atom has energy of $-E_0$. What is the wavelength of a photon with enough energy to ionize the atom when it is in the ground state, so that the ejected electron has kinetic energy equal to $2E_0$?
a. $\lambda = \dfrac{hc}{3E_0}$

b. $\lambda = \dfrac{hc}{2E_0}$

c. $\lambda = \dfrac{hc}{E_0}$

d. $\lambda = \dfrac{2hc}{E_0}$

12. While the quantum model explains many physical processes that the classical model cannot, it must be consistent with those processes that the classical model does explain. Energy and momentum conservation are fundamental principles of classical physics. Use the Compton and photoelectric effects to explain how these conservation principles carry over to the quantum model of light.

29.6 The Wave Nature of Matter

13. The least massive particle known to exist is the electron neutrino. Though scientists once believed that it had no mass, like the photon, they have now determined that this particle has an extremely low mass, equivalent to a few electron volts. Assuming a mass of 2.2 eV/c^2 (or 3.9×10^{-36} kg) and a speed of 4.4×10^6 m/s, which of the following values equals the neutrino's de Broglie wavelength?

a. 3.8×10^{-5} m

b. 4.7×10^{-7} m

c. 1.7×10^{-10} m

d. 8.9×10^{-14} m

14. Using the definition of the de Broglie wavelength, explain how wavelike properties of matter increase with a decrease in mass or decrease in speed. Use as examples an electron (mass = 9.11×10^{-31} kg) with a speed of 5.0×10^6 m/s and a proton (mass = 1.67×10^{-27} kg) with a speed of 8.0×10^6 m/s.

15. In a Davisson-Germer type of experiment, a crystal with a parallel-plane separation (d) of 9.1×10^{-2} nm produces constructive interference with an electron beam at an angle of $\theta = 50°$. Which of the following is the maximum de Broglie wavelength for these electrons?

a. 0.07nm

b. 0.09 nm

c. 0.14 nm

d. 0.21 nm

16. In a Davisson-Germer experiment, electrons with a speed of 6.5×10^6 m/s exhibit third-order ($n = 3$) constructive interference for a crystal with unknown plane separation, d. Given an angle of incidence of $\theta = 45°$, compute the value for d. Compare the de Broglie wavelength to electromagnetic radiation with the same wavelength. (Recall that the mass of the electron is 9.11×10^{-31} kg.)

29.8 The Particle-Wave Duality Reviewed

17. Which of the following describes one of the main features of wave-particle duality?

a. As speed increases, the wave nature of matter becomes more evident.

b. As momentum decreases, the particle nature of matter becomes more evident.

c. As energy increases, the wave nature of matter becomes easier to observe.

d. As mass increases, the wave nature of matter is less easy to observe.

18. Explain why Heisenberg's uncertainty principle limits the precision with which either momentum or position of a subatomic particle can be known, but becomes less applicable for matter at the macroscopic level.

30   ATOMIC PHYSICS

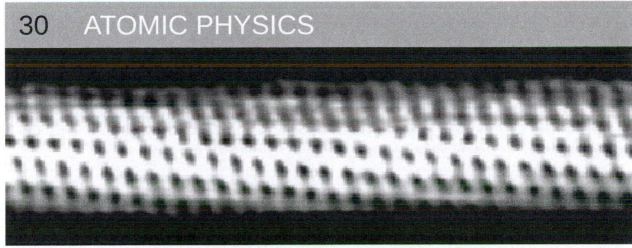

Figure 30.1 Individual carbon atoms are visible in this image of a carbon nanotube made by a scanning tunneling electron microscope. (credit: Taner Yildirim, National Institute of Standards and Technology, via Wikimedia Commons)

Chapter Outline

30.1. Discovery of the Atom

30.2. Discovery of the Parts of the Atom: Electrons and Nuclei

30.3. Bohr's Theory of the Hydrogen Atom

30.4. X Rays: Atomic Origins and Applications

30.5. Applications of Atomic Excitations and De-Excitations

30.6. The Wave Nature of Matter Causes Quantization

30.7. Patterns in Spectra Reveal More Quantization

30.8. Quantum Numbers and Rules

30.9. The Pauli Exclusion Principle

Connection for AP® Courses

Have you ever wondered how we know the composition of the Sun? After all, we cannot travel there to physically collect a sample due to the extreme conditions. Fortunately, our understanding of the internal structure of atoms gives us the tools to identify the elements in the Sun's outer layers due to an atomic "fingerprint" in the Sun's spectrum. You will learn about atoms and their substructures, as well as how these substructures determine the behavior of the atom, such as the absorption and emission of energy by electrons within an atom.

You will learn the stories of how we discovered the various properties of an atom (Essential Knowledge 1.A.4) through clever and imaginative experimentation (such as the Millikan oil drop experiment) and interpretation (such as Brownian motion). You will also learn about the probabilistic description we use to describe the nature of electrons (Essential Knowledge 7.C.1). At this scale, electrons can be thought of as discrete particles, but they also behave in a way that is consistent with a wave model of matter (Enduring Understanding 7.C). You will learn how we use the wave model to understand the energy levels in an atom (Essential Knowledge 7.C.2) and the properties of electrons.

The content in this chapter supports:

Big Idea 1 Objects and systems have properties such as mass and charge. Systems may have internal structure.

Enduring Understanding 1.A The internal structure of a system determines many properties of the system.

Essential Knowledge 1.A.4 Atoms have internal structures that determine their properties.

Essential Knowledge 1.A.5 Systems have properties determined by the properties and interactions of their constituent atomic and molecular substructures.

Enduring Understanding 1.B Electric charge is a property of an object or system that affects its interactions with other objects or systems containing charge.

Essential Knowledge 1.B.3 The smallest observed unit of charge that can be isolated is the electron charge, also known as the elementary charge.

Big Idea 5 Changes that occur as a result of interactions are constrained by conservation laws.

Enduring Understanding 5.B The energy of a system is conserved.

Essential Knowledge 5.B.8 Energy transfer occurs when photons are absorbed or emitted, for example, by atoms or nuclei.

Big Idea 7 The mathematics of probability can be used to describe the behavior of complex systems and to interpret the behavior of quantum mechanical systems.

Enduring Understanding 7.C At the quantum scale, matter is described by a wave function, which leads to a probabilistic description of the microscopic world.

Essential Knowledge 7.C.1 The probabilistic description of matter is modeled by a wave function, which can be assigned to an object and used to describe its motion and interactions. The absolute value of the wave function is related to the probability of finding a particle in some spatial region.

Essential Knowledge 7.C.2 The allowed states for an electron in an atom can be calculated from the wave model of an electron.

Essential Knowledge 7.C.4 Photon emission and absorption processes are described by probability.

30.1 Discovery of the Atom

Learning Objectives

By the end of this section, you will be able to:

- Describe the basic structure of the atom, the basic unit of all matter.

How do we know that atoms are really there if we cannot see them with our eyes? A brief account of the progression from the proposal of atoms by the Greeks to the first direct evidence of their existence follows.

People have long speculated about the structure of matter and the existence of atoms. The earliest significant ideas to survive are due to the ancient Greeks in the fifth century BCE, especially those of the philosophers Leucippus and Democritus. (There is some evidence that philosophers in both India and China made similar speculations, at about the same time.) They considered the question of whether a substance can be divided without limit into ever smaller pieces. There are only a few possible answers to this question. One is that infinitesimally small subdivision is possible. Another is what Democritus in particular believed—that there is a smallest unit that cannot be further subdivided. Democritus called this the **atom**. We now know that atoms themselves can be subdivided, but their identity is destroyed in the process, so the Greeks were correct in a respect. The Greeks also felt that atoms were in constant motion, another correct notion.

The Greeks and others speculated about the properties of atoms, proposing that only a few types existed and that all matter was formed as various combinations of these types. The famous proposal that the basic elements were earth, air, fire, and water was brilliant, but incorrect. The Greeks had identified the most common examples of the four states of matter (solid, gas, plasma, and liquid), rather than the basic elements. More than 2000 years passed before observations could be made with equipment capable of revealing the true nature of atoms.

Over the centuries, discoveries were made regarding the properties of substances and their chemical reactions. Certain systematic features were recognized, but similarities between common and rare elements resulted in efforts to transmute them (lead into gold, in particular) for financial gain. Secrecy was endemic. Alchemists discovered and rediscovered many facts but did not make them broadly available. As the Middle Ages ended, alchemy gradually faded, and the science of chemistry arose. It was no longer possible, nor considered desirable, to keep discoveries secret. Collective knowledge grew, and by the beginning of the 19th century, an important fact was well established—the masses of reactants in specific chemical reactions always have a particular mass ratio. This is very strong indirect evidence that there are basic units (atoms and molecules) that have these same mass ratios. The English chemist John Dalton (1766–1844) did much of this work, with significant contributions by the Italian physicist Amedeo Avogadro (1776–1856). It was Avogadro who developed the idea of a fixed number of atoms and molecules in a mole, and this special number is called Avogadro's number in his honor. The Austrian physicist Johann Josef Loschmidt was the first to measure the value of the constant in 1865 using the kinetic theory of gases.

Patterns and Systematics

The recognition and appreciation of patterns has enabled us to make many discoveries. The periodic table of elements was proposed as an organized summary of the known elements long before all elements had been discovered, and it led to many other discoveries. We shall see in later chapters that patterns in the properties of subatomic particles led to the proposal of quarks as their underlying structure, an idea that is still bearing fruit.

Knowledge of the properties of elements and compounds grew, culminating in the mid-19th-century development of the periodic table of the elements by Dmitri Mendeleev (1834–1907), the great Russian chemist. Mendeleev proposed an ingenious array that highlighted the periodic nature of the properties of elements. Believing in the systematics of the periodic table, he also predicted the existence of then-unknown elements to complete it. Once these elements were discovered and determined to have properties predicted by Mendeleev, his periodic table became universally accepted.

Also during the 19th century, the kinetic theory of gases was developed. Kinetic theory is based on the existence of atoms and molecules in random thermal motion and provides a microscopic explanation of the gas laws, heat transfer, and thermodynamics (see **Introduction to Temperature, Kinetic Theory, and the Gas Laws** and **Introduction to Laws of Thermodynamics**). Kinetic theory works so well that it is another strong indication of the existence of atoms. But it is still indirect evidence—individual atoms and molecules had not been observed. There were heated debates about the validity of kinetic theory until direct evidence of atoms was obtained.

The first truly direct evidence of atoms is credited to Robert Brown, a Scottish botanist. In 1827, he noticed that tiny pollen grains suspended in still water moved about in complex paths. This can be observed with a microscope for any small particles in a fluid. The motion is caused by the random thermal motions of fluid molecules colliding with particles in the fluid, and it is now called **Brownian motion**. (See **Figure 30.2**.) Statistical fluctuations in the numbers of molecules striking the sides of a visible particle cause it to move first this way, then that. Although the molecules cannot be directly observed, their effects on the particle can be. By examining Brownian motion, the size of molecules can be calculated. The smaller and more numerous they are, the smaller the fluctuations in the numbers striking different sides.

Figure 30.2 The position of a pollen grain in water, measured every few seconds under a microscope, exhibits Brownian motion. Brownian motion is due to fluctuations in the number of atoms and molecules colliding with a small mass, causing it to move about in complex paths. This is nearly direct evidence for the existence of atoms, providing a satisfactory alternative explanation cannot be found.

It was Albert Einstein who, starting in his epochal year of 1905, published several papers that explained precisely how Brownian motion could be used to measure the size of atoms and molecules. (In 1905 Einstein created special relativity, proposed photons as quanta of EM radiation, and produced a theory of Brownian motion that allowed the size of atoms to be determined. All of this was done in his spare time, since he worked days as a patent examiner. Any one of these very basic works could have been the crowning achievement of an entire career—yet Einstein did even more in later years.) Their sizes were only approximately known to be 10^{-10} m, based on a comparison of latent heat of vaporization and surface tension made in about 1805 by Thomas Young of double-slit fame and the famous astronomer and mathematician Simon Laplace.

Using Einstein's ideas, the French physicist Jean-Baptiste Perrin (1870–1942) carefully observed Brownian motion; not only did he confirm Einstein's theory, he also produced accurate sizes for atoms and molecules. Since molecular weights and densities of materials were well established, knowing atomic and molecular sizes allowed a precise value for Avogadro's number to be obtained. (If we know how big an atom is, we know how many fit into a certain volume.) Perrin also used these ideas to explain atomic and molecular agitation effects in sedimentation, and he received the 1926 Nobel Prize for his achievements. Most scientists were already convinced of the existence of atoms, but the accurate observation and analysis of Brownian motion was conclusive—it was the first truly direct evidence.

A huge array of direct and indirect evidence for the existence of atoms now exists. For example, it has become possible to accelerate ions (much as electrons are accelerated in cathode-ray tubes) and to detect them individually as well as measure their masses (see **More Applications of Magnetism** for a discussion of mass spectrometers). Other devices that observe individual atoms, such as the scanning tunneling electron microscope, will be discussed elsewhere. (See **Figure 30.3**.) All of our understanding of the properties of matter is based on and consistent with the atom. The atom's substructures, such as electron shells and the nucleus, are both interesting and important. The nucleus in turn has a substructure, as do the particles of which it is composed. These topics, and the question of whether there is a smallest basic structure to matter, will be explored in later parts of the text.

Figure 30.3 Individual atoms can be detected with devices such as the scanning tunneling electron microscope that produced this image of individual gold atoms on a graphite substrate. (credit: Erwin Rossen, Eindhoven University of Technology, via Wikimedia Commons)

30.2 Discovery of the Parts of the Atom: Electrons and Nuclei

Learning Objectives

By the end of this section, you will be able to:

- Describe how electrons were discovered.
- Explain the Millikan oil drop experiment.
- Describe Rutherford's gold foil experiment.
- Describe Rutherford's planetary model of the atom.

The information presented in this section supports the following AP® learning objectives and science practices:

- **1.B.3.1** The student is able to challenge the claim that an electric charge smaller than the elementary charge has been isolated. **(S.P. 1.5, 6.1, 7.2)**

Just as atoms are a substructure of matter, electrons and nuclei are substructures of the atom. The experiments that were used to discover electrons and nuclei reveal some of the basic properties of atoms and can be readily understood using ideas such as electrostatic and magnetic force, already covered in previous chapters.

Charges and Electromagnetic Forces

In previous discussions, we have noted that positive charge is associated with nuclei and negative charge with electrons. We have also covered many aspects of the electric and magnetic forces that affect charges. We will now explore the discovery of the electron and nucleus as substructures of the atom and examine their contributions to the properties of atoms.

The Electron

Gas discharge tubes, such as that shown in **Figure 30.4**, consist of an evacuated glass tube containing two metal electrodes and a rarefied gas. When a high voltage is applied to the electrodes, the gas glows. These tubes were the precursors to today's neon lights. They were first studied seriously by Heinrich Geissler, a German inventor and glassblower, starting in the 1860s. The English scientist William Crookes, among others, continued to study what for some time were called Crookes tubes, wherein electrons are freed from atoms and molecules in the rarefied gas inside the tube and are accelerated from the cathode (negative) to the anode (positive) by the high potential. These "*cathode rays*" collide with the gas atoms and molecules and excite them, resulting in the emission of electromagnetic (EM) radiation that makes the electrons' path visible as a ray that spreads and fades as it moves away from the cathode.

Gas discharge tubes today are most commonly called **cathode-ray tubes**, because the rays originate at the cathode. Crookes showed that the electrons carry momentum (they can make a small paddle wheel rotate). He also found that their normally straight path is bent by a magnet in the direction expected for a negative charge moving away from the cathode. These were the first direct indications of electrons and their charge.

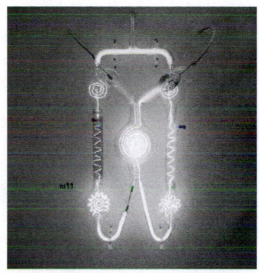

Figure 30.4 A gas discharge tube glows when a high voltage is applied to it. Electrons emitted from the cathode are accelerated toward the anode; they excite atoms and molecules in the gas, which glow in response. Once called Geissler tubes and later Crookes tubes, they are now known as cathode-ray tubes (CRTs) and are found in older TVs, computer screens, and x-ray machines. When a magnetic field is applied, the beam bends in the direction expected for negative charge. (credit: Paul Downey, Flickr)

The English physicist J. J. Thomson (1856–1940) improved and expanded the scope of experiments with gas discharge tubes. (See **Figure 30.5** and **Figure 30.6**.) He verified the negative charge of the cathode rays with both magnetic and electric fields. Additionally, he collected the rays in a metal cup and found an excess of negative charge. Thomson was also able to measure the ratio of the charge of the electron to its mass, q_e / m_e—an important step to finding the actual values of both q_e and m_e. **Figure 30.7** shows a cathode-ray tube, which produces a narrow beam of electrons that passes through charging plates connected to a high-voltage power supply. An electric field $\mathbf{E}$ is produced between the charging plates, and the cathode-ray tube is placed between the poles of a magnet so that the electric field $\mathbf{E}$ is perpendicular to the magnetic field $\mathbf{B}$ of the magnet. These fields, being perpendicular to each other, produce opposing forces on the electrons. As discussed for mass spectrometers in **More Applications of Magnetism**, if the net force due to the fields vanishes, then the velocity of the charged particle is $v = E/B$. In this manner, Thomson determined the velocity of the electrons and then moved the beam up and down by adjusting the electric field.

Figure 30.5 J. J. Thomson (credit: www.firstworldwar.com, via Wikimedia Commons)

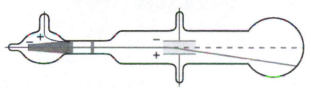

Figure 30.6 Diagram of Thomson's CRT. (credit: Kurzon, Wikimedia Commons)

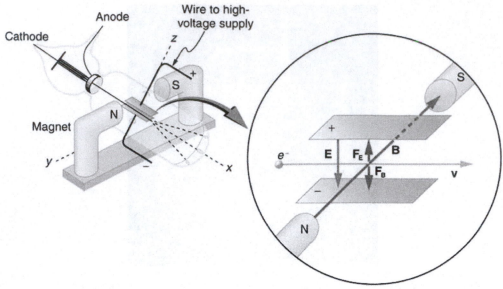

Figure 30.7 This schematic shows the electron beam in a CRT passing through crossed electric and magnetic fields and causing phosphor to glow when striking the end of the tube.

To see how the amount of deflection is used to calculate q_e/m_e, note that the deflection is proportional to the electric force on the electron:

$$F = q_e E. \tag{30.1}$$

But the vertical deflection is also related to the electron's mass, since the electron's acceleration is

$$a = \frac{F}{m_e}. \tag{30.2}$$

The value of F is not known, since q_e was not yet known. Substituting the expression for electric force into the expression for acceleration yields

$$a = \frac{F}{m_e} = \frac{q_e E}{m_e}. \tag{30.3}$$

Gathering terms, we have

$$\frac{q_e}{m_e} = \frac{a}{E}. \tag{30.4}$$

The deflection is analyzed to get a, and E is determined from the applied voltage and distance between the plates; thus, $\frac{q_e}{m_e}$ can be determined. With the velocity known, another measurement of $\frac{q_e}{m_e}$ can be obtained by bending the beam of electrons with the magnetic field. Since $F_{\text{mag}} = q_e vB = m_e a$, we have $q_e/m_e = a/vB$. Consistent results are obtained using magnetic deflection.

What is so important about q_e/m_e, the ratio of the electron's charge to its mass? The value obtained is

$$\frac{q_e}{m_e} = -1.76\times10^{11} \text{ C/kg (electron).} \tag{30.5}$$

This is a huge number, as Thomson realized, and it implies that the electron has a very small mass. It was known from electroplating that about 10^8 C/kg is needed to plate a material, a factor of about 1000 less than the charge per kilogram of electrons. Thomson went on to do the same experiment for positively charged hydrogen ions (now known to be bare protons) and found a charge per kilogram about 1000 times smaller than that for the electron, implying that the proton is about 1000 times more massive than the electron. Today, we know more precisely that

$$\frac{q_p}{m_p} = 9.58\times10^7 \text{ C/kg (proton),} \tag{30.6}$$

where q_p is the charge of the proton and m_p is its mass. This ratio (to four significant figures) is 1836 times less charge per kilogram than for the electron. Since the charges of electrons and protons are equal in magnitude, this implies $m_p = 1836 m_e$.

Thomson performed a variety of experiments using differing gases in discharge tubes and employing other methods, such as the photoelectric effect, for freeing electrons from atoms. He always found the same properties for the electron, proving it to be an independent particle. For his work, the important pieces of which he began to publish in 1897, Thomson was awarded the 1906 Nobel Prize in Physics. In retrospect, it is difficult to appreciate how astonishing it was to find that the atom has a substructure. Thomson himself said, "It was only when I was convinced that the experiment left no escape from it that I published my belief in the existence of bodies smaller than atoms."

Thomson attempted to measure the charge of individual electrons, but his method could determine its charge only to the order of magnitude expected.

Since Faraday's experiments with electroplating in the 1830s, it had been known that about 100,000 C per mole was needed to plate singly ionized ions. Dividing this by the number of ions per mole (that is, by Avogadro's number), which was approximately known, the charge per ion was calculated to be about 1.6×10^{-19} C, close to the actual value.

An American physicist, Robert Millikan (1868–1953) (see **Figure 30.8**), decided to improve upon Thomson's experiment for measuring q_e and was eventually forced to try another approach, which is now a classic experiment performed by students.

The Millikan oil drop experiment is shown in **Figure 30.9**.

Figure 30.8 Robert Millikan (credit: Unknown Author, via Wikimedia Commons)

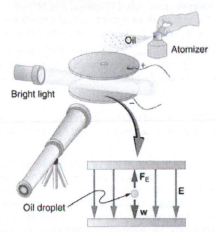

Figure 30.9 The Millikan oil drop experiment produced the first accurate direct measurement of the charge on electrons, one of the most fundamental constants in nature. Fine drops of oil become charged when sprayed. Their movement is observed between metal plates with a potential applied to oppose the gravitational force. The balance of gravitational and electric forces allows the calculation of the charge on a drop. The charge is found to be quantized in units of -1.6×10^{-19} C, thus determining directly the charge of the excess and missing electrons on the oil drops.

In the Millikan oil drop experiment, fine drops of oil are sprayed from an atomizer. Some of these are charged by the process and can then be suspended between metal plates by a voltage between the plates. In this situation, the weight of the drop is balanced by the electric force:

$$m_{\text{drop}}g = q_e E \qquad (30.7)$$

The electric field is produced by the applied voltage, hence, $E = V/d$, and V is adjusted to just balance the drop's weight. The drops can be seen as points of reflected light using a microscope, but they are too small to directly measure their size and mass. The mass of the drop is determined by observing how fast it falls when the voltage is turned off. Since air resistance is very significant for these submicroscopic drops, the more massive drops fall faster than the less massive, and sophisticated sedimentation calculations can reveal their mass. Oil is used rather than water, because it does not readily evaporate, and so mass is nearly constant. Once the mass of the drop is known, the charge of the electron is given by rearranging the previous equation:

$$q = \frac{m_{drop}\,g}{E} = \frac{m_{drop}\,g\,d}{V}, \tag{30.8}$$

where d is the separation of the plates and V is the voltage that holds the drop motionless. (The same drop can be observed for several hours to see that it really is motionless.) By 1913 Millikan had measured the charge of the electron q_e to an accuracy of 1%, and he improved this by a factor of 10 within a few years to a value of -1.60×10^{-19} C . He also observed that all charges were multiples of the basic electron charge and that sudden changes could occur in which electrons were added or removed from the drops. For this very fundamental direct measurement of q_e and for his studies of the photoelectric effect, Millikan was awarded the 1923 Nobel Prize in Physics.

With the charge of the electron known and the charge-to-mass ratio known, the electron's mass can be calculated. It is

$$m = \frac{q_e}{\left(\frac{q_e}{m_e}\right)}. \tag{30.9}$$

Substituting known values yields

$$m_e = \frac{-1.60 \times 10^{-19}\ \text{C}}{-1.76 \times 10^{11}\ \text{C/kg}} \tag{30.10}$$

or

$$m_e = 9.11 \times 10^{-31}\ \text{kg (electron's mass)}, \tag{30.11}$$

where the round-off errors have been corrected. The mass of the electron has been verified in many subsequent experiments and is now known to an accuracy of better than one part in one million. It is an incredibly small mass and remains the smallest known mass of any particle that has mass. (Some particles, such as photons, are massless and cannot be brought to rest, but travel at the speed of light.) A similar calculation gives the masses of other particles, including the proton. To three digits, the mass of the proton is now known to be

$$m_p = 1.67 \times 10^{-27}\ \text{kg (proton's mass)}, \tag{30.12}$$

which is nearly identical to the mass of a hydrogen atom. What Thomson and Millikan had done was to prove the existence of one substructure of atoms, the electron, and further to show that it had only a tiny fraction of the mass of an atom. The nucleus of an atom contains most of its mass, and the nature of the nucleus was completely unanticipated.

Another important characteristic of quantum mechanics was also beginning to emerge. All electrons are identical to one another. The charge and mass of electrons are not average values; rather, they are unique values that all electrons have. This is true of other fundamental entities at the submicroscopic level. All protons are identical to one another, and so on.

The Nucleus

Here, we examine the first direct evidence of the size and mass of the nucleus. In later chapters, we will examine many other aspects of nuclear physics, but the basic information on nuclear size and mass is so important to understanding the atom that we consider it here.

Nuclear radioactivity was discovered in 1896, and it was soon the subject of intense study by a number of the best scientists in the world. Among them was New Zealander Lord Ernest Rutherford, who made numerous fundamental discoveries and earned the title of "father of nuclear physics." Born in Nelson, Rutherford did his postgraduate studies at the Cavendish Laboratories in England before taking up a position at McGill University in Canada where he did the work that earned him a Nobel Prize in Chemistry in 1908. In the area of atomic and nuclear physics, there is much overlap between chemistry and physics, with physics providing the fundamental enabling theories. He returned to England in later years and had six future Nobel Prize winners as students. Rutherford used nuclear radiation to directly examine the size and mass of the atomic nucleus. The experiment he devised is shown in **Figure 30.10**. A radioactive source that emits alpha radiation was placed in a lead container with a hole in one side to produce a beam of alpha particles, which are a type of ionizing radiation ejected by the nuclei of a radioactive source. A thin gold foil was placed in the beam, and the scattering of the alpha particles was observed by the glow they caused when they struck a phosphor screen.

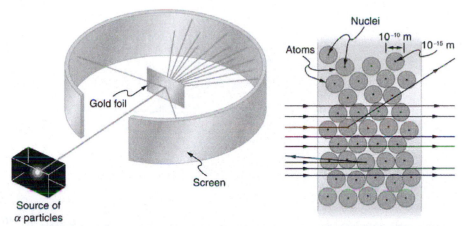

Figure 30.10 Rutherford's experiment gave direct evidence for the size and mass of the nucleus by scattering alpha particles from a thin gold foil. Alpha particles with energies of about 5 MeV are emitted from a radioactive source (which is a small metal container in which a specific amount of a radioactive material is sealed), are collimated into a beam, and fall upon the foil. The number of particles that penetrate the foil or scatter to various angles indicates that gold nuclei are very small and contain nearly all of the gold atom's mass. This is particularly indicated by the alpha particles that scatter to very large angles, much like a soccer ball bouncing off a goalie's head.

Alpha particles were known to be the doubly charged positive nuclei of helium atoms that had kinetic energies on the order of 5 MeV when emitted in nuclear decay, which is the disintegration of the nucleus of an unstable nuclide by the spontaneous emission of charged particles. These particles interact with matter mostly via the Coulomb force, and the manner in which they scatter from nuclei can reveal nuclear size and mass. This is analogous to observing how a bowling ball is scattered by an object you cannot see directly. Because the alpha particle's energy is so large compared with the typical energies associated with atoms (MeV versus eV), you would expect the alpha particles to simply crash through a thin foil much like a supersonic bowling ball would crash through a few dozen rows of bowling pins. Thomson had envisioned the atom to be a small sphere in which equal amounts of positive and negative charge were distributed evenly. The incident massive alpha particles would suffer only small deflections in such a model. Instead, Rutherford and his collaborators found that alpha particles occasionally were scattered to large angles, some even back in the direction from which they came! Detailed analysis using conservation of momentum and energy—particularly of the small number that came straight back—implied that gold nuclei are very small compared with the size of a gold atom, contain almost all of the atom's mass, and are tightly bound. Since the gold nucleus is several times more massive than the alpha particle, a head-on collision would scatter the alpha particle straight back toward the source. In addition, the smaller the nucleus, the fewer alpha particles that would hit one head on.

Although the results of the experiment were published by his colleagues in 1909, it took Rutherford two years to convince himself of their meaning. Like Thomson before him, Rutherford was reluctant to accept such radical results. Nature on a small scale is so unlike our classical world that even those at the forefront of discovery are sometimes surprised. Rutherford later wrote: "It was almost as incredible as if you fired a 15-inch shell at a piece of tissue paper and it came back and hit you. On consideration, I realized that this scattering backwards ... [meant] ... the greatest part of the mass of the atom was concentrated in a tiny nucleus." In 1911, Rutherford published his analysis together with a proposed model of the atom. The size of the nucleus was determined to be about 10^{-15} m , or 100,000 times smaller than the atom. This implies a huge density, on the order of 10^{15} g/cm^3 , vastly unlike any macroscopic matter. Also implied is the existence of previously unknown nuclear forces to counteract the huge repulsive Coulomb forces among the positive charges in the nucleus. Huge forces would also be consistent with the large energies emitted in nuclear radiation.

The small size of the nucleus also implies that the atom is mostly empty inside. In fact, in Rutherford's experiment, most alphas went straight through the gold foil with very little scattering, since electrons have such small masses and since the atom was mostly empty with nothing for the alpha to hit. There were already hints of this at the time Rutherford performed his experiments, since energetic electrons had been observed to penetrate thin foils more easily than expected. **Figure 30.11** shows a schematic of the atoms in a thin foil with circles representing the size of the atoms (about 10^{-10} m) and dots representing the nuclei. (The dots are not to scale—if they were, you would need a microscope to see them.) Most alpha particles miss the small nuclei and are only slightly scattered by electrons. Occasionally, (about once in 8000 times in Rutherford's experiment), an alpha hits a nucleus head-on and is scattered straight backward.

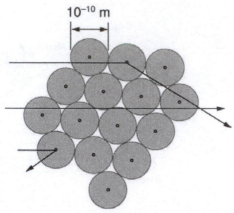

Figure 30.11 An expanded view of the atoms in the gold foil in Rutherford's experiment. Circles represent the atoms (about 10^{-10} m in diameter), while the dots represent the nuclei (about 10^{-15} m in diameter). To be visible, the dots are much larger than scale. Most alpha particles crash through but are relatively unaffected because of their high energy and the electron's small mass. Some, however, head straight toward a nucleus and are scattered straight back. A detailed analysis gives the size and mass of the nucleus.

Based on the size and mass of the nucleus revealed by his experiment, as well as the mass of electrons, Rutherford proposed the **planetary model of the atom**. The planetary model of the atom pictures low-mass electrons orbiting a large-mass nucleus. The sizes of the electron orbits are large compared with the size of the nucleus, with mostly vacuum inside the atom. This picture is analogous to how low-mass planets in our solar system orbit the large-mass Sun at distances large compared with the size of the sun. In the atom, the attractive Coulomb force is analogous to gravitation in the planetary system. (See **Figure 30.12**.) Note that a model or mental picture is needed to explain experimental results, since the atom is too small to be directly observed with visible light.

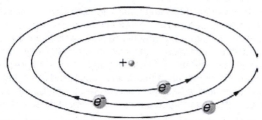

Figure 30.12 Rutherford's planetary model of the atom incorporates the characteristics of the nucleus, electrons, and the size of the atom. This model was the first to recognize the structure of atoms, in which low-mass electrons orbit a very small, massive nucleus in orbits much larger than the nucleus. The atom is mostly empty and is analogous to our planetary system.

Rutherford's planetary model of the atom was crucial to understanding the characteristics of atoms, and their interactions and energies, as we shall see in the next few sections. Also, it was an indication of how different nature is from the familiar classical world on the small, quantum mechanical scale. The discovery of a substructure to all matter in the form of atoms and molecules was now being taken a step further to reveal a substructure of atoms that was simpler than the 92 elements then known. We have continued to search for deeper substructures, such as those inside the nucleus, with some success. In later chapters, we will follow this quest in the discussion of quarks and other elementary particles, and we will look at the direction the search seems now to be heading.

PhET Explorations: Rutherford Scattering

How did Rutherford figure out the structure of the atom without being able to see it? Simulate the famous experiment in which he disproved the Plum Pudding model of the atom by observing alpha particles bouncing off atoms and determining that they must have a small core.

Figure 30.13 Rutherford Scattering (http://cnx.org/content/m54944/1.3/rutherford-scattering_en.jar)

30.3 Bohr's Theory of the Hydrogen Atom

The great Danish physicist Niels Bohr (1885–1962) made immediate use of Rutherford's planetary model of the atom. (**Figure 30.14**). Bohr became convinced of its validity and spent part of 1912 at Rutherford's laboratory. In 1913, after returning to Copenhagen, he began publishing his theory of the simplest atom, hydrogen, based on the planetary model of the atom. For decades, many questions had been asked about atomic characteristics. From their sizes to their spectra, much was known about atoms, but little had been explained in terms of the laws of physics. Bohr's theory explained the atomic spectrum of hydrogen and established new and broadly applicable principles in quantum mechanics.

Figure 30.14 Niels Bohr, Danish physicist, used the planetary model of the atom to explain the atomic spectrum and size of the hydrogen atom. His many contributions to the development of atomic physics and quantum mechanics, his personal influence on many students and colleagues, and his personal integrity, especially in the face of Nazi oppression, earned him a prominent place in history. (credit: Unknown Author, via Wikimedia Commons)

Mysteries of Atomic Spectra

As noted in **Quantization of Energy** , the energies of some small systems are quantized. Atomic and molecular emission and absorption spectra have been known for over a century to be discrete (or quantized). (See **Figure 30.15**.) Maxwell and others had realized that there must be a connection between the spectrum of an atom and its structure, something like the resonant frequencies of musical instruments. But, in spite of years of efforts by many great minds, no one had a workable theory. (It was a running joke that any theory of atomic and molecular spectra could be destroyed by throwing a book of data at it, so complex were the spectra.) Following Einstein's proposal of photons with quantized energies directly proportional to their wavelengths, it became even more evident that electrons in atoms can exist only in discrete orbits.

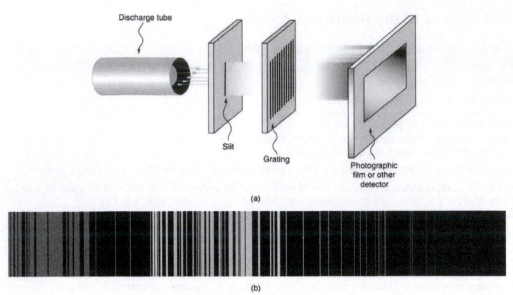

(b)

Figure 30.15 Part (a) shows, from left to right, a discharge tube, slit, and diffraction grating producing a line spectrum. Part (b) shows the emission line spectrum for iron. The discrete lines imply quantized energy states for the atoms that produce them. The line spectrum for each element is unique, providing a powerful and much used analytical tool, and many line spectra were well known for many years before they could be explained with physics. (credit for (b): Yttrium91, Wikimedia Commons)

In some cases, it had been possible to devise formulas that described the emission spectra. As you might expect, the simplest atom—hydrogen, with its single electron—has a relatively simple spectrum. The hydrogen spectrum had been observed in the infrared (IR), visible, and ultraviolet (UV), and several series of spectral lines had been observed. (See Figure 30.16.) These series are named after early researchers who studied them in particular depth.

The observed **hydrogen-spectrum wavelengths** can be calculated using the following formula:

$$\frac{1}{\lambda} = R\left(\frac{1}{n_f^2} - \frac{1}{n_i^2}\right),$$

(30.13)

where λ is the wavelength of the emitted EM radiation and R is the **Rydberg constant**, determined by the experiment to be

$$R = 1.097 \times 10^7 / \text{m (or m}^{-1}).$$

(30.14)

The constant n_f is a positive integer associated with a specific series. For the Lyman series, $n_f = 1$; for the Balmer series, $n_f = 2$; for the Paschen series, $n_f = 3$; and so on. The Lyman series is entirely in the UV, while part of the Balmer series is visible with the remainder UV. The Paschen series and all the rest are entirely IR. There are apparently an unlimited number of series, although they lie progressively farther into the infrared and become difficult to observe as n_f increases. The constant n_i is a positive integer, but it must be greater than n_f. Thus, for the Balmer series, $n_f = 2$ and $n_i = 3, 4, 5, 6, \dots$. Note that n_i can approach infinity. While the formula in the wavelengths equation was just a recipe designed to fit data and was not based on physical principles, it did imply a deeper meaning. Balmer first devised the formula for his series alone, and it was later found to describe all the other series by using different values of n_f. Bohr was the first to comprehend the deeper meaning. Again, we see the interplay between experiment and theory in physics. Experimentally, the spectra were well established, an equation was found to fit the experimental data, but the theoretical foundation was missing.

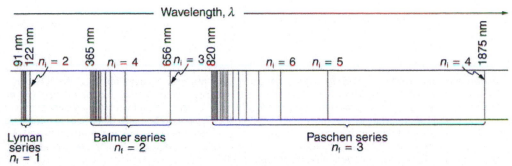

Figure 30.16 A schematic of the hydrogen spectrum shows several series named for those who contributed most to their determination. Part of the Balmer series is in the visible spectrum, while the Lyman series is entirely in the UV, and the Paschen series and others are in the IR. Values of n_f and n_i are shown for some of the lines.

Example 30.1 Calculating Wave Interference of a Hydrogen Line

What is the distance between the slits of a grating that produces a first-order maximum for the second Balmer line at an angle of $15°$?

Strategy and Concept

For an Integrated Concept problem, we must first identify the physical principles involved. In this example, we need to know (a) the wavelength of light as well as (b) conditions for an interference maximum for the pattern from a double slit. Part (a) deals with a topic of the present chapter, while part (b) considers the wave interference material of **Wave Optics**.

Solution for (a)

Hydrogen spectrum wavelength. The Balmer series requires that $n_f = 2$. The first line in the series is taken to be for $n_i = 3$, and so the second would have $n_i = 4$.

The calculation is a straightforward application of the wavelength equation. Entering the determined values for n_f and n_i yields

$$\frac{1}{\lambda} = R\left(\frac{1}{n_f^2} - \frac{1}{n_i^2}\right)$$ (30.15)

$$= \left(1.097 \times 10^7 \text{ m}^{-1}\right)\left(\frac{1}{2^2} - \frac{1}{4^2}\right)$$

$$= 2.057 \times 10^6 \text{ m}^{-1}.$$

Inverting to find λ gives

$$\lambda = \frac{1}{2.057 \times 10^6 \text{ m}^{-1}} = 486 \times 10^{-9} \text{ m}$$ (30.16)

$$= 486 \text{ nm}.$$

Discussion for (a)

This is indeed the experimentally observed wavelength, corresponding to the second (blue-green) line in the Balmer series. More impressive is the fact that the same simple recipe predicts *all* of the hydrogen spectrum lines, including new ones observed in subsequent experiments. What is nature telling us?

Solution for (b)

Double-slit interference (**Wave Optics**). To obtain constructive interference for a double slit, the path length difference from two slits must be an integral multiple of the wavelength. This condition was expressed by the equation

$$d \sin \theta = m\lambda,$$ (30.17)

where d is the distance between slits and θ is the angle from the original direction of the beam. The number m is the order of the interference; $m = 1$ in this example. Solving for d and entering known values yields

$$d = \frac{(1)(486 \text{ nm})}{\sin 15°} = 1.88 \times 10^{-6} \text{ m}.$$ (30.18)

Discussion for (b)

This number is similar to those used in the interference examples of **Introduction to Quantum Physics** (and is close to the spacing between slits in commonly used diffraction glasses).

Bohr's Solution for Hydrogen

Bohr was able to derive the formula for the hydrogen spectrum using basic physics, the planetary model of the atom, and some very important new proposals. His first proposal is that only certain orbits are allowed: we say that *the orbits of electrons in atoms are quantized*. Each orbit has a different energy, and electrons can move to a higher orbit by absorbing energy and drop to a lower orbit by emitting energy. If the orbits are quantized, the amount of energy absorbed or emitted is also quantized, producing discrete spectra. Photon absorption and emission are among the primary methods of transferring energy into and out of atoms. The energies of the photons are quantized, and their energy is explained as being equal to the change in energy of the electron when it moves from one orbit to another. In equation form, this is

$$\Delta E = hf = E_i - E_f. \tag{30.19}$$

Here, ΔE is the change in energy between the initial and final orbits, and hf is the energy of the absorbed or emitted photon. It is quite logical (that is, expected from our everyday experience) that energy is involved in changing orbits. A blast of energy is required for the space shuttle, for example, to climb to a higher orbit. What is not expected is that atomic orbits should be quantized. This is not observed for satellites or planets, which can have any orbit given the proper energy. (See **Figure 30.17**.)

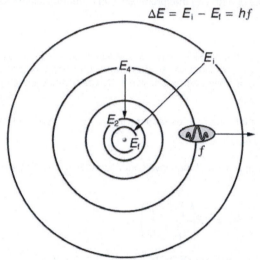

Figure 30.17 The planetary model of the atom, as modified by Bohr, has the orbits of the electrons quantized. Only certain orbits are allowed, explaining why atomic spectra are discrete (quantized). The energy carried away from an atom by a photon comes from the electron dropping from one allowed orbit to another and is thus quantized. This is likewise true for atomic absorption of photons.

Figure 30.18 shows an **energy-level diagram**, a convenient way to display energy states. In the present discussion, we take these to be the allowed energy levels of the electron. Energy is plotted vertically with the lowest or ground state at the bottom and with excited states above. Given the energies of the lines in an atomic spectrum, it is possible (although sometimes very difficult) to determine the energy levels of an atom. Energy-level diagrams are used for many systems, including molecules and nuclei. A theory of the atom or any other system must predict its energies based on the physics of the system.

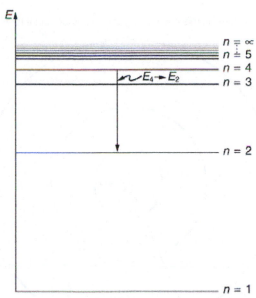

Figure 30.18 An energy-level diagram plots energy vertically and is useful in visualizing the energy states of a system and the transitions between them. This diagram is for the hydrogen-atom electrons, showing a transition between two orbits having energies E_4 and E_2.

Bohr was clever enough to find a way to calculate the electron orbital energies in hydrogen. This was an important first step that has been improved upon, but it is well worth repeating here, because it does correctly describe many characteristics of hydrogen. Assuming circular orbits, Bohr proposed that the **angular momentum L of an electron in its orbit is quantized**, that is, it has only specific, discrete values. The value for L is given by the formula

$$L = m_e v r_n = n\frac{h}{2\pi}(n = 1, 2, 3, \ldots),$$
(30.20)

where L is the angular momentum, m_e is the electron's mass, r_n is the radius of the n th orbit, and h is Planck's constant. Note that angular momentum is $L = I\omega$. For a small object at a radius r, $I = mr^2$ and $\omega = v/r$, so that

$L = (mr^2)(v/r) = mvr$. Quantization says that this value of mvr can only be equal to $h/2, 2h/2, 3h/2$, etc. At the time, Bohr himself did not know why angular momentum should be quantized, but using this assumption he was able to calculate the energies in the hydrogen spectrum, something no one else had done at the time.

From Bohr's assumptions, we will now derive a number of important properties of the hydrogen atom from the classical physics we have covered in the text. We start by noting the centripetal force causing the electron to follow a circular path is supplied by the Coulomb force. To be more general, we note that this analysis is valid for any single-electron atom. So, if a nucleus has Z protons ($Z = 1$ for hydrogen, 2 for helium, etc.) and only one electron, that atom is called a **hydrogen-like atom**. The spectra of hydrogen-like ions are similar to hydrogen, but shifted to higher energy by the greater attractive force between the electron and nucleus. The magnitude of the centripetal force is $m_e v^2 / r_n$, while the Coulomb force is $k(Zq_e)(q_e)/r_n^2$. The tacit assumption here is that the nucleus is more massive than the stationary electron, and the electron orbits about it. This is consistent with the planetary model of the atom. Equating these,

$$k\frac{Zq_e^2}{r_n^2} = \frac{m_e v^2}{r_n}\text{(Coulomb = centripetal)}.$$
(30.21)

Angular momentum quantization is stated in an earlier equation. We solve that equation for v, substitute it into the above, and rearrange the expression to obtain the radius of the orbit. This yields:

$$r_n = \frac{n^2}{Z}a_\text{B}, \text{ for allowed orbits}(n = 1, 2, 3, \ldots),$$
(30.22)

where a_B is defined to be the **Bohr radius**, since for the lowest orbit $(n = 1)$ and for hydrogen $(Z = 1)$, $r_1 = a_\text{B}$. It is left for this chapter's Problems and Exercises to show that the Bohr radius is

$$a_\text{B} = \frac{h^2}{4\pi^2 m_e k q_e^2} = 0.529 \times 10^{-10} \text{ m.}$$
(30.23)

These last two equations can be used to calculate the **radii of the allowed (quantized) electron orbits in any hydrogen-like atom**. It is impressive that the formula gives the correct size of hydrogen, which is measured experimentally to be very close to the Bohr radius. The earlier equation also tells us that the orbital radius is proportional to n^2, as illustrated in **Figure 30.19**.

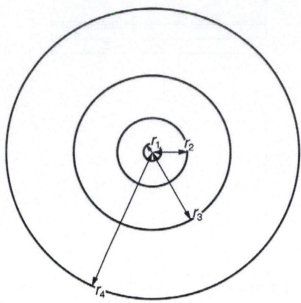

Figure 30.19 The allowed electron orbits in hydrogen have the radii shown. These radii were first calculated by Bohr and are given by the equation $r_n = \frac{n^2}{Z}a_B$. The lowest orbit has the experimentally verified diameter of a hydrogen atom.

To get the electron orbital energies, we start by noting that the electron energy is the sum of its kinetic and potential energy:

$$E_n = \text{KE} + \text{PE}. \tag{30.24}$$

Kinetic energy is the familiar $KE = (1/2)m_e v^2$, assuming the electron is not moving at relativistic speeds. Potential energy for the electron is electrical, or $PE = q_e V$, where V is the potential due to the nucleus, which looks like a point charge. The nucleus has a positive charge Zq_e; thus, $V = kZq_e / r_n$, recalling an earlier equation for the potential due to a point charge. Since the electron's charge is negative, we see that $PE = -kZq_e / r_n$. Entering the expressions for KE and PE, we find

$$E_n = \frac{1}{2}m_e v^2 - k\frac{Zq_e^2}{r_n}. \tag{30.25}$$

Now we substitute r_n and v from earlier equations into the above expression for energy. Algebraic manipulation yields

$$E_n = -\frac{Z^2}{n^2}E_0 (n = 1, 2, 3, ...) \tag{30.26}$$

for the orbital **energies of hydrogen-like atoms**. Here, E_0 is the **ground-state energy** $(n = 1)$ for hydrogen $(Z = 1)$ and is given by

$$E_0 = \frac{2\pi^2 q_e^4 m_e k^2}{h^2} = 13.6 \text{ eV}. \tag{30.27}$$

Thus, for hydrogen,

$$E_n = -\frac{13.6 \text{ eV}}{n^2}(n = 1, 2, 3, ...). \tag{30.28}$$

Figure 30.20 shows an energy-level diagram for hydrogen that also illustrates how the various spectral series for hydrogen are related to transitions between energy levels.

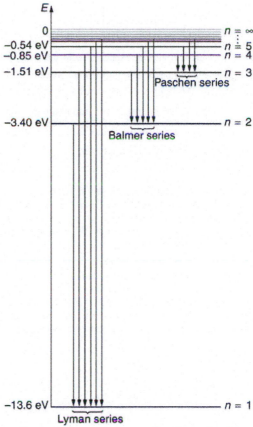

Figure 30.20 Energy-level diagram for hydrogen showing the Lyman, Balmer, and Paschen series of transitions. The orbital energies are calculated using the above equation, first derived by Bohr.

Electron total energies are negative, since the electron is bound to the nucleus, analogous to being in a hole without enough kinetic energy to escape. As n approaches infinity, the total energy becomes zero. This corresponds to a free electron with no kinetic energy, since r_n gets very large for large n, and the electric potential energy thus becomes zero. Thus, 13.6 eV is needed to ionize hydrogen (to go from –13.6 eV to 0, or unbound), an experimentally verified number. Given more energy, the electron becomes unbound with some kinetic energy. For example, giving 15.0 eV to an electron in the ground state of hydrogen strips it from the atom and leaves it with 1.4 eV of kinetic energy.

Finally, let us consider the energy of a photon emitted in a downward transition, given by the equation to be

$$\Delta E = hf = E_i - E_f. \tag{30.29}$$

Substituting $E_n = (-13.6 \text{ eV}/n^2)$, we see that

$$hf = (13.6 \text{ eV})\left(\frac{1}{n_f^2} - \frac{1}{n_i^2}\right). \tag{30.30}$$

Dividing both sides of this equation by hc gives an expression for $1/\lambda$:

$$\frac{hf}{hc} = \frac{f}{c} = \frac{1}{\lambda} = \frac{(13.6 \text{ eV})}{hc}\left(\frac{1}{n_f^2} - \frac{1}{n_i^2}\right). \tag{30.31}$$

It can be shown that

$$\left(\frac{13.6 \text{ eV}}{hc}\right) = \frac{(13.6 \text{ eV})(1.602 \times 10^{-19} \text{ J/eV})}{(6.626 \times 10^{-34} \text{ J·s})(2.998 \times 10^8 \text{ m/s})} = 1.097 \times 10^7 \text{ m}^{-1} = R \tag{30.32}$$

is the **Rydberg constant**. Thus, we have used Bohr's assumptions to derive the formula first proposed by Balmer years earlier as a recipe to fit experimental data.

$$\frac{1}{\lambda} = R\left(\frac{1}{n_f^2} - \frac{1}{n_i^2}\right)$$

(30.33)

We see that Bohr's theory of the hydrogen atom answers the question as to why this previously known formula describes the hydrogen spectrum. It is because the energy levels are proportional to $1/n^2$, where n is a non-negative integer. A downward transition releases energy, and so n_i must be greater than n_f. The various series are those where the transitions end on a certain level. For the Lyman series, $n_f = 1$ — that is, all the transitions end in the ground state (see also **Figure 30.20**). For the Balmer series, $n_f = 2$, or all the transitions end in the first excited state; and so on. What was once a recipe is now based in physics, and something new is emerging—angular momentum is quantized.

Triumphs and Limits of the Bohr Theory

Bohr did what no one had been able to do before. Not only did he explain the spectrum of hydrogen, he correctly calculated the size of the atom from basic physics. Some of his ideas are broadly applicable. Electron orbital energies are quantized in all atoms and molecules. Angular momentum is quantized. The electrons do not spiral into the nucleus, as expected classically (accelerated charges radiate, so that the electron orbits classically would decay quickly, and the electrons would sit on the nucleus—matter would collapse). These are major triumphs.

But there are limits to Bohr's theory. It cannot be applied to multielectron atoms, even one as simple as a two-electron helium atom. Bohr's model is what we call *semiclassical*. The orbits are quantized (nonclassical) but are assumed to be simple circular paths (classical). As quantum mechanics was developed, it became clear that there are no well-defined orbits; rather, there are clouds of probability. Bohr's theory also did not explain that some spectral lines are doublets (split into two) when examined closely. We shall examine many of these aspects of quantum mechanics in more detail, but it should be kept in mind that Bohr did not fail. Rather, he made very important steps along the path to greater knowledge and laid the foundation for all of atomic physics that has since evolved.

PhET Explorations: Models of the Hydrogen Atom

How did scientists figure out the structure of atoms without looking at them? Try out different models by shooting light at the atom. Check how the prediction of the model matches the experimental results.

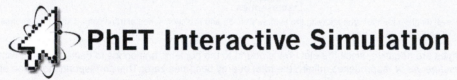

Figure 30.21 Models of the Hydrogen Atom (http://cnx.org/content/m54948/1.2/hydrogen-atom_en.jar)

30.4 X Rays: Atomic Origins and Applications

Learning Objectives

By the end of this section, you will be able to:

- Define x-ray tube and its spectrum.
- Show the x-ray characteristic energy.
- Specify the use of x rays in medical observations.
- Explain the use of x rays in CT scanners in diagnostics.

The information presented in this section supports the following AP® learning objectives and science practices:

- **5.B.8.1** The student is able to describe emission or absorption spectra associated with electronic or nuclear transitions as transitions between allowed energy states of the atom in terms of the principle of energy conservation, including characterization of the frequency of radiation emitted or absorbed. **(S.P. 1.2, 7.2)**

Each type of atom (or element) has its own characteristic electromagnetic spectrum. **X rays** lie at the high-frequency end of an atom's spectrum and are characteristic of the atom as well. In this section, we explore characteristic x rays and some of their important applications.

We have previously discussed x rays as a part of the electromagnetic spectrum in **Photon Energies and the Electromagnetic Spectrum**. That module illustrated how an x-ray tube (a specialized CRT) produces x rays. Electrons emitted from a hot filament are accelerated with a high voltage, gaining significant kinetic energy and striking the anode.

There are two processes by which x rays are produced in the anode of an x-ray tube. In one process, the deceleration of electrons produces x rays, and these x rays are called *bremsstrahlung*, or braking radiation. The second process is atomic in nature and produces *characteristic x rays*, so called because they are characteristic of the anode material. The x-ray spectrum in **Figure 30.22** is typical of what is produced by an x-ray tube, showing a broad curve of bremsstrahlung radiation with characteristic x-ray peaks on it.

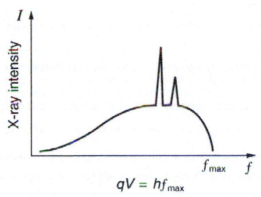

$$qV = hf_{max}$$

Figure 30.22 X-ray spectrum obtained when energetic electrons strike a material, such as in the anode of a CRT. The smooth part of the spectrum is bremsstrahlung radiation, while the peaks are characteristic of the anode material. A different anode material would have characteristic x-ray peaks at different frequencies.

The spectrum in **Figure 30.22** is collected over a period of time in which many electrons strike the anode, with a variety of possible outcomes for each hit. The broad range of x-ray energies in the bremsstrahlung radiation indicates that an incident electron's energy is not usually converted entirely into photon energy. The highest-energy x ray produced is one for which all of the electron's energy was converted to photon energy. Thus the accelerating voltage and the maximum x-ray energy are related by conservation of energy. Electric potential energy is converted to kinetic energy and then to photon energy, so that $E_{max} = hf_{max} = q_e V$. Units of electron volts are convenient. For example, a 100-kV accelerating voltage produces x-ray photons with a maximum energy of 100 keV.

Some electrons excite atoms in the anode. Part of the energy that they deposit by collision with an atom results in one or more of the atom's inner electrons being knocked into a higher orbit or the atom being ionized. When the anode's atoms de-excite, they emit characteristic electromagnetic radiation. The most energetic of these are produced when an inner-shell vacancy is filled—that is, when an $n = 1$ or $n = 2$ shell electron has been excited to a higher level, and another electron falls into the vacant spot. A *characteristic x ray* (see **Photon Energies and the Electromagnetic Spectrum**) is electromagnetic (EM) radiation emitted by an atom when an inner-shell vacancy is filled. **Figure 30.23** shows a representative energy-level diagram that illustrates the labeling of characteristic x rays. X rays created when an electron falls into an $n = 1$ shell vacancy are called K_α when they come from the next higher level; that is, an $n = 2$ to $n = 1$ transition. The labels $K, L, M,...$ come from the older alphabetical labeling of shells starting with K rather than using the principal quantum numbers 1, 2, 3, ... A more energetic K_β x ray is produced when an electron falls into an $n = 1$ shell vacancy from the $n = 3$ shell; that is, an $n = 3$ to $n = 1$ transition. Similarly, when an electron falls into the $n = 2$ shell from the $n = 3$ shell, an L_α x ray is created. The energies of these x rays depend on the energies of electron states in the particular atom and, thus, are characteristic of that element: every element has it own set of x-ray energies. This property can be used to identify elements, for example, to find trace (small) amounts of an element in an environmental or biological sample.

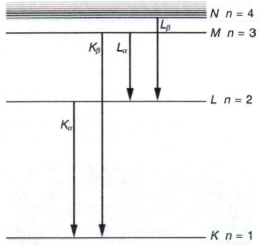

Figure 30.23 A characteristic x ray is emitted when an electron fills an inner-shell vacancy, as shown for several transitions in this approximate energy level diagram for a multiple-electron atom. Characteristic x rays are labeled according to the shell that had the vacancy and the shell from which the electron came. A K_α x ray, for example, is produced when an electron coming from the $n = 2$ shell fills the $n = 1$ shell vacancy.

Example 30.2 Characteristic X-Ray Energy

Calculate the approximate energy of a K_α x ray from a tungsten anode in an x-ray tube.

Strategy

How do we calculate energies in a multiple-electron atom? In the case of characteristic x rays, the following approximate calculation is reasonable. Characteristic x rays are produced when an inner-shell vacancy is filled. Inner-shell electrons are nearer the nucleus than others in an atom and thus feel little net effect from the others. This is similar to what happens inside a charged conductor, where its excess charge is distributed over the surface so that it produces no electric field inside. It is reasonable to assume the inner-shell electrons have hydrogen-like energies, as given by $E_n = -\frac{Z^2}{n^2}E_0$ $(n = 1, 2, 3, ...)$.

As noted, a K_α x ray is produced by an $n = 2$ to $n = 1$ transition. Since there are two electrons in a filled K shell, a vacancy would leave one electron, so that the effective charge would be $Z - 1$ rather than Z. For tungsten, $Z = 74$, so that the effective charge is 73.

Solution

$E_n = -\frac{Z^2}{n^2}E_0$ $(n = 1, 2, 3, ...)$ gives the orbital energies for hydrogen-like atoms to be $E_n = -(Z^2/n^2)E_0$, where

$E_0 = 13.6\ \text{eV}$. As noted, the effective Z is 73. Now the K_α x-ray energy is given by

$$E_{K_\alpha} = \Delta E = E_\text{i} - E_\text{f} = E_2 - E_1, \qquad (30.34)$$

where

$$E_1 = -\frac{Z^2}{1^2}E_0 = -\frac{73^2}{1}\left(13.6\ \text{eV}\right) = -72.5\ \text{keV} \qquad (30.35)$$

and

$$E_2 = -\frac{Z^2}{2^2}E_0 = -\frac{73^2}{4}\left(13.6\ \text{eV}\right) = -18.1\ \text{keV}. \qquad (30.36)$$

Thus,

$$E_{K_\alpha} = -18.1\ \text{keV} - \left(-72.5\ \text{keV}\right) = 54.4\ \text{keV}. \qquad (30.37)$$

Discussion

This large photon energy is typical of characteristic x rays from heavy elements. It is large compared with other atomic emissions because it is produced when an inner-shell vacancy is filled, and inner-shell electrons are tightly bound. Characteristic x ray energies become progressively larger for heavier elements because their energy increases approximately as Z^2. Significant accelerating voltage is needed to create these inner-shell vacancies. In the case of tungsten, at least 72.5 kV is needed, because other shells are filled and you cannot simply bump one electron to a higher filled shell. Tungsten is a common anode material in x-ray tubes; so much of the energy of the impinging electrons is absorbed, raising its temperature, that a high-melting-point material like tungsten is required.

Medical and Other Diagnostic Uses of X-rays

All of us can identify diagnostic uses of x-ray photons. Among these are the universal dental and medical x rays that have become an essential part of medical diagnostics. (See **Figure 30.25** and **Figure 30.26**.) X rays are also used to inspect our luggage at airports, as shown in **Figure 30.24**, and for early detection of cracks in crucial aircraft components. An x ray is not only a noun meaning high-energy photon, it also is an image produced by x rays, and it has been made into a familiar verb—to be x-rayed.

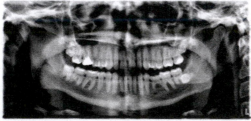

Figure 30.24 An x-ray image reveals fillings in a person's teeth. (credit: Dmitry G, Wikimedia Commons)

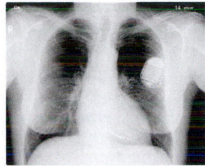

Figure 30.25 This x-ray image of a person's chest shows many details, including an artificial pacemaker. (credit: Sunzi99, Wikimedia Commons)

Figure 30.26 This x-ray image shows the contents of a piece of luggage. The denser the material, the darker the shadow. (credit: IDuke, Wikimedia Commons)

The most common x-ray images are simple shadows. Since x-ray photons have high energies, they penetrate materials that are opaque to visible light. The more energy an x-ray photon has, the more material it will penetrate. So an x-ray tube may be operated at 50.0 kV for a chest x ray, whereas it may need to be operated at 100 kV to examine a broken leg in a cast. The depth of penetration is related to the density of the material as well as to the energy of the photon. The denser the material, the fewer x-ray photons get through and the darker the shadow. Thus x rays excel at detecting breaks in bones and in imaging other physiological structures, such as some tumors, that differ in density from surrounding material. Because of their high photon energy, x rays produce significant ionization in materials and damage cells in biological organisms. Modern uses minimize exposure to the patient and eliminate exposure to others. Biological effects of x rays will be explored in the next chapter along with other types of ionizing radiation such as those produced by nuclei.

As the x-ray energy increases, the Compton effect (see Photon Momentum) becomes more important in the attenuation of the x rays. Here, the x ray scatters from an outer electron shell of the atom, giving the ejected electron some kinetic energy while losing energy itself. The probability for attenuation of the x rays depends upon the number of electrons present (the material's density) as well as the thickness of the material. Chemical composition of the medium, as characterized by its atomic number Z, is not important here. Low-energy x rays provide better contrast (sharper images). However, due to greater attenuation and less scattering, they are more absorbed by thicker materials. Greater contrast can be achieved by injecting a substance with a large atomic number, such as barium or iodine. The structure of the part of the body that contains the substance (e.g., the gastro-intestinal tract or the abdomen) can easily be seen this way.

Breast cancer is the second-leading cause of death among women worldwide. Early detection can be very effective, hence the importance of x-ray diagnostics. A mammogram cannot diagnose a malignant tumor, only give evidence of a lump or region of increased density within the breast. X-ray absorption by different types of soft tissue is very similar, so contrast is difficult; this is especially true for younger women, who typically have denser breasts. For older women who are at greater risk of developing breast cancer, the presence of more fat in the breast gives the lump or tumor more contrast. MRI (Magnetic resonance imaging) has recently been used as a supplement to conventional x rays to improve detection and eliminate false positives. The subject's radiation dose from x rays will be treated in a later chapter.

A standard x ray gives only a two-dimensional view of the object. Dense bones might hide images of soft tissue or organs. If you took another x ray from the side of the person (the first one being from the front), you would gain additional information. While shadow images are sufficient in many applications, far more sophisticated images can be produced with modern technology. Figure 30.27 shows the use of a computed tomography (CT) scanner, also called computed axial tomography (CAT) scanner. X rays are passed through a narrow section (called a slice) of the patient's body (or body part) over a range of directions. An array of many detectors on the other side of the patient registers the x rays. The system is then rotated around the patient and another image is taken, and so on. The x-ray tube and detector array are mechanically attached and so rotate together. Complex computer image processing of the relative absorption of the x rays along different directions produces a highly-detailed image. Different slices are taken as the patient moves through the scanner on a table. Multiple images of different slices can also be computer analyzed to produce three-dimensional information, sometimes enhancing specific types of tissue, as shown in Figure 30.28. G. Hounsfield (UK) and A. Cormack (US) won the Nobel Prize in Medicine in 1979 for their development of computed tomography.

Figure 30.27 A patient being positioned in a CT scanner aboard the hospital ship USNS Mercy. The CT scanner passes x rays through slices of the patient's body (or body part) over a range of directions. The relative absorption of the x rays along different directions is computer analyzed to produce highly detailed images. Three-dimensional information can be obtained from multiple slices. (credit: Rebecca Moat, U.S. Navy)

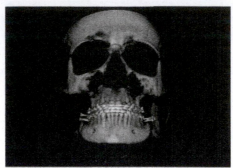

Figure 30.28 This three-dimensional image of a skull was produced by computed tomography, involving analysis of several x-ray slices of the head. (credit: Emailshankar, Wikimedia Commons)

X-Ray Diffraction and Crystallography

Since x-ray photons are very energetic, they have relatively short wavelengths. For example, the 54.4-keV K_α x ray of **Example 30.2** has a wavelength $\lambda = hc / E = 0.0228$ nm . Thus, typical x-ray photons act like rays when they encounter macroscopic objects, like teeth, and produce sharp shadows; however, since atoms are on the order of 0.1 nm in size, x rays can be used to detect the location, shape, and size of atoms and molecules. The process is called **x-ray diffraction**, because it involves the diffraction and interference of x rays to produce patterns that can be analyzed for information about the structures that scattered the x rays. Perhaps the most famous example of x-ray diffraction is the discovery of the double-helix structure of DNA in 1953 by an international team of scientists working at the Cavendish Laboratory—American James Watson, Englishman Francis Crick, and New Zealand–born Maurice Wilkins. Using x-ray diffraction data produced by Rosalind Franklin, they were the first to discern the structure of DNA that is so crucial to life. For this, Watson, Crick, and Wilkins were awarded the 1962 Nobel Prize in Physiology or Medicine. There is much debate and controversy over the issue that Rosalind Franklin was not included in the prize.

Figure 30.29 shows a diffraction pattern produced by the scattering of x rays from a crystal. This process is known as x-ray crystallography because of the information it can yield about crystal structure, and it was the type of data Rosalind Franklin supplied to Watson and Crick for DNA. Not only do x rays confirm the size and shape of atoms, they give information on the atomic arrangements in materials. For example, current research in high-temperature superconductors involves complex materials whose lattice arrangements are crucial to obtaining a superconducting material. These can be studied using x-ray crystallography.

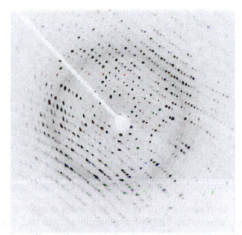

Figure 30.29 X-ray diffraction from the crystal of a protein, hen egg lysozyme, produced this interference pattern. Analysis of the pattern yields information about the structure of the protein. (credit: Del45, Wikimedia Commons)

Historically, the scattering of x rays from crystals was used to prove that x rays are energetic EM waves. This was suspected from the time of the discovery of x rays in 1895, but it was not until 1912 that the German Max von Laue (1879–1960) convinced two of his colleagues to scatter x rays from crystals. If a diffraction pattern is obtained, he reasoned, then the x rays must be waves, and their wavelength could be determined. (The spacing of atoms in various crystals was reasonably well known at the time, based on good values for Avogadro's number.) The experiments were convincing, and the 1914 Nobel Prize in Physics was given to von Laue for his suggestion leading to the proof that x rays are EM waves. In 1915, the unique father-and-son team of Sir William Henry Bragg and his son Sir William Lawrence Bragg were awarded a joint Nobel Prize for inventing the x-ray spectrometer and the then-new science of x-ray analysis. The elder Bragg had migrated to Australia from England just after graduating in mathematics. He learned physics and chemistry during his career at the University of Adelaide. The younger Bragg was born in Adelaide but went back to the Cavendish Laboratories in England to a career in x-ray and neutron crystallography; he provided support for Watson, Crick, and Wilkins for their work on unraveling the mysteries of DNA and to Max Perutz for his 1962 Nobel Prize-winning work on the structure of hemoglobin. Here again, we witness the enabling nature of physics—establishing instruments and designing experiments as well as solving mysteries in the biomedical sciences.

Certain other uses for x rays will be studied in later chapters. X rays are useful in the treatment of cancer because of the inhibiting effect they have on cell reproduction. X rays observed coming from outer space are useful in determining the nature of their sources, such as neutron stars and possibly black holes. Created in nuclear bomb explosions, x rays can also be used to detect clandestine atmospheric tests of these weapons. X rays can cause excitations of atoms, which then fluoresce (emitting characteristic EM radiation), making x-ray-induced fluorescence a valuable analytical tool in a range of fields from art to archaeology.

30.5 Applications of Atomic Excitations and De-Excitations

Learning Objectives
By the end of this section, you will be able to: • Define and discuss fluorescence.   • Define metastable.   • Describe how laser emission is produced.   • Explain population inversion.   • Define and discuss holography.   The information presented in this section supports the following AP® learning objectives and science practices: • **1.A.5.1** The student is able to model verbally or visually the properties of a system based on its substructure and to relate this to changes in the system properties over time as external variables are changed. **(S.P. 1.1, 7.1)**   • **1.A.5.2** The student is able to construct representations of how the properties of a system are determined by the interactions of its constituent substructures. **(S.P. 1.1, 1.4, 7.1)**   • **7.C.4.1** The student is able to construct or interpret representations of transitions between atomic energy states involving the emission and absorption of photons. **(S.P. 1.1, 1.2)**

Many properties of matter and phenomena in nature are directly related to atomic energy levels and their associated excitations and de-excitations. The color of a rose, the output of a laser, and the transparency of air are but a few examples. (See **Figure 30.30.**) While it may not appear that glow-in-the-dark pajamas and lasers have much in common, they are in fact different applications of similar atomic de-excitations.

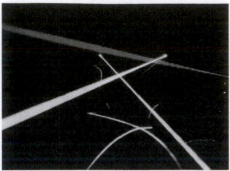

Figure 30.30 Light from a laser is based on a particular type of atomic de-excitation. (credit: Jeff Keyzer)

The color of a material is due to the ability of its atoms to absorb certain wavelengths while reflecting or reemitting others. A simple red material, for example a tomato, absorbs all visible wavelengths except red. This is because the atoms of its hydrocarbon pigment (lycopene) have levels separated by a variety of energies corresponding to all visible photon energies except red. Air is another interesting example. It is transparent to visible light, because there are few energy levels that visible photons can excite in air molecules and atoms. Visible light, thus, cannot be absorbed. Furthermore, visible light is only weakly scattered by air, because visible wavelengths are so much greater than the sizes of the air molecules and atoms. Light must pass through kilometers of air to scatter enough to cause red sunsets and blue skies.

Real World Connections: The Tomato

Let us consider the properties of a tomato from two different perspectives. When we try to explain the color of a tomato, we must consider the tomato as a system with properties that depend on its internal structure and the interactions between various parts. The internal structure of the tomato (specifically, the behavior of its pigment molecules) is very important and must be understood. Unlike a hydrogen atom, the energy level structure of a pigment molecule in a tomato is much more complicated. There are a very large number of energy levels, and the energy differences between these levels correspond to many different parts/colors of the visible spectrum, except for red.

So the photons that can be absorbed by these pigment molecules include every energy (or wavelength) in the visible spectrum except energies (or wavelengths) in the red part of the spectrum. Because these molecules absorb most of the visible photons, but reflect red photons, the color of the tomato appears red to our eyes. Without understanding the internal structure of the tomato pigment "system," we would have no way of explaining its color.

Now consider a tomato in free fall. It accelerates toward the Earth at a rate of 9.8 m/s^2, and we can say this with confidence without knowing anything about the internal structure of the tomato. In this case, we refer to the tomato as an object rather than a system. We only need to know the macroscopic properties of the tomato (its mass) in order to understand the force acting on the tomato.

Fluorescence and Phosphorescence

The ability of a material to emit various wavelengths of light is similarly related to its atomic energy levels. **Figure 30.31** shows a scorpion illuminated by a UV lamp, sometimes called a black light. Some rocks also glow in black light, the particular colors being a function of the rock's mineral composition. Black lights are also used to make certain posters glow.

Figure 30.31 Objects glow in the visible spectrum when illuminated by an ultraviolet (black) light. Emissions are characteristic of the mineral involved, since they are related to its energy levels. In the case of scorpions, proteins near the surface of their skin give off the characteristic blue glow. This is a colorful example of fluorescence in which excitation is induced by UV radiation while de-excitation occurs in the form of visible light. (credit: Ken Bosma, Flickr)

In the fluorescence process, an atom is excited to a level several steps above its ground state by the absorption of a relatively high-energy UV photon. This is called **atomic excitation**. Once it is excited, the atom can de-excite in several ways, one of which is to re-emit a photon of the same energy as excited it, a single step back to the ground state. This is called **atomic de-excitation**. All other paths of de-excitation involve smaller steps, in which lower-energy (longer wavelength) photons are emitted. Some of these may be in the visible range, such as for the scorpion in **Figure 30.31**. **Fluorescence** is defined to be any process in which an atom or molecule, excited by a photon of a given energy, and de-excites by emission of a lower-energy photon.

Fluorescence can be induced by many types of energy input. Fluorescent paint, dyes, and even soap residues in clothes make colors seem brighter in sunlight by converting some UV into visible light. X rays can induce fluorescence, as is done in x-ray fluoroscopy to make brighter visible images. Electric discharges can induce fluorescence, as in so-called neon lights and in gas-discharge tubes that produce atomic and molecular spectra. Common fluorescent lights use an electric discharge in mercury vapor to cause atomic emissions from mercury atoms. The inside of a fluorescent light is coated with a fluorescent material that emits visible light over a broad spectrum of wavelengths. By choosing an appropriate coating, fluorescent lights can be made more like sunlight or like the reddish glow of candlelight, depending on needs. Fluorescent lights are more efficient in converting electrical energy into visible light than incandescent filaments (about four times as efficient), the blackbody radiation of which is primarily in the infrared due to temperature limitations.

This atom is excited to one of its higher levels by absorbing a UV photon. It can de-excite in a single step, re-emitting a photon of the same energy, or in several steps. The process is called fluorescence if the atom de-excites in smaller steps, emitting energy different from that which excited it. Fluorescence can be induced by a variety of energy inputs, such as UV, x-rays, and electrical discharge.

The spectacular Waitomo caves on North Island in New Zealand provide a natural habitat for glow-worms. The glow-worms hang up to 70 silk threads of about 30 or 40 cm each to trap prey that fly towards them in the dark. The fluorescence process is very efficient, with nearly 100% of the energy input turning into light. (In comparison, fluorescent lights are about 20% efficient.)

Fluorescence has many uses in biology and medicine. It is commonly used to label and follow a molecule within a cell. Such tagging allows one to study the structure of DNA and proteins. Fluorescent dyes and antibodies are usually used to tag the molecules, which are then illuminated with UV light and their emission of visible light is observed. Since the fluorescence of each element is characteristic, identification of elements within a sample can be done this way.

Figure 30.32 shows a commonly used fluorescent dye called fluorescein. Below that, Figure 30.33 reveals the diffusion of a fluorescent dye in water by observing it under UV light.

Figure 30.32 Fluorescein, shown here in powder form, is used to dye laboratory samples. (credit: Benjah-bmm27, Wikimedia Commons)

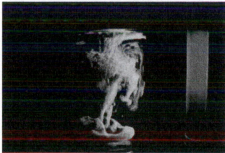

Figure 30.33 Here, fluorescent powder is added to a beaker of water. The mixture gives off a bright glow under ultraviolet light. (credit: Bricksnite, Wikimedia Commons)

Nano-Crystals

Recently, a new class of fluorescent materials has appeared—"nano-crystals." These are single-crystal molecules less than 100 nm in size. The smallest of these are called "quantum dots." These semiconductor indicators are very small (2–6 nm) and provide improved brightness. They also have the advantage that all colors can be excited with the same incident wavelength. They are brighter and more stable than organic dyes and have a longer lifetime than conventional phosphors. They have become an excellent tool for long-term studies of cells, including migration and morphology. (Figure 30.34.)

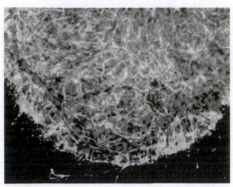

Figure 30.34 Microscopic image of chicken cells using nano-crystals of a fluorescent dye. Cell nuclei exhibit blue fluorescence while neurofilaments exhibit green. (credit: Weerapong Prasongchean, Wikimedia Commons)

Once excited, an atom or molecule will usually spontaneously de-excite quickly. (The electrons raised to higher levels are attracted to lower ones by the positive charge of the nucleus.) Spontaneous de-excitation has a very short mean lifetime of typically about 10^{-8} s . However, some levels have significantly longer lifetimes, ranging up to milliseconds to minutes or even hours. These energy levels are inhibited and are slow in de-exciting because their quantum numbers differ greatly from those of available lower levels. Although these level lifetimes are short in human terms, they are many orders of magnitude longer than is typical and, thus, are said to be **metastable**, meaning relatively stable. **Phosphorescence** is the de-excitation of a metastable state. Glow-in-the-dark materials, such as luminous dials on some watches and clocks and on children's toys and pajamas, are made of phosphorescent substances. Visible light excites the atoms or molecules to metastable states that decay slowly, releasing the stored excitation energy partially as visible light. In some ceramics, atomic excitation energy can be frozen in after the ceramic has cooled from its firing. It is very slowly released, but the ceramic can be induced to phosphoresce by heating—a process called "thermoluminescence." Since the release is slow, thermoluminescence can be used to date antiquities. The less light emitted, the older the ceramic. (See **Figure 30.35**.)

Figure 30.35 Atoms frozen in an excited state when this Chinese ceramic figure was fired can be stimulated to de-excite and emit EM radiation by heating a sample of the ceramic—a process called thermoluminescence. Since the states slowly de-excite over centuries, the amount of thermoluminescence decreases with age, making it possible to use this effect to date and authenticate antiquities. This figure dates from the 11th century. (credit: Vassil, Wikimedia Commons)

Lasers

Lasers today are commonplace. Lasers are used to read bar codes at stores and in libraries, laser shows are staged for entertainment, laser printers produce high-quality images at relatively low cost, and lasers send prodigious numbers of telephone messages through optical fibers. Among other things, lasers are also employed in surveying, weapons guidance, tumor eradication, retinal welding, and for reading music CDs and computer CD-ROMs.

Why do lasers have so many varied applications? The answer is that lasers produce single-wavelength EM radiation that is also very coherent—that is, the emitted photons are in phase. Laser output can, thus, be more precisely manipulated than incoherent mixed-wavelength EM radiation from other sources. The reason laser output is so pure and coherent is based on how it is produced, which in turn depends on a metastable state in the lasing material. Suppose a material had the energy levels shown in **Figure 30.36**. When energy is put into a large collection of these atoms, electrons are raised to all possible levels. Most return to the ground state in less than about 10^{-8} s , but those in the metastable state linger. This includes those electrons originally excited to the metastable state and those that fell into it from above. It is possible to get a majority of the atoms into the metastable state, a condition called a **population inversion**.

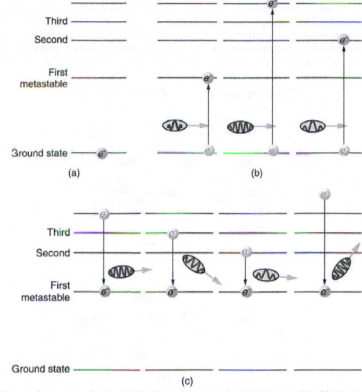

Figure 30.36 (a) Energy-level diagram for an atom showing the first few states, one of which is metastable. (b) Massive energy input excites atoms to a variety of states. (c) Most states decay quickly, leaving electrons only in the metastable and ground state. If a majority of electrons are in the metastable state, a population inversion has been achieved.

Once a population inversion is achieved, a very interesting thing can happen, as shown in **Figure 30.37**. An electron spontaneously falls from the metastable state, emitting a photon. This photon finds another atom in the metastable state and stimulates it to decay, emitting a second photon of *the same wavelength and in phase* with the first, and so on. **Stimulated emission** is the emission of electromagnetic radiation in the form of photons of a given frequency, triggered by photons of the same frequency. For example, an excited atom, with an electron in an energy orbit higher than normal, releases a photon of a specific frequency when the electron drops back to a lower energy orbit. If this photon then strikes another electron in the same high-energy orbit in another atom, another photon of the same frequency is released. The emitted photons and the triggering photons are always in phase, have the same polarization, and travel in the same direction. The probability of absorption of a photon is the same as the probability of stimulated emission, and so a majority of atoms must be in the metastable state to produce energy. Einstein (again Einstein, and back in 1917!) was one of the important contributors to the understanding of stimulated emission of radiation. Among other things, Einstein was the first to realize that stimulated emission and absorption are equally probable. The laser acts as a temporary energy storage device that subsequently produces a massive energy output of single-wavelength, in-phase photons.

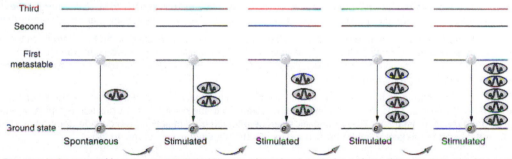

Figure 30.37 One atom in the metastable state spontaneously decays to a lower level, producing a photon that goes on to stimulate another atom to de-excite. The second photon has exactly the same energy and wavelength as the first and is in phase with it. Both go on to stimulate the emission of other photons. A population inversion is necessary for there to be a net production rather than a net absorption of the photons.

The name **laser** is an acronym for light amplification by stimulated emission of radiation, the process just described. The process was proposed and developed following the advances in quantum physics. A joint Nobel Prize was awarded in 1964 to American Charles Townes (1915–), and Nikolay Basov (1922–2001) and Aleksandr Prokhorov (1916–2002), from the Soviet Union, for the development of lasers. The Nobel Prize in 1981 went to Arthur Schawlow (1921-1999) for pioneering laser applications. The original devices were called masers, because they produced microwaves. The first working laser was created in 1960 at Hughes

Research labs (CA) by T. Maiman. It used a pulsed high-powered flash lamp and a ruby rod to produce red light. Today the name laser is used for all such devices developed to produce a variety of wavelengths, including microwave, infrared, visible, and ultraviolet radiation. Figure 30.38 shows how a laser can be constructed to enhance the stimulated emission of radiation. Energy input can be from a flash tube, electrical discharge, or other sources, in a process sometimes called optical pumping. A large percentage of the original pumping energy is dissipated in other forms, but a population inversion must be achieved. Mirrors can be used to enhance stimulated emission by multiple passes of the radiation back and forth through the lasing material. One of the mirrors is semitransparent to allow some of the light to pass through. The laser output from a laser is a mere 1% of the light passing back and forth in a laser.

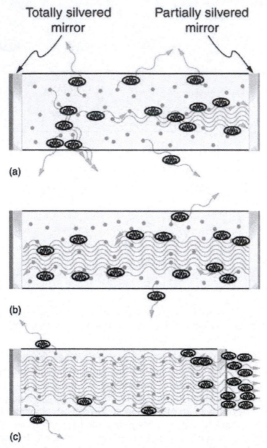

Figure 30.38 Typical laser construction has a method of pumping energy into the lasing material to produce a population inversion. (a) Spontaneous emission begins with some photons escaping and others stimulating further emissions. (b) and (c) Mirrors are used to enhance the probability of stimulated emission by passing photons through the material several times.

Real World Connections: Emission Spectrum

When observing an emission spectrum like the iron spectrum in Figure 30.15(b), you may notice the locations of the emission lines, which indicate the wavelength of each line. These wavelengths correspond to specific energy level differences for electrons in an iron atom. You may also notice that some of these emission lines are brighter than others, too.

This has to do with the probabilistic nature of emission. When an electron is in an excited state, for example in the $n = 4$ energy level of a hydrogen atom, it has a variety of possible options for emission. The electron can transition from $n = 4$ to $n = 3$, $n = 2$, or $n = 1$, but not all transitions are equally likely. Typically, transitions to lower energy states are much more probable than transitions to higher energy states.

This means photons corresponding to a transition from $n = 4$ to $n = 3$ are much less common than photons corresponding to a transition from $n = 4$ to $n = 1$. Thus, the emission line corresponding to the $n = 4$ to $n = 1$ transition is typically much brighter under ordinary circumstances. The probabilities can be affected by stimulation from outside photons, and this kind of interaction is at the heart of the laser ("light amplification by the stimulated emission of radiation").

Lasers are constructed from many types of lasing materials, including gases, liquids, solids, and semiconductors. But all lasers are based on the existence of a metastable state or a phosphorescent material. Some lasers produce continuous output; others are pulsed in bursts as brief as 10^{-14} s. Some laser outputs are fantastically powerful—some greater than 10^{12} W —but the more common, everyday lasers produce something on the order of 10^{-3} W. The helium-neon laser that produces a familiar red light is very common. Figure 30.39 shows the energy levels of helium and neon, a pair of noble gases that work well together. An electrical discharge is passed through a helium-neon gas mixture in which the number of atoms of helium is ten times that of neon. The first excited state of helium is metastable and, thus, stores energy. This energy is easily transferred by collision to neon atoms, because they have an excited state at nearly the same energy as that in helium. That state in neon is also metastable, and this is the one that produces the laser output. (The most likely transition is to the nearby state, producing 1.96 eV photons, which have a wavelength of 633 nm and appear red.) A population inversion can be produced in neon, because there are so many more helium atoms and these put energy into the neon. Helium-neon lasers often have continuous output, because the population inversion can be maintained even while lasing occurs. Probably the most common lasers in use today, including the common laser pointer, are semiconductor or diode lasers, made of silicon. Here, energy is pumped into the material by passing a current in the device to excite the electrons. Special coatings on the ends and fine cleavings of the semiconductor material allow light to bounce back and forth and a tiny fraction to emerge as laser light. Diode lasers can usually run continually and produce outputs in the milliwatt range.

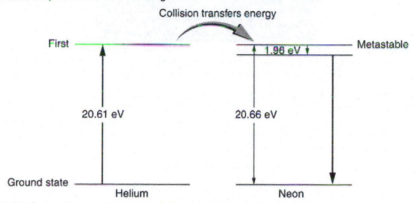

Figure 30.39 Energy levels in helium and neon. In the common helium-neon laser, an electrical discharge pumps energy into the metastable states of both atoms. The gas mixture has about ten times more helium atoms than neon atoms. Excited helium atoms easily de-excite by transferring energy to neon in a collision. A population inversion in neon is achieved, allowing lasing by the neon to occur.

There are many medical applications of lasers. Lasers have the advantage that they can be focused to a small spot. They also have a well-defined wavelength. Many types of lasers are available today that provide wavelengths from the ultraviolet to the infrared. This is important, as one needs to be able to select a wavelength that will be preferentially absorbed by the material of interest. Objects appear a certain color because they absorb all other visible colors incident upon them. What wavelengths are absorbed depends upon the energy spacing between electron orbitals in that molecule. Unlike the hydrogen atom, biological molecules are complex and have a variety of absorption wavelengths or lines. But these can be determined and used in the selection of a laser with the appropriate wavelength. Water is transparent to the visible spectrum but will absorb light in the UV and IR regions. Blood (hemoglobin) strongly reflects red but absorbs most strongly in the UV.

Laser surgery uses a wavelength that is strongly absorbed by the tissue it is focused upon. One example of a medical application of lasers is shown in Figure 30.40. A detached retina can result in total loss of vision. Burns made by a laser focused to a small spot on the retina form scar tissue that can hold the retina in place, salvaging the patient's vision. Other light sources cannot be focused as precisely as a laser due to refractive dispersion of different wavelengths. Similarly, laser surgery in the form of cutting or burning away tissue is made more accurate because laser output can be very precisely focused and is preferentially absorbed because of its single wavelength. Depending upon what part or layer of the retina needs repairing, the appropriate type of laser can be selected. For the repair of tears in the retina, a green argon laser is generally used. This light is absorbed well by tissues containing blood, so coagulation or "welding" of the tear can be done.

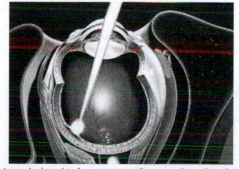

Figure 30.40 A detached retina is burned by a laser designed to focus on a small spot on the retina, the resulting scar tissue holding it in place. The lens of the eye is used to focus the light, as is the device bringing the laser output to the eye.

In dentistry, the use of lasers is rising. Lasers are most commonly used for surgery on the soft tissue of the mouth. They can be used to remove ulcers, stop bleeding, and reshape gum tissue. Their use in cutting into bones and teeth is not quite so common; here the erbium YAG (yttrium aluminum garnet) laser is used.

The massive combination of lasers shown in Figure 30.41 can be used to induce nuclear fusion, the energy source of the sun and hydrogen bombs. Since lasers can produce very high power in very brief pulses, they can be used to focus an enormous amount of energy on a small glass sphere containing fusion fuel. Not only does the incident energy increase the fuel temperature significantly so that fusion can occur, it also compresses the fuel to great density, enhancing the probability of fusion. The compression or implosion is caused by the momentum of the impinging laser photons.

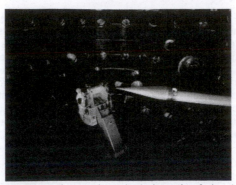

Figure 30.41 This system of lasers at Lawrence Livermore Laboratory is used to ignite nuclear fusion. A tremendous burst of energy is focused on a small fuel pellet, which is imploded to the high density and temperature needed to make the fusion reaction proceed. (credit: Lawrence Livermore National Laboratory, Lawrence Livermore National Security, LLC, and the Department of Energy)

Music CDs are now so common that vinyl records are quaint antiquities. CDs (and DVDs) store information digitally and have a much larger information-storage capacity than vinyl records. An entire encyclopedia can be stored on a single CD. **Figure 30.42** illustrates how the information is stored and read from the CD. Pits made in the CD by a laser can be tiny and very accurately spaced to record digital information. These are read by having an inexpensive solid-state infrared laser beam scatter from pits as the CD spins, revealing their digital pattern and the information encoded upon them.

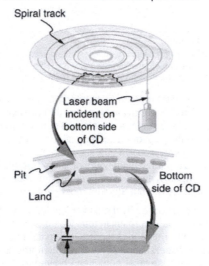

Figure 30.42 A CD has digital information stored in the form of laser-created pits on its surface. These in turn can be read by detecting the laser light scattered from the pit. Large information capacity is possible because of the precision of the laser. Shorter-wavelength lasers enable greater storage capacity.

Holograms, such as those in **Figure 30.43**, are true three-dimensional images recorded on film by lasers. Holograms are used for amusement, decoration on novelty items and magazine covers, security on credit cards and driver's licenses (a laser and other equipment is needed to reproduce them), and for serious three-dimensional information storage. You can see that a hologram is a true three-dimensional image, because objects change relative position in the image when viewed from different angles.

Figure 30.43 Credit cards commonly have holograms for logos, making them difficult to reproduce (credit: Dominic Alves, Flickr)

The name **hologram** means "entire picture" (from the Greek *holo*, as in holistic), because the image is three-dimensional. **Holography** is the process of producing holograms and, although they are recorded on photographic film, the process is quite different from normal photography. Holography uses light interference or wave optics, whereas normal photography uses geometric optics. **Figure 30.44** shows one method of producing a hologram. Coherent light from a laser is split by a mirror, with part of the light illuminating the object. The remainder, called the reference beam, shines directly on a piece of film. Light scattered from the object interferes with the reference beam, producing constructive and destructive interference. As a result, the exposed film looks foggy, but close examination reveals a complicated interference pattern stored on it. Where the interference was constructive, the film (a negative actually) is darkened. Holography is sometimes called lensless photography, because it uses the wave characteristics of light as contrasted to normal photography, which uses geometric optics and so requires lenses.

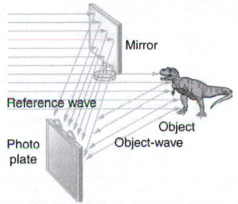

Figure 30.44 Production of a hologram. Single-wavelength coherent light from a laser produces a well-defined interference pattern on a piece of film. The laser beam is split by a partially silvered mirror, with part of the light illuminating the object and the remainder shining directly on the film.

Light falling on a hologram can form a three-dimensional image. The process is complicated in detail, but the basics can be understood as shown in **Figure 30.45**, in which a laser of the same type that exposed the film is now used to illuminate it. The myriad tiny exposed regions of the film are dark and block the light, while less exposed regions allow light to pass. The film thus acts much like a collection of diffraction gratings with various spacings. Light passing through the hologram is diffracted in various directions, producing both real and virtual images of the object used to expose the film. The interference pattern is the same as that produced by the object. Moving your eye to various places in the interference pattern gives you different perspectives, just as looking directly at the object would. The image thus looks like the object and is three-dimensional like the object.

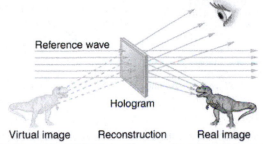

Figure 30.45 A transmission hologram is one that produces real and virtual images when a laser of the same type as that which exposed the hologram is passed through it. Diffraction from various parts of the film produces the same interference pattern as the object that was used to expose it.

The hologram illustrated in **Figure 30.45** is a transmission hologram. Holograms that are viewed with reflected light, such as the white light holograms on credit cards, are reflection holograms and are more common. White light holograms often appear a little blurry with rainbow edges, because the diffraction patterns of various colors of light are at slightly different locations due to their different wavelengths. Further uses of holography include all types of 3-D information storage, such as of statues in museums and engineering studies of structures and 3-D images of human organs. Invented in the late 1940s by Dennis Gabor (1900–1970), who won the 1971 Nobel Prize in Physics for his work, holography became far more practical with the development of the laser. Since lasers produce coherent single-wavelength light, their interference patterns are more pronounced. The precision is so great that it is even possible to record numerous holograms on a single piece of film by just changing the angle of the film for each successive image. This is how the holograms that move as you walk by them are produced—a kind of lensless movie.

In a similar way, in the medical field, holograms have allowed complete 3-D holographic displays of objects from a stack of images. Storing these images for future use is relatively easy. With the use of an endoscope, high-resolution 3-D holographic images of internal organs and tissues can be made.

30.6 The Wave Nature of Matter Causes Quantization

Learning Objectives
By the end of this section, you will be able to:
• Explain Bohr's model of the atom.

- Define and describe quantization of angular momentum.
- Calculate the angular momentum for an orbit of an atom.
- Define and describe the wave-like properties of matter.

The information presented in this section supports the following AP® learning objectives and science practices:

- **7.C.1.1** The student is able to use a graphical wave function representation of a particle to predict qualitatively the probability of finding a particle in a specific spatial region. **(S.P. 1.4)**
- **7.C.2.1** The student is able to use a standing wave model in which an electron orbit circumference is an integer multiple of the de Broglie wavelength to give a qualitative explanation that accounts for the existence of specific allowed energy states of an electron in an atom. **(S.P. 1.4)**

After visiting some of the applications of different aspects of atomic physics, we now return to the basic theory that was built upon Bohr's atom. Einstein once said it was important to keep asking the questions we eventually teach children not to ask. Why is angular momentum quantized? You already know the answer. Electrons have wave-like properties, as de Broglie later proposed. They can exist only where they interfere constructively, and only certain orbits meet proper conditions, as we shall see in the next module.

Following Bohr's initial work on the hydrogen atom, a decade was to pass before de Broglie proposed that matter has wave properties. The wave-like properties of matter were subsequently confirmed by observations of electron interference when scattered from crystals. Electrons can exist only in locations where they interfere constructively. How does this affect electrons in atomic orbits? When an electron is bound to an atom, its wavelength must fit into a small space, something like a standing wave on a string. (See **Figure 30.46**.) Allowed orbits are those orbits in which an electron constructively interferes with itself. Not all orbits produce constructive interference. Thus only certain orbits are allowed—the orbits are quantized.

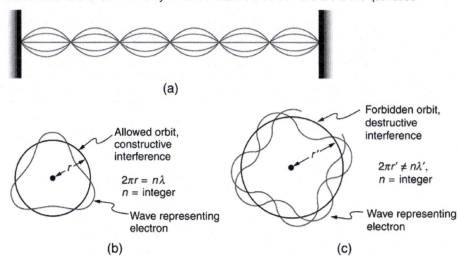

Figure 30.46 (a) Waves on a string have a wavelength related to the length of the string, allowing them to interfere constructively. (b) If we imagine the string bent into a closed circle, we get a rough idea of how electrons in circular orbits can interfere constructively. (c) If the wavelength does not fit into the circumference, the electron interferes destructively; it cannot exist in such an orbit.

For a circular orbit, constructive interference occurs when the electron's wavelength fits neatly into the circumference, so that wave crests always align with crests and wave troughs align with troughs, as shown in **Figure 30.46** (b). More precisely, when an integral multiple of the electron's wavelength equals the circumference of the orbit, constructive interference is obtained. In equation form, the *condition for constructive interference and an allowed electron orbit* is

$$n\lambda_n = 2\pi r_n (n = 1, 2, 3 \ldots),$$
(30.38)

where λ_n is the electron's wavelength and r_n is the radius of that circular orbit. The de Broglie wavelength is $\lambda = h / p = h / mv$, and so here $\lambda = h / m_e v$. Substituting this into the previous condition for constructive interference produces an interesting result:

$$\frac{nh}{m_e v} = 2\pi r_n.$$
(30.39)

Rearranging terms, and noting that $L = mvr$ for a circular orbit, we obtain the quantization of angular momentum as the condition for allowed orbits:

$$L = m_e v r_n = n\frac{h}{2\pi}(n = 1, 2, 3 \ldots).$$
(30.40)

This is what Bohr was forced to hypothesize as the rule for allowed orbits, as stated earlier. We now realize that it is the condition for constructive interference of an electron in a circular orbit. **Figure 30.47** illustrates this for $n = 3$ and $n = 4$.

Waves and Quantization

The wave nature of matter is responsible for the quantization of energy levels in bound systems. Only those states where matter interferes constructively exist, or are "allowed." Since there is a lowest orbit where this is possible in an atom, the electron cannot spiral into the nucleus. It cannot exist closer to or inside the nucleus. The wave nature of matter is what prevents matter from collapsing and gives atoms their sizes.

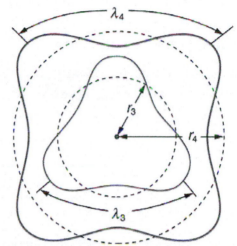

Figure 30.47 The third and fourth allowed circular orbits have three and four wavelengths, respectively, in their circumferences.

Because of the wave character of matter, the idea of well-defined orbits gives way to a model in which there is a cloud of probability, consistent with Heisenberg's uncertainty principle. **Figure 30.48** shows how this applies to the ground state of hydrogen. If you try to follow the electron in some well-defined orbit using a probe that has a small enough wavelength to get some details, you will instead knock the electron out of its orbit. Each measurement of the electron's position will find it to be in a definite location somewhere near the nucleus. Repeated measurements reveal a cloud of probability like that in the figure, with each speck the location determined by a single measurement. There is not a well-defined, circular-orbit type of distribution. Nature again proves to be different on a small scale than on a macroscopic scale.

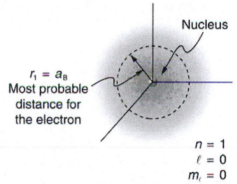

Figure 30.48 The ground state of a hydrogen atom has a probability cloud describing the position of its electron. The probability of finding the electron is proportional to the darkness of the cloud. The electron can be closer or farther than the Bohr radius, but it is very unlikely to be a great distance from the nucleus.

There are many examples in which the wave nature of matter causes quantization in bound systems such as the atom. Whenever a particle is confined or bound to a small space, its allowed wavelengths are those which fit into that space. For example, the particle in a box model describes a particle free to move in a small space surrounded by impenetrable barriers. This is true in blackbody radiators (atoms and molecules) as well as in atomic and molecular spectra. Various atoms and molecules will have different sets of electron orbits, depending on the size and complexity of the system. When a system is large, such as a grain of sand, the tiny particle waves in it can fit in so many ways that it becomes impossible to see that the allowed states are discrete. Thus the correspondence principle is satisfied. As systems become large, they gradually look less grainy, and quantization becomes less evident. Unbound systems (small or not), such as an electron freed from an atom, do not have quantized energies, since their wavelengths are not constrained to fit in a certain volume.

PhET Explorations: Quantum Wave Interference

When do photons, electrons, and atoms behave like particles and when do they behave like waves? Watch waves spread out and interfere as they pass through a double slit, then get detected on a screen as tiny dots. Use quantum detectors to explore how measurements change the waves and the patterns they produce on the screen.

Figure 30.49 Quantum Wave Interference (http://cnx.org/content/m55014/1.2/quantum-wave-interference_en.jar)

30.7 Patterns in Spectra Reveal More Quantization

Learning Objectives

By the end of this section, you will be able to:

- State and discuss the Zeeman effect.
- Define orbital magnetic field.
- Define orbital angular momentum.
- Define space quantization.

High-resolution measurements of atomic and molecular spectra show that the spectral lines are even more complex than they first appear. In this section, we will see that this complexity has yielded important new information about electrons and their orbits in atoms.

In order to explore the substructure of atoms (and knowing that magnetic fields affect moving charges), the Dutch physicist Hendrik Lorentz (1853–1930) suggested that his student Pieter Zeeman (1865–1943) study how spectra might be affected by magnetic fields. What they found became known as the **Zeeman effect**, which involved spectral lines being split into two or more separate emission lines by an external magnetic field, as shown in **Figure 30.50**. For their discoveries, Zeeman and Lorentz shared the 1902 Nobel Prize in Physics.

Zeeman splitting is complex. Some lines split into three lines, some into five, and so on. But one general feature is that the amount the split lines are separated is proportional to the applied field strength, indicating an interaction with a moving charge. The splitting means that the quantized energy of an orbit is affected by an external magnetic field, causing the orbit to have several discrete energies instead of one. Even without an external magnetic field, very precise measurements showed that spectral lines are doublets (split into two), apparently by magnetic fields within the atom itself.

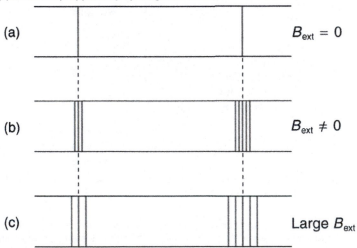

Figure 30.50 The Zeeman effect is the splitting of spectral lines when a magnetic field is applied. The number of lines formed varies, but the spread is proportional to the strength of the applied field. (a) Two spectral lines with no external magnetic field. (b) The lines split when the field is applied. (c) The splitting is greater when a stronger field is applied.

Bohr's theory of circular orbits is useful for visualizing how an electron's orbit is affected by a magnetic field. The circular orbit forms a current loop, which creates a magnetic field of its own, $\mathbf{B}_{orb}$ as seen in **Figure 30.51**. Note that the **orbital magnetic field** $\mathbf{B}_{orb}$ and the **orbital angular momentum** $\mathbf{L}_{orb}$ are along the same line. The external magnetic field and the orbital magnetic field interact; a torque is exerted to align them. A torque rotating a system through some angle does work so that there is energy associated with this interaction. Thus, orbits at different angles to the external magnetic field have different energies. What is remarkable is that the energies are quantized—the magnetic field splits the spectral lines into several discrete lines that have different energies. This means that only certain angles are allowed between the orbital angular momentum and the external field, as seen in **Figure 30.52**.

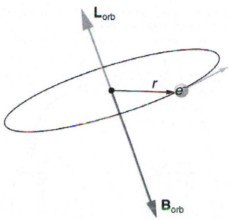

Figure 30.51 The approximate picture of an electron in a circular orbit illustrates how the current loop produces its own magnetic field, called $\mathbf{B}_{orb}$. It also shows how $\mathbf{B}_{orb}$ is along the same line as the orbital angular momentum $\mathbf{L}_{orb}$.

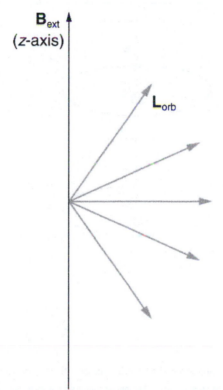

Figure 30.52 Only certain angles are allowed between the orbital angular momentum and an external magnetic field. This is implied by the fact that the Zeeman effect splits spectral lines into several discrete lines. Each line is associated with an angle between the external magnetic field and magnetic fields due to electrons and their orbits.

We already know that the magnitude of angular momentum is quantized for electron orbits in atoms. The new insight is that the *direction of the orbital angular momentum is also quantized*. The fact that the orbital angular momentum can have only certain directions is called **space quantization**. Like many aspects of quantum mechanics, this quantization of direction is totally unexpected. On the macroscopic scale, orbital angular momentum, such as that of the moon around the earth, can have any magnitude and be in any direction.

Detailed treatment of space quantization began to explain some complexities of atomic spectra, but certain patterns seemed to be caused by something else. As mentioned, spectral lines are actually closely spaced doublets, a characteristic called **fine structure**, as shown in **Figure 30.53**. The doublet changes when a magnetic field is applied, implying that whatever causes the doublet interacts with a magnetic field. In 1925, Sem Goudsmit and George Uhlenbeck, two Dutch physicists, successfully argued that electrons have properties analogous to a macroscopic charge spinning on its axis. Electrons, in fact, have an internal or intrinsic angular momentum called **intrinsic spin** $\mathbf{S}$. Since electrons are charged, their intrinsic spin creates an **intrinsic magnetic field** $\mathbf{B}_{int}$, which interacts with their orbital magnetic field $\mathbf{B}_{orb}$. Furthermore, *electron intrinsic spin is quantized in magnitude and direction*, analogous to the situation for orbital angular momentum. The spin of the electron can have only one magnitude, and its direction can be at only one of two angles relative to a magnetic field, as seen in **Figure 30.54**. We refer to this as spin up or spin down for the electron. Each spin direction has a different energy; hence, spectroscopic lines are split into two. Spectral doublets are now understood as being due to electron spin.

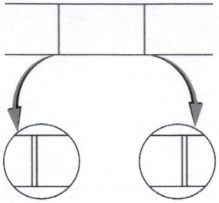

Figure 30.53 Fine structure. Upon close examination, spectral lines are doublets, even in the absence of an external magnetic field. The electron has an intrinsic magnetic field that interacts with its orbital magnetic field.

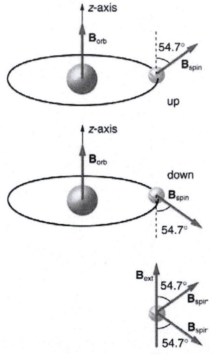

Figure 30.54 The intrinsic magnetic field $\mathbf{B}_{int}$ of an electron is attributed to its spin, $\mathbf{S}$, roughly pictured to be due to its charge spinning on its axis.

This is only a crude model, since electrons seem to have no size. The spin and intrinsic magnetic field of the electron can make only one of two angles with another magnetic field, such as that created by the electron's orbital motion. Space is quantized for spin as well as for orbital angular momentum.

These two new insights—that the direction of angular momentum, whether orbital or spin, is quantized, and that electrons have intrinsic spin—help to explain many of the complexities of atomic and molecular spectra. In magnetic resonance imaging, it is the way that the intrinsic magnetic field of hydrogen and biological atoms interact with an external field that underlies the diagnostic fundamentals.

30.8 Quantum Numbers and Rules

Learning Objectives

By the end of this section, you will be able to:

- Define quantum number.
- Calculate the angle of an angular momentum vector with an axis.
- Define spin quantum number.

Physical characteristics that are quantized—such as energy, charge, and angular momentum—are of such importance that names and symbols are given to them. The values of quantized entities are expressed in terms of **quantum numbers**, and the rules governing them are of the utmost importance in determining what nature is and does. This section covers some of the more important quantum numbers and rules—all of which apply in chemistry, material science, and far beyond the realm of atomic physics, where they were first discovered. Once again, we see how physics makes discoveries which enable other fields to grow.

The *energy states of bound systems are quantized*, because the particle wavelength can fit into the bounds of the system in only certain ways. This was elaborated for the hydrogen atom, for which the allowed energies are expressed as $E_n \propto 1/n^2$, where $n = 1, 2, 3, \dots$. We define n to be the principal quantum number that labels the basic states of a system. The lowest-energy state has $n = 1$, the first excited state has $n = 2$, and so on. Thus the allowed values for the principal quantum number are

$$n = 1, 2, 3, \dots. \tag{30.41}$$

This is more than just a numbering scheme, since the energy of the system, such as the hydrogen atom, can be expressed as some function of n, as can other characteristics (such as the orbital radii of the hydrogen atom).

The fact that the *magnitude of angular momentum is quantized* was first recognized by Bohr in relation to the hydrogen atom; it is now known to be true in general. With the development of quantum mechanics, it was found that the magnitude of angular momentum L can have only the values

$$L = \sqrt{l(l+1)}\frac{h}{2\pi} \quad (l = 0, 1, 2, \dots, n-1), \tag{30.42}$$

where l is defined to be the **angular momentum quantum number**. The rule for l in atoms is given in the parentheses. Given n, the value of l can be any integer from zero up to $n-1$. For example, if $n = 4$, then l can be 0, 1, 2, or 3.

Note that for $n = 1$, l can only be zero. This means that the ground-state angular momentum for hydrogen is actually zero, not $h/2\pi$ as Bohr proposed. The picture of circular orbits is not valid, because there would be angular momentum for any circular orbit. A more valid picture is the cloud of probability shown for the ground state of hydrogen in **Figure 30.48**. The electron actually spends time in and near the nucleus. The reason the electron does not remain in the nucleus is related to Heisenberg's uncertainty principle—the electron's energy would have to be much too large to be confined to the small space of the nucleus.

Now the first excited state of hydrogen has $n = 2$, so that l can be either 0 or 1, according to the rule in $L = \sqrt{l(l+1)}\frac{h}{2\pi}$.

Similarly, for $n = 3$, l can be 0, 1, or 2. It is often most convenient to state the value of l, a simple integer, rather than calculating the value of L from $L = \sqrt{l(l+1)}\frac{h}{2\pi}$. For example, for $l = 2$, we see that

$$L = \sqrt{2(2+1)}\frac{h}{2\pi} = \sqrt{6}\frac{h}{2\pi} = 0.390h = 2.58 \times 10^{-34} \text{ J} \cdot \text{s}. \tag{30.43}$$

It is much simpler to state $l = 2$.

As recognized in the Zeeman effect, the *direction of angular momentum is quantized*. We now know this is true in all circumstances. It is found that the component of angular momentum along one direction in space, usually called the z-axis, can have only certain values of L_z. The direction in space must be related to something physical, such as the direction of the magnetic field at that location. This is an aspect of relativity. Direction has no meaning if there is nothing that varies with direction, as does magnetic force. The allowed values of L_z are

$$L_z = m_l\frac{h}{2\pi} \quad (m_l = -l, -l+1, \dots, -1, 0, 1, \dots l-1, l), \tag{30.44}$$

where L_z is the z-**component of the angular momentum** and m_l is the angular momentum projection quantum number. The rule in parentheses for the values of m_l is that it can range from $-l$ to l in steps of one. For example, if $l = 2$, then m_l can have the five values –2, –1, 0, 1, and 2. Each m_l corresponds to a different energy in the presence of a magnetic field, so that they are related to the splitting of spectral lines into discrete parts, as discussed in the preceding section. If the z-component of angular momentum can have only certain values, then the angular momentum can have only certain directions, as illustrated in **Figure 30.55**.

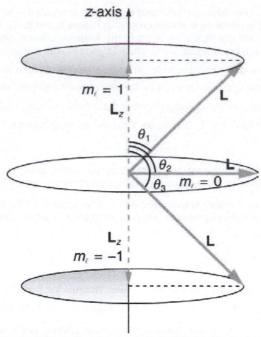

Figure 30.55 The component of a given angular momentum along the z-axis (defined by the direction of a magnetic field) can have only certain values; these are shown here for $l = 1$, for which $m_l = -1, 0,$ and $+1$. The direction of L is quantized in the sense that it can have only certain angles relative to the z-axis.

Example 30.3 What Are the Allowed Directions?

Calculate the angles that the angular momentum vector $\mathbf{L}$ can make with the z-axis for $l = 1$, as illustrated in **Figure 30.55**.

Strategy

Figure 30.55 represents the vectors $\mathbf{L}$ and $\mathbf{L}_z$ as usual, with arrows proportional to their magnitudes and pointing in the correct directions. $\mathbf{L}$ and $\mathbf{L}_z$ form a right triangle, with $\mathbf{L}$ being the hypotenuse and $\mathbf{L}_z$ the adjacent side. This means that the ratio of $\mathbf{L}_z$ to $\mathbf{L}$ is the cosine of the angle of interest. We can find $\mathbf{L}$ and $\mathbf{L}_z$ using $L = \sqrt{l(l+1)}\dfrac{h}{2\pi}$ and $L_z = m\dfrac{h}{2\pi}$.

Solution

We are given $l = 1$, so that m_l can be +1, 0, or –1. Thus L has the value given by $L = \sqrt{l(l+1)}\dfrac{h}{2\pi}$.

$$L = \frac{\sqrt{l(l+1)}h}{2\pi} = \frac{\sqrt{2}h}{2\pi} \tag{30.45}$$

L_z can have three values, given by $L_z = m_l \dfrac{h}{2\pi}$.

$$L_z = m_l\frac{h}{2\pi} = \begin{cases} \dfrac{h}{2\pi}, & m_l = +1 \\ 0, & m_l = 0 \\ -\dfrac{h}{2\pi}, & m_l = -1 \end{cases} \tag{30.46}$$

As can be seen in **Figure 30.55**, $\cos\theta = L_z/L$, and so for $m_l = +1$, we have

$$\cos\theta_1 = \frac{L_Z}{L} = \frac{\dfrac{h}{2\pi}}{\dfrac{\sqrt{2}h}{2\pi}} = \frac{1}{\sqrt{2}} = 0.707. \tag{30.47}$$

Thus,

$$\theta_1 = \cos^{-1} 0.707 = 45.0°.$$ (30.48)

Similarly, for $m_l = 0$, we find $\cos \theta_2 = 0$; thus,

$$\theta_2 = \cos^{-1} 0 = 90.0°.$$ (30.49)

And for $m_l = -1$,

$$\cos \theta_3 = \frac{L_Z}{L} = \frac{-\frac{h}{2\pi}}{\frac{\sqrt{2}h}{2\pi}} = -\frac{1}{\sqrt{2}} = -0.707,$$ (30.50)

so that

$$\theta_3 = \cos^{-1}(-0.707) = 135.0°.$$ (30.51)

Discussion

The angles are consistent with the figure. Only the angle relative to the z-axis is quantized. L can point in any direction as long as it makes the proper angle with the z-axis. Thus the angular momentum vectors lie on cones as illustrated. This behavior is not observed on the large scale. To see how the correspondence principle holds here, consider that the smallest angle (θ_1 in the example) is for the maximum value of $m_l = 0$, namely $m_l = l$. For that smallest angle,

$$\cos \theta = \frac{L_z}{L} = \frac{l}{\sqrt{l(l+1)}},$$ (30.52)

which approaches 1 as l becomes very large. If $\cos \theta = 1$, then $\theta = 0°$. Furthermore, for large l, there are many values of m_l, so that all angles become possible as l gets very large.

Intrinsic Spin Angular Momentum Is Quantized in Magnitude and Direction

There are two more quantum numbers of immediate concern. Both were first discovered for electrons in conjunction with fine structure in atomic spectra. It is now well established that electrons and other fundamental particles have *intrinsic spin*, roughly analogous to a planet spinning on its axis. This spin is a fundamental characteristic of particles, and only one magnitude of intrinsic spin is allowed for a given type of particle. Intrinsic angular momentum is quantized independently of orbital angular momentum. Additionally, the direction of the spin is also quantized. It has been found that the **magnitude of the intrinsic (internal) spin angular momentum**, S, of an electron is given by

$$S = \sqrt{s(s+1)} \frac{h}{2\pi} \quad (s = 1/2 \text{ for electrons}),$$ (30.53)

where s is defined to be the **spin quantum number**. This is very similar to the quantization of L given in $L = \sqrt{l(l+1)} \frac{h}{2\pi}$, except that the only value allowed for s for electrons is 1/2.

The *direction of intrinsic spin is quantized*, just as is the direction of orbital angular momentum. The direction of spin angular momentum along one direction in space, again called the z-axis, can have only the values

$$S_z = m_s \frac{h}{2\pi} \quad \left(m_s = -\frac{1}{2}, +\frac{1}{2} \right)$$ (30.54)

for electrons. S_z is the z-**component of spin angular momentum** and m_s is the **spin projection quantum number**. For electrons, s can only be 1/2, and m_s can be either +1/2 or –1/2. Spin projection $m_s = +1/2$ is referred to as *spin up*, whereas $m_s = -1/2$ is called *spin down*. These are illustrated in **Figure 30.54**.

Intrinsic Spin

In later chapters, we will see that intrinsic spin is a characteristic of all subatomic particles. For some particles s is half-integral, whereas for others s is integral—there are crucial differences between half-integral spin particles and integral spin particles. Protons and neutrons, like electrons, have $s = 1/2$, whereas photons have $s = 1$, and other particles called pions have $s = 0$, and so on.

To summarize, the state of a system, such as the precise nature of an electron in an atom, is determined by its particular quantum numbers. These are expressed in the form (n, l, m_l, m_s) —see **Table 30.1** *For electrons in atoms*, the principal quantum number can have the values $n = 1, 2, 3, ...$. Once n is known, the values of the angular momentum quantum number are limited to $l = 1, 2, 3, ...,n - 1$. For a given value of l, the angular momentum projection quantum number can have only the values $m_l = -l, -l + 1, ..., -1, 0, 1, ..., l - 1, l$. Electron spin is independent of n, l, and m_l, always having $s = 1/2$. The spin projection quantum number can have two values, $m_s = 1/2$ or $-1/2$.

Table 30.1 Atomic Quantum Numbers

Name	Symbol	Allowed values
Principal quantum number	n	$1, 2, 3, ...$
Angular momentum	l	$0, 1, 2, ...n - 1$
Angular momentum projection	m_l	$-l, -l + 1, ..., -1, 0, 1, ..., l - 1, l$ (or $0, \pm 1, \pm 2, ..., \pm l$)
Spin[1]	s	$1/2$(electrons)
Spin projection	m_s	$-1/2, +1/2$

Figure 30.56 shows several hydrogen states corresponding to different sets of quantum numbers. Note that these clouds of probability are the locations of electrons as determined by making repeated measurements—each measurement finds the electron in a definite location, with a greater chance of finding the electron in some places rather than others. With repeated measurements, the pattern of probability shown in the figure emerges. The clouds of probability do not look like nor do they correspond to classical orbits. The uncertainty principle actually prevents us and nature from knowing how the electron gets from one place to another, and so an orbit really does not exist as such. Nature on a small scale is again much different from that on the large scale.

---

1. The spin quantum number s is usually not stated, since it is always 1/2 for electrons

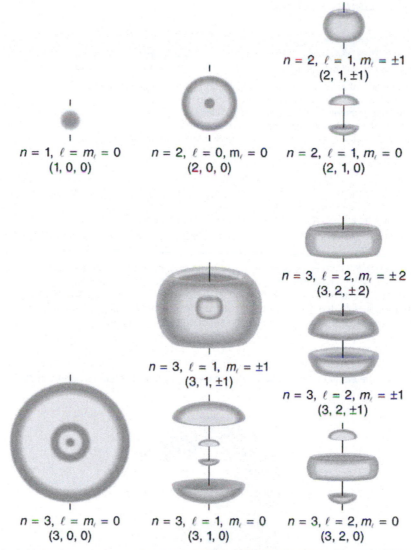

Figure 30.56 Probability clouds for the electron in the ground state and several excited states of hydrogen. The nature of these states is determined by their sets of quantum numbers, here given as $\left(n,\ l,\ m_l\right)$. The ground state is (0, 0, 0); one of the possibilities for the second excited state is (3, 2, 1).

The probability of finding the electron is indicated by the shade of color; the darker the coloring the greater the chance of finding the electron.

We will see that the quantum numbers discussed in this section are valid for a broad range of particles and other systems, such as nuclei. Some quantum numbers, such as intrinsic spin, are related to fundamental classifications of subatomic particles, and they obey laws that will give us further insight into the substructure of matter and its interactions.

PhET Explorations: Stern-Gerlach Experiment

The classic Stern-Gerlach Experiment shows that atoms have a property called spin. Spin is a kind of intrinsic angular momentum, which has no classical counterpart. When the z-component of the spin is measured, one always gets one of two values: spin up or spin down.

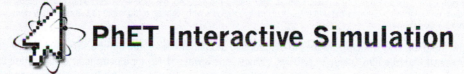

Figure 30.57 Stern-Gerlach Experiment (http://cnx.org/content/m54997/1.2/stern-gerlach_en.jar)

30.9 The Pauli Exclusion Principle

Learning Objectives

By the end of this section, you will be able to:

- Define the composition of an atom along with its electrons, neutrons, and protons.
- Explain the Pauli exclusion principle and its application to the atom.
- Specify the shell and subshell symbols and their positions.
- Define the position of electrons in different shells of an atom.
- State the position of each element in the periodic table according to shell filling.

Multiple-Electron Atoms

All atoms except hydrogen are multiple-electron atoms. The physical and chemical properties of elements are directly related to the number of electrons a neutral atom has. The periodic table of the elements groups elements with similar properties into columns. This systematic organization is related to the number of electrons in a neutral atom, called the **atomic number**, Z. We shall see in this section that the exclusion principle is key to the underlying explanations, and that it applies far beyond the realm of atomic physics.

In 1925, the Austrian physicist Wolfgang Pauli (see **Figure 30.58**) proposed the following rule: No two electrons can have the same set of quantum numbers. That is, no two electrons can be in the same state. This statement is known as the **Pauli exclusion principle**, because it excludes electrons from being in the same state. The Pauli exclusion principle is extremely powerful and very broadly applicable. It applies to any identical particles with half-integral intrinsic spin—that is, having $s = 1/2, 3/2, ...$ Thus no two electrons can have the same set of quantum numbers.

Pauli Exclusion Principle
_____

No two electrons can have the same set of quantum numbers. That is, no two electrons can be in the same state.

Figure 30.58 The Austrian physicist Wolfgang Pauli (1900–1958) played a major role in the development of quantum mechanics. He proposed the exclusion principle; hypothesized the existence of an important particle, called the neutrino, before it was directly observed; made fundamental contributions to several areas of theoretical physics; and influenced many students who went on to do important work of their own. (credit: Nobel Foundation, via Wikimedia Commons)

Let us examine how the exclusion principle applies to electrons in atoms. The quantum numbers involved were defined in **Quantum Numbers and Rules** as n, l, m_l, s, and m_s. Since s is always $1/2$ for electrons, it is redundant to list s, and so we omit it and specify the state of an electron by a set of four numbers (n, l, m_l, m_s). For example, the quantum numbers $(2, 1, 0, -1/2)$ completely specify the state of an electron in an atom.

Since no two electrons can have the same set of quantum numbers, there are limits to how many of them can be in the same energy state. Note that n determines the energy state in the absence of a magnetic field. So we first choose n, and then we see how many electrons can be in this energy state or energy level. Consider the $n = 1$ level, for example. The only value l

can have is 0 (see Table 30.1 for a list of possible values once n is known), and thus m_l can only be 0. The spin projection m_s can be either $+1/2$ or $-1/2$, and so there can be two electrons in the $n = 1$ state. One has quantum numbers $(1, 0, 0, +1/2)$, and the other has $(1, 0, 0, -1/2)$. Figure 30.59 illustrates that there can be one or two electrons having $n = 1$, but not three.

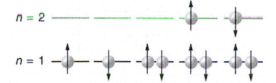

Figure 30.59 The Pauli exclusion principle explains why some configurations of electrons are allowed while others are not. Since electrons cannot have the same set of quantum numbers, a maximum of two can be in the $n = 1$ level, and a third electron must reside in the higher-energy $n = 2$ level. If there are two electrons in the $n = 1$ level, their spins must be in opposite directions. (More precisely, their spin projections must differ.)

Shells and Subshells

Because of the Pauli exclusion principle, only hydrogen and helium can have all of their electrons in the $n = 1$ state. Lithium (see the periodic table) has three electrons, and so one must be in the $n = 2$ level. This leads to the concept of shells and shell filling. As we progress up in the number of electrons, we go from hydrogen to helium, lithium, beryllium, boron, and so on, and we see that there are limits to the number of electrons for each value of n. Higher values of the shell n correspond to higher energies, and they can allow more electrons because of the various combinations of l, m_l, and m_s that are possible. Each value of the principal quantum number n thus corresponds to an atomic **shell** into which a limited number of electrons can go. Shells and the number of electrons in them determine the physical and chemical properties of atoms, since it is the outermost electrons that interact most with anything outside the atom.

The probability clouds of electrons with the lowest value of l are closest to the nucleus and, thus, more tightly bound. Thus when shells fill, they start with $l = 0$, progress to $l = 1$, and so on. Each value of l thus corresponds to a **subshell**.

The table given below lists symbols traditionally used to denote shells and subshells.

Table 30.2 Shell and
Subshell Symbols

Shell	Subshell	
n	l	Symbol
1	0	s
2	1	p
3	2	d
4	3	f
5	4	g
	5	h
	6[2]	i

To denote shells and subshells, we write nl with a number for n and a letter for l. For example, an electron in the $n = 1$ state must have $l = 0$, and it is denoted as a $1s$ electron. Two electrons in the $n = 1$ state is denoted as $1s^2$. Another example is an electron in the $n = 2$ state with $l = 1$, written as $2p$. The case of three electrons with these quantum numbers is written $2p^3$. This notation, called spectroscopic notation, is generalized as shown in **Figure 30.60**.

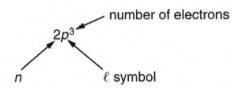

Figure 30.60

Counting the number of possible combinations of quantum numbers allowed by the exclusion principle, we can determine how many electrons it takes to fill each subshell and shell.

Example 30.4 How Many Electrons Can Be in This Shell?

List all the possible sets of quantum numbers for the $n = 2$ shell, and determine the number of electrons that can be in the shell and each of its subshells.

Strategy

Given $n = 2$ for the shell, the rules for quantum numbers limit l to be 0 or 1. The shell therefore has two subshells, labeled $2s$ and $2p$. Since the lowest l subshell fills first, we start with the $2s$ subshell possibilities and then proceed with the $2p$ subshell.

Solution

It is convenient to list the possible quantum numbers in a table, as shown below.

2. It is unusual to deal with subshells having l greater than 6, but when encountered, they continue to be labeled in alphabetical order.

n	ℓ	m_l	m_s	Subshell	Total in subshell	Total in shell
2	0	0	+1/2	2s	2	
2	0	0	−1/2			
2	1	1	+1/2	2p	6	8
2	1	1	−1/2			
2	1	0	+1/2			
2	1	0	−1/2			
2	1	−1	+1/2			
2	1	−1	−1/2			

Figure 30.61

Discussion

It is laborious to make a table like this every time we want to know how many electrons can be in a shell or subshell. There exist general rules that are easy to apply, as we shall now see.

The number of electrons that can be in a subshell depends entirely on the value of l. Once l is known, there are a fixed number of values of m_l, each of which can have two values for m_s First, since m_l goes from $-l$ to l in steps of 1, there are $2l + 1$ possibilities. This number is multiplied by 2, since each electron can be spin up or spin down. Thus the *maximum number of electrons that can be in a subshell* is $2(2l + 1)$.

For example, the $2s$ subshell in **Example 30.4** has a maximum of 2 electrons in it, since $2(2l + 1) = 2(0 + 1) = 2$ for this subshell. Similarly, the $2p$ subshell has a maximum of 6 electrons, since $2(2l + 1) = 2(2 + 1) = 6$. For a shell, the maximum number is the sum of what can fit in the subshells. Some algebra shows that the *maximum number of electrons that can be in a shell* is $2n^2$.

For example, for the first shell $n = 1$, and so $2n^2 = 2$. We have already seen that only two electrons can be in the $n = 1$ shell. Similarly, for the second shell, $n = 2$, and so $2n^2 = 8$. As found in **Example 30.4**, the total number of electrons in the $n = 2$ shell is 8.

Example 30.5 Subshells and Totals for $n = 3$

How many subshells are in the $n = 3$ shell? Identify each subshell, calculate the maximum number of electrons that will fit into each, and verify that the total is $2n^2$.

Strategy

Subshells are determined by the value of l; thus, we first determine which values of l are allowed, and then we apply the equation "maximum number of electrons that can be in a subshell $= 2(2l + 1)$" to find the number of electrons in each subshell.

Solution

Since $n = 3$, we know that l can be 0, 1, or 2; thus, there are three possible subshells. In standard notation, they are labeled the $3s$, $3p$, and $3d$ subshells. We have already seen that 2 electrons can be in an s state, and 6 in a p state, but let us use the equation "maximum number of electrons that can be in a subshell $= 2(2l + 1)$" to calculate the maximum number in each:

$$3s \text{ has } l = 0; \text{ thus, } 2(2l + 1) = 2(0 + 1) = 2 \tag{30.55}$$
$$3p \text{ has } l = 1; \text{ thus, } 2(2l + 1) = 2(2 + 1) = 6$$
$$3d \text{ has } l = 2; \text{ thus, } 2(2l + 1) = 2(4 + 1) = 10$$
$$\text{Total} = 18$$
$$(\text{in the } n = 3 \text{ shell})$$

The equation "maximum number of electrons that can be in a shell = $2n^2$" gives the maximum number in the $n = 3$ shell to be

$$\text{Maximum number of electrons} = 2n^2 = 2(3)^2 = 2(9) = 18. \tag{30.56}$$

Discussion

The total number of electrons in the three possible subshells is thus the same as the formula $2n^2$. In standard (spectroscopic) notation, a filled $n = 3$ shell is denoted as $3s^2 3p^6 3d^{10}$. Shells do not fill in a simple manner. Before the $n = 3$ shell is completely filled, for example, we begin to find electrons in the $n = 4$ shell.

Shell Filling and the Periodic Table

Table 30.3 shows electron configurations for the first 20 elements in the periodic table, starting with hydrogen and its single electron and ending with calcium. The Pauli exclusion principle determines the maximum number of electrons allowed in each shell and subshell. But the order in which the shells and subshells are filled is complicated because of the large numbers of interactions between electrons.

Table 30.3 Electron Configurations of Elements Hydrogen Through Calcium

Element	Number of electrons (Z)	Ground state configuration					
H	1	$1s^1$					
He	2	$1s^2$					
Li	3	$1s^2$	$2s^1$				
Be	4	"	$2s^2$				
B	5	"	$2s^2$	$2p^1$			
C	6	"	$2s^2$	$2p^2$			
N	7	"	$2s^2$	$2p^3$			
O	8	"	$2s^2$	$2p^4$			
F	9	"	$2s^2$	$2p^5$			
Ne	10	"	$2s^2$	$2p^6$			
Na	11	"	$2s^2$	$2p^6$	$3s^1$		
Mg	12	"	"	"	$3s^2$		
Al	13	"	"	"	$3s^2$	$3p^1$	
Si	14	"	"	"	$3s^2$	$3p^2$	
P	15	"	"	"	$3s^2$	$3p^3$	
S	16	"	"	"	$3s^2$	$3p^4$	
Cl	17	"	"	"	$3s^2$	$3p^5$	
Ar	18	"	"	"	$3s^2$	$3p^6$	
K	19	"	"	"	$3s^2$	$3p^6$	$4s^1$
Ca	20	"	"	"	"	"	$4s^2$

Examining the above table, you can see that as the number of electrons in an atom increases from 1 in hydrogen to 2 in helium and so on, the lowest-energy shell gets filled first—that is, the $n = 1$ shell fills first, and then the $n = 2$ shell begins to fill. Within a shell, the subshells fill starting with the lowest l, or with the s subshell, then the p, and so on, usually until all subshells are filled. The first exception to this occurs for potassium, where the $4s$ subshell begins to fill before any electrons go into the $3d$ subshell. The next exception is not shown in **Table 30.3**; it occurs for rubidium, where the $5s$ subshell starts to fill before the $4d$ subshell. The reason for these exceptions is that $l = 0$ electrons have probability clouds that penetrate closer to the nucleus and, thus, are more tightly bound (lower in energy).

Figure 30.62 shows the periodic table of the elements, through element 118. Of special interest are elements in the main groups, namely, those in the columns numbered 1, 2, 13, 14, 15, 16, 17, and 18.

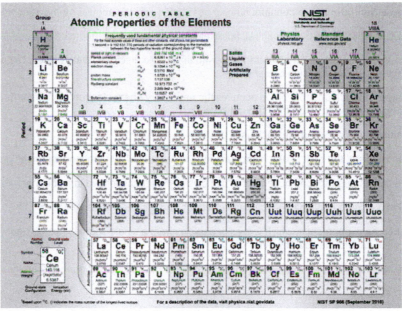

Figure 30.62 Periodic table of the elements (credit: National Institute of Standards and Technology, U.S. Department of Commerce)

The number of electrons in the outermost subshell determines the atom's chemical properties, since it is these electrons that are farthest from the nucleus and thus interact most with other atoms. If the outermost subshell can accept or give up an electron easily, then the atom will be highly reactive chemically. Each group in the periodic table is characterized by its outermost electron configuration. Perhaps the most familiar is Group 18 (Group VIII), the noble gases (helium, neon, argon, etc.). These gases are all characterized by a filled outer subshell that is particularly stable. This means that they have large ionization energies and do not readily give up an electron. Furthermore, if they were to accept an extra electron, it would be in a significantly higher level and thus loosely bound. Chemical reactions often involve sharing electrons. Noble gases can be forced into unstable chemical compounds only under high pressure and temperature.

Group 17 (Group VII) contains the halogens, such as fluorine, chlorine, iodine and bromine, each of which has one less electron than a neighboring noble gas. Each halogen has 5 p electrons (a p^5 configuration), while the p subshell can hold 6 electrons. This means the halogens have one vacancy in their outermost subshell. They thus readily accept an extra electron (it becomes tightly bound, closing the shell as in noble gases) and are highly reactive chemically. The halogens are also likely to form singly negative ions, such as Cl^-, fitting an extra electron into the vacancy in the outer subshell. In contrast, alkali metals, such as sodium and potassium, all have a single s electron in their outermost subshell (an s^1 configuration) and are members of Group 1 (Group I). These elements easily give up their extra electron and are thus highly reactive chemically. As you might expect, they also tend to form singly positive ions, such as Na^+, by losing their loosely bound outermost electron. They are metals (conductors), because the loosely bound outer electron can move freely.

Of course, other groups are also of interest. Carbon, silicon, and germanium, for example, have similar chemistries and are in Group 4 (Group IV). Carbon, in particular, is extraordinary in its ability to form many types of bonds and to be part of long chains, such as inorganic molecules. The large group of what are called transitional elements is characterized by the filling of the d subshells and crossing of energy levels. Heavier groups, such as the lanthanide series, are more complex—their shells do not fill in simple order. But the groups recognized by chemists such as Mendeleev have an explanation in the substructure of atoms.

PhET Explorations: Build an Atom

Build an atom out of protons, neutrons, and electrons, and see how the element, charge, and mass change. Then play a game to test your ideas!

Figure 30.63 Build an Atom (http://cnx.org/content/m54998/1.2/build-an-atom_en.jar)

Glossary

angular momentum quantum number: a quantum number associated with the angular momentum of electrons

atom: basic unit of matter, which consists of a central, positively charged nucleus surrounded by negatively charged electrons

atomic de-excitation: process by which an atom transfers from an excited electronic state back to the ground state electronic configuration; often occurs by emission of a photon

atomic excitation: a state in which an atom or ion acquires the necessary energy to promote one or more of its electrons to electronic states higher in energy than their ground state

atomic number: the number of protons in the nucleus of an atom

Bohr radius: the mean radius of the orbit of an electron around the nucleus of a hydrogen atom in its ground state

Brownian motion: the continuous random movement of particles of matter suspended in a liquid or gas

cathode-ray tube: a vacuum tube containing a source of electrons and a screen to view images

double-slit interference: an experiment in which waves or particles from a single source impinge upon two slits so that the resulting interference pattern may be observed

energies of hydrogen-like atoms: Bohr formula for energies of electron states in hydrogen-like atoms:

$$E_n = -\frac{Z^2}{n^2}E_0 (n = 1, 2, 3, \ldots)$$

energy-level diagram: a diagram used to analyze the energy level of electrons in the orbits of an atom

fine structure: the splitting of spectral lines of the hydrogen spectrum when the spectral lines are examined at very high resolution

fluorescence: any process in which an atom or molecule, excited by a photon of a given energy, de-excites by emission of a lower-energy photon

hologram: means *entire picture* (from the Greek word *holo*, as in holistic), because the image produced is three dimensional

holography: the process of producing holograms

hydrogen spectrum wavelengths:
the wavelengths of visible light from hydrogen; can be calculated by $\frac{1}{\lambda} = R\left(\frac{1}{n_f^2} - \frac{1}{n_i^2}\right)$

hydrogen-like atom: any atom with only a single electron

intrinsic magnetic field: the magnetic field generated due to the intrinsic spin of electrons

intrinsic spin: the internal or intrinsic angular momentum of electrons

laser: acronym for light amplification by stimulated emission of radiation

magnitude of the intrinsic (internal) spin angular momentum: given by $S = \sqrt{s(s+1)}\frac{h}{2\pi}$

metastable: a state whose lifetime is an order of magnitude longer than the most short-lived states

orbital angular momentum: an angular momentum that corresponds to the quantum analog of classical angular momentum

orbital magnetic field: the magnetic field generated due to the orbital motion of electrons

Pauli exclusion principle: a principle that states that no two electrons can have the same set of quantum numbers; that is, no two electrons can be in the same state

phosphorescence: the de-excitation of a metastable state

planetary model of the atom: the most familiar model or illustration of the structure of the atom

population inversion: the condition in which the majority of atoms in a sample are in a metastable state

quantum numbers: the values of quantized entities, such as energy and angular momentum

Rydberg constant: a physical constant related to the atomic spectra with an established value of $1.097 \times 10^7 \ \text{m}^{-1}$

shell: a probability cloud for electrons that has a single principal quantum number

space quantization: the fact that the orbital angular momentum can have only certain directions

spin projection quantum number: quantum number that can be used to calculate the intrinsic electron angular momentum along the z-axis

spin quantum number: the quantum number that parameterizes the intrinsic angular momentum (or spin angular momentum, or simply spin) of a given particle

stimulated emission: emission by atom or molecule in which an excited state is stimulated to decay, most readily caused by a photon of the same energy that is necessary to excite the state

subshell: the probability cloud for electrons that has a single angular momentum quantum number l

x rays: a form of electromagnetic radiation

x-ray diffraction: a technique that provides the detailed information about crystallographic structure of natural and manufactured materials

z-component of spin angular momentum: component of intrinsic electron spin along the z-axis

z-component of the angular momentum: component of orbital angular momentum of electron along the z-axis

Zeeman effect: the effect of external magnetic fields on spectral lines

Section Summary

30.1 Discovery of the Atom
- Atoms are the smallest unit of elements; atoms combine to form molecules, the smallest unit of compounds.
- The first direct observation of atoms was in Brownian motion.
- Analysis of Brownian motion gave accurate sizes for atoms (10^{-10} m on average) and a precise value for Avogadro's number.

30.2 Discovery of the Parts of the Atom: Electrons and Nuclei
- Atoms are composed of negatively charged electrons, first proved to exist in cathode-ray-tube experiments, and a positively charged nucleus.
- All electrons are identical and have a charge-to-mass ratio of

$$\frac{q_e}{m_e} = -1.76 \times 10^{11} \ \text{C/kg}.$$

- The positive charge in the nuclei is carried by particles called protons, which have a charge-to-mass ratio of

$$\frac{q_p}{m_p} = 9.57 \times 10^7 \ \text{C/kg}.$$

- Mass of electron,

$$m_e = 9.11 \times 10^{-31} \ \text{kg}.$$

- Mass of proton,

$$m_p = 1.67 \times 10^{-27} \ \text{kg}.$$

- The planetary model of the atom pictures electrons orbiting the nucleus in the same way that planets orbit the sun.

30.3 Bohr's Theory of the Hydrogen Atom
- The planetary model of the atom pictures electrons orbiting the nucleus in the way that planets orbit the sun. Bohr used the planetary model to develop the first reasonable theory of hydrogen, the simplest atom. Atomic and molecular spectra are

quantized, with hydrogen spectrum wavelengths given by the formula

$$\frac{1}{\lambda} = R\left(\frac{1}{n_f^2} - \frac{1}{n_i^2}\right),$$

where λ is the wavelength of the emitted EM radiation and R is the Rydberg constant, which has the value

$$R = 1.097 \times 10^7 \text{ m}^{-1}.$$

- The constants n_i and n_f are positive integers, and n_i must be greater than n_f.
- Bohr correctly proposed that the energy and radii of the orbits of electrons in atoms are quantized, with energy for transitions between orbits given by

$$\Delta E = hf = E_i - E_f,$$

where ΔE is the change in energy between the initial and final orbits and hf is the energy of an absorbed or emitted photon. It is useful to plot orbital energies on a vertical graph called an energy-level diagram.
- Bohr proposed that the allowed orbits are circular and must have quantized orbital angular momentum given by

$$L = m_e v r_n = n\frac{h}{2\pi}(n = 1, 2, 3 \ldots),$$

where L is the angular momentum, r_n is the radius of the nth orbit, and h is Planck's constant. For all one-electron (hydrogen-like) atoms, the radius of an orbit is given by

$$r_n = \frac{n^2}{Z}a_B(\text{allowed orbits } n = 1, 2, 3, \ldots),$$

Z is the atomic number of an element (the number of electrons is has when neutral) and a_B is defined to be the Bohr radius, which is

$$a_B = \frac{h^2}{4\pi^2 m_e k q_e^2} = 0.529 \times 10^{-10} \text{ m}.$$

- Furthermore, the energies of hydrogen-like atoms are given by

$$E_n = -\frac{Z^2}{n^2}E_0(n = 1, 2, 3 \ldots),$$

where E_0 is the ground-state energy and is given by

$$E_0 = \frac{2\pi^2 q_e^4 m_e k^2}{h^2} = 13.6 \text{ eV}.$$

Thus, for hydrogen,

$$E_n = -\frac{13.6 \text{ eV}}{n^2}(n, = , 1, 2, 3 \ldots).$$

- The Bohr Theory gives accurate values for the energy levels in hydrogen-like atoms, but it has been improved upon in several respects.

30.4 X Rays: Atomic Origins and Applications

- X rays are relatively high-frequency EM radiation. They are produced by transitions between inner-shell electron levels, which produce x rays characteristic of the atomic element, or by accelerating electrons.
- X rays have many uses, including medical diagnostics and x-ray diffraction.

30.5 Applications of Atomic Excitations and De-Excitations

- An important atomic process is fluorescence, defined to be any process in which an atom or molecule is excited by absorbing a photon of a given energy and de-excited by emitting a photon of a lower energy.
- Some states live much longer than others and are termed metastable.
- Phosphorescence is the de-excitation of a metastable state.
- Lasers produce coherent single-wavelength EM radiation by stimulated emission, in which a metastable state is stimulated to decay.
- Lasing requires a population inversion, in which a majority of the atoms or molecules are in their metastable state.

30.6 The Wave Nature of Matter Causes Quantization

- Quantization of orbital energy is caused by the wave nature of matter. Allowed orbits in atoms occur for constructive interference of electrons in the orbit, requiring an integral number of wavelengths to fit in an orbit's circumference; that is,

$$n\lambda_n = 2\pi r_n(n = 1, 2, 3 \ldots),$$

where λ_n is the electron's de Broglie wavelength.

- Owing to the wave nature of electrons and the Heisenberg uncertainty principle, there are no well-defined orbits; rather, there are clouds of probability.
- Bohr correctly proposed that the energy and radii of the orbits of electrons in atoms are quantized, with energy for transitions between orbits given by

$$\Delta E = hf = E_i - E_f,$$

where ΔE is the change in energy between the initial and final orbits and hf is the energy of an absorbed or emitted photon.

- It is useful to plot orbit energies on a vertical graph called an energy-level diagram.
- The allowed orbits are circular, Bohr proposed, and must have quantized orbital angular momentum given by

$$L = m_e v r_n = n\frac{h}{2\pi}(n = 1, 2, 3 \ldots),$$

where L is the angular momentum, r_n is the radius of orbit n, and h is Planck's constant.

30.7 Patterns in Spectra Reveal More Quantization

- The Zeeman effect—the splitting of lines when a magnetic field is applied—is caused by other quantized entities in atoms.
- Both the magnitude and direction of orbital angular momentum are quantized.
- The same is true for the magnitude and direction of the intrinsic spin of electrons.

30.8 Quantum Numbers and Rules

- Quantum numbers are used to express the allowed values of quantized entities. The principal quantum number n labels the basic states of a system and is given by

$$n = 1, 2, 3, \ldots.$$

- The magnitude of angular momentum is given by

$$L = \sqrt{l(l+1)}\frac{h}{2\pi} \quad (l = 0, 1, 2, \ldots, n-1),$$

where l is the angular momentum quantum number. The direction of angular momentum is quantized, in that its component along an axis defined by a magnetic field, called the z-axis is given by

$$L_z = m_l\frac{h}{2\pi} \quad (m_l = -l, -l+1, \ldots, -1, 0, 1, \ldots l-1, l),$$

where L_z is the z-component of the angular momentum and m_l is the angular momentum projection quantum number. Similarly, the electron's intrinsic spin angular momentum S is given by

$$S = \sqrt{s(s+1)}\frac{h}{2\pi} \quad (s = 1/2 \text{ for electrons}),$$

s is defined to be the spin quantum number. Finally, the direction of the electron's spin along the z-axis is given by

$$S_z = m_s\frac{h}{2\pi} \quad \left(m_s = -\frac{1}{2}, +\frac{1}{2}\right),$$

where S_z is the z-component of spin angular momentum and m_s is the spin projection quantum number. Spin projection $m_s = +1/2$ is referred to as spin up, whereas $m_s = -1/2$ is called spin down. Table 30.1 summarizes the atomic quantum numbers and their allowed values.

30.9 The Pauli Exclusion Principle

- The state of a system is completely described by a complete set of quantum numbers. This set is written as (n, l, m_l, m_s).
- The Pauli exclusion principle says that no two electrons can have the same set of quantum numbers; that is, no two electrons can be in the same state.
- This exclusion limits the number of electrons in atomic shells and subshells. Each value of n corresponds to a shell, and each value of l corresponds to a subshell.
- The maximum number of electrons that can be in a subshell is $2(2l + 1)$.
- The maximum number of electrons that can be in a shell is $2n^2$.

Conceptual Questions

30.1 Discovery of the Atom

1. Name three different types of evidence for the existence of atoms.

2. Explain why patterns observed in the periodic table of the elements are evidence for the existence of atoms, and why Brownian motion is a more direct type of evidence for their existence.

3. If atoms exist, why can't we see them with visible light?

30.2 Discovery of the Parts of the Atom: Electrons and Nuclei

4. What two pieces of evidence allowed the first calculation of m_e, the mass of the electron?

(a) The ratios q_e/m_e and q_p/m_p.

(b) The values of q_e and E_B.

(c) The ratio q_e/m_e and q_e.

Justify your response.

5. How do the allowed orbits for electrons in atoms differ from the allowed orbits for planets around the sun? Explain how the correspondence principle applies here.

30.3 Bohr's Theory of the Hydrogen Atom

6. How do the allowed orbits for electrons in atoms differ from the allowed orbits for planets around the sun? Explain how the correspondence principle applies here.

7. Explain how Bohr's rule for the quantization of electron orbital angular momentum differs from the actual rule.

8. What is a hydrogen-like atom, and how are the energies and radii of its electron orbits related to those in hydrogen?

30.4 X Rays: Atomic Origins and Applications

9. Explain why characteristic x rays are the most energetic in the EM emission spectrum of a given element.

10. Why does the energy of characteristic x rays become increasingly greater for heavier atoms?

11. Observers at a safe distance from an atmospheric test of a nuclear bomb feel its heat but receive none of its copious x rays. Why is air opaque to x rays but transparent to infrared?

12. Lasers are used to burn and read CDs. Explain why a laser that emits blue light would be capable of burning and reading more information than one that emits infrared.

13. Crystal lattices can be examined with x rays but not UV. Why?

14. CT scanners do not detect details smaller than about 0.5 mm. Is this limitation due to the wavelength of x rays? Explain.

30.5 Applications of Atomic Excitations and De-Excitations

15. How do the allowed orbits for electrons in atoms differ from the allowed orbits for planets around the sun? Explain how the correspondence principle applies here.

16. Atomic and molecular spectra are discrete. What does discrete mean, and how are discrete spectra related to the quantization of energy and electron orbits in atoms and molecules?

17. Hydrogen gas can only absorb EM radiation that has an energy corresponding to a transition in the atom, just as it can only emit these discrete energies. When a spectrum is taken of the solar corona, in which a broad range of EM wavelengths are passed through very hot hydrogen gas, the absorption spectrum shows all the features of the emission spectrum. But when such EM radiation passes through room-temperature hydrogen gas, only the Lyman series is absorbed. Explain the difference.

18. Lasers are used to burn and read CDs. Explain why a laser that emits blue light would be capable of burning and reading more information than one that emits infrared.

19. The coating on the inside of fluorescent light tubes absorbs ultraviolet light and subsequently emits visible light. An inventor claims that he is able to do the reverse process. Is the inventor's claim possible?

20. What is the difference between fluorescence and phosphorescence?

21. How can you tell that a hologram is a true three-dimensional image and that those in 3-D movies are not?

30.6 The Wave Nature of Matter Causes Quantization

22. How is the de Broglie wavelength of electrons related to the quantization of their orbits in atoms and molecules?

30.7 Patterns in Spectra Reveal More Quantization

23. What is the Zeeman effect, and what type of quantization was discovered because of this effect?

30.8 Quantum Numbers and Rules

24. Define the quantum numbers n, l, m_l, s, and m_s.

25. For a given value of n, what are the allowed values of l?

26. For a given value of l, what are the allowed values of m_l? What are the allowed values of m_l for a given value of n? Give an example in each case.

27. List all the possible values of s and m_s for an electron. Are there particles for which these values are different? The same?

30.9 The Pauli Exclusion Principle

28. Identify the shell, subshell, and number of electrons for the following: (a) $2p^3$. (b) $4d^9$. (c) $3s^1$. (d) $5g^{16}$.

29. Which of the following are not allowed? State which rule is violated for any that are not allowed. (a) $1p^3$ (b) $2p^8$ (c) $3g^{11}$ (d) $4f^2$

Problems & Exercises

30.1 Discovery of the Atom

1. Using the given charge-to-mass ratios for electrons and protons, and knowing the magnitudes of their charges are equal, what is the ratio of the proton's mass to the electron's? (Note that since the charge-to-mass ratios are given to only three-digit accuracy, your answer may differ from the accepted ratio in the fourth digit.)

2. (a) Calculate the mass of a proton using the charge-to-mass ratio given for it in this chapter and its known charge. (b) How does your result compare with the proton mass given in this chapter?

3. If someone wanted to build a scale model of the atom with a nucleus 1.00 m in diameter, how far away would the nearest electron need to be?

30.2 Discovery of the Parts of the Atom: Electrons and Nuclei

4. Rutherford found the size of the nucleus to be about 10^{-15} m. This implied a huge density. What would this density be for gold?

5. In Millikan's oil-drop experiment, one looks at a small oil drop held motionless between two plates. Take the voltage between the plates to be 2033 V, and the plate separation to be 2.00 cm. The oil drop (of density 0.81 g/cm^3) has a diameter of 4.0×10^{-6} m. Find the charge on the drop, in terms of electron units.

6. (a) An aspiring physicist wants to build a scale model of a hydrogen atom for her science fair project. If the atom is 1.00 m in diameter, how big should she try to make the nucleus?

(b) How easy will this be to do?

30.3 Bohr's Theory of the Hydrogen Atom

7. By calculating its wavelength, show that the first line in the Lyman series is UV radiation.

8. Find the wavelength of the third line in the Lyman series, and identify the type of EM radiation.

9. Look up the values of the quantities in $a_B = \dfrac{h^2}{4\pi^2 m_e kq_e^2}$, and verify that the Bohr radius a_B is 0.529×10^{-10} m.

10. Verify that the ground state energy E_0 is 13.6 eV by using $E_0 = \dfrac{2\pi^2 q_e^4 m_e k^2}{h^2}$.

11. If a hydrogen atom has its electron in the $n = 4$ state, how much energy in eV is needed to ionize it?

12. A hydrogen atom in an excited state can be ionized with less energy than when it is in its ground state. What is n for a hydrogen atom if 0.850 eV of energy can ionize it?

13. Find the radius of a hydrogen atom in the $n = 2$ state according to Bohr's theory.

14. Show that $(13.6 \text{ eV})/hc = 1.097 \times 10^7$ m $= R$ (Rydberg's constant), as discussed in the text.

15. What is the smallest-wavelength line in the Balmer series? Is it in the visible part of the spectrum?

16. Show that the entire Paschen series is in the infrared part of the spectrum. To do this, you only need to calculate the shortest wavelength in the series.

17. Do the Balmer and Lyman series overlap? To answer this, calculate the shortest-wavelength Balmer line and the longest-wavelength Lyman line.

18. (a) Which line in the Balmer series is the first one in the UV part of the spectrum?

(b) How many Balmer series lines are in the visible part of the spectrum?

(c) How many are in the UV?

19. A wavelength of 4.653 μm is observed in a hydrogen spectrum for a transition that ends in the $n_f = 5$ level. What was n_i for the initial level of the electron?

20. A singly ionized helium ion has only one electron and is denoted He^+. What is the ion's radius in the ground state compared to the Bohr radius of hydrogen atom?

21. A beryllium ion with a single electron (denoted Be^{3+}) is in an excited state with radius the same as that of the ground state of hydrogen.

(a) What is n for the Be^{3+} ion?

(b) How much energy in eV is needed to ionize the ion from this excited state?

22. Atoms can be ionized by thermal collisions, such as at the high temperatures found in the solar corona. One such ion is C^{+5}, a carbon atom with only a single electron.

(a) By what factor are the energies of its hydrogen-like levels greater than those of hydrogen?

(b) What is the wavelength of the first line in this ion's Paschen series?

(c) What type of EM radiation is this?

23. Verify Equations $r_n = \dfrac{n^2}{Z} a_B$ and $a_B = \dfrac{h^2}{4\pi^2 m_e kq_e^2} = 0.529 \times 10^{-10}$ m using the approach stated in the text. That is, equate the Coulomb and centripetal forces and then insert an expression for velocity from the condition for angular momentum quantization.

24. The wavelength of the four Balmer series lines for hydrogen are found to be 410.3, 434.2, 486.3, and 656.5 nm. What average percentage difference is found between these wavelength numbers and those predicted by $\dfrac{1}{\lambda} = R\left(\dfrac{1}{n_f^2} - \dfrac{1}{n_i^2}\right)$? It is amazing how well a simple formula (disconnected originally from theory) could duplicate this phenomenon.

30.4 X Rays: Atomic Origins and Applications

25. (a) What is the shortest-wavelength x-ray radiation that can be generated in an x-ray tube with an applied voltage of 50.0 kV? (b) Calculate the photon energy in eV. (c) Explain the relationship of the photon energy to the applied voltage.

26. A color television tube also generates some x rays when its electron beam strikes the screen. What is the shortest wavelength of these x rays, if a 30.0-kV potential is used to accelerate the electrons? (Note that TVs have shielding to prevent these x rays from exposing viewers.)

27. An x ray tube has an applied voltage of 100 kV. (a) What is the most energetic x-ray photon it can produce? Express your answer in electron volts and joules. (b) Find the wavelength of such an X–ray.

28. The maximum characteristic x-ray photon energy comes from the capture of a free electron into a K shell vacancy. What is this photon energy in keV for tungsten, assuming the free electron has no initial kinetic energy?

29. What are the approximate energies of the K_α and K_β x rays for copper?

30.5 Applications of Atomic Excitations and De-Excitations

30. Figure 30.39 shows the energy-level diagram for neon. (a) Verify that the energy of the photon emitted when neon goes from its metastable state to the one immediately below is equal to 1.96 eV. (b) Show that the wavelength of this radiation is 633 nm. (c) What wavelength is emitted when the neon makes a direct transition to its ground state?

31. A helium-neon laser is pumped by electric discharge. What wavelength electromagnetic radiation would be needed to pump it? See Figure 30.39 for energy-level information.

32. Ruby lasers have chromium atoms doped in an aluminum oxide crystal. The energy level diagram for chromium in a ruby is shown in Figure 30.64. What wavelength is emitted by a ruby laser?

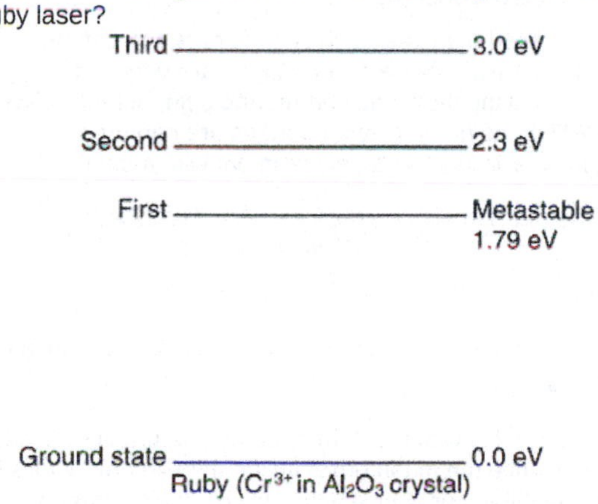

Third ──────────────── 3.0 eV

Second ──────────────── 2.3 eV

First ──────────────── Metastable
1.79 eV

Ground state ──────────────── 0.0 eV
Ruby (Cr^{3+} in Al_2O_3 crystal)

Figure 30.64 Chromium atoms in an aluminum oxide crystal have these energy levels, one of which is metastable. This is the basis of a ruby laser. Visible light can pump the atom into an excited state above the metastable state to achieve a population inversion.

33. (a) What energy photons can pump chromium atoms in a ruby laser from the ground state to its second and third excited states? (b) What are the wavelengths of these photons? Verify that they are in the visible part of the spectrum.

34. Some of the most powerful lasers are based on the energy levels of neodymium in solids, such as glass, as shown in Figure 30.65. (a) What average wavelength light can pump the neodymium into the levels above its metastable state? (b) Verify that the 1.17 eV transition produces $1.06\ \mu m$ radiation.

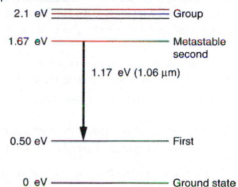

Figure 30.65 Neodymium atoms in glass have these energy levels, one of which is metastable. The group of levels above the metastable state is convenient for achieving a population inversion, since photons of many different energies can be absorbed by atoms in the ground state.

30.8 Quantum Numbers and Rules

35. If an atom has an electron in the $n = 5$ state with $m_l = 3$, what are the possible values of l?

36. An atom has an electron with $m_l = 2$. What is the smallest value of n for this electron?

37. What are the possible values of m_l for an electron in the $n = 4$ state?

38. What, if any, constraints does a value of $m_l = 1$ place on the other quantum numbers for an electron in an atom?

39. (a) Calculate the magnitude of the angular momentum for an $l = 1$ electron. (b) Compare your answer to the value Bohr proposed for the $n = 1$ state.

40. (a) What is the magnitude of the angular momentum for an $l = 1$ electron? (b) Calculate the magnitude of the electron's spin angular momentum. (c) What is the ratio of these angular momenta?

41. Repeat Exercise 30.40 for $l = 3$.

42. (a) How many angles can L make with the z-axis for an $l = 2$ electron? (b) Calculate the value of the smallest angle.

43. What angles can the spin S of an electron make with the z-axis?

30.9 The Pauli Exclusion Principle

44. (a) How many electrons can be in the $n = 4$ shell?

(b) What are its subshells, and how many electrons can be in each?

45. (a) What is the minimum value of 1 for a subshell that has 11 electrons in it?

(b) If this subshell is in the $n = 5$ shell, what is the spectroscopic notation for this atom?

46. (a) If one subshell of an atom has 9 electrons in it, what is the minimum value of l ? (b) What is the spectroscopic notation for this atom, if this subshell is part of the $n = 3$ shell?

47. (a) List all possible sets of quantum numbers (n, l, m_l, m_s) for the $n = 3$ shell, and determine the number of electrons that can be in the shell and each of its subshells.

(b) Show that the number of electrons in the shell equals $2n^2$ and that the number in each subshell is $2(2l + 1)$.

48. Which of the following spectroscopic notations are not allowed? (a) $5s^1$ (b) $1d^1$ (c) $4s^3$ (d) $3p^7$ (e) $5g^{15}$. State which rule is violated for each that is not allowed.

49. Which of the following spectroscopic notations are allowed (that is, which violate none of the rules regarding values of quantum numbers)? (a) $1s^1$ (b) $1d^3$ (c) $4s^2$ (d) $3p^7$ (e) $6h^{20}$

50. (a) Using the Pauli exclusion principle and the rules relating the allowed values of the quantum numbers (n, l, m_l, m_s), prove that the maximum number of electrons in a subshell is $2n^2$.

(b) In a similar manner, prove that the maximum number of electrons in a shell is $2n^2$.

51. Integrated Concepts

Estimate the density of a nucleus by calculating the density of a proton, taking it to be a sphere 1.2 fm in diameter. Compare your result with the value estimated in this chapter.

52. Integrated Concepts

The electric and magnetic forces on an electron in the CRT in **Figure 30.7** are supposed to be in opposite directions. Verify this by determining the direction of each force for the situation shown. Explain how you obtain the directions (that is, identify the rules used).

53. (a) What is the distance between the slits of a diffraction grating that produces a first-order maximum for the first Balmer line at an angle of $20.0°$?

(b) At what angle will the fourth line of the Balmer series appear in first order?

(c) At what angle will the second-order maximum be for the first line?

54. Integrated Concepts

A galaxy moving away from the earth has a speed of $0.0100c$. What wavelength do we observe for an $n_i = 7$ to $n_f = 2$ transition for hydrogen in that galaxy?

55. Integrated Concepts

Calculate the velocity of a star moving relative to the earth if you observe a wavelength of 91.0 nm for ionized hydrogen capturing an electron directly into the lowest orbital (that is, a $n_i = \infty$ to $n_f = 1$, or a Lyman series transition).

56. Integrated Concepts

In a Millikan oil-drop experiment using a setup like that in **Figure 30.9**, a 500-V potential difference is applied to plates separated by 2.50 cm. (a) What is the mass of an oil drop having two extra electrons that is suspended motionless by the field between the plates? (b) What is the diameter of the drop, assuming it is a sphere with the density of olive oil?

57. Integrated Concepts

What double-slit separation would produce a first-order maximum at $3.00°$ for 25.0-keV x rays? The small answer indicates that the wave character of x rays is best determined by having them interact with very small objects such as atoms and molecules.

58. Integrated Concepts

In a laboratory experiment designed to duplicate Thomson's determination of q_e / m_e , a beam of electrons having a velocity of 6.00×10^7 m/s enters a 5.00×10^{-3} T magnetic field. The beam moves perpendicular to the field in a path having a 6.80-cm radius of curvature. Determine q_e / m_e from these observations, and compare the result with the known value.

59. Integrated Concepts

Find the value of l , the orbital angular momentum quantum number, for the moon around the earth. The extremely large value obtained implies that it is impossible to tell the difference between adjacent quantized orbits for macroscopic objects.

60. Integrated Concepts

Particles called muons exist in cosmic rays and can be created in particle accelerators. Muons are very similar to electrons, having the same charge and spin, but they have a mass 207 times greater. When muons are captured by an atom, they orbit just like an electron but with a smaller radius, since the mass in $a_B = \dfrac{h^2}{4\pi^2 m_e kq_e^2} = 0.529 \times 10^{-10}$ m is $207 \, m_e$.

(a) Calculate the radius of the $n = 1$ orbit for a muon in a uranium ion ($Z = 92$).

(b) Compare this with the 7.5-fm radius of a uranium nucleus. Note that since the muon orbits inside the electron, it falls into a hydrogen-like orbit. Since your answer is less than the radius of the nucleus, you can see that the photons emitted as the muon falls into its lowest orbit can give information about the nucleus.

61. Integrated Concepts

Calculate the minimum amount of energy in joules needed to create a population inversion in a helium-neon laser containing 1.00×10^{-4} moles of neon.

62. Integrated Concepts

A carbon dioxide laser used in surgery emits infrared radiation with a wavelength of $10.6 \ \mu m$. In 1.00 ms, this laser raised the temperature of $1.00 \ cm^3$ of flesh to $100°C$ and evaporated it.

(a) How many photons were required? You may assume flesh has the same heat of vaporization as water. (b) What was the minimum power output during the flash?

63. Integrated Concepts

Suppose an MRI scanner uses 100-MHz radio waves.

(a) Calculate the photon energy.

(b) How does this compare to typical molecular binding energies?

64. Integrated Concepts

(a) An excimer laser used for vision correction emits 193-nm UV. Calculate the photon energy in eV.

(b) These photons are used to evaporate corneal tissue, which is very similar to water in its properties. Calculate the amount of energy needed per molecule of water to make the phase change from liquid to gas. That is, divide the heat of vaporization in kJ/kg by the number of water molecules in a kilogram.

(c) Convert this to eV and compare to the photon energy. Discuss the implications.

65. Integrated Concepts

A neighboring galaxy rotates on its axis so that stars on one side move toward us as fast as 200 km/s, while those on the other side move away as fast as 200 km/s. This causes the EM radiation we receive to be Doppler shifted by velocities over the entire range of ±200 km/s. What range of wavelengths will we observe for the 656.0-nm line in the Balmer series of hydrogen emitted by stars in this galaxy. (This is called line broadening.)

66. Integrated Concepts

A pulsar is a rapidly spinning remnant of a supernova. It rotates on its axis, sweeping hydrogen along with it so that hydrogen on one side moves toward us as fast as 50.0 km/s, while that on the other side moves away as fast as 50.0 km/s. This means that the EM radiation we receive will be Doppler shifted over a range of $\pm 50.0 \ km/s$. What range of wavelengths will we observe for the 91.20-nm line in the Lyman series of hydrogen? (Such line broadening is observed and actually provides part of the evidence for rapid rotation.)

67. Integrated Concepts

Prove that the velocity of charged particles moving along a straight path through perpendicular electric and magnetic fields is $v = E/B$. Thus crossed electric and magnetic fields can be used as a velocity selector independent of the charge and mass of the particle involved.

68. Unreasonable Results

(a) What voltage must be applied to an X-ray tube to obtain 0.0100-fm-wavelength X-rays for use in exploring the details of nuclei? (b) What is unreasonable about this result? (c) Which assumptions are unreasonable or inconsistent?

69. Unreasonable Results

A student in a physics laboratory observes a hydrogen spectrum with a diffraction grating for the purpose of measuring the wavelengths of the emitted radiation. In the spectrum, she observes a yellow line and finds its wavelength to be 589 nm. (a) Assuming this is part of the Balmer series, determine n_i, the principal quantum number of the initial state. (b) What is unreasonable about this result? (c) Which assumptions are unreasonable or inconsistent?

70. Construct Your Own Problem

The solar corona is so hot that most atoms in it are ionized. Consider a hydrogen-like atom in the corona that has only a single electron. Construct a problem in which you calculate selected spectral energies and wavelengths of the Lyman, Balmer, or other series of this atom that could be used to identify its presence in a very hot gas. You will need to choose the atomic number of the atom, identify the element, and choose which spectral lines to consider.

71. Construct Your Own Problem

Consider the Doppler-shifted hydrogen spectrum received from a rapidly receding galaxy. Construct a problem in which you calculate the energies of selected spectral lines in the Balmer series and examine whether they can be described with a formula like that in the equation $\frac{1}{\lambda} = R\left(\frac{1}{n_f^2} - \frac{1}{n_i^2}\right)$, but with a different constant R.

Test Prep for AP® Courses

30.2 Discovery of the Parts of the Atom: Electrons and Nuclei

1. In an experiment, three microscopic latex spheres are sprayed into a chamber and become charged with +3e, +5e, and −3e, respectively. Later, all three spheres collide simultaneously and then separate. Which of the following are possible values for the final charges on the spheres? Select two answers.
   a.   +4e, −4e, +5e
   b.   −4e, +4.5e, +4.5e
   c.   +5e, −8e, +7e
   d.   +6e, +6e, −7e

2. In Millikan's oil drop experiment, he experimented with various voltage differences between two plates to determine what voltage was necessary to hold a drop motionless. He deduced that the charge on the oil drop could be found by setting the gravitational force on the drop (pointing downward) equal to the electric force (pointing upward):

$$m_{drop} g = qE,$$

where m_{drop} is the mass of the oil drop, g is the gravitational acceleration (9.8 m/s^2), q is the net charge of the oil drop, and E is the electric field between the plates. Millikan deduced that the charge on an electron, e, is 1.6×10^{-19} C.

For a system of oil drops of equal mass (1.0×10^{-15} kilograms), describe what value or values of the electric field would hold the drops motionless.

30.3 Bohr's Theory of the Hydrogen Atom

3. A hypothetical one-electron atom in its highest excited state can only emit photons of energy 2E, 3E, and 5E before reaching the ground state. Which of the following represents the complete set of energy levels for this atom?
   a.   0, 3E, 5E
   b.   0, 2E, 3E
   c.   0, 2E, 3E, 5E
   d.   0, 5E, 8E, 10E

4. The Lyman series of photons each have an energy capable of exciting the electron of a hydrogen atom from the ground state (energy level 1) to energy levels 2, 3, 4, etc. The wavelengths of the first five photons in this series are 121.6 nm, 102.6 nm, 97.3 nm, 95.0 nm, and 93.8 nm. The ground state energy of hydrogen is −13.6 eV. Based on the wavelengths of the Lyman series, calculate the energies of the first five excited states above ground level for a hydrogen atom to the nearest 0.1 eV.

5. The ground state of a certain type of atom has energy $-E_0$. What is the wavelength of a photon with enough energy to ionize an atom in the ground state and give the ejected electron a kinetic energy of $2E_0$?
   a. $\dfrac{hc}{3E_0}$

   b. $\dfrac{hc}{2E_0}$

   c. $\dfrac{hc}{E_0}$

   d. $\dfrac{2hc}{E_0}$

6. An electron in a hydrogen atom is initially in energy level 2

(E_2 = -3.4 eV). (a) What frequency of photon must be absorbed by the atom in order for the electron to transition to energy level 3 (E_3 = -1.5 eV)? (b) What frequency of photon must be emitted by the atom in order for the electron to transition to energy level 1 (E_1 = -13.6 eV)?

30.5 Applications of Atomic Excitations and De-Excitations

7. A sample of hydrogen gas confined to a tube is initially at room temperature. As the gas is heated, the observer notices that the gas begins to glow with a pale pink color. Careful study of the spectrum shows that the light spectrum is not continuous. Instead, the hydrogen gas is only emitting visible wavelength photons of four specific colors, which combine to form the overall color to the human eye. What is the best way to explain this behavior?
   a.   As the gas heats up, atoms have more and more collisions and close approaches, so frictional heating causes the gas to glow.
   b.   As the gas heats up, the electrons within the hydrogen atoms are excited to high energy levels. As the electrons transition to lower energies, they emit light of specific colors.
   c.   As the gas heats up, more and more collisions occur, and the energy lost in these inelastic collisions is converted into light.
   d.   As the gas heats up, the turbulence of the gas within the tube causes friction between the gas and the walls of the container, causing the gas to glow.

8. A rock is illuminated with high energy ultraviolet light. This causes the rock to emit visible light. Explain what is happening in the atomic substructure of the rock that causes this effect, which we call fluorescence.

9. Which of the following is the best way of explaining why the leaves on a given tree are green?
   a.   The molecules in the leaves absorb all visible light but strongly reflect green light.
   b.   The molecules in the leaves absorb green light and reflect other visible light.
   c.   The molecules are excited by external light sources, and their electrons emit green light when they are de-excited to a lower energy level within the molecules.
   d.   The molecules glow with a characteristic green energy in order to balance the absorption of energy due to light and heat from their surroundings.

10. Explain what phosphorescence is and how it differs from fluorescence. Which process typically takes longer and why?

11. An electron is excited from the ground state of an atom (energy level 1) into a highly excited state (energy level 8). Which of the following electron behaviors represents the fluorescence effect by the atom?
   a.   The electron remains at level 8 for a very long time, then transitions up to level 9.
   b.   The electron transitions directly down from level 8 to level 1.
   c.   The electron transitions from level 8 to level 1 and then returns quickly to level 8.
   d.   The electron transitions from level 8 to level 6, then to level 5, then to level 3, then to level 1.

12. Describe the process of fluorescence in terms of the emission of photons as electron transitions between energy states. Specifically, explain how this process differs from ordinary atomic emission.

30.6 The Wave Nature of Matter Causes Quantization

13.

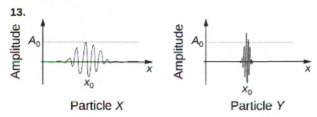

Particle X   Particle Y

Figure 30.66 This figure shows graphical representations of the wave functions of two particles, X and Y, that are moving in the positive x-direction. The maximum amplitude of particle X's wave function is A_0. Which particle has a greater probability of being located at position x_0 at this instant, and why?

   a. Particle X, because the wave function of particle X spends more time passing through x_0 than the wave function of particle Y.
   b. Particle X, because the wave function of particle X has a longer wavelength than the wave function of particle Y.
   c. Particle Y, because the wave function of particle Y is narrower than the wave function of particle X.
   d. Particle Y, because the wave function of particle Y has a greater amplitude near x_0 than the wave function of particle X.

14. In Figure 30.66, explain qualitatively the difference in the wave functions of particle X and particle Y. Which particle is more likely to be found at a larger distance from the coordinate x_0 and why? Which particle is more likely be found exactly at x_0 and why?

15. For an electron with a de Broglie wavelength λ, which of the following orbital circumferences within the atom would be disallowed? Select two answers.

   a. $0.5\ \lambda$

   b. λ

   c. $1.5\ \lambda$

   d. $2\ \lambda$

16. We have discovered that an electron's orbit must contain an integer number of de Broglie wavelengths. Explain why, under ordinary conditions, this makes it impossible for electrons to spiral in to merge with the positively charged nucleus.

31 RADIOACTIVITY AND NUCLEAR PHYSICS

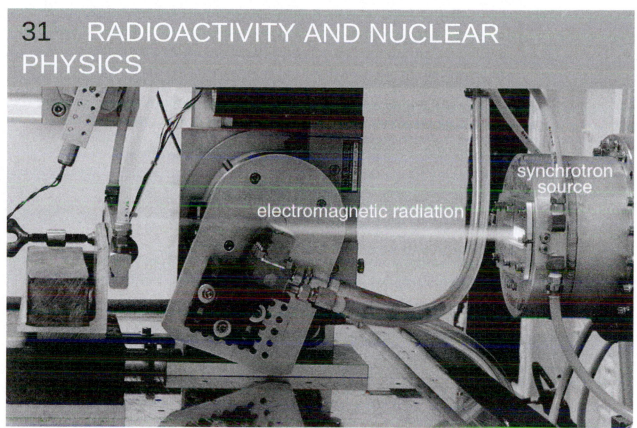

Figure 31.1 The synchrotron source produces electromagnetic radiation, as evident from the visible glow. (credit: United States Department of Energy, via Wikimedia Commons)

Chapter Outline

31.1. Nuclear Radioactivity

31.2. Radiation Detection and Detectors

31.3. Substructure of the Nucleus

31.4. Nuclear Decay and Conservation Laws

31.5. Half-Life and Activity

31.6. Binding Energy

31.7. Tunneling

Connection for AP® Courses

In this chapter, students will explore radioactivity and nuclear physics. Students will learn about the structure and properties of a nucleus (Enduring Understanding 1.A, Essential Knowledge 1.A.3), supporting Big Idea 1. Students will also study the forces that govern the behavior of the nucleus, including the weak force and the strong force (Enduring Understanding 3.G). This supports Big Idea 3 by explaining that interactions can be described by forces, such as the strong force between nucleons holding the nucleus together.

Students will also learn the conservation laws associated with nuclear physics, such as conservation of energy (Enduring Understanding 5.B), conservation of charge (Enduring Understanding 5.C) and conservation of nucleon number (Enduring Understanding 5.G). Students will study the processes that can be described using conservation laws (Big Idea 5), such as radioactive decay, nuclear absorption and emission of nuclear energy, usually regulated by photons (Essential Knowledge 5.B.8). As part of the study of conservation laws, students will explore the consequences of charge conservation (Essential Knowledge 5.C.1) during radioactive decay and during interactions between nuclei (Essential Knowledge 5.C.2). Students will also learn how conservation of nucleon number determines which nuclear reactions can occur (Essential Knowledge 5.G.1). Students will also study types of nuclear radiation, radioactivity, and the binding energy of a nucleus.

This chapter also supports Big Idea 7 by exploring how probability can describe the behavior of quantum mechanical systems. Students will study the process of radioactive decay, which can be described by probability theory. Students will also explore examples demonstrating spontaneous radioactive decay as a probabilistic statistical process (Essential Knowledge 7.C.3), thus making a connection between modeling matter with a wave function and probabilistic description of the microscopic world (Enduring Understanding 7.C).

The content in this chapter supports:

Big Idea 1 Objects and systems have properties such as mass and charge. Systems may have internal structure.

Enduring Understanding 1.A The internal structure of a system determines many properties of the system.

Essential Knowledge 1.A.3 Nuclei have internal structures that determine their properties.

Big Idea 3 The interactions of an object with other objects can be described by forces.

Enduring Understanding 3.G Certain types of forces are considered fundamental.

Essential Knowledge 3.G.3 The strong force is exerted at nuclear scales and dominates the interactions of nucleons.

Big Idea 5 Changes that occur as a result of interactions are constrained by conservation laws.

Enduring Understanding 5.B The energy of a system is conserved.

Essential Knowledge 5.B.8 Energy transfer occurs when photons are absorbed or emitted, for example, by atoms or nuclei.

Enduring Understanding 5.C The electric charge of a system is conserved.

Essential Knowledge 5.C.1 Electric charge is conserved in nuclear and elementary particle reactions, even when elementary particles are produced or destroyed. Examples should include equations representing nuclear decay.

Essential Knowledge 5.C.2 The exchange of electric charges among a set of objects in a system conserves electric charge.

Enduring Understanding 5.G Nucleon number is conserved.

Essential Knowledge 5.G.1 The possible nuclear reactions are constrained by the law of conservation of nucleon number.

Big Idea 7 The mathematics of probability can be used to describe the behavior of complex systems and to interpret the behavior of quantum mechanical systems.

Enduring Understanding 7.C At the quantum scale, matter is described by a wave function, which leads to a probabilistic description of the microscopic world.

Essential Knowledge 7.C.3 The spontaneous radioactive decay of an individual nucleus is described by probability.

31.1 Nuclear Radioactivity

Learning Objectives
By the end of this section, you will be able to:
• Explain nuclear radiation.
• Explain the types of radiation – alpha emission, beta emission, and gamma emission.
• Explain the ionization of radiation in an atom.
• Define the range of radiation.
The information presented in this section supports the following AP® learning objectives and science practices:
• **5.B.8.1** The student is able to describe emission or absorption spectra associated with electronic or nuclear transitions as transitions between allowed energy states of the atom in terms of the principle of energy conservation, including characterization of the frequency of radiation emitted or absorbed. **(S.P. 1.2, 7.2)**
• **5.C.1.1** The student is able to analyze electric charge conservation for nuclear and elementary particle reactions and make predictions related to such reactions based upon conservation of charge. **(S.P. 6.4, 7.2)**

The discovery and study of nuclear radioactivity quickly revealed evidence of revolutionary new physics. In addition, uses for nuclear radiation also emerged quickly—for example, people such as Ernest Rutherford used it to determine the size of the nucleus and devices were painted with radon-doped paint to make them glow in the dark (see **Figure 31.2**). We therefore begin our study of nuclear physics with the discovery and basic features of nuclear radioactivity.

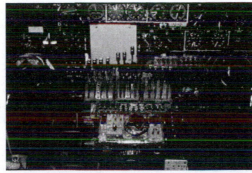

Figure 31.2 The dials of this World War II aircraft glow in the dark, because they are painted with radium-doped phosphorescent paint. It is a poignant reminder of the dual nature of radiation. Although radium paint dials are conveniently visible day and night, they emit radon, a radioactive gas that is hazardous and is not directly sensed. (credit: U.S. Air Force Photo)

Discovery of Nuclear Radioactivity

In 1896, the French physicist Antoine Henri Becquerel (1852–1908) accidentally found that a uranium-rich mineral called pitchblende emits invisible, penetrating rays that can darken a photographic plate enclosed in an opaque envelope. The rays therefore carry energy; but amazingly, the pitchblende emits them continuously without any energy input. This is an apparent violation of the law of conservation of energy, one that we now understand is due to the conversion of a small amount of mass into energy, as related in Einstein's famous equation $E = mc^2$. It was soon evident that Becquerel's rays originate in the nuclei of the atoms and have other unique characteristics. The emission of these rays is called **nuclear radioactivity** or simply **radioactivity**. The rays themselves are called **nuclear radiation**. A nucleus that spontaneously destroys part of its mass to emit radiation is said to **decay** (a term also used to describe the emission of radiation by atoms in excited states). A substance or object that emits nuclear radiation is said to be **radioactive**.

Two types of experimental evidence imply that Becquerel's rays originate deep in the heart (or nucleus) of an atom. First, the radiation is found to be associated with certain elements, such as uranium. Radiation does not vary with chemical state—that is, uranium is radioactive whether it is in the form of an element or compound. In addition, radiation does not vary with temperature, pressure, or ionization state of the uranium atom. Since all of these factors affect electrons in an atom, the radiation cannot come from electron transitions, as atomic spectra do. The huge energy emitted during each event is the second piece of evidence that the radiation cannot be atomic. Nuclear radiation has energies of the order of 10^6 eV per event, which is much greater than the typical atomic energies (a few eV), such as that observed in spectra and chemical reactions, and more than ten times as high as the most energetic characteristic x rays. Becquerel did not vigorously pursue his discovery for very long. In 1898, Marie Curie (1867–1934), then a graduate student married the already well-known French physicist Pierre Curie (1859–1906), began her doctoral study of Becquerel's rays. She and her husband soon discovered two new radioactive elements, which she named *polonium* (after her native land) and *radium* (because it radiates). These two new elements filled holes in the periodic table and, further, displayed much higher levels of radioactivity per gram of material than uranium. Over a period of four years, working under poor conditions and spending their own funds, the Curies processed more than a ton of uranium ore to isolate a gram of radium salt. Radium became highly sought after, because it was about two million times as radioactive as uranium. Curie's radium salt glowed visibly from the radiation that took its toll on them and other unaware researchers. Shortly after completing her Ph.D., both Curies and Becquerel shared the 1903 Nobel Prize in physics for their work on radioactivity. Pierre was killed in a horse cart accident in 1906, but Marie continued her study of radioactivity for nearly 30 more years. Awarded the 1911 Nobel Prize in chemistry for her discovery of two new elements, she remains the only person to win Nobel Prizes in physics and chemistry. Marie's radioactive fingerprints on some pages of her notebooks can still expose film, and she suffered from radiation-induced lesions. She died of leukemia likely caused by radiation, but she was active in research almost until her death in 1934. The following year, her daughter and son-in-law, Irene and Frederic Joliot-Curie, were awarded the Nobel Prize in chemistry for their discovery of artificially induced radiation, adding to a remarkable family legacy.

Alpha, Beta, and Gamma

Research begun by people such as New Zealander Ernest Rutherford soon after the discovery of nuclear radiation indicated that different types of rays are emitted. Eventually, three types were distinguished and named **alpha** (α), **beta** (β), and **gamma** (γ), because, like x-rays, their identities were initially unknown. **Figure 31.3** shows what happens if the rays are passed through a magnetic field. The γ s are unaffected, while the α s and β s are deflected in opposite directions, indicating the α s are positive, the β s negative, and the γ s uncharged. Rutherford used both magnetic and electric fields to show that α s have a positive charge twice the magnitude of an electron, or $+2 | q_e |$. In the process, he found the α s charge to mass ratio to be several thousand times smaller than the electron's. Later on, Rutherford collected α s from a radioactive source and passed an electric discharge through them, obtaining the spectrum of recently discovered helium gas. Among many important discoveries made by Rutherford and his collaborators was the proof that α *radiation is the emission of a helium nucleus*. Rutherford won the Nobel Prize in chemistry in 1908 for his early work. He continued to make important contributions until his death in 1934.

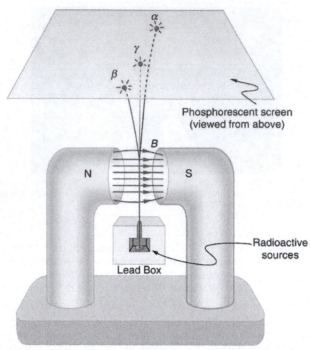

Figure 31.3 Alpha, beta, and gamma rays are passed through a magnetic field on the way to a phosphorescent screen. The α s and β s bend in opposite directions, while the γ s are unaffected, indicating a positive charge for α s, negative for β s, and neutral for γ s. Consistent results are obtained with electric fields. Collection of the radiation offers further confirmation from the direct measurement of excess charge.

Other researchers had already proved that β s are negative and have the same mass and same charge-to-mass ratio as the recently discovered electron. By 1902, it was recognized that β *radiation is the emission of an electron.* Although β s are electrons, they do not exist in the nucleus before it decays and are not ejected atomic electrons—the electron is created in the nucleus at the instant of decay.

Since γ s remain unaffected by electric and magnetic fields, it is natural to think they might be photons. Evidence for this grew, but it was not until 1914 that this was proved by Rutherford and collaborators. By scattering γ radiation from a crystal and observing interference, they demonstrated that γ *radiation is the emission of a high-energy photon by a nucleus.* In fact, γ radiation comes from the de-excitation of a nucleus, just as an x ray comes from the de-excitation of an atom. The names "γ ray" and "x ray" identify the source of the radiation. At the same energy, γ rays and x rays are otherwise identical.

Table 31.1 Properties of Nuclear Radiation

Type of Radiation	Range
α -Particles	A sheet of paper, a few cm of air, fractions of a mm of tissue
β -Particles	A thin aluminum plate, or tens of cm of tissue
γ Rays	Several cm of lead or meters of concrete

Ionization and Range

Two of the most important characteristics of α, β, and γ rays were recognized very early. All three types of nuclear radiation produce *ionization* in materials, but they penetrate different distances in materials—that is, they have different *ranges*. Let us examine why they have these characteristics and what are some of the consequences.

Like x rays, nuclear radiation in the form of α s, β s, and γ s has enough energy per event to ionize atoms and molecules in any material. The energy emitted in various nuclear decays ranges from a few keV to more than 10 MeV, while only a few eV are needed to produce ionization. The effects of x rays and nuclear radiation on biological tissues and other materials, such as solid state electronics, are directly related to the ionization they produce. All of them, for example, can damage electronics or kill cancer cells. In addition, methods for detecting x rays and nuclear radiation are based on ionization, directly or indirectly. All of them can ionize the air between the plates of a capacitor, for example, causing it to discharge. This is the basis of inexpensive personal radiation monitors, such as pictured in **Figure 31.4**. Apart from α, β, and γ, there are other forms of nuclear radiation as well, and these also produce ionization with similar effects. We define **ionizing radiation** as any form of radiation that produces ionization whether nuclear in origin or not, since the effects and detection of the radiation are related to ionization.

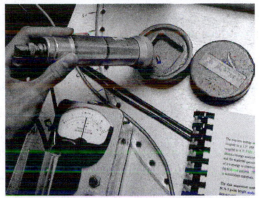

Figure 31.4 These dosimeters (literally, dose meters) are personal radiation monitors that detect the amount of radiation by the discharge of a rechargeable internal capacitor. The amount of discharge is related to the amount of ionizing radiation encountered, a measurement of dose. One dosimeter is shown in the charger. Its scale is read through an eyepiece on the top. (credit: L. Chang, Wikimedia Commons)

The **range of radiation** is defined to be the distance it can travel through a material. Range is related to several factors, including the energy of the radiation, the material encountered, and the type of radiation (see **Figure 31.5**). The higher the *energy*, the greater the range, all other factors being the same. This makes good sense, since radiation loses its energy in materials primarily by producing ionization in them, and each ionization of an atom or a molecule requires energy that is removed from the radiation. The amount of ionization is, thus, directly proportional to the energy of the particle of radiation, as is its range.

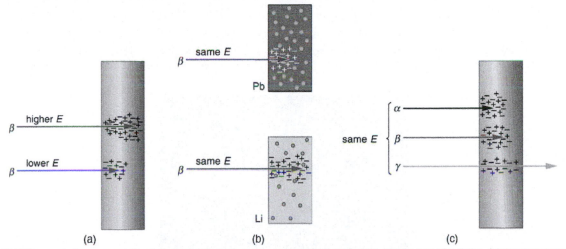

Figure 31.5 The penetration or range of radiation depends on its energy, the material it encounters, and the type of radiation. (a) Greater energy means greater range. (b) Radiation has a smaller range in materials with high electron density. (c) Alphas have the smallest range, betas have a greater range, and gammas penetrate the farthest.

Radiation can be absorbed or shielded by materials, such as the lead aprons dentists drape on us when taking x rays. Lead is a particularly effective shield compared with other materials, such as plastic or air. How does the range of radiation depend on *material*? Ionizing radiation interacts best with charged particles in a material. Since electrons have small masses, they most readily absorb the energy of the radiation in collisions. The greater the density of a material and, in particular, the greater the density of electrons within a material, the smaller the range of radiation.

Collisions

Conservation of energy and momentum often results in energy transfer to a less massive object in a collision. This was discussed in detail in **Work, Energy, and Energy Resources**, for example.

Different *types* of radiation have different ranges when compared at the same energy and in the same material. Alphas have the shortest range, betas penetrate farther, and gammas have the greatest range. This is directly related to charge and speed of the particle or type of radiation. At a given energy, each α, β, or γ will produce the same number of ionizations in a material (each ionization requires a certain amount of energy on average). The more readily the particle produces ionization, the more quickly it will lose its energy. The effect of *charge* is as follows: The α has a charge of $+2q_e$, the β has a charge of $-q_e$, and the γ is uncharged. The electromagnetic force exerted by the α is thus twice as strong as that exerted by the β and it is more likely to produce ionization. Although chargeless, the γ does interact weakly because it is an electromagnetic wave, but it is less likely to produce ionization in any encounter. More quantitatively, the change in momentum Δp given to a particle in the material is $\Delta p = F\Delta t$, where F is the force the α, β, or γ exerts over a time Δt. The smaller the charge, the smaller is F and the smaller is the momentum (and energy) lost. Since the speed of alphas is about 5% to 10% of the speed of light, classical (non-relativistic) formulas apply.

The *speed* at which they travel is the other major factor affecting the range of α s, β s, and γ s. The faster they move, the less time they spend in the vicinity of an atom or a molecule, and the less likely they are to interact. Since α s and β s are particles with mass (helium nuclei and electrons, respectively), their energy is kinetic, given classically by $\frac{1}{2}mv^2$. The mass of the β particle is thousands of times less than that of the α s, so that β s must travel much faster than α s to have the same energy. Since β s move faster (most at relativistic speeds), they have less time to interact than α s. Gamma rays are photons, which must travel at the speed of light. They are even less likely to interact than a β, since they spend even less time near a given atom (and they have no charge). The range of γ s is thus greater than the range of β s.

Alpha radiation from radioactive sources has a range much less than a millimeter of biological tissues, usually not enough to even penetrate the dead layers of our skin. On the other hand, the same α radiation can penetrate a few centimeters of air, so mere distance from a source prevents α radiation from reaching us. This makes α radiation relatively safe for our body compared to β and γ radiation. Typical β radiation can penetrate a few millimeters of tissue or about a meter of air. Beta radiation is thus hazardous even when not ingested. The range of β s in lead is about a millimeter, and so it is easy to store β sources in lead radiation-proof containers. Gamma rays have a much greater range than either α s or β s. In fact, if a given thickness of material, like a lead brick, absorbs 90% of the γ s, then a second lead brick will only absorb 90% of what got through the first. Thus, γ s do not have a well-defined range; we can only cut down the amount that gets through. Typically, γ s can penetrate many meters of air, go right through our bodies, and are effectively shielded (that is, reduced in intensity to acceptable levels) by many centimeters of lead. One benefit of γ s is that they can be used as radioactive tracers (see **Figure 31.6**).

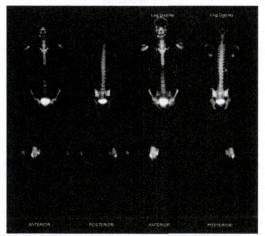

Figure 31.6 This image of the concentration of a radioactive tracer in a patient's body reveals where the most active bone cells are, an indication of bone cancer. A short-lived radioactive substance that locates itself selectively is given to the patient, and the radiation is measured with an external detector. The emitted γ radiation has a sufficient range to leave the body—the range of α s and β s is too small for them to be observed outside the patient. (credit: Kieran Maher, Wikimedia Commons)

PhET Explorations: Beta Decay

Watch beta decay occur for a collection of nuclei or for an individual nucleus.

Figure 31.7 Beta Decay (http://cnx.org/content/m54935/1.2/beta-decay_en.jar)

31.2 Radiation Detection and Detectors

Learning Objectives

By the end of this section, you will be able to:
 	
  • Explain the working principle of a Geiger tube.
  • Define and discuss radiation detectors.

It is well known that ionizing radiation affects us but does not trigger nerve impulses. Newspapers carry stories about unsuspecting victims of radiation poisoning who fall ill with radiation sickness, such as burns and blood count changes, but who never felt the radiation directly. This makes the detection of radiation by instruments more than an important research tool. This section is a brief overview of radiation detection and some of its applications.

Human Application

The first direct detection of radiation was Becquerel's fogged photographic plate. Photographic film is still the most common detector of ionizing radiation, being used routinely in medical and dental x rays. Nuclear radiation is also captured on film, such as seen in **Figure 31.8**. The mechanism for film exposure by ionizing radiation is similar to that by photons. A quantum of energy interacts with the emulsion and alters it chemically, thus exposing the film. The quantum come from an α-particle, β-particle, or photon, provided it has more than the few eV of energy needed to induce the chemical change (as does all ionizing radiation). The process is not 100% efficient, since not all incident radiation interacts and not all interactions produce the chemical change. The amount of film darkening is related to exposure, but the darkening also depends on the type of radiation, so that absorbers and other devices must be used to obtain energy, charge, and particle-identification information.

Figure 31.8 Film badges contain film similar to that used in this dental x-ray film and is sandwiched between various absorbers to determine the penetrating ability of the radiation as well as the amount. (credit: Werneuchen, Wikimedia Commons)

Another very common **radiation detector** is the **Geiger tube**. The clicking and buzzing sound we hear in dramatizations and documentaries, as well as in our own physics labs, is usually an audio output of events detected by a Geiger counter. These relatively inexpensive radiation detectors are based on the simple and sturdy Geiger tube, shown schematically in **Figure 31.9(b)**. A conducting cylinder with a wire along its axis is filled with an insulating gas so that a voltage applied between the cylinder and wire produces almost no current. Ionizing radiation passing through the tube produces free ion pairs that are attracted to the wire and cylinder, forming a current that is detected as a count. The word count implies that there is no information on energy, charge, or type of radiation with a simple Geiger counter. They do not detect every particle, since some radiation can pass through without producing enough ionization to be detected. However, Geiger counters are very useful in producing a prompt output that reveals the existence and relative intensity of ionizing radiation.

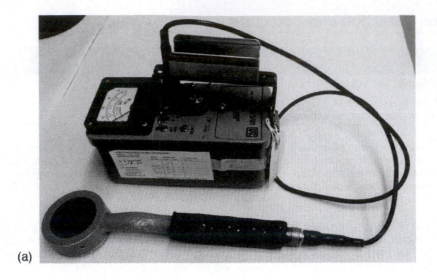

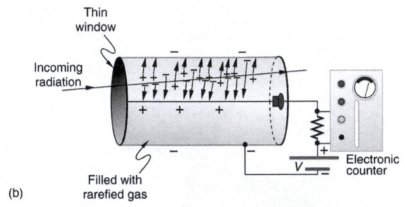

Figure 31.9 (a) Geiger counters such as this one are used for prompt monitoring of radiation levels, generally giving only relative intensity and not identifying the type or energy of the radiation. (credit: TimVickers, Wikimedia Commons) (b) Voltage applied between the cylinder and wire in a Geiger tube causes ions and electrons produced by radiation passing through the gas-filled cylinder to move towards them. The resulting current is detected and registered as a count.

Another radiation detection method records light produced when radiation interacts with materials. The energy of the radiation is sufficient to excite atoms in a material that may fluoresce, such as the phosphor used by Rutherford's group. Materials called **scintillators** use a more complex collaborative process to convert radiation energy into light. Scintillators may be liquid or solid, and they can be very efficient. Their light output can provide information about the energy, charge, and type of radiation. Scintillator light flashes are very brief in duration, enabling the detection of a huge number of particles in short periods of time. Scintillator detectors are used in a variety of research and diagnostic applications. Among these are the detection by satellite-mounted equipment of the radiation from distant galaxies, the analysis of radiation from a person indicating body burdens, and the detection of exotic particles in accelerator laboratories.

Light from a scintillator is converted into electrical signals by devices such as the **photomultiplier** tube shown schematically in **Figure 31.10**. These tubes are based on the photoelectric effect, which is multiplied in stages into a cascade of electrons, hence the name photomultiplier. Light entering the photomultiplier strikes a metal plate, ejecting an electron that is attracted by a positive potential difference to the next plate, giving it enough energy to eject two or more electrons, and so on. The final output current can be made proportional to the energy of the light entering the tube, which is in turn proportional to the energy deposited in the scintillator. Very sophisticated information can be obtained with scintillators, including energy, charge, particle identification, direction of motion, and so on.

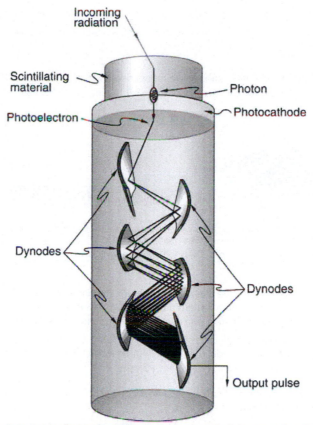

Figure 31.10 Photomultipliers use the photoelectric effect on the photocathode to convert the light output of a scintillator into an electrical signal. Each successive dynode has a more-positive potential than the last and attracts the ejected electrons, giving them more energy. The number of electrons is thus multiplied at each dynode, resulting in an easily detected output current.

Solid-state radiation detectors convert ionization produced in a semiconductor (like those found in computer chips) directly into an electrical signal. Semiconductors can be constructed that do not conduct current in one particular direction. When a voltage is applied in that direction, current flows only when ionization is produced by radiation, similar to what happens in a Geiger tube. Further, the amount of current in a solid-state detector is closely related to the energy deposited and, since the detector is solid, it can have a high efficiency (since ionizing radiation is stopped in a shorter distance in solids fewer particles escape detection). As with scintillators, very sophisticated information can be obtained from solid-state detectors.

PhET Explorations: Radioactive Dating Game

Learn about different types of radiometric dating, such as carbon dating. Understand how decay and half life work to enable radiometric dating to work. Play a game that tests your ability to match the percentage of the dating element that remains to the age of the object.

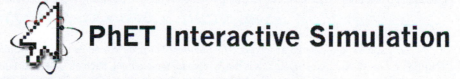

Figure 31.11 Radioactive Dating Game (http://cnx.org/content/m54926/1.2/radioactive-dating-game_en.jar)

31.3 Substructure of the Nucleus

Learning Objectives
By the end of this section, you will be able to:
• Define and discuss the nucleus in an atom.
• Define atomic number.
• Define and discuss Isotopes.
• Calculate the density of the nucleus.
• Explain nuclear force.

The information presented in this section supports the following AP® learning objectives and science practices:

- **3.G.3.1** The student is able to identify the strong force as the force that is responsible for holding the nucleus together. **(S.P. 7.2)**

What is inside the nucleus? Why are some nuclei stable while others decay? (See **Figure 31.12**.) Why are there different types of decay (α , β and γ)? Why are nuclear decay energies so large? Pursuing natural questions like these has led to far more fundamental discoveries than you might imagine.

(a)   (b)   (c)

Figure 31.12 Why is most of the carbon in this coal stable (a), while the uranium in the disk (b) slowly decays over billions of years? Why is cesium in this ampule (c) even less stable than the uranium, decaying in far less than 1/1,000,000 the time? What is the reason uranium and cesium undergo different types of decay (α and β , respectively)? (credits: (a) Bresson Thomas, Wikimedia Commons; (b) U.S. Department of Energy; (c) Tomihahndorf, Wikimedia Commons)

We have already identified **protons** as the particles that carry positive charge in the nuclei. However, there are actually *two* types of particles in the nuclei—the *proton* and the *neutron*, referred to collectively as **nucleons**, the constituents of nuclei. As its name implies, the **neutron** is a neutral particle ($q = 0$) that has nearly the same mass and intrinsic spin as the proton. **Table 31.2** compares the masses of protons, neutrons, and electrons. Note how close the proton and neutron masses are, but the neutron is slightly more massive once you look past the third digit. Both nucleons are much more massive than an electron. In fact, $m_p = 1836 m_e$ (as noted in **Medical Applications of Nuclear Physics** and $m_n = 1839 m_e$.

Table 31.2 also gives masses in terms of mass units that are more convenient than kilograms on the atomic and nuclear scale. The first of these is the *unified* **atomic mass** *unit* (u), defined as

$$1 \text{ u} = 1.6605 \times 10^{-27} \text{ kg.} \tag{31.1}$$

This unit is defined so that a neutral carbon ^{12}C atom has a mass of exactly 12 u. Masses are also expressed in units of MeV/c^2 . These units are very convenient when considering the conversion of mass into energy (and vice versa), as is so prominent in nuclear processes. Using $E = mc^2$ and units of m in MeV/c^2 , we find that c^2 cancels and E comes out conveniently in MeV. For example, if the rest mass of a proton is converted entirely into energy, then

$$E = mc^2 = (938.27 \text{ MeV}/c^2)c^2 = 938.27 \text{ MeV.} \tag{31.2}$$

It is useful to note that 1 u of mass converted to energy produces 931.5 MeV, or

$$1 \text{ u} = 931.5 \text{ MeV}/c^2. \tag{31.3}$$

All properties of a nucleus are determined by the number of protons and neutrons it has. A specific combination of protons and neutrons is called a **nuclide** and is a unique nucleus. The following notation is used to represent a particular nuclide:

$$^{A}_{Z}\text{X}_N, \tag{31.4}$$

where the symbols A , X , Z , and N are defined as follows: The *number of protons in a nucleus* is the **atomic number** Z , as defined in **Medical Applications of Nuclear Physics**. X is the *symbol for the element*, such as Ca for calcium. However, once Z is known, the element is known; hence, Z and X are redundant. For example, $Z = 20$ is always calcium, and calcium always has $Z = 20$. N is the *number of neutrons* in a nucleus. In the notation for a nuclide, the subscript N is usually omitted. The symbol A is defined as the number of nucleons or the *total number of protons and neutrons*,

$$A = N + Z, \tag{31.5}$$

where A is also called the **mass number**. This name for A is logical; the mass of an atom is nearly equal to the mass of its nucleus, since electrons have so little mass. The mass of the nucleus turns out to be nearly equal to the sum of the masses of the protons and neutrons in it, which is proportional to A . In this context, it is particularly convenient to express masses in units of u. Both protons and neutrons have masses close to 1 u, and so the mass of an atom is close to A u. For example, in an oxygen nucleus with eight protons and eight neutrons, $A = 16$, and its mass is 16 u. As noticed, the unified atomic mass unit is defined so that a neutral carbon atom (actually a ^{12}C atom) has a mass of *exactly* 12 u . Carbon was chosen as the standard, partly because of its importance in organic chemistry (see **Appendix A**.

Table 31.2 Masses of the Proton, Neutron, and Electron

Particle	Symbol	kg	u	MeVc²
Proton	p	1.67262×10^{-27}	1.007276	938.27
Neutron	n	1.67493×10^{-27}	1.008665	939.57
Electron	e	9.1094×10^{-31}	0.00054858	0.511

Let us look at a few examples of nuclides expressed in the $^A_Z X_N$ notation. The nucleus of the simplest atom, hydrogen, is a single proton, or $^1_1 H$ (the zero for no neutrons is often omitted). To check this symbol, refer to the periodic table—you see that the atomic number Z of hydrogen is 1. Since you are given that there are no neutrons, the mass number A is also 1. Suppose you are told that the helium nucleus or α particle has two protons and two neutrons. You can then see that it is written $^4_2 He_2$. There is a scarce form of hydrogen found in nature called deuterium; its nucleus has one proton and one neutron and, hence, twice the mass of common hydrogen. The symbol for deuterium is, thus, $^2_1 H_1$ (sometimes D is used, as for deuterated water $D_2 O$). An even rarer—and radioactive—form of hydrogen is called tritium, since it has a single proton and two neutrons, and it is written $^3_1 H_2$. These three varieties of hydrogen have nearly identical chemistries, but the nuclei differ greatly in mass, stability, and other characteristics. Nuclei (such as those of hydrogen) having the same Z and different N s are defined to be **isotopes** of the same element.

There is some redundancy in the symbols A, X, Z, and N. If the element X is known, then Z can be found in a periodic table and is always the same for a given element. If both A and X are known, then N can also be determined (first find Z; then, $N = A - Z$). Thus the simpler notation for nuclides is

$$^A X,\tag{31.6}$$

which is sufficient and is most commonly used. For example, in this simpler notation, the three isotopes of hydrogen are $^1 H$, $^2 H$, and $^3 H$, while the α particle is $^4 He$. We read this backward, saying helium-4 for $^4 He$, or uranium-238 for $^{238} U$. So for $^{238} U$, should we need to know, we can determine that $Z = 92$ for uranium from the periodic table, and, thus, $N = 238 - 92 = 146$.

A variety of experiments indicate that a nucleus behaves something like a tightly packed ball of nucleons, as illustrated in **Figure 31.13**. These nucleons have large kinetic energies and, thus, move rapidly in very close contact. Nucleons can be separated by a large force, such as in a collision with another nucleus, but resist strongly being pushed closer together. The most compelling evidence that nucleons are closely packed in a nucleus is that the **radius of a nucleus**, r, is found to be given approximately by

$$r = r_0 A^{1/3},\tag{31.7}$$

where $r_0 = 1.2$ fm and A is the mass number of the nucleus. Note that $r^3 \propto A$. Since many nuclei are spherical, and the volume of a sphere is $V = (4/3)\pi r^3$, we see that $V \propto A$ —that is, the volume of a nucleus is proportional to the number of nucleons in it. This is what would happen if you pack nucleons so closely that there is no empty space between them.

○ Proton

○ Neutron

Figure 31.13 A model of the nucleus.

Nucleons are held together by nuclear forces and resist both being pulled apart and pushed inside one another. The volume of the nucleus is the sum of the volumes of the nucleons in it, here shown in different colors to represent protons and neutrons.

Example 31.1 How Small and Dense Is a Nucleus?

(a) Find the radius of an iron-56 nucleus. (b) Find its approximate density in kg/m^3, approximating the mass of $^{56} Fe$ to be 56 u.

Strategy and Concept

(a) Finding the radius of ^{56}Fe is a straightforward application of $r = r_0 A^{1/3}$, given $A = 56$. (b) To find the approximate density, we assume the nucleus is spherical (this one actually is), calculate its volume using the radius found in part (a), and then find its density from $\rho = m/V$. Finally, we will need to convert density from units of $\text{u}\big/\text{fm}^3$ to $\text{kg}\big/\text{m}^3$.

Solution

(a) The radius of a nucleus is given by

$$r = r_0 A^{1/3}. \tag{31.8}$$

Substituting the values for r_0 and A yields

$$
\begin{aligned}
r &= (1.2 \text{ fm})(56)^{1/3} = (1.2 \text{ fm})(3.83) \\
&= 4.6 \text{ fm}.
\end{aligned}
\tag{31.9}
$$

(b) Density is defined to be $\rho = m/V$, which for a sphere of radius r is

$$\rho = \frac{m}{V} = \frac{m}{(4/3)\pi r^3}. \tag{31.10}$$

Substituting known values gives

$$
\begin{aligned}
\rho &= \frac{56 \text{ u}}{(1.33)(3.14)(4.6 \text{ fm})^3} \\
&= 0.138 \text{ u/fm}^3.
\end{aligned}
\tag{31.11}
$$

Converting to units of $\text{kg}\big/\text{m}^3$, we find

$$
\begin{aligned}
\rho &= (0.138 \text{ u/fm}^3)(1.66 \times 10^{-27} \text{ kg/u})\left(\frac{1 \text{ fm}}{10^{-15} \text{ m}}\right) \\
&= 2.3 \times 10^{17} \text{ kg/m}^3.
\end{aligned}
\tag{31.12}
$$

Discussion

(a) The radius of this medium-sized nucleus is found to be approximately 4.6 fm, and so its diameter is about 10 fm, or 10^{-14} m. In our discussion of Rutherford's discovery of the nucleus, we noticed that it is about 10^{-15} m in diameter (which is for lighter nuclei), consistent with this result to an order of magnitude. The nucleus is much smaller in diameter than the typical atom, which has a diameter of the order of 10^{-10} m.

(b) The density found here is so large as to cause disbelief. It is consistent with earlier discussions we have had about the nucleus being very small and containing nearly all of the mass of the atom. Nuclear densities, such as found here, are about 2×10^{14} times greater than that of water, which has a density of "only" 10^3 kg/m^3. One cubic meter of nuclear matter, such as found in a neutron star, has the same mass as a cube of water 61 km on a side.

Nuclear Forces and Stability

What forces hold a nucleus together? The nucleus is very small and its protons, being positive, exert tremendous repulsive forces on one another. (The Coulomb force increases as charges get closer, since it is proportional to $1/r^2$, even at the tiny distances found in nuclei.) The answer is that two previously unknown forces hold the nucleus together and make it into a tightly packed ball of nucleons. These forces are called the *weak and strong nuclear forces*. Nuclear forces are so short ranged that they fall to zero strength when nucleons are separated by only a few fm. However, like glue, they are strongly attracted when the nucleons get close to one another. The strong nuclear force is about 100 times more attractive than the repulsive EM force, easily holding the nucleons together. Nuclear forces become extremely repulsive if the nucleons get too close, making nucleons strongly resist being pushed inside one another, something like ball bearings.

The fact that nuclear forces are very strong is responsible for the very large energies emitted in nuclear decay. During decay, the forces do work, and since work is force times the distance ($W = Fd \cos\theta$), a large force can result in a large emitted energy. In fact, we know that there are *two* distinct nuclear forces because of the different types of nuclear decay—the strong nuclear force is responsible for α decay, while the weak nuclear force is responsible for β decay.

The many stable and unstable nuclei we have explored, and the hundreds we have not discussed, can be arranged in a table called the **chart of the nuclides**, a simplified version of which is shown in **Figure 31.14**. Nuclides are located on a plot of N versus Z. Examination of a detailed chart of the nuclides reveals patterns in the characteristics of nuclei, such as stability, abundance, and types of decay, analogous to but more complex than the systematics in the periodic table of the elements.

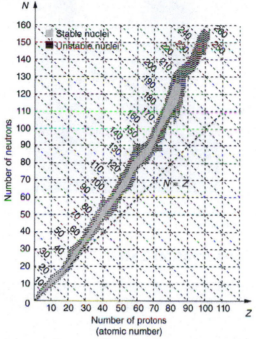

Figure 31.14 Simplified chart of the nuclides, a graph of N versus Z for known nuclides. The patterns of stable and unstable nuclides reveal characteristics of the nuclear forces. The dashed line is for $N = Z$. Numbers along diagonals are mass numbers A.

In principle, a nucleus can have any combination of protons and neutrons, but **Figure 31.14** shows a definite pattern for those that are stable. For low-mass nuclei, there is a strong tendency for N and Z to be nearly equal. This means that the nuclear force is more attractive when $N = Z$. More detailed examination reveals greater stability when N and Z are even numbers—nuclear forces are more attractive when neutrons and protons are in pairs. For increasingly higher masses, there are progressively more neutrons than protons in stable nuclei. This is due to the ever-growing repulsion between protons. Since nuclear forces are short ranged, and the Coulomb force is long ranged, an excess of neutrons keeps the protons a little farther apart, reducing Coulomb repulsion. Decay modes of nuclides out of the region of stability consistently produce nuclides closer to the region of stability. There are more stable nuclei having certain numbers of protons and neutrons, called **magic numbers**. Magic numbers indicate a shell structure for the nucleus in which closed shells are more stable. Nuclear shell theory has been very successful in explaining nuclear energy levels, nuclear decay, and the greater stability of nuclei with closed shells. We have been producing ever-heavier transuranic elements since the early 1940s, and we have now produced the element with $Z = 118$. There are theoretical predictions of an island of relative stability for nuclei with such high Z s.

Figure 31.15 The German-born American physicist Maria Goeppert Mayer (1906–1972) shared the 1963 Nobel Prize in physics with J. Jensen for the creation of the nuclear shell model. This successful nuclear model has nucleons filling shells analogous to electron shells in atoms. It was inspired by patterns observed in nuclear properties. (credit: Nobel Foundation via Wikimedia Commons)

31.4 Nuclear Decay and Conservation Laws

Learning Objectives

By the end of this section, you will be able to:

- Define and discuss nuclear decay.
- State the conservation laws.
- Explain parent and daughter nucleus.
- Calculate the energy emitted during nuclear decay.

The information presented in this section supports the following AP® learning objectives and science practices:

- **5.B.8.1** The student is able to describe emission or absorption spectra associated with electronic or nuclear transitions as transitions between allowed energy states of the atom in terms of the principle of energy conservation, including characterization of the frequency of radiation emitted or absorbed. **(S.P. 1.2, 7.2)**
- **5.C.1.1** The student is able to analyze electric charge conservation for nuclear and elementary particle reactions and make predictions related to such reactions based upon conservation of charge. **(S.P. 6.4, 7.2)**
- **5.C.2.1** The student is able to predict electric charges on objects within a system by application of the principle of charge conservation within a system. **(S.P. 6.4)**
- **5.G.1.1** The student is able to apply conservation of nucleon number and conservation of electric charge to make predictions about nuclear reactions and decays such as fission, fusion, alpha decay, beta decay, or gamma decay. **(S.P. 6.4)**

Nuclear **decay** has provided an amazing window into the realm of the very small. Nuclear decay gave the first indication of the connection between mass and energy, and it revealed the existence of two of the four basic forces in nature. In this section, we explore the major modes of nuclear decay; and, like those who first explored them, we will discover evidence of previously unknown particles and conservation laws.

Some nuclides are stable, apparently living forever. Unstable nuclides decay (that is, they are radioactive), eventually producing a stable nuclide after many decays. We call the original nuclide the **parent** and its decay products the **daughters**. Some radioactive nuclides decay in a single step to a stable nucleus. For example, ^{60}Co is unstable and decays directly to ^{60}Ni, which is stable. Others, such as ^{238}U, decay to another unstable nuclide, resulting in a **decay series** in which each subsequent nuclide decays until a stable nuclide is finally produced. The decay series that starts from ^{238}U is of particular interest, since it produces the radioactive isotopes ^{226}Ra and ^{210}Po, which the Curies first discovered (see **Figure 31.16**). Radon gas is also produced (^{222}Rn in the series), an increasingly recognized naturally occurring hazard. Since radon is a noble gas, it emanates from materials, such as soil, containing even trace amounts of ^{238}U and can be inhaled. The decay of radon and its daughters produces internal damage. The ^{238}U decay series ends with ^{206}Pb, a stable isotope of lead.

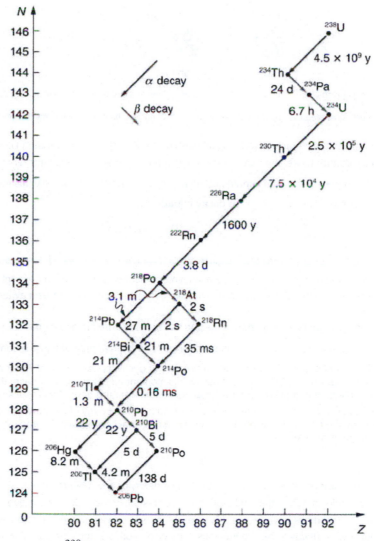

Figure 31.16 The decay series produced by ^{238}U , the most common uranium isotope. Nuclides are graphed in the same manner as in the chart of nuclides. The type of decay for each member of the series is shown, as well as the half-lives. Note that some nuclides decay by more than one mode. You can see why radium and polonium are found in uranium ore. A stable isotope of lead is the end product of the series.

Note that the daughters of α decay shown in **Figure 31.16** always have two fewer protons and two fewer neutrons than the parent. This seems reasonable, since we know that α decay is the emission of a ^{4}He nucleus, which has two protons and two neutrons. The daughters of β decay have one less neutron and one more proton than their parent. Beta decay is a little more subtle, as we shall see. No γ decays are shown in the figure, because they do not produce a daughter that differs from the parent.

Alpha Decay

In **alpha decay**, a ^{4}He nucleus simply breaks away from the parent nucleus, leaving a daughter with two fewer protons and two fewer neutrons than the parent (see **Figure 31.17**). One example of α decay is shown in **Figure 31.16** for ^{238}U . Another nuclide that undergoes α decay is ^{239}Pu . The decay equations for these two nuclides are

$$^{238}U \rightarrow {}^{234}Th_{92}^{234} + {}^4He \qquad (31.13)$$

and

$$^{239}Pu \rightarrow {}^{235}U + {}^4He. \qquad (31.14)$$

Figure 31.17 Alpha decay is the separation of a ^4He nucleus from the parent. The daughter nucleus has two fewer protons and two fewer neutrons than the parent. Alpha decay occurs spontaneously only if the daughter and ^4He nucleus have less total mass than the parent.

If you examine the periodic table of the elements, you will find that Th has $Z = 90$, two fewer than U, which has $Z = 92$. Similarly, in the second **decay equation**, we see that U has two fewer protons than Pu, which has $Z = 94$. The general rule for α decay is best written in the format $^A_Z\text{X}_N$. If a certain nuclide is known to α decay (generally this information must be looked up in a table of isotopes, such as in **Appendix B**), its α **decay equation** is

$$^A_Z\text{X}_N \rightarrow {}^{A-4}_{Z-2}\text{Y}_{N-2} + {}^4_2\text{He}_2 \quad (\alpha \text{ decay})$$

(31.15)

where Y is the nuclide that has two fewer protons than X, such as Th having two fewer than U. So if you were told that ^{239}Pu α decays and were asked to write the complete decay equation, you would first look up which element has two fewer protons (an atomic number two lower) and find that this is uranium. Then since four nucleons have broken away from the original 239, its atomic mass would be 235.

It is instructive to examine conservation laws related to α decay. You can see from the equation $^A_Z\text{X}_N \rightarrow {}^{A-4}_{Z-2}\text{Y}_{N-2} + {}^4_2\text{He}_2$ that total charge is conserved. Linear and angular momentum are conserved, too. Although conserved angular momentum is not of great consequence in this type of decay, conservation of linear momentum has interesting consequences. If the nucleus is at rest when it decays, its momentum is zero. In that case, the fragments must fly in opposite directions with equal-magnitude momenta so that total momentum remains zero. This results in the α particle carrying away most of the energy, as a bullet from a heavy rifle carries away most of the energy of the powder burned to shoot it. Total mass–energy is also conserved: the energy produced in the decay comes from conversion of a fraction of the original mass. As discussed in **Section 30.**, the general relationship is

$$E = (\Delta m)c^2.$$

(31.16)

Here, E is the **nuclear reaction energy** (the reaction can be nuclear decay or any other reaction), and Δm is the difference in mass between initial and final products. When the final products have less total mass, Δm is positive, and the reaction releases energy (is exothermic). When the products have greater total mass, the reaction is endothermic (Δm is negative) and must be induced with an energy input. For α decay to be spontaneous, the decay products must have smaller mass than the parent.

Example 31.2 Alpha Decay Energy Found from Nuclear Masses

Find the energy emitted in the α decay of ^{239}Pu.

Strategy

Nuclear reaction energy, such as released in α decay, can be found using the equation $E = (\Delta m)c^2$. We must first find Δm, the difference in mass between the parent nucleus and the products of the decay. This is easily done using masses given in **Appendix A**.

Solution

The decay equation was given earlier for ^{239}Pu; it is

$$^{239}\text{Pu} \rightarrow {}^{235}\text{U} + {}^4\text{He}.$$

(31.17)

Thus the pertinent masses are those of ^{239}Pu, ^{235}U, and the α particle or ^4He, all of which are listed in **Appendix A**. The initial mass was $m(^{239}\text{Pu}) = 239.052157 \text{ u}$. The final mass is the sum $m(^{235}\text{U}) + m(^4\text{He}) = 235.043924 \text{ u} + 4.002602 \text{ u} = 239.046526 \text{ u}$. Thus,

$$\begin{aligned} \Delta m &= m(^{239}\text{Pu}) - [m(^{235}\text{U}) + m(^4\text{He})] \\ &= 239.052157 \text{ u} - 239.046526 \text{ u} \\ &= 0.0005631 \text{ u}. \end{aligned}$$

(31.18)

Now we can find E by entering Δm into the equation:

$$E = (\Delta m)c^2 = (0.005631 \text{ u})c^2. \tag{31.19}$$

We know $1 \text{ u} = 931.5 \text{ MeV}/c^2$, and so

$$E = (0.005631)(931.5 \text{ MeV}/c^2)(c^2) = 5.25 \text{ MeV}. \tag{31.20}$$

Discussion

The energy released in this α decay is in the MeV range, about 10^6 times as great as typical chemical reaction energies, consistent with many previous discussions. Most of this energy becomes kinetic energy of the α particle (or ^{4}He nucleus), which moves away at high speed. The energy carried away by the recoil of the ^{235}U nucleus is much smaller in order to conserve momentum. The ^{235}U nucleus can be left in an excited state to later emit photons (γ rays).

This decay is spontaneous and releases energy, because the products have less mass than the parent nucleus. The question of why the products have less mass will be discussed in **Section 31.6**. Note that the masses given in **Appendix A** are atomic masses of neutral atoms, including their electrons. The mass of the electrons is the same before and after α decay, and so their masses subtract out when finding Δm. In this case, there are 94 electrons before and after the decay.

Beta Decay

There are actually *three* types of **beta decay**. The first discovered was "ordinary" beta decay and is called β^- decay or electron emission. The symbol β^- represents *an electron emitted in nuclear beta decay*. Cobalt-60 is a nuclide that β^- decays in the following manner:

$$^{60}\text{Co} \rightarrow {}^{60}\text{Ni} + \beta^- + \text{neutrino}. \tag{31.21}$$

The **neutrino** is a particle emitted in beta decay that was unanticipated and is of fundamental importance. The neutrino was not even proposed in theory until more than 20 years after beta decay was known to involve electron emissions. Neutrinos are so difficult to detect that the first direct evidence of them was not obtained until 1953. Neutrinos are nearly massless, have no charge, and do not interact with nucleons via the strong nuclear force. Traveling approximately at the speed of light, they have little time to affect any nucleus they encounter. This is, owing to the fact that they have no charge (and they are not EM waves), they do not interact through the EM force. They do interact via the relatively weak and very short range weak nuclear force. Consequently, neutrinos escape almost any detector and penetrate almost any shielding. However, neutrinos do carry energy, angular momentum (they are fermions with half-integral spin), and linear momentum away from a beta decay. When accurate measurements of beta decay were made, it became apparent that energy, angular momentum, and linear momentum were not accounted for by the daughter nucleus and electron alone. Either a previously unsuspected particle was carrying them away, or three conservation laws were being violated. Wolfgang Pauli made a formal proposal for the existence of neutrinos in 1930. The Italian-born American physicist Enrico Fermi (1901–1954) gave neutrinos their name, meaning little neutral ones, when he developed a sophisticated theory of beta decay (see **Figure 31.18**). Part of Fermi's theory was the identification of the weak nuclear force as being distinct from the strong nuclear force and in fact responsible for beta decay.

Figure 31.18 Enrico Fermi was nearly unique among 20th-century physicists—he made significant contributions both as an experimentalist and a theorist. His many contributions to theoretical physics included the identification of the weak nuclear force. The fermi (fm) is named after him, as are an entire class of subatomic particles (fermions), an element (Fermium), and a major research laboratory (Fermilab). His experimental work included studies of radioactivity, for which he won the 1938 Nobel Prize in physics, and creation of the first nuclear chain reaction. (credit: United States Department of Energy, Office of Public Affairs)

The neutrino also reveals a new conservation law. There are various families of particles, one of which is the electron family. We propose that the number of members of the electron family is constant in any process or any closed system. In our example of beta decay, there are no members of the electron family present before the decay, but after, there is an electron and a neutrino. So electrons are given an electron family number of $+1$. The neutrino in β^- decay is an **electron's antineutrino**, given the symbol $\bar{\nu}_e$, where ν is the Greek letter nu, and the subscript e means this neutrino is related to the electron. The bar indicates this is a particle of **antimatter**. (All particles have antimatter counterparts that are nearly identical except that they have the opposite charge. Antimatter is almost entirely absent on Earth, but it is found in nuclear decay and other nuclear and particle reactions as well as in outer space.) The electron's antineutrino $\bar{\nu}_e$, being antimatter, has an electron family number of -1.

The total is zero, before and after the decay. The new conservation law, obeyed in all circumstances, states that the *total electron family number is constant*. An electron cannot be created without also creating an antimatter family member. This law is analogous to the conservation of charge in a situation where total charge is originally zero, and equal amounts of positive and negative charge must be created in a reaction to keep the total zero.

If a nuclide $^A_Z X_N$ is known to β^- decay, then its β^- decay equation is

$$X_N \rightarrow Y_{N-1} + \beta^- + \bar{\nu}_e \ (\beta^- \ \text{decay}), \tag{31.22}$$

where Y is the nuclide having one more proton than X (see **Figure 31.19**). So if you know that a certain nuclide β^- decays, you can find the daughter nucleus by first looking up Z for the parent and then determining which element has atomic number $Z + 1$. In the example of the β^- decay of ^{60}Co given earlier, we see that $Z = 27$ for Co and $Z = 28$ is Ni. It is as if one of the neutrons in the parent nucleus decays into a proton, electron, and neutrino. In fact, neutrons outside of nuclei do just that—they live only an average of a few minutes and β^- decay in the following manner:

$$n \rightarrow p + \beta^- + \bar{\nu}_e. \tag{31.23}$$

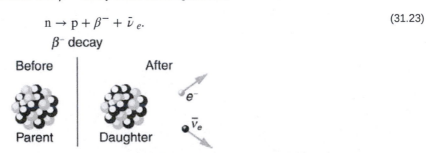

Figure 31.19 In β^- decay, the parent nucleus emits an electron and an antineutrino. The daughter nucleus has one more proton and one less neutron than its parent. Neutrinos interact so weakly that they are almost never directly observed, but they play a fundamental role in particle physics.

We see that charge is conserved in β^- decay, since the total charge is Z before and after the decay. For example, in ^{60}Co decay, total charge is 27 before decay, since cobalt has $Z = 27$. After decay, the daughter nucleus is Ni, which has $Z = 28$, and there is an electron, so that the total charge is also $28 + (-1)$ or 27. Angular momentum is conserved, but not obviously (you have to examine the spins and angular momenta of the final products in detail to verify this). Linear momentum is also conserved, again imparting most of the decay energy to the electron and the antineutrino, since they are of low and zero mass, respectively. Another new conservation law is obeyed here and elsewhere in nature. *The total number of nucleons A is conserved.* In ^{60}Co decay, for example, there are 60 nucleons before and after the decay. Note that total A is also conserved in α decay. Also note that the total number of protons changes, as does the total number of neutrons, so that total Z and total N are *not* conserved in β^- decay, as they are in α decay. Energy released in β^- decay can be calculated given the masses of the parent and products.

Example 31.3 β^- Decay Energy from Masses

Find the energy emitted in the β^- decay of ^{60}Co .

Strategy and Concept

As in the preceding example, we must first find Δm , the difference in mass between the parent nucleus and the products of the decay, using masses given in **Appendix A**. Then the emitted energy is calculated as before, using $E = (\Delta m)c^2$. The initial mass is just that of the parent nucleus, and the final mass is that of the daughter nucleus and the electron created in the decay. The neutrino is massless, or nearly so. However, since the masses given in **Appendix A** are for neutral atoms, the daughter nucleus has one more electron than the parent, and so the extra electron mass that corresponds to the β^- is included in the atomic mass of Ni. Thus,

$$\Delta m = m(^{60}\text{Co}) - m(^{60}\text{Ni}).$$ (31.24)

Solution

The β^- decay equation for ^{60}Co is

$$^{60}_{27}\text{Co}_{33} \rightarrow {}^{60}_{28}\text{Ni}_{32} + \beta^- + \bar{\nu}_e.$$ (31.25)

As noticed,

$$\Delta m = m(^{60}\text{Co}) - m(^{60}\text{Ni}).$$ (31.26)

Entering the masses found in **Appendix A** gives

$$\Delta m = 59.933820 \text{ u} - 59.930789 \text{ u} = 0.003031 \text{ u}.$$ (31.27)

Thus,

$$E = (\Delta m)c^2 = (0.003031 \text{ u})c^2.$$ (31.28)

Using $1 \text{ u} = 931.5 \text{ MeV} / c^2$, we obtain

$$E = (0.003031)(931.5 \text{ MeV} / c^2)(c^2) = 2.82 \text{ MeV}.$$ (31.29)

Discussion and Implications

Perhaps the most difficult thing about this example is convincing yourself that the β^- mass is included in the atomic mass of ^{60}Ni . Beyond that are other implications. Again the decay energy is in the MeV range. This energy is shared by all of the products of the decay. In many ^{60}Co decays, the daughter nucleus ^{60}Ni is left in an excited state and emits photons (γ rays). Most of the remaining energy goes to the electron and neutrino, since the recoil kinetic energy of the daughter nucleus is small. One final note: the electron emitted in β^- decay is created in the nucleus at the time of decay.

The second type of beta decay is less common than the first. It is β^+ decay. Certain nuclides decay by the emission of a *positive* electron. This is **antielectron** or **positron decay** (see **Figure 31.20**).

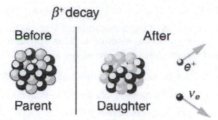

Figure 31.20 β^+ decay is the emission of a positron that eventually finds an electron to annihilate, characteristically producing gammas in opposite directions.

The antielectron is often represented by the symbol e^+, but in beta decay it is written as β^+ to indicate the antielectron was emitted in a nuclear decay. Antielectrons are the antimatter counterpart to electrons, being nearly identical, having the same mass, spin, and so on, but having a positive charge and an electron family number of -1. When a **positron** encounters an electron, there is a mutual annihilation in which all the mass of the antielectron-electron pair is converted into pure photon energy. (The reaction, $e^+ + e^- \rightarrow \gamma + \gamma$, conserves electron family number as well as all other conserved quantities.) If a nuclide $_Z^A X_N$ is known to β^+ decay, then its β^+ **decay equation** is

$$_Z^A X_N \rightarrow Y_{N+1} + \beta^+ + \nu_e \ (\beta^+ \text{ decay}), \tag{31.30}$$

where Y is the nuclide having one less proton than X (to conserve charge) and ν_e is the symbol for the **electron's neutrino**, which has an electron family number of $+1$. Since an antimatter member of the electron family (the β^+) is created in the decay, a matter member of the family (here the ν_e) must also be created. Given, for example, that ^{22}Na β^+ decays, you can write its full decay equation by first finding that $Z = 11$ for ^{22}Na, so that the daughter nuclide will have $Z = 10$, the atomic number for neon. Thus the β^+ decay equation for ^{22}Na is

$$_{11}^{22}\text{Na}_{11} \rightarrow _{10}^{22}\text{Ne}_{12} + \beta^+ + \nu_e. \tag{31.31}$$

In β^+ decay, it is as if one of the protons in the parent nucleus decays into a neutron, a positron, and a neutrino. Protons do not do this outside of the nucleus, and so the decay is due to the complexities of the nuclear force. Note again that the total number of nucleons is constant in this and any other reaction. To find the energy emitted in β^+ decay, you must again count the number of electrons in the neutral atoms, since atomic masses are used. The daughter has one less electron than the parent, and one electron mass is created in the decay. Thus, in β^+ decay,

$$\Delta m = m(\text{parent}) - [m(\text{daughter}) + 2m_e], \tag{31.32}$$

since we use the masses of neutral atoms.

Electron capture is the third type of beta decay. Here, a nucleus captures an inner-shell electron and undergoes a nuclear reaction that has the same effect as β^+ decay. Electron capture is sometimes denoted by the letters EC. We know that electrons cannot reside in the nucleus, but this is a nuclear reaction that consumes the electron and occurs spontaneously only when the products have less mass than the parent plus the electron. If a nuclide $_Z^A X_N$ is known to undergo electron capture, then its **electron capture equation** is

$$_Z^A X_N + e^- \rightarrow Y_{N+1} + \nu_e (\text{electron capture, or EC}). \tag{31.33}$$

Any nuclide that can β^+ decay can also undergo electron capture (and often does both). The same conservation laws are obeyed for EC as for β^+ decay. It is good practice to confirm these for yourself.

All forms of beta decay occur because the parent nuclide is unstable and lies outside the region of stability in the chart of nuclides. Those nuclides that have relatively more neutrons than those in the region of stability will β^- decay to produce a daughter with fewer neutrons, producing a daughter nearer the region of stability. Similarly, those nuclides having relatively more protons than those in the region of stability will β^- decay or undergo electron capture to produce a daughter with fewer protons, nearer the region of stability.

Gamma Decay

Gamma decay is the simplest form of nuclear decay—it is the emission of energetic photons by nuclei left in an excited state by some earlier process. Protons and neutrons in an excited nucleus are in higher orbitals, and they fall to lower levels by photon emission (analogous to electrons in excited atoms). Nuclear excited states have lifetimes typically of only about 10^{-14} s, an indication of the great strength of the forces pulling the nucleons to lower states. The γ decay equation is simply

$$\,^A_Z X^*_N \rightarrow X_N + \gamma_1 + \gamma_2 + \cdots \quad (\gamma \text{ decay})$$

(31.34)

where the asterisk indicates the nucleus is in an excited state. There may be one or more γ s emitted, depending on how the nuclide de-excites. In radioactive decay, γ emission is common and is preceded by γ or β decay. For example, when ^{60}Co β^- decays, it most often leaves the daughter nucleus in an excited state, written $^{60}\text{Ni}^*$. Then the nickel nucleus quickly γ decays by the emission of two penetrating γ s:

$$^{60}\text{Ni}^* \rightarrow \,^{60}\text{Ni} + \gamma_1 + \gamma_2.$$

(31.35)

These are called cobalt γ rays, although they come from nickel—they are used for cancer therapy, for example. It is again constructive to verify the conservation laws for gamma decay. Finally, since γ decay does not change the nuclide to another species, it is not prominently featured in charts of decay series, such as that in **Figure 31.16**.

There are other types of nuclear decay, but they occur less commonly than α, β, and γ decay. Spontaneous fission is the most important of the other forms of nuclear decay because of its applications in nuclear power and weapons. It is covered in the next chapter.

31.5 Half-Life and Activity

Learning Objectives
By the end of this section, you will be able to: • Define half-life. • Define dating. • Calculate the age of old objects by radioactive dating. The information presented in this section supports the following AP® learning objectives and science practices: • **7.C.3.1** The student is able to predict the number of radioactive nuclei remaining in a sample after a certain period of time, and also predict the missing species (alpha, beta, gamma) in a radioactive decay. **(S.P. 6.4)**

Unstable nuclei decay. However, some nuclides decay faster than others. For example, radium and polonium, discovered by the Curies, decay faster than uranium. This means they have shorter lifetimes, producing a greater rate of decay. In this section we explore half-life and activity, the quantitative terms for lifetime and rate of decay.

Half-Life

Why use a term like half-life rather than lifetime? The answer can be found by examining **Figure 31.21**, which shows how the number of radioactive nuclei in a sample decreases with time. The *time in which half of the original number of nuclei decay* is defined as the **half-life**, $t_{1/2}$. Half of the remaining nuclei decay in the next half-life. Further, half of that amount decays in the following half-life. Therefore, the number of radioactive nuclei decreases from N to $N/2$ in one half-life, then to $N/4$ in the next, and to $N/8$ in the next, and so on. If N is a large number, then *many* half-lives (not just two) pass before all of the nuclei decay. Nuclear decay is an example of a purely statistical process. A more precise definition of half-life is that *each nucleus has a 50% chance of living for a time equal to one half-life $t_{1/2}$*. Thus, if N is reasonably large, half of the original nuclei decay in a time of one half-life. If an individual nucleus makes it through that time, it still has a 50% chance of surviving through another half-life. Even if it happens to make it through hundreds of half-lives, it still has a 50% chance of surviving through one more. The probability of decay is the same no matter when you start counting. This is like random coin flipping. The chance of heads is 50%, no matter what has happened before.

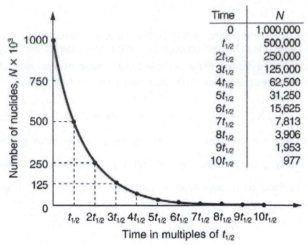

Time	N
0	1,000,000
$t_{1/2}$	500,000
$2t_{1/2}$	250,000
$3t_{1/2}$	125,000
$4t_{1/2}$	62,500
$5t_{1/2}$	31,250
$6t_{1/2}$	15,625
$7t_{1/2}$	7,813
$8t_{1/2}$	3,906
$9t_{1/2}$	1,953
$10t_{1/2}$	977

Figure 31.21 Radioactive decay reduces the number of radioactive nuclei over time. In one half-life $t_{1/2}$, the number decreases to half of its original value. Half of what remains decay in the next half-life, and half of those in the next, and so on. This is an exponential decay, as seen in the graph of the number of nuclei present as a function of time.

There is a tremendous range in the half-lives of various nuclides, from as short as 10^{-23} s for the most unstable, to more than 10^{16} y for the least unstable, or about 46 orders of magnitude. Nuclides with the shortest half-lives are those for which the nuclear forces are least attractive, an indication of the extent to which the nuclear force can depend on the particular combination of neutrons and protons. The concept of half-life is applicable to other subatomic particles, as will be discussed in **Particle Physics**. It is also applicable to the decay of excited states in atoms and nuclei. The following equation gives the quantitative relationship between the original number of nuclei present at time zero (N_0) and the number (N) at a later time t:

$$N = N_0 e^{-\lambda t},$$
(31.36)

where $e = 2.71828...$ is the base of the natural logarithm, and λ is the **decay constant** for the nuclide. The shorter the half-life, the larger is the value of λ, and the faster the exponential $e^{-\lambda t}$ decreases with time. The relationship between the decay constant λ and the half-life $t_{1/2}$ is

$$\lambda = \frac{\ln(2)}{t_{1/2}} \approx \frac{0.693}{t_{1/2}}.$$
(31.37)

To see how the number of nuclei declines to half its original value in one half-life, let $t = t_{1/2}$ in the exponential in the equation $N = N_0 e^{-\lambda t}$. This gives $N = N_0 e^{-\lambda t} = N_0 e^{-0.693} = 0.500 N_0$. For integral numbers of half-lives, you can just divide the original number by 2 over and over again, rather than using the exponential relationship. For example, if ten half-lives have passed, we divide N by 2 ten times. This reduces it to $N/1024$. For an arbitrary time, not just a multiple of the half-life, the exponential relationship must be used.

Radioactive dating is a clever use of naturally occurring radioactivity. Its most famous application is **carbon-14 dating**. Carbon-14 has a half-life of 5730 years and is produced in a nuclear reaction induced when solar neutrinos strike ^{14}N in the atmosphere. Radioactive carbon has the same chemistry as stable carbon, and so it mixes into the ecosphere, where it is consumed and becomes part of every living organism. Carbon-14 has an abundance of 1.3 parts per trillion of normal carbon. Thus, if you know the number of carbon nuclei in an object (perhaps determined by mass and Avogadro's number), you multiply that number by 1.3×10^{-12} to find the number of ^{14}C nuclei in the object. When an organism dies, carbon exchange with the environment ceases, and ^{14}C is not replenished as it decays. By comparing the abundance of ^{14}C in an artifact, such as mummy wrappings, with the normal abundance in living tissue, it is possible to determine the artifact's age (or time since death). Carbon-14 dating can be used for biological tissues as old as 50 or 60 thousand years, but is most accurate for younger samples, since the abundance of ^{14}C nuclei in them is greater. Very old biological materials contain no ^{14}C at all. There are instances in which the date of an artifact can be determined by other means, such as historical knowledge or tree-ring counting. These cross-references have confirmed the validity of carbon-14 dating and permitted us to calibrate the technique as well. Carbon-14 dating revolutionized parts of archaeology and is of such importance that it earned the 1960 Nobel Prize in chemistry for its developer, the American chemist Willard Libby (1908–1980).

One of the most famous cases of carbon-14 dating involves the Shroud of Turin, a long piece of fabric purported to be the burial shroud of Jesus (see **Figure 31.22**). This relic was first displayed in Turin in 1354 and was denounced as a fraud at that time by a French bishop. Its remarkable negative imprint of an apparently crucified body resembles the then-accepted image of Jesus, and so the shroud was never disregarded completely and remained controversial over the centuries. Carbon-14 dating was not performed on the shroud until 1988, when the process had been refined to the point where only a small amount of material needed to be destroyed. Samples were tested at three independent laboratories, each being given four pieces of cloth, with only one unidentified piece from the shroud, to avoid prejudice. All three laboratories found samples of the shroud contain 92% of the ^{14}C found in living tissues, allowing the shroud to be dated (see **Example 31.4**).

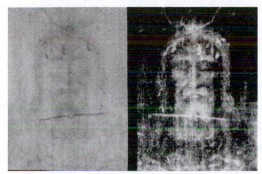

Figure 31.22 Part of the Shroud of Turin, which shows a remarkable negative imprint likeness of Jesus complete with evidence of crucifixion wounds. The shroud first surfaced in the 14th century and was only recently carbon-14 dated. It has not been determined how the image was placed on the material. (credit: Butko, Wikimedia Commons)

Example 31.4 How Old Is the Shroud of Turin?

Calculate the age of the Shroud of Turin given that the amount of ^{14}C found in it is 92% of that in living tissue.

Strategy

Knowing that 92% of the ^{14}C remains means that $N/N_0 = 0.92$. Therefore, the equation $N = N_0 e^{-\lambda t}$ can be used to find λt. We also know that the half-life of ^{14}C is 5730 y, and so once λt is known, we can use the equation $\lambda = \dfrac{0.693}{t_{1/2}}$ to find λ and then find t as requested. Here, we postulate that the decrease in ^{14}C is solely due to nuclear decay.

Solution

Solving the equation $N = N_0 e^{-\lambda t}$ for N/N_0 gives

$$\frac{N}{N_0} = e^{-\lambda t}. \tag{31.38}$$

Thus,

$$0.92 = e^{-\lambda t}. \tag{31.39}$$

Taking the natural logarithm of both sides of the equation yields

$$\ln 0.92 = -\lambda t \tag{31.40}$$

so that

$$-0.0834 = -\lambda t. \tag{31.41}$$

Rearranging to isolate t gives

$$t = \frac{0.0834}{\lambda}. \tag{31.42}$$

Now, the equation $\lambda = \dfrac{0.693}{t_{1/2}}$ can be used to find λ for ^{14}C. Solving for λ and substituting the known half-life gives

$$\lambda = \frac{0.693}{t_{1/2}} = \frac{0.693}{5730 \text{ y}}. \tag{31.43}$$

We enter this value into the previous equation to find t:

$$t = \frac{0.0834}{\dfrac{0.693}{5730 \text{ y}}} = 690 \text{ y}. \tag{31.44}$$

Discussion

This dates the material in the shroud to 1988–690 = a.d. 1300. Our calculation is only accurate to two digits, so that the year is rounded to 1300. The values obtained at the three independent laboratories gave a weighted average date of a.d. 1320 ± 60. The uncertainty is typical of carbon-14 dating and is due to the small amount of ^{14}C in living tissues, the amount of material available, and experimental uncertainties (reduced by having three independent measurements). It is meaningful that the date of the shroud is consistent with the first record of its existence and inconsistent with the period in which Jesus lived.

There are other forms of radioactive dating. Rocks, for example, can sometimes be dated based on the decay of ^{238}U. The decay series for ^{238}U ends with ^{206}Pb, so that the ratio of these nuclides in a rock is an indication of how long it has been since the rock solidified. The original composition of the rock, such as the absence of lead, must be known with some confidence. However, as with carbon-14 dating, the technique can be verified by a consistent body of knowledge. Since ^{238}U has a half-life of 4.5×10^9 y, it is useful for dating only very old materials, showing, for example, that the oldest rocks on Earth solidified about 3.5×10^9 years ago.

Activity, the Rate of Decay

What do we mean when we say a source is highly radioactive? Generally, this means the number of decays per unit time is very high. We define **activity** R to be the **rate of decay** expressed in decays per unit time. In equation form, this is

$$R = \frac{\Delta N}{\Delta t} \tag{31.45}$$

where ΔN is the number of decays that occur in time Δt. The SI unit for activity is one decay per second and is given the name **becquerel** (Bq) in honor of the discoverer of radioactivity. That is,

$$1\ Bq = 1\ decay/s. \tag{31.46}$$

Activity R is often expressed in other units, such as decays per minute or decays per year. One of the most common units for activity is the **curie** (Ci), defined to be the activity of 1 g of ^{226}Ra, in honor of Marie Curie's work with radium. The definition of curie is

$$1\ Ci = 3.70 \times 10^{10}\ Bq, \tag{31.47}$$

or 3.70×10^{10} decays per second. A curie is a large unit of activity, while a becquerel is a relatively small unit. $1\ MBq = 100$ microcuries (μCi). In countries like Australia and New Zealand that adhere more to SI units, most radioactive sources, such as those used in medical diagnostics or in physics laboratories, are labeled in Bq or megabecquerel (MBq).

Intuitively, you would expect the activity of a source to depend on two things: the amount of the radioactive substance present, and its half-life. The greater the number of radioactive nuclei present in the sample, the more will decay per unit of time. The shorter the half-life, the more decays per unit time, for a given number of nuclei. So activity R should be proportional to the number of radioactive nuclei, N, and inversely proportional to their half-life, $t_{1/2}$. In fact, your intuition is correct. It can be shown that the activity of a source is

$$R = \frac{0.693N}{t_{1/2}} \tag{31.48}$$

where N is the number of radioactive nuclei present, having half-life $t_{1/2}$. This relationship is useful in a variety of calculations, as the next two examples illustrate.

Example 31.5 How Great Is the ^{14}C Activity in Living Tissue?

Calculate the activity due to ^{14}C in 1.00 kg of carbon found in a living organism. Express the activity in units of Bq and Ci.

Strategy

To find the activity R using the equation $R = \frac{0.693N}{t_{1/2}}$, we must know N and $t_{1/2}$. The half-life of ^{14}C can be found in Appendix B, and was stated above as 5730 y. To find N, we first find the number of ^{12}C nuclei in 1.00 kg of carbon using the concept of a mole. As indicated, we then multiply by 1.3×10^{-12} (the abundance of ^{14}C in a carbon sample from a living organism) to get the number of ^{14}C nuclei in a living organism.

Solution

One mole of carbon has a mass of 12.0 g, since it is nearly pure ^{12}C. (A mole has a mass in grams equal in magnitude to A found in the periodic table.) Thus the number of carbon nuclei in a kilogram is

$$N(^{12}C) = \frac{6.02 \times 10^{23} \text{ mol}^{-1}}{12.0 \text{ g/mol}} \times (1000 \text{ g}) = 5.02 \times 10^{25}. \tag{31.49}$$

So the number of ^{14}C nuclei in 1 kg of carbon is

$$N(^{14}C) = (5.02 \times 10^{25})(1.3 \times 10^{-12}) = 6.52 \times 10^{13}. \tag{31.50}$$

Now the activity R is found using the equation $R = \frac{0.693N}{t_{1/2}}$.

Entering known values gives

$$R = \frac{0.693(6.52 \times 10^{13})}{5730 \text{ y}} = 7.89 \times 10^9 \text{ y}^{-1}, \tag{31.51}$$

or 7.89×10^9 decays per year. To convert this to the unit Bq, we simply convert years to seconds. Thus,

$$R = (7.89 \times 10^9 \text{ y}^{-1}) \frac{1.00 \text{ y}}{3.16 \times 10^7 \text{ s}} = 250 \text{ Bq}, \tag{31.52}$$

or 250 decays per second. To express R in curies, we use the definition of a curie,

$$R = \frac{250 \text{ Bq}}{3.7 \times 10^{10} \text{ Bq/Ci}} = 6.76 \times 10^{-9} \text{ Ci}. \tag{31.53}$$

Thus,

$$R = 6.76 \text{ nCi}. \tag{31.54}$$

Discussion

Our own bodies contain kilograms of carbon, and it is intriguing to think there are hundreds of ^{14}C decays per second taking place in us. Carbon-14 and other naturally occurring radioactive substances in our bodies contribute to the background radiation we receive. The small number of decays per second found for a kilogram of carbon in this example gives you some idea of how difficult it is to detect ^{14}C in a small sample of material. If there are 250 decays per second in a kilogram, then there are 0.25 decays per second in a gram of carbon in living tissue. To observe this, you must be able to distinguish decays from other forms of radiation, in order to reduce background noise. This becomes more difficult with an old tissue sample, since it contains less ^{14}C, and for samples more than 50 thousand years old, it is impossible.

Human-made (or artificial) radioactivity has been produced for decades and has many uses. Some of these include medical therapy for cancer, medical imaging and diagnostics, and food preservation by irradiation. Many applications as well as the biological effects of radiation are explored in **Medical Applications of Nuclear Physics**, but it is clear that radiation is hazardous. A number of tragic examples of this exist, one of the most disastrous being the meltdown and fire at the Chernobyl reactor complex in the Ukraine (see **Figure 31.23**). Several radioactive isotopes were released in huge quantities, contaminating many thousands of square kilometers and directly affecting hundreds of thousands of people. The most significant releases were of ^{131}I, ^{90}Sr, ^{137}Cs, ^{239}Pu, ^{238}U, and ^{235}U. Estimates are that the total amount of radiation released was about 100 million curies.

Human and Medical Applications

Figure 31.23 The Chernobyl reactor. More than 100 people died soon after its meltdown, and there will be thousands of deaths from radiation-induced cancer in the future. While the accident was due to a series of human errors, the cleanup efforts were heroic. Most of the immediate fatalities were firefighters and reactor personnel. (credit: Elena Filatova)

Example 31.6 What Mass of ^{137}Cs Escaped Chernobyl?

It is estimated that the Chernobyl disaster released 6.0 MCi of ^{137}Cs into the environment. Calculate the mass of ^{137}Cs released.

Strategy

We can calculate the mass released using Avogadro's number and the concept of a mole if we can first find the number of nuclei N released. Since the activity R is given, and the half-life of ^{137}Cs is found in **Appendix B** to be 30.2 y, we can use the equation $R = \frac{0.693N}{t_{1/2}}$ to find N.

Solution

Solving the equation $R = \frac{0.693N}{t_{1/2}}$ for N gives

$$N = \frac{Rt_{1/2}}{0.693}. \tag{31.55}$$

Entering the given values yields

$$N = \frac{(6.0 \text{ MCi})(30.2 \text{ y})}{0.693}. \tag{31.56}$$

Converting curies to becquerels and years to seconds, we get

$$\begin{aligned} N &= \frac{(6.0 \times 10^6 \text{ Ci})(3.7 \times 10^{10} \text{ Bq/Ci})(30.2 \text{ y})(3.16 \times 10^7 \text{ s/y})}{0.693} \\ &= 3.1 \times 10^{26}. \end{aligned} \tag{31.57}$$

One mole of a nuclide $^A X$ has a mass of A grams, so that one mole of ^{137}Cs has a mass of 137 g. A mole has 6.02×10^{23} nuclei. Thus the mass of ^{137}Cs released was

$$\begin{aligned} m &= \left(\frac{137 \text{ g}}{6.02 \times 10^{23}}\right)(3.1 \times 10^{26}) = 70 \times 10^3 \text{ g} \\ &= 70 \text{ kg}. \end{aligned} \tag{31.58}$$

Discussion

While 70 kg of material may not be a very large mass compared to the amount of fuel in a power plant, it is extremely radioactive, since it only has a 30-year half-life. Six megacuries (6.0 MCi) is an extraordinary amount of activity but is only a fraction of what is produced in nuclear reactors. Similar amounts of the other isotopes were also released at Chernobyl. Although the chances of such a disaster may have seemed small, the consequences were extremely severe, requiring greater caution than was used. More will be said about safe reactor design in the next chapter, but it should be noted that Western reactors have a fundamentally safer design.

Activity R decreases in time, going to half its original value in one half-life, then to one-fourth its original value in the next half-life, and so on. Since $R = \dfrac{0.693N}{t_{1/2}}$, the activity decreases as the number of radioactive nuclei decreases. The equation for R

as a function of time is found by combining the equations $N = N_0 e^{-\lambda t}$ and $R = \dfrac{0.693N}{t_{1/2}}$, yielding

$$R = R_0 e^{-\lambda t}, \tag{31.59}$$

where R_0 is the activity at $t = 0$. This equation shows exponential decay of radioactive nuclei. For example, if a source originally has a 1.00-mCi activity, it declines to 0.500 mCi in one half-life, to 0.250 mCi in two half-lives, to 0.125 mCi in three half-lives, and so on. For times other than whole half-lives, the equation $R = R_0 e^{-\lambda t}$ must be used to find R.

PhET Explorations: Alpha Decay

Watch alpha particles escape from a polonium nucleus, causing radioactive alpha decay. See how random decay times relate to the half life.

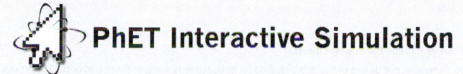

Figure 31.24 Alpha Decay (http://cnx.org/content/m54932/1.2/alpha-decay_en.jar)

31.6 Binding Energy

Learning Objectives

By the end of this section, you will be able to:

- Define and discuss binding energy.
- Calculate the binding energy per nucleon of a particle.

The information presented in this section supports the following AP® learning objectives and science practices:

- **3.G.3.1** The student is able to identify the strong force as the force that is responsible for holding the nucleus together.

The more tightly bound a system is, the stronger the forces that hold it together and the greater the energy required to pull it apart. We can therefore learn about nuclear forces by examining how tightly bound the nuclei are. We define the **binding energy** (BE) of a nucleus to be *the energy required to completely disassemble it into separate protons and neutrons*. We can determine the BE of a nucleus from its rest mass. The two are connected through Einstein's famous relationship $E = (\Delta m)c^2$. A bound system has a *smaller* mass than its separate constituents; the more tightly the nucleons are bound together, the smaller the mass of the nucleus.

Imagine pulling a nuclide apart as illustrated in Figure 31.25. Work done to overcome the nuclear forces holding the nucleus together puts energy into the system. By definition, the energy input equals the binding energy BE. The pieces are at rest when separated, and so the energy put into them increases their total rest mass compared with what it was when they were glued together as a nucleus. That mass increase is thus $\Delta m = \mathrm{BE}/c^2$. This difference in mass is known as *mass defect*. It implies that the mass of the nucleus is less than the sum of the masses of its constituent protons and neutrons. A nuclide $^A\mathrm{X}$ has Z protons and N neutrons, so that the difference in mass is

$$\Delta m = (Zm_p + Nm_n) - m_{\text{tot}}. \tag{31.60}$$

Thus,

$$\mathrm{BE} = (\Delta m)c^2 = [(Zm_p + Nm_n) - m_{\text{tot}}]c^2, \tag{31.61}$$

where m_{tot} is the mass of the nuclide $^A\mathrm{X}$, m_p is the mass of a proton, and m_n is the mass of a neutron. Traditionally, we deal with the masses of neutral atoms. To get atomic masses into the last equation, we first add Z electrons to m_{tot}, which gives $m\left(^A\mathrm{X}\right)$, the atomic mass of the nuclide. We then add Z electrons to the Z protons, which gives $Zm\left(^1\mathrm{H}\right)$, or Z times the mass of a hydrogen atom. Thus the binding energy of a nuclide $^A\mathrm{X}$ is

$$\text{BE} = \left\{[Zm(^1\text{H}) + Nm_n] - m(^A X)\right\}c^2. \tag{31.62}$$

The atomic masses can be found in **Appendix A**, most conveniently expressed in unified atomic mass units u (
$1 \text{ u} = 931.5 \text{ MeV}/c^2$). BE is thus calculated from known atomic masses.

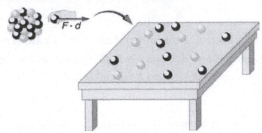

Figure 31.25 Work done to pull a nucleus apart into its constituent protons and neutrons increases the mass of the system. The work to disassemble the nucleus equals its binding energy BE. A bound system has less mass than the sum of its parts, especially noticeable in the nuclei, where forces and energies are very large.

Things Great and Small

Nuclear Decay Helps Explain Earth's Hot Interior

A puzzle created by radioactive dating of rocks is resolved by radioactive heating of Earth's interior. This intriguing story is another example of how small-scale physics can explain large-scale phenomena.

Radioactive dating plays a role in determining the approximate age of the Earth. The oldest rocks on Earth solidified about 3.5×10^9 years ago—a number determined by uranium-238 dating. These rocks could only have solidified once the surface of the Earth had cooled sufficiently. The temperature of the Earth at formation can be estimated based on gravitational potential energy of the assemblage of pieces being converted to thermal energy. Using heat transfer concepts discussed in **Thermodynamics** it is then possible to calculate how long it would take for the surface to cool to rock-formation temperatures. The result is about 10^9 years. The first rocks formed have been solid for 3.5×10^9 years, so that the age of the Earth is approximately 4.5×10^9 years. There is a large body of other types of evidence (both Earth-bound and solar system characteristics are used) that supports this age. The puzzle is that, given its age and initial temperature, the center of the Earth should be much cooler than it is today (see **Figure 31.26**).

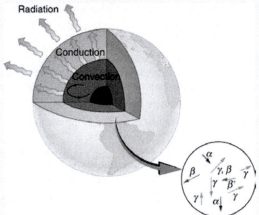

Figure 31.26 The center of the Earth cools by well-known heat transfer methods. Convection in the liquid regions and conduction move thermal energy to the surface, where it radiates into cold, dark space. Given the age of the Earth and its initial temperature, it should have cooled to a lower temperature by now. The blowup shows that nuclear decay releases energy in the Earth's interior. This energy has slowed the cooling process and is responsible for the interior still being molten.

We know from seismic waves produced by earthquakes that parts of the interior of the Earth are liquid. Shear or transverse waves cannot travel through a liquid and are not transmitted through the Earth's core. Yet compression or longitudinal waves can pass through a liquid and do go through the core. From this information, the temperature of the interior can be estimated. As noticed, the interior should have cooled more from its initial temperature in the 4.5×10^9 years since its formation. In fact, it should have taken no more than about 10^9 years to cool to its present temperature. What is keeping it hot? The answer seems to be radioactive decay of primordial elements that were part of the material that formed the Earth (see the blowup in **Figure 31.26**).

Nuclides such as ^{238}U and ^{40}K have half-lives similar to or longer than the age of the Earth, and their decay still contributes energy to the interior. Some of the primordial radioactive nuclides have unstable decay products that also release energy— ^{238}U has a long decay chain of these. Further, there were more of these primordial radioactive nuclides early in the life of the Earth, and thus the activity and energy contributed were greater then (perhaps by an order of magnitude). The amount of power created by these decays per cubic meter is very small. However, since a huge volume of material lies deep below the surface, this relatively small amount of energy cannot escape quickly. The power produced near the surface has much less distance to go to escape and has a negligible effect on surface temperatures.

A final effect of this trapped radiation merits mention. Alpha decay produces helium nuclei, which form helium atoms when they are stopped and capture electrons. Most of the helium on Earth is obtained from wells and is produced in this manner. Any helium in the atmosphere will escape in geologically short times because of its high thermal velocity.

What patterns and insights are gained from an examination of the binding energy of various nuclides? First, we find that BE is approximately proportional to the number of nucleons A in any nucleus. About twice as much energy is needed to pull apart a nucleus like ^{24}Mg compared with pulling apart ^{12}C, for example. To help us look at other effects, we divide BE by A and consider the **binding energy per nucleon**, BE/A. The graph of BE/A in **Figure 31.27** reveals some very interesting aspects of nuclei. We see that the binding energy per nucleon averages about 8 MeV, but is lower for both the lightest and heaviest nuclei. This overall trend, in which nuclei with A equal to about 60 have the greatest BE/A and are thus the most tightly bound, is due to the combined characteristics of the attractive nuclear forces and the repulsive Coulomb force. It is especially important to note two things—the strong nuclear force is about 100 times stronger than the Coulomb force, *and* the nuclear forces are shorter in range compared to the Coulomb force. So, for low-mass nuclei, the nuclear attraction dominates and each added nucleon forms bonds with all others, causing progressively heavier nuclei to have progressively greater values of BE/A. This continues up to $A \approx 60$, roughly corresponding to the mass number of iron. Beyond that, new nucleons added to a nucleus will be too far from some others to feel their nuclear attraction. Added protons, however, feel the repulsion of all other protons, since the Coulomb force is longer in range. Coulomb repulsion grows for progressively heavier nuclei, but nuclear attraction remains about the same, and so BE/A becomes smaller. This is why stable nuclei heavier than $A \approx 40$ have more neutrons than protons. Coulomb repulsion is reduced by having more neutrons to keep the protons farther apart (see **Figure 31.28**).

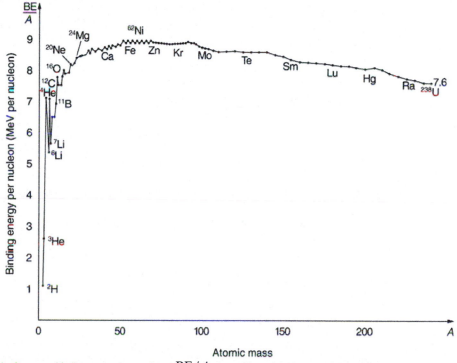

Figure 31.27 A graph of average binding energy per nucleon, BE/A, for stable nuclei. The most tightly bound nuclei are those with A near 60, where the attractive nuclear force has its greatest effect. At higher A s, the Coulomb repulsion progressively reduces the binding energy per nucleon, because the nuclear force is short ranged. The spikes on the curve are very tightly bound nuclides and indicate shell closures.

Nucleons inside range
feel nuclear force directly

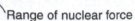

Range of nuclear force

Figure 31.28 The nuclear force is attractive and stronger than the Coulomb force, but it is short ranged. In low-mass nuclei, each nucleon feels the nuclear attraction of all others. In larger nuclei, the range of the nuclear force, shown for a single nucleon, is smaller than the size of the nucleus, but the Coulomb repulsion from all protons reaches all others. If the nucleus is large enough, the Coulomb repulsion can add to overcome the nuclear attraction.

There are some noticeable spikes on the BE/A graph, which represent particularly tightly bound nuclei. These spikes reveal further details of nuclear forces, such as confirming that closed-shell nuclei (those with magic numbers of protons or neutrons or both) are more tightly bound. The spikes also indicate that some nuclei with even numbers for Z and N, and with $Z = N$, are exceptionally tightly bound. This finding can be correlated with some of the cosmic abundances of the elements. The most common elements in the universe, as determined by observations of atomic spectra from outer space, are hydrogen, followed by ^4He, with much smaller amounts of ^{12}C and other elements. It should be noted that the heavier elements are created in supernova explosions, while the lighter ones are produced by nuclear fusion during the normal life cycles of stars, as will be discussed in subsequent chapters. The most common elements have the most tightly bound nuclei. It is also no accident that one of the most tightly bound light nuclei is ^4He, emitted in α decay.

Example 31.7 What Is BE/A for an Alpha Particle?

Calculate the binding energy per nucleon of ^4He, the α particle.

Strategy

To find BE/A, we first find BE using the Equation $BE = \{[Zm(^1\text{H}) + Nm_n] - m(^A\text{X})\}c^2$ and then divide by A. This is straightforward once we have looked up the appropriate atomic masses in Appendix A.

Solution

The binding energy for a nucleus is given by the equation

$$BE = \{[Zm(^1\text{H}) + Nm_n] - m(^A\text{X})\}c^2. \tag{31.63}$$

For ^4He, we have $Z = N = 2$; thus,

$$BE = \{[2m(^1\text{H}) + 2m_n] - m(^4\text{He})\}c^2. \tag{31.64}$$

Appendix A gives these masses as $m(^4\text{He}) = 4.002602$ u, $m(^1\text{H}) = 1.007825$ u, and $m_n = 1.008665$ u. Thus,

$$BE = (0.030378 \text{ u})c^2. \tag{31.65}$$

Noting that $1 \text{ u} = 931.5 \text{ MeV}/c^2$, we find

$$BE = (0.030378)(931.5 \text{ MeV}/c^2)c^2 = 28.3 \text{ MeV}. \tag{31.66}$$

Since $A = 4$, we see that BE/A is this number divided by 4, or

$$BE/A = 7.07 \text{ MeV/nucleon}. \tag{31.67}$$

Discussion

This is a large binding energy per nucleon compared with those for other low-mass nuclei, which have $BE/A \approx 3 \text{ MeV/nucleon}$. This indicates that ^4He is tightly bound compared with its neighbors on the chart of the nuclides. You can see the spike representing this value of BE/A for ^4He on the graph in Figure 31.27. This is why ^4He is stable. Since ^4He is tightly bound, it has less mass than other $A = 4$ nuclei and, therefore, cannot spontaneously decay into them. The large binding energy also helps to explain why some nuclei undergo α decay. Smaller mass in the decay products can mean energy release, and such decays can be spontaneous. Further, it can happen that two protons and two neutrons in a nucleus can randomly find themselves together, experience the exceptionally large nuclear force that binds this combination, and act as a ^4He unit within the nucleus, at least for a while. In some cases, the ^4He escapes, and α decay has then taken place.

There is more to be learned from nuclear binding energies. The general trend in BE/A is fundamental to energy production in stars, and to fusion and fission energy sources on Earth, for example. This is one of the applications of nuclear physics covered in **Medical Applications of Nuclear Physics**. The abundance of elements on Earth, in stars, and in the universe as a whole is related to the binding energy of nuclei and has implications for the continued expansion of the universe.

Problem-Solving Strategies

For Reaction And Binding Energies and Activity Calculations in Nuclear Physics

1. *Identify exactly what needs to be determined in the problem (identify the unknowns).* This will allow you to decide whether the energy of a decay or nuclear reaction is involved, for example, or whether the problem is primarily concerned with activity (rate of decay).

2. *Make a list of what is given or can be inferred from the problem as stated (identify the knowns).*

3. *For reaction and binding-energy problems, we use atomic rather than nuclear masses.* Since the masses of neutral atoms are used, you must count the number of electrons involved. If these do not balance (such as in β^+ decay), then an energy adjustment of 0.511 MeV per electron must be made. Also note that atomic masses may not be given in a problem; they can be found in tables.

4. *For problems involving activity, the relationship of activity to half-life, and the number of nuclei given in the equation* $R = \dfrac{0.693N}{t_{1/2}}$ *can be very useful.* Owing to the fact that number of nuclei is involved, you will also need to be familiar with moles and Avogadro's number.

5. *Perform the desired calculation; keep careful track of plus and minus signs as well as powers of 10.*

6. *Check the answer to see if it is reasonable: Does it make sense?* Compare your results with worked examples and other information in the text. (Heeding the advice in Step 5 will also help you to be certain of your result.) You must understand the problem conceptually to be able to determine whether the numerical result is reasonable.

PhET Explorations: Nuclear Fission

Start a chain reaction, or introduce non-radioactive isotopes to prevent one. Control energy production in a nuclear reactor!

Figure 31.29 Nuclear Fission (http://cnx.org/content/m54933/1.3/nuclear-fission_en.jar)

31.7 Tunneling

Learning Objectives
By the end of this section, you will be able to: • Define and discuss tunneling. • Define potential barrier. • Explain quantum tunneling.

Protons and neutrons are *bound* inside nuclei, that means energy must be supplied to break them away. The situation is analogous to a marble in a bowl that can roll around but lacks the energy to get over the rim. It is bound inside the bowl (see **Figure 31.30**). If the marble could get over the rim, it would gain kinetic energy by rolling down outside. However classically, if the marble does not have enough kinetic energy to get over the rim, it remains forever trapped in its well.

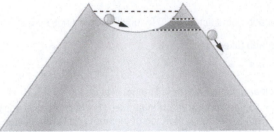

Figure 31.30 The marble in this semicircular bowl at the top of a volcano has enough kinetic energy to get to the altitude of the dashed line, but not enough to get over the rim, so that it is trapped forever. If it could find a tunnel through the barrier, it would escape, roll downhill, and gain kinetic energy.

In a nucleus, the attractive nuclear potential is analogous to the bowl at the top of a volcano (where the "volcano" refers only to the shape). Protons and neutrons have kinetic energy, but it is about 8 MeV less than that needed to get out (see Figure 31.31). That is, they are bound by an average of 8 MeV per nucleon. The slope of the hill outside the bowl is analogous to the repulsive Coulomb potential for a nucleus, such as for an α particle outside a positive nucleus. In α decay, two protons and two neutrons spontaneously break away as a ^4He unit. Yet the protons and neutrons do not have enough kinetic energy to get over the rim. So how does the α particle get out?

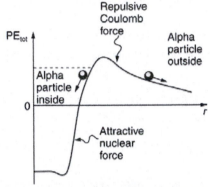

Figure 31.31 Nucleons within an atomic nucleus are bound or trapped by the attractive nuclear force, as shown in this simplified potential energy curve. An α particle outside the range of the nuclear force feels the repulsive Coulomb force. The α particle inside the nucleus does not have enough kinetic energy to get over the rim, yet it does manage to get out by quantum mechanical tunneling.

The answer was supplied in 1928 by the Russian physicist George Gamow (1904–1968). The α particle tunnels through a region of space it is forbidden to be in, and it comes out of the side of the nucleus. Like an electron making a transition between orbits around an atom, it travels from one point to another without ever having been in between. Figure 31.32 indicates how this works. The wave function of a quantum mechanical particle varies smoothly, going from within an atomic nucleus (on one side of a potential energy barrier) to outside the nucleus (on the other side of the potential energy barrier). Inside the barrier, the wave function does not become zero but decreases exponentially, and we do not observe the particle inside the barrier. The probability of finding a particle is related to the square of its wave function, and so there is a small probability of finding the particle outside the barrier, which implies that the particle can tunnel through the barrier. This process is called **barrier penetration** or **quantum mechanical tunneling**. This concept was developed in theory by J. Robert Oppenheimer (who led the development of the first nuclear bombs during World War II) and was used by Gamow and others to describe α decay.

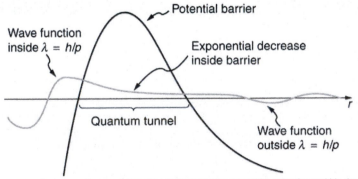

Figure 31.32 The wave function representing a quantum mechanical particle must vary smoothly, going from within the nucleus (to the left of the barrier) to outside the nucleus (to the right of the barrier). Inside the barrier, the wave function does not abruptly become zero; rather, it decreases exponentially. Outside the barrier, the wave function is small but finite, and there it smoothly becomes sinusoidal. Owing to the fact that there is a small probability of finding the particle outside the barrier, the particle can tunnel through the barrier.

Good ideas explain more than one thing. In addition to qualitatively explaining how the four nucleons in an α particle can get out of the nucleus, the detailed theory also explains quantitatively the half-life of various nuclei that undergo α decay. This description is what Gamow and others devised, and it works for α decay half-lives that vary by 17 orders of magnitude. Experiments have shown that the more energetic the α decay of a particular nuclide is, the shorter is its half-life. **Tunneling** explains this in the following manner: For the decay to be more energetic, the nucleons must have more energy in the nucleus and should be able to ascend a little closer to the rim. The barrier is therefore not as thick for more energetic decay, and the exponential decrease of the wave function inside the barrier is not as great. Thus the probability of finding the particle outside the barrier is greater, and the half-life is shorter.

Tunneling as an effect also occurs in quantum mechanical systems other than nuclei. Electrons trapped in solids can tunnel from one object to another if the barrier between the objects is thin enough. The process is the same in principle as described for α decay. It is far more likely for a thin barrier than a thick one. Scanning tunneling electron microscopes function on this principle. The current of electrons that travels between a probe and a sample tunnels through a barrier and is very sensitive to its thickness, allowing detection of individual atoms as shown in **Figure 31.33**.

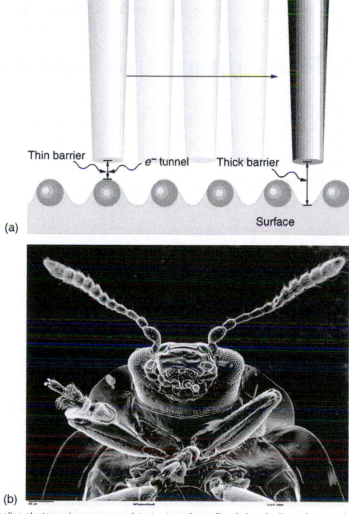

Figure 31.33 (a) A scanning tunneling electron microscope can detect extremely small variations in dimensions, such as individual atoms. Electrons tunnel quantum mechanically between the probe and the sample. The probability of tunneling is extremely sensitive to barrier thickness, so that the electron current is a sensitive indicator of surface features. (b) Head and mouthparts of *Coleoptera Chrysomelidea* as seen through an electron microscope (credit: Louisa Howard, Dartmouth College)

Making Connections: Real World Connections

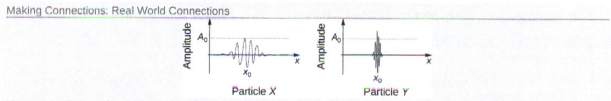

Figure 31.34 The wave function for particle X has a lower amplitude and a broader spatial distribution compared to particle Y, indicating a greater uncertainty in the position of particle X. The amplitude of the wave function is a measure of the probability of finding the particle at a precise location in x.

Recall the discussion of wave-particle duality and the uncertainty principle in **Sections 29.6** and **29.7**. At the quantum level, particles such as electrons and alpha particles can be represented by a wave function. The wave function represents the probability of finding the particle at a given precise location. Because the location of an alpha particle is not certain, at any given time, there is a small chance that it will be located far away from its original location inside the nucleus, even at a distance that would place it outside the nucleus. There is a small probability each second of the alpha particle being found at this new position, effectively resulting in alpha decay.

The greater this probability, the less time it takes to happen and thus, the shorter the half-life of the decay sequence. Particle X above has a much greater uncertainty in its position, and if it represents an alpha particle inside a particular nucleus, it would take this alpha particle very little time to tunnel out of the nucleus in this way compared to particle Y, which has a very small positional uncertainty and is not likely to be found as far from its expected location in the nucleus.

PhET Explorations: Quantum Tunneling and Wave Packets

Watch quantum "particles" tunnel through barriers. Explore the properties of the wave functions that describe these particles.

Figure 31.35 Quantum Tunneling and Wave Packets (http://cnx.org/content/m54934/1.2/quantum-tunneling_en.jar)

Glossary

activity: the rate of decay for radioactive nuclides

alpha decay: type of radioactive decay in which an atomic nucleus emits an alpha particle

alpha rays: one of the types of rays emitted from the nucleus of an atom

antielectron: another term for positron

antimatter: composed of antiparticles

atomic mass: the total mass of the protons, neutrons, and electrons in a single atom

atomic number: number of protons in a nucleus

barrier penetration: quantum mechanical effect whereby a particle has a nonzero probability to cross through a potential energy barrier despite not having sufficient energy to pass over the barrier; also called quantum mechanical tunneling

becquerel: SI unit for rate of decay of a radioactive material

beta decay: type of radioactive decay in which an atomic nucleus emits a beta particle

beta rays: one of the types of rays emitted from the nucleus of an atom

binding energy: the energy needed to separate nucleus into individual protons and neutrons

binding energy per nucleon: the binding energy calculated per nucleon; it reveals the details of the nuclear force—larger the BE/A, the more stable the nucleus

carbon-14 dating: a radioactive dating technique based on the radioactivity of carbon-14

chart of the nuclides: a table comprising stable and unstable nuclei

curie: the activity of 1g of ^{226}Ra, equal to 3.70×10^{10} Bq

daughter: the nucleus obtained when parent nucleus decays and produces another nucleus following the rules and the conservation laws

decay: the process by which an atomic nucleus of an unstable atom loses mass and energy by emitting ionizing particles

decay constant: quantity that is inversely proportional to the half-life and that is used in equation for number of nuclei as a function of time

decay equation: the equation to find out how much of a radioactive material is left after a given period of time

decay series: process whereby subsequent nuclides decay until a stable nuclide is produced

electron capture: the process in which a proton-rich nuclide absorbs an inner atomic electron and simultaneously emits a neutrino

electron capture equation: equation representing the electron capture

electron's antineutrino: antiparticle of electron's neutrino

electron's neutrino: a subatomic elementary particle which has no net electric charge

gamma decay: type of radioactive decay in which an atomic nucleus emits a gamma particle

gamma rays: one of the types of rays emitted from the nucleus of an atom

Geiger tube: a very common radiation detector that usually gives an audio output

half-life: the time in which there is a 50% chance that a nucleus will decay

ionizing radiation: radiation (whether nuclear in origin or not) that produces ionization whether nuclear in origin or not

isotopes: nuclei having the same Z and different N s

magic numbers: a number that indicates a shell structure for the nucleus in which closed shells are more stable

mass number: number of nucleons in a nucleus

neutrino: an electrically neutral, weakly interacting elementary subatomic particle

neutron: a neutral particle that is found in a nucleus

nuclear radiation: rays that originate in the nuclei of atoms, the first examples of which were discovered by Becquerel

nuclear reaction energy: the energy created in a nuclear reaction

nucleons: the particles found inside nuclei

nucleus: a region consisting of protons and neutrons at the center of an atom

nuclide: a type of atom whose nucleus has specific numbers of protons and neutrons

parent: the original state of nucleus before decay

photomultiplier: a device that converts light into electrical signals

positron: the particle that results from positive beta decay; also known as an antielectron

positron decay: type of beta decay in which a proton is converted to a neutron, releasing a positron and a neutrino

protons: the positively charged nucleons found in a nucleus

quantum mechanical tunneling: quantum mechanical effect whereby a particle has a nonzero probability to cross through a potential energy barrier despite not having sufficient energy to pass over the barrier; also called barrier penetration

radiation detector: a device that is used to detect and track the radiation from a radioactive reaction

radioactive: a substance or object that emits nuclear radiation

radioactive dating: an application of radioactive decay in which the age of a material is determined by the amount of radioactivity of a particular type that occurs

radioactivity: the emission of rays from the nuclei of atoms

radius of a nucleus: the radius of a nucleus is $r = r_0 A^{1/3}$

range of radiation: the distance that the radiation can travel through a material

rate of decay: the number of radioactive events per unit time

scintillators: a radiation detection method that records light produced when radiation interacts with materials

solid-state radiation detectors: semiconductors fabricated to directly convert incident radiation into electrical current

tunneling: a quantum mechanical process of potential energy barrier penetration

Section Summary

31.1 Nuclear Radioactivity
- Some nuclei are radioactive—they spontaneously decay destroying some part of their mass and emitting energetic rays, a process called nuclear radioactivity.
- Nuclear radiation, like x rays, is ionizing radiation, because energy sufficient to ionize matter is emitted in each decay.
- The range (or distance traveled in a material) of ionizing radiation is directly related to the charge of the emitted particle and its energy, with greater-charge and lower-energy particles having the shortest ranges.
- Radiation detectors are based directly or indirectly upon the ionization created by radiation, as are the effects of radiation on living and inert materials.

31.2 Radiation Detection and Detectors
- Radiation detectors are based directly or indirectly upon the ionization created by radiation, as are the effects of radiation on living and inert materials.

31.3 Substructure of the Nucleus
- Two particles, both called nucleons, are found inside nuclei. The two types of nucleons are protons and neutrons; they are very similar, except that the proton is positively charged while the neutron is neutral. Some of their characteristics are given in Table 31.2 and compared with those of the electron. A mass unit convenient to atomic and nuclear processes is the unified atomic mass unit (u), defined to be

$$1 \text{ u} = 1.6605 \times 10^{-27} \text{ kg} = 931.46 \text{ MeV}/c^2.$$

- A nuclide is a specific combination of protons and neutrons, denoted by

$$^A_Z X_N \text{ or simply}^A X,$$

Z is the number of protons or atomic number, X is the symbol for the element, N is the number of neutrons, and A is the mass number or the total number of protons and neutrons,

$$A = N + Z.$$

- Nuclides having the same Z but different N are isotopes of the same element.
- The radius of a nucleus, r, is approximately

$$r = r_0 A^{1/3},$$

where $r_0 = 1.2 \text{ fm}$. Nuclear volumes are proportional to A. There are two nuclear forces, the weak and the strong.

Systematics in nuclear stability seen on the chart of the nuclides indicate that there are shell closures in nuclei for values of Z and N equal to the magic numbers, which correspond to highly stable nuclei.

31.4 Nuclear Decay and Conservation Laws
- When a parent nucleus decays, it produces a daughter nucleus following rules and conservation laws. There are three major types of nuclear decay, called alpha (α), beta (β), and gamma (γ). The α decay equation is

$$^A_Z X_N \rightarrow \, ^{A-4}_{Z-2} Y_{N-2} + \, ^4_2 He_2.$$

- Nuclear decay releases an amount of energy E related to the mass destroyed Δm by

$$E = (\Delta m)c^2.$$

- There are three forms of beta decay. The β^- decay equation is

$$^A_Z X_N \rightarrow \, ^A_{Z+1} Y_{N-1} + \beta^- + \bar{\nu}_e.$$

- The β^+ decay equation is

$$^A_Z X_N \rightarrow \, ^A_{Z-1} Y_{N+1} + \beta^+ + \nu_e.$$

- The electron capture equation is

$$^A_Z X_N + e^- \rightarrow \, ^A_{Z-1} Y_{N+1} + \nu_e.$$

- β^- is an electron, β^+ is an antielectron or positron, ν_e represents an electron's neutrino, and $\bar{\nu}_e$ is an electron's antineutrino. In addition to all previously known conservation laws, two new ones arise— conservation of electron family number and conservation of the total number of nucleons. The γ decay equation is

$$X_N^* \rightarrow X_N + \gamma_1 + \gamma_2 + \cdots$$

γ is a high-energy photon originating in a nucleus.

31.5 Half-Life and Activity

- Half-life $t_{1/2}$ is the time in which there is a 50% chance that a nucleus will decay. The number of nuclei N as a function of time is

$$N = N_0 e^{-\lambda t},$$

where N_0 is the number present at $t = 0$, and λ is the decay constant, related to the half-life by

$$\lambda = \frac{0.693}{t_{1/2}}.$$

- One of the applications of radioactive decay is radioactive dating, in which the age of a material is determined by the amount of radioactive decay that occurs. The rate of decay is called the activity R:

$$R = \frac{\Delta N}{\Delta t}.$$

- The SI unit for R is the becquerel (Bq), defined by

$$1 \text{ Bq} = 1 \text{ decay/s}.$$

- R is also expressed in terms of curies (Ci), where

$$1 \text{ Ci} = 3.70 \times 10^{10} \text{ Bq}.$$

- The activity R of a source is related to N and $t_{1/2}$ by

$$R = \frac{0.693 N}{t_{1/2}}.$$

- Since N has an exponential behavior as in the equation $N = N_0 e^{-\lambda t}$, the activity also has an exponential behavior, given by

$$R = R_0 e^{-\lambda t},$$

where R_0 is the activity at $t = 0$.

31.6 Binding Energy

- The binding energy (BE) of a nucleus is the energy needed to separate it into individual protons and neutrons. In terms of atomic masses,

$$\text{BE} = \{[Zm(^1\text{H}) + Nm_n] - m(^A\text{X})\}c^2,$$

where $m(^1\text{H})$ is the mass of a hydrogen atom, $m(^A\text{X})$ is the atomic mass of the nuclide, and m_n is the mass of a neutron. Patterns in the binding energy per nucleon, BE/A, reveal details of the nuclear force. The larger the BE/A, the more stable the nucleus.

31.7 Tunneling

- Tunneling is a quantum mechanical process of potential energy barrier penetration. The concept was first applied to explain α decay, but tunneling is found to occur in other quantum mechanical systems.

Conceptual Questions

31.1 Nuclear Radioactivity

1. Suppose the range for $5.0 \text{ MeV} \alpha$ ray is known to be 2.0 mm in a certain material. Does this mean that every $5.0 \text{ MeV} \alpha$ a ray that strikes this material travels 2.0 mm, or does the range have an average value with some statistical fluctuations in the distances traveled? Explain.

2. What is the difference between γ rays and characteristic x rays? Is either necessarily more energetic than the other? Which can be the most energetic?

3. Ionizing radiation interacts with matter by scattering from electrons and nuclei in the substance. Based on the law of conservation of momentum and energy, explain why electrons tend to absorb more energy than nuclei in these interactions.

4. What characteristics of radioactivity show it to be nuclear in origin and not atomic?

5. What is the source of the energy emitted in radioactive decay? Identify an earlier conservation law, and describe how it was modified to take such processes into account.

6. Consider Figure 31.3. If an electric field is substituted for the magnetic field with positive charge instead of the north pole and negative charge instead of the south pole, in which directions will the α, β, and γ rays bend?

7. Explain how an α particle can have a larger range in air than a β particle with the same energy in lead.

8. Arrange the following according to their ability to act as radiation shields, with the best first and worst last. Explain your ordering in terms of how radiation loses its energy in matter.

(a) A solid material with low density composed of low-mass atoms.

(b) A gas composed of high-mass atoms.

(c) A gas composed of low-mass atoms.

(d) A solid with high density composed of high-mass atoms.

9. Often, when people have to work around radioactive materials spills, we see them wearing white coveralls (usually a plastic material). What types of radiation (if any) do you think these suits protect the worker from, and how?

31.2 Radiation Detection and Detectors

10. Is it possible for light emitted by a scintillator to be too low in frequency to be used in a photomultiplier tube? Explain.

31.3 Substructure of the Nucleus

11. The weak and strong nuclear forces are basic to the structure of matter. Why we do not experience them directly?

12. Define and make clear distinctions between the terms neutron, nucleon, nucleus, nuclide, and neutrino.

13. What are isotopes? Why do different isotopes of the same element have similar chemistries?

31.4 Nuclear Decay and Conservation Laws

14. Star Trek fans have often heard the term "antimatter drive." Describe how you could use a magnetic field to trap antimatter, such as produced by nuclear decay, and later combine it with matter to produce energy. Be specific about the type of antimatter, the need for vacuum storage, and the fraction of matter converted into energy.

15. What conservation law requires an electron's neutrino to be produced in electron capture? Note that the electron no longer exists after it is captured by the nucleus.

16. Neutrinos are experimentally determined to have an extremely small mass. Huge numbers of neutrinos are created in a supernova at the same time as massive amounts of light are first produced. When the 1987A supernova occurred in the Large Magellanic Cloud, visible primarily in the Southern Hemisphere and some 100,000 light-years away from Earth, neutrinos from the explosion were observed at about the same time as the light from the blast. How could the relative arrival times of neutrinos and light be used to place limits on the mass of neutrinos?

17. What do the three types of beta decay have in common that is distinctly different from alpha decay?

31.5 Half-Life and Activity

18. In a 3×10^9-year-old rock that originally contained some ^{238}U, which has a half-life of 4.5×10^9 years, we expect to find some ^{238}U remaining in it. Why are ^{226}Ra, ^{222}Rn, and ^{210}Po also found in such a rock, even though they have much shorter half-lives (1600 years, 3.8 days, and 138 days, respectively)?

19. Does the number of radioactive nuclei in a sample decrease to *exactly* half its original value in one half-life? Explain in terms of the statistical nature of radioactive decay.

20. Radioactivity depends on the nucleus and not the atom or its chemical state. Why, then, is one kilogram of uranium more radioactive than one kilogram of uranium hexafluoride?

21. Explain how a bound system can have less mass than its components. Why is this not observed classically, say for a building made of bricks?

22. Spontaneous radioactive decay occurs only when the decay products have less mass than the parent, and it tends to produce a daughter that is more stable than the parent. Explain how this is related to the fact that more tightly bound nuclei are more stable. (Consider the binding energy per nucleon.)

23. To obtain the most precise value of BE from the equation $\text{BE} = \left[ZM(^1\text{H}) + Nm_n \right]c^2 - m(^A\text{X})c^2$, we should take into account the binding energy of the electrons in the neutral atoms. Will doing this produce a larger or smaller value for BE? Why is this effect usually negligible?

24. How does the finite range of the nuclear force relate to the fact that BE/A is greatest for A near 60?

31.6 Binding Energy

25. Why is the number of neutrons greater than the number of protons in stable nuclei having A greater than about 40, and why is this effect more pronounced for the heaviest nuclei?

31.7 Tunneling

26. A physics student caught breaking conservation laws is imprisoned. She leans against the cell wall hoping to tunnel out quantum mechanically. Explain why her chances are negligible. (This is so in any classical situation.)

27. When a nucleus α decays, does the α particle move continuously from inside the nucleus to outside? That is, does it travel each point along an imaginary line from inside to out? Explain.

Problems & Exercises

31.2 Radiation Detection and Detectors

1. The energy of 30.0 eV is required to ionize a molecule of the gas inside a Geiger tube, thereby producing an ion pair. Suppose a particle of ionizing radiation deposits 0.500 MeV of energy in this Geiger tube. What maximum number of ion pairs can it create?

2. A particle of ionizing radiation creates 4000 ion pairs in the gas inside a Geiger tube as it passes through. What minimum energy was deposited, if 30.0 eV is required to create each ion pair?

3. (a) Repeat Exercise 31.2, and convert the energy to joules or calories. (b) If all of this energy is converted to thermal energy in the gas, what is its temperature increase, assuming 50.0 cm^3 of ideal gas at 0.250-atm pressure? (The small answer is consistent with the fact that the energy is large on a quantum mechanical scale but small on a macroscopic scale.)

4. Suppose a particle of ionizing radiation deposits 1.0 MeV in the gas of a Geiger tube, all of which goes to creating ion pairs. Each ion pair requires 30.0 eV of energy. (a) The applied voltage sweeps the ions out of the gas in $1.00 \text{ } \mu s$.

What is the current? (b) This current is smaller than the actual current since the applied voltage in the Geiger tube accelerates the separated ions, which then create other ion pairs in subsequent collisions. What is the current if this last effect multiplies the number of ion pairs by 900?

31.3 Substructure of the Nucleus

5. Verify that a $2.3 \times 10^{17} \text{ kg}$ mass of water at normal density would make a cube 60 km on a side, as claimed in Example 31.1. (This mass at nuclear density would make a cube 1.0 m on a side.)

6. Find the length of a side of a cube having a mass of 1.0 kg and the density of nuclear matter, taking this to be $2.3 \times 10^{17} \text{ kg/m}^3$.

7. What is the radius of an α particle?

8. Find the radius of a ^{238}Pu nucleus. ^{238}Pu is a manufactured nuclide that is used as a power source on some space probes.

9. (a) Calculate the radius of ^{58}Ni, one of the most tightly bound stable nuclei.

(b) What is the ratio of the radius of ^{58}Ni to that of ^{258}Ha, one of the largest nuclei ever made? Note that the radius of the largest nucleus is still much smaller than the size of an atom.

10. The unified atomic mass unit is defined to be $1 \text{ u} = 1.6605 \times 10^{-27} \text{ kg}$. Verify that this amount of mass converted to energy yields 931.5 MeV. Note that you must use four-digit or better values for c and $|q_e|$.

11. What is the ratio of the velocity of a β particle to that of an α particle, if they have the same nonrelativistic kinetic energy?

12. If a 1.50-cm-thick piece of lead can absorb 90.0% of the γ rays from a radioactive source, how many centimeters of lead are needed to absorb all but 0.100% of the γ rays?

13. The detail observable using a probe is limited by its wavelength. Calculate the energy of a γ-ray photon that has a wavelength of $1 \times 10^{-16} \text{ m}$, small enough to detect details about one-tenth the size of a nucleon. Note that a photon having this energy is difficult to produce and interacts poorly with the nucleus, limiting the practicability of this probe.

14. (a) Show that if you assume the average nucleus is spherical with a radius $r = r_0 A^{1/3}$, and with a mass of A u, then its density is independent of A.

(b) Calculate that density in u/fm^3 and kg/m^3, and compare your results with those found in Example 31.1 for ^{56}Fe.

15. What is the ratio of the velocity of a 5.00-MeV β ray to that of an α particle with the same kinetic energy? This should confirm that β s travel much faster than α s even when relativity is taken into consideration. (See also Exercise 31.11.)

16. (a) What is the kinetic energy in MeV of a β ray that is traveling at $0.998c$? This gives some idea of how energetic a β ray must be to travel at nearly the same speed as a γ ray. (b) What is the velocity of the γ ray relative to the β ray?

31.4 Nuclear Decay and Conservation Laws

In the following eight problems, write the complete decay equation for the given nuclide in the complete $^A_Z X_N$ notation. Refer to the periodic table for values of Z.

17. β^- decay of ^3H (tritium), a manufactured isotope of hydrogen used in some digital watch displays, and manufactured primarily for use in hydrogen bombs.

18. β^- decay of ^{40}K, a naturally occurring rare isotope of potassium responsible for some of our exposure to background radiation.

19. β^+ decay of ^{50}Mn.

20. β^+ decay of ^{52}Fe.

21. Electron capture by ^7Be.

22. Electron capture by ^{106}In.

23. α decay of ^{210}Po, the isotope of polonium in the decay series of ^{238}U that was discovered by the Curies. A favorite isotope in physics labs, since it has a short half-life and decays to a stable nuclide.

24. α decay of ^{226}Ra , another isotope in the decay series of ^{238}U , first recognized as a new element by the Curies. Poses special problems because its daughter is a radioactive noble gas.

In the following four problems, identify the parent nuclide and write the complete decay equation in the $^A_Z X_N$ notation. Refer to the periodic table for values of Z .

25. β^- decay producing ^{137}Ba . The parent nuclide is a major waste product of reactors and has chemistry similar to potassium and sodium, resulting in its concentration in your cells if ingested.

26. β^- decay producing ^{90}Y . The parent nuclide is a major waste product of reactors and has chemistry similar to calcium, so that it is concentrated in bones if ingested (^{90}Y is also radioactive.)

27. α decay producing ^{228}Ra . The parent nuclide is nearly 100% of the natural element and is found in gas lantern mantles and in metal alloys used in jets (^{228}Ra is also radioactive).

28. α decay producing ^{208}Pb . The parent nuclide is in the decay series produced by ^{232}Th , the only naturally occurring isotope of thorium.

29. When an electron and positron annihilate, both their masses are destroyed, creating two equal energy photons to preserve momentum. (a) Confirm that the annihilation equation $e^+ + e^- \rightarrow \gamma + \gamma$ conserves charge, electron family number, and total number of nucleons. To do this, identify the values of each before and after the annihilation. (b) Find the energy of each γ ray, assuming the electron and positron are initially nearly at rest. (c) Explain why the two γ rays travel in exactly opposite directions if the center of mass of the electron-positron system is initially at rest.

30. Confirm that charge, electron family number, and the total number of nucleons are all conserved by the rule for α decay given in the equation $^A_Z X_N \rightarrow ^{A-4}_{Z-2} Y_{N-2} + ^4_2 He_2$. To do this, identify the values of each before and after the decay.

31. Confirm that charge, electron family number, and the total number of nucleons are all conserved by the rule for β^- decay given in the equation $^A_Z X_N \rightarrow ^A_{Z+1} Y_{N-1} + \beta^- + \bar{\nu}_e$. To do this, identify the values of each before and after the decay.

32. Confirm that charge, electron family number, and the total number of nucleons are all conserved by the rule for β^- decay given in the equation $^A_Z X_N \rightarrow ^A_{Z-1} Y_{N-1} + \beta^- + \nu_e$. To do this, identify the values of each before and after the decay.

33. Confirm that charge, electron family number, and the total number of nucleons are all conserved by the rule for electron capture given in the equation $^A_Z X_N + e^- \rightarrow ^A_{Z-1} Y_{N+1} + \nu_e$. To do this, identify the values of each before and after the capture.

34. A rare decay mode has been observed in which ^{222}Ra emits a ^{14}C nucleus. (a) The decay equation is ^{222}Ra $\rightarrow ^A$X $+ ^{14}$C . Identify the nuclide ^{A}X . (b) Find the energy emitted in the decay. The mass of ^{222}Ra is 222.015353 u.

35. (a) Write the complete α decay equation for ^{226}Ra .

(b) Find the energy released in the decay.

36. (a) Write the complete α decay equation for ^{249}Cf .

(b) Find the energy released in the decay.

37. (a) Write the complete β^- decay equation for the neutron. (b) Find the energy released in the decay.

38. (a) Write the complete β^- decay equation for ^{90}Sr , a major waste product of nuclear reactors. (b) Find the energy released in the decay.

39. Calculate the energy released in the β^+ decay of ^{22}Na , the equation for which is given in the text. The masses of ^{22}Na and ^{22}Ne are 21.994434 and 21.991383 u, respectively.

40. (a) Write the complete β^+ decay equation for ^{11}C .

(b) Calculate the energy released in the decay. The masses of ^{11}C and ^{11}B are 11.011433 and 11.009305 u, respectively.

41. (a) Calculate the energy released in the α decay of ^{238}U .

(b) What fraction of the mass of a single ^{238}U is destroyed in the decay? The mass of ^{234}Th is 234.043593 u.

(c) Although the fractional mass loss is large for a single nucleus, it is difficult to observe for an entire macroscopic sample of uranium. Why is this?

42. (a) Write the complete reaction equation for electron capture by ^{7}Be.

(b) Calculate the energy released.

43. (a) Write the complete reaction equation for electron capture by ^{15}O .

(b) Calculate the energy released.

31.5 Half-Life and Activity

Data from the appendices and the periodic table may be needed for these problems.

44. An old campfire is uncovered during an archaeological dig. Its charcoal is found to contain less than 1/1000 the normal amount of ^{14}C. Estimate the minimum age of the charcoal, noting that $2^{10} = 1024$.

45. A ^{60}Co source is labeled 4.00 mCi, but its present activity is found to be 1.85×10^7 Bq. (a) What is the present activity in mCi? (b) How long ago did it actually have a 4.00-mCi activity?

46. (a) Calculate the activity R in curies of 1.00 g of ^{226}Ra. (b) Discuss why your answer is not exactly 1.00 Ci, given that the curie was originally supposed to be exactly the activity of a gram of radium.

47. Show that the activity of the ^{14}C in 1.00 g of ^{12}C found in living tissue is 0.250 Bq.

48. Mantles for gas lanterns contain thorium, because it forms an oxide that can survive being heated to incandescence for long periods of time. Natural thorium is almost 100% ^{232}Th, with a half-life of 1.405×10^{10} y. If an average lantern mantle contains 300 mg of thorium, what is its activity?

49. Cow's milk produced near nuclear reactors can be tested for as little as 1.00 pCi of ^{131}I per liter, to check for possible reactor leakage. What mass of ^{131}I has this activity?

50. (a) Natural potassium contains ^{40}K, which has a half-life of 1.277×10^9 y. What mass of ^{40}K in a person would have a decay rate of 4140 Bq? (b) What is the fraction of ^{40}K in natural potassium, given that the person has 140 g in his body? (These numbers are typical for a 70-kg adult.)

51. There is more than one isotope of natural uranium. If a researcher isolates 1.00 mg of the relatively scarce ^{235}U and finds this mass to have an activity of 80.0 Bq, what is its half-life in years?

52. ^{50}V has one of the longest known radioactive half-lives. In a difficult experiment, a researcher found that the activity of 1.00 kg of ^{50}V is 1.75 Bq. What is the half-life in years?

53. You can sometimes find deep red crystal vases in antique stores, called uranium glass because their color was produced by doping the glass with uranium. Look up the natural isotopes of uranium and their half-lives, and calculate the activity of such a vase assuming it has 2.00 g of uranium in it. Neglect the activity of any daughter nuclides.

54. A tree falls in a forest. How many years must pass before the ^{14}C activity in 1.00 g of the tree's carbon drops to 1.00 decay per hour?

55. What fraction of the ^{40}K that was on Earth when it formed 4.5×10^9 years ago is left today?

56. A 5000-Ci ^{60}Co source used for cancer therapy is considered too weak to be useful when its activity falls to 3500 Ci. How long after its manufacture does this happen?

57. Natural uranium is 0.7200% ^{235}U and 99.27% ^{238}U. What were the percentages of ^{235}U and ^{238}U in natural uranium when Earth formed 4.5×10^9 years ago?

58. The β^- particles emitted in the decay of ^3H (tritium) interact with matter to create light in a glow-in-the-dark exit sign. At the time of manufacture, such a sign contains 15.0 Ci of ^3H. (a) What is the mass of the tritium? (b) What is its activity 5.00 y after manufacture?

59. World War II aircraft had instruments with glowing radium-painted dials (see **Figure 31.2**). The activity of one such instrument was 1.0×10^5 Bq when new. (a) What mass of ^{226}Ra was present? (b) After some years, the phosphors on the dials deteriorated chemically, but the radium did not escape. What is the activity of this instrument 57.0 years after it was made?

60. (a) The ^{210}Po source used in a physics laboratory is labeled as having an activity of $1.0\ \mu Ci$ on the date it was prepared. A student measures the radioactivity of this source with a Geiger counter and observes 1500 counts per minute. She notices that the source was prepared 120 days before her lab. What fraction of the decays is she observing with her apparatus? (b) Identify some of the reasons that only a fraction of the α s emitted are observed by the detector.

61. Armor-piercing shells with depleted uranium cores are fired by aircraft at tanks. (The high density of the uranium makes them effective.) The uranium is called depleted because it has had its ^{235}U removed for reactor use and is nearly pure ^{238}U. Depleted uranium has been erroneously called non-radioactive. To demonstrate that this is wrong: (a) Calculate the activity of 60.0 g of pure ^{238}U. (b) Calculate the activity of 60.0 g of natural uranium, neglecting the ^{234}U and all daughter nuclides.

62. The ceramic glaze on a red-orange Fiestaware plate is U_2O_3 and contains 50.0 grams of ^{238}U, but very little ^{235}U. (a) What is the activity of the plate? (b) Calculate the total energy that will be released by the ^{238}U decay. (c) If energy is worth 12.0 cents per $kW \cdot h$, what is the monetary value of the energy emitted? (These plates went out of production some 30 years ago, but are still available as collectibles.)

63. Large amounts of depleted uranium (^{238}U) are available as a by-product of uranium processing for reactor fuel and weapons. Uranium is very dense and makes good counter weights for aircraft. Suppose you have a 4000-kg block of ^{238}U . (a) Find its activity. (b) How many calories per day are generated by thermalization of the decay energy? (c) Do you think you could detect this as heat? Explain.

64. The *Galileo* space probe was launched on its long journey past several planets in 1989, with an ultimate goal of Jupiter. Its power source is 11.0 kg of ^{238}Pu , a by-product of nuclear weapons plutonium production. Electrical energy is generated thermoelectrically from the heat produced when the 5.59-MeV α particles emitted in each decay crash to a halt inside the plutonium and its shielding. The half-life of ^{238}Pu is 87.7 years. (a) What was the original activity of the ^{238}Pu in becquerel? (b) What power was emitted in kilowatts? (c) What power was emitted 12.0 y after launch? You may neglect any extra energy from daughter nuclides and any losses from escaping γ rays.

65. Construct Your Own Problem

Consider the generation of electricity by a radioactive isotope in a space probe, such as described in **Exercise 31.64.** Construct a problem in which you calculate the mass of a radioactive isotope you need in order to supply power for a long space flight. Among the things to consider are the isotope chosen, its half-life and decay energy, the power needs of the probe and the length of the flight.

66. Unreasonable Results

A nuclear physicist finds $1.0\ \mu g$ of ^{236}U in a piece of uranium ore and assumes it is primordial since its half-life is 2.3×10^7 y . (a) Calculate the amount of ^{236}U that would had to have been on Earth when it formed 4.5×10^9 y ago for $1.0\ \mu g$ to be left today. (b) What is unreasonable about this result? (c) What assumption is responsible?

67. Unreasonable Results

(a) Repeat **Exercise 31.57** but include the 0.0055% natural abundance of ^{234}U with its 2.45×10^5 y half-life. (b) What is unreasonable about this result? (c) What assumption is responsible? (d) Where does the ^{234}U come from if it is not primordial?

68. Unreasonable Results

The manufacturer of a smoke alarm decides that the smallest current of α radiation he can detect is $1.00\ \mu A$. (a) Find the activity in curies of an α emitter that produces a $1.00\ \mu A$ current of α particles. (b) What is unreasonable about this result? (c) What assumption is responsible?

31.6 Binding Energy

69. ^{2}H is a loosely bound isotope of hydrogen. Called deuterium or heavy hydrogen, it is stable but relatively rare—it is 0.015% of natural hydrogen. Note that deuterium has $Z = N$, which should tend to make it more tightly bound, but both are odd numbers. Calculate BE/A , the binding energy per nucleon, for ^{2}H and compare it with the approximate value obtained from the graph in **Figure 31.27**.

70. ^{56}Fe is among the most tightly bound of all nuclides. It is more than 90% of natural iron. Note that ^{56}Fe has even numbers of both protons and neutrons. Calculate BE/A , the binding energy per nucleon, for ^{56}Fe and compare it with the approximate value obtained from the graph in **Figure 31.27**.

71. ^{209}Bi is the heaviest stable nuclide, and its BE/A is low compared with medium-mass nuclides. Calculate BE/A , the binding energy per nucleon, for ^{209}Bi and compare it with the approximate value obtained from the graph in **Figure 31.27**.

72. (a) Calculate BE/A for ^{235}U , the rarer of the two most common uranium isotopes. (b) Calculate BE/A for ^{238}U . (Most of uranium is ^{238}U .) Note that ^{238}U has even numbers of both protons and neutrons. Is the BE/A of ^{238}U significantly different from that of ^{235}U ?

73. (a) Calculate BE/A for ^{12}C . Stable and relatively tightly bound, this nuclide is most of natural carbon. (b) Calculate BE/A for ^{14}C . Is the difference in BE/A between ^{12}C and ^{14}C significant? One is stable and common, and the other is unstable and rare.

74. The fact that BE/A is greatest for A near 60 implies that the range of the nuclear force is about the diameter of such nuclides. (a) Calculate the diameter of an $A = 60$ nucleus. (b) Compare BE/A for ^{58}Ni and ^{90}Sr . The first is one of the most tightly bound nuclides, while the second is larger and less tightly bound.

75. The purpose of this problem is to show in three ways that the binding energy of the electron in a hydrogen atom is negligible compared with the masses of the proton and electron. (a) Calculate the mass equivalent in u of the 13.6-eV binding energy of an electron in a hydrogen atom, and compare this with the mass of the hydrogen atom obtained from **Appendix A**. (b) Subtract the mass of the proton given in **Table 31.2** from the mass of the hydrogen atom given in **Appendix A**. You will find the difference is equal to the electron's mass to three digits, implying the binding energy is small in comparison. (c) Take the ratio of the binding energy of the electron (13.6 eV) to the energy equivalent of the electron's mass (0.511 MeV). (d) Discuss how your answers confirm the stated purpose of this problem.

76. Unreasonable Results

A particle physicist discovers a neutral particle with a mass of 2.02733 u that he assumes is two neutrons bound together. (a) Find the binding energy. (b) What is unreasonable about this result? (c) What assumptions are unreasonable or inconsistent?

31.7 Tunneling

77. Derive an approximate relationship between the energy of α decay and half-life using the following data. It may be useful to graph the log of $t_{1/2}$ against E_α to find some straight-line relationship.

Table 31.3 Energy and Half-Life for α Decay

Nuclide	E_α (MeV)	$t_{1/2}$
^{216}Ra	9.5	0.18 μs
^{194}Po	7.0	0.7 s
^{240}Cm	6.4	27 d
^{226}Ra	4.91	1600 y
^{232}Th	4.1	1.4×10^{10} y

78. Integrated Concepts

A 2.00-T magnetic field is applied perpendicular to the path of charged particles in a bubble chamber. What is the radius of curvature of the path of a 10 MeV proton in this field? Neglect any slowing along its path.

79. (a) Write the decay equation for the α decay of ^{235}U.

(b) What energy is released in this decay? The mass of the daughter nuclide is 231.036298 u. (c) Assuming the residual nucleus is formed in its ground state, how much energy goes to the α particle?

80. Unreasonable Results

The relatively scarce naturally occurring calcium isotope ^{48}Ca has a half-life of about 2×10^{16} y . (a) A small sample of this isotope is labeled as having an activity of 1.0 Ci. What is the mass of the ^{48}Ca in the sample? (b) What is unreasonable about this result? (c) What assumption is responsible?

81. Unreasonable Results

A physicist scatters γ rays from a substance and sees evidence of a nucleus 7.5×10^{-13} m in radius. (a) Find the atomic mass of such a nucleus. (b) What is unreasonable about this result? (c) What is unreasonable about the assumption?

82. Unreasonable Results

A frazzled theoretical physicist reckons that all conservation laws are obeyed in the decay of a proton into a neutron, positron, and neutrino (as in β^+ decay of a nucleus) and sends a paper to a journal to announce the reaction as a possible end of the universe due to the spontaneous decay of protons. (a) What energy is released in this decay? (b) What is unreasonable about this result? (c) What assumption is responsible?

83. Construct Your Own Problem

Consider the decay of radioactive substances in the Earth's interior. The energy emitted is converted to thermal energy that reaches the earth's surface and is radiated away into cold dark space. Construct a problem in which you estimate the activity in a cubic meter of earth rock? And then calculate the power generated. Calculate how much power must cross each square meter of the Earth's surface if the power is dissipated at the same rate as it is generated. Among the things to consider are the activity per cubic meter, the energy per decay, and the size of the Earth.

Test Prep for AP® Courses

31.1 Nuclear Radioactivity

1. A nucleus is observed to emit a γ ray with a frequency of 6.3×10^{19} Hz. What must happen to the nucleus as a consequence?

a. The nucleus must gain 0.26 MeV.
b. The nucleus must also emit an α particle of energy 0.26 MeV in the opposite direction.
c. The nucleus must lose 0.26 MeV.
d. The nucleus must also emit a β particle of energy 0.26 MeV in the opposite direction.

2. A uranium nucleus emits an α particle. Assuming charge is conserved, the resulting nucleus must be
a. thorium
b. plutonium
c. radium
d. curium

31.3 Substructure of the Nucleus

3. A typical carbon nucleus contains 6 neutrons and 6 protons. The 6 protons are all positively charged and in very close proximity, with separations on the order of 10^{-15} meters, which should result in an enormous repulsive force. What prevents the nucleus from dismantling itself due to the repulsion of the electric force?
a. The attractive nature of the strong nuclear force overpowers the electric force.
b. The weak nuclear force barely offsets the electric force.
c. Magnetic forces generated by the orbiting electrons create a stable minimum in which the nuclear charged particles reside.
d. The attractive electric force of the surrounding electrons is equal in all directions and cancels out, leaving no net electric force.

31.4 Nuclear Decay and Conservation Laws

4. A nucleus in an excited state undergoes γ decay, losing 1.33 MeV when emitting a γ ray. In order to conserve energy in the reaction, what frequency must the γ ray have?

5. $^{241}_{95}\text{Am}$ is commonly used in smoke detectors because its α decay process provides a useful tool for detecting the presence of smoke particles. When $^{241}_{95}\text{Am}$ undergoes α decay, what is the resulting nucleus? If $^{241}_{95}\text{Am}$ were to undergo β decay, what would be the resulting nucleus? Explain each answer.

6. For β decay, the nucleus releases a negative charge. In order for charge to be conserved overall, the nucleus must gain a positive charge, increasing its atomic number by 1, resulting in $^{241}_{96}\text{Cm}$.

A $^{14}_{6}\text{C}$ nucleus undergoes a decay process, and the resulting nucleus is $^{14}_{7}\text{N}$. What is the value of the charge released by the original nucleus?
a. +1
b. 0
c. -1
d. -2

7. Explain why the overall charge of the nucleus is increased by +1 during the β decay process.

8. Identify the missing particle based upon conservation principles:

$$N + He \rightarrow X + O$$

a. H
b. H
c. C
d. C
e. Be

9. Are the following reactions possible? For each, explain why or why not.
a. $U \rightarrow Ra + He$
b. $Ra \rightarrow Pb + C$
c. $C \rightarrow N + e^- + \bar{v}_e$
d. $Mg \rightarrow Na + e^+ + v_e$

31.5 Half-Life and Activity

10. A radioactive sample has N atoms initially. After 3 half-lives have elapsed, how many atoms remain?
a. N/3
b. N/6
c. N/8
d. N/27

11. When Po decays, the product is Pb. The half-life of this decay process is 1.78 ms. If the initial sample contains 3.4 x 10^{17} parent nuclei, how many are remaining after 35 ms have elapsed? What kind of decay process is this (alpha, beta, or gamma)?

31.6 Binding Energy

12. Binding energy is a measure of how much work must be done against nuclear forces in order to disassemble a nucleus into its constituent parts. For example, the amount of energy in order to disassemble He into 2 protons and 2 neutrons requires 28.3 MeV of work to be done on the nuclear particles. Describe the force that makes it so difficult to pull a nucleus apart. Would it be accurate to say that the electric force plays a role in the forces within a nucleus? Explain why or why not.

32   MEDICAL APPLICATIONS OF NUCLEAR PHYSICS

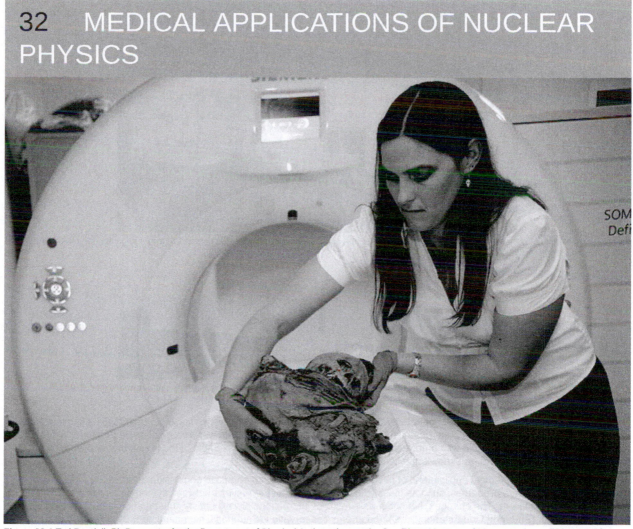

Figure 32.1 Tori Randall, Ph.D., curator for the Department of Physical Anthropology at the San Diego Museum of Man, prepares a 550-year-old Peruvian child mummy for a CT scan at Naval Medical Center San Diego. (credit: U.S. Navy photo by Mass Communication Specialist 3rd Class Samantha A. Lewis)

Chapter Outline

32.1. Medical Imaging and Diagnostics

32.2. Biological Effects of Ionizing Radiation

32.3. Therapeutic Uses of Ionizing Radiation

32.4. Food Irradiation

32.5. Fusion

32.6. Fission

32.7. Nuclear Weapons

Connection for AP® Courses

Applications of nuclear physics have become an integral part of modern life. From a bone scan that detects a cancer to a radioiodine treatment that cures another, nuclear radiation has many diagnostic and therapeutic applications in medicine. In addition nuclear radiation is used in other useful scanning applications, as seen in **Figure 32.2** and **Figure 32.3**. The fission power reactor and the hope of controlled fusion have made nuclear energy a part of our plans for the future. That said, the destructive potential of nuclear weapons haunts us, as does the possibility of nuclear reactor accidents.

Figure 32.2 Customs officers can use gamma ray-, x-ray-, or neutron-scanning devices to reveal the contents of trucks and cars. (credit: Gerald L. Nino, CBP, U.S. Dept. of Homeland Security).

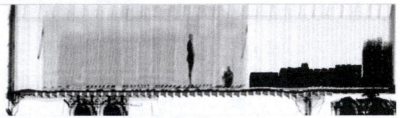

Figure 32.3 This image was obtained using gamma-ray radiography and shows two stowaways caught illegally entering the United States from Canada. (credit: U.S. Customs and Border Protection).

Nuclear physics revealed many secrets of nature, but full exploitation of the technology remains controversial as it is intertwined with human values. Because of its great potential for alleviation of suffering and its power as a giant-scale destroyer of life, nuclear physics is typically viewed with ambivalence. Nuclear physics is a classic example of the truism that applications of technology can be good or evil, but knowledge itself is neither.

This chapter focuses on medical applications of nuclear physics. The sections on fusion and fission address the ideas that objects and systems have properties, such as mass (Big Idea 1), and that interactions between systems can result in changes in those systems (Big Idea 4). The changes that occur as a result of interactions always satisfy conservation laws (Big Idea 5). The mass conservation (Enduring Understanding 1.C) and energy conservation (Enduring Understanding 5.B) are replaced by the law of conservation of mass-energy.

In nuclear fusion and fission reactions, so much potential energy is lost that the mass of the products of a reaction are measurably less than the mass of the reactants (Essential Knowledge 1.C.4, Essential Knowledge 4.C.4) in accordance with the equation $E = mc^2$. This equation explains that mass is part of the internal energy of an object or system (Essential Knowledge 5.B.11). In addition, the number of nucleons is conserved in these nuclear reactions (Enduring Understanding 5.G), and that determines which nuclear reactions are possible (Essential Knowledge 5.G.1).

Big Idea 1 Objects and systems have properties such as mass and charge. Systems may have internal structure.

Enduring Understanding 1.C Objects and systems have properties of inertial mass and gravitational mass that are experimentally verified to be the same and that satisfy conservation principles.

Essential Knowledge 1.C.4 In certain processes, mass can be converted to energy and energy can be converted to mass according to $E = mc^2$, the equation derived from the theory of special relativity.

Big Idea 4 Interactions between systems can result in changes in those systems.

Enduring Understanding 4.C Interactions with other objects or systems can change the total energy of a system.

Essential Knowledge 4.C.4 Mass can be converted into energy and energy can be converted into mass.

Big Idea 5 Changes that occur as a result of interactions are constrained by conservation laws.

Enduring Understanding 5.B The energy of a system is conserved.

Essential Knowledge 5.B.11 Beyond the classical approximation, mass is actually part of the internal energy of an object or system with $E = mc^2$.

Enduring Understanding 5.G Nucleon number is conserved.

Essential Knowledge 5.G.1 The possible nuclear reactions are constrained by the law of conservation of nucleon number.

32.1 Medical Imaging and Diagnostics

Learning Objectives
By the end of this section, you will be able to: • Explain the working principle behind an Anger camera. • Describe the SPECT and PET imaging techniques.

A host of medical imaging techniques employ nuclear radiation. What makes nuclear radiation so useful? First, γ radiation can easily penetrate tissue; hence, it is a useful probe to monitor conditions inside the body. Second, nuclear radiation depends on the nuclide and not on the chemical compound it is in, so that a radioactive nuclide can be put into a compound designed for specific purposes. The compound is said to be **tagged**. A tagged compound used for medical purposes is called a **radiopharmaceutical**. Radiation detectors external to the body can determine the location and concentration of a radiopharmaceutical to yield medically useful information. For example, certain drugs are concentrated in inflamed regions of the body, and this information can aid diagnosis and treatment as seen in Figure 32.4. Another application utilizes a radiopharmaceutical which the body sends to bone cells, particularly those that are most active, to detect cancerous tumors or healing points. Images can then be produced of such bone scans. Radioisotopes are also used to determine the functioning of body organs, such as blood flow, heart muscle activity, and iodine uptake in the thyroid gland.

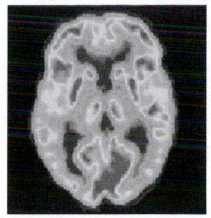

Figure 32.4 A radiopharmaceutical is used to produce this brain image of a patient with Alzheimer's disease. Certain features are computer enhanced. (credit: National Institutes of Health)

Medical Application

Table 32.1 lists certain medical diagnostic uses of radiopharmaceuticals, including isotopes and activities that are typically administered. Many organs can be imaged with a variety of nuclear isotopes replacing a stable element by a radioactive isotope. One common diagnostic employs iodine to image the thyroid, since iodine is concentrated in that organ. The most active thyroid cells, including cancerous cells, concentrate the most iodine and, therefore, emit the most radiation. Conversely, hypothyroidism is indicated by lack of iodine uptake. Note that there is more than one isotope that can be used for several types of scans. Another common nuclear diagnostic is the thallium scan for the cardiovascular system, particularly used to evaluate blockages in the coronary arteries and examine heart activity. The salt TlCl can be used, because it acts like NaCl and follows the blood. Gallium-67 accumulates where there is rapid cell growth, such as in tumors and sites of infection. Hence, it is useful in cancer imaging. Usually, the patient receives the injection one day and has a whole body scan 3 or 4 days later because it can take several days for the gallium to build up.

Table 32.1 Diagnostic Uses of Radiopharmaceuticals

Procedure, isotope	Typical activity (mCi), where $1 \text{ mCi} = 3.7 \times 10^7 \text{ Bq}$
Brain scan	
^{99m}Tc	7.5
^{113m}In	7.5
^{11}C (PET)	20
^{13}N (PET)	20
^{15}O (PET)	50
^{18}F (PET)	10
Lung scan	
^{99m}Tc	2
^{133}Xe	7.5
Cardiovascular blood pool	
^{131}I	0.2
^{99m}Tc	2
Cardiovascular arterial flow	
^{201}Tl	3
^{24}Na	7.5
Thyroid scan	
^{131}I	0.05
^{123}I	0.07
Liver scan	
^{198}Au (colloid)	0.1
^{99m}Tc (colloid)	2
Bone scan	
^{85}Sr	0.1
^{99m}Tc	10
Kidney scan	
^{197}Hg	0.1
^{99m}Tc	1.5

Note that **Table 32.1** lists many diagnostic uses for ^{99m}Tc , where "m" stands for a metastable state of the technetium nucleus. Perhaps 80 percent of all radiopharmaceutical procedures employ ^{99m}Tc because of its many advantages. One is that the decay of its metastable state produces a single, easily identified 0.142-MeV γ ray. Additionally, the radiation dose to the patient is limited by the short 6.0-h half-life of ^{99m}Tc. And, although its half-life is short, it is easily and continuously produced on site.

The basic process for production is neutron activation of molybdenum, which quickly β decays into ^{99m}Tc. Technetium-99m can be attached to many compounds to allow the imaging of the skeleton, heart, lungs, kidneys, etc.

Figure 32.5 shows one of the simpler methods of imaging the concentration of nuclear activity, employing a device called an **Anger camera** or **gamma camera**. A piece of lead with holes bored through it collimates γ rays emerging from the patient, allowing detectors to receive γ rays from specific directions only. The computer analysis of detector signals produces an image. One of the disadvantages of this detection method is that there is no depth information (i.e., it provides a two-dimensional view of the tumor as opposed to a three-dimensional view), because radiation from any location under that detector produces a signal.

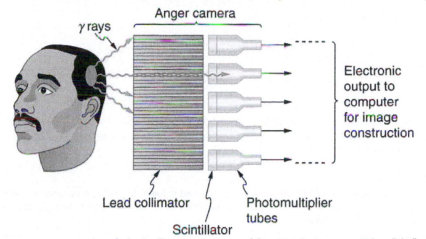

Figure 32.5 An Anger or gamma camera consists of a lead collimator and an array of detectors. Gamma rays produce light flashes in the scintillators. The light output is converted to an electrical signal by the photomultipliers. A computer constructs an image from the detector output.

Imaging techniques much like those in x-ray computed tomography (CT) scans use nuclear activity in patients to form three-dimensional images. **Figure 32.6** shows a patient in a circular array of detectors that may be stationary or rotated, with detector output used by a computer to construct a detailed image. This technique is called **single-photon-emission computed tomography(SPECT)** or sometimes simply SPET. The spatial resolution of this technique is poor, about 1 cm, but the contrast (i.e. the difference in visual properties that makes an object distinguishable from other objects and the background) is good.

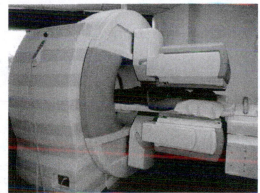

Figure 32.6 SPECT uses a geometry similar to a CT scanner to form an image of the concentration of a radiopharmaceutical compound. (credit: Woldo, Wikimedia Commons)

Images produced by β^+ emitters have become important in recent years. When the emitted positron (β^+) encounters an electron, mutual annihilation occurs, producing two γ rays. These γ rays have identical 0.511-MeV energies (the energy comes from the destruction of an electron or positron mass) and they move directly away from one another, allowing detectors to determine their point of origin accurately, as shown in **Figure 32.7**. The system is called **positron emission tomography (PET)**. It requires detectors on opposite sides to simultaneously (i.e., at the same time) detect photons of 0.511-MeV energy and utilizes computer imaging techniques similar to those in SPECT and CT scans. Examples of β^+ -emitting isotopes used in PET are

^{11}C, ^{13}N, ^{15}O, and ^{18}F, as seen in **Table 32.1**. This list includes C, N, and O, and so they have the advantage of being able to function as tags for natural body compounds. Its resolution of 0.5 cm is better than that of SPECT; the accuracy and sensitivity of PET scans make them useful for examining the brain's anatomy and function. The brain's use of oxygen and water can be monitored with ^{15}O. PET is used extensively for diagnosing brain disorders. It can note decreased metabolism in certain regions prior to a confirmation of Alzheimer's disease. PET can locate regions in the brain that become active when a person carries out specific activities, such as speaking, closing their eyes, and so on.

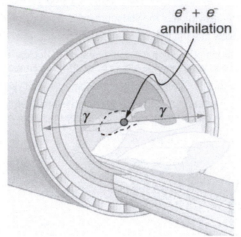

Figure 32.7 A PET system takes advantage of the two identical γ-ray photons produced by positron-electron annihilation. These γ rays are emitted in opposite directions, so that the line along which each pair is emitted is determined. Various events detected by several pairs of detectors are then analyzed by the computer to form an accurate image.

PhET Explorations: Simplified MRI

Is it a tumor? Magnetic Resonance Imaging (MRI) can tell. Your head is full of tiny radio transmitters (the nuclear spins of the hydrogen nuclei of your water molecules). In an MRI unit, these little radios can be made to broadcast their positions, giving a detailed picture of the inside of your head.

Figure 32.8 Simplified MRI (http://cnx.org/content/m54886/1.2/mri_en.jar)

32.2 Biological Effects of Ionizing Radiation

Learning Objectives

By the end of this section, you will be able to:

- Define various units of radiation.
- Describe RBE.

The information presented in this section supports the following AP® learning objectives and science practices:

- **7.C.4.1** The student is able to construct or interpret representations of transitions between atomic energy states involving the emission and absorption of photons. [For questions addressing stimulated emission, students will not be expected to recall the details of the process, such as the fact that the emitted photons have the same frequency and phase as the incident photon; but given a representation of the process, students are expected to make inferences such as figuring out from energy conservation that since the atom loses energy in the process, the emitted photons taken together must carry more energy than the incident photon.]

We hear many seemingly contradictory things about the biological effects of ionizing radiation. It can cause cancer, burns, and hair loss, yet it is used to treat and even cure cancer. How do we understand these effects? Once again, there is an underlying simplicity in nature, even in complicated biological organisms. All the effects of ionizing radiation on biological tissue can be understood by knowing that **ionizing radiation affects molecules within cells, particularly DNA molecules.**

Let us take a brief look at molecules within cells and how cells operate. Cells have long, double-helical DNA molecules containing chemical codes called genetic codes that govern the function and processes undertaken by the cell. It is for unraveling the double-helical structure of DNA that James Watson, Francis Crick, and Maurice Wilkins received the Nobel Prize. Damage to DNA consists of breaks in chemical bonds or other changes in the structural features of the DNA chain, leading to changes in the genetic code. In human cells, we can have as many as a million individual instances of damage to DNA per cell per day. It is remarkable that DNA contains codes that check whether the DNA is damaged or can repair itself. It is like an auto check and repair mechanism. This repair ability of DNA is vital for maintaining the integrity of the genetic code and for the normal functioning of the entire organism. It should be constantly active and needs to respond rapidly. The rate of DNA repair depends on various factors such as the cell type and age of the cell. A cell with a damaged ability to repair DNA, which could have been induced by ionizing radiation, can do one of the following:

- The cell can go into an irreversible state of dormancy, known as senescence.
- The cell can commit suicide, known as programmed cell death.
- The cell can go into unregulated cell division leading to tumors and cancers.

Since ionizing radiation damages the DNA, which is critical in cell reproduction, it has its greatest effect on cells that rapidly reproduce, including most types of cancer. Thus, cancer cells are more sensitive to radiation than normal cells and can be killed by it easily. Cancer is characterized by a malfunction of cell reproduction, and can also be caused by ionizing radiation. Without contradiction, ionizing radiation can be both a cure and a cause.

To discuss quantitatively the biological effects of ionizing radiation, we need a radiation dose unit that is directly related to those effects. All effects of radiation are assumed to be directly proportional to the amount of ionization produced in the biological organism. The amount of ionization is in turn proportional to the amount of deposited energy. Therefore, we define a **radiation dose unit** called the **rad**, as $1/100$ of a joule of ionizing energy deposited per kilogram of tissue, which is

$$1 \text{ rad} = 0.01 \text{ J/kg}. \tag{32.1}$$

For example, if a 50.0-kg person is exposed to ionizing radiation over her entire body and she absorbs 1.00 J, then her whole-body radiation dose is

$$(1.00 \text{ J})/ (50.0 \text{ kg}) = 0.0200 \text{ J/kg} = 2.00 \text{ rad}. \tag{32.2}$$

If the same 1.00 J of ionizing energy were absorbed in her 2.00-kg forearm alone, then the dose to the forearm would be

$$(1.00 \text{ J})/ (2.00 \text{ kg}) = 0.500 \text{ J/kg} = 50.0 \text{ rad}, \tag{32.3}$$

and the unaffected tissue would have a zero rad dose. While calculating radiation doses, you divide the energy absorbed by the mass of affected tissue. You must specify the affected region, such as the whole body or forearm in addition to giving the numerical dose in rads. The SI unit for radiation dose is the **gray (Gy)**, which is defined to be

$$1 \text{ Gy} = 1 \text{ J/kg} = 100 \text{ rad}. \tag{32.4}$$

However, the rad is still commonly used. Although the energy per kilogram in 1 rad is small, it has significant effects since the energy causes ionization. The energy needed for a single ionization is a few eV, or less than 10^{-18} J. Thus, 0.01 J of ionizing energy can create a huge number of ion pairs and have an effect at the cellular level.

The effects of ionizing radiation may be directly proportional to the dose in rads, but they also depend on the type of radiation and the type of tissue. That is, for a given dose in rads, the effects depend on whether the radiation is α, β, γ, x-ray, or some other type of ionizing radiation. In the earlier discussion of the range of ionizing radiation, it was noted that energy is deposited in a series of ionizations and not in a single interaction. Each ion pair or ionization requires a certain amount of energy, so that the number of ion pairs is directly proportional to the amount of the deposited ionizing energy. But, if the range of the radiation is small, as it is for α s, then the ionization and the damage created is more concentrated and harder for the organism to repair, as seen in **Figure 32.9**. Concentrated damage is more difficult for biological organisms to repair than damage that is spread out, so short-range particles have greater biological effects. The **relative biological effectiveness** (RBE) or **quality factor** (QF) is given in **Table 32.2** for several types of ionizing radiation—the effect of the radiation is directly proportional to the RBE. A dose unit more closely related to effects in biological tissue is called the **roentgen equivalent man** or **rem** and is defined to be the dose in rads multiplied by the relative biological effectiveness.

$$\text{rem} = \text{rad} \times \text{RBE} \tag{32.5}$$

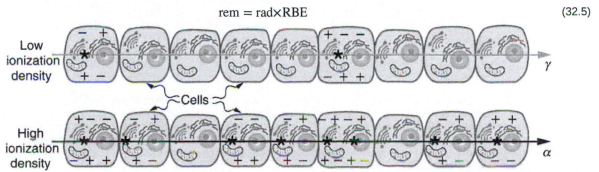

Figure 32.9 The image shows ionization created in cells by α and γ radiation. Because of its shorter range, the ionization and damage created by α is more concentrated and harder for the organism to repair. Thus, the RBE for α s is greater than the RBE for γ s, even though they create the same amount of ionization at the same energy.

So, if a person had a whole-body dose of 2.00 rad of γ radiation, the dose in rem would be

$(2.00 \text{ rad})(1) = 2.00 \text{ rem whole body}$. If the person had a whole-body dose of 2.00 rad of α radiation, then the dose in rem would be $(2.00 \text{ rad})(20) = 40.0 \text{ rem whole body}$. The α s would have 20 times the effect on the person than the γ s for the same deposited energy. The SI equivalent of the rem is the **sievert** (Sv), defined to be $Sv = Gy \times RBE$, so that

$$1 \text{ Sv} = 1 \text{ Gy} \times RBE = 100 \text{ rem}. \tag{32.6}$$

The RBEs given in **Table 32.2** are approximate, but they yield certain insights. For example, the eyes are more sensitive to radiation, because the cells of the lens do not repair themselves. Neutrons cause more damage than γ rays, although both are neutral and have large ranges, because neutrons often cause secondary radiation when they are captured. Note that the RBEs are 1 for higher-energy β s, γ s, and x-rays, three of the most common types of radiation. For those types of radiation, the numerical values of the dose in rem and rad are identical. For example, 1 rad of γ radiation is also 1 rem. For that reason, rads are still widely quoted rather than rem. **Table 32.3** summarizes the units that are used for radiation.

Misconception Alert: Activity vs. Dose
_____

"Activity" refers to the radioactive source while "dose" refers to the amount of energy from the radiation that is deposited in a person or object.

A high level of activity doesn't mean much if a person is far away from the source. The activity R of a source depends upon the quantity of material (kg) as well as the half-life. A short half-life will produce many more disintegrations per second. Recall that $R = \dfrac{0.693N}{t_{1/2}}$. Also, the activity decreases exponentially, which is seen in the equation $R = R_0 e^{-\lambda t}$.

Table 32.2 Relative Biological Effectiveness

Type and energy of radiation	RBE[1]
X-rays	1
γ rays	1
β rays greater than 32 keV	1
β rays less than 32 keV	1.7
Neutrons, thermal to slow (<20 keV)	2–5
Neutrons, fast (1–10 MeV)	10 (body), 32 (eyes)
Protons (1–10 MeV)	10 (body), 32 (eyes)
α rays from radioactive decay	10–20
Heavy ions from accelerators	10–20

Table 32.3 Units for Radiation

Quantity	SI unit name	Definition	Former unit	Conversion
Activity	Becquerel (bq)	decay/sec	Curie (Ci)	$1 \text{ Bq} = 2.7 \times 10^{-11} \text{ Ci}$
Absorbed dose	Gray (Gy)	1 J/kg	rad	$Gy = 100 \text{ rad}$
Dose Equivalent	Sievert (Sv)	1 J/kg × RBE	rem	$Sv = 100 \text{ rem}$

The large-scale effects of radiation on humans can be divided into two categories: immediate effects and long-term effects. **Table 32.4** gives the immediate effects of whole-body exposures received in less than one day. If the radiation exposure is spread out over more time, greater doses are needed to cause the effects listed. This is due to the body's ability to partially repair the damage. Any dose less than 100 mSv (10 rem) is called a **low dose**, 0.1 Sv to 1 Sv (10 to 100 rem) is called a **moderate dose**, and anything greater than 1 Sv (100 rem) is called a **high dose**. There is no known way to determine after the fact if a person has been exposed to less than 10 mSv.

---

1. Values approximate, difficult to determine.

Table 32.4 Immediate Effects of Radiation (Adults, Whole Body, Single Exposure)

Dose in Sv [2]	Effect
0–0.10	No observable effect.
0.1 – 1	Slight to moderate decrease in white blood cell counts.
0.5	Temporary sterility; 0.35 for women, 0.50 for men.
1 – 2	Significant reduction in blood cell counts, brief nausea and vomiting. Rarely fatal.
2 – 5	Nausea, vomiting, hair loss, severe blood damage, hemorrhage, fatalities.
4.5	LD50/32. Lethal to 50% of the population within 32 days after exposure if not treated.
5 – 20	Worst effects due to malfunction of small intestine and blood systems. Limited survival.
>20	Fatal within hours due to collapse of central nervous system.

Immediate effects are explained by the effects of radiation on cells and the sensitivity of rapidly reproducing cells to radiation. The first clue that a person has been exposed to radiation is a change in blood count, which is not surprising since blood cells are the most rapidly reproducing cells in the body. At higher doses, nausea and hair loss are observed, which may be due to interference with cell reproduction. Cells in the lining of the digestive system also rapidly reproduce, and their destruction causes nausea. When the growth of hair cells slows, the hair follicles become thin and break off. High doses cause significant cell death in all systems, but the lowest doses that cause fatalities do so by weakening the immune system through the loss of white blood cells.

The two known long-term effects of radiation are cancer and genetic defects. Both are directly attributable to the interference of radiation with cell reproduction. For high doses of radiation, the risk of cancer is reasonably well known from studies of exposed groups. Hiroshima and Nagasaki survivors and a smaller number of people exposed by their occupation, such as radium dial painters, have been fully documented. Chernobyl victims will be studied for many decades, with some data already available. For example, a significant increase in childhood thyroid cancer has been observed. The risk of a radiation-induced cancer for low and moderate doses is generally *assumed* to be proportional to the risk known for high doses. Under this assumption, any dose of radiation, no matter how small, involves a risk to human health. This is called the **linear hypothesis** and it may be prudent, but it *is* controversial. There is some evidence that, unlike the immediate effects of radiation, the long-term effects are cumulative and there is little self-repair. This is analogous to the risk of skin cancer from UV exposure, which is known to be cumulative.

There is a latency period for the onset of radiation-induced cancer of about 2 years for leukemia and 15 years for most other forms. The person is at risk for at least 30 years after the latency period. Omitting many details, the overall risk of a radiation-induced cancer death per year per rem of exposure is about 10 in a million, which can be written as $10/10^6 \, \text{rem} \cdot \text{y}$.

If a person receives a dose of 1 rem, his risk each year of dying from radiation-induced cancer is 10 in a million and that risk continues for about 30 years. The lifetime risk is thus 300 in a million, or 0.03 percent. Since about 20 percent of all worldwide deaths are from cancer, the increase due to a 1 rem exposure is impossible to detect demographically. But 100 rem (1 Sv), which was the dose received by the average Hiroshima and Nagasaki survivor, causes a 3 percent risk, which can be observed in the presence of a 20 percent normal or natural incidence rate.

The incidence of genetic defects induced by radiation is about one-third that of cancer deaths, but is much more poorly known.

The lifetime risk of a genetic defect due to a 1 rem exposure is about 100 in a million or $3.3/10^6 \, \text{rem} \cdot \text{y}$, but the normal incidence is 60,000 in a million. Evidence of such a small increase, tragic as it is, is nearly impossible to obtain. For example, there is no evidence of increased genetic defects among the offspring of Hiroshima and Nagasaki survivors. Animal studies do not seem to correlate well with effects on humans and are not very helpful. For both cancer and genetic defects, the approach to safety has been to use the linear hypothesis, which is likely to be an overestimate of the risks of low doses. Certain researchers even claim that low doses are *beneficial*. **Hormesis** is a term used to describe generally favorable biological responses to low exposures of toxins or radiation. Such low levels may help certain repair mechanisms to develop or enable cells to adapt to the effects of the low exposures. Positive effects may occur at low doses that could be a problem at high doses.

Even the linear hypothesis estimates of the risks are relatively small, and the average person is not exposed to large amounts of radiation. Table 32.5 lists average annual background radiation doses from natural and artificial sources for Australia, the United States, Germany, and world-wide averages. Cosmic rays are partially shielded by the atmosphere, and the dose depends upon altitude and latitude, but the average is about 0.40 mSv/y. A good example of the variation of cosmic radiation dose with altitude comes from the airline industry. Monitored personnel show an average of 2 mSv/y. A 12-hour flight might give you an exposure of 0.02 to 0.03 mSv.

Doses from the Earth itself are mainly due to the isotopes of uranium, thorium, and potassium, and vary greatly by location. Some places have great natural concentrations of uranium and thorium, yielding doses ten times as high as the average value. Internal doses come from foods and liquids that we ingest. Fertilizers containing phosphates have potassium and uranium. So we are all a little radioactive. Carbon-14 has about 66 Bq/kg radioactivity whereas fertilizers may have more than 3000 Bq/kg radioactivity. Medical and dental diagnostic exposures are mostly from x-rays. It should be noted that x-ray doses tend to be localized and are becoming much smaller with improved techniques. Table 32.6 shows typical doses received during various diagnostic x-ray examinations. Note the large dose from a CT scan. While CT scans only account for less than 20 percent of the

---

2. Multiply by 100 to obtain dose in rem.

x-ray procedures done today, they account for about 50 percent of the annual dose received.

Radon is usually more pronounced underground and in buildings with low air exchange with the outside world. Almost all soil contains some ^{226}Ra and ^{222}Rn , but radon is lower in mainly sedimentary soils and higher in granite soils. Thus, the exposure to the public can vary greatly, even within short distances. Radon can diffuse from the soil into homes, especially basements. The estimated exposure for ^{222}Rn is controversial. Recent studies indicate there is more radon in homes than had been realized, and it is speculated that radon may be responsible for 20 percent of lung cancers, being particularly hazardous to those who also smoke. Many countries have introduced limits on allowable radon concentrations in indoor air, often requiring the measurement of radon concentrations in a house prior to its sale. Ironically, it could be argued that the higher levels of radon exposure and their geographic variability, taken with the lack of demographic evidence of any effects, means that low-level radiation is *less* dangerous than previously thought.

Radiation Protection

Laws regulate radiation doses to which people can be exposed. The greatest occupational whole-body dose that is allowed depends upon the country and is about 20 to 50 mSv/y and is rarely reached by medical and nuclear power workers. Higher doses are allowed for the hands. Much lower doses are permitted for the reproductive organs and the fetuses of pregnant women. Inadvertent doses to the public are limited to $1/10$ of occupational doses, except for those caused by nuclear power, which cannot legally expose the public to more than $1/1000$ of the occupational limit or 0.05 mSv/y (5 mrem/y). This has been exceeded in the United States only at the time of the Three Mile Island (TMI) accident in 1979. Chernobyl is another story. Extensive monitoring with a variety of radiation detectors is performed to assure radiation safety. Increased ventilation in uranium mines has lowered the dose there to about 1 mSv/y.

Table 32.5 Background Radiation Sources and Average Doses

Source	Dose (mSv/y)[3]			
Source	Australia	Germany	United States	World
Natural Radiation - external				
Cosmic Rays	0.30	0.28	0.30	0.39
Soil, building materials	0.40	0.40	0.30	0.48
Radon gas	0.90	1.1	2.0	1.2
Natural Radiation - internal				
^{40}K, ^{14}C, ^{226}Ra	0.24	0.28	0.40	0.29
Medical & Dental	0.80	0.90	0.53	0.40
TOTAL	2.6	3.0	3.5	2.8

To physically limit radiation doses, we use **shielding**, increase the **distance** from a source, and limit the **time of exposure**.

Figure 32.10 illustrates how these are used to protect both the patient and the dental technician when an x-ray is taken. Shielding absorbs radiation and can be provided by any material, including sufficient air. The greater the distance from the source, the more the radiation spreads out. The less time a person is exposed to a given source, the smaller is the dose received by the person. Doses from most medical diagnostics have decreased in recent years due to faster films that require less exposure time.

---

3. Multiply by 100 to obtain dose in mrem/y.

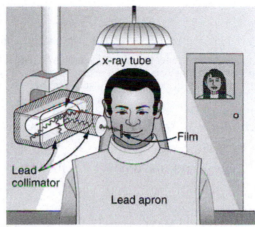

Figure 32.10 A lead apron is placed over the dental patient and shielding surrounds the x-ray tube to limit exposure to tissue other than the tissue that is being imaged. Fast films limit the time needed to obtain images, reducing exposure to the imaged tissue. The technician stands a few meters away behind a lead-lined door with a lead glass window, reducing her occupational exposure.

Table 32.6 Typical Doses Received During Diagnostic X-ray Exams

Procedure	Effective dose (mSv)
Chest	0.02
Dental	0.01
Skull	0.07
Leg	0.02
Mammogram	0.40
Barium enema	7.0
Upper GI	3.0
CT head	2.0
CT abdomen	10.0

Problem-Solving Strategy

You need to follow certain steps for dose calculations, which are

Step 1. *Examine the situation to determine that a person is exposed to ionizing radiation.*

Step 2. *Identify exactly what needs to be determined in the problem (identify the unknowns).* The most straightforward problems ask for a dose calculation.

Step 3. *Make a list of what is given or can be inferred from the problem as stated (identify the knowns).* Look for information on the type of radiation, the energy per event, the activity, and the mass of tissue affected.

Step 4. *For dose calculations, you need to determine the energy deposited.* This may take one or more steps, depending on the given information.

Step 5. *Divide the deposited energy by the mass of the affected tissue.* Use units of joules for energy and kilograms for mass. If a dose in Sv is involved, use the definition that $1 \text{ Sv} = 1 \text{ J/kg}$.

Step 6. *If a dose in mSv is involved, determine the RBE (QF) of the radiation.* Recall that $1 \text{ mSv} = 1 \text{ mGy} \times \text{RBE}$ (or $1 \text{ rem} = 1 \text{ rad} \times \text{RBE}$) .

Step 7. *Check the answer to see if it is reasonable: Does it make sense?* The dose should be consistent with the numbers given in the text for diagnostic, occupational, and therapeutic exposures.

Example 32.1 Dose from Inhaled Plutonium

Calculate the dose in rem/y for the lungs of a weapons plant employee who inhales and retains an activity of 1.00 μCi of ^{239}Pu in an accident. The mass of affected lung tissue is 2.00 kg, the plutonium decays by emission of a 5.23-MeV α particle, and you may assume the higher value of the RBE for α s from Table 32.2.

Strategy

Dose in rem is defined by $1\ \text{rad} = 0.01\ \text{J/kg}$ and $\text{rem} = \text{rad} \times \text{RBE}$. The energy deposited is divided by the mass of tissue affected and then multiplied by the RBE. The latter two quantities are given, and so the main task in this example will be to find the energy deposited in one year. Since the activity of the source is given, we can calculate the number of decays, multiply by the energy per decay, and convert MeV to joules to get the total energy.

Solution

The activity $R = 1.00\ \mu\text{Ci} = 3.70 \times 10^4\ \text{Bq} = 3.70 \times 10^4$ decays/s. So, the number of decays per year is obtained by multiplying by the number of seconds in a year:

$$\left(3.70 \times 10^4\ \text{decays/s}\right)\left(3.16 \times 10^7\ \text{s}\right) = 1.17 \times 10^{12}\ \text{decays}. \qquad (32.7)$$

Thus, the ionizing energy deposited per year is

$$E = \left(1.17 \times 10^{12}\ \text{decays}\right)\left(5.23\ \text{MeV/decay}\right) \times \left(\frac{1.60 \times 10^{-13}\ \text{J}}{\text{MeV}}\right) = 0.978\ \text{J}. \qquad (32.8)$$

Dividing by the mass of the affected tissue gives

$$\frac{E}{\text{mass}} = \frac{0.978\ \text{J}}{2.00\ \text{kg}} = 0.489\ \text{J/kg}. \qquad (32.9)$$

One Gray is 1.00 J/kg, and so the dose in Gy is

$$\text{dose in Gy} = \frac{0.489\ \text{J/kg}}{1.00\ (\text{J/kg})/\text{Gy}} = 0.489\ \text{Gy}. \qquad (32.10)$$

Now, the dose in Sv is

$$\text{dose in Sv} = \text{Gy} \times \text{RBE} \qquad (32.11)$$
$$= (0.489\ \text{Gy})(20) = 9.8\ \text{Sv}. \qquad (32.12)$$

Discussion

First note that the dose is given to two digits, because the RBE is (at best) known only to two digits. By any standard, this yearly radiation dose is high and will have a devastating effect on the health of the worker. Worse yet, plutonium has a long radioactive half-life and is not readily eliminated by the body, and so it will remain in the lungs. Being an α emitter makes the effects 10 to 20 times worse than the same ionization produced by β s, γ rays, or x-rays. An activity of $1.00\ \mu\text{Ci}$ is created by only $16\ \mu\text{g}$ of ^{239}Pu (left as an end-of-chapter problem to verify), partly justifying claims that plutonium is the most toxic substance known. Its actual hazard depends on how likely it is to be spread out among a large population and then ingested. The Chernobyl disaster's deadly legacy, for example, has nothing to do with the plutonium it put into the environment.

Risk versus Benefit

Medical doses of radiation are also limited. Diagnostic doses are generally low and have further lowered with improved techniques and faster films. With the possible exception of routine dental x-rays, radiation is used diagnostically only when needed so that the low risk is justified by the benefit of the diagnosis. Chest x-rays give the lowest doses—about 0.1 mSv to the tissue affected, with less than 5 percent scattering into tissues that are not directly imaged. Other x-ray procedures range upward to about 10 mSv in a CT scan, and about 5 mSv (0.5 rem) per dental x-ray, again both only affecting the tissue imaged. Medical images with radiopharmaceuticals give doses ranging from 1 to 5 mSv, usually localized. One exception is the thyroid scan using ^{131}I. Because of its relatively long half-life, it exposes the thyroid to about 0.75 Sv. The isotope ^{123}I is more difficult to produce, but its short half-life limits thyroid exposure to about 15 mSv.

PhET Explorations: Alpha Decay

Watch alpha particles escape from a polonium nucleus, causing radioactive alpha decay. See how random decay times relate to the half life.

Figure 32.11 Alpha Decay (http://cnx.org/content/m54890/1.2/alpha-decay_en.jar)

32.3 Therapeutic Uses of Ionizing Radiation

Learning Objectives
By the end of this section, you will be able to: • Explain the concept of radiotherapy and list typical doses for cancer therapy.

Therapeutic applications of ionizing radiation, called radiation therapy or **radiotherapy**, have existed since the discovery of x-rays and nuclear radioactivity. Today, radiotherapy is used almost exclusively for cancer therapy, where it saves thousands of lives and improves the quality of life and longevity of many it cannot save. Radiotherapy may be used alone or in combination with surgery and chemotherapy (drug treatment) depending on the type of cancer and the response of the patient. A careful examination of all available data has established that radiotherapy's beneficial effects far outweigh its long-term risks.

Medical Application

The earliest uses of ionizing radiation on humans were mostly harmful, with many at the level of snake oil as seen in **Figure 32.12**. Radium-doped cosmetics that glowed in the dark were used around the time of World War I. As recently as the 1950s, radon mine tours were promoted as healthful and rejuvenating—those who toured were exposed but gained no benefits. Radium salts were sold as health elixirs for many years. The gruesome death of a wealthy industrialist, who became psychologically addicted to the brew, alerted the unsuspecting to the dangers of radium salt elixirs. Most abuses finally ended after the legislation in the 1950s.

Figure 32.12 The properties of radiation were once touted for far more than its modern use in cancer therapy. Until 1932, radium was advertised for a variety of uses, often with tragic results. (credit: Struthious Bandersnatch.)

Radiotherapy is effective against cancer because cancer cells reproduce rapidly and, consequently, are more sensitive to radiation. The central problem in radiotherapy is to make the dose for cancer cells as high as possible while limiting the dose for normal cells. The ratio of abnormal cells killed to normal cells killed is called the **therapeutic ratio**, and all radiotherapy techniques are designed to enhance this ratio. Radiation can be concentrated in cancerous tissue by a number of techniques. One of the most prevalent techniques for well-defined tumors is a geometric technique shown in **Figure 32.13**. A narrow beam of radiation is passed through the patient from a variety of directions with a common crossing point in the tumor. This concentrates the dose in the tumor while spreading it out over a large volume of normal tissue. The external radiation can be x-rays, ^{60}Co γ rays, or ionizing-particle beams produced by accelerators. Accelerator-produced beams of neutrons, π-mesons , and heavy ions such as nitrogen nuclei have been employed, and these can be quite effective. These particles have larger QFs or RBEs and sometimes can be better localized, producing a greater therapeutic ratio. But accelerator radiotherapy is much more expensive and less frequently employed than other forms.

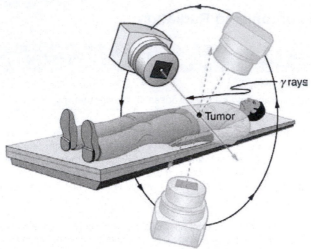

Figure 32.13 The ^{60}Co source of γ-radiation is rotated around the patient so that the common crossing point is in the tumor, concentrating the dose there. This geometric technique works for well-defined tumors.

Another form of radiotherapy uses chemically inert radioactive implants. One use is for prostate cancer. Radioactive seeds (about 40 to 100 and the size of a grain of rice) are placed in the prostate region. The isotopes used are usually ^{135}I (6-month half life) or ^{103}Pd (3-month half life). Alpha emitters have the dual advantages of a large QF and a small range for better localization.

Radiopharmaceuticals are used for cancer therapy when they can be localized well enough to produce a favorable therapeutic ratio. Thyroid cancer is commonly treated utilizing radioactive iodine. Thyroid cells concentrate iodine, and cancerous thyroid cells are more aggressive in doing this. An ingenious use of radiopharmaceuticals in cancer therapy tags antibodies with radioisotopes. Antibodies produced by a patient to combat his cancer are extracted, cultured, loaded with a radioisotope, and then returned to the patient. The antibodies are concentrated almost entirely in the tissue they developed to fight, thus localizing the radiation in abnormal tissue. The therapeutic ratio can be quite high for short-range radiation. There is, however, a significant dose for organs that eliminate radiopharmaceuticals from the body, such as the liver, kidneys, and bladder. As with most radiotherapy, the technique is limited by the tolerable amount of damage to the normal tissue.

Table 32.7 lists typical therapeutic doses of radiation used against certain cancers. The doses are large, but not fatal because they are localized and spread out in time. Protocols for treatment vary with the type of cancer and the condition and response of the patient. Three to five 200-rem treatments per week for a period of several weeks is typical. Time between treatments allows the body to repair normal tissue. This effect occurs because damage is concentrated in the abnormal tissue, and the abnormal tissue is more sensitive to radiation. Damage to normal tissue limits the doses. You will note that the greatest doses are given to any tissue that is not rapidly reproducing, such as in the adult brain. Lung cancer, on the other end of the scale, cannot ordinarily be cured with radiation because of the sensitivity of lung tissue and blood to radiation. But radiotherapy for lung cancer does alleviate symptoms and prolong life and is therefore justified in some cases.

Table 32.7 Cancer Radiotherapy

Type of Cancer	Typical dose (Sv)
Lung	10–20
Hodgkin's disease	40–45
Skin	40–50
Ovarian	50–75
Breast	50–80+
Brain	80+
Neck	80+
Bone	80+
Soft tissue	80+
Thyroid	80+

Finally, it is interesting to note that chemotherapy employs drugs that interfere with cell division and is, thus, also effective against cancer. It also has almost the same side effects, such as nausea and hair loss, and risks, such as the inducement of another cancer.

32.4 Food Irradiation

Learning Objectives

By the end of this section, you will be able to:

- Define food irradiation, low dose, and free radicals.

Ionizing radiation is widely used to sterilize medical supplies, such as bandages, and consumer products, such as tampons. Worldwide, it is also used to irradiate food, an application that promises to grow in the future. **Food irradiation** is the treatment of food with ionizing radiation. It is used to reduce pest infestation and to delay spoilage and prevent illness caused by microorganisms. Food irradiation is controversial. Proponents see it as superior to pasteurization, preservatives, and insecticides, supplanting dangerous chemicals with a more effective process. Opponents see its safety as unproven, perhaps leaving worse toxic residues as well as presenting an environmental hazard at treatment sites. In developing countries, food irradiation might increase crop production by 25.0% or more, and reduce food spoilage by a similar amount. It is used chiefly to treat spices and some fruits, and in some countries, red meat, poultry, and vegetables. Over 40 countries have approved food irradiation at some level.

Food irradiation exposes food to large doses of γ rays, x-rays, or electrons. These photons and electrons induce no nuclear reactions and thus create *no residual radioactivity*. (Some forms of ionizing radiation, such as neutron irradiation, cause residual radioactivity. These are not used for food irradiation.) The γ source is usually ^{60}Co or ^{137}Cs , the latter isotope being a major by-product of nuclear power. Cobalt-60 γ rays average 1.25 MeV, while those of ^{137}Cs are 0.67 MeV and are less penetrating.

X-rays used for food irradiation are created with voltages of up to 5 million volts and, thus, have photon energies up to 5 MeV. Electrons used for food irradiation are accelerated to energies up to 10 MeV. The higher the energy per particle, the more penetrating the radiation is and the more ionization it can create. **Figure 32.14** shows a typical γ -irradiation plant.

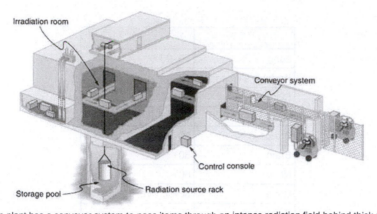

Figure 32.14 A food irradiation plant has a conveyor system to pass items through an intense radiation field behind thick shielding walls. The γ source is lowered into a deep pool of water for safe storage when not in use. Exposure times of up to an hour expose food to doses up to 10^4 Gy.

Owing to the fact that food irradiation seeks to destroy organisms such as insects and bacteria, much larger doses than those fatal to humans must be applied. Generally, the simpler the organism, the more radiation it can tolerate. (Cancer cells are a partial exception, because they are rapidly reproducing and, thus, more sensitive.) Current licensing allows up to 1000 Gy to be applied to fresh fruits and vegetables, called a *low dose* in food irradiation. Such a dose is enough to prevent or reduce the growth of many microorganisms, but about 10,000 Gy is needed to kill salmonella, and even more is needed to kill fungi. Doses greater than 10,000 Gy are considered to be high doses in food irradiation and product sterilization.

The effectiveness of food irradiation varies with the type of food. Spices and many fruits and vegetables have dramatically longer shelf lives. These also show no degradation in taste and no loss of food value or vitamins. If not for the mandatory labeling, such foods subjected to low-level irradiation (up to 1000 Gy) could not be distinguished from untreated foods in quality. However, some foods actually spoil faster after irradiation, particularly those with high water content like lettuce and peaches. Others, such as milk, are given a noticeably unpleasant taste. High-level irradiation produces significant and chemically measurable changes in foods. It produces about a 15% loss of nutrients and a 25% loss of vitamins, as well as some change in taste. Such losses are similar to those that occur in ordinary freezing and cooking.

How does food irradiation work? Ionization produces a random assortment of broken molecules and ions, some with unstable oxygen- or hydrogen-containing molecules known as **free radicals**. These undergo rapid chemical reactions, producing perhaps four or five thousand different compounds called **radiolytic products**, some of which make cell function impossible by breaking cell membranes, fracturing DNA, and so on. How safe is the food afterward? Critics argue that the radiolytic products present a lasting hazard, perhaps being carcinogenic. However, the safety of irradiated food is not known precisely. We do know that low-level food irradiation produces no compounds in amounts that can be measured chemically. This is not surprising, since trace amounts of several thousand compounds may be created. We also know that there have been no observable negative short-term effects on consumers. Long-term effects may show up if large number of people consume large quantities of irradiated food, but no effects have appeared due to the small amounts of irradiated food that are consumed regularly. The case for safety is supported by testing of animal diets that were irradiated; no transmitted genetic effects have been observed. Food irradiation (at least up to a million rad) has been endorsed by the World Health Organization and the UN Food and Agricultural Organization. Finally, the hazard to consumers, if it exists, must be weighed against the benefits in food production and preservation. It must also be weighed against the very real hazards of existing insecticides and food preservatives.

32.5 Fusion

> ### Learning Objectives
>
> By the end of this section, you will be able to:
>
> - Define nuclear fusion.
> - Discuss processes to achieve practical fusion energy generation.
>
> The information presented in this section supports the following AP® learning objectives and science practices:
>
> - **1.C.4.1** The student is able to articulate the reasons that the theory of conservation of mass was replaced by the theory of conservation of mass-energy. **(S.P. 6.3)**
> - **4.C.4.1** The student is able to apply mathematical routines to describe the relationship between mass and energy and apply this concept across domains of scale. **(S.P. 2.2, 2.3, 7.2)**
> - **5.B.11.1** The student is able to apply conservation of mass and conservation of energy concepts to a natural phenomenon and use the equation $E = mc^2$ to make a related calculation. **(S.P. 2.2, 7.2)**
> - **5.G.1.1** The student is able to apply conservation of nucleon number and conservation of electric charge to make predictions about nuclear reactions and decays such as fission, fusion, alpha decay, beta decay, or gamma decay. **(S.P. 6.4)**

While basking in the warmth of the summer sun, a student reads of the latest breakthrough in achieving sustained thermonuclear power and vaguely recalls hearing about the cold fusion controversy. The three are connected. The Sun's energy is produced by nuclear fusion (see **Figure 32.15**). Thermonuclear power is the name given to the use of controlled nuclear fusion as an energy source. While research in the area of thermonuclear power is progressing, high temperatures and containment difficulties remain. The cold fusion controversy centered around unsubstantiated claims of practical fusion power at room temperatures.

Figure 32.15 The Sun's energy is produced by nuclear fusion. (credit: Spiralz)

Nuclear fusion is a reaction in which two nuclei are combined, or *fused*, to form a larger nucleus. We know that all nuclei have less mass than the sum of the masses of the protons and neutrons that form them. The missing mass times c^2 equals the binding energy of the nucleus—the greater the binding energy, the greater the missing mass. We also know that BE/A, the binding energy per nucleon, is greater for medium-mass nuclei and has a maximum at Fe (iron). This means that if two low-mass nuclei can be fused together to form a larger nucleus, energy can be released. The larger nucleus has a greater binding energy and less mass per nucleon than the two that combined. Thus mass is destroyed in the fusion reaction, and energy is released (see **Figure 32.16**). On average, fusion of low-mass nuclei releases energy, but the details depend on the actual nuclides involved.

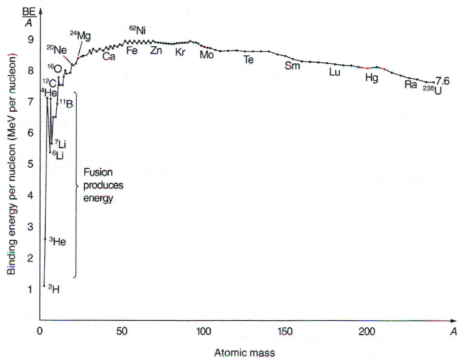

Figure 32.16 Fusion of light nuclei to form medium-mass nuclei destroys mass, because BE/A is greater for the product nuclei. The larger BE/A is, the less mass per nucleon, and so mass is converted to energy and released in these fusion reactions.

The major obstruction to fusion is the Coulomb repulsion between nuclei. Since the attractive nuclear force that can fuse nuclei together is short ranged, the repulsion of like positive charges must be overcome to get nuclei close enough to induce fusion. **Figure 32.17** shows an approximate graph of the potential energy between two nuclei as a function of the distance between their centers. The graph is analogous to a hill with a well in its center. A ball rolled from the right must have enough kinetic energy to get over the hump before it falls into the deeper well with a net gain in energy. So it is with fusion. If the nuclei are given enough kinetic energy to overcome the electric potential energy due to repulsion, then they can combine, release energy, and fall into a deep well. One way to accomplish this is to heat fusion fuel to high temperatures so that the kinetic energy of thermal motion is sufficient to get the nuclei together.

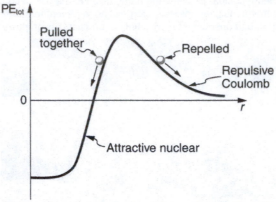

Figure 32.17 Potential energy between two light nuclei graphed as a function of distance between them. If the nuclei have enough kinetic energy to get over the Coulomb repulsion hump, they combine, release energy, and drop into a deep attractive well. Tunneling through the barrier is important in practice. The greater the kinetic energy and the higher the particles get up the barrier (or the lower the barrier), the more likely the tunneling.

You might think that, in the core of our Sun, nuclei are coming into contact and fusing. However, in fact, temperatures on the order of $10^8 \, \text{K}$ are needed to actually get the nuclei in contact, exceeding the core temperature of the Sun. Quantum mechanical tunneling is what makes fusion in the Sun possible, and tunneling is an important process in most other practical applications of fusion, too. Since the probability of tunneling is extremely sensitive to barrier height and width, increasing the temperature greatly increases the rate of fusion. The closer reactants get to one another, the more likely they are to fuse (see **Figure 32.18**). Thus most fusion in the Sun and other stars takes place at their centers, where temperatures are highest. Moreover, high temperature is needed for thermonuclear power to be a practical source of energy.

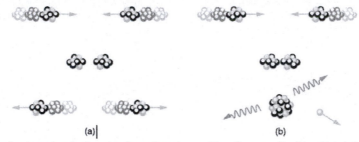

Figure 32.18 (a) Two nuclei heading toward each other slow down, then stop, and then fly away without touching or fusing. (b) At higher energies, the two nuclei approach close enough for fusion via tunneling. The probability of tunneling increases as they approach, but they do not have to touch for the reaction to occur.

The Sun produces energy by fusing protons or hydrogen nuclei ^1H (by far the Sun's most abundant nuclide) into helium nuclei ^4He. The principal sequence of fusion reactions forms what is called the **proton-proton cycle**:

$$^1\text{H} + {}^1\text{H} \rightarrow {}^2\text{H} + e^+ + \nu_e \qquad (0.42 \text{ MeV}) \tag{32.13}$$

$$^1\text{H} + {}^2\text{H} \rightarrow {}^3\text{He} + \gamma \qquad (5.49 \text{ MeV}) \tag{32.14}$$

$$^3\text{He} + {}^3\text{He} \rightarrow {}^4\text{He} + {}^1\text{H} + {}^1\text{H} \qquad (12.86 \text{ MeV}) \tag{32.15}$$

where e^+ stands for a positron and ν_e is an electron neutrino. (The energy in parentheses is *released* by the reaction.) Note that the first two reactions must occur twice for the third to be possible, so that the cycle consumes six protons (^1H) but gives back two. Furthermore, the two positrons produced will find two electrons and annihilate to form four more γ rays, for a total of six. The overall effect of the cycle is thus

$$2e^- + 4\,{}^1\text{H} \rightarrow {}^4\text{He} + 2\nu_e + 6\gamma \qquad (26.7 \text{ MeV}) \tag{32.16}$$

where the 26.7 MeV includes the annihilation energy of the positrons and electrons and is distributed among all the reaction products. The solar interior is dense, and the reactions occur deep in the Sun where temperatures are highest. It takes about 32,000 years for the energy to diffuse to the surface and radiate away. However, the neutrinos escape the Sun in less than two seconds, carrying their energy with them, because they interact so weakly that the Sun is transparent to them. Negative feedback in the Sun acts as a thermostat to regulate the overall energy output. For instance, if the interior of the Sun becomes hotter than normal, the reaction rate increases, producing energy that expands the interior. This cools it and lowers the reaction rate. Conversely, if the interior becomes too cool, it contracts, increasing the temperature and reaction rate (see **Figure 32.19**). Stars like the Sun are stable for billions of years, until a significant fraction of their hydrogen has been depleted. What happens then is discussed in **Introduction to Frontiers of Physics** .

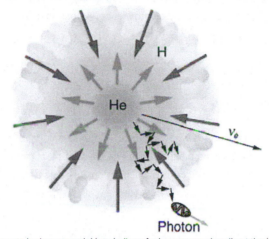

Figure 32.19 Nuclear fusion in the Sun converts hydrogen nuclei into helium; fusion occurs primarily at the boundary of the helium core, where temperature is highest and sufficient hydrogen remains. Energy released diffuses slowly to the surface, with the exception of neutrinos, which escape immediately. Energy production remains stable because of negative feedback effects.

Theories of the proton-proton cycle (and other energy-producing cycles in stars) were pioneered by the German-born, American physicist Hans Bethe (1906–2005), starting in 1938. He was awarded the 1967 Nobel Prize in physics for this work, and he has made many other contributions to physics and society. Neutrinos produced in these cycles escape so readily that they provide us an excellent means to test these theories and study stellar interiors. Detectors have been constructed and operated for more than four decades now to measure solar neutrinos (see **Figure 32.20**). Although solar neutrinos are detected and neutrinos were observed from Supernova 1987A (**Figure 32.21**), too few solar neutrinos were observed to be consistent with predictions of solar energy production. After many years, this solar neutrino problem was resolved with a blend of theory and experiment that showed that the neutrino does indeed have mass. It was also found that there are three types of neutrinos, each associated with a different type of nuclear decay.

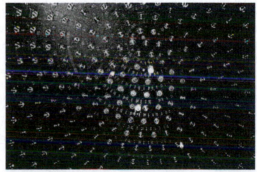

Figure 32.20 This array of photomultiplier tubes is part of the large solar neutrino detector at the Fermi National Accelerator Laboratory in Illinois. In these experiments, the neutrinos interact with heavy water and produce flashes of light, which are detected by the photomultiplier tubes. In spite of its size and the huge flux of neutrinos that strike it, very few are detected each day since they interact so weakly. This, of course, is the same reason they escape the Sun so readily. (credit: Fred Ullrich)

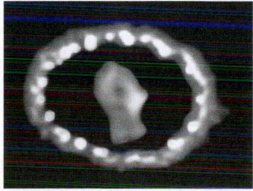

Figure 32.21 Supernovas are the source of elements heavier than iron. Energy released powers nucleosynthesis. Spectroscopic analysis of the ring of material ejected by Supernova 1987A observable in the southern hemisphere, shows evidence of heavy elements. The study of this supernova also provided indications that neutrinos might have mass. (credit: NASA, ESA, and P. Challis)

The proton-proton cycle is not a practical source of energy on Earth, in spite of the great abundance of hydrogen (^1H). The reaction $^1\text{H} + {}^1\text{H} \rightarrow {}^2\text{H} + e^+ + \nu_e$ has a very low probability of occurring. (This is why our Sun will last for about ten billion years.) However, a number of other fusion reactions are easier to induce. Among them are:

$$^2\text{H} + {}^2\text{H} \rightarrow {}^3\text{H} + {}^1\text{H} \quad (4.03 \text{ MeV}) \tag{32.17}$$

$$^2\text{H} + {}^2\text{H} \rightarrow {}^3\text{He} + n \quad (3.27 \text{ MeV}) \tag{32.18}$$

$$^2\text{H} + {}^3\text{H} \rightarrow {}^4\text{He} + n \quad (17.59 \text{ MeV}) \tag{32.19}$$

$$^2\text{H} + {}^2\text{H} \rightarrow {}^4\text{He} + \gamma \quad (23.85 \text{ MeV}). \tag{32.20}$$

Deuterium (^2H) is about 0.015% of natural hydrogen, so there is an immense amount of it in sea water alone. In addition to an abundance of deuterium fuel, these fusion reactions produce large energies per reaction (in parentheses), but they do not produce much radioactive waste. Tritium (^3H) is radioactive, but it is consumed as a fuel (the reaction $^2\text{H} + {}^3\text{H} \rightarrow {}^4\text{He} + n$), and the neutrons and γ s can be shielded. The neutrons produced can also be used to create more energy and fuel in reactions like

$$n + {}^1\text{H} \rightarrow {}^2\text{H} + \gamma \quad (20.68 \text{ MeV}) \tag{32.21}$$

and

$$n + {}^1\text{H} \rightarrow {}^2\text{H} + \gamma \quad (2.22 \text{ MeV}). \tag{32.22}$$

Note that these last two reactions, and $^2\text{H} + {}^2\text{H} \rightarrow {}^4\text{He} + \gamma$, put most of their energy output into the γ ray, and such energy is difficult to utilize.

The three keys to practical fusion energy generation are to achieve the temperatures necessary to make the reactions likely, to raise the density of the fuel, and to confine it long enough to produce large amounts of energy. These three factors—temperature, density, and time—complement one another, and so a deficiency in one can be compensated for by the others. **Ignition** is defined to occur when the reactions produce enough energy to be self-sustaining after external energy input is cut off. This goal, which must be reached before commercial plants can be a reality, has not been achieved. Another milestone, called **break-even**, occurs when the fusion power produced equals the heating power input. Break-even has nearly been reached and gives hope that ignition and commercial plants may become a reality in a few decades.

Two techniques have shown considerable promise. The first of these is called **magnetic confinement** and uses the property that charged particles have difficulty crossing magnetic field lines. The tokamak, shown in **Figure 32.22**, has shown particular promise. The tokamak's toroidal coil confines charged particles into a circular path with a helical twist due to the circulating ions themselves. In 1995, the Tokamak Fusion Test Reactor at Princeton in the US achieved world-record plasma temperatures as high as 500 million degrees Celsius. This facility operated between 1982 and 1997. A joint international effort is underway in France to build a tokamak-type reactor that will be the stepping stone to commercial power. ITER, as it is called, will be a full-scale device that aims to demonstrate the feasibility of fusion energy. It will generate 500 MW of power for extended periods of time and will achieve break-even conditions. It will study plasmas in conditions similar to those expected in a fusion power plant. Completion is scheduled for 2018.

Figure 32.22 (a) Artist's rendition of ITER, a tokamak-type fusion reactor being built in southern France. It is hoped that this gigantic machine will reach the break-even point. Completion is scheduled for 2018. (credit: Stephan Mosel, Flickr)

The second promising technique aims multiple lasers at tiny fuel pellets filled with a mixture of deuterium and tritium. Huge power input heats the fuel, evaporating the confining pellet and crushing the fuel to high density with the expanding hot plasma produced. This technique is called **inertial confinement**, because the fuel's inertia prevents it from escaping before significant fusion can take place. Higher densities have been reached than with tokamaks, but with smaller confinement times. In 2009, the Lawrence Livermore Laboratory (CA) completed a laser fusion device with 192 ultraviolet laser beams that are focused upon a

D-T pellet (see Figure 32.23).

Figure 32.23 National Ignition Facility (CA). This image shows a laser bay where 192 laser beams will focus onto a small D-T target, producing fusion. (credit: Lawrence Livermore National Laboratory, Lawrence Livermore National Security, LLC, and the Department of Energy)

Example 32.2 Calculating Energy and Power from Fusion

(a) Calculate the energy released by the fusion of a 1.00-kg mixture of deuterium and tritium, which produces helium. There are equal numbers of deuterium and tritium nuclei in the mixture.

(b) If this takes place continuously over a period of a year, what is the average power output?

Strategy

According to $^2\text{H} + {}^3\text{H} \rightarrow {}^4\text{He} + n$, the energy per reaction is 17.59 MeV. To find the total energy released, we must find the number of deuterium and tritium atoms in a kilogram. Deuterium has an atomic mass of about 2 and tritium has an atomic mass of about 3, for a total of about 5 g per mole of reactants or about 200 mol in 1.00 kg. To get a more precise figure, we will use the atomic masses from Appendix A. The power output is best expressed in watts, and so the energy output needs to be calculated in joules and then divided by the number of seconds in a year.

Solution for (a)

The atomic mass of deuterium (^2H) is 2.014102 u, while that of tritium (^3H) is 3.016049 u, for a total of 5.032151 u per reaction. So a mole of reactants has a mass of 5.03 g, and in 1.00 kg there are $(1000 \text{ g}) / (5.03 \text{ g/mol}) = 198.8 \text{ mol of reactants}$. The number of reactions that take place is therefore

$$(198.8 \text{ mol})(6.02 \times 10^{23} \text{ mol}^{-1}) = 1.20 \times 10^{26} \text{ reactions.} \tag{32.23}$$

The total energy output is the number of reactions times the energy per reaction:

$$E = (1.20 \times 10^{26} \text{ reactions})(17.59 \text{ MeV/reaction})(1.602 \times 10^{-13} \text{ J/MeV}) \tag{32.24}$$

$$= 3.37 \times 10^{14} \text{ J.}$$

Solution for (b)

Power is energy per unit time. One year has 3.16×10^7 s, so

$$P = \frac{E}{t} = \frac{3.37 \times 10^{14} \text{ J}}{3.16 \times 10^7 \text{ s}} \tag{32.25}$$

$$= 1.07 \times 10^7 \text{ W} = 10.7 \text{ MW.}$$

Discussion

By now we expect nuclear processes to yield large amounts of energy, and we are not disappointed here. The energy output of 3.37×10^{14} J from fusing 1.00 kg of deuterium and tritium is equivalent to 2.6 million gallons of gasoline and about eight times the energy output of the bomb that destroyed Hiroshima. Yet the average backyard swimming pool has about 6 kg of deuterium in it, so that fuel is plentiful if it can be utilized in a controlled manner. The average power output over a year is more than 10 MW, impressive but a bit small for a commercial power plant. About 32 times this power output would allow generation of 100 MW of electricity, assuming an efficiency of one-third in converting the fusion energy to electrical energy.

32.6 Fission

Nuclear fission is a reaction in which a nucleus is split (or *fissured*). Controlled fission is a reality, whereas controlled fusion is a hope for the future. Hundreds of nuclear fission power plants around the world attest to the fact that controlled fission is practical and, at least in the short term, economical, as seen in **Figure 32.24**. Whereas nuclear power was of little interest for decades following TMI and Chernobyl (and now Fukushima Daiichi), growing concerns over global warming has brought nuclear power back on the table as a viable energy alternative. By the end of 2009, there were 442 reactors operating in 30 countries, providing 15% of the world's electricity. France provides over 75% of its electricity with nuclear power, while the US has 104 operating reactors providing 20% of its electricity. Australia and New Zealand have none. China is building nuclear power plants at the rate of one start every month.

Figure 32.24 The people living near this nuclear power plant have no measurable exposure to radiation that is traceable to the plant. About 16% of the world's electrical power is generated by controlled nuclear fission in such plants. The cooling towers are the most prominent features but are not unique to nuclear power. The reactor is in the small domed building to the left of the towers. (credit: Kalmthouts)

Fission is the opposite of fusion and releases energy only when heavy nuclei are split. As noted in **Fusion**, energy is released if the products of a nuclear reaction have a greater binding energy per nucleon (BE/A) than the parent nuclei. **Figure 32.25** shows that BE/A is greater for medium-mass nuclei than heavy nuclei, implying that when a heavy nucleus is split, the products have less mass per nucleon, so that mass is destroyed and energy is released in the reaction. The amount of energy per fission reaction can be large, even by nuclear standards. The graph in **Figure 32.25** shows BE/A to be about 7.6 MeV/nucleon for the heaviest nuclei (A about 240), while BE/A is about 8.6 MeV/nucleon for nuclei having A about 120. Thus, if a heavy nucleus splits in half, then about 1 MeV per nucleon, or approximately 240 MeV per fission, is released. This is about 10 times the energy per fusion reaction, and about 100 times the energy of the average α , β , or γ decay.

Example 32.3 Calculating Energy Released by Fission

Calculate the energy released in the following spontaneous fission reaction:

$$^{238}\mathrm{U} \rightarrow {}^{95}\mathrm{Sr} + {}^{140}\mathrm{Xe} + 3n \tag{32.26}$$

given the atomic masses to be $m(^{238}\mathrm{U}) = 238.050784\ \mathrm{u}$, $m(^{95}\mathrm{Sr}) = 94.919388\ \mathrm{u}$, $m(^{140}\mathrm{Xe}) = 139.921610\ \mathrm{u}$, and $m(n) = 1.008665$ u .

Strategy

As always, the energy released is equal to the mass destroyed times c^2 , so we must find the difference in mass between the parent ^{238}U and the fission products.

Solution

The products have a total mass of

$$m_{products} = 94.919388 \text{ u} + 139.921610 \text{ u} + 3(1.008665 \text{ u}) \quad (32.27)$$
$$= 237.866993 \text{ u.}$$

The mass lost is the mass of ^{238}U minus $m_{products}$, or

$$\Delta m = 238.050784 \text{ u} - 237.8669933 \text{ u} = 0.183791 \text{ u,} \quad (32.28)$$

so the energy released is

$$E = (\Delta m)c^2 \quad (32.29)$$
$$= (0.183791 \text{ u})\frac{931.5 \text{ MeV}/c^2}{\text{u}}c^2 = 171.2 \text{ MeV.}$$

Discussion

A number of important things arise in this example. The 171-MeV energy released is large, but a little less than the earlier estimated 240 MeV. This is because this fission reaction produces neutrons and does not split the nucleus into two equal parts. Fission of a given nuclide, such as ^{238}U , does not always produce the same products. Fission is a statistical process in which an entire range of products are produced with various probabilities. Most fission produces neutrons, although the number varies with each fission. This is an extremely important aspect of fission, because *neutrons can induce more fission*, enabling self-sustaining chain reactions.

Spontaneous fission can occur, but this is usually not the most common decay mode for a given nuclide. For example, ^{238}U can spontaneously fission, but it decays mostly by α emission. Neutron-induced fission is crucial as seen in **Figure 32.25**. Being chargeless, even low-energy neutrons can strike a nucleus and be absorbed once they feel the attractive nuclear force. Large nuclei are described by a **liquid drop model** with surface tension and oscillation modes, because the large number of nucleons act like atoms in a drop. The neutron is attracted and thus, deposits energy, causing the nucleus to deform as a liquid drop. If stretched enough, the nucleus narrows in the middle. The number of nucleons in contact and the strength of the nuclear force binding the nucleus together are reduced. Coulomb repulsion between the two ends then succeeds in fissioning the nucleus, which pops like a water drop into two large pieces and a few neutrons. **Neutron-induced fission** can be written as

$$n + {}^{A}X \rightarrow FF_1 + FF_2 + xn, \quad (32.30)$$

where FF_1 and FF_2 are the two daughter nuclei, called **fission fragments**, and x is the number of neutrons produced. Most often, the masses of the fission fragments are not the same. Most of the released energy goes into the kinetic energy of the fission fragments, with the remainder going into the neutrons and excited states of the fragments. Since neutrons can induce fission, a self-sustaining chain reaction is possible, provided more than one neutron is produced on average — that is, if $x > 1$ in $n + {}^{A}X \rightarrow FF_1 + FF_2 + xn$. This can also be seen in **Figure 32.26**.

An example of a typical neutron-induced fission reaction is

$$n + {}^{235}_{92}U \rightarrow {}^{142}_{56}Ba + {}^{91}_{36}Kr + 3n. \quad (32.31)$$

Note that in this equation, the total charge remains the same (is conserved): $92 + 0 = 56 + 36$. Also, as far as whole numbers are concerned, the mass is constant: $1 + 235 = 142 + 91 + 3$. This is not true when we consider the masses out to 6 or 7 significant places, as in the previous example.

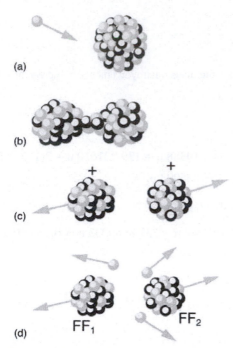

Figure 32.25 Neutron-induced fission is shown. First, energy is put into this large nucleus when it absorbs a neutron. Acting like a struck liquid drop, the nucleus deforms and begins to narrow in the middle. Since fewer nucleons are in contact, the repulsive Coulomb force is able to break the nucleus into two parts with some neutrons also flying away.

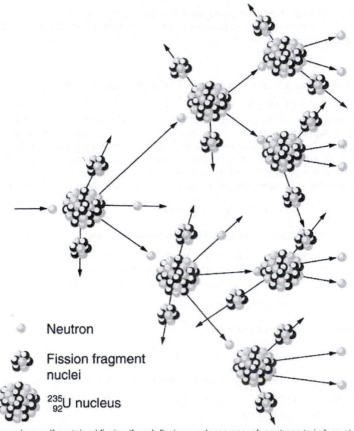

Figure 32.26 A chain reaction can produce self-sustained fission if each fission produces enough neutrons to induce at least one more fission. This depends on several factors, including how many neutrons are produced in an average fission and how easy it is to make a particular type of nuclide fission.

Not every neutron produced by fission induces fission. Some neutrons escape the fissionable material, while others interact with a nucleus without making it fission. We can enhance the number of fissions produced by neutrons by having a large amount of fissionable material. The minimum amount necessary for self-sustained fission of a given nuclide is called its **critical mass**.

Some nuclides, such as ^{239}Pu , produce more neutrons per fission than others, such as ^{235}U . Additionally, some nuclides are easier to make fission than others. In particular, ^{235}U and ^{239}Pu are easier to fission than the much more abundant ^{238}U . Both factors affect critical mass, which is smallest for ^{239}Pu .

The reason ^{235}U and ^{239}Pu are easier to fission than ^{238}U is that the nuclear force is more attractive for an even number of neutrons in a nucleus than for an odd number. Consider that $^{235}_{92}$U$_{143}$ has 143 neutrons, and $^{239}_{94}$P$_{145}$ has 145 neutrons, whereas $^{238}_{92}$U$_{146}$ has 146. When a neutron encounters a nucleus with an odd number of neutrons, the nuclear force is more attractive, because the additional neutron will make the number even. About 2-MeV more energy is deposited in the resulting nucleus than would be the case if the number of neutrons was already even. This extra energy produces greater deformation, making fission more likely. Thus, ^{235}U and ^{239}Pu are superior fission fuels. The isotope ^{235}U is only 0.72 % of natural uranium, while ^{238}U is 99.27%, and ^{239}Pu does not exist in nature. Australia has the largest deposits of uranium in the world, standing at 28% of the total. This is followed by Kazakhstan and Canada. The US has only 3% of global reserves.

Most fission reactors utilize ^{235}U , which is separated from ^{238}U at some expense. This is called enrichment. The most common separation method is gaseous diffusion of uranium hexafluoride (UF_6) through membranes. Since ^{235}U has less mass than ^{238}U , its UF_6 molecules have higher average velocity at the same temperature and diffuse faster. Another interesting characteristic of ^{235}U is that it preferentially absorbs very slow moving neutrons (with energies a fraction of an eV), whereas fission reactions produce fast neutrons with energies in the order of an MeV. To make a self-sustained fission reactor with ^{235}U , it is thus necessary to slow down ("thermalize") the neutrons. Water is very effective, since neutrons collide with protons in water molecules and lose energy. **Figure 32.27** shows a schematic of a reactor design, called the pressurized water reactor.

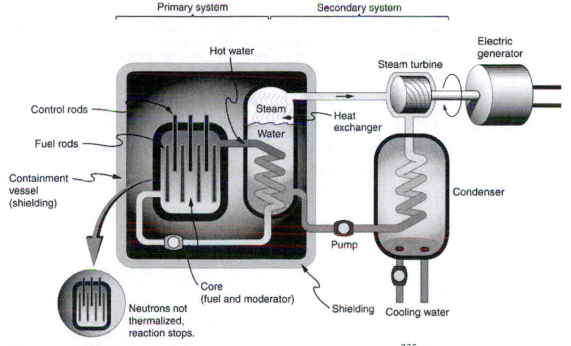

Figure 32.27 A pressurized water reactor is cleverly designed to control the fission of large amounts of ^{235}U , while using the heat produced in the fission reaction to create steam for generating electrical energy. Control rods adjust neutron flux so that criticality is obtained, but not exceeded. In case the reactor overheats and boils the water away, the chain reaction terminates, because water is needed to thermalize the neutrons. This inherent safety feature can be overwhelmed in extreme circumstances.

Control rods containing nuclides that very strongly absorb neutrons are used to adjust neutron flux. To produce large power, reactors contain hundreds to thousands of critical masses, and the chain reaction easily becomes self-sustaining, a condition called **criticality**. Neutron flux should be carefully regulated to avoid an exponential increase in fissions, a condition called **supercriticality**. Control rods help prevent overheating, perhaps even a meltdown or explosive disassembly. The water that is used to thermalize neutrons, necessary to get them to induce fission in ^{235}U , and achieve criticality, provides a negative feedback for temperature increases. In case the reactor overheats and boils the water to steam or is breached, the absence of water kills the chain reaction. Considerable heat, however, can still be generated by the reactor's radioactive fission products. Other safety features, thus, need to be incorporated in the event of a *loss of coolant* accident, including auxiliary cooling water and pumps.

Example 32.4 Calculating Energy from a Kilogram of Fissionable Fuel

Calculate the amount of energy produced by the fission of 1.00 kg of ^{235}U , given the average fission reaction of ^{235}U produces 200 MeV.

Strategy

The total energy produced is the number of ^{235}U atoms times the given energy per ^{235}U fission. We should therefore find the number of ^{235}U atoms in 1.00 kg.

Solution

The number of ^{235}U atoms in 1.00 kg is Avogadro's number times the number of moles. One mole of ^{235}U has a mass of 235.04 g; thus, there are $(1000 \text{ g})/(235.04 \text{ g/mol}) = 4.25 \text{ mol}$. The number of ^{235}U atoms is therefore,

$$(4.25 \text{ mol})(6.02 \times 10^{23} \ ^{235}\text{U/mol}) = 2.56 \times 10^{24} \ ^{235}\text{U}. \tag{32.32}$$

So the total energy released is

$$\begin{aligned} E &= (2.56 \times 10^{24} \ ^{235}\text{U})\left(\frac{200 \text{ MeV}}{^{235}\text{U}}\right)\left(\frac{1.60 \times 10^{-13} \text{ J}}{\text{MeV}}\right) \\ &= 8.21 \times 10^{13} \text{ J}. \end{aligned} \tag{32.33}$$

Discussion

This is another impressively large amount of energy, equivalent to about 14,000 barrels of crude oil or 600,000 gallons of gasoline. But, it is only one-fourth the energy produced by the fusion of a kilogram mixture of deuterium and tritium as seen in **Example 32.2**. Even though each fission reaction yields about ten times the energy of a fusion reaction, the energy per kilogram of fission fuel is less, because there are far fewer moles per kilogram of the heavy nuclides. Fission fuel is also much more scarce than fusion fuel, and less than 1% of uranium (the ^{235}U) is readily usable.

One nuclide already mentioned is ^{239}Pu , which has a 24,120-y half-life and does not exist in nature. Plutonium-239 is manufactured from ^{238}U in reactors, and it provides an opportunity to utilize the other 99% of natural uranium as an energy source. The following reaction sequence, called **breeding**, produces ^{239}Pu . Breeding begins with neutron capture by ^{238}U :

$$^{238}\text{U} + n \rightarrow \ ^{239}\text{U} + \gamma. \tag{32.34}$$

Uranium-239 then β^- decays:

$$^{239}\text{U} \rightarrow \ ^{239}\text{Np} + \beta^- + v_e(t_{1/2} = 23 \text{ min}). \tag{32.35}$$

Neptunium-239 also β^- decays:

$$^{239}\text{Np} \rightarrow \ ^{239}\text{Pu} + \beta^- + v_e(t_{1/2} = 2.4 \text{ d}). \tag{32.36}$$

Plutonium-239 builds up in reactor fuel at a rate that depends on the probability of neutron capture by ^{238}U (all reactor fuel contains more ^{238}U than ^{235}U). Reactors designed specifically to make plutonium are called **breeder reactors**. They seem to be inherently more hazardous than conventional reactors, but it remains unknown whether their hazards can be made economically acceptable. The four reactors at Chernobyl, including the one that was destroyed, were built to breed plutonium and produce electricity. These reactors had a design that was significantly different from the pressurized water reactor illustrated above.

Plutonium-239 has advantages over ^{235}U as a reactor fuel — it produces more neutrons per fission on average, and it is easier for a thermal neutron to cause it to fission. It is also chemically different from uranium, so it is inherently easier to separate from uranium ore. This means ^{239}Pu has a particularly small critical mass, an advantage for nuclear weapons.

PhET Explorations: Nuclear Fission

Start a chain reaction, or introduce non-radioactive isotopes to prevent one. Control energy production in a nuclear reactor!

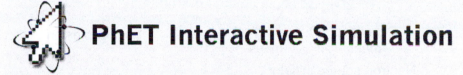

Figure 32.28 Nuclear Fission (http://cnx.org/content/m54906/1.2/nuclear-fission_en.jar)

32.7 Nuclear Weapons

Learning Objectives

By the end of this section, you will be able to:

- Discuss different types of fission and thermonuclear bombs.
- Explain the dangers and health impacts of nuclear explosions.

The world was in turmoil when fission was discovered in 1938. The discovery of fission, made by two German physicists, Otto Hahn and Fritz Strassman, was quickly verified by two Jewish refugees from Nazi Germany, Lise Meitner and her nephew Otto Frisch. Fermi, among others, soon found that not only did neutrons induce fission; more neutrons were produced during fission. The possibility of a self-sustained chain reaction was immediately recognized by leading scientists the world over. The enormous energy known to be in nuclei, but considered inaccessible, now seemed to be available on a large scale.

Within months after the announcement of the discovery of fission, Adolf Hitler banned the export of uranium from newly occupied Czechoslovakia. It seemed that the military value of uranium had been recognized in Nazi Germany, and that a serious effort to build a nuclear bomb had begun.

Alarmed scientists, many of them who fled Nazi Germany, decided to take action. None was more famous or revered than Einstein. It was felt that his help was needed to get the American government to make a serious effort at nuclear weapons as a matter of survival. Leo Szilard, an escaped Hungarian physicist, took a draft of a letter to Einstein, who, although pacifistic, signed the final version. The letter was for President Franklin Roosevelt, warning of the German potential to build extremely powerful bombs of a new type. It was sent in August of 1939, just before the German invasion of Poland that marked the start of World War II.

It was not until December 6, 1941, the day before the Japanese attack on Pearl Harbor, that the United States made a massive commitment to building a nuclear bomb. The top secret Manhattan Project was a crash program aimed at beating the Germans. It was carried out in remote locations, such as Los Alamos, New Mexico, whenever possible, and eventually came to cost billions of dollars and employ the efforts of more than 100,000 people. J. Robert Oppenheimer (1904–1967), whose talent and ambitions made him ideal, was chosen to head the project. The first major step was made by Enrico Fermi and his group in December 1942, when they achieved the first self-sustained nuclear reactor. This first "atomic pile", built in a squash court at the University of Chicago, used carbon blocks to thermalize neutrons. It not only proved that the chain reaction was possible, it began the era of nuclear reactors. Glenn Seaborg, an American chemist and physicist, received the Nobel Prize in physics in 1951 for discovery of several transuranic elements, including plutonium. Carbon-moderated reactors are relatively inexpensive and simple in design and are still used for breeding plutonium, such as at Chernobyl, where two such reactors remain in operation.

Plutonium was recognized as easier to fission with neutrons and, hence, a superior fission material very early in the Manhattan Project. Plutonium availability was uncertain, and so a uranium bomb was developed simultaneously. **Figure 32.29** shows a gun-type bomb, which takes two subcritical uranium masses and blows them together. To get an appreciable yield, the critical mass must be held together by the explosive charges inside the cannon barrel for a few microseconds. Since the buildup of the uranium chain reaction is relatively slow, the device to hold the critical mass together can be relatively simple. Owing to the fact that the rate of spontaneous fission is low, a neutron source is triggered at the same time the critical mass is assembled.

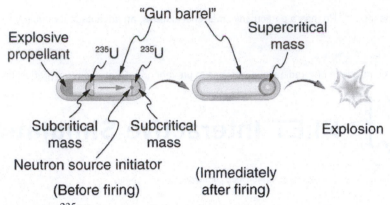

Figure 32.29 A gun-type fission bomb for ^{235}U utilizes two subcritical masses forced together by explosive charges inside a cannon barrel. The energy yield depends on the amount of uranium and the time it can be held together before it disassembles itself.

Plutonium's special properties necessitated a more sophisticated critical mass assembly, shown schematically in **Figure 32.30**. A spherical mass of plutonium is surrounded by shape charges (high explosives that release most of their blast in one direction) that implode the plutonium, crushing it into a smaller volume to form a critical mass. The implosion technique is faster and more effective, because it compresses three-dimensionally rather than one-dimensionally as in the gun-type bomb. Again, a neutron source must be triggered at just the correct time to initiate the chain reaction.

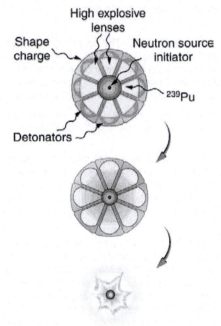

Figure 32.30 An implosion created by high explosives compresses a sphere of ^{239}Pu into a critical mass. The superior fissionability of plutonium has made it the universal bomb material.

Owing to its complexity, the plutonium bomb needed to be tested before there could be any attempt to use it. On July 16, 1945, the test named Trinity was conducted in the isolated Alamogordo Desert about 200 miles south of Los Alamos (see **Figure 32.31**). A new age had begun. The yield of this device was about 10 kilotons (kT), the equivalent of 5000 of the largest conventional bombs.

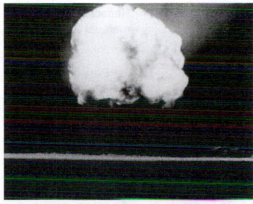

Figure 32.31 Trinity test (1945), the first nuclear bomb (credit: United States Department of Energy)

Although Germany surrendered on May 7, 1945, Japan had been steadfastly refusing to surrender for many months, forcing large casualties. Invasion plans by the Allies estimated a million casualties of their own and untold losses of Japanese lives. The bomb was viewed as a way to end the war. The first was a uranium bomb dropped on Hiroshima on August 6. Its yield of about 15 kT destroyed the city and killed an estimated 80,000 people, with 100,000 more being seriously injured (see **Figure 32.32**). The second was a plutonium bomb dropped on Nagasaki only three days later, on August 9. Its 20 kT yield killed at least 50,000 people, something less than Hiroshima because of the hilly terrain and the fact that it was a few kilometers off target. The Japanese were told that one bomb a week would be dropped until they surrendered unconditionally, which they did on August 14. In actuality, the United States had only enough plutonium for one more and as yet unassembled bomb.

Figure 32.32 Destruction in Hiroshima (credit: United States Federal Government)

Knowing that fusion produces several times more energy per kilogram of fuel than fission, some scientists pushed the idea of a fusion bomb starting very early on. Calling this bomb the Super, they realized that it could have another advantage over fission—high-energy neutrons would aid fusion, while they are ineffective in ^{239}Pu fission. Thus the fusion bomb could be virtually unlimited in energy release. The first such bomb was detonated by the United States on October 31, 1952, at Eniwetok Atoll with a yield of 10 megatons (MT), about 670 times that of the fission bomb that destroyed Hiroshima. The Soviets followed with a fusion device of their own in August 1953, and a weapons race, beyond the aim of this text to discuss, continued until the end of the Cold War.

Figure 32.33 shows a simple diagram of how a thermonuclear bomb is constructed. A fission bomb is exploded next to fusion fuel in the solid form of lithium deuteride. Before the shock wave blows it apart, γ rays heat and compress the fuel, and neutrons create tritium through the reaction $n + {}^6\text{Li} \rightarrow {}^3\text{H} + {}^4\text{He}$. Additional fusion and fission fuels are enclosed in a dense shell of ^{238}U. The shell reflects some of the neutrons back into the fuel to enhance its fusion, but at high internal temperatures fast neutrons are created that also cause the plentiful and inexpensive ^{238}U to fission, part of what allows thermonuclear bombs to be so large.

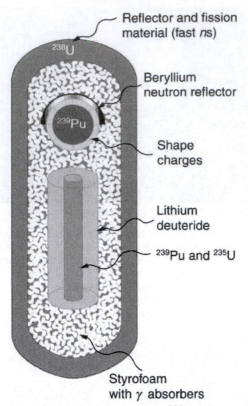

Reflector and fission
material (fast *n*s)

Beryllium
neutron reflector

Shape
charges

Lithium
deuteride

^{239}Pu and ^{235}U

Styrofoam
with γ absorbers

Figure 32.33 This schematic of a fusion bomb (H-bomb) gives some idea of how the ^{239}Pu fission trigger is used to ignite fusion fuel. Neutrons and γ rays transmit energy to the fusion fuel, create tritium from deuterium, and heat and compress the fusion fuel. The outer shell of ^{238}U serves to reflect some neutrons back into the fuel, causing more fusion, and it boosts the energy output by fissioning itself when neutron energies become high enough.

The energy yield and the types of energy produced by nuclear bombs can be varied. Energy yields in current arsenals range from about 0.1 kT to 20 MT, although the Soviets once detonated a 67 MT device. Nuclear bombs differ from conventional explosives in more than size. **Figure 32.34** shows the approximate fraction of energy output in various forms for conventional explosives and for two types of nuclear bombs. Nuclear bombs put a much larger fraction of their output into thermal energy than do conventional bombs, which tend to concentrate the energy in blast. Another difference is the immediate and residual radiation energy from nuclear weapons. This can be adjusted to put more energy into radiation (the so-called neutron bomb) so that the bomb can be used to irradiate advancing troops without killing friendly troops with blast and heat.

(a) Conventional
chemical bomb

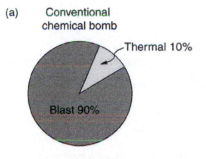

(b) Conventional
nuclear bomb

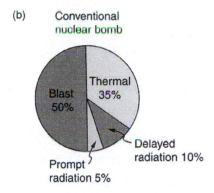

(c) Radiation-enhanced
nuclear bomb
(neutron bomb)

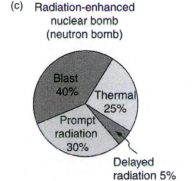

Figure 32.34 Approximate fractions of energy output by conventional and two types of nuclear weapons. In addition to yielding more energy than conventional weapons, nuclear bombs put a much larger fraction into thermal energy. This can be adjusted to enhance the radiation output to be more effective against troops. An enhanced radiation bomb is also called a neutron bomb.

At its peak in 1986, the combined arsenals of the United States and the Soviet Union totaled about 60,000 nuclear warheads. In addition, the British, French, and Chinese each have several hundred bombs of various sizes, and a few other countries have a small number. Nuclear weapons are generally divided into two categories. Strategic nuclear weapons are those intended for military targets, such as bases and missile complexes, and moderate to large cities. There were about 20,000 strategic weapons in 1988. Tactical weapons are intended for use in smaller battles. Since the collapse of the Soviet Union and the end of the Cold War in 1989, most of the 32,000 tactical weapons (including Cruise missiles, artillery shells, land mines, torpedoes, depth charges, and backpacks) have been demobilized, and parts of the strategic weapon systems are being dismantled with warheads and missiles being disassembled. According to the Treaty of Moscow of 2002, Russia and the United States have been required to reduce their strategic nuclear arsenal down to about 2000 warheads each.

A few small countries have built or are capable of building nuclear bombs, as are some terrorist groups. Two things are needed—a minimum level of technical expertise and sufficient fissionable material. The first is easy. Fissionable material is controlled but is also available. There are international agreements and organizations that attempt to control nuclear proliferation, but it is increasingly difficult given the availability of fissionable material and the small amount needed for a crude bomb. The production of fissionable fuel itself is technologically difficult. However, the presence of large amounts of such material worldwide, though in the hands of a few, makes control and accountability crucial.

Glossary

Anger camera: a common medical imaging device that uses a scintillator connected to a series of photomultipliers

break-even: when fusion power produced equals the heating power input

breeder reactors: reactors that are designed specifically to make plutonium

breeding: reaction process that produces ^{239}Pu

critical mass: minimum amount necessary for self-sustained fission of a given nuclide

criticality: condition in which a chain reaction easily becomes self-sustaining

fission fragments: a daughter nuclei

food irradiation: treatment of food with ionizing radiation

free radicals: ions with unstable oxygen- or hydrogen-containing molecules

gamma camera: another name for an Anger camera

gray (Gy): the SI unit for radiation dose which is defined to be $1 \text{ Gy} = 1 \text{ J/kg} = 100 \text{ rad}$

high dose: a dose greater than 1 Sv (100 rem)

hormesis: a term used to describe generally favorable biological responses to low exposures of toxins or radiation

ignition: when a fusion reaction produces enough energy to be self-sustaining after external energy input is cut off

inertial confinement: a technique that aims multiple lasers at tiny fuel pellets evaporating and crushing them to high density

linear hypothesis: assumption that risk is directly proportional to risk from high doses

liquid drop model: a model of nucleus (only to understand some of its features) in which nucleons in a nucleus act like atoms in a drop

low dose: a dose less than 100 mSv (10 rem)

magnetic confinement: a technique in which charged particles are trapped in a small region because of difficulty in crossing magnetic field lines

moderate dose: a dose from 0.1 Sv to 1 Sv (10 to 100 rem)

neutron-induced fission: fission that is initiated after the absorption of neutron

nuclear fission: reaction in which a nucleus splits

nuclear fusion: a reaction in which two nuclei are combined, or fused, to form a larger nucleus

positron emission tomography (PET): tomography technique that uses β^+ emitters and detects the two annihilation γ rays, aiding in source localization

proton-proton cycle: the combined reactions $^1\text{H}+^1\text{H} \rightarrow ^2\text{H}+e^+ +v_e$, $^1\text{H}+^2\text{H} \rightarrow ^3\text{He}+\gamma$, and $^3\text{He}+^3\text{He} \rightarrow ^4\text{He}+^1\text{H}+^1\text{H}$

quality factor: same as relative biological effectiveness

rad: the ionizing energy deposited per kilogram of tissue

radiolytic products: compounds produced due to chemical reactions of free radicals

radiopharmaceutical: compound used for medical imaging

radiotherapy: the use of ionizing radiation to treat ailments

relative biological effectiveness (RBE): a number that expresses the relative amount of damage that a fixed amount of ionizing radiation of a given type can inflict on biological tissues

roentgen equivalent man (rem): a dose unit more closely related to effects in biological tissue

shielding: a technique to limit radiation exposure

sievert: the SI equivalent of the rem

single-photon-emission computed tomography (SPECT): tomography performed with γ-emitting radiopharmaceuticals

supercriticality: an exponential increase in fissions

tagged: process of attaching a radioactive substance to a chemical compound

therapeutic ratio: the ratio of abnormal cells killed to normal cells killed

Section Summary

32.1 Medical Imaging and Diagnostics
- Radiopharmaceuticals are compounds that are used for medical imaging and therapeutics.
- The process of attaching a radioactive substance is called tagging.
- Table 32.1 lists certain diagnostic uses of radiopharmaceuticals including the isotope and activity typically used in diagnostics.
- One common imaging device is the Anger camera, which consists of a lead collimator, radiation detectors, and an analysis computer.
- Tomography performed with γ-emitting radiopharmaceuticals is called SPECT and has the advantages of x-ray CT scans coupled with organ- and function-specific drugs.
- PET is a similar technique that uses β^+ emitters and detects the two annihilation γ rays, which aid to localize the source.

32.2 Biological Effects of Ionizing Radiation
- The biological effects of ionizing radiation are due to two effects it has on cells: interference with cell reproduction, and destruction of cell function.
- A radiation dose unit called the rad is defined in terms of the ionizing energy deposited per kilogram of tissue:

$$1 \text{ rad} = 0.01 \text{ J/kg}.$$
- The SI unit for radiation dose is the gray (Gy), which is defined to be $1 \text{ Gy} = 1 \text{ J/kg} = 100 \text{ rad}$.
- To account for the effect of the type of particle creating the ionization, we use the relative biological effectiveness (RBE) or quality factor (QF) given in Table 32.2 and define a unit called the roentgen equivalent man (rem) as

$$\text{rem} = \text{rad} \times \text{RBE}.$$
- Particles that have short ranges or create large ionization densities have RBEs greater than unity. The SI equivalent of the rem is the sievert (Sv), defined to be

$$\text{Sv} = \text{Gy} \times \text{RBE and } 1 \text{ Sv} = 100 \text{ rem}.$$
- Whole-body, single-exposure doses of 0.1 Sv or less are low doses while those of 0.1 to 1 Sv are moderate, and those over 1 Sv are high doses. Some immediate radiation effects are given in Table 32.4. Effects due to low doses are not observed, but their risk is assumed to be directly proportional to those of high doses, an assumption known as the linear hypothesis.

  Long-term effects are cancer deaths at the rate of $10/10^6$ rem·y and genetic defects at roughly one-third this rate.

  Background radiation doses and sources are given in Table 32.5. World-wide average radiation exposure from natural sources, including radon, is about 3 mSv, or 300 mrem. Radiation protection utilizes shielding, distance, and time to limit exposure.

32.3 Therapeutic Uses of Ionizing Radiation
- Radiotherapy is the use of ionizing radiation to treat ailments, now limited to cancer therapy.
- The sensitivity of cancer cells to radiation enhances the ratio of cancer cells killed to normal cells killed, which is called the therapeutic ratio.
- Doses for various organs are limited by the tolerance of normal tissue for radiation. Treatment is localized in one region of the body and spread out in time.

32.4 Food Irradiation
- Food irradiation is the treatment of food with ionizing radiation.
- Irradiating food can destroy insects and bacteria by creating free radicals and radiolytic products that can break apart cell membranes.
- Food irradiation has produced no observable negative short-term effects for humans, but its long-term effects are unknown.

32.5 Fusion
- Nuclear fusion is a reaction in which two nuclei are combined to form a larger nucleus. It releases energy when light nuclei are fused to form medium-mass nuclei.
- Fusion is the source of energy in stars, with the proton-proton cycle,

$$^1\text{H} + {}^1\text{H} \rightarrow {}^2\text{H} + e^+ + \nu_e \quad (0.42 \text{ MeV})$$

$$^1\text{H} + {}^2\text{H} \rightarrow {}^3\text{He} + \gamma \quad (5.49 \text{ MeV})$$

$$^3\text{He} + {}^3\text{He} \rightarrow {}^4\text{He} + {}^1\text{H} + {}^1\text{H} \quad (12.86 \text{ MeV})$$

  being the principal sequence of energy-producing reactions in our Sun.

- The overall effect of the proton-proton cycle is

$$2e^- + 4\,^1\mathrm{H} \rightarrow \,^4\mathrm{He} + 2\nu_e + 6\gamma \qquad (26.7\ \mathrm{MeV}),$$

where the 26.7 MeV includes the energy of the positrons emitted and annihilated.
- Attempts to utilize controlled fusion as an energy source on Earth are related to deuterium and tritium, and the reactions play important roles.
- Ignition is the condition under which controlled fusion is self-sustaining; it has not yet been achieved. Break-even, in which the fusion energy output is as great as the external energy input, has nearly been achieved.
- Magnetic confinement and inertial confinement are the two methods being developed for heating fuel to sufficiently high temperatures, at sufficient density, and for sufficiently long times to achieve ignition. The first method uses magnetic fields and the second method uses the momentum of impinging laser beams for confinement.

32.6 Fission
- Nuclear fission is a reaction in which a nucleus is split.
- Fission releases energy when heavy nuclei are split into medium-mass nuclei.
- Self-sustained fission is possible, because neutron-induced fission also produces neutrons that can induce other fissions, $n + \,^A\mathrm{X} \rightarrow \mathrm{FF}_1 + \mathrm{FF}_2 + xn$, where FF_1 and FF_2 are the two daughter nuclei, or fission fragments, and x is the number of neutrons produced.
- A minimum mass, called the critical mass, should be present to achieve criticality.
- More than a critical mass can produce supercriticality.
- The production of new or different isotopes (especially $^{239}\mathrm{Pu}$) by nuclear transformation is called breeding, and reactors designed for this purpose are called breeder reactors.

32.7 Nuclear Weapons
- There are two types of nuclear weapons—fission bombs use fission alone, whereas thermonuclear bombs use fission to ignite fusion.
- Both types of weapons produce huge numbers of nuclear reactions in a very short time.
- Energy yields are measured in kilotons or megatons of equivalent conventional explosives and range from 0.1 kT to more than 20 MT.
- Nuclear bombs are characterized by far more thermal output and nuclear radiation output than conventional explosives.

Conceptual Questions

32.1 Medical Imaging and Diagnostics

1. In terms of radiation dose, what is the major difference between medical diagnostic uses of radiation and medical therapeutic uses?

2. One of the methods used to limit radiation dose to the patient in medical imaging is to employ isotopes with short half-lives. How would this limit the dose?

32.2 Biological Effects of Ionizing Radiation

3. Isotopes that emit α radiation are relatively safe outside the body and exceptionally hazardous inside. Yet those that emit γ radiation are hazardous outside and inside. Explain why.

4. Why is radon more closely associated with inducing lung cancer than other types of cancer?

5. The RBE for low-energy β s is 1.7, whereas that for higher-energy β s is only 1. Explain why, considering how the range of radiation depends on its energy.

6. Which methods of radiation protection were used in the device shown in the first photo in **Figure 32.35**? Which were used in the situation shown in the second photo?

(a)

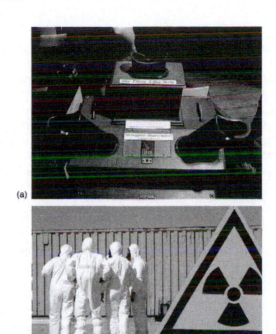

(a)

(b)

Figure 32.35 (a) This x-ray fluorescence machine is one of the thousands used in shoe stores to produce images of feet as a check on the fit of shoes. They are unshielded and remain on as long as the feet are in them, producing doses much greater than medical images. Children were fascinated with them. These machines were used in shoe stores until laws preventing such unwarranted radiation exposure were enacted in the 1950s. (credit: Andrew Kuchling) (b) Now that we know the effects of exposure to radioactive material, safety is a priority. (credit: U.S. Navy)

7. What radioisotope could be a problem in homes built of cinder blocks made from uranium mine tailings? (This is true of homes and schools in certain regions near uranium mines.)

8. Are some types of cancer more sensitive to radiation than others? If so, what makes them more sensitive?

9. Suppose a person swallows some radioactive material by accident. What information is needed to be able to assess possible damage?

32.3 Therapeutic Uses of Ionizing Radiation

10. Radiotherapy is more likely to be used to treat cancer in elderly patients than in young ones. Explain why. Why is radiotherapy used to treat young people at all?

32.4 Food Irradiation

11. Does food irradiation leave the food radioactive? To what extent is the food altered chemically for low and high doses in food irradiation?

12. Compare a low dose of radiation to a human with a low dose of radiation used in food treatment.

13. Suppose one food irradiation plant uses a ^{137}Cs source while another uses an equal activity of ^{60}Co . Assuming equal fractions of the γ rays from the sources are absorbed, why is more time needed to get the same dose using the ^{137}Cs source?

32.5 Fusion

14. Why does the fusion of light nuclei into heavier nuclei release energy?

15. Energy input is required to fuse medium-mass nuclei, such as iron or cobalt, into more massive nuclei. Explain why.

16. In considering potential fusion reactions, what is the advantage of the reaction $^2\text{H} + {}^3\text{H} \rightarrow {}^4\text{He} + n$ over the reaction

$^2\text{H} + {}^2\text{H} \rightarrow {}^3\text{He} + n$?

17. Give reasons justifying the contention made in the text that energy from the fusion reaction $^2\text{H} + {}^2\text{H} \rightarrow {}^4\text{He} + \gamma$ is relatively difficult to capture and utilize.

32.6 Fission

18. Explain why the fission of heavy nuclei releases energy. Similarly, why is it that energy input is required to fission light nuclei?

19. Explain, in terms of conservation of momentum and energy, why collisions of neutrons with protons will thermalize neutrons better than collisions with oxygen.

20. The ruins of the Chernobyl reactor are enclosed in a huge concrete structure built around it after the accident. Some rain penetrates the building in winter, and radioactivity from the building increases. What does this imply is happening inside?

21. Since the uranium or plutonium nucleus fissions into several fission fragments whose mass distribution covers a wide range of pieces, would you expect more residual radioactivity from fission than fusion? Explain.

22. The core of a nuclear reactor generates a large amount of thermal energy from the decay of fission products, even when the power-producing fission chain reaction is turned off. Would this residual heat be greatest after the reactor has run for a long time or short time? What if the reactor has been shut down for months?

23. How can a nuclear reactor contain many critical masses and not go supercritical? What methods are used to control the fission in the reactor?

24. Why can heavy nuclei with odd numbers of neutrons be induced to fission with thermal neutrons, whereas those with even numbers of neutrons require more energy input to induce fission?

25. Why is a conventional fission nuclear reactor not able to explode as a bomb?

32.7 Nuclear Weapons

26. What are some of the reasons that plutonium rather than uranium is used in all fission bombs and as the trigger in all fusion bombs?

27. Use the laws of conservation of momentum and energy to explain how a shape charge can direct most of the energy released in an explosion in a specific direction. (Note that this is similar to the situation in guns and cannons—most of the energy goes into the bullet.)

28. How does the lithium deuteride in the thermonuclear bomb shown in **Figure 32.33** supply tritium (^3H) as well as deuterium (^2H)?

29. Fallout from nuclear weapons tests in the atmosphere is mainly ^{90}Sr and ^{137}Cs, which have 28.6- and 32.2-y half-lives, respectively. Atmospheric tests were terminated in most countries in 1963, although China only did so in 1980. It has been found that environmental activities of these two isotopes are decreasing faster than their half-lives. Why might this be?

Problems & Exercises

32.1 Medical Imaging and Diagnostics

1. A neutron generator uses an α source, such as radium, to bombard beryllium, inducing the reaction $^4\text{He} + {}^9\text{Be} \rightarrow {}^{12}\text{C} + n$. Such neutron sources are called RaBe sources, or PuBe sources if they use plutonium to get the α s. Calculate the energy output of the reaction in MeV.

2. Neutrons from a source (perhaps the one discussed in the preceding problem) bombard natural molybdenum, which is 24 percent ^{98}Mo. What is the energy output of the reaction $^{98}\text{Mo} + n \rightarrow {}^{99}\text{Mo} + \gamma$? The mass of ^{98}Mo is given in Appendix A: Atomic Masses, and that of ^{99}Mo is 98.907711 u.

3. The purpose of producing ^{99}Mo (usually by neutron activation of natural molybdenum, as in the preceding problem) is to produce $^{99\text{m}}\text{Tc}$. Using the rules, verify that the β^- decay of ^{99}Mo produces $^{99\text{m}}\text{Tc}$. (Most $^{99\text{m}}\text{Tc}$ nuclei produced in this decay are left in a metastable excited state denoted $^{99\text{m}}\text{Tc}$.)

4. (a) Two annihilation γ rays in a PET scan originate at the same point and travel to detectors on either side of the patient. If the point of origin is 9.00 cm closer to one of the detectors, what is the difference in arrival times of the photons? (This could be used to give position information, but the time difference is small enough to make it difficult.)

(b) How accurately would you need to be able to measure arrival time differences to get a position resolution of 1.00 mm?

5. Table 32.1 indicates that 7.50 mCi of $^{99\text{m}}\text{Tc}$ is used in a brain scan. What is the mass of technetium?

6. The activities of ^{131}I and ^{123}I used in thyroid scans are given in Table 32.1 to be 50 and 70 μCi , respectively. Find and compare the masses of ^{131}I and ^{123}I in such scans, given their respective half-lives are 8.04 d and 13.2 h. The masses are so small that the radioiodine is usually mixed with stable iodine as a carrier to ensure normal chemistry and distribution in the body.

7. (a) Neutron activation of sodium, which is 100% ^{23}Na , produces ^{24}Na , which is used in some heart scans, as seen in Table 32.1. The equation for the reaction is $^{23}\text{Na} + n \rightarrow {}^{24}\text{Na} + \gamma$. Find its energy output, given the mass of ^{24}Na is 23.990962 u.

(b) What mass of ^{24}Na produces the needed 5.0-mCi activity, given its half-life is 15.0 h?

32.2 Biological Effects of Ionizing Radiation

8. What is the dose in mSv for: (a) a 0.1 Gy x-ray? (b) 2.5 mGy of neutron exposure to the eye? (c) 1.5 mGy of α exposure?

9. Find the radiation dose in Gy for: (a) A 10-mSv fluoroscopic x-ray series. (b) 50 mSv of skin exposure by an α emitter. (c) 160 mSv of β^- and γ rays from the ^{40}K in your body.

10. How many Gy of exposure is needed to give a cancerous tumor a dose of 40 Sv if it is exposed to α activity?

11. What is the dose in Sv in a cancer treatment that exposes the patient to 200 Gy of γ rays?

12. One half the γ rays from $^{99\text{m}}\text{Tc}$ are absorbed by a 0.170-mm-thick lead shielding. Half of the γ rays that pass through the first layer of lead are absorbed in a second layer of equal thickness. What thickness of lead will absorb all but one in 1000 of these γ rays?

13. A plumber at a nuclear power plant receives a whole-body dose of 30 mSv in 15 minutes while repairing a crucial valve. Find the radiation-induced yearly risk of death from cancer and the chance of genetic defect from this maximum allowable exposure.

14. In the 1980s, the term picowave was used to describe food irradiation in order to overcome public resistance by playing on the well-known safety of microwave radiation. Find the energy in MeV of a photon having a wavelength of a picometer.

15. Find the mass of ^{239}Pu that has an activity of 1.00 μCi .

32.3 Therapeutic Uses of Ionizing Radiation

16. A beam of 168-MeV nitrogen nuclei is used for cancer therapy. If this beam is directed onto a 0.200-kg tumor and gives it a 2.00-Sv dose, how many nitrogen nuclei were stopped? (Use an RBE of 20 for heavy ions.)

17. (a) If the average molecular mass of compounds in food is 50.0 g, how many molecules are there in 1.00 kg of food? (b) How many ion pairs are created in 1.00 kg of food, if it is exposed to 1000 Sv and it takes 32.0 eV to create an ion pair? (c) Find the ratio of ion pairs to molecules. (d) If these ion pairs recombine into a distribution of 2000 new compounds, how many parts per billion is each?

18. Calculate the dose in Sv to the chest of a patient given an x-ray under the following conditions. The x-ray beam intensity is 1.50 W/m^2 , the area of the chest exposed is 0.0750 m^2 , 35.0% of the x-rays are absorbed in 20.0 kg of tissue, and the exposure time is 0.250 s.

19. (a) A cancer patient is exposed to γ rays from a 5000-Ci ^{60}Co transillumination unit for 32.0 s. The γ rays are collimated in such a manner that only 1.00% of them strike the patient. Of those, 20.0% are absorbed in a tumor having a mass of 1.50 kg. What is the dose in rem to the tumor, if the average γ energy per decay is 1.25 MeV? None of the β s from the decay reach the patient. (b) Is the dose consistent with stated therapeutic doses?

20. What is the mass of ^{60}Co in a cancer therapy transillumination unit containing 5.00 kCi of ^{60}Co ?

21. Large amounts of ^{65}Zn are produced in copper exposed to accelerator beams. While machining contaminated copper, a physicist ingests 50.0 µCi of ^{65}Zn . Each ^{65}Zn decay emits an average γ -ray energy of 0.550 MeV, 40.0% of which is absorbed in the scientist's 75.0-kg body. What dose in mSv is caused by this in one day?

22. Naturally occurring ^{40}K is listed as responsible for 16 mrem/y of background radiation. Calculate the mass of ^{40}K that must be inside the 55-kg body of a woman to produce this dose. Each ^{40}K decay emits a 1.32-MeV β , and 50% of the energy is absorbed inside the body.

23. (a) Background radiation due to ^{226}Ra averages only 0.01 mSv/y, but it can range upward depending on where a person lives. Find the mass of ^{226}Ra in the 80.0-kg body of a man who receives a dose of 2.50-mSv/y from it, noting that each ^{226}Ra decay emits a 4.80-MeV α particle. You may neglect dose due to daughters and assume a constant amount, evenly distributed due to balanced ingestion and bodily elimination. (b) Is it surprising that such a small mass could cause a measurable radiation dose? Explain.

24. The annual radiation dose from ^{14}C in our bodies is 0.01 mSv/y. Each ^{14}C decay emits a β^- averaging 0.0750 MeV. Taking the fraction of ^{14}C to be 1.3×10^{-12} N of normal ^{12}C , and assuming the body is 13% carbon, estimate the fraction of the decay energy absorbed. (The rest escapes, exposing those close to you.)

25. If everyone in Australia received an extra 0.05 mSv per year of radiation, what would be the increase in the number of cancer deaths per year? (Assume that time had elapsed for the effects to become apparent.) Assume that there are 200×10^{-4} deaths per Sv of radiation per year. What percent of the actual number of cancer deaths recorded is this?

32.5 Fusion

26. Verify that the total number of nucleons, total charge, and electron family number are conserved for each of the fusion reactions in the proton-proton cycle in

$$^{1}H + {}^{1}H \rightarrow {}^{2}H + e^{+} + v_{e}, \quad {}^{1}H + {}^{2}H \rightarrow {}^{3}He + \gamma,$$

and

$$^{3}He + {}^{3}He \rightarrow {}^{4}He + {}^{1}H + {}^{1}H.$$

(List the value of each of the conserved quantities before and after each of the reactions.)

27. Calculate the energy output in each of the fusion reactions in the proton-proton cycle, and verify the values given in the above summary.

28. Show that the total energy released in the proton-proton cycle is 26.7 MeV, considering the overall effect in

$$^{1}H + {}^{1}H \rightarrow {}^{2}H + e^{+} + v_{e}, \quad {}^{1}H + {}^{2}H \rightarrow {}^{3}He + \gamma \text{ , and}$$

$$^{3}He + {}^{3}He \rightarrow {}^{4}He + {}^{1}H + {}^{1}H \text{ and being certain to}$$
include the annihilation energy.

29. Verify by listing the number of nucleons, total charge, and electron family number before and after the cycle that these quantities are conserved in the overall proton-proton cycle in

$$2e^{-} + 4{}^{1}H \rightarrow {}^{4}He + 2v_{e} + 6\gamma \text{ .}$$

30. The energy produced by the fusion of a 1.00-kg mixture of deuterium and tritium was found in Example **Calculating Energy and Power from Fusion**. Approximately how many kilograms would be required to supply the annual energy use in the United States?

31. Tritium is naturally rare, but can be produced by the reaction $n + {}^{2}H \rightarrow {}^{3}H + \gamma$. How much energy in MeV is released in this neutron capture?

32. Two fusion reactions mentioned in the text are

$$n + {}^{3}He \rightarrow {}^{4}He + \gamma$$

and

$$n + {}^{1}H \rightarrow {}^{2}H + \gamma \text{ .}$$

Both reactions release energy, but the second also creates more fuel. Confirm that the energies produced in the reactions are 20.58 and 2.22 MeV, respectively. Comment on which product nuclide is most tightly bound, ^{4}He or ^{2}H .

33. (a) Calculate the number of grams of deuterium in an 80,000-L swimming pool, given deuterium is 0.0150% of natural hydrogen.

(b) Find the energy released in joules if this deuterium is fused via the reaction $^{2}H + {}^{2}H \rightarrow {}^{3}He + n$.

(c) Could the neutrons be used to create more energy?

(d) Discuss the amount of this type of energy in a swimming pool as compared to that in, say, a gallon of gasoline, also taking into consideration that water is far more abundant.

34. How many kilograms of water are needed to obtain the 198.8 mol of deuterium, assuming that deuterium is 0.01500% (by number) of natural hydrogen?

35. The power output of the Sun is 4×10^{26} W .

(a) If 90% of this is supplied by the proton-proton cycle, how many protons are consumed per second?

(b) How many neutrinos per second should there be per square meter at the Earth from this process? This huge number is indicative of how rarely a neutrino interacts, since large detectors observe very few per day.

36. Another set of reactions that result in the fusing of hydrogen into helium in the Sun and especially in hotter stars is called the carbon cycle. It is

$$^{12}C + {}^1H \rightarrow {}^{13}N + \gamma,$$

$$^{13}N \rightarrow {}^{13}C + e^+ + v_e,$$

$$^{13}C + {}^1H \rightarrow {}^{14}N + \gamma,$$

$$^{14}N + {}^1H \rightarrow {}^{15}O + \gamma,$$

$$^{15}O \rightarrow {}^{15}N + e^+ + v_e,$$

$$^{15}N + {}^1H \rightarrow {}^{12}C + {}^4He.$$

Write down the overall effect of the carbon cycle (as was done for the proton-proton cycle in

$2e^- + 4\,{}^1H \rightarrow {}^4He + 2v_e + 6\gamma$). Note the number of protons (^1H) required and assume that the positrons (e^+) annihilate electrons to form more γ rays.

37. (a) Find the total energy released in MeV in each carbon cycle (elaborated in the above problem) including the annihilation energy.

(b) How does this compare with the proton-proton cycle output?

38. Verify that the total number of nucleons, total charge, and electron family number are conserved for each of the fusion reactions in the carbon cycle given in the above problem. (List the value of each of the conserved quantities before and after each of the reactions.)

39. Integrated Concepts

The laser system tested for inertial confinement can produce a 100-kJ pulse only 1.00 ns in duration. (a) What is the power output of the laser system during the brief pulse?

(b) How many photons are in the pulse, given their wavelength is $1.06\,\mu m$?

(c) What is the total momentum of all these photons?

(d) How does the total photon momentum compare with that of a single 1.00 MeV deuterium nucleus?

40. Integrated Concepts

Find the amount of energy given to the ^4He nucleus and to the γ ray in the reaction $n + {}^3He \rightarrow {}^4He + \gamma$, using the conservation of momentum principle and taking the reactants to be initially at rest. This should confirm the contention that most of the energy goes to the γ ray.

41. Integrated Concepts

(a) What temperature gas would have atoms moving fast enough to bring two ^3He nuclei into contact? Note that, because both are moving, the average kinetic energy only needs to be half the electric potential energy of these doubly charged nuclei when just in contact with one another.

(b) Does this high temperature imply practical difficulties for doing this in controlled fusion?

42. Integrated Concepts

(a) Estimate the years that the deuterium fuel in the oceans could supply the energy needs of the world. Assume world energy consumption to be ten times that of the United States which is 8×10^{19} J/y and that the deuterium in the oceans could be converted to energy with an efficiency of 32%. You must estimate or look up the amount of water in the oceans and take the deuterium content to be 0.015% of natural hydrogen to find the mass of deuterium available. Note that approximate energy yield of deuterium is 3.37×10^{14} J/kg.

(b) Comment on how much time this is by any human measure. (It is not an unreasonable result, only an impressive one.)

32.6 Fission

43. (a) Calculate the energy released in the neutron-induced fission (similar to the spontaneous fission in **Example 32.3**)

$$n + {}^{238}U \rightarrow {}^{96}Sr + {}^{140}Xe + 3n,$$

given $m(^{96}Sr) = 95.921750$ u and

$m(^{140}Xe) = 139.92164$. (b) This result is about 6 MeV greater than the result for spontaneous fission. Why? (c) Confirm that the total number of nucleons and total charge are conserved in this reaction.

44. (a) Calculate the energy released in the neutron-induced fission reaction

$$n + {}^{235}U \rightarrow {}^{92}Kr + {}^{142}Ba + 2n,$$

given $m(^{92}Kr) = 91.926269$ u and

$m(^{142}Ba) = 141.916361$ u .

(b) Confirm that the total number of nucleons and total charge are conserved in this reaction.

45. (a) Calculate the energy released in the neutron-induced fission reaction

$$n + {}^{239}Pu \rightarrow {}^{96}Sr + {}^{140}Ba + 4n,$$

given $m(^{96}Sr) = 95.921750$ u and

$m(^{140}Ba) = 139.910581$ u .

(b) Confirm that the total number of nucleons and total charge are conserved in this reaction.

46. Confirm that each of the reactions listed for plutonium breeding just following **Example 32.4** conserves the total number of nucleons, the total charge, and electron family number.

47. Breeding plutonium produces energy even before any plutonium is fissioned. (The primary purpose of the four nuclear reactors at Chernobyl was breeding plutonium for weapons. Electrical power was a by-product used by the civilian population.) Calculate the energy produced in each of the reactions listed for plutonium breeding just following **Example 32.4**. The pertinent masses are

$m(^{239}U) = 239.054289$ u , $m(^{239}Np) = 239.052932$ u

, and $m(^{239}Pu) = 239.052157$ u .

48. The naturally occurring radioactive isotope ^{232}Th does not make good fission fuel, because it has an even number of neutrons; however, it can be bred into a suitable fuel (much as ^{238}U is bred into ^{239}P).

(a) What are Z and N for ^{232}Th ?

(b) Write the reaction equation for neutron captured by ^{232}Th and identify the nuclide ^{A}X produced in

$n + {}^{232}\text{Th} \rightarrow {}^{A}X + \gamma$.

(c) The product nucleus β^{-} decays, as does its daughter. Write the decay equations for each, and identify the final nucleus.

(d) Confirm that the final nucleus has an odd number of neutrons, making it a better fission fuel.

(e) Look up the half-life of the final nucleus to see if it lives long enough to be a useful fuel.

49. The electrical power output of a large nuclear reactor facility is 900 MW. It has a 35.0% efficiency in converting nuclear power to electrical.

(a) What is the thermal nuclear power output in megawatts?

(b) How many ^{235}U nuclei fission each second, assuming the average fission produces 200 MeV?

(c) What mass of ^{235}U is fissioned in one year of full-power operation?

50. A large power reactor that has been in operation for some months is turned off, but residual activity in the core still produces 150 MW of power. If the average energy per decay of the fission products is 1.00 MeV, what is the core activity in curies?

32.7 Nuclear Weapons

51. Find the mass converted into energy by a 12.0-kT bomb.

52. What mass is converted into energy by a 1.00-MT bomb?

53. Fusion bombs use neutrons from their fission trigger to create tritium fuel in the reaction $n + {}^{6}\text{Li} \rightarrow {}^{3}\text{H} + {}^{4}\text{He}$. What is the energy released by this reaction in MeV?

54. It is estimated that the total explosive yield of all the nuclear bombs in existence currently is about 4,000 MT.

(a) Convert this amount of energy to kilowatt-hours, noting that $1 \text{ kW} \cdot \text{h} = 3.60 \times 10^{6} \text{ J}$.

(b) What would the monetary value of this energy be if it could be converted to electricity costing 10 cents per kW·h?

55. A radiation-enhanced nuclear weapon (or neutron bomb) can have a smaller total yield and still produce more prompt radiation than a conventional nuclear bomb. This allows the use of neutron bombs to kill nearby advancing enemy forces with radiation without blowing up your own forces with the blast. For a 0.500-kT radiation-enhanced weapon and a 1.00-kT conventional nuclear bomb: (a) Compare the blast yields. (b) Compare the prompt radiation yields.

56. (a) How many ^{239}Pu nuclei must fission to produce a 20.0-kT yield, assuming 200 MeV per fission? (b) What is the mass of this much ^{239}Pu ?

57. Assume one-fourth of the yield of a typical 320-kT strategic bomb comes from fission reactions averaging 200 MeV and the remainder from fusion reactions averaging 20 MeV.

(a) Calculate the number of fissions and the approximate mass of uranium and plutonium fissioned, taking the average atomic mass to be 238.

(b) Find the number of fusions and calculate the approximate mass of fusion fuel, assuming an average total atomic mass of the two nuclei in each reaction to be 5.

(c) Considering the masses found, does it seem reasonable that some missiles could carry 10 warheads? Discuss, noting that the nuclear fuel is only a part of the mass of a warhead.

58. This problem gives some idea of the magnitude of the energy yield of a small tactical bomb. Assume that half the energy of a 1.00-kT nuclear depth charge set off under an aircraft carrier goes into lifting it out of the water—that is, into gravitational potential energy. How high is the carrier lifted if its mass is 90,000 tons?

59. It is estimated that weapons tests in the atmosphere have deposited approximately 9 MCi of ^{90}Sr on the surface of the earth. Find the mass of this amount of ^{90}Sr .

60. A 1.00-MT bomb exploded a few kilometers above the ground deposits 25.0% of its energy into radiant heat.

(a) Find the calories per cm^{2} at a distance of 10.0 km by assuming a uniform distribution over a spherical surface of that radius.

(b) If this heat falls on a person's body, what temperature increase does it cause in the affected tissue, assuming it is absorbed in a layer 1.00-cm deep?

61. Integrated Concepts

One scheme to put nuclear weapons to nonmilitary use is to explode them underground in a geologically stable region and extract the geothermal energy for electricity production. There was a total yield of about 4,000 MT in the combined arsenals in 2006. If 1.00 MT per day could be converted to electricity with an efficiency of 10.0%:

(a) What would the average electrical power output be?

(b) How many years would the arsenal last at this rate?

Test Prep for AP® Courses

32.2 Biological Effects of Ionizing Radiation

1. A patient receives A rad of radiation as part of her treatment and absorbs E J of energy. The RBE of the radiation particles is R. If the RBE is increased to 1.5R, what will be the energy absorbed by the patient?

a. 1.5E J
b. E J
c. 0.75E J
d. 0.67E J

2. If a 90-kg person is exposed to 50 mrem of alpha particles (with RBE of 16), calculate the dosage (in rad) received by the person. What is the amount of energy absorbed by the person?

32.5 Fusion

3.

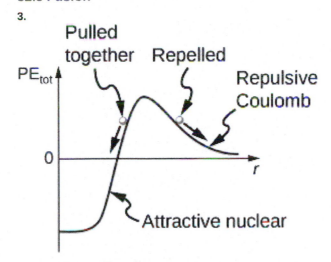

Figure 32.36 This figure shows a graph of the potential energy between two light nuclei as a function of the distance between them. Fusion can occur between the nuclei if the distance is

a. large so that kinetic energy is low.
b. large so that potential energy is low.
c. small so that nuclear attractive force can overcome Coulomb's repulsion.
d. small so that nuclear attractive force cannot overcome Coulomb's repulsion.

4. In a nuclear fusion reaction, 2 g of hydrogen is converted into 1.985 g of helium. What is the energy released?

a. 4.5×10^3 J
b. 4.5×10^6 J
c. 1.35×10^{12} J
d. 1.35×10^{15} J

5. When deuterium and tritium nuclei fuse to produce helium, what else is produced?

a. positron
b. proton
c. α-particle
d. neutron

6. Suppose two deuterium nuclei are fused to produce helium.
a. Write the equation for the fusion reaction.
b. Calculate the difference between the masses of reactants and products.
c. Using the result calculated in (b), find the energy produced in the fusion reaction.

Assume that the mass of deuterium is 2.014102 u, the mass of helium is 4.002603 u and 1 u = 1.66×10^{-27} kg.

32.6 Fission

7. Which of the following statements about nuclear fission is true?

a. No new elements can be produced in a fission reaction.
b. Energy released in fission reactions is generally less than that from fusion reactions.
c. In a fission reaction, two light nuclei are combined into a heavier one.
d. Fission reactions can be explained on the basis of the conservation of mass-energy.

8. What is the energy obtained when 10 g of mass is converted to energy with an efficiency of 70%?

a. 3.93×10^{27} MeV
b. 3.93×10^{30} MeV
c. 5.23×10^{27} MeV
d. 5.23×10^{30} MeV

9. In a neutron-induced fission reaction of ^{239}Pu, which of the following is produced along with ^{96}Sr and four neutrons?

a. $^{139}_{56}$Ba

b. $^{140}_{56}$Ba

c. $^{139}_{54}$Xe

d. $^{140}_{54}$Xe

10. When ^{235}U is bombarded with one neutron, the following fission reaction occurs: $^{235}_{92}U + n \rightarrow {}^{141}_{56}Ba + {}^{92}_{y}Kr + xn$.

a. Find the values for x and y.
b. Assuming that the mass of ^{235}U is 235.04 u, the mass of ^{141}Ba is 140.91 u, the mass of ^{92}Kr is 91.93 u, and the mass of n is 1.01 u, a student calculates the energy released in the fission reaction as 2.689×10^{-8}, but forgets to write the unit. Find the correct unit and convert the answer to MeV.

33 PARTICLE PHYSICS

Figure 33.1 Part of the Large Hadron Collider at CERN, on the border of Switzerland and France. The LHC is a particle accelerator, designed to study fundamental particles. (credit: Image Editor, Flickr)

Chapter Outline

33.1. The Yukawa Particle and the Heisenberg Uncertainty Principle Revisited

33.2. The Four Basic Forces

33.3. Accelerators Create Matter from Energy

33.4. Particles, Patterns, and Conservation Laws

33.5. Quarks: Is That All There Is?

33.6. GUTs: The Unification of Forces

Connection for AP® Courses

Continuing to use ideas that would be familiar to the ancient Greeks, we look for smaller and smaller structures in nature, hoping ultimately to find and understand the most fundamental building blocks. Atomic physics deals with the smallest units of elements and compounds. Through the study of atomic physics, we have found a relatively small number of atoms with systematic properties that explain a tremendous range of phenomena.

Nuclear physics is concerned with the nuclei of atoms and their substructures, supporting Big Idea 1, that systems have internal structure. Furthermore, the internal structure of a system determines many properties of the system (Enduring Understanding 1.A). Here, a smaller number of components—the proton and neutron—make up all nuclei. Neutrons and protons are composed of quarks. Electrons, neutrinos, photons, and quarks are examples of fundamental particles. The positive electric charge on protons and neutral charge on neutrons result from their quark compositions (Essential Knowledge 1.A.2).

This chapter divides elementary particles into fundamental particles as objects that do not have internal structure and composed particles whose properties are defined by their substructures (Essential Knowledge 1.A.2). The magnetic dipole moment, related to the properties of spin (angular momentum) and charge, is an intrinsic property of some fundamental particles such as the electron (Essential Knowledge 1.E.6). This property is the fundamental source of magnetic behavior in matter (Enduring Understanding 1.E).

Exploring the systematic behavior of interactions among particles has revealed even more about matter, forces, and energy. Mass and electric charge are properties of matter that are conserved (Enduring Understanding 1.C). The total energy of the system is also conserved (Enduring Understanding 5.B). In quantum mechanical systems, mass is actually part of the internal energy of an object or system (Essential Knowledge 5.B.11). It has been discovered experimentally that, due to certain interactions between systems, mass can be converted to energy and energy can be converted to mass (Essential Knowledge 1.C.4, Essential Knowledge 4.C.4), supporting Big Idea 4. These process can also lead to changes in the total energy of the system (Enduring Understanding 4.C).

Particle physics deals with the substructures of atoms and nuclei and is particularly aimed at finding those truly fundamental particles that have no further substructure. In general, any system can be viewed as a collection of objects, where objects do not have internal structure (Essential Knowledge 1.A.1). Just as in atomic and nuclear physics, we have found a complex array of particles and properties with systematic characteristics analogous to the periodic table and the chart of nuclides. We have discovered that changes in the systems are constrained by the conservation laws, supporting Big Idea 5. In the case of elementary particles, these conservation laws include mass-energy conservation and conservation of electric charge (Enduring Understanding 5.C). Electric charge is conserved in elementary particle reactions, even when elementary particles are produced or destroyed (Essential Knowledge 5.C.1).

The chapter revisits the ideas of fundamental forces (Enduring Understanding 3.G) and their fields in connection to elementary particles. This supports Big Ideas 2 and 3, because these particles are carriers of a specific force that provides existence of the field in space (Enduring Understanding 2.A). The field is simply the macroscopic outcome of all these force-carrying particles. The approximate relative strength and range of the gravitational force (Essential Knowledge 3.G.1), electromagnetic force (Essential Knowledge 3.G.2), strong force (Essential Knowledge 3.G.3) and weak force are considered in relation to the properties of their carrier particles. The details of these considerations go beyond AP® expectations.

An underlying structure is apparent, and there is some reason to think that we *are* finding particles that have no substructure. Of course, we have been in similar situations before. For example, atoms were once thought to be the ultimate substructure. Perhaps we will find deeper and deeper structures and never come to an ultimate substructure. We may never really know, as indicated in Figure 33.2.

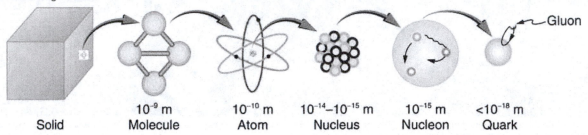

Figure 33.2 The properties of matter are based on substructures called molecules and atoms. Molecules are formed from atoms. Atoms have the substructure of a nucleus with orbiting electrons, the interactions of which explain atomic properties. Protons and neutrons, the interactions of which explain the stability and abundance of elements, form the substructure of nuclei. Protons and neutrons are not fundamental—they are composed of quarks. Like electrons and a few other particles, quarks may be fundamental building blocks, lacking any further substructure. But the story is not complete, because quarks and electrons may have substructure smaller than is presently observable.

This chapter covers the basics of particle physics as we know it today. An amazing convergence of topics is evolving in modern particle physics. We find that some particles are intimately related to forces, and that nature on the smallest scale may have a defining influence on the large-scale character of the universe. The study of particle physics is an adventure beyond even the best science fiction, because it is not only fantastic, it is real.

Big Idea 1 Objects and systems have properties such as mass and charge. Systems may have internal structure.

Enduring Understanding 1.A The internal structure of a system determines many properties of the system.

Essential Knowledge 1.A.1 A system is an object or a collection of objects. Objects are treated as having no internal structure.

Essential Knowledge 1.A.2 Fundamental particles have no internal structure.

Enduring Understanding 1.C Objects and systems have properties of inertial mass and gravitational mass that are experimentally verified to be the same and that satisfy conservation principles.

Essential Knowledge 1.C.4 In certain processes, mass can be converted to energy and energy can be converted to mass according to $E = mc^2$, the equation derived from the theory of special relativity.

Enduring Understanding 1.E Materials have many macroscopic properties that result from the arrangement and interactions of the atoms and molecules that make up the material.

Essential Knowledge 1.E.6 Matter has a property called magnetic dipole moment.

a. Magnetic dipole moment is a fundamental source of magnetic behavior of matter and an intrinsic property of some fundamental particles such as the electron.

Big Idea 2 Fields existing in space can be used to explain interactions.

Enduring Understanding 2.A A field associates a value of some physical quantity with every point in space. Field models are useful for describing interactions that occur at a distance (long-range forces) as well as a variety of other physical phenomena.

Big Idea 3 The interactions of an object with other objects can be described by forces.

Enduring Understanding 3.G Certain types of forces are considered fundamental.

Essential Knowledge 3.G.1: Gravitational forces are exerted at all scales and dominate at the largest distance and mass scales.

Essential Knowledge 3.G.2 Electromagnetic forces are exerted at all scales and can dominate at the human scale.

Essential Knowledge 3.G.3 The strong force is exerted at nuclear scales and dominates the interactions of nucleons.

Big Idea 4 Interactions between systems can result in changes in those systems.

Enduring Understanding 4.C Interactions with other objects or systems can change the total energy of a system.

Essential Knowledge 4.C.4 Mass can be converted into energy and energy can be converted into mass.

Big Idea 5. Changes that occur as a result of interactions are constrained by conservation laws.

Enduring Understanding 5.B The energy of a system is conserved.

Essential Knowledge 5.B.11 Beyond the classical approximation, mass is actually part of the internal energy of an object or system with $E = mc^2$.

Enduring Understanding 5.C The electric charge of a system is conserved.

Essential Knowledge 5.C.1 Electric charge is conserved in nuclear and elementary particle reactions, even when elementary particles are produced or destroyed. Examples should include equations representing nuclear decay.

33.1 The Yukawa Particle and the Heisenberg Uncertainty Principle Revisited

Learning Objectives
By the end of this section, you will be able to: • Define Yukawa particle.   • State the Heisenberg uncertainty principle.   • Describe a pion.   • Estimate the mass of a pion.   • Explain what a meson is.

Particle physics as we know it today began with the ideas of Hideki Yukawa in 1935. Physicists had long been concerned with how forces are transmitted, finding the concept of fields, such as electric and magnetic fields to be very useful. A field surrounds an object and carries the force exerted by the object through space. Yukawa was interested in the strong nuclear force in particular and found an ingenious way to explain its short range. His idea is a blend of particles, forces, relativity, and quantum mechanics that is applicable to all forces. Yukawa proposed that force is transmitted by the exchange of particles (called carrier particles). The field consists of these carrier particles.

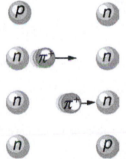

Figure 33.3 The strong nuclear force is transmitted between a proton and neutron by the creation and exchange of a pion. The pion is created through a temporary violation of conservation of mass-energy and travels from the proton to the neutron and is recaptured. It is not directly observable and is called a virtual particle. Note that the proton and neutron change identity in the process. The range of the force is limited by the fact that the pion can only exist for the short time allowed by the Heisenberg uncertainty principle. Yukawa used the finite range of the strong nuclear force to estimate the mass of the pion; the shorter the range, the larger the mass of the carrier particle.

Specifically for the strong nuclear force, Yukawa proposed that a previously unknown particle, now called a **pion**, is exchanged between nucleons, transmitting the force between them. **Figure 33.3** illustrates how a pion would carry a force between a proton and a neutron. The pion has mass and can only be created by violating the conservation of mass-energy. This is allowed by the Heisenberg uncertainty principle if it occurs for a sufficiently short period of time. As discussed in **Probability: The Heisenberg Uncertainty Principle** the Heisenberg uncertainty principle relates the uncertainties ΔE in energy and Δt in time by

$$\Delta E \Delta t \geq \frac{h}{4\pi},$$

(33.1)

where h is Planck's constant. Therefore, conservation of mass-energy can be violated by an amount ΔE for a time $\Delta t \approx \frac{h}{4\pi \Delta E}$ in which time no process can detect the violation. This allows the temporary creation of a particle of mass m, where $\Delta E = mc^2$. The larger the mass and the greater the ΔE, the shorter is the time it can exist. This means the range of the force is limited, because the particle can only travel a limited distance in a finite amount of time. In fact, the maximum distance is $d \approx c\Delta t$, where c is the speed of light. The pion must then be captured and, thus, cannot be directly observed because that would amount to a permanent violation of mass-energy conservation. Such particles (like the pion above) are called **virtual particles**, because they cannot be directly observed but their *effects* can be directly observed. Realizing all this, Yukawa used the information on the range of the strong nuclear force to estimate the mass of the pion, the particle that carries it. The steps of his reasoning are approximately retraced in the following worked example:

Example 33.1 Calculating the Mass of a Pion

Taking the range of the strong nuclear force to be about 1 fermi (10^{-15} m), calculate the approximate mass of the pion carrying the force, assuming it moves at nearly the speed of light.

Strategy

The calculation is approximate because of the assumptions made about the range of the force and the speed of the pion, but also because a more accurate calculation would require the sophisticated mathematics of quantum mechanics. Here, we use the Heisenberg uncertainty principle in the simple form stated above, as developed in **Probability: The Heisenberg Uncertainty Principle**. First, we must calculate the time Δt that the pion exists, given that the distance it travels at nearly the speed of light is about 1 fermi. Then, the Heisenberg uncertainty principle can be solved for the energy ΔE, and from that the mass of the pion can be determined. We will use the units of MeV/c^2 for mass, which are convenient since we are often considering converting mass to energy and vice versa.

Solution

The distance the pion travels is $d \approx c\Delta t$, and so the time during which it exists is approximately

$$\Delta t \approx \frac{d}{c} = \frac{10^{-15} \text{ m}}{3.0 \times 10^8 \text{ m/s}} \tag{33.2}$$

$$\approx 3.3 \times 10^{-24} \text{ s}.$$

Now, solving the Heisenberg uncertainty principle for ΔE gives

$$\Delta E \approx \frac{h}{4\pi\Delta t} \approx \frac{6.63 \times 10^{-34} \text{ J} \cdot \text{s}}{4\pi(3.3 \times 10^{-24} \text{ s})}. \tag{33.3}$$

Solving this and converting the energy to MeV gives

$$\Delta E \approx \left(1.6 \times 10^{-11} \text{ J}\right) \frac{1 \text{ MeV}}{1.6 \times 10^{-13} \text{ J}} = 100 \text{ MeV}. \tag{33.4}$$

Mass is related to energy by $\Delta E = mc^2$, so that the mass of the pion is $m = \Delta E / c^2$, or

$$m \approx 100 \text{ MeV}/c^2. \tag{33.5}$$

Discussion

This is about 200 times the mass of an electron and about one-tenth the mass of a nucleon. No such particles were known at the time Yukawa made his bold proposal.

Yukawa's proposal of particle exchange as the method of force transfer is intriguing. But how can we verify his proposal if we cannot observe the virtual pion directly? If sufficient energy is in a nucleus, it would be possible to free the pion—that is, to create its mass from external energy input. This can be accomplished by collisions of energetic particles with nuclei, but energies greater than 100 MeV are required to conserve both energy and momentum. In 1947, pions were observed in cosmic-ray experiments, which were designed to supply a small flux of high-energy protons that may collide with nuclei. Soon afterward, accelerators of sufficient energy were creating pions in the laboratory under controlled conditions. Three pions were discovered, two with charge and one neutral, and given the symbols π^+, π^-, and π^0, respectively. The masses of π^+ and π^- are identical at $139.6 \text{ MeV}/c^2$, whereas π^0 has a mass of $135.0 \text{ MeV}/c^2$. These masses are close to the predicted value of $100 \text{ MeV}/c^2$ and, since they are intermediate between electron and nucleon masses, the particles are given the name **meson** (now an entire class of particles, as we shall see in **Particles, Patterns, and Conservation Laws**).

The pions, or π-mesons as they are also called, have masses close to those predicted and feel the strong nuclear force. Another previously unknown particle, now called the muon, was discovered during cosmic-ray experiments in 1936 (one of its discoverers, Seth Neddermeyer, also originated the idea of implosion for plutonium bombs). Since the mass of a muon is around $106 \text{ MeV}/c^2$, at first it was thought to be the particle predicted by Yukawa. But it was soon realized that muons do not feel the strong nuclear force and could not be Yukawa's particle. Their role was unknown, causing the respected physicist I. I. Rabi to comment, "Who ordered that?" This remains a valid question today. We have discovered hundreds of subatomic particles; the roles of some are only partially understood. But there are various patterns and relations to forces that have led to profound insights into nature's secrets.

Summary

- Yukawa's idea of virtual particle exchange as the carrier of forces is crucial, with virtual particles being formed in temporary violation of the conservation of mass-energy as allowed by the Heisenberg uncertainty principle.

33.2 The Four Basic Forces

Learning Objectives

By the end of this section, you will be able to:

- State the four basic forces.
- Explain the Feynman diagram for the exchange of a virtual photon between two positive charges.
- Define QED.
- Describe the Feynman diagram for the exchange of a photon between a proton and a neutron.

The information presented in this section supports the following AP® learning objectives and science practices:

- **3.G.1.1** The student is able to articulate situations when the gravitational force is the dominant force and when the electromagnetic, weak, and strong forces can be ignored. **(S.P. 7.1)**
- **3.G.1.2** The student is able to connect the strength of the gravitational force between two objects to the spatial scale of the situation and the masses of the objects involved and compare that strength to other types of forces. **(S.P. 7.1)**
- **3.G.2.1** The student is able to connect the strength of electromagnetic forces with the spatial scale of the situation, the magnitude of the electric charges, and the motion of the electrically charged objects involved. **(S.P. 7.1)**
- **3.G.3.1** The student is able to identify the strong force as the force responsible for holding the nucleus together. **(S.P. 7.2)**

As first discussed in **Problem-Solving Strategies** and mentioned at various points in the text since then, there are only four distinct basic forces in all of nature. This is a remarkably small number considering the myriad phenomena they explain. Particle physics is intimately tied to these four forces. Certain fundamental particles, called carrier particles, carry these forces, and all particles can be classified according to which of the four forces they feel. The table given below summarizes important characteristics of the four basic forces.

Table 33.1 Properties of the Four Basic Forces

Force	Approximate relative strength	Range	+/−[1]	Carrier particle
Gravity	10^{-38}	∞	+ only	Graviton (conjectured)
Electromagnetic	10^{-2}	∞	+ / −	Photon (observed)
Weak force	10^{-13}	$< 10^{-18}$ m	+ / −	W^+, W^-, Z^0 (observed[2])
Strong force	1	$< 10^{-15}$ m	+ / −	Gluons (conjectured[3])

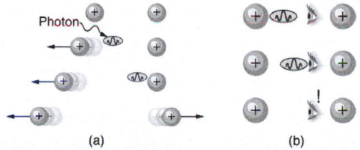

Figure 33.4 The first image shows the exchange of a virtual photon transmitting the electromagnetic force between charges, just as virtual pion exchange carries the strong nuclear force between nucleons. The second image shows that the photon cannot be directly observed in its passage, because this would disrupt it and alter the force. In this case it does not get to the other charge.

1. + attractive; - repulsive; +/− both.
2. Predicted by theory and first observed in 1983.
3. Eight proposed—indirect evidence of existence. Underlie meson exchange.

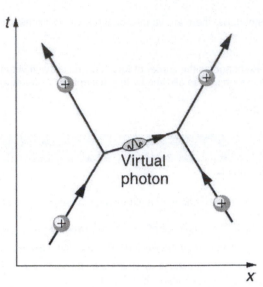

Figure 33.5 The Feynman diagram for the exchange of a virtual photon between two positive charges illustrates how the electromagnetic force is transmitted on a quantum mechanical scale. Time is graphed vertically while the distance is graphed horizontally. The two positive charges are seen to be repelled by the photon exchange.

Although these four forces are distinct and differ greatly from one another under all but the most extreme circumstances, we can see similarities among them. (In **GUTs: the Unification of Forces**, we will discuss how the four forces may be different manifestations of a single unified force.) Perhaps the most important characteristic among the forces is that they are all transmitted by the exchange of a carrier particle, exactly like what Yukawa had in mind for the strong nuclear force. Each carrier particle is a virtual particle—it cannot be directly observed while transmitting the force. **Figure 33.4** shows the exchange of a virtual photon between two positive charges. The photon cannot be directly observed in its passage, because this would disrupt it and alter the force.

Figure 33.5 shows a way of graphing the exchange of a virtual photon between two positive charges. This graph of time versus position is called a **Feynman diagram**, after the brilliant American physicist Richard Feynman (1918–1988) who developed it.

Figure 33.6 is a Feynman diagram for the exchange of a virtual pion between a proton and a neutron representing the same interaction as in **Figure 33.3**. Feynman diagrams are not only a useful tool for visualizing interactions at the quantum mechanical level, they are also used to calculate details of interactions, such as their strengths and probability of occurring. Feynman was one of the theorists who developed the field of **quantum electrodynamics** (QED), which is the quantum mechanics of electromagnetism. QED has been spectacularly successful in describing electromagnetic interactions on the submicroscopic scale. Feynman was an inspiring teacher, had a colorful personality, and made a profound impact on generations of physicists. He shared the 1965 Nobel Prize with Julian Schwinger and S. I. Tomonaga for work in QED with its deep implications for particle physics.

Why is it that particles called gluons are listed as the carrier particles for the strong nuclear force when, in **The Yukawa Particle and the Heisenberg Uncertainty Principle Revisited**, we saw that pions apparently carry that force? The answer is that pions are exchanged but they have a substructure and, as we explore it, we find that the strong force is actually related to the indirectly observed but more fundamental **gluons**. In fact, all the carrier particles are thought to be fundamental in the sense that they have no substructure. Another similarity among carrier particles is that they are all bosons (first mentioned in **Patterns in Spectra Reveal More Quantization**), having integral intrinsic spins.

There is a relationship between the mass of the carrier particle and the range of the force. The photon is massless and has energy. So, the existence of (virtual) photons is possible only by virtue of the Heisenberg uncertainty principle and can travel an unlimited distance. Thus, the range of the electromagnetic force is infinite. This is also true for gravity. It is infinite in range because its carrier particle, the graviton, has zero rest mass. (Gravity is the most difficult of the four forces to understand on a quantum scale because it affects the space and time in which the others act. But gravity is so weak that its effects are extremely difficult to observe quantum mechanically. We shall explore it further in **General Relativity and Quantum Gravity**). The

W^+, W^-, and Z^0 particles that carry the weak nuclear force have mass, accounting for the very short range of this force. In

fact, the W^+, W^-, and Z^0 are about 1000 times more massive than pions, consistent with the fact that the range of the weak nuclear force is about 1/1000 that of the strong nuclear force. Gluons are actually massless, but since they act inside massive carrier particles like pions, the strong nuclear force is also short ranged.

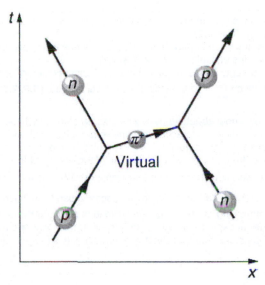

Figure 33.6 The image shows a Feynman diagram for the exchange of a π^+ between a proton and a neutron, carrying the strong nuclear force between them. This diagram represents the situation shown more pictorially in **Figure 33.4**.

The relative strengths of the forces given in the **Table 33.1** are those for the most common situations. When particles are brought very close together, the relative strengths change, and they may become identical at extremely close range. As we shall see in **GUTs: the Unification of Forces**, carrier particles may be altered by the energy required to bring particles very close together—in such a manner that they become identical.

Making Connections: Why You Stay on the Earth, but Do Not Fall Through

You are familiar with gravity pulling you towards the Earth. It's why when you jump, you come back down. In this action, and at distances and speeds that we experience in our everyday lives, gravity is the only one of the four fundamental forces that has such an obvious effect on us.

Electromagnetism is vital for our society to run, but due to your body having the same (or very nearly the same) number of positive and negative charges, it doesn't usually have as much of an effect on us. Except for one very important feature: the electrons in the bottom of your feet experience a mutually repulsive force with the electrons in the material you stand on. This is what keeps us from falling into the planet, and also allows us to push on other objects and generally interact with them.

These electromagnetic forces are dominant in the electron shells of an atom, and also the interaction of the electrons with the nucleus. However, within the nucleus, the electrostatic repulsion of the protons would break the nucleus apart if it were not for the strong force, which holds the nucleus together. At even smaller scales, within nucleons such as protons and neutrons, the weak force is responsible for nuclear decays.

The relative strengths of the forces given in the **Table 33.1** are those for the most common situations. When particles are brought very close together, the relative strengths change, and they may become identical at extremely close range. As we shall see in **GUTs: the Unification of Forces**, carrier particles may be altered by the energy required to bring particles very close together—in such a manner that they become identical.

Summary

- The four basic forces and their carrier particles are summarized in the **Table 33.1**.
- Feynman diagrams are graphs of time versus position and are highly useful pictorial representations of particle processes.
- The theory of electromagnetism on the particle scale is called quantum electrodynamics (QED).

33.3 Accelerators Create Matter from Energy

Learning Objectives

By the end of this section, you will be able to:

- State the principle of a cyclotron.
- Explain the principle of a synchrotron.
- Describe the voltage needed by an accelerator between accelerating tubes.
- State Fermilab's accelerator principle.

The information presented in this section supports the following AP® learning objectives and science practices:

- **1.C.4.1** The student is able to articulate the reasons that the theory of conservation of mass was replaced by the theory of conservation of mass–energy. **(S.P. 6.3)**
- **4.C.4.1** The student is able to apply mathematical routines to describe the relationship between mass and energy and apply this concept across domains of scale. **(S.P. 2.2, 2.3, 7.2)**
- **5.B.11.1** The student is able to apply conservation of mass and conservation of energy concepts to a natural phenomenon and use the equation $E = mc^2$ to make a related calculation. **(S.P. 2.2, 7.2)**

Before looking at all the particles we now know about, let us examine some of the machines that created them. The fundamental process in creating previously unknown particles is to accelerate known particles, such as protons or electrons, and direct a beam of them toward a target. Collisions with target nuclei provide a wealth of information, such as information obtained by Rutherford using energetic helium nuclei from natural α radiation. But if the energy of the incoming particles is large enough, new matter is sometimes created in the collision. The more energy input or ΔE, the more matter m can be created, since

$m = \Delta E / c^2$. Limitations are placed on what can occur by known conservation laws, such as conservation of mass-energy, momentum, and charge. Even more interesting are the unknown limitations provided by nature. Some expected reactions do occur, while others do not, and still other unexpected reactions may appear. New laws are revealed, and the vast majority of what we know about particle physics has come from accelerator laboratories. It is the particle physicist's favorite indoor sport, which is partly inspired by theory.

Early Accelerators

An early accelerator is a relatively simple, large-scale version of the electron gun. The **Van de Graaff** (named after the Dutch physicist), which you have likely seen in physics demonstrations, is a small version of the ones used for nuclear research since their invention for that purpose in 1932. For more, see **Figure 33.7**. These machines are electrostatic, creating potentials as great as 50 MV, and are used to accelerate a variety of nuclei for a range of experiments. Energies produced by Van de Graaffs are insufficient to produce new particles, but they have been instrumental in exploring several aspects of the nucleus. Another, equally famous, early accelerator is the **cyclotron**, invented in 1930 by the American physicist, E. O. Lawrence (1901–1958). For a visual representation with more detail, see **Figure 33.8**. Cyclotrons use fixed-frequency alternating electric fields to accelerate particles. The particles spiral outward in a magnetic field, making increasingly larger radius orbits during acceleration. This clever arrangement allows the successive addition of electric potential energy and so greater particle energies are possible than in a Van de Graaff. Lawrence was involved in many early discoveries and in the promotion of physics programs in American universities. He was awarded the 1939 Nobel Prize in Physics for the cyclotron and nuclear activations, and he has an element and two major laboratories named for him.

A **synchrotron** is a version of a cyclotron in which the frequency of the alternating voltage and the magnetic field strength are increased as the beam particles are accelerated. Particles are made to travel the same distance in a shorter time with each cycle in fixed-radius orbits. A ring of magnets and accelerating tubes, as shown in **Figure 33.9**, are the major components of synchrotrons. Accelerating voltages are synchronized (i.e., occur at the same time) with the particles to accelerate them, hence the name. Magnetic field strength is increased to keep the orbital radius constant as energy increases. High-energy particles require strong magnetic fields to steer them, so superconducting magnets are commonly employed. Still limited by achievable magnetic field strengths, synchrotrons need to be very large at very high energies, since the radius of a high-energy particle's orbit is very large. Radiation caused by a magnetic field accelerating a charged particle perpendicular to its velocity is called **synchrotron radiation** in honor of its importance in these machines. Synchrotron radiation has a characteristic spectrum and polarization, and can be recognized in cosmic rays, implying large-scale magnetic fields acting on energetic and charged particles in deep space. Synchrotron radiation produced by accelerators is sometimes used as a source of intense energetic electromagnetic radiation for research purposes.

Figure 33.7 An artist's rendition of a Van de Graaff generator.

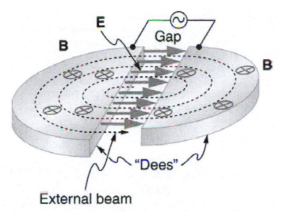

Figure 33.8 Cyclotrons use a magnetic field to cause particles to move in circular orbits. As the particles pass between the plates of the Ds, the voltage across the gap is oscillated to accelerate them twice in each orbit.

Modern Behemoths and Colliding Beams

Physicists have built ever-larger machines, first to reduce the wavelength of the probe and obtain greater detail, then to put greater energy into collisions to create new particles. Each major energy increase brought new information, sometimes producing spectacular progress, motivating the next step. One major innovation was driven by the desire to create more massive particles. Since momentum needs to be conserved in a collision, the particles created by a beam hitting a stationary target should recoil. This means that part of the energy input goes into recoil kinetic energy, significantly limiting the fraction of the beam energy that can be converted into new particles. One solution to this problem is to have head-on collisions between particles moving in opposite directions. **Colliding beams** are made to meet head-on at points where massive detectors are located. Since the total incoming momentum is zero, it is possible to create particles with momenta and kinetic energies near zero. Particles with masses equivalent to twice the beam energy can thus be created. Another innovation is to create the antimatter counterpart of the beam particle, which thus has the opposite charge and circulates in the opposite direction in the same beam pipe. For a schematic representation, see **Figure 33.10**.

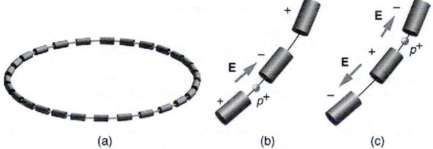

(a) (b) (c)

Figure 33.9 (a) A synchrotron has a ring of magnets and accelerating tubes. The frequency of the accelerating voltages is increased to cause the beam particles to travel the same distance in shorter time. The magnetic field should also be increased to keep each beam burst traveling in a fixed-radius path. Limits on magnetic field strength require these machines to be very large in order to accelerate particles to very high energies. (b) A positive particle is shown in the gap between accelerating tubes. (c) While the particle passes through the tube, the potentials are reversed so that there is another acceleration at the next gap. The frequency of the reversals needs to be varied as the particle is accelerated to achieve successive accelerations in each gap.

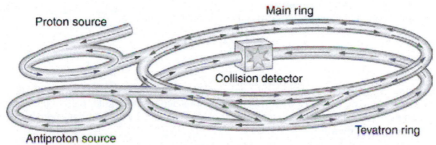

Figure 33.10 This schematic shows the two rings of Fermilab's accelerator and the scheme for colliding protons and antiprotons (not to scale).

Detectors capable of finding the new particles in the spray of material that emerges from colliding beams are as impressive as the accelerators. While the Fermilab Tevatron had proton and antiproton beam energies of about 1 TeV, so that it can create particles up to $2 \text{ TeV}/c^2$, the Large Hadron Collider (LHC) at the European Center for Nuclear Research (CERN) has achieved beam energies of 3.5 TeV, so that it has a 7-TeV collision energy; CERN hopes to double the beam energy in 2014. The now-canceled Superconducting Super Collider was being constructed in Texas with a design energy of 20 TeV to give a 40-TeV collision energy. It was to be an oval 30 km in diameter. Its cost as well as the politics of international research funding led to its demise.

In addition to the large synchrotrons that produce colliding beams of protons and antiprotons, there are other large electron-positron accelerators. The oldest of these was a straight-line or **linear accelerator**, called the Stanford Linear Accelerator (SLAC), which accelerated particles up to 50 GeV as seen in Figure 33.11. Positrons created by the accelerator were brought to the same energy and collided with electrons in specially designed detectors. Linear accelerators use accelerating tubes similar to those in synchrotrons, but aligned in a straight line. This helps eliminate synchrotron radiation losses, which are particularly severe for electrons made to follow curved paths. CERN had an electron-positron collider appropriately called the Large Electron-Positron Collider (LEP), which accelerated particles to 100 GeV and created a collision energy of 200 GeV. It was 8.5 km in diameter, while the SLAC machine was 3.2 km long.

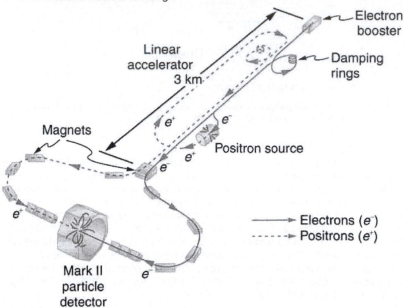

Figure 33.11 The Stanford Linear Accelerator was 3.2 km long and had the capability of colliding electron and positron beams. SLAC was also used to probe nucleons by scattering extremely short wavelength electrons from them. This produced the first convincing evidence of a quark structure inside nucleons in an experiment analogous to those performed by Rutherford long ago.

Example 33.2 Calculating the Voltage Needed by the Accelerator Between Accelerating Tubes

A linear accelerator designed to produce a beam of 800-MeV protons has 2000 accelerating tubes. What average voltage must be applied between tubes (such as in the gaps in Figure 33.9) to achieve the desired energy?

Strategy

The energy given to the proton in each gap between tubes is $\mathrm{PE}_{\mathrm{elec}} = qV$ where q is the proton's charge and V is the potential difference (voltage) across the gap. Since $q = q_e = 1.6 \times 10^{-19}\,\mathrm{C}$ and $1\,\mathrm{eV} = (1\,\mathrm{V})(1.6 \times 10^{-19}\,\mathrm{C})$, the proton gains 1 eV in energy for each volt across the gap that it passes through. The AC voltage applied to the tubes is timed so that it adds to the energy in each gap. The effective voltage is the sum of the gap voltages and equals 800 MV to give each proton an energy of 800 MeV.

Solution

There are 2000 gaps and the sum of the voltages across them is 800 MV; thus,

$$V_{\mathrm{gap}} = \frac{800\ \mathrm{MV}}{2000} = 400\ \mathrm{kV}. \tag{33.6}$$

Discussion

A voltage of this magnitude is not difficult to achieve in a vacuum. Much larger gap voltages would be required for higher energy, such as those at the 50-GeV SLAC facility. Synchrotrons are aided by the circular path of the accelerated particles, which can orbit many times, effectively multiplying the number of accelerations by the number of orbits. This makes it possible to reach energies greater than 1 TeV.

Summary

- A variety of particle accelerators have been used to explore the nature of subatomic particles and to test predictions of particle theories.
- Modern accelerators used in particle physics are either large synchrotrons or linear accelerators.
- The use of colliding beams makes much greater energy available for the creation of particles, and collisions between matter and antimatter allow a greater range of final products.

33.4 Particles, Patterns, and Conservation Laws

Learning Objectives
By the end of this section, you will be able to: • Define matter and antimatter. • Outline the differences between hadrons and leptons. • State the differences between mesons and baryons. The information presented in this section supports the following AP® learning objectives and science practices: • **5.C.1.1** The student is able to analyze electric charge conservation for nuclear and elementary particle reactions and make predictions related to such reactions based upon conservation of charge.

In the early 1930s only a small number of subatomic particles were known to exist—the proton, neutron, electron, photon and, indirectly, the neutrino. Nature seemed relatively simple in some ways, but mysterious in others. Why, for example, should the particle that carries positive charge be almost 2000 times as massive as the one carrying negative charge? Why does a neutral particle like the neutron have a magnetic moment? Does this imply an internal structure with a distribution of moving charges? Why is it that the electron seems to have no size other than its wavelength, while the proton and neutron are about 1 fermi in size? So, while the number of known particles was small and they explained a great deal of atomic and nuclear phenomena, there were many unexplained phenomena and hints of further substructures.

Things soon became more complicated, both in theory and in the prediction and discovery of new particles. In 1928, the British physicist P.A.M. Dirac (see **Figure 33.12**) developed a highly successful relativistic quantum theory that laid the foundations of quantum electrodynamics (QED). His theory, for example, explained electron spin and magnetic moment in a natural way. But Dirac's theory also predicted negative energy states for free electrons. By 1931, Dirac, along with Oppenheimer, realized this was a prediction of positively charged electrons (or positrons). In 1932, American physicist Carl Anderson discovered the positron in cosmic ray studies. The positron, or e^+, is the same particle as emitted in β^+ decay and was the first antimatter that was discovered. In 1935, Yukawa predicted pions as the carriers of the strong nuclear force, and they were eventually discovered. Muons were discovered in cosmic ray experiments in 1937, and they seemed to be heavy, unstable versions of electrons and positrons. After World War II, accelerators energetic enough to create these particles were built. Not only were predicted and known particles created, but many unexpected particles were observed. Initially called elementary particles, their numbers proliferated to dozens and then hundreds, and the term "particle zoo" became the physicist's lament at the lack of simplicity. But patterns were observed in the particle zoo that led to simplifying ideas such as quarks, as we shall soon see.

Figure 33.12 P.A.M. Dirac's theory of relativistic quantum mechanics not only explained a great deal of what was known, it also predicted antimatter. (credit: Cambridge University, Cavendish Laboratory)

Matter and Antimatter

The positron was only the first example of antimatter. Every particle in nature has an antimatter counterpart, although some particles, like the photon, are their own antiparticles. Antimatter has charge opposite to that of matter (for example, the positron is positive while the electron is negative) but is nearly identical otherwise, having the same mass, intrinsic spin, half-life, and so on. When a particle and its antimatter counterpart interact, they annihilate one another, usually totally converting their masses to pure energy in the form of photons as seen in **Figure 33.13**. Neutral particles, such as neutrons, have neutral antimatter counterparts, which also annihilate when they interact. Certain neutral particles are their own antiparticle and live correspondingly short lives. For example, the neutral pion π^0 is its own antiparticle and has a half-life about 10^{-8} shorter than π^+ and π^-, which are each other's antiparticles. Without exception, nature is symmetric—all particles have antimatter counterparts. For example, antiprotons and antineutrons were first created in accelerator experiments in 1956 and the antiproton is negative. Antihydrogen atoms, consisting of an antiproton and antielectron, were observed in 1995 at CERN, too. It is possible to contain large-scale antimatter particles such as antiprotons by using electromagnetic traps that confine the particles within a magnetic field so that they don't annihilate with other particles. However, particles of the same charge repel each other, so the more particles that are contained in a trap, the more energy is needed to power the magnetic field that contains them. It is not currently possible to store a significant quantity of antiprotons. At any rate, we now see that negative charge is associated with both low-mass (electrons) and high-mass particles (antiprotons) and the apparent asymmetry is not there. But this knowledge does raise another question—why is there such a predominance of matter and so little antimatter? Possible explanations emerge later in this and the next chapter.

Hadrons and Leptons

Particles can also be revealingly grouped according to what forces they feel between them. All particles (even those that are massless) are affected by gravity, since gravity affects the space and time in which particles exist. All charged particles are affected by the electromagnetic force, as are neutral particles that have an internal distribution of charge (such as the neutron with its magnetic moment). Special names are given to particles that feel the strong and weak nuclear forces. **Hadrons** are particles that feel the strong nuclear force, whereas **leptons** are particles that do not. The proton, neutron, and the pions are examples of hadrons. The electron, positron, muons, and neutrinos are examples of leptons, the name meaning low mass. Leptons feel the weak nuclear force. In fact, all particles feel the weak nuclear force. This means that hadrons are distinguished by being able to feel both the strong and weak nuclear forces.

Table 33.2 lists the characteristics of some of the most important subatomic particles, including the directly observed carrier particles for the electromagnetic and weak nuclear forces, all leptons, and some hadrons. Several hints related to an underlying substructure emerge from an examination of these particle characteristics. Note that the carrier particles are called **gauge bosons**. First mentioned in **Patterns in Spectra Reveal More Quantization**, a **boson** is a particle with zero or an integer value of intrinsic spin (such as $s = 0, 1, 2, ...$), whereas a **fermion** is a particle with a half-integer value of intrinsic spin (

$s = 1/2, 3/2, ...$). Fermions obey the Pauli exclusion principle whereas bosons do not. All the known and conjectured carrier particles are bosons.

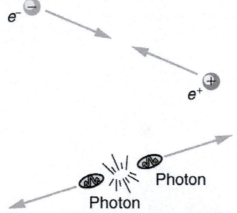

Figure 33.13 When a particle encounters its antiparticle, they annihilate, often producing pure energy in the form of photons. In this case, an electron and a positron convert all their mass into two identical energy rays, which move away in opposite directions to keep total momentum zero as it was before. Similar annihilations occur for other combinations of a particle with its antiparticle, sometimes producing more particles while obeying all conservation laws.

Making Connections: Mini-Magnets

Note that an electron has a property called spin, which implies movement in a circulatory fashion. Recall that the electron also has charge. What do you get when you have a charge moving in a circle? A current, of course, which induces a magnetic field.

Due to the combination of intrinsic spin and charge, an electron has an intrinsic magnetic dipole. This is despite the fact that there is no measureable dimension for a current loop; it is simply a fundamental property of the particle. This is why it is referred to as intrinsic spin. This property of electrons is the ultimate source of the magnetic behavior of bulk matter. Whether a material is diamagnetic, paramagnetic, or ferromagnetic depends on how the outermost layer of electrons in the atoms in the material interact with their nuclei and each other.

Table 33.2 Selected Particle Characteristics[4]

Category	Particle name	Symbol	Antiparticle	Rest mass (MeV/c^2)	B	L_e	L_μ	L_τ	S	Lifetime[5] (s)
Gauge	Photon	γ	Self	0	0	0	0	0	0	Stable
Bosons	W	W^+	W^-	80.39×10^3	0	0	0	0	0	1.6×10^{-25}
	Z	Z^0	Self	91.19×10^3	0	0	0	0	0	1.32×10^{-25}
Leptons	Electron	e^-	e^+	0.511	0	± 1	0	0	0	Stable
	Neutrino (e)	ν_e	$\bar{\nu}_e$	0(7.0eV) [6]	0	± 1	0	0	0	Stable
	Muon	μ^-	μ^+	105.7	0	0	± 1	0	0	2.20×10^{-6}
	Neutrino (μ)	ν_μ	$\bar{\nu}_\mu$	0(< 0.27)	0	0	± 1	0	0	Stable
	Tau	τ^-	τ^+	1777	0	0	0	± 1	0	2.91×10^{-13}
	Neutrino (τ)	ν_τ	$\bar{\nu}_\tau$	0(< 31)	0	0	0	± 1	0	Stable
Hadrons (selected)										
Mesons	Pion	π^+	π^-	139.6	0	0	0	0	0	2.60×10^{-8}
		π^0	Self	135.0	0	0	0	0	0	8.4×10^{-17}
	Kaon	K^+	K^-	493.7	0	0	0	0	± 1	1.24×10^{-8}
		K^0	$\bar{K}^0$	497.6	0	0	0	0	± 1	0.90×10^{-10}
	Eta	η^0	Self	547.9	0	0	0	0	0	2.53×10^{-19}
(many other mesons known)										
Baryons	Proton	p	$\bar{p}$	938.3	± 1	0	0	0	0	Stable[7]
	Neutron	n	$\bar{n}$	939.6	± 1	0	0	0	0	882
	Lambda	Λ^0	$\bar{\Lambda}^0$	1115.7	± 1	0	0	0	∓ 1	2.63×10^{-10}
	Sigma	Σ^+	$\bar{\Sigma}^-$	1189.4	± 1	0	0	0	∓ 1	0.80×10^{-10}
		Σ^0	$\bar{\Sigma}^0$	1192.6	± 1	0	0	0	∓ 1	7.4×10^{-20}
		Σ^-	$\bar{\Sigma}^+$	1197.4	± 1	0	0	0	∓ 1	1.48×10^{-10}
	Xi	Ξ^0	$\bar{\Xi}^0$	1314.9	± 1	0	0	0	∓ 2	2.90×10^{-10}
		Ξ^-	Ξ^+	1321.7	± 1	0	0	0	∓ 2	1.64×10^{-10}
	Omega	Ω^-	Ω^+	1672.5	± 1	0	0	0	∓ 3	0.82×10^{-10}
(many other baryons known)										

All known leptons are listed in the table given above. There are only six leptons (and their antiparticles), and they seem to be fundamental in that they have no apparent underlying structure. Leptons have no discernible size other than their wavelength, so

4. The lower of the $\mp$ or $\pm$ symbols are the values for antiparticles.

5. Lifetimes are traditionally given as $t_{1/2}/0.693$ (which is $1/\lambda$, the inverse of the decay constant).

6. Neutrino masses may be zero. Experimental upper limits are given in parentheses.

7. Experimental lower limit is $> 5 \times 10^{32}$ for proposed mode of decay.

that we know they are pointlike down to about 10^{-18} m . The leptons fall into three families, implying three conservation laws for three quantum numbers. One of these was known from β decay, where the existence of the electron's neutrino implied that a new quantum number, called the **electron family number** L_e is conserved. Thus, in β decay, an antielectron's neutrino $\bar{v}_e$ must be created with $L_e = -1$ when an electron with $L_e = +1$ is created, so that the total remains 0 as it was before decay.

Once the muon was discovered in cosmic rays, its decay mode was found to be

$$\mu^- \to e^- + \bar{v}_e + v_\mu, \tag{33.7}$$

which implied another "family" and associated conservation principle. The particle v_μ is a muon's neutrino, and it is created to conserve **muon family number** L_μ. So muons are leptons with a family of their own, and **conservation of total** L_μ also seems to be obeyed in many experiments.

More recently, a third lepton family was discovered when τ particles were created and observed to decay in a manner similar to muons. One principal decay mode is

$$\tau^- \to \mu^- + \bar{v}_\mu + v_\tau. \tag{33.8}$$

Conservation of total L_τ seems to be another law obeyed in many experiments. In fact, particle experiments have found that lepton family number is not universally conserved, due to neutrino "oscillations," or transformations of neutrinos from one family type to another.

Mesons and Baryons

Now, note that the hadrons in the table given above are divided into two subgroups, called mesons (originally for medium mass) and baryons (the name originally meaning large mass). The division between mesons and baryons is actually based on their observed decay modes and is not strictly associated with their masses. **Mesons** are hadrons that can decay to leptons and leave no hadrons, which implies that mesons are not conserved in number. **Baryons** are hadrons that always decay to another baryon. A new physical quantity called **baryon number** B seems to always be conserved in nature and is listed for the various particles in the table given above. Mesons and leptons have $B = 0$ so that they can decay to other particles with $B = 0$. But baryons have $B = +1$ if they are matter, and $B = -1$ if they are antimatter. The **conservation of total baryon number** is a more general rule than first noted in nuclear physics, where it was observed that the total number of nucleons was always conserved in nuclear reactions and decays. That rule in nuclear physics is just one consequence of the conservation of the total baryon number.

Forces, Reactions, and Reaction Rates

The forces that act between particles regulate how they interact with other particles. For example, pions feel the strong force and do not penetrate as far in matter as do muons, which do not feel the strong force. (This was the way those who discovered the muon knew it could not be the particle that carries the strong force—its penetration or range was too great for it to be feeling the strong force.) Similarly, reactions that create other particles, like cosmic rays interacting with nuclei in the atmosphere, have greater probability if they are caused by the strong force than if they are caused by the weak force. Such knowledge has been useful to physicists while analyzing the particles produced by various accelerators.

The forces experienced by particles also govern how particles interact with themselves if they are unstable and decay. For example, the stronger the force, the faster they decay and the shorter is their lifetime. An example of a nuclear decay via the strong force is $^8\text{Be} \to \alpha + \alpha$ with a lifetime of about 10^{-16} s . The neutron is a good example of decay via the weak force. The process $n \to p + e^- + \bar{v}_e$ has a longer lifetime of 882 s. The weak force causes this decay, as it does all β decay. An important clue that the weak force is responsible for β decay is the creation of leptons, such as e^- and $\bar{v}_e$. None would be created if the strong force was responsible, just as no leptons are created in the decay of ^8Be. The systematics of particle lifetimes is a little simpler than nuclear lifetimes when hundreds of particles are examined (not just the ones in the table given above). Particles that decay via the weak force have lifetimes mostly in the range of 10^{-16} to 10^{-12} s, whereas those that decay via the strong force have lifetimes mostly in the range of 10^{-16} to 10^{-23} s. Turning this around, if we measure the lifetime of a particle, we can tell if it decays via the weak or strong force.

Yet another quantum number emerges from decay lifetimes and patterns. Note that the particles Λ, Σ, Ξ, and Ω decay with lifetimes on the order of 10^{-10} s (the exception is Σ^0, whose short lifetime is explained by its particular quark substructure.), implying that their decay is caused by the weak force alone, although they are hadrons and feel the strong force. The decay modes of these particles also show patterns—in particular, certain decays that should be possible within all the known conservation laws do not occur. Whenever something is possible in physics, it will happen. If something does not happen, it is forbidden by a rule. All this seemed strange to those studying these particles when they were first discovered, so they named a new quantum number **strangeness**, given the symbol S in the table given above. The values of strangeness assigned to various particles are based on the decay systematics. It is found that **strangeness is conserved by the strong force**, which governs the production of most of these particles in accelerator experiments. However, **strangeness is _not conserved_ by the weak force**. This conclusion is reached from the fact that particles that have long lifetimes decay via the weak force and do not conserve strangeness. All of this also has implications for the carrier particles, since they transmit forces and are thus involved in these decays.

Example 33.3 Calculating Quantum Numbers in Two Decays

(a) The most common decay mode of the Ξ^- particle is $\Xi^- \rightarrow \Lambda^0 + \pi^-$. Using the quantum numbers in the table given above, show that strangeness changes by 1, baryon number and charge are conserved, and lepton family numbers are unaffected.

(b) Is the decay $K^+ \rightarrow \mu^+ + \nu_\mu$ allowed, given the quantum numbers in the table given above?

Strategy

In part (a), the conservation laws can be examined by adding the quantum numbers of the decay products and comparing them with the parent particle. In part (b), the same procedure can reveal if a conservation law is broken or not.

Solution for (a)

Before the decay, the Ξ^- has strangeness $S = -2$. After the decay, the total strangeness is –1 for the Λ^0, plus 0 for the π^-. Thus, total strangeness has gone from –2 to –1 or a change of +1. Baryon number for the Ξ^- is $B = +1$ before the decay, and after the decay the Λ^0 has $B = +1$ and the π^- has $B = 0$ so that the total baryon number remains +1. Charge is –1 before the decay, and the total charge after is also $0 - 1 = -1$. Lepton numbers for all the particles are zero, and so lepton numbers are conserved.

Discussion for (a)

The Ξ^- decay is caused by the weak interaction, since strangeness changes, and it is consistent with the relatively long 1.64×10^{-10}-s lifetime of the Ξ^-.

Solution for (b)

The decay $K^+ \rightarrow \mu^+ + \nu_\mu$ is allowed if charge, baryon number, mass-energy, and lepton numbers are conserved. Strangeness can change due to the weak interaction. Charge is conserved as $s \rightarrow d$. Baryon number is conserved, since all particles have $B = 0$. Mass-energy is conserved in the sense that the K^+ has a greater mass than the products, so that the decay can be spontaneous. Lepton family numbers are conserved at 0 for the electron and tau family for all particles. The muon family number is $L_\mu = 0$ before and $L_\mu = -1 + 1 = 0$ after. Strangeness changes from +1 before to 0 + 0 after, for an allowed change of 1. The decay is allowed by all these measures.

Discussion for (b)

This decay is not only allowed by our reckoning, it is, in fact, the primary decay mode of the K^+ meson and is caused by the weak force, consistent with the long 1.24×10^{-8}-s lifetime.

There are hundreds of particles, all hadrons, not listed in Table 33.2, most of which have shorter lifetimes. The systematics of those particle lifetimes, their production probabilities, and decay products are completely consistent with the conservation laws noted for lepton families, baryon number, and strangeness, but they also imply other quantum numbers and conservation laws. There are a finite, and in fact relatively small, number of these conserved quantities, however, implying a finite set of substructures. Additionally, some of these short-lived particles resemble the excited states of other particles, implying an internal structure. All of this jigsaw puzzle can be tied together and explained relatively simply by the existence of fundamental substructures. Leptons seem to be fundamental structures. Hadrons seem to have a substructure called quarks. **Quarks: Is That All There Is?** explores the basics of the underlying quark building blocks.

Figure 33.14 Murray Gell-Mann (b. 1929) proposed quarks as a substructure of hadrons in 1963 and was already known for his work on the concept of strangeness. Although quarks have never been directly observed, several predictions of the quark model were quickly confirmed, and their properties explain all known hadron characteristics. Gell-Mann was awarded the Nobel Prize in 1969. (credit: Luboš Motl)

Summary

- All particles of matter have an antimatter counterpart that has the opposite charge and certain other quantum numbers as seen in Table 33.2. These matter-antimatter pairs are otherwise very similar but will annihilate when brought together. Known particles can be divided into three major groups—leptons, hadrons, and carrier particles (gauge bosons).
- Leptons do not feel the strong nuclear force and are further divided into three groups—electron family designated by electron family number L_e; muon family designated by muon family number L_μ; and tau family designated by tau family number L_τ. The family numbers are not universally conserved due to neutrino oscillations.
- Hadrons are particles that feel the strong nuclear force and are divided into baryons, with the baryon family number B being conserved, and mesons.

33.5 Quarks: Is That All There Is?

<table>
<tr><td align="center">Learning Objectives</td></tr>
</table>

By the end of this section, you will be able to:

- Define fundamental particle.
- Describe quark and antiquark.
- List the flavors of quarks.
- Outline the quark composition of hadrons.
- Determine quantum numbers from quark composition.

The information presented in this section supports the following AP® learning objectives and science practices:

- **1.A.2.1**: The student is able to construct representations of the differences between a fundamental particle and a system composed of fundamental particles and to relate this to the properties and scales of the systems being investigated.

Quarks have been mentioned at various points in this text as fundamental building blocks and members of the exclusive club of truly elementary particles. Note that an elementary or **fundamental particle** has no substructure (it is not made of other particles) and has no finite size other than its wavelength. This does not mean that fundamental particles are stable—some decay, while others do not. Keep in mind that *all* leptons seem to be fundamental, whereas *no* hadrons are fundamental. There is strong evidence that **quarks** are the fundamental building blocks of hadrons as seen in Figure 33.15. Quarks are the second group of fundamental particles (leptons are the first). The third and perhaps final group of fundamental particles is the carrier particles for the four basic forces. Leptons, quarks, and carrier particles may be all there is. In this module we will discuss the quark substructure of hadrons and its relationship to forces as well as indicate some remaining questions and problems.

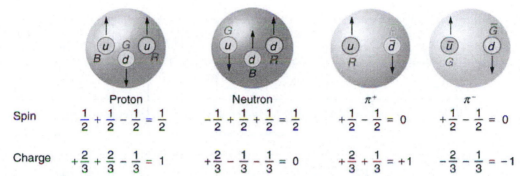

	Proton	Neutron	π^+	π^-
Spin	$\frac{1}{2} + \frac{1}{2} - \frac{1}{2} = \frac{1}{2}$	$-\frac{1}{2} + \frac{1}{2} + \frac{1}{2} = \frac{1}{2}$	$+\frac{1}{2} - \frac{1}{2} = 0$	$+\frac{1}{2} - \frac{1}{2} = 0$
Charge	$+\frac{2}{3} + \frac{2}{3} - \frac{1}{3} = 1$	$+\frac{2}{3} - \frac{1}{3} - \frac{1}{3} = 0$	$+\frac{2}{3} + \frac{1}{3} = +1$	$-\frac{2}{3} - \frac{1}{3} = -1$

Figure 33.15 All baryons, such as the proton and neutron shown here, are composed of three quarks. All mesons, such as the pions shown here, are composed of a quark-antiquark pair. Arrows represent the spins of the quarks, which, as we shall see, are also colored. The colors are such that they need to add to white for any possible combination of quarks.

Conception of Quarks

Quarks were first proposed independently by American physicists Murray Gell-Mann and George Zweig in 1963. Their quaint name was taken by Gell-Mann from a James Joyce novel—Gell-Mann was also largely responsible for the concept and name of strangeness. (Whimsical names are common in particle physics, reflecting the personalities of modern physicists.) Originally, three quark types—or **flavors**—were proposed to account for the then-known mesons and baryons. These quark flavors are named **up** (*u*), **down** (*d*), and **strange** (*s*). All quarks have half-integral spin and are thus fermions. All mesons have integral spin while all baryons have half-integral spin. Therefore, mesons should be made up of an even number of quarks while baryons need to be made up of an odd number of quarks. **Figure 33.15** shows the quark substructure of the proton, neutron, and two pions. The most radical proposal by Gell-Mann and Zweig is the fractional charges of quarks, which are $\pm\left(\frac{2}{3}\right)q_e$ and $\left(\frac{1}{3}\right)q_e$, whereas all directly observed particles have charges that are integral multiples of q_e. Note that the fractional value of the quark does not violate the fact that the e is the smallest unit of charge that is observed, because a free quark cannot exist. **Table 33.3** lists characteristics of the six quark flavors that are now thought to exist. Discoveries made since 1963 have required extra quark flavors, which are divided into three families quite analogous to leptons.

How Does it Work?

To understand how these quark substructures work, let us specifically examine the proton, neutron, and the two pions pictured in **Figure 33.15** before moving on to more general considerations. First, the proton p is composed of the three quarks *uud*, so that its total charge is $+\left(\frac{2}{3}\right)q_e + \left(\frac{2}{3}\right)q_e - \left(\frac{1}{3}\right)q_e = q_e$, as expected. With the spins aligned as in the figure, the proton's intrinsic spin is $+\left(\frac{1}{2}\right) + \left(\frac{1}{2}\right) - \left(\frac{1}{2}\right) = \left(\frac{1}{2}\right)$, also as expected. Note that the spins of the up quarks are aligned, so that they would be in the same state except that they have different colors (another quantum number to be elaborated upon a little later). Quarks obey the Pauli exclusion principle. Similar comments apply to the neutron n, which is composed of the three quarks *udd*. Note also that the neutron is made of charges that add to zero but move internally, producing its well-known magnetic moment. When the neutron β^- decays, it does so by changing the flavor of one of its quarks. Writing neutron β^- decay in terms of quarks,

$$n \rightarrow p + \beta^- + \bar{\nu}_e \text{ becomes } udd \rightarrow uud + \beta^- + \bar{\nu}_e. \tag{33.9}$$

We see that this is equivalent to a down quark changing flavor to become an up quark:

$$d \rightarrow u + \beta^- + \bar{\nu}_e \tag{33.10}$$

Table 33.3 Quarks and Antiquarks[8]

Name	Symbol	Antiparticle	Spin	Charge	B [9]	S	c	b	t	Mass (GeV/c^2) [10]
Up	u	$\bar{u}$	1/2	$\pm\frac{2}{3}q_e$	$\pm\frac{1}{3}$	0	0	0	0	0.005
Down	d	$\bar{d}$	1/2	$\mp\frac{1}{3}q_e$	$\pm\frac{1}{3}$	0	0	0	0	0.008
Strange	s	$\bar{s}$	1/2	$\mp\frac{1}{3}q_e$	$\pm\frac{1}{3}$	∓ 1	0	0	0	0.50
Charmed	c	$\bar{c}$	1/2	$\pm\frac{2}{3}q_e$	$\pm\frac{1}{3}$	0	± 1	0	0	1.6
Bottom	b	$\bar{b}$	1/2	$\mp\frac{1}{3}q_e$	$\pm\frac{1}{3}$	0	0	∓ 1	0	5
Top	t	$\bar{t}$	1/2	$\pm\frac{2}{3}q_e$	$\pm\frac{1}{3}$	0	0	0	± 1	173

8. The lower of the $\pm$ symbols are the values for antiquarks.

9. B is baryon number, S is strangeness, c is charm, b is bottomness, t is topness.

10. Values are approximate, are not directly observable, and vary with model.

Table 33.4 Quark Composition of Selected Hadrons[11]

Particle	Quark Composition
Mesons	
π^+	$u\bar{d}$
π^-	$\bar{u}d$
π^0	$u\bar{u}$, $d\bar{d}$ mixture[12]
η^0	$u\bar{u}$, $d\bar{d}$ mixture[13]
K^0	$d\bar{s}$
$\bar{K}^0$	$\bar{d}s$
K^+	$u\bar{s}$
K^-	$\bar{u}s$
J/ψ	$c\bar{c}$
Υ	$b\bar{b}$
Baryons[14],[15]	
p	uud
n	udd
Δ^0	udd
Δ^+	uud
Δ^-	ddd
Δ^{++}	uuu
Λ^0	uds
Σ^0	uds
Σ^+	uus
Σ^-	dds
Ξ^0	uss
Ξ^-	dss
Ω^-	sss

This is an example of the general fact that **the weak nuclear force can change the flavor of a quark**. By general, we mean that any quark can be converted to any other (change flavor) by the weak nuclear force. Not only can we get $d \rightarrow u$, we can also get $u \rightarrow d$. Furthermore, the strange quark can be changed by the weak force, too, making $s \rightarrow u$ and $s \rightarrow d$ possible. This explains the violation of the conservation of strangeness by the weak force noted in the preceding section. Another general fact is that **the strong nuclear force cannot change the flavor of a quark**.

---

11. These two mesons are different mixtures, but each is its own antiparticle, as indicated by its quark composition.
12. These two mesons are different mixtures, but each is its own antiparticle, as indicated by its quark composition.
13. These two mesons are different mixtures, but each is its own antiparticle, as indicated by its quark composition.

14. *Antibaryons have the antiquarks of their counterparts. The antiproton $\bar{p}$ is $\bar{u}\bar{u}\bar{d}$, for example.*

15. *Baryons composed of the same quarks are different states of the same particle. For example, the Δ^+ is an excited state of the proton.*

Again, from **Figure 33.15**, we see that the π^+ meson (one of the three pions) is composed of an up quark plus an antidown quark, or $u\bar{d}$. Its total charge is thus $+\left(\frac{2}{3}\right)q_e + \left(\frac{1}{3}\right)q_e = q_e$, as expected. Its baryon number is 0, since it has a quark and an antiquark with baryon numbers $+\left(\frac{1}{3}\right) - \left(\frac{1}{3}\right) = 0$. The π^+ half-life is relatively long since, although it is composed of matter and antimatter, the quarks are different flavors and the weak force should cause the decay by changing the flavor of one into that of the other. The spins of the u and $\bar{d}$ quarks are antiparallel, enabling the pion to have spin zero, as observed experimentally. Finally, the π^- meson shown in **Figure 33.15** is the antiparticle of the π^+ meson, and it is composed of the corresponding quark antiparticles. That is, the π^+ meson is $u\bar{d}$, while the π^- meson is $\bar{u}d$. These two pions annihilate each other quickly, because their constituent quarks are each other's antiparticles.

Two general rules for combining quarks to form hadrons are:

1. Baryons are composed of three quarks, and antibaryons are composed of three antiquarks.

2. Mesons are combinations of a quark and an antiquark.

One of the clever things about this scheme is that only integral charges result, even though the quarks have fractional charge.

All Combinations are Possible

All quark combinations are possible. **Table 33.4** lists some of these combinations. When Gell-Mann and Zweig proposed the original three quark flavors, particles corresponding to all combinations of those three had not been observed. The pattern was there, but it was incomplete—much as had been the case in the periodic table of the elements and the chart of nuclides. The Ω^- particle, in particular, had not been discovered but was predicted by quark theory. Its combination of three strange quarks, sss, gives it a strangeness of -3 (see **Table 33.2**) and other predictable characteristics, such as spin, charge, approximate mass, and lifetime. If the quark picture is complete, the Ω^- should exist. It was first observed in 1964 at Brookhaven National Laboratory and had the predicted characteristics as seen in **Figure 33.16**. The discovery of the Ω^- was convincing indirect evidence for the existence of the three original quark flavors and boosted theoretical and experimental efforts to further explore particle physics in terms of quarks.

Patterns and Puzzles: Atoms, Nuclei, and Quarks

Patterns in the properties of atoms allowed the periodic table to be developed. From it, previously unknown elements were predicted and observed. Similarly, patterns were observed in the properties of nuclei, leading to the chart of nuclides and successful predictions of previously unknown nuclides. Now with particle physics, patterns imply a quark substructure that, if taken literally, predicts previously unknown particles. These have now been observed in another triumph of underlying unity.

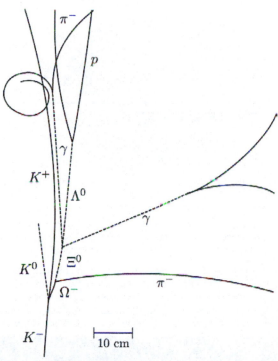

Figure 33.16 The image relates to the discovery of the Ω^-. It is a secondary reaction in which an accelerator-produced K^- collides with a proton via the strong force and conserves strangeness to produce the Ω^- with characteristics predicted by the quark model. As with other predictions of previously unobserved particles, this gave a tremendous boost to quark theory. (credit: Brookhaven National Laboratory)

Example 33.4 Quantum Numbers From Quark Composition

Verify the quantum numbers given for the Ξ^0 particle in **Table 33.2** by adding the quantum numbers for its quark composition as given in **Table 33.4**.

Strategy

The composition of the Ξ^0 is given as uss in **Table 33.4**. The quantum numbers for the constituent quarks are given in **Table 33.3**. We will not consider spin, because that is not given for the Ξ^0. But we can check on charge and the other quantum numbers given for the quarks.

Solution

The total charge of uss is $+\left(\frac{2}{3}\right)q_e - \left(\frac{1}{3}\right)q_e - \left(\frac{1}{3}\right)q_e = 0$, which is correct for the Ξ^0. The baryon number is

$$+\left(\frac{1}{3}\right) + \left(\frac{1}{3}\right) + \left(\frac{1}{3}\right) = 1$$, also correct since the Ξ^0 is a matter baryon and has $B = 1$, as listed in **Table 33.2**. Its

strangeness is $S = 0 - 1 - 1 = -2$, also as expected from **Table 33.2**. Its charm, bottomness, and topness are 0, as are its lepton family numbers (it is not a lepton).

Discussion

This procedure is similar to what the inventors of the quark hypothesis did when checking to see if their solution to the puzzle of particle patterns was correct. They also checked to see if all combinations were known, thereby predicting the previously unobserved Ω^- as the completion of a pattern.

Now, Let Us Talk About Direct Evidence

At first, physicists expected that, with sufficient energy, we should be able to free quarks and observe them directly. This has not proved possible. There is still no direct observation of a fractional charge or any isolated quark. When large energies are put into collisions, other particles are created—but no quarks emerge. There is nearly direct evidence for quarks that is quite compelling. By 1967, experiments at SLAC scattering 20-GeV electrons from protons had produced results like Rutherford had obtained for the nucleus nearly 60 years earlier. The SLAC scattering experiments showed unambiguously that there were three pointlike (meaning they had sizes considerably smaller than the probe's wavelength) charges inside the proton as seen in **Figure 33.17**. This evidence made all but the most skeptical admit that there was validity to the quark substructure of hadrons.

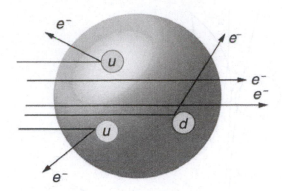

Proton

Figure 33.17 Scattering of high-energy electrons from protons at facilities like SLAC produces evidence of three point-like charges consistent with proposed quark properties. This experiment is analogous to Rutherford's discovery of the small size of the nucleus by scattering α particles. High-energy electrons are used so that the probe wavelength is small enough to see details smaller than the proton.

More recent and higher-energy experiments have produced jets of particles in collisions, highly suggestive of three quarks in a nucleon. Since the quarks are very tightly bound, energy put into separating them pulls them only so far apart before it starts being converted into other particles. More energy produces more particles, not a separation of quarks. Conservation of momentum requires that the particles come out in jets along the three paths in which the quarks were being pulled. Note that there are only three jets, and that other characteristics of the particles are consistent with the three-quark substructure.

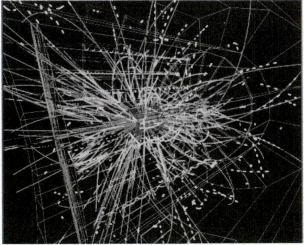

Figure 33.18 Simulation of a proton-proton collision at 14-TeV center-of-mass energy in the ALICE detector at CERN LHC. The lines follow particle trajectories and the cyan dots represent the energy depositions in the sensitive detector elements. (credit: Matevž Tadel)

Quarks Have Their Ups and Downs

The quark model actually lost some of its early popularity because the original model with three quarks had to be modified. The up and down quarks seemed to compose normal matter as seen in **Table 33.4**, while the single strange quark explained strangeness. Why didn't it have a counterpart? A fourth quark flavor called **charm** (c) was proposed as the counterpart of the strange quark to make things symmetric—there would be two normal quarks (u and d) and two exotic quarks (s and c). Furthermore, at that time only four leptons were known, two normal and two exotic. It was attractive that there would be four quarks and four leptons. The problem was that no known particles contained a charmed quark. Suddenly, in November of 1974, two groups (one headed by C. C. Ting at Brookhaven National Laboratory and the other by Burton Richter at SLAC) independently and nearly simultaneously discovered a new meson with characteristics that made it clear that its substructure is $c\bar{c}$. It was called J by one group and psi (ψ) by the other and now is known as the J/ψ meson. Since then, numerous particles have been discovered containing the charmed quark, consistent in every way with the quark model. The discovery of the J/ψ meson had such a rejuvenating effect on quark theory that it is now called the November Revolution. Ting and Richter shared the 1976 Nobel Prize.

History quickly repeated itself. In 1975, the tau (τ) was discovered, and a third family of leptons emerged as seen in **Table 33.2**). Theorists quickly proposed two more quark flavors called **top** (t) or truth and **bottom** (b) or beauty to keep the number of quarks the same as the number of leptons. And in 1976, the upsilon (Υ) meson was discovered and shown to be composed of a bottom and an antibottom quark or $b\bar{b}$, quite analogous to the J/ψ being $c\bar{c}$ as seen in **Table 33.4**. Being a single flavor, these mesons are sometimes called bare charm and bare bottom and reveal the characteristics of their quarks most clearly. Other mesons containing bottom quarks have since been observed. In 1995, two groups at Fermilab confirmed the top quark's existence, completing the picture of six quarks listed in Table 33.3. Each successive quark discovery—first c, then b, and finally t —has required higher energy because each has higher mass. Quark masses in Table 33.3 are only approximately known, because they are not directly observed. They must be inferred from the masses of the particles they combine to form.

What's Color got to do with it?—A Whiter Shade of Pale

As mentioned and shown in Figure 33.15, quarks carry another quantum number, which we call **color**. Of course, it is not the color we sense with visible light, but its properties are analogous to those of three primary and three secondary colors. Specifically, a quark can have one of three color values we call **red** (R), **green** (G), and **blue** (B) in analogy to those primary visible colors. Antiquarks have three values we call **antired or cyan** $\left(\bar{R}\right)$, **antigreen or magenta** $\left(\bar{G}\right)$, and **antiblue or yellow** $\left(\bar{B}\right)$ in analogy to those secondary visible colors. The reason for these names is that when certain visual colors are combined, the eye sees white. The analogy of the colors combining to white is used to explain why baryons are made of three quarks, why mesons are a quark and an antiquark, and why we cannot isolate a single quark. The force between the quarks is such that their combined colors produce white. This is illustrated in Figure 33.19. A baryon must have one of each primary color or RGB, which produces white. A meson must have a primary color and its anticolor, also producing white.

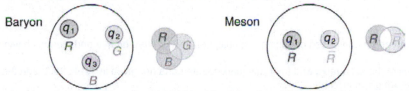

Figure 33.19 The three quarks composing a baryon must be RGB, which add to white. The quark and antiquark composing a meson must be a color and anticolor, here $R\bar{R}$ also adding to white. The force between systems that have color is so great that they can neither be separated nor exist as colored.

Why must hadrons be white? The color scheme is intentionally devised to explain why baryons have three quarks and mesons have a quark and an antiquark. Quark color is thought to be similar to charge, but with more values. An ion, by analogy, exerts much stronger forces than a neutral molecule. When the color of a combination of quarks is white, it is like a neutral atom. The forces a white particle exerts are like the polarization forces in molecules, but in hadrons these leftovers are the strong nuclear force. When a combination of quarks has color other than white, it exerts *extremely* large forces—even larger than the strong force—and perhaps cannot be stable or permanently separated. This is part of the **theory of quark confinement**, which explains how quarks can exist and yet never be isolated or directly observed. Finally, an extra quantum number with three values (like those we assign to color) is necessary for quarks to obey the Pauli exclusion principle. Particles such as the Ω^-, which is composed of three strange quarks, sss, and the Δ^{++}, which is three up quarks, uuu, can exist because the quarks have different colors and do not have the same quantum numbers. Color is consistent with all observations and is now widely accepted. Quark theory including color is called **quantum chromodynamics** (QCD), also named by Gell-Mann.

The Three Families

Fundamental particles are thought to be one of three types—leptons, quarks, or carrier particles. Each of those three types is further divided into three analogous families as illustrated in Figure 33.20. We have examined leptons and quarks in some detail. Each has six members (and their six antiparticles) divided into three analogous families. The first family is normal matter, of which most things are composed. The second is exotic, and the third more exotic and more massive than the second. The only stable particles are in the first family, which also has unstable members.

Always searching for symmetry and similarity, physicists have also divided the carrier particles into three families, omitting the graviton. Gravity is special among the four forces in that it affects the space and time in which the other forces exist and is proving most difficult to include in a Theory of Everything or TOE (to stub the pretension of such a theory). Gravity is thus often set apart. It is not certain that there is meaning in the groupings shown in Figure 33.20, but the analogies are tempting. In the past, we have been able to make significant advances by looking for analogies and patterns, and this is an example of one under current scrutiny. There are connections between the families of leptons, in that the τ decays into the μ and the μ into the e.

Similarly for quarks, the higher families eventually decay into the lowest, leaving only u and d quarks. We have long sought connections between the forces in nature. Since these are carried by particles, we will explore connections between gluons, $W^{\pm}$ and Z^0, and photons as part of the search for unification of forces discussed in GUTs: The Unification of Forces..

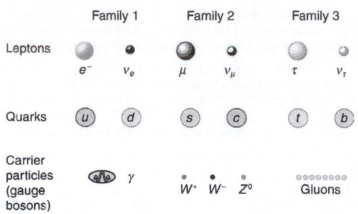

Figure 33.20 The three types of particles are leptons, quarks, and carrier particles. Each of those types is divided into three analogous families, with the graviton left out.

Summary

- Hadrons are thought to be composed of quarks, with baryons having three quarks and mesons having a quark and an antiquark.
- The characteristics of the six quarks and their antiquark counterparts are given in Table 33.3, and the quark compositions of certain hadrons are given in Table 33.4.
- Indirect evidence for quarks is very strong, explaining all known hadrons and their quantum numbers, such as strangeness, charm, topness, and bottomness.
- Quarks come in six flavors and three colors and occur only in combinations that produce white.
- Fundamental particles have no further substructure, not even a size beyond their de Broglie wavelength.
- There are three types of fundamental particles—leptons, quarks, and carrier particles. Each type is divided into three analogous families as indicated in Figure 33.20.

33.6 GUTs: The Unification of Forces

Learning Objectives

By the end of this section, you will be able to:

- State the grand unified theory.
- Explain the electroweak theory.
- Define gluons.
- Describe the principle of quantum chromodynamics.
- Define the standard model.

Present quests to show that the four basic forces are different manifestations of a single unified force follow a long tradition. In the 19th century, the distinct electric and magnetic forces were shown to be intimately connected and are now collectively called the electromagnetic force. More recently, the weak nuclear force has been shown to be connected to the electromagnetic force in a manner suggesting that a theory may be constructed in which all four forces are unified. Certainly, there are similarities in how forces are transmitted by the exchange of carrier particles, and the carrier particles themselves (the gauge bosons in Table 33.2) are also similar in important ways. The analogy to the unification of electric and magnetic forces is quite good—the four forces are distinct under normal circumstances, but there are hints of connections even on the atomic scale, and there may be conditions under which the forces are intimately related and even indistinguishable. The search for a correct theory linking the forces, called the **Grand Unified Theory (GUT)**, is explored in this section in the realm of particle physics. Frontiers of Physics expands the story in making a connection with cosmology, on the opposite end of the distance scale.

Figure 33.21 is a Feynman diagram showing how the weak nuclear force is transmitted by the carrier particle Z^0, similar to the diagrams in Figure 33.5 and Figure 33.6 for the electromagnetic and strong nuclear forces. In the 1960s, a gauge theory, called **electroweak theory**, was developed by Steven Weinberg, Sheldon Glashow, and Abdus Salam and proposed that the electromagnetic and weak forces are identical at sufficiently high energies. One of its predictions, in addition to describing both electromagnetic and weak force phenomena, was the existence of the W^+, W^-, and Z^0 carrier particles. Not only were three particles having spin 1 predicted, the mass of the W^+ and W^- was predicted to be $81 \text{ GeV}/c^2$, and that of the Z^0 was predicted to be $90 \text{ GeV}/c^2$. (Their masses had to be about 1000 times that of the pion, or about $100 \text{ GeV}/c^2$, since the range of the weak force is about 1000 times less than the strong force carried by virtual pions.) In 1983, these carrier particles were observed at CERN with the predicted characteristics, including masses having the predicted values as seen in Table 33.2. This was another triumph of particle theory and experimental effort, resulting in the 1984 Nobel Prize to the experiment's group leaders Carlo Rubbia and Simon van der Meer. Theorists Weinberg, Glashow, and Salam had already been honored with the

1979 Nobel Prize for other aspects of electroweak theory.

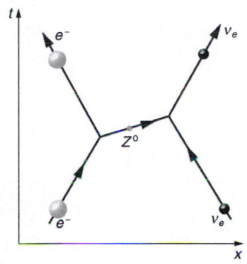

Figure 33.21 The exchange of a virtual Z^0 carries the weak nuclear force between an electron and a neutrino in this Feynman diagram. The Z^0 is one of the carrier particles for the weak nuclear force that has now been created in the laboratory with characteristics predicted by electroweak theory.

Although the weak nuclear force is very short ranged ($< 10^{-18}$ m , as indicated in Table 33.1), its effects on atomic levels can be measured given the extreme precision of modern techniques. Since electrons spend some time in the nucleus, their energies are affected, and spectra can even indicate new aspects of the weak force, such as the possibility of other carrier particles. So systems many orders of magnitude larger than the range of the weak force supply evidence of electroweak unification in addition to evidence found at the particle scale.

Gluons (g) are the proposed carrier particles for the strong nuclear force, although they are not directly observed. Like quarks, gluons may be confined to systems having a total color of white. Less is known about gluons than the fact that they are the carriers of the weak and certainly of the electromagnetic force. QCD theory calls for eight gluons, all massless and all spin 1. Six of the gluons carry a color and an anticolor, while two do not carry color, as illustrated in **Figure 33.22**(a). There is indirect evidence of the existence of gluons in nucleons. When high-energy electrons are scattered from nucleons and evidence of quarks is seen, the momenta of the quarks are smaller than they would be if there were no gluons. That means that the gluons carrying force between quarks also carry some momentum, inferred by the already indirect quark momentum measurements. At any rate, the gluons carry color charge and can change the colors of quarks when exchanged, as seen in **Figure 33.22**(b). In the figure, a red down quark interacts with a green strange quark by sending it a gluon. That gluon carries red away from the down quark and leaves it green, because it is an $R\bar{G}$ (red-antigreen) gluon. (Taking antigreen away leaves you green.) Its antigreenness kills the green in the strange quark, and its redness turns the quark red.

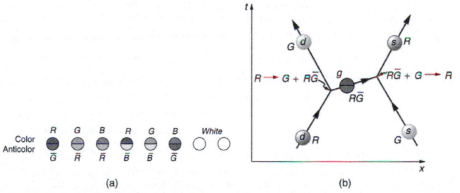

(a)   (b)

Figure 33.22 In figure (a), the eight types of gluons that carry the strong nuclear force are divided into a group of six that carry color and a group of two that do not. Figure (b) shows that the exchange of gluons between quarks carries the strong force and may change the color of a quark.

The strong force is complicated, since observable particles that feel the strong force (hadrons) contain multiple quarks. **Figure 33.23** shows the quark and gluon details of pion exchange between a proton and a neutron as illustrated earlier in **Figure 33.3** and **Figure 33.6**. The quarks within the proton and neutron move along together exchanging gluons, until the proton and neutron get close together. As the u quark leaves the proton, a gluon creates a pair of virtual particles, a d quark and a $\bar{d}$ antiquark.

The d quark stays behind and the proton turns into a neutron, while the u and $\bar{d}$ move together as a π^+ (**Table 33.4** confirms the $u\bar{d}$ composition for the π^+ .) The $\bar{d}$ annihilates a d quark in the neutron, the u joins the neutron, and the neutron becomes a proton. A pion is exchanged and a force is transmitted.

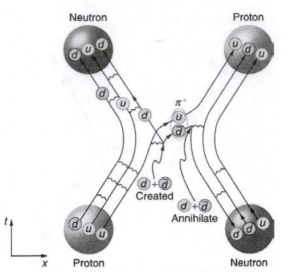

Figure 33.23 This Feynman diagram is the same interaction as shown in **Figure 33.6**, but it shows the quark and gluon details of the strong force interaction.

It is beyond the scope of this text to go into more detail on the types of quark and gluon interactions that underlie the observable particles, but the theory (**quantum chromodynamics** or QCD) is very self-consistent. So successful have QCD and the electroweak theory been that, taken together, they are called the **Standard Model**. Advances in knowledge are expected to modify, but not overthrow, the Standard Model of particle physics and forces.

Making Connections: Unification of Forces

Grand Unified Theory (GUT) is successful in describing the four forces as distinct under normal circumstances, but connected in fundamental ways. Experiments have verified that the weak and electromagnetic force become identical at very small distances and provide the GUT description of the carrier particles for the forces. GUT predicts that the other forces become identical under conditions so extreme that they cannot be tested in the laboratory, although there may be lingering evidence of them in the evolution of the universe. GUT is also successful in describing a system of carrier particles for all four forces, but there is much to be done, particularly in the realm of gravity.

How can forces be unified? They are definitely distinct under most circumstances, for example, being carried by different particles and having greatly different strengths. But experiments show that at extremely small distances, the strengths of the forces begin to become more similar. In fact, electroweak theory's prediction of the W^+ , W^- , and Z^0 carrier particles was based on the strengths of the two forces being identical at extremely small distances as seen in **Figure 33.24**. As discussed in case of the creation of virtual particles for extremely short times, the small distances or short ranges correspond to the large masses of the carrier particles and the correspondingly large energies needed to create them. Thus, the energy scale on the horizontal axis of **Figure 33.24** corresponds to smaller and smaller distances, with 100 GeV corresponding to approximately, 10^{-18} m for example. At that distance, the strengths of the EM and weak forces are the same. To test physics at that distance, energies of about 100 GeV must be put into the system, and that is sufficient to create and release the W^+ , W^- , and Z^0 carrier particles. At those and higher energies, the masses of the carrier particles becomes less and less relevant, and the Z^0 in particular resembles the massless, chargeless, spin 1 photon. In fact, there is enough energy when things are pushed to even smaller distances to transform the, and Z^0 into massless carrier particles more similar to photons and gluons. These have not been observed experimentally, but there is a prediction of an associated particle called the **Higgs boson**. The mass of this particle is not predicted with nearly the certainty with which the mass of the W^+, W^-, and Z^0 particles were predicted, but it was hoped that the Higgs boson could be observed at the now-canceled Superconducting Super Collider (SSC). Ongoing experiments at the Large Hadron Collider at CERN have presented some evidence for a Higgs boson with a mass of 125 GeV, and there is a possibility of a direct discovery during 2012. The existence of this more massive particle would give validity to the theory that the carrier particles are identical under certain circumstances.

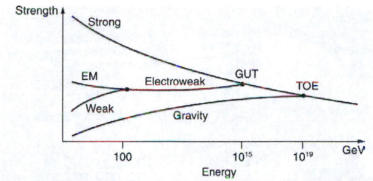

Figure 33.24 The relative strengths of the four basic forces vary with distance and, hence, energy is needed to probe small distances. At ordinary energies (a few eV or less), the forces differ greatly as indicated in Table 33.1. However, at energies available at accelerators, the weak and EM forces become identical, or unified. Unfortunately, the energies at which the strong and electroweak forces become the same are unreachable even in principle at any conceivable accelerator. The universe may provide a laboratory, and nature may show effects at ordinary energies that give us clues about the validity of this graph.

The small distances and high energies at which the electroweak force becomes identical with the strong nuclear force are not reachable with any conceivable human-built accelerator. At energies of about 10^{14} GeV (16,000 J per particle), distances of about 10^{-30} m can be probed. Such energies are needed to test theory directly, but these are about 10^{10} higher than the proposed giant SSC would have had, and the distances are about 10^{-12} smaller than any structure we have direct knowledge of. This would be the realm of various GUTs, of which there are many since there is no constraining evidence at these energies and distances. Past experience has shown that any time you probe so many orders of magnitude further (here, about 10^{12}), you find the unexpected. Even more extreme are the energies and distances at which gravity is thought to unify with the other forces in a TOE. Most speculative and least constrained by experiment are TOEs, one of which is called **Superstring theory**. Superstrings are entities that are 10^{-35} m in scale and act like one-dimensional oscillating strings and are also proposed to underlie all particles, forces, and space itself.

At the energy of GUTs, the carrier particles of the weak force would become massless and identical to gluons. If that happens, then both lepton and baryon conservation would be violated. We do not see such violations, because we do not encounter such energies. However, there is a tiny probability that, at ordinary energies, the virtual particles that violate the conservation of baryon number may exist for extremely small amounts of time (corresponding to very small ranges). All GUTs thus predict that the proton should be unstable, but would decay with an extremely long lifetime of about 10^{31} y . The predicted decay mode is

$$p \rightarrow \pi^0 + e^+, \text{ (proposed proton decay)} \qquad (33.11)$$

which violates both conservation of baryon number and electron family number. Although 10^{31} y is an extremely long time (about 10^{21} times the age of the universe), there are a lot of protons, and detectors have been constructed to look for the proposed decay mode as seen in **Figure 33.25**. It is somewhat comforting that proton decay has not been detected, and its experimental lifetime is now greater than 5×10^{32} y . This does not prove GUTs wrong, but it does place greater constraints on the theories, benefiting theorists in many ways.

From looking increasingly inward at smaller details for direct evidence of electroweak theory and GUTs, we turn around and look to the universe for evidence of the unification of forces. In the 1920s, the expansion of the universe was discovered. Thinking backward in time, the universe must once have been very small, dense, and extremely hot. At a tiny fraction of a second after the fabled Big Bang, forces would have been unified and may have left their fingerprint on the existing universe. This, one of the most exciting forefronts of physics, is the subject of **Frontiers of Physics**.

Figure 33.25 In the Tevatron accelerator at Fermilab, protons and antiprotons collide at high energies, and some of those collisions could result in the production of a Higgs boson in association with a *W* boson. When the *W* boson decays to a high-energy lepton and a neutrino, the detector triggers on the lepton, whether it is an electron or a muon. (credit: D. J. Miller)

Summary

- Attempts to show unification of the four forces are called Grand Unified Theories (GUTs) and have been partially successful, with connections proven between EM and weak forces in electroweak theory.
- The strong force is carried by eight proposed particles called gluons, which are intimately connected to a quantum number called color—their governing theory is thus called quantum chromodynamics (QCD). Taken together, QCD and the electroweak theory are widely accepted as the Standard Model of particle physics.
- Unification of the strong force is expected at such high energies that it cannot be directly tested, but it may have observable consequences in the as-yet unobserved decay of the proton and topics to be discussed in the next chapter. Although unification of forces is generally anticipated, much remains to be done to prove its validity.

Glossary

baryon number: a conserved physical quantity that is zero for mesons and leptons and ± 1 for baryons and antibaryons, respectively

baryons: hadrons that always decay to another baryon

boson: particle with zero or an integer value of intrinsic spin

bottom: a quark flavor

charm: a quark flavor, which is the counterpart of the strange quark

colliding beams: head-on collisions between particles moving in opposite directions

color: a quark flavor

conservation of total baryon number: a general rule based on the observation that the total number of nucleons was always conserved in nuclear reactions and decays

conservation of total electron family number: a general rule stating that the total electron family number stays the same through an interaction

conservation of total muon family number: a general rule stating that the total muon family number stays the same through an interaction

cyclotron: accelerator that uses fixed-frequency alternating electric fields and fixed magnets to accelerate particles in a circular spiral path

down: the second-lightest of all quarks

electron family number: the number ± 1 that is assigned to all members of the electron family, or the number 0 that is assigned to all particles not in the electron family

electroweak theory: theory showing connections between EM and weak forces

fermion: particle with a half-integer value of intrinsic spin

Feynman diagram: a graph of time versus position that describes the exchange of virtual particles between subatomic particles

flavors: quark type

fundamental particle: particle with no substructure

gauge boson: particle that carries one of the four forces

gluons: exchange particles, analogous to the exchange of photons that gives rise to the electromagnetic force between two charged particles

gluons: eight proposed particles which carry the strong force

grand unified theory: theory that shows unification of the strong and electroweak forces

hadrons: particles that feel the strong nuclear force

Higgs boson: a massive particle that, if observed, would give validity to the theory that carrier particles are identical under certain circumstances

leptons: particles that do not feel the strong nuclear force

linear accelerator: accelerator that accelerates particles in a straight line

meson: particle whose mass is intermediate between the electron and nucleon masses

meson: hadrons that can decay to leptons and leave no hadrons

muon family number: the number ± 1 that is assigned to all members of the muon family, or the number 0 that is assigned to all particles not in the muon family

particle physics: the study of and the quest for those truly fundamental particles having no substructure

pion: particle exchanged between nucleons, transmitting the force between them

quantum chromodynamics: quark theory including color

quantum chromodynamics: the governing theory of connecting quantum number color to gluons

quantum electrodynamics: the theory of electromagnetism on the particle scale

quark: an elementary particle and a fundamental constituent of matter

standard model: combination of quantum chromodynamics and electroweak theory

strange: the third lightest of all quarks

strangeness: a physical quantity assigned to various particles based on decay systematics

superstring theory: a theory of everything based on vibrating strings some 10^{-35} m in length

synchrotron: a version of a cyclotron in which the frequency of the alternating voltage and the magnetic field strength are increased as the beam particles are accelerated

synchrotron radiation: radiation caused by a magnetic field accelerating a charged particle perpendicular to its velocity

tau family number: the number ± 1 that is assigned to all members of the tau family, or the number 0 that is assigned to all particles not in the tau family

theory of quark confinement: explains how quarks can exist and yet never be isolated or directly observed

top: a quark flavor

up: the lightest of all quarks

Van de Graaff: early accelerator: simple, large-scale version of the electron gun

virtual particles: particles which cannot be directly observed but their effects can be directly observed

Conceptual Questions

33.3 Accelerators Create Matter from Energy

1. The total energy in the beam of an accelerator is far greater than the energy of the individual beam particles. Why isn't this total energy available to create a single extremely massive particle?

2. Synchrotron radiation takes energy from an accelerator beam and is related to acceleration. Why would you expect the problem to be more severe for electron accelerators than proton accelerators?

3. What two major limitations prevent us from building high-energy accelerators that are physically small?

4. What are the advantages of colliding-beam accelerators? What are the disadvantages?

33.4 Particles, Patterns, and Conservation Laws

5. Large quantities of antimatter isolated from normal matter should behave exactly like normal matter. An antiatom, for example, composed of positrons, antiprotons, and antineutrons should have the same atomic spectrum as its matter counterpart. Would you be able to tell it is antimatter by its emission of antiphotons? Explain briefly.

6. Massless particles are not only neutral, they are chargeless (unlike the neutron). Why is this so?

7. Massless particles must travel at the speed of light, while others cannot reach this speed. Why are all massless particles stable? If evidence is found that neutrinos spontaneously decay into other particles, would this imply they have mass?

8. When a star erupts in a supernova explosion, huge numbers of electron neutrinos are formed in nuclear reactions. Such neutrinos from the 1987A supernova in the relatively nearby Magellanic Cloud were observed within hours of the initial brightening, indicating they traveled to earth at approximately the speed of light. Explain how this data can be used to set an upper limit on the mass of the neutrino, noting that if the mass is small the neutrinos could travel very close to the speed of light and have a reasonable energy (on the order of MeV).

9. Theorists have had spectacular success in predicting previously unknown particles. Considering past theoretical triumphs, why should we bother to perform experiments?

10. What lifetime do you expect for an antineutron isolated from normal matter?

11. Why does the η^0 meson have such a short lifetime compared to most other mesons?

12. (a) Is a hadron always a baryon?

(b) Is a baryon always a hadron?

(c) Can an unstable baryon decay into a meson, leaving no other baryon?

13. Explain how conservation of baryon number is responsible for conservation of total atomic mass (total number of nucleons) in nuclear decay and reactions.

33.5 Quarks: Is That All There Is?

14. The quark flavor change $d \to u$ takes place in β^- decay. Does this mean that the reverse quark flavor change $u \to d$ takes place in β^+ decay? Justify your response by writing the decay in terms of the quark constituents, noting that it looks as if a proton is converted into a neutron in β^+ decay.

15. Explain how the weak force can change strangeness by changing quark flavor.

16. Beta decay is caused by the weak force, as are all reactions in which strangeness changes. Does this imply that the weak force can change quark flavor? Explain.

17. Why is it easier to see the properties of the c, b, and t quarks in mesons having composition W^- or $t\bar{t}$ rather than in baryons having a mixture of quarks, such as udb?

18. How can quarks, which are fermions, combine to form bosons? Why must an even number combine to form a boson? Give one example by stating the quark substructure of a boson.

19. What evidence is cited to support the contention that the gluon force between quarks is greater than the strong nuclear force between hadrons? How is this related to color? Is it also related to quark confinement?

20. Discuss how we know that π-mesons (π^+, π, π^0) are not fundamental particles and are not the basic carriers of the strong force.

21. An antibaryon has three antiquarks with colors $\bar{R}\bar{G}\bar{B}$. What is its color?

22. Suppose leptons are created in a reaction. Does this imply the weak force is acting? (for example, consider β decay.)

23. How can the lifetime of a particle indicate that its decay is caused by the strong nuclear force? How can a change in strangeness imply which force is responsible for a reaction? What does a change in quark flavor imply about the force that is responsible?

24. (a) Do all particles having strangeness also have at least one strange quark in them?

(b) Do all hadrons with a strange quark also have nonzero strangeness?

25. The sigma-zero particle decays mostly via the reaction $\Sigma^0 \rightarrow \Lambda^0 + \gamma$. Explain how this decay and the respective quark compositions imply that the Σ^0 is an excited state of the Λ^0.

26. What do the quark compositions and other quantum numbers imply about the relationships between the Δ^+ and the proton? The Δ^0 and the neutron?

27. Discuss the similarities and differences between the photon and the Z^0 in terms of particle properties, including forces felt.

28. Identify evidence for electroweak unification.

29. The quarks in a particle are confined, meaning individual quarks cannot be directly observed. Are gluons confined as well? Explain

33.6 GUTs: The Unification of Forces

30. If a GUT is proven, and the four forces are unified, it will still be correct to say that the orbit of the moon is determined by the gravitational force. Explain why.

31. If the Higgs boson is discovered and found to have mass, will it be considered the ultimate carrier of the weak force? Explain your response.

32. Gluons and the photon are massless. Does this imply that the W^+, W^-, and Z^0 are the ultimate carriers of the weak force?

Problems & Exercises

33.1 The Yukawa Particle and the Heisenberg Uncertainty Principle Revisited

1. A virtual particle having an approximate mass of $10^{14} \text{ GeV}/c^2$ may be associated with the unification of the strong and electroweak forces. For what length of time could this virtual particle exist (in temporary violation of the conservation of mass-energy as allowed by the Heisenberg uncertainty principle)?

2. Calculate the mass in GeV/c^2 of a virtual carrier particle that has a range limited to 10^{-30} m by the Heisenberg uncertainty principle. Such a particle might be involved in the unification of the strong and electroweak forces.

3. Another component of the strong nuclear force is transmitted by the exchange of virtual K-mesons. Taking K-mesons to have an average mass of $495 \text{ MeV}/c^2$, what is the approximate range of this component of the strong force?

33.2 The Four Basic Forces

4. (a) Find the ratio of the strengths of the weak and electromagnetic forces under ordinary circumstances.

(b) What does that ratio become under circumstances in which the forces are unified?

5. The ratio of the strong to the weak force and the ratio of the strong force to the electromagnetic force become 1 under circumstances where they are unified. What are the ratios of the strong force to those two forces under normal circumstances?

33.3 Accelerators Create Matter from Energy

6. At full energy, protons in the 2.00-km-diameter Fermilab synchrotron travel at nearly the speed of light, since their energy is about 1000 times their rest mass energy.

(a) How long does it take for a proton to complete one trip around?

(b) How many times per second will it pass through the target area?

7. Suppose a W^- created in a bubble chamber lives for 5.00×10^{-25} s. What distance does it move in this time if it is traveling at 0.900 c? Since this distance is too short to make a track, the presence of the W^- must be inferred from its decay products. Note that the time is longer than the given W^- lifetime, which can be due to the statistical nature of decay or time dilation.

8. What length track does a π^+ traveling at 0.100 c leave in a bubble chamber if it is created there and lives for 2.60×10^{-8} s ? (Those moving faster or living longer may escape the detector before decaying.)

9. The 3.20-km-long SLAC produces a beam of 50.0-GeV electrons. If there are 15,000 accelerating tubes, what average voltage must be across the gaps between them to achieve this energy?

10. Because of energy loss due to synchrotron radiation in the LHC at CERN, only 5.00 MeV is added to the energy of each proton during each revolution around the main ring. How many revolutions are needed to produce 7.00-TeV (7000 GeV) protons, if they are injected with an initial energy of 8.00 GeV?

11. A proton and an antiproton collide head-on, with each having a kinetic energy of 7.00 TeV (such as in the LHC at CERN). How much collision energy is available, taking into account the annihilation of the two masses? (Note that this is not significantly greater than the extremely relativistic kinetic energy.)

12. When an electron and positron collide at the SLAC facility, they each have 50.0 GeV kinetic energies. What is the total collision energy available, taking into account the annihilation energy? Note that the annihilation energy is insignificant, because the electrons are highly relativistic.

33.4 Particles, Patterns, and Conservation Laws

13. The π^0 is its own antiparticle and decays in the following manner: $\pi^0 \rightarrow \gamma + \gamma$. What is the energy of each γ ray if the π^0 is at rest when it decays?

14. The primary decay mode for the negative pion is $\pi^- \rightarrow \mu^- + \bar{\nu}_\mu$. What is the energy release in MeV in this decay?

15. The mass of a theoretical particle that may be associated with the unification of the electroweak and strong forces is $10^{14} \text{ GeV}/c^2$.

(a) How many proton masses is this?

(b) How many electron masses is this? (This indicates how extremely relativistic the accelerator would have to be in order to make the particle, and how large the relativistic quantity γ would have to be.)

16. The decay mode of the negative muon is
$$\mu^- \rightarrow e^- + \bar{\nu}_e + \nu_\mu.$$

(a) Find the energy released in MeV.

(b) Verify that charge and lepton family numbers are conserved.

17. The decay mode of the positive tau is
$$\tau^+ \rightarrow \mu^+ + \nu_\mu + \bar{\nu}_\tau.$$

(a) What energy is released?

(b) Verify that charge and lepton family numbers are conserved.

(c) The τ^+ is the antiparticle of the τ^-. Verify that all the decay products of the τ^+ are the antiparticles of those in the decay of the τ^- given in the text.

18. The principal decay mode of the sigma zero is
$$\Sigma^0 \to \Lambda^0 + \gamma.$$

(a) What energy is released?

(b) Considering the quark structure of the two baryons, does it appear that the Σ^0 is an excited state of the Λ^0?

(c) Verify that strangeness, charge, and baryon number are conserved in the decay.

(d) Considering the preceding and the short lifetime, can the weak force be responsible? State why or why not.

19. (a) What is the uncertainty in the energy released in the decay of a π^0 due to its short lifetime?

(b) What fraction of the decay energy is this, noting that the decay mode is $\pi^0 \to \gamma + \gamma$ (so that all the π^0 mass is destroyed)?

20. (a) What is the uncertainty in the energy released in the decay of a τ^- due to its short lifetime?

(b) Is the uncertainty in this energy greater than or less than the uncertainty in the mass of the tau neutrino? Discuss the source of the uncertainty.

33.5 Quarks: Is That All There Is?

21. (a) Verify from its quark composition that the Δ^+ particle could be an excited state of the proton.

(b) There is a spread of about 100 MeV in the decay energy of the Δ^+, interpreted as uncertainty due to its short lifetime. What is its approximate lifetime?

(c) Does its decay proceed via the strong or weak force?

22. Accelerators such as the Triangle Universities Meson Facility (TRIUMF) in British Columbia produce secondary beams of pions by having an intense primary proton beam strike a target. Such "meson factories" have been used for many years to study the interaction of pions with nuclei and, hence, the strong nuclear force. One reaction that occurs is

$\pi^+ + p \to \Delta^{++} \to \pi^+ + p$, where the Δ^{++} is a very short-lived particle. The graph in **Figure 33.26** shows the probability of this reaction as a function of energy. The width of the bump is the uncertainty in energy due to the short lifetime of the Δ^{++}.

(a) Find this lifetime.

(b) Verify from the quark composition of the particles that this reaction annihilates and then re-creates a d quark and a $\bar{d}$ antiquark by writing the reaction and decay in terms of quarks.

(c) Draw a Feynman diagram of the production and decay of the Δ^{++} showing the individual quarks involved.

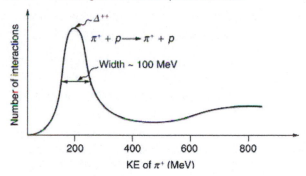

Figure 33.26 This graph shows the probability of an interaction between a π^+ and a proton as a function of energy. The bump is interpreted as a very short lived particle called a Δ^{++}. The approximately 100-MeV width of the bump is due to the short lifetime of the Δ^{++}.

23. The reaction $\pi^+ + p \to \Delta^{++}$ (described in the preceding problem) takes place via the strong force. (a) What is the baryon number of the Δ^{++} particle?

(b) Draw a Feynman diagram of the reaction showing the individual quarks involved.

24. One of the decay modes of the omega minus is
$$\Omega^- \to \Xi^0 + \pi^-.$$

(a) What is the change in strangeness?

(b) Verify that baryon number and charge are conserved, while lepton numbers are unaffected.

(c) Write the equation in terms of the constituent quarks, indicating that the weak force is responsible.

25. Repeat the previous problem for the decay mode
$$\Omega^- \to \Lambda^0 + K^-.$$

26. One decay mode for the eta-zero meson is $\eta^0 \rightarrow \gamma + \gamma$.

(a) Find the energy released.

(b) What is the uncertainty in the energy due to the short lifetime?

(c) Write the decay in terms of the constituent quarks.

(d) Verify that baryon number, lepton numbers, and charge are conserved.

27. One decay mode for the eta-zero meson is $\eta^0 \rightarrow \pi^0 + \pi^0$.

(a) Write the decay in terms of the quark constituents.

(b) How much energy is released?

(c) What is the ultimate release of energy, given the decay mode for the pi zero is $\pi^0 \rightarrow \gamma + \gamma$?

28. Is the decay $n \rightarrow e^+ + e^-$ possible considering the appropriate conservation laws? State why or why not.

29. Is the decay $\mu^- \rightarrow e^- + \nu_e + \nu_\mu$ possible considering the appropriate conservation laws? State why or why not.

30. (a) Is the decay $\Lambda^0 \rightarrow n + \pi^0$ possible considering the appropriate conservation laws? State why or why not.

(b) Write the decay in terms of the quark constituents of the particles.

31. (a) Is the decay $\Sigma^- \rightarrow n + \pi^-$ possible considering the appropriate conservation laws? State why or why not. (b) Write the decay in terms of the quark constituents of the particles.

32. The only combination of quark colors that produces a white baryon is *RGB*. Identify all the color combinations that can produce a white meson.

33. (a) Three quarks form a baryon. How many combinations of the six known quarks are there if all combinations are possible?

(b) This number is less than the number of known baryons. Explain why.

34. (a) Show that the conjectured decay of the proton, $p \rightarrow \pi^0 + e^+$, violates conservation of baryon number and conservation of lepton number.

(b) What is the analogous decay process for the antiproton?

35. Verify the quantum numbers given for the Ω^+ in **Table 33.2** by adding the quantum numbers for its quark constituents as inferred from **Table 33.4**.

36. Verify the quantum numbers given for the proton and neutron in **Table 33.2** by adding the quantum numbers for their quark constituents as given in **Table 33.4**.

37. (a) How much energy would be released if the proton did decay via the conjectured reaction $p \rightarrow \pi^0 + e^+$?

(b) Given that the π^0 decays to two γ s and that the e^+ will find an electron to annihilate, what total energy is ultimately produced in proton decay?

(c) Why is this energy greater than the proton's total mass (converted to energy)?

38. (a) Find the charge, baryon number, strangeness, charm, and bottomness of the J/Ψ particle from its quark composition.

(b) Do the same for the Υ particle.

39. There are particles called *D*-mesons. One of them is the D^+ meson, which has a single positive charge and a baryon number of zero, also the value of its strangeness, topness, and bottomness. It has a charm of $+1$. What is its quark configuration?

40. There are particles called bottom mesons or *B*-mesons. One of them is the B^- meson, which has a single negative charge; its baryon number is zero, as are its strangeness, charm, and topness. It has a bottomness of -1. What is its quark configuration?

41. (a) What particle has the quark composition $\bar{u}\,\bar{u}\,\bar{d}$?

(b) What should its decay mode be?

42. (a) Show that all combinations of three quarks produce integral charges. Thus baryons must have integral charge.

(b) Show that all combinations of a quark and an antiquark produce only integral charges. Thus mesons must have integral charge.

33.6 GUTs: The Unification of Forces

43. Integrated Concepts

The intensity of cosmic ray radiation decreases rapidly with increasing energy, but there are occasionally extremely energetic cosmic rays that create a shower of radiation from all the particles they create by striking a nucleus in the atmosphere as seen in the figure given below. Suppose a cosmic ray particle having an energy of 10^{10} GeV converts its energy into particles with masses averaging $200 \text{ MeV}/c^2$. (a) How many particles are created? (b) If the particles rain down on a 1.00-km^2 area, how many particles are there per square meter?

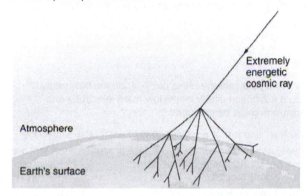

Figure 33.27 An extremely energetic cosmic ray creates a shower of particles on earth. The energy of these rare cosmic rays can approach a joule (about 10^{10} GeV) and, after multiple collisions, huge numbers of particles are created from this energy. Cosmic ray showers have been observed to extend over many square kilometers.

44. Integrated Concepts

Assuming conservation of momentum, what is the energy of each γ ray produced in the decay of a neutral at rest pion, in the reaction $\pi^0 \rightarrow \gamma + \gamma$?

45. Integrated Concepts

What is the wavelength of a 50-GeV electron, which is produced at SLAC? This provides an idea of the limit to the detail it can probe.

46. Integrated Concepts

(a) Calculate the relativistic quantity $\gamma = \dfrac{1}{\sqrt{1 - v^2/c^2}}$ for 1.00-TeV protons produced at Fermilab. (b) If such a proton created a π^+ having the same speed, how long would its life be in the laboratory? (c) How far could it travel in this time?

47. Integrated Concepts

The primary decay mode for the negative pion is $\pi^- \rightarrow \mu^- + \bar{\nu}_\mu$. (a) What is the energy release in MeV in this decay? (b) Using conservation of momentum, how much energy does each of the decay products receive, given the π^- is at rest when it decays? You may assume the muon antineutrino is massless and has momentum $p = E/c$, just like a photon.

48. Integrated Concepts

Plans for an accelerator that produces a secondary beam of K-mesons to scatter from nuclei, for the purpose of studying the strong force, call for them to have a kinetic energy of 500 MeV. (a) What would the relativistic quantity $\gamma = \dfrac{1}{\sqrt{1 - v^2/c^2}}$ be for these particles? (b) How long would their average lifetime be in the laboratory? (c) How far could they travel in this time?

49. Integrated Concepts

Suppose you are designing a proton decay experiment and you can detect 50 percent of the proton decays in a tank of water. (a) How many kilograms of water would you need to see one decay per month, assuming a lifetime of 10^{31} y ? (b) How many cubic meters of water is this? (c) If the actual lifetime is 10^{33} y , how long would you have to wait on an average to see a single proton decay?

50. Integrated Concepts

In supernovas, neutrinos are produced in huge amounts. They were detected from the 1987A supernova in the Magellanic Cloud, which is about 120,000 light years away from the Earth (relatively close to our Milky Way galaxy). If neutrinos have a mass, they cannot travel at the speed of light, but if their mass is small, they can get close. (a) Suppose a neutrino with a $7\text{-eV}/c^2$ mass has a kinetic energy of 700 keV. Find the relativistic quantity $\gamma = \dfrac{1}{\sqrt{1 - v^2/c^2}}$ for it. (b) If the neutrino leaves the 1987A supernova at the same time as a photon and both travel to Earth, how much sooner does the photon arrive? This is not a large time difference, given that it is impossible to know which neutrino left with which photon and the poor efficiency of the neutrino detectors. Thus, the fact that neutrinos were observed within hours of the brightening of the supernova only places an upper limit on the neutrino's mass. (Hint: You may need to use a series expansion to find v for the neutrino, since its γ is so large.)

51. Construct Your Own Problem

Consider an ultrahigh-energy cosmic ray entering the Earth's atmosphere (some have energies approaching a joule). Construct a problem in which you calculate the energy of the particle based on the number of particles in an observed cosmic ray shower. Among the things to consider are the average mass of the shower particles, the average number per square meter, and the extent (number of square meters covered) of the shower. Express the energy in eV and joules.

52. Construct Your Own Problem

Consider a detector needed to observe the proposed, but extremely rare, decay of an electron. Construct a problem in which you calculate the amount of matter needed in the detector to be able to observe the decay, assuming that it has a signature that is clearly identifiable. Among the things to consider are the estimated half life (long for rare events), and the number of decays per unit time that you wish to observe, as well as the number of electrons in the detector substance.

Test Prep for AP® Courses

33.2 The Four Basic Forces

1. Two intact (not ionized) hydrogen atoms are 10 cm apart. Which of the following are true?
a. Gravity, though very weak, is acting between them.
b. The neutral charge means the electromagnetic force between them can be ignored.
c. The range is too long for the strong force to be involved.
d. All of the above.

2. Explain why we only need to concern ourselves with gravitational force to describe the orbit of the Earth around the Sun.

3. Consider four forces: the gravitational force between the Earth and the Sun; the electrostatic force between the Earth and the Sun; the gravitational force between the proton and electron in a hydrogen atom, and the electrostatic force between the proton and electron in a hydrogen atom. What is the proper ordering of the magnitude of these forces, from greatest to least?
a. gravity, Earth-Sun; electrostatic, Earth-Sun; gravity, hydrogen; electrostatic, hydrogen
b. electrostatic, Earth-Sun; gravity, Earth-Sun; electrostatic, hydrogen; gravity, hydrogen
c. gravity, Earth-Sun; gravity, hydrogen; electrostatic, hydrogen; electrostatic, Earth-Sun
d. gravity, Earth-Sun; electrostatic, hydrogen; gravity, hydrogen; electrostatic, Earth-Sun

4. Deep within a nucleon, which is the stronger force between two quarks, gravity or the weak force? Why do you think so?

5. Consider the Earth-Moon system. If we were to place equal charges on the Earth and the Moon, how large would they need to be for the electrostatic repulsion to counteract the gravitational attraction?
a. 5.1×10^{13} C
b. 5.7×10^{13} C
c. 6.7×10^{13} C
d. 3.3×10^{27} C

6. What is the strength of the magnetic field created by the orbiting Moon, at the center of the orbit, in the system in the previous problem? (Treat the charge going around in orbit as a current loop.) How does this compare with the strength of the Earth's intrinsic magnetic field?

7. An atomic nucleus consists of positively charged protons and neutral neutrons, so the electrostatic repulsion should destroy it by making the protons fly apart. This doesn't happen because:
a. The strong force is ~100 times stronger than electromagnetism.
b. The weak force generates massive particles that hold it together.
c. Electromagnetism is sometimes attractive.
d. Gravity is always attractive.

8. The atomic number of an atom is the number of protons in that atom's nucleus. Make a prediction as to what happens to electromagnetic repulsion as the atomic number gets larger. Then, make a further prediction about what this implies about the number of neutrons in heavy nuclei.

33.3 Accelerators Create Matter from Energy

9. Which of the below was the first hint that conservation of mass and conservation of energy might need to be combined into one concept?
a. The Van de Graaff generator.
b. New particles showing up in accelerators.
c. Yukawa's theory.
d. They were always related.

10. How fast would two 7.0-kg bowling balls each have to be going in a collision to have enough spare energy to create a 0.10-kg tennis ball? (Ignore relativistic effects.) Can you explain why we don't see this in daily situations?

11. Use the information in Table 33.2 to answer the following questions.

Taking only energy and mass into consideration, what is the minimum amount of kinetic energy a K^- must have when colliding with a stationary proton to produce an Ω?
a. 240.5 MeV
b. 120.0 MeV
c. 15.5 MeV
d. 57.6 GeV

12. Using only energy-mass considerations, how many K^0 could a Z boson decay into? How many electrons and positrons could be produced this way?

13. A π^+ and a π^- are moving toward each other extremely slowly. When they collide, two π^0 are produced. How fast are they going? (Ignore relativistic effects.)
a. Barely moving
b. 1.0×10^7 m/s
c. 2.0×10^7 m/s
d. 7.8×10^7 m/s

14. Assume that when a free neutron decays, it transforms into a proton and an electron. Calculate the kinetic energy of the electron.

33.4 Particles, Patterns, and Conservation Laws

15. When a π^- decays, the products may include:
a. A positron.
b. A muon.
c. A proton.
d. All of the above.

16. Notice in Table 33.2 that the neutron has a half-life of 882 seconds. This is only for a free neutron, not bound with other neutrons and protons in a nucleus. Given the other particles in the table, and using both their charge and masses, what do you think the most likely decay products for a neutron are? Justify your answer.

33.5 Quarks: Is That All There Is?

17. How many pointlike particles would an experiment scattering high energy electrons from any meson discover within the meson?
a. 1
b. 2
c. 3
d. 4

18.

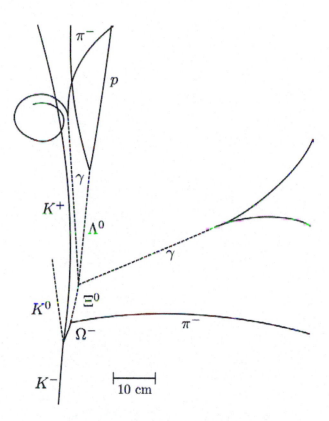

Figure 33.28 In this figure, a K⁻ initially hits a proton, and creates three new particles. Identify them, and explain how quark flavors are conserved.

34 FRONTIERS OF PHYSICS

Figure 34.1 This galaxy is ejecting huge jets of matter, powered by an immensely massive black hole at its center. (credit: X-ray: NASA/CXC/CfA/R. Kraft et al.)

Chapter Outline

34.1. Cosmology and Particle Physics

34.2. General Relativity and Quantum Gravity

34.3. Superstrings

34.4. Dark Matter and Closure

34.5. Complexity and Chaos

34.6. High-Temperature Superconductors

34.7. Some Questions We Know to Ask

Connection for AP® Courses

There is mystery, surprise, adventure, and discovery in exploring new frontiers. The search for answers is that much more intriguing because the answer to any question always leads to new questions. As our understanding of nature becomes more complete, nature still retains its sense of mystery and never loses its ability to awe us.

Looking through the lens of physics allows us to look both backward and forward in time, and we can discern marvelous patterns in nature with its myriad rules and complex connections. Moreover, we continue looking ever deeper and ever further, probing the basic structure of matter, energy, space, and time, and wondering about the scope of the universe, its beginnings, and its future.

The Big Ideas that we have been supporting and justifying throughout the previous chapters will now be used as a framework to investigate and justify new ideas. With the concepts, qualitative and quantitative problem-solving skills, the connections among topics, and all the rest of the coursework you have mastered, you will be more able to deeply appreciate the treatments that follow.

34.1 Cosmology and Particle Physics

Learning Objectives

By the end of this section, you will be able to:

- Discuss the expansion of the universe.
- Explain how the Big Bang gave rise to the universe we see today.

Look at the sky on some clear night when you are away from city lights. There you will see thousands of individual stars and a faint glowing background of millions more. The Milky Way, as it has been called since ancient times, is an arm of our galaxy of stars—the word *galaxy* coming from the Greek word *galaxias*, meaning milky. We know a great deal about our Milky Way galaxy and of the billions of other galaxies beyond its fringes. But they still provoke wonder and awe (see Figure 34.2). And there are still many questions to be answered. Most remarkable when we view the universe on the large scale is that once again explanations of its character and evolution are tied to the very small scale. Particle physics and the questions being asked about the very small scales may also have their answers in the very large scales.

Figure 34.2 Take a moment to contemplate these clusters of galaxies, photographed by the Hubble Space Telescope. Trillions of stars linked by gravity in fantastic forms, glowing with light and showing evidence of undiscovered matter. What are they like, these myriad stars? How did they evolve? What can they tell us of matter, energy, space, and time? (credit: NASA, ESA, K. Sharon (Tel Aviv University) and E. Ofek (Caltech))

As has been noted in numerous Things Great and Small vignettes, this is not the first time the large has been explained by the small and vice versa. Newton realized that the nature of gravity on Earth that pulls an apple to the ground could explain the motion of the moon and planets so much farther away. Minute atoms and molecules explain the chemistry of substances on a much larger scale. Decays of tiny nuclei explain the hot interior of the Earth. Fusion of nuclei likewise explains the energy of stars. Today, the patterns in particle physics seem to be explaining the evolution and character of the universe. And the nature of the universe has implications for unexplored regions of particle physics.

Cosmology is the study of the character and evolution of the universe. What are the major characteristics of the universe as we know them today? First, there are approximately 10^{11} galaxies in the observable part of the universe. An average galaxy contains more than 10^{11} stars, with our Milky Way galaxy being larger than average, both in its number of stars and its dimensions. Ours is a spiral-shaped galaxy with a diameter of about 100,000 light years and a thickness of about 2000 light years in the arms with a central bulge about 10,000 light years across. The Sun lies about 30,000 light years from the center near the galactic plane. There are significant clouds of gas, and there is a halo of less-dense regions of stars surrounding the main body. (See Figure 34.3.) Evidence strongly suggests the existence of a large amount of additional matter in galaxies that does not produce light—the mysterious dark matter we shall later discuss.

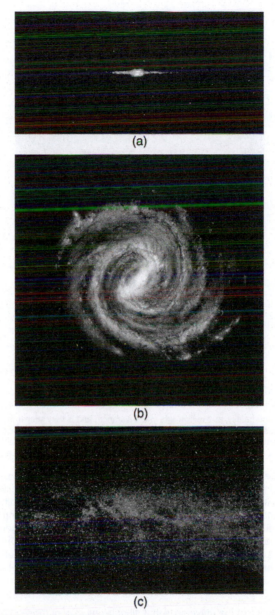

Figure 34.3 The Milky Way galaxy is typical of large spiral galaxies in its size, its shape, and the presence of gas and dust. We are fortunate to be in a location where we can see out of the galaxy and observe the vastly larger and fascinating universe around us. (a) Side view. (b) View from above. (c) The Milky Way as seen from Earth. (credits: (a) NASA, (b) Nick Risinger, (c) Andy)

Distances are great even within our galaxy and are measured in light years (the distance traveled by light in one year). The average distance between galaxies is on the order of a million light years, but it varies greatly with galaxies forming clusters such as shown in **Figure 34.2**. The Magellanic Clouds, for example, are small galaxies close to our own, some 160,000 light years from Earth. The Andromeda galaxy is a large spiral galaxy like ours and lies 2 million light years away. It is just visible to the naked eye as an extended glow in the Andromeda constellation. Andromeda is the closest large galaxy in our local group, and we can see some individual stars in it with our larger telescopes. The most distant known galaxy is 14 billion light years from Earth—a truly incredible distance. (See **Figure 34.4**.)

(a)

(b)

Figure 34.4 (a) Andromeda is the closest large galaxy, at 2 million light years distance, and is very similar to our Milky Way. The blue regions harbor young and emerging stars, while dark streaks are vast clouds of gas and dust. A smaller satellite galaxy is clearly visible. (b) The box indicates what may be the most distant known galaxy, estimated to be 13 billion light years from us. It exists in a much older part of the universe. (credit: NASA, ESA, G. Illingworth (University of California, Santa Cruz), R. Bouwens (University of California, Santa Cruz and Leiden University), and the HUDF09 Team)

Consider the fact that the light we receive from these vast distances has been on its way to us for a long time. In fact, the time in years is the same as the distance in light years. For example, the Andromeda galaxy is 2 million light years away, so that the light now reaching us left it 2 million years ago. If we could be there now, Andromeda would be different. Similarly, light from the most distant galaxy left it 14 billion years ago. We have an incredible view of the past when looking great distances. We can try to see if the universe was different then—if distant galaxies are more tightly packed or have younger-looking stars, for example, than closer galaxies, in which case there has been an evolution in time. But the problem is that the uncertainties in our data are great. Cosmology is almost typified by these large uncertainties, so that we must be especially cautious in drawing conclusions. One consequence is that there are more questions than answers, and so there are many competing theories. Another consequence is that any hard data produce a major result. Discoveries of some importance are being made on a regular basis, the hallmark of a field in its golden age.

Perhaps the most important characteristic of the universe is that all galaxies except those in our local cluster seem to be moving away from us at speeds proportional to their distance from our galaxy. It looks as if a gigantic explosion, universally called the **Big Bang**, threw matter out some billions of years ago. This amazing conclusion is based on the pioneering work of Edwin Hubble (1889–1953), the American astronomer. In the 1920s, Hubble first demonstrated conclusively that other galaxies, many previously called nebulae or clouds of stars, were outside our own. He then found that all but the closest galaxies have a red shift in their hydrogen spectra that is proportional to their distance. The explanation is that there is a **cosmological red shift** due to the expansion of space itself. The photon wavelength is stretched in transit from the source to the observer. Double the distance, and the red shift is doubled. While this cosmological red shift is often called a Doppler shift, it is not—space itself is expanding. There is no center of expansion in the universe. All observers see themselves as stationary; the other objects in space appear to be moving away from them. Hubble was directly responsible for discovering that the universe was much larger than had previously been imagined and that it had this amazing characteristic of rapid expansion.

Universal expansion on the scale of galactic clusters (that is, galaxies at smaller distances are not uniformly receding from one another) is an integral part of modern cosmology. For galaxies farther away than about 50 Mly (50 million light years), the expansion is uniform with variations due to local motions of galaxies within clusters. A representative recession velocity v can be obtained from the simple formula

$$v = H_0 d, \tag{34.1}$$

where d is the distance to the galaxy and H_0 is the **Hubble constant**. The Hubble constant is a central concept in cosmology.

Its value is determined by taking the slope of a graph of velocity versus distance, obtained from red shift measurements, such as shown in **Figure 34.5**. We shall use an approximate value of $H_0 = 20$ km/s $\cdot$ Mly. Thus, $v = H_0 d$ is an average behavior for all but the closest galaxies. For example, a galaxy 100 Mly away (as determined by its size and brightness) typically moves away from us at a speed of $v = (20 \text{ km/s} \cdot \text{Mly})(100 \text{ Mly}) = 2000 \text{ km/s}$. There can be variations in this speed due to so-called local motions or interactions with neighboring galaxies. Conversely, if a galaxy is found to be moving away from us at speed of 100,000 km/s based on its red shift, it is at a distance

$d = v / H_0 = (10,000 \text{ km/s}) / (20 \text{ km/s} \cdot \text{Mly}) = 5000 \text{ Mly} = 5 \text{ Gly}$ or $5 \times 10^9 \text{ ly}$. This last calculation is approximate, because it assumes the expansion rate was the same 5 billion years ago as now. A similar calculation in Hubble's measurement changed the notion that the universe is in a steady state.

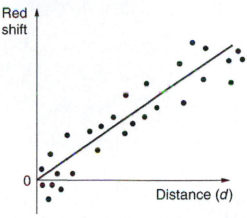

Figure 34.5 This graph of red shift versus distance for galaxies shows a linear relationship, with larger red shifts at greater distances, implying an expanding universe. The slope gives an approximate value for the expansion rate. (credit: John Cub).

One of the most intriguing developments recently has been the discovery that the expansion of the universe may be *faster now* than in the past, rather than slowing due to gravity as expected. Various groups have been looking, in particular, at supernovas in moderately distant galaxies (less than 1 Gly) to get improved distance measurements. Those distances are larger than expected for the observed galactic red shifts, implying the expansion was slower when that light was emitted. This has cosmological consequences that are discussed in **Dark Matter and Closure**. The first results, published in 1999, are only the beginning of emerging data, with astronomy now entering a data-rich era.

Figure 34.6 shows how the recession of galaxies looks like the remnants of a gigantic explosion, the famous Big Bang. Extrapolating backward in time, the Big Bang would have occurred between 13 and 15 billion years ago when all matter would have been at a point. Questions instantly arise. What caused the explosion? What happened before the Big Bang? Was there a before, or did time start then? Will the universe expand forever, or will gravity reverse it into a Big Crunch? And is there other evidence of the Big Bang besides the well-documented red shifts?

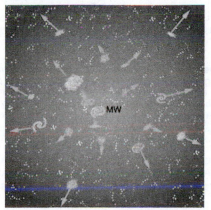

Figure 34.6 Galaxies are flying apart from one another, with the more distant moving faster as if a primordial explosion expelled the matter from which they formed. The most distant known galaxies move nearly at the speed of light relative to us.

The Russian-born American physicist George Gamow (1904–1968) was among the first to note that, if there was a Big Bang, the remnants of the primordial fireball should still be evident and should be blackbody radiation. Since the radiation from this fireball has been traveling to us since shortly after the Big Bang, its wavelengths should be greatly stretched. It will look as if the fireball has cooled in the billions of years since the Big Bang. Gamow and collaborators predicted in the late 1940s that there should be blackbody radiation from the explosion filling space with a characteristic temperature of about 7 K. Such blackbody radiation would have its peak intensity in the microwave part of the spectrum. (See **Figure 34.7**.) In 1964, Arno Penzias and Robert Wilson, two American scientists working with Bell Telephone Laboratories on a low-noise radio antenna, detected the radiation and eventually recognized it for what it is.

Figure 34.7(b) shows the spectrum of this microwave radiation that permeates space and is of cosmic origin. It is the most perfect blackbody spectrum known, and the temperature of the fireball remnant is determined from it to be $2.725 \pm 0.002 \text{K}$.

The detection of what is now called the **cosmic microwave background** (CMBR) was so important (generally considered as important as Hubble's detection that the galactic red shift is proportional to distance) that virtually every scientist has accepted the expansion of the universe as fact. Penzias and Wilson shared the 1978 Nobel Prize in Physics for their discovery.

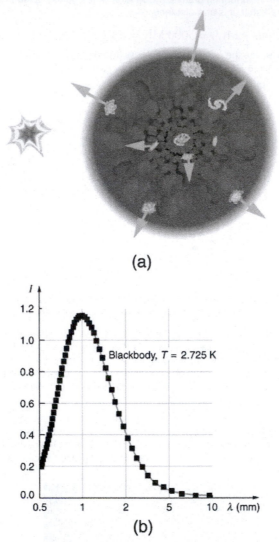

Figure 34.7 (a) The Big Bang is used to explain the present observed expansion of the universe. It was an incredibly energetic explosion some 10 to 20 billion years ago. After expanding and cooling, galaxies form inside the now-cold remnants of the primordial fireball. (b) The spectrum of cosmic microwave radiation is the most perfect blackbody spectrum ever detected. It is characteristic of a temperature of 2.725 K, the expansion-cooled temperature of the Big Bang's remnant. This radiation can be measured coming from any direction in space not obscured by some other source. It is compelling evidence of the creation of the universe in a gigantic explosion, already indicated by galactic red shifts.

Making Connections: Cosmology and Particle Physics

There are many connections of cosmology—by definition involving physics on the largest scale—with particle physics—by definition physics on the smallest scale. Among these are the dominance of matter over antimatter, the nearly perfect uniformity of the cosmic microwave background, and the mere existence of galaxies.

Matter versus antimatter We know from direct observation that antimatter is rare. The Earth and the solar system are nearly pure matter. Space probes and cosmic rays give direct evidence—the landing of the Viking probes on Mars would have been spectacular explosions of mutual annihilation energy if Mars were antimatter. We also know that most of the universe is dominated by matter. This is proven by the lack of annihilation radiation coming to us from space, particularly the relative absence of 0.511-MeV γ rays created by the mutual annihilation of electrons and positrons. It seemed possible that there could be entire solar systems or galaxies made of antimatter in perfect symmetry with our matter-dominated systems. But the interactions between stars and galaxies would sometimes bring matter and antimatter together in large amounts. The annihilation radiation they would produce is simply not observed. Antimatter in nature is created in particle collisions and in β^+ decays, but only in small amounts that quickly annihilate, leaving almost pure matter surviving.

Particle physics seems symmetric in matter and antimatter. Why isn't the cosmos? The answer is that particle physics is not quite perfectly symmetric in this regard. The decay of one of the neutral K-mesons, for example, preferentially creates more matter than antimatter. This is caused by a fundamental small asymmetry in the basic forces. This small asymmetry produced slightly more matter than antimatter in the early universe. If there was only one part in 10^9 more matter (a small asymmetry), the rest would annihilate pair for pair, leaving nearly pure matter to form the stars and galaxies we see today. So the vast number of stars we observe may be only a tiny remnant of the original matter created in the Big Bang. Here at last we see a very real and important asymmetry in nature. Rather than be disturbed by an asymmetry, most physicists are impressed by how small it is. Furthermore, if the universe were completely symmetric, the mutual annihilation would be more complete, leaving far less matter to form us and the universe we know.

How can something so old have so few wrinkles? A troubling aspect of cosmic microwave background radiation (CMBR) was soon recognized. True, the CMBR verified the Big Bang, had the correct temperature, and had a blackbody spectrum as expected. But the CMBR was *too* smooth—it looked identical in every direction. Galaxies and other similar entities could not be formed without the existence of fluctuations in the primordial stages of the universe and so there should be hot and cool spots in the CMBR, nicknamed wrinkles, corresponding to dense and sparse regions of gas caused by turbulence or early fluctuations. Over time, dense regions would contract under gravity and form stars and galaxies. Why aren't the fluctuations there? (This is a good example of an answer producing more questions.) Furthermore, galaxies are observed very far from us, so that they formed very long ago. The problem was to explain how galaxies could form so early and so quickly after the Big Bang if its remnant fingerprint is perfectly smooth. The answer is that if you look very closely, the CMBR is not perfectly smooth, only extremely smooth.

A satellite called the Cosmic Background Explorer (COBE) carried an instrument that made very sensitive and accurate measurements of the CMBR. In April of 1992, there was extraordinary publicity of COBE's first results—there were small fluctuations in the CMBR. Further measurements were carried out by experiments including NASA's Wilkinson Microwave Anisotropy Probe (WMAP), which launched in 2001. Data from WMAP provided a much more detailed picture of the CMBR fluctuations. (See **Figure 34.7**.) These amount to temperature fluctuations of only $200 \ \mu k$ out of 2.7 K, better than one part in 1000. The WMAP experiment will be followed up by the European Space Agency's Planck Surveyor, which launched in 2009.

Figure 34.8 This map of the sky uses color to show fluctuations, or wrinkles, in the cosmic microwave background observed with the WMAP spacecraft. The Milky Way has been removed for clarity. Red represents higher temperature and higher density, while blue is lower temperature and density. The fluctuations are small, less than one part in 1000, but these are still thought to be the cause of the eventual formation of galaxies. (credit: NASA/WMAP Science Team)

Let us now examine the various stages of the overall evolution of the universe from the Big Bang to the present, illustrated in **Figure 34.9**. Note that scientific notation is used to encompass the many orders of magnitude in time, energy, temperature, and size of the universe. Going back in time, the two lines approach but do not cross (there is no zero on an exponential scale). Rather, they extend indefinitely in ever-smaller time intervals to some infinitesimal point.

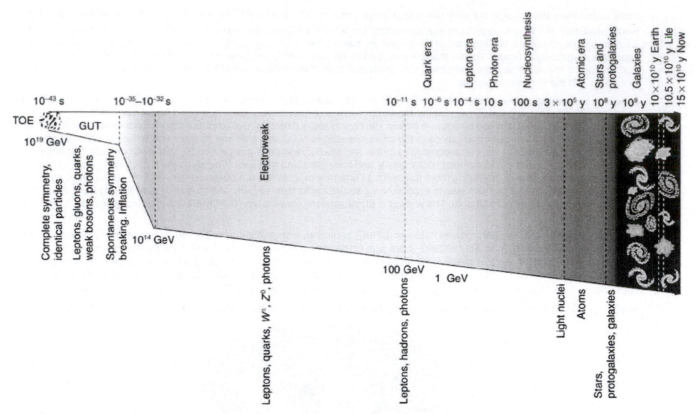

Figure 34.9 The evolution of the universe from the Big Bang onward is intimately tied to the laws of physics, especially those of particle physics at the earliest stages. The universe is relativistic throughout its history. Theories of the unification of forces at high energies may be verified by their shaping of the universe and its evolution.

Going back in time is equivalent to what would happen if expansion stopped and gravity pulled all the galaxies together, compressing and heating all matter. At a time long ago, the temperature and density were too high for stars and galaxies to exist. Before then, there was a time when the temperature was too great for atoms to exist. And farther back yet, there was a time when the temperature and density were so great that nuclei could not exist. Even farther back in time, the temperature was so high that average kinetic energy was great enough to create short-lived particles, and the density was high enough to make this likely. When we extrapolate back to the point of $W^{\pm}$ and Z^0 production (thermal energies reaching 1 TeV, or a temperature of about 10^{15} K), we reach the limits of what we know directly about particle physics. This is at a time about 10^{-12} s after the Big Bang. While 10^{-12} s may seem to be negligibly close to the instant of creation, it is not. There are important stages before this time that are tied to the unification of forces. At those stages, the universe was at extremely high energies and average particle separations were smaller than we can achieve with accelerators. What happened in the early stages before 10^{-12} s is crucial to all later stages and is possibly discerned by observing present conditions in the universe. One of these is the smoothness of the CMBR.

Names are given to early stages representing key conditions. The stage before 10^{-11} s back to 10^{-34} s is called the **electroweak epoch**, because the electromagnetic and weak forces become identical for energies above about 100 GeV. As discussed earlier, theorists expect that the strong force becomes identical to and thus unified with the electroweak force at energies of about 10^{14} GeV . The average particle energy would be this great at 10^{-34} s after the Big Bang, if there are no surprises in the unknown physics at energies above about 1 TeV. At the immense energy of 10^{14} GeV (corresponding to a temperature of about 10^{26} K), the $W^{\pm}$ and Z^0 carrier particles would be transformed into massless gauge bosons to accomplish the unification. Before 10^{-34} s back to about 10^{-43} s , we have Grand Unification in the **GUT epoch**, in which all forces except gravity are identical. At 10^{-43} s , the average energy reaches the immense 10^{19} GeV needed to unify gravity with the other forces in TOE, the Theory of Everything. Before that time is the **TOE epoch**, but we have almost no idea as to the nature of the universe then, since we have no workable theory of quantum gravity. We call the hypothetical unified force **superforce**.

Now let us imagine starting at TOE and moving forward in time to see what type of universe is created from various events along the way. As temperatures and average energies decrease with expansion, the universe reaches the stage where average particle separations are large enough to see differences between the strong and electroweak forces (at about 10^{-35} s). After this time, the forces become distinct in almost all interactions—they are no longer unified or symmetric. This transition from GUT to electroweak is an example of **spontaneous symmetry breaking**, in which conditions spontaneously evolved to a point where the forces were no longer unified, breaking that symmetry. This is analogous to a phase transition in the universe, and a clever proposal by American physicist Alan Guth in the early 1980s ties it to the smoothness of the CMBR. Guth proposed that spontaneous symmetry breaking (like a phase transition during cooling of normal matter) released an immense amount of energy that caused the universe to expand extremely rapidly for the brief time from 10^{-35} s to about 10^{-32} s. This expansion may have been by an incredible factor of 10^{50} or more in the size of the universe and is thus called the **inflationary scenario**. One result of this inflation is that it would stretch the wrinkles in the universe nearly flat, leaving an extremely smooth CMBR. While speculative, there is as yet no other plausible explanation for the smoothness of the CMBR. Unless the CMBR is not really cosmic but local in origin, the distances between regions of similar temperatures are too great for any coordination to have caused them, since any coordination mechanism must travel at the speed of light. Again, particle physics and cosmology are intimately entwined. There is little hope that we may be able to test the inflationary scenario directly, since it occurs at energies near 10^{14} GeV, vastly greater than the limits of modern accelerators. But the idea is so attractive that it is incorporated into most cosmological theories.

Characteristics of the present universe may help us determine the validity of this intriguing idea. Additionally, the recent indications that the universe's expansion rate may be *increasing* (see **Dark Matter and Closure**) could even imply that we are *in* another inflationary epoch.

It is important to note that, if conditions such as those found in the early universe could be created in the laboratory, we would see the unification of forces directly today. The forces have not changed in time, but the average energy and separation of particles in the universe have. As discussed in **The Four Basic Forces**, the four basic forces in nature are distinct under most circumstances found today. The early universe and its remnants provide evidence from times when they were unified under most circumstances.

34.2 General Relativity and Quantum Gravity

Learning Objectives

By the end of this section, you will be able to:

- Explain the effect of gravity on light.
- Discuss black holes.
- Explain quantum gravity.

When we talk of black holes or the unification of forces, we are actually discussing aspects of general relativity and quantum gravity. We know from **Special Relativity** that relativity is the study of how different observers measure the same event, particularly if they move relative to one another. Einstein's theory of **general relativity** describes all types of relative motion including accelerated motion and the effects of gravity. General relativity encompasses special relativity and classical relativity in situations where acceleration is zero and relative velocity is small compared with the speed of light. Many aspects of general relativity have been verified experimentally, some of which are better than science fiction in that they are bizarre but true. **Quantum gravity** is the theory that deals with particle exchange of gravitons as the mechanism for the force, and with extreme conditions where quantum mechanics and general relativity must both be used. A good theory of quantum gravity does not yet exist, but one will be needed to understand how all four forces may be unified. If we are successful, the theory of quantum gravity will encompass all others, from classical physics to relativity to quantum mechanics—truly a Theory of Everything (TOE).

General Relativity

Einstein first considered the case of no observer acceleration when he developed the revolutionary special theory of relativity, publishing his first work on it in 1905. By 1916, he had laid the foundation of general relativity, again almost on his own. Much of what Einstein did to develop his ideas was to mentally analyze certain carefully and clearly defined situations—doing this is to perform a **thought experiment**. **Figure 34.10** illustrates a thought experiment like the ones that convinced Einstein that light must fall in a gravitational field. Think about what a person feels in an elevator that is accelerated upward. It is identical to being in a stationary elevator in a gravitational field. The feet of a person are pressed against the floor, and objects released from hand fall with identical accelerations. In fact, it is not possible, without looking outside, to know what is happening—acceleration upward or gravity. This led Einstein to correctly postulate that acceleration and gravity will produce identical effects in all situations. So, if acceleration affects light, then gravity will, too. **Figure 34.10** shows the effect of acceleration on a beam of light shone horizontally at one wall. Since the accelerated elevator moves up during the time light travels across the elevator, the beam of light strikes low, seeming to the person to bend down. (Normally a tiny effect, since the speed of light is so great.) The same effect must occur due to gravity, Einstein reasoned, since there is no way to tell the effects of gravity acting downward from acceleration of the elevator upward. Thus gravity affects the path of light, even though we think of gravity as acting between masses and photons are massless.

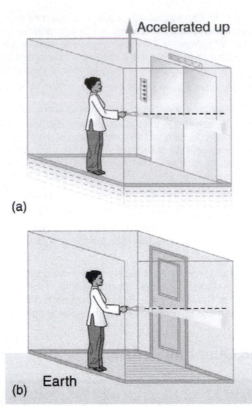

Figure 34.10 (a) A beam of light emerges from a flashlight in an upward-accelerating elevator. Since the elevator moves up during the time the light takes to reach the wall, the beam strikes lower than it would if the elevator were not accelerated. (b) Gravity has the same effect on light, since it is not possible to tell whether the elevator is accelerating upward or acted upon by gravity.

Einstein's theory of general relativity got its first verification in 1919 when starlight passing near the Sun was observed during a solar eclipse. (See **Figure 34.11**.) During an eclipse, the sky is darkened and we can briefly see stars. Those in a line of sight nearest the Sun should have a shift in their apparent positions. Not only was this shift observed, but it agreed with Einstein's predictions well within experimental uncertainties. This discovery created a scientific and public sensation. Einstein was now a folk hero as well as a very great scientist. The bending of light by matter is equivalent to a bending of space itself, with light following the curve. This is another radical change in our concept of space and time. It is also another connection that any particle with mass or energy (massless photons) is affected by gravity.

There are several current forefront efforts related to general relativity. One is the observation and analysis of gravitational lensing of light. Another is analysis of the definitive proof of the existence of black holes. Direct observation of gravitational waves or moving wrinkles in space is being searched for. Theoretical efforts are also being aimed at the possibility of time travel and wormholes into other parts of space due to black holes.

Gravitational lensing As you can see in **Figure 34.11**, light is bent toward a mass, producing an effect much like a converging lens (large masses are needed to produce observable effects). On a galactic scale, the light from a distant galaxy could be "lensed" into several images when passing close by another galaxy on its way to Earth. Einstein predicted this effect, but he considered it unlikely that we would ever observe it. A number of cases of this effect have now been observed; one is shown in **Figure 34.12**. This effect is a much larger scale verification of general relativity. But such gravitational lensing is also useful in verifying that the red shift is proportional to distance. The red shift of the intervening galaxy is always less than that of the one being lensed, and each image of the lensed galaxy has the same red shift. This verification supplies more evidence that red shift is proportional to distance. Confidence that the multiple images are not different objects is bolstered by the observations that if one image varies in brightness over time, the others also vary in the same manner.

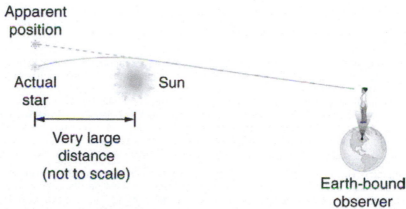

Figure 34.11 This schematic shows how light passing near a massive body like the Sun is curved toward it. The light that reaches the Earth then seems to be coming from different locations than the known positions of the originating stars. Not only was this effect observed, the amount of bending was precisely what Einstein predicted in his general theory of relativity.

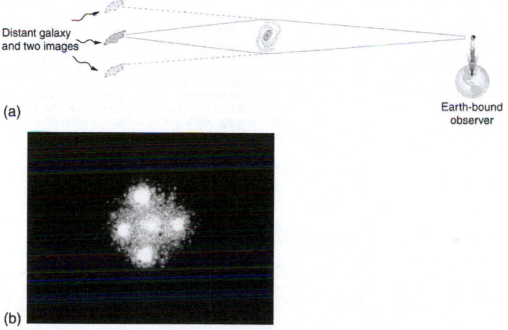

Figure 34.12 (a) Light from a distant galaxy can travel different paths to the Earth because it is bent around an intermediary galaxy by gravity. This produces several images of the more distant galaxy. (b) The images around the central galaxy are produced by gravitational lensing. Each image has the same spectrum and a larger red shift than the intermediary. (credit: NASA, ESA, and STScI)

Black holes **Black holes** are objects having such large gravitational fields that things can fall in, but nothing, not even light, can escape. Bodies, like the Earth or the Sun, have what is called an **escape velocity**. If an object moves straight up from the body, starting at the escape velocity, it will just be able to escape the gravity of the body. The greater the acceleration of gravity on the body, the greater is the escape velocity. As long ago as the late 1700s, it was proposed that if the escape velocity is greater than the speed of light, then light cannot escape. Simon Laplace (1749–1827), the French astronomer and mathematician, even incorporated this idea of a dark star into his writings. But the idea was dropped after Young's double slit experiment showed light to be a wave. For some time, light was thought not to have particle characteristics and, thus, could not be acted upon by gravity. The idea of a black hole was very quickly reincarnated in 1916 after Einstein's theory of general relativity was published. It is now thought that black holes can form in the supernova collapse of a massive star, forming an object perhaps 10 km across and having a mass greater than that of our Sun. It is interesting that several prominent physicists who worked on the concept, including Einstein, firmly believed that nature would find a way to prohibit such objects.

Black holes are difficult to observe directly, because they are small and no light comes directly from them. In fact, no light comes from inside the **event horizon**, which is defined to be at a distance from the object at which the escape velocity is exactly the speed of light. The radius of the event horizon is known as the **Schwarzschild radius** R_S and is given by

$$R_S = \frac{2GM}{c^2},$$ (34.2)

where G is the universal gravitational constant, M is the mass of the body, and c is the speed of light. The event horizon is the edge of the black hole and R_S is its radius (that is, the size of a black hole is twice R_S). Since G is small and c^2 is large, you can see that black holes are extremely small, only a few kilometers for masses a little greater than the Sun's. The object itself is inside the event horizon.

Physics near a black hole is fascinating. Gravity increases so rapidly that, as you approach a black hole, the tidal effects tear matter apart, with matter closer to the hole being pulled in with much more force than that only slightly farther away. This can pull a companion star apart and heat inflowing gases to the point of producing X rays. (See **Figure 34.13**.) We have observed X rays from certain binary star systems that are consistent with such a picture. This is not quite proof of black holes, because the X rays could also be caused by matter falling onto a neutron star. These objects were first discovered in 1967 by the British astrophysicists, Jocelyn Bell and Anthony Hewish. **Neutron stars** are literally a star composed of neutrons. They are formed by the collapse of a star's core in a supernova, during which electrons and protons are forced together to form neutrons (the reverse of neutron β decay). Neutron stars are slightly larger than a black hole of the same mass and will not collapse further because

of resistance by the strong force. However, neutron stars cannot have a mass greater than about eight solar masses or they must collapse to a black hole. With recent improvements in our ability to resolve small details, such as with the orbiting Chandra X-ray Observatory, it has become possible to measure the masses of X-ray-emitting objects by observing the motion of companion stars and other matter in their vicinity. What has emerged is a plethora of X-ray-emitting objects too massive to be neutron stars. This evidence is considered conclusive and the existence of black holes is widely accepted. These black holes are concentrated near galactic centers.

We also have evidence that supermassive black holes may exist at the cores of many galaxies, including the Milky Way. Such a black hole might have a mass millions or even billions of times that of the Sun, and it would probably have formed when matter first coalesced into a galaxy billions of years ago. Supporting this is the fact that very distant galaxies are more likely to have abnormally energetic cores. Some of the moderately distant galaxies, and hence among the younger, are known as **quasars** and emit as much or more energy than a normal galaxy but from a region less than a light year across. Quasar energy outputs may vary in times less than a year, so that the energy-emitting region must be less than a light year across. The best explanation of quasars is that they are young galaxies with a supermassive black hole forming at their core, and that they become less energetic over billions of years. In closer superactive galaxies, we observe tremendous amounts of energy being emitted from very small regions of space, consistent with stars falling into a black hole at the rate of one or more a month. The Hubble Space Telescope (1994) observed an accretion disk in the galaxy M87 rotating rapidly around a region of extreme energy emission. (See **Figure 34.13**.) A jet of material being ejected perpendicular to the plane of rotation gives further evidence of a supermassive black hole as the engine.

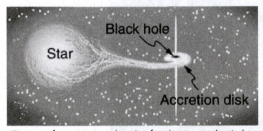

Figure 34.13 A black hole is shown pulling matter away from a companion star, forming a superheated accretion disk where X rays are emitted before the matter disappears forever into the hole. The in-fall energy also ejects some material, forming the two vertical spikes. (See also the photograph in **Introduction to Frontiers of Physics**.) There are several X-ray-emitting objects in space that are consistent with this picture and are likely to be black holes.

Gravitational waves If a massive object distorts the space around it, like the foot of a water bug on the surface of a pond, then movement of the massive object should create waves in space like those on a pond. **Gravitational waves** are mass-created distortions in space that propagate at the speed of light and are predicted by general relativity. Since gravity is by far the weakest force, extreme conditions are needed to generate significant gravitational waves. Gravity near binary neutron star systems is so great that significant gravitational wave energy is radiated as the two neutron stars orbit one another. American astronomers, Joseph Taylor and Russell Hulse, measured changes in the orbit of such a binary neutron star system. They found its orbit to change precisely as predicted by general relativity, a strong indication of gravitational waves, and were awarded the 1993 Nobel Prize. But direct detection of gravitational waves on Earth would be conclusive. For many years, various attempts have been made to detect gravitational waves by observing vibrations induced in matter distorted by these waves. American physicist Joseph Weber pioneered this field in the 1960s, but no conclusive events have been observed. (No gravity wave detectors were in operation at the time of the 1987A supernova, unfortunately.) There are now several ambitious systems of gravitational wave detectors in use around the world. These include the LIGO (Laser Interferometer Gravitational Wave Observatory) system with two laser interferometer detectors, one in the state of Washington and another in Louisiana (See **Figure 34.15**) and the VIRGO (Variability of Irradiance and Gravitational Oscillations) facility in Italy with a single detector.

Quantum Gravity

Black holes radiate Quantum gravity is important in those situations where gravity is so extremely strong that it has effects on the quantum scale, where the other forces are ordinarily much stronger. The early universe was such a place, but black holes are another. The first significant connection between gravity and quantum effects was made by the Russian physicist Yakov Zel'dovich in 1971, and other significant advances followed from the British physicist Stephen Hawking. (See **Figure 34.16**.) These two showed that black holes could radiate away energy by quantum effects just outside the event horizon (nothing can escape from inside the event horizon). Black holes are, thus, expected to radiate energy and shrink to nothing, although extremely slowly for most black holes. The mechanism is the creation of a particle-antiparticle pair from energy in the extremely strong gravitational field near the event horizon. One member of the pair falls into the hole and the other escapes, conserving momentum. (See **Figure 34.17**.) When a black hole loses energy and, hence, rest mass, its event horizon shrinks, creating an even greater gravitational field. This increases the rate of pair production so that the process grows exponentially until the black hole is nuclear in size. A final burst of particles and γ rays ensues. This is an extremely slow process for black holes about the mass of the Sun (produced by supernovas) or larger ones (like those thought to be at galactic centers), taking on the order of 10^{67} years or longer! Smaller black holes would evaporate faster, but they are only speculated to exist as remnants of the Big Bang. Searches for characteristic γ-ray bursts have produced events attributable to more mundane objects like neutron stars accreting matter.

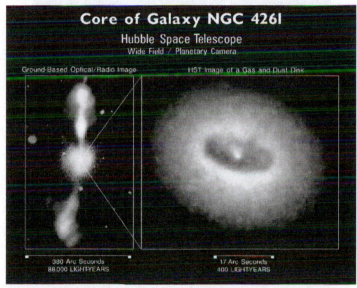

Figure 34.14 This Hubble Space Telescope photograph shows the extremely energetic core of the NGC 4261 galaxy. With the superior resolution of the orbiting telescope, it has been possible to observe the rotation of an accretion disk around the energy-producing object as well as to map jets of material being ejected from the object. A supermassive black hole is consistent with these observations, but other possibilities are not quite eliminated. (credit: NASA and ESA)

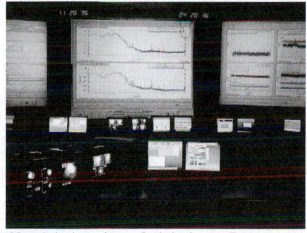

Figure 34.15 The control room of the LIGO gravitational wave detector. Gravitational waves will cause extremely small vibrations in a mass in this detector, which will be detected by laser interferometer techniques. Such detection in coincidence with other detectors and with astronomical events, such as supernovas, would provide direct evidence of gravitational waves. (credit: Tobin Fricke)

Figure 34.16 Stephen Hawking (b. 1942) has made many contributions to the theory of quantum gravity. Hawking is a long-time survivor of ALS and has produced popular books on general relativity, cosmology, and quantum gravity. (credit: Lwp Kommunikáció)

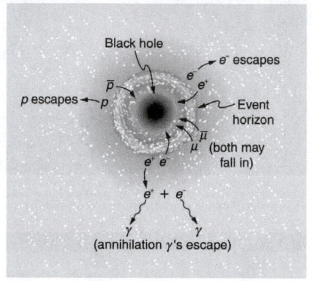

Figure 34.17 Gravity and quantum mechanics come into play when a black hole creates a particle-antiparticle pair from the energy in its gravitational field. One member of the pair falls into the hole while the other escapes, removing energy and shrinking the black hole. The search is on for the characteristic energy.

Wormholes and time travel The subject of time travel captures the imagination. Theoretical physicists, such as the American Kip Thorne, have treated the subject seriously, looking into the possibility that falling into a black hole could result in popping up in another time and place—a trip through a so-called wormhole. Time travel and wormholes appear in innumerable science fiction dramatizations, but the consensus is that time travel is not possible in theory. While still debated, it appears that quantum gravity effects inside a black hole prevent time travel due to the creation of particle pairs. Direct evidence is elusive.

The shortest time Theoretical studies indicate that, at extremely high energies and correspondingly early in the universe, quantum fluctuations may make time intervals meaningful only down to some finite time limit. Early work indicated that this might be the case for times as long as 10^{-43} s, the time at which all forces were unified. If so, then it would be meaningless to consider the universe at times earlier than this. Subsequent studies indicate that the crucial time may be as short as 10^{-95} s. But the point remains—quantum gravity seems to imply that there is no such thing as a vanishingly short time. Time may, in fact, be grainy with no meaning to time intervals shorter than some tiny but finite size.

The future of quantum gravity Not only is quantum gravity in its infancy, no one knows how to get started on a theory of gravitons and unification of forces. The energies at which TOE should be valid may be so high (at least 10^{19} GeV) and the necessary particle separation so small (less than 10^{-35} m) that only indirect evidence can provide clues. For some time, the common lament of theoretical physicists was one so familiar to struggling students—how do you even get started? But Hawking and others have made a start, and the approach many theorists have taken is called Superstring theory, the topic of the **Superstrings**.

34.3 Superstrings

Learning Objectives
By the end of this section, you will be able to:
• Define Superstring theory.
• Explain the relationship between Superstring theory and the Big Bang.

Introduced earlier in **GUTS: The Unification of Forces** **Superstring theory** is an attempt to unify gravity with the other three forces and, thus, must contain quantum gravity. The main tenet of Superstring theory is that fundamental particles, including the graviton that carries the gravitational force, act like one-dimensional vibrating strings. Since gravity affects the time and space in which all else exists, Superstring theory is an attempt at a Theory of Everything (TOE). Each independent quantum number is thought of as a separate dimension in some super space (analogous to the fact that the familiar dimensions of space are independent of one another) and is represented by a different type of Superstring. As the universe evolved after the Big Bang and forces became distinct (spontaneous symmetry breaking), some of the dimensions of superspace are imagined to have curled up and become unnoticed.

Forces are expected to be unified only at extremely high energies and at particle separations on the order of 10^{-35} m . This could mean that Superstrings must have dimensions or wavelengths of this size or smaller. Just as quantum gravity may imply that there are no time intervals shorter than some finite value, it also implies that there may be no sizes smaller than some tiny but finite value. That may be about 10^{-35} m . If so, and if Superstring theory can explain all it strives to, then the structures of Superstrings are at the lower limit of the smallest possible size and can have no further substructure. This would be the ultimate answer to the question the ancient Greeks considered. There is a finite lower limit to space.

Not only is Superstring theory in its infancy, it deals with dimensions about 17 orders of magnitude smaller than the 10^{-18} m details that we have been able to observe directly. It is thus relatively unconstrained by experiment, and there are a host of theoretical possibilities to choose from. This has led theorists to make choices subjectively (as always) on what is the most elegant theory, with less hope than usual that experiment will guide them. It has also led to speculation of alternate universes, with their Big Bangs creating each new universe with a random set of rules. These speculations may not be tested even in principle, since an alternate universe is by definition unattainable. It is something like exploring a self-consistent field of mathematics, with its axioms and rules of logic that are not consistent with nature. Such endeavors have often given insight to mathematicians and scientists alike and occasionally have been directly related to the description of new discoveries.

34.4 Dark Matter and Closure

Learning Objectives
By the end of this section, you will be able to:
• Discuss the evidence for the existence of dark matter.
• Explain neutrino oscillations and the consequences thereof.

One of the most exciting problems in physics today is the fact that there is far more matter in the universe than we can see. The motion of stars in galaxies and the motion of galaxies in clusters imply that there is about 10 times as much mass as in the luminous objects we can see. The indirectly observed non-luminous matter is called **dark matter**. Why is dark matter a problem? For one thing, we do not know what it is. It may well be 90% of all matter in the universe, yet there is a possibility that it is of a completely unknown form—a stunning discovery if verified. Dark matter has implications for particle physics. It may be possible that neutrinos actually have small masses or that there are completely unknown types of particles. Dark matter also has implications for cosmology, since there may be enough dark matter to stop the expansion of the universe. That is another problem related to dark matter—we do not know how much there is. We keep finding evidence for more matter in the universe, and we have an idea of how much it would take to eventually stop the expansion of the universe, but whether there is enough is still unknown.

Evidence

The first clues that there is more matter than meets the eye came from the Swiss-born American astronomer Fritz Zwicky in the 1930s; some initial work was also done by the American astronomer Vera Rubin. Zwicky measured the velocities of stars orbiting the galaxy, using the relativistic Doppler shift of their spectra (see **Figure 34.18**(a)). He found that velocity varied with distance from the center of the galaxy, as graphed in **Figure 34.18**(b). If the mass of the galaxy was concentrated in its center, as are its luminous stars, the velocities should decrease as the square root of the distance from the center. Instead, the velocity curve is almost flat, implying that there is a tremendous amount of matter in the galactic halo. Although not immediately recognized for its significance, such measurements have now been made for many galaxies, with similar results. Further, studies of galactic clusters have also indicated that galaxies have a mass distribution greater than that obtained from their brightness (proportional to the number of stars), which also extends into large halos surrounding the luminous parts of galaxies. Observations of other EM wavelengths, such as radio waves and X rays, have similarly confirmed the existence of dark matter. Take, for example, X rays in the relatively dark space between galaxies, which indicates the presence of previously unobserved hot, ionized gas (see

Figure 34.18(c)).

Theoretical Yearnings for Closure

Is the universe open or closed? That is, will the universe expand forever or will it stop, perhaps to contract? This, until recently, was a question of whether there is enough gravitation to stop the expansion of the universe. In the past few years, it has become a question of the combination of gravitation and what is called the **cosmological constant**. The cosmological constant was invented by Einstein to prohibit the expansion or contraction of the universe. At the time he developed general relativity, Einstein considered that an illogical possibility. The cosmological constant was discarded after Hubble discovered the expansion, but has been re-invoked in recent years.

Gravitational attraction between galaxies is slowing the expansion of the universe, but the amount of slowing down is not known directly. In fact, the cosmological constant can counteract gravity's effect. As recent measurements indicate, the universe is expanding *faster* now than in the past—perhaps a "modern inflationary era" in which the dark energy is thought to be causing the expansion of the present-day universe to accelerate. If the expansion rate were affected by gravity alone, we should be able to see that the expansion rate between distant galaxies was once greater than it is now. However, measurements show it was *less* than now. We can, however, calculate the amount of slowing based on the average density of matter we observe directly. Here we have a definite answer—there is far less visible matter than needed to stop expansion. The **critical density** ρ_c is defined to be the density needed to just halt universal expansion in a universe with no cosmological constant. It is estimated to be about

$$\rho_c \approx 10^{-26} \text{ kg/m}^3.$$
(34.3)

However, this estimate of ρ_c is only good to about a factor of two, due to uncertainties in the expansion rate of the universe.

The critical density is equivalent to an average of only a few nucleons per cubic meter, remarkably small and indicative of how truly empty intergalactic space is. Luminous matter seems to account for roughly 0.5% to 2% of the critical density, far less than that needed for closure. Taking into account the amount of dark matter we detect indirectly and all other types of indirectly observed normal matter, there is only 10% to 40% of what is needed for closure. If we are able to refine the measurements of expansion rates now and in the past, we will have our answer regarding the curvature of space and we will determine a value for the cosmological constant to justify this observation. Finally, the most recent measurements of the CMBR have implications for the cosmological constant, so it is not simply a device concocted for a single purpose.

After the recent experimental discovery of the cosmological constant, most researchers feel that the universe should be just barely open. Since matter can be thought to curve the space around it, we call an open universe **negatively curved**. This means that you can in principle travel an unlimited distance in any direction. A universe that is closed is called **positively curved**. This means that if you travel far enough in any direction, you will return to your starting point, analogous to circumnavigating the Earth. In between these two is a **flat (zero curvature) universe**. The recent discovery of the cosmological constant has shown the universe is very close to flat, and will expand forever. Why do theorists feel the universe is flat? Flatness is a part of the inflationary scenario that helps explain the flatness of the microwave background. In fact, since general relativity implies that matter creates the space in which it exists, there is a special symmetry to a flat universe.

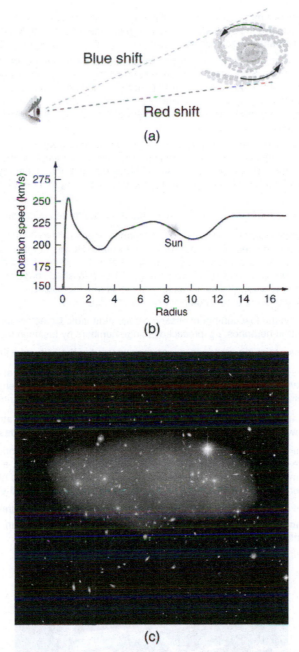

Figure 34.18 Evidence for dark matter: (a) We can measure the velocities of stars relative to their galaxies by observing the Doppler shift in emitted light, usually using the hydrogen spectrum. These measurements indicate the rotation of a spiral galaxy. (b) A graph of velocity versus distance from the galactic center shows that the velocity does not decrease as it would if the matter were concentrated in luminous stars. The flatness of the curve implies a massive galactic halo of dark matter extending beyond the visible stars. (c) This is a computer-generated image of X rays from a galactic cluster. The X rays indicate the presence of otherwise unseen hot clouds of ionized gas in the regions of space previously considered more empty. (credit: NASA, ESA, CXC, M. Bradac (University of California, Santa Barbara), and S. Allen (Stanford University))

What Is the Dark Matter We See Indirectly?

There is no doubt that dark matter exists, but its form and the amount in existence are two facts that are still being studied vigorously. As always, we seek to explain new observations in terms of known principles. However, as more discoveries are made, it is becoming more and more difficult to explain dark matter as a known type of matter.

One of the possibilities for normal matter is being explored using the Hubble Space Telescope and employing the lensing effect of gravity on light (see **Figure 34.19**). Stars glow because of nuclear fusion in them, but planets are visible primarily by reflected light. Jupiter, for example, is too small to ignite fusion in its core and become a star, but we can see sunlight reflected from it, since we are relatively close. If Jupiter orbited another star, we would not be able to see it directly. The question is open as to how many planets or other bodies smaller than about 1/1000 the mass of the Sun are there. If such bodies pass between us and a star, they will not block the star's light, being too small, but they will form a gravitational lens, as discussed in **General Relativity and Quantum Gravity**.

In a process called **microlensing**, light from the star is focused and the star appears to brighten in a characteristic manner.

Searches for dark matter in this form are particularly interested in galactic halos because of the huge amount of mass that seems to be there. Such microlensing objects are thus called **massive compact halo objects**, or **MACHOs**. To date, a few MACHOs have been observed, but not predominantly in galactic halos, nor in the numbers needed to explain dark matter.

MACHOs are among the most conventional of unseen objects proposed to explain dark matter. Others being actively pursued are red dwarfs, which are small dim stars, but too few have been seen so far, even with the Hubble Telescope, to be of significance. Old remnants of stars called white dwarfs are also under consideration, since they contain about a solar mass, but are small as the Earth and may dim to the point that we ordinarily do not observe them. While white dwarfs are known, old dim ones are not. Yet another possibility is the existence of large numbers of smaller than stellar mass black holes left from the Big Bang—here evidence is entirely absent.

There is a very real possibility that dark matter is composed of the known neutrinos, which may have small, but finite, masses. As discussed earlier, neutrinos are thought to be massless, but we only have upper limits on their masses, rather than knowing they are exactly zero. So far, these upper limits come from difficult measurements of total energy emitted in the decays and reactions in which neutrinos are involved. There is an amusing possibility of proving that neutrinos have mass in a completely different way.

We have noted in **Particles, Patterns, and Conservation Laws** that there are three flavors of neutrinos (ν_e, ν_μ, and ν_τ) and

that the weak interaction could change quark flavor. It should also change neutrino flavor—that is, any type of neutrino could change spontaneously into any other, a process called **neutrino oscillations**. However, this can occur only if neutrinos have a mass. Why? Crudely, because if neutrinos are massless, they must travel at the speed of light and time will not pass for them, so that they cannot change without an interaction. In 1999, results began to be published containing convincing evidence that neutrino oscillations do occur. Using the Super-Kamiokande detector in Japan, the oscillations have been observed and are being verified and further explored at present at the same facility and others.

Neutrino oscillations may also explain the low number of observed solar neutrinos. Detectors for observing solar neutrinos are specifically designed to detect electron neutrinos ν_e produced in huge numbers by fusion in the Sun. A large fraction of electron

neutrinos ν_e may be changing flavor to muon neutrinos ν_μ on their way out of the Sun, possibly enhanced by specific

interactions, reducing the flux of electron neutrinos to observed levels. There is also a discrepancy in observations of neutrinos produced in cosmic ray showers. While these showers of radiation produced by extremely energetic cosmic rays should contain twice as many ν_μ s as ν_e s, their numbers are nearly equal. This may be explained by neutrino oscillations from muon flavor to

electron flavor. Massive neutrinos are a particularly appealing possibility for explaining dark matter, since their existence is consistent with a large body of known information and explains more than dark matter. The question is not settled at this writing.

The most radical proposal to explain dark matter is that it consists of previously unknown leptons (sometimes obtusely referred to as non-baryonic matter). These are called **weakly interacting massive particles**, or **WIMPs**, and would also be chargeless, thus interacting negligibly with normal matter, except through gravitation. One proposed group of WIMPs would have masses several orders of magnitude greater than nucleons and are sometimes called **neutralinos**. Others are called **axions** and would

have masses about 10^{-10} that of an electron mass. Both neutralinos and axions would be gravitationally attached to galaxies, but because they are chargeless and only feel the weak force, they would be in a halo rather than interact and coalesce into spirals, and so on, like normal matter (see **Figure 34.20**).

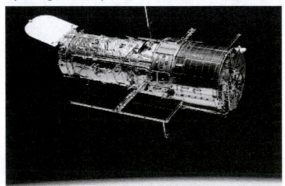

Figure 34.19 The Hubble Space Telescope is producing exciting data with its corrected optics and with the absence of atmospheric distortion. It has observed some MACHOs, disks of material around stars thought to precede planet formation, black hole candidates, and collisions of comets with Jupiter. (credit: NASA (crew of STS-125))

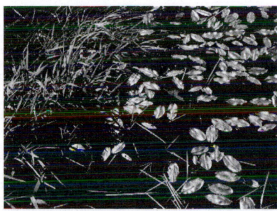

Figure 34.20 Dark matter may shepherd normal matter gravitationally in space, as this stream moves the leaves. Dark matter may be invisible and even move through the normal matter, as neutrinos penetrate us without small-scale effect. (credit: Shinichi Sugiyama)

Some particle theorists have built WIMPs into their unified force theories and into the inflationary scenario of the evolution of the universe so popular today. These particles would have been produced in just the correct numbers to make the universe flat, shortly after the Big Bang. The proposal is radical in the sense that it invokes entirely new forms of matter, in fact *two* entirely new forms, in order to explain dark matter and other phenomena. WIMPs have the extra burden of automatically being very difficult to observe directly. This is somewhat analogous to quark confinement, which guarantees that quarks are there, but they can never be seen directly. One of the primary goals of the LHC at CERN, however, is to produce and detect WIMPs. At any rate, before WIMPs are accepted as the best explanation, all other possibilities utilizing known phenomena will have to be shown inferior. Should that occur, we will be in the unanticipated position of admitting that, to date, all we know is only 10% of what exists. A far cry from the days when people firmly believed themselves to be not only the center of the universe, but also the reason for its existence.

34.5 Complexity and Chaos

Learning Objectives
By the end of this section, you will be able to: • Explain complex systems. • Discuss chaotic behavior of different systems.

Much of what impresses us about physics is related to the underlying connections and basic simplicity of the laws we have discovered. The language of physics is precise and well defined because many basic systems we study are simple enough that we can perform controlled experiments and discover unambiguous relationships. Our most spectacular successes, such as the prediction of previously unobserved particles, come from the simple underlying patterns we have been able to recognize. But there are systems of interest to physicists that are inherently complex. The simple laws of physics apply, of course, but complex systems may reveal patterns that simple systems do not. The emerging field of **complexity** is devoted to the study of complex systems, including those outside the traditional bounds of physics. Of particular interest is the ability of complex systems to adapt and evolve.

What are some examples of complex adaptive systems? One is the primordial ocean. When the oceans first formed, they were a random mix of elements and compounds that obeyed the laws of physics and chemistry. In a relatively short geological time (about 500 million years), life had emerged. Laboratory simulations indicate that the emergence of life was far too fast to have come from random combinations of compounds, even if driven by lightning and heat. There must be an underlying ability of the complex system to organize itself, resulting in the self-replication we recognize as life. Living entities, even at the unicellular level, are highly organized and systematic. Systems of living organisms are themselves complex adaptive systems. The grandest of these evolved into the biological system we have today, leaving traces in the geological record of steps taken along the way.

Complexity as a discipline examines complex systems, how they adapt and evolve, looking for similarities with other complex adaptive systems. Can, for example, parallels be drawn between biological evolution and the evolution of *economic systems*? Economic systems do emerge quickly, they show tendencies for self-organization, they are complex (in the number and types of transactions), and they adapt and evolve. Biological systems do all the same types of things. There are other examples of complex adaptive systems being studied for fundamental similarities. *Cultures* show signs of adaptation and evolution. The comparison of different cultural evolutions may bear fruit as well as comparisons to biological evolution. *Science* also is a complex system of human interactions, like culture and economics, that adapts to new information and political pressure, and evolves, usually becoming more organized rather than less. Those who study *creative thinking* also see parallels with complex systems. Humans sometimes organize almost random pieces of information, often subconsciously while doing other things, and come up with brilliant creative insights. The development of *language* is another complex adaptive system that may show similar tendencies. *Artificial intelligence* is an overt attempt to devise an adaptive system that will self-organize and evolve in the same manner as an intelligent living being learns. These are a few of the broad range of topics being studied by those who investigate complexity. There are now institutes, journals, and meetings, as well as popularizations of the emerging topic of complexity.

In traditional physics, the discipline of complexity may yield insights in certain areas. Thermodynamics treats systems on the average, while statistical mechanics deals in some detail with complex systems of atoms and molecules in random thermal motion. Yet there is organization, adaptation, and evolution in those complex systems. Non-equilibrium phenomena, such as heat transfer and phase changes, are characteristically complex in detail, and new approaches to them may evolve from complexity as a discipline. Crystal growth is another example of self-organization spontaneously emerging in a complex system. Alloys are also inherently complex mixtures that show certain simple characteristics implying some self-organization. The organization of iron atoms into magnetic domains as they cool is another. Perhaps insights into these difficult areas will emerge from complexity. But at the minimum, the discipline of complexity is another example of human effort to understand and organize the universe around us, partly rooted in the discipline of physics.

A predecessor to complexity is the topic of chaos, which has been widely publicized and has become a discipline of its own. It is also based partly in physics and treats broad classes of phenomena from many disciplines. **Chaos** is a word used to describe systems whose outcomes are extremely sensitive to initial conditions. The orbit of the planet Pluto, for example, may be chaotic in that it can change tremendously due to small interactions with other planets. This makes its long-term behavior impossible to predict with precision, just as we cannot tell precisely where a decaying Earth satellite will land or how many pieces it will break into. But the discipline of chaos has found ways to deal with such systems and has been applied to apparently unrelated systems. For example, the heartbeat of people with certain types of potentially lethal arrhythmias seems to be chaotic, and this knowledge may allow more sophisticated monitoring and recognition of the need for intervention.

Chaos is related to complexity. Some chaotic systems are also inherently complex; for example, vortices in a fluid as opposed to a double pendulum. Both are chaotic and not predictable in the same sense as other systems. But there can be organization in chaos and it can also be quantified. Examples of chaotic systems are beautiful fractal patterns such as in **Figure 34.21**. Some chaotic systems exhibit self-organization, a type of stable chaos. The orbits of the planets in our solar system, for example, may be chaotic (we are not certain yet). But they are definitely organized and systematic, with a simple formula describing the orbital radii of the first eight planets *and* the asteroid belt. Large-scale vortices in Jupiter's atmosphere are chaotic, but the Great Red Spot is a stable self-organization of rotational energy. (See **Figure 34.22**.) The Great Red Spot has been in existence for at least 400 years and is a complex self-adaptive system.

The emerging field of complexity, like the now almost traditional field of chaos, is partly rooted in physics. Both attempt to see similar systematics in a very broad range of phenomena and, hence, generate a better understanding of them. Time will tell what impact these fields have on more traditional areas of physics as well as on the other disciplines they relate to.

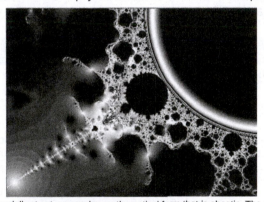

Figure 34.21 This image is related to the Mandelbrot set, a complex mathematical form that is chaotic. The patterns are infinitely fine as you look closer and closer, and they indicate order in the presence of chaos. (credit: Gilberto Santa Rosa)

Figure 34.22 The Great Red Spot on Jupiter is an example of self-organization in a complex and chaotic system. Smaller vortices in Jupiter's atmosphere behave chaotically, but the triple-Earth-size spot is self-organized and stable for at least hundreds of years. (credit: NASA)

34.6 High-Temperature Superconductors

Learning Objectives
By the end of this section, you will be able to: • Identify superconductors and their uses. • Discuss the need for a high-T_C superconductor.

Superconductors are materials with a resistivity of zero. They are familiar to the general public because of their practical applications and have been mentioned at a number of points in the text. Because the resistance of a piece of superconductor is zero, there are no heat losses for currents through them; they are used in magnets needing high currents, such as in MRI machines, and could cut energy losses in power transmission. But most superconductors must be cooled to temperatures only a few kelvin above absolute zero, a costly procedure limiting their practical applications. In the past decade, tremendous advances have been made in producing materials that become superconductors at relatively high temperatures. There is hope that room temperature superconductors may someday be manufactured.

Superconductivity was discovered accidentally in 1911 by the Dutch physicist H. Kamerlingh Onnes (1853–1926) when he used liquid helium to cool mercury. Onnes had been the first person to liquefy helium a few years earlier and was surprised to observe the resistivity of a mediocre conductor like mercury drop to zero at a temperature of 4.2 K. We define the temperature at which and below which a material becomes a superconductor to be its **critical temperature**, denoted by T_c. (See Figure 34.23.)

Progress in understanding how and why a material became a superconductor was relatively slow, with the first workable theory coming in 1957. Certain other elements were also found to become superconductors, but all had T_c s less than 10 K, which are expensive to maintain. Although Onnes received a Nobel prize in 1913, it was primarily for his work with liquid helium.

In 1986, a breakthrough was announced—a ceramic compound was found to have an unprecedented T_c of 35 K. It looked as if much higher critical temperatures could be possible, and by early 1988 another ceramic (this of thallium, calcium, barium, copper, and oxygen) had been found to have $T_c = 125 \text{ K}$ (see Figure 34.24.) The economic potential of perfect conductors saving electric energy is immense for T_c s above 77 K, since that is the temperature of liquid nitrogen. Although liquid helium has a boiling point of 4 K and can be used to make materials superconducting, it costs about $5 per liter. Liquid nitrogen boils at 77 K, but only costs about $0.30 per liter. There was general euphoria at the discovery of these complex ceramic superconductors, but this soon subsided with the sobering difficulty of forming them into usable wires. The first commercial use of a high temperature superconductor is in an electronic filter for cellular phones. High-temperature superconductors are used in experimental apparatus, and they are actively being researched, particularly in thin film applications.

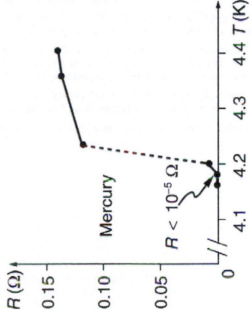

Figure 34.23 A graph of resistivity versus temperature for a superconductor shows a sharp transition to zero at the critical temperature T_C. High temperature superconductors have verifiable T_C s greater than 125 K, well above the easily achieved 77-K temperature of liquid nitrogen.

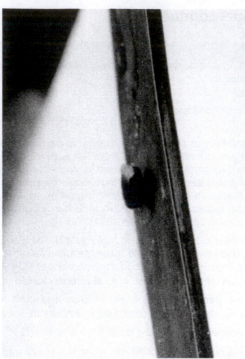

Figure 34.24 One characteristic of a superconductor is that it excludes magnetic flux and, thus, repels other magnets. The small magnet levitated above a high-temperature superconductor, which is cooled by liquid nitrogen, gives evidence that the material is superconducting. When the material warms and becomes conducting, magnetic flux can penetrate it, and the magnet will rest upon it. (credit: Saperaud)

The search is on for even higher T_c superconductors, many of complex and exotic copper oxide ceramics, sometimes including strontium, mercury, or yttrium as well as barium, calcium, and other elements. Room temperature (about 293 K) would be ideal, but any temperature close to room temperature is relatively cheap to produce and maintain. There are persistent reports of T_c s over 200 K and some in the vicinity of 270 K. Unfortunately, these observations are not routinely reproducible, with samples losing their superconducting nature once heated and recooled (cycled) a few times (see **Figure 34.25**.) They are now called USOs or unidentified superconducting objects, out of frustration and the refusal of some samples to show high T_c even though produced in the same manner as others. Reproducibility is crucial to discovery, and researchers are justifiably reluctant to claim the breakthrough they all seek. Time will tell whether USOs are real or an experimental quirk.

The theory of ordinary superconductors is difficult, involving quantum effects for widely separated electrons traveling through a material. Electrons couple in a manner that allows them to get through the material without losing energy to it, making it a superconductor. High-T_c superconductors are more difficult to understand theoretically, but theorists seem to be closing in on a workable theory. The difficulty of understanding how electrons can sneak through materials without losing energy in collisions is even greater at higher temperatures, where vibrating atoms should get in the way. Discoverers of high T_c may feel something analogous to what a politician once said upon an unexpected election victory—"I wonder what we did right?"

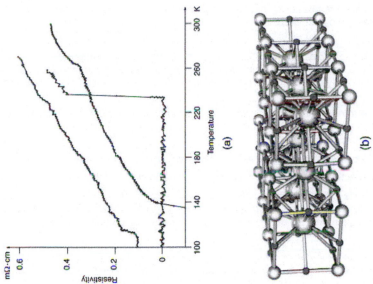

Figure 34.25 (a) This graph, adapted from an article in *Physics Today*, shows the behavior of a single sample of a high-temperature superconductor in three different trials. In one case the sample exhibited a T_c of about 230 K, whereas in the others it did not become superconducting at all. The lack of reproducibility is typical of forefront experiments and prohibits definitive conclusions. (b) This colorful diagram shows the complex but systematic nature of the lattice structure of a high-temperature superconducting ceramic. (credit: en:Cadmium, Wikimedia Commons)

34.7 Some Questions We Know to Ask

Learning Objectives

By the end of this section, you will be able to:

- Identify sample questions to be asked on the largest scales.
- Identify sample questions to be asked on the intermediate scale.
- Identify sample questions to be asked on the smallest scales.

Throughout the text we have noted how essential it is to be curious and to ask questions in order to first understand what is known, and then to go a little farther. Some questions may go unanswered for centuries; others may not have answers, but some bear delicious fruit. Part of discovery is knowing which questions to ask. You have to know something before you can even phrase a decent question. As you may have noticed, the mere act of asking a question can give you the answer. The following questions are a sample of those physicists now know to ask and are representative of the forefronts of physics. Although these questions are important, they will be replaced by others if answers are found to them. The fun continues.

On the Largest Scale

1. *Is the universe open or closed*? Theorists would like it to be just barely closed and evidence is building toward that conclusion. Recent measurements in the expansion rate of the universe and in CMBR support a flat universe. There is a connection to small-scale physics in the type and number of particles that may contribute to closing the universe.

2. *What is dark matter*? It is definitely there, but we really do not know what it is. Conventional possibilities are being ruled out, but one of them still may explain it. The answer could reveal whole new realms of physics and the disturbing possibility that most of what is out there is unknown to us, a completely different form of matter.

3. *How do galaxies form*? They exist since very early in the evolution of the universe and it remains difficult to understand how they evolved so quickly. The recent finer measurements of fluctuations in the CMBR may yet allow us to explain galaxy formation.

4. *What is the nature of various-mass black holes*? Only recently have we become confident that many black hole candidates cannot be explained by other, less exotic possibilities. But we still do not know much about how they form, what their role in the history of galactic evolution has been, and the nature of space in their vicinity. However, so many black holes are now known that correlations between black hole mass and galactic nuclei characteristics are being studied.

5. *What is the mechanism for the energy output of quasars*? These distant and extraordinarily energetic objects now seem to be early stages of galactic evolution with a supermassive black-hole-devouring material. Connections are now being made with galaxies having energetic cores, and there is evidence consistent with less consuming, supermassive black holes at the center of older galaxies. New instruments are allowing us to see deeper into our own galaxy for evidence of our own massive black hole.

6. *Where do the γ bursts come from*? We see bursts of γ rays coming from all directions in space, indicating the sources are very distant objects rather than something associated with our own galaxy. Some γ bursts finally are being correlated

with known sources so that the possibility they may originate in binary neutron star interactions or black holes eating a companion neutron star can be explored.

On the Intermediate Scale

1. *How do phase transitions take place on the microscopic scale?* We know a lot about phase transitions, such as water freezing, but the details of how they occur molecule by molecule are not well understood. Similar questions about specific heat a century ago led to early quantum mechanics. It is also an example of a complex adaptive system that may yield insights into other self-organizing systems.

2. *Is there a way to deal with nonlinear phenomena that reveals underlying connections?* Nonlinear phenomena lack a direct or linear proportionality that makes analysis and understanding a little easier. There are implications for nonlinear optics and broader topics such as chaos.

3. *How do high- T_c superconductors become resistanceless at such high temperatures?* Understanding how they work may help make them more practical or may result in surprises as unexpected as the discovery of superconductivity itself.

4. *There are magnetic effects in materials we do not understand—how do they work?* Although beyond the scope of this text, there is a great deal to learn in condensed matter physics (the physics of solids and liquids). We may find surprises analogous to lasing, the quantum Hall effect, and the quantization of magnetic flux. Complexity may play a role here, too.

On the Smallest Scale

1. *Are quarks and leptons fundamental, or do they have a substructure?* The higher energy accelerators that are just completed or being constructed may supply some answers, but there will also be input from cosmology and other systematics.

2. *Why do leptons have integral charge while quarks have fractional charge?* If both are fundamental and analogous as thought, this question deserves an answer. It is obviously related to the previous question.

3. *Why are there three families of quarks and leptons?* First, does this imply some relationship? Second, why three and only three families?

4. *Are all forces truly equal (unified) under certain circumstances?* They don't have to be equal just because we want them to be. The answer may have to be indirectly obtained because of the extreme energy at which we think they are unified.

5. *Are there other fundamental forces?* There was a flurry of activity with claims of a fifth and even a sixth force a few years ago. Interest has subsided, since those forces have not been detected consistently. Moreover, the proposed forces have strengths similar to gravity, making them extraordinarily difficult to detect in the presence of stronger forces. But the question remains; and if there are no other forces, we need to ask why only four and why these four.

6. *Is the proton stable?* We have discussed this in some detail, but the question is related to fundamental aspects of the unification of forces. We may never know from experiment that the proton is stable, only that it is very long lived.

7. *Are there magnetic monopoles?* Many particle theories call for very massive individual north- and south-pole particles—magnetic monopoles. If they exist, why are they so different in mass and elusiveness from electric charges, and if they do not exist, why not?

8. *Do neutrinos have mass?* Definitive evidence has emerged for neutrinos having mass. The implications are significant, as discussed in this chapter. There are effects on the closure of the universe and on the patterns in particle physics.

9. *What are the systematic characteristics of high- Z nuclei?* All elements with $Z = 118$ or less (with the exception of 115 and 117) have now been discovered. It has long been conjectured that there may be an island of relative stability near $Z = 114$, and the study of the most recently discovered nuclei will contribute to our understanding of nuclear forces.

These lists of questions are not meant to be complete or consistently important—you can no doubt add to it yourself. There are also important questions in topics not broached in this text, such as certain particle symmetries, that are of current interest to physicists. Hopefully, the point is clear that no matter how much we learn, there always seems to be more to know. Although we are fortunate to have the hard-won wisdom of those who preceded us, we can look forward to new enlightenment, undoubtedly sprinkled with surprise.

Glossary

axions: a type of WIMPs having masses about 10^{-10} of an electron mass

Big Bang: a gigantic explosion that threw out matter a few billion years ago

black holes: objects having such large gravitational fields that things can fall in, but nothing, not even light, can escape

chaos: word used to describe systems the outcomes of which are extremely sensitive to initial conditions

complexity: an emerging field devoted to the study of complex systems

cosmic microwave background: the spectrum of microwave radiation of cosmic origin

cosmological constant: a theoretical construct intimately related to the expansion and closure of the universe

cosmological red shift: the photon wavelength is stretched in transit from the source to the observer because of the expansion of space itself

cosmology: the study of the character and evolution of the universe

critical density: the density of matter needed to just halt universal expansion

critical temperature: the temperature at which and below which a material becomes a superconductor

dark matter: indirectly observed non-luminous matter

electroweak epoch: the stage before 10^{-11} back to 10^{-34} after the Big Bang

escape velocity: takeoff velocity when kinetic energy just cancels gravitational potential energy

event horizon: the distance from the object at which the escape velocity is exactly the speed of light

flat (zero curvature) universe: a universe that is infinite but not curved

general relativity: Einstein's theory thatdescribes all types of relative motion including accelerated motion and the effects of gravity

gravitational waves: mass-created distortions in space that propagate at the speed of light and that are predicted by general relativity

GUT epoch: the time period from 10^{-43} to 10^{-34} after the Big Bang, when Grand Unification Theory, in which all forces except gravity are identical, governed the universe

Hubble constant: a central concept in cosmology whose value is determined by taking the slope of a graph of velocity versus distance, obtained from red shift measurements

inflationary scenario: the rapid expansion of the universe by an incredible factor of 10^{-50} for the brief time from 10^{-35} to about 10^{-32}s

MACHOs: massive compact halo objects; microlensing objects of huge mass

microlensing: a process in which light from a distant star is focused and the star appears to brighten in a characteristic manner, when a small body (smaller than about 1/1000 the mass of the Sun) passes between us and the star

negatively curved: an open universe that expands forever

neutralinos: a type of WIMPs having masses several orders of magnitude greater than nucleon masses

neutrino oscillations: a process in which any type of neutrino could change spontaneously into any other

neutron stars: literally a star composed of neutrons

positively curved: a universe that is closed and eventually contracts

Quantum gravity: the theory that deals with particle exchange of gravitons as the mechanism for the force

quasars: the moderately distant galaxies that emit as much or more energy than a normal galaxy

Schwarzschild radius: the radius of the event horizon

spontaneous symmetry breaking: the transition from GUT to electroweak where the forces were no longer unified

Superconductors: materials with resistivity of zero

superforce: hypothetical unified force in TOE epoch

Superstring theory: a theory to unify gravity with the other three forces in which the fundamental particles are considered to act like one-dimensional vibrating strings

thought experiment: mental analysis of certain carefully and clearly defined situations to develop an idea

TOE epoch: before 10^{-43} after the Big Bang

WIMPs: weakly interacting massive particles; chargeless leptons (non-baryonic matter) interacting negligibly with normal matter

Section Summary

34.1 Cosmology and Particle Physics

- Cosmology is the study of the character and evolution of the universe.
- The two most important features of the universe are the cosmological red shifts of its galaxies being proportional to distance and its cosmic microwave background (CMBR). Both support the notion that there was a gigantic explosion, known as the Big Bang that created the universe.
- Galaxies farther away than our local group have, on an average, a recessional velocity given by

$$v = H_0 \, d,$$

where d is the distance to the galaxy and H_0 is the Hubble constant, taken to have the average value

$H_0 = 20 \text{km/s} \cdot \text{Mly}.$

- Explanations of the large-scale characteristics of the universe are intimately tied to particle physics.
- The dominance of matter over antimatter and the smoothness of the CMBR are two characteristics that are tied to particle physics.
- The epochs of the universe are known back to very shortly after the Big Bang, based on known laws of physics.
- The earliest epochs are tied to the unification of forces, with the electroweak epoch being partially understood, the GUT epoch being speculative, and the TOE epoch being highly speculative since it involves an unknown single superforce.
- The transition from GUT to electroweak is called spontaneous symmetry breaking. It released energy that caused the inflationary scenario, which in turn explains the smoothness of the CMBR.

34.2 General Relativity and Quantum Gravity

- Einstein's theory of general relativity includes accelerated frames and, thus, encompasses special relativity and gravity. Created by use of careful thought experiments, it has been repeatedly verified by real experiments.
- One direct result of this behavior of nature is the gravitational lensing of light by massive objects, such as galaxies, also seen in the microlensing of light by smaller bodies in our galaxy.
- Another prediction is the existence of black holes, objects for which the escape velocity is greater than the speed of light and from which nothing can escape.
- The event horizon is the distance from the object at which the escape velocity equals the speed of light c. It is called the Schwarzschild radius R_S and is given by

$$R_S = \frac{2GM}{c^2},$$

where G is the universal gravitational constant, and M is the mass of the body.

- Physics is unknown inside the event horizon, and the possibility of wormholes and time travel are being studied.
- Candidates for black holes may power the extremely energetic emissions of quasars, distant objects that seem to be early stages of galactic evolution.
- Neutron stars are stellar remnants, having the density of a nucleus, that hint that black holes could form from supernovas, too.
- Gravitational waves are wrinkles in space, predicted by general relativity but not yet observed, caused by changes in very massive objects.
- Quantum gravity is an incompletely developed theory that strives to include general relativity, quantum mechanics, and unification of forces (thus, a TOE).
- One unconfirmed connection between general relativity and quantum mechanics is the prediction of characteristic radiation from just outside black holes.

34.3 Superstrings

- Superstring theory holds that fundamental particles are one-dimensional vibrations analogous to those on strings and is an attempt at a theory of quantum gravity.

34.4 Dark Matter and Closure

- Dark matter is non-luminous matter detected in and around galaxies and galactic clusters.
- It may be 10 times the mass of the luminous matter in the universe, and its amount may determine whether the universe is open or closed (expands forever or eventually stops).
- The determining factor is the critical density of the universe and the cosmological constant, a theoretical construct intimately related to the expansion and closure of the universe.
- The critical density ρ_c is the density needed to just halt universal expansion. It is estimated to be approximately 10^{-26} kg/m^3.
- An open universe is negatively curved, a closed universe is positively curved, whereas a universe with exactly the critical density is flat.
- Dark matter's composition is a major mystery, but it may be due to the suspected mass of neutrinos or a completely unknown type of leptonic matter.
- If neutrinos have mass, they will change families, a process known as neutrino oscillations, for which there is growing

evidence.

34.5 Complexity and Chaos

- Complexity is an emerging field, rooted primarily in physics, that considers complex adaptive systems and their evolution, including self-organization.
- Complexity has applications in physics and many other disciplines, such as biological evolution.
- Chaos is a field that studies systems whose properties depend extremely sensitively on some variables and whose evolution is impossible to predict.
- Chaotic systems may be simple or complex.
- Studies of chaos have led to methods for understanding and predicting certain chaotic behaviors.

34.6 High-Temperature Superconductors

- High-temperature superconductors are materials that become superconducting at temperatures well above a few kelvin.
- The critical temperature T_c is the temperature below which a material is superconducting.
- Some high-temperature superconductors have verified T_c s above 125 K, and there are reports of T_c s as high as 250 K.

34.7 Some Questions We Know to Ask

- On the largest scale, the questions which can be asked may be about dark matter, dark energy, black holes, quasars, and other aspects of the universe.
- On the intermediate scale, we can query about gravity, phase transitions, nonlinear phenomena, high- T_c superconductors, and magnetic effects on materials.
- On the smallest scale, questions may be about quarks and leptons, fundamental forces, stability of protons, and existence of monopoles.

Conceptual Questions

34.1 Cosmology and Particle Physics

1. Explain why it only *appears* that we are at the center of expansion of the universe and why an observer in another galaxy would see the same relative motion of all but the closest galaxies away from her.

2. If there is no observable edge to the universe, can we determine where its center of expansion is? Explain.

3. If the universe is infinite, does it have a center? Discuss.

4. Another known cause of red shift in light is the source being in a high gravitational field. Discuss how this can be eliminated as the source of galactic red shifts, given that the shifts are proportional to distance and not to the size of the galaxy.

5. If some unknown cause of red shift—such as light becoming "tired" from traveling long distances through empty space—is discovered, what effect would there be on cosmology?

6. Olbers's paradox poses an interesting question: If the universe is infinite, then any line of sight should eventually fall on a star's surface. Why then is the sky dark at night? Discuss the commonly accepted evolution of the universe as a solution to this paradox.

7. If the cosmic microwave background radiation (CMBR) is the remnant of the Big Bang's fireball, we expect to see hot and cold regions in it. What are two causes of these wrinkles in the CMBR? Are the observed temperature variations greater or less than originally expected?

8. The decay of one type of K-meson is cited as evidence that nature favors matter over antimatter. Since mesons are composed of a quark and an antiquark, is it surprising that they would preferentially decay to one type over another? Is this an asymmetry in nature? Is the predominance of matter over antimatter an asymmetry?

9. Distances to local galaxies are determined by measuring the brightness of stars, called Cepheid variables, that can be observed individually and that have absolute brightnesses at a standard distance that are well known. Explain how the measured brightness would vary with distance as compared with the absolute brightness.

10. Distances to very remote galaxies are estimated based on their apparent type, which indicate the number of stars in the galaxy, and their measured brightness. Explain how the measured brightness would vary with distance. Would there be any correction necessary to compensate for the red shift of the galaxy (all distant galaxies have significant red shifts)? Discuss possible causes of uncertainties in these measurements.

11. If the smallest meaningful time interval is greater than zero, will the lines in Figure 34.9 ever meet?

34.2 General Relativity and Quantum Gravity

12. Quantum gravity, if developed, would be an improvement on both general relativity and quantum mechanics, but more mathematically difficult. Under what circumstances would it be necessary to use quantum gravity? Similarly, under what circumstances could general relativity be used? When could special relativity, quantum mechanics, or classical physics be used?

13. Does observed gravitational lensing correspond to a converging or diverging lens? Explain briefly.

14. Suppose you measure the red shifts of all the images produced by gravitational lensing, such as in **Figure 34.12**.You find that the central image has a red shift less than the outer images, and those all have the same red shift. Discuss how this not only shows that the images are of the same object, but also implies that the red shift is not affected by taking different paths through space. Does it imply that cosmological red shifts are not caused by traveling through space (light getting tired, perhaps)?

15. What are gravitational waves, and have they yet been observed either directly or indirectly?

16. Is the event horizon of a black hole the actual physical surface of the object?

17. Suppose black holes radiate their mass away and the lifetime of a black hole created by a supernova is about 10^{67} years. How does this lifetime compare with the accepted age of the universe? Is it surprising that we do not observe the predicted characteristic radiation?

34.4 Dark Matter and Closure

18. Discuss the possibility that star velocities at the edges of galaxies being greater than expected is due to unknown properties of gravity rather than to the existence of dark matter. Would this mean, for example, that gravity is greater or smaller than expected at large distances? Are there other tests that could be made of gravity at large distances, such as observing the motions of neighboring galaxies?

19. How does relativistic time dilation prohibit neutrino oscillations if they are massless?

20. If neutrino oscillations do occur, will they violate conservation of the various lepton family numbers (L_e, L_μ, and L_τ)? Will neutrino oscillations violate conservation of the total number of leptons?

21. Lacking direct evidence of WIMPs as dark matter, why must we eliminate all other possible explanations based on the known forms of matter before we invoke their existence?

34.5 Complexity and Chaos

22. Must a complex system be adaptive to be of interest in the field of complexity? Give an example to support your answer.

23. State a necessary condition for a system to be chaotic.

34.6 High-Temperature Superconductors

24. What is critical temperature T_c ? Do all materials have a critical temperature? Explain why or why not.

25. Explain how good thermal contact with liquid nitrogen can keep objects at a temperature of 77 K (liquid nitrogen's boiling point at atmospheric pressure).

26. Not only is liquid nitrogen a cheaper coolant than liquid helium, its boiling point is higher (77 K vs. 4.2 K). How does higher temperature help lower the cost of cooling a material? Explain in terms of the rate of heat transfer being related to the temperature difference between the sample and its surroundings.

34.7 Some Questions We Know to Ask

27. For experimental evidence, particularly of previously unobserved phenomena, to be taken seriously it must be reproducible or of sufficiently high quality that a single observation is meaningful. Supernova 1987A is not reproducible. How do we know observations of it were valid? The fifth force is not broadly accepted. Is this due to lack of reproducibility or poor-quality experiments (or both)? Discuss why forefront experiments are more subject to observational problems than those involving established phenomena.

28. Discuss whether you think there are limits to what humans can understand about the laws of physics. Support your arguments.

Problems & Exercises

34.1 Cosmology and Particle Physics

1. Find the approximate mass of the luminous matter in the Milky Way galaxy, given it has approximately 10^{11} stars of average mass 1.5 times that of our Sun.

2. Find the approximate mass of the dark and luminous matter in the Milky Way galaxy. Assume the luminous matter is due to approximately 10^{11} stars of average mass 1.5 times that of our Sun, and take the dark matter to be 10 times as massive as the luminous matter.

3. (a) Estimate the mass of the luminous matter in the known universe, given there are 10^{11} galaxies, each containing 10^{11} stars of average mass 1.5 times that of our Sun. (b) How many protons (the most abundant nuclide) are there in this mass? (c) Estimate the total number of particles in the observable universe by multiplying the answer to (b) by two, since there is an electron for each proton, and then by 10^9, since there are far more particles (such as photons and neutrinos) in space than in luminous matter.

4. If a galaxy is 500 Mly away from us, how fast do we expect it to be moving and in what direction?

5. On average, how far away are galaxies that are moving away from us at 2.0% of the speed of light?

6. Our solar system orbits the center of the Milky Way galaxy. Assuming a circular orbit 30,000 ly in radius and an orbital speed of 250 km/s, how many years does it take for one revolution? Note that this is approximate, assuming constant speed and circular orbit, but it is representative of the time for our system and local stars to make one revolution around the galaxy.

7. (a) What is the approximate speed relative to us of a galaxy near the edge of the known universe, some 10 Gly away? (b) What fraction of the speed of light is this? Note that we have observed galaxies moving away from us at greater than $0.9c$.

8. (a) Calculate the approximate age of the universe from the average value of the Hubble constant, $H_0 = 20 \text{km/s} \cdot \text{Mly}$. To do this, calculate the time it would take to travel 1 Mly at a constant expansion rate of 20 km/s. (b) If deceleration is taken into account, would the actual age of the universe be greater or less than that found here? Explain.

9. Assuming a circular orbit for the Sun about the center of the Milky Way galaxy, calculate its orbital speed using the following information: The mass of the galaxy is equivalent to a single mass 1.5×10^{11} times that of the Sun (or 3×10^{41} kg), located 30,000 ly away.

10. (a) What is the approximate force of gravity on a 70-kg person due to the Andromeda galaxy, assuming its total mass is 10^{13} that of our Sun and acts like a single mass 2 Mly away? (b) What is the ratio of this force to the person's weight? Note that Andromeda is the closest large galaxy.

11. Andromeda galaxy is the closest large galaxy and is visible to the naked eye. Estimate its brightness relative to the Sun, assuming it has luminosity 10^{12} times that of the Sun and lies 2 Mly away.

12. (a) A particle and its antiparticle are at rest relative to an observer and annihilate (completely destroying both masses), creating two γ rays of equal energy. What is the characteristic γ-ray energy you would look for if searching for evidence of proton-antiproton annihilation? (The fact that such radiation is rarely observed is evidence that there is very little antimatter in the universe.) (b) How does this compare with the 0.511-MeV energy associated with electron-positron annihilation?

13. The average particle energy needed to observe unification of forces is estimated to be 10^{19} GeV. (a) What is the rest mass in kilograms of a particle that has a rest mass of 10^{19} GeV/c^2? (b) How many times the mass of a hydrogen atom is this?

14. The peak intensity of the CMBR occurs at a wavelength of 1.1 mm. (a) What is the energy in eV of a 1.1-mm photon? (b) There are approximately 10^9 photons for each massive particle in deep space. Calculate the energy of 10^9 such photons. (c) If the average massive particle in space has a mass half that of a proton, what energy would be created by converting its mass to energy? (d) Does this imply that space is "matter dominated"? Explain briefly.

15. (a) What Hubble constant corresponds to an approximate age of the universe of 10^{10} y? To get an approximate value, assume the expansion rate is constant and calculate the speed at which two galaxies must move apart to be separated by 1 Mly (present average galactic separation) in a time of 10^{10} y. (b) Similarly, what Hubble constant corresponds to a universe approximately 2×10^{10}-y old?

16. Show that the velocity of a star orbiting its galaxy in a circular orbit is inversely proportional to the square root of its orbital radius, assuming the mass of the stars inside its orbit acts like a single mass at the center of the galaxy. You may use an equation from a previous chapter to support your conclusion, but you must justify its use and define all terms used.

17. The core of a star collapses during a supernova, forming a neutron star. Angular momentum of the core is conserved, and so the neutron star spins rapidly. If the initial core radius is 5.0×10^5 km and it collapses to 10.0 km, find the neutron star's angular velocity in revolutions per second, given the core's angular velocity was originally 1 revolution per 30.0 days.

18. Using data from the previous problem, find the increase in rotational kinetic energy, given the core's mass is 1.3 times that of our Sun. Where does this increase in kinetic energy come from?

19. Distances to the nearest stars (up to 500 ly away) can be measured by a technique called parallax, as shown in **Figure 34.26**. What are the angles θ_1 and θ_2 relative to the plane of the Earth's orbit for a star 4.0 ly directly above the Sun?

20. (a) Use the Heisenberg uncertainty principle to calculate the uncertainty in energy for a corresponding time interval of 10^{-43} s. (b) Compare this energy with the 10^{19} GeV unification-of-forces energy and discuss why they are similar.

21. Construct Your Own Problem

Consider a star moving in a circular orbit at the edge of a galaxy. Construct a problem in which you calculate the mass of that galaxy in kg and in multiples of the solar mass based on the velocity of the star and its distance from the center of the galaxy.

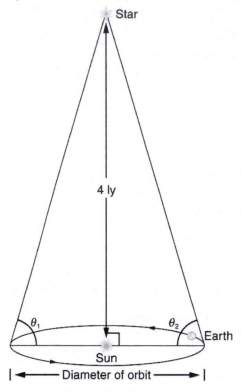

Figure 34.26 Distances to nearby stars are measured using triangulation, also called the parallax method. The angle of line of sight to the star is measured at intervals six months apart, and the distance is calculated by using the known diameter of the Earth's orbit. This can be done for stars up to about 500 ly away.

34.2 General Relativity and Quantum Gravity

22. What is the Schwarzschild radius of a black hole that has a mass eight times that of our Sun? Note that stars must be more massive than the Sun to form black holes as a result of a supernova.

23. Black holes with masses smaller than those formed in supernovas may have been created in the Big Bang. Calculate the radius of one that has a mass equal to the Earth's.

24. Supermassive black holes are thought to exist at the center of many galaxies.

(a) What is the radius of such an object if it has a mass of 10^9 Suns?

(b) What is this radius in light years?

25. Construct Your Own Problem

Consider a supermassive black hole near the center of a galaxy. Calculate the radius of such an object based on its mass. You must consider how much mass is reasonable for these large objects, and which is now nearly directly observed. (Information on black holes posted on the Web by NASA and other agencies is reliable, for example.)

34.3 Superstrings

26. The characteristic length of entities in Superstring theory is approximately 10^{-35} m .

(a) Find the energy in GeV of a photon of this wavelength.

(b) Compare this with the average particle energy of 10^{19} GeV needed for unification of forces.

34.4 Dark Matter and Closure

27. If the dark matter in the Milky Way were composed entirely of MACHOs (evidence shows it is not), approximately how many would there have to be? Assume the average mass of a MACHO is 1/1000 that of the Sun, and that dark matter has a mass 10 times that of the luminous Milky Way galaxy with its 10^{11} stars of average mass 1.5 times the Sun's mass.

28. The critical mass density needed to just halt the expansion of the universe is approximately 10^{-26} kg $/ \text{m}^3$.

(a) Convert this to $\text{eV} / c^2 \cdot \text{m}^3$.

(b) Find the number of neutrinos per cubic meter needed to close the universe if their average mass is $7 \text{ eV} / c^2$ and they have negligible kinetic energies.

29. Assume the average density of the universe is 0.1 of the critical density needed for closure. What is the average number of protons per cubic meter, assuming the universe is composed mostly of hydrogen?

30. To get an idea of how empty deep space is on the average, perform the following calculations:

(a) Find the volume our Sun would occupy if it had an average density equal to the critical density of 10^{-26} kg $/ \text{m}^3$ thought necessary to halt the expansion of the universe.

(b) Find the radius of a sphere of this volume in light years.

(c) What would this radius be if the density were that of luminous matter, which is approximately 5% that of the critical density?

(d) Compare the radius found in part (c) with the 4-ly average separation of stars in the arms of the Milky Way.

34.6 High-Temperature Superconductors

31. A section of superconducting wire carries a current of 100 A and requires 1.00 L of liquid nitrogen per hour to keep it below its critical temperature. For it to be economically advantageous to use a superconducting wire, the cost of cooling the wire must be less than the cost of energy lost to heat in the wire. Assume that the cost of liquid nitrogen is $0.30 per liter, and that electric energy costs $0.10 per kW·h. What is the resistance of a normal wire that costs as much in wasted electric energy as the cost of liquid nitrogen for the superconductor?

APPENDIX A | ATOMIC MASSES

Table A1 Atomic Masses

Atomic Number, Z	Name	Atomic Mass Number, A	Symbol	Atomic Mass (u)	Percent Abundance or Decay Mode	Half-life, $t_{1/2}$
0	neutron	1	n	1.008 665	β^-	10.37 min
1	Hydrogen	1	^{1}H	1.007 825	99.985%	
	Deuterium	2	^{2}H or D	2.014 102	0.015%	
	Tritium	3	^{3}H or T	3.016 050	β^-	12.33 y
2	Helium	3	^{3}He	3.016 030	$1.38\times10^{-4}\%$	
		4	^{4}He	4.002 603	$\approx100\%$	
3	Lithium	6	^{6}Li	6.015 121	7.5%	
		7	^{7}Li	7.016 003	92.5%	
4	Beryllium	7	^{7}Be	7.016 928	EC	53.29 d
		9	^{9}Be	9.012 182	100%	
5	Boron	10	^{10}B	10.012 937	19.9%	
		11	^{11}B	11.009 305	80.1%	
6	Carbon	11	^{11}C	11.011 432	EC, β^+	
		12	^{12}C	12.000 000	98.90%	
		13	^{13}C	13.003 355	1.10%	
		14	^{14}C	14.003 241	β^-	5730 y
7	Nitrogen	13	^{13}N	13.005 738	β^+	9.96 min
		14	^{14}N	14.003 074	99.63%	
		15	^{15}N	15.000 108	0.37%	
8	Oxygen	15	^{15}O	15.003 065	EC, β^+	122 s
		16	^{16}O	15.994 915	99.76%	
		18	^{18}O	17.999 160	0.200%	
9	Fluorine	18	^{18}F	18.000 937	EC, β^+	1.83 h
		19	^{19}F	18.998 403	100%	
10	Neon	20	^{20}Ne	19.992 435	90.51%	
		22	^{22}Ne	21.991 383	9.22%	
11	Sodium	22	^{22}Na	21.994 434	β^+	2.602 y
		23	^{23}Na	22.989 767	100%	

Atomic Number, Z	Name	Atomic Mass Number, A	Symbol	Atomic Mass (u)	Percent Abundance or Decay Mode	Half-life, $t_{1/2}$
		24	^{24}Na	23.990 961	β^-	14.96 h
12	Magnesium	24	^{24}Mg	23.985 042	78.99%	
13	Aluminum	27	^{27}Al	26.981 539	100%	
14	Silicon	28	^{28}Si	27.976 927	92.23%	2.62h
		31	^{31}Si	30.975 362	β^-	
15	Phosphorus	31	^{31}P	30.973 762	100%	
		32	^{32}P	31.973 907	β^-	14.28 d
16	Sulfur	32	^{32}S	31.972 070	95.02%	
		35	^{35}S	34.969 031	β^-	87.4 d
17	Chlorine	35	^{35}Cl	34.968 852	75.77%	
		37	^{37}Cl	36.965 903	24.23%	
18	Argon	40	^{40}Ar	39.962 384	99.60%	
19	Potassium	39	^{39}K	38.963 707	93.26%	
		40	^{40}K	39.963 999	0.0117%, EC, β^-	1.28×10^9 y
20	Calcium	40	^{40}Ca	39.962 591	96.94%	
21	Scandium	45	^{45}Sc	44.955 910	100%	
22	Titanium	48	^{48}Ti	47.947 947	73.8%	
23	Vanadium	51	^{51}V	50.943 962	99.75%	
24	Chromium	52	^{52}Cr	51.940 509	83.79%	
25	Manganese	55	^{55}Mn	54.938 047	100%	
26	Iron	56	^{56}Fe	55.934 939	91.72%	
27	Cobalt	59	^{59}Co	58.933 198	100%	
		60	^{60}Co	59.933 819	β^-	5.271 y
28	Nickel	58	^{58}Ni	57.935 346	68.27%	
		60	^{60}Ni	59.930 788	26.10%	
29	Copper	63	^{63}Cu	62.939 598	69.17%	
		65	^{65}Cu	64.927 793	30.83%	
30	Zinc	64	^{64}Zn	63.929 145	48.6%	
		66	^{66}Zn	65.926 034	27.9%	
31	Gallium	69	^{69}Ga	68.925 580	60.1%	
32	Germanium	72	^{72}Ge	71.922 079	27.4%	

Atomic Number, Z	Name	Atomic Mass Number, A	Symbol	Atomic Mass (u)	Percent Abundance or Decay Mode	Half-life, $t_{1/2}$
		74	^{74}Ge	73.921 177	36.5%	
33	Arsenic	75	^{75}As	74.921 594	100%	
34	Selenium	80	^{80}Se	79.916 520	49.7%	
35	Bromine	79	^{79}Br	78.918 336	50.69%	
36	Krypton	84	^{84}Kr	83.911 507	57.0%	
37	Rubidium	85	^{85}Rb	84.911 794	72.17%	
38	Strontium	86	^{86}Sr	85.909 267	9.86%	
		88	^{88}Sr	87.905 619	82.58%	
		90	^{90}Sr	89.907 738	β^-	28.8 y
39	Yttrium	89	^{89}Y	88.905 849	100%	
		90	^{90}Y	89.907 152	β^-	64.1 h
40	Zirconium	90	^{90}Zr	89.904 703	51.45%	
41	Niobium	93	^{93}Nb	92.906 377	100%	
42	Molybdenum	98	^{98}Mo	97.905 406	24.13%	
43	Technetium	98	^{98}Tc	97.907 215	β^-	4.2×10^6 y
44	Ruthenium	102	^{102}Ru	101.904 348	31.6%	
45	Rhodium	103	^{103}Rh	102.905 500	100%	
46	Palladium	106	^{106}Pd	105.903 478	27.33%	
47	Silver	107	^{107}Ag	106.905 092	51.84%	
		109	^{109}Ag	108.904 757	48.16%	
48	Cadmium	114	^{114}Cd	113.903 357	28.73%	
49	Indium	115	^{115}In	114.903 880	95.7%, β^-	4.4×10^{14} y
50	Tin	120	^{120}Sn	119.902 200	32.59%	
51	Antimony	121	^{121}Sb	120.903 821	57.3%	
52	Tellurium	130	^{130}Te	129.906 229	33.8%, β^-	2.5×10^{21} y
53	Iodine	127	^{127}I	126.904 473	100%	
		131	^{131}I	130.906 114	β^-	8.040 d
54	Xenon	132	^{132}Xe	131.904 144	26.9%	
		136	^{136}Xe	135.907 214	8.9%	
55	Cesium	133	^{133}Cs	132.905 429	100%	

Atomic Number, Z	Name	Atomic Mass Number, A	Symbol	Atomic Mass (u)	Percent Abundance or Decay Mode	Half-life, $t_{1/2}$
		134	^{134}Cs	133.906 696	EC, β^-	2.06 y
56	Barium	137	^{137}Ba	136.905 812	11.23%	
		138	^{138}Ba	137.905 232	71.70%	
57	Lanthanum	139	^{139}La	138.906 346	99.91%	
58	Cerium	140	^{140}Ce	139.905 433	88.48%	
59	Praseodymium	141	^{141}Pr	140.907 647	100%	
60	Neodymium	142	^{142}Nd	141.907 719	27.13%	
61	Promethium	145	^{145}Pm	144.912 743	EC, α	17.7 y
62	Samarium	152	^{152}Sm	151.919 729	26.7%	
63	Europium	153	^{153}Eu	152.921 225	52.2%	
64	Gadolinium	158	^{158}Gd	157.924 099	24.84%	
65	Terbium	159	^{159}Tb	158.925 342	100%	
66	Dysprosium	164	^{164}Dy	163.929 171	28.2%	
67	Holmium	165	^{165}Ho	164.930 319	100%	
68	Erbium	166	^{166}Er	165.930 290	33.6%	
69	Thulium	169	^{169}Tm	168.934 212	100%	
70	Ytterbium	174	^{174}Yb	173.938 859	31.8%	
71	Lutecium	175	^{175}Lu	174.940 770	97.41%	
72	Hafnium	180	^{180}Hf	179.946 545	35.10%	
73	Tantalum	181	^{181}Ta	180.947 992	99.98%	
74	Tungsten	184	^{184}W	183.950 928	30.67%	
75	Rhenium	187	^{187}Re	186.955 744	62.6%, β^-	4.6×10^{10} y
76	Osmium	191	^{191}Os	190.960 920	β^-	15.4 d
		192	^{192}Os	191.961 467	41.0%	
77	Iridium	191	^{191}Ir	190.960 584	37.3%	
		193	^{193}Ir	192.962 917	62.7%	
78	Platinum	195	^{195}Pt	194.964 766	33.8%	
79	Gold	197	^{197}Au	196.966 543	100%	
		198	^{198}Au	197.968 217	β^-	2.696 d
80	Mercury	199	^{199}Hg	198.968 253	16.87%	

Atomic Number, Z	Name	Atomic Mass Number, A	Symbol	Atomic Mass (u)	Percent Abundance or Decay Mode	Half-life, $t_{1/2}$
		202	^{202}Hg	201.970 617	29.86%	
81	Thallium	205	^{205}Tl	204.974 401	70.48%	
82	Lead	206	^{206}Pb	205.974 440	24.1%	
		207	^{207}Pb	206.975 872	22.1%	
		208	^{208}Pb	207.976 627	52.4%	
		210	^{210}Pb	209.984 163	α, β^-	22.3 y
		211	^{211}Pb	210.988 735	β^-	36.1 min
		212	^{212}Pb	211.991 871	β^-	10.64 h
83	Bismuth	209	^{209}Bi	208.980 374	100%	
		211	^{211}Bi	210.987 255	α, β^-	2.14 min
84	Polonium	210	^{210}Po	209.982 848	α	138.38 d
85	Astatine	218	^{218}At	218.008 684	α, β^-	1.6 s
86	Radon	222	^{222}Rn	222.017 570	α	3.82 d
87	Francium	223	^{223}Fr	223.019 733	α, β^-	21.8 min
88	Radium	226	^{226}Ra	226.025 402	α	1.60×10^3y
89	Actinium	227	^{227}Ac	227.027 750	α, β^-	21.8 y
90	Thorium	228	^{228}Th	228.028 715	α	1.91 y
		232	^{232}Th	232.038 054	100%, α	1.41×10^{10}y
91	Protactinium	231	^{231}Pa	231.035 880	α	3.28×10^4y
92	Uranium	233	^{233}U	233.039 628	α	1.59×10^3y
		235	^{235}U	235.043 924	0.720%, α	7.04×10^8y
		236	^{236}U	236.045 562	α	2.34×10^7y
		238	^{238}U	238.050 784	99.2745%, α	4.47×10^9y
		239	^{239}U	239.054 289	β^-	23.5 min
93	Neptunium	239	^{239}Np	239.052 933	β^-	2.355 d
94	Plutonium	239	^{239}Pu	239.052 157	α	2.41×10^4y
95	Americium	243	^{243}Am	243.061 375	α, fission	7.37×10^3y
96	Curium	245	^{245}Cm	245.065 483	α	8.50×10^3y
97	Berkelium	247	^{247}Bk	247.070 300	α	1.38×10^3y

Atomic Number, Z	Name	Atomic Mass Number, A	Symbol	Atomic Mass (u)	Percent Abundance or Decay Mode	Half-life, $t_{1/2}$
98	Californium	249	^{249}Cf	249.074 844	α	351 y
99	Einsteinium	254	^{254}Es	254.088 019	α, β^-	276 d
100	Fermium	253	^{253}Fm	253.085 173	EC, α	3.00 d
101	Mendelevium	255	^{255}Md	255.091 081	EC, α	27 min
102	Nobelium	255	^{255}No	255.093 260	EC, α	3.1 min
103	Lawrencium	257	^{257}Lr	257.099 480	EC, α	0.646 s
104	Rutherfordium	261	^{261}Rf	261.108 690	α	1.08 min
105	Dubnium	262	^{262}Db	262.113 760	α, fission	34 s
106	Seaborgium	263	^{263}Sg	263.11 86	α, fission	0.8 s
107	Bohrium	262	^{262}Bh	262.123 1	α	0.102 s
108	Hassium	264	^{264}Hs	264.128 5	α	0.08 ms
109	Meitnerium	266	^{266}Mt	266.137 8	α	3.4 ms

APPENDIX B | SELECTED RADIOACTIVE ISOTOPES

Decay modes are α, β^-, β^+, electron capture (EC) and isomeric transition (IT). EC results in the same daughter nucleus as would β^+ decay. IT is a transition from a metastable excited state. Energies for $\beta^\pm$ decays are the maxima; average energies are roughly one-half the maxima.

Table B1 Selected Radioactive Isotopes

Isotope	$t_{1/2}$	DecayMode(s)	Energy(MeV)	Percent		γ-Ray Energy(MeV)	Percent
^{3}H	12.33 y	β^-	0.0186	100%			
^{14}C	5730 y	β^-	0.156	100%			
^{13}N	9.96 min	β^+	1.20	100%			
^{22}Na	2.602 y	β^+	0.55	90%	γ	1.27	100%
^{32}P	14.28 d	β^-	1.71	100%			
^{35}S	87.4 d	β^-	0.167	100%			
^{36}Cl	3.00×10^5y	β^-	0.710	100%			
^{40}K	1.28×10^9y	β^-	1.31	89%			
^{43}K	22.3 h	β^-	0.827	87%	γ s	0.373	87%
						0.618	87%
^{45}Ca	165 d	β^-	0.257	100%			
^{51}Cr	27.70 d	EC			γ	0.320	10%
^{52}Mn	5.59d	β^+	3.69	28%	γ s	1.33	28%
						1.43	28%
^{52}Fe	8.27 h	β^+	1.80	43%		0.169	43%
						0.378	43%
^{59}Fe	44.6 d	β^- s	0.273	45%	γ s	1.10	57%
			0.466	55%		1.29	43%
^{60}Co	5.271 y	β^-	0.318	100%	γ s	1.17	100%
						1.33	100%
^{65}Zn	244.1 d	EC			γ	1.12	51%
^{67}Ga	78.3 h	EC			γ s	0.0933	70%
						0.185	35%
						0.300	19%
						others	
^{75}Se	118.5 d	EC			γ s	0.121	20%
						0.136	65%

Isotope	$t_{1/2}$	DecayMode(s)	Energy(MeV)	Percent		γ-Ray Energy(MeV)	Percent
						0.265	68%
						0.280	20%
						others	
^{86}Rb	18.8 d	β^- s	0.69	9%	γ	1.08	9%
			1.77	91%			
^{85}Sr	64.8 d	EC			γ	0.514	100%
^{90}Sr	28.8 y	β^-	0.546	100%			
^{90}Y	64.1 h	β^-	2.28	100%			
^{99m}Tc	6.02 h	IT			γ	0.142	100%
^{113m}In	99.5 min	IT			γ	0.392	100%
^{123}I	13.0 h	EC			γ	0.159	$\approx$100%
^{131}I	8.040 d	β^- s	0.248	7%	γ s	0.364	85%
			0.607	93%		others	
			others				
^{129}Cs	32.3 h	EC			γ s	0.0400	35%
						0.372	32%
						0.411	25%
						others	
^{137}Cs	30.17 y	β^- s	0.511	95%	γ	0.662	95%
			1.17	5%			
^{140}Ba	12.79 d	β^-	1.035	$\approx$100%	γ s	0.030	25%
						0.044	65%
						0.537	24%
						others	
^{198}Au	2.696 d	β^-	1.161	$\approx$100%	γ	0.412	$\approx$100%
^{197}Hg	64.1 h	EC			γ	0.0733	100%
^{210}Po	138.38 d	α	5.41	100%			
^{226}Ra	1.60×10^3 y	α s	4.68	5%	γ	0.186	100%
			4.87	95%			
^{235}U	7.038×10^8 y	α	4.68	$\approx$100%	γ s	numerous	<0.400%
^{238}U	4.468×10^9 y	α s	4.22	23%	γ	0.050	23%
			4.27	77%			
^{237}Np	2.14×10^6 y	α s	numerous		γ s	numerous	<0.250%
			4.96 (max.)				
^{239}Pu	2.41×10^4 y	α s	5.19	11%	γ s	7.5×10^{-5}	73%
			5.23	15%		0.013	15%

Isotope	$t_{1/2}$	DecayMode(s)	Energy(MeV)	Percent		γ-Ray Energy(MeV)	Percent
			5.24	73%		0.052	10%
						others	
^{243}Am	7.37×10^3y	αs	Max. 5.44		γs	0.075	
			5.37	88%		others	
			5.32	11%			
			others				

APPENDIX C | USEFUL INFORMATION

This appendix is broken into several tables.

- **Table C1**, Important Constants
- **Table C2**, Submicroscopic Masses
- **Table C3**, Solar System Data
- **Table C4**, Metric Prefixes for Powers of Ten and Their Symbols
- **Table C5**, The Greek Alphabet
- **Table C6**, SI units
- **Table C7**, Selected British Units
- **Table C8**, Other Units
- **Table C9**, Useful Formulae

Table C1 Important Constants[1]

Symbol	Meaning	Best Value	Approximate Value
c	Speed of light in vacuum	$2.99792458 \times 10^8 \, \text{m/s}$	$3.00 \times 10^8 \, \text{m/s}$
G	Gravitational constant	$6.67408(31) \times 10^{-11} \, \text{N} \cdot \text{m}^2 / \text{kg}^2$	$6.67 \times 10^{-11} \, \text{N} \cdot \text{m}^2 / \text{kg}^2$
N_A	Avogadro's number	$6.02214129(27) \times 10^{23}$	6.02×10^{23}
k	Boltzmann's constant	$1.3806488(13) \times 10^{-23} \, \text{J/K}$	$1.38 \times 10^{-23} \, \text{J/K}$
R	Gas constant	$8.3144621(75) \, \text{J/mol} \cdot \text{K}$	$8.31 \, \text{J/mol} \cdot \text{K} = 1.99 \, \text{cal/mol} \cdot \text{K} = 0.0821 \, \text{atm} \cdot \text{L/mol} \cdot \text{K}$
σ	Stefan-Boltzmann constant	$5.670373(21) \times 10^{-8} \, \text{W/m}^2 \cdot \text{K}$	$5.67 \times 10^{-8} \, \text{W/m}^2 \cdot \text{K}$
k	Coulomb force constant	$8.987551788... \times 10^9 \, \text{N} \cdot \text{m}^2 / \text{C}^2$	$8.99 \times 10^9 \, \text{N} \cdot \text{m}^2 / \text{C}^2$
q_e	Charge on electron	$-1.602176565(35) \times 10^{-19} \, \text{C}$	$-1.60 \times 10^{-19} \, \text{C}$
ε_0	Permittivity of free space	$8.854187817... \times 10^{-12} \, \text{C}^2 / \text{N} \cdot \text{m}^2$	$8.85 \times 10^{-12} \, \text{C}^2 / \text{N} \cdot \text{m}^2$
μ_0	Permeability of free space	$4\pi \times 10^{-7} \, \text{T} \cdot \text{m/A}$	$1.26 \times 10^{-6} \, \text{T} \cdot \text{m/A}$
h	Planck's constant	$6.62606957(29) \times 10^{-34} \, \text{J} \cdot \text{s}$	$6.63 \times 10^{-34} \, \text{J} \cdot \text{s}$

Table C2 Submicroscopic Masses[2]

Symbol	Meaning	Best Value	Approximate Value
m_e	Electron mass	$9.10938291(40) \times 10^{-31} \text{kg}$	$9.11 \times 10^{-31} \text{kg}$
m_p	Proton mass	$1.672621777(74) \times 10^{-27} \text{kg}$	$1.6726 \times 10^{-27} \text{kg}$
m_n	Neutron mass	$1.674927351(74) \times 10^{-27} \text{kg}$	$1.6749 \times 10^{-27} \text{kg}$

1.   Stated values are according to the National Institute of Standards and Technology Reference on Constants, Units, and Uncertainty, www.physics.nist.gov/cuu (http://www.physics.nist.gov/cuu) (accessed May 18, 2012). Values in parentheses are the uncertainties in the last digits. Numbers without uncertainties are exact as defined.
2.   Stated values are according to the National Institute of Standards and Technology Reference on Constants, Units, and Uncertainty, www.physics.nist.gov/cuu (http://www.physics.nist.gov/cuu) (accessed May 18, 2012). Values in parentheses are the uncertainties in the last digits. Numbers without uncertainties are exact as defined.

Symbol	Meaning	Best Value	Approximate Value
u	Atomic mass unit	$1.660538921(73) \times 10^{-27}$ kg	1.6605×10^{-27} kg

Table C3 Solar System Data

Sun	mass	1.99×10^{30} kg
	average radius	6.96×10^8 m
	Earth-sun distance (average)	1.496×10^{11} m
Earth	mass	5.9736×10^{24} kg
	average radius	6.376×10^6 m
	orbital period	3.16×10^7 s
Moon	mass	7.35×10^{22} kg
	average radius	1.74×10^6 m
	orbital period (average)	2.36×10^6 s
	Earth-moon distance (average)	3.84×10^8 m

Table C4 Metric Prefixes for Powers of Ten and Their Symbols

Prefix	Symbol	Value	Prefix	Symbol	Value
tera	T	10^{12}	deci	d	10^{-1}
giga	G	10^9	centi	c	10^{-2}
mega	M	10^6	milli	m	10^{-3}
kilo	k	10^3	micro	μ	10^{-6}
hecto	h	10^2	nano	n	10^{-9}
deka	da	10^1	pico	p	10^{-12}
—	—	$10^0 (= 1)$	femto	f	10^{-15}

Table C5 The Greek Alphabet

Alpha	A	α	Eta	H	η	Nu	N	ν	Tau	T	τ
Beta	B	β	Theta	Θ	θ	Xi	Ξ	ξ	Upsilon	Υ	υ
Gamma	Γ	γ	Iota	I	ι	Omicron	O	o	Phi	Φ	ϕ
Delta	Δ	δ	Kappa	K	κ	Pi	Π	π	Chi	X	χ
Epsilon	E	ε	Lambda	Λ	λ	Rho	P	ρ	Psi	Ψ	ψ
Zeta	Z	ζ	Mu	M	μ	Sigma	Σ	σ	Omega	Ω	ω

Table C6 SI Units

	Entity	Abbreviation	Name
Fundamental units	Length	m	meter
	Mass	kg	kilogram

Download for free at https://openstax.org/details/books/college-physics-ap-course

	Entity	Abbreviation	Name
	Time	s	second
	Current	A	ampere
Supplementary unit	Angle	rad	radian
Derived units	Force	$N = kg \cdot m/s^2$	newton
	Energy	$J = kg \cdot m^2/s^2$	joule
	Power	$W = J/s$	watt
	Pressure	$Pa = N/m^2$	pascal
	Frequency	$Hz = 1/s$	hertz
	Electronic potential	$V = J/C$	volt
	Capacitance	$F = C/V$	farad
	Charge	$C = s \cdot A$	coulomb
	Resistance	$\Omega = V/A$	ohm
	Magnetic field	$T = N/(A \cdot m)$	tesla
	Nuclear decay rate	$Bq = 1/s$	becquerel

Table C7 Selected British Units

Length	1 inch (in.) = 2.54 cm (exactly)
	1 foot (ft) = 0.3048 m
	1 mile (mi) = 1.609 km
Force	1 pound (lb) = 4.448 N
Energy	1 British thermal unit (Btu) = 1.055×10^3 J
Power	1 horsepower (hp) = 746 W
Pressure	$1 \, lb/in^2 = 6.895 \times 10^3$ Pa

Table C8 Other Units

Length	1 light year (ly) = 9.46×10^{15} m
	1 astronomical unit (au) = 1.50×10^{11} m
	1 nautical mile = 1.852 km
	1 angstrom($\mathring{A}$) $= 10^{-10}$ m
Area	1 acre (ac) = $4.05 \times 10^3 \, m^2$
	1 square foot (ft^2) = $9.29 \times 10^{-2} \, m^2$
	1 barn (b) $= 10^{-28} \, m^2$
Volume	1 liter (L) $= 10^{-3} \, m^3$
	1 U.S. gallon (gal) = $3.785 \times 10^{-3} \, m^3$

Mass	1 solar mass $= 1.99\times10^{30}\,\text{kg}$
	1 metric ton $= 10^3\,\text{kg}$
	1 atomic mass unit $(u) = 1.6605\times10^{-27}\,\text{kg}$
Time	1 year $(y) = 3.16\times10^7\,\text{s}$
	1 day $(d) = 86,400\,\text{s}$
Speed	1 mile per hour (mph) $= 1.609\,\text{km}/\text{h}$
	1 nautical mile per hour (naut) $= 1.852\,\text{km}/\text{h}$
Angle	1 degree $(°) = 1.745\times10^{-2}\,\text{rad}$
	1 minute of arc $(') = 1/60$ degree
	1 second of arc $('') = 1/60$ minute of arc
	1 grad $= 1.571\times10^{-2}\,\text{rad}$
Energy	1 kiloton TNT (kT) $= 4.2\times10^{12}\,\text{J}$
	1 kilowatt hour $(\text{kW}\cdot h) = 3.60\times10^6\,\text{J}$
	1 food calorie (kcal) $= 4186\,\text{J}$
	1 calorie (cal) $= 4.186\,\text{J}$
	1 electron volt (eV) $= 1.60\times10^{-19}\,\text{J}$
Pressure	1 atmosphere (atm) $= 1.013\times10^5\,\text{Pa}$
	1 millimeter of mercury (mm Hg) $= 133.3\,\text{Pa}$
	1 torricelli (torr) $= 1\,\text{mm Hg} = 133.3\,\text{Pa}$
Nuclear decay rate	1 curie (Ci) $= 3.70\times10^{10}\,\text{Bq}$

Table C9 Useful Formulae

Circumference of a circle with radius r or diameter d	$C = 2\pi r = \pi d$
Area of a circle with radius r or diameter d	$A = \pi r^2 = \pi d^2/4$
Area of a sphere with radius r	$A = 4\pi r^2$
Volume of a sphere with radius r	$V = (4/3)\left(\pi r^3\right)$

APPENDIX D | GLOSSARY OF KEY SYMBOLS AND NOTATION

In this glossary, key symbols and notation are briefly defined.

Table D1

Symbol	Definition
$\overline{\text{any symbol}}$	average (indicated by a bar over a symbol—e.g., $\bar{v}$ is average velocity)
°C	Celsius degree
°F	Fahrenheit degree
//	parallel
⊥	perpendicular
∝	proportional to
±	plus or minus
0	zero as a subscript denotes an initial value
α	alpha rays
α	angular acceleration
α	temperature coefficient(s) of resistivity
β	beta rays
β	sound level
β	volume coefficient of expansion
β^-	electron emitted in nuclear beta decay
β^+	positron decay
γ	gamma rays
γ	surface tension
$\gamma = 1/\sqrt{1 - v^2/c^2}$	a constant used in relativity
Δ	change in whatever quantity follows
δ	uncertainty in whatever quantity follows
ΔE	change in energy between the initial and final orbits of an electron in an atom
ΔE	uncertainty in energy
Δm	difference in mass between initial and final products
ΔN	number of decays that occur
Δp	change in momentum
Δp	uncertainty in momentum
ΔPE_g	change in gravitational potential energy
$\Delta \theta$	rotation angle

Symbol	Definition
Δs	distance traveled along a circular path
Δt	uncertainty in time
Δt_0	proper time as measured by an observer at rest relative to the process
ΔV	potential difference
Δx	uncertainty in position
ε_0	permittivity of free space
η	viscosity
θ	angle between the force vector and the displacement vector
θ	angle between two lines
θ	contact angle
θ	direction of the resultant
θ_b	Brewster's angle
θ_c	critical angle
κ	dielectric constant
λ	decay constant of a nuclide
λ	wavelength
λ_n	wavelength in a medium
μ_0	permeability of free space
μ_k	coefficient of kinetic friction
μ_s	coefficient of static friction
ν_e	electron neutrino
π^+	positive pion
π^-	negative pion
π^0	neutral pion
ρ	density
ρ_c	critical density, the density needed to just halt universal expansion
ρ_{fl}	fluid density
$\bar{\rho}_{obj}$	average density of an object
ρ / ρ_w	specific gravity
τ	characteristic time constant for a resistance and inductance (RL) or resistance and capacitance (RC) circuit
τ	characteristic time for a resistor and capacitor (RC) circuit
τ	torque
Υ	upsilon meson
Φ	magnetic flux

Symbol	Definition
ϕ	phase angle
Ω	ohm (unit)
ω	angular velocity
A	ampere (current unit)
A	area
A	cross-sectional area
A	total number of nucleons
a	acceleration
a_B	Bohr radius
a_c	centripetal acceleration
a_t	tangential acceleration
AC	alternating current
AM	amplitude modulation
atm	atmosphere
B	baryon number
B	blue quark color
$\bar{B}$	antiblue (yellow) antiquark color
b	quark flavor bottom or beauty
B	bulk modulus
B	magnetic field strength
B_{int}	electron's intrinsic magnetic field
B_{orb}	orbital magnetic field
BE	binding energy of a nucleus—it is the energy required to completely disassemble it into separate protons and neutrons
BE$/A$	binding energy per nucleon
Bq	becquerel—one decay per second
C	capacitance (amount of charge stored per volt)
C	coulomb (a fundamental SI unit of charge)
C_p	total capacitance in parallel
C_s	total capacitance in series
CG	center of gravity
CM	center of mass
c	quark flavor charm
c	specific heat
c	speed of light
Cal	kilocalorie

Symbol	Definition
cal	calorie
COP_{hp}	heat pump's coefficient of performance
COP_{ref}	coefficient of performance for refrigerators and air conditioners
$\cos\theta$	cosine
$\cot\theta$	cotangent
$\csc\theta$	cosecant
D	diffusion constant
d	displacement
d	quark flavor down
dB	decibel
d_i	distance of an image from the center of a lens
d_o	distance of an object from the center of a lens
DC	direct current
E	electric field strength
ε	emf (voltage) or Hall electromotive force
emf	electromotive force
E	energy of a single photon
E	nuclear reaction energy
E	relativistic total energy
E	total energy
E_0	ground state energy for hydrogen
E_0	rest energy
EC	electron capture
E_{cap}	energy stored in a capacitor
Eff	efficiency—the useful work output divided by the energy input
Eff_C	Carnot efficiency
E_{in}	energy consumed (food digested in humans)
E_{ind}	energy stored in an inductor
E_{out}	energy output
e	emissivity of an object
e^+	antielectron or positron
eV	electron volt
F	farad (unit of capacitance, a coulomb per volt)
F	focal point of a lens
$\mathbf{F}$	force

Symbol	Definition
F	magnitude of a force
F	restoring force
F_B	buoyant force
F_c	centripetal force
F_i	force input
$\mathbf{F}_{net}$	net force
F_o	force output
FM	frequency modulation
f	focal length
f	frequency
f_0	resonant frequency of a resistance, inductance, and capacitance (RLC) series circuit
f_0	threshold frequency for a particular material (photoelectric effect)
f_1	fundamental
f_2	first overtone
f_3	second overtone
f_B	beat frequency
f_k	magnitude of kinetic friction
f_s	magnitude of static friction
G	gravitational constant
G	green quark color
$\bar{G}$	antigreen (magenta) antiquark color
g	acceleration due to gravity
g	gluons (carrier particles for strong nuclear force)
h	change in vertical position
h	height above some reference point
h	maximum height of a projectile
h	Planck's constant
hf	photon energy
h_i	height of the image
h_o	height of the object
I	electric current
I	intensity
I	intensity of a transmitted wave
I	moment of inertia (also called rotational inertia)

Symbol	Definition
I_0	intensity of a polarized wave before passing through a filter
I_{ave}	average intensity for a continuous sinusoidal electromagnetic wave
I_{rms}	average current
J	joule
J/Ψ	Joules/psi meson
K	kelvin
k	Boltzmann constant
k	force constant of a spring
K_α	x rays created when an electron falls into an $n = 1$ shell vacancy from the $n = 3$ shell
K_β	x rays created when an electron falls into an $n = 2$ shell vacancy from the $n = 3$ shell
kcal	kilocalorie
KE	translational kinetic energy
KE + PE	mechanical energy
KE_e	kinetic energy of an ejected electron
KE_{rel}	relativistic kinetic energy
KE_{rot}	rotational kinetic energy
$\overline{KE}$	thermal energy
kg	kilogram (a fundamental SI unit of mass)
L	angular momentum
L	liter
L	magnitude of angular momentum
L	self-inductance
ℓ	angular momentum quantum number
L_α	x rays created when an electron falls into an $n = 2$ shell from the $n = 3$ shell
L_e	electron total family number
L_μ	muon family total number
L_τ	tau family total number
L_f	heat of fusion
L_f and L_v	latent heat coefficients
L_{orb}	orbital angular momentum
L_s	heat of sublimation
L_v	heat of vaporization
L_z	z - component of the angular momentum
M	angular magnification

Symbol	Definition
M	mutual inductance
m	indicates metastable state
m	magnification
m	mass
m	mass of an object as measured by a person at rest relative to the object
m	meter (a fundamental SI unit of length)
m	order of interference
m	overall magnification (product of the individual magnifications)
$m\left(^A\mathrm{X}\right)$	atomic mass of a nuclide
MA	mechanical advantage
m_e	magnification of the eyepiece
m_e	mass of the electron
m_ℓ	angular momentum projection quantum number
m_n	mass of a neutron
m_o	magnification of the objective lens
mol	mole
m_p	mass of a proton
m_s	spin projection quantum number
N	magnitude of the normal force
N	newton
$\mathbf{N}$	normal force
N	number of neutrons
n	index of refraction
n	number of free charges per unit volume
N_A	Avogadro's number
N_r	Reynolds number
$\mathrm{N} \cdot \mathrm{m}$	newton-meter (work-energy unit)
$\mathrm{N} \cdot \mathrm{m}$	newtons times meters (SI unit of torque)
OE	other energy
P	power
P	power of a lens
P	pressure
$\mathbf{p}$	momentum
p	momentum magnitude
p	relativistic momentum
$\mathbf{p}_{\text{tot}}$	total momentum

Symbol	Definition
$\mathbf{p}'_{tot}$	total momentum some time later
P_{abs}	absolute pressure
P_{atm}	atmospheric pressure
P_{atm}	standard atmospheric pressure
PE	potential energy
PE_{el}	elastic potential energy
PE_{elec}	electric potential energy
PE_s	potential energy of a spring
P_g	gauge pressure
P_{in}	power consumption or input
P_{out}	useful power output going into useful work or a desired, form of energy
Q	latent heat
Q	net heat transferred into a system
Q	flow rate—volume per unit time flowing past a point
$+Q$	positive charge
$-Q$	negative charge
q	electron charge
q_p	charge of a proton
q	test charge
QF	quality factor
R	activity, the rate of decay
R	radius of curvature of a spherical mirror
R	red quark color
$\bar{R}$	antired (cyan) quark color
R	resistance
R	resultant or total displacement
R	Rydberg constant
R	universal gas constant
r	distance from pivot point to the point where a force is applied
r	internal resistance
$r_\perp$	perpendicular lever arm
r	radius of a nucleus
r	radius of curvature
r	resistivity

Symbol	Definition
r or rad	radiation dose unit
rem	roentgen equivalent man
rad	radian
RBE	relative biological effectiveness
RC	resistor and capacitor circuit
rms	root mean square
r_n	radius of the nth H-atom orbit
R_p	total resistance of a parallel connection
R_s	total resistance of a series connection
R_s	Schwarzschild radius
S	entropy
S	intrinsic spin (intrinsic angular momentum)
S	magnitude of the intrinsic (internal) spin angular momentum
S	shear modulus
S	strangeness quantum number
s	quark flavor strange
s	second (fundamental SI unit of time)
s	spin quantum number
s	total displacement
$\sec \theta$	secant
$\sin \theta$	sine
s_z	z-component of spin angular momentum
T	period—time to complete one oscillation
T	temperature
T_c	critical temperature—temperature below which a material becomes a superconductor
T	tension
T	tesla (magnetic field strength B)
t	quark flavor top or truth
t	time
$t_{1/2}$	half-life—the time in which half of the original nuclei decay
$\tan \theta$	tangent
U	internal energy
u	quark flavor up
u	unified atomic mass unit
u	velocity of an object relative to an observer
u'	velocity relative to another observer

Symbol	Definition
V	electric potential
V	terminal voltage
V	volt (unit)
V	volume
$\mathbf{v}$	relative velocity between two observers
v	speed of light in a material
$\mathbf{v}$	velocity
$\bar{\mathbf{v}}$	average fluid velocity
$V_B - V_A$	change in potential
$\mathbf{v}_d$	drift velocity
V_p	transformer input voltage
V_{rms}	rms voltage
V_s	transformer output voltage
$\mathbf{v}_{tot}$	total velocity
v_w	propagation speed of sound or other wave
$\mathbf{v}_w$	wave velocity
W	work
W	net work done by a system
W	watt
w	weight
w_{fl}	weight of the fluid displaced by an object
W_c	total work done by all conservative forces
W_{nc}	total work done by all nonconservative forces
W_{out}	useful work output
X	amplitude
X	symbol for an element
$^Z X_N$	notation for a particular nuclide
x	deformation or displacement from equilibrium
x	displacement of a spring from its undeformed position
x	horizontal axis
X_C	capacitive reactance
X_L	inductive reactance
x_{rms}	root mean square diffusion distance
y	vertical axis
Y	elastic modulus or Young's modulus

Symbol	Definition
Z	atomic number (number of protons in a nucleus)
Z	impedance

ANSWER KEY

Chapter 18

Problems & Exercises

1

(a) 1.25×10^{10}

(b) 3.13×10^{12}

3
-600 C

5
1.03×10^{12}

7
9.09×10^{-13}

9
1.48×10^{8} C

15

(a) $E_{x = 1.00 \text{ cm}} = -\infty$

(b) 2.12×10^{5} N/C

(c) one charge of $+q$

17

(a) 0.252 N to the left

(b) $x = 6.07$ cm

19

(a)The electric field at the center of the square will be straight up, since q_a and q_b are positive and q_c and q_d are negative and all have the same magnitude.

(b) 2.04×10^{7} N/C (upward)

21

0.102 N, in the $-y$ direction

23

(a) $\vec{E} = 4.36 \times 10^3$ N/C, 35.0°, below the horizontal.

(b) No

25

(a) 0.263 N

(b) If the charges are distributed over some area, there will be a concentration of charge along the side closest to the oppositely charged object. This effect will increase the net force.

27

The separation decreased by a factor of 5.

31

$$F = k\frac{|q_1 q_2|}{r^2} = ma \qquad a = \frac{kq^2}{mr^2} \Longrightarrow$$

$$= \frac{\left(9.00 \times 10^9 \text{ N} \cdot \text{m}^2 \big/ \text{C}^2\right)\left(1.60 \times 10^{-19} \text{ m}\right)^2}{\left(1.67 \times 10^{-27} \text{ kg}\right)\left(2.00 \times 10^{-9} \text{ m}\right)^2}$$

$$= 3.45 \times 10^{16} \text{ m/s}^2$$

32

(a) 3.2

(b) If the distance increases by 3.2, then the force will decrease by a factor of 10 ; if the distance decreases by 3.2, then the force will increase by a factor of 10. Either way, the force changes by a factor of 10.

34

(a) 1.04×10^{-9} C

(b) This charge is approximately 1 nC, which is consistent with the magnitude of charge typical for static electricity

37

1.02×10^{-11}

39

  a.   0.859 m beyond negative charge on line connecting two charges

  b.   0.109 m from lesser charge on line connecting two charges

42

8.75×10^{-4} N

44

(a) 6.94×10^{-8} C

(b) 6.25 N/C

46

(a) 300 N/C (east)

(b) 4.80×10^{-17} N (east)

52

(a) 5.58×10^{-11} N/C

(b) the coulomb force is extraordinarily stronger than gravity

54

(a) -6.76×10^5 C

(b) 2.63×10^{13} m/s^2 (upward)

(c) 2.45×10^{-18} kg

56
The charge q_2 is 9 times greater than q_1.

Test Prep for AP® Courses

1
(b)
3
(c)
5
(a)
7
(b)
9
(a) -0.1 C, (b) 1.1 C, (c) Both charges will be equal to 1 C, law of conservation of charge, (d) 0.9 C
11
W is negative, X is positive, Y is negative, Z is neutral.
13
(c)
15
(c)
17
(b)
19
a) Ball 1 will have positive charge and Ball 2 will have negative charge. b) The negatively charged rod attracts positive charge of Ball 1. The electrons of Ball 1 are transferred to Ball 2, making it negatively charged. c) If Ball 2 is grounded while the rod is still there, it will lose its negative charge to the ground. d) Yes, Ball 1 will be positively charged and Ball 2 will be negatively charge.
21
(c)
23
decrease by 77.78%.
25
(a)
27
(d)
29
(a) 3.60×10^{10} N, (b) It will become 1/4 of the original value; hence it will be equal to 8.99×10^9 N
31
(c)
33
(a)
35
(b)
37
(a) 350 N/C, (b) west, (c) 5.6×10−17 N, (d) west.
39
(b)
41
(a) i) Field vectors near objects point toward negatively charged objects and away from positively charged objects.

(a) ii) The vectors closest to R and T are about the same length and start at about the same distance. We have that $q_R\big/d^2 = q_T\big/d^2$, so the charge on R is about the same as the charge on T. The closest vectors around S are about the same length as those around R and T. The vectors near S start at about 6 units away, while vectors near R and T start at about 4 units. We have that $q_R\big/d^2 = q_S\big/D^2$, so $q_S\big/q_R = D^2\big/d^2 = 36\big/16 = 2.25$, and so the charge on S is about twice that on R and T.

(b)

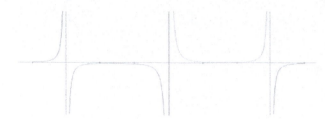

Figure 18.35. A vector diagram.

(c)

$$E = k\left[-\frac{q}{(d+x)^2} + \frac{2q}{(x)^2} + \frac{q}{(d-x)^2}\right]$$

(d) The statement is not true. The vector diagram shows field vectors in this region with nonzero length, and the vectors not shown have even greater lengths. The equation in part (c) shows that, when $0 < x < d$, the denominator of the negative term is always greater than the denominator of the third term, but the numerator is the same. So the negative term always has a smaller magnitude than the third term and since the second term is positive the sum of the terms is always positive.

Chapter 19

Problems & Exercises

1
42.8
4

1.00×10^5 K

6

(a) 4×10^4 W

(b) A defibrillator does not cause serious burns because the skin conducts electricity well at high voltages, like those used in defibrillators. The gel used aids in the transfer of energy to the body, and the skin doesn't absorb the energy, but rather lets it pass through to the heart.

8

(a) 7.40×10^3 C

(b) 1.54×10^{20} electrons per second

9

3.89×10^6 C

11

(a) 1.44×10^{12} V

(b) This voltage is very high. A 10.0 cm diameter sphere could never maintain this voltage; it would discharge.

(c) An 8.00 C charge is more charge than can reasonably be accumulated on a sphere of that size.

15

(a) 3.00 kV

(b) 750 V

17

(a) No. The electric field strength between the plates is 2.5×10^6 V/m, which is lower than the breakdown strength for air (3.0×10^6 V/m).

(b) 1.7 mm

19

44.0 mV
21
15 kV
23

(a) 800 KeV

(b) 25.0 km

24
144 V
26

(a) 1.80 km

(b) A charge of 1 C is a very large amount of charge; a sphere of radius 1.80 km is not practical.

28
-2.22×10^{-13} C
30

(a) 3.31×10^6 V

(b) 152 MeV

32

(a) 2.78×10^{-7} C

(b) 2.00×10^{-10} C

35

(a) 2.96×10^9 m/s

(b) This velocity is far too great. It is faster than the speed of light.

(c) The assumption that the speed of the electron is far less than that of light and that the problem does not require a relativistic treatment produces an answer greater than the speed of light.

46
21.6 mC
48
80.0 mC
50
20.0 kV
52
667 pF
54

(a) 4.4 μF

(b) 4.0×10^{-5} C

56

(a) 14.2 kV

(b) The voltage is unreasonably large, more than 100 times the breakdown voltage of nylon.

(c) The assumed charge is unreasonably large and cannot be stored in a capacitor of these dimensions.

57
0.293 μF
59
3.08 μF in series combination, 13.0 μF in parallel combination
60
2.79 μF
62

(a) −3.00 μF

(b) You cannot have a negative value of capacitance.

(c) The assumption that the capacitors were hooked up in parallel, rather than in series, was incorrect. A parallel connection always produces a greater capacitance, while here a smaller capacitance was assumed. This could happen only if the capacitors are connected in series.

63

(a) 405 J

(b) 90.0 mC

64

(a) 3.16 kV

(b) 25.3 mC

66

(a) 1.42×10^{-5} C, 6.38×10^{-5} J

(b) 8.46×10^{-5} C, 3.81×10^{-4} J

67

(a) 4.43×10^{-12} F

(b) 452 V

(c) 4.52×10^{-7} J

70

(a) 133 F

(b) Such a capacitor would be too large to carry with a truck. The size of the capacitor would be enormous.

(c) It is unreasonable to assume that a capacitor can store the amount of energy needed.

Test Prep for AP® Courses

1
(a)
3
(b)
5
(c)
7
(a)
9
(b)
11
(b)
13
(a)
15
(c)
17
(b)
19
(a)
21
(d)
23
(d)
25
(a)
27

(b)
29
(c)
31
(d)
33
(a)
35
(c)
37
(c)
39
(b)
41
(a)
43
(d)

Chapter 20

Problems & Exercises

1
0.278 mA

3
0.250 A

5
1.50ms

7

(a) $1.67k\,\Omega$

(b) If a 50 times larger resistance existed, keeping the current about the same, the power would be increased by a factor of about 50 (based on the equation $P = I^2R$), causing much more energy to be transferred to the skin, which could cause serious burns. The gel used reduces the resistance, and therefore reduces the power transferred to the skin.

9

(a) 0.120 C

(b) 7.50×10^{17} electrons

11
96.3 s

13

(a) 7.81×10^{14} He^{++} nuclei/s

(b) 4.00×10^3 s

(c) 7.71×10^8 s

15
-1.13×10^{-4} m/s

17
9.42×10^{13} electrons

18
0.833 A

20
$7.33 \times 10^{-2}\ \Omega$

22

(a) 0.300 V

(b) 1.50 V

(c) The voltage supplied to whatever appliance is being used is reduced because the total voltage drop from the wall to the final output of the appliance is fixed. Thus, if the voltage drop across the extension cord is large, the voltage drop across the appliance is significantly decreased, so the power output by the appliance can be significantly decreased, reducing the ability of the appliance to work properly.

24

$0.104 \ \Omega$

26

2.8×10^{-2} m

28

1.10×10^{-3} A

30

$-5°C$ to $45°C$

32

1.03

34

0.06%

36

$-17°C$

38

(a) $4.7 \ \Omega$ (total)

(b) 3.0% decrease

40

2.00×10^{12} W

44

(a) 1.50 W

(b) 7.50 W

46

$$\frac{V^2}{\Omega} = \frac{V^2}{V/A} = AV = \left(\frac{C}{s}\right)\left(\frac{J}{C}\right) = \frac{J}{s} = 1 \ W$$

48

$$1 \ kW \cdot h = \left(\frac{1 \times 10^3 \ J}{1 \ s}\right)(1 \ h)\left(\frac{3600 \ s}{1 \ h}\right) = 3.60 \times 10^6 \ J$$

50

$438/y

52

$6.25

54

1.58 h

56

$3.94 billion/year

58

25.5 W

60

(a) 2.00×10^9 J

(b) 769 kg

62

45.0 s

64

(a) 343 A

(b) 2.17×10^3 A

(c) 1.10×10^3 A

66

(a) 1.23×10^3 kg

(b) 2.64×10^3 kg

69

(a) 2.08×10^5 A

(b) 4.33×10^4 MW

(c) The transmission lines dissipate more power than they are supposed to transmit.

(d) A voltage of 480 V is unreasonably low for a transmission voltage. Long-distance transmission lines are kept at much higher voltages (often hundreds of kilovolts) to reduce power losses.

73
480 V
75
2.50 ms
77

(a) 4.00 kA

(b) 16.0 MW

(c) 16.0%

79
2.40 kW
81

(a) 4.0

(b) 0.50

(c) 4.0

83

(a) 1.39 ms

(b) 4.17 ms

(c) 8.33 ms

85

(a) 230 kW

(b) 960 A

87

(a) 0.400 mA, no effect

(b) 26.7 mA, muscular contraction for duration of the shock (can't let go)

89
1.20×10^5 Ω
91

(a) 1.00 Ω

(b) 14.4 kW

93
Temperature increases $860°$ C . It is very likely to be damaging.
95
80 beats/minute

Test Prep for AP® Courses

1
(a)
3
10 A

5
(a)
7
3.2 Ω, 2.19 A
9
(b), (d)
11
9.72×10^{-8} Ω·m
13
18 Ω
15
10:3 or 3.33

Chapter 21

Problems & Exercises

1

(a) 2.75 k Ω

(b) 27.5 Ω

3

(a) 786 Ω

(b) 20.3 Ω

5

 29.6 W

7

(a) 0.74 A

(b) 0.742 A

9

(a) 60.8 W

(b) 3.18 kW

11

(a)
$$R_s = R_1 + R_2$$
$$\Rightarrow R_s \approx R_1 (R_1 >> R_2)$$

(b) $\dfrac{1}{R_p} = \dfrac{1}{R_1} + \dfrac{1}{R_2} = \dfrac{R_1 + R_2}{R_1 R_2},$

so that

$$R_p = \frac{R_1 R_2}{R_1 + R_2} \approx \frac{R_1 R_2}{R_1} = R_2 (R_1 >> R_2).$$

13

(a) -400 k Ω

(b) Resistance cannot be negative.

(c) Series resistance is said to be less than one of the resistors, but it must be greater than any of the resistors.

14
2.00 V
16
2.9994 V
18
 0.375 Ω
21

(a) 0.658 A

(b) 0.997 W

(c) 0.997 W; yes

23

(a) 200 A

(b) 10.0 V

(c) 2.00 kW

(d) $0.1000 \; \Omega \; ; 80.0$ A, 4.0 V, 320 W

25

(a) $0.400 \; \Omega$

(b) No, there is only one independent equation, so only r can be found.

29

(a) −0.120 V

(b) $-1.41 \times 10^{-2} \; \Omega$

(c) Negative terminal voltage; negative load resistance.

(d) The assumption that such a cell could provide 8.50 A is inconsistent with its internal resistance.

31

$$-I_2 R_2 + emf_1 - I_2 r_1 + I_3 R_3 + I_3 r_2 - emf_2 = 0$$

35

$$I_3 = I_1 + I_2$$

37

$$emf_2 - I_2 r_2 - I_2 R_2 + I_1 R_5 + I_1 r_1 - emf_1 + I_1 R_1 = 0$$

39

(a) $I_1 = 4.75$ A

(b) $I_2 = -3.5$ A

(c) $I_3 = 8.25$ A

41

(a) No, you would get inconsistent equations to solve.

(b) $I_1 \neq I_2 + I_3$. The assumed currents violate the junction rule.

42

$30 \; \mu A$

44

$1.98 \; k\Omega$

46

$1.25 \times 10^{-4} \; \Omega$

48

(a) $3.00 \; M\Omega$

(b) $2.99 \; k\Omega$

50

(a) 1.58 mA

(b) 1.5848 V (need four digits to see the difference)

(c) 0.99990 (need five digits to see the difference from unity)

52

$15.0 \; \mu A$

54

(a)

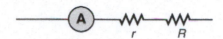

Figure 21.39.

(b) $10.02\ \Omega$

(c) 0.9980, or a 2.0×10^{-1} percent decrease

(d) 1.002, or a 2.0×10^{-1} percent increase

(e) Not significant.

56

(a) $-66.7\ \Omega$

(b) You can't have negative resistance.

(c) It is unreasonable that I_{G} is greater than I_{tot} (see **Figure 21.36**). You cannot achieve a full-scale deflection using a current less than the sensitivity of the galvanometer.

57
24.0 V
59
$1.56\ \mathrm{k}\,\Omega$
61

(a) 2.00 V

(b) $9.68\ \Omega$

62

$$\text{Range} = 5.00\ \Omega\ \text{to}\ 5.00\ \mathrm{k}\,\Omega \qquad\qquad (21.78)$$

63
range 4.00 to 30.0 M Ω
65

(a) $2.50\ \mu F$

(b) 2.00 s
67
86.5%
69

(a) $1.25\ \mathrm{k}\,\Omega$

(b) 30.0 ms

71

(a) 20.0 s

(b) 120 s

(c) 16.0 ms

73
$1.73\times10^{-2}\ \mathrm{s}$
74
$3.33\times10^{-3}\ \Omega$
76

(a) 4.99 s

(b) 3.87°C

(c) $31.1\ \mathrm{k}\,\Omega$

(d) No

Test Prep for AP® Courses

1

(a), (b)

3

(b)

5

(a) 4-Ω resistor; (b) combination of 20-Ω, 20-Ω, and 10-Ω resistors; (c) 20 W in each 20-Ω resistor, 40 W in 10-Ω resistor, 64 W in 4-Ω resistor, total 144W total in resistors, output power is 144 W, yes they are equal (law of conservation of energy); (d) 4 Ω and 3 Ω for part (a) and no change for part (b); (e) no effect, it will remain the same.

7

0.25 Ω, 0.50 Ω, no change

9

a. (c)

b. (c)

c. (d)

d. (d)

11

a. $I_1 + I_3 = I_2$

b. $E_1 - I_1R_1 - I_2R_2 - I_1r_1 = 0$; $- E_2 + I_1R_1 - I_3R_3 - I_3r_2 = 0$

c. $I_1 = 8/15$ A, $I_2 = 7/15$ A and $I_3 = -1/15$ A

d. $I_1 = 2/5$ A, $I_2 = 3/5$ A and $I_3 = 1/5$ A

e. $P_{E1} = 18/5$ W and $P_{R1} = 24/25$ W, $P_{R2} = 54/25$ W, $P_{R3} = 12/25$ W. Yes, $P_{E1} = P_{R1} + P_{R2} + P_{R3}$

f. R_3, losses in the circuit

13

(a) 20 mA, **Figure 21.44**, 5.5 s; (b) 24 mA, **Figure 21.35**, 2 s

Chapter 22

Problems & Exercises

1

(a) Left (West)

(b) Into the page

(c) Up (North)

(d) No force

(e) Right (East)

(f) Down (South)

3

(a) East (right)

(b) Into page

(c) South (down)

5

(a) Into page

(b) West (left)

(c) Out of page

7

7.50×10^{-7} N perpendicular to both the magnetic field lines and the velocity

9

(a) 3.01×10^{-5} T

(b) This is slightly less then the magnetic field strength of 5×10^{-5} T at the surface of the Earth, so it is consistent.

11

(a) 6.67×10^{-10} C (taking the Earth's field to be 5.00×10^{-5} T)

(b) Less than typical static, therefore difficult

12

4.27 m

14

(a) 0.261 T

(b) This strength is definitely obtainable with today's technology. Magnetic field strengths of 0.500 T are obtainable with permanent magnets.

16

4.36×10^{-4} m

18

(a) 3.00 kV/m

(b) 30.0 V

20

0.173 m

22

7.50×10^{-4} V

24

(a) 1.18×10^3 m/s

(b) Once established, the Hall emf pushes charges one direction and the magnetic force acts in the opposite direction resulting in no net force on the charges. Therefore, no current flows in the direction of the Hall emf. This is the same as in a current-carrying conductor—current does not flow in the direction of the Hall emf.

26

11.3 mV

28

$1.16 \, \mu$V

30

2.00 T

31

(a) west (left)

(b) into page

(c) north (up)

(d) no force

(e) east (right)

(f) south (down)

33

(a) into page

(b) west (left)

(c) out of page

35

(a) 2.50 N

(b) This is about half a pound of force per 100 m of wire, which is much less than the weight of the wire itself. Therefore, it does not cause any special concerns.

37

1.80 T

39

(a) $30°$

(b) 4.80 N

41

(a) τ decreases by 5.00% if B decreases by 5.00%

(b) 5.26% increase

43
10.0 A

45
$$A \cdot m^2 \cdot T = A \cdot m^2 \left(\frac{N}{A \cdot m} \right) = N \cdot m \, .$$

47
$3.48 \times 10^{-26}\, N \cdot m$

49

(a) $0.666\, N \cdot m$ west

(b) This is not a very significant torque, so practical use would be limited. Also, the current would need to be alternated to make the loop rotate (otherwise it would oscillate).

50

(a) 8.53 N, repulsive

(b) This force is repulsive and therefore there is never a risk that the two wires will touch and short circuit.

52
400 A in the opposite direction

54

(a) $1.67 \times 10^{-3}\, N/m$

(b) $3.33 \times 10^{-3}\, N/m$

(c) Repulsive

(d) No, these are very small forces

56

(a) Top wire: $2.65 \times 10^{-4}\, N/m$ s, $10.9°$ to left of up

(b) Lower left wire: $3.61 \times 10^{-4}\, N/m$, $13.9°$ down from right

(c) Lower right wire: $3.46 \times 10^{-4}\, N/m$, $30.0°$ down from left

58

(a) right-into page, left-out of page

(b) right-out of page, left-into page

(c) right-out of page, left-into page

60

(a) clockwise

(b) clockwise as seen from the left

(c) clockwise as seen from the right

61
$1.01 \times 10^{13}\, T$

63

(a) $4.80 \times 10^{-4}\, T$

(b) Zero

(c) If the wires are not paired, the field is about 10 times stronger than Earth's magnetic field and so could severely disrupt the use of a compass.

65

39.8 A
67

(a) 3.14×10^{-5} T

(b) 0.314 T

69

7.55×10^{-5} T , 23.4°

71
10.0 A
73

(a) 9.09×10^{-7} N upward

(b) 3.03×10^{-5} m/s^2

75
60.2 cm
77

(a) 1.02×10^3 N/m^2

(b) Not a significant fraction of an atmosphere

79

17.0×10^{-4}%/°C

81
18.3 MHz
83

(a) Straight up

(b) 6.00×10^{-4} N/m

(c) 94.1 μm

(d)2.47 Ω/m, 49.4 V/m

85

(a) 571 C

(b) Impossible to have such a large separated charge on such a small object.

(c) The 1.00-N force is much too great to be realistic in the Earth's field.

87

(a) 2.40×10^6 m/s

(b) The speed is too high to be practical $\leq$ 1% speed of light

(c) The assumption that you could reasonably generate such a voltage with a single wire in the Earth's field is unreasonable

89

(a) 25.0 kA

(b) This current is unreasonably high. It implies a total power delivery in the line of 50.0x10^9 W, which is much too high for standard transmission lines.

(c)100 meters is a long distance to obtain the required field strength. Also coaxial cables are used for transmission lines so that there is virtually no field for DC power lines, because of cancellation from opposing currents. The surveyor's concerns are not a problem for his magnetic field measurements.

Test Prep for AP® Courses

1
(a)
3
(b)

5
(b)
7
(a)
9
(b)
11
(e)
13
(c)
15
(c)

Chapter 23

Problems & Exercises

1
Zero
3

(a) CCW

(b) CW

(c) No current induced

5

(a) 1 CCW, 2 CCW, 3 CW

(b) 1, 2, and 3 no current induced

(c) 1 CW, 2 CW, 3 CCW

9
(a) 3.04 mV
(b) As a lower limit on the ring, estimate R = 1.00 mΩ. The heat transferred will be 2.31 mJ. This is not a significant amount of heat.
11
0.157 V
13

proportional to $\frac{1}{r}$

17

(a) 0.630 V

(b) No, this is a very small emf.

19
2.22 m/s
25

(a) 10.0 N

(b) 2.81×10^8 J

(c) 0.36 m/s

(d) For a week-long mission (168 hours), the change in velocity will be 60 m/s, or approximately 1%. In general, a decrease in velocity would cause the orbit to start spiraling inward because the velocity would no longer be sufficient to keep the circular orbit. The long-term consequences are that the shuttle would require a little more fuel to maintain the desired speed, otherwise the orbit would spiral slightly inward.

28
474 V
30
0.247 V
32

(a) 50

(b) yes

34

(a) 0.477 T

(b) This field strength is small enough that it can be obtained using either a permanent magnet or an electromagnet.

36

(a) 5.89 V

(b) At t=0

(c) 0.393 s

(d) 0.785 s

38

(a) 1.92×10^{6} rad/s

(b) This angular velocity is unreasonably high, higher than can be obtained for any mechanical system.

(c) The assumption that a voltage as great as 12.0 kV could be obtained is unreasonable.

39

(a) $12.00\ \Omega$

(b) 1.67 A

41
72.0 V

43
$0.100\ \Omega$

44

(a) 30.0

(b) 9.75×10^{-2} A

46

(a) 20.0 mA

(b) 2.40 W

(c) Yes, this amount of power is quite reasonable for a small appliance.

48

(a) 0.063 A

(b) Greater input current needed.

50

(a) 2.2

(b) 0.45

(c) 0.20, or 20.0%

52

(a) 335 MV

(b) way too high, well beyond the breakdown voltage of air over reasonable distances

(c) input voltage is too high

54

(a) 15.0 V

(b) 75.0 A

(c) yes

55
1.80 mH

57
3.60 V

61

(a) 31.3 kV

(b) 125 kJ

(c) 1.56 MW

(d) No, it is not surprising since this power is very high.

63

(a) 1.39 mH

(b) 3.33 V

(c) Zero

65

60.0 mH

67

(a) 200 H

(b) 5.00°C

69

500 H

71

 50.0 Ω

73

 1.00×10^{-18} s to 0.100 s

75

95.0%

77

(a) 24.6 ms

(b) 26.7 ms

(c) 9% difference, which is greater than the inherent uncertainty in the given parameters.

79

531 Hz

81

1.33 nF

83

(a) 2.55 A

(b) 1.53 mA

85

 63.7 μH

87

(a) 21.2 mH

(b) 8.00 Ω

89

(a) 3.18 mF

(b) 16.7 Ω

92

(a) 40.02 Ω at 60.0 Hz, 193 Ω at 10.0 kHz

(b) At 60 Hz, with a capacitor, $Z = 531 \ \Omega$, over 13 times as high as without the capacitor. The capacitor makes a large difference at low frequencies. At 10 kHz, with a capacitor $Z = 190 \ \Omega$, about the same as without the capacitor. The capacitor has a smaller effect at high frequencies.

94

(a) $529 \, \Omega$ at 60.0 Hz, $185 \, \Omega$ at 10.0 kHz

(b) These values are close to those obtained in **Example 23.12** because at low frequency the capacitor dominates and at high frequency the inductor dominates. So in both cases the resistor makes little contribution to the total impedance.

96
9.30 nF to 101 nF
98
3.17 pF
100

(a) $1.31 \, \mu H$

(b) 1.66 pF

102

(a) $12.8 \, k\Omega$

(b) $1.31 \, k\Omega$

(c) 31.9 mA at 500 Hz, 312 mA at 7.50 kHz

(d) 82.2 kHz

(e) 0.408 A

104

(a) 0.159

(b) $80.9°$

(c) 26.4 W

(d) 166 W

106
16.0 W

Test Prep for AP® Courses

1

(c)

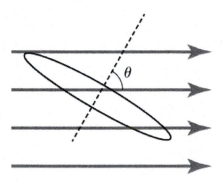

Figure 23.6.

3
(c)
5
(a), (d)
7
(c)

Chapter 24

Problems & Exercises

3
150 kV/m
6

(a) 33.3 cm (900 MHz) 11.7 cm (2560 MHz)

(b) The microwave oven with the smaller wavelength would produce smaller hot spots in foods, corresponding to the one with the frequency 2560 MHz.

8

26.96 MHz

10

5.0×10^{14} Hz

12

$$\lambda = \frac{c}{f} = \frac{3.00 \times 10^8 \text{ m/s}}{1.20 \times 10^{15} \text{ Hz}} = 2.50 \times 10^{-7} \text{ m}$$

14

0.600 m

16

(a) $f = \frac{c}{\lambda} = \frac{3.00 \times 10^8 \text{ m/s}}{1 \times 10^{-10} \text{ m}} = 3 \times 10^{18}$ Hz

(b) X-rays

19

(a) 6.00×10^6 m

(b) 4.33×10^{-5} T

21

(a) 1.50×10^6 Hz, AM band

(b) The resonance of currents on an antenna that is 1/4 their wavelength is analogous to the fundamental resonant mode of an air column closed at one end, since the tube also has a length equal to 1/4 the wavelength of the fundamental oscillation.

23

(a) 1.55×10^{15} Hz

(b) The shortest wavelength of visible light is 380 nm, so that

$$\frac{\lambda_{\text{visible}}}{\lambda_{\text{UV}}}$$

$$= \frac{380 \text{ nm}}{193 \text{ nm}}$$

$$= 1.97.$$

In other words, the UV radiation is 97% more accurate than the shortest wavelength of visible light, or almost twice as accurate!

25

3.90×10^8 m

27

(a) 1.50×10^{11} m

(b) $0.500 \, \mu\text{s}$

(c) 66.7 ns

29

(a) -3.5×10^2 W/m^2

(b) 88%

(c) $1.7 \, \mu\text{T}$

30

$$I = \frac{c\varepsilon_0 E_0^2}{2}$$

$$= \frac{(3.00\times10^8 \text{ m/s})(8.85\times10^{-12}\text{ C}^2/\text{N}\cdot\text{m}^2)(125 \text{ V/m})^2}{2}$$

$$= 20.7 \text{ W/m}^2$$

32

(a) $I = \dfrac{P}{A} = \dfrac{P}{\pi r^2} = \dfrac{0.250\times10^{-3} \text{ W}}{\pi(0.500\times10^{-3} \text{ m})^2} = 318 \text{ W/m}^2$

$$I_{\text{ave}} = \frac{cB_0^2}{2\mu_0} \Rightarrow B_0 = \left(\frac{2\mu_0 I}{c}\right)^{1/2}$$

(b)
$$= \left(\frac{2(4\pi\times10^{-7} \text{ T}\cdot\text{m/A})(318.3 \text{ W/m}^2)}{3.00\times10^8 \text{ m/s}}\right)^{1/2}$$

$$= 1.63\times10^{-6} \text{ T}$$

(c)
$$E_0 = cB_0 = (3.00\times10^8 \text{ m/s})(1.633\times10^{-6} \text{ T})$$

$$= 4.90\times10^2 \text{ V/m}$$

34

(a) 89.2 cm

(b) 27.4 V/m

36

(a) 333 T

(b) $1.33\times10^{19} \text{ W/m}^2$

(c) 13.3 kJ

38

(a) $I = \dfrac{P}{A} = \dfrac{P}{4\pi r^2} \propto \dfrac{1}{r^2}$

(b) $I \propto E_0^2, B_0^2 \Rightarrow E_0^2, B_0^2 \propto \dfrac{1}{r^2} \Rightarrow E_0, B_0 \propto \dfrac{1}{r}$

40
13.5 pF
42

(a) 4.07 kW/m^2

(b) 1.75 kV/m

(c) 5.84 μT

(d) 2 min 19 s

44

(a) $5.00\times10^3 \text{ W/m}^2$

(b) $3.88\times10^{-6} \text{ N}$

(c) $5.18\times10^{-12} \text{ N}$

46

(a) $t = 0$

(b) 7.50×10^{-10} s

(c) 1.00×10^{-9} s

48

(a) 1.01×10^{6} W/m^{2}

(b) Much too great for an oven.

(c) The assumed magnetic field is unreasonably large.

50

(a) 2.53×10^{-20} H

(b) L is much too small.

(c) The wavelength is unreasonably small.

Test Prep for AP® Courses

1
(b)
3
(a)
5
(d)
7
(d)
9
(d)
11
(a)

Chapter 25

Problems & Exercises

1
Top 1.715 m from floor, bottom 0.825 m from floor. Height of mirror is 0.890 m , or precisely one-half the height of the person.
5

2.25×10^{8} m/s in water

2.04×10^{8} m/s in glycerine

7
1.490 , polystyrene
9
1.28 s
11
1.03 ns
13
$n = 1.46$, fused quartz
17

(a) 0.898

(b) Can't have $n < 1.00$ since this would imply a speed greater than c .

(c) Refracted angle is too big relative to the angle of incidence.

19

(a) $\dfrac{c}{5.00}$

(b) Speed of light too slow, since index is much greater than that of diamond.

(c) Angle of refraction is unreasonable relative to the angle of incidence.

22

66.3°

24

> 1.414

26

1.50, benzene

29

46.5°, red; 46.0°, violet

31

(a) 0.043°

(b) 1.33 m

33

71.3°

35

53.5°, red; 55.2°, violet

37

5.00 to 12.5 D

39

−0.222 m

41

(a) 3.43 m

(b) 0.800 by 1.20 m

42

(a) -1.35 m (on the object side of the lens).

(b) $+10.0$

(c) 5.00 cm

43

44.4 cm

45

(a) 6.60 cm

(b) −0.333

47

(a) $+7.50$ cm

(b) 13.3 D

(c) Much greater

49

(a) +6.67

(b) +20.0

(c) The magnification increases without limit (to infinity) as the object distance increases to the limit of the focal distance.

51

−0.933 mm

53

+0.667 m

55

(a) -1.5×10^{-2} m

(b) −66.7 D

57

+0.360 m (concave)

59

(a) +0.111

(b) -0.334 cm (behind "mirror")

(c) 0.752cm

61

$$m = \frac{h_i}{h_o} = -\frac{d_i}{d_o} = -\frac{-d_o}{d_o} = \frac{d_o}{d_o} = 1 \Rightarrow h_i = h_o \qquad (25.61)$$

63

6.82 kW/m^2

Test Prep for AP® Courses

1
(c)
3
(c)
5
(a)
7
Since light bends toward the normal upon entering a medium with a higher index of refraction, the upper path is a more accurate representation of a light ray moving from A to B.
9

First, measure the angle of incidence and the angle of refraction for light entering the plastic from air. Since the two angles can be measured and the index of refraction of air is known, the student can solve for the index of refraction of the plastic.

Next, measure the angle of incidence and the angle of refraction for light entering the gas from the plastic. Since the two angles can be measured and the index of refraction of the plastic is known, the student can solve for the index of refraction of the gas.

11

The speed of light in a medium is simply c/n, so the speed of light in water is 2.25×10^8 m/s. From Snell's law, the angle of incidence is 44°.
13
(d)
15
(a)
17
(a)
19
(b)

Chapter 26

Problems & Exercises

1
52.0 D
3

(a) −0.233 mm

(b) The size of the rods and the cones is smaller than the image height, so we can distinguish letters on a page.

5

(a) +62.5 D

(b) −0.250 mm

(c) −0.0800 mm

6
2.00 m

8

(a) ± 0.45 D

(b) The person was nearsighted because the patient was myopic and the power was reduced.

10

0.143 m

12

1.00 m

14

20.0 cm

16

-5.00 D

18

25.0 cm

20

-0.198 D

22

30.8 cm

24

-0.444 D

26

(a) 4.00

(b) 1600

28

(a) 0.501 cm

(b) Eyepiece should be 204 cm behind the objective lens.

30

(a) +18.3 cm (on the eyepiece side of the objective lens)

(b) -60.0

(c) -11.3 cm (on the objective side of the eyepiece)

(d) +6.67

(e) -400

33

-40.0

35

-1.67

37

$+10.0$ cm

39

(a) 0.251 μm

(b) Yes, this thickness implies that the shape of the cornea can be very finely controlled, producing normal distant vision in more than 90% of patients.

Test Prep for AP® Courses

1

(a)

3

(c)

5

(a)

7

(b)

9

(d)

11

(c)

Chapter 27

Problems & Exercises

1

$1 / 1.333 = 0.750$

3

1.49, Polystyrene

5

0.877 glass to water

6

$0.516°$

8

$1.22×10^{-6}$ m

10

600 nm

12

$2.06°$

14

1200 nm (not visible)

16

(a) 760 nm

(b) 1520 nm

18

For small angles $\sin\theta - \tan\theta \approx \theta$ (in radians).

For two adjacent fringes we have,

$d \sin\theta_m = m\lambda$

and

$d \sin\theta_{m+1} = (m+1)\lambda$

Subtracting these equations gives

$$d(\sin\theta_{m+1} - \sin\theta_m) = [(m+1) - m]\lambda$$
$$d(\theta_{m+1} - \theta_m) = \lambda$$
$$\tan\theta_m = \frac{y_m}{x} \approx \theta_m \Rightarrow d\left(\frac{y_{m+1}}{x} - \frac{y_m}{x}\right) = \lambda$$
$$d\frac{\Delta y}{x} = \lambda \Rightarrow \Delta y = \frac{x\lambda}{d}$$

20

450 nm

21

$5.97°$

23

$8.99×10^3$

25

707 nm

27

(a) $11.8°$, $12.5°$, $14.1°$, $19.2°$

(b) $24.2°$, $25.7°$, $29.1°$, $41.0°$

(c) Decreasing the number of lines per centimeter by a factor of x means that the angle for the x-order maximum is the same as the original angle for the first- order maximum.

29

589.1 nm and 589.6 nm

31

28.7°

33

43.2°

35

90.0°

37

(a) The longest wavelength is 333.3 nm, which is not visible.

(b) 333 nm (UV)

(c) 6.58×10^3 cm

39

1.13×10^{-2} m

41

(a) 42.3 nm

(b) Not a visible wavelength

The number of slits in this diffraction grating is too large. Etching in integrated circuits can be done to a resolution of 50 nm, so slit separations of 400 nm are at the limit of what we can do today. This line spacing is too small to produce diffraction of light.

43

(a) 33.4°

(b) No

45

(a) 1.35×10^{-6} m

(b) 69.9°

47

750 nm

49

(a) 9.04°

(b) 12

51

(a) 0.0150°

(b) 0.262 mm

(c) This distance is not easily measured by human eye, but under a microscope or magnifying glass it is quite easily measurable.

53

(a) 30.1°

(b) 48.7°

(c) No

(d) $2\theta_1 = (2)(14.5°) = 29°$, $\theta_2 - \theta_1 = 30.05° - 14.5° = 15.56°$. Thus, $29° \approx (2)(15.56°) = 31.1°$.

55

23.6° and 53.1°

57

(a) 1.63×10^{-4} rad

(b) 326 ly

59

1.46×10^{-5} rad

61

(a) 3.04×10^{-7} rad

(b) Diameter of 235 m

63

5.15 cm

65

(a) Yes. Should easily be able to discern.

(b) The fact that it is just barely possible to discern that these are separate bodies indicates the severity of atmospheric aberrations.

70

532 nm (green)

72

83.9 nm

74

620 nm (orange)

76

380 nm

78

33.9 nm

80

4.42×10^{-5} m

82

The oil film will appear black, since the reflected light is not in the visible part of the spectrum.

84

45.0°

86

45.7 mW/m^2

88

90.0%

90

I_0

92

48.8°

94

41.2°

96

(a) 1.92, not diamond (Zircon)

(b) 55.2°

98

$B_2 = 0.707\, B_1$

100

(a) 2.07×10^{-2} °C/s

(b) Yes, the polarizing filters get hot because they absorb some of the lost energy from the sunlight.

Test Prep for AP® Courses

1

(b)

3

(b) and (c)

5

(b)

7

(b)

9

(b)

11

(d)

13

(b)

15

(d)

17

(b)

Chapter 28

Problems & Exercises

1

(a) 1.0328

(b) 1.15

3

5.96×10^{-8} s

5

$0.800c$

7

$0.140c$

9

(a) $0.745c$

(b) $0.99995c$ (to five digits to show effect)

11

(a) 0.996

(b) γ cannot be less than 1.

(c) Assumption that time is longer in moving ship is unreasonable.

12

48.6 m

14

(a) 1.387 km = 1.39 km

(b) 0.433 km

(c) $\quad =$

Thus, the distances in parts (a) and (b) are related when $\gamma = 3.20$.

16

(a) 4.303 y (to four digits to show any effect)

(b) 0.1434 y

(c) $\Delta t = \gamma \Delta t_0 \Rightarrow \gamma = \dfrac{\Delta t}{\Delta t_0} = \dfrac{4.303 \text{ y}}{0.1434 \text{ y}} = 30.0$

Thus, the two times are related when $\gamma = 30.00$.

18

(a) 0.250

(b) γ must be ≥ 1

(c) The Earth-bound observer must measure a shorter length, so it is unreasonable to assume a longer length.

20

(a) $0.909c$

(b) $0.400c$

22

$0.198c$

24

a) 658 nm

b) red

c) $v/c = 9.92 \times 10^{-5}$ (negligible)

26

$0.991c$

28

$-0.696c$

30

$0.01324c$

32

$u' = c$, so

$$u = \frac{v+u'}{1 + (vu'/c^2)} = \frac{v+c}{1 + (vc/c^2)} = \frac{v+c}{1 + (v/c)}$$
$$= \frac{c(v+c)}{c+v} = c$$

34

a) $0.99947c$

b) 1.2064×10^{11} y

c) 1.2058×10^{11} y (all to sufficient digits to show effects)

35

4.09×10^{-19} kg $\cdot$ m/s

37

(a) $3.000000015 \times 10^{13}$ kg $\cdot$ m/s .

(b) Ratio of relativistic to classical momenta equals 1.000000005 (extra digits to show small effects)

39

2.9957×10^8 m/s

41

(a) 1.121×10^{-8} m/s

(b) The small speed tells us that the mass of a proton is substantially smaller than that of even a tiny amount of macroscopic matter!

43

8.20×10^{-14} J

0.512 MeV

45

2.3×10^{-30} kg

47

(a) 1.11×10^{27} kg

(b) 5.56×10^{-5}

49

7.1×10^{-3} kg

7.1×10^{-3}

The ratio is greater for hydrogen.

51

208

$0.999988c$

53

6.92×10^5 J

1.54

55

(a) $0.914c$

(b) The rest mass energy of an electron is 0.511 MeV, so the kinetic energy is approximately 150% of the rest mass energy. The electron should be traveling close to the speed of light.

57
90.0 MeV
59

(a)
$E^2 = p^2 c^2 + m^2 c^4 = \gamma^2 m^2 c^4$, so that
$p^2 c^2 = \left(\gamma^2 - 1\right) m^2 c^4$, and therefore
$$\frac{(pc)^2}{\left(mc^2\right)^2} = \gamma^2 - 1$$

(b) yes

61

1.07×10^3

63

6.56×10^{-8} kg

4.37×10^{-10}

65

$0.314c$

$0.99995c$

67

(a) 1.00 kg

(b) This much mass would be measurable, but probably not observable just by looking because it is 0.01% of the total mass.

69

(a) 6.3×10^{11} kg/s

(b) 4.5×10^{10} y

(c) 4.44×10^9 kg

(d) 0.32%

Test Prep for AP® Courses

1

(a)

3

The relativistic Doppler effect takes into account the special relativity concept of time dilation and also does not require a medium of propagation to be used as a point of reference (light does not require a medium for propagation).

5

Relativistic kinetic energy is given as $\mathrm{KE}_{rel} = (\gamma - 1)mc^2$

where $\gamma = \dfrac{1}{\sqrt{1 - \dfrac{v^2}{c^2}}}$

Classical kinetic energy is given as $\mathrm{KE}_{class} = \frac{1}{2}mv^2$

At low velocities $v = 0$, a binomial expansion and subsequent approximation of γ gives:

$\gamma = 1 + \dfrac{1v^2}{2c^2}$ or $\gamma - 1 = \dfrac{1v^2}{2c^2}$

Substituting $\gamma - 1$ in the expression for KE_{rel} gives

$$\mathrm{KE}_{rel} = \left[\dfrac{1v^2}{2c^2}\right]mc^2 = \frac{1}{2}mv^2 = \mathrm{KE}_{class}$$

Hence, relativistic kinetic energy becomes classical kinetic energy when $v \ll c$.

Chapter 29

Problems & Exercises

1

(a) 0.070 eV

(b) 14

3

(a) 2.21×10^{34} J

(b) 2.26×10^{34}

(c) No

4

263 nm

6

3.69 eV

8

0.483 eV

10

2.25 eV

12

(a) 264 nm

(b) Ultraviolet

14

1.95×10^6 m/s

16

(a) 4.02×10^{15} /s

(b) 0.256 mW

18

(a) −1.90 eV

(b) Negative kinetic energy

(c) That the electrons would be knocked free.

20

6.34×10^{-9} eV, 1.01×10^{-27} J

22

2.42×10^{20} Hz

24

$$hc = \left(6.62607 \times 10^{-34} \text{ J} \cdot \text{s}\right)\left(2.99792 \times 10^8 \text{ m/s}\right)\left(\frac{10^9 \text{ nm}}{1 \text{ m}}\right)\left(\frac{1.00000 \text{ eV}}{1.60218 \times 10^{-19} \text{ J}}\right)$$
$$= 1239.84 \text{ eV} \cdot \text{nm}$$
$$\approx 1240 \text{ eV} \cdot \text{nm}$$

26

(a) 0.0829 eV

(b) 121

(c) 1.24 MeV

(d) 1.24×10^5

28

(a) 25.0×10^3 eV

(b) 6.04×10^{18} Hz

30

(a) 2.69

(b) 0.371

32

(a) 1.25×10^{13} photons/s

(b) 997 km

34

8.33×10^{13} photons/s

36

181 km

38

(a) 1.66×10^{-32} kg · m/s

(b) The wavelength of microwave photons is large, so the momentum they carry is very small.

40

(a) 13.3 μm

(b) 9.38×10^{-2} eV

42

(a) 2.65×10^{-28} kg · m/s

(b) 291 m/s

(c) electron 3.86×10^{-26} J, photon 7.96×10^{-20} J, ratio 2.06×10^6

44

(a) 1.32×10^{-13} m

(b) 9.39 MeV

(c) 4.70×10^{-2} MeV

46

$E = \gamma mc^2$ and $P = \gamma mu$, so

$$\frac{E}{P} = \frac{\gamma mc^2}{\gamma mu} = \frac{c^2}{u}.$$

As the mass of particle approaches zero, its velocity u will approach c , so that the ratio of energy to momentum in this limit is

$$\lim_{m \to 0} \frac{E}{P} = \frac{c^2}{c} = c$$

which is consistent with the equation for photon energy.

48

(a) 3.00×10^6 W

(b) Headlights are way too bright.

(c) Force is too large.

49

7.28×10^{-4} m

51

6.62×10^7 m/s

53

1.32×10^{-13} m

55

(a) 6.62×10^7 m/s

(b) 22.9 MeV

57
15.1 keV
59

(a) 5.29 fm

(b) 4.70×10^{-12} J

(c) 29.4 MV

61

(a) 7.28×10^{12} m/s

(b) This is thousands of times the speed of light (an impossibility).

(c) The assumption that the electron is non-relativistic is unreasonable at this wavelength.

62

(a) 57.9 m/s

(b) 9.55×10^{-9} eV

(c) From **Table 29.1**, we see that typical molecular binding energies range from about 1eV to 10 eV, therefore the result in part (b) is approximately 9 orders of magnitude smaller than typical molecular binding energies.

64

29 nm,

290 times greater

66
1.10×10^{-13} eV

68
3.3×10^{-22} s

70
2.66×10^{-46} kg

72
0.395 nm

74

(a) 1.3×10^{-19} J

(b) 2.1×10^{23}

(c) 1.4×10^{2} s

76

(a) 3.35×10^{5} J

(b) 1.12×10^{-3} kg $\cdot$ m/s

(c) 1.12×10^{-3} m/s

(d) 6.23×10^{-7} J

78

(a) 1.06×10^{3}

(b) 5.33×10^{-16} kg $\cdot$ m/s

(c) 1.24×10^{-18} m

80

(a) 1.62×10^{3} m/s

(b) 4.42×10^{-19} J for photon, 1.19×10^{-24} J for electron, photon energy is 3.71×10^{5} times greater

(c) The light is easier to make because 450-nm light is blue light and therefore easy to make. Creating electrons with 7.43 µeV of energy would not be difficult, but would require a vacuum.

81

(a) 2.30×10^{-6} m

(b) 3.20×10^{-12} m

83
3.69×10^{-4} °C

85

(a) 2.00 kJ

(b) 1.33×10^{-5} kg $\cdot$ m/s

(c) 1.33×10^{-5} N

(d) yes

Test Prep for AP® Courses

1
(b)
3

(c)

5

(b)

7

(c)

9

(c)

11

(a)

13

(a)

15

(c)

17

(d)

Chapter 30

Problems & Exercises

1

1.84×10^3

3

50 km

4

$6 \times 10^{20} \text{ kg/m}^3$

6

(a) 10.0 μm

(b) It isn't hard to make one of approximately this size. It would be harder to make it exactly 10.0 μm .

7

$$\frac{1}{\lambda} = R\left(\frac{1}{n_f^2} - \frac{1}{n_i^2}\right) \Rightarrow \lambda = \frac{1}{R}\left[\frac{(n_i \cdot n_f)^2}{n_i^2 - n_f^2}\right]; \; n_i = 2, n_f = 1, \;\; \text{so that}$$

$$\lambda = \left(\frac{\text{m}}{1.097 \times 10^7}\right)\left[\frac{(2 \times 1)^2}{2^2 - 1^2}\right] = 1.22 \times 10^{-7} \text{ m} = 122 \text{ nm} \;\;\text{, which is UV radiation.}$$

9

$$a_B = \frac{h^2}{4\pi^2 m_e kZq_e^2} = \frac{(6.626 \times 10^{-34} \text{ J·s})^2}{4\pi^2 (9.109 \times 10^{-31} \text{ kg})(8.988 \times 10^9 \text{ N·m}^2/\text{C}^2)(1)(1.602 \times 10^{-19} \text{ C})^2} = 0.529 \times 10^{-10} \text{ m}$$

11

0.850 eV

13

$2.12 \times 10^{-10} \text{ m}$

15

365 nm

It is in the ultraviolet.

17

No overlap

365 nm

122 nm

19

7

21

(a) 2

(b) 54.4 eV

23

$\frac{kZq_e^2}{r_n^2} = \frac{m_e V^2}{r_n}$, so that $r_n = \frac{kZq_e^2}{m_e V^2} = \frac{kZq_e^2}{m_e} \frac{1}{V^2}$. From the equation $m_e v r_n = n\frac{h}{2\pi}$, we can substitute for the

velocity, giving: $r_n = \frac{kZq_e^2}{m_e} \cdot \frac{4\pi^2 m_e^2 r_n^2}{n^2 h^2}$ so that $r_n = \frac{n^2}{Z} \frac{h^2}{4\pi^2 m_e kq_e^2} = \frac{n^2}{Z} a_B$, where $a_B = \frac{h^2}{4\pi^2 m_e kq_e^2}$.

25

(a) 0.248×10^{-10} m

(b) 50.0 keV

(c) The photon energy is simply the applied voltage times the electron charge, so the value of the voltage in volts is the same as the value of the energy in electron volts.

27

(a) 100×10^3 eV, 1.60×10^{-14} J

(b) 0.124×10^{-10} m

29

(a) 8.00 keV

(b) 9.48 keV

30

(a) 1.96 eV

(b) $(1240 \text{ eV} \cdot \text{nm}) / (1.96 \text{ eV}) = 633$ nm

(c) 60.0 nm

32
693 nm
34

(a) 590 nm

(b) $(1240 \text{ eV} \cdot \text{nm}) / (1.17 \text{ eV}) = 1.06$ μm

35
$l = 4, 3$ are possible since $l < n$ and $|m_l| \leq l$.

37
$n = 4 \Rightarrow l = 3, 2, 1, 0 \Rightarrow m_l = \pm 3, \pm 2, \pm 1, 0$ are possible.

39

(a) 1.49×10^{-34} J · s

(b) 1.06×10^{-34} J · s

41

(a) 3.66×10^{-34} J · s

(b) $s = 9.13 \times 10^{-35}$ J · s

(c) $\frac{L}{S} = \frac{\sqrt{12}}{\sqrt{3/4}} = 4$

43
$\theta = 54.7°, 125.3°$

44
(a) 32. (b) 2 in s, 6 in p, 10 in d, and 14 in f, for a total of 32.

46

(a) 2

(b) $3d^9$

48

(b) $n \geq l$ is violated,

(c) cannot have 3 electrons in s subshell since $3 > (2l + 1) = 2$

(d) cannot have 7 electrons in p subshell since $7 > (2l + 1) = 2(2 + 1) = 6$

50

(a) The number of different values of m_l is $\pm l, \pm (l - 1), ...,0$ for each $l > 0$ and one for $l = 0 \Rightarrow (2l + 1)$. Also an overall factor of 2 since each m_l can have m_s equal to either $+1/2$ or $-1/2 \Rightarrow 2(2l + 1)$.

(b) for each value of l, you get $2(2l + 1)$

$$= 0, 1, 2, ...,(n-1) \Rightarrow 2\{[(2)(0) + 1] + [(2)(1) + 1] + ... + [(2)(n - 1) + 1]\} = 2[1 + 3 + ... + (2n - 3) + (2n - 1)] \text{ to see}$$
n terms

that the expression in the box is $= n^2$, imagine taking $(n - 1)$ from the last term and adding it to first term $= 2[1 + (n-1) + 3 + ... + (2n - 3) + (2n - 1)-(n - 1)] = 2[n + 3 + ... + (2n - 3) + n]$. Now take $(n - 3)$ from penultimate term and add to the second term $2[n + n + ... + n + n] = 2n^2$.

n terms

52

The electric force on the electron is up (toward the positively charged plate). The magnetic force is down (by the RHR).

54

401 nm

56

(a) 6.54×10^{-16} kg

(b) 5.54×10^{-7} m

58

1.76×10^{11} C/kg , which agrees with the known value of 1.759×10^{11} C/kg to within the precision of the measurement

60

(a) 2.78 fm

(b) 0.37 of the nuclear radius.

62

(a) 1.34×10^{23}

(b) 2.52 MW

64

(a) 6.42 eV

(b) 7.27×10^{-20} J/molecule

(c) 0.454 eV, 14.1 times less than a single UV photon. Therefore, each photon will evaporate approximately 14 molecules of tissue. This gives the surgeon a rather precise method of removing corneal tissue from the surface of the eye.

66

91.18 nm to 91.22 nm

68

(a) 1.24×10^{11} V

(b) The voltage is extremely large compared with any practical value.

(c) The assumption of such a short wavelength by this method is unreasonable.

Test Prep for AP® Courses

1

(a), (d)

3

(a)

5

(a)

7

(b)

9

(a)

11

(d)

13

(d)

15

(a), (c)

Chapter 31

Problems & Exercises

1

1.67×10^4

5

$$m = \rho V = \rho d^3 \quad \Rightarrow \quad a = \left(\frac{m}{\rho}\right)^{1/3} = \left(\frac{2.3 \times 10^{17} \text{ kg}}{1000 \text{ kg/m}^3}\right)^{\frac{1}{3}}$$

$$= \quad 61 \times 10^3 \text{ m} = 61 \text{ km}$$

7

1.9 fm

9

(a) 4.6 fm

(b) 0.61 to 1

11

85.4 to 1

13

12.4 GeV

15

19.3 to 1

17

$^3_1\text{H}_2 \rightarrow \, ^3_2\text{He}_1 + \beta^- + \bar{\nu}_e$

19

$^{50}_{25}\text{M}_{25} \rightarrow \, ^{50}_{24}\text{Cr}_{26} + \beta^+ + \nu_e$

21

$^7_4\text{Be}_3 + e^- \rightarrow \, ^7_3\text{Li}_4 + \nu_e$

23

$^{210}_{84}\text{Po}_{126} \rightarrow \, ^{206}_{82}\text{Pb}_{124} + \, ^4_2\text{He}_2$

25

$^{137}_{55}\text{Cs}_{82} \rightarrow \, ^{137}_{56}\text{Ba}_{81} + \beta^- + \bar{\nu}_e$

27

$^{232}_{90}\text{Th}_{142} \rightarrow \, ^{228}_{88}\text{Ra}_{140} + \, ^4_2\text{He}_2$

29

(a) charge: $(+1) + (-1) = 0$; electron family number: $(+1) + (-1) = 0$; $A: 0 + 0 = 0$

(b) 0.511 MeV

(c) The two γ rays must travel in exactly opposite directions in order to conserve momentum, since initially there is zero momentum if the center of mass is initially at rest.

31

$Z = (Z + 1) - 1$; $A = A$; efn : $0 = (+1) + (-1)$

33

$Z - 1 = Z - 1$; $A = A$; efn : $(+1) = (+1)$

35

(a) $^{226}_{88}\text{Ra}_{138} \rightarrow\, ^{222}_{86}\text{Rn}_{136} +\, ^{4}_{2}\text{He}_2$

(b) 4.87 MeV

37

(a) $\text{n} \rightarrow \text{p} + \beta^- + \bar{\nu}_e$

(b)) 0.783 MeV

39

1.82 MeV

41

(a) 4.274 MeV

(b) 1.927×10^{-5}

(c) Since U-238 is a slowly decaying substance, only a very small number of nuclei decay on human timescales; therefore, although those nuclei that decay lose a noticeable fraction of their mass, the change in the total mass of the sample is not detectable for a macroscopic sample.

43

(a) $^{15}_{8}\text{O}_7 + e^- \rightarrow\, ^{15}_{7}\text{N}_8 + \nu_e$

(b) 2.754 MeV

44

57,300 y

46

(a) 0.988 Ci

(b) The half-life of ^{226}Ra is now better known.

48

1.22×10^3 Bq

50

(a) 16.0 mg

(b) 0.0114%

52

1.48×10^{17} y

54

5.6×10^4 y

56

2.71 y

58

(a) 1.56 mg

(b) 11.3 Ci

60

(a) 1.23×10^{-3}

(b) Only part of the emitted radiation goes in the direction of the detector. Only a fraction of that causes a response in the detector. Some of the emitted radiation (mostly α particles) is observed within the source. Some is absorbed within the source, some is absorbed by the detector, and some does not penetrate the detector.

62

(a) 1.68×10^{-5} Ci

(b) 8.65×10^{10} J

(c) $ \$2.9 \times 10^3$

64

(a) 6.97×10^{15} Bq

(b) 6.24 kW

(c) 5.67 kW

68

(a) 84.5 Ci

(b) An extremely large activity, many orders of magnitude greater than permitted for home use.

(c) The assumption of $1.00 \, \mu A$ is unreasonably large. Other methods can detect much smaller decay rates.

69
1.112 MeV, consistent with graph
71
7.848 MeV, consistent with graph
73

(a) 7.680 MeV, consistent with graph

(b) 7.520 MeV, consistent with graph. Not significantly different from value for ^{12}C, but sufficiently lower to allow decay into another nuclide that is more tightly bound.

75

(a) 1.46×10^{-8} u vs. 1.007825 u for ^1H

(b) 0.000549 u

(c) 2.66×10^{-5}

76

(a) −9.315 MeV

(b) The negative binding energy implies an unbound system.

(c) This assumption that it is two bound neutrons is incorrect.

78
22.8 cm
79

(a) $^{235}_{92}U_{143} \rightarrow \, ^{231}_{90}Th_{141} + \, ^4_2He_2$

(b) 4.679 MeV
(c) 4.599 MeV

81

a) 2.4×10^8 u

(b) The greatest known atomic masses are about 260. This result found in (a) is extremely large.

(c) The assumed radius is much too large to be reasonable.

82

(a) -1.805 MeV

(b) Negative energy implies energy input is necessary and the reaction cannot be spontaneous.

(c) Although all conversation laws are obeyed, energy must be supplied, so the assumption of spontaneous decay is incorrect.

Test Prep for AP® Courses

1
(c)
3
(a)
5

When $^{241}_{95}\text{Am}$ undergoes α decay, it loses 2 neutrons and 2 protons. The resulting nucleus is therefore $^{237}_{93}\text{Np}$.

7
During this process, the nucleus emits a particle with -1 charge. In order for the overall charge of the system to remain constant, the charge of the nucleus must therefore increase by +1.

9
  a. No. Nucleon number is conserved (238 = 234 + 4), but the atomic number or charge is NOT conserved (92 ≠ 88+2).

  b. Yes. Nucleon number is conserved (223 = 209 + 14), and atomic number is conserved (88 = 82 + 6).

  c. Yes. Nucleon number is conserved (14 = 14), and charge is conserved if the electron's charge is properly counted (6 = 7 + (-1)).

  d. No. Nucleon number is not conserved (24 ≠ 23). The positron released counts as a charge to conserve charge, but it doesn't count as a nucleon.

11

This must be alpha decay since 4 nucleons (2 positive charges) are lost from the parent nucleus. The number remaining is found from:

$$\text{N}(t) = \text{N}_0\, e\!\left(\frac{-0.693t}{t_{\frac{1}{2}}}\right) = 3.4{\times}10^{17}\, e\!\left(\frac{-(0.693)(0.035)}{0.00173}\right)$$

$$\text{N}(t) = 4.1{\times}10^{11}\ \text{nuclei}$$

Chapter 32

Problems & Exercises

1
5.701 MeV
3

$$^{99}_{42}\text{Mo}_{57} \rightarrow\ ^{99}_{43}\text{Tc}_{56} + \beta^- + \bar{v}_e$$

5

$1.43{\times}10^{-9}$ g

7

(a) 6.958 MeV

(b) $5.7{\times}10^{-10}$ g

8

(a) 100 mSv

(b) 80 mSv

(c) ~30 mSv

10
~2 Gy

12
1.69 mm

14
1.24 MeV

16
7.44×10^8

18
4.92×10^{-4} Sv

20
4.43 g

22
0.010 g

24
95%

26

(a) $A=1+1=2$, $Z=1+1=1+1$, efn $= 0 = -1 + 1$

(b) $A=1+2=3$, $Z=1+1=2$, efn$=0=0$

(c) $A=3+3=4+1+1$, $Z=2+2=2+1+1$, efn$=0=0$

28

$$
\begin{aligned}
E &= (m_i - m_f)c^2 \\
&= \left[4m(^1\text{H}) - m(^4\text{He})\right]c^2 \\
&= [4(1.007825) - 4.002603](931.5 \text{ MeV}) \\
&= 26.73 \text{ MeV}
\end{aligned}
$$

30
3.12×10^5 kg (about 200 tons)

32

$$
\begin{aligned}
E &= (m_i - m_f)c^2 \\
E_1 &= (1.008665 + 3.016030 - 4.002603)(931.5 \text{ MeV}) \\
&= 20.58 \text{ MeV} \\
E_2 &= (1.008665 + 1.007825 - 2.014102)(931.5 \text{ MeV}) \\
&= 2.224 \text{ MeV}
\end{aligned}
$$

^4He is more tightly bound, since this reaction gives off mo e energy per nucleon.

34
1.19×10^4 kg

36
$2e^- + 4\,^1\text{H} \rightarrow\ ^4\text{He} + 7\gamma + 2\nu_e$

38

(a) $A=12+1=13$, $Z=6+1=7$, efn $= 0 = 0$

(b) $A=13=13$, $Z=7=6+1$, efn $= 0 = -1 + 1$

(c) $A=13 + 1=14$, $Z=6+1=7$, efn $= 0 = 0$

(d) $A=14 + 1=15$, $Z=7+1=8$, efn $= 0 = 0$

(e) $A=15=15$, $Z=8=7+1$, efn $= 0 = -1 + 1$

(f) $A=15 + 1=12 + 4$, $Z=7+1=6 + 2$, efn $= 0 = 0$

40

$E_\gamma = 20.6$ MeV

$E_{4_{He}} = 5.68 \times 10^{-2}$ MeV

42

(a) 3×10^9 y

(b) This is approximately half the lifetime of the Earth.

43

(a) 177.1 MeV

(b) Because the gain of an external neutron yields about 6 MeV, which is the average BE/A for heavy nuclei.

(c) $A = 1 + 238 = 96 + 140 + 1 + 1 + 1, Z = 92 = 38 + 53$, efn $= 0 = 0$

45

(a) 180.6 MeV

(b) $A = 1 + 239 = 96 + 140 + 1 + 1 + 1 + 1, Z = 94 = 38 + 56$, efn $= 0 = 0$

47

$^{238}U + n \rightarrow \ ^{239}U + \gamma$ 4.81 MeV

$^{239}U \rightarrow \ ^{239}Np + \beta^- + \nu_e$ 0.753 MeV

$Np \rightarrow Pu + \beta^- + \nu_e$ 0.211 MeV

49

(a) 2.57×10^3 MW

(b) 8.03×10^{19} fission/

(c) 991 kg

51
0.56 g
53
4.781 MeV
55

(a) Blast yields 2.1×10^{12} J to 8.4×10^{11} J, or 2.5 to 1, conventional to radiation enhanced.

(b) Prompt radiation yields 6.3×10^{11} J to 2.1×10^{11} J, or 3 to 1, radiation enhanced to conventional.

57

(a) 1.1×10^{25} fission , 4.4 kg

(b) 3.2×10^{26} fusions , 2.7 kg

(c) The nuclear fuel totals only 6 kg, so it is quite reasonable that some missiles carry 10 overheads. The mass of the fuel would only be 60 kg and therefore the mass of the 10 warheads, weighing about 10 times the nuclear fuel, would be only 1500 lbs. If the fuel for the missiles weighs 5 times the total weight of the warheads, the missile would weigh about 9000 lbs or 4.5 tons. This is not an unreasonable weight for a missile.

59
7×10^4 g
61

(a) 4.86×10^9 W

(b) 11.0 y

Test Prep for AP® Courses

1

(b)

3

(c)

5

(d)

7

(d)

9

(b)

Chapter 33

Problems & Exercises

1

3×10^{-39} s

3

1.99×10^{-16} m (0.2 fm)

4

(a) 10^{-11} to 1, weak to EM

(b) 1 to 1

6

(a) 2.09×10^{-5} s

(b) 4.77×10^4 Hz

8

78.0 cm

10

1.40×10^6

12

100 GeV

13

67.5 MeV

15

(a) 1×10^{14}

(b) 2×10^{17}

17

(a) 1671 MeV

(b) $Q = 1, Q' = 1 + 0 + 0 = 1.\ L_\tau = -1;\ L'\tau = -1;\ L\mu = 0;\ L'\mu = -1 + 1 = 0$

(c) $\tau^- \rightarrow \mu^- + \nu_\mu + \bar{\nu}_\tau$

$\Rightarrow \mu^-$ antiparticle of μ^+; ν_μ of $\bar{\nu}_\mu$; $\bar{\nu}_\tau$ of ν_τ

19

(a) 3.9 eV

(b) 2.9×10^{-8}

21

(a) The uud composition is the same as for a proton.

(b) 3.3×10^{-24} s

(c) Strong (short lifetime)

23

a) $\Delta^{++}(uuu)$; $B = \frac{1}{3} + \frac{1}{3} + \frac{1}{3} = 1$

b)

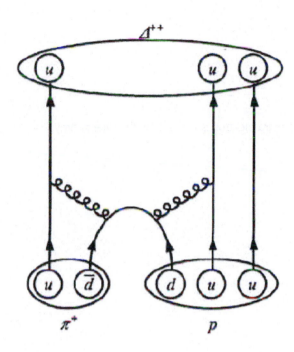

Figure 33.20.

25

(a) $+1$

(b) $B = 1 = 1 + 0$, $Z = = 0 + (-1)$, all lepton numbers are 0 before and after

(c) $(sss) \rightarrow (uds) + (\bar{u}s)$

27

(a) $(u\bar{u} + d\bar{d}) \rightarrow (u\bar{u} + d\bar{d}) + (u\bar{u} + d\bar{d})$

(b) 277.9 MeV

(c) 547.9 MeV

29

No. Charge $= -1$ is conserved. $L_{e_i} = 0 \neq L_{e_f} = 2$ is not conserved. $L_\mu = 1$ is conserved.

31

(a)Yes. $Z = -1 = 0 + (-1)$, $B = 1 = 1 + 0$, all lepton family numbers are 0 before and after, spontaneous since mass greater before reaction.

(b) $dds \rightarrow udd + \bar{u}d$

33

(a) 216

(b) There are more baryons observed because we have the 6 antiquarks and various mixtures of quarks (as for the π-meson) as well.

35

$\Omega^+ (\bar{s}\,\bar{s}\,\bar{s})$

$B = -\frac{1}{3} - \frac{1}{3} - \frac{1}{3} = -1,$

$L_e, \mu, \tau = 0 + 0 + 0 = 0,$

$Q = \frac{1}{3} + \frac{1}{3} + \frac{1}{3} = 1,$

$S = 1 + 1 + 1 = 3.$

37

(a)803 MeV

(b) 938.8 MeV

(c) The annihilation energy of an extra electron is included in the total energy.

39

$\bar{c}\,d$

41

a)The antiproton

b) $\bar{p} \rightarrow \pi^0 + e^-$

43

(a) 5×10^{10}

(b) 5×10^4 particles/m^2

45

2.5×10^{-17} m

47

(a) 33.9 MeV

(b) Muon antineutrino 29.8 MeV, muon 4.1 MeV (kinetic energy)

49

(a) 7.2×10^5 kg

(b) 7.2×10^2 m^3

(c) 100 months

Test Prep for AP® Courses

1
(d)
3
(d)
5
(b)
7
(a)
9
(c), though this comes from Einstein's special relativity
11
(a)
13
(d)
15
(b)
17
(b)

Chapter 34

Problems & Exercises

1

3×10^{41} kg

3

(a) 3×10^{52} kg

(b) 2×10^{79}

(c) 4×10^{88}

5

0.30 Gly

7

(a) 2.0×10^5 km/s

(b) $0.67c$

9

2.7×10^5 m/s

11

6×10^{-11} (an overestimate, since some of the light from Andromeda is blocked by gas and dust within that galaxy)

13

(a) 2×10^{-8} kg

(b) 1×10^{19}

15

(a) $30 \text{km/s} \cdot \text{Mly}$

(b) $15 \text{km/s} \cdot \text{Mly}$

17

960 rev/s

19

$89.999773°$ (many digits are used to show the difference between $90°$)

22

23.6 km

24

(a) 2.95×10^{12} m

(b) 3.12×10^{-4} ly

26

(a) 1×10^{20}

(b) 10 times greater

27

1.5×10^{15}

29

0.6 m^{-3}

31

$0.30 \, \Omega$

Some items in the index may be found in Chapters 1-17, which are not in this book.

The page numbers for Chapters 18-34 are the same as in the original version of the book.
The page numbers for the Appendices and Index have been recalculated to reflect the removal of the Answers pages for Chapters 1-17.

Table of Contents

PREFACE

Welcome to *College Physics for AP® Courses*, an OpenStax resource. This textbook was written to increase student access to high-quality learning materials, maintaining highest standards of academic rigor at little to no cost.

About OpenStax

OpenStax is a nonprofit based at Rice University, and it's our mission to improve student access to education. Our first openly

licensed college textbook was published in 2012, and our library has since scaled to over 25 books for college and AP® courses used by hundreds of thousands of students. Our adaptive learning technology, designed to improve learning outcomes through personalized educational paths, is being piloted in college courses throughout the country. Through our partnerships with philanthropic foundations and our alliance with other educational resource organizations, OpenStax is breaking down the most common barriers to learning and empowering students and instructors to succeed.

About OpenStax Resources

Customization

College Physics for AP® Courses is licensed under a Creative Commons Attribution 4.0 International (CC BY) license, which means that you can distribute, remix, and build upon the content, as long as you provide attribution to OpenStax and its content contributors.

Because our books are openly licensed, you are free to use the entire book or pick and choose the sections that are most relevant to the needs of your course. Feel free to remix the content by assigning your students certain chapters and sections in your syllabus, in the order that you prefer. You can even provide a direct link in your syllabus to the sections in the web view of your book.

Instructors also have the option of creating a customized version of their OpenStax book through the OpenStax Custom platform. The custom version can be made available to students in low-cost print or digital form through their campus bookstore. Visit your book page on openstax.org for a link to your book on OpenStax Custom.

Errata

All OpenStax textbooks undergo a rigorous review process. However, like any professional-grade textbook, errors sometimes occur. Since our books are web based, we can make updates periodically when deemed pedagogically necessary. If you have a correction to suggest, submit it through the link on your book page on openstax.org. Subject matter experts review all errata suggestions. OpenStax is committed to remaining transparent about all updates, so you will also find a list of past errata changes on your book page on openstax.org.

Format

You can access this textbook for free in web view or PDF through openstax.org, and in low-cost print and iBooks editions.

About *College Physics for AP® Courses*

College Physics for AP® Courses is designed to engage students in their exploration of physics and help them apply these concepts to the Advanced Placement® test. Because physics is integral to modern technology and other sciences, the book also includes content that goes beyond the scope of the AP® course to further student understanding. The AP® Connection in each chapter directs students to the material they should focus on for the AP® exam, and what content — although interesting — is not necessarily part of the AP® curriculum. This book is Learning List-approved for AP® Physics courses.

Coverage, Scope, and Alignment to the AP® Curriculum

The current AP® Physics curriculum framework outlines the two full-year physics courses AP® Physics 1: Algebra-Based and AP® Physics 2: Algebra-Based. These two courses replaced the one-year AP® Physics B course, which over the years had become a fast-paced survey of physics facts and formulas that did not provide in-depth conceptual understanding of major physics ideas and the connections between them.

AP® Physics 1 and 2 courses focus on the big ideas typically included in the first and second semesters of an algebra-based, introductory college-level physics course, providing students with the essential knowledge and skills required to support future advanced course work in physics. The AP® Physics 1 curriculum includes mechanics, mechanical waves, sound, and

electrostatics. The AP® Physics 2 curriculum focuses on thermodynamics, fluid statics, dynamics, electromagnetism, geometric and physical optics, quantum physics, atomic physics, and nuclear physics. Seven unifying themes of physics called the Big Ideas each include three to seven enduring understandings (EU), which are themselves composed of essential knowledge (EK) that provides details and context for students as they explore physics.

AP® science practices emphasize inquiry-based learning and development of critical thinking and reasoning skills. Inquiry usually uses a series of steps to gain new knowledge, beginning with an observation and following with a hypothesis to explain the observation; then experiments are conducted to test the hypothesis, gather results, and draw conclusions from data. The AP$^®$ framework has identified seven major science practices, which can be described by short phrases: using representations and models to communicate information and solve problems; using mathematics appropriately; engaging in questioning; planning and implementing data collection strategies; analyzing and evaluating data; justifying scientific explanations; and connecting concepts. The framework's Learning Objectives merge content (EU and EK) with one or more of the seven science practices that students should develop as they prepare for the AP$^®$ Physics exam.

College Physics for AP$^®$ Courses is based on the OpenStax *College Physics* text, adapted to focus on the AP curriculum's concepts and practices. Each chapter of OpenStax *College Physics for AP$^®$ Courses* begins with a *Connection for AP$^®$ Courses* introduction that explains how the content in the chapter sections align to the Big Ideas, enduring understandings, and essential knowledge in the AP® framework. This textbook contains a wealth of information, and the *Connection for AP$^®$ Courses* sections will help you distill the required AP$^®$ content from material that, although interesting, exceeds the scope of an introductory-level course.

Each section opens with the program's learning objectives as well as the AP$^®$ learning objectives and science practices addressed. We have also developed *Real World Connections* features and *Applying the Science Practices* features that highlight concepts, examples, and practices in the framework.

- 1 Introduction: The Nature of Science and Physics
- 2 Kinematics
- 3 Two-Dimensional Kinematics
- 4 Dynamics: Force and Newton's Laws of Motion
- 5 Further Applications of Newton's Laws: Friction, Drag, and Elasticity
- 6 Gravitation and Uniform Circular Motion
- 7 Work, Energy, and Energy Resources
- 8 Linear Momentum and Collisions
- 9 Statics and Torque
- 10 Rotational Motion and Angular Momentum
- 11 Fluid Statics
- 12 Fluid Dynamics and Its Biological and Medical Applications
- 13 Temperature, Kinetic Theory, and the Gas Laws
- 14 Heat and Heat Transfer Methods
- 15 Thermodynamics
- 16 Oscillatory Motion and Waves
- 17 Physics of Hearing
- 18 Electric Charge and Electric Field
- 19 Electric Potential and Electric Field
- 20 Electric Current, Resistance, and Ohm's Law
- 21 Circuits, Bioelectricity, and DC Instruments
- 22 Magnetism
- 23 Electromagnetic Induction, AC Circuits, and Electrical Technologies
- 24 Electromagnetic Waves
- 25 Geometric Optics
- 26 Vision and Optical Instruments
- 27 Wave Optics
- 28 Special Relativity
- 29 Introduction to Quantum Physics
- 30 Atomic Physics
- 31 Radioactivity and Nuclear Physics
- 32 Medical Applications of Nuclear Physics
- 33 Particle Physics
- 34 Frontiers of Physics
- Appendix A: Atomic Masses
- Appendix B: Selected Radioactive Isotopes
- Appendix C: Useful Information
- Appendix D: Glossary of Key Symbols and Notation

Pedagogical Foundation and Features

College Physics for AP$^®$ Courses is organized so that topics are introduced conceptually with a steady progression to precise definitions and analytical applications. The analytical, problem-solving aspect is tied back to the conceptual before moving on to another topic. Each introductory chapter, for example, opens with an engaging photograph relevant to the subject of the chapter and interesting applications that are easy for most students to visualize.

- **Connections for AP$^®$ Courses** introduce each chapter and explain how its content addresses the AP$^®$ curriculum.
- **Worked Examples** Examples start with problems based on real-life situations, then describe a strategy for solving the problem that emphasizes key concepts. The subsequent detailed mathematical solution also includes a follow-up

discussion.

- **Problem-solving Strategies** are presented independently and subsequently appear at crucial points in the text where students can benefit most from them.
- **Misconception Alerts** address common misconceptions that students may bring to class.
- **Take-Home Investigations** provide the opportunity for students to apply or explore what they have learned with a hands-on activity.
- **Real World Connections** highlight important concepts and examples in the AP® framework.
- **Applying the Science Practices** includes activities and challenging questions that engage students while they apply the AP® science practices.
- **Things Great and Small** explains macroscopic phenomena (such as air pressure) with submicroscopic phenomena (such as atoms bouncing off of walls).
- **PhET Explorations** link students to interactive PHeT physics simulations, developed by the University of Colorado, to help them further explore the physics concepts they have learned about in their book module.

Assessment

College Physics for AP® Courses offers a wealth of assessment options, including the following end-of-module problems:

- **Integrated Concept Problems** challenge students to apply both conceptual knowledge and skills to solve a problem.
- **Unreasonable Results** encourage students to solve a problem and then evaluate why the premise or answer to the problem are unrealistic.
- **Construct Your Own Problem** requires students to construct how to solve a particular problem, justify their starting assumptions, show their steps to find the solution to the problem, and finally discuss the meaning of the result.
- **Test Prep for AP® Courses** includes assessment items with the format and rigor found in the AP® exam to help prepare students for the exam.

AP Physics Collection

College Physics for AP® Courses is a part of the AP Physics Collection. The AP Physics Collection is a free, turnkey solution for your AP® Physics course, brought to you through a collaboration between OpenStax and Rice Online Learning. The integrated collection pairs the OpenStax College Physics for AP® Courses text with Concept Trailer videos, instructional videos, problem solution videos, and a correlation guide to help you align all of your content. The instructional videos and problem solution videos were developed by Rice Professor Jason Hafner and AP® Physics teachers Gigi Nevils-Noe and Matt Wilson through Rice Online Learning. You can access all of this free material through the College Physics for AP® Courses page on openstax.org.

Additional Resources

Student and Instructor Resources

We've compiled additional resources for both students and instructors, including Getting Started Guides, an instructor solution manual, and instructional videos. Instructor resources require a verified instructor account, which you can apply for when you log in or create your account on openstax.org. Take advantage of these resources to supplement your OpenStax book.

Partner Resources

OpenStax Partners are our allies in the mission to make high-quality learning materials affordable and accessible to students and instructors everywhere. Their tools integrate seamlessly with our OpenStax titles at a low cost. To access the partner resources for your text, visit your book page on openstax.org.

About the Authors

Senior Contributing Authors

Irina Lyublinskaya, CUNY College of Staten Island
Gregg Wolfe, Avonworth High School
Douglas Ingram, Trinity Christian University
Liza Pujji, Manukau Institute of Technology, New Zealand
Sudhi Oberoi, Visiting Research Student, QuIC Lab, Raman Research Institute, India
Nathan Czuba, Sabio Academy
Julie Kretchman, Science Writer, BS, University of Toronto
John Stoke, Science Writer, MS, University of Chicago
David Anderson, Science Writer, PhD, College of William and Mary
Erika Gasper, Science Writer, MA, University of California, Santa Cruz

Advanced Placement Teacher Reviewers

John Boehringer, Prosper High School
Victor Brazil, Petaluma High School
Michelle Burgess, Avon Lake High School
Bryan Callow, Lindenwold High School
Brian Hastings, Spring Grove Area School District

Alexander Lavy, Xavier High School
Jerome Mass, Glastonbury Public Schools

Faculty Reviewers

John Aiken, Georgia Institute of Technology
Robert Arts, University of Pikeville
Anand Batra, Howard University
Michael Ottinger, Missouri Western State University
James Smith, Caldwell University
Ulrich Zurcher, Cleveland State University

To the AP® Physics Student

The fundamental goal of physics is to discover and understand the "laws" that govern observed phenomena in the world around us. Why study physics? If you plan to become a physicist, the answer is obvious—introductory physics provides the foundation for your career; or if you want to become an engineer, physics provides the basis for the engineering principles used to solve applied and practical problems. For example, after the discovery of the photoelectric effect by physicists, engineers developed photocells that are used in solar panels to convert sunlight to electricity. What if you are an aspiring medical doctor? Although the applications of the laws of physics may not be obvious, their understanding is tremendously valuable. Physics is involved in medical diagnostics, such as x-rays, magnetic resonance imaging (MRI), and ultrasonic blood flow measurements. Medical therapy sometimes directly involves physics; cancer radiotherapy uses ionizing radiation. What if you are planning a nonscience career? Learning physics provides you with a well-rounded education and the ability to make important decisions, such as evaluating the pros and cons of energy production sources or voting on decisions about nuclear waste disposal.

This AP® Physics 1 course begins with **kinematics,** the study of motion without considering its causes. Motion is everywhere: from the vibration of atoms to the planetary revolutions around the Sun. Understanding motion is key to understanding other concepts in physics. You will then study **dynamics,** which considers the forces that affect the motion of moving objects and systems. Newton's laws of motion are the foundation of dynamics. These laws provide an example of the breadth and simplicity of the principles under which nature functions. One of the most remarkable simplifications in physics is that only four distinct forces account for all known phenomena. Your journey will continue as you learn about energy. Energy plays an essential role both in everyday events and in scientific phenomena. You can likely name many forms of energy, from that provided by our foods, to the energy we use to run our cars, to the sunlight that warms us on the beach. The next stop is learning about oscillatory motion and waves. All oscillations involve force and energy: you push a child in a swing to get the motion started and you put energy into a guitar string when you pluck it. Some oscillations create waves. For example, a guitar creates sound waves. You will conclude this first physics course with the study of static electricity and electric currents. Many of the characteristics of static electricity can be explored by rubbing things together. Rubbing creates the spark you get from walking across a wool carpet, for example. Similarly, lightning results from air movements under certain weather conditions.

In the AP® Physics 2 course, you will continue your journey by studying **fluid dynamics,** which explains why rising smoke curls and twists and how the body regulates blood flow. The next stop is **thermodynamics,** the study of heat transfer—energy in transit—that can be used to do work. Basic physical laws govern how heat transfers and its efficiency. Then you will learn more about electric phenomena as you delve into **electromagnetism.** An electric current produces a magnetic field; similarly, a magnetic field produces a current. This phenomenon, known as **magnetic induction,** is essential to our technological society. The generators in cars and nuclear plants use magnetism to generate a current. Other devices that use magnetism to induce currents include pickup coils in electric guitars, transformers of every size, certain microphones, airport security gates, and damping mechanisms on sensitive chemical balances. From electromagnetism you will continue your journey to **optics,** the study of light. You already know that visible light is the type of electromagnetic waves to which our eyes respond. Through vision, light can evoke deep emotions, such as when we view a magnificent sunset or glimpse a rainbow breaking through the clouds. Optics is concerned with the generation and propagation of light. The **quantum mechanics, atomic physics,** and **nuclear physics** are at the end of your journey. These areas of physics have been developed at the end of the 19th and early 20th centuries and deal with submicroscopic objects. Because these objects are smaller than we can observe directly with our senses and generally must be observed with the aid of instruments, parts of these physics areas may seem foreign and bizarre to you at first. However, we have experimentally confirmed most of the ideas in these areas of physics.

AP® Physics is a challenging course. After all, you are taking physics at the introductory college level. You will discover that some concepts are more difficult to understand than others; most students, for example, struggle to understand rotational motion and angular momentum or particle-wave duality. The AP® curriculum promotes depth of understanding over breadth of content, and to make your exploration of topics more manageable, concepts are organized around seven major themes called the **Big Ideas** that apply to all levels of physical systems and interactions between them (see web diagram below). Each Big Idea identifies **enduring understandings** (EU), **essential knowledge** (EK), and **illustrative examples** that support key concepts and content. Simple descriptions define the focus of each Big Idea.

- Big Idea 1: Objects and systems have properties.
- Big Idea 2: Fields explain interactions.
- Big Idea 3: The interactions are described by forces.
- Big Idea 4: Interactions result in changes.
- Big Idea 5: Changes are constrained by conservation laws.
- Big Idea 6: Waves can transfer energy and momentum.
- Big Idea 7: The mathematics of probability can to describe the behavior of complex and quantum mechanical systems.

Doing college work is not easy, but completion of AP® classes is a reliable predictor of college success and prepares you for subsequent courses. The more you engage in the subject, the easier your journey through the curriculum will be. Bring your enthusiasm to class every day along with your notebook, pencil, and calculator. Prepare for class the day before, and review

concepts daily. Form a peer study group and ask your teacher for extra help if necessary. The AP® lab program focuses on more open-ended, student-directed, and inquiry-based lab investigations designed to make you think, ask questions, and analyze data like scientists. You will develop critical thinking and reasoning skills and apply different means of communicating information. By

the time you sit for the AP® exam in May, you will be fluent in the language of physics; because you have been doing real science, you will be ready to show what you have learned. Along the way, you will find the study of the world around us to be one of the most relevant and enjoyable experiences of your high school career.

Irina Lyublinskaya, PhD
Professor of Science Education

To the AP® Physics Teacher

The AP® curriculum was designed to allow instructors flexibility in their approach to teaching the physics courses. *College Physics for AP® Courses* helps you orient students as they delve deeper into the world of physics. Each chapter includes a Connection for AP® Courses introduction that describes the AP® Physics Big Ideas, enduring understandings, and essential knowledge addressed in that chapter.

Each section starts with specific AP® learning objectives and includes essential concepts, illustrative examples, and science practices, along with suggestions for applying the learning objectives through take-home experiments, virtual lab investigations, and activities and questions for preparation and review. At the end of each section, students will find the Test Prep for AP® courses with multiple-choice and open-response questions addressing AP® learning objectives to help them prepare for the AP® exam.

College Physics for AP® Courses has been written to engage students in their exploration of physics and help them relate what they learn in the classroom to their lives outside of it. Physics underlies much of what is happening today in other sciences and in technology. Thus, the book content includes interesting facts and ideas that go beyond the scope of the AP® course. The AP® Connection in each chapter directs students to the material they should focus on for the AP® exam, and what content—although interesting—is not part of the AP® curriculum. Physics is a beautiful and fascinating science. It is in your hands to engage and inspire your students to dive into an amazing world of physics, so they can enjoy it beyond just preparation for the AP® exam.

Irina Lyublinskaya, PhD
Professor of Science Education

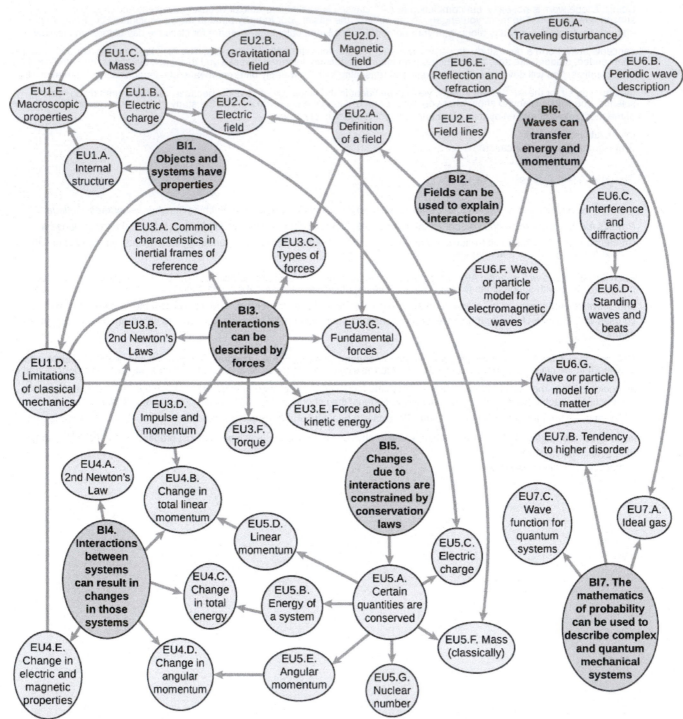

The concept map showing major links between Big Ideas and Enduring Understandings is provided below for visual reference.

18 ELECTRIC CHARGE AND ELECTRIC FIELD

Figure 18.1 Static electricity from this plastic slide causes the child's hair to stand on end. The sliding motion stripped electrons away from the child's body, leaving an excess of positive charges, which repel each other along each strand of hair. (credit: Ken Bosma/Wikimedia Commons)

Chapter Outline

18.1. **Static Electricity and Charge: Conservation of Charge**

18.2. **Conductors and Insulators**

18.3. **Conductors and Electric Fields in Static Equilibrium**

18.4. **Coulomb's Law**

18.5. **Electric Field: Concept of a Field Revisited**

18.6. **Electric Field Lines: Multiple Charges**

18.7. **Electric Forces in Biology**

18.8. **Applications of Electrostatics**

Connection for AP® Courses

The image of American politician and scientist Benjamin Franklin (1706–1790) flying a kite in a thunderstorm (shown in Figure 18.2) is familiar to every schoolchild. In this experiment, Franklin demonstrated a connection between lightning and static electricity. Sparks were drawn from a key hung on a kite string during an electrical storm. These sparks were like those produced by static electricity, such as the spark that jumps from your finger to a metal doorknob after you walk across a wool carpet. Much has been written about Franklin. His experiments were only part of the life of a man who was a scientist, inventor, revolutionary, statesman, and writer. Franklin's experiments were not performed in isolation, nor were they the only ones to reveal connections.

Figure 18.2 Benjamin Franklin, his kite, and electricity.

When Benjamin Franklin demonstrated that lightning was related to static electricity, he made a connection that is now part of the evidence that all directly experienced forces (except gravitational force) are manifestations of the electromagnetic force. For example, the Italian scientist Luigi Galvani (1737-1798) performed a series of experiments in which static electricity was used to stimulate contractions of leg muscles of dead frogs, an effect already known in humans subjected to static discharges. But Galvani also found that if he joined one end of two metal wires (say copper and zinc) and touched the other ends of the wires to muscles; he produced the same effect in frogs as static discharge. Alessandro Volta (1745-1827), partly inspired by Galvani's work, experimented with various combinations of metals and developed the battery.

During the same era, other scientists made progress in discovering fundamental connections. The periodic table was developed as systematic properties of the elements were discovered. This influenced the development and refinement of the concept of atoms as the basis of matter. Such submicroscopic descriptions of matter also help explain a great deal more. Atomic and molecular interactions, such as the forces of friction, cohesion, and adhesion, are now known to be manifestations of the electromagnetic force.

Static electricity is just one aspect of the electromagnetic force, which also includes moving electricity and magnetism. All the macroscopic forces that we experience directly, such as the sensations of touch and the tension in a rope, are due to the electromagnetic force, one of the four fundamental forces in nature. The gravitational force, another fundamental force, is actually sensed through the electromagnetic interaction of molecules, such as between those in our feet and those on the top of a bathroom scale. (The other two fundamental forces, the strong nuclear force and the weak nuclear force, cannot be sensed on the human scale.)

This chapter begins the study of electromagnetic phenomena at a fundamental level. The next several chapters will cover static electricity, moving electricity, and magnetism – collectively known as electromagnetism. In this chapter, we begin with the study of electric phenomena due to charges that are at least temporarily stationary, called electrostatics, or static electricity.

The chapter introduces several very important concepts of charge, electric force, and electric field, as well as defining the relationships between these concepts. The charge is defined as a property of a system (Big Idea 1) that can affect its interaction with other charged systems (Enduring Understanding 1.B). The law of conservation of electric charge is also discussed (Essential Knowledge 1.B.1). The two kinds of electric charge are defined as positive and negative, providing an explanation for having positively charged, negatively charged, or neutral objects (containing equal quantities of positive and negative charges) (Essential Knowledge 1.B.2). The discrete nature of the electric charge is introduced in this chapter by defining the elementary charge as the smallest observed unit of charge that can be isolated, which is the electron charge (Essential Knowledge 1.B.3). The concepts of a system (having internal structure) and of an object (having no internal structure) are implicitly introduced to explain charges carried by the electron and proton (Enduring Understanding 1.A, Essential Knowledge 1.A.1).

An electric field is caused by the presence of charged objects (Enduring Understanding 2.C) and can be used to explain interactions between electrically charged objects (Big Idea 2). The electric force represents the effect of an electric field on a charge placed in the field. The magnitude and direction of the electric force are defined by the magnitude and direction of the electric field and magnitude and sign of the charge (Essential Knowledge 2.C.1). The magnitude of the electric field is proportional to the net charge of the objects that created that field (Essential Knowledge 2.C.2). For the special case of a spherically symmetric charged object, the electric field outside the object is radial, and its magnitude varies as the inverse square of the radial distance from the center of that object (Essential Knowledge 2.C.3). The chapter provides examples of vector field maps for various charged systems, including point charges, spherically symmetric charge distributions, and uniformly charged parallel plates (Essential Knowledge 2.C.1, Essential Knowledge 2.C.2). For multiple point charges, the chapter explains how to find the vector field map by adding the electric field vectors of each individual object, including the special case of two equal charges having opposite signs, known as an electric dipole (Essential Knowledge 2.C.4). The special case of two oppositely charged parallel plates with uniformly distributed electric charge when the electric field is perpendicular to the plates and is constant in both magnitude and direction is described in detail, providing many opportunities for problem solving and applications (Essential Knowledge 2.C.5).

The idea that interactions can be described by forces is also reinforced in this chapter (Big Idea 3). Like all other forces that you have learned about so far, electric force is a vector that affects the motion according to Newton's laws (Enduring Understanding 3.A). It is clearly stated in the chapter that electric force appears as a result of interactions between two charged objects (Essential Knowledge 3.A.3, Essential Knowledge 3.C.2). At the macroscopic level the electric force is a long-range force (Enduring Understanding 3.C); however, at the microscopic level many contact forces, such as friction, can be explained by interatomic electric forces (Essential Knowledge 3.C.4). This understanding of friction is helpful when considering properties of conductors and insulators and the transfer of charge by conduction.

Interactions between systems can result in changes in those systems (Big Idea 4). In the case of charged systems, such interactions can lead to changes of electric properties (Enduring Understanding 4.E), such as charge distribution (Essential Knowledge 4.E.3). Any changes are governed by conservation laws (Big Idea 5). Depending on whether the system is closed or open, certain quantities of the system remain the same or changes in those quantities are equal to the amount of transfer of this quantity from or to the system (Enduring Understanding 5.A). The electric charge is one of these quantities (Essential Knowledge 5.A.2). Therefore, the electric charge of a system is conserved (Enduring Understanding 5.C) and the exchange of electric charge between objects in a system does not change the total electric charge of the system (Essential Knowledge 5.C.2).

Big Idea 1 Objects and systems have properties such as mass and charge. Systems may have internal structure.

Enduring Understanding 1.A The internal structure of a system determines many properties of the system.

Essential Knowledge 1.A.1 A system is an object or a collection of objects. Objects are treated as having no internal structure.

Enduring Understanding 1.B Electric charge is a property of an object or system that affects its interactions with other objects or systems containing charge.

Essential Knowledge 1.B.1 Electric charge is conserved. The net charge of a system is equal to the sum of the charges of all the objects in the system.

Essential Knowledge 1.B.2 There are only two kinds of electric charge. Neutral objects or systems contain equal quantities of positive and negative charge, with the exception of some fundamental particles that have no electric charge.

Essential Knowledge 1.B.3 The smallest observed unit of charge that can be isolated is the electron charge, also known as the elementary charge.

Big Idea 2 Fields existing in space can be used to explain interactions.

Enduring Understanding 2.C An electric field is caused by an object with electric charge.

Essential Knowledge 2.C.1 The magnitude of the electric force F exerted on an object with electric charge q by an electric field ($\vec{E}$ is $\vec{F} = q\,\vec{E}$. The direction of the force is determined by the direction of the field and the sign of the charge, with positively charged objects accelerating in the direction of the field and negatively charged objects accelerating in the direction opposite the field. This should include a vector field map for positive point charges, negative point charges, spherically symmetric charge distribution, and uniformly charged parallel plates.

Essential Knowledge 2.C.2 The magnitude of the electric field vector is proportional to the net electric charge of the object(s) creating that field. This includes positive point charges, negative point charges, spherically symmetric charge distributions, and uniformly charged parallel plates.

Essential Knowledge 2.C.3 The electric field outside a spherically symmetric charged object is radial, and its magnitude varies as the inverse square of the radial distance from the center of that object. Electric field lines are not in the curriculum. Students will be expected to rely only on the rough intuitive sense underlying field lines, wherein the field is viewed as analogous to something emanating uniformly from a source.

Essential Knowledge 2.C.4 The electric field around dipoles and other systems of electrically charged objects (that can be modeled as point objects) is found by vector addition of the field of each individual object. Electric dipoles are treated qualitatively in this course as a teaching analogy to facilitate student understanding of magnetic dipoles.

Essential Knowledge 2.C.5 Between two oppositely charged parallel plates with uniformly distributed electric charge, at points far from the edges of the plates, the electric field is perpendicular to the plates and is constant in both magnitude and direction.

Big Idea 3 The interactions of an object with other objects can be described by forces.

Enduring Understanding 3.A All forces share certain common characteristics when considered by observers in inertial reference frames.

Essential Knowledge 3.A.3 A force exerted on an object is always due to the interaction of that object with another object.

Enduring Understanding 3.C At the macroscopic level, forces can be categorized as either long-range (action-at-a-distance) forces or contact forces.

Essential Knowledge 3.C.2 Electric force results from the interaction of one object that has an electric charge with another object that has an electric charge.

Essential Knowledge 3.C.4 Contact forces result from the interaction of one object touching another object, and they arise from interatomic electric forces. These forces include tension, friction, normal, spring (Physics 1), and buoyant (Physics 2).

Big Idea 4 Interactions between systems can result in changes in those systems.

Enduring Understanding 4.E The electric and magnetic properties of a system can change in response to the presence of, or changes in, other objects or systems.

Essential Knowledge 4.E.3 The charge distribution in a system can be altered by the effects of electric forces produced by a charged object.

Big Idea 5 Changes that occur as a result of interactions are constrained by conservation laws.

Enduring Understanding 5.A Certain quantities are conserved, in the sense that the changes of those quantities in a given system are always equal to the transfer of that quantity to or from the system by all possible interactions with other systems.

Essential Knowledge 5.A.2 For all systems under all circumstances, energy, charge, linear momentum, and angular momentum are conserved.

Enduring Understanding 5.C The electric charge of a system is conserved.

Essential Knowledge 5.C.2 The exchange of electric charges among a set of objects in a system conserves electric charge.

18.1 Static Electricity and Charge: Conservation of Charge

Learning Objectives

By the end of this section, you will be able to:

- Define electric charge, and describe how the two types of charge interact.
- Describe three common situations that generate static electricity.
- State the law of conservation of charge.

The information presented in this section supports the following AP® learning objectives and science practices:

- **1.B.1.1** The student is able to make claims about natural phenomena based on conservation of electric charge. **(S.P. 6.4)**
- **1.B.1.2** The student is able to make predictions, using the conservation of electric charge, about the sign and relative quantity of net charge of objects or systems after various charging processes, including conservation of charge in simple circuits. **(S.P. 6.4, 7.2)**
- **1.B.2.1** The student is able to construct an explanation of the two-charge model of electric charge based on evidence produced through scientific practices. **(S.P. 6.4)**
- **1.B.3.1** The student is able to challenge the claim that an electric charge smaller than the elementary charge has been isolated. **(S.P. 1.5, 6.1, 7.2)**
- **5.A.2.1** The student is able to define open and closed systems for everyday situations and apply conservation concepts for energy, charge, and linear momentum to those situations. **(S.P. 6.4, 7.2)**
- **5.C.2.1** The student is able to predict electric charges on objects within a system by application of the principle of charge conservation within a system. **(S.P. 6.4)**
- **5.C.2.2** The student is able to design a plan to collect data on the electrical charging of objects and electric charge induction on neutral objects and qualitatively analyze that data. **(S.P. 4.2, 5.1)**
- **5.C.2.3** The student is able to justify the selection of data relevant to an investigation of the electrical charging of objects and electric charge induction on neutral objects. **(S.P. 4.1)**

Figure 18.3 Borneo amber was mined in Sabah, Malaysia, from shale-sandstone-mudstone veins. When a piece of amber is rubbed with a piece of silk, the amber gains more electrons, giving it a net negative charge. At the same time, the silk, having lost electrons, becomes positively charged. (credit: Sebakoamber, Wikimedia Commons)

What makes plastic wrap cling? Static electricity. Not only are applications of static electricity common these days, its existence has been known since ancient times. The first record of its effects dates to ancient Greeks who noted more than 500 years B.C. that polishing amber temporarily enabled it to attract bits of straw (see **Figure 18.3**). The very word *electric* derives from the Greek word for amber (*electron*).

Many of the characteristics of static electricity can be explored by rubbing things together. Rubbing creates the spark you get from walking across a wool carpet, for example. Static cling generated in a clothes dryer and the attraction of straw to recently polished amber also result from rubbing. Similarly, lightning results from air movements under certain weather conditions. You can also rub a balloon on your hair, and the static electricity created can then make the balloon cling to a wall. We also have to be cautious of static electricity, especially in dry climates. When we pump gasoline, we are warned to discharge ourselves (after sliding across the seat) on a metal surface before grabbing the gas nozzle. Attendants in hospital operating rooms must wear booties with aluminum foil on the bottoms to avoid creating sparks which may ignite the oxygen being used.

Some of the most basic characteristics of static electricity include:

- The effects of static electricity are explained by a physical quantity not previously introduced, called electric charge.
- There are only two types of charge, one called positive and the other called negative.
- Like charges repel, whereas unlike charges attract.
- The force between charges decreases with distance.

How do we know there are two types of **electric charge**? When various materials are rubbed together in controlled ways, certain combinations of materials always produce one type of charge on one material and the opposite type on the other. By convention, we call one type of charge "positive", and the other type "negative." For example, when glass is rubbed with silk, the glass becomes positively charged and the silk negatively charged. Since the glass and silk have opposite charges, they attract one another like clothes that have rubbed together in a dryer. Two glass rods rubbed with silk in this manner will repel one another, since each rod has positive charge on it. Similarly, two silk cloths so rubbed will repel, since both cloths have negative charge. **Figure 18.4** shows how these simple materials can be used to explore the nature of the force between charges.

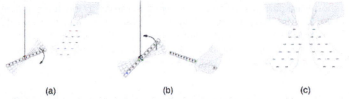

(a)   (b)   (c)

Figure 18.4 A glass rod becomes positively charged when rubbed with silk, while the silk becomes negatively charged. (a) The glass rod is attracted to the silk because their charges are opposite. (b) Two similarly charged glass rods repel. (c) Two similarly charged silk cloths repel.

More sophisticated questions arise. Where do these charges come from? Can you create or destroy charge? Is there a smallest unit of charge? Exactly how does the force depend on the amount of charge and the distance between charges? Such questions obviously occurred to Benjamin Franklin and other early researchers, and they interest us even today.

Charge Carried by Electrons and Protons

Franklin wrote in his letters and books that he could see the effects of electric charge but did not understand what caused the phenomenon. Today we have the advantage of knowing that normal matter is made of atoms, and that atoms contain positive and negative charges, usually in equal amounts.

Figure 18.5 shows a simple model of an atom with negative **electrons** orbiting its positive nucleus. The nucleus is positive due to the presence of positively charged **protons**. Nearly all charge in nature is due to electrons and protons, which are two of the three building blocks of most matter. (The third is the neutron, which is neutral, carrying no charge.) Other charge-carrying particles are observed in cosmic rays and nuclear decay, and are created in particle accelerators. All but the electron and proton survive only a short time and are quite rare by comparison.

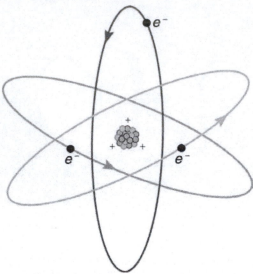

Figure 18.5 This simplified (and not to scale) view of an atom is called the planetary model of the atom. Negative electrons orbit a much heavier positive nucleus, as the planets orbit the much heavier sun. There the similarity ends, because forces in the atom are electromagnetic, whereas those in the planetary system are gravitational. Normal macroscopic amounts of matter contain immense numbers of atoms and molecules and, hence, even greater numbers of individual negative and positive charges.

The charges of electrons and protons are identical in magnitude but opposite in sign. Furthermore, all charged objects in nature are integral multiples of this basic quantity of charge, meaning that all charges are made of combinations of a basic unit of charge. Usually, charges are formed by combinations of electrons and protons. The magnitude of this basic charge is

$$|q_e| = 1.60 \times 10^{-19}\,\text{C}. \tag{18.1}$$

The symbol q is commonly used for charge and the subscript e indicates the charge of a single electron (or proton).

The SI unit of charge is the coulomb (C). The number of protons needed to make a charge of 1.00 C is

$$1.00\,\text{C} \times \frac{1\,\text{proton}}{1.60 \times 10^{-19}\,\text{C}} = 6.25 \times 10^{18}\,\text{protons.} \tag{18.2}$$

Similarly, 6.25×10^{18} electrons have a combined charge of −1.00 coulomb. Just as there is a smallest bit of an element (an atom), there is a smallest bit of charge. There is no directly observed charge smaller than $|q_e|$ (see **Things Great and Small: The Submicroscopic Origin of Charge**), and all observed charges are integral multiples of $|q_e|$.

Things Great and Small: The Submicroscopic Origin of Charge

With the exception of exotic, short-lived particles, all charge in nature is carried by electrons and protons. Electrons carry the charge we have named negative. Protons carry an equal-magnitude charge that we call positive. (See **Figure 18.6**.) Electron and proton charges are considered fundamental building blocks, since all other charges are integral multiples of those carried by electrons and protons. Electrons and protons are also two of the three fundamental building blocks of ordinary matter. The neutron is the third and has zero total charge.

Figure 18.6 shows a person touching a Van de Graaff generator and receiving excess positive charge. The expanded view of a hair shows the existence of both types of charges but an excess of positive. The repulsion of these positive like charges causes the strands of hair to repel other strands of hair and to stand up. The further blowup shows an artist's conception of an electron and a proton perhaps found in an atom in a strand of hair.

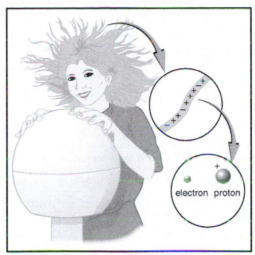

Figure 18.6 When this person touches a Van de Graaff generator, she receives an excess of positive charge, causing her hair to stand on end. The charges in one hair are shown. An artist's conception of an electron and a proton illustrate the particles carrying the negative and positive charges. We cannot really see these particles with visible light because they are so small (the electron seems to be an infinitesimal point), but we know a great deal about their measurable properties, such as the charges they carry.

The electron seems to have no substructure; in contrast, when the substructure of protons is explored by scattering extremely energetic electrons from them, it appears that there are point-like particles inside the proton. These sub-particles, named quarks, have never been directly observed, but they are believed to carry fractional charges as seen in **Figure 18.7**. Charges on electrons and protons and all other directly observable particles are unitary, but these quark substructures carry charges of either $-\frac{1}{3}$ or $+\frac{2}{3}$. There are continuing attempts to observe fractional charge directly and to learn of the properties of quarks, which are perhaps the ultimate substructure of matter.

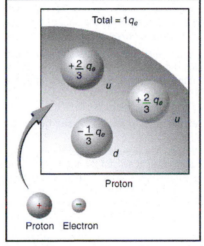

Figure 18.7 Artist's conception of fractional quark charges inside a proton. A group of three quark charges add up to the single positive charge on the proton: $-\frac{1}{3}q_e + \frac{2}{3}q_e + \frac{2}{3}q_e = +1q_e$.

Separation of Charge in Atoms

Charges in atoms and molecules can be separated—for example, by rubbing materials together. Some atoms and molecules have a greater affinity for electrons than others and will become negatively charged by close contact in rubbing, leaving the other material positively charged. (See **Figure 18.8**.) Positive charge can similarly be induced by rubbing. Methods other than rubbing can also separate charges. Batteries, for example, use combinations of substances that interact in such a way as to separate charges. Chemical interactions may transfer negative charge from one substance to the other, making one battery terminal negative and leaving the first one positive.

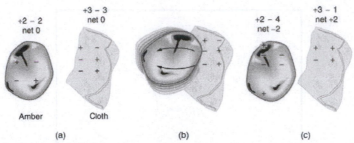

Figure 18.8 When materials are rubbed together, charges can be separated, particularly if one material has a greater affinity for electrons than another. (a) Both the amber and cloth are originally neutral, with equal positive and negative charges. Only a tiny fraction of the charges are involved, and only a few of them are shown here. (b) When rubbed together, some negative charge is transferred to the amber, leaving the cloth with a net positive charge. (c) When separated, the amber and cloth now have net charges, but the absolute value of the net positive and negative charges will be equal.

No charge is actually created or destroyed when charges are separated as we have been discussing. Rather, existing charges are moved about. In fact, in all situations the total amount of charge is always constant. This universally obeyed law of nature is called the **law of conservation of charge**.

Law of Conservation of Charge

Total charge is constant in any process.

Making Connections: Net Charge

Hence if a closed system is neutral, it will remain neutral. Similarly, if a closed system has a charge, say, −10e, it will always have that charge. The only way to change the charge of a system is to transfer charge outside, either by bringing in charge or removing charge. If it is possible to transfer charge outside, the system is no longer closed/isolated and is known as an open system. However, charge is always conserved, for both open and closed systems. Consequently, the charge transferred to/from an open system is equal to the change in the system's charge.

For example, each of the two materials (amber and cloth) discussed in **Figure 18.8** have no net charge initially. The only way to change their charge is to transfer charge from outside each object. When they are rubbed together, negative charge is transferred to the amber and the final charge of the amber is the sum of the initial charge and the charge transferred to it. On the other hand, the final charge on the cloth is equal to its initial charge minus the charge transferred out.

Similarly when glass is rubbed with silk, the net charge on the silk is its initial charge plus the incoming charge and the charge on the glass is the initial charge minus the outgoing charge. Also the charge gained by the silk will be equal to the charge lost by the glass, which means that if the silk gains −5e charge, the glass would have lost −5e charge.

In more exotic situations, such as in particle accelerators, mass, Δm, can be created from energy in the amount $\Delta m = \frac{E}{c^2}$.

Sometimes, the created mass is charged, such as when an electron is created. Whenever a charged particle is created, another having an opposite charge is always created along with it, so that the total charge created is zero. Usually, the two particles are "matter-antimatter" counterparts. For example, an antielectron would usually be created at the same time as an electron. The antielectron has a positive charge (it is called a positron), and so the total charge created is zero. (See **Figure 18.9**.) All particles have antimatter counterparts with opposite signs. When matter and antimatter counterparts are brought together, they completely annihilate one another. By annihilate, we mean that the mass of the two particles is converted to energy E, again obeying the relationship $\Delta m = \frac{E}{c^2}$. Since the two particles have equal and opposite charge, the total charge is zero before and after the annihilation; thus, total charge is conserved.

Making Connections: Conservation Laws

Only a limited number of physical quantities are universally conserved. Charge is one—energy, momentum, and angular momentum are others. Because they are conserved, these physical quantities are used to explain more phenomena and form more connections than other, less basic quantities. We find that conserved quantities give us great insight into the rules followed by nature and hints to the organization of nature. Discoveries of conservation laws have led to further discoveries, such as the weak nuclear force and the quark substructure of protons and other particles.

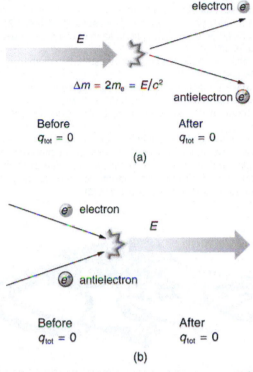

$$\Delta m = 2m_e = E/c^2$$

Before
$q_{tot} = 0$

After
$q_{tot} = 0$

(a)

Before
$q_{tot} = 0$

After
$q_{tot} = 0$

(b)

Figure 18.9 (a) When enough energy is present, it can be converted into matter. Here the matter created is an electron–antielectron pair. (m_e is the electron's mass.) The total charge before and after this event is zero. (b) When matter and antimatter collide, they annihilate each other; the total charge is conserved at zero before and after the annihilation.

The law of conservation of charge is absolute—it has never been observed to be violated. Charge, then, is a special physical quantity, joining a very short list of other quantities in nature that are always conserved. Other conserved quantities include energy, momentum, and angular momentum.

PhET Explorations: Balloons and Static Electricity

Why does a balloon stick to your sweater? Rub a balloon on a sweater, then let go of the balloon and it flies over and sticks to the sweater. View the charges in the sweater, balloons, and the wall.

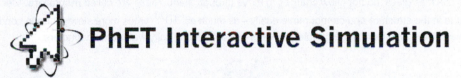

Figure 18.10 Balloons and Static Electricity (http://cnx.org/content/m55300/1.2/balloons_en.jar)

Applying the Science Practices: Electrical Charging

Design an experiment to demonstrate the electrical charging of objects, by using a glass rod, a balloon, small bits of paper, and different pieces of cloth (like silk, wool, or nylon). Also show that like charges repel each other whereas unlike charges attract each other.

18.2 Conductors and Insulators

Learning Objectives

By the end of this section, you will be able to:

- Define conductor and insulator, explain the difference, and give examples of each.
- Describe three methods for charging an object.
- Explain what happens to an electric force as you move farther from the source.
- Define polarization.

The information presented in this section supports the following AP® learning objectives and science practices:

- **1.B.2.2** The student is able to make a qualitative prediction about the distribution of positive and negative electric charges within neutral systems as they undergo various processes. **(S.P. 6.4, 7.2)**
- **1.B.2.3** The student is able to challenge claims that polarization of electric charge or separation of charge must result in a net charge on the object. **(S.P. 6.1)**
- **4.E.3.1** The student is able to make predictions about the redistribution of charge during charging by friction, conduction, and induction. **(S.P. 6.4)**
- **4.E.3.2** The student is able to make predictions about the redistribution of charge caused by the electric field due to other systems, resulting in charged or polarized objects. **(S.P. 6.4, 7.2)**
- **4.E.3.3** The student is able to construct a representation of the distribution of fixed and mobile charge in insulators and conductors. **(S.P. 1.1, 1.4, 6.4)**
- **4.E.3.4** The student is able to construct a representation of the distribution of fixed and mobile charge in insulators and conductors that predicts charge distribution in processes involving induction or conduction. **(S.P. 1.1, 1.4, 6.4)**
- **4.E.3.5** The student is able to plan and/or analyze the results of experiments in which electric charge rearrangement occurs by electrostatic induction, or is able to refine a scientific question relating to such an experiment by identifying anomalies in a data set or procedure. **(S.P. 3.2, 4.1, 4.2, 5.1, 5.3)**

Figure 18.11 This power adapter uses metal wires and connectors to conduct electricity from the wall socket to a laptop computer. The conducting wires allow electrons to move freely through the cables, which are shielded by rubber and plastic. These materials act as insulators that don't allow electric charge to escape outward. (credit: Evan-Amos, Wikimedia Commons)

Some substances, such as metals and salty water, allow charges to move through them with relative ease. Some of the electrons in metals and similar conductors are not bound to individual atoms or sites in the material. These **free electrons** can move through the material much as air moves through loose sand. Any substance that has free electrons and allows charge to move relatively freely through it is called a **conductor**. The moving electrons may collide with fixed atoms and molecules, losing some energy, but they can move in a conductor. Superconductors allow the movement of charge without any loss of energy. Salty water and other similar conducting materials contain free ions that can move through them. An ion is an atom or molecule having a positive or negative (nonzero) total charge. In other words, the total number of electrons is not equal to the total number of protons.

Other substances, such as glass, do not allow charges to move through them. These are called **insulators**. Electrons and ions in insulators are bound in the structure and cannot move easily—as much as 10^{23} times more slowly than in conductors. Pure water and dry table salt are insulators, for example, whereas molten salt and salty water are conductors.

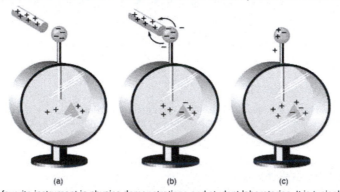

Figure 18.12 An electroscope is a favorite instrument in physics demonstrations and student laboratories. It is typically made with gold foil leaves hung from a (conducting) metal stem and is insulated from the room air in a glass-walled container. (a) A positively charged glass rod is brought near the tip of the electroscope, attracting electrons to the top and leaving a net positive charge on the leaves. Like charges in the light flexible gold leaves repel, separating them. (b) When the rod is touched against the ball, electrons are attracted and transferred, reducing the net charge on the glass rod but leaving the electroscope positively charged. (c) The excess charges are evenly distributed in the stem and leaves of the electroscope once the glass rod is removed.

Charging by Contact

Figure 18.12 shows an electroscope being charged by touching it with a positively charged glass rod. Because the glass rod is an insulator, it must actually touch the electroscope to transfer charge to or from it. (Note that the extra positive charges reside on the surface of the glass rod as a result of rubbing it with silk before starting the experiment.) Since only electrons move in metals, we see that they are attracted to the top of the electroscope. There, some are transferred to the positive rod by touch, leaving the electroscope with a net positive charge.

Electrostatic repulsion in the leaves of the charged electroscope separates them. The electrostatic force has a horizontal component that results in the leaves moving apart as well as a vertical component that is balanced by the gravitational force. Similarly, the electroscope can be negatively charged by contact with a negatively charged object.

Charging by Induction

It is not necessary to transfer excess charge directly to an object in order to charge it. **Figure 18.13** shows a method of **induction** wherein a charge is created in a nearby object, without direct contact. Here we see two neutral metal spheres in contact with one another but insulated from the rest of the world. A positively charged rod is brought near one of them, attracting negative charge to that side, leaving the other sphere positively charged.

This is an example of induced **polarization** of neutral objects. Polarization is the separation of charges in an object that remains neutral. If the spheres are now separated (before the rod is pulled away), each sphere will have a net charge. Note that the object closest to the charged rod receives an opposite charge when charged by induction. Note also that no charge is removed from the charged rod, so that this process can be repeated without depleting the supply of excess charge.

Another method of charging by induction is shown in **Figure 18.14**. The neutral metal sphere is polarized when a charged rod is brought near it. The sphere is then grounded, meaning that a conducting wire is run from the sphere to the ground. Since the earth is large and most ground is a good conductor, it can supply or accept excess charge easily. In this case, electrons are attracted to the sphere through a wire called the ground wire, because it supplies a conducting path to the ground. The ground connection is broken before the charged rod is removed, leaving the sphere with an excess charge opposite to that of the rod. Again, an opposite charge is achieved when charging by induction and the charged rod loses none of its excess charge.

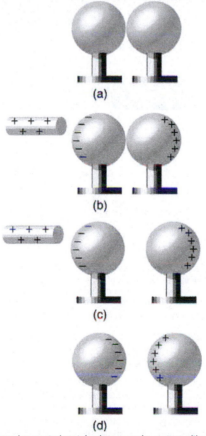

Figure 18.13 Charging by induction. (a) Two uncharged or neutral metal spheres are in contact with each other but insulated from the rest of the world. (b) A positively charged glass rod is brought near the sphere on the left, attracting negative charge and leaving the other sphere positively charged. (c) The spheres are separated before the rod is removed, thus separating negative and positive charge. (d) The spheres retain net charges after the inducing rod is removed—without ever having been touched by a charged object.

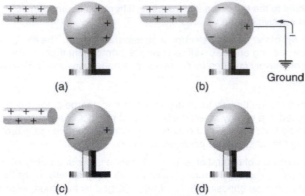

Figure 18.14 Charging by induction, using a ground connection. (a) A positively charged rod is brought near a neutral metal sphere, polarizing it. (b) The sphere is grounded, allowing electrons to be attracted from the earth's ample supply. (c) The ground connection is broken. (d) The positive rod is removed, leaving the sphere with an induced negative charge.

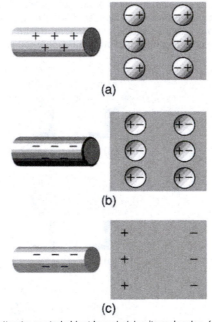

Figure 18.15 Both positive and negative objects attract a neutral object by polarizing its molecules. (a) A positive object brought near a neutral insulator polarizes its molecules. There is a slight shift in the distribution of the electrons orbiting the molecule, with unlike charges being brought nearer and like charges moved away. Since the electrostatic force decreases with distance, there is a net attraction. (b) A negative object produces the opposite polarization, but again attracts the neutral object. (c) The same effect occurs for a conductor; since the unlike charges are closer, there is a net attraction.

Neutral objects can be attracted to any charged object. The pieces of straw attracted to polished amber are neutral, for example. If you run a plastic comb through your hair, the charged comb can pick up neutral pieces of paper. **Figure 18.15** shows how the polarization of atoms and molecules in neutral objects results in their attraction to a charged object.

When a charged rod is brought near a neutral substance, an insulator in this case, the distribution of charge in atoms and molecules is shifted slightly. Opposite charge is attracted nearer the external charged rod, while like charge is repelled. Since the electrostatic force decreases with distance, the repulsion of like charges is weaker than the attraction of unlike charges, and so there is a net attraction. Thus a positively charged glass rod attracts neutral pieces of paper, as will a negatively charged rubber rod. Some molecules, like water, are polar molecules. Polar molecules have a natural or inherent separation of charge, although they are neutral overall. Polar molecules are particularly affected by other charged objects and show greater polarization effects than molecules with naturally uniform charge distributions.

Check Your Understanding

Can you explain the attraction of water to the charged rod in the figure below?

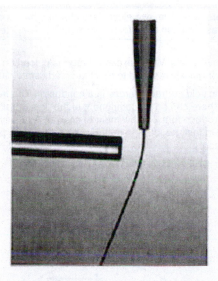

Figure 18.16

Solution

Water molecules are polarized, giving them slightly positive and slightly negative sides. This makes water even more susceptible to a charged rod's attraction. As the water flows downward, due to the force of gravity, the charged conductor exerts a net attraction to the opposite charges in the stream of water, pulling it closer.

Applying the Science Practices: Electrostatic Induction

Plan an experiment to demonstrate electrostatic induction using household items, like balloons, woolen cloth, aluminum drink cans, or foam cups. Explain the process of induction in your experiment by discussing details of (and making diagrams relating to) the movement and alignment of charges.

PhET Explorations: John Travoltage

Make sparks fly with John Travoltage. Wiggle Johnnie's foot and he picks up charges from the carpet. Bring his hand close to the door knob and get rid of the excess charge.

Figure 18.17 John Travoltage (http://cnx.org/content/m55301/1.2/travoltage_en.jar)

18.3 Conductors and Electric Fields in Static Equilibrium

Learning Objectives
By the end of this section, you will be able to: • List the three properties of a conductor in electrostatic equilibrium. • Explain the effect of an electric field on free charges in a conductor. • Explain why no electric field may exist inside a conductor. • Describe the electric field surrounding Earth. • Explain what happens to an electric field applied to an irregular conductor. • Describe how a lightning rod works. • Explain how a metal car may protect passengers inside from the dangerous electric fields caused by a downed line touching the car. The information presented in this section supports the following AP learning objectives: • **2.C.3.1** The student is able to explain the inverse square dependence of the electric field surrounding a spherically symmetric electrically charged object. • **2.C.5.1** The student is able to create representations of the magnitude and direction of the electric field at various

distances (small compared to plate size) from two electrically charged plates of equal magnitude and opposite signs and is able to recognize that the assumption of uniform field is not appropriate near edges of plates.

Conductors contain **free charges** that move easily. When excess charge is placed on a conductor or the conductor is put into a static electric field, charges in the conductor quickly respond to reach a steady state called **electrostatic equilibrium**.

Figure 18.18 shows the effect of an electric field on free charges in a conductor. The free charges move until the field is perpendicular to the conductor's surface. There can be no component of the field parallel to the surface in electrostatic equilibrium, since, if there were, it would produce further movement of charge. A positive free charge is shown, but free charges can be either positive or negative and are, in fact, negative in metals. The motion of a positive charge is equivalent to the motion of a negative charge in the opposite direction.

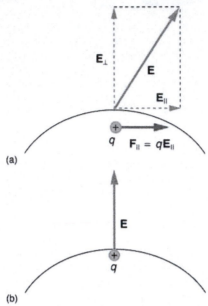

(a)

(b)

Figure 18.18 When an electric field $\mathbf{E}$ is applied to a conductor, free charges inside the conductor move until the field is perpendicular to the surface. (a) The electric field is a vector quantity, with both parallel and perpendicular components. The parallel component ($\mathbf{E}_{\parallel}$) exerts a force ($\mathbf{F}_{\parallel}$) on the free charge q, which moves the charge until $\mathbf{F}_{\parallel} = 0$. (b) The resulting field is perpendicular to the surface. The free charge has been brought to the conductor's surface, leaving electrostatic forces in equilibrium.

A conductor placed in an **electric field** will be **polarized**. **Figure 18.19** shows the result of placing a neutral conductor in an originally uniform electric field. The field becomes stronger near the conductor but entirely disappears inside it.

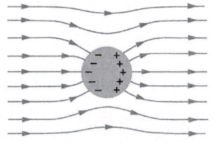

Figure 18.19 This illustration shows a spherical conductor in static equilibrium with an originally uniform electric field. Free charges move within the conductor, polarizing it, until the electric field lines are perpendicular to the surface. The field lines end on excess negative charge on one section of the surface and begin again on excess positive charge on the opposite side. No electric field exists inside the conductor, since free charges in the conductor would continue moving in response to any field until it was neutralized.

Misconception Alert: Electric Field inside a Conductor

Excess charges placed on a spherical conductor repel and move until they are evenly distributed, as shown in **Figure 18.20**. Excess charge is forced to the surface until the field inside the conductor is zero. Outside the conductor, the field is exactly the same as if the conductor were replaced by a point charge at its center equal to the excess charge.

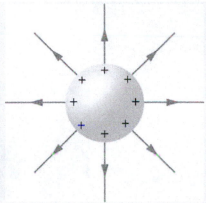

Figure 18.20 The mutual repulsion of excess positive charges on a spherical conductor distributes them uniformly on its surface. The resulting electric field is perpendicular to the surface and zero inside. Outside the conductor, the field is identical to that of a point charge at the center equal to the excess charge.

Properties of a Conductor in Electrostatic Equilibrium

1. The electric field is zero inside a conductor.

2. Just outside a conductor, the electric field lines are perpendicular to its surface, ending or beginning on charges on the surface.

3. Any excess charge resides entirely on the surface or surfaces of a conductor.

The properties of a conductor are consistent with the situations already discussed and can be used to analyze any conductor in electrostatic equilibrium. This can lead to some interesting new insights, such as described below.

How can a very uniform electric field be created? Consider a system of two metal plates with opposite charges on them, as shown in **Figure 18.21**. The properties of conductors in electrostatic equilibrium indicate that the electric field between the plates will be uniform in strength and direction. Except near the edges, the excess charges distribute themselves uniformly, producing field lines that are uniformly spaced (hence uniform in strength) and perpendicular to the surfaces (hence uniform in direction, since the plates are flat). The edge effects are less important when the plates are close together.

Figure 18.21 Two metal plates with equal, but opposite, excess charges. The field between them is uniform in strength and direction except near the edges. One use of such a field is to produce uniform acceleration of charges between the plates, such as in the electron gun of a TV tube.

Earth's Electric Field

A near uniform electric field of approximately 150 N/C, directed downward, surrounds Earth, with the magnitude increasing slightly as we get closer to the surface. What causes the electric field? At around 100 km above the surface of Earth we have a layer of charged particles, called the **ionosphere**. The ionosphere is responsible for a range of phenomena including the electric field surrounding Earth. In fair weather the ionosphere is positive and the Earth largely negative, maintaining the electric field (**Figure 18.22**(a)).

In storm conditions clouds form and localized electric fields can be larger and reversed in direction (**Figure 18.22**(b)). The exact charge distributions depend on the local conditions, and variations of **Figure 18.22**(b) are possible.

If the electric field is sufficiently large, the insulating properties of the surrounding material break down and it becomes conducting. For air this occurs at around 3×10^6 N/C. Air ionizes ions and electrons recombine, and we get discharge in the form of lightning sparks and corona discharge.

Figure 18.22 Earth's electric field. (a) Fair weather field. Earth and the ionosphere (a layer of charged particles) are both conductors. They produce a uniform electric field of about 150 N/C. (credit: D. H. Parks) (b) Storm fields. In the presence of storm clouds, the local electric fields can be larger. At very high fields, the insulating properties of the air break down and lightning can occur. (credit: Jan-Joost Verhoef)

Electric Fields on Uneven Surfaces

So far we have considered excess charges on a smooth, symmetrical conductor surface. What happens if a conductor has sharp corners or is pointed? Excess charges on a nonuniform conductor become concentrated at the sharpest points. Additionally, excess charge may move on or off the conductor at the sharpest points.

To see how and why this happens, consider the charged conductor in **Figure 18.23**. The electrostatic repulsion of like charges is most effective in moving them apart on the flattest surface, and so they become least concentrated there. This is because the forces between identical pairs of charges at either end of the conductor are identical, but the components of the forces parallel to the surfaces are different. The component parallel to the surface is greatest on the flattest surface and, hence, more effective in moving the charge.

The same effect is produced on a conductor by an externally applied electric field, as seen in **Figure 18.23** (c). Since the field lines must be perpendicular to the surface, more of them are concentrated on the most curved parts.

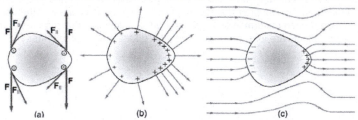

Figure 18.23 Excess charge on a nonuniform conductor becomes most concentrated at the location of greatest curvature. (a) The forces between identical pairs of charges at either end of the conductor are identical, but the components of the forces parallel to the surface are different. It is $\mathbf{F}_{\parallel}$ that moves the charges apart once they have reached the surface. (b) $\mathbf{F}_{\parallel}$ is smallest at the more pointed end, the charges are left closer together, producing the electric field shown. (c) An uncharged conductor in an originally uniform electric field is polarized, with the most concentrated charge at its most pointed end.

Applications of Conductors

On a very sharply curved surface, such as shown in **Figure 18.24**, the charges are so concentrated at the point that the resulting electric field can be great enough to remove them from the surface. This can be useful.

Lightning rods work best when they are most pointed. The large charges created in storm clouds induce an opposite charge on a building that can result in a lightning bolt hitting the building. The induced charge is bled away continually by a lightning rod, preventing the more dramatic lightning strike.

Of course, we sometimes wish to prevent the transfer of charge rather than to facilitate it. In that case, the conductor should be very smooth and have as large a radius of curvature as possible. (See **Figure 18.25**.) Smooth surfaces are used on high-voltage transmission lines, for example, to avoid leakage of charge into the air.

Another device that makes use of some of these principles is a **Faraday cage**. This is a metal shield that encloses a volume. All electrical charges will reside on the outside surface of this shield, and there will be no electrical field inside. A Faraday cage is used to prohibit stray electrical fields in the environment from interfering with sensitive measurements, such as the electrical signals inside a nerve cell.

During electrical storms if you are driving a car, it is best to stay inside the car as its metal body acts as a Faraday cage with zero electrical field inside. If in the vicinity of a lightning strike, its effect is felt on the outside of the car and the inside is unaffected, provided you remain totally inside. This is also true if an active ("hot") electrical wire was broken (in a storm or an accident) and fell on your car.

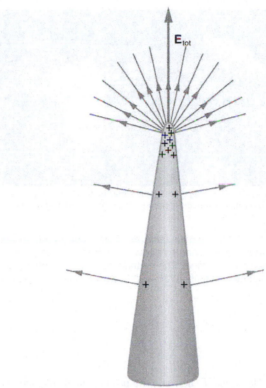

Figure 18.24 A very pointed conductor has a large charge concentration at the point. The electric field is very strong at the point and can exert a force large enough to transfer charge on or off the conductor. Lightning rods are used to prevent the buildup of large excess charges on structures and, thus, are pointed.

Figure 18.25 (a) A lightning rod is pointed to facilitate the transfer of charge. (credit: Romaine, Wikimedia Commons) (b) This Van de Graaff generator has a smooth surface with a large radius of curvature to prevent the transfer of charge and allow a large voltage to be generated. The mutual repulsion of like charges is evident in the person's hair while touching the metal sphere. (credit: Jon 'ShakataGaNai' Davis/Wikimedia Commons).

18.4 Coulomb's Law

Learning Objectives

By the end of this section, you will be able to:

- State Coulomb's law in terms of how the electrostatic force changes with the distance between two objects.
- Calculate the electrostatic force between two point charges, such as electrons or protons.
- Compare the electrostatic force to the gravitational attraction for a proton and an electron; for a human and the Earth.

The information presented in this section supports the following AP® learning objectives and science practices:

- **3.A.3.3** The student is able to describe a force as an interaction between two objects and identify both objects for any force. **(S.P. 1.4)**
- **3.A.3.4** The student is able to make claims about the force on an object due to the presence of other objects with the same property: mass, electric charge. **(S.P. 6.1, 6.4)**
- **3.C.2.1** The student is able to use Coulomb's law qualitatively and quantitatively to make predictions about the interaction between two electric point charges (interactions between collections of electric point charges are not covered in Physics 1 and instead are restricted to Physics 2). **(S.P. 2.2, 6.4)**
- **3.C.2.2** The student is able to connect the concepts of gravitational force and electric force to compare similarities and differences between the forces. **(S.P. 7.2)**

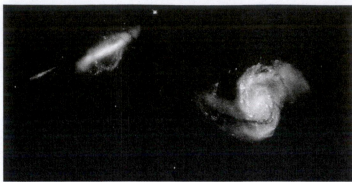

Figure 18.26 This NASA image of Arp 87 shows the result of a strong gravitational attraction between two galaxies. In contrast, at the subatomic level, the electrostatic attraction between two objects, such as an electron and a proton, is far greater than their mutual attraction due to gravity. (credit: NASA/HST)

Through the work of scientists in the late 18th century, the main features of the **electrostatic force**—the existence of two types of charge, the observation that like charges repel, unlike charges attract, and the decrease of force with distance—were eventually refined, and expressed as a mathematical formula. The mathematical formula for the electrostatic force is called **Coulomb's law** after the French physicist Charles Coulomb (1736–1806), who performed experiments and first proposed a formula to calculate it.

Coulomb's Law

$$F = k\frac{|q_1 q_2|}{r^2}.$$

(18.3)

Coulomb's law calculates the magnitude of the force F between two point charges, q_1 and q_2, separated by a distance r. In SI units, the constant k is equal to

$$k = 8.988 \times 10^9 \frac{\text{N} \cdot \text{m}^2}{\text{C}^2} \approx 8.99 \times 10^9 \frac{\text{N} \cdot \text{m}^2}{\text{C}^2}.$$

(18.4)

The electrostatic force is a vector quantity and is expressed in units of newtons. The force is understood to be along the line joining the two charges. (See **Figure 18.27**.)

Although the formula for Coulomb's law is simple, it was no mean task to prove it. The experiments Coulomb did, with the primitive equipment then available, were difficult. Modern experiments have verified Coulomb's law to great precision. For example, it has been shown that the force is inversely proportional to distance between two objects squared $\left(F \propto 1/r^2\right)$ to an accuracy of 1 part in 10^{16}. No exceptions have ever been found, even at the small distances within the atom.

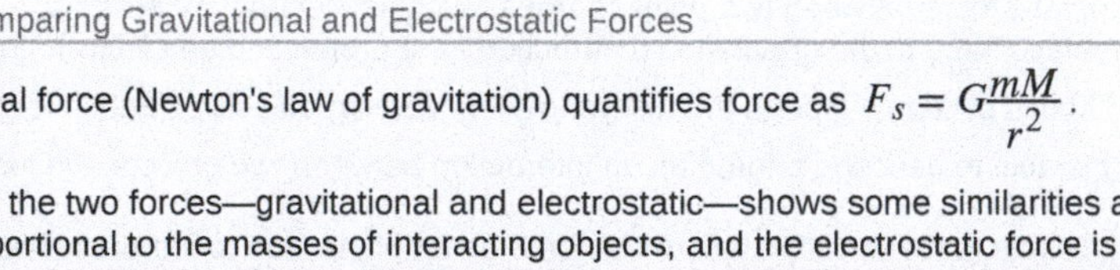

(a)

F_{21} ← r → F_{12}

q_1 q_2

(b)

F_{21} ——— r ——— F_{12}

q_1 q_2

Figure 18.27 The magnitude of the electrostatic force F between point charges q_1 and q_2 separated by a distance r is given by Coulomb's law. Note that Newton's third law (every force exerted creates an equal and opposite force) applies as usual—the force on q_1 is equal in magnitude and opposite in direction to the force it exerts on q_2. (a) Like charges. (b) Unlike charges.

Making Connections: Comparing Gravitational and Electrostatic Forces

Recall that the gravitational force (Newton's law of gravitation) quantifies force as $F_s = G\frac{mM}{r^2}$.

The comparison between the two forces—gravitational and electrostatic—shows some similarities and differences. Gravitational force is proportional to the masses of interacting objects, and the electrostatic force is proportional to the magnitudes of the charges of interacting objects. Hence both forces are proportional to a property that represents the strength of interaction for a given field. In addition, both forces are inversely proportional to the square of the distances between them. It may seem that the two forces are related but that is not the case. In fact, there are huge variations in the magnitudes of the two forces as they depend on different parameters and different mechanisms. For electrons (or protons), electrostatic force is dominant and is much greater than the gravitational force. On the other hand, gravitational force is generally dominant for objects with large masses. Another major difference between the two forces is that gravitational force can only be attractive, whereas electrostatic could be attractive or repulsive (depending on the sign of charges; unlike charges attract and like charges repel).

Example 18.1 How Strong is the Coulomb Force Relative to the Gravitational Force?

Compare the electrostatic force between an electron and proton separated by 0.530×10^{-10} m with the gravitational force between them. This distance is their average separation in a hydrogen atom.

Strategy

To compare the two forces, we first compute the electrostatic force using Coulomb's law, $F = k\frac{|q_1 q_2|}{r^2}$. We then calculate the gravitational force using Newton's universal law of gravitation. Finally, we take a ratio to see how the forces compare in magnitude.

Solution

Entering the given and known information about the charges and separation of the electron and proton into the expression of Coulomb's law yields

$$F = k\frac{|q_1 q_2|}{r^2} \tag{18.5}$$

$$= \left(8.99 \times 10^9 \text{ N} \cdot \text{m}^2/\text{C}^2\right) \times \frac{(1.60 \times 10^{-19} \text{ C})(1.60 \times 10^{-19} \text{ C})}{(0.530 \times 10^{-10} \text{ m})^2} \tag{18.6}$$

Thus the Coulomb force is

$$F = 8.19 \times 10^{-8} \text{ N}. \tag{18.7}$$

The charges are opposite in sign, so this is an attractive force. This is a very large force for an electron—it would cause an acceleration of 8.99×10^{22} m / s^2 (verification is left as an end-of-section problem).The gravitational force is given by Newton's law of gravitation as:

$$F_G = G\frac{mM}{r^2}, \tag{18.8}$$

where $G = 6.67 \times 10^{-11}$ N $\cdot$ m^2/kg^2. Here m and M represent the electron and proton masses, which can be found in the appendices. Entering values for the knowns yields

$$F_G = (6.67 \times 10^{-11} \text{ N} \cdot \text{m}^2/\text{kg}^2) \times \frac{(9.11 \times 10^{-31} \text{ kg})(1.67 \times 10^{-27} \text{ kg})}{(0.530 \times 10^{-10} \text{ m})^2} = 3.61 \times 10^{-47} \text{ N} \tag{18.9}$$

This is also an attractive force, although it is traditionally shown as positive since gravitational force is always attractive. The ratio of the magnitude of the electrostatic force to gravitational force in this case is, thus,

$$\frac{F}{F_G} = 2.27 \times 10^{39}. \tag{18.10}$$

Discussion

This is a remarkably large ratio! Note that this will be the ratio of electrostatic force to gravitational force for an electron and a proton at any distance (taking the ratio before entering numerical values shows that the distance cancels). This ratio gives some indication of just how much larger the Coulomb force is than the gravitational force between two of the most common particles in nature.

As the example implies, gravitational force is completely negligible on a small scale, where the interactions of individual charged particles are important. On a large scale, such as between the Earth and a person, the reverse is true. Most objects are nearly electrically neutral, and so attractive and repulsive **Coulomb forces** nearly cancel. Gravitational force on a large scale dominates interactions between large objects because it is always attractive, while Coulomb forces tend to cancel.

18.5 Electric Field: Concept of a Field Revisited

Learning Objectives

By the end of this section, you will be able to:

- Describe a force field and calculate the strength of an electric field due to a point charge.
- Calculate the force exerted on a test charge by an electric field.
- Explain the relationship between electrical force (F) on a test charge and electrical field strength (E).

The information presented in this section supports the following AP® learning objectives and science practices:

- **2.C.1.1** The student is able to predict the direction and the magnitude of the force exerted on an object with an electric charge q placed in an electric field E using the mathematical model of the relation between an electric force and an electric field: $\overrightarrow{F} = q\overrightarrow{E}$, a vector relation. **(S.P. 2.2)**
- **2.C.1.2** The student is able to calculate any one of the variables – electric force, electric charge, and electric field – at a point given the values and sign or direction of the other two quantities. **(S.P. 2.2)**
- **2.C.2.1** The student is able to qualitatively and semiquantitatively apply the vector relationship between the electric field and the net electric charge creating that field. **(S.P. 2.2, 6.4)**
- **3.C.4.1** The student is able to make claims about various contact forces between objects based on the microscopic cause of those forces. **(S.P. 6.1)**
- **3.C.4.2** The student is able to explain contact forces (tension, friction, normal, buoyant, spring) as arising from interatomic electric forces and that they therefore have certain directions. **(S.P. 6.2)**

Contact forces, such as between a baseball and a bat, are explained on the small scale by the interaction of the charges in atoms and molecules in close proximity. They interact through forces that include the **Coulomb force**. Action at a distance is a force between objects that are not close enough for their atoms to "touch." That is, they are separated by more than a few atomic diameters.

For example, a charged rubber comb attracts neutral bits of paper from a distance via the Coulomb force. It is very useful to think of an object being surrounded in space by a **force field**. The force field carries the force to another object (called a test object) some distance away.

Concept of a Field

A field is a way of conceptualizing and mapping the force that surrounds any object and acts on another object at a distance without apparent physical connection. For example, the gravitational field surrounding the earth (and all other masses) represents the gravitational force that would be experienced if another mass were placed at a given point within the field.

In the same way, the Coulomb force field surrounding any charge extends throughout space. Using Coulomb's law, $F = k|q_1 q_2|/r^2$, its magnitude is given by the equation $F = k|qQ|/r^2$, for a **point charge** (a particle having a charge Q) acting on a **test charge** q at a distance r (see **Figure 18.28**). Both the magnitude and direction of the Coulomb force field depend on Q and the test charge q .

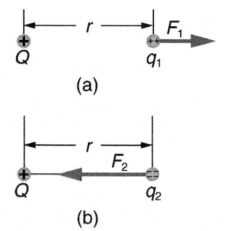

Figure 18.28 The Coulomb force field due to a positive charge Q is shown acting on two different charges. Both charges are the same distance from Q . (a) Since q_1 is positive, the force F_1 acting on it is repulsive. (b) The charge q_2 is negative and greater in magnitude than q_1 , and so the force F_2 acting on it is attractive and stronger than F_1 . The Coulomb force field is thus not unique at any point in space, because it depends on the test charges q_1 and q_2 as well as the charge Q .

To simplify things, we would prefer to have a field that depends only on Q and not on the test charge q . The electric field is defined in such a manner that it represents only the charge creating it and is unique at every point in space. Specifically, the electric field E is defined to be the ratio of the Coulomb force to the test charge:

$$\mathbf{E} = \frac{\mathbf{F}}{q},$$

(18.11)

where $\mathbf{F}$ is the electrostatic force (or Coulomb force) exerted on a positive test charge q. It is understood that $\mathbf{E}$ is in the same direction as $\mathbf{F}$. It is also assumed that q is so small that it does not alter the charge distribution creating the electric field. The units of electric field are newtons per coulomb (N/C). If the electric field is known, then the electrostatic force on any charge q is simply obtained by multiplying charge times electric field, or $\mathbf{F} = q\mathbf{E}$. Consider the electric field due to a point charge Q. According to Coulomb's law, the force it exerts on a test charge q is $F = k|qQ|/r^2$. Thus the magnitude of the electric field, E, for a point charge is

$$E = \left|\frac{F}{q}\right| = k\left|\frac{qQ}{qr^2}\right| = k\frac{|Q|}{r^2}.$$

(18.12)

Since the test charge cancels, we see that

$$E = k\frac{|Q|}{r^2}.$$

(18.13)

The electric field is thus seen to depend only on the charge Q and the distance r; it is completely independent of the test charge q.

Example 18.2 Calculating the Electric Field of a Point Charge

Calculate the strength and direction of the electric field E due to a point charge of 2.00 nC (nano-Coulombs) at a distance of 5.00 mm from the charge.

Strategy

We can find the electric field created by a point charge by using the equation $E = kQ/r^2$.

Solution

Here $Q = 2.00 \times 10^{-9}$ C and $r = 5.00 \times 10^{-3}$ m. Entering those values into the above equation gives

$$
\begin{aligned}
E &= k\frac{Q}{r^2} \\
&= (8.99 \times 10^9 \text{ N} \cdot \text{m}^2/\text{C}^2) \times \frac{(2.00 \times 10^{-9} \text{ C})}{(5.00 \times 10^{-3} \text{ m})^2} \\
&= 7.19 \times 10^5 \text{ N/C}.
\end{aligned}
$$

(18.14)

Discussion

This **electric field strength** is the same at any point 5.00 mm away from the charge Q that creates the field. It is positive, meaning that it has a direction pointing away from the charge Q.

Example 18.3 Calculating the Force Exerted on a Point Charge by an Electric Field

What force does the electric field found in the previous example exert on a point charge of -0.250 µC ?

Strategy

Since we know the electric field strength and the charge in the field, the force on that charge can be calculated using the definition of electric field $\mathbf{E} = \mathbf{F}/q$ rearranged to $\mathbf{F} = q\mathbf{E}$.

Solution

The magnitude of the force on a charge $q = -0.250$ µC exerted by a field of strength $E = 7.20 \times 10^5$ N/C is thus,

$$
\begin{aligned}
F &= -qE \\
&= (0.250 \times 10^{-6} \text{ C})(7.20 \times 10^5 \text{ N/C}) \\
&= 0.180 \text{ N}.
\end{aligned}
$$

(18.15)

Because q is negative, the force is directed opposite to the direction of the field.

Discussion

The force is attractive, as expected for unlike charges. (The field was created by a positive charge and here acts on a negative charge.) The charges in this example are typical of common static electricity, and the modest attractive force obtained is similar to forces experienced in static cling and similar situations.

PhET Explorations: Electric Field of Dreams

Play ball! Add charges to the Field of Dreams and see how they react to the electric field. Turn on a background electric field and adjust the direction and magnitude.

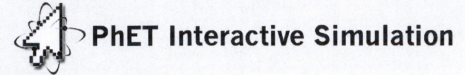

Figure 18.29 Electric Field of Dreams (http://cnx.org/content/m55304/1.2/efield_en.jar)

18.6 Electric Field Lines: Multiple Charges

Drawings using lines to represent **electric fields** around charged objects are very useful in visualizing field strength and direction. Since the electric field has both magnitude and direction, it is a vector. Like all **vectors**, the electric field can be represented by an arrow that has length proportional to its magnitude and that points in the correct direction. (We have used arrows extensively to represent force vectors, for example.)

Figure 18.30 shows two pictorial representations of the same electric field created by a positive point charge Q. **Figure 18.30** (b) shows the standard representation using continuous lines. **Figure 18.30** (b) shows numerous individual arrows with each arrow representing the force on a test charge q. Field lines are essentially a map of infinitesimal force vectors.

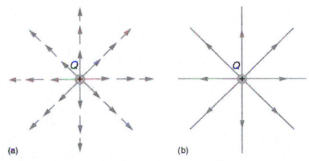

Figure 18.30 Two equivalent representations of the electric field due to a positive charge Q. (a) Arrows representing the electric field's magnitude and direction. (b) In the standard representation, the arrows are replaced by continuous field lines having the same direction at any point as the electric field. The closeness of the lines is directly related to the strength of the electric field. A test charge placed anywhere will feel a force in the direction of the field line; this force will have a strength proportional to the density of the lines (being greater near the charge, for example).

Note that the electric field is defined for a positive test charge q, so that the field lines point away from a positive charge and toward a negative charge. (See **Figure 18.31**.) The electric field strength is exactly proportional to the number of field lines per unit area, since the magnitude of the electric field for a point charge is $E = k|Q|/r^2$ and area is proportional to r^2. This pictorial representation, in which field lines represent the direction and their closeness (that is, their areal density or the number of lines crossing a unit area) represents strength, is used for all fields: electrostatic, gravitational, magnetic, and others.

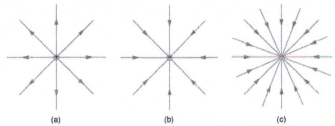

Figure 18.31 The electric field surrounding three different point charges. (a) A positive charge. (b) A negative charge of equal magnitude. (c) A larger negative charge.

In many situations, there are multiple charges. The total electric field created by multiple charges is the vector sum of the individual fields created by each charge. The following example shows how to add electric field vectors.

Example 18.4 Adding Electric Fields

Find the magnitude and direction of the total electric field due to the two point charges, q_1 and q_2, at the origin of the coordinate system as shown in **Figure 18.32**.

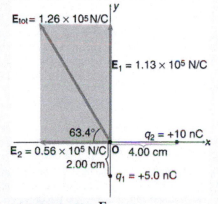

Figure 18.32 The electric fields $\mathbf{E}_1$ and $\mathbf{E}_2$ at the origin O add to $\mathbf{E}_{\text{tot}}$.

Strategy

Since the electric field is a vector (having magnitude and direction), we add electric fields with the same vector techniques used for other types of vectors. We first must find the electric field due to each charge at the point of interest, which is the origin of the coordinate system (O) in this instance. We pretend that there is a positive test charge, q, at point O, which allows us to determine the direction of the fields $\mathbf{E}_1$ and $\mathbf{E}_2$. Once those fields are found, the total field can be determined using **vector addition**.

Solution

The electric field strength at the origin due to q_1 is labeled E_1 and is calculated:

$$E_1 = k\frac{q_1}{r_1^2} = \left(8.99\times10^9 \text{ N} \cdot \text{m}^2/\text{C}^2\right)\frac{\left(5.00\times10^{-9} \text{ C}\right)}{\left(2.00\times10^{-2} \text{ m}\right)^2} \tag{18.16}$$

$$E_1 = 1.124\times10^5 \text{ N/C}.$$

Similarly, E_2 is

$$E_2 = k\frac{q_2}{r_2^2} = \left(8.99\times10^9 \text{ N} \cdot \text{m}^2/\text{C}^2\right)\frac{\left(10.0\times10^{-9} \text{ C}\right)}{\left(4.00\times10^{-2} \text{ m}\right)^2} \tag{18.17}$$

$$E_2 = 0.5619\times10^5 \text{ N/C}.$$

Four digits have been retained in this solution to illustrate that E_1 is exactly twice the magnitude of E_2. Now arrows are drawn to represent the magnitudes and directions of $\mathbf{E}_1$ and $\mathbf{E}_2$. (See Figure 18.32.) The direction of the electric field is that of the force on a positive charge so both arrows point directly away from the positive charges that create them. The arrow for $\mathbf{E}_1$ is exactly twice the length of that for $\mathbf{E}_2$. The arrows form a right triangle in this case and can be added using the Pythagorean theorem. The magnitude of the total field E_{tot} is

$$\begin{aligned} E_{\text{tot}} &= (E_1^2 + E_2^2)^{1/2} \\ &= \{(1.124\times10^5 \text{ N/C})^2 + (0.5619\times10^5 \text{ N/C})^2\}^{1/2} \\ &= 1.26\times10^5 \text{ N/C}. \end{aligned} \tag{18.18}$$

The direction is

$$\begin{aligned} \theta &= \tan^{-1}\left(\frac{E_1}{E_2}\right) \\ &= \tan^{-1}\left(\frac{1.124\times10^5 \text{ N/C}}{0.5619\times10^5 \text{ N/C}}\right) \\ &= 63.4°, \end{aligned} \tag{18.19}$$

or $63.4°$ above the x-axis.

Discussion

In cases where the electric field vectors to be added are not perpendicular, vector components or graphical techniques can be used. The total electric field found in this example is the total electric field at only one point in space. To find the total electric field due to these two charges over an entire region, the same technique must be repeated for each point in the region. This impossibly lengthy task (there are an infinite number of points in space) can be avoided by calculating the total field at representative points and using some of the unifying features noted next.

Figure 18.33 shows how the electric field from two point charges can be drawn by finding the total field at representative points and drawing electric field lines consistent with those points. While the electric fields from multiple charges are more complex than those of single charges, some simple features are easily noticed.

For example, the field is weaker between like charges, as shown by the lines being farther apart in that region. (This is because the fields from each charge exert opposing forces on any charge placed between them.) (See Figure 18.33 and Figure 18.34(a).) Furthermore, at a great distance from two like charges, the field becomes identical to the field from a single, larger charge.

Figure 18.34(b) shows the electric field of two unlike charges.

Making Connections: Electric Dipole

As the two unlike charges are also equal in magnitude, the pair of charges is also known as an electric dipole.

The field is stronger between the charges. In that region, the fields from each charge are in the same direction, and so their strengths add. The field of two unlike charges is weak at large distances, because the fields of the individual charges are in opposite directions and so their strengths subtract. At very large distances, the field of two unlike charges looks like that of a smaller single charge.

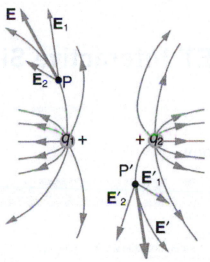

Figure 18.33 Two positive point charges q_1 and q_2 produce the resultant electric field shown. The field is calculated at representative points and then smooth field lines drawn following the rules outlined in the text.

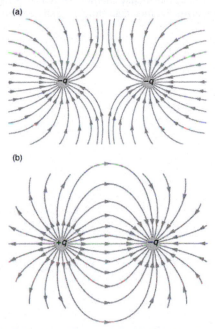

Figure 18.34 (a) Two negative charges produce the fields shown. It is very similar to the field produced by two positive charges, except that the directions are reversed. The field is clearly weaker between the charges. The individual forces on a test charge in that region are in opposite directions. (b) Two opposite charges produce the field shown, which is stronger in the region between the charges.

We use electric field lines to visualize and analyze electric fields (the lines are a pictorial tool, not a physical entity in themselves). The properties of electric field lines for any charge distribution can be summarized as follows:

1.   Field lines must begin on positive charges and terminate on negative charges, or at infinity in the hypothetical case of isolated charges.

2.   The number of field lines leaving a positive charge or entering a negative charge is proportional to the magnitude of the charge.

3.   The strength of the field is proportional to the closeness of the field lines—more precisely, it is proportional to the number of lines per unit area perpendicular to the lines.

4.   The direction of the electric field is tangent to the field line at any point in space.

5.   Field lines can never cross.

The last property means that the field is unique at any point. The field line represents the direction of the field; so if they crossed, the field would have two directions at that location (an impossibility if the field is unique).

Move point charges around on the playing field and then view the electric field, voltages, equipotential lines, and more. It's colorful, it's dynamic, it's free.

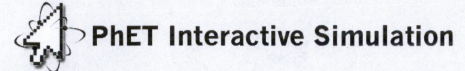

Figure 18.35 Charges and Fields (http://cnx.org/content/m55321/1.2/charges-and-fields_en.jar)

18.7 Electric Forces in Biology

Learning Objectives
By the end of this section, you will be able to: • Describe how a water molecule is polar. • Explain electrostatic screening by a water molecule within a living cell.

Classical electrostatics has an important role to play in modern molecular biology. Large molecules such as proteins, nucleic acids, and so on—so important to life—are usually electrically charged. DNA itself is highly charged; it is the electrostatic force that not only holds the molecule together but gives the molecule structure and strength. **Figure 18.36** is a schematic of the DNA double helix.

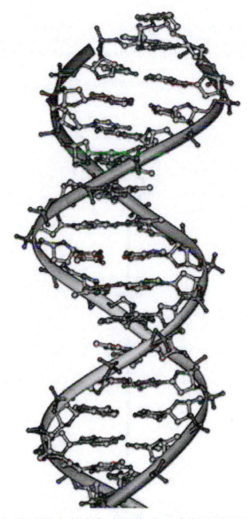

Figure 18.36 DNA is a highly charged molecule. The DNA double helix shows the two coiled strands each containing a row of nitrogenous bases, which "code" the genetic information needed by a living organism. The strands are connected by bonds between pairs of bases. While pairing combinations between certain bases are fixed (C-G and A-T), the sequence of nucleotides in the strand varies. (credit: Jerome Walker)

The four nucleotide bases are given the symbols A (adenine), C (cytosine), G (guanine), and T (thymine). The order of the four bases varies in each strand, but the pairing between bases is always the same. C and G are always paired and A and T are always paired, which helps to preserve the order of bases in cell division (mitosis) so as to pass on the correct genetic information. Since the Coulomb force drops with distance ($F \propto 1/r^2$), the distances between the base pairs must be small enough that the electrostatic force is sufficient to hold them together.

DNA is a highly charged molecule, with about $2q_e$ (fundamental charge) per 0.3×10^{-9} m. The distance separating the two strands that make up the DNA structure is about 1 nm, while the distance separating the individual atoms within each base is about 0.3 nm.

One might wonder why electrostatic forces do not play a larger role in biology than they do if we have so many charged molecules. The reason is that the electrostatic force is "diluted" due to **screening** between molecules. This is due to the presence of other charges in the cell.

Polarity of Water Molecules

The best example of this charge screening is the water molecule, represented as H_2O . Water is a strongly **polar molecule**. Its 10 electrons (8 from the oxygen atom and 2 from the two hydrogen atoms) tend to remain closer to the oxygen nucleus than the hydrogen nuclei. This creates two centers of equal and opposite charges—what is called a **dipole**, as illustrated in **Figure 18.37**. The magnitude of the dipole is called the dipole moment.

These two centers of charge will terminate some of the electric field lines coming from a free charge, as on a DNA molecule. This results in a reduction in the strength of the **Coulomb interaction**. One might say that screening makes the Coulomb force a short range force rather than long range.

Other ions of importance in biology that can reduce or screen Coulomb interactions are Na^+ , and K^+ , and Cl^- . These ions are located both inside and outside of living cells. The movement of these ions through cell membranes is crucial to the motion of nerve impulses through nerve axons.

Recent studies of electrostatics in biology seem to show that electric fields in cells can be extended over larger distances, in spite of screening, by "microtubules" within the cell. These microtubules are hollow tubes composed of proteins that guide the movement of chromosomes when cells divide, the motion of other organisms within the cell, and provide mechanisms for motion of some cells (as motors).

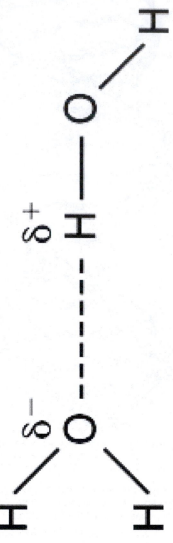

Figure 18.37 This schematic shows water (H_2O) as a polar molecule. Unequal sharing of electrons between the oxygen (O) and hydrogen (H) atoms leads to a net separation of positive and negative charge—forming a dipole. The symbols δ^- and δ^+ indicate that the oxygen side of the H_2O molecule tends to be more negative, while the hydrogen ends tend to be more positive. This leads to an attraction of opposite charges between molecules.

18.8 Applications of Electrostatics

Learning Objectives

By the end of this section, you will be able to:

- Name several real-world applications of the study of electrostatics.

The study of **electrostatics** has proven useful in many areas. This module covers just a few of the many applications of electrostatics.

The Van de Graaff Generator

Van de Graaff generators (or Van de Graaffs) are not only spectacular devices used to demonstrate high voltage due to static electricity—they are also used for serious research. The first was built by Robert Van de Graaff in 1931 (based on original suggestions by Lord Kelvin) for use in nuclear physics research. **Figure 18.38** shows a schematic of a large research version. Van de Graaffs utilize both smooth and pointed surfaces, and conductors and insulators to generate large static charges and, hence, large voltages.

A very large excess charge can be deposited on the sphere, because it moves quickly to the outer surface. Practical limits arise because the large electric fields polarize and eventually ionize surrounding materials, creating free charges that neutralize excess charge or allow it to escape. Nevertheless, voltages of 15 million volts are well within practical limits.

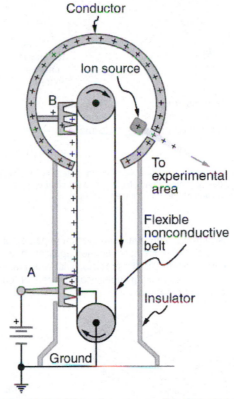

Figure 18.38 Schematic of Van de Graaff generator. A battery (A) supplies excess positive charge to a pointed conductor, the points of which spray the charge onto a moving insulating belt near the bottom. The pointed conductor (B) on top in the large sphere picks up the charge. (The induced electric field at the points is so large that it removes the charge from the belt.) This can be done because the charge does not remain inside the conducting sphere but moves to its outside surface. An ion source inside the sphere produces positive ions, which are accelerated away from the positive sphere to high velocities.

Take-Home Experiment: Electrostatics and Humidity

Rub a comb through your hair and use it to lift pieces of paper. It may help to tear the pieces of paper rather than cut them neatly. Repeat the exercise in your bathroom after you have had a long shower and the air in the bathroom is moist. Is it easier to get electrostatic effects in dry or moist air? Why would torn paper be more attractive to the comb than cut paper? Explain your observations.

Xerography

Most copy machines use an electrostatic process called **xerography**—a word coined from the Greek words *xeros* for dry and *graphos* for writing. The heart of the process is shown in simplified form in **Figure 18.39**.

A selenium-coated aluminum drum is sprayed with positive charge from points on a device called a corotron. Selenium is a substance with an interesting property—it is a **photoconductor**. That is, selenium is an insulator when in the dark and a conductor when exposed to light.

In the first stage of the xerography process, the conducting aluminum drum is **grounded** so that a negative charge is induced under the thin layer of uniformly positively charged selenium. In the second stage, the surface of the drum is exposed to the image of whatever is to be copied. Where the image is light, the selenium becomes conducting, and the positive charge is neutralized. In dark areas, the positive charge remains, and so the image has been transferred to the drum.

The third stage takes a dry black powder, called toner, and sprays it with a negative charge so that it will be attracted to the positive regions of the drum. Next, a blank piece of paper is given a greater positive charge than on the drum so that it will pull the toner from the drum. Finally, the paper and electrostatically held toner are passed through heated pressure rollers, which melt and permanently adhere the toner within the fibers of the paper.

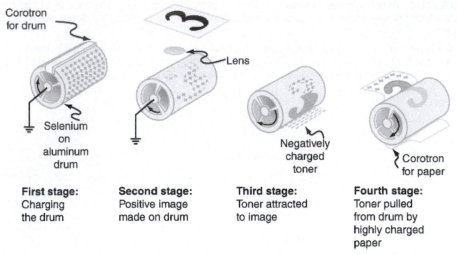

Figure 18.39 Xerography is a dry copying process based on electrostatics. The major steps in the process are the charging of the photoconducting drum, transfer of an image creating a positive charge duplicate, attraction of toner to the charged parts of the drum, and transfer of toner to the paper. Not shown are heat treatment of the paper and cleansing of the drum for the next copy.

Laser Printers

Laser printers use the xerographic process to make high-quality images on paper, employing a laser to produce an image on the photoconducting drum as shown in **Figure 18.40**. In its most common application, the laser printer receives output from a computer, and it can achieve high-quality output because of the precision with which laser light can be controlled. Many laser printers do significant information processing, such as making sophisticated letters or fonts, and may contain a computer more powerful than the one giving them the raw data to be printed.

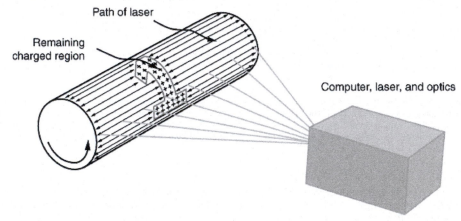

Figure 18.40 In a laser printer, a laser beam is scanned across a photoconducting drum, leaving a positive charge image. The other steps for charging the drum and transferring the image to paper are the same as in xerography. Laser light can be very precisely controlled, enabling laser printers to produce high-quality images.

Ink Jet Printers and Electrostatic Painting

The **ink jet printer**, commonly used to print computer-generated text and graphics, also employs electrostatics. A nozzle makes a fine spray of tiny ink droplets, which are then given an electrostatic charge. (See **Figure 18.41**.)

Once charged, the droplets can be directed, using pairs of charged plates, with great precision to form letters and images on paper. Ink jet printers can produce color images by using a black jet and three other jets with primary colors, usually cyan, magenta, and yellow, much as a color television produces color. (This is more difficult with xerography, requiring multiple drums and toners.)

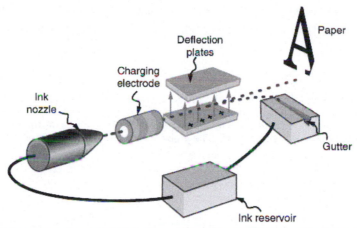

Figure 18.41 The nozzle of an ink-jet printer produces small ink droplets, which are sprayed with electrostatic charge. Various computer-driven devices are then used to direct the droplets to the correct positions on a page.

Electrostatic painting employs electrostatic charge to spray paint onto odd-shaped surfaces. Mutual repulsion of like charges causes the paint to fly away from its source. Surface tension forms drops, which are then attracted by unlike charges to the surface to be painted. Electrostatic painting can reach those hard-to-get at places, applying an even coat in a controlled manner. If the object is a conductor, the electric field is perpendicular to the surface, tending to bring the drops in perpendicularly. Corners and points on conductors will receive extra paint. Felt can similarly be applied.

Smoke Precipitators and Electrostatic Air Cleaning

Another important application of electrostatics is found in air cleaners, both large and small. The electrostatic part of the process places excess (usually positive) charge on smoke, dust, pollen, and other particles in the air and then passes the air through an oppositely charged grid that attracts and retains the charged particles. (See Figure 18.42.)

Large **electrostatic precipitators** are used industrially to remove over 99% of the particles from stack gas emissions associated with the burning of coal and oil. Home precipitators, often in conjunction with the home heating and air conditioning system, are very effective in removing polluting particles, irritants, and allergens.

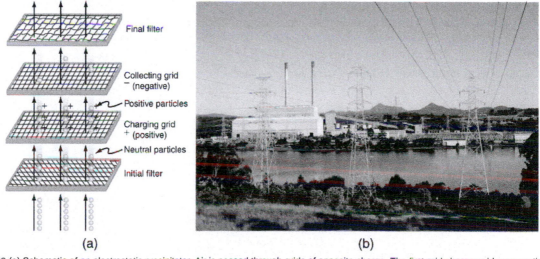

(a)   **(b)**

Figure 18.42 (a) Schematic of an electrostatic precipitator. Air is passed through grids of opposite charge. The first grid charges airborne particles, while the second attracts and collects them. (b) The dramatic effect of electrostatic precipitators is seen by the absence of smoke from this power plant. (credit: Cmdalgleish, Wikimedia Commons)

Problem-Solving Strategies for Electrostatics

1. Examine the situation to determine if static electricity is involved. This may concern separated stationary charges, the forces among them, and the electric fields they create.

2. Identify the system of interest. This includes noting the number, locations, and types of charges involved.

3. Identify exactly what needs to be determined in the problem (identify the unknowns). A written list is useful. Determine whether the Coulomb force is to be considered directly—if so, it may be useful to draw a free-body diagram, using electric field lines.

4. Make a list of what is given or can be inferred from the problem as stated (identify the knowns). It is important to distinguish the Coulomb force F from the electric field E, for example.

5. Solve the appropriate equation for the quantity to be determined (the unknown) or draw the field lines as requested.

6. Examine the answer to see if it is reasonable: Does it make sense? Are units correct and the numbers involved reasonable?

Integrated Concepts

The Integrated Concepts exercises for this module involve concepts such as electric charges, electric fields, and several other topics. Physics is most interesting when applied to general situations involving more than a narrow set of physical principles. The electric field exerts force on charges, for example, and hence the relevance of **Dynamics: Force and Newton's Laws of Motion**. The following topics are involved in some or all of the problems labeled "Integrated Concepts":

- **Kinematics**
- **Two-Dimensional Kinematics**
- **Dynamics: Force and Newton's Laws of Motion**
- **Uniform Circular Motion and Gravitation**
- **Statics and Torque**
- **Fluid Statics**

The following worked example illustrates how this strategy is applied to an Integrated Concept problem:

Example 18.5 Acceleration of a Charged Drop of Gasoline

If steps are not taken to ground a gasoline pump, static electricity can be placed on gasoline when filling your car's tank. Suppose a tiny drop of gasoline has a mass of 4.00×10^{-15} kg and is given a positive charge of 3.20×10^{-19} C. (a) Find the weight of the drop. (b) Calculate the electric force on the drop if there is an upward electric field of strength 3.00×10^5 N/C due to other static electricity in the vicinity. (c) Calculate the drop's acceleration.

Strategy

To solve an integrated concept problem, we must first identify the physical principles involved and identify the chapters in which they are found. Part (a) of this example asks for weight. This is a topic of dynamics and is defined in **Dynamics: Force and Newton's Laws of Motion**. Part (b) deals with electric force on a charge, a topic of **Electric Charge and Electric Field**. Part (c) asks for acceleration, knowing forces and mass. These are part of Newton's laws, also found in **Dynamics: Force and Newton's Laws of Motion**.

The following solutions to each part of the example illustrate how the specific problem-solving strategies are applied. These involve identifying knowns and unknowns, checking to see if the answer is reasonable, and so on.

Solution for (a)

Weight is mass times the acceleration due to gravity, as first expressed in

$$w = mg. \tag{18.20}$$

Entering the given mass and the average acceleration due to gravity yields

$$w = (4.00 \times 10^{-15} \text{ kg})(9.80 \text{ m/s}^2) = 3.92 \times 10^{-14} \text{ N}. \tag{18.21}$$

Discussion for (a)

This is a small weight, consistent with the small mass of the drop.

Solution for (b)

The force an electric field exerts on a charge is given by rearranging the following equation:

$$F = qE. \tag{18.22}$$

Here we are given the charge (3.20×10^{-19} C is twice the fundamental unit of charge) and the electric field strength, and so the electric force is found to be

$$F = (3.20 \times 10^{-19} \text{ C})(3.00 \times 10^5 \text{ N/C}) = 9.60 \times 10^{-14} \text{ N}. \tag{18.23}$$

Discussion for (b)

While this is a small force, it is greater than the weight of the drop.

Solution for (c)

The acceleration can be found using Newton's second law, provided we can identify all of the external forces acting on the drop. We assume only the drop's weight and the electric force are significant. Since the drop has a positive charge and the electric field is given to be upward, the electric force is upward. We thus have a one-dimensional (vertical direction) problem, and we can state Newton's second law as

$$a = \frac{F_{net}}{m}. \tag{18.24}$$

where $F_{net} = F - w$. Entering this and the known values into the expression for Newton's second law yields

$$
\begin{aligned}
a &= \frac{F - w}{m} \tag{18.25}\\
&= \frac{9.60 \times 10^{-14}\,\text{N} - 3.92 \times 10^{-14}\,\text{N}}{4.00 \times 10^{-15}\,\text{kg}}\\
&= 14.2\,\text{m/s}^2.
\end{aligned}
$$

Discussion for (c)

This is an upward acceleration great enough to carry the drop to places where you might not wish to have gasoline.

This worked example illustrates how to apply problem-solving strategies to situations that include topics in different chapters. The first step is to identify the physical principles involved in the problem. The second step is to solve for the unknown using familiar problem-solving strategies. These are found throughout the text, and many worked examples show how to use them for single topics. In this integrated concepts example, you can see how to apply them across several topics. You will find these techniques useful in applications of physics outside a physics course, such as in your profession, in other science disciplines, and in everyday life. The following problems will build your skills in the broad application of physical principles.

Unreasonable Results

The Unreasonable Results exercises for this module have results that are unreasonable because some premise is unreasonable or because certain of the premises are inconsistent with one another. Physical principles applied correctly then produce unreasonable results. The purpose of these problems is to give practice in assessing whether nature is being accurately described, and if it is not to trace the source of difficulty.

Problem-Solving Strategy

To determine if an answer is reasonable, and to determine the cause if it is not, do the following.

1. Solve the problem using strategies as outlined above. Use the format followed in the worked examples in the text to solve the problem as usual.

2. Check to see if the answer is reasonable. Is it too large or too small, or does it have the wrong sign, improper units, and so on?

3. If the answer is unreasonable, look for what specifically could cause the identified difficulty. Usually, the manner in which the answer is unreasonable is an indication of the difficulty. For example, an extremely large Coulomb force could be due to the assumption of an excessively large separated charge.

Glossary

conductor: a material that allows electrons to move separately from their atomic orbits

conductor: an object with properties that allow charges to move about freely within it

Coulomb force: another term for the electrostatic force

Coulomb interaction: the interaction between two charged particles generated by the Coulomb forces they exert on one another

Coulomb's law: the mathematical equation calculating the electrostatic force vector between two charged particles

dipole: a molecule's lack of symmetrical charge distribution, causing one side to be more positive and another to be more negative

electric charge: a physical property of an object that causes it to be attracted toward or repelled from another charged object; each charged object generates and is influenced by a force called an electromagnetic force

electric field: a three-dimensional map of the electric force extended out into space from a point charge

electric field lines: a series of lines drawn from a point charge representing the magnitude and direction of force exerted by that charge

electromagnetic force: one of the four fundamental forces of nature; the electromagnetic force consists of static electricity, moving electricity and magnetism

electron: a particle orbiting the nucleus of an atom and carrying the smallest unit of negative charge

electrostatic equilibrium: an electrostatically balanced state in which all free electrical charges have stopped moving about

electrostatic force: the amount and direction of attraction or repulsion between two charged bodies

electrostatic precipitators: filters that apply charges to particles in the air, then attract those charges to a filter, removing them from the airstream

electrostatic repulsion: the phenomenon of two objects with like charges repelling each other

electrostatics: the study of electric forces that are static or slow-moving

Faraday cage: a metal shield which prevents electric charge from penetrating its surface

field: a map of the amount and direction of a force acting on other objects, extending out into space

free charge: an electrical charge (either positive or negative) which can move about separately from its base molecule

free electron: an electron that is free to move away from its atomic orbit

grounded: when a conductor is connected to the Earth, allowing charge to freely flow to and from Earth's unlimited reservoir

grounded: connected to the ground with a conductor, so that charge flows freely to and from the Earth to the grounded object

induction: the process by which an electrically charged object brought near a neutral object creates a charge in that object

ink-jet printer: small ink droplets sprayed with an electric charge are controlled by electrostatic plates to create images on paper

insulator: a material that holds electrons securely within their atomic orbits

ionosphere: a layer of charged particles located around 100 km above the surface of Earth, which is responsible for a range of phenomena including the electric field surrounding Earth

laser printer: uses a laser to create a photoconductive image on a drum, which attracts dry ink particles that are then rolled onto a sheet of paper to print a high-quality copy of the image

law of conservation of charge: states that whenever a charge is created, an equal amount of charge with the opposite sign is created simultaneously

photoconductor: a substance that is an insulator until it is exposed to light, when it becomes a conductor

point charge: A charged particle, designated Q, generating an electric field

polar molecule: a molecule with an asymmetrical distribution of positive and negative charge

polarization: slight shifting of positive and negative charges to opposite sides of an atom or molecule

polarized: a state in which the positive and negative charges within an object have collected in separate locations

proton: a particle in the nucleus of an atom and carrying a positive charge equal in magnitude and opposite in sign to the amount of negative charge carried by an electron

screening: the dilution or blocking of an electrostatic force on a charged object by the presence of other charges nearby

static electricity: a buildup of electric charge on the surface of an object

test charge: A particle (designated q) with either a positive or negative charge set down within an electric field generated by a point charge

Van de Graaff generator: a machine that produces a large amount of excess charge, used for experiments with high voltage

vector: a quantity with both magnitude and direction

vector addition: mathematical combination of two or more vectors, including their magnitudes, directions, and positions

xerography: a dry copying process based on electrostatics

Section Summary

18.1 Static Electricity and Charge: Conservation of Charge

- There are only two types of charge, which we call positive and negative.
- Like charges repel, unlike charges attract, and the force between charges decreases with the square of the distance.
- The vast majority of positive charge in nature is carried by protons, while the vast majority of negative charge is carried by electrons.
- The electric charge of one electron is equal in magnitude and opposite in sign to the charge of one proton.
- An ion is an atom or molecule that has nonzero total charge due to having unequal numbers of electrons and protons.
- The SI unit for charge is the coulomb (C), with protons and electrons having charges of opposite sign but equal magnitude; the magnitude of this basic charge $|q_e|$ is

$$|q_e| = 1.60 \times 10^{-19}\,\text{C}.$$

- Whenever charge is created or destroyed, equal amounts of positive and negative are involved.
- Most often, existing charges are separated from neutral objects to obtain some net charge.
- Both positive and negative charges exist in neutral objects and can be separated by rubbing one object with another. For macroscopic objects, negatively charged means an excess of electrons and positively charged means a depletion of electrons.
- The law of conservation of charge ensures that whenever a charge is created, an equal charge of the opposite sign is created at the same time.

18.2 Conductors and Insulators

- Polarization is the separation of positive and negative charges in a neutral object.
- A conductor is a substance that allows charge to flow freely through its atomic structure.
- An insulator holds charge within its atomic structure.
- Objects with like charges repel each other, while those with unlike charges attract each other.
- A conducting object is said to be grounded if it is connected to the Earth through a conductor. Grounding allows transfer of charge to and from the earth's large reservoir.
- Objects can be charged by contact with another charged object and obtain the same sign charge.
- If an object is temporarily grounded, it can be charged by induction, and obtains the opposite sign charge.
- Polarized objects have their positive and negative charges concentrated in different areas, giving them a non-symmetrical charge.
- Polar molecules have an inherent separation of charge.

18.3 Conductors and Electric Fields in Static Equilibrium

- A conductor allows free charges to move about within it.
- The electrical forces around a conductor will cause free charges to move around inside the conductor until static equilibrium is reached.
- Any excess charge will collect along the surface of a conductor.
- Conductors with sharp corners or points will collect more charge at those points.
- A lightning rod is a conductor with sharply pointed ends that collect excess charge on the building caused by an electrical storm and allow it to dissipate back into the air.
- Electrical storms result when the electrical field of Earth's surface in certain locations becomes more strongly charged, due to changes in the insulating effect of the air.
- A Faraday cage acts like a shield around an object, preventing electric charge from penetrating inside.

18.4 Coulomb's Law

- Frenchman Charles Coulomb was the first to publish the mathematical equation that describes the electrostatic force between two objects.
- Coulomb's law gives the magnitude of the force between point charges. It is

$$F = k\frac{|q_1 q_2|}{r^2},$$

where q_1 and q_2 are two point charges separated by a distance r, and $k \approx 8.99 \times 10^9\,\text{N} \cdot \text{m}^2 / \text{C}^2$

- This Coulomb force is extremely basic, since most charges are due to point-like particles. It is responsible for all electrostatic effects and underlies most macroscopic forces.
- The Coulomb force is extraordinarily strong compared with the gravitational force, another basic force—but unlike gravitational force it can cancel, since it can be either attractive or repulsive.
- The electrostatic force between two subatomic particles is far greater than the gravitational force between the same two particles.

18.5 Electric Field: Concept of a Field Revisited

- The electrostatic force field surrounding a charged object extends out into space in all directions.

- The electrostatic force exerted by a point charge on a test charge at a distance r depends on the charge of both charges, as well as the distance between the two.
- The electric field $\mathbf{E}$ is defined to be

$$\mathbf{E} = \frac{\mathbf{F}}{q},$$

where $\mathbf{F}$ is the Coulomb or electrostatic force exerted on a small positive test charge q. $\mathbf{E}$ has units of N/C.

- The magnitude of the electric field $\mathbf{E}$ created by a point charge Q is

$$\mathbf{E} = k\frac{|Q|}{r^2}.$$

where r is the distance from Q. The electric field $\mathbf{E}$ is a vector and fields due to multiple charges add like vectors.

18.6 Electric Field Lines: Multiple Charges

- Drawings of electric field lines are useful visual tools. The properties of electric field lines for any charge distribution are that:
- Field lines must begin on positive charges and terminate on negative charges, or at infinity in the hypothetical case of isolated charges.
- The number of field lines leaving a positive charge or entering a negative charge is proportional to the magnitude of the charge.
- The strength of the field is proportional to the closeness of the field lines—more precisely, it is proportional to the number of lines per unit area perpendicular to the lines.
- The direction of the electric field is tangent to the field line at any point in space.
- Field lines can never cross.

18.7 Electric Forces in Biology

- Many molecules in living organisms, such as DNA, carry a charge.
- An uneven distribution of the positive and negative charges within a polar molecule produces a dipole.
- The effect of a Coulomb field generated by a charged object may be reduced or blocked by other nearby charged objects.
- Biological systems contain water, and because water molecules are polar, they have a strong effect on other molecules in living systems.

18.8 Applications of Electrostatics

- Electrostatics is the study of electric fields in static equilibrium.
- In addition to research using equipment such as a Van de Graaff generator, many practical applications of electrostatics exist, including photocopiers, laser printers, ink-jet printers and electrostatic air filters.

Conceptual Questions

18.1 Static Electricity and Charge: Conservation of Charge

1. There are very large numbers of charged particles in most objects. Why, then, don't most objects exhibit static electricity?

2. Why do most objects tend to contain nearly equal numbers of positive and negative charges?

18.2 Conductors and Insulators

3. An eccentric inventor attempts to levitate by first placing a large negative charge on himself and then putting a large positive charge on the ceiling of his workshop. Instead, while attempting to place a large negative charge on himself, his clothes fly off. Explain.

4. If you have charged an electroscope by contact with a positively charged object, describe how you could use it to determine the charge of other objects. Specifically, what would the leaves of the electroscope do if other charged objects were brought near its knob?

5. When a glass rod is rubbed with silk, it becomes positive and the silk becomes negative—yet both attract dust. Does the dust have a third type of charge that is attracted to both positive and negative? Explain.

6. Why does a car always attract dust right after it is polished? (Note that car wax and car tires are insulators.)

7. Describe how a positively charged object can be used to give another object a negative charge. What is the name of this process?

8. What is grounding? What effect does it have on a charged conductor? On a charged insulator?

18.3 Conductors and Electric Fields in Static Equilibrium

9. Is the object in a conductor or an insulator? Justify your answer.

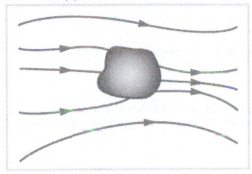

Figure 18.43

10. If the electric field lines in the figure above were perpendicular to the object, would it necessarily be a conductor? Explain.

11. The discussion of the electric field between two parallel conducting plates, in this module states that edge effects are less important if the plates are close together. What does close mean? That is, is the actual plate separation crucial, or is the ratio of plate separation to plate area crucial?

12. Would the self-created electric field at the end of a pointed conductor, such as a lightning rod, remove positive or negative charge from the conductor? Would the same sign charge be removed from a neutral pointed conductor by the application of a similar externally created electric field? (The answers to both questions have implications for charge transfer utilizing points.)

13. Why is a golfer with a metal club over her shoulder vulnerable to lightning in an open fairway? Would she be any safer under a tree?

14. Can the belt of a Van de Graaff accelerator be a conductor? Explain.

15. Are you relatively safe from lightning inside an automobile? Give two reasons.

16. Discuss pros and cons of a lightning rod being grounded versus simply being attached to a building.

17. Using the symmetry of the arrangement, show that the net Coulomb force on the charge q at the center of the square below (**Figure 18.44**) is zero if the charges on the four corners are exactly equal.

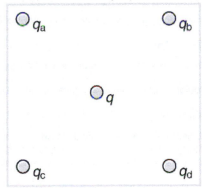

Figure 18.44 Four point charges q_a, q_b, q_c, and q_d lie on the corners of a square and q is located at its center.

18. (a) Using the symmetry of the arrangement, show that the electric field at the center of the square in **Figure 18.44** is zero if the charges on the four corners are exactly equal. (b) Show that this is also true for any combination of charges in which $q_a = q_b$ and $q_b = q_c$

19. (a) What is the direction of the total Coulomb force on q in **Figure 18.44** if q is negative, $q_a = q_c$ and both are negative, and $q_b = q_c$ and both are positive? (b) What is the direction of the electric field at the center of the square in this situation?

20. Considering **Figure 18.44**, suppose that $q_a = q_d$ and $q_b = q_c$. First show that q is in static equilibrium. (You may neglect the gravitational force.) Then discuss whether the equilibrium is stable or unstable, noting that this may depend on the signs of the charges and the direction of displacement of q from the center of the square.

21. If $q_a = 0$ in **Figure 18.44**, under what conditions will there be no net Coulomb force on q?

22. In regions of low humidity, one develops a special "grip" when opening car doors, or touching metal door knobs. This involves placing as much of the hand on the device as possible, not just the ends of one's fingers. Discuss the induced charge and explain why this is done.

23. Tollbooth stations on roadways and bridges usually have a piece of wire stuck in the pavement before them that will touch a car as it approaches. Why is this done?

24. Suppose a woman carries an excess charge. To maintain her charged status can she be standing on ground wearing just any pair of shoes? How would you discharge her? What are the consequences if she simply walks away?

18.4 Coulomb's Law

25. Figure 18.45 shows the charge distribution in a water molecule, which is called a polar molecule because it has an inherent separation of charge. Given water's polar character, explain what effect humidity has on removing excess charge from objects.

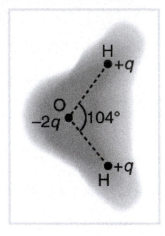

Figure 18.45 Schematic representation of the outer electron cloud of a neutral water molecule. The electrons spend more time near the oxygen than the hydrogens, giving a permanent charge separation as shown. Water is thus a *polar molecule*. It is more easily affected by electrostatic forces than molecules with uniform charge distributions.

26. Using Figure 18.45, explain, in terms of Coulomb's law, why a polar molecule (such as in Figure 18.45) is attracted by both positive and negative charges.

27. Given the polar character of water molecules, explain how ions in the air form nucleation centers for rain droplets.

18.5 Electric Field: Concept of a Field Revisited

28. Why must the test charge q in the definition of the electric field be vanishingly small?

29. Are the direction and magnitude of the Coulomb force unique at a given point in space? What about the electric field?

18.6 Electric Field Lines: Multiple Charges

30. Compare and contrast the Coulomb force field and the electric field. To do this, make a list of five properties for the Coulomb force field analogous to the five properties listed for electric field lines. Compare each item in your list of Coulomb force field properties with those of the electric field—are they the same or different? (For example, electric field lines cannot cross. Is the same true for Coulomb field lines?)

31. Figure 18.46 shows an electric field extending over three regions, labeled I, II, and III. Answer the following questions. (a) Are there any isolated charges? If so, in what region and what are their signs? (b) Where is the field strongest? (c) Where is it weakest? (d) Where is the field the most uniform?

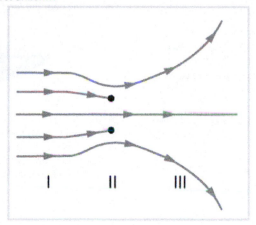

Figure 18.46

18.7 Electric Forces in Biology

32. A cell membrane is a thin layer enveloping a cell. The thickness of the membrane is much less than the size of the cell. In a static situation the membrane has a charge distribution of -2.5×10^{-6} C/m^2 on its inner surface and $+2.5 \times 10^{-6}$ C/m^2 on its outer surface. Draw a diagram of the cell and the surrounding cell membrane. Include on this diagram the charge distribution and the corresponding electric field. Is there any electric field inside the cell? Is there any electric field outside the cell?

Problems & Exercises

18.1 Static Electricity and Charge: Conservation of Charge

1. Common static electricity involves charges ranging from nanocoulombs to microcoulombs. (a) How many electrons are needed to form a charge of −2.00 nC (b) How many electrons must be removed from a neutral object to leave a net charge of $0.500\ \mu C$?

2. If 1.80×10^{20} electrons move through a pocket calculator during a full day's operation, how many coulombs of charge moved through it?

3. To start a car engine, the car battery moves 3.75×10^{21} electrons through the starter motor. How many coulombs of charge were moved?

4. A certain lightning bolt moves 40.0 C of charge. How many fundamental units of charge $|q_e|$ is this?

18.2 Conductors and Insulators

5. Suppose a speck of dust in an electrostatic precipitator has 1.0000×10^{12} protons in it and has a net charge of −5.00 nC (a very large charge for a small speck). How many electrons does it have?

6. An amoeba has 1.00×10^{16} protons and a net charge of 0.300 pC. (a) How many fewer electrons are there than protons? (b) If you paired them up, what fraction of the protons would have no electrons?

7. A 50.0 g ball of copper has a net charge of $2.00\ \mu C$.

What fraction of the copper's electrons has been removed? (Each copper atom has 29 protons, and copper has an atomic mass of 63.5.)

8. What net charge would you place on a 100 g piece of sulfur if you put an extra electron on 1 in 10^{12} of its atoms? (Sulfur has an atomic mass of 32.1.)

9. How many coulombs of positive charge are there in 4.00 kg of plutonium, given its atomic mass is 244 and that each plutonium atom has 94 protons?

18.3 Conductors and Electric Fields in Static Equilibrium

10. Sketch the electric field lines in the vicinity of the conductor in **Figure 18.47** given the field was originally uniform and parallel to the object's long axis. Is the resulting field small near the long side of the object?

Figure 18.47

11. Sketch the electric field lines in the vicinity of the conductor in **Figure 18.48** given the field was originally uniform and parallel to the object's long axis. Is the resulting field small near the long side of the object?

Figure 18.48

12. Sketch the electric field between the two conducting plates shown in **Figure 18.49**, given the top plate is positive and an equal amount of negative charge is on the bottom plate. Be certain to indicate the distribution of charge on the plates.

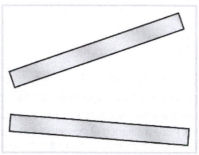

Figure 18.49

13. Sketch the electric field lines in the vicinity of the charged insulator in **Figure 18.50** noting its nonuniform charge distribution.

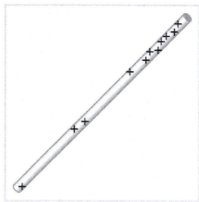

Figure 18.50 A charged insulating rod such as might be used in a classroom demonstration.

14. What is the force on the charge located at $x = 8.00$ cm in **Figure 18.51**(a) given that $q = 1.00\ \mu C$?

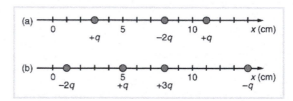

Figure 18.51 (a) Point charges located at 3.00, 8.00, and 11.0 cm along the x-axis. (b) Point charges located at 1.00, 5.00, 8.00, and 14.0 cm along the x-axis.

15. (a) Find the total electric field at $x = 1.00$ cm in **Figure 18.51(b)** given that $q = 5.00$ nC. (b) Find the total electric field at $x = 11.00$ cm in **Figure 18.51(b)**. (c) If the charges are allowed to move and eventually be brought to rest by friction, what will the final charge configuration be? (That is, will there be a single charge, double charge, etc., and what will its value(s) be?)

16. (a) Find the electric field at $x = 5.00$ cm in **Figure 18.51(a)**, given that $q = 1.00$ μC. (b) At what position between 3.00 and 8.00 cm is the total electric field the same as that for $-2q$ alone? (c) Can the electric field be zero anywhere between 0.00 and 8.00 cm? (d) At very large positive or negative values of x, the electric field approaches zero in both (a) and (b). In which does it most rapidly approach zero and why? (e) At what position to the right of 11.0 cm is the total electric field zero, other than at infinity? (Hint: A graphing calculator can yield considerable insight in this problem.)

17. (a) Find the total Coulomb force on a charge of 2.00 nC located at $x = 4.00$ cm in **Figure 18.51 (b)**, given that $q = 1.00$ μC. (b) Find the x-position at which the electric field is zero in **Figure 18.51 (b)**.

18. Using the symmetry of the arrangement, determine the direction of the force on q in the figure below, given that $q_a = q_b = +7.50$ μC and $q_c = q_d = -7.50$ μC. (b) Calculate the magnitude of the force on the charge q, given that the square is 10.0 cm on a side and $q = 2.00$ μC.

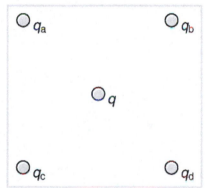

Figure 18.52

19. (a) Using the symmetry of the arrangement, determine the direction of the electric field at the center of the square in **Figure 18.52**, given that $q_a = q_b = -1.00$ μC and $q_c = q_d = +1.00$ μC. (b) Calculate the magnitude of the electric field at the location of q, given that the square is 5.00 cm on a side.

20. Find the electric field at the location of q_a in **Figure 18.52** given that $q_b = q_c = q_d = +2.00$ nC, $q = -1.00$ nC, and the square is 20.0 cm on a side.

21. Find the total Coulomb force on the charge q in **Figure 18.52**, given that $q = 1.00$ μC, $q_a = 2.00$ μC, $q_b = -3.00$ μC, $q_c = -4.00$ μC, and $q_d = +1.00$ μC. The square is 50.0 cm on a side.

22. (a) Find the electric field at the location of q_a in **Figure 18.53**, given that $q_b = +10.00$ μC and $q_c = -5.00$ μC. (b) What is the force on q_a, given that $q_a = +1.50$ nC?

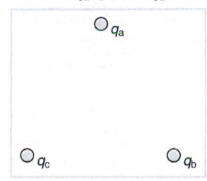

Figure 18.53 Point charges located at the corners of an equilateral triangle 25.0 cm on a side.

23. (a) Find the electric field at the center of the triangular configuration of charges in **Figure 18.53**, given that $q_a = +2.50$ nC, $q_b = -8.00$ nC, and $q_c = +1.50$ nC. (b) Is there any combination of charges, other than $q_a = q_b = q_c$, that will produce a zero strength electric field at the center of the triangular configuration?

18.4 Coulomb's Law

24. What is the repulsive force between two pith balls that are 8.00 cm apart and have equal charges of -30.0 nC?

25. (a) How strong is the attractive force between a glass rod with a 0.700 μC charge and a silk cloth with a -0.600 μC charge, which are 12.0 cm apart, using the approximation that they act like point charges? (b) Discuss how the answer to this problem might be affected if the charges are distributed over some area and do not act like point charges.

26. Two point charges exert a 5.00 N force on each other. What will the force become if the distance between them is increased by a factor of three?

27. Two point charges are brought closer together, increasing the force between them by a factor of 25. By what factor was their separation decreased?

28. How far apart must two point charges of 75.0 nC (typical of static electricity) be to have a force of 1.00 N between them?

29. If two equal charges each of 1 C each are separated in air by a distance of 1 km, what is the magnitude of the force acting between them? You will see that even at a distance as large as 1 km, the repulsive force is substantial because 1 C is a very significant amount of charge.

30. A test charge of $+2\ \mu C$ is placed halfway between a charge of $+6\ \mu C$ and another of $+4\ \mu C$ separated by 10 cm. (a) What is the magnitude of the force on the test charge? (b) What is the direction of this force (away from or toward the $+6\ \mu C$ charge)?

31. Bare free charges do not remain stationary when close together. To illustrate this, calculate the acceleration of two isolated protons separated by 2.00 nm (a typical distance between gas atoms). Explicitly show how you follow the steps in the Problem-Solving Strategy for electrostatics.

32. (a) By what factor must you change the distance between two point charges to change the force between them by a factor of 10? (b) Explain how the distance can either increase or decrease by this factor and still cause a factor of 10 change in the force.

33. Suppose you have a total charge q_{tot} that you can split in any manner. Once split, the separation distance is fixed. How do you split the charge to achieve the greatest force?

34. (a) Common transparent tape becomes charged when pulled from a dispenser. If one piece is placed above another, the repulsive force can be great enough to support the top piece's weight. Assuming equal point charges (only an approximation), calculate the magnitude of the charge if electrostatic force is great enough to support the weight of a 10.0 mg piece of tape held 1.00 cm above another. (b) Discuss whether the magnitude of this charge is consistent with what is typical of static electricity.

35. (a) Find the ratio of the electrostatic to gravitational force between two electrons. (b) What is this ratio for two protons? (c) Why is the ratio different for electrons and protons?

36. At what distance is the electrostatic force between two protons equal to the weight of one proton?

37. A certain five cent coin contains 5.00 g of nickel. What fraction of the nickel atoms' electrons, removed and placed 1.00 m above it, would support the weight of this coin? The atomic mass of nickel is 58.7, and each nickel atom contains 28 electrons and 28 protons.

38. (a) Two point charges totaling $8.00\ \mu C$ exert a repulsive force of 0.150 N on one another when separated by 0.500 m. What is the charge on each? (b) What is the charge on each if the force is attractive?

39. Point charges of $5.00\ \mu C$ and $-3.00\ \mu C$ are placed 0.250 m apart. (a) Where can a third charge be placed so that the net force on it is zero? (b) What if both charges are positive?

40. Two point charges q_1 and q_2 are 3.00 m apart, and their total charge is $20\ \mu C$. (a) If the force of repulsion between them is 0.075N, what are magnitudes of the two charges? (b) If one charge attracts the other with a force of 0.525N, what are the magnitudes of the two charges? Note that you may need to solve a quadratic equation to reach your answer.

18.5 Electric Field: Concept of a Field Revisited

41. What is the magnitude and direction of an electric field that exerts a $2.00 \times 10^{-5}\ \text{N}$ upward force on a $-1.75\ \mu C$ charge?

42. What is the magnitude and direction of the force exerted on a $3.50\ \mu C$ charge by a 250 N/C electric field that points due east?

43. Calculate the magnitude of the electric field 2.00 m from a point charge of 5.00 mC (such as found on the terminal of a Van de Graaff).

44. (a) What magnitude point charge creates a 10,000 N/C electric field at a distance of 0.250 m? (b) How large is the field at 10.0 m?

45. Calculate the initial (from rest) acceleration of a proton in a 5.00×10^6 N/C electric field (such as created by a research Van de Graaff). Explicitly show how you follow the steps in the Problem-Solving Strategy for electrostatics.

46. (a) Find the direction and magnitude of an electric field that exerts a 4.80×10^{-17} N westward force on an electron. (b) What magnitude and direction force does this field exert on a proton?

18.6 Electric Field Lines: Multiple Charges

47. (a) Sketch the electric field lines near a point charge $+q$.

(b) Do the same for a point charge $-3.00q$.

48. Sketch the electric field lines a long distance from the charge distributions shown in **Figure 18.34** (a) and (b)

49. **Figure 18.54** shows the electric field lines near two charges q_1 and q_2. What is the ratio of their magnitudes? (b) Sketch the electric field lines a long distance from the charges shown in the figure.

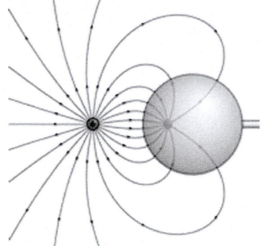

Figure 18.54 The electric field near two charges.

50. Sketch the electric field lines in the vicinity of two opposite charges, where the negative charge is three times greater in magnitude than the positive. (See **Figure 18.54** for a similar situation).

18.8 Applications of Electrostatics

51. (a) What is the electric field 5.00 m from the center of the terminal of a Van de Graaff with a 3.00 mC charge, noting that the field is equivalent to that of a point charge at the center of the terminal? (b) At this distance, what force does the field exert on a $2.00\ \mu C$ charge on the Van de Graaff's belt?

52. (a) What is the direction and magnitude of an electric field that supports the weight of a free electron near the surface of Earth? (b) Discuss what the small value for this field implies regarding the relative strength of the gravitational and electrostatic forces.

53. A simple and common technique for accelerating electrons is shown in **Figure 18.55**, where there is a uniform electric field between two plates. Electrons are released, usually from a hot filament, near the negative plate, and there is a small hole in the positive plate that allows the electrons to continue moving. (a) Calculate the acceleration of the electron if the field strength is 2.50×10^4 N/C. (b) Explain why the electron will not be pulled back to the positive plate once it moves through the hole.

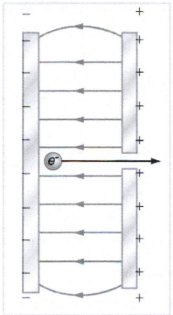

Figure 18.55 Parallel conducting plates with opposite charges on them create a relatively uniform electric field used to accelerate electrons to the right. Those that go through the hole can be used to make a TV or computer screen glow or to produce X-rays.

54. Earth has a net charge that produces an electric field of approximately 150 N/C downward at its surface. (a) What is the magnitude and sign of the excess charge, noting the electric field of a conducting sphere is equivalent to a point charge at its center? (b) What acceleration will the field produce on a free electron near Earth's surface? (c) What mass object with a single extra electron will have its weight supported by this field?

55. Point charges of $25.0\ \mu C$ and $45.0\ \mu C$ are placed 0.500 m apart. (a) At what point along the line between them is the electric field zero? (b) What is the electric field halfway between them?

56. What can you say about two charges q_1 and q_2, if the electric field one-fourth of the way from q_1 to q_2 is zero?

57. Integrated Concepts

Calculate the angular velocity ω of an electron orbiting a proton in the hydrogen atom, given the radius of the orbit is 0.530×10^{-10} m. You may assume that the proton is stationary and the centripetal force is supplied by Coulomb attraction.

58. Integrated Concepts

An electron has an initial velocity of 5.00×10^6 m/s in a uniform 2.00×10^5 N/C strength electric field. The field accelerates the electron in the direction opposite to its initial velocity. (a) What is the direction of the electric field? (b) How far does the electron travel before coming to rest? (c) How long does it take the electron to come to rest? (d) What is the electron's velocity when it returns to its starting point?

59. Integrated Concepts

The practical limit to an electric field in air is about 3.00×10^6 N/C. Above this strength, sparking takes place because air begins to ionize and charges flow, reducing the field. (a) Calculate the distance a free proton must travel in this field to reach 3.00% of the speed of light, starting from rest. (b) Is this practical in air, or must it occur in a vacuum?

60. Integrated Concepts

A 5.00 g charged insulating ball hangs on a 30.0 cm long string in a uniform horizontal electric field as shown in **Figure 18.56**. Given the charge on the ball is $1.00\ \mu C$, find the strength of the field.

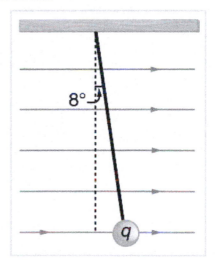

Figure 18.56 A horizontal electric field causes the charged ball to hang at an angle of $8.00°$.

61. Integrated Concepts

Figure 18.57 shows an electron passing between two charged metal plates that create an 100 N/C vertical electric field perpendicular to the electron's original horizontal velocity. (These can be used to change the electron's direction, such as in an oscilloscope.) The initial speed of the electron is 3.00×10^6 m/s , and the horizontal distance it travels in the uniform field is 4.00 cm. (a) What is its vertical deflection? (b) What is the vertical component of its final velocity? (c) At what angle does it exit? Neglect any edge effects.

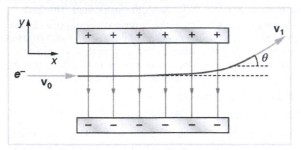

Figure 18.57

62. Integrated Concepts

The classic Millikan oil drop experiment was the first to obtain an accurate measurement of the charge on an electron. In it, oil drops were suspended against the gravitational force by a vertical electric field. (See **Figure 18.58**.) Given the oil drop to be $1.00 \ \mu m$ in radius and have a density of $920 \ \text{kg/m}^3$:

(a) Find the weight of the drop. (b) If the drop has a single excess electron, find the electric field strength needed to balance its weight.

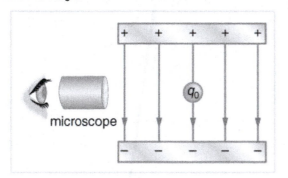

Figure 18.58 In the Millikan oil drop experiment, small drops can be suspended in an electric field by the force exerted on a single excess electron. Classically, this experiment was used to determine the electron charge q_e by measuring the electric field and mass of the drop.

63. Integrated Concepts

(a) In **Figure 18.59**, four equal charges q lie on the corners of a square. A fifth charge Q is on a mass m directly above the center of the square, at a height equal to the length d of one side of the square. Determine the magnitude of q in terms of Q, m, and d, if the Coulomb force is to equal the weight of m. (b) Is this equilibrium stable or unstable? Discuss.

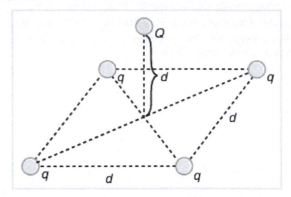

Figure 18.59 Four equal charges on the corners of a horizontal square support the weight of a fifth charge located directly above the center of the square.

64. Unreasonable Results

(a) Calculate the electric field strength near a 10.0 cm diameter conducting sphere that has 1.00 C of excess charge on it. (b) What is unreasonable about this result? (c) Which assumptions are responsible?

65. Unreasonable Results

(a) Two 0.500 g raindrops in a thunderhead are 1.00 cm apart when they each acquire 1.00 mC charges. Find their acceleration. (b) What is unreasonable about this result? (c) Which premise or assumption is responsible?

66. Unreasonable Results

A wrecking yard inventor wants to pick up cars by charging a 0.400 m diameter ball and inducing an equal and opposite charge on the car. If a car has a 1000 kg mass and the ball is to be able to lift it from a distance of 1.00 m: (a) What minimum charge must be used? (b) What is the electric field near the surface of the ball? (c) Why are these results unreasonable? (d) Which premise or assumption is responsible?

67. Construct Your Own Problem

Consider two insulating balls with evenly distributed equal and opposite charges on their surfaces, held with a certain distance between the centers of the balls. Construct a problem in which you calculate the electric field (magnitude and direction) due to the balls at various points along a line running through the centers of the balls and extending to infinity on either side. Choose interesting points and comment on the meaning of the field at those points. For example, at what points might the field be just that due to one ball and where does the field become negligibly small? Among the things to be considered are the magnitudes of the charges and the distance between the centers of the balls. Your instructor may wish for you to consider the electric field off axis or for a more complex array of charges, such as those in a water molecule.

68. Construct Your Own Problem

Consider identical spherical conducting space ships in deep space where gravitational fields from other bodies are negligible compared to the gravitational attraction between the ships. Construct a problem in which you place identical excess charges on the space ships to exactly counter their gravitational attraction. Calculate the amount of excess charge needed. Examine whether that charge depends on the distance between the centers of the ships, the masses of the ships, or any other factors. Discuss whether this would be an easy, difficult, or even impossible thing to do in practice.

Test Prep for AP® Courses

18.1 Static Electricity and Charge: Conservation of Charge

1. When a glass rod is rubbed against silk, which of the following statements is true?
   a. Electrons are removed from the silk.
   b. Electrons are removed from the rod.
   c. Protons are removed from the silk.
   d. Protons are removed from the rod.

2. In an experiment, three microscopic latex spheres are sprayed into a chamber and become charged with $+3e$, $+5e$, and $-3e$, respectively. Later, all three spheres collide simultaneously and then separate. Which of the following are possible values for the final charges on the spheres? Select *two* answers.

	X	Y	Z
(a)	+4e	−4e	+5e
(b)	−4e	+4.5e	+5.5e
(c)	+5e	−8e	+7e
(d)	+6e	+6e	−7e

3. If objects X and Y attract each other, which of the following will be false?
   a. X has positive charge and Y has negative charge.
   b. X has negative charge and Y has positive charge.
   c. X and Y both have positive charge.
   d. X is neutral and Y has a charge.

4. Suppose a positively charged object A is brought in contact with an uncharged object B in a closed system. What type of charge will be left on object B?

   a. negative
   b. positive
   c. neutral
   d. cannot be determined

5. What will be the net charge on an object which attracts neutral pieces of paper but repels a negatively charged balloon?
   a. negative
   b. positive
   c. neutral
   d. cannot be determined

6. When two neutral objects are rubbed against each other, the first one gains a net charge of 3e. Which of the following statements is true?
   a. The second object gains 3e and is negatively charged.
   b. The second object loses 3e and is negatively charged.
   c. The second object gains 3e and is positively charged.
   d. The second object loses 3e and is positively charged.

7. In an experiment, a student runs a comb through his hair several times and brings it close to small pieces of paper. Which of the following will he observe?
   a. Pieces of paper repel the comb.
   b. Pieces of paper are attracted to the comb.
   c. Some pieces of paper are attracted and some repel the comb.
   d. There is no attraction or repulsion between the pieces of paper and the comb.

8. In an experiment a negatively charged balloon (balloon X) is repelled by another charged balloon Y. However, an object Z is attracted to balloon Y. Which of the following can be the charge on Z? Select *two* answers.
   a. negative
   b. positive
   c. neutral
   d. cannot be determined

9. Suppose an object has a charge of 1 C and gains 6.88×10^{18} electrons.

a. What will be the net charge of the object?
b. If the object has gained electrons from a neutral object, what will be the charge on the neutral object?
c. Find and explain the relationship between the total charges of the two objects before and after the transfer.
d. When a third object is brought in contact with the first object (after it gains the electrons), the resulting charge on the third object is 0.4 C. What was its initial charge?

10. The charges on two identical metal spheres (placed in a closed system) are -2.4×10^{-17} C and -4.8×10^{-17} C.

a. How many electrons will be equivalent to the charge on each sphere?
b. If the two spheres are brought in contact and then separated, find the charge on each sphere.
c. Calculate the number of electrons that would be equivalent to the resulting charge on each sphere.

11. In an experiment the following observations are made by a student for four charged objects W, X, Y, and Z:

- A glass rod rubbed with silk attracts W.
- W attracts Z but repels X.
- X attracts Z but repels Y.
- Y attracts W and Z.

Estimate whether the charges on each of the four objects are positive, negative, or neutral.

18.2 Conductors and Insulators

12. Some students experimenting with an uncharged metal sphere want to give the sphere a net charge using a charged aluminum pie plate. Which of the following steps would give the sphere a net charge of the same sign as the pie plate?

a. bringing the pie plate close to, but not touching, the metal sphere, then moving the pie plate away.
b. bringing the pie plate close to, but not touching, the metal sphere, then momentarily touching a grounding wire to the metal sphere.
c. bringing the pie plate close to, but not touching, the metal sphere, then momentarily touching a grounding wire to the pie plate.
d. touching the pie plate to the metal sphere.

13.

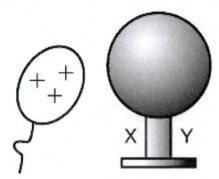

Figure 18.60 Balloon and sphere.

When the balloon is brought closer to the sphere, there will be a redistribution of charges. What is this phenomenon called?

a. electrostatic repulsion
b. conduction
c. polarization
d. none of the above

14. What will be the charge at Y (i.e., the part of the sphere furthest from the balloon)?

a. positive
b. negative
c. zero
d. It can be positive or negative depending on the material.

15. What will be the net charge on the sphere?

a. positive
b. negative
c. zero
d. It can be positive or negative depending on the material.

16. If Y is grounded while the balloon is still close to X, which of the following will be true?

a. Electrons will flow from the sphere to the ground.
b. Electrons will flow from the ground to the sphere.
c. Protons will flow from the sphere to the ground.
d. Protons will flow from the ground to the sphere.

17. If the balloon is moved away after grounding, what will be the net charge on the sphere?

a. positive
b. negative
c. zero
d. It can be positive or negative depending on the material.

18. A positively charged rod is used to charge a sphere by induction. Which of the following is true?

a. The sphere must be a conductor.
b. The sphere must be an insulator.
c. The sphere can be a conductor or insulator but must be connected to ground.
d. The sphere can be a conductor or insulator but must be already charged.

19.

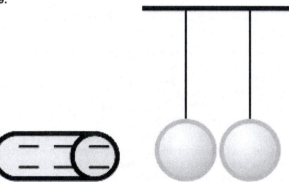
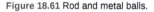

Figure 18.61 Rod and metal balls.

As shown in the figure above, two metal balls are suspended and a negatively charged rod is brought close to them.

a. If the two balls are in contact with each other what will be the charges on each ball?
b. Explain how the balls get these charges.
c. What will happen to the charge on the second ball (i.e., the ball further away from the rod) if it is momentarily grounded while the rod is still there?
d. If (instead of grounding) the second ball is moved away and then the rod is removed from the first ball, will the two balls have induced charges? If yes, what will be the charges? If no, why not?

20. Two experiments are performed using positively charged glass rods and neutral electroscopes. In the first experiment the rod is brought in contact with the electroscope. In the second experiment the rod is only brought close to the electroscope but not in contact. However, while the rod is close, the electroscope is momentarily grounded and then the rod is removed. In both experiments the needles of the

electroscopes deflect, which indicates the presence of charges.
  a. What is the charging method in each of the two experiments?
  b. What is the net charge on the electroscope in the first experiment? Explain how the electroscope obtains that charge.
  c. Is the net charge on the electroscope in the second experiment different from that of the first experiment? Explain why.

18.3 Conductors and Electric Fields in Static Equilibrium

21.

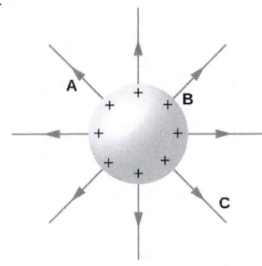

Figure 18.62 A sphere conductor.

An electric field due to a positively charged spherical conductor is shown above. Where will the electric field be weakest?

  a. Point A
  b. Point B
  c. Point C
  d. Same at all points

22.

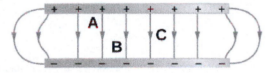

Figure 18.63 Electric field between two parallel metal plates.

The electric field created by two parallel metal plates is shown above. Where will the electric field be strongest?

  a. Point A
  b. Point B
  c. Point C
  d. Same at all points

23. Suppose that the electric field experienced due to a positively charged small spherical conductor at a certain distance is E. What will be the percentage change in electric field experienced at thrice the distance if the charge on the conductor is doubled?

24.

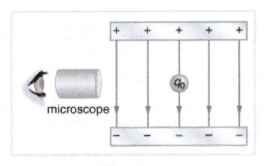

Figure 18.64 Millikan oil drop experiment.

The classic Millikan oil drop experiment setup is shown above. In this experiment oil drops are suspended in a vertical electric field against the gravitational force to measure their charge. If the mass of a negatively charged drop suspended in an electric field of 1.18×10^{-4} N/C strength is 3.85×10^{-21} g, find the number of excess electrons in the drop.

18.4 Coulomb's Law

25. For questions 25–27, suppose that the electrostatics force between two charges is F.

What will be the force if the distance between them is halved?

  a. $4F$
  b. $2F$
  c. $F/4$
  d. $F/2$

26. Which of the following is false?
  a. If the charge of one of the particles is doubled and that of the second is unchanged, the force will become $2F$.
  b. If the charge of one of the particles is doubled and that of the second is halved, the force will remain F.
  c. If the charge of both the particles is doubled, the force will become $4F$.
  d. None of the above.

27. Which of the following is true about the gravitational force between the particles?
  a. It will be 3.25×10^{-38} F.
  b. It will be 3.25×10^{38} F.
  c. It will be equal to F.
  d. It is not possible to determine the gravitational force as the masses of the particles are not given.

28. Two massive, positively charged particles are initially held a fixed distance apart. When they are moved farther apart, the magnitude of their mutual gravitational force changes by a factor of n. Which of the following indicates the factor by which the magnitude of their mutual electrostatic force changes?
  a. $1/n^2$
  b. $1/n$
  c. n
  d. n^2

29.
  a. What is the electrostatic force between two charges of 1 C each, separated by a distance of 0.5 m?
  b. How will this force change if the distance is increased to 1 m?

30.

a. Find the ratio of the electrostatic force to the gravitational force between two electrons.
b. Will this ratio change if the two electrons are replaced by protons? If yes, find the new ratio.

18.5 Electric Field: Concept of a Field Revisited

31. Two particles with charges $+2q$ and $+q$ are separated by a distance r. The $+2q$ particle has an electric field E at distance r and exerts a force F on the $+q$ particle. Use this information to answer questions 31–32.

What is the electric field of the $+q$ particle at the same distance and what force does it exert on the $+2q$ particle?

a. $E/2$, $F/2$
b. E, $F/2$
c. $E/2$, F
d. E, F

32. When the $+q$ particle is replaced by a $+3q$ particle, what will be the electric field and force from the $+2q$ particle experienced by the $+3q$ particle?
a. $E/3$, $3F$
b. E, $3F$
c. $E/3$, F
d. E, F

33. The direction of the electric field of a negative charge is
a. inward for both positive and negative charges.
b. outward for both positive and negative charges.
c. inward for other positive charges and outward for other negative charges.
d. outward for other positive charges and inward for other negative charges.

34. The force responsible for holding an atom together is
a. frictional
b. electric
c. gravitational
d. magnetic

35. When a positively charged particle exerts an inward force on another particle P, what will be the charge of P?
a. positive
b. negative
c. neutral
d. cannot be determined

36. Find the force exerted due to a particle having a charge of 3.2×10^{-19} C on another identical particle 5 cm away.

37. Suppose that the force exerted on an electron is 5.6×10^{-17} N, directed to the east.
a. Find the magnitude of the electric field that exerts the force.
b. What will be the direction of the electric field?
c. If the electron is replaced by a proton, what will be the magnitude of force exerted?
d. What will be the direction of force on the proton?

18.6 Electric Field Lines: Multiple Charges

38.

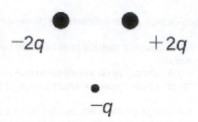

Figure 18.65 An electric dipole (with $+2q$ and $-2q$ as the two charges) is shown in the figure above. A third charge, $-q$ is placed equidistant from the dipole charges. What will be the direction of the net force on the third charge?

a. →
b. ←
c. ↓
d. ↑

39.

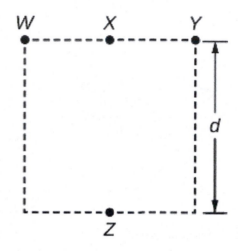

Figure 18.66

Four objects, each with charge $+q$, are held fixed on a square with sides of length d, as shown in the figure. Objects X and Z are at the midpoints of the sides of the square. The electrostatic force exerted by object W on object X is F. Use this information to answer questions 39–40.

What is the magnitude of force exerted by object W on Z?

a. $F/7$
b. $F/5$
c. $F/3$
d. $F/2$

40. What is the magnitude of the net force exerted on object X by objects W, Y, and Z?
a. $F/4$
b. $F/2$
c. $9F/4$
d. $3F$

41.

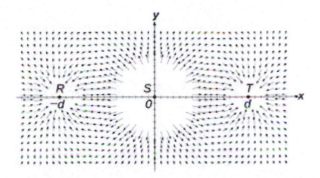

Figure 18.67 Electric field with three charged objects.

The figure above represents the electric field in the vicinity of three small charged objects, R, S, and T. The objects have charges −q, +2q, and −q, respectively, and are located on the x-axis at −d, 0, and d. Field vectors of very large magnitude are omitted for clarity.

(a) i) Briefly describe the characteristics of the field diagram that indicate that the sign of the charges of objects R and T is negative and that the sign of the charge of object S is positive.

ii) Briefly describe the characteristics of the field diagram that indicate that the magnitudes of the charges of objects R and T are equal and that the magnitude of the charge of object S is about twice that of objects R and T.

For the following parts, an electric field directed to the right is defined to be positive.

(b) On the axes below, sketch a graph of the electric field E along the x-axis as a function of position x.

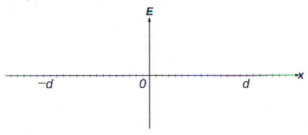

Figure 18.68 An Electric field (E) axis and Position (x) axis.

(c) Write an expression for the electric field E along the x-axis as a function of position x in the region between objects S and T in terms of q, d, and fundamental constants, as appropriate.

(d) Your classmate tells you there is a point between S and T where the electric field is zero. Determine whether this statement is true, and explain your reasoning using two of the representations from parts (a), (b), or (c).

19   ELECTRIC POTENTIAL AND ELECTRIC FIELD

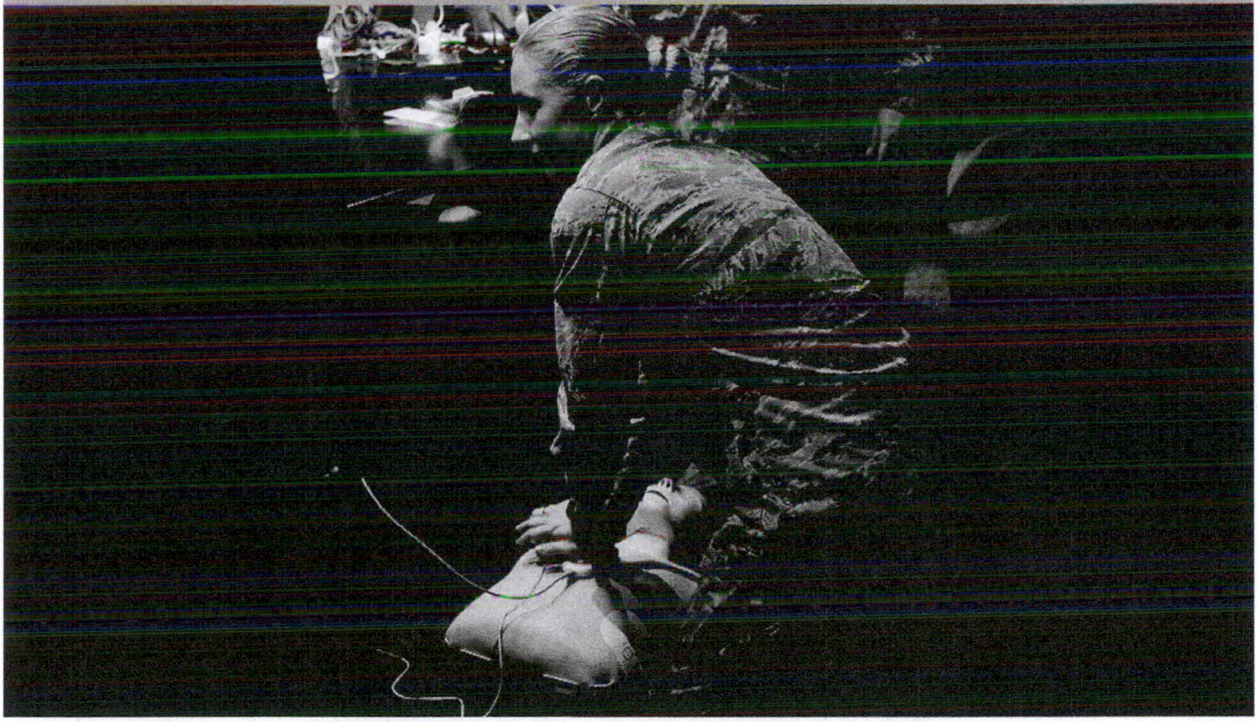

Figure 19.1 Automated external defibrillator unit (AED) (credit: U.S. Defense Department photo/Tech. Sgt. Suzanne M. Day)

Chapter Outline

19.1. Electric Potential Energy: Potential Difference

19.2. Electric Potential in a Uniform Electric Field

19.3. Electrical Potential Due to a Point Charge

19.4. Equipotential Lines

19.5. Capacitors and Dielectrics

19.6. Capacitors in Series and Parallel

19.7. Energy Stored in Capacitors

Connection for AP® Courses

In Electric Charge and Electric Field, we just scratched the surface (or at least rubbed it) of electrical phenomena. Two of the most familiar aspects of electricity are its energy and *voltage*. We know, for example, that great amounts of electrical energy can be stored in batteries, are transmitted cross-country through power lines, and may jump from clouds to explode the sap of trees. In a similar manner, at molecular levels, *ions* cross cell membranes and transfer information. We also know about voltages associated with electricity. Batteries are typically a few volts, the outlets in your home produce 120 volts, and power lines can be as high as hundreds of thousands of volts. But energy and voltage are not the same thing. A motorcycle battery, for example, is small and would not be very successful in replacing the much larger battery in a car, yet each has the same voltage. In this chapter, we shall examine the relationship between voltage and electrical energy and begin to explore some of the many applications of electricity. We do so by introducing the concept of electric potential and describing the relationship between electric field and electric potential.

This chapter presents the concept of equipotential lines (lines of equal potential) as a way to visualize the electric field (Enduring Understanding 2.E, Essential Knowledge 2.E.2). An analogy between the isolines on topographic maps for gravitational field and equipotential lines for the electric field is used to develop a conceptual understanding of equipotential lines (Essential Knowledge 2.E.1). The relationship between the magnitude of an electric field, change in electric potential, and displacement is stated for a uniform field and extended for the more general case using the concept of the "average value" of the electric field (Essential Knowledge 2.E.3).

The concept that an electric field is caused by charged objects (Enduring Understanding 2.C) supports Big Idea 2, that fields exist in space and can be used to explain interactions. The relationship between the electric field, electric charge, and electric force (Essential Knowledge 2.C.1) is used to describe the behavior of charged particles. The uniformity of the electric field between two oppositely charged parallel plates with uniformly distributed electric charge (Essential Knowledge 2.C.5), as well as the properties of materials and their geometry, are used to develop understanding of the capacitance of a capacitor (Essential Knowledge 4.E.4).

This chapter also supports Big Idea 4, that interactions between systems result in changes in those systems. This idea is applied to electric properties of various systems of charged objects, demonstrating the effect of electric interactions on electric properties of systems (Enduring Understanding 4.E). This fact in turn supports Big Idea 5, that changes due to interactions are governed by conservation laws. In particular, the energy of a system is conserved (Enduring Understanding 5.B). Any system that has internal structure can have internal energy. For a system of charged objects, internal energy can change as a result of changes in the arrangement of charges and their geometric configuration as long as work is done on, or by, the system (Essential Knowledge 5.B.2). When objects within the system interact with conservative forces, such as electric forces, the internal energy is defined by the potential energy of that interaction (Essential Knowledge 5.B.3). In general, the internal energy of a system is the sum of the kinetic energies of all its objects and the potential energy of interaction between the objects within the system (Essential Knowledge 5.B.4).

The concepts in this chapter support:

Big Idea 1 Objects and systems have properties such as mass and charge. Systems may have internal structure.

Enduring Understanding 1.E Materials have many macroscopic properties that result from the arrangement and interactions of the atoms and molecules that make up the material.

Essential Knowledge 1.E.4 Matter has a property called electric permittivity.

Big Idea 2 Fields existing in space can be used to explain interactions.

Enduring Understanding 2.C An electric field is caused by an object with electric charge.

Essential Knowledge 2.C.1 The magnitude of the electric force F exerted on an object with electric charge q by an electric field F= qE. The direction of the force is determined by the direction of the field and the sign of the charge, with positively charged objects accelerating in the direction of the field and negatively charged objects accelerating in the direction opposite the field. This should include a vector field map for positive point charges, negative point charges, spherically symmetric charge distribution, and uniformly charged parallel plates.

Essential Knowledge 2.C.5 Between two oppositely charged parallel plates with uniformly distributed electric charge, at points far from the edges of the plates, the electric field is perpendicular to the plates and is constant in both magnitude and direction.

Enduring Understanding 2.E Physicists often construct a map of isolines connecting points of equal value for some quantity related to a field and use these maps to help visualize the field.

Essential Knowledge 2.E.1 Isolines on a topographic (elevation) map describe lines of approximately equal gravitational potential energy per unit mass (gravitational equipotential). As the distance between two different isolines decreases, the steepness of the surface increases. [Contour lines on topographic maps are useful teaching tools for introducing the concept of equipotential lines. Students are encouraged to use the analogy in their answers when explaining gravitational and electrical potential and potential differences.]

Essential Knowledge 2.E.2 Isolines in a region where an electric field exists represent lines of equal electric potential, referred to as equipotential lines.

Essential Knowledge 2.E.3 The average value of the electric field in a region equals the change in electric potential across that region divided by the change in position (displacement) in the relevant direction.

Big Idea 4 Interactions between systems can result in changes in those systems.

Enduring Understanding 4.E The electric and magnetic properties of a system can change in response to the presence of, or changes in, other objects or systems.

Essential Knowledge 4.E.4 The resistance of a resistor, and the capacitance of a capacitor, can be understood from the basic properties of electric fields and forces, as well as the properties of materials and their geometry.

Big Idea 5 Changes that occur as a result of interactions are constrained by conservation laws.

Enduring Understanding 5.B The energy of a system is conserved.

Essential Knowledge 5.B.2 A system with internal structure can have internal energy, and changes in a system's internal structure can result in changes in internal energy. [Physics 1: includes mass–spring oscillators and simple pendulums. Physics 2: charged object in electric fields and examining changes in internal energy with changes in configuration.]

Essential Knowledge 5.B.3 A system with internal structure can have potential energy. Potential energy exists within a system if the objects within that system interact with conservative forces.

Essential Knowledge 5.B.4 The internal energy of a system includes the kinetic energy of the objects that make up the system and the potential energy of the configuration of the objects that make up the system.

19.1 Electric Potential Energy: Potential Difference

Learning Objectives

By the end of this section you will be able to:

- Define electric potential and electric potential energy.
- Describe the relationship between potential difference and electrical potential energy.
- Explain electron volt and its usage in submicroscopic processes.
- Determine electric potential energy given potential difference and amount of charge.

The information presented in this section supports the following AP® learning objectives and science practices:

- **2.C.1.1** The student is able to predict the direction and the magnitude of the force exerted on an object with an electric charge q placed in an electric field **E** using the mathematical model of the relation between an electric force and an electric field: **F** = q**E**; a vector relation. **(S.P. 6.4, 7.2)**
- **2.C.1.2** The student is able to calculate any one of the variables—electric force, electric charge, and electric field—at a point given the values and sign or direction of the other two quantities. **(S.P. 2.2)**
- **5.B.2.1** The student is able to calculate the expected behavior of a system using the object model (i.e., by ignoring changes in internal structure) to analyze a situation. Then, when the model fails, the student can justify the use of conservation of energy principles to calculate the change in internal energy due to changes in internal structure because the object is actually a system. **(S.P. 1.4, 2.1)**
- **5.B.3.1** The student is able to describe and make qualitative and/or quantitative predictions about everyday examples of systems with internal potential energy. **(S.P. 2.2, 6.4, 7.2)**
- **5.B.3.2** The student is able to make quantitative calculations of the internal potential energy of a system from a description or diagram of that system. **(S.P. 1.4, 2.2)**
- **5.B.3.3** The student is able to apply mathematical reasoning to create a description of the internal potential energy of a system from a description or diagram of the objects and interactions in that system. **(S.P. 1.4, 2.2)**
- **5.B.4.1** The student is able to describe and make predictions about the internal energy of systems. **(S.P. 6.4, 7.2)**
- **5.B.4.2** The student is able to calculate changes in kinetic energy and potential energy of a system, using information from representations of that system. **(S.P. 1.4, 2.1, 2.2)**

When a free positive charge q is accelerated by an electric field, such as shown in **Figure 19.2**, it is given kinetic energy. The process is analogous to an object being accelerated by a gravitational field. It is as if the charge is going down an electrical hill where its electric potential energy is converted to kinetic energy. Let us explore the work done on a charge q by the electric field in this process, so that we may develop a definition of electric potential energy.

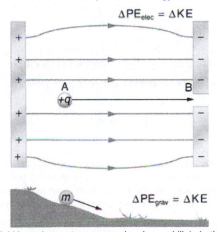

Figure 19.2 A charge accelerated by an electric field is analogous to a mass going down a hill. In both cases potential energy is converted to another form. Work is done by a force, but since this force is conservative, we can write $W = -\Delta PE$.

The electrostatic or Coulomb force is conservative, which means that the work done on q is independent of the path taken. This is exactly analogous to the gravitational force in the absence of dissipative forces such as friction. When a force is conservative, it is possible to define a potential energy associated with the force, and it is usually easier to deal with the potential energy (because it depends only on position) than to calculate the work directly.

We use the letters PE to denote electric potential energy, which has units of joules (J). The change in potential energy, ΔPE, is crucial, since the work done by a conservative force is the negative of the change in potential energy; that is, $W = -\Delta PE$. For example, work W done to accelerate a positive charge from rest is positive and results from a loss in PE, or a negative ΔPE. There must be a minus sign in front of ΔPE to make W positive. PE can be found at any point by taking one point as a reference and calculating the work needed to move a charge to the other point.

Potential Energy

$W = -\Delta \text{PE}$. For example, work W done to accelerate a positive charge from rest is positive and results from a loss in PE, or a negative ΔPE. There must be a minus sign in front of ΔPE to make W positive. PE can be found at any point by taking one point as a reference and calculating the work needed to move a charge to the other point.

Gravitational potential energy and electric potential energy are quite analogous. Potential energy accounts for work done by a conservative force and gives added insight regarding energy and energy transformation without the necessity of dealing with the force directly. It is much more common, for example, to use the concept of voltage (related to electric potential energy) than to deal with the Coulomb force directly.

Calculating the work directly is generally difficult, since $W = Fd \cos \theta$ and the direction and magnitude of F can be complex for multiple charges, for odd-shaped objects, and along arbitrary paths. But we do know that, since $F = qE$, the work, and hence ΔPE, is proportional to the test charge q. To have a physical quantity that is independent of test charge, we define **electric potential** V (or simply potential, since electric is understood) to be the potential energy per unit charge:

$$V = \frac{\text{PE}}{q}. \tag{19.1}$$

Electric Potential

This is the electric potential energy per unit charge.

$$V = \frac{\text{PE}}{q} \tag{19.2}$$

Since PE is proportional to q, the dependence on q cancels. Thus V does not depend on q. The change in potential energy ΔPE is crucial, and so we are concerned with the difference in potential or potential difference ΔV between two points, where

$$\Delta V = V_{\text{B}} - V_{\text{A}} = \frac{\Delta \text{PE}}{q}. \tag{19.3}$$

The **potential difference** between points A and B, $V_{\text{B}} - V_{\text{A}}$, is thus defined to be the change in potential energy of a charge q moved from A to B, divided by the charge. Units of potential difference are joules per coulomb, given the name volt (V) after Alessandro Volta.

$$1 \text{ V} = 1 \frac{\text{J}}{\text{C}} \tag{19.4}$$

Potential Difference

The potential difference between points A and B, $V_{\text{B}} - V_{\text{A}}$, is defined to be the change in potential energy of a charge q moved from A to B, divided by the charge. Units of potential difference are joules per coulomb, given the name volt (V) after Alessandro Volta.

$$1 \text{ V} = 1 \frac{\text{J}}{\text{C}} \tag{19.5}$$

The familiar term **voltage** is the common name for potential difference. Keep in mind that whenever a voltage is quoted, it is understood to be the potential difference between two points. For example, every battery has two terminals, and its voltage is the potential difference between them. More fundamentally, the point you choose to be zero volts is arbitrary. This is analogous to the fact that gravitational potential energy has an arbitrary zero, such as sea level or perhaps a lecture hall floor.

In summary, the relationship between potential difference (or voltage) and electrical potential energy is given by

$$\Delta V = \frac{\Delta \text{PE}}{q} \text{ and } \Delta \text{PE} = q\Delta V. \tag{19.6}$$

Potential Difference and Electrical Potential Energy

The relationship between potential difference (or voltage) and electrical potential energy is given by

$$\Delta V = \frac{\Delta \text{PE}}{q} \text{ and } \Delta \text{PE} = q\Delta V. \tag{19.7}$$

The second equation is equivalent to the first.

Real World Connections: Electric Potential in Electronic Devices

You probably use devices with stored electric potential daily. Do you own or use any electronic devices that do not have to be attached to a wall socket? What happens if you use these items long enough? Do they cease functioning? What do you do in that case? Choose one of these types of electronic devices and determine how much electric potential (measured in volts) the item requires for proper functioning. Then estimate the amount of time between replenishments of potential. Describe how the time between replenishments of potential depends on use.

Answer

Ready examples include calculators and cell phones. The former will either be solar powered, or have replaceable batteries, probably four 1.5 V for a total of 6 V. The latter will need to be recharged with a specialized charger, which probably puts out 5 V. Times will be highly dependent on which item is used, but should be less with more intense use.

Voltage is not the same as energy. Voltage is the energy per unit charge. Thus a motorcycle battery and a car battery can both have the same voltage (more precisely, the same potential difference between battery terminals), yet one stores much more energy than the other since $\Delta PE = q\Delta V$. The car battery can move more charge than the motorcycle battery, although both are 12 V batteries.

Example 19.1 Calculating Energy

Suppose you have a 12.0 V motorcycle battery that can move 5000 C of charge, and a 12.0 V car battery that can move 60,000 C of charge. How much energy does each deliver? (Assume that the numerical value of each charge is accurate to three significant figures.)

Strategy

To say we have a 12.0 V battery means that its terminals have a 12.0 V potential difference. When such a battery moves charge, it puts the charge through a potential difference of 12.0 V, and the charge is given a change in potential energy equal to $\Delta PE = q\Delta V$.

So to find the energy output, we multiply the charge moved by the potential difference.

Solution

For the motorcycle battery, $q = 5000 \text{ C}$ and $\Delta V = 12.0 \text{ V}$. The total energy delivered by the motorcycle battery is

$$
\begin{aligned}
\Delta PE_{\text{cycle}} &= (5000 \text{ C})(12.0 \text{ V}) \\
&= (5000 \text{ C})(12.0 \text{ J/C}) \\
&= 6.00 \times 10^4 \text{ J}.
\end{aligned}
\tag{19.8}
$$

Similarly, for the car battery, $q = 60,000 \text{ C}$ and

$$
\begin{aligned}
\Delta PE_{\text{car}} &= (60,000 \text{ C})(12.0 \text{ V}) \\
&= 7.20 \times 10^5 \text{ J}.
\end{aligned}
\tag{19.9}
$$

Discussion

While voltage and energy are related, they are not the same thing. The voltages of the batteries are identical, but the energy supplied by each is quite different. Note also that as a battery is discharged, some of its energy is used internally and its terminal voltage drops, such as when headlights dim because of a low car battery. The energy supplied by the battery is still calculated as in this example, but not all of the energy is available for external use.

Note that the energies calculated in the previous example are absolute values. The change in potential energy for the battery is negative, since it loses energy. These batteries, like many electrical systems, actually move negative charge—electrons in particular. The batteries repel electrons from their negative terminals (A) through whatever circuitry is involved and attract them to their positive terminals (B) as shown in **Figure 19.3**. The change in potential is $\Delta V = V_B - V_A = +12 \text{ V}$ and the charge q is negative, so that $\Delta PE = q\Delta V$ is negative, meaning the potential energy of the battery has decreased when q has moved from A to B.

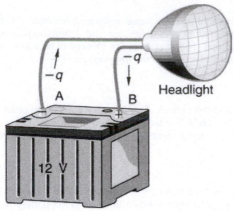

Figure 19.3 A battery moves negative charge from its negative terminal through a headlight to its positive terminal. Appropriate combinations of chemicals in the battery separate charges so that the negative terminal has an excess of negative charge, which is repelled by it and attracted to the excess positive charge on the other terminal. In terms of potential, the positive terminal is at a higher voltage than the negative. Inside the battery, both positive and negative charges move.

Making Connections: Potential Energy in a Battery

The previous example stated that the potential energy of a battery decreased with each electron it pushed out. However, shouldn't this reduced internal energy reduce the potential, as well? Yes, it should. So why don't we notice this?

Part of the answer is that the amount of energy taken by any one electron is extremely small, and therefore it doesn't reduce the potential much. But the main reason is that the energy is stored in the battery as a chemical reaction waiting to happen, not as electric potential. This reaction only runs when a load is attached to both terminals of the battery. Any one set of chemical reactants has a certain maximum potential that it can provide; this is why larger batteries consist of cells attached in series, so that the overall potential increases additively. As these reactants get used up, each cell gives less potential to the electrons it is moving; eventually this potential falls below a useful threshold. Then the battery either needs to be charged, which reverses the chemical reaction and reconstitutes the original reactants; or changed.

Example 19.2 How Many Electrons Move through a Headlight Each Second?

When a 12.0 V car battery runs a single 30.0 W headlight, how many electrons pass through it each second?

Strategy

To find the number of electrons, we must first find the charge that moved in 1.00 s. The charge moved is related to voltage and energy through the equation $\Delta\text{PE} = q\Delta V$. A 30.0 W lamp uses 30.0 joules per second. Since the battery loses energy, we have $\Delta\text{PE} = -30.0\,\text{J}$ and, since the electrons are going from the negative terminal to the positive, we see that $\Delta V = +12.0\,\text{V}$.

Solution

To find the charge q moved, we solve the equation $\Delta\text{PE} = q\Delta V$:

$$q = \frac{\Delta\text{PE}}{\Delta V}. \tag{19.10}$$

Entering the values for ΔPE and ΔV, we get

$$q = \frac{-30.0\,\text{J}}{+12.0\,\text{V}} = \frac{-30.0\,\text{J}}{+12.0\,\text{J/C}} = -2.50\,\text{C}. \tag{19.11}$$

The number of electrons n_e is the total charge divided by the charge per electron. That is,

$$n_e = \frac{-2.50\,\text{C}}{-1.60\times10^{-19}\,\text{C/e}^-} = 1.56\times10^{19}\ \text{electrons}. \tag{19.12}$$

Discussion

This is a very large number. It is no wonder that we do not ordinarily observe individual electrons with so many being present in ordinary systems. In fact, electricity had been in use for many decades before it was determined that the moving charges in many circumstances were negative. Positive charge moving in the opposite direction of negative charge often produces identical effects; this makes it difficult to determine which is moving or whether both are moving.

Download for free at https://openstax.org/details/books/college-physics-ap-courses

Applying the Science Practices: Work and Potential Energy in Point Charges

Consider a system consisting of two positive point charges, each 2.0 μC, placed 1.0 m away from each other. We can calculate the potential (i.e. internal) energy of this configuration by computing the potential due to one of the charges, and then calculating the potential energy of the second charge in the potential of the first. Applying Equations (19.38) and (19.2) gives us a potential energy of 3.6×10^{-2} J. If we move the charges closer to each other, say, to 0.50 m apart, the potential energy doubles. Note that, to create this second case, some outside force would have had to do work on this system to change the configuration, and hence it was not a closed system. However, because the electric force is conservative, we can use the work-energy theorem to state that, since there was no change in kinetic energy, all of the work done went into increasing the internal energy of the system. Also note that if the point charges had different signs they would be attracted to each other, so they would be capable of doing work on an outside system when the distance between them decreased. As work is done on the outside system, the internal energy in the two-charge system decreases.

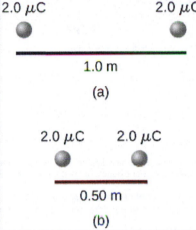

Figure 19.4 Work is done by moving two charges with the same sign closer to each other, increasing the internal energy of the two-charge system.

The Electron Volt

The energy per electron is very small in macroscopic situations like that in the previous example—a tiny fraction of a joule. But on a submicroscopic scale, such energy per particle (electron, proton, or ion) can be of great importance. For example, even a tiny fraction of a joule can be great enough for these particles to destroy organic molecules and harm living tissue. The particle may do its damage by direct collision, or it may create harmful x rays, which can also inflict damage. It is useful to have an energy unit related to submicroscopic effects. **Figure 19.5** shows a situation related to the definition of such an energy unit. An electron is accelerated between two charged metal plates as it might be in an old-model television tube or oscilloscope. The electron is given kinetic energy that is later converted to another form—light in the television tube, for example. (Note that downhill for the electron is uphill for a positive charge.) Since energy is related to voltage by $\Delta \mathrm{PE} = q\Delta V,$ we can think of the joule as a coulomb-volt.

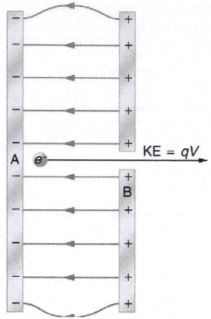

Figure 19.5 A typical electron gun accelerates electrons using a potential difference between two metal plates. The energy of the electron in electron volts is numerically the same as the voltage between the plates. For example, a 5000 V potential difference produces 5000 eV electrons.

On the submicroscopic scale, it is more convenient to define an energy unit called the **electron volt** (eV), which is the energy given to a fundamental charge accelerated through a potential difference of 1 V. In equation form,

$$1 \text{ eV} = \left(1.60 \times 10^{-19} \text{ C}\right)(1 \text{ V}) = \left(1.60 \times 10^{-19} \text{ C}\right)(1 \text{ J/C}) \qquad (19.13)$$
$$= 1.60 \times 10^{-19} \text{ J}.$$

Electron Volt

On the submicroscopic scale, it is more convenient to define an energy unit called the electron volt (eV), which is the energy given to a fundamental charge accelerated through a potential difference of 1 V. In equation form,

$$1 \text{ eV} = \left(1.60 \times 10^{-19} \text{ C}\right)(1 \text{ V}) = \left(1.60 \times 10^{-19} \text{ C}\right)(1 \text{ J/C}) \qquad (19.14)$$
$$= 1.60 \times 10^{-19} \text{ J}.$$

An electron accelerated through a potential difference of 1 V is given an energy of 1 eV. It follows that an electron accelerated through 50 V is given 50 eV. A potential difference of 100,000 V (100 kV) will give an electron an energy of 100,000 eV (100 keV), and so on. Similarly, an ion with a double positive charge accelerated through 100 V will be given 200 eV of energy. These simple relationships between accelerating voltage and particle charges make the electron volt a simple and convenient energy unit in such circumstances.

Connections: Energy Units

The electron volt (eV) is the most common energy unit for submicroscopic processes. This will be particularly noticeable in the chapters on modern physics. Energy is so important to so many subjects that there is a tendency to define a special energy unit for each major topic. There are, for example, calories for food energy, kilowatt-hours for electrical energy, and therms for natural gas energy.

The electron volt is commonly employed in submicroscopic processes—chemical valence energies and molecular and nuclear binding energies are among the quantities often expressed in electron volts. For example, about 5 eV of energy is required to break up certain organic molecules. If a proton is accelerated from rest through a potential difference of 30 kV, it is given an energy of 30 keV (30,000 eV) and it can break up as many as 6000 of these molecules (
$30,000 \text{ eV} \div 5 \text{ eV per molecule} = 6000 \text{ molecules}$). Nuclear decay energies are on the order of 1 MeV (1,000,000 eV) per event and can, thus, produce significant biological damage.

Conservation of Energy

The total energy of a system is conserved if there is no net addition (or subtraction) of work or heat transfer. For conservative forces, such as the electrostatic force, conservation of energy states that mechanical energy is a constant.

Mechanical energy is the sum of the kinetic energy and potential energy of a system; that is, $\text{KE} + \text{PE} = \text{constant}$. A loss of

PE of a charged particle becomes an increase in its KE. Here PE is the electric potential energy. Conservation of energy is stated in equation form as

$$KE + PE = \text{constant} \tag{19.15}$$

or

$$KE_i + PE_i = KE_f + PE_f \,, \tag{19.16}$$

where i and f stand for initial and final conditions. As we have found many times before, considering energy can give us insights and facilitate problem solving.

Example 19.3 Electrical Potential Energy Converted to Kinetic Energy

Calculate the final speed of a free electron accelerated from rest through a potential difference of 100 V. (Assume that this numerical value is accurate to three significant figures.)

Strategy

We have a system with only conservative forces. Assuming the electron is accelerated in a vacuum, and neglecting the gravitational force (we will check on this assumption later), all of the electrical potential energy is converted into kinetic energy. We can identify the initial and final forms of energy to be $KE_i = 0$, $KE_f = \frac{1}{2}mv^2$, $PE_i = qV$, and $PE_f = 0$.

Solution

Conservation of energy states that

$$KE_i + PE_i = KE_f + PE_f. \tag{19.17}$$

Entering the forms identified above, we obtain

$$qV = \frac{mv^2}{2}. \tag{19.18}$$

We solve this for v:

$$v = \sqrt{\frac{2qV}{m}}. \tag{19.19}$$

Entering values for q, V, and m gives

$$\begin{aligned} v &= \sqrt{\frac{2(-1.60\times10^{-19}\ \text{C})(-100\ \text{J/C})}{9.11\times10^{-31}\ \text{kg}}} \\ &= 5.93\times10^6\ \text{m/s}. \end{aligned} \tag{19.20}$$

Discussion

Note that both the charge and the initial voltage are negative, as in **Figure 19.5**. From the discussions in **Electric Charge and Electric Field**, we know that electrostatic forces on small particles are generally very large compared with the gravitational force. The large final speed confirms that the gravitational force is indeed negligible here. The large speed also indicates how easy it is to accelerate electrons with small voltages because of their very small mass. Voltages much higher than the 100 V in this problem are typically used in electron guns. Those higher voltages produce electron speeds so great that relativistic effects must be taken into account. That is why a low voltage is considered (accurately) in this example.

Making Connections: Kinetic and Potential Energy in Point Charges

Now consider another system of two point charges. One has a mass of 1000 kg and a charge of 50.0 μC, and is initially stationary. The other has a mass of 1.00 kg, a charge of 10.0 μC, and is initially traveling directly at the first point charge at 10.0 m/s from very far away. What will be the closest approach of these two objects to each other?

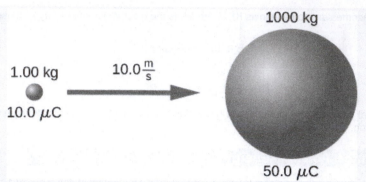

Figure 19.6 A system consisting of two point charges initially has the smaller charge moving toward the larger charge

Note that the internal energy of this two-charge system will not change, due to an absence of external forces acting on the system. Initially, the internal energy is equal to the kinetic energy of the smaller charge, and the potential energy is effectively zero due to the enormous distance between the two objects. Conservation of energy tells us that the internal energy of this system will not change. Hence the distance of closest approach will be when the internal energy is equal to the potential energy between the two charges, and there is no kinetic energy in this system.

The initial kinetic energy may be calculated as 50.0 J. Applying Equations (19.38) and (19.2), we find a distance of 9.00 cm. After this, the mutual repulsion will send the lighter object off to infinity again. Note that we did not include potential energy due to gravity, as the masses concerned are so small compared to the charges that the result will never come close to showing up in significant digits. Furthermore, the first object is much more massive than the second. As a result, any motion induced in it will also be too small to show up in the significant digits.

19.2 Electric Potential in a Uniform Electric Field

Learning Objectives

By the end of this section, you will be able to:

- Describe the relationship between voltage and electric field.
- Derive an expression for the electric potential and electric field.
- Calculate electric field strength given distance and voltage.

The information presented in this section supports the following AP® learning objectives and science practices:

- **2.C.5.2** The student is able to calculate the magnitude and determine the direction of the electric field between two electrically charged parallel plates, given the charge of each plate, or the electric potential difference and plate separation. **(S.P. 2.2)**
- **2.C.5.3** The student is able to represent the motion of an electrically charged particle in the uniform field between two oppositely charged plates and express the connection of this motion to projectile motion of an object with mass in the Earth's gravitational field. **(S.P. 1.1, 2.2, 7.1)**
- **2.E.3.1** The student is able to apply mathematical routines to calculate the average value of the magnitude of the electric field in a region from a description of the electric potential in that region using the displacement along the line on which the difference in potential is evaluated. **(S.P. 2.2)**
- **2.E.3.2** The student is able to apply the concept of the isoline representation of electric potential for a given electric charge distribution to predict the average value of the electric field in the region. **(S.P. 1.4, 6.4)**

In the previous section, we explored the relationship between voltage and energy. In this section, we will explore the relationship between voltage and electric field. For example, a uniform electric field $\mathbf{E}$ is produced by placing a potential difference (or voltage) ΔV across two parallel metal plates, labeled A and B. (See **Figure 19.7**.) Examining this will tell us what voltage is needed to produce a certain electric field strength; it will also reveal a more fundamental relationship between electric potential and electric field. From a physicist's point of view, either ΔV or $\mathbf{E}$ can be used to describe any charge distribution. ΔV is most closely tied to energy, whereas $\mathbf{E}$ is most closely related to force. ΔV is a **scalar** quantity and has no direction, while $\mathbf{E}$ is a **vector** quantity, having both magnitude and direction. (Note that the magnitude of the electric field strength, a scalar quantity, is represented by E below.) The relationship between ΔV and $\mathbf{E}$ is revealed by calculating the work done by the force in moving a charge from point A to point B. But, as noted in **Electric Potential Energy: Potential Difference**, this is complex for arbitrary charge distributions, requiring calculus. We therefore look at a uniform electric field as an interesting special case.

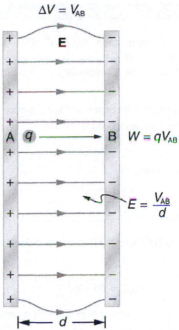

Figure 19.7 The relationship between V and E for parallel conducting plates is $E = V/d$. (Note that $\Delta V = V_{AB}$ in magnitude. For a charge that is moved from plate A at higher potential to plate B at lower potential, a minus sign needs to be included as follows: $-\Delta V = V_A - V_B = V_{AB}$. See the text for details.)

The work done by the electric field in **Figure 19.7** to move a positive charge q from A, the positive plate, higher potential, to B, the negative plate, lower potential, is

$$W = -\Delta\text{PE} = -q\Delta V. \tag{19.21}$$

The potential difference between points A and B is

$$-\Delta V = -(V_B - V_A) = V_A - V_B = V_{AB}. \tag{19.22}$$

Entering this into the expression for work yields

$$W = qV_{AB}. \tag{19.23}$$

Work is $W = Fd\cos\theta$; here $\cos\theta = 1$, since the path is parallel to the field, and so $W = Fd$. Since $F = qE$, we see that $W = qEd$. Substituting this expression for work into the previous equation gives

$$qEd = qV_{AB}. \tag{19.24}$$

The charge cancels, and so the voltage between points A and B is seen to be

$$\left.\begin{aligned} V_{AB} &= Ed \\ E &= \frac{V_{AB}}{d} \end{aligned}\right\}\text{(uniform } E\text{ - field on y)}, \tag{19.25}$$

where d is the distance from A to B, or the distance between the plates in **Figure 19.7**. Note that the above equation implies the units for electric field are volts per meter. We already know the units for electric field are newtons per coulomb; thus the following relation among units is valid:

$$1\,\text{N/C} = 1\,\text{V/m}. \tag{19.26}$$

Voltage between Points A and B

$$\left.\begin{aligned} V_{AB} &= Ed \\ E &= \frac{V_{AB}}{d} \end{aligned}\right\}\text{(uniform } E\text{ - field on y)}, \tag{19.27}$$

where d is the distance from A to B, or the distance between the plates.

Example 19.4 What Is the Highest Voltage Possible between Two Plates?

Dry air will support a maximum electric field strength of about 3.0×10^6 V/m . Above that value, the field creates enough ionization in the air to make the air a conductor. This allows a discharge or spark that reduces the field. What, then, is the maximum voltage between two parallel conducting plates separated by 2.5 cm of dry air?

Strategy

We are given the maximum electric field E between the plates and the distance d between them. The equation $V_{AB} = Ed$ can thus be used to calculate the maximum voltage.

Solution

The potential difference or voltage between the plates is

$$V_{AB} = Ed. \tag{19.28}$$

Entering the given values for E and d gives

$$V_{AB} = (3.0 \times 10^6 \text{ V/m})(0.025 \text{ m}) = 7.5 \times 10^4 \text{ V} \tag{19.29}$$

or

$$V_{AB} = 75 \text{ kV}. \tag{19.30}$$

(The answer is quoted to only two digits, since the maximum field strength is approximate.)

Discussion

One of the implications of this result is that it takes about 75 kV to make a spark jump across a 2.5 cm (1 in.) gap, or 150 kV for a 5 cm spark. This limits the voltages that can exist between conductors, perhaps on a power transmission line. A smaller voltage will cause a spark if there are points on the surface, since points create greater fields than smooth surfaces. Humid air breaks down at a lower field strength, meaning that a smaller voltage will make a spark jump through humid air. The largest voltages can be built up, say with static electricity, on dry days.

Figure 19.8 A spark chamber is used to trace the paths of high-energy particles. Ionization created by the particles as they pass through the gas between the plates allows a spark to jump. The sparks are perpendicular to the plates, following electric field lines between them. The potential difference between adjacent plates is not high enough to cause sparks without the ionization produced by particles from accelerator experiments (or cosmic rays). (credit: Daderot, Wikimedia Commons)

Making Connections: Uniform Fields

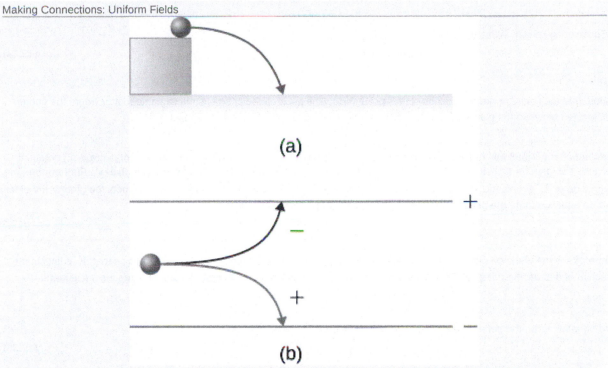

(a)

(b)

Figure 19.9 (a) A massive particle launched horizontally in a downward gravitational field will fall to the ground. (b) A positively charged particle launched horizontally in a downward electric field will fall toward the negative potential; a negatively charged particle will move in the opposite direction.

Recall from Projectile Motion (Section 3.4) that a massive projectile launched horizontally (for example, from a cliff) in a uniform downward gravitational field (as we find near the surface of the Earth) will follow a parabolic trajectory downward until it hits the ground, as shown in Figure 19.9(a).

An identical outcome occurs for a positively charged particle in a uniform electric field (Figure 19.9(b)); it follows the electric field "downhill" until it runs into something. The difference between the two cases is that the gravitational force is always attractive; the electric force has two kinds of charges, and therefore may be either attractive or repulsive. Therefore, a negatively charged particle launched into the same field will fall "uphill".

Example 19.5 Field and Force inside an Electron Gun

(a) An electron gun has parallel plates separated by 4.00 cm and gives electrons 25.0 keV of energy. What is the electric field strength between the plates? (b) What force would this field exert on a piece of plastic with a $0.500 \ \mu C$ charge that gets between the plates?

Strategy

Since the voltage and plate separation are given, the electric field strength can be calculated directly from the expression $E = \dfrac{V_{AB}}{d}$. Once the electric field strength is known, the force on a charge is found using $\mathbf{F} = q\,\mathbf{E}$. Since the electric field is in only one direction, we can write this equation in terms of the magnitudes, $F = q\,E$.

Solution for (a)

The expression for the magnitude of the electric field between two uniform metal plates is

$$E = \frac{V_{AB}}{d}. \tag{19.31}$$

Since the electron is a single charge and is given 25.0 keV of energy, the potential difference must be 25.0 kV. Entering this value for V_{AB} and the plate separation of 0.0400 m, we obtain

$$E = \frac{25.0 \text{ kV}}{0.0400 \text{ m}} = 6.25 \times 10^5 \text{ V/m}. \tag{19.32}$$

Solution for (b)

The magnitude of the force on a charge in an electric field is obtained from the equation

$$F = qE.\tag{19.33}$$

Substituting known values gives

$$F = (0.500 \times 10^{-6}\ \text{C})(6.25 \times 10^5\ \text{V/m}) = 0.313\ \text{N}.\tag{19.34}$$

Discussion

Note that the units are newtons, since $1\ \text{V/m} = 1\ \text{N/C}$. The force on the charge is the same no matter where the charge is located between the plates. This is because the electric field is uniform between the plates.

In more general situations, regardless of whether the electric field is uniform, it points in the direction of decreasing potential, because the force on a positive charge is in the direction of $\mathbf{E}$ and also in the direction of lower potential V. Furthermore, the magnitude of $\mathbf{E}$ equals the rate of decrease of V with distance. The faster V decreases over distance, the greater the electric field. In equation form, the general relationship between voltage and electric field is

$$E = -\frac{\Delta V}{\Delta s},\tag{19.35}$$

where Δs is the distance over which the change in potential, ΔV, takes place. The minus sign tells us that $\mathbf{E}$ points in the direction of decreasing potential. The electric field is said to be the *gradient* (as in grade or slope) of the electric potential.

Relationship between Voltage and Electric Field

In equation form, the general relationship between voltage and electric field is

$$E = -\frac{\Delta V}{\Delta s},\tag{19.36}$$

where Δs is the distance over which the change in potential, ΔV, takes place. The minus sign tells us that $\mathbf{E}$ points in the direction of decreasing potential. The electric field is said to be the *gradient* (as in grade or slope) of the electric potential.

Note that Equation (19.36) is defining the <u>average</u> electric field over the given region.

For continually changing potentials, ΔV and Δs become infinitesimals and differential calculus must be employed to determine the electric field.

Making Connections: Non-Parallel Conducting Plates

Consider two conducting plates, placed as shown in **Figure 19.10**. If the plates are held at a fixed potential difference ΔV, the average electric field is strongest between the near edges of the plates, and weakest between the two far edges of the plates.

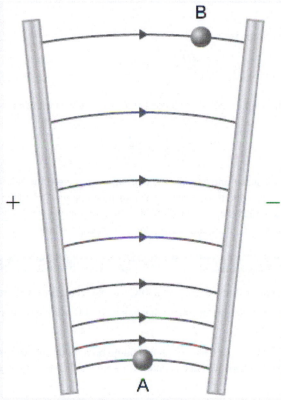

Figure 19.10 Two nonparallel plates, held at a fixed potential difference.

Now assume that the potential difference is 60 V. If the arc length along the field line labeled by A is 10 cm, what is the electric field at point A? How about point B, if the arc length along that field line is 17 cm? How would the density of the electric potential isolines, if they were drawn on the figure, compare at these two points? Can you use this concept to estimate what the electric field strength would be at a point midway between A and B?

Answer

Applying Equation. 19.36, we find that the electric fields at A and B are 600 V/m and 350 V/m respectively. The isolines would be denser at A than at B, and would spread out evenly from A to B. Therefore, the electric field at a point halfway between the two would have an arc length of 13.5 cm and be approximately 440 V/m.

19.3 Electrical Potential Due to a Point Charge

Learning Objectives
By the end of this section, you will be able to: • Explain point charges and express the equation for electric potential of a point charge. • Distinguish between electric potential and electric field. • Determine the electric potential of a point charge given charge and distance.

Point charges, such as electrons, are among the fundamental building blocks of matter. Furthermore, spherical charge distributions (like on a metal sphere) create external electric fields exactly like a point charge. The electric potential due to a point charge is, thus, a case we need to consider. Using calculus to find the work needed to move a test charge q from a large distance away to a distance of r from a point charge Q, and noting the connection between work and potential $(W = -q\Delta V)$, it can be shown that the *electric potential* V *of a point charge* is

$$V = \frac{kQ}{r} \text{ (Point Charge)},$$

(19.37)

where k is a constant equal to $9.0 \times 10^9 \text{ N} \cdot \text{m}^2/\text{C}^2$.

Electric Potential V of a Point Charge

The electric potential V of a point charge is given by

$$V = \frac{kQ}{r} \text{ (Point Charge)}. \tag{19.38}$$

The potential at infinity is chosen to be zero. Thus V for a point charge decreases with distance, whereas E for a point charge decreases with distance squared:

$$E = \frac{F}{q} = \frac{kQ}{r^2}. \tag{19.39}$$

Recall that the electric potential V is a scalar and has no direction, whereas the electric field E is a vector. To find the voltage due to a combination of point charges, you add the individual voltages as numbers. To find the total electric field, you must add the individual fields as *vectors*, taking magnitude and direction into account. This is consistent with the fact that V is closely associated with energy, a scalar, whereas E is closely associated with force, a vector.

Example 19.6 What Voltage Is Produced by a Small Charge on a Metal Sphere?

Charges in static electricity are typically in the nanocoulomb (nC) to microcoulomb (μC) range. What is the voltage 5.00 cm away from the center of a 1-cm diameter metal sphere that has a -3.00 nC static charge?

Strategy

As we have discussed in **Electric Charge and Electric Field**, charge on a metal sphere spreads out uniformly and produces a field like that of a point charge located at its center. Thus we can find the voltage using the equation $V = kQ/r$
.

Solution

Entering known values into the expression for the potential of a point charge, we obtain

$$\begin{aligned} V &= k\frac{Q}{r} \\ &= \left(8.99 \times 10^9 \text{ N} \cdot \text{m}^2/\text{C}^2\right)\left(\frac{-3.00 \times 10^{-9} \text{ C}}{5.00 \times 10^{-2} \text{ m}}\right) \\ &= -539 \text{ V}. \end{aligned} \tag{19.40}$$

Discussion

The negative value for voltage means a positive charge would be attracted from a larger distance, since the potential is lower (more negative) than at larger distances. Conversely, a negative charge would be repelled, as expected.

Example 19.7 What Is the Excess Charge on a Van de Graaff Generator

A demonstration Van de Graaff generator has a 25.0 cm diameter metal sphere that produces a voltage of 100 kV near its surface. (See **Figure 19.11**.) What excess charge resides on the sphere? (Assume that each numerical value here is shown with three significant figures.)

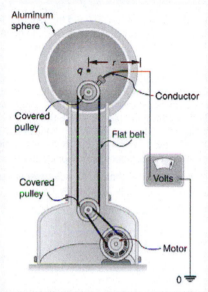

Figure 19.11 The voltage of this demonstration Van de Graaff generator is measured between the charged sphere and ground. Earth's potential is taken to be zero as a reference. The potential of the charged conducting sphere is the same as that of an equal point charge at its center.

Strategy

The potential on the surface will be the same as that of a point charge at the center of the sphere, 12.5 cm away. (The radius of the sphere is 12.5 cm.) We can thus determine the excess charge using the equation

$$V = \frac{kQ}{r}. \tag{19.41}$$

Solution

Solving for Q and entering known values gives

$$
\begin{aligned}
Q &= \frac{rV}{k} \\
&= \frac{(0.125 \text{ m})\left(100 \times 10^3 \text{ V}\right)}{8.99 \times 10^9 \text{ N} \cdot \text{m}^2 / \text{C}^2} \\
&= 1.39 \times 10^{-6} \text{ C} = 1.39 \text{ μC}.
\end{aligned}
\tag{19.42}
$$

Discussion

This is a relatively small charge, but it produces a rather large voltage. We have another indication here that it is difficult to store isolated charges.

The voltages in both of these examples could be measured with a meter that compares the measured potential with ground potential. Ground potential is often taken to be zero (instead of taking the potential at infinity to be zero). It is the potential difference between two points that is of importance, and very often there is a tacit assumption that some reference point, such as Earth or a very distant point, is at zero potential. As noted in **Electric Potential Energy: Potential Difference**, this is analogous to taking sea level as $h = 0$ when considering gravitational potential energy, $\text{PE}_g = mgh$.

19.4 Equipotential Lines

Learning Objectives

By the end of this section, you will be able to:

- Explain equipotential lines (also called isolines of electric potential) and equipotential surfaces.
- Describe the action of grounding an electrical appliance.
- Compare electric field and equipotential lines.

The information presented in this section supports the following AP® learning objectives and science practices:

- **2.E.2.1** The student is able to determine the structure of isolines of electric potential by constructing them in a given electric field. **(S.P. 6.4, 7.2)**

- **2.E.2.2** The student is able to predict the structure of isolines of electric potential by constructing them in a given electric field and make connections between these isolines and those found in a gravitational field. **(S.P. 6.4, 7.2)**
- **2.E.2.3** The student is able to qualitatively use the concept of isolines to construct isolines of electric potential in an electric field and determine the effect of that field on electrically charged objects. **(S.P. 1.4)**

We can represent electric potentials (voltages) pictorially, just as we drew pictures to illustrate electric fields. Of course, the two are related. Consider **Figure 19.12**, which shows an isolated positive point charge and its electric field lines. Electric field lines radiate out from a positive charge and terminate on negative charges. While we use blue arrows to represent the magnitude and direction of the electric field, we use green lines to represent places where the electric potential is constant. These are called **equipotential lines** in two dimensions, or *equipotential surfaces* in three dimensions. The term *equipotential* is also used as a noun, referring to an equipotential line or surface. The potential for a point charge is the same anywhere on an imaginary sphere of radius r surrounding the charge. This is true since the potential for a point charge is given by $V = kQ/r$ and, thus, has the same value at any point that is a given distance r from the charge. An equipotential sphere is a circle in the two-dimensional view of **Figure 19.12**. Since the electric field lines point radially away from the charge, they are perpendicular to the equipotential lines.

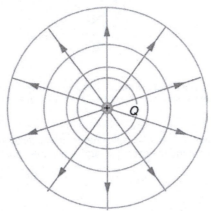

Figure 19.12 An isolated point charge Q with its electric field lines in blue and equipotential lines in green. The potential is the same along each equipotential line, meaning that no work is required to move a charge anywhere along one of those lines. Work is needed to move a charge from one equipotential line to another. Equipotential lines are perpendicular to electric field lines in every case.

Applying the Science Practices: Electric Potential and Peaks

Starting with **Figure 19.13** as a rough example, draw diagrams of isolines for both positive and negative isolated point charges. Be sure to take care with what happens to the spacing of the isolines as you get closer to the charge. Then copy both of these sets of lines, but relabel them as gravitational equipotential lines. Then try to draw the sort of hill or hole or other shape that would have equipotential lines of this form. Does this shape exist in nature?

Figure 19.13 An example of a topographical map.

You should notice that the lines get closer together the closer you get to the point charge. The hill (or sinkhole, for the equivalent from a negative charge) should have a 1/*r* sort of form, which is not a very common topographical feature.

It is important to note that *equipotential lines are always perpendicular to electric field lines.* No work is required to move a charge along an equipotential, since $\Delta V = 0$. Thus the work is

$$W = -\Delta \text{PE} = -q\Delta V = 0. \tag{19.43}$$

Work is zero if force is perpendicular to motion. Force is in the same direction as $\mathbf{E}$, so that motion along an equipotential must be perpendicular to $\mathbf{E}$. More precisely, work is related to the electric field by

$$W = Fd \cos \theta = qEd \cos \theta = 0. \tag{19.44}$$

Note that in the above equation, E and F symbolize the magnitudes of the electric field strength and force, respectively. Neither q nor $\mathbf{E}$ nor d is zero, and so $\cos \theta$ must be 0, meaning θ must be $90°$. In other words, motion along an equipotential is perpendicular to $\mathbf{E}$.

One of the rules for static electric fields and conductors is that the electric field must be perpendicular to the surface of any conductor. This implies that *a conductor is an equipotential surface in static situations*. There can be no voltage difference across the surface of a conductor, or charges will flow. One of the uses of this fact is that a conductor can be fixed at zero volts by connecting it to the earth with a good conductor—a process called **grounding**. Grounding can be a useful safety tool. For example, grounding the metal case of an electrical appliance ensures that it is at zero volts relative to the earth.

Grounding

A conductor can be fixed at zero volts by connecting it to the earth with a good conductor—a process called grounding.

Because a conductor is an equipotential, it can replace any equipotential surface. For example, in **Figure 19.12** a charged spherical conductor can replace the point charge, and the electric field and potential surfaces outside of it will be unchanged, confirming the contention that a spherical charge distribution is equivalent to a point charge at its center.

Figure 19.14 shows the electric field and equipotential lines for two equal and opposite charges. Given the electric field lines, the equipotential lines can be drawn simply by making them perpendicular to the electric field lines. Conversely, given the equipotential lines, as in **Figure 19.15**(a), the electric field lines can be drawn by making them perpendicular to the equipotentials, as in **Figure 19.15**(b).

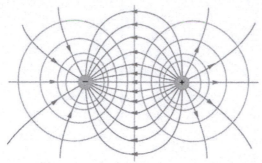

Figure 19.14 The electric field lines and equipotential lines for two equal but opposite charges. The equipotential lines can be drawn by making them perpendicular to the electric field lines, if those are known. Note that the potential is greatest (most positive) near the positive charge and least (most negative) near the negative charge.

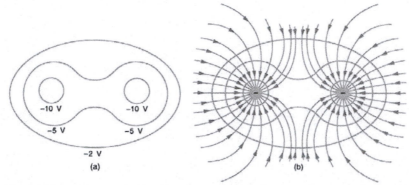

Figure 19.15 (a) These equipotential lines might be measured with a voltmeter in a laboratory experiment. (b) The corresponding electric field lines are found by drawing them perpendicular to the equipotentials. Note that these fields are consistent with two equal negative charges.

One of the most important cases is that of the familiar parallel conducting plates shown in **Figure 19.16**. Between the plates, the equipotentials are evenly spaced and parallel. The same field could be maintained by placing conducting plates at the equipotential lines at the potentials shown.

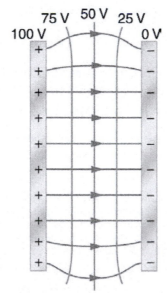

Figure 19.16 The electric field and equipotential lines between two metal plates.

Making Connections: Slopes and Parallel Plates

Consider the parallel plates in **Figure 19.2**. These have equipotential lines that are parallel to the plates in the space between, and evenly spaced. An example of this (with sample values) is given in **Figure 19.16**. One could draw a similar set of equipotential isolines for gravity on the hill shown in **Figure 19.2**. If the hill has any extent at the same slope, the isolines along that extent would be parallel to each other. Furthermore, in regions of constant slope, the isolines would be evenly spaced.

Figure 19.17 Note that a topographical map along a ridge has roughly parallel elevation lines, similar to the equipotential lines in Figure 19.16.

An important application of electric fields and equipotential lines involves the heart. The heart relies on electrical signals to maintain its rhythm. The movement of electrical signals causes the chambers of the heart to contract and relax. When a person has a heart attack, the movement of these electrical signals may be disturbed. An artificial pacemaker and a defibrillator can be used to initiate the rhythm of electrical signals. The equipotential lines around the heart, the thoracic region, and the axis of the heart are useful ways of monitoring the structure and functions of the heart. An electrocardiogram (ECG) measures the small electric signals being generated during the activity of the heart. More about the relationship between electric fields and the heart is discussed in **Energy Stored in Capacitors**.

PhET Explorations: Charges and Fields

Move point charges around on the playing field and then view the electric field, voltages, equipotential lines, and more. It's colorful, it's dynamic, it's free.

Figure 19.18 Charges and Fields (http://cnx.org/content/m55336/1.3/charges-and-fields_en.jar)

19.5 Capacitors and Dielectrics

Learning Objectives
By the end of this section, you will be able to: • Describe the action of a capacitor and define capacitance. • Explain parallel plate capacitors and their capacitances. • Discuss the process of increasing the capacitance of a capacitor with a dielectric. • Determine capacitance given charge and voltage. The information presented in this section supports the following AP® learning objectives and science pracitces: • **4.E.4.1** The student is able to make predictions about the properties of resistors and/or capacitors when placed in a simple circuit based on the geometry of the circuit element and supported by scientific theories and mathematical relationships. **(S.P. 2.2, 6.4)** • **4.E.4.2** The student is able to design a plan for the collection of data to determine the effect of changing the geometry and/or materials on the resistance or capacitance of a circuit element and relate results to the basic properties of resistors and capacitors. **(S.P. 4.1, 4.2)** • **4.E.4.3** The student is able to analyze data to determine the effect of changing the geometry and/or materials on the resistance or capacitance of a circuit element and relate results to the basic properties of resistors and capacitors. **(S.P. 5.1)**

A **capacitor** is a device used to store electric charge. Capacitors have applications ranging from filtering static out of radio reception to energy storage in heart defibrillators. Typically, commercial capacitors have two conducting parts close to one another, but not touching, such as those in **Figure 19.19**. (Most of the time an insulator is used between the two plates to provide separation—see the discussion on dielectrics below.) When battery terminals are connected to an initially uncharged capacitor, equal amounts of positive and negative charge, $+Q$ and $-Q$, are separated into its two plates. The capacitor remains neutral overall, but we refer to it as storing a charge Q in this circumstance.

Capacitor

A capacitor is a device used to store electric charge.

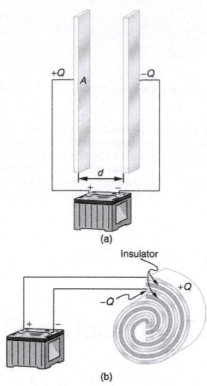

Figure 19.19 Both capacitors shown here were initially uncharged before being connected to a battery. They now have separated charges of $+Q$ and $-Q$ on their two halves. (a) A parallel plate capacitor. (b) A rolled capacitor with an insulating material between its two conducting sheets.

The amount of charge Q a *capacitor* can store depends on two major factors—the voltage applied and the capacitor's physical characteristics, such as its size.

The Amount of Charge Q a Capacitor Can Store

The amount of charge Q a *capacitor* can store depends on two major factors—the voltage applied and the capacitor's physical characteristics, such as its size.

A system composed of two identical, parallel conducting plates separated by a distance, as in **Figure 19.20**, is called a **parallel plate capacitor**. It is easy to see the relationship between the voltage and the stored charge for a parallel plate capacitor, as shown in **Figure 19.20**. Each electric field line starts on an individual positive charge and ends on a negative one, so that there will be more field lines if there is more charge. (Drawing a single field line per charge is a convenience, only. We can draw many field lines for each charge, but the total number is proportional to the number of charges.) The electric field strength is, thus, directly proportional to Q.

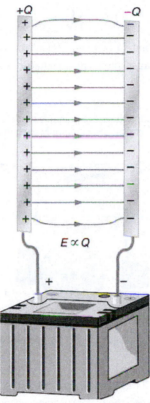

Figure 19.20 Electric field lines in this parallel plate capacitor, as always, start on positive charges and end on negative charges. Since the electric field strength is proportional to the density of field lines, it is also proportional to the amount of charge on the capacitor.

The field is proportional to the charge:

$$E \propto Q, \tag{19.45}$$

where the symbol $\propto$ means "proportional to." From the discussion in **Electric Potential in a Uniform Electric Field**, we know that the voltage across parallel plates is $V = Ed$. Thus,

$$V \propto E. \tag{19.46}$$

It follows, then, that $V \propto Q$, and conversely,

$$Q \propto V. \tag{19.47}$$

This is true in general: The greater the voltage applied to any capacitor, the greater the charge stored in it.

Different capacitors will store different amounts of charge for the same applied voltage, depending on their physical characteristics. We define their **capacitance** C to be such that the charge Q stored in a capacitor is proportional to C. The charge stored in a capacitor is given by

$$Q = CV. \tag{19.48}$$

This equation expresses the two major factors affecting the amount of charge stored. Those factors are the physical characteristics of the capacitor, C, and the voltage, V. Rearranging the equation, we see that *capacitance C is the amount of charge stored per volt, or*

$$C = \frac{Q}{V}. \tag{19.49}$$

Capacitance

Capacitance C is the amount of charge stored per volt, or

$$C = \frac{Q}{V}. \tag{19.50}$$

The unit of capacitance is the farad (F), named for Michael Faraday (1791–1867), an English scientist who contributed to the fields of electromagnetism and electrochemistry. Since capacitance is charge per unit voltage, we see that a farad is a coulomb per volt, or

$$1 \text{ F} = \frac{1 \text{ C}}{1 \text{ V}}. \qquad (19.51)$$

A 1-farad capacitor would be able to store 1 coulomb (a very large amount of charge) with the application of only 1 volt. One farad is, thus, a very large capacitance. Typical capacitors range from fractions of a picofarad $\left(1 \text{ pF} = 10^{-12} \text{ F}\right)$ to millifarads $\left(1 \text{ mF} = 10^{-3} \text{ F}\right)$.

Figure 19.21 shows some common capacitors. Capacitors are primarily made of ceramic, glass, or plastic, depending upon purpose and size. Insulating materials, called dielectrics, are commonly used in their construction, as discussed below.

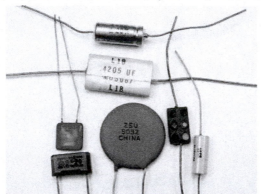

Figure 19.21 Some typical capacitors. Size and value of capacitance are not necessarily related. (credit: Windell Oskay)

Parallel Plate Capacitor

The parallel plate capacitor shown in **Figure 19.22** has two identical conducting plates, each having a surface area A, separated by a distance d (with no material between the plates). When a voltage V is applied to the capacitor, it stores a charge Q, as shown. We can see how its capacitance depends on A and d by considering the characteristics of the Coulomb force. We know that like charges repel, unlike charges attract, and the force between charges decreases with distance. So it seems quite reasonable that the bigger the plates are, the more charge they can store—because the charges can spread out more. Thus C should be greater for larger A. Similarly, the closer the plates are together, the greater the attraction of the opposite charges on them. So C should be greater for smaller d.

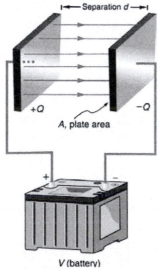

V (battery)

Figure 19.22 Parallel plate capacitor with plates separated by a distance d. Each plate has an area A.

It can be shown that for a parallel plate capacitor there are only two factors (A and d) that affect its capacitance C. The capacitance of a parallel plate capacitor in equation form is given by

$$C = \varepsilon_0 \frac{A}{d}. \qquad (19.52)$$

Capacitance of a Parallel Plate Capacitor

$$C = \varepsilon_0 \frac{A}{d} \qquad (19.53)$$

A is the area of one plate in square meters, and d is the distance between the plates in meters. The constant ε_0 is the permittivity of free space; its numerical value in SI units is $\varepsilon_0 = 8.85 \times 10^{-12}$ F/m . The units of F/m are equivalent to $C^2/N \cdot m^2$. The small numerical value of ε_0 is related to the large size of the farad. A parallel plate capacitor must have a large area to have a capacitance approaching a farad. (Note that the above equation is valid when the parallel plates are separated by air or free space. When another material is placed between the plates, the equation is modified, as discussed below.)

Example 19.8 Capacitance and Charge Stored in a Parallel Plate Capacitor

(a) What is the capacitance of a parallel plate capacitor with metal plates, each of area 1.00 m^2 , separated by 1.00 mm?

(b) What charge is stored in this capacitor if a voltage of 3.00×10^3 V is applied to it?

Strategy

Finding the capacitance C is a straightforward application of the equation $C = \varepsilon_0 A / d$. Once C is found, the charge stored can be found using the equation $Q = CV$.

Solution for (a)

Entering the given values into the equation for the capacitance of a parallel plate capacitor yields

$$
\begin{aligned}
C &= \varepsilon_0 \frac{A}{d} = \left(8.85 \times 10^{-12} \frac{\text{F}}{\text{m}}\right) \frac{1.00 \text{ m}^2}{1.00 \times 10^{-3} \text{ m}} \\
&= 8.85 \times 10^{-9} \text{ F} = 8.85 \text{ nF}.
\end{aligned}
\tag{19.54}
$$

Discussion for (a)

This small value for the capacitance indicates how difficult it is to make a device with a large capacitance. Special techniques help, such as using very large area thin foils placed close together.

Solution for (b)

The charge stored in any capacitor is given by the equation $Q = CV$. Entering the known values into this equation gives

$$
\begin{aligned}
Q &= CV = \left(8.85 \times 10^{-9} \text{ F}\right)\left(3.00 \times 10^3 \text{ V}\right) \\
&= 26.6 \text{ } \mu\text{C}.
\end{aligned}
\tag{19.55}
$$

Discussion for (b)

This charge is only slightly greater than those found in typical static electricity. Since air breaks down at about 3.00×10^6 V/m , more charge cannot be stored on this capacitor by increasing the voltage.

Another interesting biological example dealing with electric potential is found in the cell's plasma membrane. The membrane sets a cell off from its surroundings and also allows ions to selectively pass in and out of the cell. There is a potential difference across the membrane of about -70 mV . This is due to the mainly negatively charged ions in the cell and the predominance of positively charged sodium (Na^+) ions outside. Things change when a nerve cell is stimulated. Na^+ ions are allowed to pass through the membrane into the cell, producing a positive membrane potential—the nerve signal. The cell membrane is about 7 to 10 nm thick. An approximate value of the electric field across it is given by

$$
E = \frac{V}{d} = \frac{-70 \times 10^{-3} \text{ V}}{8 \times 10^{-9} \text{ m}} = -9 \times 10^6 \text{ V/m}.
\tag{19.56}
$$

This electric field is enough to cause a breakdown in air.

Dielectric

The previous example highlights the difficulty of storing a large amount of charge in capacitors. If d is made smaller to produce a larger capacitance, then the maximum voltage must be reduced proportionally to avoid breakdown (since $E = V / d$). An important solution to this difficulty is to put an insulating material, called a **dielectric**, between the plates of a capacitor and allow d to be as small as possible. Not only does the smaller d make the capacitance greater, but many insulators can withstand greater electric fields than air before breaking down.

There is another benefit to using a dielectric in a capacitor. Depending on the material used, the capacitance is greater than that given by the equation $C = \varepsilon_0 \frac{A}{d}$ by a factor κ , called the *relative permittivity*.[1] A parallel plate capacitor with a dielectric between its plates has a capacitance given by

$$C = \kappa \varepsilon_0 \frac{A}{d} \text{ (parallel plate capacitor with dielectric)}. \tag{19.57}$$

Values of the dielectric constant κ for various materials are given in **Table 19.1**. Note that κ for vacuum is exactly 1, and so the above equation is valid in that case, too. If a dielectric is used, perhaps by placing Teflon between the plates of the capacitor in **Example 19.8**, then the capacitance is greater by the factor κ, which for Teflon is 2.1.

Take-Home Experiment: Building a Capacitor

How large a capacitor can you make using a chewing gum wrapper? The plates will be the aluminum foil, and the separation (dielectric) in between will be the paper.

Table 19.1 Dielectric Constants and Dielectric Strengths for Various Materials at 20ºC

Material	Dielectric constant κ	Dielectric strength (V/m)
Vacuum	1.00000	—
Air	1.00059	3×10^6
Bakelite	4.9	24×10^6
Fused quartz	3.78	8×10^6
Neoprene rubber	6.7	12×10^6
Nylon	3.4	14×10^6
Paper	3.7	16×10^6
Polystyrene	2.56	24×10^6
Pyrex glass	5.6	14×10^6
Silicon oil	2.5	15×10^6
Strontium titanate	233	8×10^6
Teflon	2.1	60×10^6
Water	80	—

Note also that the dielectric constant for air is very close to 1, so that air-filled capacitors act much like those with vacuum between their plates *except* that the air can become conductive if the electric field strength becomes too great. (Recall that $E = V/d$ for a parallel plate capacitor.) Also shown in **Table 19.1** are maximum electric field strengths in V/m, called **dielectric strengths**, for several materials. These are the fields above which the material begins to break down and conduct. The dielectric strength imposes a limit on the voltage that can be applied for a given plate separation. For instance, in **Example 19.8**, the separation is 1.00 mm, and so the voltage limit for air is

$$\begin{aligned} V &= E \cdot d \\ &= (3\times10^6 \text{ V/m})(1.00\times10^{-3} \text{ m}) \\ &= 3000 \text{ V}. \end{aligned} \tag{19.58}$$

However, the limit for a 1.00 mm separation filled with Teflon is 60,000 V, since the dielectric strength of Teflon is 60×10^6 V/m. So the same capacitor filled with Teflon has a greater capacitance and can be subjected to a much greater voltage. Using the capacitance we calculated in the above example for the air-filled parallel plate capacitor, we find that the Teflon-filled capacitor can store a maximum charge of

---

1. Historically, the term *dielectric constant* was used. However, it is currently deprecated by standards organizations because this term was used for both relative and absolute permittivity, creating unfortunate and unnecessary ambiguity.

$$
\begin{aligned}
Q &= CV \\
&= \kappa C_{\text{air}} V \\
&= (2.1)(8.85 \text{ nF})(6.0 \times 10^4 \text{ V}) \\
&= 1.1 \text{ mC}.
\end{aligned}
$$

(19.59)

This is 42 times the charge of the same air-filled capacitor.

Dielectric Strength

The maximum electric field strength above which an insulating material begins to break down and conduct is called its **dielectric strength**.

Microscopically, how does a dielectric increase capacitance? Polarization of the insulator is responsible. The more easily it is polarized, the greater its dielectric constant κ. Water, for example, is a **polar molecule** because one end of the molecule has a slight positive charge and the other end has a slight negative charge. The polarity of water causes it to have a relatively large dielectric constant of 80. The effect of polarization can be best explained in terms of the characteristics of the Coulomb force. **Figure 19.23** shows the separation of charge schematically in the molecules of a dielectric material placed between the charged plates of a capacitor. The Coulomb force between the closest ends of the molecules and the charge on the plates is attractive and very strong, since they are very close together. This attracts more charge onto the plates than if the space were empty and the opposite charges were a distance d away.

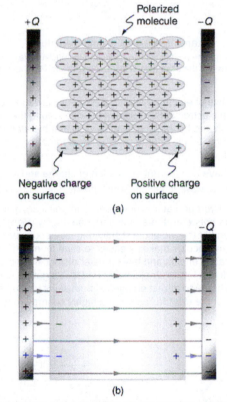

Figure 19.23 (a) The molecules in the insulating material between the plates of a capacitor are polarized by the charged plates. This produces a layer of opposite charge on the surface of the dielectric that attracts more charge onto the plate, increasing its capacitance. (b) The dielectric reduces the electric field strength inside the capacitor, resulting in a smaller voltage between the plates for the same charge. The capacitor stores the same charge for a smaller voltage, implying that it has a larger capacitance because of the dielectric.

Another way to understand how a dielectric increases capacitance is to consider its effect on the electric field inside the capacitor. **Figure 19.23(b)** shows the electric field lines with a dielectric in place. Since the field lines end on charges in the dielectric, there are fewer of them going from one side of the capacitor to the other. So the electric field strength is less than if there were a vacuum between the plates, even though the same charge is on the plates. The voltage between the plates is $V = Ed$, so it too is reduced by the dielectric. Thus there is a smaller voltage V for the same charge Q; since $C = Q/V$, the capacitance C is greater.

The dielectric constant is generally defined to be $\kappa = E_0 / E$, or the ratio of the electric field in a vacuum to that in the dielectric material, and is intimately related to the polarizability of the material.

Things Great and Small

The Submicroscopic Origin of Polarization

Polarization is a separation of charge within an atom or molecule. As has been noted, the planetary model of the atom pictures it as having a positive nucleus orbited by negative electrons, analogous to the planets orbiting the Sun. Although this model is not completely accurate, it is very helpful in explaining a vast range of phenomena and will be refined elsewhere, such as in **Atomic Physics**. The submicroscopic origin of polarization can be modeled as shown in **Figure 19.24**.

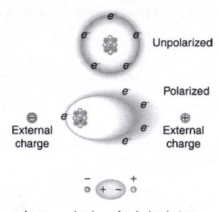

Large-scale view of polarized atom

Figure 19.24 Artist's conception of a polarized atom. The orbits of electrons around the nucleus are shifted slightly by the external charges (shown exaggerated). The resulting separation of charge within the atom means that it is polarized. Note that the unlike charge is now closer to the external charges, causing the polarization.

We will find in **Atomic Physics** that the orbits of electrons are more properly viewed as electron clouds with the density of the cloud related to the probability of finding an electron in that location (as opposed to the definite locations and paths of planets in their orbits around the Sun). This cloud is shifted by the Coulomb force so that the atom on average has a separation of charge. Although the atom remains neutral, it can now be the source of a Coulomb force, since a charge brought near the atom will be closer to one type of charge than the other.

Some molecules, such as those of water, have an inherent separation of charge and are thus called polar molecules. **Figure 19.25** illustrates the separation of charge in a water molecule, which has two hydrogen atoms and one oxygen atom (H_2O).

The water molecule is not symmetric—the hydrogen atoms are repelled to one side, giving the molecule a boomerang shape. The electrons in a water molecule are more concentrated around the more highly charged oxygen nucleus than around the hydrogen nuclei. This makes the oxygen end of the molecule slightly negative and leaves the hydrogen ends slightly positive. The inherent separation of charge in polar molecules makes it easier to align them with external fields and charges. Polar molecules therefore exhibit greater polarization effects and have greater dielectric constants. Those who study chemistry will find that the polar nature of water has many effects. For example, water molecules gather ions much more effectively because they have an electric field and a separation of charge to attract charges of both signs. Also, as brought out in the previous chapter, polar water provides a shield or screening of the electric fields in the highly charged molecules of interest in biological systems.

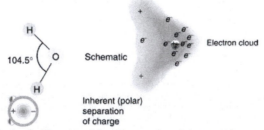

Figure 19.25 Artist's conception of a water molecule. There is an inherent separation of charge, and so water is a polar molecule. Electrons in the molecule are attracted to the oxygen nucleus and leave an excess of positive charge near the two hydrogen nuclei. (Note that the schematic on the right is a rough illustration of the distribution of electrons in the water molecule. It does not show the actual numbers of protons and electrons involved in the structure.)

PhET Explorations: Capacitor Lab

Explore how a capacitor works! Change the size of the plates and add a dielectric to see the effect on capacitance. Change the voltage and see charges built up on the plates. Observe the electric field in the capacitor. Measure the voltage and the electric field.

Figure 19.26 Capacitor Lab (http://cnx.org/content/m55340/1.2/capacitor-lab_en.jar)

19.6 Capacitors in Series and Parallel

Learning Objectives

By the end of this section, you will be able to:
 	
  • Derive expressions for total capacitance in series and in parallel.
  • Identify series and parallel parts in the combination of connection of capacitors.
  • Calculate the effective capacitance in series and parallel given individual capacitances.

Several capacitors may be connected together in a variety of applications. Multiple connections of capacitors act like a single equivalent capacitor. The total capacitance of this equivalent single capacitor depends both on the individual capacitors and how they are connected. There are two simple and common types of connections, called *series* and *parallel*, for which we can easily calculate the total capacitance. Certain more complicated connections can also be related to combinations of series and parallel.

Capacitance in Series

Figure 19.27(a) shows a series connection of three capacitors with a voltage applied. As for any capacitor, the capacitance of the combination is related to charge and voltage by $C = \frac{Q}{V}$.

Note in Figure 19.27 that opposite charges of magnitude Q flow to either side of the originally uncharged combination of capacitors when the voltage V is applied. Conservation of charge requires that equal-magnitude charges be created on the plates of the individual capacitors, since charge is only being separated in these originally neutral devices. The end result is that the combination resembles a single capacitor with an effective plate separation greater than that of the individual capacitors alone. (See Figure 19.27(b).) Larger plate separation means smaller capacitance. It is a general feature of series connections of capacitors that the total capacitance is less than any of the individual capacitances.

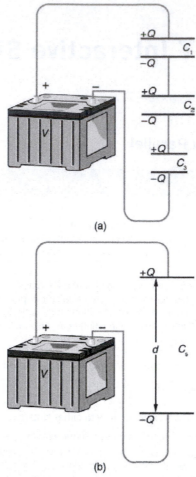

(a)

(b)

Figure 19.27 (a) Capacitors connected in series. The magnitude of the charge on each plate is Q. (b) An equivalent capacitor has a larger plate separation d. Series connections produce a total capacitance that is less than that of any of the individual capacitors.

We can find an expression for the total capacitance by considering the voltage across the individual capacitors shown in **Figure 19.27**. Solving $C = \frac{Q}{V}$ for V gives $V = \frac{Q}{C}$. The voltages across the individual capacitors are thus $V_1 = \frac{Q}{C_1}$, $V_2 = \frac{Q}{C_2}$, and $V_3 = \frac{Q}{C_3}$. The total voltage is the sum of the individual voltages:

$$V = V_1 + V_2 + V_3. \tag{19.60}$$

Now, calling the total capacitance C_S for series capacitance, consider that

$$V = \frac{Q}{C_S} = V_1 + V_2 + V_3. \tag{19.61}$$

Entering the expressions for V_1, V_2, and V_3, we get

$$\frac{Q}{C_S} = \frac{Q}{C_1} + \frac{Q}{C_2} + \frac{Q}{C_3}. \tag{19.62}$$

Canceling the Qs, we obtain the equation for the total capacitance in series C_S to be

$$\frac{1}{C_S} = \frac{1}{C_1} + \frac{1}{C_2} + \frac{1}{C_3} + ..., \tag{19.63}$$

where "..." indicates that the expression is valid for any number of capacitors connected in series. An expression of this form always results in a total capacitance C_S that is less than any of the individual capacitances C_1, C_2, ..., as the next example illustrates.

Total Capacitance in Series, C_s

Total capacitance in series: $\frac{1}{C_S} = \frac{1}{C_1} + \frac{1}{C_2} + \frac{1}{C_3} + \ldots$

Example 19.9 What Is the Series Capacitance?

Find the total capacitance for three capacitors connected in series, given their individual capacitances are 1.000, 5.000, and 8.000 μF.

Strategy

With the given information, the total capacitance can be found using the equation for capacitance in series.

Solution

Entering the given capacitances into the expression for $\frac{1}{C_S}$ gives $\frac{1}{C_S} = \frac{1}{C_1} + \frac{1}{C_2} + \frac{1}{C_3}$.

$$\frac{1}{C_S} = \frac{1}{1.000 \ \mu F} + \frac{1}{5.000 \ \mu F} + \frac{1}{8.000 \ \mu F} = \frac{1.325}{\mu F} \tag{19.64}$$

Inverting to find C_S yields $C_S = \frac{\mu F}{1.325} = 0.755 \ \mu F$.

Discussion

The total series capacitance C_s is less than the smallest individual capacitance, as promised. In series connections of capacitors, the sum is less than the parts. In fact, it is less than any individual. Note that it is sometimes possible, and more convenient, to solve an equation like the above by finding the least common denominator, which in this case (showing only whole-number calculations) is 40. Thus,

$$\frac{1}{C_S} = \frac{40}{40 \ \mu F} + \frac{8}{40 \ \mu F} + \frac{5}{40 \ \mu F} = \frac{53}{40 \ \mu F}, \tag{19.65}$$

so that

$$C_S = \frac{40 \ \mu F}{53} = 0.755 \ \mu F. \tag{19.66}$$

Capacitors in Parallel

Figure 19.28(a) shows a parallel connection of three capacitors with a voltage applied. Here the total capacitance is easier to find than in the series case. To find the equivalent total capacitance C_p, we first note that the voltage across each capacitor is V, the same as that of the source, since they are connected directly to it through a conductor. (Conductors are equipotentials, and so the voltage across the capacitors is the same as that across the voltage source.) Thus the capacitors have the same charges on them as they would have if connected individually to the voltage source. The total charge Q is the sum of the individual charges:

$$Q = Q_1 + Q_2 + Q_3. \tag{19.67}$$

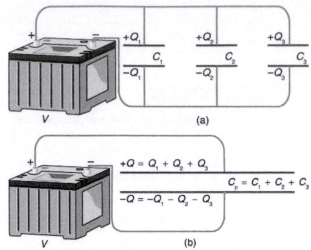

(a)

(b)

Figure 19.28 (a) Capacitors in parallel. Each is connected directly to the voltage source just as if it were all alone, and so the total capacitance in parallel is just the sum of the individual capacitances. (b) The equivalent capacitor has a larger plate area and can therefore hold more charge than the individual capacitors.

Using the relationship $Q = CV$, we see that the total charge is $Q = C_p V$, and the individual charges are $Q_1 = C_1 V$, $Q_2 = C_2 V$, and $Q_3 = C_3 V$. Entering these into the previous equation gives

$$C_p V = C_1 V + C_2 V + C_3 V. \tag{19.68}$$

Canceling V from the equation, we obtain the equation for the total capacitance in parallel C_p:

$$C_p = C_1 + C_2 + C_3 + ... \tag{19.69}$$

Total capacitance in parallel is simply the sum of the individual capacitances. (Again the "..." indicates the expression is valid for any number of capacitors connected in parallel.) So, for example, if the capacitors in the example above were connected in parallel, their capacitance would be

$$C_p = 1.000 \ \mu\text{F} + 5.000 \ \mu\text{F} + 8.000 \ \mu\text{F} = 14.000 \ \mu\text{F}. \tag{19.70}$$

The equivalent capacitor for a parallel connection has an effectively larger plate area and, thus, a larger capacitance, as illustrated in **Figure 19.28**(b).

Total Capacitance in Parallel, C_p

Total capacitance in parallel $C_p = C_1 + C_2 + C_3 + ...$

More complicated connections of capacitors can sometimes be combinations of series and parallel. (See **Figure 19.29**.) To find the total capacitance of such combinations, we identify series and parallel parts, compute their capacitances, and then find the total.

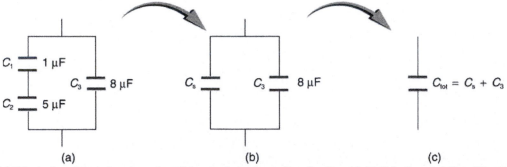

(a)   (b)   (c)

Figure 19.29 (a) This circuit contains both series and parallel connections of capacitors. See **Example 19.10** for the calculation of the overall capacitance of the circuit. (b) C_1 and C_2 are in series; their equivalent capacitance C_s is less than either of them. (c) Note that C_s is in parallel with C_3. The total capacitance is, thus, the sum of C_s and C_3.

Example 19.10 A Mixture of Series and Parallel Capacitance

Find the total capacitance of the combination of capacitors shown in **Figure 19.29**. Assume the capacitances in **Figure 19.29** are known to three decimal places ($C_1 = 1.000\ \mu F$, $C_2 = 5.000\ \mu F$, and $C_3 = 8.000\ \mu F$), and round your answer to three decimal places.

Strategy

To find the total capacitance, we first identify which capacitors are in series and which are in parallel. Capacitors C_1 and C_2 are in series. Their combination, labeled C_S in the figure, is in parallel with C_3 .

Solution

Since C_1 and C_2 are in series, their total capacitance is given by $\frac{1}{C_S} = \frac{1}{C_1} + \frac{1}{C_2} + \frac{1}{C_3}$. Entering their values into the equation gives

$$\frac{1}{C_S} = \frac{1}{C_1} + \frac{1}{C_2} = \frac{1}{1.000\ \mu F} + \frac{1}{5.000\ \mu F} = \frac{1.200}{\mu F}. \tag{19.71}$$

Inverting gives

$$C_S = 0.833\ \mu F. \tag{19.72}$$

This equivalent series capacitance is in parallel with the third capacitor; thus, the total is the sum

$$\begin{aligned} C_{\text{tot}} &= C_S + C_S \\ &= 0.833\ \mu F + 8.000\ \mu F \\ &= 8.833\ \mu F. \end{aligned} \tag{19.73}$$

Discussion

This technique of analyzing the combinations of capacitors piece by piece until a total is obtained can be applied to larger combinations of capacitors.

19.7 Energy Stored in Capacitors

Learning Objectives

By the end of this section, you will be able to:

- List some uses of capacitors.
- Express in equation form the energy stored in a capacitor.
- Explain the function of a defibrillator.

The information presented in this section supports the following AP® learning objectives and science practices:

- **5.B.2.1** The student is able to calculate the expected behavior of a system using the object model (i.e., by ignoring changes in internal structure) to analyze a situation. Then, when the model fails, the student can justify the use of conservation of energy principles to calculate the change in internal energy due to changes in internal structure because the object is actually a system. **(S.P. 1.4, 2.1)**
- **5.B.3.1** The student is able to describe and make qualitative and/or quantitative predictions about everyday examples of systems with internal potential energy. **(S.P. 2.2, 6.4, 7.2)**
- **5.B.3.2** The student is able to make quantitative calculations of the internal potential energy of a system from a description or diagram of that system. **(S.P. 1.4, 2.2)**
- **5.B.3.3** The student is able to apply mathematical reasoning to create a description of the internal potential energy of a system from a description or diagram of the objects and interactions in that system. **(S.P. 1.4, 2.2)**

Most of us have seen dramatizations in which medical personnel use a **defibrillator** to pass an electric current through a patient's heart to get it to beat normally. (Review **Figure 19.30**.) Often realistic in detail, the person applying the shock directs another person to "make it 400 joules this time." The energy delivered by the defibrillator is stored in a capacitor and can be adjusted to fit the situation. SI units of joules are often employed. Less dramatic is the use of capacitors in microelectronics, such as certain handheld calculators, to supply energy when batteries are charged. (See **Figure 19.30**.) Capacitors are also used to supply energy for flash lamps on cameras.

Figure 19.30 Energy stored in the large capacitor is used to preserve the memory of an electronic calculator when its batteries are charged. (credit: Kucharek, Wikimedia Commons)

Energy stored in a capacitor is electrical potential energy, and it is thus related to the charge Q and voltage V on the capacitor. We must be careful when applying the equation for electrical potential energy $\Delta\text{PE} = q\Delta V$ to a capacitor. Remember that ΔPE is the potential energy of a charge q going through a voltage ΔV. But the capacitor starts with zero voltage and gradually comes up to its full voltage as it is charged. The first charge placed on a capacitor experiences a change in voltage $\Delta V = 0$, since the capacitor has zero voltage when uncharged. The final charge placed on a capacitor experiences $\Delta V = V$, since the capacitor now has its full voltage V on it. The average voltage on the capacitor during the charging process is $V/2$, and so the average voltage experienced by the full charge q is $V/2$. Thus the energy stored in a capacitor, E_{cap}, is

$$E_{\text{cap}} = \frac{QV}{2},$$

(19.74)

where Q is the charge on a capacitor with a voltage V applied. (Note that the energy is not QV, but $QV/2$.) Charge and voltage are related to the capacitance C of a capacitor by $Q = CV$, and so the expression for E_{cap} can be algebraically manipulated into three equivalent expressions:

$$E_{\text{cap}} = \frac{QV}{2} = \frac{CV^2}{2} = \frac{Q^2}{2C},$$

(19.75)

where Q is the charge and V the voltage on a capacitor C. The energy is in joules for a charge in coulombs, voltage in volts, and capacitance in farads.

Energy Stored in Capacitors

The energy stored in a capacitor can be expressed in three ways:

$$E_{\text{cap}} = \frac{QV}{2} = \frac{CV^2}{2} = \frac{Q^2}{2C},$$

(19.76)

where Q is the charge, V is the voltage, and C is the capacitance of the capacitor. The energy is in joules for a charge in coulombs, voltage in volts, and capacitance in farads. Energy stored in the capacitor is internal potential energy.

Making Connections: Point Charges and Capacitors

Recall that we were able to calculate the stored potential energy of a configuration of point charges, and how the energy changed when the configuration changed in **Applying the Science Practices: Work and Potential Energy in Point Charges**. Since the charges in a capacitor are, ultimately, all point charges, we can do the same with capacitors. However, we write it down in terms of the macroscopic quantities of (total) charge, voltage, and capacitance; hence Equation (19.76).

For example, consider a parallel plate capacitor with a variable distance between the plates connected to a battery (fixed voltage). When you move the plates closer together, the voltage still doesn't change. However, this increases the capacitance, and hence the internal energy stored in this system (the capacitor) increases. It turns out that the increase in capacitance for a fixed voltage results in an increased charge. The work you did moving the plates closer together ultimately went into moving more electrons from the positive plate to the negative plate.

In a defibrillator, the delivery of a large charge in a short burst to a set of paddles across a person's chest can be a lifesaver. The person's heart attack might have arisen from the onset of fast, irregular beating of the heart—cardiac or ventricular fibrillation. The application of a large shock of electrical energy can terminate the arrhythmia and allow the body's pacemaker to resume normal patterns. Today it is common for ambulances to carry a defibrillator, which also uses an electrocardiogram to analyze the patient's heartbeat pattern. Automated external defibrillators (AED) are found in many public places (**Figure 19.31**). These are designed to be used by lay persons. The device automatically diagnoses the patient's heart condition and then applies the shock with appropriate energy and waveform. CPR is recommended in many cases before use of an AED.

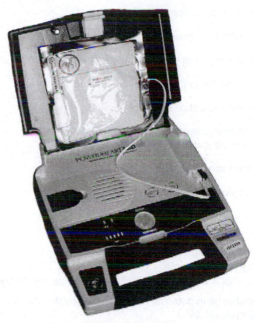

Figure 19.31 Automated external defibrillators are found in many public places. These portable units provide verbal instructions for use in the important first few minutes for a person suffering a cardiac attack. (credit: Owain Davies, Wikimedia Commons)

Example 19.11 Capacitance in a Heart Defibrillator

A heart defibrillator delivers $4.00{\times}10^2$ J of energy by discharging a capacitor initially at $1.00{\times}10^4$ V . What is its capacitance?

Strategy

We are given E_{cap} and V , and we are asked to find the capacitance C . Of the three expressions in the equation for E_{cap} , the most convenient relationship is

$$E_{cap} = \frac{CV^2}{2}. \tag{19.77}$$

Solution

Solving this expression for C and entering the given values yields

$$C = \frac{2E_{cap}}{V^2} = \frac{2(4.00{\times}10^2 \text{ J})}{(1.00{\times}10^4 \text{ V})^2} = 8.00{\times}10^{-6} \text{ F} \tag{19.78}$$

$$= 8.00 \text{ μF}.$$

Discussion

This is a fairly large, but manageable, capacitance at $1.00{\times}10^4$ V .

Glossary

capacitance: amount of charge stored per unit volt

capacitor: a device that stores electric charge

defibrillator: a machine used to provide an electrical shock to a heart attack victim's heart in order to restore the heart's normal rhythmic pattern

dielectric: an insulating material

dielectric strength: the maximum electric field above which an insulating material begins to break down and conduct

electric potential: potential energy per unit charge

electron volt: the energy given to a fundamental charge accelerated through a potential difference of one volt

equipotential line: a line along which the electric potential is constant

grounding: fixing a conductor at zero volts by connecting it to the earth or ground

mechanical energy: sum of the kinetic energy and potential energy of a system; this sum is a constant

parallel plate capacitor: two identical conducting plates separated by a distance

polar molecule: a molecule with inherent separation of charge

potential difference (or voltage): change in potential energy of a charge moved from one point to another, divided by the charge; units of potential difference are joules per coulomb, known as volt

scalar: physical quantity with magnitude but no direction

vector: physical quantity with both magnitude and direction

Section Summary

19.1 Electric Potential Energy: Potential Difference
- Electric potential is potential energy per unit charge.
- The potential difference between points A and B, $V_B - V_A$, defined to be the change in potential energy of a charge q moved from A to B, is equal to the change in potential energy divided by the charge, Potential difference is commonly called voltage, represented by the symbol ΔV.

$$\Delta V = \frac{\Delta \text{PE}}{q} \text{ and } \Delta \text{PE} = q\Delta V.$$

- An electron volt is the energy given to a fundamental charge accelerated through a potential difference of 1 V. In equation form,

$$1 \text{ eV} = \left(1.60 \times 10^{-19} \text{ C}\right)(1 \text{ V}) = \left(1.60 \times 10^{-19} \text{ C}\right)(1 \text{ J/C})$$
$$= 1.60 \times 10^{-19} \text{ J}.$$

- Mechanical energy is the sum of the kinetic energy and potential energy of a system, that is, $\text{KE} + \text{PE}$. This sum is a constant.

19.2 Electric Potential in a Uniform Electric Field
- The voltage between points A and B is

$$\left.\begin{array}{c} V_{AB} = Ed \\ E = \dfrac{V_{AB}}{d} \end{array}\right\} (\text{uniform } E \text{ - field on y}),$$

where d is the distance from A to B, or the distance between the plates.
- In equation form, the general relationship between voltage and electric field is

$$E = -\frac{\Delta V}{\Delta s},$$

where Δs is the distance over which the change in potential, ΔV, takes place. The minus sign tells us that $\mathbf{E}$ points in the direction of decreasing potential.) The electric field is said to be the *gradient* (as in grade or slope) of the electric potential.

19.3 Electrical Potential Due to a Point Charge
- Electric potential of a point charge is $V = kQ/r$.
- Electric potential is a scalar, and electric field is a vector. Addition of voltages as numbers gives the voltage due to a combination of point charges, whereas addition of individual fields as vectors gives the total electric field.

19.4 Equipotential Lines
- An equipotential line is a line along which the electric potential is constant.
- An equipotential surface is a three-dimensional version of equipotential lines.
- Equipotential lines are always perpendicular to electric field lines.
- The process by which a conductor can be fixed at zero volts by connecting it to the earth with a good conductor is called grounding.

19.5 Capacitors and Dielectrics
- A capacitor is a device used to store charge.

- The amount of charge Q a capacitor can store depends on two major factors—the voltage applied and the capacitor's physical characteristics, such as its size.
- The capacitance C is the amount of charge stored per volt, or

$$C = \frac{Q}{V}.$$

- The capacitance of a parallel plate capacitor is $C = \varepsilon_0 \frac{A}{d}$, when the plates are separated by air or free space. ε_0 is called the permittivity of free space.
- A parallel plate capacitor with a dielectric between its plates has a capacitance given by

$$C = \kappa \varepsilon_0 \frac{A}{d},$$

where κ is the dielectric constant of the material.
- The maximum electric field strength above which an insulating material begins to break down and conduct is called dielectric strength.

19.6 Capacitors in Series and Parallel

- Total capacitance in series $\frac{1}{C_S} = \frac{1}{C_1} + \frac{1}{C_2} + \frac{1}{C_3} + ...$
- Total capacitance in parallel $C_p = C_1 + C_2 + C_3 + ...$
- If a circuit contains a combination of capacitors in series and parallel, identify series and parallel parts, compute their capacitances, and then find the total.

19.7 Energy Stored in Capacitors

- Capacitors are used in a variety of devices, including defibrillators, microelectronics such as calculators, and flash lamps, to supply energy.
- The energy stored in a capacitor can be expressed in three ways:

$$E_{cap} = \frac{QV}{2} = \frac{CV^2}{2} = \frac{Q^2}{2C},$$

where Q is the charge, V is the voltage, and C is the capacitance of the capacitor. The energy is in joules when the charge is in coulombs, voltage is in volts, and capacitance is in farads.

Conceptual Questions

19.1 Electric Potential Energy: Potential Difference

1. Voltage is the common word for potential difference. Which term is more descriptive, voltage or potential difference?

2. If the voltage between two points is zero, can a test charge be moved between them with zero net work being done? Can this necessarily be done without exerting a force? Explain.

3. What is the relationship between voltage and energy? More precisely, what is the relationship between potential difference and electric potential energy?

4. Voltages are always measured between two points. Why?

5. How are units of volts and electron volts related? How do they differ?

19.2 Electric Potential in a Uniform Electric Field

6. Discuss how potential difference and electric field strength are related. Give an example.

7. What is the strength of the electric field in a region where the electric potential is constant?

8. Will a negative charge, initially at rest, move toward higher or lower potential? Explain why.

19.3 Electrical Potential Due to a Point Charge

9. In what region of space is the potential due to a uniformly charged sphere the same as that of a point charge? In what region does it differ from that of a point charge?

10. Can the potential of a non-uniformly charged sphere be the same as that of a point charge? Explain.

19.4 Equipotential Lines

11. What is an equipotential line? What is an equipotential surface?

12. Explain in your own words why equipotential lines and surfaces must be perpendicular to electric field lines.

13. Can different equipotential lines cross? Explain.

19.5 Capacitors and Dielectrics

14. Does the capacitance of a device depend on the applied voltage? What about the charge stored in it?

15. Use the characteristics of the Coulomb force to explain why capacitance should be proportional to the plate area of a capacitor. Similarly, explain why capacitance should be inversely proportional to the separation between plates.

16. Give the reason why a dielectric material increases capacitance compared with what it would be with air between the plates of a capacitor. What is the independent reason that a dielectric material also allows a greater voltage to be applied to a capacitor? (The dielectric thus increases C and permits a greater V.)

17. How does the polar character of water molecules help to explain water's relatively large dielectric constant? (**Figure 19.25**)

18. Sparks will occur between the plates of an air-filled capacitor at lower voltage when the air is humid than when dry. Explain why, considering the polar character of water molecules.

19. Water has a large dielectric constant, but it is rarely used in capacitors. Explain why.

20. Membranes in living cells, including those in humans, are characterized by a separation of charge across the membrane. Effectively, the membranes are thus charged capacitors with important functions related to the potential difference across the membrane. Is energy required to separate these charges in living membranes and, if so, is its source the metabolization of food energy or some other source?

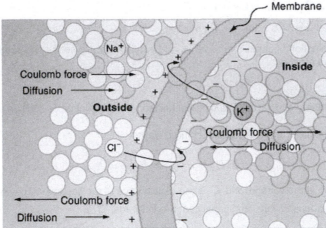

Figure 19.32 The semipermeable membrane of a cell has different concentrations of ions inside and out. Diffusion moves the K^+ (potassium) and Cl^- (chloride) ions in the directions shown, until the Coulomb force halts further transfer. This results in a layer of positive charge on the outside, a layer of negative charge on the inside, and thus a voltage across the cell membrane. The membrane is normally impermeable to Na^+ (sodium ions).

19.6 Capacitors in Series and Parallel

21. If you wish to store a large amount of energy in a capacitor bank, would you connect capacitors in series or parallel? Explain.

19.7 Energy Stored in Capacitors

22. How does the energy contained in a charged capacitor change when a dielectric is inserted, assuming the capacitor is isolated and its charge is constant? Does this imply that work was done?

23. What happens to the energy stored in a capacitor connected to a battery when a dielectric is inserted? Was work done in the process?

Problems & Exercises

19.1 Electric Potential Energy: Potential Difference

1. Find the ratio of speeds of an electron and a negative hydrogen ion (one having an extra electron) accelerated through the same voltage, assuming non-relativistic final speeds. Take the mass of the hydrogen ion to be 1.67×10^{-27} kg.

2. An evacuated tube uses an accelerating voltage of 40 kV to accelerate electrons to hit a copper plate and produce x rays. Non-relativistically, what would be the maximum speed of these electrons?

3. A bare helium nucleus has two positive charges and a mass of 6.64×10^{-27} kg. (a) Calculate its kinetic energy in joules at 2.00% of the speed of light. (b) What is this in electron volts? (c) What voltage would be needed to obtain this energy?

4. Integrated Concepts

Singly charged gas ions are accelerated from rest through a voltage of 13.0 V. At what temperature will the average kinetic energy of gas molecules be the same as that given these ions?

5. Integrated Concepts

The temperature near the center of the Sun is thought to be 15 million degrees Celsius $\left(1.5 \times 10^7 \ ^\circ\text{C}\right)$. Through what voltage must a singly charged ion be accelerated to have the same energy as the average kinetic energy of ions at this temperature?

6. Integrated Concepts

(a) What is the average power output of a heart defibrillator that dissipates 400 J of energy in 10.0 ms? (b) Considering the high-power output, why doesn't the defibrillator produce serious burns?

7. Integrated Concepts

A lightning bolt strikes a tree, moving 20.0 C of charge through a potential difference of 1.00×10^2 MV. (a) What energy was dissipated? (b) What mass of water could be raised from 15°C to the boiling point and then boiled by this energy? (c) Discuss the damage that could be caused to the tree by the expansion of the boiling steam.

8. Integrated Concepts

A 12.0 V battery-operated bottle warmer heats 50.0 g of glass, 2.50×10^2 g of baby formula, and 2.00×10^2 g of aluminum from 20.0°C to 90.0°C. (a) How much charge is moved by the battery? (b) How many electrons per second flow if it takes 5.00 min to warm the formula? (Hint: Assume that the specific heat of baby formula is about the same as the specific heat of water.)

9. Integrated Concepts

A battery-operated car utilizes a 12.0 V system. Find the charge the batteries must be able to move in order to accelerate the 750 kg car from rest to 25.0 m/s, make it climb a 2.00×10^2 m high hill, and then cause it to travel at a constant 25.0 m/s by exerting a 5.00×10^2 N force for an hour.

10. Integrated Concepts

Fusion probability is greatly enhanced when appropriate nuclei are brought close together, but mutual Coulomb repulsion must be overcome. This can be done using the kinetic energy of high-temperature gas ions or by accelerating the nuclei toward one another. (a) Calculate the potential energy of two singly charged nuclei separated by 1.00×10^{-12} m by finding the voltage of one at that distance and multiplying by the charge of the other. (b) At what temperature will atoms of a gas have an average kinetic energy equal to this needed electrical potential energy?

11. Unreasonable Results

(a) Find the voltage near a 10.0 cm diameter metal sphere that has 8.00 C of excess positive charge on it. (b) What is unreasonable about this result? (c) Which assumptions are responsible?

12. Construct Your Own Problem

Consider a battery used to supply energy to a cellular phone. Construct a problem in which you determine the energy that must be supplied by the battery, and then calculate the amount of charge it must be able to move in order to supply this energy. Among the things to be considered are the energy needs and battery voltage. You may need to look ahead to interpret manufacturer's battery ratings in ampere-hours as energy in joules.

19.2 Electric Potential in a Uniform Electric Field

13. Show that units of V/m and N/C for electric field strength are indeed equivalent.

14. What is the strength of the electric field between two parallel conducting plates separated by 1.00 cm and having a potential difference (voltage) between them of 1.50×10^4 V?

15. The electric field strength between two parallel conducting plates separated by 4.00 cm is 7.50×10^4 V/m. (a) What is the potential difference between the plates? (b) The plate with the lowest potential is taken to be at zero volts. What is the potential 1.00 cm from that plate (and 3.00 cm from the other)?

16. How far apart are two conducting plates that have an electric field strength of 4.50×10^3 V/m between them, if their potential difference is 15.0 kV?

17. (a) Will the electric field strength between two parallel conducting plates exceed the breakdown strength for air (3.0×10^6 V/m) if the plates are separated by 2.00 mm and a potential difference of 5.0×10^3 V is applied? (b) How close together can the plates be with this applied voltage?

18. The voltage across a membrane forming a cell wall is 80.0 mV and the membrane is 9.00 nm thick. What is the electric field strength? (The value is surprisingly large, but correct. Membranes are discussed in **Capacitors and Dielectrics** and **Nerve Conduction—Electrocardiograms**.) You may assume a uniform electric field.

19. Membrane walls of living cells have surprisingly large electric fields across them due to separation of ions. (Membranes are discussed in some detail in **Nerve Conduction—Electrocardiograms**.) What is the voltage across an 8.00 nm–thick membrane if the electric field strength across it is 5.50 MV/m? You may assume a uniform electric field.

20. Two parallel conducting plates are separated by 10.0 cm, and one of them is taken to be at zero volts. (a) What is the electric field strength between them, if the potential 8.00 cm from the zero volt plate (and 2.00 cm from the other) is 450 V? (b) What is the voltage between the plates?

21. Find the maximum potential difference between two parallel conducting plates separated by 0.500 cm of air, given the maximum sustainable electric field strength in air to be 3.0×10^6 V/m.

22. A doubly charged ion is accelerated to an energy of 32.0 keV by the electric field between two parallel conducting plates separated by 2.00 cm. What is the electric field strength between the plates?

23. An electron is to be accelerated in a uniform electric field having a strength of 2.00×10^6 V/m. (a) What energy in keV is given to the electron if it is accelerated through 0.400 m? (b) Over what distance would it have to be accelerated to increase its energy by 50.0 GeV?

19.3 Electrical Potential Due to a Point Charge

24. A 0.500 cm diameter plastic sphere, used in a static electricity demonstration, has a uniformly distributed 40.0 pC charge on its surface. What is the potential near its surface?

25. What is the potential 0.530×10^{-10} m from a proton (the average distance between the proton and electron in a hydrogen atom)?

26. (a) A sphere has a surface uniformly charged with 1.00 C. At what distance from its center is the potential 5.00 MV? (b) What does your answer imply about the practical aspect of isolating such a large charge?

27. How far from a $1.00 \ \mu C$ point charge will the potential be 100 V? At what distance will it be 2.00×10^2 V?

28. What are the sign and magnitude of a point charge that produces a potential of -2.00 V at a distance of 1.00 mm?

29. If the potential due to a point charge is 5.00×10^2 V at a distance of 15.0 m, what are the sign and magnitude of the charge?

30. In nuclear fission, a nucleus splits roughly in half. (a) What is the potential 2.00×10^{-14} m from a fragment that has 46 protons in it? (b) What is the potential energy in MeV of a similarly charged fragment at this distance?

31. A research Van de Graaff generator has a 2.00-m-diameter metal sphere with a charge of 5.00 mC on it. (a) What is the potential near its surface? (b) At what distance from its center is the potential 1.00 MV? (c) An oxygen atom with three missing electrons is released near the Van de Graaff generator. What is its energy in MeV at this distance?

32. An electrostatic paint sprayer has a 0.200-m-diameter metal sphere at a potential of 25.0 kV that repels paint droplets onto a grounded object. (a) What charge is on the sphere? (b) What charge must a 0.100-mg drop of paint have to arrive at the object with a speed of 10.0 m/s?

33. In one of the classic nuclear physics experiments at the beginning of the 20th century, an alpha particle was accelerated toward a gold nucleus, and its path was substantially deflected by the Coulomb interaction. If the energy of the doubly charged alpha nucleus was 5.00 MeV, how close to the gold nucleus (79 protons) could it come before being deflected?

34. (a) What is the potential between two points situated 10 cm and 20 cm from a $3.0 \ \mu C$ point charge? (b) To what location should the point at 20 cm be moved to increase this potential difference by a factor of two?

35. Unreasonable Results

(a) What is the final speed of an electron accelerated from rest through a voltage of 25.0 MV by a negatively charged Van de Graaff terminal?

(b) What is unreasonable about this result?

(c) Which assumptions are responsible?

19.4 Equipotential Lines

36. (a) Sketch the equipotential lines near a point charge $+ \ q$. Indicate the direction of increasing potential. (b) Do the same for a point charge $- 3 \ q$.

37. Sketch the equipotential lines for the two equal positive charges shown in **Figure 19.33**. Indicate the direction of increasing potential.

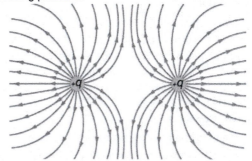

Figure 19.33 The electric field near two equal positive charges is directed away from each of the charges.

38. **Figure 19.34** shows the electric field lines near two charges q_1 and q_2, the first having a magnitude four times that of the second. Sketch the equipotential lines for these two charges, and indicate the direction of increasing potential.

39. Sketch the equipotential lines a long distance from the charges shown in **Figure 19.34**. Indicate the direction of increasing potential.

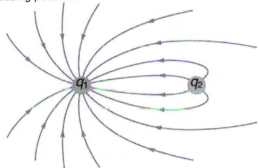

Figure 19.34 The electric field near two charges.

40. Sketch the equipotential lines in the vicinity of two opposite charges, where the negative charge is three times as great in magnitude as the positive. See **Figure 19.34** for a similar situation. Indicate the direction of increasing potential.

41. Sketch the equipotential lines in the vicinity of the negatively charged conductor in **Figure 19.35**. How will these equipotentials look a long distance from the object?

Figure 19.35 A negatively charged conductor.

42. Sketch the equipotential lines surrounding the two conducting plates shown in **Figure 19.36**, given the top plate is positive and the bottom plate has an equal amount of negative charge. Be certain to indicate the distribution of charge on the plates. Is the field strongest where the plates are closest? Why should it be?

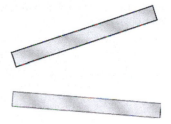

Figure 19.36

43. (a) Sketch the electric field lines in the vicinity of the charged insulator in **Figure 19.37**. Note its non-uniform charge distribution. (b) Sketch equipotential lines surrounding the insulator. Indicate the direction of increasing potential.

Figure 19.37 A charged insulating rod such as might be used in a classroom demonstration.

44. The naturally occurring charge on the ground on a fine day out in the open country is -1.00 nC/m^2. (a) What is the electric field relative to ground at a height of 3.00 m? (b) Calculate the electric potential at this height. (c) Sketch electric field and equipotential lines for this scenario.

45. The lesser electric ray (*Narcine bancroftii*) maintains an incredible charge on its head and a charge equal in magnitude but opposite in sign on its tail (**Figure 19.38**). (a) Sketch the equipotential lines surrounding the ray. (b) Sketch the equipotentials when the ray is near a ship with a conducting surface. (c) How could this charge distribution be of use to the ray?

Figure 19.38 Lesser electric ray (*Narcine bancroftii*) (credit: National Oceanic and Atmospheric Administration, NOAA's Fisheries Collection).

19.5 Capacitors and Dielectrics

46. What charge is stored in a $180 \text{ } \mu\text{F}$ capacitor when 120 V is applied to it?

47. Find the charge stored when 5.50 V is applied to an 8.00 pF capacitor.

48. What charge is stored in the capacitor in **Example 19.8**?

49. Calculate the voltage applied to a $2.00 \text{ } \mu\text{F}$ capacitor when it holds $3.10 \text{ } \mu\text{C}$ of charge.

50. What voltage must be applied to an 8.00 nF capacitor to store 0.160 mC of charge?

51. What capacitance is needed to store $3.00\ \mu C$ of charge at a voltage of 120 V?

52. What is the capacitance of a large Van de Graaff generator's terminal, given that it stores 8.00 mC of charge at a voltage of 12.0 MV?

53. Find the capacitance of a parallel plate capacitor having plates of area $5.00\ m^2$ that are separated by 0.100 mm of Teflon.

54. (a)What is the capacitance of a parallel plate capacitor having plates of area $1.50\ m^2$ that are separated by 0.0200 mm of neoprene rubber? (b) What charge does it hold when 9.00 V is applied to it?

55. Integrated Concepts

A prankster applies 450 V to an $80.0\ \mu F$ capacitor and then tosses it to an unsuspecting victim. The victim's finger is burned by the discharge of the capacitor through 0.200 g of flesh. What is the temperature increase of the flesh? Is it reasonable to assume no phase change?

56. Unreasonable Results

(a) A certain parallel plate capacitor has plates of area $4.00\ m^2$, separated by 0.0100 mm of nylon, and stores 0.170 C of charge. What is the applied voltage? (b) What is unreasonable about this result? (c) Which assumptions are responsible or inconsistent?

19.6 Capacitors in Series and Parallel

57. Find the total capacitance of the combination of capacitors in **Figure 19.39**.

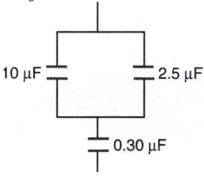

Figure 19.39 A combination of series and parallel connections of capacitors.

58. Suppose you want a capacitor bank with a total capacitance of 0.750 F and you possess numerous 1.50 mF capacitors. What is the smallest number you could hook together to achieve your goal, and how would you connect them?

59. What total capacitances can you make by connecting a $5.00\ \mu F$ and an $8.00\ \mu F$ capacitor together?

60. Find the total capacitance of the combination of capacitors shown in **Figure 19.40**.

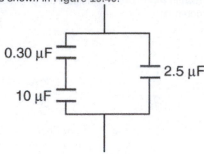

Figure 19.40 A combination of series and parallel connections of capacitors.

61. Find the total capacitance of the combination of capacitors shown in **Figure 19.41**.

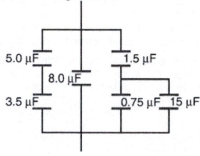

Figure 19.41 A combination of series and parallel connections of capacitors.

62. Unreasonable Results

(a) An $8.00\ \mu F$ capacitor is connected in parallel to another capacitor, producing a total capacitance of $5.00\ \mu F$. What is the capacitance of the second capacitor? (b) What is unreasonable about this result? (c) Which assumptions are unreasonable or inconsistent?

19.7 Energy Stored in Capacitors

63. (a) What is the energy stored in the $10.0\ \mu F$ capacitor of a heart defibrillator charged to 9.00×10^3 V ? (b) Find the amount of stored charge.

64. In open heart surgery, a much smaller amount of energy will defibrillate the heart. (a) What voltage is applied to the $8.00\ \mu F$ capacitor of a heart defibrillator that stores 40.0 J of energy? (b) Find the amount of stored charge.

65. A $165\ \mu F$ capacitor is used in conjunction with a motor. How much energy is stored in it when 119 V is applied?

66. Suppose you have a 9.00 V battery, a $2.00\ \mu F$ capacitor, and a $7.40\ \mu F$ capacitor. (a) Find the charge and energy stored if the capacitors are connected to the battery in series. (b) Do the same for a parallel connection.

67. A nervous physicist worries that the two metal shelves of his wood frame bookcase might obtain a high voltage if charged by static electricity, perhaps produced by friction. (a) What is the capacitance of the empty shelves if they have area 1.00×10^2 m^2 and are 0.200 m apart? (b) What is the voltage between them if opposite charges of magnitude 2.00 nC are placed on them? (c) To show that this voltage poses a small hazard, calculate the energy stored.

68. Show that for a given dielectric material the maximum energy a parallel plate capacitor can store is directly proportional to the volume of dielectric ($\text{Volume} = A \cdot d$). Note that the applied voltage is limited by the dielectric strength.

69. Construct Your Own Problem

Consider a heart defibrillator similar to that discussed in Example 19.11. Construct a problem in which you examine the charge stored in the capacitor of a defibrillator as a function of stored energy. Among the things to be considered are the applied voltage and whether it should vary with energy to be delivered, the range of energies involved, and the capacitance of the defibrillator. You may also wish to consider the much smaller energy needed for defibrillation during open-heart surgery as a variation on this problem.

70. Unreasonable Results

(a) On a particular day, it takes 9.60×10^3 J of electric energy to start a truck's engine. Calculate the capacitance of a capacitor that could store that amount of energy at 12.0 V. (b) What is unreasonable about this result? (c) Which assumptions are responsible?

Test Prep for AP® Courses

19.1 Electric Potential Energy: Potential Difference

1. An electron is placed in an electric field of 12.0 N/C to the right. What is the resulting force on the electron?
a. 1.33×10^{-20} N right
b. 1.33×10^{-20} N left
c. 1.92×10^{-18} N right
d. 1.92×10^{-18} N left

2. A positively charged object in a certain electric field is currently being pushed west by the resulting force. How will the force change if the charge grows? What if it becomes negative?

3. A −5.0 C charge is being forced south by a 60 N force. What are the magnitude and direction of the local electric field?
a. 12 N/C south
b. 12 N/C north
c. 300 N/C south
d. 300 N/C north

4. A charged object has a net force of 100 N east acting on it due to an electric field of 50 N/C pointing north. How is this possible? If not, why not?

5. How many electrons have to be moved by a car battery containing 7.20×10^5 J at 12 V to reduce the energy by 1%?

a. 4.80×10^{27}
b. 4.00×10^{26}
c. 3.75×10^{21}
d. 3.13×10^{20}

6. Most of the electricity in the power grid is generated by powerful turbines spinning around. Why don't these turbines slow down from the work they do moving electrons?

7. A typical AAA battery can move 2000 C of charge at 1.5 V. How long will this run a 50 mW LED?
a. 1000 minutes
b. 120,000 seconds
c. 15 hours
d. 250 minutes

8. Find an example car (or other vehicle) battery, and compute how many of the AAA batteries in the previous problem it would take to equal the energy stored in it. Which is more compact?

9. What is the internal energy of a system consisting of two point charges, one 2.0 μC, and the other −3.0 μC, placed 1.2 m away from each other?
a. -3.8×10^{-2} J
b. -4.5×10^{-2} J
c. 4.5×10^{-2} J
d. 3.8×10^{-2} J

10. A system of three point charges has a 1.00 μC charge at the origin, a −2.00 μC charge at x=30 cm, and a 3.00 μC charge at x=70 cm. What is the total stored potential energy of this configuration?

11. A system has 2.00 µC charges at (50 cm, 0) and (−50 cm, 0) and a −1.00 µC charge at (0, 70 cm). As the y-coordinate of the −1.00 µC charge increases, the potential energy ___. As the y-coordinate of the −1.00 µC charge decreases, the potential energy ___.
   a.   increases, increases
   b.   increases, decreases
   c.   decreases, increases
   d.   decreases, decreases

12. A system of three point charges has a 1.00 µC charge at the origin, a −2.00 µC charge at x=30 cm, and a 3.00 µC charge at x=70 cm. What happens to the total potential energy of this system if the −2.00 µC charge and the 3.00 µC charge trade places?

13. Take a square configuration of point charges, two positive and two negative, all of the same magnitude, with like charges sharing diagonals. What will happen to the internal energy of this system if one of the negative charges becomes a positive charge of the same magnitude?
   a.   increase
   b.   decrease
   c.   no change
   d.   not enough information

14. Take a square configuration of point charges, two positive and two negative, all of the same magnitude, with like charges sharing diagonals. What will happen to the internal energy of this system if the sides of the square decrease in length?

15. A system has 2.00 µC charges at (50 cm, 0) and (−50 cm, 0) and a −1.00 µC charge at (0, 70 cm), with a velocity in the −y-direction. When the −1.00 µC charge is at (0, 0) the potential energy is at a ___ and the kinetic energy is ___.
   a.   maximum, maximum
   b.   maximum, minimum
   c.   minimum, maximum
   d.   minimum, minimum

16. What is the velocity of an electron that goes through a 10 V potential after initially being at rest?

19.2 Electric Potential in a Uniform Electric Field

17. A negatively charged massive particle is dropped from above the two plates in **Figure 19.7** into the space between them. Which best describes the trajectory it takes?
   a.   A rightward-curving parabola
   b.   A leftward-curving parabola
   c.   A rightward-curving section of a circle
   d.   A leftward-curving section of a circle

18. Two massive particles with identical charge are launched into the uniform field between two plates from the same launch point with the same velocity. They both impact the positively charged plate, but the second one does so four times as far as the first. What sign is the charge? What physical difference would give them different impact points (quantify as a relative percent)? How does this compare to the gravitational projectile motion case?

19. Two plates are lying horizontally, but stacked with one 10.0 cm above the other. If the upper plate is held at +100 V, what is the magnitude and direction of the electric field between the plates if the lower is held at +50.0 V? -50.0 V?
   a.   500 V/m, 1500 V/m, down
   b.   500 V/m, 1500 V/m, up
   c.   1500 V/m, 500 V/m, down
   d.   1500 V/m, 500 V/m, up

20. Two parallel conducting plates are 15 cm apart, each with an area of 0.75 m². The left one has a charge of -0.225 C placed on it, while the right has a charge of 0.225 C. What is the magnitude and direction of the electric field between the two?

21. Consider three parallel conducting plates, with a space of 3.0 cm between them. The leftmost one is at a potential of +45 V, the middle one is held at ground, and the rightmost is at a potential of -75 V. What is the magnitude of the average electric field on an electron traveling between the plates? (Assume that the middle one has holes for the electron to go through.)
   a.   1500 V/m
   b.   2500 V/m
   c.   4000 V/m
   d.   2000 V/m

22. A new kind of electron gun has a rear plate at −25.0 kV, a grounded plate 2.00 cm in front of that, and a +25.0 kV plate 4.00 cm in front of that. What is the magnitude of the average electric field?

23. A certain electric potential isoline graph has isolines every 5.0 V. If six of these lines cross a 40 cm path drawn between two points of interest, what is the (magnitude of the average) electric field along this path?
   a.   750 V/m
   b.   150 V/m
   c.   38 V/m
   d.   75 V/m

24. Given a system of two parallel conducting plates held at a fixed potential difference, describe what happens to the isolines of the electric potential between them as the distance between them is changed. How does this relate to the electric field strength?

19.4 Equipotential Lines

25. How would **Figure 19.15** be different with two positive charges replacing the two negative charges?
   a.   The equipotential lines would have positive values.
   b.   It would actually resemble **Figure 19.14**.
   c.   no change
   d.   not enough information

26. Consider two conducting plates, placed on adjacent sides of a square, but with a 1-m space between the corner of the square and the plate. These plates are not touching, not centered on each other, but are at right angles. Each plate is 1 m wide. If the plates are held at a fixed potential difference ΔV, draw the equipotential lines for this system.

27. As isolines of electric potential get closer together, the electric field gets stronger. What shape would a hill have as the isolines of gravitational potential get closer together?
   a.   constant slope
   b.   steeper slope
   c.   shallower slope
   d.   a U-shape

28. Between **Figure 19.14** and **Figure 19.15**, which more closely resembles the gravitational field between two equal masses, and why?

29. How much work is necessary to keep a positive point charge in orbit around a negative point charge?

a. A lot; this system is unstable.
b. Just a little; the isolines are far enough apart that crossing them doesn't take much work.
c. None; we're traveling along an isoline, which requires no work.
d. There's not enough information to tell.

30. Consider two conducting plates, placed on adjacent sides of a square, but with a 1-m space between the corner of the square and the plate. These plates are not touching, not centered on each other, but are at right angles. Each plate is 1 m wide. If the plates are held at a fixed potential difference ΔV, sketch the path of both a positively charged object placed between the near ends, and a negatively charged object placed near the open ends.

19.5 Capacitors and Dielectrics

31. Two parallel plate capacitors are otherwise identical, except the second one has twice the distance between the plates of the first. If placed in otherwise identical circuits, how much charge will the second plate have on it compared to the first?
a. four times as much
b. twice as much
c. the same
d. half as much

32. In a very simple circuit consisting of a battery and a capacitor with an adjustable distance between the plates, how does the voltage vary as the distance is altered?

33. A parallel plate capacitor with adjustable-size square plates is placed in a circuit. How does the charge on the capacitor change as the length of the sides of the plates is increased?
a. it grows proportional to length^2
b. it grows proportional to length
c. it shrinks proportional to length
d. it shrinks proportional to length^2

34. Design an experiment to test the relative permittivities of various materials, and briefly describe some basic features of the results.

35. A student was changing one of the dimensions of a square parallel plate capacitor and measuring the resultant charge in a circuit with a battery. However, the student forgot which dimension was being varied, and didn't write it or any units down. Given the table, which dimension was it?

Table 19.2

Dimension	1.00	1.10	1.20	1.30
Charge(μC)	0.50	0.61	0.71	0.86

a. The distance between the plates
b. The area
c. The length of a side
d. Both the area and the length of a side

36. In an experiment in which a circular parallel plate capacitor in a circuit with a battery has the radius and plate separation grow at the same relative rate, what will happen to the total charge on the capacitor?

19.7 Energy Stored in Capacitors

37. Consider a parallel plate capacitor, with no dielectric material, attached to a battery with a fixed voltage. What happens when a dielectric is inserted into the capacitor?

a. Nothing changes, except now there is a dielectric in the capacitor.
b. The energy in the system decreases, making it very easy to move the dielectric in.
c. You have to do work to move the dielectric, increasing the energy in the system.
d. The reversed polarity destroys the battery.

38. Consider a parallel plate capacitor with no dielectric material. It was attached to a battery with a fixed voltage to charge up, but now the battery has been disconnected. What happens to the energy of the system and the dielectric material when a dielectric is inserted into the capacitor?

39. What happens to the energy stored in a circuit as you increase the number of capacitors connected in parallel? Series?
a. increases, increases
b. increases, decreases
c. decreases, increases
d. decreases, decreases

40. What would the capacitance of a capacitor with the same total internal energy as the car battery in Example 19.1 have to be? Can you explain why we use batteries instead of capacitors for this application?

41. Consider a parallel plate capacitor with metal plates, each of square shape of 1.00 m on a side, separated by 1.00 mm. What is the energy of this capacitor with 3.00×10^3 V applied to it?
a. 3.98×10^{-2} J
b. 5.08×10^{14} J
c. 1.33×10^{-5} J
d. 1.69×10^{11} J

42. Consider a parallel plate capacitor with metal plates, each of square shape of 1.00 m on a side, separated by 1.00 mm. What is the internal energy stored in this system if the charge on the capacitor is 30.0 μC?

43. Consider a parallel plate capacitor with metal plates, each of square shape of 1.00 m on a side, separated by 1.00 mm. If the plates grow in area while the voltage is held fixed, the capacitance ___ and the stored energy ___.
a. decreases, decreases
b. decreases, increases
c. increases, decreases
d. increases, increases

44. Consider a parallel plate capacitor with metal plates, each of square shape of 1.00 m on a side, separated by 1.00 mm. What happens to the energy of this system if the area of the plates increases while the charge remains fixed?

20 ELECTRIC CURRENT, RESISTANCE, AND OHM'S LAW

Figure 20.1 Electric energy in massive quantities is transmitted from this hydroelectric facility, the **Srisailam power station located along the Krishna River in India (http://en.wikipedia.org/wiki/Srisailam_Dam)** , by the movement of charge—that is, by electric current. (credit: Chintohere, Wikimedia Commons)

Chapter Outline

20.1. Current

20.2. Ohm's Law: Resistance and Simple Circuits

20.3. Resistance and Resistivity

20.4. Electric Power and Energy

20.5. Alternating Current versus Direct Current

20.6. Electric Hazards and the Human Body

20.7. Nerve Conduction–Electrocardiograms

Connection for AP® Courses

In our daily lives, we see and experience many examples of electricity which involve electric current, the movement of charge. These include the flicker of numbers on a handheld calculator, nerve impulses carrying signals of vision to the brain, an ultrasound device sending a signal to a computer screen, the brain sending a message for a baby to twitch its toes, an electric train pulling its load over a mountain pass, and a hydroelectric plant sending energy to metropolitan and rural users.

Humankind has indeed harnessed electricity, the basis of technology, to improve the quality of life. While the previous two chapters concentrated on static electricity and the fundamental force underlying its behavior, the next few chapters will be devoted to electric and magnetic phenomena involving electric current. In addition to exploring applications of electricity, we shall gain new insights into its nature – in particular, the fact that all magnetism results from electric current.

This chapter supports learning objectives covered under Big Ideas 1, 4, and 5 of the AP Physics Curriculum Framework. Electric charge is a property of a system (Big Idea 1) that affects its interaction with other charged systems (Enduring Understanding 1.B), whereas electric current is fundamentally the movement of charge through a conductor and is based on the fact that electric charge is conserved within a system (Essential Knowledge 1.B.1). The conservation of charge also leads to the concept of an electric circuit as a closed loop of electrical current. In addition, this chapter discusses examples showing that the current in a circuit is resisted by the elements of the circuit and the strength of the resistance depends on the material of the elements. The macroscopic properties of materials, including resistivity, depend on their molecular and atomic structure (Enduring Understanding 1.E). In addition, resistivity depends on the temperature of the material (Essential Knowledge 1.E.2).

The chapter also describes how the interaction of systems of objects can result in changes in those systems (Big Idea 4). For example, electric properties of a system of charged objects can change in response to the presence of, or changes in, other charged objects or systems (Enduring Understanding 4.E). A simple circuit with a resistor and an energy source is an example of such a system. The current through the resistor in the circuit is equal to the difference of potentials across the resistor divided by its resistance (Essential Knowledge 4.E.4).

The unifying theme of the physics curriculum is that any changes in the systems due to interactions are governed by laws of conservation (Big Idea 5). This chapter applies the idea of energy conservation (Enduring Understanding 5.B) to electric circuits and connects concepts of electric energy and electric power as rates of energy use (Essential Knowledge 5.B.5). While the laws of conservation of energy in electric circuits are fully described by Kirchoff's rules, which are introduced in the next chapter (Essential Knowledge 5.B.9), the specific definition of power (based on Essential Knowledge 5.B.9) is that it is the rate at which energy is transferred from a resistor as the product of the electric potential difference across the resistor and the current through the resistor.

Big Idea 1 Objects and systems have properties such as mass and charge. Systems may have internal structure.

Enduring Understanding 1.B Electric charge is a property of an object or system that affects its interactions with other objects or systems containing charge.

Essential Knowledge 1.B.1 Electric charge is conserved. The net charge of a system is equal to the sum of the charges of all the objects in the system.

Enduring Understanding 1.E Materials have many macroscopic properties that result from the arrangement and interactions of the atoms and molecules that make up the material.

Essential Knowledge 1.E.2 Matter has a property called resistivity.

Big Idea 4 Interactions between systems can result in changes in those systems.

Enduring Understanding 4.E The electric and magnetic properties of a system can change in response to the presence of, or changes in, other objects or systems.

Essential Knowledge 4.E.4 The resistance of a resistor, and the capacitance of a capacitor, can be understood from the basic properties of electric fields and forces, as well as the properties of materials and their geometry.

Big Idea 5: Changes that occur as a result of interactions are constrained by conservation laws.

Enduring Understanding 5.B The energy of a system is conserved.

Essential Knowledge 5.B.5 Energy can be transferred by an external force exerted on an object or system that moves the object or system through a distance; this energy transfer is called work. Energy transfer in mechanical or electrical systems may occur at different rates. Power is defined as the rate of energy transfer into, out of, or within a system. [A piston filled with gas getting compressed or expanded is treated in Physics 2 as a part of thermodynamics.]

Essential Knowledge 5.B.9 Kirchhoff's loop rule describes conservation of energy in electrical circuits. [The application of Kirchhoff's laws to circuits is introduced in Physics 1 and further developed in Physics 2 in the context of more complex circuits, including those with capacitors.]

20.1 Current

Learning Objectives

By the end of this section, you will be able to:

- Define electric current, ampere, and drift velocity.
- Describe the direction of charge flow in conventional current.
- Use drift velocity to calculate current and vice versa.

The information presented in this section supports the following AP® learning objectives and science practices:

- **1.B.1.1** The student is able to make claims about natural phenomena based on conservation of electric charge. **(S.P. 6.4)**
- **1.B.1.2** The student is able to make predictions, using the conservation of electric charge, about the sign and relative quantity of net charge of objects or systems after various charging processes, including conservation of charge in simple circuits. **(S.P. 6.4, 7.2)**

Electric Current

Electric current is defined to be the rate at which charge flows. A large current, such as that used to start a truck engine, moves a large amount of charge in a small time, whereas a small current, such as that used to operate a hand-held calculator, moves a small amount of charge over a long period of time. In equation form, **electric current** I is defined to be

$$I = \frac{\Delta Q}{\Delta t},$$ (20.1)

where ΔQ is the amount of charge passing through a given area in time Δt. (As in previous chapters, initial time is often taken to be zero, in which case $\Delta t = t$.) (See **Figure 20.2**.) The SI unit for current is the **ampere** (A), named for the French physicist André-Marie Ampère (1775–1836). Since $I = \Delta Q / \Delta t$, we see that an ampere is one coulomb per second:

$$1 \text{ A} = 1 \text{ C/s}$$ (20.2)

Not only are fuses and circuit breakers rated in amperes (or amps), so are many electrical appliances.

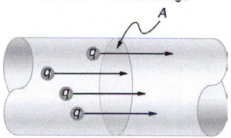

Figure 20.2 The rate of flow of charge is current. An ampere is the flow of one coulomb through an area in one second.

Example 20.1 Calculating Currents: Current in a Truck Battery and a Handheld Calculator

(a) What is the current involved when a truck battery sets in motion 720 C of charge in 4.00 s while starting an engine? (b) How long does it take 1.00 C of charge to flow through a handheld calculator if a 0.300-mA current is flowing?

Strategy

We can use the definition of current in the equation $I = \Delta Q / \Delta t$ to find the current in part (a), since charge and time are given. In part (b), we rearrange the definition of current and use the given values of charge and current to find the time required.

Solution for (a)

Entering the given values for charge and time into the definition of current gives

$$\begin{aligned} I &= \frac{\Delta Q}{\Delta t} = \frac{720 \text{ C}}{4.00 \text{ s}} = 180 \text{ C/s} \\ &= 180 \text{ A}. \end{aligned}$$ (20.3)

Discussion for (a)

This large value for current illustrates the fact that a large charge is moved in a small amount of time. The currents in these "starter motors" are fairly large because large frictional forces need to be overcome when setting something in motion.

Solution for (b)

Solving the relationship $I = \Delta Q / \Delta t$ for time Δt, and entering the known values for charge and current gives

$$\begin{aligned} \Delta t &= \frac{\Delta Q}{I} = \frac{1.00 \text{ C}}{0.300 \times 10^{-3} \text{ C/s}} \\ &= 3.33 \times 10^3 \text{ s}. \end{aligned}$$ (20.4)

Discussion for (b)

This time is slightly less than an hour. The small current used by the hand-held calculator takes a much longer time to move a smaller charge than the large current of the truck starter. So why can we operate our calculators only seconds after turning them on? It's because calculators require very little energy. Such small current and energy demands allow handheld calculators to operate from solar cells or to get many hours of use out of small batteries. Remember, calculators do not have moving parts in the same way that a truck engine has with cylinders and pistons, so the technology requires smaller currents.

Figure 20.3 shows a simple circuit and the standard schematic representation of a battery, conducting path, and load (a resistor).

Schematics are very useful in visualizing the main features of a circuit. A single schematic can represent a wide variety of situations. The schematic in **Figure 20.3** (b), for example, can represent anything from a truck battery connected to a headlight lighting the street in front of the truck to a small battery connected to a penlight lighting a keyhole in a door. Such schematics are useful because the analysis is the same for a wide variety of situations. We need to understand a few schematics to apply the concepts and analysis to many more situations.

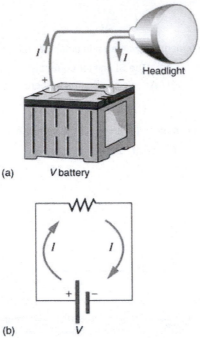

Figure 20.3 (a) A simple electric circuit. A closed path for current to flow through is supplied by conducting wires connecting a load to the terminals of a battery. (b) In this schematic, the battery is represented by the two parallel red lines, conducting wires are shown as straight lines, and the zigzag represents the load. The schematic represents a wide variety of similar circuits.

Note that the direction of current in **Figure 20.3** is from positive to negative. *The direction of conventional current is the direction that positive charge would flow.* In a single loop circuit (as shown in **Figure 20.3**), the value for current at all points of the circuit should be the same if there are no losses. This is because current is the flow of charge and charge is conserved, i.e., the charge flowing out from the battery will be the same as the charge flowing into the battery. Depending on the situation, positive charges, negative charges, or both may move. In metal wires, for example, current is carried by electrons—that is, negative charges move. In ionic solutions, such as salt water, both positive and negative charges move. This is also true in nerve cells. A Van de Graaff generator used for nuclear research can produce a current of pure positive charges, such as protons. **Figure 20.4** illustrates the movement of charged particles that compose a current. The fact that conventional current is taken to be in the direction that positive charge would flow can be traced back to American politician and scientist Benjamin Franklin in the 1700s. He named the type of charge associated with electrons negative, long before they were known to carry current in so many situations. Franklin, in fact, was totally unaware of the small-scale structure of electricity.

It is important to realize that there is an electric field in conductors responsible for producing the current, as illustrated in **Figure 20.4**. Unlike static electricity, where a conductor in equilibrium cannot have an electric field in it, conductors carrying a current have an electric field and are not in static equilibrium. An electric field is needed to supply energy to move the charges.

Making Connections: Take-Home Investigation—Electric Current Illustration

Find a straw and little peas that can move freely in the straw. Place the straw flat on a table and fill the straw with peas. When you pop one pea in at one end, a different pea should pop out the other end. This demonstration is an analogy for an electric current. Identify what compares to the electrons and what compares to the supply of energy. What other analogies can you find for an electric current?

Note that the flow of peas is based on the peas physically bumping into each other; electrons flow due to mutually repulsive electrostatic forces.

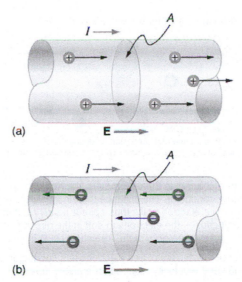

(a)

(b)

Figure 20.4 Current I is the rate at which charge moves through an area A, such as the cross-section of a wire. Conventional current is defined to move in the direction of the electric field. (a) Positive charges move in the direction of the electric field and the same direction as conventional current. (b) Negative charges move in the direction opposite to the electric field. Conventional current is in the direction opposite to the movement of negative charge. The flow of electrons is sometimes referred to as electronic flow.

Example 20.2 Calculating the Number of Electrons that Move through a Calculator

If the 0.300-mA current through the calculator mentioned in the **Example 20.1** example is carried by electrons, how many electrons per second pass through it?

Strategy

The current calculated in the previous example was defined for the flow of positive charge. For electrons, the magnitude is the same, but the sign is opposite, $I_{electrons} = -0.300 \times 10^{-3} \text{ C/s}$. Since each electron (e^-) has a charge of $-1.60 \times 10^{-19} \text{ C}$, we can convert the current in coulombs per second to electrons per second.

Solution

Starting with the definition of current, we have

$$I_{electrons} = \frac{\Delta Q_{electrons}}{\Delta t} = \frac{-0.300 \times 10^{-3} \text{ C}}{\text{s}}. \tag{20.5}$$

We divide this by the charge per electron, so that

$$\frac{e^-}{\text{s}} = \frac{-0.300 \times 10^{-3} \text{ C}}{\text{s}} \times \frac{1 \, e^-}{-1.60 \times 10^{-19} \text{ C}} \tag{20.6}$$

$$= 1.88 \times 10^{15} \frac{e^-}{\text{s}}.$$

Discussion

There are so many charged particles moving, even in small currents, that individual charges are not noticed, just as individual water molecules are not noticed in water flow. Even more amazing is that they do not always keep moving forward like soldiers in a parade. Rather they are like a crowd of people with movement in different directions but a general trend to move forward. There are lots of collisions with atoms in the metal wire and, of course, with other electrons.

Drift Velocity

Electrical signals are known to move very rapidly. Telephone conversations carried by currents in wires cover large distances without noticeable delays. Lights come on as soon as a switch is flicked. Most electrical signals carried by currents travel at speeds on the order of 10^8 m/s, a significant fraction of the speed of light. Interestingly, the individual charges that make up the current move *much* more slowly on average, typically drifting at speeds on the order of 10^{-4} m/s. How do we reconcile these two speeds, and what does it tell us about standard conductors?

The high speed of electrical signals results from the fact that the force between charges acts rapidly at a distance. Thus, when a free charge is forced into a wire, as in **Figure 20.5**, the incoming charge pushes other charges ahead of it, which in turn push on charges farther down the line. The density of charge in a system cannot easily be increased, and so the signal is passed on rapidly. The resulting electrical shock wave moves through the system at nearly the speed of light. To be precise, this rapidly moving signal or shock wave is a rapidly propagating change in electric field.

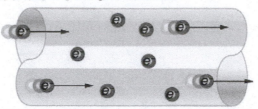

Figure 20.5 When charged particles are forced into this volume of a conductor, an equal number are quickly forced to leave. The repulsion between like charges makes it difficult to increase the number of charges in a volume. Thus, as one charge enters, another leaves almost immediately, carrying the signal rapidly forward.

Good conductors have large numbers of free charges in them. In metals, the free charges are free electrons. **Figure 20.6** shows how free electrons move through an ordinary conductor. The distance that an individual electron can move between collisions with atoms or other electrons is quite small. The electron paths thus appear nearly random, like the motion of atoms in a gas. But there is an electric field in the conductor that causes the electrons to drift in the direction shown (opposite to the field, since they are negative). The **drift velocity** v_d is the average velocity of the free charges. Drift velocity is quite small, since there are so many free charges. If we have an estimate of the density of free electrons in a conductor, we can calculate the drift velocity for a given current. The larger the density, the lower the velocity required for a given current.

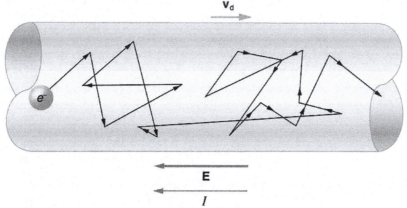

Figure 20.6 Free electrons moving in a conductor make many collisions with other electrons and atoms. The path of one electron is shown. The average velocity of the free charges is called the drift velocity, v_d, and it is in the direction opposite to the electric field for electrons. The collisions normally transfer energy to the conductor, requiring a constant supply of energy to maintain a steady current.

Conduction of Electricity and Heat

Good electrical conductors are often good heat conductors, too. This is because large numbers of free electrons can carry electrical current and can transport thermal energy.

The free-electron collisions transfer energy to the atoms of the conductor. The electric field does work in moving the electrons through a distance, but that work does not increase the kinetic energy (nor speed, therefore) of the electrons. The work is transferred to the conductor's atoms, possibly increasing temperature. Thus a continuous power input is required to maintain current. An exception, of course, is found in superconductors, for reasons we shall explore in a later chapter. Superconductors can have a steady current without a continual supply of energy—a great energy savings. In contrast, the supply of energy can be useful, such as in a lightbulb filament. The supply of energy is necessary to increase the temperature of the tungsten filament, so that the filament glows.

Making Connections: Take-Home Investigation—Filament Observations

Find a lightbulb with a filament. Look carefully at the filament and describe its structure. To what points is the filament connected?

We can obtain an expression for the relationship between current and drift velocity by considering the number of free charges in a segment of wire, as illustrated in **Figure 20.7**. *The number of free charges per unit volume* is given the symbol n and depends on the material. The shaded segment has a volume Ax, so that the number of free charges in it is nAx. The charge ΔQ in this segment is thus $qnAx$, where q is the amount of charge on each carrier. (Recall that for electrons, q is

-1.60×10^{-19} C .) Current is charge moved per unit time; thus, if all the original charges move out of this segment in time Δt, the current is

Download for free at https://openstax.org/details/books/college-physics-ap-courses

$$I = \frac{\Delta Q}{\Delta t} = \frac{qnAx}{\Delta t}. \tag{20.7}$$

Note that $x / \Delta t$ is the magnitude of the drift velocity, v_d, since the charges move an average distance x in a time Δt. Rearranging terms gives

$$I = nqAv_d, \tag{20.8}$$

where I is the current through a wire of cross-sectional area A made of a material with a free charge density n. The carriers of the current each have charge q and move with a drift velocity of magnitude v_d.

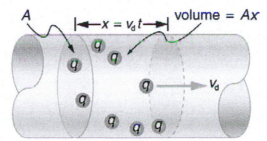

Figure 20.7 All the charges in the shaded volume of this wire move out in a time t, having a drift velocity of magnitude $v_d = x / t$. See text for further discussion.

Note that simple drift velocity is not the entire story. The speed of an electron is much greater than its drift velocity. In addition, not all of the electrons in a conductor can move freely, and those that do might move somewhat faster or slower than the drift velocity. So what do we mean by free electrons? Atoms in a metallic conductor are packed in the form of a lattice structure. Some electrons are far enough away from the atomic nuclei that they do not experience the attraction of the nuclei as much as the inner electrons do. These are the free electrons. They are not bound to a single atom but can instead move freely among the atoms in a "sea" of electrons. These free electrons respond by accelerating when an electric field is applied. Of course as they move they collide with the atoms in the lattice and other electrons, generating thermal energy, and the conductor gets warmer. In an insulator, the organization of the atoms and the structure do not allow for such free electrons.

Example 20.3 Calculating Drift Velocity in a Common Wire

Calculate the drift velocity of electrons in a 12-gauge copper wire (which has a diameter of 2.053 mm) carrying a 20.0-A current, given that there is one free electron per copper atom. (Household wiring often contains 12-gauge copper wire, and the maximum current allowed in such wire is usually 20 A.) The density of copper is 8.80×10^3 kg/m^3.

Strategy

We can calculate the drift velocity using the equation $I = nqAv_d$. The current $I = 20.0$ A is given, and $q = -1.60 \times 10^{-19}$ C is the charge of an electron. We can calculate the area of a cross-section of the wire using the formula $A = \pi r^2$, where r is one-half the given diameter, 2.053 mm. We are given the density of copper, 8.80×10^3 kg/m^3, and the periodic table shows that the atomic mass of copper is 63.54 g/mol. We can use these two quantities along with Avogadro's number, 6.02×10^{23} atoms/mol, to determine n, the number of free electrons per cubic meter.

Solution

First, calculate the density of free electrons in copper. There is one free electron per copper atom. Therefore, is the same as the number of copper atoms per m^3. We can now find n as follows:

$$\begin{aligned} n &= \frac{1\ e^-}{\text{atom}} \times \frac{6.02 \times 10^{23}\ \text{atoms}}{\text{mol}} \times \frac{1\ \text{mol}}{63.54\ \text{g}} \times \frac{1000\ \text{g}}{\text{kg}} \times \frac{8.80 \times 10^3\ \text{kg}}{1\ \text{m}^3} \\ &= 8.342 \times 10^{28}\ e^-/\text{m}^3. \end{aligned} \tag{20.9}$$

The cross-sectional area of the wire is

$$A = \pi r^2 \tag{20.10}$$

$$= \pi\left(\frac{2.053\times10^{-3}\text{ m}}{2}\right)^2$$

$$= 3.310\times10^{-6}\text{ m}^2.$$

Rearranging $I = nqAv_d$ to isolate drift velocity gives

$$v_d = \frac{I}{nqA} \tag{20.11}$$

$$= \frac{20.0\text{ A}}{(8.342\times10^{28}/\text{m}^3)(-1.60\times10^{-19}\text{ C})(3.310\times10^{-6}\text{ m}^2)}$$

$$= -4.53\times10^{-4}\text{ m/s}.$$

Discussion

The minus sign indicates that the negative charges are moving in the direction opposite to conventional current. The small value for drift velocity (on the order of 10^{-4} m/s) confirms that the signal moves on the order of 10^{12} times faster (about 10^8 m/s) than the charges that carry it.

20.2 Ohm's Law: Resistance and Simple Circuits

Learning Objectives

By the end of this section, you will be able to:

- Explain the origin of Ohm's law.
- Calculate voltages, currents, and resistances with Ohm's law.
- Explain the difference between ohmic and non-ohmic materials.
- Describe a simple circuit.

The information presented in this section supports the following AP® learning objectives and science practices:

- **4.E.4.1** The student is able to make predictions about the properties of resistors and/or capacitors when placed in a simple circuit based on the geometry of the circuit element and supported by scientific theories and mathematical relationships. **(S.P. 2.2, 6.4)**

What drives current? We can think of various devices—such as batteries, generators, wall outlets, and so on—which are necessary to maintain a current. All such devices create a potential difference and are loosely referred to as voltage sources. When a voltage source is connected to a conductor, it applies a potential difference V that creates an electric field. The electric field in turn exerts force on charges, causing current.

Ohm's Law

The current that flows through most substances is directly proportional to the voltage V applied to it. The German physicist Georg Simon Ohm (1787–1854) was the first to demonstrate experimentally that the current in a metal wire is *directly proportional to the voltage applied*:

$$I \propto V. \tag{20.12}$$

This important relationship is known as **Ohm's law**. It can be viewed as a cause-and-effect relationship, with voltage the cause and current the effect. This is an empirical law like that for friction—an experimentally observed phenomenon. Such a linear relationship doesn't always occur.

Resistance and Simple Circuits

If voltage drives current, what impedes it? The electric property that impedes current (crudely similar to friction and air resistance) is called **resistance** R. Collisions of moving charges with atoms and molecules in a substance transfer energy to the substance and limit current. Resistance is defined as inversely proportional to current, or

$$I \propto \frac{1}{R}. \tag{20.13}$$

Thus, for example, current is cut in half if resistance doubles. Combining the relationships of current to voltage and current to resistance gives

$$I = \frac{V}{R}. \tag{20.14}$$

This relationship is also called Ohm's law. Ohm's law in this form really defines resistance for certain materials. Ohm's law (like Hooke's law) is not universally valid. The many substances for which Ohm's law holds are called **ohmic**. These include good conductors like copper and aluminum, and some poor conductors under certain circumstances. Ohmic materials have a resistance R that is independent of voltage V and current I. An object that has simple resistance is called a *resistor*, even if its resistance is small. The unit for resistance is an **ohm** and is given the symbol Ω (upper case Greek omega). Rearranging $I = V/R$ gives $R = V/I$, and so the units of resistance are 1 ohm = 1 volt per ampere:

$$1\ \Omega = 1\frac{V}{A}.$$

(20.15)

Figure 20.8 shows the schematic for a simple circuit. A **simple circuit** has a single voltage source and a single resistor. The wires connecting the voltage source to the resistor can be assumed to have negligible resistance, or their resistance can be included in R.

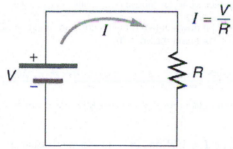

Figure 20.8 A simple electric circuit in which a closed path for current to flow is supplied by conductors (usually metal wires) connecting a load to the terminals of a battery, represented by the red parallel lines. The zigzag symbol represents the single resistor and includes any resistance in the connections to the voltage source.

Making Connections: Real World Connections

Ohm's law ($V = IR$) is a fundamental relationship that could be presented by a linear function with the slope of the line being the resistance. The resistance represents the voltage that needs to be applied to the resistor to create a current of 1 A through the circuit. The graph (in the figure below) shows this representation for two simple circuits with resistors that have different resistances and thus different slopes.

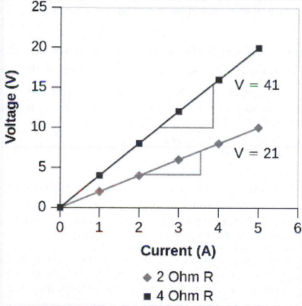

Figure 20.9 The figure illustrates the relationship between current and voltage for two different resistors. The slope of the graph represents the resistance value, which is 2Ω and 4Ω for the two lines shown.

Making Connections: Real World Connections

The materials which follow Ohm's law by having a linear relationship between voltage and current are known as ohmic materials. On the other hand, some materials exhibit a nonlinear voltage-current relationship and hence are known as non-ohmic materials. The figure below shows current voltage relationships for the two types of materials.

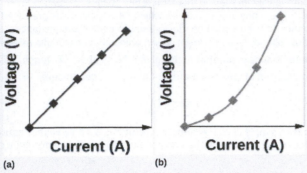

Figure 20.10 The relationship between voltage and current for ohmic and non-ohmic materials are shown.

Clearly the resistance of an ohmic material (shown in (a)) remains constant and can be calculated by finding the slope of the graph but that is not true for a non-ohmic material (shown in (b)).

Example 20.4 Calculating Resistance: An Automobile Headlight

What is the resistance of an automobile headlight through which 2.50 A flows when 12.0 V is applied to it?

Strategy

We can rearrange Ohm's law as stated by $I = V/R$ and use it to find the resistance.

Solution

Rearranging $I = V/R$ and substituting known values gives

$$R = \frac{V}{I} = \frac{12.0 \text{ V}}{2.50 \text{ A}} = 4.80 \ \Omega. \tag{20.16}$$

Discussion

This is a relatively small resistance, but it is larger than the cold resistance of the headlight. As we shall see in **Resistance and Resistivity**, resistance usually increases with temperature, and so the bulb has a lower resistance when it is first switched on and will draw considerably more current during its brief warm-up period.

Resistances range over many orders of magnitude. Some ceramic insulators, such as those used to support power lines, have resistances of $10^{12} \ \Omega$ or more. A dry person may have a hand-to-foot resistance of $10^5 \ \Omega$, whereas the resistance of the human heart is about $10^3 \ \Omega$. A meter-long piece of large-diameter copper wire may have a resistance of $10^{-5} \ \Omega$, and superconductors have no resistance at all (they are non-ohmic). Resistance is related to the shape of an object and the material of which it is composed, as will be seen in **Resistance and Resistivity**.

Additional insight is gained by solving $I = V/R$ for V, yielding

$$V = IR. \tag{20.17}$$

This expression for V can be interpreted as the *voltage drop across a resistor produced by the current I*. The phrase IR *drop* is often used for this voltage. For instance, the headlight in **Example 20.4** has an IR drop of 12.0 V. If voltage is measured at various points in a circuit, it will be seen to increase at the voltage source and decrease at the resistor. Voltage is similar to fluid pressure. The voltage source is like a pump, creating a pressure difference, causing current—the flow of charge. The resistor is like a pipe that reduces pressure and limits flow because of its resistance. Conservation of energy has important consequences here. The voltage source supplies energy (causing an electric field and a current), and the resistor converts it to another form (such as thermal energy). In a simple circuit (one with a single simple resistor), the voltage supplied by the source equals the voltage drop across the resistor, since $\text{PE} = q\Delta V$, and the same q flows through each. Thus the energy supplied by the voltage source and the energy converted by the resistor are equal. (See **Figure 20.11**.)

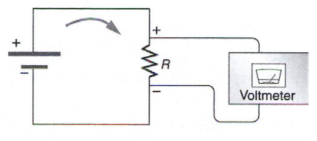

$$V = IR = 18 \text{ V}$$

Figure 20.11 The voltage drop across a resistor in a simple circuit equals the voltage output of the battery.

Making Connections: Conservation of Energy

In a simple electrical circuit, the sole resistor converts energy supplied by the source into another form. Conservation of energy is evidenced here by the fact that all of the energy supplied by the source is converted to another form by the resistor alone. We will find that conservation of energy has other important applications in circuits and is a powerful tool in circuit analysis.

PhET Explorations: Ohm's Law

See how the equation form of Ohm's law relates to a simple circuit. Adjust the voltage and resistance, and see the current change according to Ohm's law. The sizes of the symbols in the equation change to match the circuit diagram.

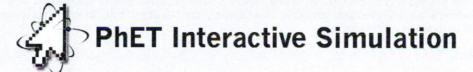

Figure 20.12 Ohm's Law (http://cnx.org/content/m55356/1.2/ohms-law_en.jar)

20.3 Resistance and Resistivity

Learning Objectives

By the end of this section, you will be able to:

- Explain the concept of resistivity.
- Use resistivity to calculate the resistance of specified configurations of material.
- Use the thermal coefficient of resistivity to calculate the change of resistance with temperature.

The information presented in this section supports the following AP® learning objectives and science practices:

- **1.E.2.1** The student is able to choose and justify the selection of data needed to determine resistivity for a given material. **(S.P. 4.1)**
- **4.E.4.2** The student is able to design a plan for the collection of data to determine the effect of changing the geometry and/or materials on the resistance or capacitance of a circuit element and relate results to the basic properties of resistors and capacitors. **(S.P. 4.1, 4.2)**
- **4.E.4.3** The student is able to analyze data to determine the effect of changing the geometry and/or materials on the resistance or capacitance of a circuit element and relate results to the basic properties of resistors and capacitors. **(S.P. 5.1)**

Material and Shape Dependence of Resistance

The resistance of an object depends on its shape and the material of which it is composed. The cylindrical resistor in **Figure 20.13** is easy to analyze, and, by so doing, we can gain insight into the resistance of more complicated shapes. As you might expect, the cylinder's electric resistance R is directly proportional to its length L, similar to the resistance of a pipe to fluid flow. The longer the cylinder, the more collisions charges will make with its atoms. The greater the diameter of the cylinder, the more current it can carry (again similar to the flow of fluid through a pipe). In fact, R is inversely proportional to the cylinder's cross-sectional area A.

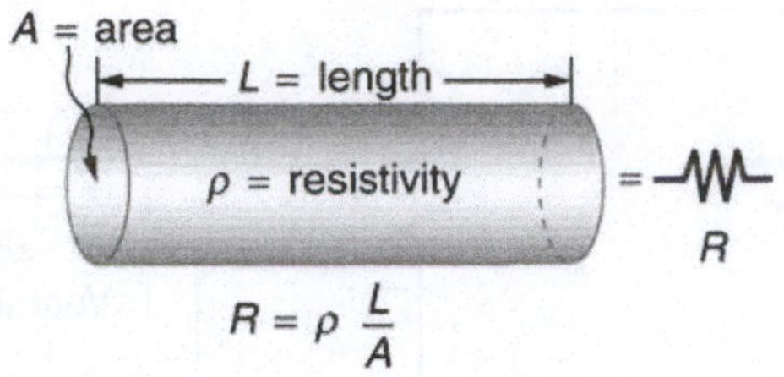

Figure 20.13 A uniform cylinder of length L and cross-sectional area A. Its resistance to the flow of current is similar to the resistance posed by a pipe to fluid flow. The longer the cylinder, the greater its resistance. The larger its cross-sectional area A, the smaller its resistance.

For a given shape, the resistance depends on the material of which the object is composed. Different materials offer different resistance to the flow of charge. We define the **resistivity** ρ of a substance so that the **resistance** R of an object is directly proportional to ρ. Resistivity ρ is an *intrinsic* property of a material, independent of its shape or size. The resistance R of a uniform cylinder of length L, of cross-sectional area A, and made of a material with resistivity ρ, is

$$R = \frac{\rho L}{A}.$$ (20.18)

Table 20.1 gives representative values of ρ. The materials listed in the table are separated into categories of conductors, semiconductors, and insulators, based on broad groupings of resistivities. Conductors have the smallest resistivities, and insulators have the largest; semiconductors have intermediate resistivities. Conductors have varying but large free charge densities, whereas most charges in insulators are bound to atoms and are not free to move. Semiconductors are intermediate, having far fewer free charges than conductors, but having properties that make the number of free charges depend strongly on the type and amount of impurities in the semiconductor. These unique properties of semiconductors are put to use in modern electronics, as will be explored in later chapters.

Table 20.1 Resistivities ρ of Various materials at $20°C$

Material	Resistivity ρ ($\Omega \cdot m$)
Conductors	
Silver	1.59×10^{-8}
Copper	1.72×10^{-8}
Gold	2.44×10^{-8}
Aluminum	2.65×10^{-8}
Tungsten	5.6×10^{-8}
Iron	9.71×10^{-8}
Platinum	10.6×10^{-8}
Steel	20×10^{-8}
Lead	22×10^{-8}
Manganin (Cu, Mn, Ni alloy)	44×10^{-8}
Constantan (Cu, Ni alloy)	49×10^{-8}
Mercury	96×10^{-8}
Nichrome (Ni, Fe, Cr alloy)	100×10^{-8}
Semiconductors[1]	
Carbon (pure)	3.5×10^{5}
Carbon	$(3.5 - 60) \times 10^{5}$
Germanium (pure)	600×10^{-3}
Germanium	$(1 - 600) \times 10^{-3}$
Silicon (pure)	2300
Silicon	$0.1 – 2300$
Insulators	
Amber	5×10^{14}
Glass	$10^{9} - 10^{14}$
Lucite	$> 10^{13}$
Mica	$10^{11} - 10^{15}$
Quartz (fused)	75×10^{16}
Rubber (hard)	$10^{13} - 10^{16}$
Sulfur	10^{15}
Teflon	$> 10^{13}$
Wood	$10^{8} - 10^{14}$

Example 20.5 Calculating Resistor Diameter: A Headlight Filament

A car headlight filament is made of tungsten and has a cold resistance of $0.350 \ \Omega$. If the filament is a cylinder 4.00 cm long (it may be coiled to save space), what is its diameter?

Strategy

We can rearrange the equation $R = \dfrac{\rho L}{A}$ to find the cross-sectional area A of the filament from the given information. Then its diameter can be found by assuming it has a circular cross-section.

Solution

The cross-sectional area, found by rearranging the expression for the resistance of a cylinder given in $R = \dfrac{\rho L}{A}$, is

$$A = \frac{\rho L}{R}. \tag{20.19}$$

Substituting the given values, and taking ρ from **Table 20.1**, yields

$$
\begin{aligned}
A &= \frac{(5.6 \times 10^{-8} \ \Omega \cdot \text{m})(4.00 \times 10^{-2} \ \text{m})}{0.350 \ \Omega} \\
&= 6.40 \times 10^{-9} \ \text{m}^2 .
\end{aligned} \tag{20.20}
$$

The area of a circle is related to its diameter D by

$$A = \frac{\pi D^2}{4}. \tag{20.21}$$

Solving for the diameter D , and substituting the value found for A , gives

$$
\begin{aligned}
D &= 2\left(\frac{A}{p}\right)^{\frac{1}{2}} = 2\left(\frac{6.40 \times 10^{-9} \ \text{m}^2}{3.14}\right)^{\frac{1}{2}} \\
&= 9.0 \times 10^{-5} \ \text{m}.
\end{aligned} \tag{20.22}
$$

Discussion

The diameter is just under a tenth of a millimeter. It is quoted to only two digits, because ρ is known to only two digits.

Temperature Variation of Resistance

The resistivity of all materials depends on temperature. Some even become superconductors (zero resistivity) at very low temperatures. (See **Figure 20.14**.) Conversely, the resistivity of conductors increases with increasing temperature. Since the atoms vibrate more rapidly and over larger distances at higher temperatures, the electrons moving through a metal make more collisions, effectively making the resistivity higher. Over relatively small temperature changes (about $100°C$ or less), resistivity ρ varies with temperature change ΔT as expressed in the following equation

$$\rho = \rho_0(1 + \alpha \Delta T), \tag{20.23}$$

where ρ_0 is the original resistivity and α is the **temperature coefficient of resistivity**. (See the values of α in **Table 20.2** below.) For larger temperature changes, α may vary or a nonlinear equation may be needed to find ρ . Note that α is positive for metals, meaning their resistivity increases with temperature. Some alloys have been developed specifically to have a small temperature dependence. Manganin (which is made of copper, manganese and nickel), for example, has α close to zero (to three digits on the scale in **Table 20.2**), and so its resistivity varies only slightly with temperature. This is useful for making a temperature-independent resistance standard, for example.

---

1. Values depend strongly on amounts and types of impurities

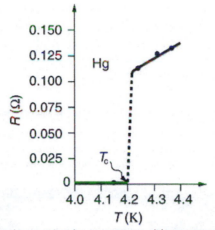

Figure 20.14 The resistance of a sample of mercury is zero at very low temperatures—it is a superconductor up to about 4.2 K. Above that critical temperature, its resistance makes a sudden jump and then increases nearly linearly with temperature.

Table 20.2 Temperature Coefficients of Resistivity α

Material	Coefficient α (1/°C) [2]
Conductors	
Silver	3.8×10^{-3}
Copper	3.9×10^{-3}
Gold	3.4×10^{-3}
Aluminum	3.9×10^{-3}
Tungsten	4.5×10^{-3}
Iron	5.0×10^{-3}
Platinum	3.93×10^{-3}
Lead	4.3×10^{-3}
Manganin (Cu, Mn, Ni alloy)	0.000×10^{-3}
Constantan (Cu, Ni alloy)	0.002×10^{-3}
Mercury	0.89×10^{-3}
Nichrome (Ni, Fe, Cr alloy)	0.4×10^{-3}
Semiconductors	
Carbon (pure)	-0.5×10^{-3}
Germanium (pure)	-50×10^{-3}
Silicon (pure)	-70×10^{-3}

Note also that α is negative for the semiconductors listed in **Table 20.2**, meaning that their resistivity decreases with increasing temperature. They become better conductors at higher temperature, because increased thermal agitation increases the number of free charges available to carry current. This property of decreasing ρ with temperature is also related to the type and amount of impurities present in the semiconductors.

The resistance of an object also depends on temperature, since R_0 is directly proportional to ρ. For a cylinder we know

2. Values at 20°C.

$R = \rho L / A$, and so, if L and A do not change greatly with temperature, R will have the same temperature dependence as ρ . (Examination of the coefficients of linear expansion shows them to be about two orders of magnitude less than typical temperature coefficients of resistivity, and so the effect of temperature on L and A is about two orders of magnitude less than on ρ .) Thus,

$$R = R_0(1 + \alpha\Delta T) \tag{20.24}$$

is the temperature dependence of the resistance of an object, where R_0 is the original resistance and R is the resistance after a temperature change ΔT . Numerous thermometers are based on the effect of temperature on resistance. (See **Figure 20.15**.) One of the most common is the thermistor, a semiconductor crystal with a strong temperature dependence, the resistance of which is measured to obtain its temperature. The device is small, so that it quickly comes into thermal equilibrium with the part of a person it touches.

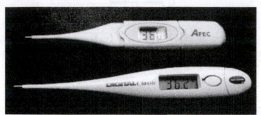

Figure 20.15 These familiar thermometers are based on the automated measurement of a thermistor's temperature-dependent resistance. (credit: Biol, Wikimedia Commons)

Example 20.6 Calculating Resistance: Hot-Filament Resistance

Although caution must be used in applying $\rho = \rho_0(1 + \alpha\Delta T)$ and $R = R_0(1 + \alpha\Delta T)$ for temperature changes greater than $100°C$, for tungsten the equations work reasonably well for very large temperature changes. What, then, is the resistance of the tungsten filament in the previous example if its temperature is increased from room temperature ($20°C$) to a typical operating temperature of $2850°C$?

Strategy

This is a straightforward application of $R = R_0(1 + \alpha\Delta T)$, since the original resistance of the filament was given to be $R_0 = 0.350 \ \Omega$, and the temperature change is $\Delta T = 2830°C$.

Solution

The hot resistance R is obtained by entering known values into the above equation:

$$\begin{aligned} R &= R_0(1 + \alpha\Delta T) \\ &= (0.350 \ \Omega)[1 + (4.5\times10^{-3}/°C)(2830°C)] \\ &= 4.8 \ \Omega. \end{aligned} \tag{20.25}$$

Discussion

This value is consistent with the headlight resistance example in **Ohm's Law: Resistance and Simple Circuits.**

PhET Explorations: Resistance in a Wire

Learn about the physics of resistance in a wire. Change its resistivity, length, and area to see how they affect the wire's resistance. The sizes of the symbols in the equation change along with the diagram of a wire.

Figure 20.16 Resistance in a Wire (http://cnx.org/content/m55357/1.2/resistance-in-a-wire_en.jar)

Applying the Science Practices: Examining Resistance

Using the PhET Simulation "Resistance in a Wire", design an experiment to determine how different variables – resistivity, length, and area – affect the resistance of a resistor. For each variable, you should record your results in a table and then create a graph to determine the relationship.

20.4 Electric Power and Energy

Learning Objectives

By the end of this section, you will be able to:

- Calculate the power dissipated by a resistor and the power supplied by a power supply.
- Calculate the cost of electricity under various circumstances.

The information presented in this section supports the following AP® learning objectives and science practices:

- **5.B.9.8** The student is able to translate between graphical and symbolic representations of experimental data describing relationships among power, current, and potential difference across a resistor. **(S.P. 1.5)**

Power in Electric Circuits

Power is associated by many people with electricity. Knowing that power is the rate of energy use or energy conversion, what is the expression for **electric power**? Power transmission lines might come to mind. We also think of lightbulbs in terms of their power ratings in watts. Let us compare a 25-W bulb with a 60-W bulb. (See **Figure 20.17**(a).) Since both operate on the same voltage, the 60-W bulb must draw more current to have a greater power rating. Thus the 60-W bulb's resistance must be lower than that of a 25-W bulb. If we increase voltage, we also increase power. For example, when a 25-W bulb that is designed to operate on 120 V is connected to 240 V, it briefly glows very brightly and then burns out. Precisely how are voltage, current, and resistance related to electric power?

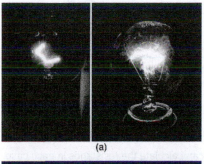

(a)

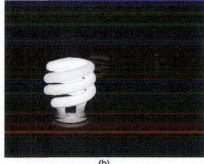

(b)

Figure 20.17 (a) Which of these lightbulbs, the 25-W bulb (upper left) or the 60-W bulb (upper right), has the higher resistance? Which draws more current? Which uses the most energy? Can you tell from the color that the 25-W filament is cooler? Is the brighter bulb a different color and if so why? (credits: Dickbauch, Wikimedia Commons; Greg Westfall, Flickr) (b) This compact fluorescent light (CFL) puts out the same intensity of light as the 60-W bulb, but at 1/4 to 1/10 the input power. (credit: dbgg1979, Flickr)

Electric energy depends on both the voltage involved and the charge moved. This is expressed most simply as $\text{PE} = qV$, where q is the charge moved and V is the voltage (or more precisely, the potential difference the charge moves through). Power is the rate at which energy is moved, and so electric power is

$$P = \frac{PE}{t} = \frac{qV}{t}.$$

(20.26)

Recognizing that current is $I = q/t$ (note that $\Delta t = t$ here), the expression for power becomes

$$P = IV.$$

(20.27)

Electric power (P) is simply the product of current times voltage. Power has familiar units of watts. Since the SI unit for potential energy (PE) is the joule, power has units of joules per second, or watts. Thus, $1 \text{ A} \cdot \text{V} = 1 \text{ W}$. For example, cars often have one or more auxiliary power outlets with which you can charge a cell phone or other electronic devices. These outlets may be rated at 20 A, so that the circuit can deliver a maximum power $P = IV = (20 \text{ A})(12 \text{ V}) = 240 \text{ W}$. In some applications, electric power may be expressed as volt-amperes or even kilovolt-amperes ($1 \text{ kA} \cdot \text{V} = 1 \text{ kW}$).

To see the relationship of power to resistance, we combine Ohm's law with $P = IV$. Substituting $I = V/R$ gives $P = (V/R)V = V^2/R$. Similarly, substituting $V = IR$ gives $P = I(IR) = I^2R$. Three expressions for electric power are listed together here for convenience:

$$P = IV \tag{20.28}$$

$$P = \frac{V^2}{R} \tag{20.29}$$

$$P = I^2R. \tag{20.30}$$

Note that the first equation is always valid, whereas the other two can be used only for resistors. In a simple circuit, with one voltage source and a single resistor, the power supplied by the voltage source and that dissipated by the resistor are identical. (In more complicated circuits, P can be the power dissipated by a single device and not the total power in the circuit.)

Making Connections: Using Graphs to Calculate Resistance

As $p \propto I^2$ and $p \propto V^2$, the graph for power versus current or voltage is quadratic. An example is shown in the figure below.

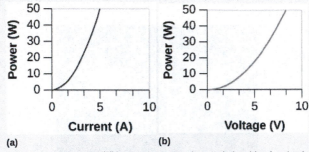

Figure 20.18 The figure shows (a) power versus current and (b) power versus voltage relationships for simple resistor circuits.

Using equations (20.29) and (20.30), we can calculate the resistance in each case. In graph (a), the power is 50 W when current is 5 A; hence, the resistance can be calculated as $R = P \big/ I^2 = 50 \big/ 5^2 = 2 \; \Omega$. Similarly, the resistance value can be calculated in graph (b) as $R = V^2 \big/ P = 10^2 \big/ 50 = 2 \; \Omega$

Different insights can be gained from the three different expressions for electric power. For example, $P = V^2/R$ implies that the lower the resistance connected to a given voltage source, the greater the power delivered. Furthermore, since voltage is squared in $P = V^2/R$, the effect of applying a higher voltage is perhaps greater than expected. Thus, when the voltage is doubled to a 25-W bulb, its power nearly quadruples to about 100 W, burning it out. If the bulb's resistance remained constant, its power would be exactly 100 W, but at the higher temperature its resistance is higher, too.

Example 20.7 Calculating Power Dissipation and Current: Hot and Cold Power

(a) Consider the examples given in **Ohm's Law: Resistance and Simple Circuits** and **Resistance and Resistivity**. Then find the power dissipated by the car headlight in these examples, both when it is hot and when it is cold. (b) What current does it draw when cold?

Strategy for (a)

For the hot headlight, we know voltage and current, so we can use $P = IV$ to find the power. For the cold headlight, we know the voltage and resistance, so we can use $P = V^2/R$ to find the power.

Solution for (a)

Entering the known values of current and voltage for the hot headlight, we obtain

$$P = IV = (2.50 \text{ A})(12.0 \text{ V}) = 30.0 \text{ W}. \tag{20.31}$$

The cold resistance was $0.350 \ \Omega$, and so the power it uses when first switched on is

$$P = \frac{V^2}{R} = \frac{(12.0 \ \text{V})^2}{0.350 \ \Omega} = 411 \ \text{W}. \tag{20.32}$$

Discussion for (a)

The 30 W dissipated by the hot headlight is typical. But the 411 W when cold is surprisingly higher. The initial power quickly decreases as the bulb's temperature increases and its resistance increases.

Strategy and Solution for (b)

The current when the bulb is cold can be found several different ways. We rearrange one of the power equations, $P = I^2 R$, and enter known values, obtaining

$$I = \sqrt{\frac{P}{R}} = \sqrt{\frac{411 \ \text{W}}{0.350 \ \Omega}} = 34.3 \ \text{A}. \tag{20.33}$$

Discussion for (b)

The cold current is remarkably higher than the steady-state value of 2.50 A, but the current will quickly decline to that value as the bulb's temperature increases. Most fuses and circuit breakers (used to limit the current in a circuit) are designed to tolerate very high currents briefly as a device comes on. In some cases, such as with electric motors, the current remains high for several seconds, necessitating special "slow blow" fuses.

The Cost of Electricity

The more electric appliances you use and the longer they are left on, the higher your electric bill. This familiar fact is based on the relationship between energy and power. You pay for the energy used. Since $P = E/t$, we see that

$$E = Pt \tag{20.34}$$

is the energy used by a device using power P for a time interval t . For example, the more lightbulbs burning, the greater P used; the longer they are on, the greater t is. The energy unit on electric bills is the kilowatt-hour ($\text{kW} \cdot \text{h}$), consistent with the relationship $E = Pt$. It is easy to estimate the cost of operating electric appliances if you have some idea of their power consumption rate in watts or kilowatts, the time they are on in hours, and the cost per kilowatt-hour for your electric utility. Kilowatt-hours, like all other specialized energy units such as food calories, can be converted to joules. You can prove to yourself that $1 \ \text{kW} \cdot \text{h} = 3.6 \times 10^6 \ \text{J}$.

The electrical energy (E) used can be reduced either by reducing the time of use or by reducing the power consumption of that appliance or fixture. This will not only reduce the cost, but it will also result in a reduced impact on the environment. Improvements to lighting are some of the fastest ways to reduce the electrical energy used in a home or business. About 20% of a home's use of energy goes to lighting, while the number for commercial establishments is closer to 40%. Fluorescent lights are about four times more efficient than incandescent lights—this is true for both the long tubes and the compact fluorescent lights (CFL). (See **Figure 20.17**(b).) Thus, a 60-W incandescent bulb can be replaced by a 15-W CFL, which has the same brightness and color. CFLs have a bent tube inside a globe or a spiral-shaped tube, all connected to a standard screw-in base that fits standard incandescent light sockets. (Original problems with color, flicker, shape, and high initial investment for CFLs have been addressed in recent years.) The heat transfer from these CFLs is less, and they last up to 10 times longer. The significance of an investment in such bulbs is addressed in the next example. New white LED lights (which are clusters of small LED bulbs) are even more efficient (twice that of CFLs) and last 5 times longer than CFLs. However, their cost is still high.

Making Connections: Energy, Power, and Time

The relationship $E = Pt$ is one that you will find useful in many different contexts. The energy your body uses in exercise is related to the power level and duration of your activity, for example. The amount of heating by a power source is related to the power level and time it is applied. Even the radiation dose of an X-ray image is related to the power and time of exposure.

Example 20.8 Calculating the Cost Effectiveness of Compact Fluorescent Lights (CFL)

If the cost of electricity in your area is 12 cents per kWh, what is the total cost (capital plus operation) of using a 60-W incandescent bulb for 1000 hours (the lifetime of that bulb) if the bulb cost 25 cents? (b) If we replace this bulb with a compact fluorescent light that provides the same light output, but at one-quarter the wattage, and which costs $1.50 but lasts 10 times longer (10,000 hours), what will that total cost be?

Strategy

To find the operating cost, we first find the energy used in kilowatt-hours and then multiply by the cost per kilowatt-hour.

Solution for (a)

The energy used in kilowatt-hours is found by entering the power and time into the expression for energy:

$$E = Pt = (60 \text{ W})(1000 \text{ h}) = 60,000 \text{ W} \cdot \text{h}. \tag{20.35}$$

In kilowatt-hours, this is

$$E = 60.0 \text{ kW} \cdot \text{h}. \tag{20.36}$$

Now the electricity cost is

$$\text{cost} = (60.0 \text{ kW} \cdot \text{h})(\$0.12/\text{kW} \cdot \text{h}) = \$7.20. \tag{20.37}$$

The total cost will be $7.20 for 1000 hours (about one-half year at 5 hours per day).

Solution for (b)

Since the CFL uses only 15 W and not 60 W, the electricity cost will be $7.20/4 = $1.80. The CFL will last 10 times longer than the incandescent, so that the investment cost will be 1/10 of the bulb cost for that time period of use, or 0.1($1.50) = $0.15. Therefore, the total cost will be $1.95 for 1000 hours.

Discussion

Therefore, it is much cheaper to use the CFLs, even though the initial investment is higher. The increased cost of labor that a business must include for replacing the incandescent bulbs more often has not been figured in here.

Making Connections: Take-Home Experiment—Electrical Energy Use Inventory

1) Make a list of the power ratings on a range of appliances in your home or room. Explain why something like a toaster has a higher rating than a digital clock. Estimate the energy consumed by these appliances in an average day (by estimating their time of use). Some appliances might only state the operating current. If the household voltage is 120 V, then use $P = IV$. 2) Check out the total wattage used in the rest rooms of your school's floor or building. (You might need to assume the long fluorescent lights in use are rated at 32 W.) Suppose that the building was closed all weekend and that these lights were left on from 6 p.m. Friday until 8 a.m. Monday. What would this oversight cost? How about for an entire year of weekends?

20.5 Alternating Current versus Direct Current

Learning Objectives

By the end of this section, you will be able to:

- Explain the differences and similarities between AC and DC current.
- Calculate *rms* voltage, current, and average power.
- Explain why AC current is used for power transmission.

Alternating Current

Most of the examples dealt with so far, and particularly those utilizing batteries, have constant voltage sources. Once the current is established, it is thus also a constant. **Direct current** (DC) is the flow of electric charge in only one direction. It is the steady state of a constant-voltage circuit. Most well-known applications, however, use a time-varying voltage source. **Alternating current** (AC) is the flow of electric charge that periodically reverses direction. If the source varies periodically, particularly sinusoidally, the circuit is known as an alternating current circuit. Examples include the commercial and residential power that serves so many of our needs. **Figure 20.19** shows graphs of voltage and current versus time for typical DC and AC power. The AC voltages and frequencies commonly used in homes and businesses vary around the world.

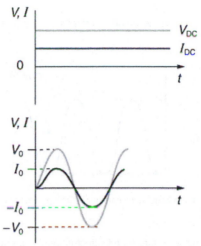

Figure 20.19 (a) DC voltage and current are constant in time, once the current is established. (b) A graph of voltage and current versus time for 60-Hz AC power. The voltage and current are sinusoidal and are in phase for a simple resistance circuit. The frequencies and peak voltages of AC sources differ greatly.

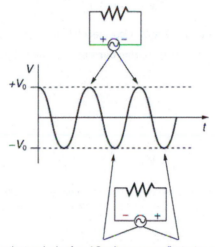

Figure 20.20 The potential difference V between the terminals of an AC voltage source fluctuates as shown. The mathematical expression for V is given by $V = V_0 \sin 2\pi ft$.

Figure 20.20 shows a schematic of a simple circuit with an AC voltage source. The voltage between the terminals fluctuates as shown, with the **AC voltage** given by

$$V = V_0 \sin 2\pi ft, \tag{20.38}$$

where V is the voltage at time t, V_0 is the peak voltage, and f is the frequency in hertz. For this simple resistance circuit, $I = V/R$, and so the **AC current** is

$$I = I_0 \sin 2\pi ft, \tag{20.39}$$

where I is the current at time t, and $I_0 = V_0/R$ is the peak current. For this example, the voltage and current are said to be in phase, as seen in **Figure 20.19(b)**.

Current in the resistor alternates back and forth just like the driving voltage, since $I = V/R$. If the resistor is a fluorescent light bulb, for example, it brightens and dims 120 times per second as the current repeatedly goes through zero. A 120-Hz flicker is too rapid for your eyes to detect, but if you wave your hand back and forth between your face and a fluorescent light, you will see a stroboscopic effect evidencing AC. The fact that the light output fluctuates means that the power is fluctuating. The power supplied is $P = IV$. Using the expressions for I and V above, we see that the time dependence of power is

$P = I_0 V_0 \sin^2 2\pi ft$, as shown in **Figure 20.21**.

Making Connections: Take-Home Experiment—AC/DC Lights

Wave your hand back and forth between your face and a fluorescent light bulb. Do you observe the same thing with the headlights on your car? Explain what you observe. *Warning: Do not look directly at very bright light.*

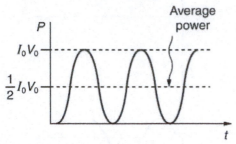

Figure 20.21 AC power as a function of time. Since the voltage and current are in phase here, their product is non-negative and fluctuates between zero and $I_0 V_0$. Average power is $(1/2)I_0 V_0$.

We are most often concerned with average power rather than its fluctuations—that 60-W light bulb in your desk lamp has an average power consumption of 60 W, for example. As illustrated in **Figure 20.21**, the average power P_{ave} is

$$P_{\text{ave}} = \frac{1}{2}I_0 V_0. \tag{20.40}$$

This is evident from the graph, since the areas above and below the $(1/2)I_0 V_0$ line are equal, but it can also be proven using trigonometric identities. Similarly, we define an average or **rms current** I_{rms} and average or **rms voltage** V_{rms} to be, respectively,

$$I_{\text{rms}} = \frac{I_0}{\sqrt{2}} \tag{20.41}$$

and

$$V_{\text{rms}} = \frac{V_0}{\sqrt{2}}. \tag{20.42}$$

where rms stands for root mean square, a particular kind of average. In general, to obtain a root mean square, the particular quantity is squared, its mean (or average) is found, and the square root is taken. This is useful for AC, since the average value is zero. Now,

$$P_{\text{ave}} = I_{\text{rms}}V_{\text{rms}}, \tag{20.43}$$

which gives

$$P_{\text{ave}} = \frac{I_0}{\sqrt{2}} \cdot \frac{V_0}{\sqrt{2}} = \frac{1}{2}I_0 V_0, \tag{20.44}$$

as stated above. It is standard practice to quote I_{rms}, V_{rms}, and P_{ave} rather than the peak values. For example, most household electricity is 120 V AC, which means that V_{rms} is 120 V. The common 10-A circuit breaker will interrupt a sustained I_{rms} greater than 10 A. Your 1.0-kW microwave oven consumes $P_{\text{ave}} = 1.0\ \text{kW}$, and so on. You can think of these rms and average values as the equivalent DC values for a simple resistive circuit.

To summarize, when dealing with AC, Ohm's law and the equations for power are completely analogous to those for DC, but rms and average values are used for AC. Thus, for AC, Ohm's law is written

$$I_{\text{rms}} = \frac{V_{\text{rms}}}{R}. \tag{20.45}$$

The various expressions for AC power P_{ave} are

$$P_{\text{ave}} = I_{\text{rms}}V_{\text{rms}}, \tag{20.46}$$

$$P_{\text{ave}} = \frac{V_{\text{rms}}^2}{R}, \tag{20.47}$$

and

$$P_{\text{ave}} = I_{\text{rms}}^2 R. \tag{20.48}$$

Example 20.9 Peak Voltage and Power for AC

(a) What is the value of the peak voltage for 120-V AC power? (b) What is the peak power consumption rate of a 60.0-W AC light bulb?

Strategy

We are told that V_{rms} is 120 V and P_{ave} is 60.0 W. We can use $V_{rms} = \dfrac{V_0}{\sqrt{2}}$ to find the peak voltage, and we can manipulate the definition of power to find the peak power from the given average power.

Solution for (a)

Solving the equation $V_{rms} = \dfrac{V_0}{\sqrt{2}}$ for the peak voltage V_0 and substituting the known value for V_{rms} gives

$$V_0 = \sqrt{2}V_{rms} = 1.414(120 \text{ V}) = 170 \text{ V}. \tag{20.49}$$

Discussion for (a)

This means that the AC voltage swings from 170 V to -170 V and back 60 times every second. An equivalent DC voltage is a constant 120 V.

Solution for (b)

Peak power is peak current times peak voltage. Thus,

$$P_0 = I_0 V_0 = 2\left(\frac{1}{2}I_0 V_0\right) = 2P_{ave}. \tag{20.50}$$

We know the average power is 60.0 W, and so

$$P_0 = 2(60.0 \text{ W}) = 120 \text{ W}. \tag{20.51}$$

Discussion

So the power swings from zero to 120 W one hundred twenty times per second (twice each cycle), and the power averages 60 W.

Why Use AC for Power Distribution?

Most large power-distribution systems are AC. Moreover, the power is transmitted at much higher voltages than the 120-V AC (240 V in most parts of the world) we use in homes and on the job. Economies of scale make it cheaper to build a few very large electric power-generation plants than to build numerous small ones. This necessitates sending power long distances, and it is obviously important that energy losses en route be minimized. High voltages can be transmitted with much smaller power losses than low voltages, as we shall see. (See **Figure 20.22**.) For safety reasons, the voltage at the user is reduced to familiar values. The crucial factor is that it is much easier to increase and decrease AC voltages than DC, so AC is used in most large power distribution systems.

Figure 20.22 Power is distributed over large distances at high voltage to reduce power loss in the transmission lines. The voltages generated at the power plant are stepped up by passive devices called transformers (see **Transformers**) to 330,000 volts (or more in some places worldwide). At the point of use, the transformers reduce the voltage transmitted for safe residential and commercial use. (Credit: GeorgHH, Wikimedia Commons)

Example 20.10 Power Losses Are Less for High-Voltage Transmission

(a) What current is needed to transmit 100 MW of power at 200 kV? (b) What is the power dissipated by the transmission lines if they have a resistance of $1.00 \ \Omega$? (c) What percentage of the power is lost in the transmission lines?

Strategy

We are given $P_{ave} = 100 \text{ MW}$, $V_{rms} = 200 \text{ kV}$, and the resistance of the lines is $R = 1.00 \ \Omega$. Using these givens, we can find the current (from $P = IV$) and then the power dissipated in the lines ($P = I^2 R$), and we take the ratio to the total power transmitted.

Solution

To find the current, we rearrange the relationship $P_{ave} = I_{rms} V_{rms}$ and substitute known values. This gives

$$I_{rms} = \frac{P_{ave}}{V_{rms}} = \frac{100 \times 10^6 \text{ W}}{200 \times 10^3 \text{ V}} = 500 \text{ A}. \tag{20.52}$$

Solution

Knowing the current and given the resistance of the lines, the power dissipated in them is found from $P_{ave} = I_{rms}^2 R$. Substituting the known values gives

$$P_{ave} = I_{rms}^2 R = (500 \text{ A})^2 (1.00 \ \Omega) = 250 \text{ kW}. \tag{20.53}$$

Solution

The percent loss is the ratio of this lost power to the total or input power, multiplied by 100:

$$\% \text{ loss} = \frac{250 \text{ kW}}{100 \text{ MW}} \times 100 = 0.250 \ \%. \tag{20.54}$$

Discussion

One-fourth of a percent is an acceptable loss. Note that if 100 MW of power had been transmitted at 25 kV, then a current of 4000 A would have been needed. This would result in a power loss in the lines of 16.0 MW, or 16.0% rather than 0.250%. The lower the voltage, the more current is needed, and the greater the power loss in the fixed-resistance transmission lines. Of course, lower-resistance lines can be built, but this requires larger and more expensive wires. If superconducting lines could be economically produced, there would be no loss in the transmission lines at all. But, as we shall see in a later chapter, there is a limit to current in superconductors, too. In short, high voltages are more economical for transmitting power, and AC voltage is much easier to raise and lower, so that AC is used in most large-scale power distribution systems.

It is widely recognized that high voltages pose greater hazards than low voltages. But, in fact, some high voltages, such as those associated with common static electricity, can be harmless. So it is not voltage alone that determines a hazard. It is not so widely recognized that AC shocks are often more harmful than similar DC shocks. Thomas Edison thought that AC shocks were more harmful and set up a DC power-distribution system in New York City in the late 1800s. There were bitter fights, in particular between Edison and George Westinghouse and Nikola Tesla, who were advocating the use of AC in early power-distribution systems. AC has prevailed largely due to transformers and lower power losses with high-voltage transmission.

PhET Explorations: Generator

Generate electricity with a bar magnet! Discover the physics behind the phenomena by exploring magnets and how you can use them to make a bulb light.

Figure 20.23 Generator (http://cnx.org/content/m55359/1.2/generator_en.jar)

20.6 Electric Hazards and the Human Body

Learning Objectives

By the end of this section, you will be able to:

- Define thermal hazard, shock hazard, and short circuit.
- Explain what effects various levels of current have on the human body.

There are two known hazards of electricity—thermal and shock. A **thermal hazard** is one where excessive electric power causes undesired thermal effects, such as starting a fire in the wall of a house. A **shock hazard** occurs when electric current passes through a person. Shocks range in severity from painful, but otherwise harmless, to heart-stopping lethality. This section considers these hazards and the various factors affecting them in a quantitative manner. **Electrical Safety: Systems and Devices** will consider systems and devices for preventing electrical hazards.

Thermal Hazards

Electric power causes undesired heating effects whenever electric energy is converted to thermal energy at a rate faster than it can be safely dissipated. A classic example of this is the **short circuit**, a low-resistance path between terminals of a voltage source. An example of a short circuit is shown in **Figure 20.24**. Insulation on wires leading to an appliance has worn through, allowing the two wires to come into contact. Such an undesired contact with a high voltage is called a *short*. Since the resistance of the short, r, is very small, the power dissipated in the short, $P = V^2/r$, is very large. For example, if V is 120 V and r is $0.100\ \Omega$, then the power is 144 kW, *much* greater than that used by a typical household appliance. Thermal energy delivered at this rate will very quickly raise the temperature of surrounding materials, melting or perhaps igniting them.

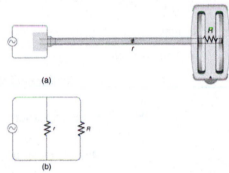

(a)

(b)

Figure 20.24 A short circuit is an undesired low-resistance path across a voltage source. (a) Worn insulation on the wires of a toaster allow them to come into contact with a low resistance r. Since $P = V^2/r$, thermal power is created so rapidly that the cord melts or burns. (b) A schematic of the short circuit.

One particularly insidious aspect of a short circuit is that its resistance may actually be decreased due to the increase in temperature. This can happen if the short creates ionization. These charged atoms and molecules are free to move and, thus, lower the resistance r. Since $P = V^2/r$, the power dissipated in the short rises, possibly causing more ionization, more power, and so on. High voltages, such as the 480-V AC used in some industrial applications, lend themselves to this hazard, because higher voltages create higher initial power production in a short.

Another serious, but less dramatic, thermal hazard occurs when wires supplying power to a user are overloaded with too great a current. As discussed in the previous section, the power dissipated in the supply wires is $P = I^2 R_\mathrm{w}$, where R_w is the resistance of the wires and I the current flowing through them. If either I or R_w is too large, the wires overheat. For example, a worn appliance cord (with some of its braided wires broken) may have $R_\mathrm{w} = 2.00\ \Omega$ rather than the $0.100\ \Omega$ it should be. If 10.0 A of current passes through the cord, then $P = I^2 R_\mathrm{w} = 200\ \mathrm{W}$ is dissipated in the cord—much more than is safe. Similarly, if a wire with a 0.100 - Ω resistance is meant to carry a few amps, but is instead carrying 100 A, it will severely overheat. The power dissipated in the wire will in that case be $P = 1000\ \mathrm{W}$. Fuses and circuit breakers are used to limit excessive currents. (See **Figure 20.25** and **Figure 20.26**.) Each device opens the circuit automatically when a sustained current exceeds safe limits.

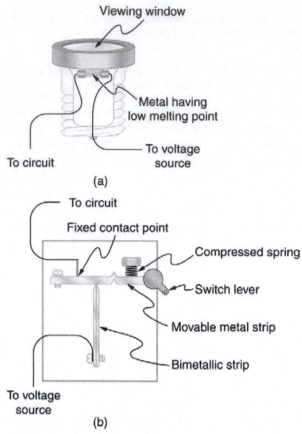

Figure 20.25 (a) A fuse has a metal strip with a low melting point that, when overheated by an excessive current, permanently breaks the connection of a circuit to a voltage source. (b) A circuit breaker is an automatic but restorable electric switch. The one shown here has a bimetallic strip that bends to the right and into the notch if overheated. The spring then forces the metal strip downward, breaking the electrical connection at the points.

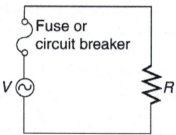

Figure 20.26 Schematic of a circuit with a fuse or circuit breaker in it. Fuses and circuit breakers act like automatic switches that open when sustained current exceeds desired limits.

Fuses and circuit breakers for typical household voltages and currents are relatively simple to produce, but those for large voltages and currents experience special problems. For example, when a circuit breaker tries to interrupt the flow of high-voltage electricity, a spark can jump across its points that ionizes the air in the gap and allows the current to continue. Large circuit breakers found in power-distribution systems employ insulating gas and even use jets of gas to blow out such sparks. Here AC is safer than DC, since AC current goes through zero 120 times per second, giving a quick opportunity to extinguish these arcs.

Shock Hazards

Electrical currents through people produce tremendously varied effects. An electrical current can be used to block back pain. The possibility of using electrical current to stimulate muscle action in paralyzed limbs, perhaps allowing paraplegics to walk, is under study. TV dramatizations in which electrical shocks are used to bring a heart attack victim out of ventricular fibrillation (a massively irregular, often fatal, beating of the heart) are more than common. Yet most electrical shock fatalities occur because a current put the heart into fibrillation. A pacemaker uses electrical shocks to stimulate the heart to beat properly. Some fatal shocks do not produce burns, but warts can be safely burned off with electric current (though freezing using liquid nitrogen is now more common). Of course, there are consistent explanations for these disparate effects. The major factors upon which the effects of electrical shock depend are

1. The amount of current I

2. The path taken by the current

3. The duration of the shock

4. The frequency f of the current ($f = 0$ for DC)

Table 20.3 gives the effects of electrical shocks as a function of current for a typical accidental shock. The effects are for a shock that passes through the trunk of the body, has a duration of 1 s, and is caused by 60-Hz power.

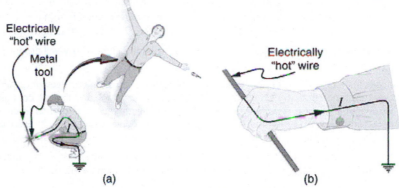

Figure 20.27 An electric current can cause muscular contractions with varying effects. (a) The victim is "thrown" backward by involuntary muscle contractions that extend the legs and torso. (b) The victim can't let go of the wire that is stimulating all the muscles in the hand. Those that close the fingers are stronger than those that open them.

Table 20.3 Effects of Electrical Shock as a Function of Current[3]

Current (mA)	Effect
1	Threshold of sensation
5	Maximum harmless current
10–20	Onset of sustained muscular contraction; cannot let go for duration of shock; contraction of chest muscles may stop breathing during shock
50	Onset of pain
100–300+	Ventricular fibrillation possible; often fatal
300	Onset of burns depending on concentration of current
6000 (6 A)	Onset of sustained ventricular contraction and respiratory paralysis; both cease when shock ends; heartbeat may return to normal; used to defibrillate the heart

Our bodies are relatively good conductors due to the water in our bodies. Given that larger currents will flow through sections with lower resistance (to be further discussed in the next chapter), electric currents preferentially flow through paths in the human body that have a minimum resistance in a direct path to Earth. The Earth is a natural electron sink. Wearing insulating shoes, a requirement in many professions, prohibits a pathway for electrons by providing a large resistance in that path. Whenever working with high-power tools (drills), or in risky situations, ensure that you do not provide a pathway for current flow (especially through the heart).

Very small currents pass harmlessly and unfelt through the body. This happens to you regularly without your knowledge. The threshold of sensation is only 1 mA and, although unpleasant, shocks are apparently harmless for currents less than 5 mA. A great number of safety rules take the 5-mA value for the maximum allowed shock. At 10 to 20 mA and above, the current can stimulate sustained muscular contractions much as regular nerve impulses do. People sometimes say they were knocked across the room by a shock, but what really happened was that certain muscles contracted, propelling them in a manner not of their own choosing. (See **Figure 20.27**(a).) More frightening, and potentially more dangerous, is the "can't let go" effect illustrated in **Figure 20.27**(b). The muscles that close the fingers are stronger than those that open them, so the hand closes involuntarily on the wire shocking it. This can prolong the shock indefinitely. It can also be a danger to a person trying to rescue the victim, because the rescuer's hand may close about the victim's wrist. Usually the best way to help the victim is to give the fist a hard knock/blow/jar with an insulator or to throw an insulator at the fist. Modern electric fences, used in animal enclosures, are now pulsed on and off to allow people who touch them to get free, rendering them less lethal than in the past.

Greater currents may affect the heart. Its electrical patterns can be disrupted, so that it beats irregularly and ineffectively in a condition called "ventricular fibrillation." This condition often lingers after the shock and is fatal due to a lack of blood circulation. The threshold for ventricular fibrillation is between 100 and 300 mA. At about 300 mA and above, the shock can cause burns, depending on the concentration of current—the more concentrated, the greater the likelihood of burns.

Very large currents cause the heart and diaphragm to contract for the duration of the shock. Both the heart and breathing stop. Interestingly, both often return to normal following the shock. The electrical patterns on the heart are completely erased in a manner that the heart can start afresh with normal beating, as opposed to the permanent disruption caused by smaller currents

3. For an average male shocked through trunk of body for 1 s by 60-Hz AC. Values for females are 60–80% of those listed.

that can put the heart into ventricular fibrillation. The latter is something like scribbling on a blackboard, whereas the former completely erases it. TV dramatizations of electric shock used to bring a heart attack victim out of ventricular fibrillation also show large paddles. These are used to spread out current passed through the victim to reduce the likelihood of burns.

Current is the major factor determining shock severity (given that other conditions such as path, duration, and frequency are fixed, such as in the table and preceding discussion). A larger voltage is more hazardous, but since $I = V/R$, the severity of the shock depends on the combination of voltage and resistance. For example, a person with dry skin has a resistance of about $200 \text{ k} \Omega$. If he comes into contact with 120-V AC, a current $I = (120 \text{ V}) / (200 \text{ k} \Omega) = 0.6 \text{ mA}$ passes harmlessly through him. The same person soaking wet may have a resistance of $10.0 \text{ k} \Omega$ and the same 120 V will produce a current of 12 mA—above the "can't let go" threshold and potentially dangerous.

Most of the body's resistance is in its dry skin. When wet, salts go into ion form, lowering the resistance significantly. The interior of the body has a much lower resistance than dry skin because of all the ionic solutions and fluids it contains. If skin resistance is bypassed, such as by an intravenous infusion, a catheter, or exposed pacemaker leads, a person is rendered **microshock sensitive**. In this condition, currents about 1/1000 those listed in **Table 20.3** produce similar effects. During open-heart surgery, currents as small as 20 μA can be used to still the heart. Stringent electrical safety requirements in hospitals, particularly in surgery and intensive care, are related to the doubly disadvantaged microshock-sensitive patient. The break in the skin has reduced his resistance, and so the same voltage causes a greater current, and a much smaller current has a greater effect.

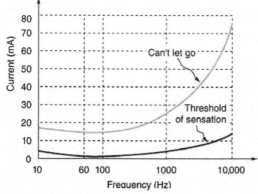

Figure 20.28 Graph of average values for the threshold of sensation and the "can't let go" current as a function of frequency. The lower the value, the more sensitive the body is at that frequency.

Factors other than current that affect the severity of a shock are its path, duration, and AC frequency. Path has obvious consequences. For example, the heart is unaffected by an electric shock through the brain, such as may be used to treat manic depression. And it is a general truth that the longer the duration of a shock, the greater its effects. **Figure 20.28** presents a graph that illustrates the effects of frequency on a shock. The curves show the minimum current for two different effects, as a function of frequency. The lower the current needed, the more sensitive the body is at that frequency. Ironically, the body is most sensitive to frequencies near the 50- or 60-Hz frequencies in common use. The body is slightly less sensitive for DC ($f = 0$), mildly confirming Edison's claims that AC presents a greater hazard. At higher and higher frequencies, the body becomes progressively less sensitive to any effects that involve nerves. This is related to the maximum rates at which nerves can fire or be stimulated. At very high frequencies, electrical current travels only on the surface of a person. Thus a wart can be burned off with very high frequency current without causing the heart to stop. (Do not try this at home with 60-Hz AC!) Some of the spectacular demonstrations of electricity, in which high-voltage arcs are passed through the air and over people's bodies, employ high frequencies and low currents. (See **Figure 20.29**.) Electrical safety devices and techniques are discussed in detail in **Electrical Safety: Systems and Devices**.

Figure 20.29 Is this electric arc dangerous? The answer depends on the AC frequency and the power involved. (credit: Khimich Alex, Wikimedia Commons)

20.7 Nerve Conduction–Electrocardiograms

Learning Objectives
By the end of this section, you will be able to: • Explain the process by which electric signals are transmitted along a neuron.   • Explain the effects myelin sheaths have on signal propagation.   • Explain what the features of an ECG signal indicate.

Nerve Conduction

Electric currents in the vastly complex system of billions of nerves in our body allow us to sense the world, control parts of our body, and think. These are representative of the three major functions of nerves. First, nerves carry messages from our sensory organs and others to the central nervous system, consisting of the brain and spinal cord. Second, nerves carry messages from the central nervous system to muscles and other organs. Third, nerves transmit and process signals within the central nervous system. The sheer number of nerve cells and the incredibly greater number of connections between them makes this system the subtle wonder that it is. **Nerve conduction** is a general term for electrical signals carried by nerve cells. It is one aspect of **bioelectricity**, or electrical effects in and created by biological systems.

Nerve cells, properly called *neurons*, look different from other cells—they have tendrils, some of them many centimeters long, connecting them with other cells. (See **Figure 20.30**.) Signals arrive at the cell body across *synapses* or through *dendrites*, stimulating the neuron to generate its own signal, sent along its long *axon* to other nerve or muscle cells. Signals may arrive from many other locations and be transmitted to yet others, conditioning the synapses by use, giving the system its complexity and its ability to learn.

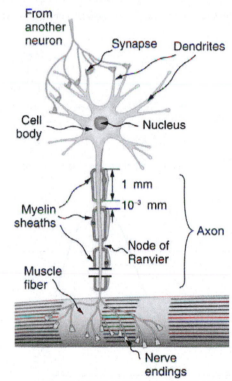

Figure 20.30 A neuron with its dendrites and long axon. Signals in the form of electric currents reach the cell body through dendrites and across synapses, stimulating the neuron to generate its own signal sent down the axon. The number of interconnections can be far greater than shown here.

The method by which these electric currents are generated and transmitted is more complex than the simple movement of free charges in a conductor, but it can be understood with principles already discussed in this text. The most important of these are the Coulomb force and diffusion.

Figure 20.31 illustrates how a voltage (potential difference) is created across the cell membrane of a neuron in its resting state. This thin membrane separates electrically neutral fluids having differing concentrations of ions, the most important varieties being Na^+, K^+, and Cl^- (these are sodium, potassium, and chlorine ions with single plus or minus charges as indicated). As discussed in **Molecular Transport Phenomena: Diffusion, Osmosis, and Related Processes**, free ions will diffuse from a region of high concentration to one of low concentration. But the cell membrane is **semipermeable**, meaning that some ions may cross it while others cannot. In its resting state, the cell membrane is permeable to K^+ and Cl^-, and impermeable to Na^+.

Diffusion of K^+ and Cl^- thus creates the layers of positive and negative charge on the outside and inside of the membrane.

The Coulomb force prevents the ions from diffusing across in their entirety. Once the charge layer has built up, the repulsion of like charges prevents more from moving across, and the attraction of unlike charges prevents more from leaving either side. The result is two layers of charge right on the membrane, with diffusion being balanced by the Coulomb force. A tiny fraction of the charges move across and the fluids remain neutral (other ions are present), while a separation of charge and a voltage have been created across the membrane.

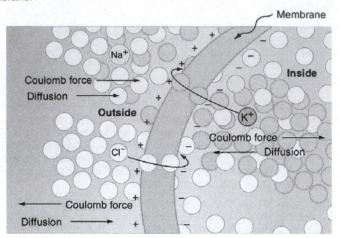

Figure 20.31 The semipermeable membrane of a cell has different concentrations of ions inside and out. Diffusion moves the K^+ and Cl^- ions in the direction shown, until the Coulomb force halts further transfer. This results in a layer of positive charge on the outside, a layer of negative charge on the inside, and thus a voltage across the cell membrane. The membrane is normally impermeable to Na^+.

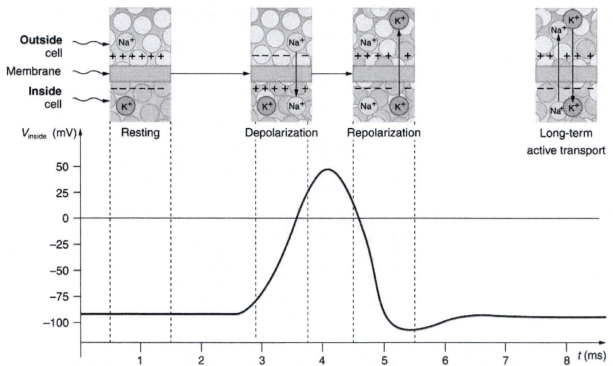

Figure 20.32 An action potential is the pulse of voltage inside a nerve cell graphed here. It is caused by movements of ions across the cell membrane as shown. Depolarization occurs when a stimulus makes the membrane permeable to Na^+ ions. Repolarization follows as the membrane again becomes impermeable to Na^+, and K^+ moves from high to low concentration. In the long term, active transport slowly maintains the concentration differences, but the cell may fire hundreds of times in rapid succession without seriously depleting them.

The separation of charge creates a potential difference of 70 to 90 mV across the cell membrane. While this is a small voltage, the resulting electric field ($E = V/d$) across the only 8-nm-thick membrane is immense (on the order of 11 MV/m!) and has fundamental effects on its structure and permeability. Now, if the exterior of a neuron is taken to be at 0 V, then the interior has a *resting potential* of about –90 mV. Such voltages are created across the membranes of almost all types of animal cells but are largest in nerve and muscle cells. In fact, fully 25% of the energy used by cells goes toward creating and maintaining these potentials.

Electric currents along the cell membrane are created by any stimulus that changes the membrane's permeability. The membrane thus temporarily becomes permeable to Na^+, which then rushes in, driven both by diffusion and the Coulomb force. This inrush of Na^+ first neutralizes the inside membrane, or *depolarizes* it, and then makes it slightly positive. The depolarization causes the membrane to again become impermeable to Na^+, and the movement of K^+ quickly returns the cell to its resting potential, or *repolarizes* it. This sequence of events results in a voltage pulse, called the *action potential*. (See Figure 20.32.) Only small fractions of the ions move, so that the cell can fire many hundreds of times without depleting the excess concentrations of Na^+ and K^+. Eventually, the cell must replenish these ions to maintain the concentration differences that create bioelectricity. This sodium-potassium pump is an example of *active transport*, wherein cell energy is used to move ions across membranes against diffusion gradients and the Coulomb force.

The action potential is a voltage pulse at one location on a cell membrane. How does it get transmitted along the cell membrane, and in particular down an axon, as a nerve impulse? The answer is that the changing voltage and electric fields affect the permeability of the adjacent cell membrane, so that the same process takes place there. The adjacent membrane depolarizes, affecting the membrane further down, and so on, as illustrated in Figure 20.33. Thus the action potential stimulated at one location triggers a *nerve impulse* that moves slowly (about 1 m/s) along the cell membrane.

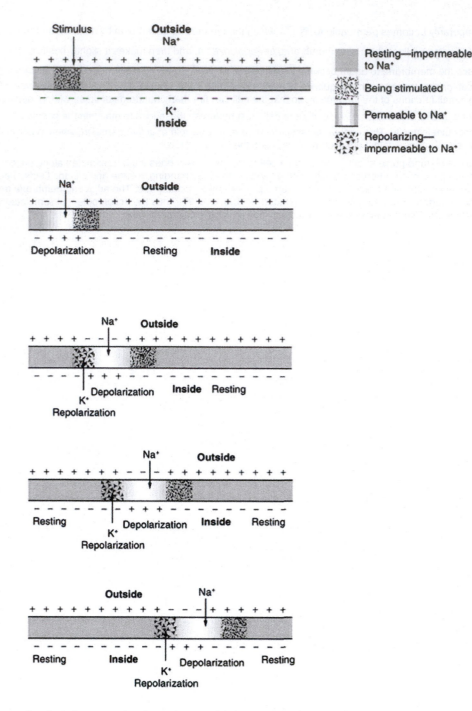

Figure 20.33 A nerve impulse is the propagation of an action potential along a cell membrane. A stimulus causes an action potential at one location, which changes the permeability of the adjacent membrane, causing an action potential there. This in turn affects the membrane further down, so that the action potential moves slowly (in electrical terms) along the cell membrane. Although the impulse is due to Na^+ and K^+ going across the membrane, it is equivalent to a wave of charge moving along the outside and inside of the membrane.

Some axons, like that in **Figure 20.30**, are sheathed with *myelin*, consisting of fat-containing cells. **Figure 20.34** shows an enlarged view of an axon having myelin sheaths characteristically separated by unmyelinated gaps (called nodes of Ranvier). This arrangement gives the axon a number of interesting properties. Since myelin is an insulator, it prevents signals from jumping between adjacent nerves (cross talk). Additionally, the myelinated regions transmit electrical signals at a very high speed, as an ordinary conductor or resistor would. There is no action potential in the myelinated regions, so that no cell energy is used in them. There is an IR signal loss in the myelin, but the signal is regenerated in the gaps, where the voltage pulse triggers the action potential at full voltage. So a myelinated axon transmits a nerve impulse faster, with less energy consumption, and is better protected from cross talk than an unmyelinated one. Not all axons are myelinated, so that cross talk and slow signal transmission are a characteristic of the normal operation of these axons, another variable in the nervous system.

The degeneration or destruction of the myelin sheaths that surround the nerve fibers impairs signal transmission and can lead to numerous neurological effects. One of the most prominent of these diseases comes from the body's own immune system attacking the myelin in the central nervous system—multiple sclerosis. MS symptoms include fatigue, vision problems, weakness of arms and legs, loss of balance, and tingling or numbness in one's extremities (neuropathy). It is more apt to strike younger adults, especially females. Causes might come from infection, environmental or geographic affects, or genetics. At the moment there is no known cure for MS.

Most animal cells can fire or create their own action potential. Muscle cells contract when they fire and are often induced to do so by a nerve impulse. In fact, nerve and muscle cells are physiologically similar, and there are even hybrid cells, such as in the heart, that have characteristics of both nerves and muscles. Some animals, like the infamous electric eel (see **Figure 20.35**), use muscles ganged so that their voltages add in order to create a shock great enough to stun prey.

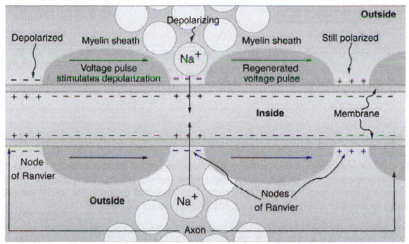

Figure 20.34 Propagation of a nerve impulse down a myelinated axon, from left to right. The signal travels very fast and without energy input in the myelinated regions, but it loses voltage. It is regenerated in the gaps. The signal moves faster than in unmyelinated axons and is insulated from signals in other nerves, limiting cross talk.

Figure 20.35 An electric eel flexes its muscles to create a voltage that stuns prey. (credit: chrisbb, Flickr)

Electrocardiograms

Just as nerve impulses are transmitted by depolarization and repolarization of adjacent membrane, the depolarization that causes muscle contraction can also stimulate adjacent muscle cells to depolarize (fire) and contract. Thus, a depolarization wave can be sent across the heart, coordinating its rhythmic contractions and enabling it to perform its vital function of propelling blood through the circulatory system. **Figure 20.36** is a simplified graphic of a depolarization wave spreading across the heart from the *sinoarterial (SA) node*, the heart's natural pacemaker.

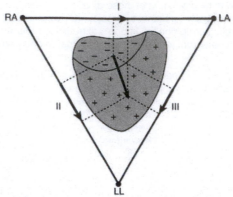

Figure 20.36 The outer surface of the heart changes from positive to negative during depolarization. This wave of depolarization is spreading from the top of the heart and is represented by a vector pointing in the direction of the wave. This vector is a voltage (potential difference) vector. Three electrodes, labeled RA, LA, and LL, are placed on the patient. Each pair (called leads I, II, and III) measures a component of the depolarization vector and is graphed in an ECG.

An **electrocardiogram (ECG)** is a record of the voltages created by the wave of depolarization and subsequent repolarization in the heart. Voltages between pairs of electrodes placed on the chest are vector components of the voltage wave on the heart. Standard ECGs have 12 or more electrodes, but only three are shown in **Figure 20.36** for clarity. Decades ago, three-electrode ECGs were performed by placing electrodes on the left and right arms and the left leg. The voltage between the right arm and the left leg is called the *lead II potential* and is the most often graphed. We shall examine the lead II potential as an indicator of heart-muscle function and see that it is coordinated with arterial blood pressure as well.

Heart function and its four-chamber action are explored in **Viscosity and Laminar Flow; Poiseuille's Law**. Basically, the right and left atria receive blood from the body and lungs, respectively, and pump the blood into the ventricles. The right and left ventricles, in turn, pump blood through the lungs and the rest of the body, respectively. Depolarization of the heart muscle causes it to contract. After contraction it is repolarized to ready it for the next beat. The ECG measures components of depolarization and repolarization of the heart muscle and can yield significant information on the functioning and malfunctioning of the heart.

Figure 20.37 shows an ECG of the lead II potential and a graph of the corresponding arterial blood pressure. The major features are labeled P, Q, R, S, and T. The *P wave* is generated by the depolarization and contraction of the atria as they pump blood into the ventricles. The *QRS complex* is created by the depolarization of the ventricles as they pump blood to the lungs and body. Since the shape of the heart and the path of the depolarization wave are not simple, the QRS complex has this typical shape and time span. The lead II QRS signal also masks the repolarization of the atria, which occur at the same time. Finally, the *T wave* is generated by the repolarization of the ventricles and is followed by the next P wave in the next heartbeat. Arterial blood pressure varies with each part of the heartbeat, with systolic (maximum) pressure occurring closely after the QRS complex, which signals contraction of the ventricles.

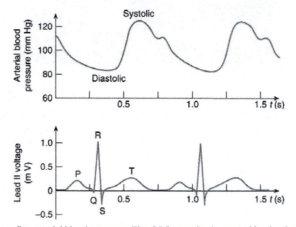

Figure 20.37 A lead II ECG with corresponding arterial blood pressure. The QRS complex is created by the depolarization and contraction of the ventricles and is followed shortly by the maximum or systolic blood pressure. See text for further description.

Taken together, the 12 leads of a state-of-the-art ECG can yield a wealth of information about the heart. For example, regions of damaged heart tissue, called infarcts, reflect electrical waves and are apparent in one or more lead potentials. Subtle changes due to slight or gradual damage to the heart are most readily detected by comparing a recent ECG to an older one. This is particularly the case since individual heart shape, size, and orientation can cause variations in ECGs from one individual to another. ECG technology has advanced to the point where a portable ECG monitor with a liquid crystal instant display and a printer can be carried to patients' homes or used in emergency vehicles. See **Figure 20.38**.

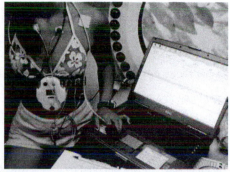

Figure 20.38 This NASA scientist and NEEMO 5 aquanaut's heart rate and other vital signs are being recorded by a portable device while living in an underwater habitat. (credit: NASA, Life Sciences Data Archive at Johnson Space Center, Houston, Texas)

PhET Explorations: Neuron

Figure 20.39 Neuron (http://cnx.org/content/m55361/1.2/neuron_en.jar)

Stimulate a neuron and monitor what happens. Pause, rewind, and move forward in time in order to observe the ions as they move across the neuron membrane.

Glossary

AC current: current that fluctuates sinusoidally with time, expressed as $I = I_0 \sin 2\pi ft$, where I is the current at time t, I_0 is the peak current, and f is the frequency in hertz

AC voltage: voltage that fluctuates sinusoidally with time, expressed as $V = V_0 \sin 2\pi ft$, where V is the voltage at time t, V_0 is the peak voltage, and f is the frequency in hertz

alternating current: (AC) the flow of electric charge that periodically reverses direction

ampere: (amp) the SI unit for current; 1 A = 1 C/s

bioelectricity: electrical effects in and created by biological systems

direct current: (DC) the flow of electric charge in only one direction

drift velocity: the average velocity at which free charges flow in response to an electric field

electric current: the rate at which charge flows, $I = \Delta Q/\Delta t$

electric power: the rate at which electrical energy is supplied by a source or dissipated by a device; it is the product of current times voltage

electrocardiogram (ECG): usually abbreviated ECG, a record of voltages created by depolarization and repolarization, especially in the heart

microshock sensitive: a condition in which a person's skin resistance is bypassed, possibly by a medical procedure, rendering the person vulnerable to electrical shock at currents about 1/1000 the normally required level

nerve conduction: the transport of electrical signals by nerve cells

ohm: the unit of resistance, given by $1\Omega = 1$ V/A

Ohm's law: an empirical relation stating that the current I is proportional to the potential difference V, $\propto V$; it is often written as $I = V/R$, where R is the resistance

ohmic: a type of material for which Ohm's law is valid

resistance: the electric property that impedes current; for ohmic materials, it is the ratio of voltage to current, $R = V/I$

resistivity: an intrinsic property of a material, independent of its shape or size, directly proportional to the resistance, denoted by ρ

rms current: the root mean square of the current, $I_{rms} = I_0 / \sqrt{2}$, where I_0 is the peak current, in an AC system

rms voltage: the root mean square of the voltage, $V_{rms} = V_0 / \sqrt{2}$, where V_0 is the peak voltage, in an AC system

semipermeable: property of a membrane that allows only certain types of ions to cross it

shock hazard: when electric current passes through a person

short circuit: also known as a "short," a low-resistance path between terminals of a voltage source

simple circuit: a circuit with a single voltage source and a single resistor

temperature coefficient of resistivity: an empirical quantity, denoted by α, which describes the change in resistance or resistivity of a material with temperature

thermal hazard: a hazard in which electric current causes undesired thermal effects

Section Summary

20.1 Current
- Electric current I is the rate at which charge flows, given by

$$I = \frac{\Delta Q}{\Delta t},$$

  where ΔQ is the amount of charge passing through an area in time Δt .
- The direction of conventional current is taken as the direction in which positive charge moves.
- The SI unit for current is the ampere (A), where $1 \text{ A} = 1 \text{ C/s}$.
- Current is the flow of free charges, such as electrons and ions.
- Drift velocity v_d is the average speed at which these charges move.
- Current I is proportional to drift velocity v_d, as expressed in the relationship $I = nqAv_d$. Here, I is the current through a wire of cross-sectional area A. The wire's material has a free-charge density n, and each carrier has charge q and a drift velocity v_d .
- Electrical signals travel at speeds about 10^{12} times greater than the drift velocity of free electrons.

20.2 Ohm's Law: Resistance and Simple Circuits
- A simple circuit *is* one in which there is a single voltage source and a single resistance.
- One statement of Ohm's law gives the relationship between current I, voltage V, and resistance R in a simple circuit to be $I = \frac{V}{R}$.
- Resistance has units of ohms (Ω), related to volts and amperes by $1 \ \Omega = 1 \text{ V/A}$.
- There is a voltage or IR drop across a resistor, caused by the current flowing through it, given by $V = IR$.

20.3 Resistance and Resistivity
- The resistance R of a cylinder of length L and cross-sectional area A is $R = \frac{\rho L}{A}$, where ρ is the resistivity of the material.
- Values of ρ in **Table 20.1** show that materials fall into three groups—*conductors, semiconductors, and insulators*.
- Temperature affects resistivity; for relatively small temperature changes ΔT, resistivity is $\rho = \rho_0(1 + \alpha \Delta T)$, where ρ_0 is the original resistivity and α is the temperature coefficient of resistivity.
- **Table 20.2** gives values for α, the temperature coefficient of resistivity.
- The resistance R of an object also varies with temperature: $R = R_0(1 + \alpha \Delta T)$, where R_0 is the original resistance, and R is the resistance after the temperature change.

20.4 Electric Power and Energy
- Electric power P is the rate (in watts) that energy is supplied by a source or dissipated by a device.
- Three expressions for electrical power are

$$P = IV,$$

$$P = \frac{V^2}{R},$$

and

$$P = I^2 R.$$

- The energy used by a device with a power P over a time t is $E = Pt$.

20.5 Alternating Current versus Direct Current

- Direct current (DC) is the flow of electric current in only one direction. It refers to systems where the source voltage is constant.
- The voltage source of an alternating current (AC) system puts out $V = V_0 \sin 2\pi ft$, where V is the voltage at time t, V_0 is the peak voltage, and f is the frequency in hertz.
- In a simple circuit, $I = V/R$ and AC current is $I = I_0 \sin 2\pi ft$, where I is the current at time t, and $I_0 = V_0/R$ is the peak current.
- The average AC power is $P_{ave} = \frac{1}{2}I_0 V_0$.
- Average (rms) current I_{rms} and average (rms) voltage V_{rms} are $I_{rms} = \frac{I_0}{\sqrt{2}}$ and $V_{rms} = \frac{V_0}{\sqrt{2}}$, where rms stands for root mean square.
- Thus, $P_{ave} = I_{rms}V_{rms}$.
- Ohm's law for AC is $I_{rms} = \frac{V_{rms}}{R}$.
- Expressions for the average power of an AC circuit are $P_{ave} = I_{rms}V_{rms}$, $P_{ave} = \frac{V_{rms}^2}{R}$, and $P_{ave} = I_{rms}^2 R$, analogous to the expressions for DC circuits.

20.6 Electric Hazards and the Human Body

- The two types of electric hazards are thermal (excessive power) and shock (current through a person).
- Shock severity is determined by current, path, duration, and AC frequency.
- Table 20.3 lists shock hazards as a function of current.
- Figure 20.28 graphs the threshold current for two hazards as a function of frequency.

20.7 Nerve Conduction–Electrocardiograms

- Electric potentials in neurons and other cells are created by ionic concentration differences across semipermeable membranes.
- Stimuli change the permeability and create action potentials that propagate along neurons.
- Myelin sheaths speed this process and reduce the needed energy input.
- This process in the heart can be measured with an electrocardiogram (ECG).

Conceptual Questions

20.1 Current

1. Can a wire carry a current and still be neutral—that is, have a total charge of zero? Explain.

2. Car batteries are rated in ampere-hours ($A \cdot h$). To what physical quantity do ampere-hours correspond (voltage, charge, . . .), and what relationship do ampere-hours have to energy content?

3. If two different wires having identical cross-sectional areas carry the same current, will the drift velocity be higher or lower in the better conductor? Explain in terms of the equation $v_d = \frac{I}{nqA}$, by considering how the density of charge carriers n relates to whether or not a material is a good conductor.

4. Why are two conducting paths from a voltage source to an electrical device needed to operate the device?

5. In cars, one battery terminal is connected to the metal body. How does this allow a single wire to supply current to electrical devices rather than two wires?

6. Why isn't a bird sitting on a high-voltage power line electrocuted? Contrast this with the situation in which a large bird hits two wires simultaneously with its wings.

20.2 Ohm's Law: Resistance and Simple Circuits

7. The IR drop across a resistor means that there is a change in potential or voltage across the resistor. Is there any change in current as it passes through a resistor? Explain.

8. How is the IR drop in a resistor similar to the pressure drop in a fluid flowing through a pipe?

20.3 Resistance and Resistivity

9. In which of the three semiconducting materials listed in Table 20.1 do impurities supply free charges? (Hint: Examine the range of resistivity for each and determine whether the pure semiconductor has the higher or lower conductivity.)

10. Does the resistance of an object depend on the path current takes through it? Consider, for example, a rectangular bar—is its resistance the same along its length as across its width? (See Figure 20.40.)

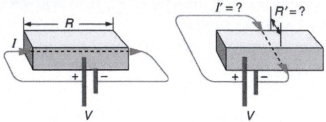

Figure 20.40 Does current taking two different paths through the same object encounter different resistance?

11. If aluminum and copper wires of the same length have the same resistance, which has the larger diameter? Why?

12. Explain why $R = R_0(1 + \alpha\Delta T)$ for the temperature variation of the resistance R of an object is not as accurate as $\rho = \rho_0(1 + \alpha\Delta T)$, which gives the temperature variation of resistivity ρ.

20.4 Electric Power and Energy

13. Why do incandescent lightbulbs grow dim late in their lives, particularly just before their filaments break?

14. The power dissipated in a resistor is given by $P = V^2/R$, which means power decreases if resistance increases. Yet this power is also given by $P = I^2 R$, which means power increases if resistance increases. Explain why there is no contradiction here.

20.5 Alternating Current versus Direct Current

15. Give an example of a use of AC power other than in the household. Similarly, give an example of a use of DC power other than that supplied by batteries.

16. Why do voltage, current, and power go through zero 120 times per second for 60-Hz AC electricity?

17. You are riding in a train, gazing into the distance through its window. As close objects streak by, you notice that the nearby fluorescent lights make *dashed* streaks. Explain.

20.6 Electric Hazards and the Human Body

18. Using an ohmmeter, a student measures the resistance between various points on his body. He finds that the resistance between two points on the same finger is about the same as the resistance between two points on opposite hands—both are several hundred thousand ohms. Furthermore, the resistance decreases when more skin is brought into contact with the probes of the ohmmeter. Finally, there is a dramatic drop in resistance (to a few thousand ohms) when the skin is wet. Explain these observations and their implications regarding skin and internal resistance of the human body.

19. What are the two major hazards of electricity?

20. Why isn't a short circuit a shock hazard?

21. What determines the severity of a shock? Can you say that a certain voltage is hazardous without further information?

22. An electrified needle is used to burn off warts, with the circuit being completed by having the patient sit on a large butt plate. Why is this plate large?

23. Some surgery is performed with high-voltage electricity passing from a metal scalpel through the tissue being cut. Considering the nature of electric fields at the surface of conductors, why would you expect most of the current to flow from the sharp edge of the scalpel? Do you think high- or low-frequency AC is used?

24. Some devices often used in bathrooms, such as hairdryers, often have safety messages saying "Do not use when the bathtub or basin is full of water." Why is this so?

25. We are often advised to not flick electric switches with wet hands, dry your hand first. We are also advised to never throw water on an electric fire. Why is this so?

26. Before working on a power transmission line, linemen will touch the line with the back of the hand as a final check that the voltage is zero. Why the back of the hand?

27. Why is the resistance of wet skin so much smaller than dry, and why do blood and other bodily fluids have low resistances?

28. Could a person on intravenous infusion (an IV) be microshock sensitive?

29. In view of the small currents that cause shock hazards and the larger currents that circuit breakers and fuses interrupt, how do they play a role in preventing shock hazards?

20.7 Nerve Conduction–Electrocardiograms

30. Note that in Figure 20.31, both the concentration gradient and the Coulomb force tend to move Na^+ ions into the cell. What prevents this?

31. Define depolarization, repolarization, and the action potential.

32. Explain the properties of myelinated nerves in terms of the insulating properties of myelin.

Problems & Exercises

20.1 Current

1. What is the current in milliamperes produced by the solar cells of a pocket calculator through which 4.00 C of charge passes in 4.00 h?

2. A total of 600 C of charge passes through a flashlight in 0.500 h. What is the average current?

3. What is the current when a typical static charge of $0.250\ \mu C$ moves from your finger to a metal doorknob in $1.00\ \mu s$?

4. Find the current when 2.00 nC jumps between your comb and hair over a $0.500 - \mu s$ time interval.

5. A large lightning bolt had a 20,000-A current and moved 30.0 C of charge. What was its duration?

6. The 200-A current through a spark plug moves 0.300 mC of charge. How long does the spark last?

7. (a) A defibrillator sends a 6.00-A current through the chest of a patient by applying a 10,000-V potential as in the figure below. What is the resistance of the path? (b) The defibrillator paddles make contact with the patient through a conducting gel that greatly reduces the path resistance. Discuss the difficulties that would ensue if a larger voltage were used to produce the same current through the patient, but with the path having perhaps 50 times the resistance. (Hint: The current must be about the same, so a higher voltage would imply greater power. Use this equation for power: $P = I^2 R$.)

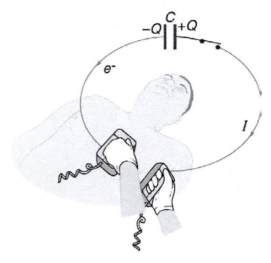

Figure 20.41 The capacitor in a defibrillation unit drives a current through the heart of a patient.

8. During open-heart surgery, a defibrillator can be used to bring a patient out of cardiac arrest. The resistance of the path is $500\ \Omega$ and a 10.0-mA current is needed. What voltage should be applied?

9. (a) A defibrillator passes 12.0 A of current through the torso of a person for 0.0100 s. How much charge moves? (b) How many electrons pass through the wires connected to the patient? (See figure two problems earlier.)

10. A clock battery wears out after moving 10,000 C of charge through the clock at a rate of 0.500 mA. (a) How long did the clock run? (b) How many electrons per second flowed?

11. The batteries of a submerged non-nuclear submarine supply 1000 A at full speed ahead. How long does it take to move Avogadro's number (6.02×10^{23}) of electrons at this rate?

12. Electron guns are used in X-ray tubes. The electrons are accelerated through a relatively large voltage and directed onto a metal target, producing X-rays. (a) How many electrons per second strike the target if the current is 0.500 mA? (b) What charge strikes the target in 0.750 s?

13. A large cyclotron directs a beam of He^{++} nuclei onto a target with a beam current of 0.250 mA. (a) How many He^{++} nuclei per second is this? (b) How long does it take for 1.00 C to strike the target? (c) How long before 1.00 mol of He^{++} nuclei strike the target?

14. Repeat the above example on **Example 20.3**, but for a wire made of silver and given there is one free electron per silver atom.

15. Using the results of the above example on **Example 20.3**, find the drift velocity in a copper wire of twice the diameter and carrying 20.0 A.

16. A 14-gauge copper wire has a diameter of 1.628 mm. What magnitude current flows when the drift velocity is 1.00 mm/s? (See above example on **Example 20.3** for useful information.)

17. SPEAR, a storage ring about 72.0 m in diameter at the Stanford Linear Accelerator (closed in 2009), has a 20.0-A circulating beam of electrons that are moving at nearly the speed of light. (See **Figure 20.42**.) How many electrons are in the beam?

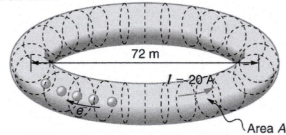

Figure 20.42 Electrons circulating in the storage ring called SPEAR constitute a 20.0-A current. Because they travel close to the speed of light, each electron completes many orbits in each second.

20.2 Ohm's Law: Resistance and Simple Circuits

18. What current flows through the bulb of a 3.00-V flashlight when its hot resistance is $3.60\ \Omega$?

19. Calculate the effective resistance of a pocket calculator that has a 1.35-V battery and through which 0.200 mA flows.

20. What is the effective resistance of a car's starter motor when 150 A flows through it as the car battery applies 11.0 V to the motor?

21. How many volts are supplied to operate an indicator light on a DVD player that has a resistance of $140\ \Omega$, given that 25.0 mA passes through it?

22. (a) Find the voltage drop in an extension cord having a $0.0600\text{-}\ \Omega$ resistance and through which 5.00 A is flowing. (b) A cheaper cord utilizes thinner wire and has a resistance of $0.300\ \Omega$. What is the voltage drop in it when 5.00 A flows? (c) Why is the voltage to whatever appliance is being used reduced by this amount? What is the effect on the appliance?

23. A power transmission line is hung from metal towers with glass insulators having a resistance of $1.00\times10^9\ \Omega$. What current flows through the insulator if the voltage is 200 kV? (Some high-voltage lines are DC.)

20.3 Resistance and Resistivity

24. What is the resistance of a 20.0-m-long piece of 12-gauge copper wire having a 2.053-mm diameter?

25. The diameter of 0-gauge copper wire is 8.252 mm. Find the resistance of a 1.00-km length of such wire used for power transmission.

26. If the 0.100-mm diameter tungsten filament in a light bulb is to have a resistance of $0.200\ \Omega$ at $20.0°C$, how long should it be?

27. Find the ratio of the diameter of aluminum to copper wire, if they have the same resistance per unit length (as they might in household wiring).

28. What current flows through a 2.54-cm-diameter rod of pure silicon that is 20.0 cm long, when $1.00\times10^3\ V$ is applied to it? (Such a rod may be used to make nuclear-particle detectors, for example.)

29. (a) To what temperature must you raise a copper wire, originally at $20.0°C$, to double its resistance, neglecting any changes in dimensions? (b) Does this happen in household wiring under ordinary circumstances?

30. A resistor made of Nichrome wire is used in an application where its resistance cannot change more than 1.00% from its value at $20.0°C$. Over what temperature range can it be used?

31. Of what material is a resistor made if its resistance is 40.0% greater at $100°C$ than at $20.0°C$?

32. An electronic device designed to operate at any temperature in the range from $-10.0°C$ to $55.0°C$ contains pure carbon resistors. By what factor does their resistance increase over this range?

33. (a) Of what material is a wire made, if it is 25.0 m long with a 0.100 mm diameter and has a resistance of $77.7\ \Omega$ at $20.0°C$? (b) What is its resistance at $150°C$?

34. Assuming a constant temperature coefficient of resistivity, what is the maximum percent decrease in the resistance of a constantan wire starting at $20.0°C$?

35. A wire is drawn through a die, stretching it to four times its original length. By what factor does its resistance increase?

36. A copper wire has a resistance of $0.500\ \Omega$ at $20.0°C$, and an iron wire has a resistance of $0.525\ \Omega$ at the same temperature. At what temperature are their resistances equal?

37. (a) Digital medical thermometers determine temperature by measuring the resistance of a semiconductor device called a thermistor (which has $\alpha = -0.0600/°C$) when it is at the same temperature as the patient. What is a patient's temperature if the thermistor's resistance at that temperature is 82.0% of its value at $37.0°C$ (normal body temperature)? (b) The negative value for α may not be maintained for very low temperatures. Discuss why and whether this is the case here. (Hint: Resistance can't become negative.)

38. Integrated Concepts

(a) Redo **Exercise 20.25** taking into account the thermal expansion of the tungsten filament. You may assume a thermal expansion coefficient of $12\times10^{-6}/°C$. (b) By what percentage does your answer differ from that in the example?

39. Unreasonable Results

(a) To what temperature must you raise a resistor made of constantan to double its resistance, assuming a constant temperature coefficient of resistivity? (b) To cut it in half? (c) What is unreasonable about these results? (d) Which assumptions are unreasonable, or which premises are inconsistent?

20.4 Electric Power and Energy

40. What is the power of a $1.00\times10^2\ MV$ lightning bolt having a current of $2.00\times10^4\ A$?

41. What power is supplied to the starter motor of a large truck that draws 250 A of current from a 24.0-V battery hookup?

42. A charge of 4.00 C of charge passes through a pocket calculator's solar cells in 4.00 h. What is the power output, given the calculator's voltage output is 3.00 V? (See **Figure 20.43**.)

Figure 20.43 The strip of solar cells just above the keys of this calculator convert light to electricity to supply its energy needs. (credit: Evan-Amos, Wikimedia Commons)

43. How many watts does a flashlight that has $6.00\times10^2\ C$ pass through it in 0.500 h use if its voltage is 3.00 V?

44. Find the power dissipated in each of these extension cords: (a) an extension cord having a $0.0600\text{ - }\Omega$ resistance and through which 5.00 A is flowing; (b) a cheaper cord utilizing thinner wire and with a resistance of $0.300\ \Omega$.

45. Verify that the units of a volt-ampere are watts, as implied by the equation $P = IV$.

46. Show that the units $1 \text{ V}^2 / \Omega = 1 \text{W}$, as implied by the equation $P = V^2/R$.

47. Show that the units $1 \text{ A}^2 \cdot \Omega = 1 \text{ W}$, as implied by the equation $P = I^2R$.

48. Verify the energy unit equivalence that $1 \text{ kW} \cdot \text{h} = 3.60 \times 10^6 \text{ J}$.

49. Electrons in an X-ray tube are accelerated through $1.00 \times 10^2 \text{ kV}$ and directed toward a target to produce X-rays. Calculate the power of the electron beam in this tube if it has a current of 15.0 mA.

50. An electric water heater consumes 5.00 kW for 2.00 h per day. What is the cost of running it for one year if electricity costs $12.0 \text{ cents/kW} \cdot \text{h}$? See **Figure 20.44**.

Figure 20.44 On-demand electric hot water heater. Heat is supplied to water only when needed. (credit: aviddavid, Flickr)

51. With a 1200-W toaster, how much electrical energy is needed to make a slice of toast (cooking time = 1 minute)? At $9.0 \text{ cents/kW} \cdot \text{h}$, how much does this cost?

52. What would be the maximum cost of a CFL such that the total cost (investment plus operating) would be the same for both CFL and incandescent 60-W bulbs? Assume the cost of the incandescent bulb is 25 cents and that electricity costs 10 cents/kWh. Calculate the cost for 1000 hours, as in the cost effectiveness of CFL example.

53. Some makes of older cars have 6.00-V electrical systems. (a) What is the hot resistance of a 30.0-W headlight in such a car? (b) What current flows through it?

54. Alkaline batteries have the advantage of putting out constant voltage until very nearly the end of their life. How long will an alkaline battery rated at $1.00 \text{ A} \cdot \text{h}$ and 1.58 V keep a 1.00-W flashlight bulb burning?

55. A cauterizer, used to stop bleeding in surgery, puts out 2.00 mA at 15.0 kV. (a) What is its power output? (b) What is the resistance of the path?

56. The average television is said to be on 6 hours per day. Estimate the yearly cost of electricity to operate 100 million TVs, assuming their power consumption averages 150 W and the cost of electricity averages $12.0 \text{ cents/kW} \cdot \text{h}$.

57. An old lightbulb draws only 50.0 W, rather than its original 60.0 W, due to evaporative thinning of its filament. By what factor is its diameter reduced, assuming uniform thinning along its length? Neglect any effects caused by temperature differences.

58. 00-gauge copper wire has a diameter of 9.266 mm. Calculate the power loss in a kilometer of such wire when it carries $1.00 \times 10^2 \text{ A}$.

59. Integrated Concepts

Cold vaporizers pass a current through water, evaporating it with only a small increase in temperature. One such home device is rated at 3.50 A and utilizes 120 V AC with 95.0% efficiency. (a) What is the vaporization rate in grams per minute? (b) How much water must you put into the vaporizer for 8.00 h of overnight operation? (See **Figure 20.45**.)

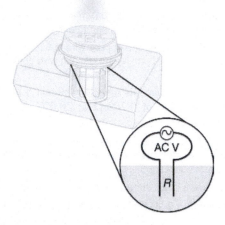

Figure 20.45 This cold vaporizer passes current directly through water, vaporizing it directly with relatively little temperature increase.

60. Integrated Concepts

(a) What energy is dissipated by a lightning bolt having a 20,000-A current, a voltage of $1.00 \times 10^2 \text{ MV}$, and a length of 1.00 ms? (b) What mass of tree sap could be raised from $18.0°C$ to its boiling point and then evaporated by this energy, assuming sap has the same thermal characteristics as water?

61. Integrated Concepts

What current must be produced by a 12.0-V battery-operated bottle warmer in order to heat 75.0 g of glass, 250 g of baby formula, and $3.00 \times 10^2 \text{ g}$ of aluminum from $20.0°C$ to $90.0°C$ in 5.00 min?

62. Integrated Concepts

How much time is needed for a surgical cauterizer to raise the temperature of 1.00 g of tissue from $37.0°C$ to $100°C$ and then boil away 0.500 g of water, if it puts out 2.00 mA at 15.0 kV? Ignore heat transfer to the surroundings.

63. Integrated Concepts

Hydroelectric generators (see **Figure 20.46**) at Hoover Dam produce a maximum current of 8.00×10^3 A at 250 kV. (a) What is the power output? (b) The water that powers the generators enters and leaves the system at low speed (thus its kinetic energy does not change) but loses 160 m in altitude. How many cubic meters per second are needed, assuming 85.0% efficiency?

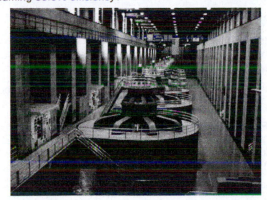

Figure 20.46 Hydroelectric generators at the Hoover dam. (credit: Jon Sullivan)

64. Integrated Concepts

(a) Assuming 95.0% efficiency for the conversion of electrical power by the motor, what current must the 12.0-V batteries of a 750-kg electric car be able to supply: (a) To accelerate from rest to 25.0 m/s in 1.00 min? (b) To climb a 2.00×10^2-m -high hill in 2.00 min at a constant 25.0-m/s speed while exerting 5.00×10^2 N of force to overcome air resistance and friction? (c) To travel at a constant 25.0-m/s speed, exerting a 5.00×10^2 N force to overcome air resistance and friction? See **Figure 20.47**.

Figure 20.47 This REVAi, an electric car, gets recharged on a street in London. (credit: Frank Hebbert)

65. Integrated Concepts

A light-rail commuter train draws 630 A of 650-V DC electricity when accelerating. (a) What is its power consumption rate in kilowatts? (b) How long does it take to reach 20.0 m/s starting from rest if its loaded mass is 5.30×10^4 kg , assuming 95.0% efficiency and constant power? (c) Find its average acceleration. (d) Discuss how the acceleration you found for the light-rail train compares to what might be typical for an automobile.

66. Integrated Concepts

(a) An aluminum power transmission line has a resistance of $0.0580 \ \Omega \ / \text{km}$. What is its mass per kilometer? (b) What is the mass per kilometer of a copper line having the same resistance? A lower resistance would shorten the heating time. Discuss the practical limits to speeding the heating by lowering the resistance.

67. Integrated Concepts

(a) An immersion heater utilizing 120 V can raise the temperature of a 1.00×10^2-g aluminum cup containing 350 g of water from $20.0°C$ to $95.0°C$ in 2.00 min. Find its resistance, assuming it is constant during the process. (b) A lower resistance would shorten the heating time. Discuss the practical limits to speeding the heating by lowering the resistance.

68. Integrated Concepts

(a) What is the cost of heating a hot tub containing 1500 kg of water from $10.0°C$ to $40.0°C$, assuming 75.0% efficiency to account for heat transfer to the surroundings? The cost of electricity is $9 \ \text{cents/kW} \cdot \text{h}$. (b) What current was used by the 220-V AC electric heater, if this took 4.00 h?

69. Unreasonable Results

(a) What current is needed to transmit 1.00×10^2 MW of power at 480 V? (b) What power is dissipated by the transmission lines if they have a $1.00 - \Omega$ resistance? (c) What is unreasonable about this result? (d) Which assumptions are unreasonable, or which premises are inconsistent?

70. Unreasonable Results

(a) What current is needed to transmit 1.00×10^2 MW of power at 10.0 kV? (b) Find the resistance of 1.00 km of wire that would cause a 0.0100% power loss. (c) What is the diameter of a 1.00-km-long copper wire having this resistance? (d) What is unreasonable about these results? (e) Which assumptions are unreasonable, or which premises are inconsistent?

71. Construct Your Own Problem

Consider an electric immersion heater used to heat a cup of water to make tea. Construct a problem in which you calculate the needed resistance of the heater so that it increases the temperature of the water and cup in a reasonable amount of time. Also calculate the cost of the electrical energy used in your process. Among the things to be considered are the voltage used, the masses and heat capacities involved, heat losses, and the time over which the heating takes place. Your instructor may wish for you to consider a thermal safety switch (perhaps bimetallic) that will halt the process before damaging temperatures are reached in the immersion unit.

20.5 Alternating Current versus Direct Current

72. (a) What is the hot resistance of a 25-W light bulb that runs on 120-V AC? (b) If the bulb's operating temperature is $2700°C$, what is its resistance at $2600°C$?

73. Certain heavy industrial equipment uses AC power that has a peak voltage of 679 V. What is the rms voltage?

74. A certain circuit breaker trips when the rms current is 15.0 A. What is the corresponding peak current?

75. Military aircraft use 400-Hz AC power, because it is possible to design lighter-weight equipment at this higher frequency. What is the time for one complete cycle of this power?

76. A North American tourist takes his 25.0-W, 120-V AC razor to Europe, finds a special adapter, and plugs it into 240 V AC. Assuming constant resistance, what power does the razor consume as it is ruined?

77. In this problem, you will verify statements made at the end of the power losses for **Example 20.10.** (a) What current is needed to transmit 100 MW of power at a voltage of 25.0 kV? (b) Find the power loss in a $1.00 - \Omega$ transmission line. (c) What percent loss does this represent?

78. A small office-building air conditioner operates on 408-V AC and consumes 50.0 kW. (a) What is its effective resistance? (b) What is the cost of running the air conditioner during a hot summer month when it is on 8.00 h per day for 30 days and electricity costs $9.00 \text{ cents/kW} \cdot \text{h}$?

79. What is the peak power consumption of a 120-V AC microwave oven that draws 10.0 A?

80. What is the peak current through a 500-W room heater that operates on 120-V AC power?

81. Two different electrical devices have the same power consumption, but one is meant to be operated on 120-V AC and the other on 240-V AC. (a) What is the ratio of their resistances? (b) What is the ratio of their currents? (c) Assuming its resistance is unaffected, by what factor will the power increase if a 120-V AC device is connected to 240-V AC?

82. Nichrome wire is used in some radiative heaters. (a) Find the resistance needed if the average power output is to be 1.00 kW utilizing 120-V AC. (b) What length of Nichrome wire, having a cross-sectional area of 5.00mm^2, is needed if the operating temperature is $500° C$? (c) What power will it draw when first switched on?

83. Find the time after $t = 0$ when the instantaneous voltage of 60-Hz AC first reaches the following values: (a) $V_0/2$ (b) V_0 (c) 0.

84. (a) At what two times in the first period following $t = 0$ does the instantaneous voltage in 60-Hz AC equal V_{rms}? (b) $-V_{rms}$?

20.6 Electric Hazards and the Human Body

85. (a) How much power is dissipated in a short circuit of 240-V AC through a resistance of $0.250 \ \Omega$? (b) What current flows?

86. What voltage is involved in a 1.44-kW short circuit through a $0.100 - \Omega$ resistance?

87. Find the current through a person and identify the likely effect on her if she touches a 120-V AC source: (a) if she is standing on a rubber mat and offers a total resistance of $300 \text{ k} \Omega$; (b) if she is standing barefoot on wet grass and has a resistance of only $4000 \text{ k} \Omega$.

88. While taking a bath, a person touches the metal case of a radio. The path through the person to the drainpipe and ground has a resistance of $4000 \ \Omega$. What is the smallest voltage on the case of the radio that could cause ventricular fibrillation?

89. Foolishly trying to fish a burning piece of bread from a toaster with a metal butter knife, a man comes into contact with 120-V AC. He does not even feel it since, luckily, he is wearing rubber-soled shoes. What is the minimum resistance of the path the current follows through the person?

90. (a) During surgery, a current as small as $20.0 \ \mu A$ applied directly to the heart may cause ventricular fibrillation. If the resistance of the exposed heart is $300 \ \Omega$, what is the smallest voltage that poses this danger? (b) Does your answer imply that special electrical safety precautions are needed?

91. (a) What is the resistance of a 220-V AC short circuit that generates a peak power of 96.8 kW? (b) What would the average power be if the voltage was 120 V AC?

92. A heart defibrillator passes 10.0 A through a patient's torso for 5.00 ms in an attempt to restore normal beating. (a) How much charge passed? (b) What voltage was applied if 500 J of energy was dissipated? (c) What was the path's resistance? (d) Find the temperature increase caused in the 8.00 kg of affected tissue.

93. Integrated Concepts

A short circuit in a 120-V appliance cord has a $0.500- \Omega$ resistance. Calculate the temperature rise of the 2.00 g of surrounding materials, assuming their specific heat capacity is $0.200 \text{ cal/g} \cdot °C$ and that it takes 0.0500 s for a circuit breaker to interrupt the current. Is this likely to be damaging?

94. Construct Your Own Problem

Consider a person working in an environment where electric currents might pass through her body. Construct a problem in which you calculate the resistance of insulation needed to protect the person from harm. Among the things to be considered are the voltage to which the person might be exposed, likely body resistance (dry, wet, ...), and acceptable currents (safe but sensed, safe and unfelt, ...).

20.7 Nerve Conduction–Electrocardiograms

95. Integrated Concepts

Use the ECG in **Figure 20.37** to determine the heart rate in beats per minute assuming a constant time between beats.

96. Integrated Concepts

(a) Referring to **Figure 20.37**, find the time systolic pressure lags behind the middle of the QRS complex. (b) Discuss the reasons for the time lag.

Test Prep for AP® Courses

20.1 Current

1. Which of the following can be explained on the basis of conservation of charge in a closed circuit consisting of a battery, resistor, and metal wires?
  a. The number of electrons leaving the battery will be equal to the number of electrons entering the battery.
  b. The number of electrons leaving the battery will be less than the number of electrons entering the battery.
  c. The number of protons leaving the battery will be equal to the number of protons entering the battery.
  d. The number of protons leaving the battery will be less than the number of protons entering the battery.

2. When a battery is connected to a bulb, there is 2.5 A of current in the circuit. What amount of charge will flow though the circuit in a time of 0.5 s?
  a. 0.5 C
  b. 1 C
  c. 1.25 C
  d. 1.5 C

3. If 0.625×10^{20} electrons flow through a circuit each second, what is the current in the circuit?

4. Two students calculate the charge flowing through a circuit. The first student concludes that 300 C of charge flows in 1 minute. The second student concludes that 3.125×10^{19} electrons flow per second. If the current measured in the circuit is 5 A, which of the two students (if any) have performed the calculations correctly?

20.2 Ohm's Law: Resistance and Simple Circuits

5. If the voltage across a fixed resistance is doubled, what happens to the current?
  a. It doubles.
  b. It halves.
  c. It stays the same.
  d. The current cannot be determined.

6. The table below gives the voltages and currents recorded across a resistor.

Table 20.4

Voltage (V)	2.50	5.00	7.50	10.00	12.50
Current (A)	0.69	1.38	2.09	2.76	3.49

  a. Plot the graph and comment on the shape.
  b. Calculate the value of the resistor.

7. What is the resistance of a bulb if the current in it is 1.25 A when a 4 V voltage supply is connected to it? If the voltage supply is increased to 7 V, what will be the current in the bulb?

20.3 Resistance and Resistivity

8. Which of the following affect the resistivity of a wire?
  a. length
  b. area of cross section
  c. material
  d. all of the above

9. The lengths and diameters of four wires are given as shown.

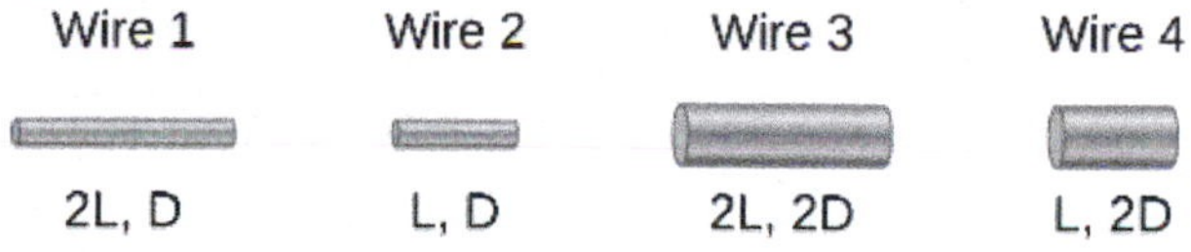

Figure 20.48

If the four wires are made from the same material, which of the following is true? Select *two* answers.

  a. Resistance of Wire 3 > Resistance of Wire 2
  b. Resistance of Wire 1 > Resistance of Wire 2
  c. Resistance of Wire 1 < Resistance of Wire 4
  d. Resistance of Wire 4 < Resistance of Wire 3

10. Suppose the resistance of a wire is R Ω. What will be the resistance of another wire of the same material having the same length but double the diameter?
  a. $R/2$
  b. $2R$
  c. $R/4$
  d. $4R$

11. The resistances of two wires having the same lengths and cross section areas are 3 Ω and 11 Ω. If the resistivity of the 3 Ω wire is 2.65×10^{-8} Ω·m, find the resistivity of the 1 Ω wire.

12. The lengths and diameters of three wires are given below. If they all have the same resistance, find the ratio of their resistivities.

Table 20.5

Wire	Length	Diameter
Wire 1	2 m	1 cm
Wire 2	1 m	0.5 cm
Wire 3	1 m	1 cm

13. Suppose the resistance of a wire is 2 Ω. If the wire is stretched to three times its length, what will be its resistance? Assume that the volume does not change.

20.4 Electric Power and Energy

14.

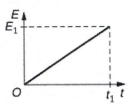

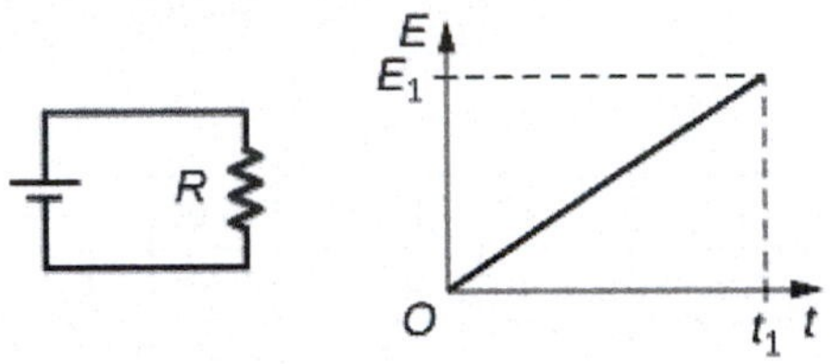

Figure 20.49 The circuit shown contains a resistor R connected to a voltage supply. The graph shows the total energy E dissipated by the resistance as a function of time. Which of the following shows the corresponding graph for double resistance, i.e., if R is replaced by $2R$?

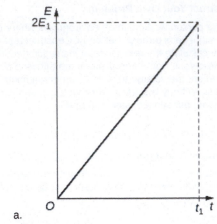

a.

Figure 20.50

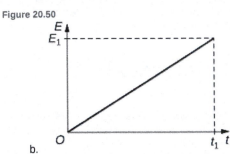

b.

Figure 20.51

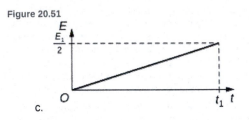

c.

Figure 20.52

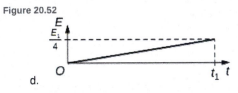

d.

Figure 20.53

15. What will be the ratio of the resistance of a 120 W, 220 V lamp to that of a 100 W, 110 V lamp?

21 CIRCUITS, BIOELECTRICITY, AND DC INSTRUMENTS

Figure 21.1 Electric circuits in a computer allow large amounts of data to be quickly and accurately analyzed.. (credit: Airman 1st Class Mike Meares, United States Air Force)

Chapter Outline

21.1. Resistors in Series and Parallel

21.2. Electromotive Force: Terminal Voltage

21.3. Kirchhoff's Rules

21.4. DC Voltmeters and Ammeters

21.5. Null Measurements

21.6. DC Circuits Containing Resistors and Capacitors

Connection for AP® Courses

Electric circuits are commonplace in our everyday lives. Some circuits are simple, such as those in flashlights while others are extremely complex, such as those used in supercomputers. This chapter takes the topic of electric circuits a step beyond simple circuits by addressing both changes that result from interactions between systems (Big Idea 4) and constraints on such changes due to laws of conservation (Big Idea 5). When the circuit is purely resistive, everything in this chapter applies to both DC and AC. However, matters become more complex when capacitance is involved. We do consider what happens when capacitors are connected to DC voltage sources, but the interaction of capacitors (and other nonresistive devices) with AC sources is left for a later chapter. In addition, a number of important DC instruments, such as meters that measure voltage and current, are covered in this chapter.

Information and examples presented in the chapter examine cause-effect relationships inherent in interactions involving electrical systems. The electrical properties of an electric circuit can change due to other systems (Enduring Understanding 4.E). More specifically, values of currents and potential differences in electric circuits depend on arrangements of individual circuit components (Essential Knowledge 4.E.5). In this chapter several series and parallel combinations of resistors are discussed and their effects on currents and potential differences are analyzed.

In electric circuits the total energy (Enduring Understanding 5.B) and the total electric charge (Enduring Understanding 5.C) are conserved. Kirchoff's rules describe both, energy conservation (Essential Knowledge 5.B.9) and charge conservation (Essential

Knowledge 5.C.3). Energy conservation is discussed in terms of the loop rule which specifies that the potential around any closed circuit path must be zero. Charge conservation is applied as conservation of current by equating the sum of all currents entering a junction to the sum of all currents leaving the junction (also known as the junction rule). Kirchoff's rules are used to calculate currents and potential differences in circuits that combine resistors in series and parallel, and resistors and capacitors.

The concepts in this chapter support:

Big Idea 4 Interactions between systems can result in changes in those systems.

Enduring Understanding 4.E The electric and magnetic properties of a system can change in response to the presence of, or changes in, other objects or systems.

Essential Knowledge 4.E.5 The values of currents and electric potential differences in an electric circuit are determined by the properties and arrangement of the individual circuit elements such as sources of emf, resistors, and capacitors.

Big Idea 5 Changes that occur as a result of interactions are constrained by conservation laws.

Enduring Understanding 5.B The energy of a system is conserved.

Essential Knowledge 5.B.9 Kirchhoff's loop rule describes conservation of energy in electrical circuits.

Enduring Understanding 5.C The electric charge of a system is conserved.

Essential Knowledge 5.C.3 Kirchhoff's junction rule describes the conservation of electric charge in electrical circuits. Since charge is conserved, current must be conserved at each junction in the circuit. Examples should include circuits that combine resistors in series and parallel.

21.1 Resistors in Series and Parallel

Learning Objectives

By the end of this section, you will be able to:

- Draw a circuit with resistors in parallel and in series.
- Use Ohm's law to calculate the voltage drop across a resistor when current passes through it.
- Contrast the way total resistance is calculated for resistors in series and in parallel.
- Explain why total resistance of a parallel circuit is less than the smallest resistance of any of the resistors in that circuit.
- Calculate total resistance of a circuit that contains a mixture of resistors connected in series and in parallel.

The information presented in this section supports the following AP® learning objectives and science practices:

- **4.E.5.1** The student is able to make and justify a quantitative prediction of the effect of a change in values or arrangements of one or two circuit elements on the currents and potential differences in a circuit containing a small number of sources of emf, resistors, capacitors, and switches in series and/or parallel. **(S.P. 2.2, 6.4)**
- **4.E.5.2** The student is able to make and justify a qualitative prediction of the effect of a change in values or arrangements of one or two circuit elements on currents and potential differences in a circuit containing a small number of sources of emf, resistors, capacitors, and switches in series and/or parallel. **(S.P. 6.1, 6.4)**
- **4.E.5.3** The student is able to plan data collection strategies and perform data analysis to examine the values of currents and potential differences in an electric circuit that is modified by changing or rearranging circuit elements, including sources of emf, resistors, and capacitors. **(S.P. 2.2, 4.2, 5.1)**
- **5.B.9.3** The student is able to apply conservation of energy (Kirchhoff's loop rule) in calculations involving the total electric potential difference for complete circuit loops with only a single battery and resistors in series and/or in, at most, one parallel branch. **(S.P. 2.2, 6.4, 7.2)**

Most circuits have more than one component, called a **resistor** that limits the flow of charge in the circuit. A measure of this limit on charge flow is called **resistance**. The simplest combinations of resistors are the series and parallel connections illustrated in **Figure 21.2**. The total resistance of a combination of resistors depends on both their individual values and how they are connected.

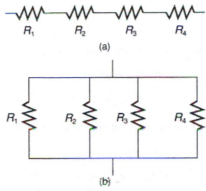

(a)

(b)

Figure 21.2 (a) A series connection of resistors. (b) A parallel connection of resistors.

Resistors in Series

When are resistors in **series**? Resistors are in series whenever the flow of charge, called the **current**, must flow through devices sequentially. For example, if current flows through a person holding a screwdriver and into the Earth, then R_1 in **Figure 21.2(a)** could be the resistance of the screwdriver's shaft, R_2 the resistance of its handle, R_3 the person's body resistance, and R_4 the resistance of her shoes.

Figure 21.3 shows resistors in series connected to a **voltage** source. It seems reasonable that the total resistance is the sum of the individual resistances, considering that the current has to pass through each resistor in sequence. (This fact would be an advantage to a person wishing to avoid an electrical shock, who could reduce the current by wearing high-resistance rubber-soled shoes. It could be a disadvantage if one of the resistances were a faulty high-resistance cord to an appliance that would reduce the operating current.)

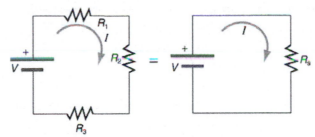

Figure 21.3 Three resistors connected in series to a battery (left) and the equivalent single or series resistance (right).

To verify that resistances in series do indeed add, let us consider the loss of electrical power, called a **voltage drop**, in each resistor in **Figure 21.3**.

According to **Ohm's law**, the voltage drop, V, across a resistor when a current flows through it is calculated using the equation $V = IR$, where I equals the current in amps (A) and R is the resistance in ohms (Ω). Another way to think of this is that V is the voltage necessary to make a current I flow through a resistance R.

So the voltage drop across R_1 is $V_1 = IR_1$, that across R_2 is $V_2 = IR_2$, and that across R_3 is $V_3 = IR_3$. The sum of these voltages equals the voltage output of the source; that is,

$$V = V_1 + V_2 + V_3.\tag{21.1}$$

This equation is based on the conservation of energy and conservation of charge. Electrical potential energy can be described by the equation $PE = qV$, where q is the electric charge and V is the voltage. Thus the energy supplied by the source is qV, while that dissipated by the resistors is

$$qV_1 + qV_2 + qV_3.\tag{21.2}$$

Connections: Conservation Laws

The derivations of the expressions for series and parallel resistance are based on the laws of conservation of energy and conservation of charge, which state that total charge and total energy are constant in any process. These two laws are directly involved in all electrical phenomena and will be invoked repeatedly to explain both specific effects and the general behavior of electricity.

These energies must be equal, because there is no other source and no other destination for energy in the circuit. Thus, $qV = qV_1 + qV_2 + qV_3$. The charge q cancels, yielding $V = V_1 + V_2 + V_3$, as stated. (Note that the same amount of charge passes through the battery and each resistor in a given amount of time, since there is no capacitance to store charge, there is no place for charge to leak, and charge is conserved.)

Now substituting the values for the individual voltages gives

$$V = IR_1 + IR_2 + IR_3 = I(R_1 + R_2 + R_3). \tag{21.3}$$

Note that for the equivalent single series resistance R_s, we have

$$V = IR_s. \tag{21.4}$$

This implies that the total or equivalent series resistance R_s of three resistors is $R_s = R_1 + R_2 + R_3$.

This logic is valid in general for any number of resistors in series; thus, the total resistance R_s of a series connection is

$$R_s = R_1 + R_2 + R_3 + ..., \tag{21.5}$$

as proposed. Since all of the current must pass through each resistor, it experiences the resistance of each, and resistances in series simply add up.

Example 21.1 Calculating Resistance, Current, Voltage Drop, and Power Dissipation: Analysis of a Series Circuit

Suppose the voltage output of the battery in **Figure 21.3** is 12.0 V, and the resistances are $R_1 = 1.00 \ \Omega$,

$R_2 = 6.00 \ \Omega$, and $R_3 = 13.0 \ \Omega$. (a) What is the total resistance? (b) Find the current. (c) Calculate the voltage drop in each resistor, and show these add to equal the voltage output of the source. (d) Calculate the power dissipated by each resistor. (e) Find the power output of the source, and show that it equals the total power dissipated by the resistors.

Strategy and Solution for (a)

The total resistance is simply the sum of the individual resistances, as given by this equation:

$$
\begin{aligned}
R_s &= R_1 + R_2 + R_3 \\
&= 1.00 \ \Omega + 6.00 \ \Omega + 13.0 \ \Omega \\
&= 20.0 \ \Omega.
\end{aligned}
\tag{21.6}
$$

Strategy and Solution for (b)

The current is found using Ohm's law, $V = IR$. Entering the value of the applied voltage and the total resistance yields the current for the circuit:

$$I = \frac{V}{R_s} = \frac{12.0 \text{ V}}{20.0 \ \Omega} = 0.600 \text{ A}. \tag{21.7}$$

Strategy and Solution for (c)

The voltage—or IR drop—in a resistor is given by Ohm's law. Entering the current and the value of the first resistance yields

$$V_1 = IR_1 = (0.600 \text{ A})(1.0 \ \Omega) = 0.600 \text{ V}. \tag{21.8}$$

Similarly,

$$V_2 = IR_2 = (0.600 \text{ A})(6.0 \ \Omega) = 3.60 \text{ V} \tag{21.9}$$

and

$$V_3 = IR_3 = (0.600 \text{ A})(13.0 \ \Omega) = 7.80 \text{ V}. \tag{21.10}$$

Discussion for (c)

The three IR drops add to 12.0 V, as predicted:

$$V_1 + V_2 + V_3 = (0.600 + 3.60 + 7.80) \text{ V} = 12.0 \text{ V}. \tag{21.11}$$

Strategy and Solution for (d)

The easiest way to calculate power in watts (W) dissipated by a resistor in a DC circuit is to use **Joule's law**, $P = IV$, where P is electric power. In this case, each resistor has the same full current flowing through it. By substituting Ohm's law $V = IR$ into Joule's law, we get the power dissipated by the first resistor as

$$P_1 = I^2 R_1 = (0.600 \text{ A})^2 (1.00 \ \Omega) = 0.360 \text{ W}. \tag{21.12}$$

Similarly,

$$P_2 = I^2 R_2 = (0.600 \text{ A})^2 (6.00 \ \Omega) = 2.16 \text{ W} \tag{21.13}$$

and

$$P_3 = I^2 R_3 = (0.600 \text{ A})^2 (13.0 \ \Omega) = 4.68 \text{ W}. \tag{21.14}$$

Discussion for (d)

Power can also be calculated using either $P = IV$ or $P = \dfrac{V^2}{R}$, where V is the voltage drop across the resistor (not the full voltage of the source). The same values will be obtained.

Strategy and Solution for (e)

The easiest way to calculate power output of the source is to use $P = IV$, where V is the source voltage. This gives

$$P = (0.600 \text{ A})(12.0 \text{ V}) = 7.20 \text{ W}. \tag{21.15}$$

Discussion for (e)

Note, coincidentally, that the total power dissipated by the resistors is also 7.20 W, the same as the power put out by the source. That is,

$$P_1 + P_2 + P_3 = (0.360 + 2.16 + 4.68) \text{ W} = 7.20 \text{ W}. \tag{21.16}$$

Power is energy per unit time (watts), and so conservation of energy requires the power output of the source to be equal to the total power dissipated by the resistors.

Major Features of Resistors in Series

1. Series resistances add: $R_s = R_1 + R_2 + R_3 + ...$

2. The same current flows through each resistor in series.

3. Individual resistors in series do not get the total source voltage, but divide it.

Resistors in Parallel

Figure 21.4 shows resistors in **parallel**, wired to a voltage source. Resistors are in parallel when each resistor is connected directly to the voltage source by connecting wires having negligible resistance. Each resistor thus has the full voltage of the source applied to it.

Each resistor draws the same current it would if it alone were connected to the voltage source (provided the voltage source is not overloaded). For example, an automobile's headlights, radio, and so on, are wired in parallel, so that they utilize the full voltage of the source and can operate completely independently. The same is true in your house, or any building. (See **Figure 21.4**(b).)

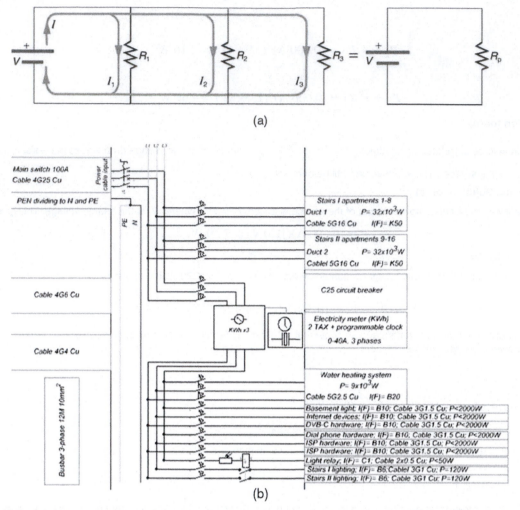

Figure 21.4 (a) Three resistors connected in parallel to a battery and the equivalent single or parallel resistance. (b) Electrical power setup in a house. (credit: Dmitry G, Wikimedia Commons)

To find an expression for the equivalent parallel resistance R_p, let us consider the currents that flow and how they are related to resistance. Since each resistor in the circuit has the full voltage, the currents flowing through the individual resistors are $I_1 = \frac{V}{R_1}$, $I_2 = \frac{V}{R_2}$, and $I_3 = \frac{V}{R_3}$. Conservation of charge implies that the total current I produced by the source is the sum of these currents:

$$I = I_1 + I_2 + I_3. \tag{21.17}$$

Substituting the expressions for the individual currents gives

$$I = \frac{V}{R_1} + \frac{V}{R_2} + \frac{V}{R_3} = V\left(\frac{1}{R_1} + \frac{1}{R_2} + \frac{1}{R_3}\right). \tag{21.18}$$

Note that Ohm's law for the equivalent single resistance gives

$$I = \frac{V}{R_p} = V\left(\frac{1}{R_p}\right). \tag{21.19}$$

The terms inside the parentheses in the last two equations must be equal. Generalizing to any number of resistors, the total resistance R_p of a parallel connection is related to the individual resistances by

$$\frac{1}{R_p} = \frac{1}{R_1} + \frac{1}{R_2} + \frac{1}{R_{.3}} + \dots \tag{21.20}$$

This relationship results in a total resistance R_p that is less than the smallest of the individual resistances. (This is seen in the next example.) When resistors are connected in parallel, more current flows from the source than would flow for any of them individually, and so the total resistance is lower.

Example 21.2 Calculating Resistance, Current, Power Dissipation, and Power Output: Analysis of a Parallel Circuit

Let the voltage output of the battery and resistances in the parallel connection in **Figure 21.4** be the same as the previously considered series connection: $V = 12.0 \text{ V}$, $R_1 = 1.00 \ \Omega$, $R_2 = 6.00 \ \Omega$, and $R_3 = 13.0 \ \Omega$. (a) What is the total resistance? (b) Find the total current. (c) Calculate the currents in each resistor, and show these add to equal the total current output of the source. (d) Calculate the power dissipated by each resistor. (e) Find the power output of the source, and show that it equals the total power dissipated by the resistors.

Strategy and Solution for (a)

The total resistance for a parallel combination of resistors is found using the equation below. Entering known values gives

$$\frac{1}{R_p} = \frac{1}{R_1} + \frac{1}{R_2} + \frac{1}{R_3} = \frac{1}{1.00 \ \Omega} + \frac{1}{6.00 \ \Omega} + \frac{1}{13.0 \ \Omega}. \tag{21.21}$$

Thus,

$$\frac{1}{R_p} = \frac{1.00}{\Omega} + \frac{0.1667}{\Omega} + \frac{0.07692}{\Omega} = \frac{1.2436}{\Omega}. \tag{21.22}$$

(Note that in these calculations, each intermediate answer is shown with an extra digit.)

We must invert this to find the total resistance R_p. This yields

$$R_p = \frac{1}{1.2436} \ \Omega = 0.8041 \ \Omega. \tag{21.23}$$

The total resistance with the correct number of significant digits is $R_p = 0.804 \ \Omega$.

Discussion for (a)

R_p is, as predicted, less than the smallest individual resistance.

Strategy and Solution for (b)

The total current can be found from Ohm's law, substituting R_p for the total resistance. This gives

$$I = \frac{V}{R_p} = \frac{12.0 \text{ V}}{0.8041 \ \Omega} = 14.92 \text{ A}. \tag{21.24}$$

Discussion for (b)

Current I for each device is much larger than for the same devices connected in series (see the previous example). A circuit with parallel connections has a smaller total resistance than the resistors connected in series.

Strategy and Solution for (c)

The individual currents are easily calculated from Ohm's law, since each resistor gets the full voltage. Thus,

$$I_1 = \frac{V}{R_1} = \frac{12.0 \text{ V}}{1.00 \ \Omega} = 12.0 \text{ A}. \tag{21.25}$$

Similarly,

$$I_2 = \frac{V}{R_2} = \frac{12.0 \text{ V}}{6.00 \ \Omega} = 2.00 \text{ A} \tag{21.26}$$

and

$$I_3 = \frac{V}{R_3} = \frac{12.0 \text{ V}}{13.0 \ \Omega} = 0.92 \text{ A}. \tag{21.27}$$

Discussion for (c)

The total current is the sum of the individual currents:

$$I_1 + I_2 + I_3 = 14.92 \text{ A}. \tag{21.28}$$

This is consistent with conservation of charge.

Strategy and Solution for (d)

The power dissipated by each resistor can be found using any of the equations relating power to current, voltage, and resistance, since all three are known. Let us use $P = \frac{V^2}{R}$, since each resistor gets full voltage. Thus,

$$P_1 = \frac{V^2}{R_1} = \frac{(12.0 \text{ V})^2}{1.00 \ \Omega} = 144 \text{ W}.$$ (21.29)

Similarly,

$$P_2 = \frac{V^2}{R_2} = \frac{(12.0 \text{ V})^2}{6.00 \ \Omega} = 24.0 \text{ W}$$ (21.30)

and

$$P_3 = \frac{V^2}{R_3} = \frac{(12.0 \text{ V})^2}{13.0 \ \Omega} = 11.1 \text{ W}.$$ (21.31)

Discussion for (d)

The power dissipated by each resistor is considerably higher in parallel than when connected in series to the same voltage source.

Strategy and Solution for (e)

The total power can also be calculated in several ways. Choosing $P = IV$, and entering the total current, yields

$$P = IV = (14.92 \text{ A})(12.0 \text{ V}) = 179 \text{ W}.$$ (21.32)

Discussion for (e)

Total power dissipated by the resistors is also 179 W:

$$P_1 + P_2 + P_3 = 144 \text{ W} + 24.0 \text{ W} + 11.1 \text{ W} = 179 \text{ W}.$$ (21.33)

This is consistent with the law of conservation of energy.

Overall Discussion

Note that both the currents and powers in parallel connections are greater than for the same devices in series.

Major Features of Resistors in Parallel

1. Parallel resistance is found from $\frac{1}{R_p} = \frac{1}{R_1} + \frac{1}{R_2} + \frac{1}{R_3} + \dots$, and it is smaller than any individual resistance in the combination.

2. Each resistor in parallel has the same full voltage of the source applied to it. (Power distribution systems most often use parallel connections to supply the myriad devices served with the same voltage and to allow them to operate independently.)

3. Parallel resistors do not each get the total current; they divide it.

Combinations of Series and Parallel

More complex connections of resistors are sometimes just combinations of series and parallel. These are commonly encountered, especially when wire resistance is considered. In that case, wire resistance is in series with other resistances that are in parallel.

Combinations of series and parallel can be reduced to a single equivalent resistance using the technique illustrated in **Figure 21.5**. Various parts are identified as either series or parallel, reduced to their equivalents, and further reduced until a single resistance is left. The process is more time consuming than difficult.

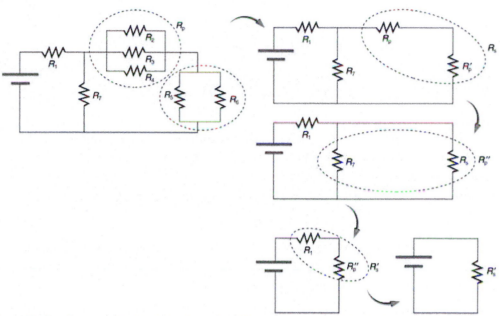

Figure 21.5 This combination of seven resistors has both series and parallel parts. Each is identified and reduced to an equivalent resistance, and these are further reduced until a single equivalent resistance is reached.

The simplest combination of series and parallel resistance, shown in **Figure 21.6**, is also the most instructive, since it is found in many applications. For example, R_1 could be the resistance of wires from a car battery to its electrical devices, which are in parallel. R_2 and R_3 could be the starter motor and a passenger compartment light. We have previously assumed that wire resistance is negligible, but, when it is not, it has important effects, as the next example indicates.

Example 21.3 Calculating Resistance, IR Drop, Current, and Power Dissipation: Combining Series and Parallel Circuits

Figure 21.6 shows the resistors from the previous two examples wired in a different way—a combination of series and parallel. We can consider R_1 to be the resistance of wires leading to R_2 and R_3. (a) Find the total resistance. (b) What is the IR drop in R_1? (c) Find the current I_2 through R_2. (d) What power is dissipated by R_2?

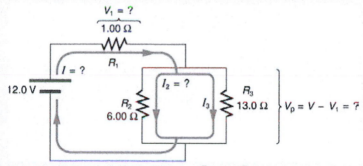

Figure 21.6 These three resistors are connected to a voltage source so that R_2 and R_3 are in parallel with one another and that combination is in series with R_1.

Strategy and Solution for (a)

To find the total resistance, we note that R_2 and R_3 are in parallel and their combination R_p is in series with R_1. Thus the total (equivalent) resistance of this combination is

$$R_{\text{tot}} = R_1 + R_p. \tag{21.34}$$

First, we find R_p using the equation for resistors in parallel and entering known values:

$$\frac{1}{R_p} = \frac{1}{R_2} + \frac{1}{R_3} = \frac{1}{6.00 \ \Omega} + \frac{1}{13.0 \ \Omega} = \frac{0.2436}{\Omega}. \tag{21.35}$$

Inverting gives

$$R_p = \frac{1}{0.2436} \ \Omega \ = 4.11 \ \Omega \ . \tag{21.36}$$

So the total resistance is

$$R_{tot} = R_1 + R_p = 1.00 \ \Omega \ + 4.11 \ \Omega \ = 5.11 \ \Omega \ . \tag{21.37}$$

Discussion for (a)

The total resistance of this combination is intermediate between the pure series and pure parallel values ($20.0 \ \Omega$ and $0.804 \ \Omega$, respectively) found for the same resistors in the two previous examples.

Strategy and Solution for (b)

To find the IR drop in R_1 , we note that the full current I flows through R_1 . Thus its IR drop is

$$V_1 = IR_1. \tag{21.38}$$

We must find I before we can calculate V_1 . The total current I is found using Ohm's law for the circuit. That is,

$$I = \frac{V}{R_{tot}} = \frac{12.0 \ V}{5.11 \ \Omega} = 2.35 \ A. \tag{21.39}$$

Entering this into the expression above, we get

$$V_1 = IR_1 = (2.35 \ A)(1.00 \ \Omega \ ) = 2.35 \ V. \tag{21.40}$$

Discussion for (b)

The voltage applied to R_2 and R_3 is less than the total voltage by an amount V_1 . When wire resistance is large, it can significantly affect the operation of the devices represented by R_2 and R_3 .

Strategy and Solution for (c)

To find the current through R_2 , we must first find the voltage applied to it. We call this voltage V_p , because it is applied to a parallel combination of resistors. The voltage applied to both R_2 and R_3 is reduced by the amount V_1 , and so it is

$$V_p = V - V_1 = 12.0 \ V - 2.35 \ V = 9.65 \ V. \tag{21.41}$$

Now the current I_2 through resistance R_2 is found using Ohm's law:

$$I_2 = \frac{V_p}{R_2} = \frac{9.65 \ V}{6.00 \ \Omega} = 1.61 \ A. \tag{21.42}$$

Discussion for (c)

The current is less than the 2.00 A that flowed through R_2 when it was connected in parallel to the battery in the previous parallel circuit example.

Strategy and Solution for (d)

The power dissipated by R_2 is given by

$$P_2 = (I_2)^2 R_2 = (1.61 \ A)^2 (6.00 \ \Omega \ ) = 15.5 \ W. \tag{21.43}$$

Discussion for (d)

The power is less than the 24.0 W this resistor dissipated when connected in parallel to the 12.0-V source.

Applying the Science Practices: Circuit Construction Kit (DC only)

Plan an experiment to analyze the effect on currents and potential differences due to rearrangement of resistors and variations in voltage sources. Your experimental investigation should include data collection for at least five different scenarios of rearranged resistors (i.e., several combinations of series and parallel) and three scenarios of different voltage sources.

Practical Implications

One implication of this last example is that resistance in wires reduces the current and power delivered to a resistor. If wire resistance is relatively large, as in a worn (or a very long) extension cord, then this loss can be significant. If a large current is drawn, the IR drop in the wires can also be significant.

For example, when you are rummaging in the refrigerator and the motor comes on, the refrigerator light dims momentarily. Similarly, you can see the passenger compartment light dim when you start the engine of your car (although this may be due to resistance inside the battery itself).

What is happening in these high-current situations is illustrated in **Figure 21.7**. The device represented by R_3 has a very low resistance, and so when it is switched on, a large current flows. This increased current causes a larger IR drop in the wires represented by R_1, reducing the voltage across the light bulb (which is R_2), which then dims noticeably.

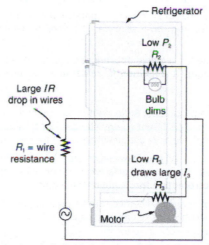

Figure 21.7 Why do lights dim when a large appliance is switched on? The answer is that the large current the appliance motor draws causes a significant IR drop in the wires and reduces the voltage across the light.

Check Your Understanding

Can any arbitrary combination of resistors be broken down into series and parallel combinations? See if you can draw a circuit diagram of resistors that cannot be broken down into combinations of series and parallel.

Solution
No, there are many ways to connect resistors that are not combinations of series and parallel, including loops and junctions. In such cases Kirchhoff's rules, to be introduced in **Kirchhoff's Rules**, will allow you to analyze the circuit.

Problem-Solving Strategies for Series and Parallel Resistors

1. Draw a clear circuit diagram, labeling all resistors and voltage sources. This step includes a list of the knowns for the problem, since they are labeled in your circuit diagram.

2. Identify exactly what needs to be determined in the problem (identify the unknowns). A written list is useful.

3. Determine whether resistors are in series, parallel, or a combination of both series and parallel. Examine the circuit diagram to make this assessment. Resistors are in series if the same current must pass sequentially through them.

4. Use the appropriate list of major features for series or parallel connections to solve for the unknowns. There is one list for series and another for parallel. If your problem has a combination of series and parallel, reduce it in steps by considering individual groups of series or parallel connections, as done in this module and the examples. Special note: When finding R, the reciprocal must be taken with care.

5. Check to see whether the answers are reasonable and consistent. Units and numerical results must be reasonable. Total series resistance should be greater, whereas total parallel resistance should be smaller, for example. Power should be greater for the same devices in parallel compared with series, and so on.

21.2 Electromotive Force: Terminal Voltage

Learning Objectives

By the end of this section, you will be able to:

- Compare and contrast the voltage and the electromagnetic force of an electric power source.
- Describe what happens to the terminal voltage, current, and power delivered to a load as internal resistance of the

voltage source increases.
- Explain why it is beneficial to use more than one voltage source connected in parallel.

The information presented in this section supports the following AP® learning objectives and science practices:

- **5.B.9.7** The student is able to refine and analyze a scientific question for an experiment using Kirchhoff's loop rule for circuits that includes determination of internal resistance of the battery and analysis of a nonohmic resistor. **(S.P. 4.1, 4.2, 5.1, 5.3)**

When you forget to turn off your car lights, they slowly dim as the battery runs down. Why don't they simply blink off when the battery's energy is gone? Their gradual dimming implies that battery output voltage decreases as the battery is depleted.

Furthermore, if you connect an excessive number of 12-V lights in parallel to a car battery, they will be dim even when the battery is fresh and even if the wires to the lights have very low resistance. This implies that the battery's output voltage is reduced by the overload.

The reason for the decrease in output voltage for depleted or overloaded batteries is that all voltage sources have two fundamental parts—a source of electrical energy and an **internal resistance**. Let us examine both.

Electromotive Force

You can think of many different types of voltage sources. Batteries themselves come in many varieties. There are many types of mechanical/electrical generators, driven by many different energy sources, ranging from nuclear to wind. Solar cells create voltages directly from light, while thermoelectric devices create voltage from temperature differences.

A few voltage sources are shown in **Figure 21.8**. All such devices create a **potential difference** and can supply current if connected to a resistance. On the small scale, the potential difference creates an electric field that exerts force on charges, causing current. We thus use the name **electromotive force**, abbreviated emf.

Emf is not a force at all; it is a special type of potential difference. To be precise, the electromotive force (emf) is the potential difference of a source when no current is flowing. Units of emf are volts.

Figure 21.8 A variety of voltage sources (clockwise from top left): the Brazos Wind Farm in Fluvanna, Texas (credit: Leaflet, Wikimedia Commons); the Krasnoyarsk Dam in Russia (credit: Alex Polezhaev); a solar farm (credit: U.S. Department of Energy); and a group of nickel metal hydride batteries (credit: Tiaa Monto). The voltage output of each depends on its construction and load, and equals emf only if there is no load.

Electromotive force is directly related to the source of potential difference, such as the particular combination of chemicals in a battery. However, emf differs from the voltage output of the device when current flows. The voltage across the terminals of a battery, for example, is less than the emf when the battery supplies current, and it declines further as the battery is depleted or loaded down. However, if the device's output voltage can be measured without drawing current, then output voltage will equal emf (even for a very depleted battery).

Internal Resistance

As noted before, a 12-V truck battery is physically larger, contains more charge and energy, and can deliver a larger current than a 12-V motorcycle battery. Both are lead-acid batteries with identical emf, but, because of its size, the truck battery has a smaller internal resistance r. Internal resistance is the inherent resistance to the flow of current within the source itself.

Figure 21.9 is a schematic representation of the two fundamental parts of any voltage source. The emf (represented by a script E in the figure) and internal resistance r are in series. The smaller the internal resistance for a given emf, the more current and the more power the source can supply.

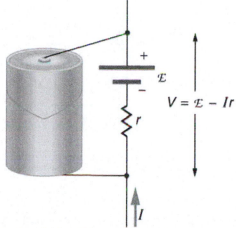

$$V = \mathcal{E} - Ir$$

Figure 21.9 Any voltage source (in this case, a carbon-zinc dry cell) has an emf related to its source of potential difference, and an internal resistance r related to its construction. (Note that the script E stands for emf.). Also shown are the output terminals across which the terminal voltage V is measured. Since $V = \text{emf} - Ir$, terminal voltage equals emf only if there is no current flowing.

The internal resistance r can behave in complex ways. As noted, r increases as a battery is depleted. But internal resistance may also depend on the magnitude and direction of the current through a voltage source, its temperature, and even its history. The internal resistance of rechargeable nickel-cadmium cells, for example, depends on how many times and how deeply they have been depleted.

Things Great and Small: The Submicroscopic Origin of Battery Potential

Various types of batteries are available, with emfs determined by the combination of chemicals involved. We can view this as a molecular reaction (what much of chemistry is about) that separates charge.

The lead-acid battery used in cars and other vehicles is one of the most common types. A single cell (one of six) of this battery is seen in **Figure 21.10**. The cathode (positive) terminal of the cell is connected to a lead oxide plate, while the anode (negative) terminal is connected to a lead plate. Both plates are immersed in sulfuric acid, the electrolyte for the system.

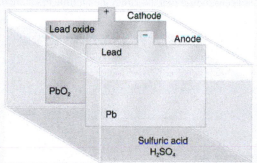

Figure 21.10 Artist's conception of a lead-acid cell. Chemical reactions in a lead-acid cell separate charge, sending negative charge to the anode, which is connected to the lead plates. The lead oxide plates are connected to the positive or cathode terminal of the cell. Sulfuric acid conducts the charge as well as participating in the chemical reaction.

The details of the chemical reaction are left to the reader to pursue in a chemistry text, but their results at the molecular level help explain the potential created by the battery. **Figure 21.11** shows the result of a single chemical reaction. Two electrons are placed on the anode, making it negative, provided that the cathode supplied two electrons. This leaves the cathode positively charged, because it has lost two electrons. In short, a separation of charge has been driven by a chemical reaction.

Note that the reaction will not take place unless there is a complete circuit to allow two electrons to be supplied to the cathode. Under many circumstances, these electrons come from the anode, flow through a resistance, and return to the cathode. Note also that since the chemical reactions involve substances with resistance, it is not possible to create the emf without an internal resistance.

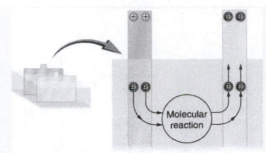

Figure 21.11 Artist's conception of two electrons being forced onto the anode of a cell and two electrons being removed from the cathode of the cell. The chemical reaction in a lead-acid battery places two electrons on the anode and removes two from the cathode. It requires a closed circuit to proceed, since the two electrons must be supplied to the cathode.

Why are the chemicals able to produce a unique potential difference? Quantum mechanical descriptions of molecules, which take into account the types of atoms and numbers of electrons in them, are able to predict the energy states they can have and the energies of reactions between them.

In the case of a lead-acid battery, an energy of 2 eV is given to each electron sent to the anode. Voltage is defined as the electrical potential energy divided by charge: $V = \frac{P_E}{q}$. An electron volt is the energy given to a single electron by a voltage of 1 V. So the voltage here is 2 V, since 2 eV is given to each electron. It is the energy produced in each molecular reaction that produces the voltage. A different reaction produces a different energy and, hence, a different voltage.

Terminal Voltage

The voltage output of a device is measured across its terminals and, thus, is called its **terminal voltage** V. Terminal voltage is given by

$$V = \text{emf} - Ir, \tag{21.44}$$

where r is the internal resistance and I is the current flowing at the time of the measurement.

I is positive if current flows away from the positive terminal, as shown in **Figure 21.9**. You can see that the larger the current, the smaller the terminal voltage. And it is likewise true that the larger the internal resistance, the smaller the terminal voltage.

Suppose a load resistance R_{load} is connected to a voltage source, as in **Figure 21.12**. Since the resistances are in series, the total resistance in the circuit is $R_{\text{load}} + r$. Thus the current is given by Ohm's law to be

$$I = \frac{\text{emf}}{R_{\text{load}} + r}. \tag{21.45}$$

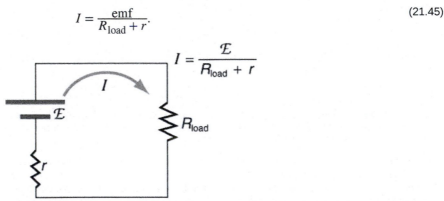

Figure 21.12 Schematic of a voltage source and its load R_{load}. Since the internal resistance r is in series with the load, it can significantly affect the terminal voltage and current delivered to the load. (Note that the script E stands for emf.)

We see from this expression that the smaller the internal resistance r, the greater the current the voltage source supplies to its load R_{load}. As batteries are depleted, r increases. If r becomes a significant fraction of the load resistance, then the current is significantly reduced, as the following example illustrates.

Example 21.4 Calculating Terminal Voltage, Power Dissipation, Current, and Resistance: Terminal Voltage and Load

A certain battery has a 12.0-V emf and an internal resistance of $0.100\ \Omega$. (a) Calculate its terminal voltage when connected to a $10.0\text{-}\Omega$ load. (b) What is the terminal voltage when connected to a $0.500\text{-}\Omega$ load? (c) What power does the $0.500\text{-}\Omega$ load dissipate? (d) If the internal resistance grows to $0.500\ \Omega$, find the current, terminal voltage, and power dissipated by a $0.500\text{-}\Omega$ load.

Strategy

The analysis above gave an expression for current when internal resistance is taken into account. Once the current is found, the terminal voltage can be calculated using the equation $V = \text{emf} - Ir$. Once current is found, the power dissipated by a resistor can also be found.

Solution for (a)

Entering the given values for the emf, load resistance, and internal resistance into the expression above yields

$$I = \frac{\text{emf}}{R_{\text{load}} + r} = \frac{12.0\text{ V}}{10.1\ \Omega} = 1.188\text{ A}. \tag{21.46}$$

Enter the known values into the equation $V = \text{emf} - Ir$ to get the terminal voltage:

$$\begin{aligned} V &= \text{emf} - Ir = 12.0\text{ V} - (1.188\text{ A})(0.100\ \Omega) \\ &= 11.9\text{ V}. \end{aligned} \tag{21.47}$$

Discussion for (a)

The terminal voltage here is only slightly lower than the emf, implying that $10.0\ \Omega$ is a light load for this particular battery.

Solution for (b)

Similarly, with $R_{\text{load}} = 0.500\ \Omega$, the current is

$$I = \frac{\text{emf}}{R_{\text{load}} + r} = \frac{12.0\text{ V}}{0.600\ \Omega} = 20.0\text{ A}. \tag{21.48}$$

The terminal voltage is now

$$\begin{aligned} V &= \text{emf} - Ir = 12.0\text{ V} - (20.0\text{ A})(0.100\ \Omega) \\ &= 10.0\text{ V}. \end{aligned} \tag{21.49}$$

Discussion for (b)

This terminal voltage exhibits a more significant reduction compared with emf, implying $0.500\ \Omega$ is a heavy load for this battery.

Solution for (c)

The power dissipated by the $0.500\text{-}\Omega$ load can be found using the formula $P = I^2 R$. Entering the known values gives

$$P_{\text{load}} = I^2 R_{\text{load}} = (20.0\text{ A})^2(0.500\ \Omega) = 2.00 \times 10^2\text{ W}. \tag{21.50}$$

Discussion for (c)

Note that this power can also be obtained using the expressions $\frac{V^2}{R}$ or IV, where V is the terminal voltage (10.0 V in this case).

Solution for (d)

Here the internal resistance has increased, perhaps due to the depletion of the battery, to the point where it is as great as the load resistance. As before, we first find the current by entering the known values into the expression, yielding

$$I = \frac{\text{emf}}{R_{\text{load}} + r} = \frac{12.0\text{ V}}{1.00\ \Omega} = 12.0\text{ A}. \tag{21.51}$$

Now the terminal voltage is

$$\begin{aligned} V &= \text{emf} - Ir = 12.0\text{ V} - (12.0\text{ A})(0.500\ \Omega) \\ &= 6.00\text{ V}, \end{aligned} \tag{21.52}$$

and the power dissipated by the load is

$$P_{\text{load}} = I^2 R_{\text{load}} = (12.0\text{ A})^2(0.500\ \Omega) = 72.0\text{ W}. \tag{21.53}$$

Discussion for (d)

We see that the increased internal resistance has significantly decreased terminal voltage, current, and power delivered to a load.

Applying the Science Practices: Internal Resistance

The internal resistance of a battery can be estimated using a simple activity. The circuit shown in the figure below includes a resistor R in series with a battery along with an ammeter and voltmeter to measure the current and voltage respectively.

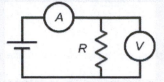

Figure 21.13

The currents and voltages measured for several R values are shown in the table below. Using the data given in the table and applying graphical analysis, determine the emf and internal resistance of the battery. Your response should clearly explain the method used to obtain the result.

Table 21.1

Resistance	Current (A)	Voltage (V)
R_1	3.53	4.24
R_2	2.07	4.97
R_3	1.46	5.27
R_4	1.13	5.43

Answer

Plot the measured currents and voltages on a graph. The terminal voltage of a battery is equal to the emf of the battery minus the voltage drop across the internal resistance of the battery or $V = emf - Ir$. Using this linear relationship and the plotted graph, the internal resistance and emf of the battery can be found. The graph for this case is shown below. The equation is $V = -0.50I + 6.0$ and hence the internal resistance will be equal to 0.5 Ω and emf will be equal to 6 V.

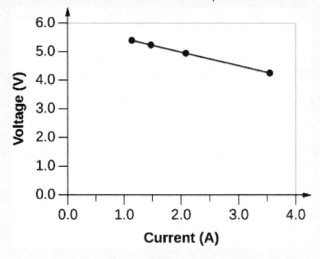

Figure 21.14

Battery testers, such as those in **Figure 21.15**, use small load resistors to intentionally draw current to determine whether the terminal voltage drops below an acceptable level. They really test the internal resistance of the battery. If internal resistance is high, the battery is weak, as evidenced by its low terminal voltage.

Figure 21.15 These two battery testers measure terminal voltage under a load to determine the condition of a battery. The large device is being used by a U.S. Navy electronics technician to test large batteries aboard the aircraft carrier USS *Nimitz* and has a small resistance that can dissipate large amounts of power. (credit: U.S. Navy photo by Photographer's Mate Airman Jason A. Johnston) The small device is used on small batteries and has a digital display to indicate the acceptability of their terminal voltage. (credit: Keith Williamson)

Some batteries can be recharged by passing a current through them in the direction opposite to the current they supply to a resistance. This is done routinely in cars and batteries for small electrical appliances and electronic devices, and is represented pictorially in **Figure 21.16**. The voltage output of the battery charger must be greater than the emf of the battery to reverse current through it. This will cause the terminal voltage of the battery to be greater than the emf, since $V = \text{emf} - Ir$, and I is now negative.

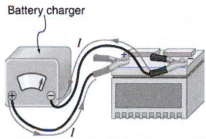

Figure 21.16 A car battery charger reverses the normal direction of current through a battery, reversing its chemical reaction and replenishing its chemical potential.

Multiple Voltage Sources

There are two voltage sources when a battery charger is used. Voltage sources connected in series are relatively simple. When voltage sources are in series, their internal resistances add and their emfs add algebraically. (See **Figure 21.17**.) Series connections of voltage sources are common—for example, in flashlights, toys, and other appliances. Usually, the cells are in series in order to produce a larger total emf.

But if the cells oppose one another, such as when one is put into an appliance backward, the total emf is less, since it is the algebraic sum of the individual emfs.

A battery is a multiple connection of voltaic cells, as shown in **Figure 21.18**. The disadvantage of series connections of cells is that their internal resistances add. One of the authors once owned a 1957 MGA that had two 6-V batteries in series, rather than a single 12-V battery. This arrangement produced a large internal resistance that caused him many problems in starting the engine.

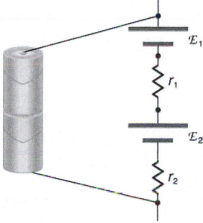

Figure 21.17 A series connection of two voltage sources. The emfs (each labeled with a script E) and internal resistances add, giving a total emf of $\text{emf}_1 + \text{emf}_2$ and a total internal resistance of $r_1 + r_2$.

Figure 21.18 Batteries are multiple connections of individual cells, as shown in this modern rendition of an old print. Single cells, such as AA or C cells, are commonly called batteries, although this is technically incorrect.

If the *series* connection of two voltage sources is made into a complete circuit with the emfs in opposition, then a current of

magnitude $I = \dfrac{(\text{emf}_1 - \text{emf}_2)}{r_1 + r_2}$ flows. See **Figure 21.19**, for example, which shows a circuit exactly analogous to the battery

charger discussed above. If two voltage sources in series with emfs in the same sense are connected to a load R_{load}, as in

Figure 21.20, then $I = \dfrac{(\text{emf}_1 + \text{emf}_2)}{r_1 + r_2 + R_{\text{load}}}$ flows.

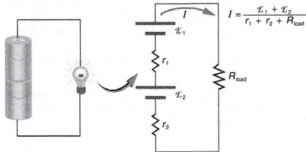

Figure 21.19 These two voltage sources are connected in series with their emfs in opposition. Current flows in the direction of the greater emf and is

limited to $I = \dfrac{(\text{emf}_1 - \text{emf}_2)}{r_1 + r_2}$ by the sum of the internal resistances. (Note that each emf is represented by script E in the figure.) A battery

charger connected to a battery is an example of such a connection. The charger must have a larger emf than the battery to reverse current through it.

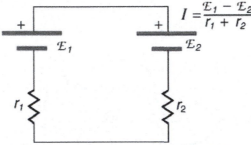

Figure 21.20 This schematic represents a flashlight with two cells (voltage sources) and a single bulb (load resistance) in series. The current that flows

is $I = \dfrac{(\text{emf}_1 + \text{emf}_2)}{r_1 + r_2 + R_{\text{load}}}$. (Note that each emf is represented by script E in the figure.)

Take-Home Experiment: Flashlight Batteries

Find a flashlight that uses several batteries and find new and old batteries. Based on the discussions in this module, predict the brightness of the flashlight when different combinations of batteries are used. Do your predictions match what you observe? Now place new batteries in the flashlight and leave the flashlight switched on for several hours. Is the flashlight still quite bright? Do the same with the old batteries. Is the flashlight as bright when left on for the same length of time with old and new batteries? What does this say for the case when you are limited in the number of available new batteries?

Figure 21.21 shows two voltage sources with identical emfs in parallel and connected to a load resistance. In this simple case, the total emf is the same as the individual emfs. But the total internal resistance is reduced, since the internal resistances are in parallel. The parallel connection thus can produce a larger current.

Here, $I = \dfrac{\text{emf}}{(r_{\text{tot}} + R_{\text{load}})}$ flows through the load, and r_{tot} is less than those of the individual batteries. For example, some

diesel-powered cars use two 12-V batteries in parallel; they produce a total emf of 12 V but can deliver the larger current needed to start a diesel engine.

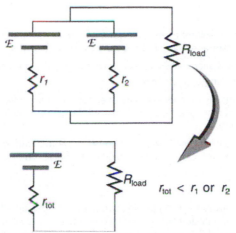

Figure 21.21 Two voltage sources with identical emfs (each labeled by script E) connected in parallel produce the same emf but have a smaller total internal resistance than the individual sources. Parallel combinations are often used to deliver more current. Here $I = \dfrac{\text{emf}}{\left(r_{\text{tot}} + R_{\text{load}}\right)}$ flows through the load.

Animals as Electrical Detectors

A number of animals both produce and detect electrical signals. Fish, sharks, platypuses, and echidnas (spiny anteaters) all detect electric fields generated by nerve activity in prey. Electric eels produce their own emf through biological cells (electric organs) called electroplaques, which are arranged in both series and parallel as a set of batteries.

Electroplaques are flat, disk-like cells; those of the electric eel have a voltage of 0.15 V across each one. These cells are usually located toward the head or tail of the animal, although in the case of the electric eel, they are found along the entire body. The electroplaques in the South American eel are arranged in 140 rows, with each row stretching horizontally along the body and containing 5,000 electroplaques. This can yield an emf of approximately 600 V, and a current of 1 A—deadly.

The mechanism for detection of external electric fields is similar to that for producing nerve signals in the cell through depolarization and repolarization—the movement of ions across the cell membrane. Within the fish, weak electric fields in the water produce a current in a gel-filled canal that runs from the skin to sensing cells, producing a nerve signal. The Australian platypus, one of the very few mammals that lay eggs, can detect fields of 30 $\dfrac{\text{mV}}{\text{m}}$, while sharks have been found to be able to sense a field in their snouts as small as 100 $\dfrac{\text{mV}}{\text{m}}$ (**Figure 21.22**). Electric eels use their own electric fields produced by the electroplaques to stun their prey or enemies.

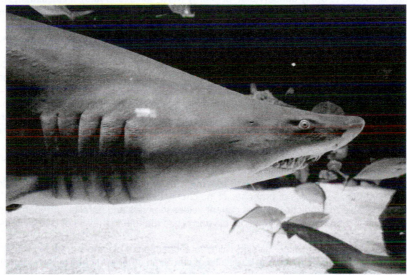

Figure 21.22 Sand tiger sharks (*Carcharias taurus*), like this one at the Minnesota Zoo, use electroreceptors in their snouts to locate prey. (credit: Jim Winstead, Flickr)

Solar Cell Arrays

Another example dealing with multiple voltage sources is that of combinations of solar cells—wired in both series and parallel combinations to yield a desired voltage and current. Photovoltaic generation (PV), the conversion of sunlight directly into electricity, is based upon the photoelectric effect, in which photons hitting the surface of a solar cell create an electric current in the cell.

Most solar cells are made from pure silicon—either as single-crystal silicon, or as a thin film of silicon deposited upon a glass or metal backing. Most single cells have a voltage output of about 0.5 V, while the current output is a function of the amount of sunlight upon the cell (the incident solar radiation—the insolation). Under bright noon sunlight, a current of about 100 mA/cm^2 of cell surface area is produced by typical single-crystal cells.

Individual solar cells are connected electrically in modules to meet electrical-energy needs. They can be wired together in series or in parallel—connected like the batteries discussed earlier. A solar-cell array or module usually consists of between 36 and 72 cells, with a power output of 50 W to 140 W.

The output of the solar cells is direct current. For most uses in a home, AC is required, so a device called an inverter must be used to convert the DC to AC. Any extra output can then be passed on to the outside electrical grid for sale to the utility.

Take-Home Experiment: Virtual Solar Cells

One can assemble a "virtual" solar cell array by using playing cards, or business or index cards, to represent a solar cell. Combinations of these cards in series and/or parallel can model the required array output. Assume each card has an output of 0.5 V and a current (under bright light) of 2 A. Using your cards, how would you arrange them to produce an output of 6 A at 3 V (18 W)?

Suppose you were told that you needed only 18 W (but no required voltage). Would you need more cards to make this arrangement?

21.3 Kirchhoff's Rules

Learning Objectives

By the end of this section, you will be able to:

- Analyze a complex circuit using Kirchhoff's rules, applying the conventions for determining the correct signs of various terms.

The information presented in this section supports the following AP® learning objectives and science practices:

- **5.B.9.1** The student is able to construct or interpret a graph of the energy changes within an electrical circuit with only a single battery and resistors in series and/or in, at most, one parallel branch as an application of the conservation of energy (Kirchhoff's loop rule). **(S.P. 1.1, 1.4)**
- **5.B.9.2** The student is able to apply conservation of energy concepts to the design of an experiment that will demonstrate the validity of Kirchhoff's loop rule in a circuit with only a battery and resistors either in series or in, at most, one pair of parallel branches. **(S.P. 4.2, 6.4, 7.2)**
- **5.B.9.3** The student is able to apply conservation of energy (Kirchhoff's loop rule) in calculations involving the total electric potential difference for complete circuit loops with only a single battery and resistors in series and/or in, at most, one parallel branch. **(S.P. 2.2, 6.4, 7.2)**
- **5.B.9.4** The student is able to analyze experimental data including an analysis of experimental uncertainty that will demonstrate the validity of Kirchhoff's loop rule. **(S.P. 5.1)**
- **5.B.9.5** The student is able to use conservation of energy principles (Kirchhoff's loop rule) to describe and make predictions regarding electrical potential difference, charge, and current in steady-state circuits composed of various combinations of resistors and capacitors. **(S.P. 6.4)**
- **5.C.3.1** The student is able to apply conservation of electric charge (Kirchhoff's junction rule) to the comparison of electric current in various segments of an electrical circuit with a single battery and resistors in series and in, at most, one parallel branch and predict how those values would change if configurations of the circuit are changed. **(S.P. 6.4, 7.2)**
- **5.C.3.2** The student is able to design an investigation of an electrical circuit with one or more resistors in which evidence of conservation of electric charge can be collected and analyzed. **(S.P. 4.1, 4.2, 5.1)**
- **5.C.3.3** The student is able to use a description or schematic diagram of an electrical circuit to calculate unknown values of current in various segments or branches of the circuit. **(S.P. 1.4, 2.2)**
- **5.C.3.4** The student is able to predict or explain current values in series and parallel arrangements of resistors and other branching circuits using Kirchhoff's junction rule and relate the rule to the law of charge conservation. **(S.P. 6.4, 7.2)**
- **5.C.3.5** The student is able to determine missing values and direction of electric current in branches of a circuit with resistors and NO capacitors from values and directions of current in other branches of the circuit through appropriate selection of nodes and application of the junction rule. **(S.P. 1.4, 2.2)**

Many complex circuits, such as the one in Figure 21.23, cannot be analyzed with the series-parallel techniques developed in **Resistors in Series and Parallel** and **Electromotive Force: Terminal Voltage**. There are, however, two circuit analysis rules that can be used to analyze any circuit, simple or complex. These rules are special cases of the laws of conservation of charge and conservation of energy. The rules are known as **Kirchhoff's rules**, after their inventor Gustav Kirchhoff (1824–1887).

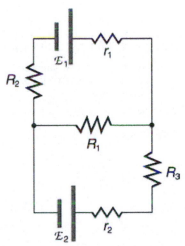

Figure 21.23 This circuit cannot be reduced to a combination of series and parallel connections. Kirchhoff's rules, special applications of the laws of conservation of charge and energy, can be used to analyze it. (Note: The script E in the figure represents electromotive force, emf.)

Kirchhoff's Rules

- Kirchhoff's first rule—the junction rule. The sum of all currents entering a junction must equal the sum of all currents leaving the junction.
- Kirchhoff's second rule—the loop rule. The algebraic sum of changes in potential around any closed circuit path (loop) must be zero.

Explanations of the two rules will now be given, followed by problem-solving hints for applying Kirchhoff's rules, and a worked example that uses them.

Kirchhoff's First Rule

Kirchhoff's first rule (the **junction rule**) is an application of the conservation of charge to a junction; it is illustrated in **Figure 21.24**. Current is the flow of charge, and charge is conserved; thus, whatever charge flows into the junction must flow out. Kirchhoff's first rule requires that $I_1 = I_2 + I_3$ (see figure). Equations like this can and will be used to analyze circuits and to solve circuit problems.

Making Connections: Conservation Laws

Kirchhoff's rules for circuit analysis are applications of **conservation laws** to circuits. The first rule is the application of conservation of charge, while the second rule is the application of conservation of energy. Conservation laws, even used in a specific application, such as circuit analysis, are so basic as to form the foundation of that application.

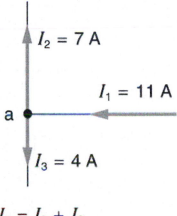

$$I_1 = I_2 + I_3$$

Figure 21.24 The junction rule. The diagram shows an example of Kirchhoff's first rule where the sum of the currents into a junction equals the sum of the currents out of a junction. In this case, the current going into the junction splits and comes out as two currents, so that $I_1 = I_2 + I_3$. Here I_1 must be 11 A, since I_2 is 7 A and I_3 is 4 A.

Kirchhoff's Second Rule

Kirchhoff's second rule (the **loop rule**) is an application of conservation of energy. The loop rule is stated in terms of potential, V, rather than potential energy, but the two are related since $PE_{elec} = qV$. Recall that **emf** is the potential difference of a source when no current is flowing. In a closed loop, whatever energy is supplied by emf must be transferred into other forms by devices in the loop, since there are no other ways in which energy can be transferred into or out of the circuit. **Figure 21.25** illustrates the changes in potential in a simple series circuit loop.

Kirchhoff's second rule requires $emf - Ir - IR_1 - IR_2 = 0$. Rearranged, this is $emf = Ir + IR_1 + IR_2$, which means the emf equals the sum of the IR (voltage) drops in the loop.

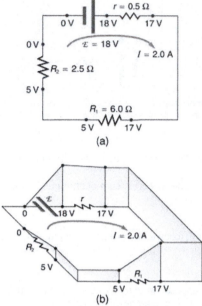

(a)

(b)

Figure 21.25 The loop rule. An example of Kirchhoff's second rule where the sum of the changes in potential around a closed loop must be zero. (a) In this standard schematic of a simple series circuit, the emf supplies 18 V, which is reduced to zero by the resistances, with 1 V across the internal resistance, and 12 V and 5 V across the two load resistances, for a total of 18 V. (b) This perspective view represents the potential as something like a roller coaster, where charge is raised in potential by the emf and lowered by the resistances. (Note that the script E stands for emf.)

Applying Kirchhoff's Rules

By applying Kirchhoff's rules, we generate equations that allow us to find the unknowns in circuits. The unknowns may be currents, emfs, or resistances. Each time a rule is applied, an equation is produced. If there are as many independent equations as unknowns, then the problem can be solved. There are two decisions you must make when applying Kirchhoff's rules. These decisions determine the signs of various quantities in the equations you obtain from applying the rules.

1. When applying Kirchhoff's first rule, the junction rule, you must label the current in each branch and decide in what direction it is going. For example, in **Figure 21.23**, **Figure 21.24**, and **Figure 21.25**, currents are labeled I_1, I_2, I_3, and I, and arrows indicate their directions. There is no risk here, for if you choose the wrong direction, the current will be of the correct magnitude but negative.

2. When applying Kirchhoff's second rule, the loop rule, you must identify a closed loop and decide in which direction to go around it, clockwise or counterclockwise. For example, in **Figure 21.25** the loop was traversed in the same direction as the current (clockwise). Again, there is no risk; going around the circuit in the opposite direction reverses the sign of every term in the equation, which is like multiplying both sides of the equation by -1.

Figure 21.26 and the following points will help you get the plus or minus signs right when applying the loop rule. Note that the resistors and emfs are traversed by going from a to b. In many circuits, it will be necessary to construct more than one loop. In traversing each loop, one needs to be consistent for the sign of the change in potential. (See **Example 21.5**.)

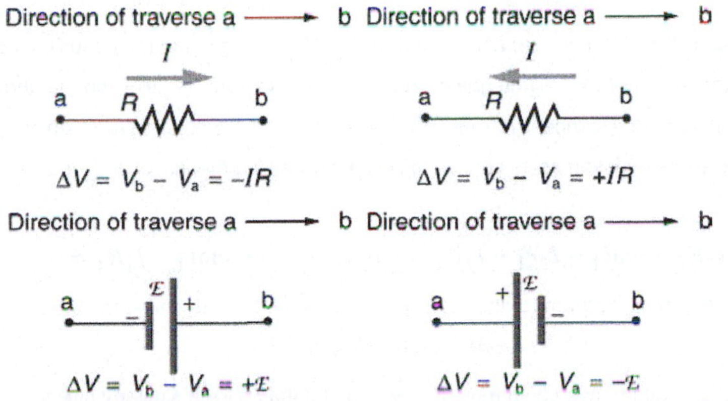

Direction of traverse a ——▶ b   Direction of traverse a ——▶ b

$$\Delta V = V_b - V_a = -IR \qquad \Delta V = V_b - V_a = +IR$$

Direction of traverse a ——▶ b   Direction of traverse a ——▶ b

$$\Delta V = V_b - V_a = +\mathcal{E} \qquad \Delta V = V_b - V_a = -\mathcal{E}$$

Figure 21.26 Each of these resistors and voltage sources is traversed from a to b. The potential changes are shown beneath each element and are explained in the text. (Note that the script E stands for emf.)

- When a resistor is traversed in the same direction as the current, the change in potential is $-IR$. (See **Figure 21.26**.)
- When a resistor is traversed in the direction opposite to the current, the change in potential is $+IR$. (See **Figure 21.26**.)
- When an emf is traversed from $-$ to $+$ (the same direction it moves positive charge), the change in potential is +emf. (See **Figure 21.26**.)
- When an emf is traversed from + to $-$ (opposite to the direction it moves positive charge), the change in potential is $-$emf. (See **Figure 21.26**.)

Example 21.5 Calculating Current: Using Kirchhoff's Rules

Find the currents flowing in the circuit in **Figure 21.27**.

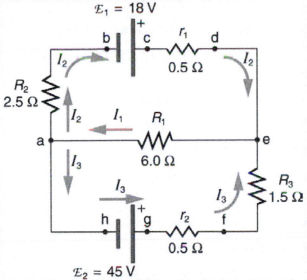

Figure 21.27 This circuit is similar to that in **Figure 21.23**, but the resistances and emfs are specified. (Each emf is denoted by script E.) The currents in each branch are labeled and assumed to move in the directions shown. This example uses Kirchhoff's rules to find the currents.

Strategy

This circuit is sufficiently complex that the currents cannot be found using Ohm's law and the series-parallel techniques—it is necessary to use Kirchhoff's rules. Currents have been labeled I_1, I_2, and I_3 in the figure and assumptions have been made about their directions. Locations on the diagram have been labeled with letters a through h. In the solution we will apply the junction and loop rules, seeking three independent equations to allow us to solve for the three unknown currents.

Solution

We begin by applying Kirchhoff's first or junction rule at point a. This gives

$$I_1 = I_2 + I_3, \tag{21.54}$$

since I_1 flows into the junction, while I_2 and I_3 flow out. Applying the junction rule at e produces exactly the same equation, so that no new information is obtained. This is a single equation with three unknowns—three independent equations are needed, and so the loop rule must be applied.

Now we consider the loop abcdea. Going from a to b, we traverse R_2 in the same (assumed) direction of the current I_2, and so the change in potential is $-I_2 R_2$. Then going from b to c, we go from $-$ to +, so that the change in potential is $+\text{emf}_1$. Traversing the internal resistance r_1 from c to d gives $-I_2 r_1$. Completing the loop by going from d to a again traverses a resistor in the same direction as its current, giving a change in potential of $-I_1 R_1$.

The loop rule states that the changes in potential sum to zero. Thus,

$$-I_2 R_2 + \text{emf}_1 - I_2 r_1 - I_1 R_1 = -I_2(R_2 + r_1) + \text{emf}_1 - I_1 R_1 = 0. \tag{21.55}$$

Substituting values from the circuit diagram for the resistances and emf, and canceling the ampere unit gives

$$-3I_2 + 18 - 6I_1 = 0. \tag{21.56}$$

Now applying the loop rule to aefgha (we could have chosen abcdefgha as well) similarly gives

$$+ I_1 R_1 + I_3 R_3 + I_3 r_2 - \text{emf}_2 = +I_1 R_1 + I_3(R_3 + r_2) - \text{emf}_2 = 0. \tag{21.57}$$

Note that the signs are reversed compared with the other loop, because elements are traversed in the opposite direction. With values entered, this becomes

$$+ 6I_1 + 2I_3 - 45 = 0. \tag{21.58}$$

These three equations are sufficient to solve for the three unknown currents. First, solve the second equation for I_2:

$$I_2 = 6 - 2I_1. \tag{21.59}$$

Now solve the third equation for I_3:

$$I_3 = 22.5 - 3I_1. \tag{21.60}$$

Substituting these two new equations into the first one allows us to find a value for I_1:

$$I_1 = I_2 + I_3 = (6 - 2I_1) + (22.5 - 3I_1) = 28.5 - 5I_1. \tag{21.61}$$

Combining terms gives

$$6I_1 = 28.5, \text{ and} \tag{21.62}$$

$$I_1 = 4.75 \text{ A}. \tag{21.63}$$

Substituting this value for I_1 back into the fourth equation gives

$$I_2 = 6 - 2I_1 = 6 - 9.50 \tag{21.64}$$

$$I_2 = -3.50 \text{ A}. \tag{21.65}$$

The minus sign means I_2 flows in the direction opposite to that assumed in **Figure 21.27**.

Finally, substituting the value for I_1 into the fifth equation gives

$$I_3 = 22.5 - 3I_1 = 22.5 - 14.25 \tag{21.66}$$

$$I_3 = 8.25 \text{ A}. \tag{21.67}$$

Discussion

Just as a check, we note that indeed $I_1 = I_2 + I_3$. The results could also have been checked by entering all of the values into the equation for the abcdefgha loop.

Problem-Solving Strategies for Kirchhoff's Rules

1. Make certain there is a clear circuit diagram on which you can label all known and unknown resistances, emfs, and currents. If a current is unknown, you must assign it a direction. This is necessary for determining the signs of potential changes. If you assign the direction incorrectly, the current will be found to have a negative value—no harm done.

2. Apply the junction rule to any junction in the circuit. Each time the junction rule is applied, you should get an equation with a current that does not appear in a previous application—if not, then the equation is redundant.

3. Apply the loop rule to as many loops as needed to solve for the unknowns in the problem. (There must be as many independent equations as unknowns.) To apply the loop rule, you must choose a direction to go around the loop. Then

carefully and consistently determine the signs of the potential changes for each element using the four bulleted points discussed above in conjunction with **Figure 21.26**.

4. Solve the simultaneous equations for the unknowns. This may involve many algebraic steps, requiring careful checking and rechecking.

5. Check to see whether the answers are reasonable and consistent. The numbers should be of the correct order of magnitude, neither exceedingly large nor vanishingly small. The signs should be reasonable—for example, no resistance should be negative. Check to see that the values obtained satisfy the various equations obtained from applying the rules. The currents should satisfy the junction rule, for example.

The material in this section is correct in theory. We should be able to verify it by making measurements of current and voltage. In fact, some of the devices used to make such measurements are straightforward applications of the principles covered so far and are explored in the next modules. As we shall see, a very basic, even profound, fact results—making a measurement alters the quantity being measured.

Check Your Understanding

Can Kirchhoff's rules be applied to simple series and parallel circuits or are they restricted for use in more complicated circuits that are not combinations of series and parallel?

Solution
Kirchhoff's rules can be applied to any circuit since they are applications to circuits of two conservation laws. Conservation laws are the most broadly applicable principles in physics. It is usually mathematically simpler to use the rules for series and parallel in simpler circuits so we emphasize Kirchhoff's rules for use in more complicated situations. But the rules for series and parallel can be derived from Kirchhoff's rules. Moreover, Kirchhoff's rules can be expanded to devices other than resistors and emfs, such as capacitors, and are one of the basic analysis devices in circuit analysis.

Making Connections: Parallel Resistors

A simple circuit shown below – with two parallel resistors and a voltage source – is implemented in a laboratory experiment with $\varepsilon = 6.00 \pm 0.02$ V and $R_1 = 4.8 \pm 0.1$ Ω and $R_2 = 9.6 \pm 0.1$ Ω. The values include an allowance for experimental uncertainties as they cannot be measured with perfect certainty. For example if you measure the value for a resistor a few times, you may get slightly different results. Hence values are expressed with some level of uncertainty.

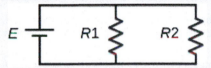

Figure 21.28

In the laboratory experiment the currents measured in the two resistors are $I_1 = 1.27$ A and $I_2 = 0.62$ A respectively. Let us examine these values using Kirchhoff's laws.

For the two loops,

$E - I_1 R_1 = 0$ or $I_1 = E/R_1$

$E - I_2 R_2 = 0$ or $I_2 = E/R_2$

Converting the given uncertainties for voltage and resistances into percentages, we get

$E = 6.00$ V $\pm 0.33\%$

$R_1 = 4.8$ $\Omega \pm 2.08\%$

$R_2 = 9.6$ $\Omega \pm 1.04\%$

We now find the currents for the two loops. While the voltage is divided by the resistance to find the current, uncertainties in voltage and resistance are directly added to find the uncertainty in the current value.

$I_1 = (6.00/4.8) \pm (0.33\% + 2.08\%)$

$= 1.25 \pm 2.4\%$

$= 1.25 \pm 0.03$ A

$I_2 = (6.00/9.6) \pm (0.33\% + 1.04\%)$

$= 0.63 \pm 1.4\%$

$= 0.63 \pm 0.01$ A

Finally you can check that the two measured values in this case are within the uncertainty ranges found for the currents. However there can also be additional experimental uncertainty in the measurements of currents.

21.4 DC Voltmeters and Ammeters

Voltmeters measure voltage, whereas **ammeters** measure current. Some of the meters in automobile dashboards, digital cameras, cell phones, and tuner-amplifiers are voltmeters or ammeters. (See **Figure 21.29**.) The internal construction of the simplest of these meters and how they are connected to the system they monitor give further insight into applications of series and parallel connections.

Figure 21.29 The fuel and temperature gauges (far right and far left, respectively) in this 1996 Volkswagen are voltmeters that register the voltage output of "sender" units, which are hopefully proportional to the amount of gasoline in the tank and the engine temperature. (credit: Christian Giersing)

Voltmeters are connected in parallel with whatever device's voltage is to be measured. A parallel connection is used because objects in parallel experience the same potential difference. (See **Figure 21.30**, where the voltmeter is represented by the symbol V.)

Ammeters are connected in series with whatever device's current is to be measured. A series connection is used because objects in series have the same current passing through them. (See **Figure 21.31**, where the ammeter is represented by the symbol A.)

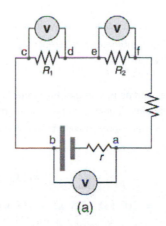

(a)

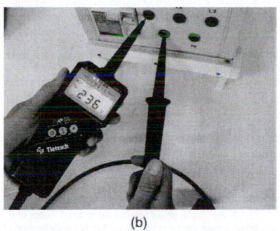

(b)

Figure 21.30 (a) To measure potential differences in this series circuit, the voltmeter (V) is placed in parallel with the voltage source or either of the resistors. Note that terminal voltage is measured between points a and b. It is not possible to connect the voltmeter directly across the emf without including its internal resistance, r . (b) A digital voltmeter in use. (credit: Messtechniker, Wikimedia Commons)

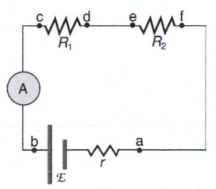

Figure 21.31 An ammeter (A) is placed in series to measure current. All of the current in this circuit flows through the meter. The ammeter would have the same reading if located between points d and e or between points f and a as it does in the position shown. (Note that the script capital E stands for emf, and r stands for the internal resistance of the source of potential difference.)

Analog Meters: Galvanometers

Analog meters have a needle that swivels to point at numbers on a scale, as opposed to **digital meters**, which have numerical readouts similar to a hand-held calculator. The heart of most analog meters is a device called a **galvanometer**, denoted by G. Current flow through a galvanometer, I_G , produces a proportional needle deflection. (This deflection is due to the force of a magnetic field upon a current-carrying wire.)

The two crucial characteristics of a given galvanometer are its resistance and current sensitivity. **Current sensitivity** is the current that gives a **full-scale deflection** of the galvanometer's needle, the maximum current that the instrument can measure. For example, a galvanometer with a current sensitivity of $50\ \mu A$ has a maximum deflection of its needle when $50\ \mu A$ flows through it, reads half-scale when $25\ \mu A$ flows through it, and so on.

If such a galvanometer has a $25\text{-}\Omega$ resistance, then a voltage of only $V = IR = (50 \text{ }\mu\text{A})(25 \text{ }\Omega) = 1.25 \text{ mV}$ produces a full-scale reading. By connecting resistors to this galvanometer in different ways, you can use it as either a voltmeter or ammeter that can measure a broad range of voltages or currents.

Galvanometer as Voltmeter

Figure 21.32 shows how a galvanometer can be used as a voltmeter by connecting it in series with a large resistance, R. The value of the resistance R is determined by the maximum voltage to be measured. Suppose you want 10 V to produce a full-scale deflection of a voltmeter containing a $25\text{-}\Omega$ galvanometer with a $50\text{-}\mu\text{A}$ sensitivity. Then 10 V applied to the meter must produce a current of $50 \text{ }\mu\text{A}$. The total resistance must be

$$R_{\text{tot}} = R + r = \frac{V}{I} = \frac{10 \text{ V}}{50 \text{ }\mu\text{A}} = 200 \text{ k}\Omega, \text{ or} \qquad (21.68)$$

$$R = R_{\text{tot}} - r = 200 \text{ k}\Omega - 25 \text{ }\Omega \approx 200 \text{ k }\Omega. \qquad (21.69)$$

(R is so large that the galvanometer resistance, r, is nearly negligible.) Note that 5 V applied to this voltmeter produces a half-scale deflection by producing a $25\text{-}\mu\text{A}$ current through the meter, and so the voltmeter's reading is proportional to voltage as desired.

This voltmeter would not be useful for voltages less than about half a volt, because the meter deflection would be small and difficult to read accurately. For other voltage ranges, other resistances are placed in series with the galvanometer. Many meters have a choice of scales. That choice involves switching an appropriate resistance into series with the galvanometer.

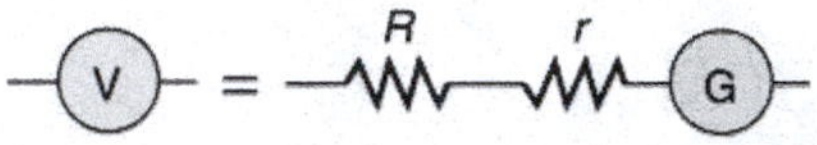

Figure 21.32 A large resistance R placed in series with a galvanometer G produces a voltmeter, the full-scale deflection of which depends on the choice of R. The larger the voltage to be measured, the larger R must be. (Note that r represents the internal resistance of the galvanometer.)

Galvanometer as Ammeter

The same galvanometer can also be made into an ammeter by placing it in parallel with a small resistance R, often called the **shunt resistance**, as shown in **Figure 21.33**. Since the shunt resistance is small, most of the current passes through it, allowing an ammeter to measure currents much greater than those producing a full-scale deflection of the galvanometer.

Suppose, for example, an ammeter is needed that gives a full-scale deflection for 1.0 A, and contains the same $25\text{-}\Omega$ galvanometer with its $50\text{-}\mu\text{A}$ sensitivity. Since R and r are in parallel, the voltage across them is the same.

These IR drops are $IR = I_G r$ so that $IR = \frac{I_G}{I} = \frac{R}{r}$. Solving for R, and noting that I_G is $50 \text{ }\mu\text{A}$ and I is 0.999950 A, we have

$$R = r\frac{I_G}{I} = (25 \text{ }\Omega)\frac{50 \text{ }\mu\text{A}}{0.999950 \text{ A}} = 1.25 \times 10^{-3} \text{ }\Omega. \qquad (21.70)$$

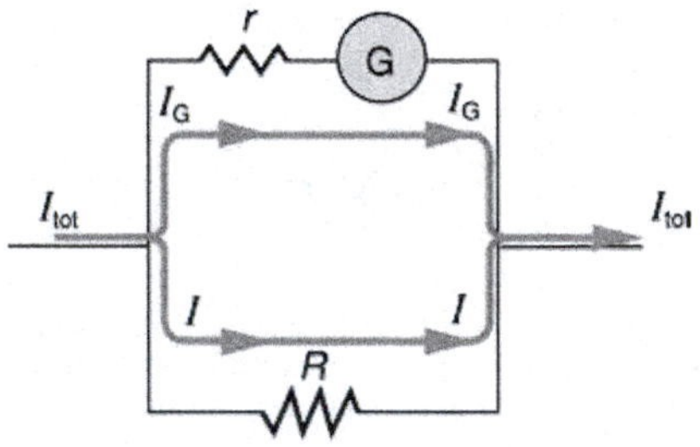

Figure 21.33 A small shunt resistance R placed in parallel with a galvanometer G produces an ammeter, the full-scale deflection of which depends on the choice of R. The larger the current to be measured, the smaller R must be. Most of the current (I) flowing through the meter is shunted through R to protect the galvanometer. (Note that r represents the internal resistance of the galvanometer.) Ammeters may also have multiple scales for greater flexibility in application. The various scales are achieved by switching various shunt resistances in parallel with the galvanometer—the greater the maximum current to be measured, the smaller the shunt resistance must be.

Taking Measurements Alters the Circuit

When you use a voltmeter or ammeter, you are connecting another resistor to an existing circuit and, thus, altering the circuit. Ideally, voltmeters and ammeters do not appreciably affect the circuit, but it is instructive to examine the circumstances under which they do or do not interfere.

First, consider the voltmeter, which is always placed in parallel with the device being measured. Very little current flows through the voltmeter if its resistance is a few orders of magnitude greater than the device, and so the circuit is not appreciably affected. (See **Figure 21.34**(a).) (A large resistance in parallel with a small one has a combined resistance essentially equal to the small one.) If, however, the voltmeter's resistance is comparable to that of the device being measured, then the two in parallel have a smaller resistance, appreciably affecting the circuit. (See Figure 21.34(b).) The voltage across the device is not the same as when the voltmeter is out of the circuit.

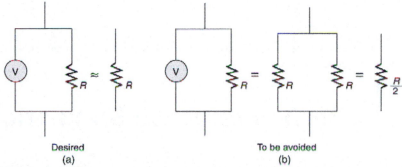

Desired
(a)

To be avoided
(b)

Figure 21.34 (a) A voltmeter having a resistance much larger than the device ($R_{\text{Voltmeter}} >> R$) with which it is in parallel produces a parallel resistance essentially the same as the device and does not appreciably affect the circuit being measured. (b) Here the voltmeter has the same resistance as the device ($R_{\text{Voltmeter}} \cong R$), so that the parallel resistance is half of what it is when the voltmeter is not connected. This is an example of a significant alteration of the circuit and is to be avoided.

An ammeter is placed in series in the branch of the circuit being measured, so that its resistance adds to that branch. Normally, the ammeter's resistance is very small compared with the resistances of the devices in the circuit, and so the extra resistance is negligible. (See Figure 21.35(a).) However, if very small load resistances are involved, or if the ammeter is not as low in resistance as it should be, then the total series resistance is significantly greater, and the current in the branch being measured is reduced. (See Figure 21.35(b).)

A practical problem can occur if the ammeter is connected incorrectly. If it was put in parallel with the resistor to measure the current in it, you could possibly damage the meter; the low resistance of the ammeter would allow most of the current in the circuit to go through the galvanometer, and this current would be larger since the effective resistance is smaller.

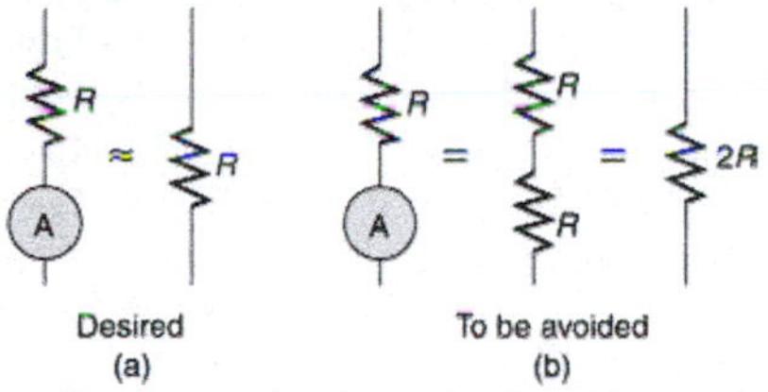

Desired
(a)

To be avoided
(b)

Figure 21.35 (a) An ammeter normally has such a small resistance that the total series resistance in the branch being measured is not appreciably increased. The circuit is essentially unaltered compared with when the ammeter is absent. (b) Here the ammeter's resistance is the same as that of the branch, so that the total resistance is doubled and the current is half what it is without the ammeter. This significant alteration of the circuit is to be avoided.

One solution to the problem of voltmeters and ammeters interfering with the circuits being measured is to use galvanometers with greater sensitivity. This allows construction of voltmeters with greater resistance and ammeters with smaller resistance than when less sensitive galvanometers are used.

There are practical limits to galvanometer sensitivity, but it is possible to get analog meters that make measurements accurate to a few percent. Note that the inaccuracy comes from altering the circuit, not from a fault in the meter.

Connections: Limits to Knowledge

Making a measurement alters the system being measured in a manner that produces uncertainty in the measurement. For macroscopic systems, such as the circuits discussed in this module, the alteration can usually be made negligibly small, but it cannot be eliminated entirely. For submicroscopic systems, such as atoms, nuclei, and smaller particles, measurement alters the system in a manner that cannot be made arbitrarily small. This actually limits knowledge of the system—even limiting what nature can know about itself. We shall see profound implications of this when the Heisenberg uncertainty principle is discussed in the modules on quantum mechanics.

There is another measurement technique based on drawing no current at all and, hence, not altering the circuit at all. These are called null measurements and are the topic of **Null Measurements**. Digital meters that employ solid-state electronics and null measurements can attain accuracies of one part in 10^6 .

Check Your Understanding

Digital meters are able to detect smaller currents than analog meters employing galvanometers. How does this explain their ability to measure voltage and current more accurately than analog meters?

Solution

Since digital meters require less current than analog meters, they alter the circuit less than analog meters. Their resistance as a voltmeter can be far greater than an analog meter, and their resistance as an ammeter can be far less than an analog meter. Consult **Figure 21.30** and **Figure 21.31** and their discussion in the text.

PhET Explorations: Circuit Construction Kit (DC Only), Virtual Lab

Stimulate a neuron and monitor what happens. Pause, rewind, and move forward in time in order to observe the ions as they move across the neuron membrane.

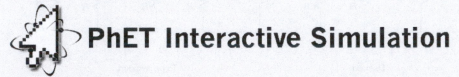

Figure 21.36 Circuit Construction Kit (DC Only), Virtual Lab (http://cnx.org/content/m55368/1.3/circuit-construction-kit-dc-virtual-lab_en.jar)

21.5 Null Measurements

Learning Objectives

By the end of this section, you will be able to:

- Explain why a null measurement device is more accurate than a standard voltmeter or ammeter.
- Demonstrate how a Wheatstone bridge can be used to accurately calculate the resistance in a circuit.

Standard measurements of voltage and current alter the circuit being measured, introducing uncertainties in the measurements. Voltmeters draw some extra current, whereas ammeters reduce current flow. **Null measurements** balance voltages so that there is no current flowing through the measuring device and, therefore, no alteration of the circuit being measured.

Null measurements are generally more accurate but are also more complex than the use of standard voltmeters and ammeters, and they still have limits to their precision. In this module, we shall consider a few specific types of null measurements, because they are common and interesting, and they further illuminate principles of electric circuits.

The Potentiometer

Suppose you wish to measure the emf of a battery. Consider what happens if you connect the battery directly to a standard voltmeter as shown in **Figure 21.37**. (Once we note the problems with this measurement, we will examine a null measurement that improves accuracy.) As discussed before, the actual quantity measured is the terminal voltage V, which is related to the emf of the battery by $V = \text{emf} - Ir$, where I is the current that flows and r is the internal resistance of the battery.

The emf could be accurately calculated if r were very accurately known, but it is usually not. If the current I could be made zero, then $V = \text{emf}$, and so emf could be directly measured. However, standard voltmeters need a current to operate; thus, another technique is needed.

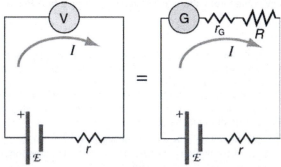

Figure 21.37 An analog voltmeter attached to a battery draws a small but nonzero current and measures a terminal voltage that differs from the emf of the battery. (Note that the script capital E symbolizes electromotive force, or emf.) Since the internal resistance of the battery is not known precisely, it is not possible to calculate the emf precisely.

A **potentiometer** is a null measurement device for measuring potentials (voltages). (See **Figure 21.38**.) A voltage source is connected to a resistor R, say, a long wire, and passes a constant current through it. There is a steady drop in potential (an IR drop) along the wire, so that a variable potential can be obtained by making contact at varying locations along the wire.

Figure 21.38(b) shows an unknown emf_x (represented by script E_x in the figure) connected in series with a galvanometer.

Note that emf_{x} opposes the other voltage source. The location of the contact point (see the arrow on the drawing) is adjusted until the galvanometer reads zero. When the galvanometer reads zero, $\text{emf}_{\text{x}} = IR_{\text{x}}$, where R_{x} is the resistance of the section of wire up to the contact point. Since no current flows through the galvanometer, none flows through the unknown emf, and so emf_{x} is directly sensed.

Now, a very precisely known standard emf_{s} is substituted for emf_{x}, and the contact point is adjusted until the galvanometer again reads zero, so that $\text{emf}_{\text{s}} = IR_{\text{s}}$. In both cases, no current passes through the galvanometer, and so the current I through the long wire is the same. Upon taking the ratio $\dfrac{\text{emf}_{\text{x}}}{\text{emf}_{\text{s}}}$, I cancels, giving

$$\frac{\text{emf}_{\text{x}}}{\text{emf}_{\text{s}}} = \frac{IR_{\text{x}}}{IR_{\text{s}}} = \frac{R_{\text{x}}}{R_{\text{s}}}. \tag{21.71}$$

Solving for emf_{x} gives

$$\text{emf}_{\text{x}} = \text{emf}_{\text{s}}\frac{R_{\text{x}}}{R_{\text{s}}}. \tag{21.72}$$

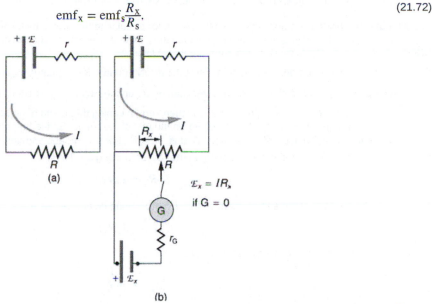

Figure 21.38 The potentiometer, a null measurement device. (a) A voltage source connected to a long wire resistor passes a constant current I through it. (b) An unknown emf (labeled script E_{x} in the figure) is connected as shown, and the point of contact along R is adjusted until the galvanometer reads zero. The segment of wire has a resistance R_{x} and script $E_{\text{x}} = IR_{\text{x}}$, where I is unaffected by the connection since no current flows through the galvanometer. The unknown emf is thus proportional to the resistance of the wire segment.

Because a long uniform wire is used for R, the ratio of resistances $R_{\text{x}}/R_{\text{s}}$ is the same as the ratio of the lengths of wire that zero the galvanometer for each emf. The three quantities on the right-hand side of the equation are now known or measured, and emf_{x} can be calculated. The uncertainty in this calculation can be considerably smaller than when using a voltmeter directly, but it is not zero. There is always some uncertainty in the ratio of resistances $R_{\text{x}}/R_{\text{s}}$ and in the standard emf_{s}. Furthermore, it is not possible to tell when the galvanometer reads exactly zero, which introduces error into both R_{x} and R_{s}, and may also affect the current I.

Resistance Measurements and the Wheatstone Bridge

There is a variety of so-called **ohmmeters** that purport to measure resistance. What the most common ohmmeters actually do is to apply a voltage to a resistance, measure the current, and calculate the resistance using Ohm's law. Their readout is this calculated resistance. Two configurations for ohmmeters using standard voltmeters and ammeters are shown in **Figure 21.39**. Such configurations are limited in accuracy, because the meters alter both the voltage applied to the resistor and the current that flows through it.

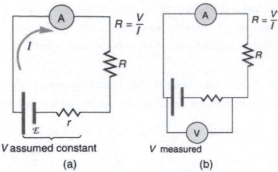

V assumed constant   V measured

(a)   (b)

Figure 21.39 Two methods for measuring resistance with standard meters. (a) Assuming a known voltage for the source, an ammeter measures current, and resistance is calculated as $R = \dfrac{V}{I}$. (b) Since the terminal voltage V varies with current, it is better to measure it. V is most accurately known when I is small, but I itself is most accurately known when it is large.

The **Wheatstone bridge** is a null measurement device for calculating resistance by balancing potential drops in a circuit. (See Figure 21.40.) The device is called a bridge because the galvanometer forms a bridge between two branches. A variety of **bridge devices** are used to make null measurements in circuits.

Resistors R_1 and R_2 are precisely known, while the arrow through R_3 indicates that it is a variable resistance. The value of R_3 can be precisely read. With the unknown resistance R_x in the circuit, R_3 is adjusted until the galvanometer reads zero.

The potential difference between points b and d is then zero, meaning that b and d are at the same potential. With no current running through the galvanometer, it has no effect on the rest of the circuit. So the branches abc and adc are in parallel, and each branch has the full voltage of the source. That is, the IR drops along abc and adc are the same. Since b and d are at the same potential, the IR drop along ad must equal the IR drop along ab. Thus,

$$I_1 R_1 = I_2 R_3. \tag{21.73}$$

Again, since b and d are at the same potential, the IR drop along dc must equal the IR drop along bc. Thus,

$$I_1 R_2 = I_2 R_x. \tag{21.74}$$

Taking the ratio of these last two expressions gives

$$\frac{I_1 R_1}{I_1 R_2} = \frac{I_2 R_3}{I_2 R_x}. \tag{21.75}$$

Canceling the currents and solving for R$_x$ yields

$$R_x = R_3 \frac{R_2}{R_1}. \tag{21.76}$$

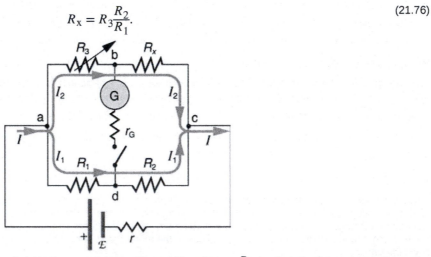

Figure 21.40 The Wheatstone bridge is used to calculate unknown resistances. The variable resistance R_3 is adjusted until the galvanometer reads zero with the switch closed. This simplifies the circuit, allowing R_x to be calculated based on the IR drops as discussed in the text.

This equation is used to calculate the unknown resistance when current through the galvanometer is zero. This method can be very accurate (often to four significant digits), but it is limited by two factors. First, it is not possible to get the current through the galvanometer to be exactly zero. Second, there are always uncertainties in R_1, R_2, and R_3, which contribute to the uncertainty in R_x.

Identify other factors that might limit the accuracy of null measurements. Would the use of a digital device that is more sensitive than a galvanometer improve the accuracy of null measurements?

Solution
One factor would be resistance in the wires and connections in a null measurement. These are impossible to make zero, and they can change over time. Another factor would be temperature variations in resistance, which can be reduced but not completely eliminated by choice of material. Digital devices sensitive to smaller currents than analog devices do improve the accuracy of null measurements because they allow you to get the current closer to zero.

21.6 DC Circuits Containing Resistors and Capacitors

Learning Objectives

By the end of this section, you will be able to:

- Explain the importance of the time constant τ, and calculate the time constant for a given resistance and capacitance.
- Explain why batteries in a flashlight gradually lose power and the light dims over time.
- Describe what happens to a graph of the voltage across a capacitor over time as it charges.
- Explain how a timing circuit works and list some applications.
- Calculate the necessary speed of a strobe flash needed to "stop" the movement of an object over a particular length.

The information presented in this section supports the following AP® learning objectives and science practices:

- **5.C.3.6** The student is able to determine missing values and direction of electric current in branches of a circuit with both resistors and capacitors from values and directions of current in other branches of the circuit through appropriate selection of nodes and application of the junction rule. **(S.P. 1.4, 2.2)**
- **5.C.3.7** The student is able to determine missing values, direction of electric current, charge of capacitors at steady state, and potential differences within a circuit with resistors and capacitors from values and directions of current in other branches of the circuit. **(S.P. 1.4, 2.2)**

When you use a flash camera, it takes a few seconds to charge the capacitor that powers the flash. The light flash discharges the capacitor in a tiny fraction of a second. Why does charging take longer than discharging? This question and a number of other phenomena that involve charging and discharging capacitors are discussed in this module.

RC Circuits

An RC **circuit** is one containing a **resistor** R and a **capacitor** C. The capacitor is an electrical component that stores electric charge.

Figure 21.41 shows a simple RC circuit that employs a DC (direct current) voltage source. The capacitor is initially uncharged. As soon as the switch is closed, current flows to and from the initially uncharged capacitor. As charge increases on the capacitor plates, there is increasing opposition to the flow of charge by the repulsion of like charges on each plate.

In terms of voltage, this is because voltage across the capacitor is given by $V_c = Q/C$, where Q is the amount of charge stored on each plate and C is the **capacitance**. This voltage opposes the battery, growing from zero to the maximum emf when fully charged. The current thus decreases from its initial value of $I_0 = \frac{\text{emf}}{R}$ to zero as the voltage on the capacitor reaches the same value as the emf. When there is no current, there is no IR drop, and so the voltage on the capacitor must then equal the emf of the voltage source. This can also be explained with Kirchhoff's second rule (the loop rule), discussed in **Kirchhoff's Rules**, which says that the algebraic sum of changes in potential around any closed loop must be zero.

The initial current is $I_0 = \frac{\text{emf}}{R}$, because all of the IR drop is in the resistance. Therefore, the smaller the resistance, the faster a given capacitor will be charged. Note that the internal resistance of the voltage source is included in R, as are the resistances of the capacitor and the connecting wires. In the flash camera scenario above, when the batteries powering the camera begin to wear out, their internal resistance rises, reducing the current and lengthening the time it takes to get ready for the next flash.

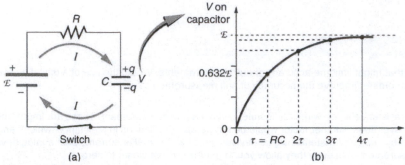

Figure 21.41 (a) An RC circuit with an initially uncharged capacitor. Current flows in the direction shown (opposite of electron flow) as soon as the switch is closed. Mutual repulsion of like charges in the capacitor progressively slows the flow as the capacitor is charged, stopping the current when the capacitor is fully charged and $Q = C \cdot \text{emf}$. (b) A graph of voltage across the capacitor versus time, with the switch closing at time $t = 0$. (Note that in the two parts of the figure, the capital script E stands for emf, q stands for the charge stored on the capacitor, and τ is the RC time constant.)

Voltage on the capacitor is initially zero and rises rapidly at first, since the initial current is a maximum. **Figure 21.41**(b) shows a graph of capacitor voltage versus time (t) starting when the switch is closed at $t = 0$. The voltage approaches emf asymptotically, since the closer it gets to emf the less current flows. The equation for voltage versus time when charging a capacitor C through a resistor R, derived using calculus, is

$$V = \text{emf}(1 - e^{-t/RC}) \text{ (charging)},$$ (21.77)

where V is the voltage across the capacitor, emf is equal to the emf of the DC voltage source, and the exponential e = 2.718 ... is the base of the natural logarithm. Note that the units of RC are seconds. We define

$$\tau = RC,$$ (21.78)

where τ (the Greek letter tau) is called the time constant for an RC circuit. As noted before, a small resistance R allows the capacitor to charge faster. This is reasonable, since a larger current flows through a smaller resistance. It is also reasonable that the smaller the capacitor C, the less time needed to charge it. Both factors are contained in $\tau = RC$.

More quantitatively, consider what happens when $t = \tau = RC$. Then the voltage on the capacitor is

$$V = \text{emf}\left(1 - e^{-1}\right) = \text{emf}(1 - 0.368) = 0.632 \cdot \text{emf}.$$ (21.79)

This means that in the time $\tau = RC$, the voltage rises to 0.632 of its final value. The voltage will rise 0.632 of the remainder in the next time τ. It is a characteristic of the exponential function that the final value is never reached, but 0.632 of the remainder to that value is achieved in every time, τ. In just a few multiples of the time constant τ, then, the final value is very nearly achieved, as the graph in **Figure 21.41**(b) illustrates.

Discharging a Capacitor

Discharging a capacitor through a resistor proceeds in a similar fashion, as **Figure 21.42** illustrates. Initially, the current is $I_0 = \dfrac{V_0}{R}$, driven by the initial voltage V_0 on the capacitor. As the voltage decreases, the current and hence the rate of discharge decreases, implying another exponential formula for V. Using calculus, the voltage V on a capacitor C being discharged through a resistor R is found to be

$$V = V \, e^{-t/RC} \text{(discharging).}$$ (21.80)

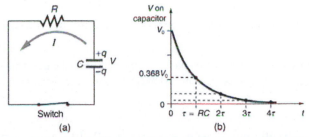

Figure 21.42 (a) Closing the switch discharges the capacitor C through the resistor R. Mutual repulsion of like charges on each plate drives the current. (b) A graph of voltage across the capacitor versus time, with $V = V_0$ at $t = 0$. The voltage decreases exponentially, falling a fixed fraction of the way to zero in each subsequent time constant τ.

The graph in Figure 21.42(b) is an example of this exponential decay. Again, the time constant is $\tau = RC$. A small resistance R allows the capacitor to discharge in a small time, since the current is larger. Similarly, a small capacitance requires less time to discharge, since less charge is stored. In the first time interval $\tau = RC$ after the switch is closed, the voltage falls to 0.368 of its initial value, since $V = V_0 \cdot e^{-1} = 0.368 V_0$.

During each successive time τ, the voltage falls to 0.368 of its preceding value. In a few multiples of τ, the voltage becomes very close to zero, as indicated by the graph in Figure 21.42(b).

Now we can explain why the flash camera in our scenario takes so much longer to charge than discharge; the resistance while charging is significantly greater than while discharging. The internal resistance of the battery accounts for most of the resistance while charging. As the battery ages, the increasing internal resistance makes the charging process even slower. (You may have noticed this.)

The flash discharge is through a low-resistance ionized gas in the flash tube and proceeds very rapidly. Flash photographs, such as in Figure 21.43, can capture a brief instant of a rapid motion because the flash can be less than a microsecond in duration. Such flashes can be made extremely intense.

During World War II, nighttime reconnaissance photographs were made from the air with a single flash illuminating more than a square kilometer of enemy territory. The brevity of the flash eliminated blurring due to the surveillance aircraft's motion. Today, an important use of intense flash lamps is to pump energy into a laser. The short intense flash can rapidly energize a laser and allow it to reemit the energy in another form.

Figure 21.43 This stop-motion photograph of a rufous hummingbird (*Selasphorus rufus*) feeding on a flower was obtained with an extremely brief and intense flash of light powered by the discharge of a capacitor through a gas. (credit: Dean E. Biggins, U.S. Fish and Wildlife Service)

Example 21.6 Integrated Concept Problem: Calculating Capacitor Size—Strobe Lights

High-speed flash photography was pioneered by Doc Edgerton in the 1930s, while he was a professor of electrical engineering at MIT. You might have seen examples of his work in the amazing shots of hummingbirds in motion, a drop of milk splattering on a table, or a bullet penetrating an apple (see Figure 21.43). To stop the motion and capture these pictures, one needs a high-intensity, very short pulsed flash, as mentioned earlier in this module.

Suppose one wished to capture the picture of a bullet (moving at 5.0×10^2 m/s) that was passing through an apple. The duration of the flash is related to the RC time constant, τ. What size capacitor would one need in the RC circuit to succeed, if the resistance of the flash tube was $10.0 \ \Omega$? Assume the apple is a sphere with a diameter of 8.0×10^{-2} m.

Strategy

We begin by identifying the physical principles involved. This example deals with the strobe light, as discussed above.

Figure 21.42 shows the circuit for this probe. The characteristic time τ of the strobe is given as $\tau = RC$.

Solution

We wish to find C, but we don't know τ. We want the flash to be on only while the bullet traverses the apple. So we need to use the kinematic equations that describe the relationship between distance x, velocity v, and time t:

$$x = vt \text{ or } t = \tfrac{x}{v}. \tag{21.81}$$

The bullet's velocity is given as 5.0×10^2 m/s, and the distance x is 8.0×10^{-2} m. The traverse time, then, is

$$t = \tfrac{x}{v} = \frac{8.0 \times 10^{-2} \text{ m}}{5.0 \times 10^2 \text{ m/s}} = 1.6 \times 10^{-4} \text{ s}. \tag{21.82}$$

We set this value for the crossing time t equal to τ. Therefore,

$$C = \tfrac{t}{R} = \frac{1.6 \times 10^{-4} \text{ s}}{10.0 \; \Omega} = 16 \; \mu\text{F}. \tag{21.83}$$

(Note: Capacitance C is typically measured in farads, F, defined as Coulombs per volt. From the equation, we see that C can also be stated in units of seconds per ohm.)

Discussion

The flash interval of $160 \; \mu\text{s}$ (the traverse time of the bullet) is relatively easy to obtain today. Strobe lights have opened up new worlds from science to entertainment. The information from the picture of the apple and bullet was used in the Warren Commission Report on the assassination of President John F. Kennedy in 1963 to confirm that only one bullet was fired.

RC Circuits for Timing

RC circuits are commonly used for timing purposes. A mundane example of this is found in the ubiquitous intermittent wiper systems of modern cars. The time between wipes is varied by adjusting the resistance in an RC circuit. Another example of an RC circuit is found in novelty jewelry, Halloween costumes, and various toys that have battery-powered flashing lights. (See **Figure 21.44** for a timing circuit.)

A more crucial use of RC circuits for timing purposes is in the artificial pacemaker, used to control heart rate. The heart rate is normally controlled by electrical signals generated by the sino-atrial (SA) node, which is on the wall of the right atrium chamber. This causes the muscles to contract and pump blood. Sometimes the heart rhythm is abnormal and the heartbeat is too high or too low.

The artificial pacemaker is inserted near the heart to provide electrical signals to the heart when needed with the appropriate time constant. Pacemakers have sensors that detect body motion and breathing to increase the heart rate during exercise to meet the body's increased needs for blood and oxygen.

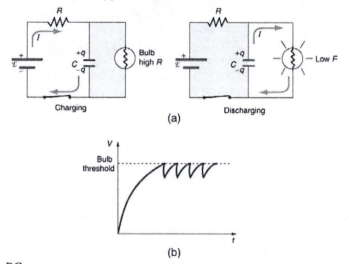

Figure 21.44 (a) The lamp in this RC circuit ordinarily has a very high resistance, so that the battery charges the capacitor as if the lamp were not there. When the voltage reaches a threshold value, a current flows through the lamp that dramatically reduces its resistance, and the capacitor discharges through the lamp as if the battery and charging resistor were not there. Once discharged, the process starts again, with the flash period determined by the RC constant τ. (b) A graph of voltage versus time for this circuit.

Example 21.7 Calculating Time: *RC* Circuit in a Heart Defibrillator

A heart defibrillator is used to resuscitate an accident victim by discharging a capacitor through the trunk of her body. A simplified version of the circuit is seen in **Figure 21.42**. (a) What is the time constant if an $8.00\text{-}\mu F$ capacitor is used and the path resistance through her body is 1.00×10^3 Ω ? (b) If the initial voltage is 10.0 kV, how long does it take to decline to 5.00×10^2 V ?

Strategy

Since the resistance and capacitance are given, it is straightforward to multiply them to give the time constant asked for in part (a). To find the time for the voltage to decline to 5.00×10^2 V , we repeatedly multiply the initial voltage by 0.368 until a voltage less than or equal to 5.00×10^2 V is obtained. Each multiplication corresponds to a time of τ seconds.

Solution for (a)

The time constant τ is given by the equation $\tau = RC$. Entering the given values for resistance and capacitance (and remembering that units for a farad can be expressed as s / Ω) gives

$$\tau = RC = (1.00 \times 10^3 \ \Omega)(8.00 \ \mu F) = 8.00 \text{ ms.} \tag{21.84}$$

Solution for (b)

In the first 8.00 ms, the voltage (10.0 kV) declines to 0.368 of its initial value. That is:

$$V = 0.368V_0 = 3.680 \times 10^3 \text{ V at } t = 8.00 \text{ ms.} \tag{21.85}$$

(Notice that we carry an extra digit for each intermediate calculation.) After another 8.00 ms, we multiply by 0.368 again, and the voltage is

$$
\begin{aligned}
V' &= 0.368V \\
&= (0.368)\!\left(3.680 \times 10^3 \text{ V}\right) \\
&= 1.354 \times 10^3 \text{ V at } t = 16.0 \text{ ms.}
\end{aligned}
\tag{21.86}
$$

Similarly, after another 8.00 ms, the voltage is

$$
\begin{aligned}
V'' &= 0.368V' = (0.368)(1.354 \times 10^3 \text{ V}) \\
&= 498 \text{ V at } t = 24.0 \text{ ms.}
\end{aligned}
\tag{21.87}
$$

Discussion

So after only 24.0 ms, the voltage is down to 498 V, or 4.98% of its original value.Such brief times are useful in heart defibrillation, because the brief but intense current causes a brief but effective contraction of the heart. The actual circuit in a heart defibrillator is slightly more complex than the one in **Figure 21.42**, to compensate for magnetic and AC effects that will be covered in **Magnetism**.

Check Your Understanding

When is the potential difference across a capacitor an emf?

Solution

Only when the current being drawn from or put into the capacitor is zero. Capacitors, like batteries, have internal resistance, so their output voltage is not an emf unless current is zero. This is difficult to measure in practice so we refer to a capacitor's voltage rather than its emf. But the source of potential difference in a capacitor is fundamental and it is an emf.

PhET Explorations: Circuit Construction Kit (DC only)

An electronics kit in your computer! Build circuits with resistors, light bulbs, batteries, and switches. Take measurements with the realistic ammeter and voltmeter. View the circuit as a schematic diagram, or switch to a life-like view.

Figure 21.45 Circuit Construction Kit (DC only) (http://cnx.org/content/m55370/1.3/circuit-construction-kit-dc_en.jar)

Glossary

ammeter: an instrument that measures current

analog meter: a measuring instrument that gives a readout in the form of a needle movement over a marked gauge

bridge device: a device that forms a bridge between two branches of a circuit; some bridge devices are used to make null measurements in circuits

capacitance: the maximum amount of electric potential energy that can be stored (or separated) for a given electric potential

capacitor: an electrical component used to store energy by separating electric charge on two opposing plates

conservation laws: require that energy and charge be conserved in a system

current: the flow of charge through an electric circuit past a given point of measurement

current sensitivity: the maximum current that a galvanometer can read

digital meter: a measuring instrument that gives a readout in a digital form

electromotive force (emf): the potential difference of a source of electricity when no current is flowing; measured in volts

full-scale deflection: the maximum deflection of a galvanometer needle, also known as current sensitivity; a galvanometer with a full-scale deflection of $50 \, \mu A$ has a maximum deflection of its needle when $50 \, \mu A$ flows through it

galvanometer: an analog measuring device, denoted by G, that measures current flow using a needle deflection caused by a magnetic field force acting upon a current-carrying wire

internal resistance: the amount of resistance within the voltage source

Joule's law: the relationship between potential electrical power, voltage, and resistance in an electrical circuit, given by:
$$P_e = IV$$

junction rule: Kirchhoff's first rule, which applies the conservation of charge to a junction; current is the flow of charge; thus, whatever charge flows into the junction must flow out; the rule can be stated $I_1 = I_2 + I_3$

Kirchhoff's rules: a set of two rules, based on conservation of charge and energy, governing current and changes in potential in an electric circuit

loop rule: Kirchhoff's second rule, which states that in a closed loop, whatever energy is supplied by emf must be transferred into other forms by devices in the loop, since there are no other ways in which energy can be transferred into or out of the circuit. Thus, the emf equals the sum of the IR (voltage) drops in the loop and can be stated:
$$\mathrm{emf} = Ir + IR_1 + IR_2$$

null measurements: methods of measuring current and voltage more accurately by balancing the circuit so that no current flows through the measurement device

ohmmeter: an instrument that applies a voltage to a resistance, measures the current, calculates the resistance using Ohm's law, and provides a readout of this calculated resistance

Ohm's law: the relationship between current, voltage, and resistance within an electrical circuit: $V = IR$

parallel: the wiring of resistors or other components in an electrical circuit such that each component receives an equal voltage from the power source; often pictured in a ladder-shaped diagram, with each component on a rung of the ladder

potential difference: the difference in electric potential between two points in an electric circuit, measured in volts

potentiometer: a null measurement device for measuring potentials (voltages)

RC circuit: a circuit that contains both a resistor and a capacitor

resistance: causing a loss of electrical power in a circuit

resistor: a component that provides resistance to the current flowing through an electrical circuit

series: a sequence of resistors or other components wired into a circuit one after the other

shunt resistance: a small resistance R placed in parallel with a galvanometer G to produce an ammeter; the larger the current to be measured, the smaller R must be; most of the current flowing through the meter is shunted through R to protect the galvanometer

terminal voltage: the voltage measured across the terminals of a source of potential difference

voltage: the electrical potential energy per unit charge; electric pressure created by a power source, such as a battery

voltage drop: the loss of electrical power as a current travels through a resistor, wire or other component

voltmeter: an instrument that measures voltage

Wheatstone bridge: a null measurement device for calculating resistance by balancing potential drops in a circuit

Section Summary

21.1 Resistors in Series and Parallel

- The total resistance of an electrical circuit with resistors wired in a series is the sum of the individual resistances:
  $R_s = R_1 + R_2 + R_3 + ...$
- Each resistor in a series circuit has the same amount of current flowing through it.
- The voltage drop, or power dissipation, across each individual resistor in a series is different, and their combined total adds up to the power source input.
- The total resistance of an electrical circuit with resistors wired in parallel is less than the lowest resistance of any of the components and can be determined using the formula:

$$\frac{1}{R_p} = \frac{1}{R_1} + \frac{1}{R_2} + \frac{1}{R_3} + ...$$

- Each resistor in a parallel circuit has the same full voltage of the source applied to it.
- The current flowing through each resistor in a parallel circuit is different, depending on the resistance.
- If a more complex connection of resistors is a combination of series and parallel, it can be reduced to a single equivalent resistance by identifying its various parts as series or parallel, reducing each to its equivalent, and continuing until a single resistance is eventually reached.

21.2 Electromotive Force: Terminal Voltage

- All voltage sources have two fundamental parts—a source of electrical energy that has a characteristic electromotive force (emf), and an internal resistance r.
- The emf is the potential difference of a source when no current is flowing.
- The numerical value of the emf depends on the source of potential difference.
- The internal resistance r of a voltage source affects the output voltage when a current flows.
- The voltage output of a device is called its terminal voltage V and is given by $V = \text{emf} - Ir$, where I is the electric current and is positive when flowing away from the positive terminal of the voltage source.
- When multiple voltage sources are in series, their internal resistances add and their emfs add algebraically.
- Solar cells can be wired in series or parallel to provide increased voltage or current, respectively.

21.3 Kirchhoff's Rules

- Kirchhoff's rules can be used to analyze any circuit, simple or complex.
- Kirchhoff's first rule—the junction rule: The sum of all currents entering a junction must equal the sum of all currents leaving the junction.
- Kirchhoff's second rule—the loop rule: The algebraic sum of changes in potential around any closed circuit path (loop) must be zero.
- The two rules are based, respectively, on the laws of conservation of charge and energy.
- When calculating potential and current using Kirchhoff's rules, a set of conventions must be followed for determining the correct signs of various terms.
- The simpler series and parallel rules are special cases of Kirchhoff's rules.

21.4 DC Voltmeters and Ammeters

- Voltmeters measure voltage, and ammeters measure current.
- A voltmeter is placed in parallel with the voltage source to receive full voltage and must have a large resistance to limit its effect on the circuit.
- An ammeter is placed in series to get the full current flowing through a branch and must have a small resistance to limit its effect on the circuit.
- Both can be based on the combination of a resistor and a galvanometer, a device that gives an analog reading of current.
- Standard voltmeters and ammeters alter the circuit being measured and are thus limited in accuracy.

21.5 Null Measurements

- Null measurement techniques achieve greater accuracy by balancing a circuit so that no current flows through the

measuring device.
- One such device, for determining voltage, is a potentiometer.
- Another null measurement device, for determining resistance, is the Wheatstone bridge.
- Other physical quantities can also be measured with null measurement techniques.

21.6 DC Circuits Containing Resistors and Capacitors

- An RC circuit is one that has both a resistor and a capacitor.
- The time constant τ for an RC circuit is $\tau = RC$.
- When an initially uncharged ($V_0 = 0$ at $t = 0$) capacitor in series with a resistor is charged by a DC voltage source, the voltage rises, asymptotically approaching the emf of the voltage source; as a function of time,

$$V = \text{emf}(1 - e^{-t/RC})(\text{charging}).$$

- Within the span of each time constant τ, the voltage rises by 0.632 of the remaining value, approaching the final voltage asymptotically.
- If a capacitor with an initial voltage V_0 is discharged through a resistor starting at $t = 0$, then its voltage decreases exponentially as given by

$$V = V_0 e^{-t/RC}(\text{discharging}).$$

- In each time constant τ, the voltage falls by 0.368 of its remaining initial value, approaching zero asymptotically.

Conceptual Questions

21.1 Resistors in Series and Parallel

1. A switch has a variable resistance that is nearly zero when closed and extremely large when open, and it is placed in series with the device it controls. Explain the effect the switch in **Figure 21.46** has on current when open and when closed.

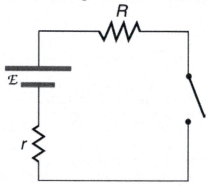

Figure 21.46 A switch is ordinarily in series with a resistance and voltage source. Ideally, the switch has nearly zero resistance when closed but has an extremely large resistance when open. (Note that in this diagram, the script E represents the voltage (or electromotive force) of the battery.)

2. What is the voltage across the open switch in **Figure 21.46**?

3. There is a voltage across an open switch, such as in **Figure 21.46**. Why, then, is the power dissipated by the open switch small?

4. Why is the power dissipated by a closed switch, such as in **Figure 21.46**, small?

5. A student in a physics lab mistakenly wired a light bulb, battery, and switch as shown in **Figure 21.47**. Explain why the bulb is on when the switch is open, and off when the switch is closed. (Do not try this—it is hard on the battery!)

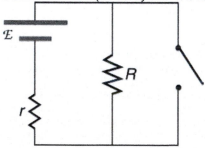

Figure 21.47 A wiring mistake put this switch in parallel with the device represented by R. (Note that in this diagram, the script E represents the voltage (or electromotive force) of the battery.)

6. Knowing that the severity of a shock depends on the magnitude of the current through your body, would you prefer to be in series or parallel with a resistance, such as the heating element of a toaster, if shocked by it? Explain.

7. Would your headlights dim when you start your car's engine if the wires in your automobile were superconductors? (Do not neglect the battery's internal resistance.) Explain.

8. Some strings of holiday lights are wired in series to save wiring costs. An old version utilized bulbs that break the electrical connection, like an open switch, when they burn out. If one such bulb burns out, what happens to the others? If such a string operates on 120 V and has 40 identical bulbs, what is the normal operating voltage of each? Newer versions use bulbs that short circuit, like a closed switch, when they burn out. If one such bulb burns out, what happens to the others? If such a string operates on 120 V and has 39 remaining identical bulbs, what is then the operating voltage of each?

9. If two household lightbulbs rated 60 W and 100 W are connected in series to household power, which will be brighter? Explain.

10. Suppose you are doing a physics lab that asks you to put a resistor into a circuit, but all the resistors supplied have a larger resistance than the requested value. How would you connect the available resistances to attempt to get the smaller value asked for?

11. Before World War II, some radios got power through a "resistance cord" that had a significant resistance. Such a resistance cord reduces the voltage to a desired level for the radio's tubes and the like, and it saves the expense of a transformer. Explain why resistance cords become warm and waste energy when the radio is on.

12. Some light bulbs have three power settings (not including zero), obtained from multiple filaments that are individually switched and wired in parallel. What is the minimum number of filaments needed for three power settings?

21.2 Electromotive Force: Terminal Voltage

13. Is every emf a potential difference? Is every potential difference an emf? Explain.

14. Explain which battery is doing the charging and which is being charged in Figure 21.48.

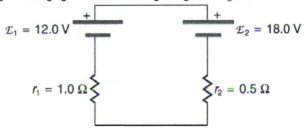

$\mathcal{E}_1 = 12.0$ V $\mathcal{E}_2 = 18.0$ V

$r_1 = 1.0\ \Omega$ $r_2 = 0.5\ \Omega$

Figure 21.48

15. Given a battery, an assortment of resistors, and a variety of voltage and current measuring devices, describe how you would determine the internal resistance of the battery.

16. Two different 12-V automobile batteries on a store shelf are rated at 600 and 850 "cold cranking amps." Which has the smallest internal resistance?

17. What are the advantages and disadvantages of connecting batteries in series? In parallel?

18. Semitractor trucks use four large 12-V batteries. The starter system requires 24 V, while normal operation of the truck's other electrical components utilizes 12 V. How could the four batteries be connected to produce 24 V? To produce 12 V? Why is 24 V better than 12 V for starting the truck's engine (a very heavy load)?

21.3 Kirchhoff's Rules

19. Can all of the currents going into the junction in Figure 21.49 be positive? Explain.

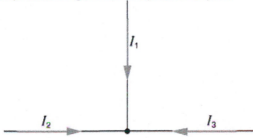

I_1

I_2 I_3

Figure 21.49

20. Apply the junction rule to junction b in Figure 21.50. Is any new information gained by applying the junction rule at e? (In the figure, each emf is represented by script E.)

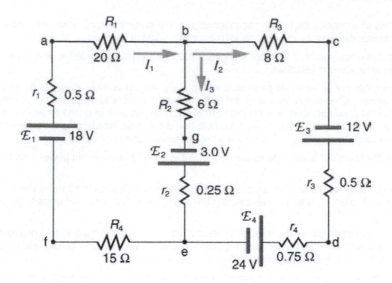

Figure 21.50

21. (a) What is the potential difference going from point a to point b in **Figure 21.50**? (b) What is the potential difference going from c to b? (c) From e to g? (d) From e to d?

22. Apply the loop rule to loop afedcba in **Figure 21.50**.

23. Apply the loop rule to loops abgefa and cbgedc in **Figure 21.50**.

21.4 DC Voltmeters and Ammeters

24. Why should you not connect an ammeter directly across a voltage source as shown in **Figure 21.51**? (Note that script E in the figure stands for emf.)

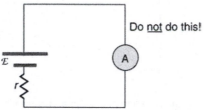

Figure 21.51

25. Suppose you are using a multimeter (one designed to measure a range of voltages, currents, and resistances) to measure current in a circuit and you inadvertently leave it in a voltmeter mode. What effect will the meter have on the circuit? What would happen if you were measuring voltage but accidentally put the meter in the ammeter mode?

26. Specify the points to which you could connect a voltmeter to measure the following potential differences in **Figure 21.52**: (a) the potential difference of the voltage source; (b) the potential difference across R_1; (c) across R_2; (d) across R_3; (e) across R_2 and R_3. Note that there may be more than one answer to each part.

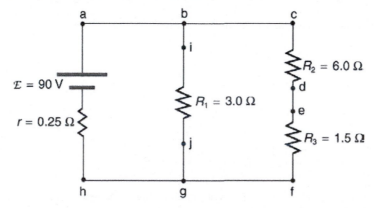

Figure 21.52

27. To measure currents in **Figure 21.52**, you would replace a wire between two points with an ammeter. Specify the points between which you would place an ammeter to measure the following: (a) the total current; (b) the current flowing through R_1; (c) through R_2; (d) through R_3. Note that there may be more than one answer to each part.

21.5 Null Measurements

28. Why can a null measurement be more accurate than one using standard voltmeters and ammeters? What factors limit the accuracy of null measurements?

29. If a potentiometer is used to measure cell emfs on the order of a few volts, why is it most accurate for the standard emf_s to be the same order of magnitude and the resistances to be in the range of a few ohms?

21.6 DC Circuits Containing Resistors and Capacitors

30. Regarding the units involved in the relationship $\tau = RC$, verify that the units of resistance times capacitance are time, that is, $\Omega \cdot \text{F} = \text{s}$.

31. The RC time constant in heart defibrillation is crucial to limiting the time the current flows. If the capacitance in the defibrillation unit is fixed, how would you manipulate resistance in the circuit to adjust the RC constant τ? Would an adjustment of the applied voltage also be needed to ensure that the current delivered has an appropriate value?

32. When making an ECG measurement, it is important to measure voltage variations over small time intervals. The time is limited by the RC constant of the circuit—it is not possible to measure time variations shorter than RC. How would you manipulate R and C in the circuit to allow the necessary measurements?

33. Draw two graphs of charge versus time on a capacitor. Draw one for charging an initially uncharged capacitor in series with a resistor, as in the circuit in **Figure 21.41**, starting from $t = 0$. Draw the other for discharging a capacitor through a resistor, as in the circuit in **Figure 21.42**, starting at $t = 0$, with an initial charge Q_0. Show at least two intervals of τ.

34. When charging a capacitor, as discussed in conjunction with **Figure 21.41**, how long does it take for the voltage on the capacitor to reach emf? Is this a problem?

35. When discharging a capacitor, as discussed in conjunction with **Figure 21.42**, how long does it take for the voltage on the capacitor to reach zero? Is this a problem?

36. Referring to **Figure 21.41**, draw a graph of potential difference across the resistor versus time, showing at least two intervals of τ. Also draw a graph of current versus time for this situation.

37. A long, inexpensive extension cord is connected from inside the house to a refrigerator outside. The refrigerator doesn't run as it should. What might be the problem?

38. In **Figure 21.44**, does the graph indicate the time constant is shorter for discharging than for charging? Would you expect ionized gas to have low resistance? How would you adjust R to get a longer time between flashes? Would adjusting R affect the discharge time?

39. An electronic apparatus may have large capacitors at high voltage in the power supply section, presenting a shock hazard even when the apparatus is switched off. A "bleeder resistor" is therefore placed across such a capacitor, as shown schematically in **Figure 21.53**, to bleed the charge from it after the apparatus is off. Why must the bleeder resistance be much greater than the effective resistance of the rest of the circuit? How does this affect the time constant for discharging the capacitor?

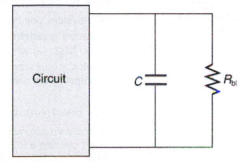

Figure 21.53 A bleeder resistor R_{bl} discharges the capacitor in this electronic device once it is switched off.

Problems & Exercises

21.1 Resistors in Series and Parallel

Note: Data taken from figures can be assumed to be accurate to three significant digits.

1. (a) What is the resistance of ten 275-Ω resistors connected in series? (b) In parallel?

2. (a) What is the resistance of a 1.00×10^{2}-Ω, a 2.50-$k\Omega$, and a 4.00-$k\,\Omega$ resistor connected in series? (b) In parallel?

3. What are the largest and smallest resistances you can obtain by connecting a 36.0-Ω, a 50.0-Ω, and a 700-Ω resistor together?

4. An 1800-W toaster, a 1400-W electric frying pan, and a 75-W lamp are plugged into the same outlet in a 15-A, 120-V circuit. (The three devices are in parallel when plugged into the same socket.). (a) What current is drawn by each device? (b) Will this combination blow the 15-A fuse?

5. Your car's 30.0-W headlight and 2.40-kW starter are ordinarily connected in parallel in a 12.0-V system. What power would one headlight and the starter consume if connected in series to a 12.0-V battery? (Neglect any other resistance in the circuit and any change in resistance in the two devices.)

6. (a) Given a 48.0-V battery and 24.0-Ω and 96.0-Ω resistors, find the current and power for each when connected in series. (b) Repeat when the resistances are in parallel.

7. Referring to the example combining series and parallel circuits and **Figure 21.6**, calculate I_3 in the following two different ways: (a) from the known values of I and I_2; (b) using Ohm's law for R_3. In both parts explicitly show how you follow the steps in the **Problem-Solving Strategies for Series and Parallel Resistors**.

8. Referring to **Figure 21.6**: (a) Calculate P_3 and note how it compares with P_3 found in the first two example problems in this module. (b) Find the total power supplied by the source and compare it with the sum of the powers dissipated by the resistors.

9. Refer to **Figure 21.7** and the discussion of lights dimming when a heavy appliance comes on. (a) Given the voltage source is 120 V, the wire resistance is 0.400 Ω, and the bulb is nominally 75.0 W, what power will the bulb dissipate if a total of 15.0 A passes through the wires when the motor comes on? Assume negligible change in bulb resistance. (b) What power is consumed by the motor?

10. A 240-kV power transmission line carrying 5.00×10^{2} A is hung from grounded metal towers by ceramic insulators, each having a 1.00×10^{9}-Ω resistance. **Figure 21.54**. (a) What is the resistance to ground of 100 of these insulators? (b) Calculate the power dissipated by 100 of them. (c) What fraction of the power carried by the line is this? Explicitly show how you follow the steps in the **Problem-Solving Strategies for Series and Parallel Resistors**.

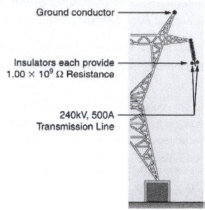

Ground conductor

Insulators each provide 1.00×10^9 Ω Resistance

240kV, 500A Transmission Line

Figure 21.54 High-voltage (240-kV) transmission line carrying 5.00×10^{2} A is hung from a grounded metal transmission tower. The row of ceramic insulators provide 1.00×10^{9} Ω of resistance each.

11. Show that if two resistors R_1 and R_2 are combined and one is much greater than the other ($R_1 >> R_2$): (a) Their series resistance is very nearly equal to the greater resistance R_1. (b) Their parallel resistance is very nearly equal to smaller resistance R_2.

12. Unreasonable Results

Two resistors, one having a resistance of 145 Ω, are connected in parallel to produce a total resistance of 150 Ω. (a) What is the value of the second resistance? (b) What is unreasonable about this result? (c) Which assumptions are unreasonable or inconsistent?

13. Unreasonable Results

Two resistors, one having a resistance of 900 $k\Omega$, are connected in series to produce a total resistance of 0.500 $M\Omega$. (a) What is the value of the second resistance? (b) What is unreasonable about this result? (c) Which assumptions are unreasonable or inconsistent?

21.2 Electromotive Force: Terminal Voltage

14. Standard automobile batteries have six lead-acid cells in series, creating a total emf of 12.0 V. What is the emf of an individual lead-acid cell?

15. Carbon-zinc dry cells (sometimes referred to as non-alkaline cells) have an emf of 1.54 V, and they are produced as single cells or in various combinations to form other voltages. (a) How many 1.54-V cells are needed to make the common 9-V battery used in many small electronic devices? (b) What is the actual emf of the approximately 9-V battery? (c) Discuss how internal resistance in the series connection of cells will affect the terminal voltage of this approximately 9-V battery.

16. What is the output voltage of a 3.0000-V lithium cell in a digital wristwatch that draws 0.300 mA, if the cell's internal resistance is $2.00\ \Omega$?

17. (a) What is the terminal voltage of a large 1.54-V carbon-zinc dry cell used in a physics lab to supply 2.00 A to a circuit, if the cell's internal resistance is $0.100\ \Omega$? (b) How much electrical power does the cell produce? (c) What power goes to its load?

18. What is the internal resistance of an automobile battery that has an emf of 12.0 V and a terminal voltage of 15.0 V while a current of 8.00 A is charging it?

19. (a) Find the terminal voltage of a 12.0-V motorcycle battery having a $0.600\text{-}\Omega$ internal resistance, if it is being charged by a current of 10.0 A. (b) What is the output voltage of the battery charger?

20. A car battery with a 12-V emf and an internal resistance of $0.050\ \Omega$ is being charged with a current of 60 A. Note that in this process the battery is being charged. (a) What is the potential difference across its terminals? (b) At what rate is thermal energy being dissipated in the battery? (c) At what rate is electric energy being converted to chemical energy? (d) What are the answers to (a) and (b) when the battery is used to supply 60 A to the starter motor?

21. The hot resistance of a flashlight bulb is $2.30\ \Omega$, and it is run by a 1.58-V alkaline cell having a $0.100\text{-}\Omega$ internal resistance. (a) What current flows? (b) Calculate the power supplied to the bulb using $I^2 R_{\text{bulb}}$. (c) Is this power the same as calculated using $\dfrac{V^2}{R_{\text{bulb}}}$?

22. The label on a portable radio recommends the use of rechargeable nickel-cadmium cells (nicads), although they have a 1.25-V emf while alkaline cells have a 1.58-V emf. The radio has a $3.20\text{-}\Omega$ resistance. (a) Draw a circuit diagram of the radio and its batteries. Now, calculate the power delivered to the radio. (b) When using Nicad cells each having an internal resistance of $0.0400\ \Omega$. (c) When using alkaline cells each having an internal resistance of $0.200\ \Omega$. (d) Does this difference seem significant, considering that the radio's effective resistance is lowered when its volume is turned up?

23. An automobile starter motor has an equivalent resistance of $0.0500\ \Omega$ and is supplied by a 12.0-V battery with a $0.0100\text{-}\Omega$ internal resistance. (a) What is the current to the motor? (b) What voltage is applied to it? (c) What power is supplied to the motor? (d) Repeat these calculations for when the battery connections are corroded and add $0.0900\ \Omega$ to the circuit. (Significant problems are caused by even small amounts of unwanted resistance in low-voltage, high-current applications.)

24. A child's electronic toy is supplied by three 1.58-V alkaline cells having internal resistances of $0.0200\ \Omega$ in series with a 1.53-V carbon-zinc dry cell having a $0.100\text{-}\Omega$ internal resistance. The load resistance is $10.0\ \Omega$. (a) Draw a circuit diagram of the toy and its batteries. (b) What current flows? (c) How much power is supplied to the load? (d) What is the internal resistance of the dry cell if it goes bad, resulting in only 0.500 W being supplied to the load?

25. (a) What is the internal resistance of a voltage source if its terminal voltage drops by 2.00 V when the current supplied increases by 5.00 A? (b) Can the emf of the voltage source be found with the information supplied?

26. A person with body resistance between his hands of $10.0\ \text{k}\Omega$ accidentally grasps the terminals of a 20.0-kV power supply. (Do NOT do this!) (a) Draw a circuit diagram to represent the situation. (b) If the internal resistance of the power supply is $2000\ \Omega$, what is the current through his body? (c) What is the power dissipated in his body? (d) If the power supply is to be made safe by increasing its internal resistance, what should the internal resistance be for the maximum current in this situation to be 1.00 mA or less? (e) Will this modification compromise the effectiveness of the power supply for driving low-resistance devices? Explain your reasoning.

27. Electric fish generate current with biological cells called electroplaques, which are physiological emf devices. The electroplaques in the South American eel are arranged in 140 rows, each row stretching horizontally along the body and each containing 5000 electroplaques. Each electroplaque has an emf of 0.15 V and internal resistance of $0.25\ \Omega$. If the water surrounding the fish has resistance of $800\ \Omega$, how much current can the eel produce in water from near its head to near its tail?

28. Integrated Concepts

A 12.0-V emf automobile battery has a terminal voltage of 16.0 V when being charged by a current of 10.0 A. (a) What is the battery's internal resistance? (b) What power is dissipated inside the battery? (c) At what rate (in $^\circ\text{C/min}$) will its temperature increase if its mass is 20.0 kg and it has a specific heat of $0.300\ \text{kcal/kg} \cdot {^\circ\text{C}}$, assuming no heat escapes?

29. Unreasonable Results

A 1.58-V alkaline cell with a $0.200\text{-}\Omega$ internal resistance is supplying 8.50 A to a load. (a) What is its terminal voltage? (b) What is the value of the load resistance? (c) What is unreasonable about these results? (d) Which assumptions are unreasonable or inconsistent?

30. Unreasonable Results

(a) What is the internal resistance of a 1.54-V dry cell that supplies 1.00 W of power to a $15.0\text{-}\Omega$ bulb? (b) What is unreasonable about this result? (c) Which assumptions are unreasonable or inconsistent?

21.3 Kirchhoff's Rules

31. Apply the loop rule to loop abcdefgha in **Figure 21.27**.

32. Apply the loop rule to loop aedcba in **Figure 21.27**.

33. Verify the second equation in **Example 21.5** by substituting the values found for the currents I_1 and I_2 .

34. Verify the third equation in **Example 21.5** by substituting the values found for the currents I_1 and I_3.

35. Apply the junction rule at point a in **Figure 21.55**.

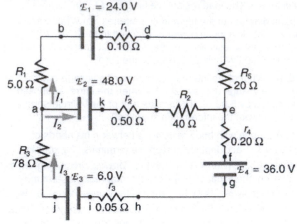

Figure 21.55

36. Apply the loop rule to loop abcdefghija in **Figure 21.55**.

37. Apply the loop rule to loop akledcba in **Figure 21.55**.

38. Find the currents flowing in the circuit in **Figure 21.55**. Explicitly show how you follow the steps in the **Problem-Solving Strategies for Series and Parallel Resistors**.

39. Solve **Example 21.5**, but use loop abcdefgha instead of loop akledcba. Explicitly show how you follow the steps in the **Problem-Solving Strategies for Series and Parallel Resistors**.

40. Find the currents flowing in the circuit in **Figure 21.50**.

41. Unreasonable Results

Consider the circuit in **Figure 21.56**, and suppose that the emfs are unknown and the currents are given to be $I_1 = 5.00 \text{ A}$, $I_2 = 3.0 \text{ A}$, and $I_3 = -2.00 \text{ A}$. (a) Could you find the emfs? (b) What is wrong with the assumptions?

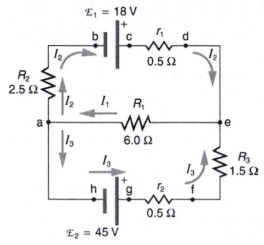

Figure 21.56

21.4 DC Voltmeters and Ammeters

42. What is the sensitivity of the galvanometer (that is, what current gives a full-scale deflection) inside a voltmeter that has a $1.00\text{-M}\,\Omega$ resistance on its 30.0-V scale?

43. What is the sensitivity of the galvanometer (that is, what current gives a full-scale deflection) inside a voltmeter that has a $25.0\text{-k}\,\Omega$ resistance on its 100-V scale?

44. Find the resistance that must be placed in series with a $25.0\text{-}\Omega$ galvanometer having a $50.0\text{-}\mu\text{A}$ sensitivity (the same as the one discussed in the text) to allow it to be used as a voltmeter with a 0.100-V full-scale reading.

45. Find the resistance that must be placed in series with a $25.0\text{-}\Omega$ galvanometer having a $50.0\text{-}\mu\text{A}$ sensitivity (the same as the one discussed in the text) to allow it to be used as a voltmeter with a 3000-V full-scale reading. Include a circuit diagram with your solution.

46. Find the resistance that must be placed in parallel with a $25.0\text{-}\Omega$ galvanometer having a $50.0\text{-}\mu\text{A}$ sensitivity (the same as the one discussed in the text) to allow it to be used as an ammeter with a 10.0-A full-scale reading. Include a circuit diagram with your solution.

47. Find the resistance that must be placed in parallel with a $25.0\text{-}\Omega$ galvanometer having a $50.0\text{-}\mu\text{A}$ sensitivity (the same as the one discussed in the text) to allow it to be used as an ammeter with a 300-mA full-scale reading.

48. Find the resistance that must be placed in series with a $10.0\text{-}\Omega$ galvanometer having a $100\text{-}\mu\text{A}$ sensitivity to allow it to be used as a voltmeter with: (a) a 300-V full-scale reading, and (b) a 0.300-V full-scale reading.

49. Find the resistance that must be placed in parallel with a $10.0\text{-}\Omega$ galvanometer having a $100\text{-}\mu\text{A}$ sensitivity to allow it to be used as an ammeter with: (a) a 20.0-A full-scale reading, and (b) a 100-mA full-scale reading.

50. Suppose you measure the terminal voltage of a 1.585-V alkaline cell having an internal resistance of $0.100\ \Omega$ by placing a $1.00\text{-k}\,\Omega$ voltmeter across its terminals. (See **Figure 21.57**.) (a) What current flows? (b) Find the terminal voltage. (c) To see how close the measured terminal voltage is to the emf, calculate their ratio.

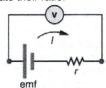

Figure 21.57

51. Suppose you measure the terminal voltage of a 3.200-V lithium cell having an internal resistance of $5.00\ \Omega$ by placing a $1.00\text{-k}\,\Omega$ voltmeter across its terminals. (a) What current flows? (b) Find the terminal voltage. (c) To see how close the measured terminal voltage is to the emf, calculate their ratio.

52. A certain ammeter has a resistance of $5.00 \times 10^{-5}\ \Omega$ on its 3.00-A scale and contains a $10.0\text{-}\Omega$ galvanometer. What is the sensitivity of the galvanometer?

53. A $1.00\text{-}M\Omega$ voltmeter is placed in parallel with a $75.0\text{-}k\,\Omega$ resistor in a circuit. (a) Draw a circuit diagram of the connection. (b) What is the resistance of the combination? (c) If the voltage across the combination is kept the same as it was across the $75.0\text{-}k\,\Omega$ resistor alone, what is the percent increase in current? (d) If the current through the combination is kept the same as it was through the $75.0\text{-}k\,\Omega$ resistor alone, what is the percentage decrease in voltage? (e) Are the changes found in parts (c) and (d) significant? Discuss.

54. A $0.0200\text{-}\Omega$ ammeter is placed in series with a $10.00\text{-}\Omega$ resistor in a circuit. (a) Draw a circuit diagram of the connection. (b) Calculate the resistance of the combination. (c) If the voltage is kept the same across the combination as it was through the $10.00\text{-}\Omega$ resistor alone, what is the percent decrease in current? (d) If the current is kept the same through the combination as it was through the $10.00\text{-}\Omega$ resistor alone, what is the percent increase in voltage? (e) Are the changes found in parts (c) and (d) significant? Discuss.

55. Unreasonable Results

Suppose you have a $40.0\text{-}\Omega$ galvanometer with a $25.0\text{-}\mu A$ sensitivity. (a) What resistance would you put in series with it to allow it to be used as a voltmeter that has a full-scale deflection for 0.500 mV? (b) What is unreasonable about this result? (c) Which assumptions are responsible?

56. Unreasonable Results

(a) What resistance would you put in parallel with a $40.0\text{-}\Omega$ galvanometer having a $25.0\text{-}\mu A$ sensitivity to allow it to be used as an ammeter that has a full-scale deflection for $10.0\text{-}\mu A$? (b) What is unreasonable about this result? (c) Which assumptions are responsible?

21.5 Null Measurements

57. What is the emf_x of a cell being measured in a potentiometer, if the standard cell's emf is 12.0 V and the potentiometer balances for $R_x = 5.000\ \Omega$ and $R_s = 2.500\ \Omega$?

58. Calculate the emf_x of a dry cell for which a potentiometer is balanced when $R_x = 1.200\ \Omega$, while an alkaline standard cell with an emf of 1.600 V requires $R_s = 1.247\ \Omega$ to balance the potentiometer.

59. When an unknown resistance R_x is placed in a Wheatstone bridge, it is possible to balance the bridge by adjusting R_3 to be $2500\ \Omega$. What is R_x if $\frac{R_2}{R_1} = 0.625$?

60. To what value must you adjust R_3 to balance a Wheatstone bridge, if the unknown resistance R_x is $100\ \Omega$, R_1 is $50.0\ \Omega$, and R_2 is $175\ \Omega$?

61. (a) What is the unknown emf_x in a potentiometer that balances when R_x is $10.0\ \Omega$, and balances when R_s is $15.0\ \Omega$ for a standard 3.000-V emf? (b) The same emf_x is placed in the same potentiometer, which now balances when R_s is $15.0\ \Omega$ for a standard emf of 3.100 V. At what resistance R_x will the potentiometer balance?

62. Suppose you want to measure resistances in the range from $10.0\ \Omega$ to $10.0\ k\Omega$ using a Wheatstone bridge that has $\frac{R_2}{R_1} = 2.000$. Over what range should R_3 be adjustable?

21.6 DC Circuits Containing Resistors and Capacitors

63. The timing device in an automobile's intermittent wiper system is based on an RC time constant and utilizes a $0.500\text{-}\mu F$ capacitor and a variable resistor. Over what range must R be made to vary to achieve time constants from 2.00 to 15.0 s?

64. A heart pacemaker fires 72 times a minute, each time a 25.0-nF capacitor is charged (by a battery in series with a resistor) to 0.632 of its full voltage. What is the value of the resistance?

65. The duration of a photographic flash is related to an RC time constant, which is $0.100\ \mu s$ for a certain camera. (a) If the resistance of the flash lamp is $0.0400\ \Omega$ during discharge, what is the size of the capacitor supplying its energy? (b) What is the time constant for charging the capacitor, if the charging resistance is $800\ k\Omega$?

66. A 2.00- and a $7.50\text{-}\mu F$ capacitor can be connected in series or parallel, as can a 25.0- and a $100\text{-}k\Omega$ resistor. Calculate the four RC time constants possible from connecting the resulting capacitance and resistance in series.

67. After two time constants, what percentage of the final voltage, emf, is on an initially uncharged capacitor C , charged through a resistance R ?

68. A $500\text{-}\Omega$ resistor, an uncharged $1.50\text{-}\mu F$ capacitor, and a 6.16-V emf are connected in series. (a) What is the initial current? (b) What is the RC time constant? (c) What is the current after one time constant? (d) What is the voltage on the capacitor after one time constant?

69. A heart defibrillator being used on a patient has an RC time constant of 10.0 ms due to the resistance of the patient and the capacitance of the defibrillator. (a) If the defibrillator has an $8.00\text{-}\mu F$ capacitance, what is the resistance of the path through the patient? (You may neglect the capacitance of the patient and the resistance of the defibrillator.) (b) If the initial voltage is 12.0 kV, how long does it take to decline to $6.00 \times 10^2\ V$?

70. An ECG monitor must have an RC time constant less than 1.00×10^2 µs to be able to measure variations in voltage over small time intervals. (a) If the resistance of the circuit (due mostly to that of the patient's chest) is $1.00 \text{ k}\Omega$, what is the maximum capacitance of the circuit? (b) Would it be difficult in practice to limit the capacitance to less than the value found in (a)?

71. Figure 21.58 shows how a bleeder resistor is used to discharge a capacitor after an electronic device is shut off, allowing a person to work on the electronics with less risk of shock. (a) What is the time constant? (b) How long will it take to reduce the voltage on the capacitor to 0.250% (5% of 5%) of its full value once discharge begins? (c) If the capacitor is charged to a voltage V_0 through a $100\text{-}\Omega$ resistance, calculate the time it takes to rise to $0.865V_0$ (This is about two time constants.)

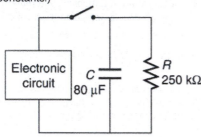

Figure 21.58

72. Using the exact exponential treatment, find how much time is required to discharge a $250\text{-}\mu\text{F}$ capacitor through a $500\text{-}\Omega$ resistor down to 1.00% of its original voltage.

73. Using the exact exponential treatment, find how much time is required to charge an initially uncharged 100-pF capacitor through a $75.0\text{-M }\Omega$ resistor to 90.0% of its final voltage.

74. Integrated Concepts

If you wish to take a picture of a bullet traveling at 500 m/s, then a very brief flash of light produced by an RC discharge through a flash tube can limit blurring. Assuming 1.00 mm of motion during one RC constant is acceptable, and given that the flash is driven by a $600\text{-}\mu\text{F}$ capacitor, what is the resistance in the flash tube?

75. Integrated Concepts

A flashing lamp in a Christmas earring is based on an RC discharge of a capacitor through its resistance. The effective duration of the flash is 0.250 s, during which it produces an average 0.500 W from an average 3.00 V. (a) What energy does it dissipate? (b) How much charge moves through the lamp? (c) Find the capacitance. (d) What is the resistance of the lamp?

76. Integrated Concepts

A $160\text{-}\mu\text{F}$ capacitor charged to 450 V is discharged through a $31.2\text{-k }\Omega$ resistor. (a) Find the time constant. (b) Calculate the temperature increase of the resistor, given that its mass is 2.50 g and its specific heat is $1.67\dfrac{\text{kJ}}{\text{kg} \cdot \text{°C}}$, noting that most of the thermal energy is retained in the short time of the discharge. (c) Calculate the new resistance, assuming it is pure carbon. (d) Does this change in resistance seem significant?

77. Unreasonable Results

(a) Calculate the capacitance needed to get an RC time constant of 1.00×10^3 s with a $0.100\text{-}\Omega$ resistor. (b) What is unreasonable about this result? (c) Which assumptions are responsible?

78. Construct Your Own Problem

Consider a camera's flash unit. Construct a problem in which you calculate the size of the capacitor that stores energy for the flash lamp. Among the things to be considered are the voltage applied to the capacitor, the energy needed in the flash and the associated charge needed on the capacitor, the resistance of the flash lamp during discharge, and the desired RC time constant.

79. Construct Your Own Problem

Consider a rechargeable lithium cell that is to be used to power a camcorder. Construct a problem in which you calculate the internal resistance of the cell during normal operation. Also, calculate the minimum voltage output of a battery charger to be used to recharge your lithium cell. Among the things to be considered are the emf and useful terminal voltage of a lithium cell and the current it should be able to supply to a camcorder.

Test Prep for AP® Courses

21.1 Resistors in Series and Parallel

1.

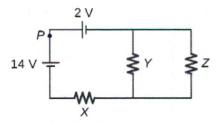

Figure 21.59 The figure above shows a circuit containing two batteries and three identical resistors with resistance R. Which of the following changes to the circuit will result in an increase in the current at point P? Select *two* answers.
   a. Reversing the connections to the 14 V battery.
   b. Removing the 2 V battery and connecting the wires to close the left loop.
   c. Rearranging the resistors so all three are in series.
   d. Removing the branch containing resistor Z.

2. In a circuit, a parallel combination of six 1.6-kΩ resistors is connected in series with a parallel combination of four 2.4-kΩ resistors. If the source voltage is 24 V, what will be the percentage of total current in one of the 2.4-kΩ resistors?
   a. 10%
   b. 12%
   c. 20%
   d. 25%

3. If the circuit in the previous question is modified by removing some of the 1.6 kΩ resistors, the total current in the circuit is 24 mA. How many resistors were removed?
   a. 1
   b. 2
   c. 3
   d. 4

4.

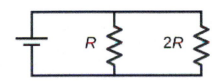

Figure 21.60 Two resistors, with resistances R and $2R$ are connected to a voltage source as shown in this figure. If the power dissipated in R is 10 W, what is the power dissipated in $2R$?
   a. 1 W
   b. 2.5 W
   c. 5 W
   d. 10 W

5. In a circuit, a parallel combination of two 20-Ω and one 10-Ω resistors is connected in series with a 4-Ω resistor. The source voltage is 36 V.
   a. Find the resistor(s) with the maximum current.
   b. Find the resistor(s) with the maximum voltage drop.
   c. Find the power dissipated in each resistor and hence the total power dissipated in all the resistors. Also find the power output of the source. Are they equal or not? Justify your answer.
   d. Will the answers for questions (a) and (b) differ if a 3 Ω resistor is added in series to the 4 Ω resistor? If yes, repeat the question(s) for the new resistor combination.
   e. If the values of all the resistors and the source voltage are doubled, what will be the effect on the current?

21.2 Electromotive Force: Terminal Voltage

6. Suppose there are two voltage sources – Sources A and B – with the same emfs but different internal resistances, i.e., the internal resistance of Source A is lower than Source B. If they both supply the same current in their circuits, which of the following statements is true?
   a. External resistance in Source A's circuit is more than Source B's circuit.
   b. External resistance in Source A's circuit is less than Source B's circuit.
   c. External resistance in Source A's circuit is the same as Source B's circuit.
   d. The relationship between external resistances in the two circuits can't be determined.

7. Calculate the internal resistance of a voltage source if the terminal voltage of the source increases by 1 V when the current supplied decreases by 4 A? Suppose this source is connected in series (in the same direction) to another source with a different voltage but same internal resistance. What will be the total internal resistance? How will the total internal resistance change if the sources are connected in the opposite direction?

21.3 Kirchhoff's Rules

8. An experiment was set up with the circuit diagram shown. Assume $R_1 = 10 \ \Omega$, $R_2 = R_3 = 5 \ \Omega$, $r = 0 \ \Omega$ and $E = 6$ V.

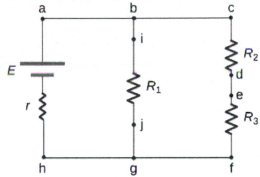

Figure 21.61

a.   One of the steps to examine the set-up is to test points
     with the same potential. Which of the following points
     can be tested?
     a.   Points b, c and d.
     b.   Points d, e and f.
     c.   Points f, h and j.
     d.   Points a, h and i.
b.   At which three points should the currents be measured
     so that Kirchhoff's junction rule can be directly
     confirmed?
     a.   Points b, c and d.
     b.   Points d, e and f.
     c.   Points f, h and j.
     d.   Points a, h and i.
c.   If the current in the branch with the voltage source is
     upward and currents in the other two branches are
     downward, i.e. $I_a = I_i + I_c$, identify which of the following
     can be true? Select *two* answers.
     a. $I_i = I_j - I_f$
     b. $I_e = I_h - I_i$
     c. $I_c = I_j - I_a$
     d. $I_d = I_h - I_j$
d.   The measurements reveal that the current through R_1 is
     0.5 A and R_3 is 0.6 A. Based on your knowledge of
     Kirchoff's laws, confirm which of the following
     statements are true.
     a.   The measured current for R_1 is correct but for R_3
          is incorrect.
     b.   The measured current for R_3 is correct but for R_1
          is incorrect.
     c.   Both the measured currents are correct.
     d.   Both the measured currents are incorrect.

e.   The graph shown in the following figure is the energy
     dissipated at R_1 as a function of time.

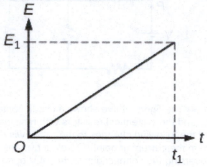

Figure 21.62
Which of the following shows the graph for energy
dissipated at R_2 as a function of time?

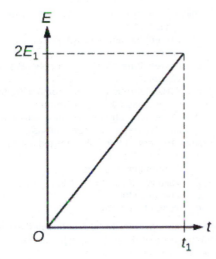

a.

Figure 21.63

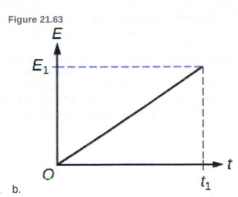

b.

Figure 21.64

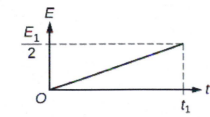

c.

Figure 21.65

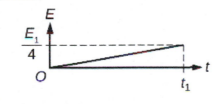

d.

Figure 21.66

9. For this question, consider the circuit shown in the following figure.

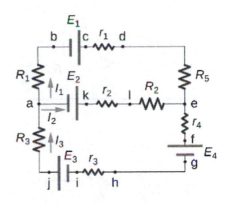

Figure 21.67

a. Assuming that none of the three currents (I_1, I_2, and I_3) are equal to zero, which of the following statements is false?
   a. $I_3 = I_1 + I_2$ at point a.
   b. $I_2 = I_3 - I_1$ at point e.
   c. The current through R_3 is equal to the current through R_5.
   d. The current through R_1 is equal to the current through R_5.

b. Which of the following statements is true?
   a. $E_1 + E_2 + I_1R_1 - I_2R_2 + I_1r_1 - I_2r_2 + I_1R_5 = 0$
   b. $-E_1 + E_2 + I_1R_1 - I_2R_2 + I_1r_1 - I_2r_2 - I_1R_5 = 0$
   c. $E_1 - E_2 - I_1R_1 + I_2R_2 - I_1r_1 + I_2r_2 - I_1R_5 = 0$
   d. $E_1 + E_2 - I_1R_1 + I_2R_2 - I_1r_1 + I_2r_2 + I_1R_5 = 0$

c. If $I_1 = 5$ A and $I_3 = -2$ A, which of the following statements is false?
   a. The current through R_1 will flow from a to b and will be equal to 5 A.
   b. The current through R_3 will flow from a to j and will be equal to 2 A.
   c. The current through R_5 will flow from d to e and will be equal to 5 A.
   d. None of the above.

d. If $I_1 = 5$ A and $I_3 = -2$ A, I_2 will be equal to
   a. 3 A
   b. -3 A
   c. 7 A
   d. -7 A

10.

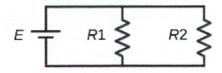

Figure 21.68 In an experiment this circuit is set up. Three ammeters are used to record the currents in the three vertical branches (with R_1, R_2, and E). The readings of the ammeters in the resistor branches (i.e. currents in R_1 and R_2) are 2 A and 3 A respectively.

a. Find the equation obtained by applying Kirchhoff's loop rule in the loop involving R_1 and R_2.

b. What will be the reading of the third ammeter (i.e. the branch with E)? If E were replaced by $3E$, how would this reading change?

c. If the original circuit is modified by adding another voltage source (as shown in the following circuit), find the readings of the three ammeters.

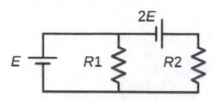

Figure 21.69

11.

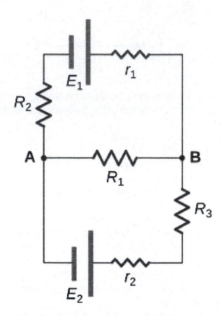

Figure 21.70 In this circuit, assume the currents through R_1, R_2 and R_3 are I_1, I_2 and I_3 respectively and all are flowing in the clockwise direction.

a. Find the equation obtained by applying Kirchhoff's junction rule at point A.
b. Find the equations obtained by applying Kirchhoff's loop rule in the upper and lower loops.
c. Assume $R_1 = R_2 = 6$ Ω, $R_3 = 12$ Ω, $r_1 = r_2 = 0$ Ω, $E_1 = 6$ V and $E_2 = 4$ V. Calculate I_1, I_2 and I_3.
d. For the situation in which E_2 is replaced by a closed switch, repeat parts (a) and (b). Using the values for R_1, R_2, R_3, r_1 and E_1 from part (c) calculate the currents through the three resistors.
e. For the circuit in part (d) calculate the output power of the voltage source and across all the resistors. Examine if energy is conserved in the circuit.
f. A student implemented the circuit of part (d) in the lab and measured the current though one of the resistors as 0.19 A. According to the results calculated in part (d) identify the resistor(s). Justify any difference in measured and calculated value.

21.6 DC Circuits Containing Resistors and Capacitors

12. A battery is connected to a resistor and an uncharged capacitor. The switch for the circuit is closed at $t = 0$ s.

a. While the capacitor is being charged, which of the following is true?
   a. Current through and voltage across the resistor increase.
   b. Current through and voltage across the resistor decrease.
   c. Current through and voltage across the resistor first increase and then decrease.
   d. Current through and voltage across the resistor first decrease and then increase.
b. When the capacitor is fully charged, which of the following is NOT zero?
   a. Current in the resistor.
   b. Voltage across the resistor.
   c. Current in the capacitor.
   d. None of the above.

13. An uncharged capacitor C is connected in series (with a switch) to a resistor R_1 and a voltage source E. Assume $E = 24$ V, $R_1 = 1.2$ kΩ and $C = 1$ mF.

a. What will be the current through the circuit as the switch is closed? Draw a circuit diagram and show the direction of current after the switch is closed. How long will it take for the capacitor to be 99% charged?
b. After full charging, this capacitor is connected in series to another resistor, $R_2 = 1$ kΩ. What will be the current in the circuit as soon as it's connected? Draw a circuit diagram and show the direction of current. How long will it take for the capacitor voltage to reach 3.24 V?

22 MAGNETISM

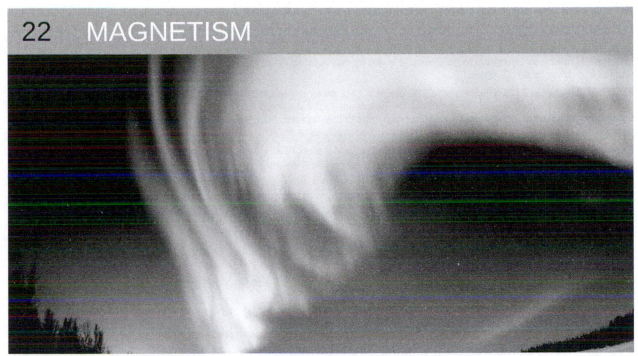

Figure 22.1 The magnificent spectacle of the Aurora Borealis, or northern lights, glows in the northern sky above Bear Lake near Eielson Air Force Base, Alaska. Shaped by the Earth's magnetic field, this light is produced by radiation spewed from solar storms. (credit: Senior Airman Joshua Strang, via Flickr)

Chapter Outline

Connection for AP® Courses

Magnetism plays a major role in your everyday life. All electric motors, with uses as diverse as powering refrigerators, starting cars, and moving elevators, contain magnets. Magnetic resonance imaging (MRI) has become an important diagnostic tool in the field of medicine, and the use of magnetism to explore brain activity is a subject of contemporary research and development. Other applications of magnetism include computer memory, levitation of high-speed trains, the aurora borealis, and, of course, the first important historical use of magnetism: navigation. You will find all of these applications of magnetism linked by a small number of underlying principles.

In this chapter, you will learn that both the internal properties of an object and the movement of charged particles can generate a magnetic field, and you will learn why all magnetic fields have a north and south pole. You will also learn how magnetic fields exert forces on objects, resulting in the magnetic alignment that makes a compass work. You will learn how we use this principle to weigh the smallest of subatomic particles with precision and contain superheated plasma to facilitate nuclear fusion.

Big Idea 1 Objects and systems have properties such as mass and charge. Systems may have internal structure.

Enduring Understanding 1.E Materials have many macroscopic properties that result from the arrangement and interactions of the atoms and molecules that make up the material.

Essential Knowledge 1.E.5 Matter has a property called magnetic permeability.

Essential Knowledge 1.E.6 Matter has a property called magnetic dipole moment.

Big Idea 2 Fields existing in space can be used to explain interactions.

Enduring Understanding 2.D A magnetic field is caused by a magnet or a moving electrically charged object. Magnetic fields observed in nature always seem to be produced either by moving charged objects or by magnetic dipoles or combinations of dipoles and never by single poles.

Essential Knowledge 2.D.1 The magnetic field exerts a force on a moving electrically charged object. That magnetic force is perpendicular to the direction of the velocity of the object and to the magnetic field and is proportional to the magnitude of the charge, the magnitude of the velocity, and the magnitude of the magnetic field. It also depends on the angle between the velocity and the magnetic field vectors. Treatment is quantitative for angles of 0°, 90°, or 180° and qualitative for other angles.

Essential Knowledge 2.D.2 The magnetic field vectors around a straight wire that carries electric current are tangent to concentric circles centered on that wire. The field has no component toward the current-carrying wire.

Essential Knowledge 2.D.3 A magnetic dipole placed in a magnetic field, such as the ones created by a magnet or the Earth, will tend to align with the magnetic field vector.

Essential Knowledge 2.D.4 Ferromagnetic materials contain magnetic domains that are themselves magnets.

Big Idea 3 The interactions of an object with other objects can be described by forces.

Enduring Understanding 3.C At the macroscopic level, forces can be categorized as either long-range (action-at-a-distance) forces or contact forces.

Essential Knowledge 3.C.3 A magnetic force results from the interaction of a moving charged object or a magnet with other moving charged objects or another magnet.

Big Idea 4 Interactions between systems can result in changes in those systems.

Enduring Understanding 4.E The electric and magnetic properties of a system can change in response to the presence of, or changes in, other objects or systems.

Essential Knowledge 4.E.1 The magnetic properties of some materials can be affected by magnetic fields at the system. Students should focus on the underlying concepts and not the use of the vocabulary.

Figure 22.2 Engineering of technology like iPods would not be possible without a deep understanding of magnetism. (credit: Jesse! S?, Flickr)

22.1 Magnets

Learning Objectives
By the end of this section, you will be able to:
• Describe the difference between the north and south poles of a magnet. • Describe how magnetic poles interact with each other.

Figure 22.3 Magnets come in various shapes, sizes, and strengths. All have both a north pole and a south pole. There is never an isolated pole (a monopole).

All magnets attract iron, such as that in a refrigerator door. However, magnets may attract or repel other magnets. Experimentation shows that all magnets have two poles. If freely suspended, one pole will point toward the north. The two poles are thus named the **north magnetic pole** and the **south magnetic pole** (or more properly, north-seeking and south-seeking poles, for the attractions in those directions).

Universal Characteristics of Magnets and Magnetic Poles

It is a universal characteristic of all magnets that *like poles repel and unlike poles attract*. (Note the similarity with electrostatics: unlike charges attract and like charges repel.)

Further experimentation shows that it is *impossible to separate north and south poles* in the manner that + and − charges can be separated.

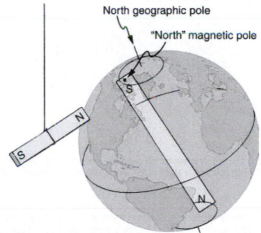

Figure 22.4 One end of a bar magnet is suspended from a thread that points toward north. The magnet's two poles are labeled N and S for north-seeking and south-seeking poles, respectively.

Misconception Alert: Earth's Geographic North Pole Hides an S

The Earth acts like a very large bar magnet with its south-seeking pole near the geographic North Pole. That is why the north pole of your compass is attracted toward the geographic north pole of the Earth—because the magnetic pole that is near the geographic North Pole is actually a south magnetic pole! Confusion arises because the geographic term "North Pole" has come to be used (incorrectly) for the magnetic pole that is near the North Pole. Thus, "North magnetic pole" is actually a misnomer—it should be called the South magnetic pole.

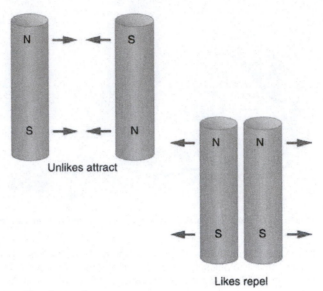

Figure 22.5 Unlike poles attract, whereas like poles repel.

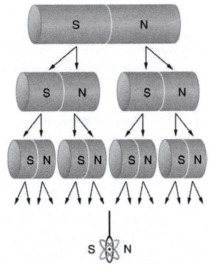

Figure 22.6 North and south poles always occur in pairs. Attempts to separate them result in more pairs of poles. If we continue to split the magnet, we will eventually get down to an iron atom with a north pole and a south pole—these, too, cannot be separated.

Real World Connections: Dipoles and Monopoles

Figure 22.6 shows that no matter how many times you divide a magnet the resulting objects are always magnetic dipoles. Formally, a magnetic dipole is an object (usually very small) with a north and south magnetic pole. Magnetic dipoles have a vector property called magnetic momentum. The magnitude of this vector is equal to the strength of its poles and the distance between the poles, and the direction points from the south pole to the north pole.

A magnetic dipole can also be thought of as a very small closed current loop. There is no way to isolate north and south magnetic poles like you can isolate positive and negative charges. Another way of saying this is that magnetic fields of a magnetic object always make closed loops, starting at a north pole and ending at a south pole.

With a positive charge, you might imagine drawing a spherical surface enclosing that charge, and there would be a net flux of electric field lines flowing outward through that surface. In fact, Gauss's law states that the electric flux through a surface is proportional to the amount of charge enclosed.

With a magnetic object, every surface you can imagine that encloses all or part of the magnet ultimately has zero net flux of magnetic field lines flowing through the surface. Just as many outward-flowing lines from the north pole of the magnet pass through the surface as inward-flowing lines from the south pole of the magnet.

Some physicists have theorized that magnetic monopoles exist. These would be isolated magnetic "charges" that would only generate field lines that flow outward or inward (not loops). Despite many searches, we have yet to experimentally verify the existence of magnetic monopoles.

The fact that magnetic poles always occur in pairs of north and south is true from the very large scale—for example, sunspots always occur in pairs that are north and south magnetic poles—all the way down to the very small scale. Magnetic atoms have both a north pole and a south pole, as do many types of subatomic particles, such as electrons, protons, and neutrons.

We know that like magnetic poles repel and unlike poles attract. See if you can show this for two refrigerator magnets. Will the magnets stick if you turn them over? Why do they stick to the door anyway? What can you say about the magnetic properties of the door next to the magnet? Do refrigerator magnets stick to metal or plastic spoons? Do they stick to all types of metal?

22.2 Ferromagnets and Electromagnets

Learning Objectives

By the end of this section, you will be able to:

- Define ferromagnet.
- Describe the role of magnetic domains in magnetization.
- Explain the significance of the Curie temperature.
- Describe the relationship between electricity and magnetism.

The information presented in this section supports the following AP® learning objectives and science practices:

- **2.D.3.1** The student is able to describe the orientation of a magnetic dipole placed in a magnetic field in general and the particular cases of a compass in the magnetic field of the Earth and iron filings surrounding a bar magnet. **(S.P. 1.2)**
- **2.D.4.1** The student is able to use the representation of magnetic domains to qualitatively analyze the magnetic behavior of a bar magnet composed of ferromagnetic material. **(S.P. 1.4)**
- **4.E.1.1** The student is able to use representations and models to qualitatively describe the magnetic properties of some materials that can be affected by magnetic properties of other objects in the system. **(S.P. 1.1, 1.4, 2.2)**

Ferromagnets

Only certain materials, such as iron, cobalt, nickel, and gadolinium, exhibit strong magnetic effects. Such materials are called **ferromagnetic**, after the Latin word for iron, *ferrum*. A group of materials made from the alloys of the rare earth elements are also used as strong and permanent magnets; a popular one is neodymium. Other materials exhibit weak magnetic effects, which are detectable only with sensitive instruments. Not only do ferromagnetic materials respond strongly to magnets (the way iron is attracted to magnets), they can also be **magnetized** themselves—that is, they can be induced to be magnetic or made into permanent magnets.

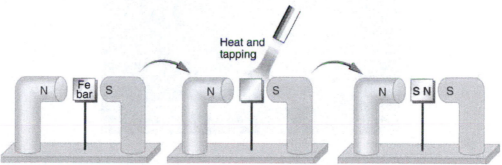

Figure 22.7 An unmagnetized piece of iron is placed between two magnets, heated, and then cooled, or simply tapped when cold. The iron becomes a permanent magnet with the poles aligned as shown: its south pole is adjacent to the north pole of the original magnet, and its north pole is adjacent to the south pole of the original magnet. Note that there are attractive forces between the magnets.

When a magnet is brought near a previously unmagnetized ferromagnetic material, it causes local magnetization of the material with unlike poles closest, as in **Figure 22.7**. (This results in the attraction of the previously unmagnetized material to the magnet.) What happens on a microscopic scale is illustrated in **Figure 22.8**. The regions within the material called **domains** act like small bar magnets. Within domains, the poles of individual atoms are aligned. Each atom acts like a tiny bar magnet. Domains are small and randomly oriented in an unmagnetized ferromagnetic object. In response to an external magnetic field, the domains may grow to millimeter size, aligning themselves as shown in **Figure 22.8**(b). This induced magnetization can be made permanent if the material is heated and then cooled, or simply tapped in the presence of other magnets.

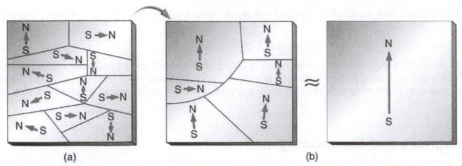

(a)   (b)

Figure 22.8 (a) An unmagnetized piece of iron (or other ferromagnetic material) has randomly oriented domains. (b) When magnetized by an external field, the domains show greater alignment, and some grow at the expense of others. Individual atoms are aligned within domains; each atom acts like a tiny bar magnet.

Conversely, a permanent magnet can be demagnetized by hard blows or by heating it in the absence of another magnet. Increased thermal motion at higher temperature can disrupt and randomize the orientation and the size of the domains. There is a well-defined temperature for ferromagnetic materials, which is called the **Curie temperature**, above which they cannot be magnetized. The Curie temperature for iron is 1043 K $(770°C)$, which is well above room temperature. There are several elements and alloys that have Curie temperatures much lower than room temperature and are ferromagnetic only below those temperatures.

Electromagnets

Early in the 19th century, it was discovered that electrical currents cause magnetic effects. The first significant observation was by the Danish scientist Hans Christian Oersted (1777–1851), who found that a compass needle was deflected by a current-carrying wire. This was the first significant evidence that the movement of charges had any connection with magnets. **Electromagnetism** is the use of electric current to make magnets. These temporarily induced magnets are called **electromagnets**. Electromagnets are employed for everything from a wrecking yard crane that lifts scrapped cars to controlling the beam of a 90-km-circumference particle accelerator to the magnets in medical imaging machines (See **Figure 22.9**).

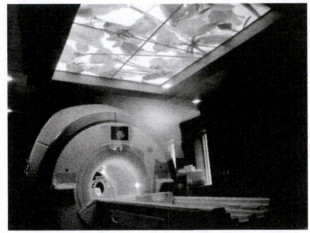

Figure 22.9 Instrument for magnetic resonance imaging (MRI). The device uses a superconducting cylindrical coil for the main magnetic field. The patient goes into this "tunnel" on the gurney. (credit: Bill McChesney, Flickr)

Figure 22.10 shows that the response of iron filings to a current-carrying coil and to a permanent bar magnet. The patterns are similar. In fact, electromagnets and ferromagnets have the same basic characteristics—for example, they have north and south poles that cannot be separated and for which like poles repel and unlike poles attract.

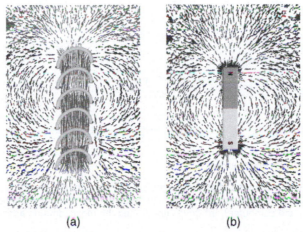

(a)   (b)

Figure 22.10 Iron filings near (a) a current-carrying coil and (b) a magnet act like tiny compass needles, showing the shape of their fields. Their response to a current-carrying coil and a permanent magnet is seen to be very similar, especially near the ends of the coil and the magnet.

Making Connections: Compasses

Figure 22.11 Compass needles and a bar magnet.

In **Figure 22.11**, a series of tiny compass needles are placed in an external magnetic field. These dipoles respond exactly like the iron filings in **Figure 22.10**. For needles close to one pole of the magnet, the needle is aligned so that the opposite pole of the needle points at the bar magnet. For example, close to the north pole of the bar magnet, the south pole of the compass needle is aligned to be closest to the bar magnet.

As an experimenter moves each compass around, the needle will rotate in such a way as to orient itself with the field of the bar magnet at that location. In this way, the magnetic field lines may be mapped out precisely.

The strength of the magnetic field depends on the medium in which the magnetic field exists. Some substances (like iron) respond to external magnetic fields in a way that amplifies the external magnetic field. The magnetic permeability of a substance is a measure of the substance's ability to support or amplify an already existing external magnetic field. Ferromagnets, in which magnetic domains in the substance align with and amplify an external magnetic field, are examples of objects with high permeability.

Combining a ferromagnet with an electromagnet can produce particularly strong magnetic effects. (See **Figure 22.12**.) Whenever strong magnetic effects are needed, such as lifting scrap metal, or in particle accelerators, electromagnets are enhanced by ferromagnetic materials. Limits to how strong the magnets can be made are imposed by coil resistance (it will overheat and melt at sufficiently high current), and so superconducting magnets may be employed. These are still limited, because superconducting properties are destroyed by too great a magnetic field.

Figure 22.12 An electromagnet with a ferromagnetic core can produce very strong magnetic effects. Alignment of domains in the core produces a magnet, the poles of which are aligned with the electromagnet.

Figure 22.13 shows a few uses of combinations of electromagnets and ferromagnets. Ferromagnetic materials can act as memory devices, because the orientation of the magnetic fields of small domains can be reversed or erased. Magnetic information storage on videotapes and computer hard drives are among the most common applications. This property is vital in our digital world.

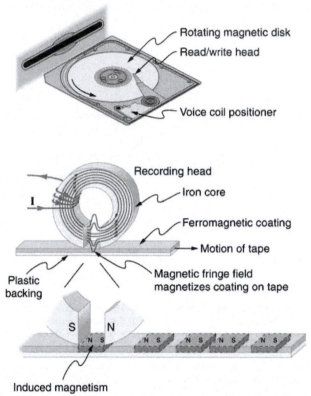

Figure 22.13 An electromagnet induces regions of permanent magnetism on a floppy disk coated with a ferromagnetic material. The information stored here is digital (a region is either magnetic or not); in other applications, it can be analog (with a varying strength), such as on audiotapes.

Current: The Source of All Magnetism

An electromagnet creates magnetism with an electric current. In later sections we explore this more quantitatively, finding the strength and direction of magnetic fields created by various currents. But what about ferromagnets? Figure 22.14 shows models of how electric currents create magnetism at the submicroscopic level. (Note that we cannot directly observe the paths of individual electrons about atoms, and so a model or visual image, consistent with all direct observations, is made. We can directly observe the electron's orbital angular momentum, its spin momentum, and subsequent magnetic moments, all of which are explained with electric-current-creating subatomic magnetism.) Currents, including those associated with other submicroscopic particles like protons, allow us to explain ferromagnetism and all other magnetic effects. Ferromagnetism, for example, results from an internal cooperative alignment of electron spins, possible in some materials but not in others.

Crucial to the statement that electric current is the source of all magnetism is the fact that it is impossible to separate north and south magnetic poles. (This is far different from the case of positive and negative charges, which are easily separated.) A current loop always produces a magnetic dipole—that is, a magnetic field that acts like a north pole and south pole pair. Since isolated north and south magnetic poles, called **magnetic monopoles**, are not observed, currents are used to explain all magnetic effects. If magnetic monopoles did exist, then we would have to modify this underlying connection that all magnetism is due to electrical current. There is no known reason that magnetic monopoles should not exist—they are simply never observed—and so searches at the subnuclear level continue. If they do *not* exist, we would like to find out why not. If they *do* exist, we would like to see evidence of them.

Electric Currents and Magnetism

Electric current is the source of all magnetism.

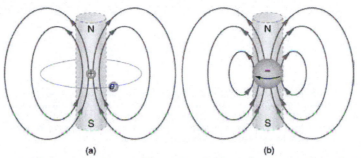

(a)   (b)

Figure 22.14 (a) In the planetary model of the atom, an electron orbits a nucleus, forming a closed-current loop and producing a magnetic field with a north pole and a south pole. (b) Electrons have spin and can be crudely pictured as rotating charge, forming a current that produces a magnetic field with a north pole and a south pole. Neither the planetary model nor the image of a spinning electron is completely consistent with modern physics. However, they do provide a useful way of understanding phenomena.

PhET Explorations: Magnets and Electromagnets

Explore the interactions between a compass and bar magnet. Discover how you can use a battery and wire to make a magnet! Can you make it a stronger magnet? Can you make the magnetic field reverse?

Figure 22.15 Magnets and Electromagnets (http://cnx.org/content/m55375/1.2/magnets-and-electromagnets_en.jar)

22.3 Magnetic Fields and Magnetic Field Lines

Learning Objectives
By the end of this section, you will be able to: • Define magnetic field and describe the magnetic field lines of various magnetic fields.

Einstein is said to have been fascinated by a compass as a child, perhaps musing on how the needle felt a force without direct physical contact. His ability to think deeply and clearly about action at a distance, particularly for gravitational, electric, and magnetic forces, later enabled him to create his revolutionary theory of relativity. Since magnetic forces act at a distance, we define a **magnetic field** to represent magnetic forces. The pictorial representation of **magnetic field lines** is very useful in visualizing the strength and direction of the magnetic field. As shown in **Figure 22.16**, the **direction of magnetic field lines** is defined to be the direction in which the north end of a compass needle points. The magnetic field is traditionally called the **B-field**.

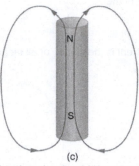

(a)   (b)   (c)

Figure 22.16 Magnetic field lines are defined to have the direction that a small compass points when placed at a location. (a) If small compasses are used to map the magnetic field around a bar magnet, they will point in the directions shown: away from the north pole of the magnet, toward the south pole of the magnet. (Recall that the Earth's north magnetic pole is really a south pole in terms of definitions of poles on a bar magnet.) (b) Connecting the arrows gives continuous magnetic field lines. The strength of the field is proportional to the closeness (or density) of the lines. (c) If the interior of the magnet could be probed, the field lines would be found to form continuous closed loops.

Small compasses used to test a magnetic field will not disturb it. (This is analogous to the way we tested electric fields with a small test charge. In both cases, the fields represent only the object creating them and not the probe testing them.) **Figure 22.17** shows how the magnetic field appears for a current loop and a long straight wire, as could be explored with small compasses. A small compass placed in these fields will align itself parallel to the field line at its location, with its north pole pointing in the direction of B. Note the symbols used for field into and out of the paper.

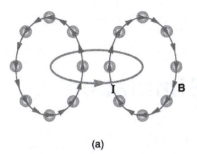
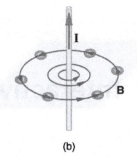
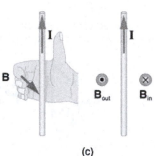

(a)   (b)   (c)

Figure 22.17 Small compasses could be used to map the fields shown here. (a) The magnetic field of a circular current loop is similar to that of a bar magnet. (b) A long and straight wire creates a field with magnetic field lines forming circular loops. (c) When the wire is in the plane of the paper, the field is perpendicular to the paper. Note that the symbols used for the field pointing inward (like the tail of an arrow) and the field pointing outward (like the tip of an arrow).

Making Connections: Concept of a Field

A field is a way of mapping forces surrounding any object that can act on another object at a distance without apparent physical connection. The field represents the object generating it. Gravitational fields map gravitational forces, electric fields map electrical forces, and magnetic fields map magnetic forces.

Extensive exploration of magnetic fields has revealed a number of hard-and-fast rules. We use magnetic field lines to represent the field (the lines are a pictorial tool, not a physical entity in and of themselves). The properties of magnetic field lines can be summarized by these rules:

1. The direction of the magnetic field is tangent to the field line at any point in space. A small compass will point in the direction of the field line.

2. The strength of the field is proportional to the closeness of the lines. It is exactly proportional to the number of lines per unit area perpendicular to the lines (called the areal density).

3. Magnetic field lines can never cross, meaning that the field is unique at any point in space.

4. Magnetic field lines are continuous, forming closed loops without beginning or end. They go from the north pole to the south pole.

The last property is related to the fact that the north and south poles cannot be separated. It is a distinct difference from electric field lines, which begin and end on the positive and negative charges. If magnetic monopoles existed, then magnetic field lines would begin and end on them.

22.4 Magnetic Field Strength: Force on a Moving Charge in a Magnetic Field

Learning Objectives

By the end of this section, you will be able to:

- Describe the effects of magnetic fields on moving charges.
- Use the right-hand rule 1 to determine the velocity of a charge, the direction of the magnetic field, and the direction of magnetic force on a moving charge.
- Calculate the magnetic force on a moving charge.

The information presented in this section supports the following AP® learning objectives and science practices:

- **2.D.1.1** The student is able to apply mathematical routines to express the force exerted on a moving charged object by a magnetic field. **(S.P. 2.2)**
- **3.C.3.1** The student is able to use right-hand rules to analyze a situation involving a current-carrying conductor and a moving electrically charged object to determine the direction of the magnetic force exerted on the charged object due to the magnetic field created by the current-carrying conductor. **(S.P. 1.4)**

What is the mechanism by which one magnet exerts a force on another? The answer is related to the fact that all magnetism is caused by current, the flow of charge. *Magnetic fields exert forces on moving charges*, and so they exert forces on other magnets, all of which have moving charges.

Right Hand Rule 1

The magnetic force on a moving charge is one of the most fundamental known. Magnetic force is as important as the electrostatic or Coulomb force. Yet the magnetic force is more complex, in both the number of factors that affects it and in its direction, than the relatively simple Coulomb force. The magnitude of the **magnetic force** F on a charge q moving at a speed v in a magnetic field of strength B is given by

$$F = qvB \sin \theta, \tag{22.1}$$

where θ is the angle between the directions of $\mathbf{v}$ and $\mathbf{B}$. This force is often called the **Lorentz force**. In fact, this is how we define the magnetic field strength B—in terms of the force on a charged particle moving in a magnetic field. The SI unit for magnetic field strength B is called the **tesla** (T) after the eccentric but brilliant inventor Nikola Tesla (1856–1943). To determine how the tesla relates to other SI units, we solve $F = qvB \sin \theta$ for B.

$$B = \frac{F}{qv \sin \theta} \tag{22.2}$$

Because $\sin \theta$ is unitless, the tesla is

$$1 \text{ T} = \frac{1 \text{ N}}{\text{C} \cdot \text{m/s}} = \frac{1 \text{ N}}{\text{A} \cdot \text{m}} \tag{22.3}$$

(note that C/s = A).

Another smaller unit, called the **gauss** (G), where $1 \text{ G} = 10^{-4} \text{ T}$, is sometimes used. The strongest permanent magnets have fields near 2 T; superconducting electromagnets may attain 10 T or more. The Earth's magnetic field on its surface is only about $5 \times 10^{-5} \text{ T}$, or 0.5 G.

The *direction* of the magnetic force $\mathbf{F}$ is perpendicular to the plane formed by $\mathbf{v}$ and $\mathbf{B}$, as determined by the **right hand rule 1** (or RHR-1), which is illustrated in **Figure 22.18**. RHR-1 states that, to determine the direction of the magnetic force on a positive moving charge, you point the thumb of the right hand in the direction of $\mathbf{v}$, the fingers in the direction of $\mathbf{B}$, and a perpendicular to the palm points in the direction of $\mathbf{F}$. One way to remember this is that there is one velocity, and so the thumb represents it. There are many field lines, and so the fingers represent them. The force is in the direction you would push with your palm. The force on a negative charge is in exactly the opposite direction to that on a positive charge.

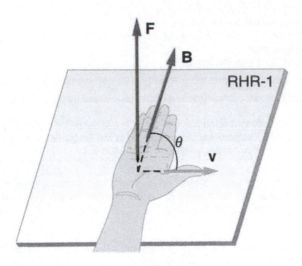

$$F = qvB \sin \theta$$

$$\mathbf{F} \perp \text{ plane of } \mathbf{v} \text{ and } \mathbf{B}$$

Figure 22.18 Magnetic fields exert forces on moving charges. This force is one of the most basic known. The direction of the magnetic force on a moving charge is perpendicular to the plane formed by $\mathbf{v}$ and $\mathbf{B}$ and follows right hand rule–1 (RHR-1) as shown. The magnitude of the force is proportional to q, v, B, and the sine of the angle between $\mathbf{v}$ and $\mathbf{B}$.

Making Connections: Charges and Magnets

There is no magnetic force on static charges. However, there is a magnetic force on moving charges. When charges are stationary, their electric fields do not affect magnets. But, when charges move, they produce magnetic fields that exert forces on other magnets. When there is relative motion, a connection between electric and magnetic fields emerges—each affects the other.

Example 22.1 Calculating Magnetic Force: Earth's Magnetic Field on a Charged Glass Rod

With the exception of compasses, you seldom see or personally experience forces due to the Earth's small magnetic field. To illustrate this, suppose that in a physics lab you rub a glass rod with silk, placing a 20-nC positive charge on it. Calculate the force on the rod due to the Earth's magnetic field, if you throw it with a horizontal velocity of 10 m/s due west in a place where the Earth's field is due north parallel to the ground. (The direction of the force is determined with right hand rule 1 as shown in **Figure 22.19**.)

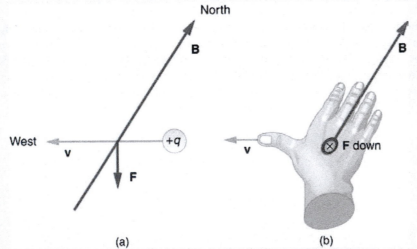

Figure 22.19 A positively charged object moving due west in a region where the Earth's magnetic field is due north experiences a force that is straight down as shown. A negative charge moving in the same direction would feel a force straight up.

Strategy

We are given the charge, its velocity, and the magnetic field strength and direction. We can thus use the equation $F = qvB \sin \theta$ to find the force.

Solution

The magnetic force is

$$F = qvb \sin \theta. \tag{22.4}$$

We see that $\sin \theta = 1$, since the angle between the velocity and the direction of the field is $90°$. Entering the other given quantities yields

$$
\begin{aligned}
F &= \left(20 \times 10^{-9} \text{ C}\right)\left(10 \text{ m/s}\right)\left(5 \times 10^{-5} \text{ T}\right) \\
&= 1 \times 10^{-11} \text{ (C} \cdot \text{m/s)}\left(\frac{\text{N}}{\text{C} \cdot \text{m/s}}\right) = 1 \times 10^{-11} \text{ N.}
\end{aligned}
\tag{22.5}
$$

Discussion

This force is completely negligible on any macroscopic object, consistent with experience. (It is calculated to only one digit, since the Earth's field varies with location and is given to only one digit.) The Earth's magnetic field, however, does produce very important effects, particularly on submicroscopic particles. Some of these are explored in **Force on a Moving Charge in a Magnetic Field: Examples and Applications.**

22.5 Force on a Moving Charge in a Magnetic Field: Examples and Applications

Learning Objectives
By the end of this section, you will be able to:
• Describe the effects of a magnetic field on a moving charge.
• Calculate the radius of curvature of the path of a charge that is moving in a magnetic field.
The information presented in this section supports the following AP® learning objectives and science practices:
• **3.C.3.1** The student is able to use right-hand rules to analyze a situation involving a current-carrying conductor and a moving electrically charged object to determine the direction of the magnetic force exerted on the charged object due to the magnetic field created by the current-carrying conductor. **(S.P. 1.4)**

Magnetic force can cause a charged particle to move in a circular or spiral path. Cosmic rays are energetic charged particles in outer space, some of which approach the Earth. They can be forced into spiral paths by the Earth's magnetic field. Protons in giant accelerators are kept in a circular path by magnetic force. The bubble chamber photograph in **Figure 22.20** shows charged particles moving in such curved paths. The curved paths of charged particles in magnetic fields are the basis of a number of phenomena and can even be used analytically, such as in a mass spectrometer.

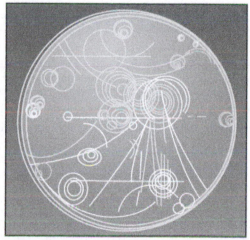

Figure 22.20 Trails of bubbles are produced by high-energy charged particles moving through the superheated liquid hydrogen in this artist's rendition of a bubble chamber. There is a strong magnetic field perpendicular to the page that causes the curved paths of the particles. The radius of the path can be used to find the mass, charge, and energy of the particle.

So does the magnetic force cause circular motion? Magnetic force is always perpendicular to velocity, so that it does no work on the charged particle. The particle's kinetic energy and speed thus remain constant. The direction of motion is affected, but not the speed. This is typical of uniform circular motion. The simplest case occurs when a charged particle moves perpendicular to a uniform B-field, such as shown in **Figure 22.21**. (If this takes place in a vacuum, the magnetic field is the dominant factor determining the motion.) Here, the magnetic force supplies the centripetal force $F_c = mv^2/r$. Noting that $\sin \theta = 1$, we see that $F = qvB$.

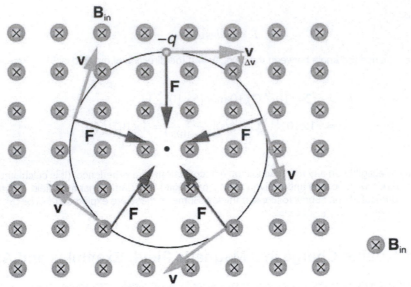

Figure 22.21 A negatively charged particle moves in the plane of the page in a region where the magnetic field is perpendicular into the page (represented by the small circles with x's—like the tails of arrows). The magnetic force is perpendicular to the velocity, and so velocity changes in direction but not magnitude. Uniform circular motion results.

Because the magnetic force F supplies the centripetal force F_c, we have

$$qvB = \frac{mv^2}{r}.$$

(22.6)

Solving for r yields

$$r = \frac{mv}{qB}.$$

(22.7)

Here, r is the radius of curvature of the path of a charged particle with mass m and charge q, moving at a speed v perpendicular to a magnetic field of strength B. If the velocity is not perpendicular to the magnetic field, then v is the component of the velocity perpendicular to the field. The component of the velocity parallel to the field is unaffected, since the magnetic force is zero for motion parallel to the field. This produces a spiral motion rather than a circular one.

Example 22.2 Calculating the Curvature of the Path of an Electron Moving in a Magnetic Field: A Magnet on a TV Screen

A magnet brought near an old-fashioned TV screen such as in **Figure 22.22** (TV sets with cathode ray tubes instead of LCD screens) severely distorts its picture by altering the path of the electrons that make its phosphors glow. *(Don't try this at home, as it will permanently magnetize and ruin the TV.)* To illustrate this, calculate the radius of curvature of the path of an electron having a velocity of 6.00×10^7 m/s (corresponding to the accelerating voltage of about 10.0 kV used in some TVs) perpendicular to a magnetic field of strength $B = 0.500$ T (obtainable with permanent magnets).

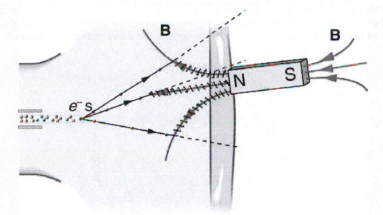

Figure 22.22 Side view showing what happens when a magnet comes in contact with a computer monitor or TV screen. Electrons moving toward the screen spiral about magnetic field lines, maintaining the component of their velocity parallel to the field lines. This distorts the image on the screen.

Strategy

We can find the radius of curvature r directly from the equation $r = \frac{mv}{qB}$, since all other quantities in it are given or known.

Solution

Using known values for the mass and charge of an electron, along with the given values of v and B gives us

$$r = \frac{mv}{qB} = \frac{(9.11 \times 10^{-31}\ \text{kg})(6.00 \times 10^{7}\ \text{m/s})}{(1.60 \times 10^{-19}\ \text{C})(0.500\ \text{T})}$$ (22.8)

$$= 6.83 \times 10^{-4}\ \text{m}$$

or

$$r = 0.683\ \text{mm.}$$ (22.9)

Discussion

The small radius indicates a large effect. The electrons in the TV picture tube are made to move in very tight circles, greatly altering their paths and distorting the image.

Figure 22.23 shows how electrons not moving perpendicular to magnetic field lines follow the field lines. The component of velocity parallel to the lines is unaffected, and so the charges spiral along the field lines. If field strength increases in the direction of motion, the field will exert a force to slow the charges, forming a kind of magnetic mirror, as shown below.

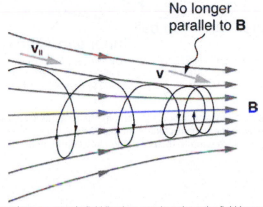

Figure 22.23 When a charged particle moves along a magnetic field line into a region where the field becomes stronger, the particle experiences a force that reduces the component of velocity parallel to the field. This force slows the motion along the field line and here reverses it, forming a "magnetic mirror."

The properties of charged particles in magnetic fields are related to such different things as the Aurora Australis or Aurora Borealis and particle accelerators. *Charged particles approaching magnetic field lines may get trapped in spiral orbits about the lines rather than crossing them*, as seen above. Some cosmic rays, for example, follow the Earth's magnetic field lines, entering the atmosphere near the magnetic poles and causing the southern or northern lights through their ionization of molecules in the atmosphere. This glow of energized atoms and molecules is seen in **Figure 22.1**. Those particles that approach middle latitudes must cross magnetic field lines, and many are prevented from penetrating the atmosphere. Cosmic rays are a component of background radiation; consequently, they give a higher radiation dose at the poles than at the equator.

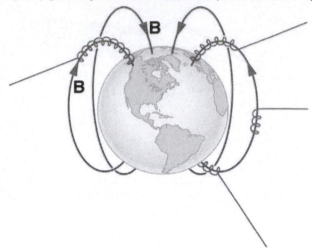

Figure 22.24 Energetic electrons and protons, components of cosmic rays, from the Sun and deep outer space often follow the Earth's magnetic field lines rather than cross them. (Recall that the Earth's north magnetic pole is really a south pole in terms of a bar magnet.)

Some incoming charged particles become trapped in the Earth's magnetic field, forming two belts above the atmosphere known as the Van Allen radiation belts after the discoverer James A. Van Allen, an American astrophysicist. (See **Figure 22.25**.) Particles trapped in these belts form radiation fields (similar to nuclear radiation) so intense that manned space flights avoid them and satellites with sensitive electronics are kept out of them. In the few minutes it took lunar missions to cross the Van Allen radiation belts, astronauts received radiation doses more than twice the allowed annual exposure for radiation workers. Other planets have similar belts, especially those having strong magnetic fields like Jupiter.

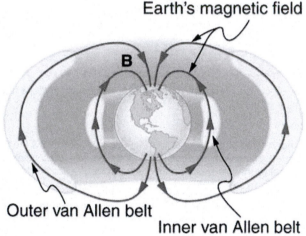

Figure 22.25 The Van Allen radiation belts are two regions in which energetic charged particles are trapped in the Earth's magnetic field. One belt lies about 300 km above the Earth's surface, the other about 16,000 km. Charged particles in these belts migrate along magnetic field lines and are partially reflected away from the poles by the stronger fields there. The charged particles that enter the atmosphere are replenished by the Sun and sources in deep outer space.

Back on Earth, we have devices that employ magnetic fields to contain charged particles. Among them are the giant particle accelerators that have been used to explore the substructure of matter. (See **Figure 22.26**.) Magnetic fields not only control the direction of the charged particles, they also are used to focus particles into beams and overcome the repulsion of like charges in these beams.

Figure 22.26 The Fermilab facility in Illinois has a large particle accelerator (the most powerful in the world until 2008) that employs magnetic fields (magnets seen here in orange) to contain and direct its beam. This and other accelerators have been in use for several decades and have allowed us to discover some of the laws underlying all matter. (credit: ammcrim, Flickr)

Thermonuclear fusion (like that occurring in the Sun) is a hope for a future clean energy source. One of the most promising devices is the *tokamak*, which uses magnetic fields to contain (or trap) and direct the reactive charged particles. (See **Figure 22.27**.) Less exotic, but more immediately practical, amplifiers in microwave ovens use a magnetic field to contain oscillating electrons. These oscillating electrons generate the microwaves sent into the oven.

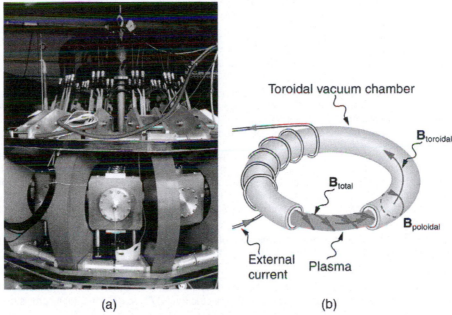

Figure 22.27 Tokamaks such as the one shown in the figure are being studied with the goal of economical production of energy by nuclear fusion. Magnetic fields in the doughnut-shaped device contain and direct the reactive charged particles. (credit: David Mellis, Flickr)

Mass spectrometers have a variety of designs, and many use magnetic fields to measure mass. The curvature of a charged particle's path in the field is related to its mass and is measured to obtain mass information. (See **More Applications of Magnetism**.) Historically, such techniques were employed in the first direct observations of electron charge and mass. Today, mass spectrometers (sometimes coupled with gas chromatographs) are used to determine the make-up and sequencing of large biological molecules.

22.6 The Hall Effect

Learning Objectives

By the end of this section, you will be able to:

- Describe the Hall effect.
- Calculate the Hall emf across a current-carrying conductor.

The information presented in this section supports the following AP® learning objectives and science practices:

- **3.C.3.1** The student is able to use right-hand rules to analyze a situation involving a current-carrying conductor and a moving electrically charged object to determine the direction of the magnetic force exerted on the charged object due to the magnetic field created by the current-carrying conductor. **(S.P. 1.4)**

We have seen effects of a magnetic field on free-moving charges. The magnetic field also affects charges moving in a conductor.

One result is the Hall effect, which has important implications and applications.

Figure 22.28 shows what happens to charges moving through a conductor in a magnetic field. The field is perpendicular to the electron drift velocity and to the width of the conductor. Note that conventional current is to the right in both parts of the figure. In part (a), electrons carry the current and move to the left. In part (b), positive charges carry the current and move to the right. Moving electrons feel a magnetic force toward one side of the conductor, leaving a net positive charge on the other side. This separation of charge *creates a voltage* ε, known as the **Hall emf**, *across* the conductor. The creation of a voltage *across* a current-carrying conductor by a magnetic field is known as the **Hall effect**, after Edwin Hall, the American physicist who discovered it in 1879.

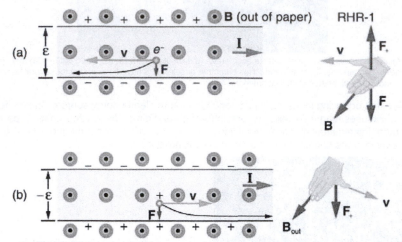

Figure 22.28 The Hall effect. (a) Electrons move to the left in this flat conductor (conventional current to the right). The magnetic field is directly out of the page, represented by circled dots; it exerts a force on the moving charges, causing a voltage ε, the Hall emf, across the conductor. (b) Positive charges moving to the right (conventional current also to the right) are moved to the side, producing a Hall emf of the opposite sign, $-\varepsilon$. Thus, if the direction of the field and current are known, the sign of the charge carriers can be determined from the Hall effect.

One very important use of the Hall effect is to determine whether positive or negative charges carries the current. Note that in **Figure 22.28(b)**, where positive charges carry the current, the Hall emf has the sign opposite to when negative charges carry the current. Historically, the Hall effect was used to show that electrons carry current in metals and it also shows that positive charges carry current in some semiconductors. The Hall effect is used today as a research tool to probe the movement of charges, their drift velocities and densities, and so on, in materials. In 1980, it was discovered that the Hall effect is quantized, an example of quantum behavior in a macroscopic object.

The Hall effect has other uses that range from the determination of blood flow rate to precision measurement of magnetic field strength. To examine these quantitatively, we need an expression for the Hall emf, ε, across a conductor. Consider the balance of forces on a moving charge in a situation where B, v, and l are mutually perpendicular, such as shown in **Figure 22.29**. Although the magnetic force moves negative charges to one side, they cannot build up without limit. The electric field caused by their separation opposes the magnetic force, $F = qvB$, and the electric force, $F_e = qE$, eventually grows to equal it. That is,

$$qE = qvB \tag{22.10}$$

or

$$E = vB. \tag{22.11}$$

Note that the electric field E is uniform across the conductor because the magnetic field B is uniform, as is the conductor. For a uniform electric field, the relationship between electric field and voltage is $E = \varepsilon / l$, where l is the width of the conductor and ε is the Hall emf. Entering this into the last expression gives

$$\frac{\varepsilon}{l} = vB. \tag{22.12}$$

Solving this for the Hall emf yields

$$\varepsilon = Blv \ (B, \ v, \text{ and } l, \text{ mutually perpendicular}), \tag{22.13}$$

where ε is the Hall effect voltage across a conductor of width l through which charges move at a speed v.

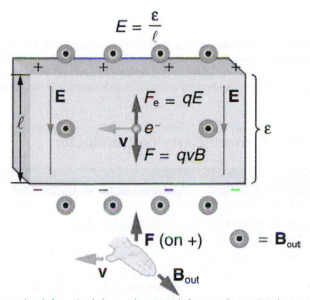

$$E = \frac{\varepsilon}{\ell}$$

Figure 22.29 The Hall emf ε produces an electric force that balances the magnetic force on the moving charges. The magnetic force produces charge separation, which builds up until it is balanced by the electric force, an equilibrium that is quickly reached.

One of the most common uses of the Hall effect is in the measurement of magnetic field strength B. Such devices, called *Hall probes*, can be made very small, allowing fine position mapping. Hall probes can also be made very accurate, usually accomplished by careful calibration. Another application of the Hall effect is to measure fluid flow in any fluid that has free charges (most do). (See **Figure 22.30**.) A magnetic field applied perpendicular to the flow direction produces a Hall emf ε as shown. Note that the sign of ε depends not on the sign of the charges, but only on the directions of B and v. The magnitude of the Hall emf is $\varepsilon = Blv$, where l is the pipe diameter, so that the average velocity v can be determined from ε providing the other factors are known.

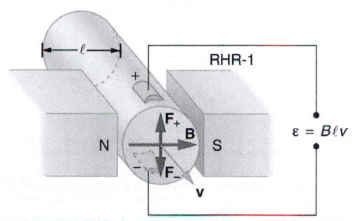

Figure 22.30 The Hall effect can be used to measure fluid flow in any fluid having free charges, such as blood. The Hall emf ε is measured across the tube perpendicular to the applied magnetic field and is proportional to the average velocity v.

Example 22.3 Calculating the Hall emf: Hall Effect for Blood Flow

A Hall effect flow probe is placed on an artery, applying a 0.100-T magnetic field across it, in a setup similar to that in **Figure 22.30**. What is the Hall emf, given the vessel's inside diameter is 4.00 mm and the average blood velocity is 20.0 cm/s?

Strategy

Because B, v, and l are mutually perpendicular, the equation $\varepsilon = Blv$ can be used to find ε.

Solution

Entering the given values for B, v, and l gives

$$\begin{aligned}
\varepsilon &= Blv = (0.100 \text{ T})(4.00 \times 10^{-3} \text{ m})(0.200 \text{ m/s}) \\
&= 80.0 \text{ }\mu\text{V}
\end{aligned}$$
(22.14)

Discussion

This is the average voltage output. Instantaneous voltage varies with pulsating blood flow. The voltage is small in this type of measurement. ε is particularly difficult to measure, because there are voltages associated with heart action (ECG voltages) that are on the order of millivolts. In practice, this difficulty is overcome by applying an AC magnetic field, so that the Hall emf is AC with the same frequency. An amplifier can be very selective in picking out only the appropriate frequency, eliminating signals and noise at other frequencies.

22.7 Magnetic Force on a Current-Carrying Conductor

Learning Objectives

By the end of this section, you will be able to:

- Describe the effects of a magnetic force on a current-carrying conductor.
- Calculate the magnetic force on a current-carrying conductor.

The information presented in this section supports the following AP® learning objectives and science practices:

- **3.C.3.1** The student is able to use right-hand rules to analyze a situation involving a current-carrying conductor and a moving electrically charged object to determine the direction of the magnetic force exerted on the charged object due to the magnetic field created by the current-carrying conductor. **(S.P. 1.4)**

Because charges ordinarily cannot escape a conductor, the magnetic force on charges moving in a conductor is transmitted to the conductor itself.

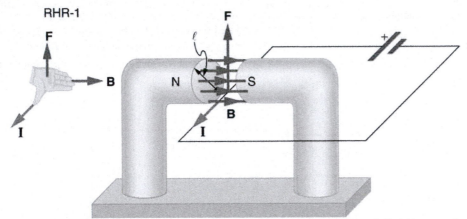

Figure 22.31 The magnetic field exerts a force on a current-carrying wire in a direction given by the right hand rule 1 (the same direction as that on the individual moving charges). This force can easily be large enough to move the wire, since typical currents consist of very large numbers of moving charges.

We can derive an expression for the magnetic force on a current by taking a sum of the magnetic forces on individual charges. (The forces add because they are in the same direction.) The force on an individual charge moving at the drift velocity v_d is given by $F = qv_dB \sin \theta$. Taking B to be uniform over a length of wire l and zero elsewhere, the total magnetic force on the wire is then $F = (qv_dB \sin \theta)(N)$, where N is the number of charge carriers in the section of wire of length l. Now, $N = nV$, where n is the number of charge carriers per unit volume and V is the volume of wire in the field. Noting that $V = Al$, where A is the cross-sectional area of the wire, then the force on the wire is $F = (qv_dB \sin \theta)(nAl)$. Gathering terms,

$$F = (nqAv_d)lB \sin \theta. \tag{22.15}$$

Because $nqAv_d = I$ (see **Current**),

$$F = IlB \sin \theta \tag{22.16}$$

is the equation for *magnetic force on a length l of wire carrying a current I in a uniform magnetic field B*, as shown in **Figure 22.32**. If we divide both sides of this expression by l, we find that the magnetic force per unit length of wire in a uniform field is $\frac{F}{l} = IB \sin \theta$. The direction of this force is given by RHR-1, with the thumb in the direction of the current I. Then, with the fingers in the direction of B, a perpendicular to the palm points in the direction of F, as in **Figure 22.32**.

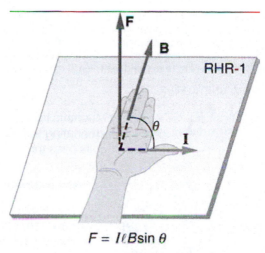

$$F = IlB\sin\theta$$

F ⊥ plane of **I** and **B**

Figure 22.32 The force on a current-carrying wire in a magnetic field is $F = IlB\sin\theta$. Its direction is given by RHR-1.

Example 22.4 Calculating Magnetic Force on a Current-Carrying Wire: A Strong Magnetic Field

Calculate the force on the wire shown in **Figure 22.31**, given $B = 1.50$ T, $l = 5.00$ cm, and $I = 20.0$ A.

Strategy

The force can be found with the given information by using $F = IlB\sin\theta$ and noting that the angle θ between I and B is $90°$, so that $\sin\theta = 1$.

Solution

Entering the given values into $F = IlB\sin\theta$ yields

$$F = IlB\sin\theta = (20.0\text{ A})(0.0500\text{ m})(1.50\text{ T})(1). \tag{22.17}$$

The units for tesla are $1\text{ T} = \dfrac{\text{N}}{\text{A}\cdot\text{m}}$; thus,

$$F = 1.50\text{ N}. \tag{22.18}$$

Discussion

This large magnetic field creates a significant force on a small length of wire.

Magnetic force on current-carrying conductors is used to convert electric energy to work. (Motors are a prime example—they employ loops of wire and are considered in the next section.) Magnetohydrodynamics (MHD) is the technical name given to a clever application where magnetic force pumps fluids without moving mechanical parts. (See **Figure 22.33**.)

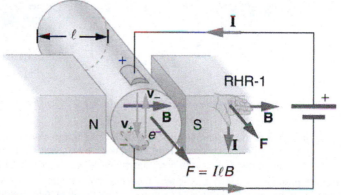

Figure 22.33 Magnetohydrodynamics. The magnetic force on the current passed through this fluid can be used as a nonmechanical pump.

A strong magnetic field is applied across a tube and a current is passed through the fluid at right angles to the field, resulting in a force on the fluid parallel to the tube axis as shown. The absence of moving parts makes this attractive for moving a hot, chemically active substance, such as the liquid sodium employed in some nuclear reactors. Experimental artificial hearts are testing with this technique for pumping blood, perhaps circumventing the adverse effects of mechanical pumps. (Cell membranes, however, are affected by the large fields needed in MHD, delaying its practical application in humans.) MHD propulsion for nuclear submarines has been proposed, because it could be considerably quieter than conventional propeller drives. The deterrent value of nuclear submarines is based on their ability to hide and survive a first or second nuclear strike. As we slowly disassemble our nuclear weapons arsenals, the submarine branch will be the last to be decommissioned because of this ability (See **Figure 22.34**.) Existing MHD drives are heavy and inefficient—much development work is needed.

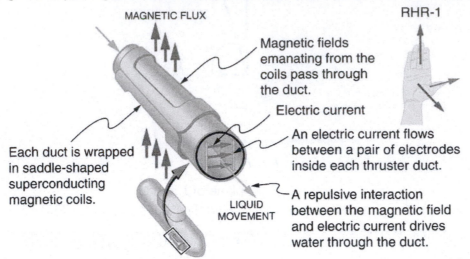

Figure 22.34 An MHD propulsion system in a nuclear submarine could produce significantly less turbulence than propellers and allow it to run more silently. The development of a silent drive submarine was dramatized in the book and the film *The Hunt for Red October*.

22.8 Torque on a Current Loop: Motors and Meters

Learning Objectives
By the end of this section, you will be able to: • Describe how motors and meters work in terms of torque on a current loop. • Calculate the torque on a current-carrying loop in a magnetic field.

Motors are the most common application of magnetic force on current-carrying wires. Motors have loops of wire in a magnetic field. When current is passed through the loops, the magnetic field exerts torque on the loops, which rotates a shaft. Electrical energy is converted to mechanical work in the process. (See **Figure 22.35**.)

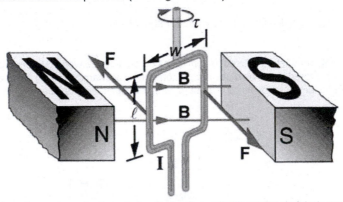

Figure 22.35 Torque on a current loop. A current-carrying loop of wire attached to a vertically rotating shaft feels magnetic forces that produce a clockwise torque as viewed from above.

Let us examine the force on each segment of the loop in **Figure 22.35** to find the torques produced about the axis of the vertical shaft. (This will lead to a useful equation for the torque on the loop.) We take the magnetic field to be uniform over the rectangular loop, which has width w and height l. First, we note that the forces on the top and bottom segments are vertical and, therefore, parallel to the shaft, producing no torque. Those vertical forces are equal in magnitude and opposite in direction, so that they also produce no net force on the loop. **Figure 22.36** shows views of the loop from above. Torque is defined as $\tau = rF \sin \theta$, where F is the force, r is the distance from the pivot that the force is applied, and θ is the angle between r and F. As seen in **Figure 22.36**(a), right hand rule 1 gives the forces on the sides to be equal in magnitude and opposite in direction, so that the net force is again zero. However, each force produces a clockwise torque. Since $r = w/2$, the torque on each vertical segment is $(w/2)F \sin \theta$, and the two add to give a total torque.

$$\tau = \frac{w}{2}F \sin \theta + \frac{w}{2}F \sin \theta = wF \sin \theta \qquad (22.19)$$

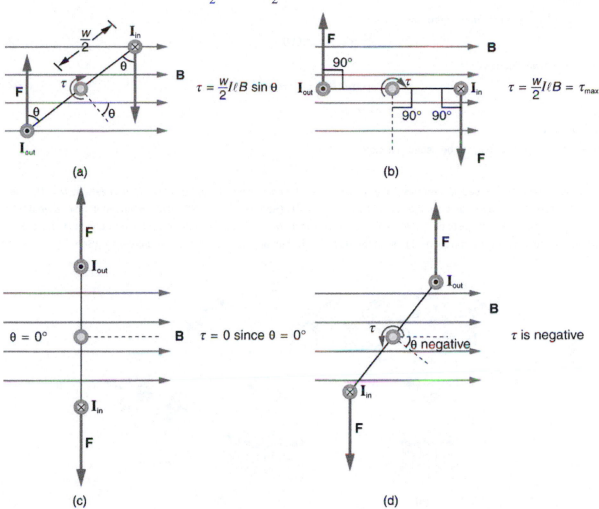

(a)

(b)

(c)

(d)

Figure 22.36 Top views of a current-carrying loop in a magnetic field. (a) The equation for torque is derived using this view. Note that the perpendicular to the loop makes an angle θ with the field that is the same as the angle between $w/2$ and $\mathbf{F}$. (b) The maximum torque occurs when θ is a right angle and $\sin \theta = 1$. (c) Zero (minimum) torque occurs when θ is zero and $\sin \theta = 0$. (d) The torque reverses once the loop rotates past $\theta = 0$.

Now, each vertical segment has a length l that is perpendicular to B, so that the force on each is $F = IlB$. Entering F into the expression for torque yields

$$\tau = wIlB \sin \theta. \qquad (22.20)$$

If we have a multiple loop of N turns, we get N times the torque of one loop. Finally, note that the area of the loop is $A = wl$; the expression for the torque becomes

$$\tau = NIAB \sin \theta. \qquad (22.21)$$

This is the torque on a current-carrying loop in a uniform magnetic field. This equation can be shown to be valid for a loop of any shape. The loop carries a current I, has N turns, each of area A, and the perpendicular to the loop makes an angle θ with the field B. The net force on the loop is zero.

Example 22.5 Calculating Torque on a Current-Carrying Loop in a Strong Magnetic Field

Find the maximum torque on a 100-turn square loop of a wire of 10.0 cm on a side that carries 15.0 A of current in a 2.00-T field.

Strategy

Torque on the loop can be found using $\tau = NIAB \sin \theta$. Maximum torque occurs when $\theta = 90°$ and $\sin \theta = 1$.

Solution

For $\sin \theta = 1$, the maximum torque is

$$\tau_{\max} = NIAB. \qquad (22.22)$$

Entering known values yields

$$
\begin{aligned}
\tau_{\max} &= (100)(15.0 \text{ A})(0.100 \text{ m}^2)(2.00 \text{ T}) \qquad (22.23) \\
&= 30.0 \text{ N} \cdot \text{m}.
\end{aligned}
$$

Discussion

This torque is large enough to be useful in a motor.

The torque found in the preceding example is the maximum. As the coil rotates, the torque decreases to zero at $\theta = 0$. The torque then *reverses* its direction once the coil rotates past $\theta = 0$. (See **Figure 22.36**(d).) This means that, unless we do something, the coil will oscillate back and forth about equilibrium at $\theta = 0$. To get the coil to continue rotating in the same direction, we can reverse the current as it passes through $\theta = 0$ with automatic switches called *brushes*. (See **Figure 22.37**.)

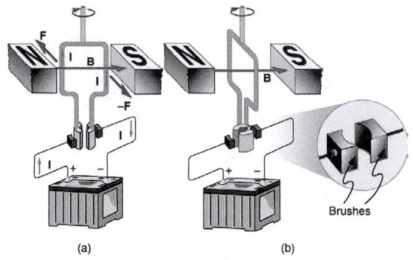

Figure 22.37 (a) As the angular momentum of the coil carries it through $\theta = 0$, the brushes reverse the current to keep the torque clockwise. (b) The coil will rotate continuously in the clockwise direction, with the current reversing each half revolution to maintain the clockwise torque.

Meters, such as those in analog fuel gauges on a car, are another common application of magnetic torque on a current-carrying loop. **Figure 22.38** shows that a meter is very similar in construction to a motor. The meter in the figure has its magnets shaped to limit the effect of θ by making B perpendicular to the loop over a large angular range. Thus the torque is proportional to I and not θ. A linear spring exerts a counter-torque that balances the current-produced torque. This makes the needle deflection proportional to I. If an exact proportionality cannot be achieved, the gauge reading can be calibrated. To produce a galvanometer for use in analog voltmeters and ammeters that have a low resistance and respond to small currents, we use a large loop area A, high magnetic field B, and low-resistance coils.

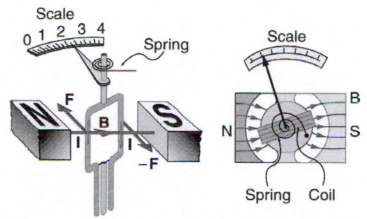

Figure 22.38 Meters are very similar to motors but only rotate through a part of a revolution. The magnetic poles of this meter are shaped to keep the component of B perpendicular to the loop constant, so that the torque does not depend on θ and the deflection against the return spring is proportional only to the current I.

22.9 Magnetic Fields Produced by Currents: Ampere's Law

Learning Objectives

By the end of this section, you will be able to:

- Calculate current that produces a magnetic field.
- Use the right-hand rule 2 to determine the direction of current or the direction of magnetic field loops.

The information presented in this section supports the following AP® learning objectives and science practices:

- **2.D.2.1** The student is able to create a verbal or visual representation of a magnetic field around a long straight wire or a pair of parallel wires. **(S.P. 1.1)**
- **3.C.3.1** The student is able to use right-hand rules to analyze a situation involving a current-carrying conductor and a moving electrically charged object to determine the direction of the magnetic force exerted on the charged object due to the magnetic field created by the current-carrying conductor. **(S.P. 1.4)**
- **3.C.3.2** The student is able to plan a data collection strategy appropriate to an investigation of the direction of the force on a moving electrically charged object caused by a current in a wire in the context of a specific set of equipment and instruments and analyze the resulting data to arrive at a conclusion. **(S.P. 4.2, 5.1)**

How much current is needed to produce a significant magnetic field, perhaps as strong as the Earth's field? Surveyors will tell you that overhead electric power lines create magnetic fields that interfere with their compass readings. Indeed, when Oersted discovered in 1820 that a current in a wire affected a compass needle, he was not dealing with extremely large currents. How does the shape of wires carrying current affect the shape of the magnetic field created? We noted earlier that a current loop created a magnetic field similar to that of a bar magnet, but what about a straight wire or a toroid (doughnut)? How is the direction of a current-created field related to the direction of the current? Answers to these questions are explored in this section, together with a brief discussion of the law governing the fields created by currents.

Magnetic Field Created by a Long Straight Current-Carrying Wire: Right Hand Rule 2

Magnetic fields have both direction and magnitude. As noted before, one way to explore the direction of a magnetic field is with compasses, as shown for a long straight current-carrying wire in **Figure 22.39**. Hall probes can determine the magnitude of the field. The field around a long straight wire is found to be in circular loops. The **right hand rule 2** (RHR-2) emerges from this exploration and is valid for any current segment—*point the thumb in the direction of the current, and the fingers curl in the direction of the magnetic field loops* created by it.

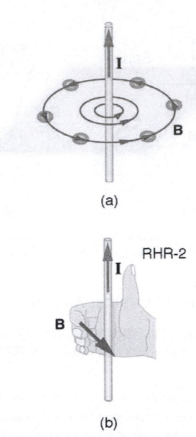

(a)

(b)

Figure 22.39 (a) Compasses placed near a long straight current-carrying wire indicate that field lines form circular loops centered on the wire. (b) Right hand rule 2 states that, if the right hand thumb points in the direction of the current, the fingers curl in the direction of the field. This rule is consistent with the field mapped for the long straight wire and is valid for any current segment.

Making Connections: Notation

For a wire oriented perpendicular to the page, if the current in the wire is directed out of the page, the right-hand rule tells us that the magnetic field lines will be oriented in a counterclockwise direction around the wire. If the current in the wire is directed into the page, the magnetic field lines will be oriented in a clockwise direction around the wire. We use $\odot$ to indicate that the direction of the current in the wire is out of the page, and $\otimes$ for the direction into the page.

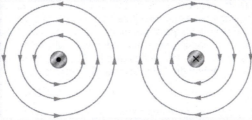

Figure 22.40 Two parallel wires have currents pointing into or out of the page as shown. The direction of the magnetic field in the vicinity of the two wires is shown.

The **magnetic field strength (magnitude) produced by a long straight current-carrying wire** is found by experiment to be

$$B = \frac{\mu_0 I}{2\pi r} \text{ (long straight wire)}, \tag{22.24}$$

where I is the current, r is the shortest distance to the wire, and the constant $\mu_0 = 4\pi \times 10^{-7} \text{ T} \cdot \text{m/A}$ is the **permeability of free space**. (μ_0 is one of the basic constants in nature. We will see later that μ_0 is related to the speed of light.) Since the wire is very long, the magnitude of the field depends only on distance from the wire r, not on position along the wire.

Example 22.6 Calculating Current that Produces a Magnetic Field

Find the current in a long straight wire that would produce a magnetic field twice the strength of the Earth's at a distance of 5.0 cm from the wire.

Strategy

The Earth's field is about 5.0×10^{-5} T , and so here B due to the wire is taken to be 1.0×10^{-4} T . The equation $B = \frac{\mu_0 I}{2\pi r}$ can be used to find I , since all other quantities are known.

Solution

Solving for I and entering known values gives

$$\begin{aligned} I &= \frac{2\pi r B}{\mu_0} = \frac{2\pi \left(5.0 \times 10^{-2} \text{ m}\right)\left(1.0 \times 10^{-4} \text{ T}\right)}{4\pi \times 10^{-7} \text{ T} \cdot \text{m/A}} \\ &= 25 \text{ A}. \end{aligned} \qquad (22.25)$$

Discussion

So a moderately large current produces a significant magnetic field at a distance of 5.0 cm from a long straight wire. Note that the answer is stated to only two digits, since the Earth's field is specified to only two digits in this example.

Ampere's Law and Others

The magnetic field of a long straight wire has more implications than you might at first suspect. *Each segment of current produces a magnetic field like that of a long straight wire, and the total field of any shape current is the vector sum of the fields due to each segment.* The formal statement of the direction and magnitude of the field due to each segment is called the **Biot-Savart law**. Integral calculus is needed to sum the field for an arbitrary shape current. This results in a more complete law, called **Ampere's law**, which relates magnetic field and current in a general way. Ampere's law in turn is a part of **Maxwell's equations**, which give a complete theory of all electromagnetic phenomena. Considerations of how Maxwell's equations appear to different observers led to the modern theory of relativity, and the realization that electric and magnetic fields are different manifestations of the same thing. Most of this is beyond the scope of this text in both mathematical level, requiring calculus, and in the amount of space that can be devoted to it. But for the interested student, and particularly for those who continue in physics, engineering, or similar pursuits, delving into these matters further will reveal descriptions of nature that are elegant as well as profound. In this text, we shall keep the general features in mind, such as RHR-2 and the rules for magnetic field lines listed in **Magnetic Fields and Magnetic Field Lines**, while concentrating on the fields created in certain important situations.

Making Connections: Relativity

Hearing all we do about Einstein, we sometimes get the impression that he invented relativity out of nothing. On the contrary, one of Einstein's motivations was to solve difficulties in knowing how different observers see magnetic and electric fields.

Magnetic Field Produced by a Current-Carrying Circular Loop

The magnetic field near a current-carrying loop of wire is shown in **Figure 22.41**. Both the direction and the magnitude of the magnetic field produced by a current-carrying loop are complex. RHR-2 can be used to give the direction of the field near the loop, but mapping with compasses and the rules about field lines given in **Magnetic Fields and Magnetic Field Lines** are needed for more detail. There is a simple formula for the **magnetic field strength at the center of a circular loop**. It is

$$B = \frac{\mu_0 I}{2R} \text{ (at center of loop)}, \qquad (22.26)$$

where R is the radius of the loop. This equation is very similar to that for a straight wire, but it is valid *only* at the center of a circular loop of wire. The similarity of the equations does indicate that similar field strength can be obtained at the center of a loop. One way to get a larger field is to have N loops; then, the field is $B = N\mu_0 I / (2R)$. Note that the larger the loop, the smaller the field at its center, because the current is farther away.

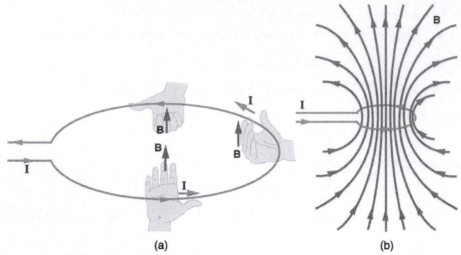

Figure 22.41 (a) RHR-2 gives the direction of the magnetic field inside and outside a current-carrying loop. (b) More detailed mapping with compasses or with a Hall probe completes the picture. The field is similar to that of a bar magnet.

Magnetic Field Produced by a Current-Carrying Solenoid

A **solenoid** is a long coil of wire (with many turns or loops, as opposed to a flat loop). Because of its shape, the field inside a solenoid can be very uniform, and also very strong. The field just outside the coils is nearly zero. **Figure 22.42** shows how the field looks and how its direction is given by RHR-2.

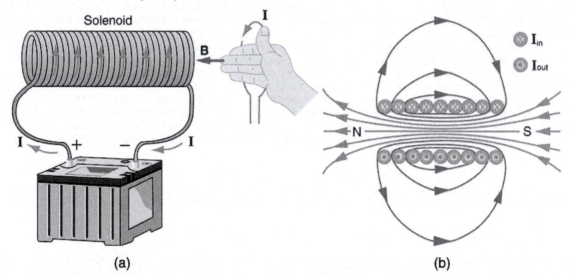

Figure 22.42 (a) Because of its shape, the field inside a solenoid of length l is remarkably uniform in magnitude and direction, as indicated by the straight and uniformly spaced field lines. The field outside the coils is nearly zero. (b) This cutaway shows the magnetic field generated by the current in the solenoid.

The magnetic field inside of a current-carrying solenoid is very uniform in direction and magnitude. Only near the ends does it begin to weaken and change direction. The field outside has similar complexities to flat loops and bar magnets, but the **magnetic field strength inside a solenoid** is simply

$$B = \mu_0 nI \quad \text{(inside a solenoid)}, \tag{22.27}$$

where n is the number of loops per unit length of the solenoid ($n = N/l$, with N being the number of loops and l the length). Note that B is the field strength anywhere in the uniform region of the interior and not just at the center. Large uniform fields spread over a large volume are possible with solenoids, as **Example 22.7** implies.

Example 22.7 Calculating Field Strength inside a Solenoid

What is the field inside a 2.00-m-long solenoid that has 2000 loops and carries a 1600-A current?

Strategy

To find the field strength inside a solenoid, we use $B = \mu_0 nI$. First, we note the number of loops per unit length is

$$n = \frac{N}{l} = \frac{2000}{2.00 \text{ m}} = 1000 \text{ m}^{-1} = 10 \text{ cm}^{-1}. \tag{22.28}$$

Solution

Substituting known values gives

$$
\begin{aligned}
B &= \mu_0 nI = \left(4\pi \times 10^{-7} \text{ T} \cdot \text{m/A}\right)\left(1000 \text{ m}^{-1}\right)(1600 \text{ A}) \\
&= 2.01 \text{ T}.
\end{aligned}
\tag{22.29}
$$

Discussion

This is a large field strength that could be established over a large-diameter solenoid, such as in medical uses of magnetic resonance imaging (MRI). The very large current is an indication that the fields of this strength are not easily achieved, however. Such a large current through 1000 loops squeezed into a meter's length would produce significant heating. Higher currents can be achieved by using superconducting wires, although this is expensive. There is an upper limit to the current, since the superconducting state is disrupted by very large magnetic fields.

Applying the Science Practices: Charged Particle in a Magnetic Field
---

Visit **here (http://openstaxcollege.org/l/31particlemagnetic)** and start the simulation applet "Particle in a Magnetic Field (2D)" in order to explore the magnetic force that acts on a charged particle in a magnetic field. Experiment with the simulation to see how it works and what parameters you can change; then construct a plan to methodically investigate how magnetic fields affect charged particles. Some questions you may want to answer as part of your experiment are:

- Are the paths of charged particles in magnetic fields always similar in two dimensions? Why or why not?
- How would the path of a neutral particle in the magnetic field compare to the path of a charged particle?
- How would the path of a positive particle differ from the path of a negative particle in a magnetic field?
- What quantities dictate the properties of the particle's path?
- If you were attempting to measure the mass of a charged particle moving through a magnetic field, what would you need to measure about its path? Would you need to see it moving at many different velocities or through different field strengths, or would one trial be sufficient if your measurements were correct?
- Would doubling the charge change the path through the field? Predict an answer to this question, and then test your hypothesis.
- Would doubling the velocity change the path through the field? Predict an answer to this question, and then test your hypothesis.
- Would doubling the magnetic field strength change the path through the field? Predict an answer to this question, and then test your hypothesis.
- Would increasing the mass change the path? Predict an answer to this question, and then test your hypothesis.

There are interesting variations of the flat coil and solenoid. For example, the toroidal coil used to confine the reactive particles in tokamaks is much like a solenoid bent into a circle. The field inside a toroid is very strong but circular. Charged particles travel in circles, following the field lines, and collide with one another, perhaps inducing fusion. But the charged particles do not cross field lines and escape the toroid. A whole range of coil shapes are used to produce all sorts of magnetic field shapes. Adding ferromagnetic materials produces greater field strengths and can have a significant effect on the shape of the field. Ferromagnetic materials tend to trap magnetic fields (the field lines bend into the ferromagnetic material, leaving weaker fields outside it) and are used as shields for devices that are adversely affected by magnetic fields, including the Earth's magnetic field.

PhET Explorations: Generator

Generate electricity with a bar magnet! Discover the physics behind the phenomena by exploring magnets and how you can use them to make a bulb light.

Figure 22.43 Generator (http://cnx.org/content/m55399/1.4/generator_en.jar)

22.10 Magnetic Force between Two Parallel Conductors

You might expect that there are significant forces between current-carrying wires, since ordinary currents produce significant magnetic fields and these fields exert significant forces on ordinary currents. But you might not expect that the force between wires is used to *define* the ampere. It might also surprise you to learn that this force has something to do with why large circuit breakers burn up when they attempt to interrupt large currents.

The force between two long straight and parallel conductors separated by a distance r can be found by applying what we have developed in preceding sections. **Figure 22.44** shows the wires, their currents, the fields they create, and the subsequent forces they exert on one another. Let us consider the field produced by wire 1 and the force it exerts on wire 2 (call the force F_2). The field due to I_1 at a distance r is given to be

$$B_1 = \frac{\mu_0 I_1}{2\pi r}.$$

(22.30)

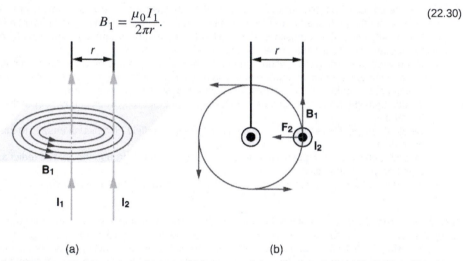

(a)   (b)

Figure 22.44 (a) The magnetic field produced by a long straight conductor is perpendicular to a parallel conductor, as indicated by RHR-2. (b) A view from above of the two wires shown in (a), with one magnetic field line shown for each wire. RHR-1 shows that the force between the parallel conductors is attractive when the currents are in the same direction. A similar analysis shows that the force is repulsive between currents in opposite directions.

This field is uniform along wire 2 and perpendicular to it, and so the force F_2 it exerts on wire 2 is given by $F = IlB \sin \theta$ with $\sin \theta = 1$:

$$F_2 = I_2 l B_1.$$

(22.31)

By Newton's third law, the forces on the wires are equal in magnitude, and so we just write F for the magnitude of F_2. (Note that $F_1 = -F_2$.) Since the wires are very long, it is convenient to think in terms of F/l, the force per unit length. Substituting the expression for B_1 into the last equation and rearranging terms gives

$$\frac{F}{l} = \frac{\mu_0 I_1 I_2}{2\pi r}.$$

(22.32)

F/l is the force per unit length between two parallel currents I_1 and I_2 separated by a distance r. The force is attractive if the currents are in the same direction and repulsive if they are in opposite directions.

Making Connections: Field Canceling

For two parallel wires, the fields will tend to cancel out in the area between the wires.

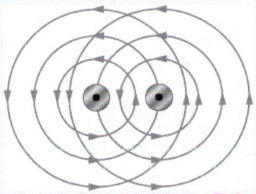

Figure 22.45 Two parallel wires have currents pointing in the same direction, out of the page. The direction of the magnetic fields induced by each wire is shown.

Note that the magnetic influence of the wire on the left-hand side extends beyond the wire on the right-hand side. To the right of both wires, the total magnetic field is directed toward the top of the page and is the result of the sum of the fields of both wires. Obviously, the closer wire has a greater effect on the overall magnetic field, but the more distant wire also contributes. One wire cannot block the magnetic field of another wire any more than a massive stone floor beneath you can block the gravitational field of the Earth.

Parallel wires with currents in the same direction attract, as you can see if we isolate the magnetic field lines of wire 2 influencing the current in wire 1. Right-hand rule 1 tells us the direction of the resulting magnetic force.

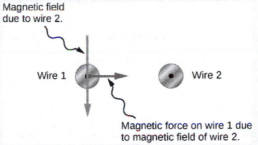

Figure 22.46 The same two wires are shown, but now only a part of the magnetic field due to wire 2 is shown in order to demonstrate how the magnetic force from wire 2 affects wire 1.

When the currents point in opposite directions as shown, the magnetic field in between the two wires is augmented. In the region outside of the two wires, along the horizontal line connecting the wires, the magnetic fields partially cancel.

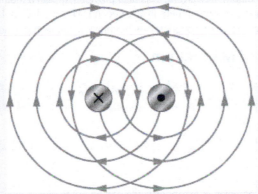

Figure 22.47 Two wires with parallel currents pointing in opposite directions are shown. The direction of the magnetic field due to each wire is indicated.

Parallel wires with currents in opposite directions repel, as you can see if we isolate the magnetic field lines of wire 2 influencing the current in wire 1. Right-hand rule 1 tells us the direction of the resulting magnetic force.

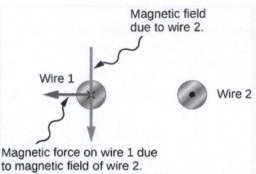

Figure 22.48 The same two wires with opposite currents are shown, but now only a part of the magnetic field due to wire 2 is shown in order to demonstrate how the magnetic force from wire 2 affects wire 1.

This force is responsible for the *pinch effect* in electric arcs and plasmas. The force exists whether the currents are in wires or not. In an electric arc, where currents are moving parallel to one another, there is an attraction that squeezes currents into a smaller tube. In large circuit breakers, like those used in neighborhood power distribution systems, the pinch effect can concentrate an arc between plates of a switch trying to break a large current, burn holes, and even ignite the equipment. Another example of the pinch effect is found in the solar plasma, where jets of ionized material, such as solar flares, are shaped by magnetic forces.

The *operational definition of the ampere* is based on the force between current-carrying wires. Note that for parallel wires separated by 1 meter with each carrying 1 ampere, the force per meter is

$$\frac{F}{l} = \frac{\left(4\pi \times 10^{-7} \text{ T} \cdot \text{m/A}\right)(1 \text{ A})^2}{(2\pi)(1 \text{ m})} = 2 \times 10^{-7} \text{ N/m}. \tag{22.33}$$

Since μ_0 is exactly $4\pi \times 10^{-7}$ T · m/A by definition, and because $1 \text{ T} = 1 \text{ N/(A} \cdot \text{m)}$, the force per meter is exactly 2×10^{-7} N/m . This is the basis of the operational definition of the ampere.

The Ampere

The official definition of the ampere is:

One ampere of current through each of two parallel conductors of infinite length, separated by one meter in empty space free of other magnetic fields, causes a force of exactly 2×10^{-7} N/m on each conductor.

Infinite-length straight wires are impractical and so, in practice, a current balance is constructed with coils of wire separated by a few centimeters. Force is measured to determine current. This also provides us with a method for measuring the coulomb. We measure the charge that flows for a current of one ampere in one second. That is, $1 \text{ C} = 1 \text{ A} \cdot \text{s}$. For both the ampere and the coulomb, the method of measuring force between conductors is the most accurate in practice.

22.11 More Applications of Magnetism

Learning Objectives

By the end of this section, you will be able to:

 - Describe some applications of magnetism.

Mass Spectrometry

The curved paths followed by charged particles in magnetic fields can be put to use. A charged particle moving perpendicular to a magnetic field travels in a circular path having a radius r.

$$r = \frac{mv}{qB} \tag{22.34}$$

It was noted that this relationship could be used to measure the mass of charged particles such as ions. A mass spectrometer is a device that measures such masses. Most mass spectrometers use magnetic fields for this purpose, although some of them have extremely sophisticated designs. Since there are five variables in the relationship, there are many possibilities. However, if v, q, and B can be fixed, then the radius of the path r is simply proportional to the mass m of the charged particle. Let us examine one such mass spectrometer that has a relatively simple design. (See **Figure 22.49**.) The process begins with an ion source, a device like an electron gun. The ion source gives ions their charge, accelerates them to some velocity v, and directs a beam of them into the next stage of the spectrometer. This next region is a *velocity selector* that only allows particles with a particular value of v to get through.

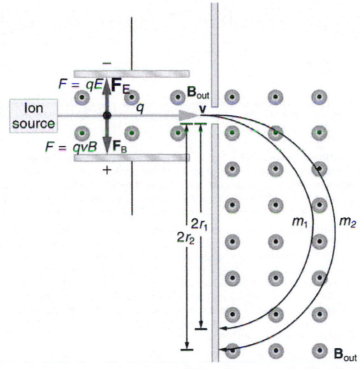

Figure 22.49 This mass spectrometer uses a velocity selector to fix v so that the radius of the path is proportional to mass.

The velocity selector has both an electric field and a magnetic field, perpendicular to one another, producing forces in opposite directions on the ions. Only those ions for which the forces balance travel in a straight line into the next region. If the forces balance, then the electric force $F = qE$ equals the magnetic force $F = qvB$, so that $qE = qvB$. Noting that q cancels, we see that

$$v = \frac{E}{B}$$

(22.35)

is the velocity particles must have to make it through the velocity selector, and further, that v can be selected by varying E and B. In the final region, there is only a uniform magnetic field, and so the charged particles move in circular arcs with radii proportional to particle mass. The paths also depend on charge q, but since q is in multiples of electron charges, it is easy to determine and to discriminate between ions in different charge states.

Mass spectrometry today is used extensively in chemistry and biology laboratories to identify chemical and biological substances according to their mass-to-charge ratios. In medicine, mass spectrometers are used to measure the concentration of isotopes used as tracers. Usually, biological molecules such as proteins are very large, so they are broken down into smaller fragments before analyzing. Recently, large virus particles have been analyzed as a whole on mass spectrometers. Sometimes a gas chromatograph or high-performance liquid chromatograph provides an initial separation of the large molecules, which are then input into the mass spectrometer.

Cathode Ray Tubes—CRTs—and the Like

What do non-flat-screen TVs, old computer monitors, x-ray machines, and the 2-mile-long Stanford Linear Accelerator have in common? All of them accelerate electrons, making them different versions of the electron gun. Many of these devices use magnetic fields to steer the accelerated electrons. **Figure 22.50** shows the construction of the type of cathode ray tube (CRT) found in some TVs, oscilloscopes, and old computer monitors. Two pairs of coils are used to steer the electrons, one vertically and the other horizontally, to their desired destination.

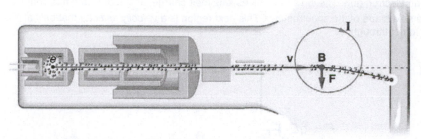

Figure 22.50 The cathode ray tube (CRT) is so named because rays of electrons originate at the cathode in the electron gun. Magnetic coils are used to steer the beam in many CRTs. In this case, the beam is moved down. Another pair of horizontal coils would steer the beam horizontally.

Magnetic Resonance Imaging

Magnetic resonance imaging (MRI) is one of the most useful and rapidly growing medical imaging tools. It non-invasively produces two-dimensional and three-dimensional images of the body that provide important medical information with none of the hazards of x-rays. MRI is based on an effect called **nuclear magnetic resonance (NMR)** in which an externally applied magnetic field interacts with the nuclei of certain atoms, particularly those of hydrogen (protons). These nuclei possess their own small magnetic fields, similar to those of electrons and the current loops discussed earlier in this chapter.

When placed in an external magnetic field, such nuclei experience a torque that pushes or aligns the nuclei into one of two new energy states—depending on the orientation of its spin (analogous to the N pole and S pole in a bar magnet). Transitions from the lower to higher energy state can be achieved by using an external radio frequency signal to "flip" the orientation of the small magnets. (This is actually a quantum mechanical process. The direction of the nuclear magnetic field is quantized as is energy in the radio waves. We will return to these topics in later chapters.) The specific frequency of the radio waves that are absorbed and reemitted depends sensitively on the type of nucleus, the chemical environment, and the external magnetic field strength. Therefore, this is a *resonance* phenomenon in which *nuclei* in a *magnetic* field act like resonators (analogous to those discussed in the treatment of sound in **Oscillatory Motion and Waves**) that absorb and reemit only certain frequencies. Hence, the phenomenon is named *nuclear magnetic resonance (NMR)*.

NMR has been used for more than 50 years as an analytical tool. It was formulated in 1946 by F. Bloch and E. Purcell, with the 1952 Nobel Prize in Physics going to them for their work. Over the past two decades, NMR has been developed to produce detailed images in a process now called magnetic resonance imaging (MRI), a name coined to avoid the use of the word "nuclear" and the concomitant implication that nuclear radiation is involved. (It is not.) The 2003 Nobel Prize in Medicine went to P. Lauterbur and P. Mansfield for their work with MRI applications.

The largest part of the MRI unit is a superconducting magnet that creates a magnetic field, typically between 1 and 2 T in strength, over a relatively large volume. MRI images can be both highly detailed and informative about structures and organ functions. It is helpful that normal and non-normal tissues respond differently for slight changes in the magnetic field. In most medical images, the protons that are hydrogen nuclei are imaged. (About 2/3 of the atoms in the body are hydrogen.) Their location and density give a variety of medically useful information, such as organ function, the condition of tissue (as in the brain), and the shape of structures, such as vertebral disks and knee-joint surfaces. MRI can also be used to follow the movement of certain ions across membranes, yielding information on active transport, osmosis, dialysis, and other phenomena. With excellent spatial resolution, MRI can provide information about tumors, strokes, shoulder injuries, infections, etc.

An image requires position information as well as the density of a nuclear type (usually protons). By varying the magnetic field slightly over the volume to be imaged, the resonant frequency of the protons is made to vary with position. Broadcast radio frequencies are swept over an appropriate range and nuclei absorb and reemit them only if the nuclei are in a magnetic field with the correct strength. The imaging receiver gathers information through the body almost point by point, building up a tissue map. The reception of reemitted radio waves as a function of frequency thus gives position information. These "slices" or cross sections through the body are only several mm thick. The intensity of the reemitted radio waves is proportional to the concentration of the nuclear type being flipped, as well as information on the chemical environment in that area of the body. Various techniques are available for enhancing contrast in images and for obtaining more information. Scans called T1, T2, or proton density scans rely on different relaxation mechanisms of nuclei. Relaxation refers to the time it takes for the protons to return to equilibrium after the external field is turned off. This time depends upon tissue type and status (such as inflammation).

While MRI images are superior to x rays for certain types of tissue and have none of the hazards of x rays, they do not completely supplant x-ray images. MRI is less effective than x rays for detecting breaks in bone, for example, and in imaging breast tissue, so the two diagnostic tools complement each other. MRI images are also expensive compared to simple x-ray images and tend to be used most often where they supply information not readily obtained from x rays. Another disadvantage of MRI is that the patient is totally enclosed with detectors close to the body for about 30 minutes or more, leading to claustrophobia. It is also difficult for the obese patient to be in the magnet tunnel. New "open-MRI" machines are now available in which the magnet does not completely surround the patient.

Over the last decade, the development of much faster scans, called "functional MRI" (fMRI), has allowed us to map the functioning of various regions in the brain responsible for thought and motor control. This technique measures the change in blood flow for activities (thought, experiences, action) in the brain. The nerve cells increase their consumption of oxygen when active. Blood hemoglobin releases oxygen to active nerve cells and has somewhat different magnetic properties when oxygenated than when deoxygenated. With MRI, we can measure this and detect a blood oxygen-dependent signal. Most of the brain scans today use fMRI.

Other Medical Uses of Magnetic Fields

Currents in nerve cells and the heart create magnetic fields like any other currents. These can be measured but with some difficulty since their strengths are about 10^{-6} to 10^{-8} *less* than the Earth's magnetic field. Recording of the heart's magnetic field as it beats is called a **magnetocardiogram (MCG)**, while measurements of the brain's magnetic field is called a **magnetoencephalogram (MEG)**. Both give information that differs from that obtained by measuring the electric fields of these organs (ECGs and EEGs), but they are not yet of sufficient importance to make these difficult measurements common.

In both of these techniques, the sensors do not touch the body. MCG can be used in fetal studies, and is probably more sensitive than echocardiography. MCG also looks at the heart's electrical activity whose voltage output is too small to be recorded by surface electrodes as in EKG. It has the potential of being a rapid scan for early diagnosis of cardiac ischemia (obstruction of blood flow to the heart) or problems with the fetus.

MEG can be used to identify abnormal electrical discharges in the brain that produce weak magnetic signals. Therefore, it looks at brain activity, not just brain structure. It has been used for studies of Alzheimer's disease and epilepsy. Advances in instrumentation to measure very small magnetic fields have allowed these two techniques to be used more in recent years. What is used is a sensor called a SQUID, for superconducting quantum interference device. This operates at liquid helium temperatures and can measure magnetic fields thousands of times smaller than the Earth's.

Finally, there is a burgeoning market for magnetic cures in which magnets are applied in a variety of ways to the body, from magnetic bracelets to magnetic mattresses. The best that can be said for such practices is that they are apparently harmless, unless the magnets get close to the patient's computer or magnetic storage disks. Claims are made for a broad spectrum of benefits from cleansing the blood to giving the patient more energy, but clinical studies have not verified these claims, nor is there an identifiable mechanism by which such benefits might occur.

PhET Explorations: Magnet and Compass

Ever wonder how a compass worked to point you to the Arctic? Explore the interactions between a compass and bar magnet, and then add the Earth and find the surprising answer! Vary the magnet's strength, and see how things change both inside and outside. Use the field meter to measure how the magnetic field changes.

Figure 22.51 Magnet and Compass (http://cnx.org/content/m55403/1.2/magnet-and-compass_en.jar)

Glossary

Ampere's law: the physical law that states that the magnetic field around an electric current is proportional to the current; each segment of current produces a magnetic field like that of a long straight wire, and the total field of any shape current is the vector sum of the fields due to each segment

B-field: another term for magnetic field

Biot-Savart law: a physical law that describes the magnetic field generated by an electric current in terms of a specific equation

Curie temperature: the temperature above which a ferromagnetic material cannot be magnetized

direction of magnetic field lines: the direction that the north end of a compass needle points

domains: regions within a material that behave like small bar magnets

electromagnet: an object that is temporarily magnetic when an electrical current is passed through it

electromagnetism: the use of electrical currents to induce magnetism

ferromagnetic: materials, such as iron, cobalt, nickel, and gadolinium, that exhibit strong magnetic effects

gauss: G, the unit of the magnetic field strength; $1 \text{ G} = 10^{-4} \text{ T}$

Hall effect: the creation of voltage across a current-carrying conductor by a magnetic field

Hall emf: the electromotive force created by a current-carrying conductor by a magnetic field, $\varepsilon = Blv$

Lorentz force: the force on a charge moving in a magnetic field

magnetic field: the representation of magnetic forces

magnetic field lines: the pictorial representation of the strength and the direction of a magnetic field

magnetic field strength (magnitude) produced by a long straight current-carrying wire: defined as $B = \frac{\mu_0 I}{2\pi r}$, where I is the current, r is the shortest distance to the wire, and μ_0 is the permeability of free space

magnetic field strength at the center of a circular loop: defined as $B = \frac{\mu_0 I}{2R}$ where R is the radius of the loop

magnetic field strength inside a solenoid: defined as $B = \mu_0 nI$ where n is the number of loops per unit length of the solenoid ($n = N/l$, with N being the number of loops and l the length)

magnetic force: the force on a charge produced by its motion through a magnetic field; the Lorentz force

magnetic monopoles: an isolated magnetic pole; a south pole without a north pole, or vice versa (no magnetic monopole has ever been observed)

magnetic resonance imaging (MRI): a medical imaging technique that uses magnetic fields create detailed images of internal tissues and organs

magnetized: to be turned into a magnet; to be induced to be magnetic

magnetocardiogram (MCG): a recording of the heart's magnetic field as it beats

magnetoencephalogram (MEG): a measurement of the brain's magnetic field

Maxwell's equations: a set of four equations that describe electromagnetic phenomena

meter: common application of magnetic torque on a current-carrying loop that is very similar in construction to a motor; by design, the torque is proportional to I and not θ, so the needle deflection is proportional to the current

motor: loop of wire in a magnetic field; when current is passed through the loops, the magnetic field exerts torque on the loops, which rotates a shaft; electrical energy is converted to mechanical work in the process

north magnetic pole: the end or the side of a magnet that is attracted toward Earth's geographic north pole

nuclear magnetic resonance (NMR): a phenomenon in which an externally applied magnetic field interacts with the nuclei of certain atoms

permeability of free space: the measure of the ability of a material, in this case free space, to support a magnetic field; the constant $\mu_0 = 4\pi \times 10^{-7}$ T · m/A

right hand rule 1 (RHR-1): the rule to determine the direction of the magnetic force on a positive moving charge: when the thumb of the right hand points in the direction of the charge's velocity **v** and the fingers point in the direction of the magnetic field **B**, then the force on the charge is perpendicular and away from the palm; the force on a negative charge is perpendicular and into the palm

right hand rule 2 (RHR-2): a rule to determine the direction of the magnetic field induced by a current-carrying wire: Point the thumb of the right hand in the direction of current, and the fingers curl in the direction of the magnetic field loops

solenoid: a thin wire wound into a coil that produces a magnetic field when an electric current is passed through it

south magnetic pole: the end or the side of a magnet that is attracted toward Earth's geographic south pole

tesla: T, the SI unit of the magnetic field strength; $1\,\text{T} = \frac{1\,\text{N}}{\text{A} \cdot \text{m}}$

Section Summary

22.1 Magnets

- Magnetism is a subject that includes the properties of magnets, the effect of the magnetic force on moving charges and currents, and the creation of magnetic fields by currents.
- There are two types of magnetic poles, called the north magnetic pole and south magnetic pole.
- North magnetic poles are those that are attracted toward the Earth's geographic north pole.

- Like poles repel and unlike poles attract.
- Magnetic poles always occur in pairs of north and south—it is not possible to isolate north and south poles.

22.2 Ferromagnets and Electromagnets

- Magnetic poles always occur in pairs of north and south—it is not possible to isolate north and south poles.
- All magnetism is created by electric current.
- Ferromagnetic materials, such as iron, are those that exhibit strong magnetic effects.
- The atoms in ferromagnetic materials act like small magnets (due to currents within the atoms) and can be aligned, usually in millimeter-sized regions called domains.
- Domains can grow and align on a larger scale, producing permanent magnets. Such a material is magnetized, or induced to be magnetic.
- Above a material's Curie temperature, thermal agitation destroys the alignment of atoms, and ferromagnetism disappears.
- Electromagnets employ electric currents to make magnetic fields, often aided by induced fields in ferromagnetic materials.

22.3 Magnetic Fields and Magnetic Field Lines

- Magnetic fields can be pictorially represented by magnetic field lines, the properties of which are as follows:

1. The field is tangent to the magnetic field line.

2. Field strength is proportional to the line density.

3. Field lines cannot cross.

4. Field lines are continuous loops.

22.4 Magnetic Field Strength: Force on a Moving Charge in a Magnetic Field

- Magnetic fields exert a force on a moving charge q, the magnitude of which is

$$F = qvB \sin \theta,$$

where θ is the angle between the directions of v and B.
- The SI unit for magnetic field strength B is the tesla (T), which is related to other units by

$$1 \text{ T} = \frac{1 \text{ N}}{\text{C} \cdot \text{m/s}} = \frac{1 \text{ N}}{\text{A} \cdot \text{m}}.$$

- The *direction* of the force on a moving charge is given by right hand rule 1 (RHR-1): Point the thumb of the right hand in the direction of v, the fingers in the direction of B, and a perpendicular to the palm points in the direction of F.
- The force is perpendicular to the plane formed by $\mathbf{v}$ and $\mathbf{B}$. Since the force is zero if $\mathbf{v}$ is parallel to $\mathbf{B}$, charged particles often follow magnetic field lines rather than cross them.

22.5 Force on a Moving Charge in a Magnetic Field: Examples and Applications

- Magnetic force can supply centripetal force and cause a charged particle to move in a circular path of radius

$$r = \frac{mv}{qB},$$

where v is the component of the velocity perpendicular to B for a charged particle with mass m and charge q.

22.6 The Hall Effect

- The Hall effect is the creation of voltage ε, known as the Hall emf, across a current-carrying conductor by a magnetic field.
- The Hall emf is given by

$$\varepsilon = Blv \ (B, v, \text{ and } l, \text{ mutually perpendicular})$$

for a conductor of width l through which charges move at a speed v.

22.7 Magnetic Force on a Current-Carrying Conductor

- The magnetic force on current-carrying conductors is given by

$$F = IlB \sin \theta,$$

where I is the current, l is the length of a straight conductor in a uniform magnetic field B, and θ is the angle between I and B. The force follows RHR-1 with the thumb in the direction of I.

22.8 Torque on a Current Loop: Motors and Meters

- The torque τ on a current-carrying loop of any shape in a uniform magnetic field. is

$$\tau = NIAB \sin \theta,$$

where N is the number of turns, I is the current, A is the area of the loop, B is the magnetic field strength, and θ is the angle between the perpendicular to the loop and the magnetic field.

22.9 Magnetic Fields Produced by Currents: Ampere's Law

- The strength of the magnetic field created by current in a long straight wire is given by

$$B = \frac{\mu_0 I}{2\pi r} \text{ (long straight wire)},$$

I is the current, r is the shortest distance to the wire, and the constant $\mu_0 = 4\pi \times 10^{-7} \text{ T} \cdot \text{m/A}$ is the permeability of free space.

- The direction of the magnetic field created by a long straight wire is given by right hand rule 2 (RHR-2): *Point the thumb of the right hand in the direction of current, and the fingers curl in the direction of the magnetic field loops* created by it.
- The magnetic field created by current following any path is the sum (or integral) of the fields due to segments along the path (magnitude and direction as for a straight wire), resulting in a general relationship between current and field known as Ampere's law.
- The magnetic field strength at the center of a circular loop is given by

$$B = \frac{\mu_0 I}{2R} \text{ (at center of loop)},$$

R is the radius of the loop. This equation becomes $B = \mu_0 n I / (2R)$ for a flat coil of N loops. RHR-2 gives the direction of the field about the loop. A long coil is called a solenoid.

- The magnetic field strength inside a solenoid is

$$B = \mu_0 n I \text{ (inside a solenoid)},$$

where n is the number of loops per unit length of the solenoid. The field inside is very uniform in magnitude and direction.

22.10 Magnetic Force between Two Parallel Conductors

- The force between two parallel currents I_1 and I_2, separated by a distance r, has a magnitude per unit length given by

$$\frac{F}{l} = \frac{\mu_0 I_1 I_2}{2\pi r}.$$

- The force is attractive if the currents are in the same direction, repulsive if they are in opposite directions.

22.11 More Applications of Magnetism

- Crossed (perpendicular) electric and magnetic fields act as a velocity filter, giving equal and opposite forces on any charge with velocity perpendicular to the fields and of magnitude

$$v = \frac{E}{B}.$$

Conceptual Questions

22.1 Magnets

1. Volcanic and other such activity at the mid-Atlantic ridge extrudes material to fill the gap between separating tectonic plates associated with continental drift. The magnetization of rocks is found to reverse in a coordinated manner with distance from the ridge. What does this imply about the Earth's magnetic field and how could the knowledge of the spreading rate be used to give its historical record?

22.3 Magnetic Fields and Magnetic Field Lines

2. Explain why the magnetic field would not be unique (that is, not have a single value) at a point in space where magnetic field lines might cross. (Consider the direction of the field at such a point.)

3. List the ways in which magnetic field lines and electric field lines are similar. For example, the field direction is tangent to the line at any point in space. Also list the ways in which they differ. For example, electric force is parallel to electric field lines, whereas magnetic force on moving charges is perpendicular to magnetic field lines.

4. Noting that the magnetic field lines of a bar magnet resemble the electric field lines of a pair of equal and opposite charges, do you expect the magnetic field to rapidly decrease in strength with distance from the magnet? Is this consistent with your experience with magnets?

5. Is the Earth's magnetic field parallel to the ground at all locations? If not, where is it parallel to the surface? Is its strength the same at all locations? If not, where is it greatest?

22.4 Magnetic Field Strength: Force on a Moving Charge in a Magnetic Field

6. If a charged particle moves in a straight line through some region of space, can you say that the magnetic field in that region is necessarily zero?

22.5 Force on a Moving Charge in a Magnetic Field: Examples and Applications

7. How can the motion of a charged particle be used to distinguish between a magnetic and an electric field?

8. High-velocity charged particles can damage biological cells and are a component of radiation exposure in a variety of locations ranging from research facilities to natural background. Describe how you could use a magnetic field to shield yourself.

9. If a cosmic ray proton approaches the Earth from outer space along a line toward the center of the Earth that lies in the plane of the equator, in what direction will it be deflected by the Earth's magnetic field? What about an electron? A neutron?

10. What are the signs of the charges on the particles in **Figure 22.52**?

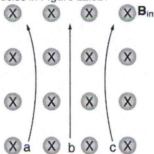

Figure 22.52

11. Which of the particles in **Figure 22.53** has the greatest velocity, assuming they have identical charges and masses?

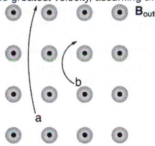

Figure 22.53

12. Which of the particles in **Figure 22.53** has the greatest mass, assuming all have identical charges and velocities?

13. While operating, a high-precision TV monitor is placed on its side during maintenance. The image on the monitor changes color and blurs slightly. Discuss the possible relation of these effects to the Earth's magnetic field.

22.6 The Hall Effect

14. Discuss how the Hall effect could be used to obtain information on free charge density in a conductor. (Hint: Consider how drift velocity and current are related.)

22.7 Magnetic Force on a Current-Carrying Conductor

15. Draw a sketch of the situation in **Figure 22.31** showing the direction of electrons carrying the current, and use RHR-1 to verify the direction of the force on the wire.

16. Verify that the direction of the force in an MHD drive, such as that in **Figure 22.33**, does not depend on the sign of the charges carrying the current across the fluid.

17. Why would a magnetohydrodynamic drive work better in ocean water than in fresh water? Also, why would superconducting magnets be desirable?

18. Which is more likely to interfere with compass readings, AC current in your refrigerator or DC current when you start your car? Explain.

22.8 Torque on a Current Loop: Motors and Meters

19. Draw a diagram and use RHR-1 to show that the forces on the top and bottom segments of the motor's current loop in **Figure 22.35** are vertical and produce no torque about the axis of rotation.

22.9 Magnetic Fields Produced by Currents: Ampere's Law

20. Make a drawing and use RHR-2 to find the direction of the magnetic field of a current loop in a motor (such as in **Figure 22.35**). Then show that the direction of the torque on the loop is the same as produced by like poles repelling and unlike poles attracting.

22.10 Magnetic Force between Two Parallel Conductors

21. Is the force attractive or repulsive between the hot and neutral lines hung from power poles? Why?

22. If you have three parallel wires in the same plane, as in **Figure 22.54**, with currents in the outer two running in opposite directions, is it possible for the middle wire to be repelled by both? Attracted by both? Explain.

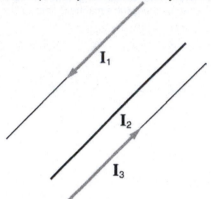

Figure 22.54 Three parallel coplanar wires with currents in the outer two in opposite directions.

23. Suppose two long straight wires run perpendicular to one another without touching. Does one exert a net force on the other? If so, what is its direction? Does one exert a net torque on the other? If so, what is its direction? Justify your responses by using the right hand rules.

24. Use the right hand rules to show that the force between the two loops in **Figure 22.55** is attractive if the currents are in the same direction and repulsive if they are in opposite directions. Is this consistent with like poles of the loops repelling and unlike poles of the loops attracting? Draw sketches to justify your answers.

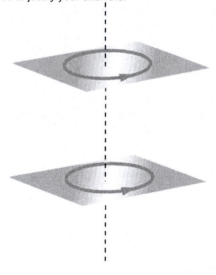

Figure 22.55 Two loops of wire carrying currents can exert forces and torques on one another.

25. If one of the loops in **Figure 22.55** is tilted slightly relative to the other and their currents are in the same direction, what are the directions of the torques they exert on each other? Does this imply that the poles of the bar magnet-like fields they create will line up with each other if the loops are allowed to rotate?

26. Electric field lines can be shielded by the Faraday cage effect. Can we have magnetic shielding? Can we have gravitational shielding?

22.11 More Applications of Magnetism

27. Measurements of the weak and fluctuating magnetic fields associated with brain activity are called magnetoencephalograms (MEGs). Do the brain's magnetic fields imply coordinated or uncoordinated nerve impulses? Explain.

28. Discuss the possibility that a Hall voltage would be generated on the moving heart of a patient during MRI imaging. Also discuss the same effect on the wires of a pacemaker. (The fact that patients with pacemakers are not given MRIs is significant.)

29. A patient in an MRI unit turns his head quickly to one side and experiences momentary dizziness and a strange taste in his mouth. Discuss the possible causes.

30. You are told that in a certain region there is either a uniform electric or magnetic field. What measurement or observation could you make to determine the type? (Ignore the Earth's magnetic field.)

31. An example of magnetohydrodynamics (MHD) comes from the flow of a river (salty water). This fluid interacts with the Earth's magnetic field to produce a potential difference between the two river banks. How would you go about calculating the potential difference?

32. Draw gravitational field lines between 2 masses, electric field lines between a positive and a negative charge, electric field lines between 2 positive charges and magnetic field lines around a magnet. Qualitatively describe the differences between the fields and the entities responsible for the field lines.

Problems & Exercises

22.4 Magnetic Field Strength: Force on a Moving Charge in a Magnetic Field

1. What is the direction of the magnetic force on a positive charge that moves as shown in each of the six cases shown in **Figure 22.56**?

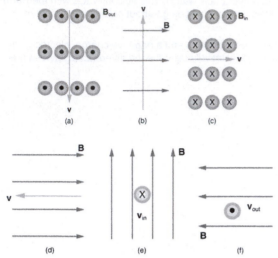

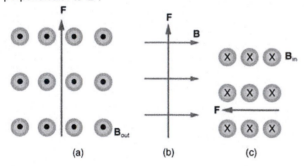

Figure 22.56

2. Repeat **Exercise 22.1** for a negative charge.

3. What is the direction of the velocity of a negative charge that experiences the magnetic force shown in each of the three cases in **Figure 22.57**, assuming it moves perpendicular to $\mathbf{B}$?

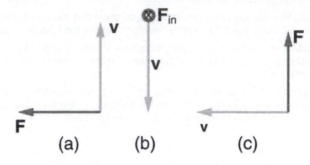

Figure 22.57

4. Repeat **Exercise 22.3** for a positive charge.

5. What is the direction of the magnetic field that produces the magnetic force on a positive charge as shown in each of the three cases in the figure below, assuming $\mathbf{B}$ is perpendicular to $\mathbf{v}$?

Figure 22.58

6. Repeat **Exercise 22.5** for a negative charge.

7. What is the maximum force on an aluminum rod with a $0.100\text{-}\mu C$ charge that you pass between the poles of a 1.50-T permanent magnet at a speed of 5.00 m/s? In what direction is the force?

8. (a) Aircraft sometimes acquire small static charges. Suppose a supersonic jet has a $0.500\text{-}\mu C$ charge and flies due west at a speed of 660 m/s over the Earth's south magnetic pole, where the $8.00\times10^{-5}\text{-T}$ magnetic field points straight up. What are the direction and the magnitude of the magnetic force on the plane? (b) Discuss whether the value obtained in part (a) implies this is a significant or negligible effect.

9. (a) A cosmic ray proton moving toward the Earth at 5.00×10^{7} m/s experiences a magnetic force of 1.70×10^{-16} N. What is the strength of the magnetic field if there is a $45°$ angle between it and the proton's velocity? (b) Is the value obtained in part (a) consistent with the known strength of the Earth's magnetic field on its surface? Discuss.

10. An electron moving at 4.00×10^{3} m/s in a 1.25-T magnetic field experiences a magnetic force of 1.40×10^{-16} N. What angle does the velocity of the electron make with the magnetic field? There are two answers.

11. (a) A physicist performing a sensitive measurement wants to limit the magnetic force on a moving charge in her equipment to less than 1.00×10^{-12} N. What is the greatest the charge can be if it moves at a maximum speed of 30.0 m/s in the Earth's field? (b) Discuss whether it would be difficult to limit the charge to less than the value found in (a) by comparing it with typical static electricity and noting that static is often absent.

22.5 Force on a Moving Charge in a Magnetic Field: Examples and Applications

If you need additional support for these problems, see **More Applications of Magnetism**.

12. A cosmic ray electron moves at 7.50×10^6 m/s perpendicular to the Earth's magnetic field at an altitude where field strength is 1.00×10^{-5} T. What is the radius of the circular path the electron follows?

13. A proton moves at 7.50×10^7 m/s perpendicular to a magnetic field. The field causes the proton to travel in a circular path of radius 0.800 m. What is the field strength?

14. (a) Viewers of *Star Trek* hear of an antimatter drive on the Starship *Enterprise*. One possibility for such a futuristic energy source is to store antimatter charged particles in a vacuum chamber, circulating in a magnetic field, and then extract them as needed. Antimatter annihilates with normal matter, producing pure energy. What strength magnetic field is needed to hold antiprotons, moving at 5.00×10^7 m/s in a circular path 2.00 m in radius? Antiprotons have the same mass as protons but the opposite (negative) charge. (b) Is this field strength obtainable with today's technology or is it a futuristic possibility?

15. (a) An oxygen-16 ion with a mass of 2.66×10^{-26} kg travels at 5.00×10^6 m/s perpendicular to a 1.20-T magnetic field, which makes it move in a circular arc with a 0.231-m radius. What positive charge is on the ion? (b) What is the ratio of this charge to the charge of an electron? (c) Discuss why the ratio found in (b) should be an integer.

16. What radius circular path does an electron travel if it moves at the same speed and in the same magnetic field as the proton in **Exercise 22.13**?

17. A velocity selector in a mass spectrometer uses a 0.100-T magnetic field. (a) What electric field strength is needed to select a speed of 4.00×10^6 m/s? (b) What is the voltage between the plates if they are separated by 1.00 cm?

18. An electron in a TV CRT moves with a speed of 6.00×10^7 m/s, in a direction perpendicular to the Earth's field, which has a strength of 5.00×10^{-5} T. (a) What strength electric field must be applied perpendicular to the Earth's field to make the electron moves in a straight line? (b) If this is done between plates separated by 1.00 cm, what is the voltage applied? (Note that TVs are usually surrounded by a ferromagnetic material to shield against external magnetic fields and avoid the need for such a correction.)

19. (a) At what speed will a proton move in a circular path of the same radius as the electron in **Exercise 22.12**? (b) What would the radius of the path be if the proton had the same speed as the electron? (c) What would the radius be if the proton had the same kinetic energy as the electron? (d) The same momentum?

20. A mass spectrometer is being used to separate common oxygen-16 from the much rarer oxygen-18, taken from a sample of old glacial ice. (The relative abundance of these oxygen isotopes is related to climatic temperature at the time the ice was deposited.) The ratio of the masses of these two ions is 16 to 18, the mass of oxygen-16 is 2.66×10^{-26} kg, and they are singly charged and travel at 5.00×10^6 m/s in a 1.20-T magnetic field. What is the separation between their paths when they hit a target after traversing a semicircle?

21. (a) Triply charged uranium-235 and uranium-238 ions are being separated in a mass spectrometer. (The much rarer uranium-235 is used as reactor fuel.) The masses of the ions are 3.90×10^{-25} kg and 3.95×10^{-25} kg, respectively, and they travel at 3.00×10^5 m/s in a 0.250-T field. What is the separation between their paths when they hit a target after traversing a semicircle? (b) Discuss whether this distance between their paths seems to be big enough to be practical in the separation of uranium-235 from uranium-238.

22.6 The Hall Effect

22. A large water main is 2.50 m in diameter and the average water velocity is 6.00 m/s. Find the Hall voltage produced if the pipe runs perpendicular to the Earth's 5.00×10^{-5}-T field.

23. What Hall voltage is produced by a 0.200-T field applied across a 2.60-cm-diameter aorta when blood velocity is 60.0 cm/s?

24. (a) What is the speed of a supersonic aircraft with a 17.0-m wingspan, if it experiences a 1.60-V Hall voltage between its wing tips when in level flight over the north magnetic pole, where the Earth's field strength is 8.00×10^{-5} T? (b) Explain why very little current flows as a result of this Hall voltage.

25. A nonmechanical water meter could utilize the Hall effect by applying a magnetic field across a metal pipe and measuring the Hall voltage produced. What is the average fluid velocity in a 3.00-cm-diameter pipe, if a 0.500-T field across it creates a 60.0-mV Hall voltage?

26. Calculate the Hall voltage induced on a patient's heart while being scanned by an MRI unit. Approximate the conducting path on the heart wall by a wire 7.50 cm long that moves at 10.0 cm/s perpendicular to a 1.50-T magnetic field.

27. A Hall probe calibrated to read $1.00\ \mu V$ when placed in a 2.00-T field is placed in a 0.150-T field. What is its output voltage?

28. Using information in **Example 20.6**, what would the Hall voltage be if a 2.00-T field is applied across a 10-gauge copper wire (2.588 mm in diameter) carrying a 20.0-A current?

29. Show that the Hall voltage across wires made of the same material, carrying identical currents, and subjected to the same magnetic field is inversely proportional to their diameters. (Hint: Consider how drift velocity depends on wire diameter.)

30. A patient with a pacemaker is mistakenly being scanned for an MRI image. A 10.0-cm-long section of pacemaker wire moves at a speed of 10.0 cm/s perpendicular to the MRI unit's magnetic field and a 20.0-mV Hall voltage is induced. What is the magnetic field strength?

22.7 Magnetic Force on a Current-Carrying Conductor

31. What is the direction of the magnetic force on the current in each of the six cases in **Figure 22.59**?

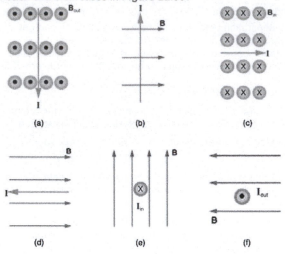

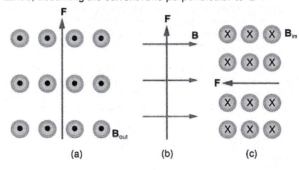

Figure 22.59

32. What is the direction of a current that experiences the magnetic force shown in each of the three cases in **Figure 22.60**, assuming the current runs perpendicular to B ?

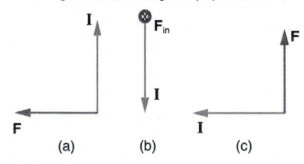

Figure 22.60

33. What is the direction of the magnetic field that produces the magnetic force shown on the currents in each of the three cases in **Figure 22.61**, assuming $\mathbf{B}$ is perpendicular to $\mathbf{I}$?

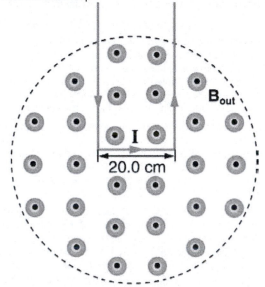

Figure 22.61

34. (a) What is the force per meter on a lightning bolt at the equator that carries 20,000 A perpendicular to the Earth's 3.00×10^{-5}-T field? (b) What is the direction of the force if the current is straight up and the Earth's field direction is due north, parallel to the ground?

35. (a) A DC power line for a light-rail system carries 1000 A at an angle of $30.0°$ to the Earth's 5.00×10^{-5} -T field. What is the force on a 100-m section of this line? (b) Discuss practical concerns this presents, if any.

36. What force is exerted on the water in an MHD drive utilizing a 25.0-cm-diameter tube, if 100-A current is passed across the tube that is perpendicular to a 2.00-T magnetic field? (The relatively small size of this force indicates the need for very large currents and magnetic fields to make practical MHD drives.)

37. A wire carrying a 30.0-A current passes between the poles of a strong magnet that is perpendicular to its field and experiences a 2.16-N force on the 4.00 cm of wire in the field. What is the average field strength?

38. (a) A 0.750-m-long section of cable carrying current to a car starter motor makes an angle of $60°$ with the Earth's 5.50×10^{-5} T field. What is the current when the wire experiences a force of 7.00×10^{-3} N ? (b) If you run the wire between the poles of a strong horseshoe magnet, subjecting 5.00 cm of it to a 1.75-T field, what force is exerted on this segment of wire?

39. (a) What is the angle between a wire carrying an 8.00-A current and the 1.20-T field it is in if 50.0 cm of the wire experiences a magnetic force of 2.40 N? (b) What is the force on the wire if it is rotated to make an angle of $90°$ with the field?

40. The force on the rectangular loop of wire in the magnetic field in **Figure 22.62** can be used to measure field strength. The field is uniform, and the plane of the loop is perpendicular to the field. (a) What is the direction of the magnetic force on the loop? Justify the claim that the forces on the sides of the loop are equal and opposite, independent of how much of the loop is in the field and do not affect the net force on the loop. (b) If a current of 5.00 A is used, what is the force per tesla on the 20.0-cm-wide loop?

Figure 22.62 A rectangular loop of wire carrying a current is perpendicular to a magnetic field. The field is uniform in the region shown and is zero outside that region.

22.8 Torque on a Current Loop: Motors and

Meters

41. (a) By how many percent is the torque of a motor decreased if its permanent magnets lose 5.0% of their strength? (b) How many percent would the current need to be increased to return the torque to original values?

42. (a) What is the maximum torque on a 150-turn square loop of wire 18.0 cm on a side that carries a 50.0-A current in a 1.60-T field? (b) What is the torque when θ is $10.9°$?

43. Find the current through a loop needed to create a maximum torque of $9.00 \, \text{N} \cdot \text{m}$. The loop has 50 square turns that are 15.0 cm on a side and is in a uniform 0.800-T magnetic field.

44. Calculate the magnetic field strength needed on a 200-turn square loop 20.0 cm on a side to create a maximum torque of $300 \, \text{N} \cdot \text{m}$ if the loop is carrying 25.0 A.

45. Since the equation for torque on a current-carrying loop is $\tau = NIAB \sin \theta$, the units of $\text{N} \cdot \text{m}$ must equal units of $\text{A} \cdot \text{m}^2 \, \text{T}$. Verify this.

46. (a) At what angle θ is the torque on a current loop 90.0% of maximum? (b) 50.0% of maximum? (c) 10.0% of maximum?

47. A proton has a magnetic field due to its spin on its axis. The field is similar to that created by a circular current loop 0.650×10^{-15} m in radius with a current of 1.05×10^4 A (no kidding). Find the maximum torque on a proton in a 2.50-T field. (This is a significant torque on a small particle.)

48. (a) A 200-turn circular loop of radius 50.0 cm is vertical, with its axis on an east-west line. A current of 100 A circulates clockwise in the loop when viewed from the east. The Earth's field here is due north, parallel to the ground, with a strength of 3.00×10^{-5} T. What are the direction and magnitude of the torque on the loop? (b) Does this device have any practical applications as a motor?

49. Repeat **Exercise 22.41**, but with the loop lying flat on the ground with its current circulating counterclockwise (when viewed from above) in a location where the Earth's field is north, but at an angle $45.0°$ below the horizontal and with a strength of 6.00×10^{-5} T.

22.10 Magnetic Force between Two Parallel Conductors

50. (a) The hot and neutral wires supplying DC power to a light-rail commuter train carry 800 A and are separated by 75.0 cm. What is the magnitude and direction of the force between 50.0 m of these wires? (b) Discuss the practical consequences of this force, if any.

51. The force per meter between the two wires of a jumper cable being used to start a stalled car is 0.225 N/m. (a) What is the current in the wires, given they are separated by 2.00 cm? (b) Is the force attractive or repulsive?

52. A 2.50-m segment of wire supplying current to the motor of a submerged submarine carries 1000 A and feels a 4.00-N repulsive force from a parallel wire 5.00 cm away. What is the direction and magnitude of the current in the other wire?

53. The wire carrying 400 A to the motor of a commuter train feels an attractive force of 4.00×10^{-3} N/m due to a parallel wire carrying 5.00 A to a headlight. (a) How far apart are the wires? (b) Are the currents in the same direction?

54. An AC appliance cord has its hot and neutral wires separated by 3.00 mm and carries a 5.00-A current. (a) What is the average force per meter between the wires in the cord? (b) What is the maximum force per meter between the wires? (c) Are the forces attractive or repulsive? (d) Do appliance cords need any special design features to compensate for these forces?

55. **Figure 22.63** shows a long straight wire near a rectangular current loop. What is the direction and magnitude of the total force on the loop?

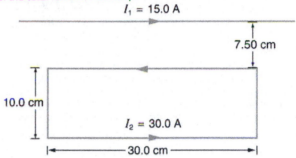

Figure 22.63

56. Find the direction and magnitude of the force that each wire experiences in **Figure 22.64**(a) by, using vector addition.

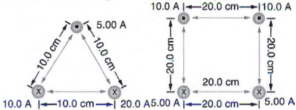

Figure 22.64

57. Find the direction and magnitude of the force that each wire experiences in **Figure 22.64**(b), using vector addition.

22.11 More Applications of Magnetism

58. Indicate whether the magnetic field created in each of the three situations shown in **Figure 22.65** is into or out of the page on the left and right of the current.

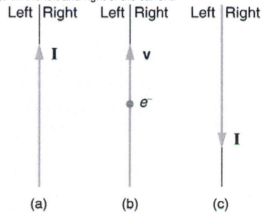

Figure 22.65

59. What are the directions of the fields in the center of the loop and coils shown in **Figure 22.66**?

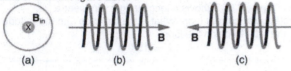

(a)   (b)   (c)

Figure 22.66

60. What are the directions of the currents in the loop and coils shown in **Figure 22.67**?

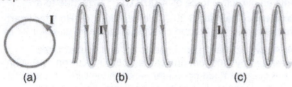

(a)   (b)   (c)

Figure 22.67

61. To see why an MRI utilizes iron to increase the magnetic field created by a coil, calculate the current needed in a 400-loop-per-meter circular coil 0.660 m in radius to create a 1.20-T field (typical of an MRI instrument) at its center with no iron present. The magnetic field of a proton is approximately like that of a circular current loop 0.650×10^{-15} m in radius carrying 1.05×10^4 A. What is the field at the center of such a loop?

62. Inside a motor, 30.0 A passes through a 250-turn circular loop that is 10.0 cm in radius. What is the magnetic field strength created at its center?

63. Nonnuclear submarines use batteries for power when submerged. (a) Find the magnetic field 50.0 cm from a straight wire carrying 1200 A from the batteries to the drive mechanism of a submarine. (b) What is the field if the wires to and from the drive mechanism are side by side? (c) Discuss the effects this could have for a compass on the submarine that is not shielded.

64. How strong is the magnetic field inside a solenoid with 10,000 turns per meter that carries 20.0 A?

65. What current is needed in the solenoid described in **Exercise 22.58** to produce a magnetic field 10^4 times the Earth's magnetic field of 5.00×10^{-5} T?

66. How far from the starter cable of a car, carrying 150 A, must you be to experience a field less than the Earth's $(5.00 \times 10^{-5}$ T)? Assume a long straight wire carries the current. (In practice, the body of your car shields the dashboard compass.)

67. Measurements affect the system being measured, such as the current loop in **Figure 22.62**. (a) Estimate the field the loop creates by calculating the field at the center of a circular loop 20.0 cm in diameter carrying 5.00 A. (b) What is the smallest field strength this loop can be used to measure, if its field must alter the measured field by less than 0.0100%?

68. **Figure 22.68** shows a long straight wire just touching a loop carrying a current I_1. Both lie in the same plane. (a) What direction must the current I_2 in the straight wire have to create a field at the center of the loop in the direction opposite to that created by the loop? (b) What is the ratio of I_1/I_2 that gives zero field strength at the center of the loop? (c) What is the direction of the field directly above the loop under this circumstance?

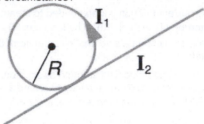

Figure 22.68

69. Find the magnitude and direction of the magnetic field at the point equidistant from the wires in **Figure 22.64(a)**, using the rules of vector addition to sum the contributions from each wire.

70. Find the magnitude and direction of the magnetic field at the point equidistant from the wires in **Figure 22.64(b)**, using the rules of vector addition to sum the contributions from each wire.

71. What current is needed in the top wire in **Figure 22.64(a)** to produce a field of zero at the point equidistant from the wires, if the currents in the bottom two wires are both 10.0 A into the page?

72. Calculate the size of the magnetic field 20 m below a high voltage power line. The line carries 450 MW at a voltage of 300,000 V.

73. Integrated Concepts

(a) A pendulum is set up so that its bob (a thin copper disk) swings between the poles of a permanent magnet as shown in **Figure 22.69**. What is the magnitude and direction of the magnetic force on the bob at the lowest point in its path, if it has a positive $0.250\ \mu C$ charge and is released from a height of 30.0 cm above its lowest point? The magnetic field strength is 1.50 T. (b) What is the acceleration of the bob at the bottom of its swing if its mass is 30.0 grams and it is hung from a flexible string? Be certain to include a free-body diagram as part of your analysis.

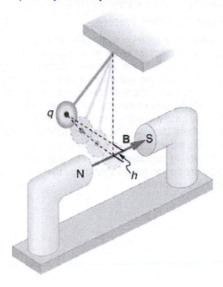

Figure 22.69

74. Integrated Concepts

(a) What voltage will accelerate electrons to a speed of 6.00×10^{-7} m/s ? (b) Find the radius of curvature of the path of a *proton* accelerated through this potential in a 0.500-T field and compare this with the radius of curvature of an electron accelerated through the same potential.

75. Integrated Concepts

Find the radius of curvature of the path of a 25.0-MeV proton moving perpendicularly to the 1.20-T field of a cyclotron.

76. Integrated Concepts

To construct a nonmechanical water meter, a 0.500-T magnetic field is placed across the supply water pipe to a home and the Hall voltage is recorded. (a) Find the flow rate in liters per second through a 3.00-cm-diameter pipe if the Hall voltage is 60.0 mV. (b) What would the Hall voltage be for the same flow rate through a 10.0-cm-diameter pipe with the same field applied?

77. Integrated Concepts

(a) Using the values given for an MHD drive in **Exercise 22.59**, and assuming the force is uniformly applied to the fluid, calculate the pressure created in N/m^2. (b) Is this a significant fraction of an atmosphere?

78. Integrated Concepts

(a) Calculate the maximum torque on a 50-turn, 1.50 cm radius circular current loop carrying $50\ \mu A$ in a 0.500-T field.

(b) If this coil is to be used in a galvanometer that reads $50\ \mu A$ full scale, what force constant spring must be used, if it is attached 1.00 cm from the axis of rotation and is stretched by the $60°$ arc moved?

79. Integrated Concepts

A current balance used to define the ampere is designed so that the current through it is constant, as is the distance between wires. Even so, if the wires change length with temperature, the force between them will change. What percent change in force per degree will occur if the wires are copper?

80. Integrated Concepts

(a) Show that the period of the circular orbit of a charged particle moving perpendicularly to a uniform magnetic field is $T = 2\pi m / (qB)$. (b) What is the frequency f ? (c) What is the angular velocity ω ? Note that these results are independent of the velocity and radius of the orbit and, hence, of the energy of the particle. (**Figure 22.70**.)

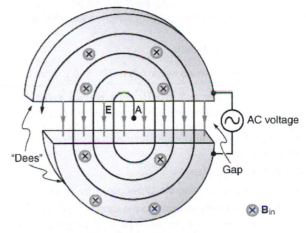

Figure 22.70 Cyclotrons accelerate charged particles orbiting in a magnetic field by placing an AC voltage on the metal Dees, between which the particles move, so that energy is added twice each orbit. The frequency is constant, since it is independent of the particle energy—the radius of the orbit simply increases with energy until the particles approach the edge and are extracted for various experiments and applications.

81. Integrated Concepts

A cyclotron accelerates charged particles as shown in **Figure 22.70**. Using the results of the previous problem, calculate the frequency of the accelerating voltage needed for a proton in a 1.20-T field.

82. Integrated Concepts

(a) A 0.140-kg baseball, pitched at 40.0 m/s horizontally and perpendicular to the Earth's horizontal 5.00×10^{-5} T field, has a 100-nC charge on it. What distance is it deflected from its path by the magnetic force, after traveling 30.0 m horizontally? (b) Would you suggest this as a secret technique for a pitcher to throw curve balls?

83. Integrated Concepts

(a) What is the direction of the force on a wire carrying a current due east in a location where the Earth's field is due north? Both are parallel to the ground. (b) Calculate the force per meter if the wire carries 20.0 A and the field strength is 3.00×10^{-5} T . (c) What diameter copper wire would have its weight supported by this force? (d) Calculate the resistance per meter and the voltage per meter needed.

84. Integrated Concepts

One long straight wire is to be held directly above another by repulsion between their currents. The lower wire carries 100 A and the wire 7.50 cm above it is 10-gauge (2.588 mm diameter) copper wire. (a) What current must flow in the upper wire, neglecting the Earth's field? (b) What is the smallest current if the Earth's 3.00×10^{-5} T field is parallel to the ground and is not neglected? (c) Is the supported wire in a stable or unstable equilibrium if displaced vertically? If displaced horizontally?

85. Unreasonable Results

(a) Find the charge on a baseball, thrown at 35.0 m/s perpendicular to the Earth's 5.00×10^{-5} T field, that experiences a 1.00-N magnetic force. (b) What is unreasonable about this result? (c) Which assumption or premise is responsible?

86. Unreasonable Results

A charged particle having mass 6.64×10^{-27} kg (that of a helium atom) moving at 8.70×10^{5} m/s perpendicular to a 1.50-T magnetic field travels in a circular path of radius 16.0 mm. (a) What is the charge of the particle? (b) What is unreasonable about this result? (c) Which assumptions are responsible?

87. Unreasonable Results

An inventor wants to generate 120-V power by moving a 1.00-m-long wire perpendicular to the Earth's 5.00×10^{-5} T field. (a) Find the speed with which the wire must move. (b) What is unreasonable about this result? (c) Which assumption is responsible?

88. Unreasonable Results

Frustrated by the small Hall voltage obtained in blood flow measurements, a medical physicist decides to increase the applied magnetic field strength to get a 0.500-V output for blood moving at 30.0 cm/s in a 1.50-cm-diameter vessel. (a) What magnetic field strength is needed? (b) What is unreasonable about this result? (c) Which premise is responsible?

89. Unreasonable Results

A surveyor 100 m from a long straight 200-kV DC power line suspects that its magnetic field may equal that of the Earth and affect compass readings. (a) Calculate the current in the wire needed to create a 5.00×10^{-5} T field at this distance. (b) What is unreasonable about this result? (c) Which assumption or premise is responsible?

90. Construct Your Own Problem

Consider a mass separator that applies a magnetic field perpendicular to the velocity of ions and separates the ions based on the radius of curvature of their paths in the field. Construct a problem in which you calculate the magnetic field strength needed to separate two ions that differ in mass, but not charge, and have the same initial velocity. Among the things to consider are the types of ions, the velocities they can be given before entering the magnetic field, and a reasonable value for the radius of curvature of the paths they follow. In addition, calculate the separation distance between the ions at the point where they are detected.

91. Construct Your Own Problem

Consider using the torque on a current-carrying coil in a magnetic field to detect relatively small magnetic fields (less than the field of the Earth, for example). Construct a problem in which you calculate the maximum torque on a current-carrying loop in a magnetic field. Among the things to be considered are the size of the coil, the number of loops it has, the current you pass through the coil, and the size of the field you wish to detect. Discuss whether the torque produced is large enough to be effectively measured. Your instructor may also wish for you to consider the effects, if any, of the field produced by the coil on the surroundings that could affect detection of the small field.

Test Prep for AP® Courses

22.2 Ferromagnets and Electromagnets

1. A bar magnet is oriented so that the north pole of the bar magnet points north. A compass needle is placed to the north of the bar magnet. In which direction does the north pole of the compass needle point?

a. North
b. East
c. South
d. West

2. Assume for simplicity that the Earth's magnetic north pole is at the same location as its geographic north pole. If you are in an airplane flying due west along the equator, as you cross the prime meridian (0° longitude) facing west and look down at a compass you are carrying, you see that the compass needle is perpendicular to your direction of motion, and the north pole of the needle dipole points to your right. As you continue flying due west, describe how and why the orientation of the needle will (or will not) change.

3. Describe how the magnetic domains in an unmagnetized iron rod will respond to the presence of a strong external magnetic field.

a. The domains will split into monopoles.
b. The domains will tend to align with the external field.
c. The domains will tend to orient themselves perpendicular to the external field.
d. The domains will tend to align so as to cancel out the external field.

4. Describe what steps must be undertaken in order to convert an unmagnetized iron rod into a permanently magnetized state. As part of your answer, explain what a magnetic domain is and how it responds to the steps described.

5. Iron is ferromagnetic and lead is diamagnetic, which means its magnetic domains respond in the opposite direction of ferromagnets but many orders of magnitude more weakly. The two blocks are placed in a magnetic field that points to the right. Which of the following best represents the orientations of the dipoles when the field is present?

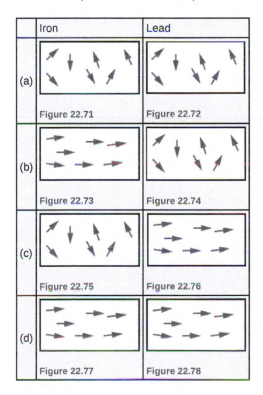

	Iron	Lead
(a)	Figure 22.71	Figure 22.72
(b)	Figure 22.73	Figure 22.74
(c)	Figure 22.75	Figure 22.76
(d)	Figure 22.77	Figure 22.78

6. A weather vane is some sort of directional arrow parallel to

the ground that may rotate freely in a horizontal plane. A typical weather vane has a large cross-sectional area perpendicular to the direction the arrow is pointing, like a "One Way" street sign. The purpose of the weather vane is to indicate the direction of the wind. As wind blows past the weather vane, it tends to orient the arrow in the same direction as the wind. Consider a weather vane's response to a strong wind. Explain how this is both similar to and different from a magnetic domain's response to an external magnetic field. How does each affect its surroundings?

22.4 Magnetic Field Strength: Force on a Moving Charge in a Magnetic Field

7. A proton moves in the –x-direction and encounters a uniform magnetic field pointing in the +x-direction. In what direction is the resulting magnetic force on the proton?

a. The proton experiences no magnetic force.
b. +x-direction
c. −y-direction
d. +y-direction

8. A proton moves with a speed of 240 m/s in the +x-direction into a region of a 4.5-T uniform magnetic field directed 62° above the +x-direction in the x,y-plane. Calculate the magnitude of the magnetic force on the proton.

22.5 Force on a Moving Charge in a Magnetic Field: Examples and Applications

9. A wire oriented north-south carries current south. The wire is immersed in the Earth's magnetic field, which is also oriented north-south (with a horizontal component pointing north). The Earth's magnetic field also has a vertical component pointing down. What is the direction of the magnetic force felt by the wire?

a. West
b. East
c. Up
d. North

22.6 The Hall Effect

10. An airplane wingspan can be approximated as a conducting rod of length 35 m. As the airplane flies due north, it is flying at a rate of 82 m/s through the Earth's magnetic field, which has a magnitude of 45 μT toward the north in a direction 57° below the horizontal plane. (a) Which end of the wingspan is positively charged, the east or west end? Explain. (b) What is the Hall emf along the wingspan?

22.9 Magnetic Fields Produced by Currents: Ampere's Law

11. An experimentalist fires a beam of electrons, creating a visible path in the air that can be measured. The beam is fired along a direction parallel to a current-carrying wire, and the electrons travel in a circular path in response to the wire's magnetic field. Assuming the mass and charge of the electrons is known, what quantities would you need to measure in order to deduce the current in the wire?

a. the radius of the circular path
b. the average distance between the electrons and the wire
c. the velocity of the electrons
d. two of the above
e. all of the above

12. Electrons starting from rest are accelerated through a potential difference of 240 V and fired into a region of uniform

3.5-mT magnetic field generated by a large solenoid. The electrons are initially moving in the +x-direction upon entering the field, and the field is directed into the page. Determine (a) the radius of the circle in which the electrons will move in this uniform magnetic field and (b) the initial direction of the magnetic force the electrons feel upon entering the uniform field of the solenoid.

13. In terms of the direction of force, we use the left-hand rule. Pointing your thumb in the +x-direction with the velocity and fingers of the left hand into the page reveals that the magnetic force points down toward the bottom of the page in the −y-direction.

A wire along the y-axis carries current in the +y-direction. In what direction is the magnetic field at a point on the +x-axis near the wire?

   a.   away from the wire
   b.   vertically upward
   c.   into the page
   d.   out of the page

14. Imagine the *xy* coordinate plane is the plane of the page. A wire along the *z*-axis carries current in the +z-direction (out of the page, or $\odot$). Draw a diagram of the magnetic field in the vicinity of this wire indicating the direction of the field. Also, describe how the strength of the magnetic field varies according to the distance from the *z*-axis.

22.10 Magnetic Force between Two Parallel Conductors

15. Two parallel wires carry equal currents in the same direction and are separated by a small distance. What is the direction of the magnetic force exerted by the two wires on each other?

   a.   No force since the wires are parallel.
   b.   No force since the currents are in the same direction.
   c.   The force is attractive.
   d.   The force is repulsive.

16. A wire along the y-axis carries current in the +y-direction. An experimenter would like to arrange a second wire parallel to the first wire and crossing the x-axis at the coordinate $x = 2.0 \text{ m}$ so that the total magnetic field at the coordinate $x = 1.0 \text{ m}$ is zero. In what direction must the current flow in the second wire, assuming it is equal in magnitude to the current in the first wire? Explain.

23 ELECTROMAGNETIC INDUCTION, AC CIRCUITS, AND ELECTRICAL TECHNOLOGIES

Figure 23.1 This wind turbine in the Thames Estuary in the UK is an example of induction at work. Wind pushes the blades of the turbine, spinning a shaft attached to magnets. The magnets spin around a conductive coil, inducing an electric current in the coil, and eventually feeding the electrical grid. (credit: phault, Flickr)

Chapter Outline

23.1. Induced Emf and Magnetic Flux

23.2. Faraday's Law of Induction: Lenz's Law

23.3. Motional Emf

23.4. Eddy Currents and Magnetic Damping

23.5. Electric Generators

23.6. Back Emf

23.7. Transformers

23.8. Electrical Safety: Systems and Devices

23.9. Inductance

23.10. RL Circuits

23.11. Reactance, Inductive and Capacitive

> ### 23.12. RLC Series AC Circuits

Connection for AP® Courses

Nature's displays of symmetry are beautiful and alluring. As shown in **Figure 23.2**, a butterfly's wings exhibit an appealing symmetry in a complex system. The laws of physics display symmetries at the most basic level – these symmetries are a source of wonder and imply deeper meaning. Since we place a high value on symmetry, we look for it when we explore nature. The remarkable thing is that we find it.

Figure 23.2 A butterfly with symmetrically patterned wings is resting on a flower. Physics, like this butterfly, has inherent symmetries. (credit: Thomas Bresson).

This chapter supports Big Idea 4, illustrating how electric and magnetic changes can take place in a system due to interactions with other systems. The hint of symmetry between electricity and magnetism found in the preceding chapter will be elaborated upon in this chapter. Specifically, we know that a current creates a magnetic field. If nature is symmetric in this case, then perhaps a magnetic field can create a current. Historically, it was very shortly after Oersted discovered that currents cause magnetic fields that other scientists asked the following question: Can magnetic fields cause currents? The answer was soon found by experiment to be yes. In 1831, some 12 years after Oersted's discovery, the English scientist Michael Faraday (1791–1862) and the American scientist Joseph Henry (1797–1878) independently demonstrated that magnetic fields can produce currents. The basic process of generating emfs (electromotive forces), and hence currents, with magnetic fields is known as induction; this process is also called "magnetic induction" to distinguish it from charging by induction, which utilizes the Coulomb force.

Today, currents induced by magnetic fields are essential to our technological society. The ubiquitous generator – found in automobiles, on bicycles, in nuclear power plants, and so on – uses magnetism to generate current. Other devices that use magnetism to induce currents include pickup coils in electric guitars, transformers of every size, certain microphones, airport security gates, and damping mechanisms on sensitive chemical balances. Explanations and examples in this chapter will help you understand current induction via magnetic interactions in mechanical systems (Enduring Understanding 4.E, Essential Knowledge 4.E.2). You will also learn how the behavior of AC circuits depends strongly on the effect of magnetic fields on currents.

The content of this chapter supports:

Big Idea 4 Interactions between systems can result in changes in those systems.

Enduring Understanding 4.E The electric and magnetic properties of a system can change in response to the presence of, or changes in, other objects or systems.

Essential Knowledge 4.E.2 Changing magnetic flux induces an electric field that can establish an induced emf in a system.

23.1 Induced Emf and Magnetic Flux

Learning Objectives

By the end of this section, you will be able to:

- Calculate the flux of a uniform magnetic field through a loop of arbitrary orientation.
- Describe methods to produce an electromotive force (emf) with a magnetic field or a magnet and a loop of wire.

The information presented in this section supports the following AP® learning objectives and science practices:

- **4.E.2.1** The student is able to construct an explanation of the function of a simple electromagnetic device in which an induced emf is produced by a changing magnetic flux through an area defined by a current loop (i.e., a simple microphone or generator) or of the effect on behavior of a device in which an induced emf is produced by a constant magnetic field through a changing area. **(S.P. 6.4)**

The apparatus used by Faraday to demonstrate that magnetic fields can create currents is illustrated in **Figure 23.3**. When the switch is closed, a magnetic field is produced in the coil on the top part of the iron ring and transmitted to the coil on the bottom part of the ring. The galvanometer is used to detect any current induced in the coil on the bottom. It was found that each time the switch is closed, the galvanometer detects a current in one direction in the coil on the bottom. (You can also observe this in a physics lab.) Each time the switch is opened, the galvanometer detects a current in the opposite direction. Interestingly, if the switch remains closed or open for any length of time, there is no current through the galvanometer. *Closing and opening the switch* induces the current. It is the *change* in magnetic field that creates the current. More basic than the current that flows is the emfthat causes it. The current is a result of an *emf induced by a changing magnetic field*, whether or not there is a path for current to flow.

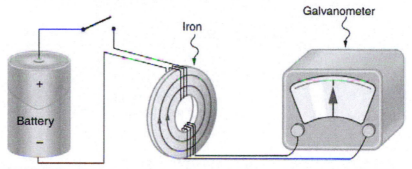

Figure 23.3 Faraday's apparatus for demonstrating that a magnetic field can produce a current. A change in the field produced by the top coil induces an emf and, hence, a current in the bottom coil. When the switch is opened and closed, the galvanometer registers currents in opposite directions. No current flows through the galvanometer when the switch remains closed or open.

An experiment easily performed and often done in physics labs is illustrated in **Figure 23.4**. An emf is induced in the coil when a bar magnet is pushed in and out of it. Emfs of opposite signs are produced by motion in opposite directions, and the emfs are also reversed by reversing poles. The same results are produced if the coil is moved rather than the magnet—it is the relative motion that is important. The faster the motion, the greater the emf, and there is no emf when the magnet is stationary relative to the coil.

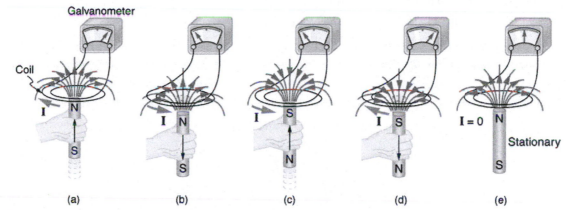

Figure 23.4 Movement of a magnet relative to a coil produces emfs as shown. The same emfs are produced if the coil is moved relative to the magnet. The greater the speed, the greater the magnitude of the emf, and the emf is zero when there is no motion.

The method of inducing an emf used in most electric generators is shown in **Figure 23.5**. A coil is rotated in a magnetic field, producing an alternating current emf, which depends on rotation rate and other factors that will be explored in later sections. Note that the generator is remarkably similar in construction to a motor (another symmetry).

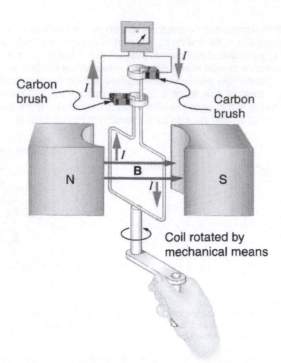

Figure 23.5 Rotation of a coil in a magnetic field produces an emf. This is the basic construction of a generator, where work done to turn the coil is converted to electric energy. Note the generator is very similar in construction to a motor.

So we see that changing the magnitude or direction of a magnetic field produces an emf. Experiments revealed that there is a crucial quantity called the **magnetic flux**, Φ , given by

$$\Phi = BA \cos \theta, \tag{23.1}$$

where B is the magnetic field strength over an area A , at an angle θ with the perpendicular to the area as shown in **Figure 23.6**. **Any change in magnetic flux Φ induces an emf.** This process is defined to be **electromagnetic induction**. Units of magnetic flux Φ are $\mathrm{T} \cdot \mathrm{m}^2$. As seen in **Figure 23.6**, $B \cos \theta = B_\perp$, which is the component of B perpendicular to the area A . Thus magnetic flux is $\Phi = B_\perp A$, the product of the area and the component of the magnetic field perpendicular to it.

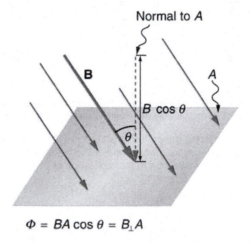

$$\Phi = BA \cos \theta = B_\perp A$$

Figure 23.6 Magnetic flux Φ is related to the magnetic field and the area over which it exists. The flux $\Phi = BA \cos \theta$ is related to induction; any change in Φ induces an emf.

All induction, including the examples given so far, arises from some change in magnetic flux Φ . For example, Faraday changed B and hence Φ when opening and closing the switch in his apparatus (shown in **Figure 23.3**). This is also true for the bar magnet and coil shown in **Figure 23.4**. When rotating the coil of a generator, the angle θ and, hence, Φ is changed. Just how great an emf and what direction it takes depend on the change in Φ and how rapidly the change is made, as examined in the next section.

23.2 Faraday's Law of Induction: Lenz's Law

Faraday's and Lenz's Law

Faraday's experiments showed that the emf induced by a change in magnetic flux depends on only a few factors. First, emf is directly proportional to the change in flux $\Delta\Phi$. Second, emf is greatest when the change in time Δt is smallest—that is, emf is inversely proportional to Δt. Finally, if a coil has N turns, an emf will be produced that is N times greater than for a single coil, so that emf is directly proportional to N. The equation for the emf induced by a change in magnetic flux is

$$\text{emf} = -N\frac{\Delta\Phi}{\Delta t}.$$

(23.2)

This relationship is known as **Faraday's law of induction**. The units for emf are volts, as is usual.

The minus sign in Faraday's law of induction is very important. The minus means that *the emf creates a current I and magnetic field B that oppose the change in flux* $\Delta\Phi$ *—this is known as Lenz's law*. The direction (given by the minus sign) of the emf is so important that it is called **Lenz's law** after the Russian Heinrich Lenz (1804–1865), who, like Faraday and Henry, independently investigated aspects of induction. Faraday was aware of the direction, but Lenz stated it so clearly that he is credited for its discovery. (See **Figure 23.7**.)

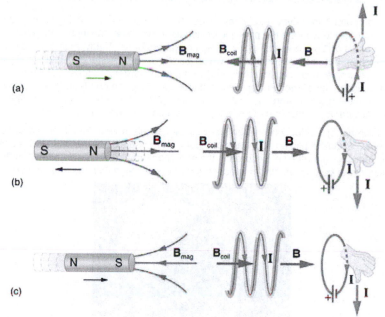

Figure 23.7 (a) When this bar magnet is thrust into the coil, the strength of the magnetic field increases in the coil. The current induced in the coil creates another field, in the opposite direction of the bar magnet's to oppose the increase. This is one aspect of *Lenz's law—induction opposes any change in flux*. (b) and (c) are two other situations. Verify for yourself that the direction of the induced B_{coil} shown indeed opposes the change in flux and that the current direction shown is consistent with RHR-2.

Problem-Solving Strategy for Lenz's Law

To use Lenz's law to determine the directions of the induced magnetic fields, currents, and emfs:

1. Make a sketch of the situation for use in visualizing and recording directions.
2. Determine the direction of the magnetic field B.
3. Determine whether the flux is increasing or decreasing.
4. Now determine the direction of the induced magnetic field B. It opposes the *change* in flux by adding or subtracting from the original field.
5. Use RHR-2 to determine the direction of the induced current I that is responsible for the induced magnetic field B.
6. The direction (or polarity) of the induced emf will now drive a current in this direction and can be represented as current

emerging from the positive terminal of the emf and returning to its negative terminal.

For practice, apply these steps to the situations shown in **Figure 23.7** and to others that are part of the following text material.

Applications of Electromagnetic Induction

There are many applications of Faraday's Law of induction, as we will explore in this chapter and others. At this juncture, let us mention several that have to do with data storage and magnetic fields. A very important application has to do with audio and video *recording tapes*. A plastic tape, coated with iron oxide, moves past a recording head. This recording head is basically a round iron ring about which is wrapped a coil of wire—an electromagnet (**Figure 23.8**). A signal in the form of a varying input current from a microphone or camera goes to the recording head. These signals (which are a function of the signal amplitude and frequency) produce varying magnetic fields at the recording head. As the tape moves past the recording head, the magnetic field orientations of the iron oxide molecules on the tape are changed thus recording the signal. In the playback mode, the magnetized tape is run past another head, similar in structure to the recording head. The different magnetic field orientations of the iron oxide molecules on the tape induces an emf in the coil of wire in the playback head. This signal then is sent to a loudspeaker or video player.

Figure 23.8 Recording and playback heads used with audio and video magnetic tapes. (credit: Steve Jurvetson)

Similar principles apply to computer hard drives, except at a much faster rate. Here recordings are on a coated, spinning disk. Read heads historically were made to work on the principle of induction. However, the input information is carried in digital rather than analog form – a series of 0's or 1's are written upon the spinning hard drive. Today, most hard drive readout devices do not work on the principle of induction, but use a technique known as *giant magnetoresistance*. (The discovery that weak changes in a magnetic field in a thin film of iron and chromium could bring about much larger changes in electrical resistance was one of the first large successes of nanotechnology.) Another application of induction is found on the magnetic stripe on the back of your personal credit card as used at the grocery store or the ATM machine. This works on the same principle as the audio or video tape mentioned in the last paragraph in which a head reads personal information from your card.

Another application of electromagnetic induction is when electrical signals need to be transmitted across a barrier. Consider the *cochlear implant* shown below. Sound is picked up by a microphone on the outside of the skull and is used to set up a varying magnetic field. A current is induced in a receiver secured in the bone beneath the skin and transmitted to electrodes in the inner ear. Electromagnetic induction can be used in other instances where electric signals need to be conveyed across various media.

Figure 23.9 Electromagnetic induction used in transmitting electric currents across mediums. The device on the baby's head induces an electrical current in a receiver secured in the bone beneath the skin. (credit: Bjorn Knetsch)

Another contemporary area of research in which electromagnetic induction is being successfully implemented (and with substantial potential) is transcranial magnetic simulation. A host of disorders, including depression and hallucinations can be traced to irregular localized electrical activity in the brain. In *transcranial magnetic stimulation*, a rapidly varying and very localized magnetic field is placed close to certain sites identified in the brain. Weak electric currents are induced in the identified sites and can result in recovery of electrical functioning in the brain tissue.

Sleep apnea ("the cessation of breath") affects both adults and infants (especially premature babies and it may be a cause of sudden infant deaths [SID]). In such individuals, breath can stop repeatedly during their sleep. A cessation of more than 20 seconds can be very dangerous. Stroke, heart failure, and tiredness are just some of the possible consequences for a person having sleep apnea. The concern in infants is the stopping of breath for these longer times. One type of monitor to alert parents when a child is not breathing uses electromagnetic induction. A wire wrapped around the infant's chest has an alternating current running through it. The expansion and contraction of the infant's chest as the infant breathes changes the area through the coil. A pickup coil located nearby has an alternating current induced in it due to the changing magnetic field of the initial wire. If the child stops breathing, there will be a change in the induced current, and so a parent can be alerted.

Making Connections: Conservation of Energy

Lenz's law is a manifestation of the conservation of energy. The induced emf produces a current that opposes the change in flux, because a change in flux means a change in energy. Energy can enter or leave, but not instantaneously. Lenz's law is a consequence. As the change begins, the law says induction opposes and, thus, slows the change. In fact, if the induced emf were in the same direction as the change in flux, there would be a positive feedback that would give us free energy from no apparent source—conservation of energy would be violated.

Example 23.1 Calculating Emf: How Great Is the Induced Emf?

Calculate the magnitude of the induced emf when the magnet in **Figure 23.7**(a) is thrust into the coil, given the following information: the single loop coil has a radius of 6.00 cm and the average value of $B \cos \theta$ (this is given, since the bar magnet's field is complex) increases from 0.0500 T to 0.250 T in 0.100 s.

Strategy

To find the *magnitude* of emf, we use Faraday's law of induction as stated by $\text{emf} = -N\dfrac{\Delta \Phi}{\Delta t}$, but without the minus sign that indicates direction:

$$\text{emf} = N\frac{\Delta \Phi}{\Delta t}. \tag{23.3}$$

Solution

We are given that $N = 1$ and $\Delta t = 0.100$ s, but we must determine the change in flux $\Delta \Phi$ before we can find emf. Since the area of the loop is fixed, we see that

$$\Delta \Phi = \Delta(BA \cos \theta) = A\Delta(B \cos \theta). \tag{23.4}$$

Now $\Delta(B \cos \theta) = 0.200$ T, since it was given that $B \cos \theta$ changes from 0.0500 to 0.250 T. The area of the loop is $A = \pi r^2 = (3.14...)(0.060 \text{ m})^2 = 1.13 \times 10^{-2} \text{ m}^2$. Thus,

$$\Delta \Phi = (1.13 \times 10^{-2} \text{ m}^2)(0.200 \text{ T}). \tag{23.5}$$

Entering the determined values into the expression for emf gives

$$\text{Emf} = N\frac{\Delta \Phi}{\Delta t} = \frac{(1.13 \times 10^{-2} \text{ m}^2)(0.200 \text{ T})}{0.100 \text{ s}} = 22.6 \text{ mV}. \tag{23.6}$$

Discussion

While this is an easily measured voltage, it is certainly not large enough for most practical applications. More loops in the coil, a stronger magnet, and faster movement make induction the practical source of voltages that it is.

PhET Explorations: Faraday's Electromagnetic Lab

Play with a bar magnet and coils to learn about Faraday's law. Move a bar magnet near one or two coils to make a light bulb glow. View the magnetic field lines. A meter shows the direction and magnitude of the current. View the magnetic field lines or use a meter to show the direction and magnitude of the current. You can also play with electromagnets, generators and transformers!

Figure 23.10 Faraday's Electromagnetic Lab (http://cnx.org/content/m55407/1.2/faraday_en.jar)

23.3 Motional Emf

Learning Objectives

By the end of this section, you will be able to:

- Calculate emf, force, magnetic field, and work due to the motion of an object in a magnetic field.

As we have seen, any change in magnetic flux induces an emf opposing that change—a process known as induction. Motion is one of the major causes of induction. For example, a magnet moved toward a coil induces an emf, and a coil moved toward a magnet produces a similar emf. In this section, we concentrate on motion in a magnetic field that is stationary relative to the Earth, producing what is loosely called *motional emf*.

One situation where motional emf occurs is known as the Hall effect and has already been examined. Charges moving in a magnetic field experience the magnetic force $F = qvB \sin \theta$, which moves opposite charges in opposite directions and produces an $\mathrm{emf} = B\ell v$. We saw that the Hall effect has applications, including measurements of B and v. We will now see that the Hall effect is one aspect of the broader phenomenon of induction, and we will find that motional emf can be used as a power source.

Consider the situation shown in **Figure 23.11**. A rod is moved at a speed v along a pair of conducting rails separated by a distance ℓ in a uniform magnetic field B. The rails are stationary relative to B and are connected to a stationary resistor R. The resistor could be anything from a light bulb to a voltmeter. Consider the area enclosed by the moving rod, rails, and resistor. B is perpendicular to this area, and the area is increasing as the rod moves. Thus the magnetic flux enclosed by the rails, rod, and resistor is increasing. When flux changes, an emf is induced according to Faraday's law of induction.

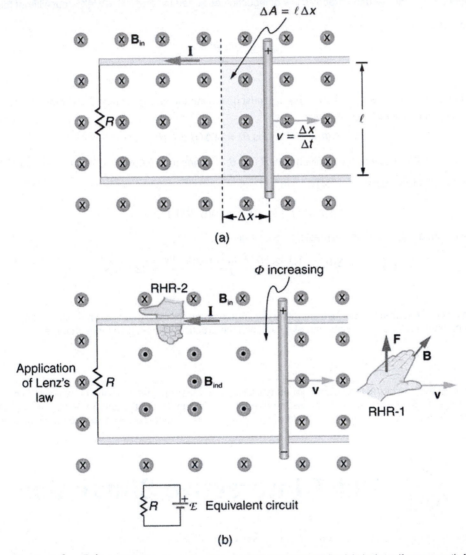

(a)

(b)

Figure 23.11 (a) A motional $\mathrm{emf} = B\ell v$ is induced between the rails when this rod moves to the right in the uniform magnetic field. The magnetic field B is into the page, perpendicular to the moving rod and rails and, hence, to the area enclosed by them. (b) Lenz's law gives the directions of the induced field and current, and the polarity of the induced emf. Since the flux is increasing, the induced field is in the opposite direction, or out of the page. RHR-2 gives the current direction shown, and the polarity of the rod will drive such a current. RHR-1 also indicates the same polarity for the rod. (Note that the script E symbol used in the equivalent circuit at the bottom of part (b) represents emf.)

To find the magnitude of emf induced along the moving rod, we use Faraday's law of induction without the sign:

$$\mathrm{emf} = N\frac{\Delta \Phi}{\Delta t}. \tag{23.7}$$

Here and below, "emf" implies the magnitude of the emf. In this equation, $N = 1$ and the flux $\Phi = BA \cos \theta$. We have $\theta = 0°$ and $\cos \theta = 1$, since B is perpendicular to A. Now $\Delta\Phi = \Delta(BA) = B\Delta A$, since B is uniform. Note that the area swept out by the rod is $\Delta A = \ell \, \Delta \, x$. Entering these quantities into the expression for emf yields

$$\text{emf} = \frac{B\Delta A}{\Delta t} = B\frac{\ell\Delta x}{\Delta t}. \qquad (23.8)$$

Finally, note that $\Delta x / \Delta t = v$, the velocity of the rod. Entering this into the last expression shows that

$$\text{emf} = B\ell v \qquad (B, \ell, \text{ and } v \text{ perpendicular}) \qquad (23.9)$$

is the motional emf. This is the same expression given for the Hall effect previously.

Making Connections: Unification of Forces

There are many connections between the electric force and the magnetic force. The fact that a moving electric field produces a magnetic field and, conversely, a moving magnetic field produces an electric field is part of why electric and magnetic forces are now considered to be different manifestations of the same force. This classic unification of electric and magnetic forces into what is called the electromagnetic force is the inspiration for contemporary efforts to unify other basic forces.

To find the direction of the induced field, the direction of the current, and the polarity of the induced emf, we apply Lenz's law as explained in **Faraday's Law of Induction: Lenz's Law.** (See **Figure 23.11**(b).) Flux is increasing, since the area enclosed is increasing. Thus the induced field must oppose the existing one and be out of the page. And so the RHR-2 requires that I be counterclockwise, which in turn means the top of the rod is positive as shown.

Motional emf also occurs if the magnetic field moves and the rod (or other object) is stationary relative to the Earth (or some observer). We have seen an example of this in the situation where a moving magnet induces an emf in a stationary coil. It is the relative motion that is important. What is emerging in these observations is a connection between magnetic and electric fields. A moving magnetic field produces an electric field through its induced emf. We already have seen that a moving electric field produces a magnetic field—moving charge implies moving electric field and moving charge produces a magnetic field.

Motional emfs in the Earth's weak magnetic field are not ordinarily very large, or we would notice voltage along metal rods, such as a screwdriver, during ordinary motions. For example, a simple calculation of the motional emf of a 1 m rod moving at 3.0 m/s perpendicular to the Earth's field gives $\text{emf} = B\ell v = (5.0 \times 10^{-5} \text{ T})(1.0 \text{ m})(3.0 \text{ m/s}) = 150 \ \mu\text{V}$. This small value is consistent with experience. There is a spectacular exception, however. In 1992 and 1996, attempts were made with the space shuttle to create large motional emfs. The Tethered Satellite was to be let out on a 20 km length of wire as shown in **Figure 23.12**, to create a 5 kV emf by moving at orbital speed through the Earth's field. This emf could be used to convert some of the shuttle's kinetic and potential energy into electrical energy if a complete circuit could be made. To complete the circuit, the stationary ionosphere was to supply a return path for the current to flow. (The ionosphere is the rarefied and partially ionized atmosphere at orbital altitudes. It conducts because of the ionization. The ionosphere serves the same function as the stationary rails and connecting resistor in **Figure 23.11**, without which there would not be a complete circuit.) Drag on the current in the cable due to the magnetic force $F = I\ell B \sin \theta$ does the work that reduces the shuttle's kinetic and potential energy and allows it to be converted to electrical energy. The tests were both unsuccessful. In the first, the cable hung up and could only be extended a couple of hundred meters; in the second, the cable broke when almost fully extended. **Example 23.2** indicates feasibility in principle.

Example 23.2 Calculating the Large Motional Emf of an Object in Orbit

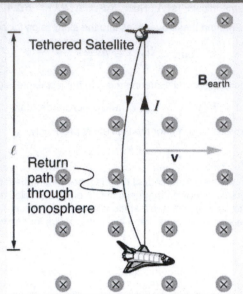

Figure 23.12 Motional emf as electrical power conversion for the space shuttle is the motivation for the Tethered Satellite experiment. A 5 kV emf was predicted to be induced in the 20 km long tether while moving at orbital speed in the Earth's magnetic field. The circuit is completed by a return path through the stationary ionosphere.

Calculate the motional emf induced along a 20.0 km long conductor moving at an orbital speed of 7.80 km/s perpendicular to the Earth's 5.00×10^{-5} T magnetic field.

Strategy

This is a straightforward application of the expression for motional emf— $\mathrm{emf} = B\ell v$.

Solution

Entering the given values into $\mathrm{emf} = B\ell v$ gives

$$
\begin{aligned}
\mathrm{emf} &= B\ell v \\
&= (5.00 \times 10^{-5}\ \mathrm{T})(2.0 \times 10^{4}\ \mathrm{m})(7.80 \times 10^{3}\ \mathrm{m/s}) \\
&= 7.80 \times 10^{3}\ \mathrm{V}.
\end{aligned}
\tag{23.10}
$$

Discussion

The value obtained is greater than the 5 kV measured voltage for the shuttle experiment, since the actual orbital motion of the tether is not perpendicular to the Earth's field. The 7.80 kV value is the maximum emf obtained when $\theta = 90°$ and $\sin \theta = 1$.

23.4 Eddy Currents and Magnetic Damping

Learning Objectives

By the end of this section, you will be able to:

- Explain the magnitude and direction of an induced eddy current, and the effect this will have on the object it is induced in.
- Describe several applications of magnetic damping.

Eddy Currents and Magnetic Damping

As discussed in **Motional Emf**, motional emf is induced when a conductor moves in a magnetic field or when a magnetic field moves relative to a conductor. If motional emf can cause a current loop in the conductor, we refer to that current as an **eddy current**. Eddy currents can produce significant drag, called **magnetic damping**, on the motion involved. Consider the apparatus shown in **Figure 23.13**, which swings a pendulum bob between the poles of a strong magnet. (This is another favorite physics lab activity.) If the bob is metal, there is significant drag on the bob as it enters and leaves the field, quickly damping the motion. If, however, the bob is a slotted metal plate, as shown in **Figure 23.13**(b), there is a much smaller effect due to the magnet. There is no discernible effect on a bob made of an insulator. Why is there drag in both directions, and are there any uses for magnetic drag?

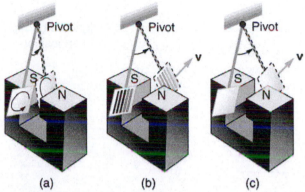

(a)   (b)   (c)

Figure 23.13 A common physics demonstration device for exploring eddy currents and magnetic damping. (a) The motion of a metal pendulum bob swinging between the poles of a magnet is quickly damped by the action of eddy currents. (b) There is little effect on the motion of a slotted metal bob, implying that eddy currents are made less effective. (c) There is also no magnetic damping on a nonconducting bob, since the eddy currents are extremely small.

Figure 23.14 shows what happens to the metal plate as it enters and leaves the magnetic field. In both cases, it experiences a force opposing its motion. As it enters from the left, flux increases, and so an eddy current is set up (Faraday's law) in the counterclockwise direction (Lenz's law), as shown. Only the right-hand side of the current loop is in the field, so that there is an unopposed force on it to the left (RHR-1). When the metal plate is completely inside the field, there is no eddy current if the field is uniform, since the flux remains constant in this region. But when the plate leaves the field on the right, flux decreases, causing an eddy current in the clockwise direction that, again, experiences a force to the left, further slowing the motion. A similar analysis of what happens when the plate swings from the right toward the left shows that its motion is also damped when entering and leaving the field.

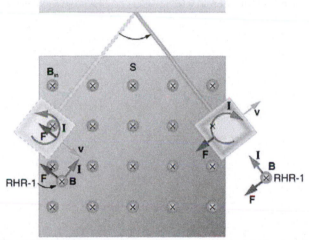

Figure 23.14 A more detailed look at the conducting plate passing between the poles of a magnet. As it enters and leaves the field, the change in flux produces an eddy current. Magnetic force on the current loop opposes the motion. There is no current and no magnetic drag when the plate is completely inside the uniform field.

When a slotted metal plate enters the field, as shown in **Figure 23.15**, an emf is induced by the change in flux, but it is less effective because the slots limit the size of the current loops. Moreover, adjacent loops have currents in opposite directions, and their effects cancel. When an insulating material is used, the eddy current is extremely small, and so magnetic damping on insulators is negligible. If eddy currents are to be avoided in conductors, then they can be slotted or constructed of thin layers of conducting material separated by insulating sheets.

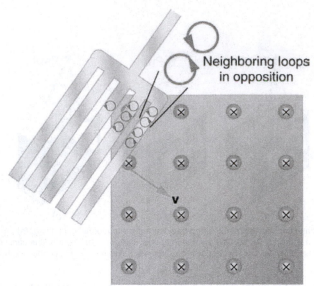

Figure 23.15 Eddy currents induced in a slotted metal plate entering a magnetic field form small loops, and the forces on them tend to cancel, thereby making magnetic drag almost zero.

Applications of Magnetic Damping

One use of magnetic damping is found in sensitive laboratory balances. To have maximum sensitivity and accuracy, the balance must be as friction-free as possible. But if it is friction-free, then it will oscillate for a very long time. Magnetic damping is a simple and ideal solution. With magnetic damping, drag is proportional to speed and becomes zero at zero velocity. Thus the oscillations are quickly damped, after which the damping force disappears, allowing the balance to be very sensitive. (See **Figure 23.16**.) In most balances, magnetic damping is accomplished with a conducting disc that rotates in a fixed field.

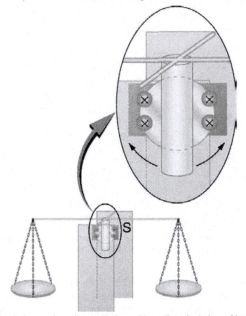

Figure 23.16 Magnetic damping of this sensitive balance slows its oscillations. Since Faraday's law of induction gives the greatest effect for the most rapid change, damping is greatest for large oscillations and goes to zero as the motion stops.

Since eddy currents and magnetic damping occur only in conductors, recycling centers can use magnets to separate metals from other materials. Trash is dumped in batches down a ramp, beneath which lies a powerful magnet. Conductors in the trash are slowed by magnetic damping while nonmetals in the trash move on, separating from the metals. (See **Figure 23.17**.) This works for all metals, not just ferromagnetic ones. A magnet can separate out the ferromagnetic materials alone by acting on stationary trash.

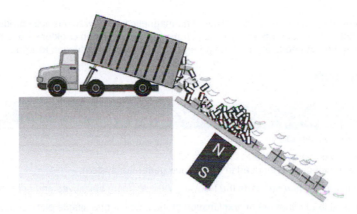

Figure 23.17 Metals can be separated from other trash by magnetic drag. Eddy currents and magnetic drag are created in the metals sent down this ramp by the powerful magnet beneath it. Nonmetals move on.

Other major applications of eddy currents are in metal detectors and braking systems in trains and roller coasters. Portable metal detectors (Figure 23.18) consist of a primary coil carrying an alternating current and a secondary coil in which a current is induced. An eddy current will be induced in a piece of metal close to the detector which will cause a change in the induced current within the secondary coil, leading to some sort of signal like a shrill noise. Braking using eddy currents is safer because factors such as rain do not affect the braking and the braking is smoother. However, eddy currents cannot bring the motion to a complete stop, since the force produced decreases with speed. Thus, speed can be reduced from say 20 m/s to 5 m/s, but another form of braking is needed to completely stop the vehicle. Generally, powerful rare earth magnets such as neodymium magnets are used in roller coasters. Figure 23.19 shows rows of magnets in such an application. The vehicle has metal fins (normally containing copper) which pass through the magnetic field slowing the vehicle down in much the same way as with the pendulum bob shown in Figure 23.13.

Figure 23.18 A soldier in Iraq uses a metal detector to search for explosives and weapons. (credit: U.S. Army)

Figure 23.19 The rows of rare earth magnets (protruding horizontally) are used for magnetic braking in roller coasters. (credit: Stefan Scheer, Wikimedia Commons)

Induction cooktops have electromagnets under their surface. The magnetic field is varied rapidly producing eddy currents in the base of the pot, causing the pot and its contents to increase in temperature. Induction cooktops have high efficiencies and good response times but the base of the pot needs to be ferromagnetic, iron or steel for induction to work.

23.5 Electric Generators

Electric generators induce an emf by rotating a coil in a magnetic field, as briefly discussed in **Induced Emf and Magnetic Flux**. We will now explore generators in more detail. Consider the following example.

Example 23.3 Calculating the Emf Induced in a Generator Coil

The generator coil shown in **Figure 23.20** is rotated through one-fourth of a revolution (from $\theta = 0°$ to $\theta = 90°$) in 15.0 ms. The 200-turn circular coil has a 5.00 cm radius and is in a uniform 1.25 T magnetic field. What is the average emf induced?

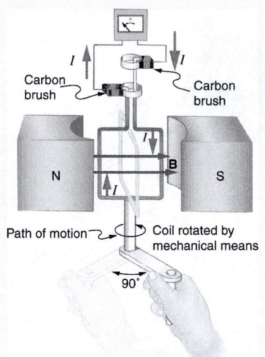

Figure 23.20 When this generator coil is rotated through one-fourth of a revolution, the magnetic flux Φ changes from its maximum to zero, inducing an emf.

Strategy

We use Faraday's law of induction to find the average emf induced over a time Δt :

$$\text{emf} = -N\frac{\Delta \Phi}{\Delta t}. \tag{23.11}$$

We know that $N = 200$ and $\Delta t = 15.0 \text{ ms}$, and so we must determine the change in flux $\Delta \Phi$ to find emf.

Solution

Since the area of the loop and the magnetic field strength are constant, we see that

$$\Delta\Phi = \Delta(BA\cos\theta) = AB\Delta(\cos\theta). \tag{23.12}$$

Now, $\Delta(\cos\theta) = -1.0$, since it was given that θ goes from $0°$ to $90°$. Thus $\Delta\Phi = -AB$, and

$$\text{emf} = N\frac{AB}{\Delta t}. \tag{23.13}$$

The area of the loop is $A = \pi r^2 = (3.14...)(0.0500\text{ m})^2 = 7.85\times10^{-3}\text{ m}^2$. Entering this value gives

$$\text{emf} = 200\frac{(7.85\times10^{-3}\text{ m}^2)(1.25\text{ T})}{15.0\times10^{-3}\text{ s}} = 131\text{ V}. \tag{23.14}$$

Discussion

This is a practical average value, similar to the 120 V used in household power.

The emf calculated in **Example 23.3** is the average over one-fourth of a revolution. What is the emf at any given instant? It varies with the angle between the magnetic field and a perpendicular to the coil. We can get an expression for emf as a function of time by considering the motional emf on a rotating rectangular coil of width w and height ℓ in a uniform magnetic field, as illustrated in **Figure 23.21**.

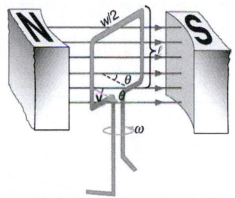

Figure 23.21 A generator with a single rectangular coil rotated at constant angular velocity in a uniform magnetic field produces an emf that varies sinusoidally in time. Note the generator is similar to a motor, except the shaft is rotated to produce a current rather than the other way around.

Charges in the wires of the loop experience the magnetic force, because they are moving in a magnetic field. Charges in the vertical wires experience forces parallel to the wire, causing currents. But those in the top and bottom segments feel a force perpendicular to the wire, which does not cause a current. We can thus find the induced emf by considering only the side wires. Motional emf is given to be $\text{emf} = B\ell v$, where the velocity v is perpendicular to the magnetic field B. Here the velocity is at an angle θ with B, so that its component perpendicular to B is $v\sin\theta$ (see **Figure 23.21**). Thus in this case the emf induced on each side is $\text{emf} = B\ell v\sin\theta$, and they are in the same direction. The total emf around the loop is then

$$\text{emf} = 2B\ell v\sin\theta. \tag{23.15}$$

This expression is valid, but it does not give emf as a function of time. To find the time dependence of emf, we assume the coil rotates at a constant angular velocity ω. The angle θ is related to angular velocity by $\theta = \omega t$, so that

$$\text{emf} = 2B\ell v\sin\omega t. \tag{23.16}$$

Now, linear velocity v is related to angular velocity ω by $v = r\omega$. Here $r = w/2$, so that $v = (w/2)\omega$, and

$$\text{emf} = 2B\ell\frac{w}{2}\omega\sin\omega t = (\ell w)B\omega\sin\omega t. \tag{23.17}$$

Noting that the area of the loop is $A = \ell w$, and allowing for N loops, we find that

$$\text{emf} = NAB\omega\sin\omega t \tag{23.18}$$

is the **emf induced in a generator coil** of N turns and area A rotating at a constant angular velocity ω in a uniform magnetic field B. This can also be expressed as

$$\text{emf} = \text{emf}_0\sin\omega t, \tag{23.19}$$

where

$$\text{emf}_0 = NAB\omega \tag{23.20}$$

is the maximum **(peak) emf**. Note that the frequency of the oscillation is $f = \omega/2\pi$, and the period is $T = 1/f = 2\pi/\omega$.

Figure 23.22 shows a graph of emf as a function of time, and it now seems reasonable that AC voltage is sinusoidal.

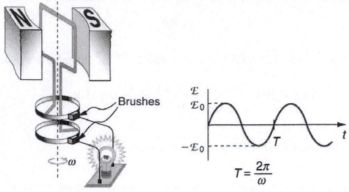

Figure 23.22 The emf of a generator is sent to a light bulb with the system of rings and brushes shown. The graph gives the emf of the generator as a function of time. emf_0 is the peak emf. The period is $T = 1/f = 2\pi/\omega$, where f is the frequency. Note that the script E stands for emf.

The fact that the peak emf, $\text{emf}_0 = NAB\omega$, makes good sense. The greater the number of coils, the larger their area, and the stronger the field, the greater the output voltage. It is interesting that the faster the generator is spun (greater ω), the greater the emf. This is noticeable on bicycle generators—at least the cheaper varieties. One of the authors as a juvenile found it amusing to ride his bicycle fast enough to burn out his lights, until he had to ride home lightless one dark night.

Figure 23.23 shows a scheme by which a generator can be made to produce pulsed DC. More elaborate arrangements of multiple coils and split rings can produce smoother DC, although electronic rather than mechanical means are usually used to make ripple-free DC.

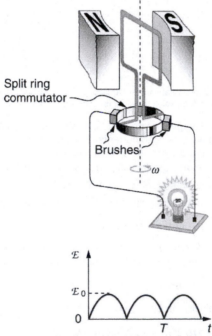

Figure 23.23 Split rings, called commutators, produce a pulsed DC emf output in this configuration.

Example 23.4 Calculating the Maximum Emf of a Generator

Calculate the maximum emf, emf_0, of the generator that was the subject of **Example 23.3**.

Strategy

Once ω, the angular velocity, is determined, $\text{emf}_0 = NAB\omega$ can be used to find emf_0. All other quantities are known.

Solution

Angular velocity is defined to be the change in angle per unit time:

$$\omega = \frac{\Delta\theta}{\Delta t}. \tag{23.21}$$

One-fourth of a revolution is $\pi/2$ radians, and the time is 0.0150 s; thus,

$$\begin{aligned} \omega &= \frac{\pi/2 \text{ rad}}{0.0150 \text{ s}} \\ &= 104.7 \text{ rad/s}. \end{aligned} \tag{23.22}$$

104.7 rad/s is exactly 1000 rpm. We substitute this value for ω and the information from the previous example into $\text{emf}_0 = NAB\omega$, yielding

$$\begin{aligned} \text{emf}_0 &= NAB\omega \\ &= 200(7.85\times10^{-3} \text{ m}^2)(1.25 \text{ T})(104.7 \text{ rad/s}) \cdot \\ &= 206 \text{ V} \end{aligned} \tag{23.23}$$

Discussion

The maximum emf is greater than the average emf of 131 V found in the previous example, as it should be.

In real life, electric generators look a lot different than the figures in this section, but the principles are the same. The source of mechanical energy that turns the coil can be falling water (hydropower), steam produced by the burning of fossil fuels, or the kinetic energy of wind. **Figure 23.24** shows a cutaway view of a steam turbine; steam moves over the blades connected to the shaft, which rotates the coil within the generator.

Figure 23.24 Steam turbine/generator. The steam produced by burning coal impacts the turbine blades, turning the shaft which is connected to the generator. (credit: Nabonaco, Wikimedia Commons)

Generators illustrated in this section look very much like the motors illustrated previously. This is not coincidental. In fact, a motor becomes a generator when its shaft rotates. Certain early automobiles used their starter motor as a generator. In **Back Emf**, we shall further explore the action of a motor as a generator.

23.6 Back Emf

Learning Objectives
By the end of this section, you will be able to: • Explain what back emf is and how it is induced.

It has been noted that motors and generators are very similar. Generators convert mechanical energy into electrical energy, whereas motors convert electrical energy into mechanical energy. Furthermore, motors and generators have the same construction. When the coil of a motor is turned, magnetic flux changes, and an emf (consistent with Faraday's law of induction) is induced. The motor thus acts as a generator whenever its coil rotates. This will happen whether the shaft is turned by an external input, like a belt drive, or by the action of the motor itself. That is, when a motor is doing work and its shaft is turning, an emf is generated. Lenz's law tells us the emf opposes any change, so that the input emf that powers the motor will be opposed by the motor's self-generated emf, called the **back emf** of the motor. (See **Figure 23.25**.)

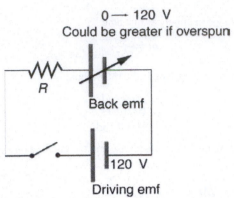

Figure 23.25 The coil of a DC motor is represented as a resistor in this schematic. The back emf is represented as a variable emf that opposes the one driving the motor. Back emf is zero when the motor is not turning, and it increases proportionally to the motor's angular velocity.

Back emf is the generator output of a motor, and so it is proportional to the motor's angular velocity ω. It is zero when the motor is first turned on, meaning that the coil receives the full driving voltage and the motor draws maximum current when it is on but not turning. As the motor turns faster and faster, the back emf grows, always opposing the driving emf, and reduces the voltage across the coil and the amount of current it draws. This effect is noticeable in a number of situations. When a vacuum cleaner, refrigerator, or washing machine is first turned on, lights in the same circuit dim briefly due to the IR drop produced in feeder lines by the large current drawn by the motor. When a motor first comes on, it draws more current than when it runs at its normal operating speed. When a mechanical load is placed on the motor, like an electric wheelchair going up a hill, the motor slows, the back emf drops, more current flows, and more work can be done. If the motor runs at too low a speed, the larger current can overheat it (via resistive power in the coil, $P = I^2 R$), perhaps even burning it out. On the other hand, if there is no mechanical load on the motor, it will increase its angular velocity ω until the back emf is nearly equal to the driving emf. Then the motor uses only enough energy to overcome friction.

Consider, for example, the motor coils represented in **Figure 23.25**. The coils have a $0.400\ \Omega$ equivalent resistance and are driven by a 48.0 V emf. Shortly after being turned on, they draw a current $I = V/R = (48.0\ \text{V})/(0.400\ \Omega) = 120\ \text{A}$ and,

thus, dissipate $P = I^2 R = 5.76\ \text{kW}$ of energy as heat transfer. Under normal operating conditions for this motor, suppose the back emf is 40.0 V. Then at operating speed, the total voltage across the coils is 8.0 V (48.0 V minus the 40.0 V back emf), and the current drawn is $I = V/R = (8.0\ \text{V})/(0.400\ \Omega) = 20\ \text{A}$. Under normal load, then, the power dissipated is

$P = IV = (20\ \text{A})/(8.0\ \text{V}) = 160\ \text{W}$. The latter will not cause a problem for this motor, whereas the former 5.76 kW would burn out the coils if sustained.

23.7 Transformers

Learning Objectives

By the end of this section, you will be able to:

- Explain how a transformer works.
- Calculate voltage, current, and/or number of turns given the other quantities.

The information presented in this section supports the following AP® learning objectives and science practices:

- **4.E.2.1** The student is able to construct an explanation of the function of a simple electromagnetic device in which an induced emf is produced by a changing magnetic flux through an area defined by a current loop (i.e., a simple microphone or generator) or of the effect on behavior of a device in which an induced emf is produced by a constant magnetic field through a changing area. **(S.P. 6.4)**

Transformers do what their name implies—they transform voltages from one value to another (The term voltage is used rather than emf, because transformers have internal resistance). For example, many cell phones, laptops, video games, and power tools and small appliances have a transformer built into their plug-in unit (like that in **Figure 23.26**) that changes 120 V or 240 V AC into whatever voltage the device uses. Transformers are also used at several points in the power distribution systems, such as illustrated in **Figure 23.27**. Power is sent long distances at high voltages, because less current is required for a given amount of power, and this means less line loss, as was discussed previously. But high voltages pose greater hazards, so that transformers are employed to produce lower voltage at the user's location.

Figure 23.26 The plug-in transformer has become increasingly familiar with the proliferation of electronic devices that operate on voltages other than common 120 V AC. Most are in the 3 to 12 V range. (credit: Shop Xtreme)

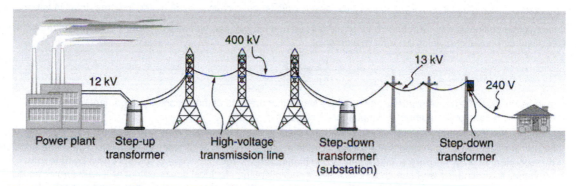

Figure 23.27 Transformers change voltages at several points in a power distribution system. Electric power is usually generated at greater than 10 kV, and transmitted long distances at voltages over 200 kV—sometimes as great as 700 kV—to limit energy losses. Local power distribution to neighborhoods or industries goes through a substation and is sent short distances at voltages ranging from 5 to 13 kV. This is reduced to 120, 240, or 480 V for safety at the individual user site.

The type of transformer considered in this text—see **Figure 23.28**—is based on Faraday's law of induction and is very similar in construction to the apparatus Faraday used to demonstrate magnetic fields could cause currents. The two coils are called the *primary* and *secondary coils*. In normal use, the input voltage is placed on the primary, and the secondary produces the transformed output voltage. Not only does the iron core trap the magnetic field created by the primary coil, its magnetization increases the field strength. Since the input voltage is AC, a time-varying magnetic flux is sent to the secondary, inducing its AC output voltage.

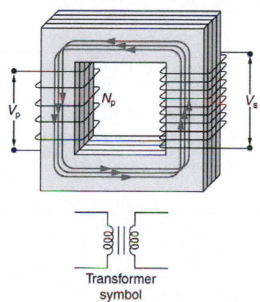

Transformer
symbol

Figure 23.28 A typical construction of a simple transformer has two coils wound on a ferromagnetic core that is laminated to minimize eddy currents. The magnetic field created by the primary is mostly confined to and increased by the core, which transmits it to the secondary coil. Any change in current in the primary induces a current in the secondary.

For the simple transformer shown in **Figure 23.28**, the output voltage V_s depends almost entirely on the input voltage V_p and the ratio of the number of loops in the primary and secondary coils. Faraday's law of induction for the secondary coil gives its induced output voltage V_s to be

$$V_s = -N_s \frac{\Delta \Phi}{\Delta t},$$ (23.24)

where N_s is the number of loops in the secondary coil and $\Delta \Phi / \Delta t$ is the rate of change of magnetic flux. Note that the output voltage equals the induced emf ($V_s = \text{emf}_s$), provided coil resistance is small (a reasonable assumption for transformers). The cross-sectional area of the coils is the same on either side, as is the magnetic field strength, and so $\Delta \Phi / \Delta t$ is the same on either side. The input primary voltage V_p is also related to changing flux by

$$V_p = -N_p \frac{\Delta \Phi}{\Delta t}.$$ (23.25)

The reason for this is a little more subtle. Lenz's law tells us that the primary coil opposes the change in flux caused by the input voltage V_p, hence the minus sign (This is an example of *self-inductance*, a topic to be explored in some detail in later sections).

Assuming negligible coil resistance, Kirchhoff's loop rule tells us that the induced emf exactly equals the input voltage. Taking the ratio of these last two equations yields a useful relationship:

$$\frac{V_s}{V_p} = \frac{N_s}{N_p}.$$ (23.26)

This is known as the **transformer equation**, and it simply states that the ratio of the secondary to primary voltages in a transformer equals the ratio of the number of loops in their coils.

The output voltage of a transformer can be less than, greater than, or equal to the input voltage, depending on the ratio of the number of loops in their coils. Some transformers even provide a variable output by allowing connection to be made at different points on the secondary coil. A **step-up transformer** is one that increases voltage, whereas a **step-down transformer** decreases voltage. Assuming, as we have, that resistance is negligible, the electrical power output of a transformer equals its input. This is nearly true in practice—transformer efficiency often exceeds 99%. Equating the power input and output,

$$P_p = I_p V_p = I_s V_s = P_s.$$ (23.27)

Rearranging terms gives

$$\frac{V_s}{V_p} = \frac{I_p}{I_s}.$$ (23.28)

Combining this with $\dfrac{V_s}{V_p} = \dfrac{N_s}{N_p}$, we find that

$$\frac{I_s}{I_p} = \frac{N_p}{N_s}$$ (23.29)

is the relationship between the output and input currents of a transformer. So if voltage increases, current decreases. Conversely, if voltage decreases, current increases.

Example 23.5 Calculating Characteristics of a Step-Up Transformer

A portable x-ray unit has a step-up transformer, the 120 V input of which is transformed to the 100 kV output needed by the x-ray tube. The primary has 50 loops and draws a current of 10.00 A when in use. (a) What is the number of loops in the secondary? (b) Find the current output of the secondary.

Strategy and Solution for (a)

We solve $\dfrac{V_s}{V_p} = \dfrac{N_s}{N_p}$ for N_s, the number of loops in the secondary, and enter the known values. This gives

$$\begin{aligned} N_s &= N_p \frac{V_s}{V_p} \\ &= (50)\frac{100,000 \text{ V}}{120 \text{ V}} = 4.17 \times 10^4. \end{aligned}$$ (23.30)

Discussion for (a)

A large number of loops in the secondary (compared with the primary) is required to produce such a large voltage. This would be true for neon sign transformers and those supplying high voltage inside TVs and CRTs.

Strategy and Solution for (b)

We can similarly find the output current of the secondary by solving $\frac{I_s}{I_p} = \frac{N_p}{N_s}$ for I_s and entering known values. This gives

$$
\begin{aligned}
I_s &= I_p \frac{N_p}{N_s} \\
&= (10.00 \text{ A}) \frac{50}{4.17 \times 10^4} = 12.0 \text{ mA}.
\end{aligned}
$$

(23.31)

Discussion for (b)

As expected, the current output is significantly less than the input. In certain spectacular demonstrations, very large voltages are used to produce long arcs, but they are relatively safe because the transformer output does not supply a large current. Note that the power input here is $P_p = I_p V_p = (10.00 \text{ A})(120 \text{ V}) = 1.20 \text{ kW}$. This equals the power output

$P_p = I_s V_s = (12.0 \text{ mA})(100 \text{ kV}) = 1.20 \text{ kW}$, as we assumed in the derivation of the equations used.

The fact that transformers are based on Faraday's law of induction makes it clear why we cannot use transformers to change DC voltages. If there is no change in primary voltage, there is no voltage induced in the secondary. One possibility is to connect DC to the primary coil through a switch. As the switch is opened and closed, the secondary produces a voltage like that in **Figure 23.29**. This is not really a practical alternative, and AC is in common use wherever it is necessary to increase or decrease voltages.

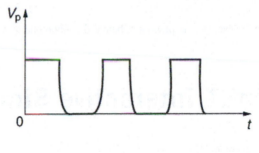

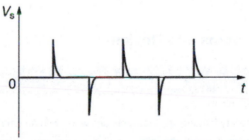

Figure 23.29 Transformers do not work for pure DC voltage input, but if it is switched on and off as on the top graph, the output will look something like that on the bottom graph. This is not the sinusoidal AC most AC appliances need.

Example 23.6 Calculating Characteristics of a Step-Down Transformer

A battery charger meant for a series connection of ten nickel-cadmium batteries (total emf of 12.5 V DC) needs to have a 15.0 V output to charge the batteries. It uses a step-down transformer with a 200-loop primary and a 120 V input. (a) How many loops should there be in the secondary coil? (b) If the charging current is 16.0 A, what is the input current?

Strategy and Solution for (a)

You would expect the secondary to have a small number of loops. Solving $\frac{V_s}{V_p} = \frac{N_s}{N_p}$ for N_s and entering known values gives

$$
\begin{aligned}
N_s &= N_p \frac{V_s}{V_p} \\
&= (200) \frac{15.0 \text{ V}}{120 \text{ V}} = 25.
\end{aligned}
$$

(23.32)

Strategy and Solution for (b)

The current input can be obtained by solving $\frac{I_s}{I_p} = \frac{N_p}{N_s}$ for I_p and entering known values. This gives

$$I_p = I_s \frac{N_s}{N_p} \tag{23.33}$$

$$= (16.0 \text{ A})\frac{25}{200} = 2.00 \text{ A}.$$

Discussion

The number of loops in the secondary is small, as expected for a step-down transformer. We also see that a small input current produces a larger output current in a step-down transformer. When transformers are used to operate large magnets, they sometimes have a small number of very heavy loops in the secondary. This allows the secondary to have low internal resistance and produce large currents. Note again that this solution is based on the assumption of 100% efficiency—or power out equals power in ($P_p = P_s$)—reasonable for good transformers. In this case the primary and secondary power is

240 W. (Verify this for yourself as a consistency check.) Note that the Ni-Cd batteries need to be charged from a DC power source (as would a 12 V battery). So the AC output of the secondary coil needs to be converted into DC. This is done using something called a rectifier, which uses devices called diodes that allow only a one-way flow of current.

Transformers have many applications in electrical safety systems, which are discussed in **Electrical Safety: Systems and Devices**.

PhET Explorations: Generator

Generate electricity with a bar magnet! Discover the physics behind the phenomena by exploring magnets and how you can use them to make a bulb light.

Figure 23.30 Generator (http://cnx.org/content/m55414/1.2/generator_en.jar)

23.8 Electrical Safety: Systems and Devices

Learning Objectives

By the end of this section, you will be able to:

- Explain how various modern safety features in electric circuits work, with an emphasis on how induction is employed.

The information presented in this section supports the following AP® learning objectives and science practices:

- **4.E.2.1** The student is able to construct an explanation of the function of a simple electromagnetic device in which an induced emf is produced by a changing magnetic flux through an area defined by a current loop (i.e., a simple microphone or generator) or of the effect on behavior of a device in which an induced emf is produced by a constant magnetic field through a changing area. **(S.P. 6.4)**

Electricity has two hazards. A **thermal hazard** occurs when there is electrical overheating. A **shock hazard** occurs when electric current passes through a person. Both hazards have already been discussed. Here we will concentrate on systems and devices that prevent electrical hazards.

Figure 23.31 shows the schematic for a simple AC circuit with no safety features. This is not how power is distributed in practice. Modern household and industrial wiring requires the **three-wire system**, shown schematically in **Figure 23.32**, which has several safety features. First is the familiar *circuit breaker* (or *fuse*) to prevent thermal overload. Second, there is a protective *case* around the appliance, such as a toaster or refrigerator. The case's safety feature is that it prevents a person from touching exposed wires and coming into electrical contact with the circuit, helping prevent shocks.

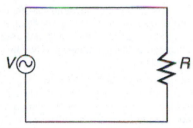

Figure 23.31 Schematic of a simple AC circuit with a voltage source and a single appliance represented by the resistance R. There are no safety features in this circuit.

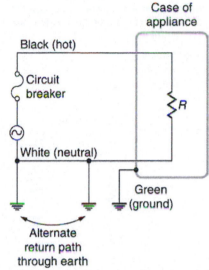

Figure 23.32 The three-wire system connects the neutral wire to the earth at the voltage source and user location, forcing it to be at zero volts and supplying an alternative return path for the current through the earth. Also grounded to zero volts is the case of the appliance. A circuit breaker or fuse protects against thermal overload and is in series on the active (live/hot) wire. Note that wire insulation colors vary with region and it is essential to check locally to determine which color codes are in use (and even if they were followed in the particular installation).

There are *three connections to earth or ground* (hereafter referred to as "earth/ground") shown in **Figure 23.32**. Recall that an earth/ground connection is a low-resistance path directly to the earth. The two earth/ground connections on the *neutral wire* force it to be at zero volts relative to the earth, giving the wire its name. This wire is therefore safe to touch even if its insulation, usually white, is missing. The neutral wire is the return path for the current to follow to complete the circuit. Furthermore, the two earth/ground connections supply an alternative path through the earth, a good conductor, to complete the circuit. The earth/ground connection closest to the power source could be at the generating plant, while the other is at the user's location. The third earth/ground is to the case of the appliance, through the green *earth/ground wire*, forcing the case, too, to be at zero volts. The *live* or *hot wire* (hereafter referred to as "live/hot") supplies voltage and current to operate the appliance. **Figure 23.33** shows a more pictorial version of how the three-wire system is connected through a three-prong plug to an appliance.

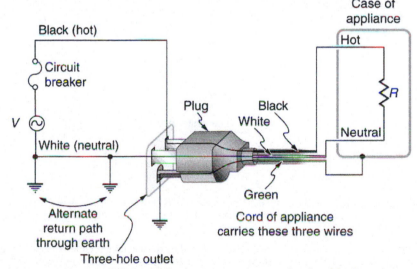

Figure 23.33 The standard three-prong plug can only be inserted in one way, to assure proper function of the three-wire system.

A note on insulation color-coding: Insulating plastic is color-coded to identify live/hot, neutral and ground wires but these codes vary around the world. Live/hot wires may be brown, red, black, blue or grey. Neutral wire may be blue, black or white. Since the same color may be used for live/hot or neutral in different parts of the world, it is essential to determine the color code in your region. The only exception is the earth/ground wire which is often green but may be yellow or just bare wire. Striped coatings are sometimes used for the benefit of those who are colorblind.

The three-wire system replaced the older two-wire system, which lacks an earth/ground wire. Under ordinary circumstances, insulation on the live/hot and neutral wires prevents the case from being directly in the circuit, so that the earth/ground wire may seem like double protection. Grounding the case solves more than one problem, however. The simplest problem is worn insulation on the live/hot wire that allows it to contact the case, as shown in **Figure 23.34**. Lacking an earth/ground connection (some people cut the third prong off the plug because they only have outdated two hole receptacles), a severe shock is possible. This is particularly dangerous in the kitchen, where a good connection to earth/ground is available through water on the floor or a water faucet. With the earth/ground connection intact, the circuit breaker will trip, forcing repair of the appliance. Why are some appliances still sold with two-prong plugs? These have nonconducting cases, such as power tools with impact resistant plastic cases, and are called *doubly insulated*. Modern two-prong plugs can be inserted into the asymmetric standard outlet in only one way, to ensure proper connection of live/hot and neutral wires.

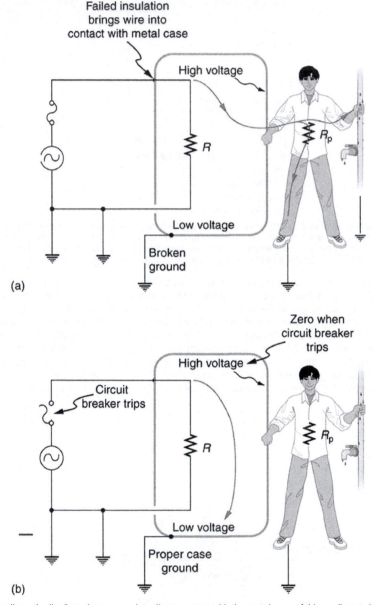

Figure 23.34 Worn insulation allows the live/hot wire to come into direct contact with the metal case of this appliance. (a) The earth/ground connection being broken, the person is severely shocked. The appliance may operate normally in this situation. (b) With a proper earth/ground, the circuit breaker trips, forcing repair of the appliance.

Electromagnetic induction causes a more subtle problem that is solved by grounding the case. The AC current in appliances can induce an emf on the case. If grounded, the case voltage is kept near zero, but if the case is not grounded, a shock can occur as pictured in **Figure 23.35**. Current driven by the induced case emf is called a *leakage current*, although current does not necessarily pass from the resistor to the case.

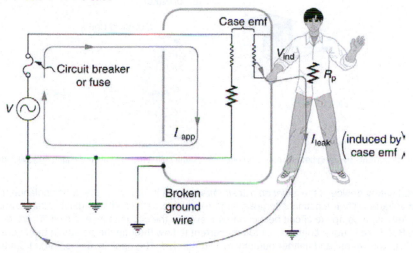

Figure 23.35 AC currents can induce an emf on the case of an appliance. The voltage can be large enough to cause a shock. If the case is grounded, the induced emf is kept near zero.

A *ground fault interrupter* (GFI) is a safety device found in updated kitchen and bathroom wiring that works based on electromagnetic induction. GFIs compare the currents in the live/hot and neutral wires. When live/hot and neutral currents are not equal, it is almost always because current in the neutral is less than in the live/hot wire. Then some of the current, again called a leakage current, is returning to the voltage source by a path other than through the neutral wire. It is assumed that this path presents a hazard, such as shown in **Figure 23.36**. GFIs are usually set to interrupt the circuit if the leakage current is greater than 5 mA, the accepted maximum harmless shock. Even if the leakage current goes safely to earth/ground through an intact earth/ground wire, the GFI will trip, forcing repair of the leakage.

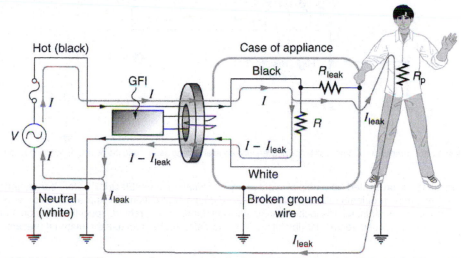

Figure 23.36 A ground fault interrupter (GFI) compares the currents in the live/hot and neutral wires and will trip if their difference exceeds a safe value. The leakage current here follows a hazardous path that could have been prevented by an intact earth/ground wire.

Figure 23.37 shows how a GFI works. If the currents in the live/hot and neutral wires are equal, then they induce equal and opposite emfs in the coil. If not, then the circuit breaker will trip.

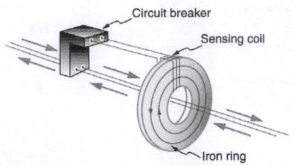

Figure 23.37 A GFI compares currents by using both to induce an emf in the same coil. If the currents are equal, they will induce equal but opposite emfs.

Another induction-based safety device is the *isolation transformer*, shown in **Figure 23.38**. Most isolation transformers have equal input and output voltages. Their function is to put a large resistance between the original voltage source and the device being operated. This prevents a complete circuit between them, even in the circumstance shown. There is a complete circuit through the appliance. But there is not a complete circuit for current to flow through the person in the figure, who is touching only one of the transformer's output wires, and neither output wire is grounded. The appliance is isolated from the original voltage source by the high resistance of the material between the transformer coils, hence the name isolation transformer. For current to flow through the person, it must pass through the high-resistance material between the coils, through the wire, the person, and back through the earth—a path with such a large resistance that the current is negligible.

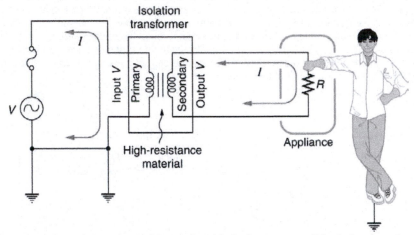

Figure 23.38 An isolation transformer puts a large resistance between the original voltage source and the device, preventing a complete circuit between them.

The basics of electrical safety presented here help prevent many electrical hazards. Electrical safety can be pursued to greater depths. There are, for example, problems related to different earth/ground connections for appliances in close proximity. Many other examples are found in hospitals. Microshock-sensitive patients, for instance, require special protection. For these people, currents as low as 0.1 mA may cause ventricular fibrillation. The interested reader can use the material presented here as a basis for further study.

23.9 Inductance

Learning Objectives
By the end of this section, you will be able to:
• Calculate the inductance of an inductor. • Calculate the energy stored in an inductor. • Calculate the emf generated in an inductor.

Inductors

Induction is the process in which an emf is induced by changing magnetic flux. Many examples have been discussed so far, some more effective than others. Transformers, for example, are designed to be particularly effective at inducing a desired voltage and current with very little loss of energy to other forms. Is there a useful physical quantity related to how "effective" a given device is? The answer is yes, and that physical quantity is called **inductance**.

Mutual inductance is the effect of Faraday's law of induction for one device upon another, such as the primary coil in transmitting energy to the secondary in a transformer. See **Figure 23.39**, where simple coils induce emfs in one another.

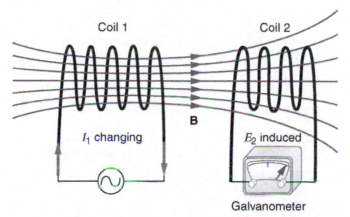

Figure 23.39 These coils can induce emfs in one another like an inefficient transformer. Their mutual inductance M indicates the effectiveness of the coupling between them. Here a change in current in coil 1 is seen to induce an emf in coil 2. (Note that "E_2 induced" represents the induced emf in coil 2.)

In the many cases where the geometry of the devices is fixed, flux is changed by varying current. We therefore concentrate on the rate of change of current, $\Delta I/\Delta t$, as the cause of induction. A change in the current I_1 in one device, coil 1 in the figure, induces an emf_2 in the other. We express this in equation form as

$$\text{emf}_2 = -M\frac{\Delta I_1}{\Delta t},$$

(23.34)

where M is defined to be the mutual inductance between the two devices. The minus sign is an expression of Lenz's law. The larger the mutual inductance M, the more effective the coupling. For example, the coils in **Figure 23.39** have a small M compared with the transformer coils in **Figure 23.28**. Units for M are $(\text{V} \cdot \text{s})/\text{A} = \Omega \cdot \text{s}$, which is named a **henry** (H), after Joseph Henry. That is, $1\,\text{H} = 1\,\Omega \cdot \text{s}$.

Nature is symmetric here. If we change the current I_2 in coil 2, we induce an emf_1 in coil 1, which is given by

$$\text{emf}_1 = -M\frac{\Delta I_2}{\Delta t},$$

(23.35)

where M is the same as for the reverse process. Transformers run backward with the same effectiveness, or mutual inductance M.

A large mutual inductance M may or may not be desirable. We want a transformer to have a large mutual inductance. But an appliance, such as an electric clothes dryer, can induce a dangerous emf on its case if the mutual inductance between its coils and the case is large. One way to reduce mutual inductance M is to counterwind coils to cancel the magnetic field produced. (See **Figure 23.40**.)

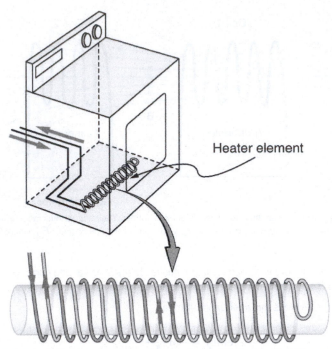

Heater element

Figure 23.40 The heating coils of an electric clothes dryer can be counter-wound so that their magnetic fields cancel one another, greatly reducing the mutual inductance with the case of the dryer.

Self-inductance, the effect of Faraday's law of induction of a device on itself, also exists. When, for example, current through a coil is increased, the magnetic field and flux also increase, inducing a counter emf, as required by Lenz's law. Conversely, if the current is decreased, an emf is induced that opposes the decrease. Most devices have a fixed geometry, and so the change in flux is due entirely to the change in current ΔI through the device. The induced emf is related to the physical geometry of the device and the rate of change of current. It is given by

$$\text{emf} = -L\frac{\Delta I}{\Delta t},$$

(23.36)

where L is the self-inductance of the device. A device that exhibits significant self-inductance is called an **inductor**, and given the symbol in **Figure 23.41**.

Figure 23.41

The minus sign is an expression of Lenz's law, indicating that emf opposes the change in current. Units of self-inductance are henries (H) just as for mutual inductance. The larger the self-inductance L of a device, the greater its opposition to any change in current through it. For example, a large coil with many turns and an iron core has a large L and will not allow current to change quickly. To avoid this effect, a small L must be achieved, such as by counterwinding coils as in **Figure 23.40**.

A 1 H inductor is a large inductor. To illustrate this, consider a device with $L = 1.0\text{ H}$ that has a 10 A current flowing through it. What happens if we try to shut off the current rapidly, perhaps in only 1.0 ms? An emf, given by $\text{emf} = -L(\Delta I / \Delta t)$, will oppose the change. Thus an emf will be induced given by $\text{emf} = -L(\Delta I / \Delta t) = (1.0\text{ H})[(10\text{ A}) / (1.0\text{ ms})] = 10{,}000\text{ V}$.

The positive sign means this large voltage is in the same direction as the current, opposing its decrease. Such large emfs can cause arcs, damaging switching equipment, and so it may be necessary to change current more slowly.

There are uses for such a large induced voltage. Camera flashes use a battery, two inductors that function as a transformer, and a switching system or oscillator to induce large voltages. (Remember that we need a changing magnetic field, brought about by a changing current, to induce a voltage in another coil.) The oscillator system will do this many times as the battery voltage is boosted to over one thousand volts. (You may hear the high pitched whine from the transformer as the capacitor is being charged.) A capacitor stores the high voltage for later use in powering the flash. (See **Figure 23.42**.)

Download for free at https://openstax.org/details/books/college-physics-ap-courses

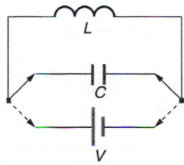

Figure 23.42 Through rapid switching of an inductor, 1.5 V batteries can be used to induce emfs of several thousand volts. This voltage can be used to store charge in a capacitor for later use, such as in a camera flash attachment.

It is possible to calculate L for an inductor given its geometry (size and shape) and knowing the magnetic field that it produces. This is difficult in most cases, because of the complexity of the field created. So in this text the inductance L is usually a given quantity. One exception is the solenoid, because it has a very uniform field inside, a nearly zero field outside, and a simple shape. It is instructive to derive an equation for its inductance. We start by noting that the induced emf is given by Faraday's law of induction as $\text{emf} = -N(\Delta\Phi/\Delta t)$ and, by the definition of self-inductance, as $\text{emf} = -L(\Delta I/\Delta t)$. Equating these yields

$$\text{emf} = -N\frac{\Delta\Phi}{\Delta t} = -L\frac{\Delta I}{\Delta t}. \tag{23.37}$$

Solving for L gives

$$L = N\frac{\Delta\Phi}{\Delta I}. \tag{23.38}$$

This equation for the self-inductance L of a device is always valid. It means that self-inductance L depends on how effective the current is in creating flux; the more effective, the greater $\Delta\Phi/\Delta I$ is.

Let us use this last equation to find an expression for the inductance of a solenoid. Since the area A of a solenoid is fixed, the change in flux is $\Delta\Phi = \Delta(BA) = A\Delta B$. To find ΔB, we note that the magnetic field of a solenoid is given by $B = \mu_0 nI = \mu_0\frac{NI}{\ell}$. (Here $n = N/\ell$, where N is the number of coils and ℓ is the solenoid's length.) Only the current changes, so that $\Delta\Phi = A\Delta B = \mu_0 NA\frac{\Delta I}{\ell}$. Substituting $\Delta\Phi$ into $L = N\frac{\Delta\Phi}{\Delta I}$ gives

$$L = N\frac{\Delta\Phi}{\Delta I} = N\frac{\mu_0 NA\frac{\Delta I}{\ell}}{\Delta I}. \tag{23.39}$$

This simplifies to

$$L = \frac{\mu_0 N^2 A}{\ell}(\text{solenoid}). \tag{23.40}$$

This is the self-inductance of a solenoid of cross-sectional area A and length ℓ. Note that the inductance depends only on the physical characteristics of the solenoid, consistent with its definition.

Example 23.7 Calculating the Self-inductance of a Moderate Size Solenoid

Calculate the self-inductance of a 10.0 cm long, 4.00 cm diameter solenoid that has 200 coils.

Strategy

This is a straightforward application of $L = \frac{\mu_0 N^2 A}{\ell}$, since all quantities in the equation except L are known.

Solution

Use the following expression for the self-inductance of a solenoid:

$$L = \frac{\mu_0 N^2 A}{\ell}. \tag{23.41}$$

The cross-sectional area in this example is $A = \pi r^2 = (3.14...)(0.0200\ \text{m})^2 = 1.26\times10^{-3}\ \text{m}^2$, N is given to be 200, and the length ℓ is 0.100 m. We know the permeability of free space is $\mu_0 = 4\pi\times10^{-7}\ \text{T} \cdot \text{m/A}$. Substituting these into the expression for L gives

$$
\begin{aligned}
L &= \frac{(4\pi\times10^{-7}\ \text{T} \cdot \text{m/A})(200)^2(1.26\times10^{-3}\ \text{m}^2)}{0.100\ \text{m}} \\
&= 0.632\ \text{mH}.
\end{aligned}
\tag{23.42}
$$

Discussion

This solenoid is moderate in size. Its inductance of nearly a millihenry is also considered moderate.

One common application of inductance is used in traffic lights that can tell when vehicles are waiting at the intersection. An electrical circuit with an inductor is placed in the road under the place a waiting car will stop over. The body of the car increases the inductance and the circuit changes sending a signal to the traffic lights to change colors. Similarly, metal detectors used for airport security employ the same technique. A coil or inductor in the metal detector frame acts as both a transmitter and a receiver. The pulsed signal in the transmitter coil induces a signal in the receiver. The self-inductance of the circuit is affected by any metal object in the path. Such detectors can be adjusted for sensitivity and also can indicate the approximate location of metal found on a person. (But they will not be able to detect any plastic explosive such as that found on the "underwear bomber.") See **Figure 23.43**.

Figure 23.43 The familiar security gate at an airport can not only detect metals but also indicate their approximate height above the floor. (credit: Alexbuirds, Wikimedia Commons)

Energy Stored in an Inductor

We know from Lenz's law that inductances oppose changes in current. There is an alternative way to look at this opposition that is based on energy. Energy is stored in a magnetic field. It takes time to build up energy, and it also takes time to deplete energy; hence, there is an opposition to rapid change. In an inductor, the magnetic field is directly proportional to current and to the inductance of the device. It can be shown that the **energy stored in an inductor** E_{ind} is given by

$$
E_{\text{ind}} = \frac{1}{2}LI^2.
\tag{23.43}
$$

This expression is similar to that for the energy stored in a capacitor.

Example 23.8 Calculating the Energy Stored in the Field of a Solenoid

How much energy is stored in the 0.632 mH inductor of the preceding example when a 30.0 A current flows through it?

Strategy

The energy is given by the equation $E_{\text{ind}} = \frac{1}{2}LI^2$, and all quantities except E_{ind} are known.

Solution

Substituting the value for L found in the previous example and the given current into $E_{\text{ind}} = \frac{1}{2}LI^2$ gives

$$
\begin{aligned}
E_{\text{ind}} &= \frac{1}{2}LI^2 \\
&= 0.5(0.632\times10^{-3}\ \text{H})(30.0\ \text{A})^2 = 0.284\ \text{J}.
\end{aligned}
\tag{23.44}
$$

Discussion

This amount of energy is certainly enough to cause a spark if the current is suddenly switched off. It cannot be built up instantaneously unless the power input is infinite.

23.10 RL Circuits

Learning Objectives

By the end of this section, you will be able to:

- Calculate the current in an RL circuit after a specified number of characteristic time steps.
- Calculate the characteristic time of an RL circuit.
- Sketch the current in an RL circuit over time.

We know that the current through an inductor L cannot be turned on or off instantaneously. The change in current changes flux, inducing an emf opposing the change (Lenz's law). How long does the opposition last? Current *will* flow and *can* be turned off, but how long does it take? **Figure 23.44** shows a switching circuit that can be used to examine current through an inductor as a function of time.

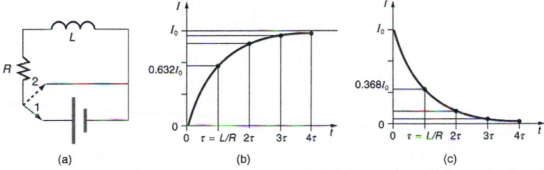

Figure 23.44 (a) An RL circuit with a switch to turn current on and off. When in position 1, the battery, resistor, and inductor are in series and a current is established. In position 2, the battery is removed and the current eventually stops because of energy loss in the resistor. (b) A graph of current growth versus time when the switch is moved to position 1. (c) A graph of current decay when the switch is moved to position 2.

When the switch is first moved to position 1 (at $t = 0$), the current is zero and it eventually rises to $I_0 = V/R$, where R is the total resistance of the circuit. The opposition of the inductor L is greatest at the beginning, because the amount of change is greatest. The opposition it poses is in the form of an induced emf, which decreases to zero as the current approaches its final value. The opposing emf is proportional to the amount of change left. This is the hallmark of an exponential behavior, and it can be shown with calculus that

$$I = I_0(1 - e^{-t/\tau}) \quad \text{(turning on)}, \tag{23.45}$$

is the current in an RL circuit when switched on (Note the similarity to the exponential behavior of the voltage on a charging capacitor). The initial current is zero and approaches $I_0 = V/R$ with a **characteristic time constant** τ for an RL circuit, given by

$$\tau = \frac{L}{R}, \tag{23.46}$$

where τ has units of seconds, since $1 \text{ H} = 1 \ \Omega \cdot \text{s}$. In the first period of time τ, the current rises from zero to $0.632I_0$, since $I = I_0(1 - e^{-1}) = I_0(1 - 0.368) = 0.632I_0$. The current will go 0.632 of the remainder in the next time τ. A well-known property of the exponential is that the final value is never exactly reached, but 0.632 of the remainder to that value is achieved in every characteristic time τ. In just a few multiples of the time τ, the final value is very nearly achieved, as the graph in **Figure 23.44(b)** illustrates.

The characteristic time τ depends on only two factors, the inductance L and the resistance R. The greater the inductance L, the greater τ is, which makes sense since a large inductance is very effective in opposing change. The smaller the resistance R, the greater τ is. Again this makes sense, since a small resistance means a large final current and a greater change to get there. In both cases—large L and small R —more energy is stored in the inductor and more time is required to get it in and out.

When the switch in **Figure 23.44(a)** is moved to position 2 and cuts the battery out of the circuit, the current drops because of energy dissipation by the resistor. But this is also not instantaneous, since the inductor opposes the decrease in current by inducing an emf in the same direction as the battery that drove the current. Furthermore, there is a certain amount of energy, $(1/2)LI_0^2$, stored in the inductor, and it is dissipated at a finite rate. As the current approaches zero, the rate of decrease slows, since the energy dissipation rate is I^2R. Once again the behavior is exponential, and I is found to be

$$I = I_0 e^{-t/\tau} \quad \text{(turning off)} \tag{23.47}$$

(See Figure 23.44(c).) In the first period of time $\tau = L/R$ after the switch is closed, the current falls to 0.368 of its initial value, since $I = I_0 e^{-1} = 0.368 I_0$. In each successive time τ, the current falls to 0.368 of the preceding value, and in a few multiples of τ, the current becomes very close to zero, as seen in the graph in Figure 23.44(c).

Example 23.9 Calculating Characteristic Time and Current in an RL Circuit

(a) What is the characteristic time constant for a 7.50 mH inductor in series with a $3.00\ \Omega$ resistor? (b) Find the current 5.00 ms after the switch is moved to position 2 to disconnect the battery, if it is initially 10.0 A.

Strategy for (a)

The time constant for an RL circuit is defined by $\tau = L/R$.

Solution for (a)

Entering known values into the expression for τ given in $\tau = L/R$ yields

$$\tau = \frac{L}{R} = \frac{7.50\ \text{mH}}{3.00\ \Omega} = 2.50\ \text{ms}. \tag{23.48}$$

Discussion for (a)

This is a small but definitely finite time. The coil will be very close to its full current in about ten time constants, or about 25 ms.

Strategy for (b)

We can find the current by using $I = I_0 e^{-t/\tau}$, or by considering the decline in steps. Since the time is twice the characteristic time, we consider the process in steps.

Solution for (b)

In the first 2.50 ms, the current declines to 0.368 of its initial value, which is

$$\begin{aligned} I &= 0.368 I_0 = (0.368)(10.0\ \text{A}) \\ &= 3.68\ \text{A at } t = 2.50\ \text{ms}. \end{aligned} \tag{23.49}$$

After another 2.50 ms, or a total of 5.00 ms, the current declines to 0.368 of the value just found. That is,

$$\begin{aligned} I' &= 0.368 I = (0.368)(3.68\ \text{A}) \\ &= 1.35\ \text{A at } t = 5.00\ \text{ms}. \end{aligned} \tag{23.50}$$

Discussion for (b)

After another 5.00 ms has passed, the current will be 0.183 A (see Exercise 23.69); so, although it does die out, the current certainly does not go to zero instantaneously.

In summary, when the voltage applied to an inductor is changed, the current also changes, *but the change in current lags the change in voltage in an RL circuit*. In Reactance, Inductive and Capacitive, we explore how an RL circuit behaves when a sinusoidal AC voltage is applied.

23.11 Reactance, Inductive and Capacitive

Learning Objectives

By the end of this section, you will be able to:

- Sketch voltage and current versus time in simple inductive, capacitive, and resistive circuits.
- Calculate inductive and capacitive reactance.
- Calculate current and/or voltage in simple inductive, capacitive, and resistive circuits.

Many circuits also contain capacitors and inductors, in addition to resistors and an AC voltage source. We have seen how capacitors and inductors respond to DC voltage when it is switched on and off. We will now explore how inductors and capacitors react to sinusoidal AC voltage.

Inductors and Inductive Reactance

Suppose an inductor is connected directly to an AC voltage source, as shown in **Figure 23.45**. It is reasonable to assume negligible resistance, since in practice we can make the resistance of an inductor so small that it has a negligible effect on the circuit. Also shown is a graph of voltage and current as functions of time.

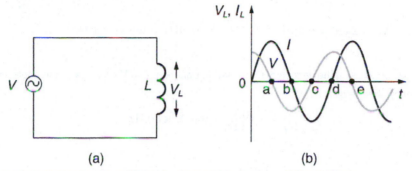

(a)   (b)

Figure 23.45 (a) An AC voltage source in series with an inductor having negligible resistance. (b) Graph of current and voltage across the inductor as functions of time.

The graph in **Figure 23.45**(b) starts with voltage at a maximum. Note that the current starts at zero and rises to its peak *after* the voltage that drives it, just as was the case when DC voltage was switched on in the preceding section. When the voltage becomes negative at point a, the current begins to decrease; it becomes zero at point b, where voltage is its most negative. The current then becomes negative, again following the voltage. The voltage becomes positive at point c and begins to make the current less negative. At point d, the current goes through zero just as the voltage reaches its positive peak to start another cycle. This behavior is summarized as follows:

AC Voltage in an Inductor

When a sinusoidal voltage is applied to an inductor, the voltage leads the current by one-fourth of a cycle, or by a $90°$ phase angle.

Current lags behind voltage, since inductors oppose change in current. Changing current induces a back emf $V = -L(\Delta I / \Delta t)$. This is considered to be an effective resistance of the inductor to AC. The rms current I through an inductor L is given by a version of Ohm's law:

$$I = \frac{V}{X_L},$$ (23.51)

where V is the rms voltage across the inductor and X_L is defined to be

$$X_L = 2\pi f L,$$ (23.52)

with f the frequency of the AC voltage source in hertz (An analysis of the circuit using Kirchhoff's loop rule and calculus actually produces this expression). X_L is called the **inductive reactance**, because the inductor reacts to impede the current. X_L has units of ohms ($1 \text{ H} = 1 \Omega \cdot \text{s}$, so that frequency times inductance has units of $(\text{cycles/s})(\Omega \cdot \text{s}) = \Omega$), consistent with its role as an effective resistance. It makes sense that X_L is proportional to L , since the greater the induction the greater its resistance to change. It is also reasonable that X_L is proportional to frequency f , since greater frequency means greater change in current. That is, $\Delta I/\Delta t$ is large for large frequencies (large f , small Δt). The greater the change, the greater the opposition of an inductor.

Example 23.10 Calculating Inductive Reactance and then Current

(a) Calculate the inductive reactance of a 3.00 mH inductor when 60.0 Hz and 10.0 kHz AC voltages are applied. (b) What is the rms current at each frequency if the applied rms voltage is 120 V?

Strategy

The inductive reactance is found directly from the expression $X_L = 2\pi f L$. Once X_L has been found at each frequency, Ohm's law as stated in the Equation $I = V/X_L$ can be used to find the current at each frequency.

Solution for (a)

Entering the frequency and inductance into Equation $X_L = 2\pi fL$ gives

$$X_L = 2\pi fL = 6.28(60.0/\text{s})(3.00 \text{ mH}) = 1.13 \ \Omega \text{ at 60 Hz.} \tag{23.53}$$

Similarly, at 10 kHz,

$$X_L = 2\pi fL = 6.28(1.00 \times 10^4/\text{s})(3.00 \text{ mH}) = 188 \ \Omega \text{ at 10 kHz.} \tag{23.54}$$

Solution for (b)

The rms current is now found using the version of Ohm's law in Equation $I = V/X_L$, given the applied rms voltage is 120 V. For the first frequency, this yields

$$I = \frac{V}{X_L} = \frac{120 \text{ V}}{1.13 \ \Omega} = 106 \text{ A at 60 Hz.} \tag{23.55}$$

Similarly, at 10 kHz,

$$I = \frac{V}{X_L} = \frac{120 \text{ V}}{188 \ \Omega} = 0.637 \text{ A at 10 kHz.} \tag{23.56}$$

Discussion

The inductor reacts very differently at the two different frequencies. At the higher frequency, its reactance is large and the current is small, consistent with how an inductor impedes rapid change. Thus high frequencies are impeded the most. Inductors can be used to filter out high frequencies; for example, a large inductor can be put in series with a sound reproduction system or in series with your home computer to reduce high-frequency sound output from your speakers or high-frequency power spikes into your computer.

Note that although the resistance in the circuit considered is negligible, the AC current is not extremely large because inductive reactance impedes its flow. With AC, there is no time for the current to become extremely large.

Capacitors and Capacitive Reactance

Consider the capacitor connected directly to an AC voltage source as shown in **Figure 23.46**. The resistance of a circuit like this can be made so small that it has a negligible effect compared with the capacitor, and so we can assume negligible resistance. Voltage across the capacitor and current are graphed as functions of time in the figure.

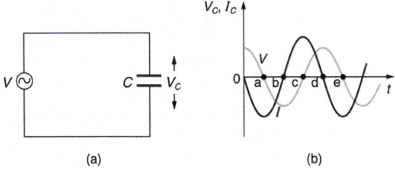

(a)   (b)

Figure 23.46 (a) An AC voltage source in series with a capacitor C having negligible resistance. (b) Graph of current and voltage across the capacitor as functions of time.

The graph in **Figure 23.46** starts with voltage across the capacitor at a maximum. The current is zero at this point, because the capacitor is fully charged and halts the flow. Then voltage drops and the current becomes negative as the capacitor discharges. At point a, the capacitor has fully discharged ($Q = 0$ on it) and the voltage across it is zero. The current remains negative

between points a and b, causing the voltage on the capacitor to reverse. This is complete at point b, where the current is zero and the voltage has its most negative value. The current becomes positive after point b, neutralizing the charge on the capacitor and bringing the voltage to zero at point c, which allows the current to reach its maximum. Between points c and d, the current drops to zero as the voltage rises to its peak, and the process starts to repeat. Throughout the cycle, the voltage follows what the current is doing by one-fourth of a cycle:

AC Voltage in a Capacitor

When a sinusoidal voltage is applied to a capacitor, the voltage follows the current by one-fourth of a cycle, or by a $90°$ phase angle.

The capacitor is affecting the current, having the ability to stop it altogether when fully charged. Since an AC voltage is applied, there is an rms current, but it is limited by the capacitor. This is considered to be an effective resistance of the capacitor to AC, and so the rms current I in the circuit containing only a capacitor C is given by another version of Ohm's law to be

$$I = \frac{V}{X_C},$$ (23.57)

where V is the rms voltage and X_C is defined (As with X_L, this expression for X_C results from an analysis of the circuit using Kirchhoff's rules and calculus) to be

$$X_C = \frac{1}{2\pi fC},$$ (23.58)

where X_C is called the **capacitive reactance**, because the capacitor reacts to impede the current. X_C has units of ohms (verification left as an exercise for the reader). X_C is inversely proportional to the capacitance C; the larger the capacitor, the greater the charge it can store and the greater the current that can flow. It is also inversely proportional to the frequency f; the greater the frequency, the less time there is to fully charge the capacitor, and so it impedes current less.

Example 23.11 Calculating Capacitive Reactance and then Current

(a) Calculate the capacitive reactance of a 5.00 mF capacitor when 60.0 Hz and 10.0 kHz AC voltages are applied. (b) What is the rms current if the applied rms voltage is 120 V?

Strategy

The capacitive reactance is found directly from the expression in $X_C = \frac{1}{2\pi fC}$. Once X_C has been found at each frequency, Ohm's law stated as $I = V / X_C$ can be used to find the current at each frequency.

Solution for (a)

Entering the frequency and capacitance into $X_C = \frac{1}{2\pi fC}$ gives

$$
\begin{aligned}
X_C &= \frac{1}{2\pi fC} \\
&= \frac{1}{6.28(60.0\,/\,\text{s})(5.00\,\mu\text{F})} = 531\ \Omega \text{ at 60 Hz.}
\end{aligned}
$$ (23.59)

Similarly, at 10 kHz,

$$
\begin{aligned}
X_C &= \frac{1}{2\pi fC} = \frac{1}{6.28(1.00\times10^4\,/\,\text{s})(5.00\,\mu\text{F}).} \\
&= 3.18\ \Omega \text{ at 10 kHz}
\end{aligned}
$$ (23.60)

Solution for (b)

The rms current is now found using the version of Ohm's law in $I = V / X_C$, given the applied rms voltage is 120 V. For the first frequency, this yields

$$I = \frac{V}{X_C} = \frac{120\ \text{V}}{531\ \Omega} = 0.226\ \text{A at 60 Hz.}$$ (23.61)

Similarly, at 10 kHz,

$$I = \frac{V}{X_C} = \frac{120\ \text{V}}{3.18\ \Omega} = 37.7\ \text{A at 10 kHz.}$$ (23.62)

Discussion

The capacitor reacts very differently at the two different frequencies, and in exactly the opposite way an inductor reacts. At the higher frequency, its reactance is small and the current is large. Capacitors favor change, whereas inductors oppose change. Capacitors impede low frequencies the most, since low frequency allows them time to become charged and stop the current. Capacitors can be used to filter out low frequencies. For example, a capacitor in series with a sound reproduction system rids it of the 60 Hz hum.

Although a capacitor is basically an open circuit, there is an rms current in a circuit with an AC voltage applied to a capacitor. This is because the voltage is continually reversing, charging and discharging the capacitor. If the frequency goes to zero (DC),

X_C tends to infinity, and the current is zero once the capacitor is charged. At very high frequencies, the capacitor's reactance tends to zero—it has a negligible reactance and does not impede the current (it acts like a simple wire). *Capacitors have the opposite effect on AC circuits that inductors have.*

Resistors in an AC Circuit

Just as a reminder, consider **Figure 23.47**, which shows an AC voltage applied to a resistor and a graph of voltage and current versus time. The voltage and current are exactly *in phase* in a resistor. There is no frequency dependence to the behavior of plain resistance in a circuit:

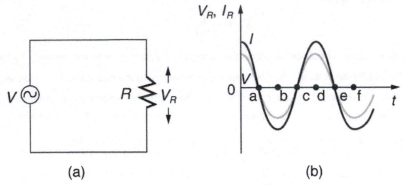

(a)   (b)

Figure 23.47 (a) An AC voltage source in series with a resistor. (b) Graph of current and voltage across the resistor as functions of time, showing them to be exactly in phase.

AC Voltage in a Resistor

When a sinusoidal voltage is applied to a resistor, the voltage is exactly in phase with the current—they have a $0°$ phase angle.

23.12 RLC Series AC Circuits

Learning Objectives
By the end of this section, you will be able to: • Calculate the impedance, phase angle, resonant frequency, power, power factor, voltage, and/or current in an RLC series circuit. • Draw the circuit diagram for an RLC series circuit. • Explain the significance of the resonant frequency.

Impedance

When alone in an AC circuit, inductors, capacitors, and resistors all impede current. How do they behave when all three occur together? Interestingly, their individual resistances in ohms do not simply add. Because inductors and capacitors behave in opposite ways, they partially to totally cancel each other's effect. **Figure 23.48** shows an *RLC* series circuit with an AC voltage source, the behavior of which is the subject of this section. The crux of the analysis of an *RLC* circuit is the frequency dependence of X_L and X_C, and the effect they have on the phase of voltage versus current (established in the preceding section). These give rise to the frequency dependence of the circuit, with important "resonance" features that are the basis of many applications, such as radio tuners.

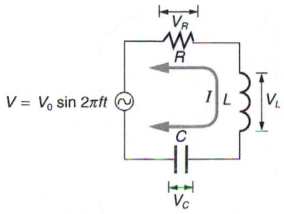

Figure 23.48 An *RLC* series circuit with an AC voltage source.

The combined effect of resistance R, inductive reactance X_L, and capacitive reactance X_C is defined to be **impedance**, an AC analogue to resistance in a DC circuit. Current, voltage, and impedance in an *RLC* circuit are related by an AC version of Ohm's law:

$$I_0 = \frac{V_0}{Z} \text{ or } I_{\text{rms}} = \frac{V_{\text{rms}}}{Z}. \tag{23.63}$$

Here I_0 is the peak current, V_0 the peak source voltage, and Z is the impedance of the circuit. The units of impedance are ohms, and its effect on the circuit is as you might expect: the greater the impedance, the smaller the current. To get an expression for Z in terms of R, X_L, and X_C, we will now examine how the voltages across the various components are related to the source voltage. Those voltages are labeled V_R, V_L, and V_C in **Figure 23.48**.

Conservation of charge requires current to be the same in each part of the circuit at all times, so that we can say the currents in R, L, and C are equal and in phase. But we know from the preceding section that the voltage across the inductor V_L leads the current by one-fourth of a cycle, the voltage across the capacitor V_C follows the current by one-fourth of a cycle, and the voltage across the resistor V_R is exactly in phase with the current. **Figure 23.49** shows these relationships in one graph, as well as showing the total voltage around the circuit $V = V_R + V_L + V_C$, where all four voltages are the instantaneous values. According to Kirchhoff's loop rule, the total voltage around the circuit V is also the voltage of the source.

You can see from **Figure 23.49** that while V_R is in phase with the current, V_L leads by $90°$, and V_C follows by $90°$. Thus V_L and V_C are $180°$ out of phase (crest to trough) and tend to cancel, although not completely unless they have the same magnitude. Since the peak voltages are not aligned (not in phase), the peak voltage V_0 of the source does *not* equal the sum of the peak voltages across R, L, and C. The actual relationship is

$$V_0 = \sqrt{V_{0R}^2 + (V_{0L} - V_{0C})^2}, \tag{23.64}$$

where V_{0R}, V_{0L}, and V_{0C} are the peak voltages across R, L, and C, respectively. Now, using Ohm's law and definitions from **Reactance, Inductive and Capacitive**, we substitute $V_0 = I_0 Z$ into the above, as well as $V_{0R} = I_0 R$, $V_{0L} = I_0 X_L$, and $V_{0C} = I_0 X_C$, yielding

$$I_0 Z = \sqrt{I_0^2 R^2 + (I_0 X_L - I_0 X_C)^2} = I_0 \sqrt{R^2 + (X_L - X_C)^2}. \tag{23.65}$$

I_0 cancels to yield an expression for Z:

$$Z = \sqrt{R^2 + (X_L - X_C)^2}, \tag{23.66}$$

which is the impedance of an *RLC* series AC circuit. For circuits without a resistor, take $R = 0$; for those without an inductor, take $X_L = 0$; and for those without a capacitor, take $X_C = 0$.

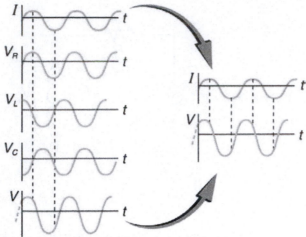

Figure 23.49 This graph shows the relationships of the voltages in an *RLC* circuit to the current. The voltages across the circuit elements add to equal the voltage of the source, which is seen to be out of phase with the current.

Example 23.12 Calculating Impedance and Current

An *RLC* series circuit has a $40.0 \ \Omega$ resistor, a 3.00 mH inductor, and a $5.00 \ \mu F$ capacitor. (a) Find the circuit's impedance at 60.0 Hz and 10.0 kHz, noting that these frequencies and the values for L and C are the same as in **Example 23.10** and **Example 23.11**. (b) If the voltage source has $V_{rms} = 120 \ V$, what is I_{rms} at each frequency?

Strategy

For each frequency, we use $Z = \sqrt{R^2 + (X_L - X_C)^2}$ to find the impedance and then Ohm's law to find current. We can take advantage of the results of the previous two examples rather than calculate the reactances again.

Solution for (a)

At 60.0 Hz, the values of the reactances were found in **Example 23.10** to be $X_L = 1.13 \ \Omega$ and in **Example 23.11** to be

$X_C = 531 \ \Omega$. Entering these and the given $40.0 \ \Omega$ for resistance into $Z = \sqrt{R^2 + (X_L - X_C)^2}$ yields

$$
\begin{aligned}
Z &= \sqrt{R^2 + (X_L - X_C)^2} \\
&= \sqrt{(40.0 \ \Omega)^2 + (1.13 \ \Omega - 531 \ \Omega)^2} \\
&= 531 \ \Omega \text{ at } 60.0 \text{ Hz.}
\end{aligned}
\tag{23.67}
$$

Similarly, at 10.0 kHz, $X_L = 188 \ \Omega$ and $X_C = 3.18 \ \Omega$, so that

$$
\begin{aligned}
Z &= \sqrt{(40.0 \ \Omega)^2 + (188 \ \Omega - 3.18 \ \Omega)^2} \\
&= 190 \ \Omega \text{ at } 10.0 \text{ kHz.}
\end{aligned}
\tag{23.68}
$$

Discussion for (a)

In both cases, the result is nearly the same as the largest value, and the impedance is definitely not the sum of the individual values. It is clear that X_L dominates at high frequency and X_C dominates at low frequency.

Solution for (b)

The current I_{rms} can be found using the AC version of Ohm's law in Equation $I_{rms} = V_{rms} / Z$:

$$
I_{rms} = \frac{V_{rms}}{Z} = \frac{120 \ V}{531 \ \Omega} = 0.226 \ A \text{ at } 60.0 \text{ Hz}
$$

Finally, at 10.0 kHz, we find

$$
I_{rms} = \frac{V_{rms}}{Z} = \frac{120 \ V}{190 \ \Omega} = 0.633 \ A \text{ at } 10.0 \text{ kHz}
$$

Discussion for (a)

The current at 60.0 Hz is the same (to three digits) as found for the capacitor alone in **Example 23.11**. The capacitor dominates at low frequency. The current at 10.0 kHz is only slightly different from that found for the inductor alone in

Example 23.10. The inductor dominates at high frequency.

Resonance in *RLC* Series AC Circuits

How does an *RLC* circuit behave as a function of the frequency of the driving voltage source? Combining Ohm's law, $I_{rms} = V_{rms}/Z$, and the expression for impedance Z from $Z = \sqrt{R^2 + (X_L - X_C)^2}$ gives

$$I_{rms} = \frac{V_{rms}}{\sqrt{R^2 + (X_L - X_C)^2}}. \tag{23.69}$$

The reactances vary with frequency, with X_L large at high frequencies and X_C large at low frequencies, as we have seen in three previous examples. At some intermediate frequency f_0, the reactances will be equal and cancel, giving $Z = R$ —this is a minimum value for impedance, and a maximum value for I_{rms} results. We can get an expression for f_0 by taking

$$X_L = X_C. \tag{23.70}$$

Substituting the definitions of X_L and X_C,

$$2\pi f_0 L = \frac{1}{2\pi f_0 C}. \tag{23.71}$$

Solving this expression for f_0 yields

$$f_0 = \frac{1}{2\pi\sqrt{LC}}, \tag{23.72}$$

where f_0 is the **resonant frequency** of an *RLC* series circuit. This is also the *natural frequency* at which the circuit would oscillate if not driven by the voltage source. At f_0, the effects of the inductor and capacitor cancel, so that $Z = R$, and I_{rms} is a maximum.

Resonance in AC circuits is analogous to mechanical resonance, where resonance is defined to be a forced oscillation—in this case, forced by the voltage source—at the natural frequency of the system. The receiver in a radio is an *RLC* circuit that oscillates best at its f_0. A variable capacitor is often used to adjust f_0 to receive a desired frequency and to reject others.

Figure 23.50 is a graph of current as a function of frequency, illustrating a resonant peak in I_{rms} at f_0. The two curves are for two different circuits, which differ only in the amount of resistance in them. The peak is lower and broader for the higher-resistance circuit. Thus the higher-resistance circuit does not resonate as strongly and would not be as selective in a radio receiver, for example.

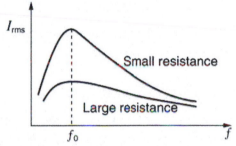

Figure 23.50 A graph of current versus frequency for two *RLC* series circuits differing only in the amount of resistance. Both have a resonance at f_0, but that for the higher resistance is lower and broader. The driving AC voltage source has a fixed amplitude V_0.

Example 23.13 Calculating Resonant Frequency and Current

For the same *RLC* series circuit having a $40.0\ \Omega$ resistor, a 3.00 mH inductor, and a $5.00\ \mu F$ capacitor: (a) Find the resonant frequency. (b) Calculate I_{rms} at resonance if V_{rms} is 120 V.

Strategy

The resonant frequency is found by using the expression in $f_0 = \frac{1}{2\pi\sqrt{LC}}$. The current at that frequency is the same as if the resistor alone were in the circuit.

Solution for (a)

Entering the given values for L and C into the expression given for f_0 in $f_0 = \dfrac{1}{2\pi\sqrt{LC}}$ yields

$$f_0 = \frac{1}{2\pi\sqrt{LC}} \tag{23.73}$$

$$= \frac{1}{2\pi\sqrt{(3.00\times10^{-3}\text{ H})(5.00\times10^{-6}\text{ F})}} = 1.30\text{ kHz}.$$

Discussion for (a)

We see that the resonant frequency is between 60.0 Hz and 10.0 kHz, the two frequencies chosen in earlier examples. This was to be expected, since the capacitor dominated at the low frequency and the inductor dominated at the high frequency. Their effects are the same at this intermediate frequency.

Solution for (b)

The current is given by Ohm's law. At resonance, the two reactances are equal and cancel, so that the impedance equals the resistance alone. Thus,

$$I_{\text{rms}} = \frac{V_{\text{rms}}}{Z} = \frac{120\text{ V}}{40.0\ \Omega} = 3.00\text{ A}. \tag{23.74}$$

Discussion for (b)

At resonance, the current is greater than at the higher and lower frequencies considered for the same circuit in the preceding example.

Power in *RLC* Series AC Circuits

If current varies with frequency in an *RLC* circuit, then the power delivered to it also varies with frequency. But the average power is not simply current times voltage, as it is in purely resistive circuits. As was seen in **Figure 23.49**, voltage and current are out of phase in an *RLC* circuit. There is a **phase angle** ϕ between the source voltage V and the current I, which can be found from

$$\cos\phi = \frac{R}{Z}. \tag{23.75}$$

For example, at the resonant frequency or in a purely resistive circuit $Z = R$, so that $\cos\phi = 1$. This implies that $\phi = 0\degree$ and that voltage and current are in phase, as expected for resistors. At other frequencies, average power is less than at resonance. This is both because voltage and current are out of phase and because I_{rms} is lower. The fact that source voltage and current are out of phase affects the power delivered to the circuit. It can be shown that the *average power* is

$$P_{\text{ave}} = I_{\text{rms}}V_{\text{rms}}\cos\phi, \tag{23.76}$$

Thus $\cos\phi$ is called the **power factor**, which can range from 0 to 1. Power factors near 1 are desirable when designing an efficient motor, for example. At the resonant frequency, $\cos\phi = 1$.

Example 23.14 Calculating the Power Factor and Power

For the same *RLC* series circuit having a $40.0\ \Omega$ resistor, a 3.00 mH inductor, a $5.00\ \mu\text{F}$ capacitor, and a voltage source with a V_{rms} of 120 V: (a) Calculate the power factor and phase angle for $f = 60.0\text{Hz}$. (b) What is the average power at 50.0 Hz? (c) Find the average power at the circuit's resonant frequency.

Strategy and Solution for (a)

The power factor at 60.0 Hz is found from

$$\cos\phi = \frac{R}{Z}. \tag{23.77}$$

We know $Z = 531\ \Omega$ from **Example 23.12**, so that

$$\cos\phi = \frac{40.0\ \Omega}{531\ \Omega} = 0.0753 \text{ at } 60.0\text{ Hz}. \tag{23.78}$$

This small value indicates the voltage and current are significantly out of phase. In fact, the phase angle is

$$\phi = \cos^{-1} 0.0753 = 85.7° \text{ at } 60.0 \text{ Hz.} \tag{23.79}$$

Discussion for (a)

The phase angle is close to $90°$, consistent with the fact that the capacitor dominates the circuit at this low frequency (a pure RC circuit has its voltage and current $90°$ out of phase).

Strategy and Solution for (b)

The average power at 60.0 Hz is

$$P_{ave} = I_{rms}V_{rms}\cos\phi. \tag{23.80}$$

I_{rms} was found to be 0.226 A in **Example 23.12**. Entering the known values gives

$$P_{ave} = (0.226 \text{ A})(120 \text{ V})(0.0753) = 2.04 \text{ W at } 60.0 \text{ Hz.} \tag{23.81}$$

Strategy and Solution for (c)

At the resonant frequency, we know $\cos\phi = 1$, and I_{rms} was found to be 6.00 A in **Example 23.13**. Thus,

$$P_{ave} = (3.00 \text{ A})(120 \text{ V})(1) = 360 \text{ W} \text{ at resonance (1.30 kHz)}$$

Discussion

Both the current and the power factor are greater at resonance, producing significantly greater power than at higher and lower frequencies.

Power delivered to an RLC series AC circuit is dissipated by the resistance alone. The inductor and capacitor have energy input and output but do not dissipate it out of the circuit. Rather they transfer energy back and forth to one another, with the resistor dissipating exactly what the voltage source puts into the circuit. This assumes no significant electromagnetic radiation from the inductor and capacitor, such as radio waves. Such radiation can happen and may even be desired, as we will see in the next chapter on electromagnetic radiation, but it can also be suppressed as is the case in this chapter. The circuit is analogous to the wheel of a car driven over a corrugated road as shown in **Figure 23.51**. The regularly spaced bumps in the road are analogous to the voltage source, driving the wheel up and down. The shock absorber is analogous to the resistance damping and limiting the amplitude of the oscillation. Energy within the system goes back and forth between kinetic (analogous to maximum current, and energy stored in an inductor) and potential energy stored in the car spring (analogous to no current, and energy stored in the electric field of a capacitor). The amplitude of the wheels' motion is a maximum if the bumps in the road are hit at the resonant frequency.

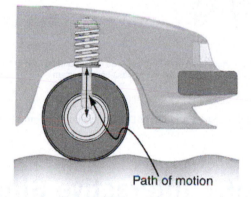

Path of motion

Figure 23.51 The forced but damped motion of the wheel on the car spring is analogous to an RLC series AC circuit. The shock absorber damps the motion and dissipates energy, analogous to the resistance in an RLC circuit. The mass and spring determine the resonant frequency.

A pure LC circuit with negligible resistance oscillates at f_0, the same resonant frequency as an RLC circuit. It can serve as a frequency standard or clock circuit—for example, in a digital wristwatch. With a very small resistance, only a very small energy input is necessary to maintain the oscillations. The circuit is analogous to a car with no shock absorbers. Once it starts oscillating, it continues at its natural frequency for some time. **Figure 23.52** shows the analogy between an LC circuit and a mass on a spring.

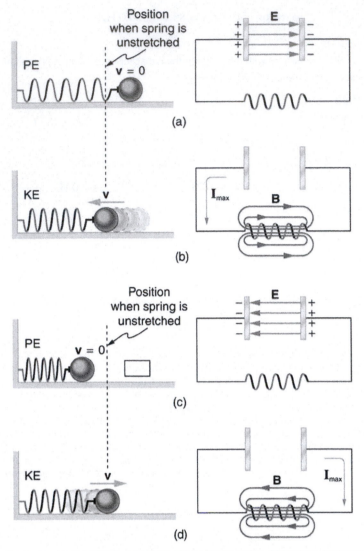

Figure 23.52 An *LC* circuit is analogous to a mass oscillating on a spring with no friction and no driving force. Energy moves back and forth between the inductor and capacitor, just as it moves from kinetic to potential in the mass-spring system.

PhET Explorations: Circuit Construction Kit (AC+DC), Virtual Lab

Build circuits with capacitors, inductors, resistors and AC or DC voltage sources, and inspect them using lab instruments such as voltmeters and ammeters.

Figure 23.53 Circuit Construction Kit (AC+DC), Virtual Lab (http://cnx.org/content/m55420/1.2/circuit-construction-kit-ac-virtual-lab_en.jar)

Glossary

back emf: the emf generated by a running motor, because it consists of a coil turning in a magnetic field; it opposes the voltage powering the motor

capacitive reactance: the opposition of a capacitor to a change in current; calculated by $X_C = \frac{1}{2\pi fC}$

characteristic time constant: denoted by τ, of a particular series *RL* circuit is calculated by $\tau = \frac{L}{R}$, where L is the inductance and R is the resistance

eddy current: a current loop in a conductor caused by motional emf

electric generator: a device for converting mechanical work into electric energy; it induces an emf by rotating a coil in a magnetic field

electromagnetic induction: the process of inducing an emf (voltage) with a change in magnetic flux

emf induced in a generator coil: $\text{emf} = NAB\omega \sin \omega t$, where A is the area of an N-turn coil rotated at a constant angular velocity ω in a uniform magnetic field B, over a period of time t

energy stored in an inductor: self-explanatory; calculated by $E_{\text{ind}} = \frac{1}{2}LI^2$

Faraday's law of induction: the means of calculating the emf in a coil due to changing magnetic flux, given by
$$\text{emf} = -N\frac{\Delta\Phi}{\Delta t}$$

henry: the unit of inductance; $1\ \text{H} = 1\ \Omega \cdot \text{s}$

impedance: the AC analogue to resistance in a DC circuit; it is the combined effect of resistance, inductive reactance, and capacitive reactance in the form $Z = \sqrt{R^2 + (X_L - X_C)^2}$

inductance: a property of a device describing how efficient it is at inducing emf in another device

induction: (magnetic induction) the creation of emfs and hence currents by magnetic fields

inductive reactance: the opposition of an inductor to a change in current; calculated by $X_L = 2\pi fL$

inductor: a device that exhibits significant self-inductance

Lenz's law: the minus sign in Faraday's law, signifying that the emf induced in a coil opposes the change in magnetic flux

magnetic damping: the drag produced by eddy currents

magnetic flux: the amount of magnetic field going through a particular area, calculated with $\Phi = BA\cos\theta$ where B is the magnetic field strength over an area A at an angle θ with the perpendicular to the area

mutual inductance: how effective a pair of devices are at inducing emfs in each other

peak emf: $\text{emf}_0 = NAB\omega$

phase angle: denoted by ϕ, the amount by which the voltage and current are out of phase with each other in a circuit

power factor: the amount by which the power delivered in the circuit is less than the theoretical maximum of the circuit due to voltage and current being out of phase; calculated by $\cos\phi$

resonant frequency: the frequency at which the impedance in a circuit is at a minimum, and also the frequency at which the circuit would oscillate if not driven by a voltage source; calculated by $f_0 = \frac{1}{2\pi\sqrt{LC}}$

self-inductance: how effective a device is at inducing emf in itself

shock hazard: the term for electrical hazards due to current passing through a human

step-down transformer: a transformer that decreases voltage

step-up transformer: a transformer that increases voltage

thermal hazard: the term for electrical hazards due to overheating

three-wire system: the wiring system used at present for safety reasons, with live, neutral, and ground wires

transformer: a device that transforms voltages from one value to another using induction

transformer equation: the equation showing that the ratio of the secondary to primary voltages in a transformer equals the ratio of the number of loops in their coils; $\frac{V_s}{V_p} = \frac{N_s}{N_p}$

Section Summary

23.1 Induced Emf and Magnetic Flux

- The crucial quantity in induction is magnetic flux Φ, defined to be $\Phi = BA \cos \theta$, where B is the magnetic field strength over an area A at an angle θ with the perpendicular to the area.
- Units of magnetic flux Φ are $T \cdot m^2$.
- Any change in magnetic flux Φ induces an emf—the process is defined to be electromagnetic induction.

23.2 Faraday's Law of Induction: Lenz's Law

- Faraday's law of induction states that the emf induced by a change in magnetic flux is

$$\text{emf} = -N\frac{\Delta\Phi}{\Delta t}$$

  when flux changes by $\Delta\Phi$ in a time Δt.
- If emf is induced in a coil, N is its number of turns.
- The minus sign means that the emf creates a current I and magnetic field B that *oppose the change in flux* $\Delta\Phi$ —this opposition is known as Lenz's law.

23.3 Motional Emf

- An emf induced by motion relative to a magnetic field B is called a *motional emf* and is given by

$$\text{emf} = B\ell v \qquad (B, \ell, \text{ and } v \text{ perpendicular}),$$

  where ℓ is the length of the object moving at speed v relative to the field.

23.4 Eddy Currents and Magnetic Damping

- Current loops induced in moving conductors are called eddy currents.
- They can create significant drag, called magnetic damping.

23.5 Electric Generators

- An electric generator rotates a coil in a magnetic field, inducing an emf given as a function of time by

$$\text{emf} = NAB\omega \sin \omega t,$$

  where A is the area of an N-turn coil rotated at a constant angular velocity ω in a uniform magnetic field B.
- The peak emf emf_0 of a generator is

$$\text{emf}_0 = NAB\omega.$$

23.6 Back Emf

- Any rotating coil will have an induced emf—in motors, this is called back emf, since it opposes the emf input to the motor.

23.7 Transformers

- Transformers use induction to transform voltages from one value to another.
- For a transformer, the voltages across the primary and secondary coils are related by

$$\frac{V_s}{V_p} = \frac{N_s}{N_p},$$

  where V_p and V_s are the voltages across primary and secondary coils having N_p and N_s turns.
- The currents I_p and I_s in the primary and secondary coils are related by $\frac{I_s}{I_p} = \frac{N_p}{N_s}$.
- A step-up transformer increases voltage and decreases current, whereas a step-down transformer decreases voltage and increases current.

23.8 Electrical Safety: Systems and Devices

- Electrical safety systems and devices are employed to prevent thermal and shock hazards.

Download for free at https://openstax.org/details/books/college-physics-ap-courses

- Circuit breakers and fuses interrupt excessive currents to prevent thermal hazards.
- The three-wire system guards against thermal and shock hazards, utilizing live/hot, neutral, and earth/ground wires, and grounding the neutral wire and case of the appliance.
- A ground fault interrupter (GFI) prevents shock by detecting the loss of current to unintentional paths.
- An isolation transformer insulates the device being powered from the original source, also to prevent shock.
- Many of these devices use induction to perform their basic function.

23.9 Inductance

- Inductance is the property of a device that tells how effectively it induces an emf in another device.
- Mutual inductance is the effect of two devices in inducing emfs in each other.
- A change in current $\Delta I_1 / \Delta t$ in one induces an emf emf_2 in the second:

$$\text{emf}_2 = -M\frac{\Delta I_1}{\Delta t},$$

  where M is defined to be the mutual inductance between the two devices, and the minus sign is due to Lenz's law.
- Symmetrically, a change in current $\Delta I_2 / \Delta t$ through the second device induces an emf emf_1 in the first:

$$\text{emf}_1 = -M\frac{\Delta I_2}{\Delta t},$$

  where M is the same mutual inductance as in the reverse process.
- Current changes in a device induce an emf in the device itself.
- Self-inductance is the effect of the device inducing emf in itself.
- The device is called an inductor, and the emf induced in it by a change in current through it is

$$\text{emf} = -L\frac{\Delta I}{\Delta t},$$

  where L is the self-inductance of the inductor, and $\Delta I / \Delta t$ is the rate of change of current through it. The minus sign indicates that emf opposes the change in current, as required by Lenz's law.
- The unit of self- and mutual inductance is the henry (H), where $1\text{ H} = 1\,\Omega \cdot \text{s}$.
- The self-inductance L of an inductor is proportional to how much flux changes with current. For an N-turn inductor,

$$L = N\frac{\Delta \Phi}{\Delta I}.$$

- The self-inductance of a solenoid is

$$L = \frac{\mu_0 N^2 A}{\ell}\text{(solenoid)},$$

  where N is its number of turns in the solenoid, A is its cross-sectional area, ℓ is its length, and $\mu_0 = 4\pi \times 10^{-7}\text{ T} \cdot \text{m/A}$ is the permeability of free space.
- The energy stored in an inductor E_{ind} is

$$E_{\text{ind}} = \frac{1}{2}LI^2.$$

23.10 RL Circuits

- When a series connection of a resistor and an inductor—an *RL* circuit—is connected to a voltage source, the time variation of the current is

$$I = I_0(1 - e^{-t/\tau}) \quad \text{(turning on)}.$$

  where $I_0 = V / R$ is the final current.

- The characteristic time constant τ is $\tau = \frac{L}{R}$, where L is the inductance and R is the resistance.

- In the first time constant τ, the current rises from zero to $0.632 I_0$, and 0.632 of the remainder in every subsequent time interval τ.

- When the inductor is shorted through a resistor, current decreases as

$$I = I_0 e^{-t/\tau} \quad \text{(turning off)}$$

  Here I_0 is the initial current.

- Current falls to $0.368I_0$ in the first time interval τ, and 0.368 of the remainder toward zero in each subsequent time τ.

23.11 Reactance, Inductive and Capacitive

- For inductors in AC circuits, we find that when a sinusoidal voltage is applied to an inductor, the voltage leads the current by one-fourth of a cycle, or by a $90°$ phase angle.
- The opposition of an inductor to a change in current is expressed as a type of AC resistance.
- Ohm's law for an inductor is

$$I = \frac{V}{X_L},$$

where V is the rms voltage across the inductor.

- X_L is defined to be the inductive reactance, given by

$$X_L = 2\pi f L,$$

with f the frequency of the AC voltage source in hertz.

- Inductive reactance X_L has units of ohms and is greatest at high frequencies.

- For capacitors, we find that when a sinusoidal voltage is applied to a capacitor, the voltage follows the current by one-fourth of a cycle, or by a $90°$ phase angle.
- Since a capacitor can stop current when fully charged, it limits current and offers another form of AC resistance; Ohm's law for a capacitor is

$$I = \frac{V}{X_C},$$

where V is the rms voltage across the capacitor.

- X_C is defined to be the capacitive reactance, given by

$$X_C = \frac{1}{2\pi f C}.$$

- X_C has units of ohms and is greatest at low frequencies.

23.12 RLC Series AC Circuits

- The AC analogy to resistance is impedance Z, the combined effect of resistors, inductors, and capacitors, defined by the AC version of Ohm's law:

$$I_0 = \frac{V_0}{Z} \text{ or } I_{rms} = \frac{V_{rms}}{Z},$$

where I_0 is the peak current and V_0 is the peak source voltage.

- Impedance has units of ohms and is given by $Z = \sqrt{R^2 + (X_L - X_C)^2}$.
- The resonant frequency f_0, at which $X_L = X_C$, is

$$f_0 = \frac{1}{2\pi\sqrt{LC}}.$$

- In an AC circuit, there is a phase angle ϕ between source voltage V and the current I, which can be found from

$$\cos \phi = \frac{R}{Z},$$

- $\phi = 0°$ for a purely resistive circuit or an *RLC* circuit at resonance.
- The average power delivered to an *RLC* circuit is affected by the phase angle and is given by

$$P_{ave} = I_{rms}V_{rms} \cos \phi,$$

$\cos \phi$ is called the power factor, which ranges from 0 to 1.

Conceptual Questions

23.1 Induced Emf and Magnetic Flux

1. How do the multiple-loop coils and iron ring in the version of Faraday's apparatus shown in **Figure 23.3** enhance the observation of induced emf?

2. When a magnet is thrust into a coil as in **Figure 23.4**(a), what is the direction of the force exerted by the coil on the magnet? Draw a diagram showing the direction of the current induced in the coil and the magnetic field it produces, to justify your response. How does the magnitude of the force depend on the resistance of the galvanometer?

3. Explain how magnetic flux can be zero when the magnetic field is not zero.

4. Is an emf induced in the coil in **Figure 23.54** when it is stretched? If so, state why and give the direction of the induced current.

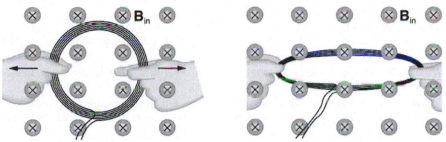

Figure 23.54 A circular coil of wire is stretched in a magnetic field.

23.2 Faraday's Law of Induction: Lenz's Law

5. A person who works with large magnets sometimes places her head inside a strong field. She reports feeling dizzy as she quickly turns her head. How might this be associated with induction?

6. A particle accelerator sends high-velocity charged particles down an evacuated pipe. Explain how a coil of wire wrapped around the pipe could detect the passage of individual particles. Sketch a graph of the voltage output of the coil as a single particle passes through it.

23.3 Motional Emf

7. Why must part of the circuit be moving relative to other parts, to have usable motional emf? Consider, for example, that the rails in **Figure 23.11** are stationary relative to the magnetic field, while the rod moves.

8. A powerful induction cannon can be made by placing a metal cylinder inside a solenoid coil. The cylinder is forcefully expelled when solenoid current is turned on rapidly. Use Faraday's and Lenz's laws to explain how this works. Why might the cylinder get live/hot when the cannon is fired?

9. An induction stove heats a pot with a coil carrying an alternating current located beneath the pot (and without a hot surface). Can the stove surface be a conductor? Why won't a coil carrying a direct current work?

10. Explain how you could thaw out a frozen water pipe by wrapping a coil carrying an alternating current around it. Does it matter whether or not the pipe is a conductor? Explain.

23.4 Eddy Currents and Magnetic Damping

11. Explain why magnetic damping might not be effective on an object made of several thin conducting layers separated by insulation.

12. Explain how electromagnetic induction can be used to detect metals? This technique is particularly important in detecting buried landmines for disposal, geophysical prospecting and at airports.

23.5 Electric Generators

13. Using RHR-1, show that the emfs in the sides of the generator loop in **Figure 23.23** are in the same sense and thus add.

14. The source of a generator's electrical energy output is the work done to turn its coils. How is the work needed to turn the generator related to Lenz's law?

23.6 Back Emf

15. Suppose you find that the belt drive connecting a powerful motor to an air conditioning unit is broken and the motor is running freely. Should you be worried that the motor is consuming a great deal of energy for no useful purpose? Explain why or why not.

23.7 Transformers

16. Explain what causes physical vibrations in transformers at twice the frequency of the AC power involved.

23.8 Electrical Safety: Systems and Devices

17. Does plastic insulation on live/hot wires prevent shock hazards, thermal hazards, or both?

18. Why are ordinary circuit breakers and fuses ineffective in preventing shocks?

19. A GFI may trip just because the live/hot and neutral wires connected to it are significantly different in length. Explain why.

23.9 Inductance

20. How would you place two identical flat coils in contact so that they had the greatest mutual inductance? The least?

21. How would you shape a given length of wire to give it the greatest self-inductance? The least?

22. Verify, as was concluded without proof in **Example 23.7**, that units of $T \cdot m^2 / A = \Omega \cdot s = H$.

23.11 Reactance, Inductive and Capacitive

23. Presbycusis is a hearing loss due to age that progressively affects higher frequencies. A hearing aid amplifier is designed to amplify all frequencies equally. To adjust its output for presbycusis, would you put a capacitor in series or parallel with the hearing aid's speaker? Explain.

24. Would you use a large inductance or a large capacitance in series with a system to filter out low frequencies, such as the 100 Hz hum in a sound system? Explain.

25. High-frequency noise in AC power can damage computers. Does the plug-in unit designed to prevent this damage use a large inductance or a large capacitance (in series with the computer) to filter out such high frequencies? Explain.

26. Does inductance depend on current, frequency, or both? What about inductive reactance?

27. Explain why the capacitor in **Figure 23.55**(a) acts as a low-frequency filter between the two circuits, whereas that in **Figure 23.55**(b) acts as a high-frequency filter.

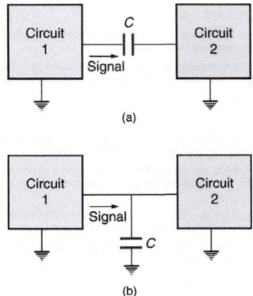

Figure 23.55 Capacitors and inductors. Capacitor with high frequency and low frequency.

28. If the capacitors in **Figure 23.55** are replaced by inductors, which acts as a low-frequency filter and which as a high-frequency filter?

23.12 RLC Series AC Circuits

29. Does the resonant frequency of an AC circuit depend on the peak voltage of the AC source? Explain why or why not.

30. Suppose you have a motor with a power factor significantly less than 1. Explain why it would be better to improve the power factor as a method of improving the motor's output, rather than to increase the voltage input.

Problems & Exercises

23.1 Induced Emf and Magnetic Flux

1. What is the value of the magnetic flux at coil 2 in **Figure 23.56** due to coil 1?

Figure 23.56 (a) The planes of the two coils are perpendicular. (b) The wire is perpendicular to the plane of the coil.

2. What is the value of the magnetic flux through the coil in **Figure 23.56(b)** due to the wire?

23.2 Faraday's Law of Induction: Lenz's Law

3. Referring to **Figure 23.57**(a), what is the direction of the current induced in coil 2: (a) If the current in coil 1 increases? (b) If the current in coil 1 decreases? (c) If the current in coil 1 is constant? Explicitly show how you follow the steps in the **Problem-Solving Strategy for Lenz's Law.**

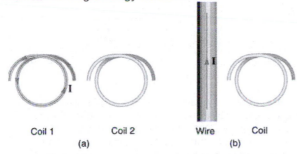

Figure 23.57 (a) The coils lie in the same plane. (b) The wire is in the plane of the coil

4. Referring to **Figure 23.57**(b), what is the direction of the current induced in the coil: (a) If the current in the wire increases? (b) If the current in the wire decreases? (c) If the current in the wire suddenly changes direction? Explicitly show how you follow the steps in the **Problem-Solving Strategy for Lenz's Law.**

5. Referring to **Figure 23.58**, what are the directions of the currents in coils 1, 2, and 3 (assume that the coils are lying in the plane of the circuit): (a) When the switch is first closed? (b) When the switch has been closed for a long time? (c) Just after the switch is opened?

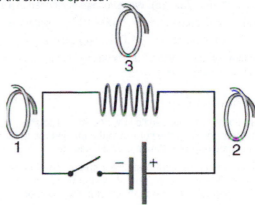

Figure 23.58

6. Repeat the previous problem with the battery reversed.

7. Verify that the units of $\Delta \Phi / \Delta t$ are volts. That is, show that $1 \text{ T} \cdot \text{m}^2 / \text{s} = 1 \text{ V}$.

8. Suppose a 50-turn coil lies in the plane of the page in a uniform magnetic field that is directed into the page. The coil originally has an area of 0.250 m^2. It is stretched to have no area in 0.100 s. What is the direction and magnitude of the induced emf if the uniform magnetic field has a strength of 1.50 T?

9. (a) An MRI technician moves his hand from a region of very low magnetic field strength into an MRI scanner's 2.00 T field with his fingers pointing in the direction of the field. Find the average emf induced in his wedding ring, given its diameter is 2.20 cm and assuming it takes 0.250 s to move it into the field. (b) Discuss whether this current would significantly change the temperature of the ring.

10. Integrated Concepts

Referring to the situation in the previous problem: (a) What current is induced in the ring if its resistance is 0.0100 Ω ? (b) What average power is dissipated? (c) What magnetic field is induced at the center of the ring? (d) What is the direction of the induced magnetic field relative to the MRI's field?

11. An emf is induced by rotating a 1000-turn, 20.0 cm diameter coil in the Earth's $5.00 \times 10^{-5} \text{ T}$ magnetic field. What average emf is induced, given the plane of the coil is originally perpendicular to the Earth's field and is rotated to be parallel to the field in 10.0 ms?

12. A 0.250 m radius, 500-turn coil is rotated one-fourth of a revolution in 4.17 ms, originally having its plane perpendicular to a uniform magnetic field. (This is 60 rev/s.) Find the magnetic field strength needed to induce an average emf of 10,000 V.

13. Integrated Concepts

Approximately how does the emf induced in the loop in **Figure 23.57**(b) depend on the distance of the center of the loop from the wire?

14. Integrated Concepts

(a) A lightning bolt produces a rapidly varying magnetic field. If the bolt strikes the earth vertically and acts like a current in a long straight wire, it will induce a voltage in a loop aligned like that in **Figure 23.57**(b). What voltage is induced in a 1.00 m diameter loop 50.0 m from a 2.00×10^6 A lightning strike, if the current falls to zero in 25.0 μs ? (b) Discuss circumstances under which such a voltage would produce noticeable consequences.

23.3 Motional Emf

15. Use Faraday's law, Lenz's law, and RHR-1 to show that the magnetic force on the current in the moving rod in **Figure 23.11** is in the opposite direction of its velocity.

16. If a current flows in the Satellite Tether shown in **Figure 23.12**, use Faraday's law, Lenz's law, and RHR-1 to show that there is a magnetic force on the tether in the direction opposite to its velocity.

17. (a) A jet airplane with a 75.0 m wingspan is flying at 280 m/s. What emf is induced between wing tips if the vertical component of the Earth's field is 3.00×10^{-5} T ? (b) Is an emf of this magnitude likely to have any consequences? Explain.

18. (a) A nonferrous screwdriver is being used in a 2.00 T magnetic field. What maximum emf can be induced along its 12.0 cm length when it moves at 6.00 m/s? (b) Is it likely that this emf will have any consequences or even be noticed?

19. At what speed must the sliding rod in **Figure 23.11** move to produce an emf of 1.00 V in a 1.50 T field, given the rod's length is 30.0 cm?

20. The 12.0 cm long rod in **Figure 23.11** moves at 4.00 m/s. What is the strength of the magnetic field if a 95.0 V emf is induced?

21. Prove that when B, ℓ, and v are not mutually perpendicular, motional emf is given by $\text{emf} = B\ell v \sin\theta$. If v is perpendicular to B, then θ is the angle between ℓ and B. If ℓ is perpendicular to B, then θ is the angle between v and B.

22. In the August 1992 space shuttle flight, only 250 m of the conducting tether considered in **Example 23.2** could be let out. A 40.0 V motional emf was generated in the Earth's 5.00×10^{-5} T field, while moving at 7.80×10^3 m/s . What was the angle between the shuttle's velocity and the Earth's field, assuming the conductor was perpendicular to the field?

23. Integrated Concepts

Derive an expression for the current in a system like that in **Figure 23.11**, under the following conditions. The resistance between the rails is R, the rails and the moving rod are identical in cross section A and have the same resistivity ρ. The distance between the rails is l, and the rod moves at constant speed v perpendicular to the uniform field B. At time zero, the moving rod is next to the resistance R.

24. Integrated Concepts

The Tethered Satellite in **Figure 23.12** has a mass of 525 kg and is at the end of a 20.0 km long, 2.50 mm diameter cable with the tensile strength of steel. (a) How much does the cable stretch if a 100 N force is exerted to pull the satellite in? (Assume the satellite and shuttle are at the same altitude above the Earth.) (b) What is the effective force constant of the cable? (c) How much energy is stored in it when stretched by the 100 N force?

25. Integrated Concepts

The Tethered Satellite discussed in this module is producing 5.00 kV, and a current of 10.0 A flows. (a) What magnetic drag force does this produce if the system is moving at 7.80 km/s? (b) How much kinetic energy is removed from the system in 1.00 h, neglecting any change in altitude or velocity during that time? (c) What is the change in velocity if the mass of the system is 100,000 kg? (d) Discuss the long term consequences (say, a week-long mission) on the space shuttle's orbit, noting what effect a decrease in velocity has and assessing the magnitude of the effect.

23.4 Eddy Currents and Magnetic Damping

26. Make a drawing similar to **Figure 23.14**, but with the pendulum moving in the opposite direction. Then use Faraday's law, Lenz's law, and RHR-1 to show that magnetic force opposes motion.

27.

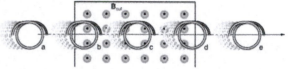

Figure 23.59 A coil is moved into and out of a region of uniform magnetic field. A coil is moved through a magnetic field as shown in **Figure 23.59**. The field is uniform inside the rectangle and zero outside. What is the direction of the induced current and what is the direction of the magnetic force on the coil at each position shown?

23.5 Electric Generators

28. Calculate the peak voltage of a generator that rotates its 200-turn, 0.100 m diameter coil at 3600 rpm in a 0.800 T field.

29. At what angular velocity in rpm will the peak voltage of a generator be 480 V, if its 500-turn, 8.00 cm diameter coil rotates in a 0.250 T field?

30. What is the peak emf generated by rotating a 1000-turn, 20.0 cm diameter coil in the Earth's 5.00×10^{-5} T magnetic field, given the plane of the coil is originally perpendicular to the Earth's field and is rotated to be parallel to the field in 10.0 ms?

31. What is the peak emf generated by a 0.250 m radius, 500-turn coil is rotated one-fourth of a revolution in 4.17 ms, originally having its plane perpendicular to a uniform magnetic field. (This is 60 rev/s.)

32. (a) A bicycle generator rotates at 1875 rad/s, producing an 18.0 V peak emf. It has a 1.00 by 3.00 cm rectangular coil in a 0.640 T field. How many turns are in the coil? (b) Is this number of turns of wire practical for a 1.00 by 3.00 cm coil?

33. Integrated Concepts

This problem refers to the bicycle generator considered in the previous problem. It is driven by a 1.60 cm diameter wheel that rolls on the outside rim of the bicycle tire. (a) What is the velocity of the bicycle if the generator's angular velocity is 1875 rad/s? (b) What is the maximum emf of the generator when the bicycle moves at 10.0 m/s, noting that it was 18.0 V under the original conditions? (c) If the sophisticated generator can vary its own magnetic field, what field strength will it need at 5.00 m/s to produce a 9.00 V maximum emf?

34. (a) A car generator turns at 400 rpm when the engine is idling. Its 300-turn, 5.00 by 8.00 cm rectangular coil rotates in an adjustable magnetic field so that it can produce sufficient voltage even at low rpms. What is the field strength needed to produce a 24.0 V peak emf? (b) Discuss how this required field strength compares to those available in permanent and electromagnets.

35. Show that if a coil rotates at an angular velocity ω, the period of its AC output is $2\pi/\omega$.

36. A 75-turn, 10.0 cm diameter coil rotates at an angular velocity of 8.00 rad/s in a 1.25 T field, starting with the plane of the coil parallel to the field. (a) What is the peak emf? (b) At what time is the peak emf first reached? (c) At what time is the emf first at its most negative? (d) What is the period of the AC voltage output?

37. (a) If the emf of a coil rotating in a magnetic field is zero at $t = 0$, and increases to its first peak at $t = 0.100 \text{ ms}$, what is the angular velocity of the coil? (b) At what time will its next maximum occur? (c) What is the period of the output? (d) When is the output first one-fourth of its maximum? (e) When is it next one-fourth of its maximum?

38. Unreasonable Results

A 500-turn coil with a 0.250 m^2 area is spun in the Earth's $5.00 \times 10^{-5} \text{ T}$ field, producing a 12.0 kV maximum emf. (a) At what angular velocity must the coil be spun? (b) What is unreasonable about this result? (c) Which assumption or premise is responsible?

23.6 Back Emf

39. Suppose a motor connected to a 120 V source draws 10.0 A when it first starts. (a) What is its resistance? (b) What current does it draw at its normal operating speed when it develops a 100 V back emf?

40. A motor operating on 240 V electricity has a 180 V back emf at operating speed and draws a 12.0 A current. (a) What is its resistance? (b) What current does it draw when it is first started?

41. What is the back emf of a 120 V motor that draws 8.00 A at its normal speed and 20.0 A when first starting?

42. The motor in a toy car operates on 6.00 V, developing a 4.50 V back emf at normal speed. If it draws 3.00 A at normal speed, what current does it draw when starting?

43. Integrated Concepts

The motor in a toy car is powered by four batteries in series, which produce a total emf of 6.00 V. The motor draws 3.00 A and develops a 4.50 V back emf at normal speed. Each battery has a $0.100 \ \Omega$ internal resistance. What is the resistance of the motor?

23.7 Transformers

44. A plug-in transformer, like that in **Figure 23.29**, supplies 9.00 V to a video game system. (a) How many turns are in its secondary coil, if its input voltage is 120 V and the primary coil has 400 turns? (b) What is its input current when its output is 1.30 A?

45. An American traveler in New Zealand carries a transformer to convert New Zealand's standard 240 V to 120 V so that she can use some small appliances on her trip. (a) What is the ratio of turns in the primary and secondary coils of her transformer? (b) What is the ratio of input to output current? (c) How could a New Zealander traveling in the United States use this same transformer to power her 240 V appliances from 120 V?

46. A cassette recorder uses a plug-in transformer to convert 120 V to 12.0 V, with a maximum current output of 200 mA. (a) What is the current input? (b) What is the power input? (c) Is this amount of power reasonable for a small appliance?

47. (a) What is the voltage output of a transformer used for rechargeable flashlight batteries, if its primary has 500 turns, its secondary 4 turns, and the input voltage is 120 V? (b) What input current is required to produce a 4.00 A output? (c) What is the power input?

48. (a) The plug-in transformer for a laptop computer puts out 7.50 V and can supply a maximum current of 2.00 A. What is the maximum input current if the input voltage is 240 V? Assume 100% efficiency. (b) If the actual efficiency is less than 100%, would the input current need to be greater or smaller? Explain.

49. A multipurpose transformer has a secondary coil with several points at which a voltage can be extracted, giving outputs of 5.60, 12.0, and 480 V. (a) The input voltage is 240 V to a primary coil of 280 turns. What are the numbers of turns in the parts of the secondary used to produce the output voltages? (b) If the maximum input current is 5.00 A, what are the maximum output currents (each used alone)?

50. A large power plant generates electricity at 12.0 kV. Its old transformer once converted the voltage to 335 kV. The secondary of this transformer is being replaced so that its output can be 750 kV for more efficient cross-country transmission on upgraded transmission lines. (a) What is the ratio of turns in the new secondary compared with the old secondary? (b) What is the ratio of new current output to old output (at 335 kV) for the same power? (c) If the upgraded transmission lines have the same resistance, what is the ratio of new line power loss to old?

51. If the power output in the previous problem is 1000 MW and line resistance is $2.00 \ \Omega$, what were the old and new line losses?

52. Unreasonable Results

The 335 kV AC electricity from a power transmission line is fed into the primary coil of a transformer. The ratio of the number of turns in the secondary to the number in the primary is $N_s / N_p = 1000$. (a) What voltage is induced in the secondary? (b) What is unreasonable about this result? (c) Which assumption or premise is responsible?

53. Construct Your Own Problem

Consider a double transformer to be used to create very large voltages. The device consists of two stages. The first is a transformer that produces a much larger output voltage than its input. The output of the first transformer is used as input to a second transformer that further increases the voltage. Construct a problem in which you calculate the output voltage of the final stage based on the input voltage of the first stage and the number of turns or loops in both parts of both transformers (four coils in all). Also calculate the maximum output current of the final stage based on the input current. Discuss the possibility of power losses in the devices and the effect on the output current and power.

23.8 Electrical Safety: Systems and Devices

54. Integrated Concepts

A short circuit to the grounded metal case of an appliance occurs as shown in **Figure 23.60**. The person touching the case is wet and only has a $3.00 \text{ k}\Omega$ resistance to earth/ground. (a) What is the voltage on the case if 5.00 mA flows through the person? (b) What is the current in the short circuit if the resistance of the earth/ground wire is $0.200 \ \Omega$? (c) Will this trigger the 20.0 A circuit breaker supplying the appliance?

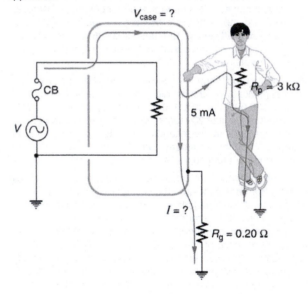

Figure 23.60 A person can be shocked even when the case of an appliance is grounded. The large short circuit current produces a voltage on the case of the appliance, since the resistance of the earth/ground wire is not zero.

23.9 Inductance

55. Two coils are placed close together in a physics lab to demonstrate Faraday's law of induction. A current of 5.00 A in one is switched off in 1.00 ms, inducing a 9.00 V emf in the other. What is their mutual inductance?

56. If two coils placed next to one another have a mutual inductance of 5.00 mH, what voltage is induced in one when the 2.00 A current in the other is switched off in 30.0 ms?

57. The 4.00 A current through a 7.50 mH inductor is switched off in 8.33 ms. What is the emf induced opposing this?

58. A device is turned on and 3.00 A flows through it 0.100 ms later. What is the self-inductance of the device if an induced 150 V emf opposes this?

59. Starting with $\text{emf}_2 = -M\dfrac{\Delta I_1}{\Delta t}$, show that the units of inductance are $(\text{V} \cdot \text{s})/\text{A} = \Omega \cdot \text{s}$.

60. Camera flashes charge a capacitor to high voltage by switching the current through an inductor on and off rapidly. In what time must the 0.100 A current through a 2.00 mH inductor be switched on or off to induce a 500 V emf?

61. A large research solenoid has a self-inductance of 25.0 H. (a) What induced emf opposes shutting it off when 100 A of current through it is switched off in 80.0 ms? (b) How much energy is stored in the inductor at full current? (c) At what rate in watts must energy be dissipated to switch the current off in 80.0 ms? (d) In view of the answer to the last part, is it surprising that shutting it down this quickly is difficult?

62. (a) Calculate the self-inductance of a 50.0 cm long, 10.0 cm diameter solenoid having 1000 loops. (b) How much energy is stored in this inductor when 20.0 A of current flows through it? (c) How fast can it be turned off if the induced emf cannot exceed 3.00 V?

63. A precision laboratory resistor is made of a coil of wire 1.50 cm in diameter and 4.00 cm long, and it has 500 turns. (a) What is its self-inductance? (b) What average emf is induced if the 12.0 A current through it is turned on in 5.00 ms (one-fourth of a cycle for 50 Hz AC)? (c) What is its inductance if it is shortened to half its length and counter-wound (two layers of 250 turns in opposite directions)?

64. The heating coils in a hair dryer are 0.800 cm in diameter, have a combined length of 1.00 m, and a total of 400 turns. (a) What is their total self-inductance assuming they act like a single solenoid? (b) How much energy is stored in them when 6.00 A flows? (c) What average emf opposes shutting them off if this is done in 5.00 ms (one-fourth of a cycle for 50 Hz AC)?

65. When the 20.0 A current through an inductor is turned off in 1.50 ms, an 800 V emf is induced, opposing the change. What is the value of the self-inductance?

66. How fast can the 150 A current through a 0.250 H inductor be shut off if the induced emf cannot exceed 75.0 V?

67. Integrated Concepts

A very large, superconducting solenoid such as one used in MRI scans, stores 1.00 MJ of energy in its magnetic field when 100 A flows. (a) Find its self-inductance. (b) If the coils "go normal," they gain resistance and start to dissipate thermal energy. What temperature increase is produced if all the stored energy goes into heating the 1000 kg magnet, given its average specific heat is 200 J/kg·°C ?

68. Unreasonable Results

A 25.0 H inductor has 100 A of current turned off in 1.00 ms. (a) What voltage is induced to oppose this? (b) What is unreasonable about this result? (c) Which assumption or premise is responsible?

23.10 RL Circuits

69. If you want a characteristic RL time constant of 1.00 s, and you have a $500 \ \Omega$ resistor, what value of self-inductance is needed?

70. Your *RL* circuit has a characteristic time constant of 20.0 ns, and a resistance of $5.00 \text{ M}\Omega$. (a) What is the inductance of the circuit? (b) What resistance would give you a 1.00 ns time constant, perhaps needed for quick response in an oscilloscope?

71. A large superconducting magnet, used for magnetic resonance imaging, has a 50.0 H inductance. If you want current through it to be adjustable with a 1.00 s characteristic time constant, what is the minimum resistance of system?

72. Verify that after a time of 10.0 ms, the current for the situation considered in **Example 23.9** will be 0.183 A as stated.

73. Suppose you have a supply of inductors ranging from 1.00 nH to 10.0 H, and resistors ranging from $0.100 \ \Omega$ to $1.00 \text{ M}\Omega$. What is the range of characteristic *RL* time constants you can produce by connecting a single resistor to a single inductor?

74. (a) What is the characteristic time constant of a 25.0 mH inductor that has a resistance of $4.00 \ \Omega$? (b) If it is connected to a 12.0 V battery, what is the current after 12.5 ms?

75. What percentage of the final current I_0 flows through an inductor L in series with a resistor R, three time constants after the circuit is completed?

76. The 5.00 A current through a 1.50 H inductor is dissipated by a $2.00 \ \Omega$ resistor in a circuit like that in **Figure 23.44** with the switch in position 2. (a) What is the initial energy in the inductor? (b) How long will it take the current to decline to 5.00% of its initial value? (c) Calculate the average power dissipated, and compare it with the initial power dissipated by the resistor.

77. (a) Use the exact exponential treatment to find how much time is required to bring the current through an 80.0 mH inductor in series with a $15.0 \ \Omega$ resistor to 99.0% of its final value, starting from zero. (b) Compare your answer to the approximate treatment using integral numbers of τ. (c) Discuss how significant the difference is.

78. (a) Using the exact exponential treatment, find the time required for the current through a 2.00 H inductor in series with a $0.500 \ \Omega$ resistor to be reduced to 0.100% of its original value. (b) Compare your answer to the approximate treatment using integral numbers of τ. (c) Discuss how significant the difference is.

23.11 Reactance, Inductive and Capacitive

79. At what frequency will a 30.0 mH inductor have a reactance of $100 \ \Omega$?

80. What value of inductance should be used if a $20.0 \text{ k}\Omega$ reactance is needed at a frequency of 500 Hz?

81. What capacitance should be used to produce a $2.00 \text{ M}\Omega$ reactance at 60.0 Hz?

82. At what frequency will an 80.0 mF capacitor have a reactance of $0.250 \ \Omega$?

83. (a) Find the current through a 0.500 H inductor connected to a 60.0 Hz, 480 V AC source. (b) What would the current be at 100 kHz?

84. (a) What current flows when a 60.0 Hz, 480 V AC source is connected to a $0.250 \ \mu\text{F}$ capacitor? (b) What would the current be at 25.0 kHz?

85. A 20.0 kHz, 16.0 V source connected to an inductor produces a 2.00 A current. What is the inductance?

86. A 20.0 Hz, 16.0 V source produces a 2.00 mA current when connected to a capacitor. What is the capacitance?

87. (a) An inductor designed to filter high-frequency noise from power supplied to a personal computer is placed in series with the computer. What minimum inductance should it have to produce a $2.00 \text{ k}\Omega$ reactance for 15.0 kHz noise? (b) What is its reactance at 60.0 Hz?

88. The capacitor in **Figure 23.55**(a) is designed to filter low-frequency signals, impeding their transmission between circuits. (a) What capacitance is needed to produce a $100 \text{ k}\Omega$ reactance at a frequency of 120 Hz? (b) What would its reactance be at 1.00 MHz? (c) Discuss the implications of your answers to (a) and (b).

89. The capacitor in **Figure 23.55**(b) will filter high-frequency signals by shorting them to earth/ground. (a) What capacitance is needed to produce a reactance of $10.0 \text{ m}\Omega$ for a 5.00 kHz signal? (b) What would its reactance be at 3.00 Hz? (c) Discuss the implications of your answers to (a) and (b).

90. Unreasonable Results

In a recording of voltages due to brain activity (an EEG), a 10.0 mV signal with a 0.500 Hz frequency is applied to a capacitor, producing a current of 100 mA. Resistance is negligible. (a) What is the capacitance? (b) What is unreasonable about this result? (c) Which assumption or premise is responsible?

91. Construct Your Own Problem

Consider the use of an inductor in series with a computer operating on 60 Hz electricity. Construct a problem in which you calculate the relative reduction in voltage of incoming high frequency noise compared to 60 Hz voltage. Among the things to consider are the acceptable series reactance of the inductor for 60 Hz power and the likely frequencies of noise coming through the power lines.

23.12 RLC Series AC Circuits

92. An *RL* circuit consists of a $40.0 \ \Omega$ resistor and a 3.00 mH inductor. (a) Find its impedance Z at 60.0 Hz and 10.0 kHz. (b) Compare these values of Z with those found in **Example 23.12** in which there was also a capacitor.

93. An *RC* circuit consists of a $40.0 \ \Omega$ resistor and a $5.00 \ \mu\text{F}$ capacitor. (a) Find its impedance at 60.0 Hz and 10.0 kHz. (b) Compare these values of Z with those found in **Example 23.12**, in which there was also an inductor.

94. An *LC* circuit consists of a 3.00 mH inductor and a $5.00 \ \mu\text{F}$ capacitor. (a) Find its impedance at 60.0 Hz and 10.0 kHz. (b) Compare these values of Z with those found in **Example 23.12** in which there was also a resistor.

95. What is the resonant frequency of a 0.500 mH inductor connected to a $40.0 \ \mu\text{F}$ capacitor?

96. To receive AM radio, you want an *RLC* circuit that can be made to resonate at any frequency between 500 and 1650 kHz. This is accomplished with a fixed $1.00\ \mu H$ inductor connected to a variable capacitor. What range of capacitance is needed?

97. Suppose you have a supply of inductors ranging from 1.00 nH to 10.0 H, and capacitors ranging from 1.00 pF to 0.100 F. What is the range of resonant frequencies that can be achieved from combinations of a single inductor and a single capacitor?

98. What capacitance do you need to produce a resonant frequency of 1.00 GHz, when using an 8.00 nH inductor?

99. What inductance do you need to produce a resonant frequency of 60.0 Hz, when using a $2.00\ \mu F$ capacitor?

100. The lowest frequency in the FM radio band is 88.0 MHz. (a) What inductance is needed to produce this resonant frequency if it is connected to a 2.50 pF capacitor? (b) The capacitor is variable, to allow the resonant frequency to be adjusted to as high as 108 MHz. What must the capacitance be at this frequency?

101. An *RLC* series circuit has a $2.50\ \Omega$ resistor, a $100\ \mu H$ inductor, and an $80.0\ \mu F$ capacitor.(a) Find the circuit's impedance at 120 Hz. (b) Find the circuit's impedance at 5.00 kHz. (c) If the voltage source has $V_{rms} = 5.60\ V$, what is I_{rms} at each frequency? (d) What is the resonant frequency of the circuit? (e) What is I_{rms} at resonance?

102. An *RLC* series circuit has a $1.00\ k\Omega$ resistor, a $150\ \mu H$ inductor, and a 25.0 nF capacitor. (a) Find the circuit's impedance at 500 Hz. (b) Find the circuit's impedance at 7.50 kHz. (c) If the voltage source has $V_{rms} = 408\ V$, what is I_{rms} at each frequency? (d) What is the resonant frequency of the circuit? (e) What is I_{rms} at resonance?

103. An *RLC* series circuit has a $2.50\ \Omega$ resistor, a $100\ \mu H$ inductor, and an $80.0\ \mu F$ capacitor. (a) Find the power factor at $f = 120\ Hz$. (b) What is the phase angle at 120 Hz? (c) What is the average power at 120 Hz? (d) Find the average power at the circuit's resonant frequency.

104. An *RLC* series circuit has a $1.00\ k\Omega$ resistor, a $150\ \mu H$ inductor, and a 25.0 nF capacitor. (a) Find the power factor at $f = 7.50\ Hz$. (b) What is the phase angle at this frequency? (c) What is the average power at this frequency? (d) Find the average power at the circuit's resonant frequency.

105. An *RLC* series circuit has a $200\ \Omega$ resistor and a 25.0 mH inductor. At 8000 Hz, the phase angle is $45.0°$. (a) What is the impedance? (b) Find the circuit's capacitance. (c) If $V_{rms} = 408\ V$ is applied, what is the average power supplied?

106. Referring to **Example 23.14**, find the average power at 10.0 kHz.

Download for free at https://openstax.org/details/books/college-physics-ap-courses

23.1 Induced Emf and Magnetic Flux

1. To produce current with a coil and bar magnet you can:
   a.   move the coil but not the magnet.
   b.   move the magnet but not the coil.
   c.   move either the coil or the magnet.
   d.   It is not possible to produce current.

2. Calculate the magnetic flux for a coil of area 0.2 m² placed at an angle of $\theta=60°$ (as shown in the figure above) to a magnetic field of strength 1.5×10^{-3} T. At what angle will the flux be at its maximum?

23.5 Electric Generators

3. The emf induced in a coil that is rotating in a magnetic field will be at a maximum when
   a.   the magnetic flux is at a maximum.
   b.   the magnetic flux is at a minimum.
   c.   the change in magnetic flux is at a maximum.
   d.   the change in magnetic flux is at a minimum.

4. A coil with circular cross section and 20 turns is rotating at a rate of 400 rpm between the poles of a magnet. If the magnetic field strength is 0.6 T and peak voltage is 0.2 V, what is the radius of the coil? If the emf of the coil is zero at t = 0 s, when will it reach its peak emf?

23.7 Transformers

5. Which of the following statements is true for a step-down transformer? Select *two* answers.
   a.   Primary voltage is higher than secondary voltage.
   b.   Primary voltage is lower than secondary voltage.
   c.   Primary current is higher than secondary current.
   d.   Primary current is lower than secondary current.

6. An ideal step-up transformer with turn ratio 1:30 is supplied with an input power of 120 W. If the output voltage is 210 V, calculate the output power and input current.

23.8 Electrical Safety: Systems and Devices

7. Which of the following statements is true for an isolation transformer?
   a.   It has more primary turns than secondary turns.
   b.   It has fewer primary turns than secondary turns.
   c.   It has an equal number of primary and secondary turns.
   d.   It can have more, fewer, or an equal number of primary and secondary turns.

8. Explain the working of a ground fault interrupter (GFI).

24 ELECTROMAGNETIC WAVES

Figure 24.1 Human eyes detect these orange "sea goldie" fish swimming over a coral reef in the blue waters of the Gulf of Eilat (Red Sea) using visible light. (credit: Daviddarom, Wikimedia Commons)

Chapter Outline

24.1. Maxwell's Equations: Electromagnetic Waves Predicted and Observed

24.2. Production of Electromagnetic Waves

24.3. The Electromagnetic Spectrum

24.4. Energy in Electromagnetic Waves

Connection for AP® Courses

Electromagnetic waves are all around us. The beauty of a coral reef, the warmth of sunshine, sunburn, an X-ray image revealing a broken bone, even microwave popcorn—all involve **electromagnetic waves**. The list of the various types of electromagnetic waves, ranging from radio transmission waves to nuclear gamma-rays (γ-rays), is interesting in itself. Even more intriguing is that all of these widely varied phenomena are different manifestations of the same thing—electromagnetic waves. (See **Figure 24.2**.)

What are electromagnetic waves? How are they created, and how do they travel? How can we understand and conceptualize their widely varying properties? What is their relationship to electric and magnetic effects? These and other questions will be explored in this chapter.

Electromagnetic waves support Big Idea 6 that waves can transport energy and momentum. In general, electromagnetic waves behave like any other wave, as they are traveling disturbances (Enduring Understanding 6.A). They consist of oscillating electric and magnetic fields, which can be conceived of as transverse waves (Essential Knowledge 6.A.1). They are periodic and can be described by their amplitude, frequency, wavelength, speed, and energy (Enduring Understanding 6.B).

Simple waves can be modeled mathematically using sine or cosine functions involving the wavelength, amplitude, and frequency of the wave. (Essential Knowledge 6.B.3). However, electromagnetic waves also have some unique properties compared to other waves. They can travel through both matter and a vacuum (Essential Knowledge 6.F.2), unlike mechanical waves, including sound, that require a medium (Essential Knowledge 6.A.2).

Maxwell's equations define the relationship between electric permittivity, the magnetic permeability of free space (vacuum), and the speed of light, which is the speed of propagation of all electromagnetic waves in a vacuum. This chapter uses the properties electric permittivity (Essential Knowledge 1.E.4) and magnetic permeability (Essential Knowledge 1.E.5) to support Big Idea 1 that objects and systems have certain properties and may have internal structure.

The particular properties mentioned are the macroscopic results of the atomic and molecular structure of materials (Enduring Understanding 1.E). Electromagnetic radiation can be modeled as a wave or as fundamental particles (Enduring Understanding 6.F). This chapter also introduces different types of electromagnetic radiation that are characterized by their wavelengths (Essential Knowledge 6.F.1) and have been given specific names (see **Figure 24.2**).

Big Idea 1 Objects and systems have properties such as mass and charge. Systems may have internal structure.

Enduring Understanding 1.E Materials have many macroscopic properties that result from the arrangement and interactions of the atoms and molecules that make up the material.

Essential Knowledge 1.E.4 Matter has a property called electric permittivity.

Essential Knowledge 1.E.5 Matter has a property called magnetic permeability.

Big Idea 6. Waves can transfer energy and momentum from one location to another without the permanent transfer of mass and serve as a mathematical model for the description of other phenomena.

Enduring Understanding 6.A A wave is a traveling disturbance that transfers energy and momentum.

Essential Knowledge 6.A.1 Waves can propagate via different oscillation modes such as transverse and longitudinal.

Essential Knowledge 6.A.2 For propagation, mechanical waves require a medium, while electromagnetic waves do not require a physical medium. Examples include light traveling through a vacuum and sound not traveling through a vacuum.

Enduring Understanding 6.B A periodic wave is one that repeats as a function of both time and position and can be described by its amplitude, frequency, wavelength, speed, and energy.

Essential Knowledge 6.B.3 A simple wave can be described by an equation involving one sine or cosine function involving the wavelength, amplitude, and frequency of the wave.

Enduring Understanding 6.F Electromagnetic radiation can be modeled as waves or as fundamental particles.

Essential Knowledge 6.F.1 Types of electromagnetic radiation are characterized by their wavelengths, and certain ranges of wavelength have been given specific names. These include (in order of increasing wavelength spanning a range from picometers to kilometers) gamma rays, x-rays, ultraviolet, visible light, infrared, microwaves, and radio waves.

Essential Knowledge 6.F.2 Electromagnetic waves can transmit energy through a medium and through a vacuum.

Misconception Alert: Sound Waves vs. Radio Waves

Many people confuse sound waves with **radio waves**, one type of electromagnetic (EM) wave. However, sound and radio waves are completely different phenomena. Sound creates pressure variations (waves) in matter, such as air or water, or your eardrum. Conversely, radio waves are *electromagnetic waves*, like visible light, infrared, ultraviolet, X-rays, and gamma rays. EM waves don't need a medium in which to propagate; they can travel through a vacuum, such as outer space.

A radio works because sound waves played by the D.J. at the radio station are converted into electromagnetic waves, then encoded and transmitted in the radio-frequency range. The radio in your car receives the radio waves, decodes the information, and uses a speaker to change it back into a sound wave, bringing sweet music to your ears.

Discovering a New Phenomenon

It is worth noting at the outset that the general phenomenon of electromagnetic waves was predicted by theory before it was realized that light is a form of electromagnetic wave. The prediction was made by James Clerk Maxwell in the mid-19th century when he formulated a single theory combining all the electric and magnetic effects known by scientists at that time. "Electromagnetic waves" was the name he gave to the phenomena his theory predicted.

Such a theoretical prediction followed by experimental verification is an indication of the power of science in general, and physics in particular. The underlying connections and unity of physics allow certain great minds to solve puzzles without having all the pieces. The prediction of electromagnetic waves is one of the most spectacular examples of this power. Certain others, such as the prediction of antimatter, will be discussed in later modules.

Figure 24.2 The electromagnetic waves sent and received by this 50-foot radar dish antenna at Kennedy Space Center in Florida are not visible, but help track expendable launch vehicles with high-definition imagery. The first use of this C-band radar dish was for the launch of the Atlas V rocket sending the New Horizons probe toward Pluto. (credit: NASA)

24.1 Maxwell's Equations: Electromagnetic Waves Predicted and Observed

Learning Objectives
By the end of this section, you will be able to:
• Restate Maxwell's equations.

The Scotsman James Clerk Maxwell (1831–1879) is regarded as the greatest theoretical physicist of the 19th century. (See **Figure 24.3**.) Although he died young, Maxwell not only formulated a complete electromagnetic theory, represented by **Maxwell's equations**, he also developed the kinetic theory of gases and made significant contributions to the understanding of color vision and the nature of Saturn's rings.

Figure 24.3 James Clerk Maxwell, a 19th-century physicist, developed a theory that explained the relationship between electricity and magnetism and correctly predicted that visible light is caused by electromagnetic waves. (credit: G. J. Stodart)

Maxwell brought together all the work that had been done by brilliant physicists such as Oersted, Coulomb, Gauss, and Faraday, and added his own insights to develop the overarching theory of electromagnetism. Maxwell's equations are paraphrased here in words because their mathematical statement is beyond the level of this text. However, the equations illustrate how apparently simple mathematical statements can elegantly unite and express a multitude of concepts—why mathematics is the language of science.

Maxwell's Equations

1. **Electric field lines** originate on positive charges and terminate on negative charges. The electric field is defined as the force per unit charge on a test charge, and the strength of the force is related to the electric constant ε_0, also known as the permittivity of free space. From Maxwell's first equation we obtain a special form of Coulomb's law known as Gauss's law for electricity.

2. **Magnetic field lines** are continuous, having no beginning or end. No magnetic monopoles are known to exist. The

strength of the magnetic force is related to the magnetic constant μ_0, also known as the permeability of free space. This second of Maxwell's equations is known as Gauss's law for magnetism.

3. A changing magnetic field induces an electromotive force (emf) and, hence, an electric field. The direction of the emf opposes the change. This third of Maxwell's equations is Faraday's law of induction, and includes Lenz's law.

4. Magnetic fields are generated by moving charges or by changing electric fields. This fourth of Maxwell's equations encompasses Ampere's law and adds another source of magnetism—changing electric fields.

Maxwell's equations encompass the major laws of electricity and magnetism. What is not so apparent is the symmetry that Maxwell introduced in his mathematical framework. Especially important is his addition of the hypothesis that changing electric fields create magnetic fields. This is exactly analogous (and symmetric) to Faraday's law of induction and had been suspected for some time, but fits beautifully into Maxwell's equations.

Symmetry is apparent in nature in a wide range of situations. In contemporary research, symmetry plays a major part in the search for sub-atomic particles using massive multinational particle accelerators such as the new Large Hadron Collider at CERN.

Making Connections: Unification of Forces

Maxwell's complete and symmetric theory showed that electric and magnetic forces are not separate, but different manifestations of the same thing—the electromagnetic force. This classical unification of forces is one motivation for current attempts to unify the four basic forces in nature—the gravitational, electrical, strong, and weak nuclear forces.

Since changing electric fields create relatively weak magnetic fields, they could not be easily detected at the time of Maxwell's hypothesis. Maxwell realized, however, that oscillating charges, like those in AC circuits, produce changing electric fields. He predicted that these changing fields would propagate from the source like waves generated on a lake by a jumping fish.

The waves predicted by Maxwell would consist of oscillating electric and magnetic fields—defined to be an electromagnetic wave (EM wave). Electromagnetic waves would be capable of exerting forces on charges great distances from their source, and they might thus be detectable. Maxwell calculated that electromagnetic waves would propagate at a speed given by the equation

$$c = \frac{1}{\sqrt{\mu_0 \varepsilon_0}}.$$ (24.1)

When the values for μ_0 and ε_0 are entered into the equation for c, we find that

$$c = \frac{1}{\sqrt{(8.85 \times 10^{-12} \frac{C^2}{N \cdot m^2})(4\pi \times 10^{-7} \frac{T \cdot m}{A})}} = 3.00 \times 10^8 \text{ m/s},$$ (24.2)

which is the speed of light. In fact, Maxwell concluded that light is an electromagnetic wave having such wavelengths that it can be detected by the eye.

Other wavelengths should exist—it remained to be seen if they did. If so, Maxwell's theory and remarkable predictions would be verified, the greatest triumph of physics since Newton. Experimental verification came within a few years, but not before Maxwell's death.

Hertz's Observations

The German physicist Heinrich Hertz (1857–1894) was the first to generate and detect certain types of electromagnetic waves in the laboratory. Starting in 1887, he performed a series of experiments that not only confirmed the existence of electromagnetic waves, but also verified that they travel at the speed of light.

Hertz used an AC RLC (resistor-inductor-capacitor) circuit that resonates at a known frequency $f_0 = \frac{1}{2\pi\sqrt{LC}}$ and connected it to a loop of wire as shown in **Figure 24.4**. High voltages induced across the gap in the loop produced sparks that were visible evidence of the current in the circuit and that helped generate electromagnetic waves.

Across the laboratory, Hertz had another loop attached to another RLC circuit, which could be tuned (as the dial on a radio) to the same resonant frequency as the first and could, thus, be made to receive electromagnetic waves. This loop also had a gap across which sparks were generated, giving solid evidence that electromagnetic waves had been received.

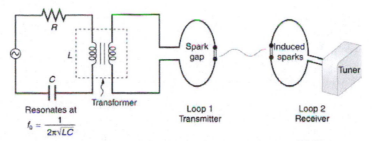

Figure 24.4 The apparatus used by Hertz in 1887 to generate and detect electromagnetic waves. An RLC circuit connected to the first loop caused sparks across a gap in the wire loop and generated electromagnetic waves. Sparks across a gap in the second loop located across the laboratory gave evidence that the waves had been received.

Hertz also studied the reflection, refraction, and interference patterns of the electromagnetic waves he generated, verifying their wave character. He was able to determine wavelength from the interference patterns, and knowing their frequency, he could calculate the propagation speed using the equation $v = f\lambda$ (velocity—or speed—equals frequency times wavelength). Hertz was thus able to prove that electromagnetic waves travel at the speed of light. The SI unit for frequency, the hertz ($1\ \mathrm{Hz} = 1\ \mathrm{cycle/sec}$), is named in his honor.

24.2 Production of Electromagnetic Waves

Learning Objectives

By the end of this section, you will be able to:

- Describe the electric and magnetic waves as they move out from a source, such as an AC generator.
- Explain the mathematical relationship between the magnetic field strength and the electrical field strength.
- Calculate the maximum strength of the magnetic field in an electromagnetic wave, given the maximum electric field strength.

The information presented in this section supports the following AP® learning objectives and science practices:

- **6.A.1.1** The student is able to use a visual representation to construct an explanation of the distinction between transverse and longitudinal waves by focusing on the vibration that generates the wave. **(S.P. 6.2)**
- **6.A.1.2** The student is able to describe representations of transverse and longitudinal waves. **(S.P. 1.2)**
- **6.A.2.2** The student is able to contrast mechanical and electromagnetic waves in terms of the need for a medium in wave propagation. **(S.P. 6.4, 7.2)**
- **6.B.3.1** The student is able to construct an equation relating the wavelength and amplitude of a wave from a graphical representation of the electric or magnetic field value as a function of position at a given time instant and vice versa, or construct an equation relating the frequency or period and amplitude of a wave from a graphical representation of the electric or magnetic field value at a given position as a function of time and vice versa. **(S.P. 1.4)**
- **6.F.2.1** The student is able to describe representations and models of electromagnetic waves that explain the transmission of energy when no medium is present. **(S.P. 1.1)**

We can get a good understanding of **electromagnetic waves** (EM) by considering how they are produced. Whenever a current varies, associated electric and magnetic fields vary, moving out from the source like waves. Perhaps the easiest situation to visualize is a varying current in a long straight wire, produced by an AC generator at its center, as illustrated in **Figure 24.5**.

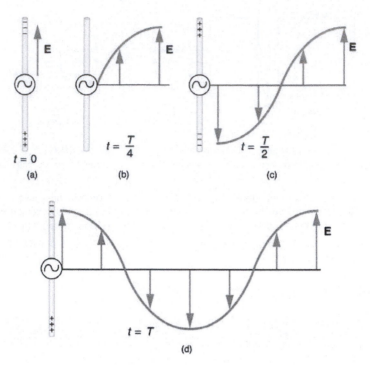

Figure 24.5 This long straight gray wire with an AC generator at its center becomes a broadcast antenna for electromagnetic waves. Shown here are the charge distributions at four different times. The electric field ($\mathbf{E}$) propagates away from the antenna at the speed of light, forming part of an electromagnetic wave.

The **electric field** ($\mathbf{E}$) shown surrounding the wire is produced by the charge distribution on the wire. Both the $\mathbf{E}$ and the charge distribution vary as the current changes. The changing field propagates outward at the speed of light.

There is an associated **magnetic field** ($\mathbf{B}$) which propagates outward as well (see **Figure 24.6**). The electric and magnetic fields are closely related and propagate as an electromagnetic wave. This is what happens in broadcast antennae such as those in radio and TV stations.

Closer examination of the one complete cycle shown in **Figure 24.5** reveals the periodic nature of the generator-driven charges oscillating up and down in the antenna and the electric field produced. At time $t = 0$, there is the maximum separation of charge, with negative charges at the top and positive charges at the bottom, producing the maximum magnitude of the electric field (or E -field) in the upward direction. One-fourth of a cycle later, there is no charge separation and the field next to the antenna is zero, while the maximum E -field has moved away at speed c .

As the process continues, the charge separation reverses and the field reaches its maximum downward value, returns to zero, and rises to its maximum upward value at the end of one complete cycle. The outgoing wave has an **amplitude** proportional to the maximum separation of charge. Its **wavelength** (λ) is proportional to the period of the oscillation and, hence, is smaller for short periods or high frequencies. (As usual, wavelength and **frequency** (f) are inversely proportional.)

Electric and Magnetic Waves: Moving Together

Following Ampere's law, current in the antenna produces a magnetic field, as shown in **Figure 24.6**. The relationship between $\mathbf{E}$ and $\mathbf{B}$ is shown at one instant in **Figure 24.6** (a). As the current varies, the magnetic field varies in magnitude and direction.

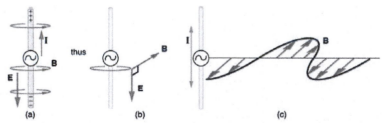

Figure 24.6 (a) The current in the antenna produces the circular magnetic field lines. The current ($\mathbf{I}$) produces the separation of charge along the wire, which in turn creates the electric field as shown. (b) The electric and magnetic fields ($\mathbf{E}$ and $\mathbf{B}$) near the wire are perpendicular; they are shown here for one point in space. (c) The magnetic field varies with current and propagates away from the antenna at the speed of light.

The magnetic field lines also propagate away from the antenna at the speed of light, forming the other part of the electromagnetic wave, as seen in **Figure 24.6** (b). The magnetic part of the wave has the same period and wavelength as the electric part, since they are both produced by the same movement and separation of charges in the antenna.

The electric and magnetic waves are shown together at one instant in time in **Figure 24.7**. The electric and magnetic fields produced by a long straight wire antenna are exactly in phase. Note that they are perpendicular to one another and to the direction of propagation, making this a **transverse wave**.

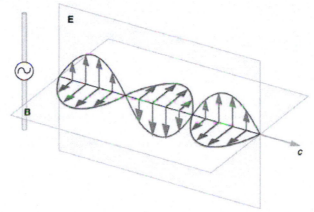

Figure 24.7 A part of the electromagnetic wave sent out from the antenna at one instant in time. The electric and magnetic fields (**E** and **B**) are in phase, and they are perpendicular to one another and the direction of propagation. For clarity, the waves are shown only along one direction, but they propagate out in other directions too.

Electromagnetic waves generally propagate out from a source in all directions, sometimes forming a complex radiation pattern. A linear antenna like this one will not radiate parallel to its length, for example. The wave is shown in one direction from the antenna in **Figure 24.7** to illustrate its basic characteristics.

Making Connections: Self-Propagating Wave

Note that an electromagnetic wave, as shown in **Figure 24.7**, is the result of a changing electric field causing a changing magnetic field, which causes a changing electric field, and so on. Therefore, unlike other waves, an electromagnetic wave is self-propagating, even in a vacuum (empty space). It does not need a medium to travel through. This is unlike mechanical waves, which do need a medium. The classic standing wave on a string, for example, does not exist without the string. Similarly, sound waves travel by molecules colliding with their neighbors. If there is no matter, sound waves cannot travel.

Instead of the AC generator, the antenna can also be driven by an AC circuit. In fact, charges radiate whenever they are accelerated. But while a current in a circuit needs a complete path, an antenna has a varying charge distribution forming a **standing wave**, driven by the AC. The dimensions of the antenna are critical for determining the frequency of the radiated electromagnetic waves. This is a **resonant** phenomenon and when we tune radios or TV, we vary electrical properties to achieve appropriate resonant conditions in the antenna.

Applying the Science Practices: Wave Properties and Graphs

Exercise 24.1

From the illustration of the electric field given in **Figure 24.8**(a) of an electromagnetic wave at some instant in time, please state what the amplitude and wavelength of the given waveform are. Then write down the equation for this particular wave.

Solution

The amplitude is 60 V/m, while the wavelength in this case is 1.5 m. The equation is $E = (60\frac{V}{m})\cos\left(\frac{4}{3}\pi x\right)$.

Exercise 24.2

Now, consider another electromagnetic wave for which the electric field at a particular location is given over time by $E = (30\frac{V}{m})\sin\left(4.0\pi \times 10^6\, t\right)$. What are the amplitude, frequency, and period? Finally, draw and label an appropriate graph for this electric field.

Solution

The amplitude is 30 V/m, while the frequency is 2.0 MHz and hence the period is 5.0×10^{-7} s. The graph should be similar to that in **Figure 24.8**(b).

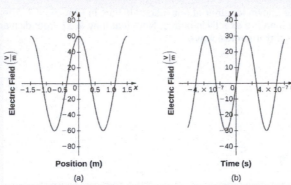

Figure 24.8 Two waveforms which describe two electromagnetic waves. Notice that (a) describes a wave in space, while (b) describes a wave at a particular point in space over time.

Receiving Electromagnetic Waves

Electromagnetic waves carry energy away from their source, similar to a sound wave carrying energy away from a standing wave on a guitar string. An antenna for receiving EM signals works in reverse. And like antennas that produce EM waves, receiver antennas are specially designed to resonate at particular frequencies.

An incoming electromagnetic wave accelerates electrons in the antenna, setting up a standing wave. If the radio or TV is switched on, electrical components pick up and amplify the signal formed by the accelerating electrons. The signal is then converted to audio and/or video format. Sometimes big receiver dishes are used to focus the signal onto an antenna.

In fact, charges radiate whenever they are accelerated. When designing circuits, we often assume that energy does not quickly escape AC circuits, and mostly this is true. A broadcast antenna is specially designed to enhance the rate of electromagnetic radiation, and shielding is necessary to keep the radiation close to zero. Some familiar phenomena are based on the production of electromagnetic waves by varying currents. Your microwave oven, for example, sends electromagnetic waves, called microwaves, from a concealed antenna that has an oscillating current imposed on it.

Relating E-Field and B-Field Strengths

There is a relationship between the E- and B-field strengths in an electromagnetic wave. This can be understood by again considering the antenna just described. The stronger the E-field created by a separation of charge, the greater the current and, hence, the greater the B-field created.

Since current is directly proportional to voltage (Ohm's law) and voltage is directly proportional to E-field strength, the two should be directly proportional. It can be shown that the magnitudes of the fields do have a constant ratio, equal to the speed of light. That is,

$$\frac{E}{B} = c \tag{24.3}$$

is the ratio of E-field strength to B-field strength in any electromagnetic wave. This is true at all times and at all locations in space. A simple and elegant result.

Example 24.1 Calculating B-Field Strength in an Electromagnetic Wave

What is the maximum strength of the B-field in an electromagnetic wave that has a maximum E-field strength of 1000 V/m ?

Strategy

To find the B-field strength, we rearrange the above equation to solve for B, yielding

$$B = \frac{E}{c}. \tag{24.4}$$

Solution

We are given E, and c is the speed of light. Entering these into the expression for B yields

$$B = \frac{1000 \text{ V/m}}{3.00 \times 10^8 \text{ m/s}} = 3.33 \times 10^{-6} \text{ T}, \tag{24.5}$$

Where T stands for Tesla, a measure of magnetic field strength.

Discussion

The B-field strength is less than a tenth of the Earth's admittedly weak magnetic field. This means that a relatively strong electric field of 1000 V/m is accompanied by a relatively weak magnetic field. Note that as this wave spreads out, say with distance from an antenna, its field strengths become progressively weaker.

The result of this example is consistent with the statement made in the module **Maxwell's Equations: Electromagnetic Waves Predicted and Observed** that changing electric fields create relatively weak magnetic fields. They can be detected in electromagnetic waves, however, by taking advantage of the phenomenon of resonance, as Hertz did. A system with the same natural frequency as the electromagnetic wave can be made to oscillate. All radio and TV receivers use this principle to pick up and then amplify weak electromagnetic waves, while rejecting all others not at their resonant frequency.

Take-Home Experiment: Antennas

For your TV or radio at home, identify the antenna, and sketch its shape. If you don't have cable, you might have an outdoor or indoor TV antenna. Estimate its size. If the TV signal is between 60 and 216 MHz for basic channels, then what is the wavelength of those EM waves?

Try tuning the radio and note the small range of frequencies at which a reasonable signal for that station is received. (This is easier with digital readout.) If you have a car with a radio and extendable antenna, note the quality of reception as the length of the antenna is changed.

PhET Explorations: Radio Waves and Electromagnetic Fields

Broadcast radio waves from KPhET. Wiggle the transmitter electron manually or have it oscillate automatically. Display the field as a curve or vectors. The strip chart shows the electron positions at the transmitter and at the receiver.

Figure 24.9 Radio Waves and Electromagnetic Fields (http://cnx.org/content/m55427/1.5/radio-waves_en.jar)

24.3 The Electromagnetic Spectrum

Learning Objectives
By the end of this section, you will be able to:

- List three "rules of thumb" that apply to the different frequencies along the electromagnetic spectrum.
- Explain why the higher the frequency, the shorter the wavelength of an electromagnetic wave.
- Draw a simplified electromagnetic spectrum, indicating the relative positions, frequencies, and spacing of the different types of radiation bands.
- List and explain the different methods by which electromagnetic waves are produced across the spectrum.

The information presented in this section supports the following AP® learning objectives and science practices:

- **6.F.1.1** The student is able to make qualitative comparisons of the wavelengths of types of electromagnetic radiation. **(S.P. 6.4, 7.2)**

In this module we examine how electromagnetic waves are classified into categories such as radio, infrared, ultraviolet, and so on, so that we can understand some of their similarities as well as some of their differences. We will also find that there are many connections with previously discussed topics, such as wavelength and resonance. A brief overview of the production and utilization of electromagnetic waves is found in **Table 24.1**.

Table 24.1 Electromagnetic Waves

Type of EM wave	Production	Applications	Life sciences aspect	Issues
Radio & TV	Accelerating charges	Communications Remote controls	MRI	Requires controls for band use
Microwaves	Accelerating charges & thermal agitation	Communications Ovens Radar	Deep heating	Cell phone use
Infrared	Thermal agitations & electronic transitions	Thermal imaging Heating	Absorbed by atmosphere	Greenhouse effect
Visible light	Thermal agitations & electronic transitions	All pervasive	Photosynthesis Human vision	
Ultraviolet	Thermal agitations & electronic transitions	Sterilization Cancer control	Vitamin D production	Ozone depletion Cancer causing
X-rays	Inner electronic transitions and fast collisions	Medical Security	Medical diagnosis Cancer therapy	Cancer causing
Gamma rays	Nuclear decay	Nuclear medicineSecurity	Medical diagnosis Cancer therapy	Cancer causing Radiation damage

Connections: Waves

There are many types of waves, such as water waves and even earthquakes. Among the many shared attributes of waves are propagation speed, frequency, and wavelength. These are always related by the expression $v_W = f\lambda$. This module concentrates on EM waves, but other modules contain examples of all of these characteristics for sound waves and submicroscopic particles.

As noted before, an electromagnetic wave has a frequency and a wavelength associated with it and travels at the speed of light, or c. The relationship among these wave characteristics can be described by $v_W = f\lambda$, where v_W is the propagation speed of the wave, f is the frequency, and λ is the wavelength. Here $v_W = c$, so that for all electromagnetic waves,

$$c = f\lambda. \tag{24.6}$$

Thus, for all electromagnetic waves, the greater the frequency, the smaller the wavelength.

Figure 24.10 shows how the various types of electromagnetic waves are categorized according to their wavelengths and frequencies—that is, it shows the electromagnetic spectrum. Many of the characteristics of the various types of electromagnetic waves are related to their frequencies and wavelengths, as we shall see.

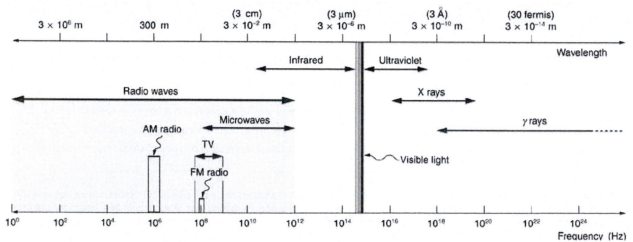

Figure 24.10 The electromagnetic spectrum, showing the major categories of electromagnetic waves. The range of frequencies and wavelengths is remarkable. The dividing line between some categories is distinct, whereas other categories overlap.

Electromagnetic Spectrum: Rules of Thumb

Three rules that apply to electromagnetic waves in general are as follows:

- High-frequency electromagnetic waves are more energetic and are more able to penetrate than low-frequency waves.
- High-frequency electromagnetic waves can carry more information per unit time than low-frequency waves.

- The shorter the wavelength of any electromagnetic wave probing a material, the smaller the detail it is possible to resolve.

Note that there are exceptions to these rules of thumb.

Transmission, Reflection, and Absorption

What happens when an electromagnetic wave impinges on a material? If the material is transparent to the particular frequency, then the wave can largely be transmitted. If the material is opaque to the frequency, then the wave can be totally reflected. The wave can also be absorbed by the material, indicating that there is some interaction between the wave and the material, such as the thermal agitation of molecules.

Of course it is possible to have partial transmission, reflection, and absorption. We normally associate these properties with visible light, but they do apply to all electromagnetic waves. What is not obvious is that something that is transparent to light may be opaque at other frequencies. For example, ordinary glass is transparent to visible light but largely opaque to ultraviolet radiation. Human skin is opaque to visible light—we cannot see through people—but transparent to X-rays.

Radio and TV Waves

The broad category of **radio waves** is defined to contain any electromagnetic wave produced by currents in wires and circuits. Its name derives from their most common use as a carrier of audio information (i.e., radio). The name is applied to electromagnetic waves of similar frequencies regardless of source. Radio waves from outer space, for example, do not come from alien radio stations. They are created by many astronomical phenomena, and their study has revealed much about nature on the largest scales.

There are many uses for radio waves, and so the category is divided into many subcategories, including microwaves and those electromagnetic waves used for AM and FM radio, cellular telephones, and TV.

The lowest commonly encountered radio frequencies are produced by high-voltage AC power transmission lines at frequencies of 50 or 60 Hz. (See Figure 24.11.) These extremely long wavelength electromagnetic waves (about 6000 km!) are one means of energy loss in long-distance power transmission.

Figure 24.11 This high-voltage traction power line running to Eutingen Railway Substation in Germany radiates electromagnetic waves with very long wavelengths. (credit: Zonk43, Wikimedia Commons)

There is an ongoing controversy regarding potential health hazards associated with exposure to these electromagnetic fields (E-fields). Some people suspect that living near such transmission lines may cause a variety of illnesses, including cancer. But demographic data are either inconclusive or simply do not support the hazard theory. Recent reports that have looked at many European and American epidemiological studies have found no increase in risk for cancer due to exposure to E-fields.

Extremely low frequency (ELF) radio waves of about 1 kHz are used to communicate with submerged submarines. The ability of radio waves to penetrate salt water is related to their wavelength (much like ultrasound penetrating tissue)—the longer the wavelength, the farther they penetrate. Since salt water is a good conductor, radio waves are strongly absorbed by it, and very long wavelengths are needed to reach a submarine under the surface. (See Figure 24.12.)

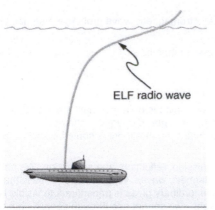

Figure 24.12 Very long wavelength radio waves are needed to reach this submarine, requiring extremely low frequency signals (ELF). Shorter wavelengths do not penetrate to any significant depth.

AM radio waves are used to carry commercial radio signals in the frequency range from 540 to 1600 kHz. The abbreviation AM stands for **amplitude modulation**, which is the method for placing information on these waves. (See **Figure 24.13**.) A **carrier wave** having the basic frequency of the radio station, say 1530 kHz, is varied or modulated in amplitude by an audio signal. The resulting wave has a constant frequency, but a varying amplitude.

A radio receiver tuned to have the same resonant frequency as the carrier wave can pick up the signal, while rejecting the many other frequencies impinging on its antenna. The receiver's circuitry is designed to respond to variations in amplitude of the carrier wave to replicate the original audio signal. That audio signal is amplified to drive a speaker or perhaps to be recorded.

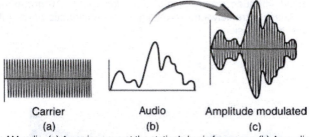

Figure 24.13 Amplitude modulation for AM radio. (a) A carrier wave at the station's basic frequency. (b) An audio signal at much lower audible frequencies. (c) The amplitude of the carrier is modulated by the audio signal without changing its basic frequency.

FM Radio Waves

FM radio waves are also used for commercial radio transmission, but in the frequency range of 88 to 108 MHz. FM stands for **frequency modulation**, another method of carrying information. (See **Figure 24.14**.) Here a carrier wave having the basic frequency of the radio station, perhaps 105.1 MHz, is modulated in frequency by the audio signal, producing a wave of constant amplitude but varying frequency.

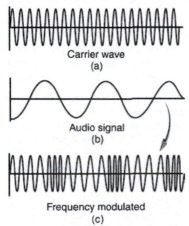

Figure 24.14 Frequency modulation for FM radio. (a) A carrier wave at the station's basic frequency. (b) An audio signal at much lower audible frequencies. (c) The frequency of the carrier is modulated by the audio signal without changing its amplitude.

Since audible frequencies range up to 20 kHz (or 0.020 MHz) at most, the frequency of the FM radio wave can vary from the carrier by as much as 0.020 MHz. Thus the carrier frequencies of two different radio stations cannot be closer than 0.020 MHz. An FM receiver is tuned to resonate at the carrier frequency and has circuitry that responds to variations in frequency, reproducing the audio information.

FM radio is inherently less subject to noise from stray radio sources than AM radio. The reason is that amplitudes of waves add. So an AM receiver would interpret noise added onto the amplitude of its carrier wave as part of the information. An FM receiver can be made to reject amplitudes other than that of the basic carrier wave and only look for variations in frequency. It is thus easier to reject noise from FM, since noise produces a variation in amplitude.

Television is also broadcast on electromagnetic waves. Since the waves must carry a great deal of visual as well as audio information, each channel requires a larger range of frequencies than simple radio transmission. TV channels utilize frequencies in the range of 54 to 88 MHz and 174 to 222 MHz. (The entire FM radio band lies between channels 88 MHz and 174 MHz.) These TV channels are called VHF (for **very high frequency**). Other channels called UHF (for **ultra high frequency**) utilize an even higher frequency range of 470 to 1000 MHz.

The TV video signal is AM, while the TV audio is FM. Note that these frequencies are those of free transmission with the user utilizing an old-fashioned roof antenna. Satellite dishes and cable transmission of TV occurs at significantly higher frequencies and is rapidly evolving with the use of the high-definition or HD format.

Example 24.2 Calculating Wavelengths of Radio Waves

Calculate the wavelengths of a 1530-kHz AM radio signal, a 105.1-MHz FM radio signal, and a 1.90-GHz cell phone signal.

Strategy

The relationship between wavelength and frequency is $c = f\lambda$, where $c = 3.00\times10^8$ m / s is the speed of light (the speed of light is only very slightly smaller in air than it is in a vacuum). We can rearrange this equation to find the wavelength for all three frequencies.

Solution

Rearranging gives

$$\lambda = \frac{c}{f}. \tag{24.7}$$

(a) For the $f = 1530$ kHz AM radio signal, then,

$$\begin{aligned} \lambda &= \frac{3.00\times10^8 \text{ m/s}}{1530\times10^3 \text{ cycles/s}} \\ &= 196 \text{ m.} \end{aligned} \tag{24.8}$$

(b) For the $f = 105.1$ MHz FM radio signal,

$$\begin{aligned} \lambda &= \frac{3.00\times10^8 \text{ m/s}}{105.1\times10^6 \text{ cycles/s}} \\ &= 2.85 \text{ m.} \end{aligned} \tag{24.9}$$

(c) And for the $f = 1.90$ GHz cell phone,

$$\begin{aligned} \lambda &= \frac{3.00\times10^8 \text{ m/s}}{1.90\times10^9 \text{ cycles/s}} \\ &= 0.158 \text{ m.} \end{aligned} \tag{24.10}$$

Discussion

These wavelengths are consistent with the spectrum in **Figure 24.10**. The wavelengths are also related to other properties of these electromagnetic waves, as we shall see.

The wavelengths found in the preceding example are representative of AM, FM, and cell phones, and account for some of the differences in how they are broadcast and how well they travel. The most efficient length for a linear antenna, such as discussed in **Production of Electromagnetic Waves**, is $\lambda / 2$, half the wavelength of the electromagnetic wave. Thus a very large antenna is needed to efficiently broadcast typical AM radio with its carrier wavelengths on the order of hundreds of meters.

One benefit to these long AM wavelengths is that they can go over and around rather large obstacles (like buildings and hills), just as ocean waves can go around large rocks. FM and TV are best received when there is a line of sight between the broadcast antenna and receiver, and they are often sent from very tall structures. FM, TV, and mobile phone antennas themselves are much smaller than those used for AM, but they are elevated to achieve an unobstructed line of sight. (See **Figure 24.15**.)

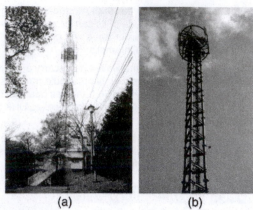

Figure 24.15 (a) A large tower is used to broadcast TV signals. The actual antennas are small structures on top of the tower—they are placed at great heights to have a clear line of sight over a large broadcast area. (credit: Ozizo, Wikimedia Commons) (b) The NTT Dokomo mobile phone tower at Tokorozawa City, Japan. (credit: tokoroten, Wikimedia Commons)

Radio Wave Interference

Astronomers and astrophysicists collect signals from outer space using electromagnetic waves. A common problem for astrophysicists is the "pollution" from electromagnetic radiation pervading our surroundings from communication systems in general. Even everyday gadgets like our car keys having the facility to lock car doors remotely and being able to turn TVs on and off using remotes involve radio-wave frequencies. In order to prevent interference between all these electromagnetic signals, strict regulations are drawn up for different organizations to utilize different radio frequency bands.

One reason why we are sometimes asked to switch off our mobile phones (operating in the range of 1.9 GHz) on airplanes and in hospitals is that important communications or medical equipment often uses similar radio frequencies and their operation can be affected by frequencies used in the communication devices.

For example, radio waves used in magnetic resonance imaging (MRI) have frequencies on the order of 100 MHz, although this varies significantly depending on the strength of the magnetic field used and the nuclear type being scanned. MRI is an important medical imaging and research tool, producing highly detailed two- and three-dimensional images. Radio waves are broadcast, absorbed, and reemitted in a resonance process that is sensitive to the density of nuclei (usually protons or hydrogen nuclei).

The wavelength of 100-MHz radio waves is 3 m, yet using the sensitivity of the resonant frequency to the magnetic field strength, details smaller than a millimeter can be imaged. This is a good example of an exception to a rule of thumb (in this case, the rubric that details much smaller than the probe's wavelength cannot be detected). The intensity of the radio waves used in MRI presents little or no hazard to human health.

Microwaves

Microwaves are the highest-frequency electromagnetic waves that can be produced by currents in macroscopic circuits and devices. Microwave frequencies range from about 10^9 Hz to the highest practical LC resonance at nearly 10^{12} Hz. Since they have high frequencies, their wavelengths are short compared with those of other radio waves—hence the name "microwave."

Microwaves can also be produced by atoms and molecules. They are, for example, a component of electromagnetic radiation generated by **thermal agitation**. The thermal motion of atoms and molecules in any object at a temperature above absolute zero causes them to emit and absorb radiation.

Since it is possible to carry more information per unit time on high frequencies, microwaves are quite suitable for communications. Most satellite-transmitted information is carried on microwaves, as are land-based long-distance transmissions. A clear line of sight between transmitter and receiver is needed because of the short wavelengths involved.

Radar is a common application of microwaves that was first developed in World War II. By detecting and timing microwave echoes, radar systems can determine the distance to objects as diverse as clouds and aircraft. A Doppler shift in the radar echo can be used to determine the speed of a car or the intensity of a rainstorm. Sophisticated radar systems are used to map the Earth and other planets, with a resolution limited by wavelength. (See **Figure 24.16.**) The shorter the wavelength of any probe, the smaller the detail it is possible to observe.